SI Prefixes

Prefix	Symbol	Power of 10	Prefix	Symbol	Power of 10
tera	T	10^{12}	centi	c	10^{-2}
giga	G	10^{9}	milli	m	10^{-3}
mega	M	10^{6}	micro	μ	10^{-6}
kilo	k	10^{3}	nano	n	10^{-9}
hecto	h	10^{2}	pico	p	10^{-12}
deca	da	10^{1}	femto	f	10^{-15}
deci	d	10^{-1}	atto	a	10^{-18}

Greek Alphabet

α	Alpha	A	ν	Nu	N
β	Beta	B	ξ	Xi	Ξ
γ	Gamma	Γ	o	Omicron	O
δ	Delta	Δ	π	Pi	Π
ϵ	Epsilon	E	ρ	Rho	P
ζ	Zeta	Z	σ	Sigma	Σ
η	Eta	H	τ	Tau	T
θ	Theta	Θ	υ	Upsilon	Υ
ι	Iota	I	ϕ	Phi	Φ
κ	Kappa	K	χ	Chi	X
λ	Lambda	Λ	ψ	Psi	Ψ
μ	Mu	M	ω	Omega	Ω

ELECTRIC CIRCUITS

McGRAW-HILL SERIES IN ELECTRICAL AND COMPUTER ENGINEERING

SENIOR CONSULTING EDITOR
Stephen W. Director, *Carnegie Mellon University*

- Circuits and Systems
- Communications and Signal Processing
- Computer Engineering
- Control Theory
- Electromagnetics
- Electronics and VLSI Circuits
- Introductory
- Power and Energy
- Radar and Antennas

Previous Consulting Editors
Ronald N. Bracewell, Colin Cherry, James F. Gibbons, Willis W. Harman, Hubert Heffner, Edward W. Herold, John G. Linvill, Simon Ramo, Ronald A. Rohrer, Anthony E. Siegman, Charles Susskind, Frederick E. Terman, John G. Truxal, Ernst Weber, and John R. Whinnery

CIRCUITS AND SYSTEMS

SENIOR CONSULTING EDITOR
Stephen W. Director, *Carnegie Mellon University*

- Antoniou: Digital Filters: Analysis and Design
- Balabanian: Electric Circuits
- Bracewell: The Fourier Transform and Its Applications
- Cannon: Dynamics of Physical Systems
- Chua, Desoer, and Kuh: Linear and Non-Linear Circuits
- Desoer and Kuh: Basic Circuit Theory
- Glisson: Introduction to System Analysis
- Hayt and Kemmerly: Engineering Circuit Analysis
- Papoulis: The Fourier Integral and Its Application
- Reid: Linear System Fundamentals: Continuous and Discrete, Classic and Modern
- Scott: An Introduction to Circuit Analysis: A Systems Approach
- Temes and LaPatra: Introduction to Circuit Synthesis
- Truxal: Introductory System Engineering

ELECTRIC CIRCUITS

Norman Balabanian

Professor Emeritus
Electrical and Computer Engineering
Syracuse University

Visiting Scholar
Electrical Engineering
Tufts University

Visiting Scholar
Science, Technology and Society
Massachusetts Institute of Technology

McGraw-Hill, Inc.

New York St. Louis San Francisco Auckland Bogotá Caracas
Lisbon London Madrid Mexico City Milan Montreal New Delhi
San Juan Singapore Sydney Tokyo Toronto

ELECTRIC CIRCUITS

This book is printed on recycled, acid-free paper containing 10% postconsumer waste.

234567890 DOH DOH 90987654

ISBN 0-07-004804-5

This book was set in Times Roman.
The editors were Anne T. Brown and Margery Luhrs;
the text and cover designer was Keithley and Associates;
the production supervisor was Leroy A. Young.
R. R. Donnelley & Sons Company was printer and binder.

Library of Congress Cataloging-in-Publication Data

Balabanian, Norman, (date).
Electric circuits/Norman Balabanian.
p. cm.—(McGraw-Hill series in electrical and computer engineering. Circuits and systems)
Includes index.
ISBN 0-07-004804-5
1. Electric circuit analysis. I. Title. II. Series.
TK454.B32 1994
621.319'2—dc20 93–26186

When ordering this title, use ISBN 0-07-1138471.

ABOUT THE AUTHOR

Norman Balabanian retired as Professor Emeritus from the Department of Electrical & Computer Engineering at Syracuse University after long years of teaching and research, culminating in seven years as Department Chair. Currently, he is a Visiting Scholar both in the Department of Electrical Engineering at Tufts University and in the Program in Science, Technology and Society at the Massachusetts Institute of Technology. He is the author or coauthor of eight previous books on network analysis or synthesis at various levels and is the author or editor of three previous books on engineering education. He has taught as visiting professor at the University of Colorado at Boulder and the University of California at Berkeley; also at the Instituto Politecnico Nacional in Mexico as a member of UNESCO, at the University of Zagreb in the former Yugoslavia as a Fulbright Fellow, and at the Institut National d'Électricité et d'Électronique in Algeria as Academic Advisor for Engineering.

Dr. Balabanian is a Life Fellow of the Institute of Electrical and Electronic Engineers and of the American Association for the Advancement of Science. He received the IEEE Centennial Award in 1984. He has been Editor of the IEEE Transactions on Circuit Theory and the IEEE Technology and Society Magazine. He has also served as President of the IEEE Society on Social Implications of Technology and of the Electrical Engineering Division of the American Society for Engineering Education.

To Wilbur R. LePage

Profound scholar, dedicated teacher,
colleague, and friend

CONTENTS

CHAPTER 6

SIGNALS AND FIRST-ORDER CIRCUITS 241

PREFACE

Pedagogical Issues

By now, circuit theory as a college subject is more than a century old. Nevertheless, textbooks on the subject still make their appearance regularly. They vary in level, in profundity, in degree of detail, in organization, in reliance on software, and in pedagogical approach.

In seeking to explain and predict electrical phenomena in the physical world of circuits, this book follows the aproach that is generally described as the "scientific method." Over the years, observations and experiments in the physical world have led to a set of laws (including Ohm's law, Kirchhoff's laws, Faraday's law, and Ampère's law) that are enunciated in terms of a hypothetical model of the part of the physical world under discussion. The model consists of a number of components. These elements do not actually exist in the physical world, although the behavior of *some* physical devices, under *some* conditions of operation, may approximate that of one or another of the hypothetical components, or some combination of them. When practical circuits are under consideration in the book, they are replaced by models appropriate to the specific conditions of operation.

The deductive process—applying general principles to specific cases—is usually well illustrated in textbooks. In this book, I have also tried to emphasize the inductive process—the development of a generally valid law or procedure from an examination of specific cases. Sometimes, the study of one or more specific cases leads to a conjecture about something generally valid. The student might then find that this same conjecture can be deduced from a reexamination of previously established principles. This text follows such a pattern in several places.

When a subject like circuit theory reaches a degree of maturity, there is a tendency for a textbook to acquire some of the characteristics of an encyclopedia; every conceivable topic is "covered." This robs the learner of all the joys of discovery. The learner is given the complete story and told to learn it, mainly by practicing on exercises and problems just like detailed examples provided. I have tried to avoid the pitfall of cataloging for students all that we know on a subject and have left for them the pleasure of developing (with guidance) some results which, though known to those familiar with circuit theory, are not essential for going on with the subject matter being developed.

At key points in the exposition, I have tried to raise questions about a conclusion just reached, to explore the alternatives it presents for going on and the reasonableness of pursuing a particular line of attack. *Why* we are pursuing a particular thread and *how* a particular component in our model is created are just as important to explain as the details of manipulating this model, of following that specific procedure, or of applying some algorithm. The tenor of the text is that we are conducting an investigative exploration, almost like a research project, for the purpose of discovering and assimilating knowledge about the subject. When a topic is introduced, I make an effort to help the student understand why we ought to devote time to it. Why study the sinusoidal steady state, for example? When some step has to be taken, alternatives are explored. "We could do this or we could do that," the commentary might go; "let's first try this, for the following reasons."

Students learn best if they are engaged. If teachers (or authors) could ensure that, their day would be made. Therefore, I try to remind readers now and then that they should participate in the derivation of an equation by performing missing steps, observe the relevant features of a diagram by describing it to themselves, or think through a proposed plan before going on.

Level of Presentation

The subject of circuits is taught in engineering schools at different levels. At some universities, the subject is delayed until the third year, when more background in physics and mathematics is expected. This will influence the level at which the subject is approached. At most other schools, the subject is introduced sometimes in the second year. Students have taken a course in differential and integral calculus but perhaps not yet in differential equations. They might have studied electricity and magnetism in physics (or might be doing so concurrently), as well as statics and dynamics in mechanical engineering. This book is aimed at this level. Regardless of the level, I have not compromised in the rigor of presentation of any topic.

The book is suitable for use in courses spanning an academic year, either two semesters or three quarters. By proper selection of chapters and of topics within chapters, a one-semester course can also be accommodated.

Topic Selection

This book deals mainly with linear circuits, although nonlinear diode circuits are treated briefly. Because nonlinear circuits are treated to such a limited extent and no general propositions are discussed in this area, the wisdom of including them at all might be questioned. Normally, if one covers only linear circuits, then no matter how often one cautions students that the rules and procedures being studied apply only to linear circuits, they tend to believe that what they are studying is quite general—unless they see cases of nonlinear circuits where the methods don't apply. Furthermore, even the linear models we use have a limited dynamic range. This is especially true for such things as controlled sources and op amps. It is important to ensure that the conditions of a problem are such that the model is truly operating in the linear region. Sensitivity to these matters cannot be raised if only linear circuits are treated.

Topics have been selected and ordered so that the book can be used at institutions with different calendars and a variety of emphases. Several "enrichment" topics are introduced in sections that can be omitted without incurring a subsequent penalty. Later sections or problems based on this material can be similarly omitted. The most salient are a section on transistors, the material on gyrators and negative converters, the proof of Thévenin's theorem, the coverage of stagger-tuned circuits, the Fourier transforms of certain functions, and the derivation of the convolution integral. Including such material permits students with more time or curiosity to benefit without penalizing others.

Auxiliary material is provided in two forms. Inside the book covers are four tables of reference information. One gives the SI prefix names, symbols, and corresponding powers of 10. The second is a short list of useful trigonometric identities. The third is a brief table of commonly encountered integrals. A table listing the Greek alphabet is also provided. Because they are easy to locate without riffling through pages, these tables are easy to use. The remaining auxiliary material is in the form of two appendices on a number of topics in mathematics that students have probably encountered but don't necessarily have at their fingertips. They serve as memory aids, not as thorough expositions. Appendix A treats matrix algebra, the solution of linear algebraic equations, Cramer's rule, and pivotal condensation. Appendix B treats the arithmetic and algebra of complex numbers; it also provides a brief look at complex variables, Euler's theorem, and the law of exponents.

Numbering Scheme for Equations and Figures

Some schemes for numbering equations and figures, and the manner of referring to them, can cause students to be distracted as they engage unproductively in reading the numbers and searching for them. In this book a sequential numbering system, starting fresh within each chapter, for both equations and figures is used, without attached prefixes consisting of chapter numbers or section numbers. (On those few occasions when reference is made to an equation in an earlier chapter, the chapter number does have to be stated.) Not all equations are numbered, only significant ones or those to which reference might be made later. No special identifying "tag," such as *Eq.,* is used when referring to an equation, but when referring to a figure, the word *Figure* is used.

Illustrations, Examples, Exercises, and Problems

When a particular topic is first being developed, illustrations are used to illuminate it. Indeed, an illustration might precede the theoretical development of the topic as part of the process of induction. *Illustrations*, thus, are incorporated in the development of the material. There are also numbered *examples,* separated from the text and easily distinguished, which are worked out using the concepts just developed, together with other recently assimilated ideas. Both literal and numerical illustrations and examples are used.

Scattered throughout, but in a format that distinguishes them from the text, are numbered *exercises* for students to work out at the time they are studying the relevant sections. The purpose of these exercises is to provide reinforcement for the concepts under study by having students carry out some simple calculations applying results just encountered.

At the end of each chapter there is a set of *problems.* The problems in each set range from a simple application of procedures developed in the book to a challenging solution of a practical problem. Sometimes a problem requires students to apply a specified technique for its solution. At other times they are asked to solve a problem by two or more specified approaches and to compare the ease or difficulty. This is so that students can practice specific techniques and reinforce their understanding of them. Sometimes the problem is open-ended, so that students have to make decisions about the methods to use and then apply them. Certain types of problems specify a circuit structure and component and excitation values and then ask for the solution for certain values of voltage or current, power dissipated or supplied, or energy stored. In other problems, not all element values are specified but certain other desired outcomes are, such as the voltages or currents somewhere. The unknown component values are to be determined in order to achieve the specified goals. Such problems have a design flavor.

Unrealistic circuits that throw together an arbitrary collection of circuit elements, even though the resulting circuit model does not remotely correspond to a practical circuit, are avoided. Among such culprits are resistors across voltage sources or in series with current sources, and contrived circuits that include control sources in unrealistic configurations that do not correspond to models of real devices; their effect is often nothing more than that of a resistor.

From Chapter 3 on, there is, in addition, a category called *design problems.* The objective in these problems is the design of an engineering circuit to achieve specified goals—say, a sawtooth wave. At the level of this book it may be necessary to specify the structure of a specific design and require the students to determine the parameter values that will achieve specified values of peak and valley voltage, repetition rate, or return time, for example. The design problems are all realistic. (In the context of this book, however, it is not possible to include all the factors that come into play in a real design, such as cost, for example.)

Use of Software

I am ambivalent about students' use of circuit analysis software during the study of circuits. For professionals, such software is indispensable, and familiarity with it is a requirement for being a professional. The real question for the author of a textbook is how valuable such software is during the early stages of study. My decision has been not to incorporate the use of circuit analysis software directly into the text but to orient specific topics in a way that is consistent with the availability and use of PSpice and MATLAB. Thus, for example, the usual approach in mesh and node analysis of *RLC* circuits that emphasizes the writing of the coefficient matrix by inspection is downplayed, explicitly because such a skill is of no value when using PSpice, besides being applicable only to *RLC* circuits. (On the other hand, note the section on supplements that follows.)

Text Supplements

There are two varieties of supplements. One variety is provided to instructors who adopt this book for use in their courses and is not available to students. It includes a *Solutions Manual,* which contains full solutions of all the problems in the book. It also includes a set of *transparency masters* of appropriate figures from the book. The figures are enlarged so that instructors can use them in the classroom.

The second is a separate *Study Guide* intended for student use. The Study Guide is keyed to the chapters in the book and provides instruction in problem solving and solution checking, using a step-by-step, programmed approach. It also includes a tutorial in the use of PSpice and MATLAB in circuit analysis, giving examples of problem solutions using this software. The greatest part of the Study Guide consists of worked-out problems covering the entire range of topics in the book. Many of the problems are explicitly phrased for use of PSpice or MATLAB. Instructors adopting the book for their courses will be provided a copy of the Study Guide for their use.

Acknowledgments

Countless individuals have contributed indirectly to the development of this book. They include my teachers, my students over a long career, and the many users of my earlier books who provided helpful comments. McGraw-Hill and I would also like to thank the following reviewers of the present manuscript, who provided useful comments during its development: Stephen Director (Consulting Editor), Carnegie Mellon University; James G. Gottling, The Ohio State University; DeVerl S. Humpherys, Brigham Young University; Erman Kudeki, Cornell University; Yusef Leblebici, University of Illinois at Urbana-Champaign; Fu Li, Portland State University; Frank J. Lofy, The University of Wisconsin at Platteville; George W. Lucky, New Mexico State University; Pradeep Msira, Wright State University; James G. Smith, Southern Illinois University at Carbondale; Renjeng Su, University of Colorado at Boulder; James Svoboda, Clarkson University; John E. Tyler, Jr., Texas A&M University; and Bogdan M. Wilamowski, University of Wyoming.

Special thanks are due to Dikran Meliksetian, Assistant Professor of Electrical Engineering at South Dakota School of Mines and Technology, who contributed many design problems to the text, single-handedly produced the solutions manual, and contributed to the preparation of the study guide.

I owe a debt of gratitude to the Department of Electrical and Computer Engineering at Syracuse University for making available to me a year's free time to devote to this book. I am grateful also to the faculty of the Department of Electrical Engineering at Tufts University and to its Chairman, Prof. Dennis Fermental, for their generosity in providing me, for several years, an office, facilities, and collegiality while this book was being completed.

The one individual who has influenced me the most is my teacher and mentor, Wilbur R. LePage. We have spent countless days and nights discussing this or that esoteric topic in circuit theory, arguing about interpretations and nuances and how best to think of, and to present to students, this or that topic.

Norman Balabanian

ELECTRIC CIRCUITS

1 The Fundamental Laws

1 PRELIMINARY THOUGHTS

An electrical engineer is concerned with energy and information—their generation, their conversion from one form to another, their distribution to users, and their subsequent utilization. You can give countless examples of these various tasks from your own experience. Electricity is generated by converting solar energy (by means of photovoltaic cells) and energy from water flow (hydropower), chemical reactions (in a battery), organic and nuclear fuels (in power stations), and other sources. It is then transmitted, over long distances or short distances, to distribution centers from which it is sent out to operate hundreds of varieties of different electrical devices, large and small.

The information contained in the TV images and radio programs we receive at home originates in a studio; is converted to electrical form for transmission through the atmosphere, by microwave relay, or by optical cable; and is eventually reconverted to sound or visual images by such devices as radio and TV receivers. The telephone system converts sound waves to electrical form and transmits the electrical signals via wires, microwave links, or optical fibers to a destination where they are reconverted into sound. The same system can convert the information on a printed page into electrical signals, transmit these to a remote fax machine, and regenerate the printed page. These are just some familiar examples.

In performing these four operations, or functions—*generation, conversion, transmission,* and *utilization* of energy and information—the electrical engineer must deal with many aspects of the physical world. These operations are performed by interconnecting many physical devices and pieces of equipment.

The engineer, then, must be able to understand how interconnections of electrical devices function, to describe how they behave electrically, and, ultimately, to design an interconnection of such devices to perform a desirable function, such as producing music. The electrical behavior must be described in terms of electrical quantities, or *variables*: voltage, current, power, electric charge, energy, magnetic flux, electric field intensity, magnetic field intensity, and others. This description of the behavior of an electrical device involves a statement (or equation) about the relationships among the variables.

To give an example from another field, the behavior of a body of matter subjected to an external force is described in terms of the force and the velocity (or acceleration). Following Newton, we say that the force is equal to the mass of the body times the acceleration, which is the time rate of change of the velocity.

In describing the electrical behavior of devices, we would be hopelessly lost if we had to discover the relationships among the variables separately for every conceivable device. Instead, as in all science, we seek a *model*; not a physical model, but an abstract mathematical model, a conceptual scheme, a *theory*. If we are successful, the behavior of the model will be similar to the behavior of the actual device or the interconnection of devices.

The way in which we develop this model, of course, is through observation and experimentation in the physical world. We make many repeated observations of the same or of similar phenomena, and we note the trends that our results follow. We try to vary one of the factors that might affect the result, holding other factors constant, and we note how the results change.

A classical example of this procedure is Galileo's experiment to determine how long it takes a body to fall from a given height. He dropped bodies of various sizes, shapes, and weights and measured the time it took for each body to fall a given distance. He found that, for the most part, all the bodies took about the same length of time, except the lightest ones, such as pieces of paper or feathers, which took longer. Even these light bodies, however, tended to fall faster if the air through which they were falling was not turbulent or windy. On the basis of these experiments, Galileo contended that all bodies, even the lightest ones, would take the same length of time to fall a given distance, provided they were in a vacuum so the air resistance to the fall was removed.

This conclusion appears perfectly reasonable if it is assumed that the acceleration of gravity g is a constant. With such an assumption, if a body is dropped with no initial velocity, the distance it falls in a time t is $x = gt^2/2$. This expression is independent of the weight, size, and shape of the body. If the distance of fall is fixed, then the time it takes will be constant independent of the properties of the body—so long as g is constant.

Based on such considerations, we can *postulate,* or hypothesize, that the acceleration of gravity is constant. We have just seen that, on the basis of this postulate, we can predict something which is quite closely confirmed by experimentation. Armed with this information, we can now go on to predict other results concerning motions of bodies under the influence of gravity. Perhaps, as we go along, we will need to make additional hypotheses. If, at any stage of the development, a result predicted from our assumptions fails to agree with our observations (assuming that the error made in the measurement is less than the observed discrepancy), then we must reexamine our postulates and modify them, in order that the observed results can again be predicted from the new postulates.

This is what happened to the classical theory of mechanics, the view built up by Galileo and Newton and called *Newtonian mechanics.* Early in the twentieth century Einstein observed that, for rapidly moving bodies, conclusions drawn from classical mechanics deviated more and more from observations. He formulated the theory of (special) relativity to explain the observed results. Relativistic mechanics gives results that agree with observations whenever the speed of an object is an appreciable fraction of the speed of light. Furthermore, it reduces to the classical theory when the speed of an object is small compared with the speed of light.

What we have described is the scientific method. The first step is to *make observations,* to experiment. The next step is to *postulate a model of the physical world* (that portion of it which is under observation) and to generalize the results of the observations to this model. This is done by *stating a relationship among the variables* used to describe the phenomenon under consideration. This relationship comes to be known as a *law.* Newton's law relating force and acceleration and Ohm's law relating voltage and current (to be discussed in the next chapter) are examples.

On the basis of the postulated laws, we next try to invent other relationships among the variables in the model. In other words, we derive additional results based on the assumed laws. Finally, we test the validity of these results by making more observations in the physical world. If these observations confirm our predictions, we gain confidence in our original "laws" and in our model. On the other hand, if the observations fail to confirm our predictions, we seek modifications of the model, just as Einstein successfully did with Newton's model of mechanics.

The conservation laws of science are perhaps the most fundamental and universal postulates in physical science. The law or *principle of conservation of energy* states that the total energy in a closed system must remain constant. People have tried to find situations in which these principles are violated, but they have not succeeded. Many "perpetual-motion" machines have been invented, but nature has always foiled the inventor. Any additional postulates we may introduce in developing a theory about the interconnections of electrical devices must be consistent with the conservation laws.

In addition to the conservation laws, any theory we develop must be consistent with the *atomic theory of matter.* In accordance with this theory, all matter consists of units called *atoms.* An atom is viewed as possessing a nucleus carrying a positive charge surrounded by a "cloud" of electrons each carrying a negative charge. Each element of matter (for example, hydrogen, copper, or silicon) has a specific atomic number which equals the number of electrons surrounding the nucleus of each atom of the element. As a unit, each atom is normally electrically neutral, the charge of the electrons just balancing the charge of the nucleus. Atomic dimensions are of the order of 10^{-10} meters. With certain exceptions (such as microelectronics), the distances with which we are concerned in engineering are usually millions of times greater than atomic dimensions.

It is possible to remove one or more electrons from their home nuclei, thus creating bodies which are electrically charged. Over time, it has been observed that such electrically charged bodies exert forces (of attraction and repulsion) on each other, even when the charges are stationary (relative to each other). Coulomb first gave a quantitative relationship (known as *Coulomb's law*) for the force of attraction or repulsion of electric charges on each other. Additional kinds of forces on charges occur when charges are not stationary but in motion.

Systems of Units

In order to carry out scientific investigations and engineering analysis and design, it is essential to define terms clearly and accurately and to use a consistent system of units in terms of which to conduct measurements. Such a system of units is the *Système International* (SI). Although several other systems have been used in the past, SI is now the worldwide standard. A few quantities are adopted as basic and all other physical quantities are derived from them. The six basic quantities in SI are *length, mass, time, electric current, temperature,* and *luminance*; the last two of these are not needed for studying the interconnection of electrical devices. The units of the first four quantities in SI are the *meter, kilogram, second,* and *ampere.*

One of the advantages of the SI units is that a system of decimal prefixes can be used with any unit, so that there is no necessity of using measures like 0.000025 second or 53,000 meters; they can be referred to as 25 *micro*seconds and 53 *kilo*meters, respectively. A list of the prefixes in SI units is given in the inside front cover.

Whenever mention is made of the value of a physical quantity, the unit of measurement must be specified. To say "This book is 3 in thickness" would be meaningless unless we also mentioned the units: 3 centimeters. In this book we will always use the SI units, with prefixes where convenient. Units should always be specified. However, when the context makes clear that the SI unit is implied, it will sometimes be omitted in the book. When the unit is something other than the SI unit, it must *always* be specified, as in our example of 3 centimeters. If we say a certain length is 3 without specifying units, we *must* mean 3 meters, the SI unit. However, you must get into the habit of *always* specifying units. Only when you have earned the right (by no longer making any mistakes in this area) should you take liberties with units.

2 ELECTRIC CURRENT

The smallest known amount of charge is that of an electron. However, this is so small compared with the amount of charge involved in practical situations that it is not commonly chosen as a unit of measure. In the SI system of measurement, the unit of charge is called the *coulomb* (C).[1] One coulomb equals 6.25×10^{18} times the magnitude of the charge of an electron; or, stated alternatively, an electronic charge, which is negative, equals 1.6×10^{-19} coulomb.

In many situations, electric charges can move when subjected to forces. In metallic conductors, the moving charges are electrons, and hence negative. On the other hand, the plasma in a neon lamp contains both moving electrons and positive ions. In semiconductor materials, likewise, both electrons and positively charged "holes" contribute to the current. Thus, current can be due to the motion of either positive or negative charge, or both.

Definition of Current

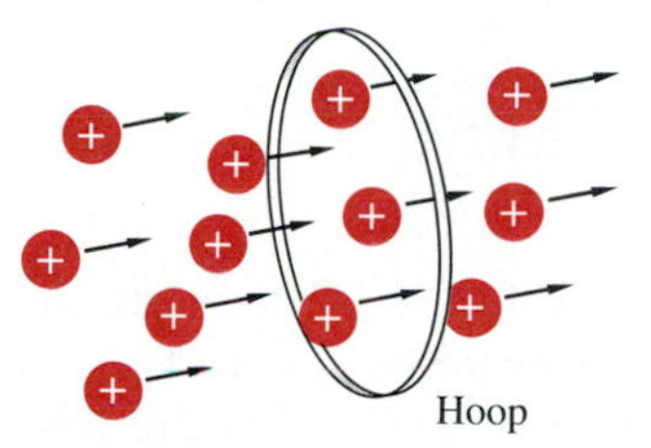

Figure 1
Charges moving through a hoop.

To develop a definition of current, suppose there is a cloud of positive charges in which all charges are moving along parallel paths as in Figure 1. Furthermore, imagine a hoop placed in such a way that charges pass through the hoop as time progresses; and also imagine you are sitting on the hoop and counting positive charges as they pass through the hoop. (This is essentially the same as sitting beside the road and counting cars as they go by.)

The average current equals $\Delta q/\Delta t$, where Δq is the amount of charge that passes by in a time interval Δt; in this expression, charge and time are, respectively, in coulombs and seconds. The physical dimension of current, hence, is coulombs per second (C/s), but current has been accorded its own unit, the *ampere* (abbreviated A).[2] As an example, if charges are counted for half an hour and found to total 3600 C, then, since half an hour is 1800 seconds (s), the average current over this time interval will be:

$$\text{Average current} = \frac{3600}{1800} = 2\ \text{A}$$

An average taken over such a long interval as this will not tell in detail what happens during different parts of that interval. For example, it might be that during the first 5 minutes 1200 C would pass; during the next 10 minutes no

[1] This unit is named after Charles Augustin de Coulomb (1736–1806). A prolific experimental scientist, Coulomb combined accuracy of measurement and mathematical powers to discover fundamental relationships in mechanics, magnetism, and, most important for us, electricity. There are two "Coulomb's laws." The one of interest to us is the law of force between two electric charges: it is directly proportional to the charges and inversely proportional to the square of the distance between them.

[2] This unit was named after the French mathematician André Marie Ampère (1775–1836), who, through a set of ingenious experiments, developed the fundamental laws of electrodynamics, including what is now called Ampère's law.

charge would pass; during the next 5 minutes charge would actually go the other way in the amount of 800 C; and during the last 10 minutes 3200 C would pass. In the half-hour interval this would give:

$$q = 1200 + 0 - 800 + 3200 = 3600$$

Consequently, the average current would again be 2A, as computed previously. (Note that 1200, 800, and 3200 were all given in coulombs; hence the 3600 is obviously in coulombs, from the context. Thus, we did not specify it.)

A larger amount of information can be obtained for this half-hour interval by counting for smaller intervals of time, say, 15 minutes in this example. For the first and second 15-min (900-s) intervals, the average currents would be:

First 15 minutes: Average current = 1200/900 = 1.33 A

Second 15 minutes: Average current = 2400/900 = 2.67 A

The average over the two intervals is still 2 A. Further refinements are possible by using shorter measurement intervals, for example, 1 s, 0.001 s, or 10^{-6} s. This leads to the notion of *instantaneous* current, defined as the limit approached by the average current as the time interval over which the average is taken approaches zero. Thus, the instantaneous current is:

$$i = \lim_{\Delta t \to 0} \frac{\Delta q}{\Delta t} = \frac{dq}{dt} \tag{1}$$

At any given time t, then, the current at that instant is the time derivative of charge. In this expression, all variables are in their SI units: charge in coulombs, time in seconds, and current in amperes.

Since current is the time derivative of charge transferred, the amount of charge transferred over a period of time must be equal to the integral of the current over this time interval. Let $q(t)$ be the charge transferred from an initial time t_0 up to time t. Using x as a dummy variable of integration, the expression for the charge transferred over this time interval will be:

$$q(t) = \int_{t_0}^{t} i(x)\, dx \tag{2}$$

In speaking of charges "moving through the hoop," something must be said of the direction of motion. This motion can be in either one direction or the other. Therefore we arbitrarily choose a particular direction as the *reference.* If positive charges go through in that (the reference) direction, we say that the current is positive; if they go through in the opposite direction, we say that the current is negative. But how about negative charges?

Suppose a positive and a negative charge of equal magnitude move at the same rate in the same direction. The *net* flow of charge will be zero. This means that a negative charge moving in one direction is equivalent to a positive charge of equal magnitude moving in the opposite direction. As long as the amount of charge transferred can be found, it is unimportant in defining current whether positive or negative charges move.

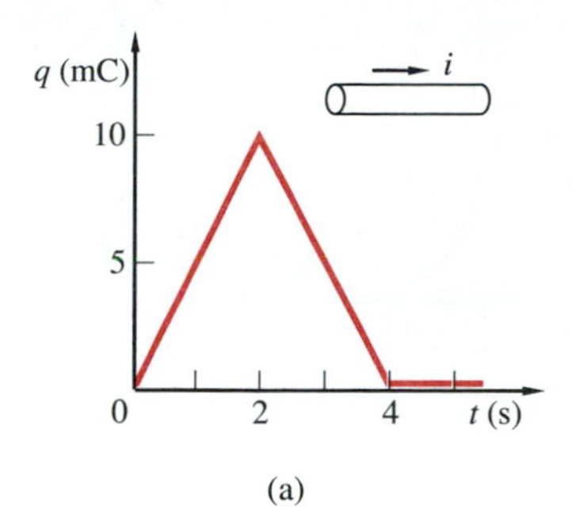

(a)

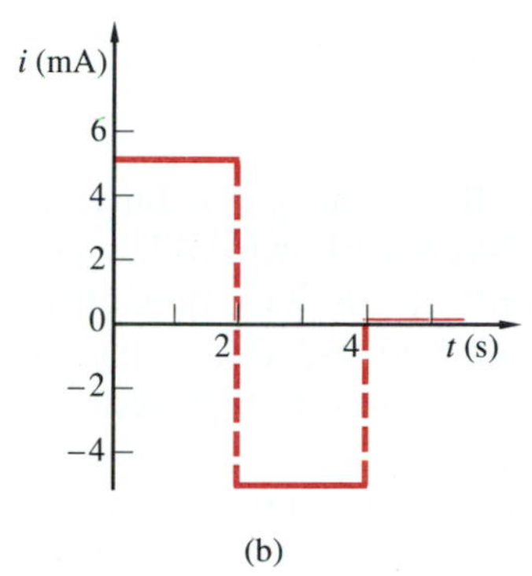

(b)

Figure 2
Current determined from a given charge.

Suppose that a positive charge q is transferred along a path in the direction indicated in Figure 2. The variation of q with time is as shown by the curve in Figure 2a. Then the current will be represented by the curve in Figure 2b, which, at every point (except $t = 0$, 2, and 4 s), is the derivative of the charge curve. At $t = 2$, the derivative of q is undefined; the slope is positive when approaching the point $t = 2$ from the right but negative when approaching $t = 2$ from the left. The current is discontinuous at this point. A similar description applies for $t = 4$ s.

In the real physical world, an abrupt change in a physical quantity like this may not be possible exactly; but it can be approximated so closely that nothing is lost by using the approximation. The discontinuity in current arises because of the sharp point in the charge curve. Suppose, instead, the tip of the triangle was slightly rounded. The derivative would then change smoothly but rapidly at $t = 2$. The current curve would resemble Figure 2b, but its mathematical description would be more complicated.

In Figure 2, suppose that it is the current curve that is given and we wish to find the charge transferred from $t = 0$ up to any time t. Thus, from 0 to 1 we find the charge transferred to be:

$$q = \int_0^1 5\,dt = 5t\big|_0^1 = 5 \text{ mC}$$

EXERCISE 1

Find the charge transferred from 0 to any time t lying in the two ranges (a) up to 2 s and (b) from 2 to 4 s.

ANSWER: (a) $q(t) = 5t$; (b) $q(t) = 20 - 5t$. Verify for $t = 2$. □

Defining a Current Reference

When charges are flowing in a discrete path, we arbitrarily assign a particular orientation along the path to be the reference orientation. The average current is called positive over a time interval if, during this time, the net charge transported along the reference orientation is positive, or, equivalently, if the net charge transported opposite to the reference orientation is negative. The reference orientation is indicated by an arrow drawn beside the path of charge flow.

The box shown in Figure 3a represents any electrical device with two terminals through which charges are flowing, thus constituting a current i. The arrow beside this box is the reference orientation for the current. We could

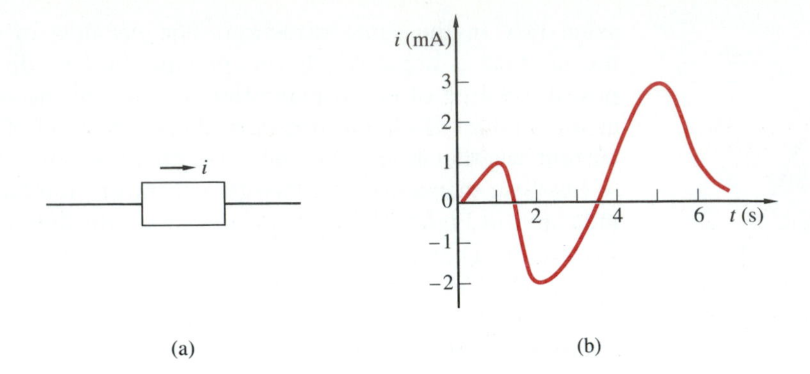

Figure 3
Representation of electrical device and variation of current with time.

just as easily have chosen the opposite reference. If, at any given time, the actual current is to the right, then i with the reference shown will have a positive value at this time. Later, if the actual current is to the left, then at that time i will have a negative value. In either case, from a knowledge of the sign of i and its reference orientation, we will know whether the actual current is to the right (i positive) or to the left (i negative).

The term *orientation* will be used instead of *direction* for current. This is a more accurate description, because the current path can be curved and accordingly not have a single direction. For example, a horizontal conductor might have a bend causing part of the conductor to be vertical. Although the *direction* of charge flow is horizontal over part of the path and vertical over another part, its *orientation* along the path is the same. When there is no possibility of ambiguity, instead of using the full description *current reference orientation* we shall simply say *current reference.*

How we choose the reference for current in any particular path is unimportant. What is important is to recognize that a reference *must* be chosen. The arrow along a path just means that we have agreed to call the current positive when positive charges are flowing in the arrow direction, and otherwise, negative. For example, Figure 3b shows the variation of a current i with time. Whenever the curve is above the horizontal axis, the current is flowing to the right in Figure 3a, and whenever the curve is below the horizontal axis, it is flowing to the left.

If we wish to measure the current, we should have a meter that tells us both the amount of the current and its orientation. Let us think of a meter (not digital) with a needle that can swing in either direction about its equilibrium (center) point, which corresponds to zero current. A sketch is shown in Figure 4.

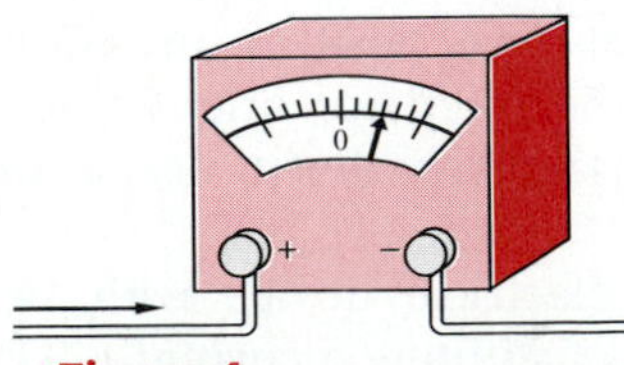

Figure 4
Current meter.

The meter can be connected in two ways. When it is connected one way, the needle will swing in one direction for a given current. If the connections of the meter are reversed, the needle will swing in the other direction for the same current. Again, we have a choice of how to connect the meter. How we do it is immaterial, as long as we agree on the correspondence between the direction of the current and the direction of swing of the needle. This agreement is

made by placing some mark on the terminals of the meter, say, a + on one of the terminals. This indicates the direction of swing of the needle when, at a given time, the current enters the + terminal. Let us agree to connect the meter so that the needle swings to the right (clockwise) when the current enters the + terminal. The current reference for a given path is then toward the + terminal of the meter from the outside (or from the + terminal to the unmarked terminal through the meter) when it is connected in this manner.

3 VOLTAGE

The existence of electrical phenomena depends on electrical charges becoming separated, with some objects being positively charged and others negatively charged. Any of the various devices that act as sources (such as batteries, generators, or thermocouples) have the fundamental property of being able to accumulate positive charge at one terminal and negative charge at the other terminal. These separated electrical charges exert forces on each other in accordance with Coulomb's law—so-called *Coulomb forces.* These forces are balanced by the forces internal to the generating devices, which cause the charge separation in the first place (e.g., forces that result from chemical, thermodynamic, or quantum mechanical phenomena, as in an electrolytic cell, a thermocouple, or a photocell; or from a time-varying magnetic field or the motion of a charge in a magnetic field, as in an alternator). These internal forces act against the Coulomb forces and thereby cause energy to be stored in the configuration of charges so created. This stored energy is (*electric*) *potential energy.*

Potential Energy

Now if an external conducting path is provided between the terminals of a generating source, as in Figure 5, the Coulomb force exerted on the separated charges will cause them to flow back together through that conducting path, as illustrated in the figure. When the charges flow from one terminal to the other, they do work and thereby lose potential energy.

The potential energy that inheres in a particular configuration of charges depends only on that specific configuration, that is, on the specific positions of the charge. That means that, if a different distribution of these charges is brought about by causing one or more charges to move to new positions, the potential energy will change to a value inherent in the new distribution. It

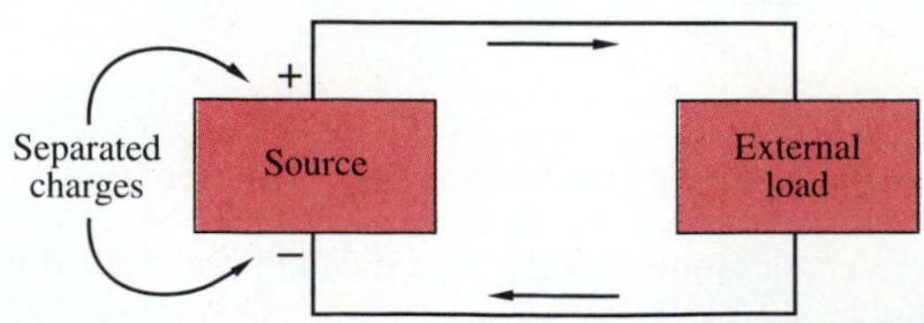

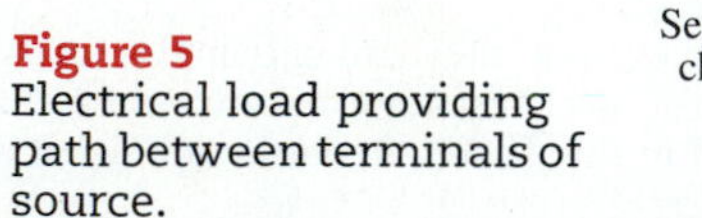
Figure 5
Electrical load providing path between terminals of source.

makes no difference to the change in potential energy of a charged particle how—by what path—it has reached its position, only that it occupies that position among all the other charges. In this respect, electric potential energy is like gravitational potential energy.

For example, if you walk from one point on the side of a hill to another point at a different elevation, the work you do against the force of gravity (which equals the change in gravitational potential energy) will be independent of your route, even if you go to the bottom of the hill on the way. Since the change in potential energy is independent of the path between the two points, it is uniquely a function of those endpoints. In particular, if those endpoints are identical (that is, the path is closed), the change in potential energy "between" them must be zero. In other words, the work done by *electrostatic* (Coulomb) forces acting on a charged particle traversing a closed path is always zero.[3]

Definition of Voltage

Now consider an electrical device represented by the box shown in Figure 6a. It is part of an interconnection of physical devices, the remainder of which is not shown. (An interconnection of electrical devices will be called a *circuit.* Shortly, we will define this term more carefully; for the moment just consider it shorthand for an interconnection of electrical devices.) A state of charge separation exists in the circuit, and so electrostatic forces exist.

To be specific, assume these forces act in the direction from A to B. Imagine a particle, having positive charge Δq, located at point A. It could have arrived there by having been pushed through a source of charges or moved in some way through the air from B to A. By virtue of the property of the electrostatic force just noted, the potential energy of the charge is the same, whatever the route by which q has reached A. Now imagine the charge moving through the box. Its potential energy will *decrease* by an amount that we'll call Δw_{AB}. This is also the amount of work done by the electrostatic force in moving q from A to B and is equal to work previously done by some unspecified agent when moving the charge from B to A.

Just as the decrease of potential energy of a falling object located in a gravitational field is proportional to the mass of the object, so also the decrease of electrical potential energy Δw_{AB} of the charge just described is proportional to the charge Δq. (Strictly speaking, Δq must be small so that its motion does not disturb the distribution of other charges.) Then, the quantity $\Delta w_{AB}/\Delta q$ is unique; that is, it is independent of Δq (so long as Δq is very small). In the

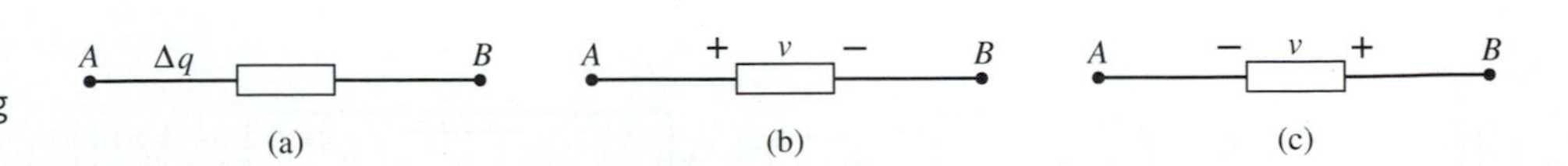

Figure 6 Definition and designation of voltage, including reference.

[3] This conclusion is true if charges are stationary or even if they are moving at constant velocity, that is, in a steady current. It is no longer true if charges not only have a velocity but are accelerating. For the circuit conditions studied in this book, the latter condition does not hold.

limit as Δq becomes infinitesimal, this quantity is given a name:

$$v_{AB} = \frac{dw_{AB}}{dq} \tag{3}$$

From the form of the equation, it is seen that v_{AB} is the *decrease, from A to B, of potential energy per unit charge,* but it is customary to drop the words *energy* and *per unit charge* and call it merely *potential decrease* (or *drop*) from A to B. This potential decrease is more commonly called the *voltage* from A to B. The unit of voltage is the *volt* (V).[4]

The order of the subscripts (A to B, for example) is related to the direction the "test" charge Δq is imagined to move. This point is emphasized because, in the example given here, the direction of the electrostatic force is also from A to B and Δw_{AB} and v_{AB} are positive numbers. In the event the electrostatic force is directed from B to A, we can still speak of the voltage from A to B, but Δw_{AB} and v_{AB} will be negative numbers in this case.

Voltage Reference

Since the definition of voltage involves two points, there must be some method of designating *from* which point *to* which point there is a potential *decrease.* This can be done with the use of double subscripts. Thus, for the electrical device in Figure 6b, v_{AB} specifies the potential decrease *from A,* the first subscript, *to B,* the second subscript. In a particular situation, v_{AB} may turn out to be positive or it may turn out to be negative. If v_{AB} is positive, it means that there is *actually* a decrease in potential in going from A to B. In this case, point A is at a *higher* potential. If v_{AB} (the decrease in potential energy) is negative, it means that there is *actually* an increase in potential in moving from A to B, and so point A is at a lower potential.

Even though—in the definition of voltage—there is the idea of motion of a *test charge* from one point to another, unlike current, voltage *does not* "flow." We say "the current *through*" such and such a path, but we cannot use the same word for voltage, since it does not "flow through." We say instead "the voltage *across*" the two points. For voltage, the word *polarity* corresponds to what is called the "direction of flow" in the case of current. When we were speaking of current, however, we introduced the term *orientation* to replace *direction* because it describes the situation more accurately. We shall use the same word *orientation* to stand for polarity. In this way, we can refer to the orientation of both current and voltage, even though for current its meaning is direction and for voltage its meaning is polarity.[5]

[4] This unit is named after Alessandro Volta (1745–1827), who in 1800 invented the electric battery (formerly called a *voltaic pile,* after him). This device, for the first time, provided a source of electric current that scientists could use for studies of electricity. However, Volta himself did very little scientific work to explain the generation of voltage by his pile or to determine the nature of the electricity so produced.

[5] The same concepts of "through" and "across" variables are valid in other systems—for example, mechanical, thermal, acoustical—as well. In acoustical systems, pressure is an "across" and flow is a "through" variable. In mechanical systems, force is an "across" and velocity is a "through" variable.

The double-subscript notation is cumbersome. Another way of designating the initial and final points in the definition of voltage is to use + and − markings, as in the diagram in Figure 6b, together with a symbol for voltage, say, v. If we write v and put a plus sign at point A and a minus sign at point B, it is the same as writing v_{AB}.[6]

The plus and minus signs (or just the plus sign) constitute a *voltage reference orientation* (or simply voltage reference). The reference is chosen *arbitrarily.* To specify the voltage reference for two terminals is to specify at which terminal of the device the plus sign is placed. If v is to be the voltage of a branch, it is necessary to show both the symbol v and its reference, namely, the plus and minus (or only plus) signs, as shown in the diagram of Figure 6c. In this diagram, with the reference shown, v is the same as v_{BA}. In Figure 6b, because the reference was different, v was the same as v_{AB}. Of course, v is an algebraic quantity that may be positive or negative. When v is positive, the *actual* polarity of the voltage is the same as its reference orientation. When v is negative, the *actual* voltage polarity is opposite to its reference.[7]

4 ENERGY AND POWER

Our interest in electrical circuits lies in the fact that they can transmit energy and process information, or signals. So the question arises how voltage and current are related to energy and information. We will not take up the theory of information in this book.[8]

As for energy, a common definition is that *energy is the ability to do work*; but this statement lacks quantitative measure. Perhaps your earliest exposure to the idea of energy is with the form called mechanical energy. Here we find two types: *kinetic energy,* associated with motion, and *potential energy,* associated with position (or location). You should also be familiar with the idea that heat is a form of energy. The counterparts of these types are also found in electrical energy. In SI units, the unit of energy is the *joule*; its

[6] Of course, the plus sign alone is sufficient, since the other point will carry a minus sign by implication. Similarly, any other mark, or pair of marks, could be used. It is customary, however, for a DC voltmeter to carry a plus sign on one terminal. If a voltmeter is connected across a branch so that the plus sign on the voltmeter and the plus sign on the branch coincide, then the voltmeter will read up scale when v is positive.

[7] In many books, you will see the terms *voltage drop* and *voltage rise* used. These are useful terms if properly employed. As voltage has been defined here, with its polarity reference, voltage *is* a "drop," a *decrease* in potential. Discussing a voltage *rise* must mean that the reference has been reversed. Sometimes the terms are used to describe the *actual* polarity of the voltage in a particular case. For example, one might say that "the voltage from point B to point A is a rise of 10 volts." In the terminology used here, this would mean that if the voltage reference polarity is + at point A and − at point B, then, for the given case, the value of the voltage is -10 V.

[8] The concept of information is complex; however, it can be made quantitative, and the "information" aspects of voltage and current as signals have been treated in advanced works which we, needless to say, will not consider. However, some aspects of the processing of signals (by amplifying, filtering, modulating, or shaping) will be taken up subsequently.

dimensions are *newton-meters,* as determined by the fact that the mechanical energy supplied to a body is the product of the force exerted (newtons) times the distance moved (meters).

Energy in Terms of Voltage

The concept of voltage between two points was introduced earlier through a consideration of the work done by an electrostatic field when a positive charge moves between the two points. Specifically, the voltage from point A to point B is the energy expended per unit charge by an electrostatic field when the charge moves from A to B. If work is done by the field as the charge moves, the potential energy will decrease.

Now, instead of a single charge, suppose there is a stream of positive charges, constituting a current, moving from A to B through an electrical device that is part of a circuit, as in Figure 7. The voltage and current reference orientations are as shown. (Note these carefully.)

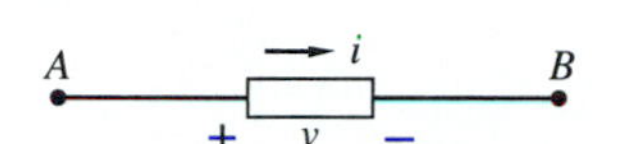

Figure 7
Charges constituting a current moving from A to B.

If a positive charge moves through a device from one point to another of lower potential, the electrostatic field created by the rest of the circuit to which the device is connected has expended energy. This energy is *delivered to* the device, where it is transformed to some other form: for example, stored as chemical energy (as in charging a battery), or changed to heat.

If, instead, the positive charge moves in a device from a point of lower to a point of higher potential, the device will be delivering energy to the electrostatic field (that is, to the rest of the circuit).

Suppose that a positive charge of $q = 3$ C goes through the device from A to B in 2 s. The amount of energy expended by the electrostatic field would be $w_{AB} = 3v$ joules, where v is the voltage from A to B. The average current during this time will be $\frac{3}{2}$ A.

Now suppose the charge moving from A to B is infinitesimal, say, Δq, and the time it takes for it to go from A to B is Δt. The voltage is still v. The energy expended by the rest of the circuit will be $\Delta w_{AB} = v\,\Delta q$. During the interval Δt the *average current* equals $\Delta q/\Delta t$. If the expression for the energy expended, $\Delta w_{AB} = v_{AB}\,\Delta q$, from which (3) was obtained is divided by Δt, we get:

$$\frac{\Delta w_{AB}}{\Delta t} = v\frac{\Delta q}{\Delta t} \tag{4}$$

In the limit as Δt approaches 0, this becomes:

$$\frac{dw_{AB}}{dt} = v\frac{dq}{dt} = vi \tag{5}$$

In other words, the rate at which energy is being expended by other parts of the circuit in which this device is connected (which means the rate at which energy is delivered to the device) is the product of the voltage across the device and the current through it, assuming the references for v and i are those shown in Figure 7.

Power in Terms of Voltage and Current

The term *power* is defined as the *rate of energy transfer,* or the *rate of doing work,* because, when work is being done, energy must be transferred. We should be careful to specify the *sense* of the energy transfer, that is, whether the energy is being expended by the rest of the circuit and being transferred *to* the box in the preceding diagram or the opposite. We shall define *power delivered to* a device as the rate at which energy is expended *by* the circuit to which the device is connected. It will be represented by p. The relationship between power and energy, therefore, is:

$$p(t) = \frac{dw(t)}{dt}$$
$$w(t) = \int_0^t p(x)\,dx \tag{6}$$

It is assumed that the origin of time is chosen so that the energy delivered to the device up to time $t = 0$ is zero. Hence, the previous expression $dw_{AB}/dt = vi$ becomes:

$$p = vi \tag{7}$$

Reexamine the diagram to note the particular combination of references for v and i for which $p = vi$. The voltage reference plus is at the tail of the current reference arrow.[9] $p = vi$ is a fundamental expression relating power to voltage and current. The unit of power is the *watt* (abbreviated W).[10] From the definition of power as the rate of energy expenditure, 1 watt = 1 joule per second (1 W = 1 J/s). In terms of the units of voltage and current, 1 watt = 1 volt-ampere.

In Figure 7 suppose i and v are simultaneously positive at a particular time. The power actually delivered to the branch will then be positive at that time. At another time, suppose v is numerically negative while i is positive. The power delivered to the branch will then be negative. Thus, the power p delivered to (or absorbed by) a branch can be either positive or negative. If p is negative, this means that power is actually being supplied *by* the branch *to* its embedding circuit.

The expression $p = vi$ refers to the power delivered to a branch for a

[9] This description might lead to confusion if care is not taken in the placement of the reference symbols. It will be assumed that the current reference arrow is *always* placed alongside the branch to which it pertains, not in the leads connecting that branch to others. Thus, the voltage plus sign will always be at one end of the arrow and the minus sign at the other end. Even if the current reference arrow is actually placed in the lead, it must be mentally slid along to its proper resting place when considering the current reference relative to that of the voltage.

[10] Named after James Watt (1736–1819), considered by some the greatest British engineer. While still young, he learned the craft of mathematical instrument making. Although not the first to conceive of a steam engine, he invented the improvements at the heart of the steam engine we know now. His many other inventions include a document-copying machine which remained the office standard for a century. When water was still considered an element in 1783, Watt, the instrument maker and engineer, proposed in a paper to the Royal Society that water was a compound of what are now known as hydrogen and oxygen.

particular set of reference orientations for v and i. Suppose one of the references, v or i but not both, is reversed, say, v. The voltage with the new reference is the negative of what it was. Hence, the expression for power delivered to this branch will be $p = -vi$.

Reference for Power

A distinction must be made between the *expression* relating p to v and i and the actual *sign* of the numerical value of p. We have defined p as the *power delivered to* (or the power *into*) a branch under consideration (by whatever else the branch is connected to). Suppose the references of v and i are chosen as given in Figure 7, with the voltage + at the tail of the current arrow. At a particular time, suppose the values of v and i are $v = 10$ V and $i = -3$ A. Is the expression for power delivered to the branch $+vi$ or $-vi$? Look at equation (7) and the diagram showing the v and i references. The voltage and current are functions of time. Even though the *numerical values* of v or i can become negative as time goes on, if the *references* remain as shown, the expression for power will be $p = +vi$. For the preceding numerical values of v and i, the *numerical value* of the power delivered to the branch is $p = -30$ W. This means that, at the time in question, since the power delivered to the branch is negative, the branch itself is not absorbing power delivered to it, but is actually delivering power.

As current flows through a branch, either the branch is *absorbing* the power delivered to it or it is *delivering* power. The power *absorbed by* the branch is the same as p, the power delivered *to* the branch.

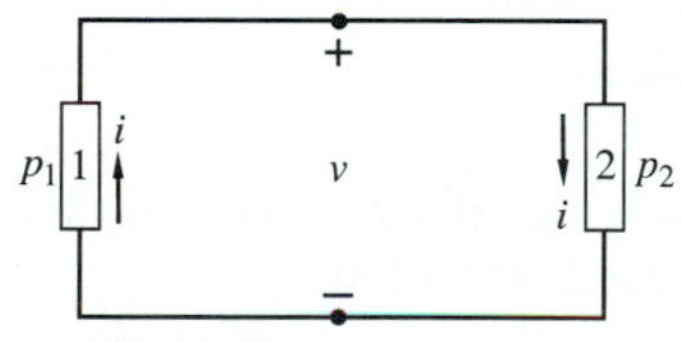

Figure 8
Power related to voltage and current.

Figure 8 shows two branches connected together. They have a common voltage v and a common current i. For emphasis, two reference arrows are shown for i, one beside each branch. Both arrows have the same orientation along the path of the current. Let p_1 and p_2 be the power delivered to branches 1 and 2, respectively. In terms of v and i, the expressions for p_1 and p_2 are:

$$\begin{aligned} p_1 &= -vi \\ p_2 &= vi \end{aligned} \tag{8}$$

(You should assure yourself that you agree with this before proceeding. Review the previous few pages if necessary.) When v and i are numerically both positive or both negative, p_2 is positive. The power is actually being delivered to branch 2.

Standard References

In common terminology, when power is actually being delivered to a branch, the branch is said to be a *load*. Thus, when v and i are positive in Figure 8, branch 2 can be considered a load.

For convenience, the set of v and i references for which $p = +vi$ is called a set of *standard references*. Referring back to the diagram, for a set of standard references, the voltage reference + sign is located at the tail of the current reference arrow. For a standard set of references (for short, we say standard references), p will be positive if v and i have the same sign. A distinction must be made between (1) the voltage and current references of a branch and (2) whether or not the branch is actually absorbing power at a particular time. Standard references might be chosen for a branch, but either v or i could be negative at some time. The power delivered to the branch at that time is negative; hence the branch is not then acting like a load.

Return now to Figure 8 and focus attention on p_1, which we had found to be $-vi$. At a given time, if v and i are positive, then p_1 is negative; hence, it is not a load at that time. Is power actually being delivered to branch 1 at that time? No, it is not; branch 1 is itself delivering power. When a branch is actually delivering power, it is commonly called a *source.* Thus, under the preceding conditions, branch 1 is acting as a source and $p = -vi$. The v and i references for which $p = -vi$ are *nonstandard references.* Looking back at branch 1 in Figure 8, nonstandard references can be described by saying the voltage plus is located at the tip of the current arrow. (See footnote 6.)

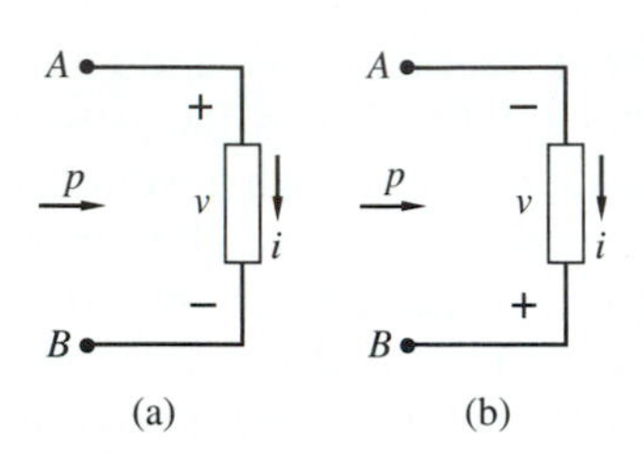

Figure 9
Power delivered to a device; (a) standard references ($p = +vi$); (b) nonstandard references ($p = -vi$).

The relationship between power and v and i is summarized in Figure 9. The arrow designating the direction of power flow is the *reference* for power, namely, *power delivered to* the box. For standard voltage and current references, $p = vi$; when the references are not standard, $p = -vi$.[11]

EXERCISE 2

Each of the branches shown in Figure 10 is connected to other branches (not shown). The voltage and current references and numerical values at particular times are as shown. Write the appropriate expressions for power p in terms of v and i and determine the numerical value of p in each case. Specify whether or not the references are standard and whether the branch is actually absorbing power or supplying it.

ANSWER: (a) $p = -vi = -60$ W; nonstandard; supplying. (b) $p = vi = 60$ W; standard; absorbing. (c) $p = -vi = 60$ W; nonstandard; absorbing. □

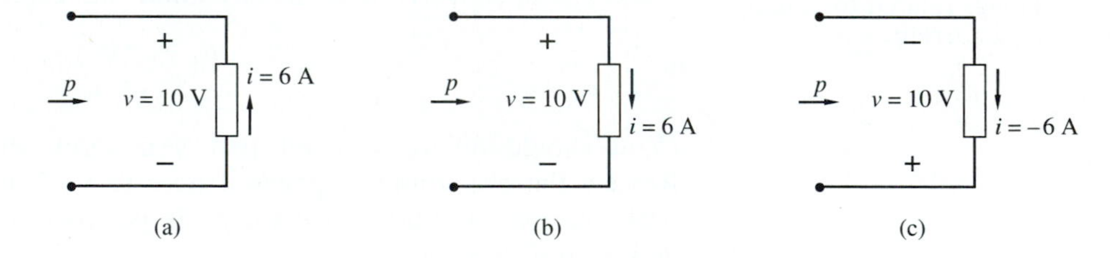

Figure 10
Power calculation.

5 ELECTRIC CIRCUITS AND MODELS

As stated in Section 1, in order to explain and predict the electrical behavior of an interconnection of actual (physical) electrical devices, we must construct a

[11] Since standard references were defined using the concept of a load, this set of references might also be called *load references.* Similarly, being based on the idea of a source, nonstandard references might also be called *source references.* But we will not use these designations. Some authors use the term *associated,* instead of standard, references.

mathematical model to represent the real world. Fortunately, we are not the pioneers; the models have already been constructed. In this book we will deal with the models.

The physical "electrical devices" we have mentioned a few times include such things as batteries, transformers, transistors, physical capacitors, integrated circuits, operational amplifiers, relays, motors, power generators, and many others. Such devices are interconnected by conductors (usually wires). An interconnection of physical electrical devices is referred to as a *physical circuit.* An example, representing an automobile ignition system, is shown in Figure 11a. Electrical properties of such devices have been identified after substantial study. Each elementary property (such as the ability to dissipate heat or to store energy) has been represented by an idealized model, called an *element.* The elements are idealized (hypothetical) devices; an interconnection of electrical elements is called a *circuit.*

When this term was first used, it referred to a single closed path. We still talk of "closing a circuit," that is, making a connection that completes a closed path.[12] In present usage, the term *circuit* still carries the connotation, if not of just a single closed path, then of just a few closed paths—a rather simple interconnection of electrical components. When we think of a more substantial interconnection, we use the term *network.* In this book, we use the two terms more or less interchangeably, but when emphasis is needed to designate something large or more general, we will use the term *network.*

Under certain specific conditions, a single physical device may be modeled by a certain combination of circuit elements, but under other conditions, it may require a different model because the first model is not accurate enough under the new conditions.

A model of the automobile ignition system is shown in Figure 11b. The boxes represent elements. Note that all but one of the elements have two terminals while the model depicting the most important property of the transformer has three. (A transformer consists of two adjacent coils of wire each having two terminals, as shown, but in the diagram one terminal of each coil is connected to a terminal of the other, reducing the total to three terminals. Chapter 10 is devoted to the subject of transformers.)

Note also that there are two elements not obviously present in the physical circuit. These two elements (coil and wire resistances) are models of properties of the physical circuit which may have to be taken into account. In the diagrams representing both the physical circuit and the model, the devices are interconnected by what appear to be the same kinds of lines. In the physical circuit, the lines represent physical wires. Physical wires are never perfect

[12] In the United States there is a judicial system called the *circuit court.* It was established early in the life of the country when the population was thin and the small towns scattered around did not warrant a full-time judge to be in residence. An itinerant judge would "ride the circuit" on his horse. He (yes, the judge was a male) would be in one town one day and ride off to the next town the next day, and so on, until he traversed a complete loop and completed the circuit. The judges no longer travel but the territories they used to cover on horseback are still called circuits and the judges are still *circuit-court judges*.

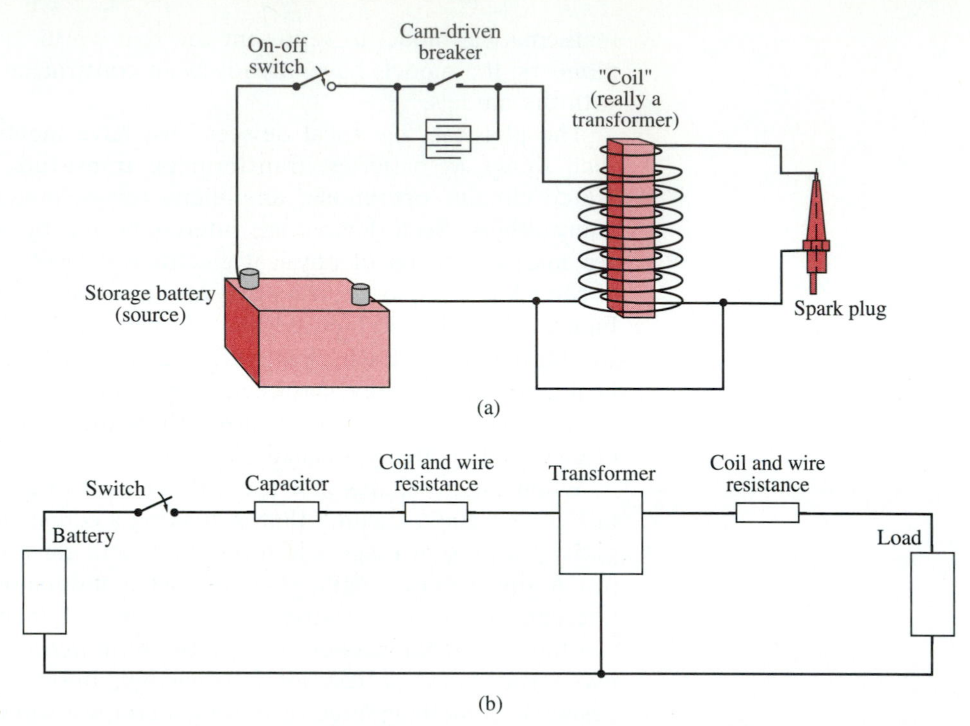

Figure 11
(a) Automobile ignition system and (b) its model.

conductors. (If you ever iron your clothes, feel the cord sometimes; it will be warm, showing that the cord dissipates heat and is not a perfect conductor.) However, most of the time, a model in which this imperfection is neglected will be sufficiently accurate.

The lines in the ideal model, on the other hand, do not represent wires; they are simply convenient ways for showing the connections. They have no other electrical effects. In Figure 11b, for example, the elements can be rearranged so that the bottom terminals of the models of the battery, the transformer, and the load are all connected directly together without any intervening lines, but the diagram would look awkward. (Do it.) We will subsequently introduce a number of specific circuit elements and their representation in a circuit diagram. For the immediate purpose here, it is not important what specific elements are contained in a circuit, and so we shall represent them as "boxes" with two or more terminals.

In the circuits to be considered in this book, we assume that all electrical effects of physical electrical devices—including those of the connecting wires—are localized, concentrated, or *lumped* in the ideal devices. Another assumption made is that the effect of a change in current or voltage in the circuit is felt instantaneously anywhere else in the circuit. Thus, in the automobile ignition system, when the switch is turned on, we assume that the effect of the battery voltage is felt throughout the circuit immediately. In this particular example, the "lumped" assumption is an excellent approximation to reality. You will need more study of both circuits and electromagnetic fields to determine in specific cases the conditions under which the lumped approximation holds.[13]

6 KIRCHHOFF'S CURRENT LAW

We have just defined a (lumped) electric circuit to be an interconnection of idealized models of physical electrical devices called circuit elements. Each type of element is pictured with a particular symbol. For the present purpose, however, it is not important what the particular types of elements might be, just that there are some elements interconnected in a circuit. As previously described, we show each element by a box with as many terminals as the element has.

One such circuit is shown in Figure 12a. A point at which two or more elements are connected is called a *node* and is represented in the circuit by a heavy dot (clear or darkened). The node labeled A and the elements connected to it are shown in Figure 12b. We assume that the circuit is energized and currents exist in the elements. The currents associated with

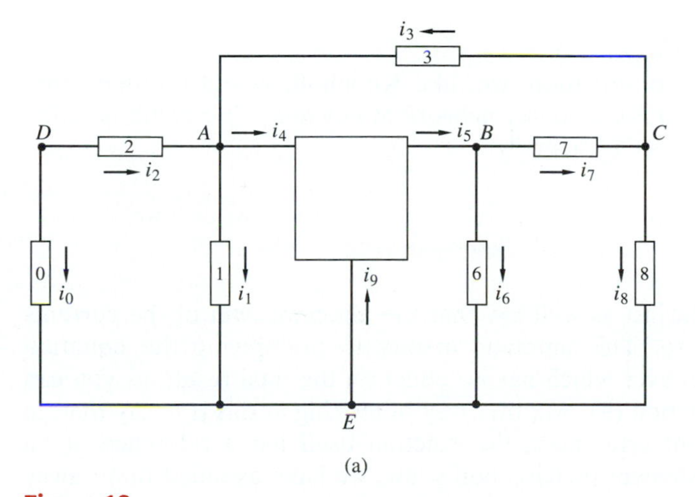

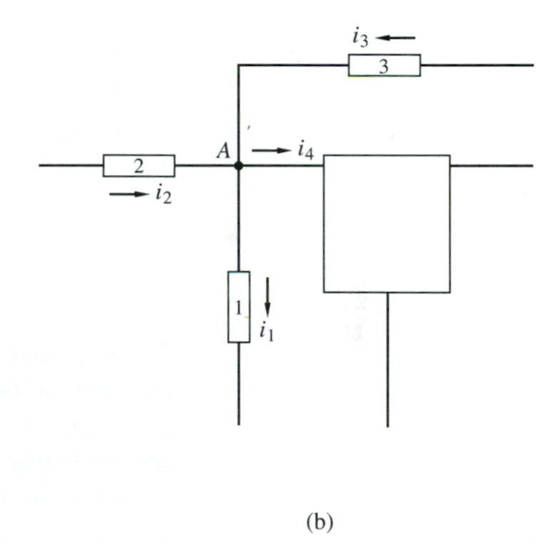

Figure 12
An electric circuit and one of its nodes, showing all connected branches.

[13] Electromagnetic waves travel at the speed of light: $c = 3 \times 10^8$ meters per second. The frequencies in the AM broadcast band lie between about 600 and 1500 kilohertz (kHz). The shortest wavelength in this band is, therefore, $3 \times 10^8/1.5 \times 10^6 = 200$ meters. The circuits in a modern radio receiver have dimensions of less than 20 cm (0.2 m); that's more than 1000 times smaller than the shortest wavelength. Thus, the electromagnetic wave has almost the same value over all parts of the circuit. The lumped approximation is valid whenever the dimensions of a circuit are much smaller than the relevant wavelength. (If this explanation makes sense to you, fine; but don't worry if it's still a little hazy. The haziness will clear up after more study.)

node A are labeled i_1 through i_4; their values will generally be varying with time. As previously discussed, the arrow associated with each current is its reference orientation and has nothing to do with the actual direction of current flow at any given time. In what sense we choose a current orientation is unimportant; it is chosen arbitrarily. What is important is that a reference *must* be chosen.

The foundations of circuit theory were laid in the 1840s by Gustav Kirchhoff after repeated observations of the phenomena that resulted when the simple electrical devices of his day were interconnected.[14] If we were to make measurements in the physical circuit of which Figure 12a is a model, with instruments connected according to the references shown, we would find, as did Kirchhoff, that:

$$i_1 - i_2 - i_3 + i_4 \approx 0 \tag{9}$$

at all instants of time, within the limits of accuracy of the instruments. Note that the references of i_1 and i_4 are directed away from the node and those of i_2 and i_3 are directed toward the node. If we were to repeat the experiment at any other node, we would find each time a similar result, namely, that the algebraic sum of the currents is approximately zero.

One Form of KCL

On the basis of this observation, we, like Kirchhoff, would postulate that such a result is true at *all nodes* in *any network* at *any time.* This result is called *Kirchhoff's current law* (KCL), namely:

At each node in any electrical network and at each instant of time, the algebraic sum of all currents leaving a node is zero.

Of course, we might just as well say that the algebraic sum of the currents *entering* a node is zero. This amounts merely to multiplying the equation through by -1, a maneuver which has no effect on the final result, as you can see by trying it on equation (9). Another way of looking at this is to say that, in addition to the element references, the junction itself has a reference or an orientation. In our statement of Kirchhoff's law, we have assumed that "away from the junction" is the junction reference. If an element reference coincides with the junction reference, the symbol for the corresponding current in an application of KCL is preceded by a plus sign; otherwise, a minus sign. This scheme was used in writing equation (9). Note that there is nothing basic about this junction reference; there is no rule to be memorized here.

[14] Gustav Robert Kirchhoff (1824–1887) was a German professor of physics. His first research work was in the conduction of electricity. His circuit laws were presented when he was just 21 years old. He went on to make major contributions in many areas of physics, including spectrum analysis and the diffraction of light, analytical dynamics, and wave phenomena.

Conservation of Charge

If (9) is true, it must remain true if we integrate it over some time interval, say, t_1 to t_2. The intergration result is:

$$\int_{t_1}^{t_2} i_1\, dx + \int_{t_1}^{t_2} i_4\, dx - \left(\int_{t_1}^{t_2} i_2\, dx + \int_{t_1}^{t_2} i_3\, dx\right) = 0 \tag{10}$$

From the relationship between current and charge, each of the integrals represents the charge transported through the corresponding elements toward or away from the node (depending on the current references) in the time interval $t_2 - t_1$. Hence, the equation states that the net charge transported away from (or toward) the node, over any interval of time, is zero. That is, no charge can accumulate at a node, or be removed therefrom; this is the *principle of conservation of charge.* This conclusion is valid no matter how long the time interval is, *even if it is infinitesimally small.* This means that charge must be a continuous function of time. This is a fundamental principle of science. Indeed, we could have started with this principle and derived KCL from it as a consequence. (Do it.)

Let us return to the circuit of Figure 12 and write expressions for Kirchhoff's current law at each of the other nodes in the circuit. (The equation for node A is repeated.)[15]

Node A:	$i_1 - i_2 - i_3 + i_4$	$= 0$
Node B:	$-i_5 - i_6 + i_7$	$= 0$
Node C:	$i_3 - i_7 + i_8$	$= 0$
Node D:	$i_0 + i_2$	$= 0$
Node E:	$-i_0 - i_1 + i_6 - i_8 + i_9$	$= 0$

If all these equations are true, then their sum is also true. Adding them all leads to

$$i_4 - i_5 + i_9 = 0$$

What does this equation represent? A glance at Figure 12a discloses that this equation can be read as follows: the net current entering (or leaving) the three-terminal element is zero. This is certainly not unexpected; it is a consequence of the principle of conservation of charge.

This result can be generalized to elements having any number of terminals:

[15] There will be many occasions in this book when a set of equations will be obtained from a circuit diagram by applying fundamental laws or derived relationships. If you are to follow the development—and ultimately become capable yourself of generating similar equations for other circuits—you must understand each step of the process. One approach is to examine each term in the given equation and confirm in your own mind how it was obtained, and whether it has the correct sign and the correct dimensions. Another approach is to independently carry out the stipulated process (in the present example, applying KCL) and then compare your result with the one given in the book. If you don't do this, you will be giving yourself a severe handicap.

At any instant of time, the net current leaving (or entering) an electrical element having any number of terminals must be zero.

For a two-terminal element, this is self-evident, since the current enters one terminal and leaves the other.

EXERCISE 3

Using boxes to represent the elements, draw a circuit having four nodes and as many two-terminal elements as needed to connect each node to every other node. How many elements will there be? Assign current references to each element and write KCL equations at each node. □

Another result is evident from the preceding set of equations. Since adding the KCL equations at all the nodes yields a sum which is zero, the KCL equation at any one node is the negative sum of the KCL equations at all the other nodes. This conclusion is a general result applicable to all circuits, as we will demonstrate in Chapter 5. It means that the KCL equations are not all independent, since one equation can be deduced from the others. Confirm this conclusion with the circuit you constructed in the preceding exercise.

Alternative Form of KCL

The "syntax" of KCL as described earlier in this section is, "An algebraic sum of certain quantities equals zero." (Some wit in the class was overheard to say that, since KCL adds up to nothing, why bother with it in the first place?) Simple algebra permits transposing terms from one side of an equation to another. Thus, KCL at node A, for example, can be written:

$$i_1 + i_4 = i_2 + i_3$$

We observe from Figure 12 that the reference orientations of i_1 and i_4 are away from the node and those of i_2 and i_3 are toward the node. Thus, another way of stating KCL is the following:

At each node in an electrical network and at all times, the sum of all currents with references oriented away from a node equals the sum of all currents with references oriented toward the node.

As an exercise, use this alternative form to write KCL equations at the other nodes in Figure 12. Then transpose terms with negative signs in the previous KCL equations at nodes B, C, D, and E to confirm your results.

Double-Subscript Notation

It is sometimes convenient to use an alternative system for indicating references for current (as it is for voltage also). In Figure 12, for example, the reference arrow for i_7 is directed from node B to node C. The same information about this current and its reference can be given by using a name for the current which includes information about its reference. Thus, the double-subscript notation i_{BC} specifies the current in the branch lying between nodes B and C, with reference directed from B (the first subscript) to C (the second); $i_{BC} = i_7$. The reversal of the two subscripts means reversing the reference orientation; thus $i_{CB} = -i_{BC} = -i_7$. This notation is sometimes handy.

7 KIRCHHOFF'S VOLTAGE LAW

The concept of voltage was defined in Section 3. We observed there that the decrease in potential energy of an electric charge in going between two points is independent of the path taken. If the two points in question are the same point, then there is no net decrease in potential. This idea is the basis of a second fundamental law proposed by Kirchhoff.

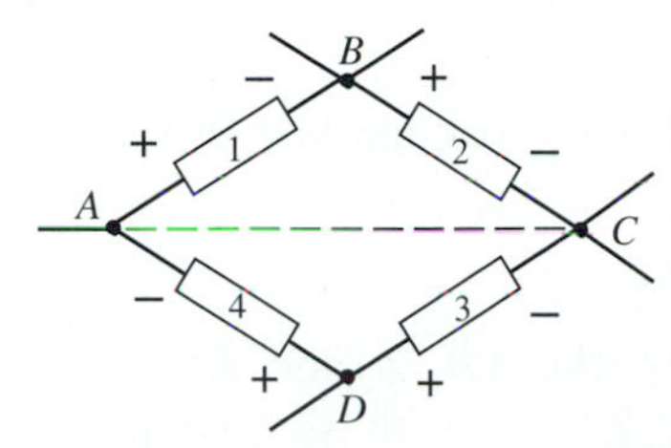

Figure 13
Portion of a circuit.

Consider the portion of a circuit—which may be part of a larger circuit—shown in Figure 13. The voltage from A to C is the decrease in potential energy of a test charge in moving from A to C along any path. Using double-subscript notation, along the upper path, through B, we can write:

$$v_{AC} = v_{AB} + v_{BC} \tag{11}$$

That is, the decrease in potential energy from A to C is the decrease from A to B plus the decrease B to C. Similarly, along the lower path:

$$v_{AC} = v_{AD} + v_{DC} \tag{12}$$

Since voltage is independent of path, these two expressions must be equal. Thus:

$$v_{AB} + v_{BC} = v_{AD} + v_{DC} = -v_{DA} - v_{CD}$$

On the right-hand side, the order of subscripts in each term has been reversed, so the signs must be changed. Transposing terms from right to left leads to:

$$v_{AB} + v_{BC} + v_{CD} + v_{DA} = 0$$

That is, the sum of voltages of all branches encountered on a closed path equals zero.

The order of the subscripts shows that the traversal of the closed path in this case is made in the clockwise direction. However, multiplying the preceding equation by -1 and reversing the order of subscripts will give the same result (namely, that the sum of voltages equals zero), but now the traversal of the path is counterclockwise. (Do it.)

While using double-subscript notation facilitates understanding the preceding result, this notation is cumbersome. It is more convenient in the analysis of networks to number the branches and to identify the voltages of the branches themselves rather than identify voltages by the branch terminals. Thus, in Figure 13 the branch voltages would be labeled $v_1, \ldots, v_4$, with voltage references chosen arbitrarily. Let us again add potential decreases in traversing the closed path starting at node A, passing through B, C, and D and returning to node A. The potential decrease from A to B, for example, is $-v_1$, in view of the chosen reference for v_1.

Completing the summation of voltages yields

$$-v_1 + v_2 - v_3 + v_4 = 0 \tag{13}$$

If the path is traversed in the counterclockwise direction instead of clockwise, the only difference would be that the equation would be multiplied by -1.

What has been done for any one closed path in a circuit holds true for any other closed path. Clearly, though, such closed paths must exist. The circuit must be *connected,* by which we mean that there must exist a path from one node in a circuit to any other node through elements of the circuit.

One Form of KVL

We can now make a general statement of Kirchhoff's voltage law (KVL):

The algebraic sum of all voltages encountered in traversing any closed path in a lumped connected circuit is zero at any instant of time.

For example, KVL would apply to the closed paths *ABCA* and *ADCA* in Figure 13. Indeed, (11) and (12) would express KVL for these two paths. However, the closed path *ABCDA* has a different character. All segments of this closed path consist of two-terminal branches of the circuit, whereas in the preceding two closed paths, this isn't the case; there is no two-terminal branch connecting *A* to *C* directly. On this basis, we define a *loop* in a circuit as *a closed path all segments of which consist of two-terminal branches.* If KVL applies to all closed paths in a circuit, it certainly applies to those closed paths which are loops. Back in Figure 12, for example, the closed path *AEDA* through branches 1, 0, 2 is a loop; so is the closed path *ACBEA* through branches 3, 7, 6, and 1. (See Chapter 5 for more extensive discussion of loops.)

EXERCISE 4

Write equations expressing KVL around the loops *AEDA* and *ACBEA* in Figure 12. Assume standard references for the voltages.

ANSWER: $v_1 - v_0 + v_2 = 0$ and $-v_3 - v_7 - v_6 - v_1 = 0$. □

Alternative Form of KVL

An alternative statement of KVL can be obtained in the following way. Equation (13) was obtained by applying KVL to the closed path shown in Figure 13. The expression can be rewritten by transposing terms, as follows:

$$v_2 + v_4 = v_1 + v_3$$

We note that, in a clockwise traversal of the closed path, the references for v_2 and v_4 are oriented consistent with the path orientation, while v_1 and v_3 are opposite to it—or consistent with the orientation of a counterclockwise traversal of the path. Hence, the following alternative statement of KVL can be made:

In a lumped connected circuit, the sum of all voltages along a closed path with references that agree with a clockwise traversal of the path equals the sum of all voltages whose references agree with a counterclockwise traversal of the path.

In different circuits, one or another way of viewing KVL may be simplest in arriving at the appropriate equations. Whenever we wish to find the voltage across a specific pair of nodes, for example, it is simplest to proceed as follows. Find a path between the two nodes and equate the sum of the voltages encountered along this path with the desired voltage, with due attention to references.

Illustrations

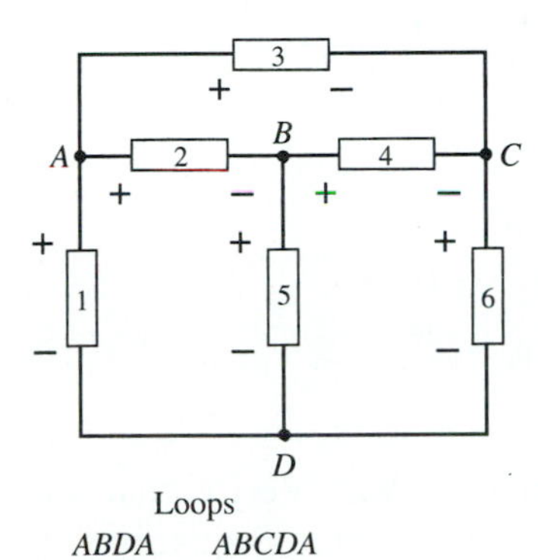

Loops
ABDA *ABCDA*
ABCA *ABDCA*
ACDA *ACBDA*
BCDB

Figure 14
Circuit with seven loops.

We will illustrate the KVL idea in the circuit of Figure 14. The branches have been numbered but, to avoid clutter, voltage symbols have not been shown. Voltage references have been chosen arbitrarily; v_1 is the voltage v_{AD}. One approach that might be entertained is to find all the closed paths that include branch 1 and to write a KVL equation around each one, and to do this for the other branches as well.

Besides the direct path, there are four other paths from node A to node D: *ABD, ACD, ABCD, ACBD*. Thus, four different equations expressing voltage v_1 in terms of voltages along these four paths can be written, as follows:

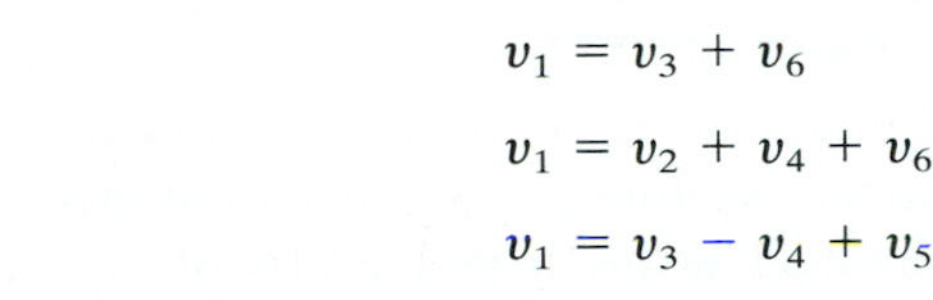

$$v_1 = v_2 + v_5$$
$$v_1 = v_3 + v_6$$
$$v_1 = v_2 + v_4 + v_6$$
$$v_1 = v_3 - v_4 + v_5$$

(Make sure you confirm these independently yourself.)

Similarly, voltage $v_{BD} = v_5$. Again, there are four other paths from B to D (*BAD, BCD, BCAD, BACD*), so four other expressions can be obtained stating that the voltage is independent of path:

$$v_5 = -v_2 + v_1$$
$$v_5 = v_4 + v_6$$
$$v_5 = v_4 - v_3 + v_1$$
$$v_5 = -v_2 + v_3 + v_6$$

Note that the first of these is identical with the first equation in the previous set. This is not surprising, since they both come from KVL around the same loop: *ABDA*. Similarly, the third of these is identical with the fourth in the previous set.

To continue, $v_6 = v_{CD}$. But from node C to node D there are four other paths: *CBD, CAD, CBAD, CABD*. Hence, a set of four equations similar to the preceding two sets can be written. (Do it. You will find that each of the equations is identical to one of the previous equations.)

Enough of this! This approach does not seem to be a useful avenue. We ought to be more systematic about this. Let us return to Figure 14, identify all the loops, and then apply one of the forms of KVL, say, the first form. Identifying three of the loops is easy: the inside "windows" consist of branches

{1, 2, 5}, {4, 5, 6}, and {2, 3, 4}. Another loop is also easily visualized: the one that encloses all the "windows," consisting of branches {1, 3, 6}. Finally, somewhat more difficult, we can find loops enclosing all pairs of the three windows. They consist of branches {1, 2, 4, 6}, {1, 3, 4, 5}, and {2, 3, 6, 5}. Traversing each of these loops in a clockwise sense leads to the following seven KVL equations:[16]

$$
\begin{array}{rrrrrrl}
-v_1 & + v_2 & & & + v_5 & & = 0 \\
 & & & v_4 & - v_5 & + v_6 & = 0 \\
 & -v_2 & + v_3 & - v_4 & & & = 0 \\
-v_1 & & + v_3 & & & + v_6 & = 0 \\
-v_1 & + v_2 & & + v_4 & & + v_6 & = 0 \\
-v_1 & & + v_3 & - v_4 & + v_5 & & = 0 \\
 & -v_2 & + v_3 & & - v_5 & + v_6 & = 0
\end{array}
$$

EXERCISE 5

For the same loops, apply the alternative form of KVL. Verify that the resulting equations can be obtained from the preceding set of equations by transposing terms. □

When we were considering the KCL equations written at the nodes of a circuit, we found that not all are independent, that any one equation can be obtained as the negative sum of all the others. It is also true that the KVL equations around all the loops in a circuit are not independent. In this case, though, the relationship among the equations is not as simple as it is for KCL. We shall consider this question more fully in Chapter 5.

[16] Don't forget: confirm each term in each equation. To help, you might draw partial diagrams showing each closed path isolated.

SUMMARY

Topics covered in this chapter include the following:

- The international system of units (SI)
- Electric current and its reference
- Voltage and its reference
- Electrical power and its relationship to voltage and current
- Electrical energy and its relationship to power
- Standard references
- Kirchhoff's current law (KCL)
- Principle of conservation of charge
- Kirchhoff's voltage law (KVL)

PROBLEMS

1. The current i in a circuit branch, with a particular reference, is given by the curve in Figure P1.

a. Find the positive charge q transported along the current orientation from $t = 0$ to $t = 2$ s.

b. Repeat part (a) for $t = 4, 6, 8$, and 10 s.
c. Sketch the function $q(t)$ up to $t = 10$ s.

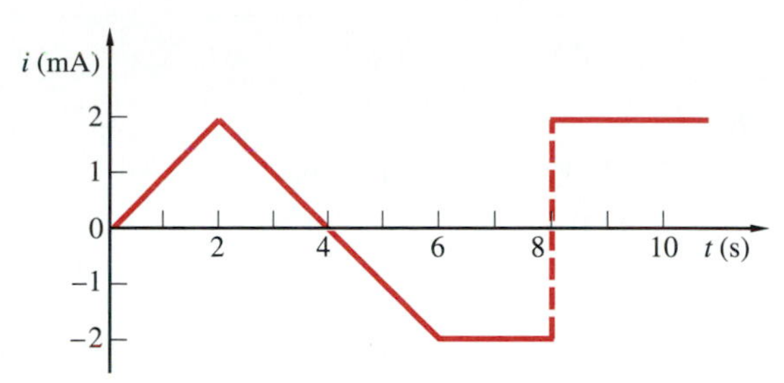

Figure P1

2. In each of the branches in Figure P2, find the value of power into the branch, including units.

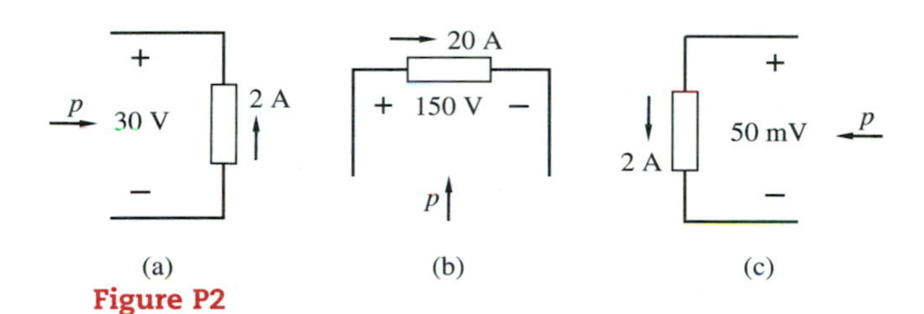

Figure P2

3. The current in a two-terminal circuit component is found to be $i(t) = 20te^{-2t}$ A.
a. Find an expression for the positive charge as a function of time flowing in the direction of the current reference.
b. Repeat for $i(t) = 10e^{-t/2} \cos 400t$.

4. The power delivered to a branch in a circuit is given in Figure P4a.
a. Sketch the curve of the energy delivered to the component from $t = 0$ up to time t, where t is any time up to 3 s.
b. Let the curve in Figure P4b be the voltage across the branch. Sketch the current of the branch, assuming standard references.

5. The switch connecting a circuit component to a source of some kind is closed at $t = 0$, as shown in Figure P5. After that, it is found that $i(t) = 2t$ mA and $v(t) = (15t - 40)$ V.
a. Find a function of time representing the energy delivered to the component.

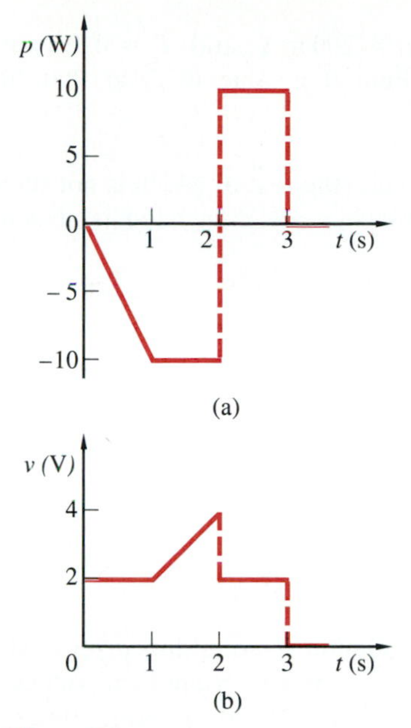

Figure P4

b. Also find the net energy delivered to the component (with units) from $t = 0$ to $t = 5$ s.

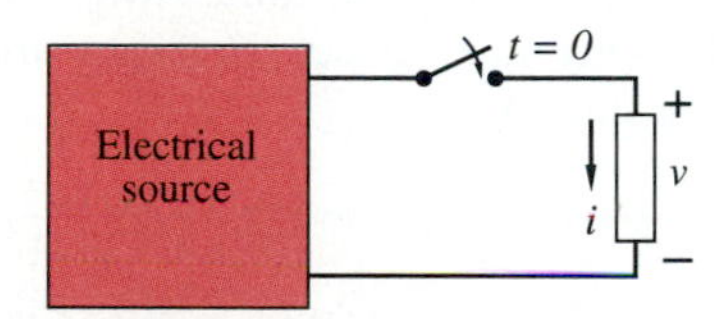

Figure P5

6. Part of a circuit is shown in Figure P6. Student A proposed the set of references for the currents shown in the figure. Student B objected, saying that these references are not possible because then there would be a net charge leaving the node, thus violating KCL. What do you say?

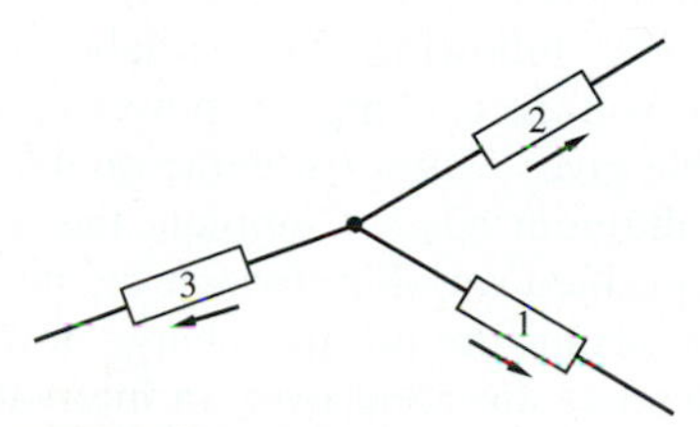

Figure P6

7. In Figure P6, $i_1 = 200$ mA and $i_2 = 0.3$ A at a particular time. Find the value of i_3 at that time, including units.

8. One node of a circuit (the rest of which is not shown) to which four branches are connected is shown in Figure P8.

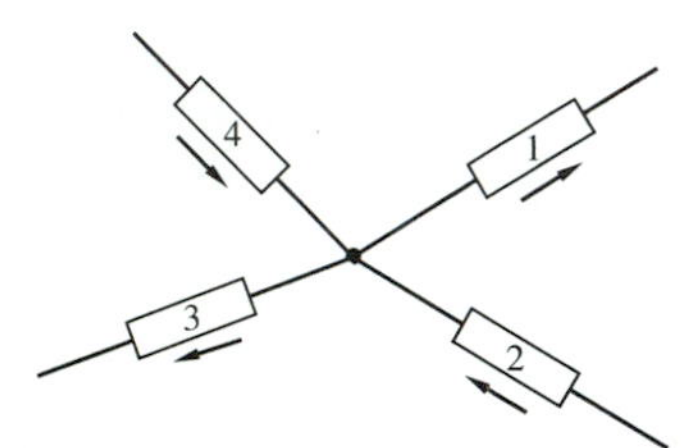

Figure P8

a. Write a KCL equation relating the currents whose references are as shown. Numerical values (in amperes) of the currents at different times are given in the accompanying table.
b. Complete the table by finding the value of the missing term in each case.
c. Also, for each row, fill in the two columns which give the total positive charge *actually leaving* and *actually entering* the node per unit time. Note that these *are not* the currents whose *references* are leaving or entering the node.

i_1	i_2	i_3	i_4	Sum of currents actually leaving	Sum of currents actually entering
2	3	5	4	7	7
−1		0.5	3		
	−4	1	−3		
−5	−2	−1			

9. Standard references are chosen for the voltage and current of a circuit branch. (Draw a diagram showing the references.) The following five variables are relevant: current i, voltage v, charge q, power p, and energy w. The table gives values (or expressions) for some of these in different cases. Complete the table by finding the unspecified variables (watch the units). In all cases, charge means the positive charge flowing in the current reference direction over an interval of time from 0 to t, power is delivered *to* the branch, and energy means the energy delivered to the branch over an interval of time from 0 to t.

i (mA)	v (V)	p (W)	w (J)	q (mC)
20		1		
	$120(t - 1)$	$6t^2 - 6t$		
	$25t$			$40e^{-2t}$
			te^{-t}	$25t(2 - t)$
$100t$	$30t + 40$			

10. For each circuit in Figure P10, select branch current references and write KCL equations at each of the nodes, using either form of KCL. Add the equations; if you do not get $0 = 0$, you must have an error.

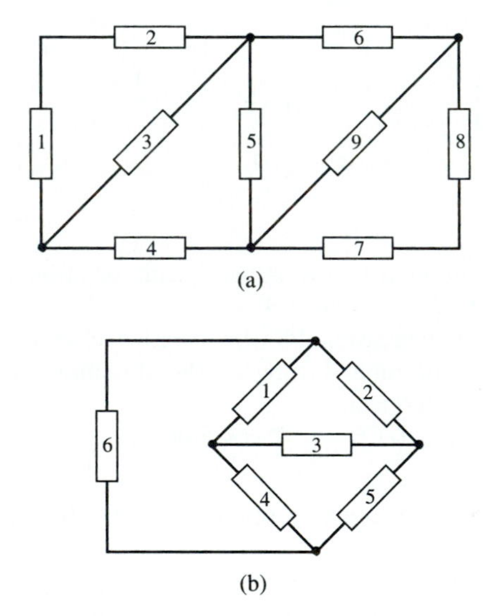

Figure P10

11. For each circuit in Figure P10, select branch voltage references and write KVL equations around each of the loops, using either form of KVL. (There are 10 loops in Figure P10a and 7 in Figure P10b.)

12. One loop of a circuit (the rest of which is not shown) that contains four branches is given in Figure P12.
a. Write a KVL equation relating the four branch voltages.

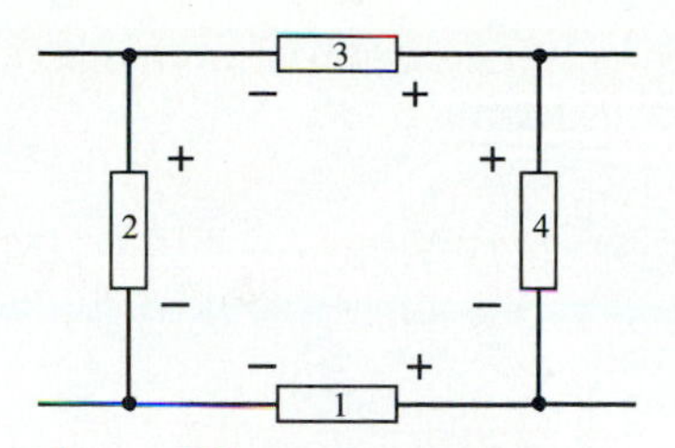

Figure P12

b. Numerical values (in volts) of the voltages at different times are given in the accompanying table. Complete the table by finding the value of the missing term in each row.

c. For each row, complete the two columns which specify the sum of voltages whose *actual* values coincide with a *clockwise* traversal and a counterclockwise traversal of the loop, respectively.

v_1	v_2	v_3	v_4	Sum of voltages actually clockwise	Sum of voltages actually counterclockwise
50	30	10	−10	50	50
−40	−20	50			
10	30		40		
	−35	−25	−40		
$10t$		$5(t-1)$	20		

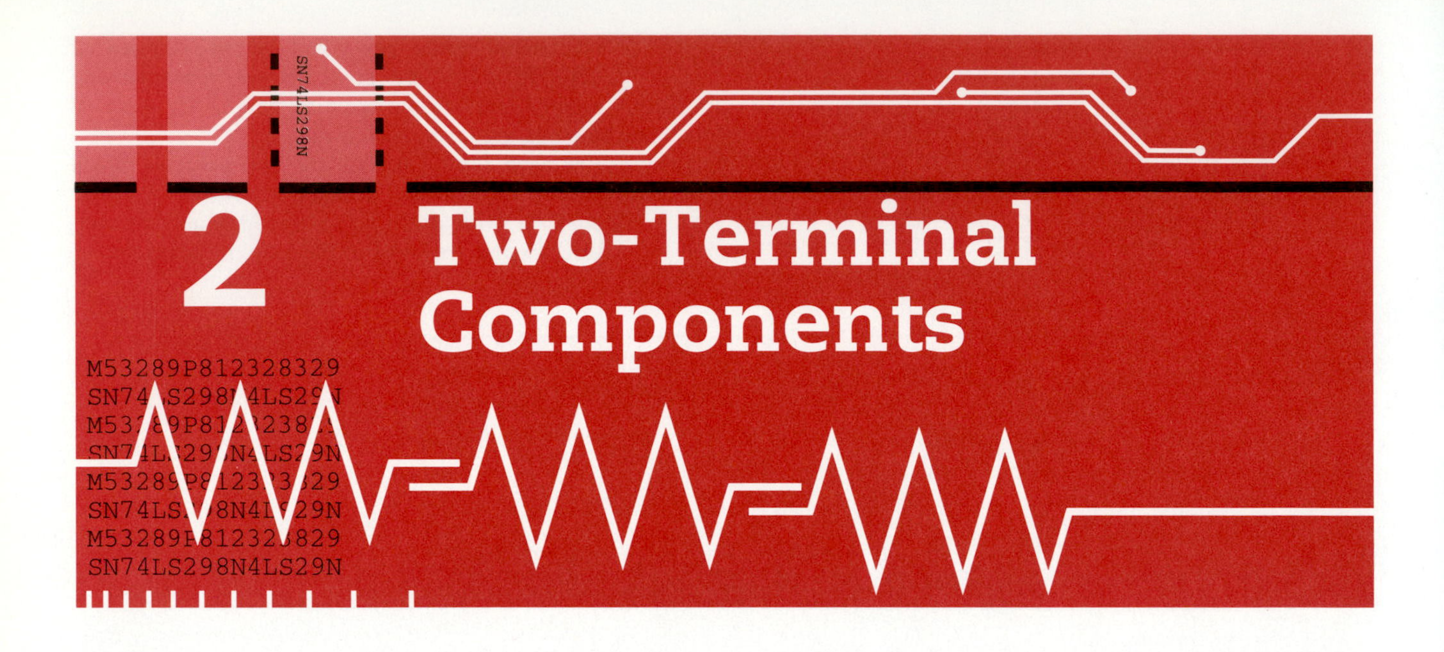

2 Two-Terminal Components

In the previous chapter we introduced the fundamental laws related to the interconnection of electrical devices into circuits. The two laws of Kirchhoff apply to circuits quite independently of the actual components that make up the circuit. However, actual circuits are made up of specific components. Thus, the next order of business is to introduce models of specific electrical devices and describe how the electrical variables are related to each other for these models.

Many practical electrical devices (such as batteries, heaters, diodes, and neon lamps) have two terminals. Each device, when connected in a circuit, will have associated with it a current *through,* and a voltage *across,* its terminals. In the ideal representation, currents will satisfy Kirchhoff's current law at each node of the circuit. Similarly, the voltages of the branches around each loop will satisfy Kirchhoff's voltage law. These two laws are quite independent of each other. Nothing need be said of a branch voltage when applying the current law, and nothing need be said of a branch current when applying the voltage law.

For each electrical device, however, there is a characteristic way in which the voltage and current of the device are related. For actual physical devices, these relationships can be discovered by simultaneous measurement of the voltage and current. In another approach, mathematical formulas relating the voltage and current of a device can be obtained by carrying out an analysis of the physical properties of the device (such as density and mobility of electrons, electronic charge, dimensions, and temperature) and relating the voltage and current to these physical properties. If this latter approach is to succeed, a great deal must be known about the physics of a device. Since such knowledge is not assumed here, this approach will not be used.

Figure 1 shows what we shall call a *voltage-current tester,* an imaginary device that could be constructed in a laboratory. Any sort of current-carrying two-terminal device can be connected between terminals A and B for testing. The essential features are a center-scale voltmeter, a center-scale ammeter, and a means for varying the voltage, represented in the figure by a knob. What is inside the box with the knob is not really important. It could be a set of dry cells connected to a rotary switch, as in Figure 1a, or something else.

The details of the operation of the meter are not important now. We assume that the indicator is centered when there is no voltage or current. The design of the meters is such that when an actual current goes through the ammeter *from* the plus-marked terminal *to* the other terminal, the indicating needle swings up scale (to the right); when the actual current is the reverse of this, the needle swings down scale (counterclockwise). The voltmeter operates in a similar fashion. Both meters are assumed to be ideal in the sense that across the ammeter there is zero voltage and through the voltmeter there is zero current. Actual meters are made to approximate this ideal.

In making the measurements, the control knob is adjusted to a sequence of positions, and simultaneous readings of the voltage and current are made. The resulting data can be plotted as a graph of voltage against current (or current against voltage). Different curves will be obtained depending on the device under test.

Three such curves are shown in Figure 2. The second curve is typical of a device called a *pn junction diode.* The third curve is typical of a device called a *tunnel diode.* We shall devote attention to diodes in a later section.

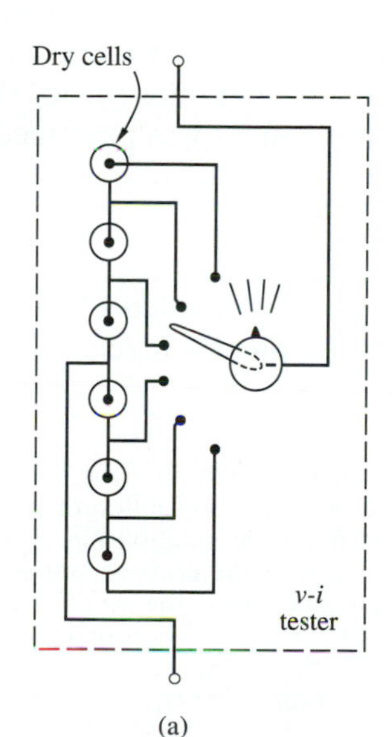

(a)

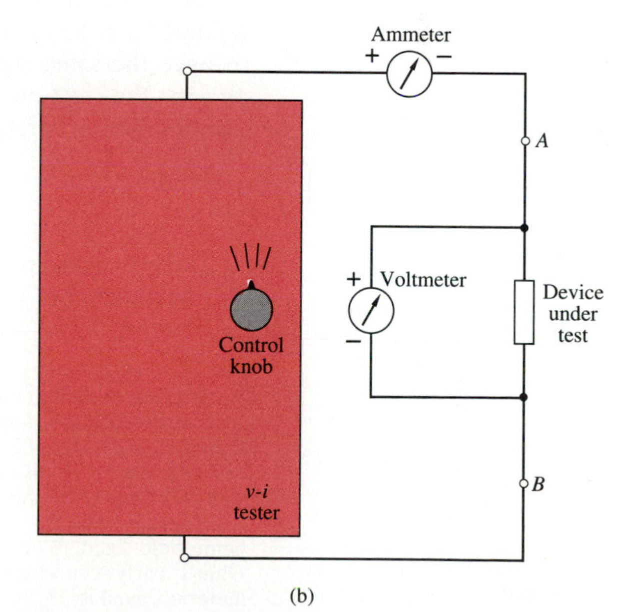

(b)

Figure 1
Voltage-current tester.

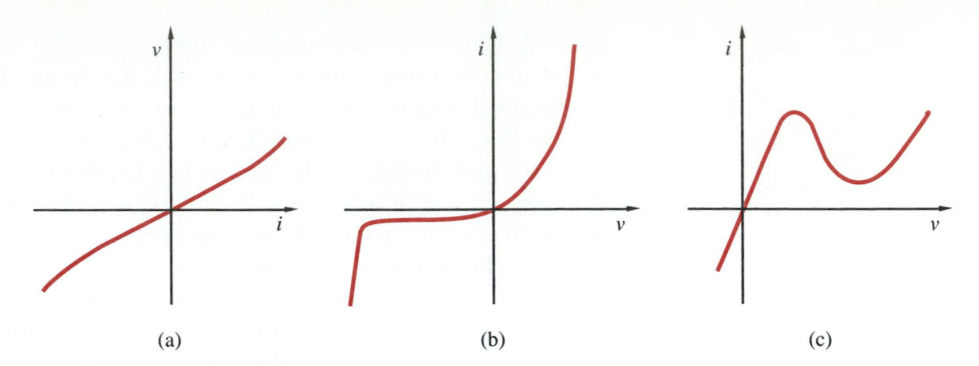

Figure 2
Plot of meter readings: v against i.

1 THE LINEAR RESISTOR

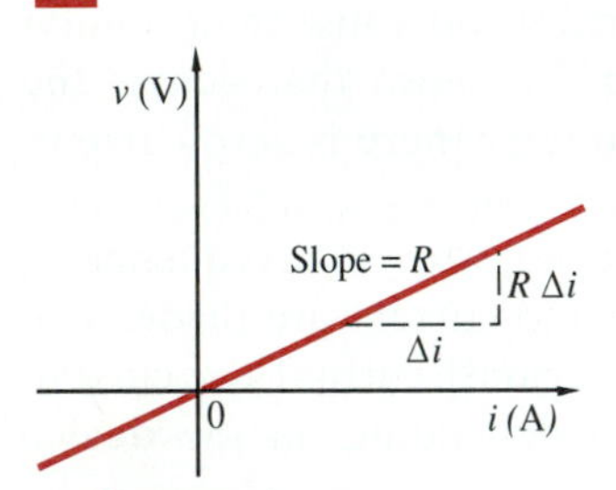

Figure 3
A resistor whose v-i relationship is linear: a linear resistor.

The curve in Figure 2a is almost a straight line, symmetrical about the origin. It deviates from linearity as the current increases in either direction but is quite straight for small values of current. On this basis, we create a hypothetical component (a model) having a linear relationship between its voltage and current over *all* values of current, as in Figure 3. Since the line, whose slope is called R, passes through the origin, its equation is:

$$v = Ri \tag{1}$$

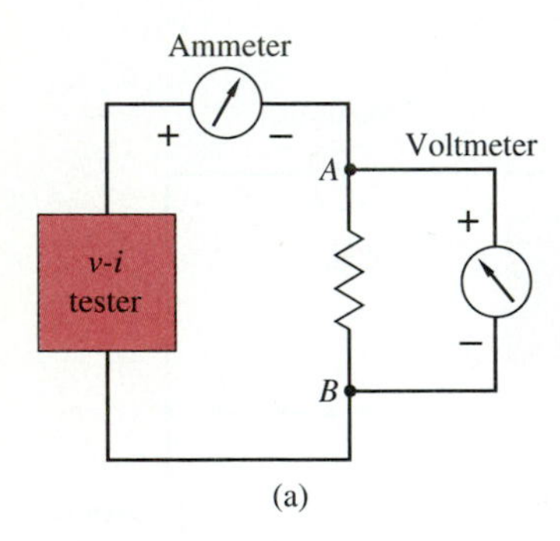

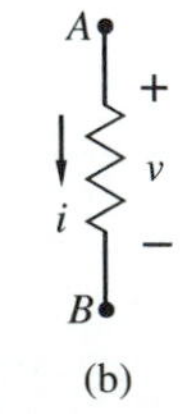

Figure 4
Measurement of resistance.

In Figure 4a we have replaced the actual physical device having the v-i relation of Figure 2a with its model, having the ideal linear relation of Figure 3. The experimental result previously obtained in Figure 2a showed that the voltmeter reading and the ammeter reading have the same sign when connected as shown.

As discussed in the preceding chapter, the same information given by meter readings can be obtained by using reference orientations. In order for v and i to have the same sign, the references for v and i must be as shown in Figure 4b. In the section on power and energy in the preceding chapter this combination of references for v and i was described as *standard* references.

Ohm's Law

The relationship in (1), $v = Ri$, is called *Ohm's law.*[1] This expression holds at every instant of time. The parameter R relating i to v is called the *resistance*;

[1] Georg Simon Ohm (1789–1854), a German physicist, made extensive measurements of the current and voltage of many different metals of different cross-sectional shapes and lengths in the early 1820s. By 1826 he had found the relationship between voltage and current to vary with the length of his conductors, with the cross-sectional area, with the temperature, and with a property of the material now called the *conductivity,* but the linear (straight-line) relationship between voltage and current was approximated very well, all other parameters being held fixed. Who knows whether circuit analysis was advanced or retarded by Georg Ohm's early concentration on *linear* circuits, partly because he didn't have nonlinear materials available?

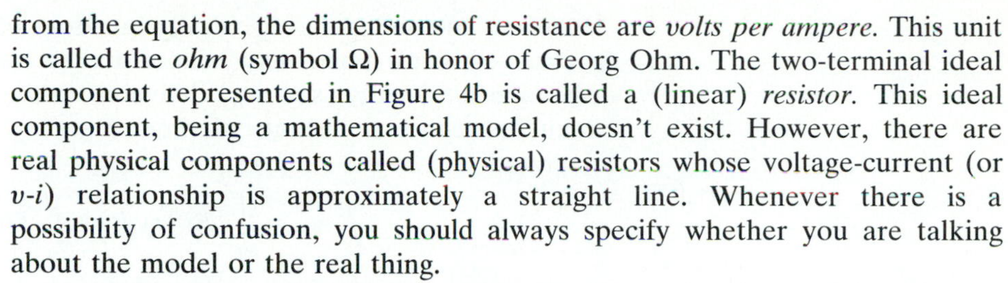

from the equation, the dimensions of resistance are *volts per ampere*. This unit is called the *ohm* (symbol Ω) in honor of Georg Ohm. The two-terminal ideal component represented in Figure 4b is called a (linear) *resistor*. This ideal component, being a mathematical model, doesn't exist. However, there are real physical components called (physical) resistors whose voltage-current (or v-i) relationship is approximately a straight line. Whenever there is a possibility of confusion, you should always specify whether you are talking about the model or the real thing.

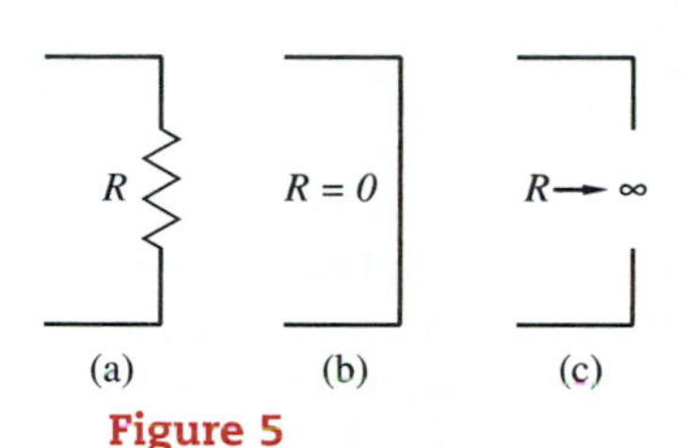

Figure 5
General resistance (a) and its limiting values: (b) short circuit, (c) open circuit.

Practical values of resistance can range from fractions of an ohm to thousands of ohms (kilohms, or kΩ) and millions of ohms (megohms, or MΩ). The limiting cases when R approaches 0 and when R approaches infinity are interesting enough to deserve special names. The case $R = 0$ is called a *short circuit*, while the case $R \rightarrow$ infinity is called an *open circuit*. The two cases are illustrated in Figure 5.

Suppose either the voltage reference in Figure 4b or the current reference (but not both) is reversed. That is, the new v or the new i is the negative of the previous one. Then, for these nonstandard references, the correct expression for Ohm's law would be $v = -Ri$. The two sets of references are shown in Figure 6. As outlined in Chapter 1, we describe the standard references by saying that the voltage plus is at the tail of the current arrow. Conversely, for nonstandard references, the voltage plus is at the tip of the current arrow. (Caution: The current reference should be placed alongside the component for this description to be valid, not placed in the connecting lead. Test it out.)

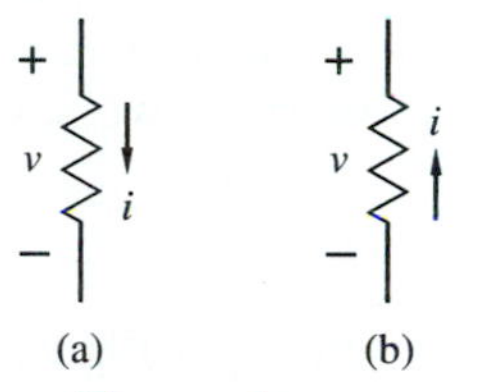

Figure 6
Joint v and i references: (a) standard, (b) nonstandard.

Note that the form of Ohm's law in (1) expresses v in terms of i. The equation can be inverted and written:

$$i = \left(\frac{1}{R}\right)v \qquad \text{or} \qquad i = Gv \tag{2}$$

The parameter G is the reciprocal of R and is called the *conductance*. The unit of conductance, instead of being named after a worthy person, was earlier called the *mho* which is *ohm* spelled backward. More recently, it has been renamed the *siemens* (S) after the name of a German scientist.

Power and Energy in a Linear Resistor

An expression for the power delivered to a device in terms of the voltage and current of the device was obtained in the preceding chapter. We have now identified a specific device—a resistor—and should determine what the expression for power becomes in this case.

The diagram in Figure 7 shows a resistor as part of a circuit. The rest of the circuit is contained in the box on the left. By virtue of this connection, there is a voltage v across the resistor and a current i through it. The references chosen for the resistor are the standard references. Hence, the power delivered to the resistor is $p = vi$. When this is combined with Ohm's law ($v = Ri$ or $i = Gv$), either v or i can be eliminated from the expression for power. Thus:

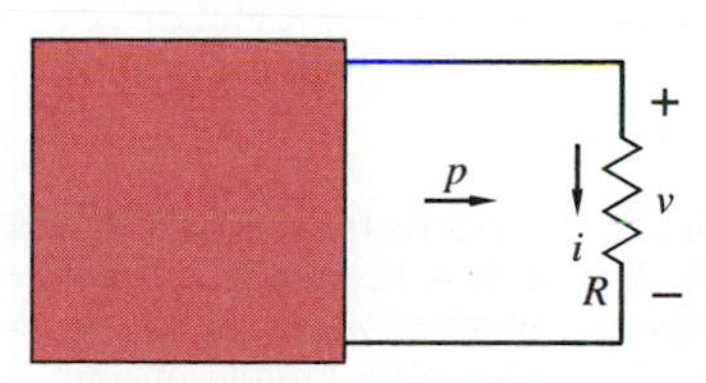

Figure 7
A resistor connected in a circuit.

$$\begin{aligned} p = vi &= Ri^2 \\ &= Gv^2 \end{aligned} \tag{3}$$

Current and voltage are generally functions of time (a constant is a specific function of time). Each may be positive or negative at different times. To determine the energy (say, w_R) delivered *to* the resistor from the rest of the circuit between two instants of time t_1 and t_2, the expression for power is integrated:

$$w_R = \int_{t_1}^{t_2} Ri^2(t)\, dt = \int_{t_1}^{t_2} Gv^2(t)\, dt \tag{4}$$

The question naturally arises as to the disposition of this energy. What happens to it? Can we get it back from the resistor? The answers are clarified by noting that, even though the current or voltage may be negative or positive at different times, their square is always positive. Hence, assuming a positive value of R, the energy supplied to a (linear) resistor can never be negative. It cannot even be zero except in the trivial case when the current is zero identically (for all time). Saying that the energy delivered to a (linear) resistor can never be negative is another way of saying that electrical energy can never be supplied by a (linear) resistor.

This answer is corroborated by observations in the physical world. Whenever current is passed through a conducting medium, there is an attendant appearance of heat. (The electric toaster operates on this principle.) The conversion of the electrical energy into heat is an irreversible process; we say that the energy is dissipated in the resistor.[2]

EXAMPLE 1 An arbitrary current through a 1000-Ω resistor is given by the curve in Figure 8. Find the energy supplied to the resistor over each of the following time

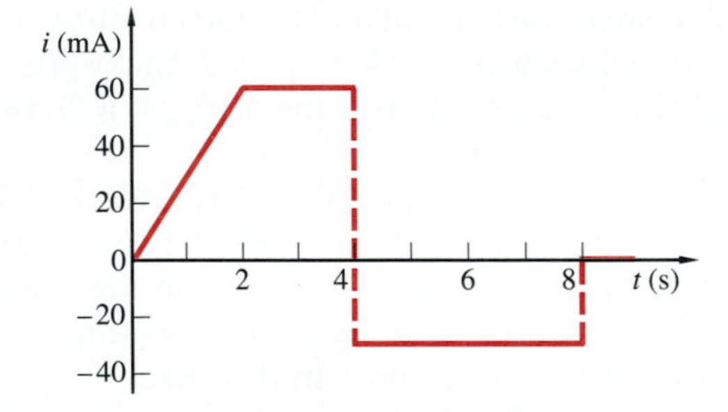

Figure 8
Current waveform.

[2] Of course, the heat may be used to generate steam, which can drive a turbine, which drives an electric generator, thus yielding electrical energy again. But such a sequence of energy conversions is not what is meant by reversibility. Quantitative measurements of the heat produced by the passage of an electric current through a conductor were first made by James Joule (1818–1889) in England in the 1840s. At the age of 21 he wrote his paper *On the Production of Heat by Voltaic Electricity,* and he began the series of experiments that eventually led to Joule's mechanical theory of heat. In fact, equation (4) deserves to be called *Joule's law,* and it often is.

intervals:

$$0 \leqslant t \leqslant 2\,\text{s}$$
$$2\,\text{s} \leqslant t \leqslant 4\,\text{s}$$
$$4\,\text{s} \leqslant t \leqslant 8\,\text{s}$$
$$8\,\text{s} < t$$

SOLUTION We first need expressions for the current over each of the specific intervals of time. Then we insert these expressions into Joule's law, equation (4):

$0 \leqslant t \leqslant 2$:

$$i(t) = 30 \times 10^{-3} t$$
$$w_R = \int_0^2 0.9t^2\,dt = \frac{(0.9t^3)}{3}\bigg|_0^2 = 2.4\,\text{J}$$

$2 \leqslant t \leqslant 4$:

$$i(t) = 60 \times 10^{-3}$$
$$w_R = \int_2^4 3.6\,dt = 3.6t\big|_2^4 = 7.2\,\text{J}$$

$4 \leqslant t \leqslant 8$:

$$i(t) = -30 \times 10^{-3}$$
$$w_R = \int_4^8 0.9\,dt = 0.9t\big|_4^8 = 3.6\,\text{J}$$

$t > 8$:

$$i(t) = 0 \qquad w_R = 0$$

POWER RATING AND TOLERANCE OF PHYSICAL RESISTORS In the preceding example, we found that a certain amount of energy was supplied to a resistor over a certain time period. What happens to this energy? In the real world, it is converted into heat, thereby raising the temperature of the resistor. Again in the real world, the resistance of a physical resistor is temperature-dependent, and so, as the temperature rises, the resistance will change. Too much change will mean that the circuit in which the resistor is located will not operate as it was designed to—assuming it does not burn up altogether!

Physical resistors are designed to dissipate a certain amount of energy per unit time without an excessive rise in temperature. Safe operation of resistors requires that the energy supplied to a resistor per unit time—namely, the power—not exceed a specified value. Resistors are, thus, rated in accordance with this maximum power they can dissipate. This is their *power rating.*

In discussing resistors so far, in addition to the power rating of physical resistors just mentioned, we have specified a parameter, the resistance R (or its reciprocal G). In the real world, when physical resistors are manufactured, although it is intended that each batch have a specific value (say, $10\,\text{k}\Omega$), the manufacturing process cannot guarantee that every resistor in the batch will have exactly this value. Normally, a deviation from this nominal value is acceptable. Hence, a *tolerance* (some percent) is specified. A 1% tolerance is

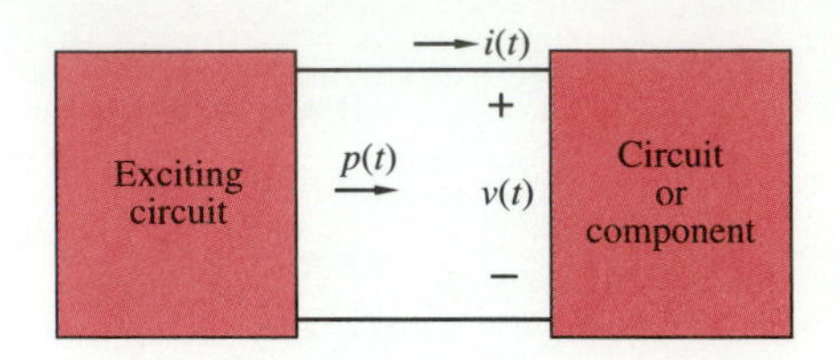

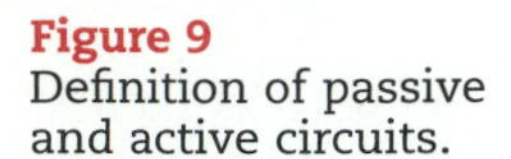
Figure 9
Definition of passive and active circuits.

normally viewed as excellent; 1% resistors will require more manufacturing care, and, therefore, will cost more than 5% or 10% resistors, which are quite common.

Thus, in the real world, not only do resistors have nominal values of resistance, but two other numbers would be associated with them: the power rating in watts and the tolerance in percent.

Passive and Active Components

What we have learned about the energy supplied to a resistor is the basis for defining a general class of components or circuits.

Figure 9 shows a two-terminal circuit or circuit component on the right being excited by a two-terminal circuit on the left, resulting in a voltage $v(t)$ across the component and current $i(t)$ through it. The power supplied from the left is $p(t) = v(t)i(t)$. The fact that v, i, and p are time-dependent variables has been explicitly expressed. The total energy input from time immemorial to the present time t is obtained by integrating the power. If this energy is never negative, for any time t, we say the circuit or the circuit component is *passive*. That is, if:

$$w(t) = \int_{-\infty}^{t} v(x)i(x)\,dx \geqslant 0 \qquad \text{for all } t$$
$$w(t) = w(t_0) + \int_{t_0}^{t} v(x)i(x)\,dx \geqslant 0 \tag{5}$$

then the circuit is passive. The second form is obtained by choosing some initial time t_0, the energy supplied up to that time being labeled $w(t_0)$. The total energy supplied is the amount supplied up to the initial time plus the amount supplied from then on, until time t.

If the circuit or component is not passive, it is said to be *active*. A two-terminal circuit is active if:

$$w(t) = \int_{-\infty}^{t} v(x)i(x)\,dx < 0 \qquad \text{for some time } t \tag{6}$$

That is to say, although the energy supplied may remain positive for some time, there will come a time when the net energy supplied to the circuit on the right is negative. This means the circuit itself is supplying energy, and so it is active.

E X A M P L E 2 Suppose the current and voltage of a two-terminal device whose references are standard are $i(t) = t - 2$ and $v(t) = 5t$ between $t = 0$ and 5, and 0 otherwise.

The expression for instantaneous power will be $p(t) = vi = 5t(t - 2)$. The energy *into* the component is:

$$E(t) = \int_0^t 5x(x - 2)\,dx = \left(\frac{5x^3}{3} - 5x^2\right)\Big|_0^t = \tfrac{5}{3}t^2(t - 3) \qquad (7)$$

Clearly, $E(t) < 0$ for $t < 3$. Hence, this component is not passive; it is active.

2 SOURCES

In a resistor, energy is irreversibly transformed from an electrical form to heat. The process is irreversible because cooling the resistor in a refrigerator, for example, will not produce electrical energy.

In the physical world there exist many devices that seem to generate electrical energy. Of course, the electrical energy is not really "generated"; it is converted from some other form. There is the rotating electrical generator, the workhorse that supplies electrical energy "in bulk"; there are the storage battery and various dry cells which convert chemical energy; there is the photoelectric cell, or electric eye, in which light energy is converted to electric current; there is the thermocouple that converts a difference in temperature to an electric voltage; and others. A large part of electrical technology is concerned with the design of such energy-converting devices. This requires an extensive knowledge of the physical processes whereby the energy is converted. But in this text we are concerned only with the use of such devices in a circuit. Therefore, we are interested only in the voltage or current at the terminals of these devices, not in their internal operation.

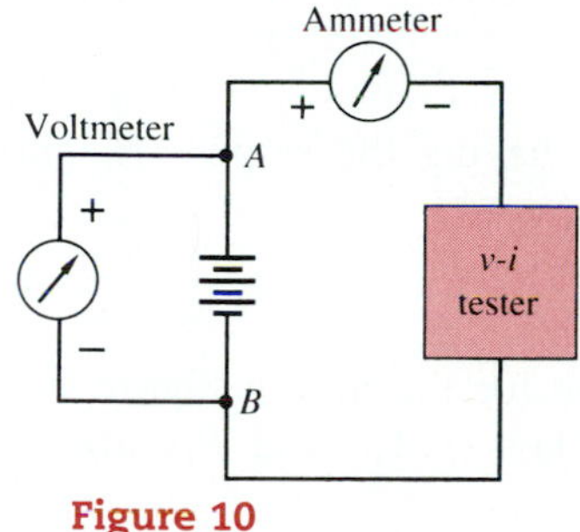

Figure 10
Battery measurements.

We shall use a battery as the prototype from which models of electric generators will be derived. Figure 10 shows the previously described voltage-current tester connected to a battery, represented in the diagram by a stack of alternate long and short lines. As before, using the ideal instruments described earlier, simultaneous measurements of voltage and current at the terminals are made. A plot of v against i is found to give an approximately straight line with negative slope $-R_0$, as shown in Figure 11. The intercept of the line on the

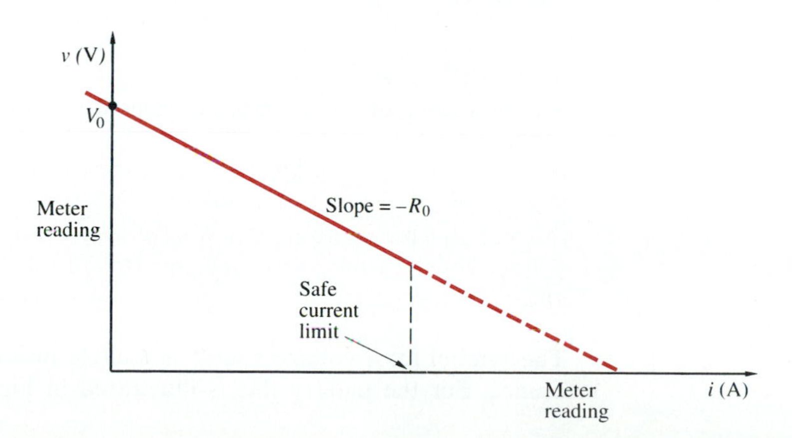

Figure 11
Plot of battery meter readings.

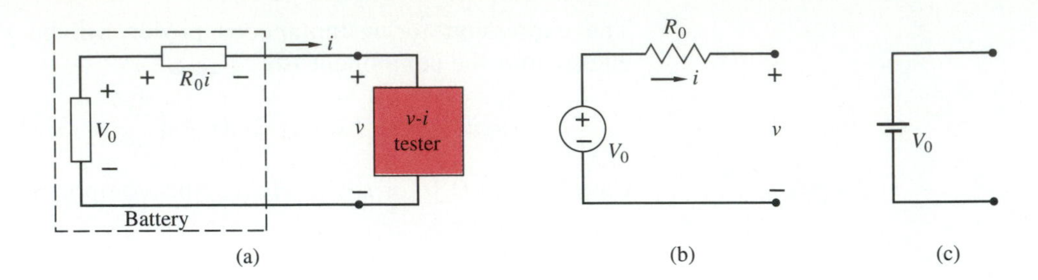

Figure 12
Model of battery.

voltage axis is labeled V_0. (The slope of the line is exaggerated in this diagram over the actual slope in order to show the intercept on the current axis.)

A model representing the battery is obtained from this empirical result in the following way:

1. Even though there may actually be a slight curvature, the plot in Figure 11 is assumed to be a straight line. That is, the v-i curve is assumed to be linear.

2. Suitable ideal components are postulated to account for the empirical result.

The equation of the straight line in Figure 11, having the slope $-R_0$ and v-axis intercept V_0, is:

$$v = -R_0 i + V_0 \tag{8}$$

(You should verify that this is the correct equation for the line in Figure 11.) This equation reminds us of Kirchhoff's voltage law (v, V_0, and $R_0 i$ are all dimensionally voltage).

The same equation would result from an application of KVL in the circuit shown in Figure 12a consisting of two boxes labeled with voltages V_0 and $R_0 i$ and connected to the v-i tester. The box with voltage $R_0 i$ and current i is clearly a resistor having a resistance R_0. The other box has a constant voltage V_0 whose value does not change, no matter how we change the current i by changing the dial on the v-i tester.

The Voltage Source

The preceding observation is the basis for defining an ideal component:

A *voltage source* is an ideal two-terminal component with the property that the voltage across its terminals does not depend on the current flowing through it. That is, no matter what else is connected at the terminals of a voltage source, causing various amounts of current, the voltage will not be affected.

The symbol for a voltage source is a circle inside which appears the voltage reference. For the battery this is illustrated in Figure 12b, where the voltage source has a constant voltage, V_0. For the case of a voltage source that does not vary with time, another symbol is also used, as shown in Figure 12c. The single long and short parallel lines are reminiscent of the stack of such lines which is the symbol for a physical battery. In general, the voltage of a source is a function of time, $v_s(t)$ or $v_g(t)$ (s for source, g for generator).

For example, the voltage supplied by the electric utilities to an outlet in a house can be considered a voltage source whose variation with time is sinusoidal, as shown in Figure 13a. Similarly, the voltage that causes the "spot" to scan the screen of a cathode-ray tube varies with time as a "sawtooth," as in Figure 13b.

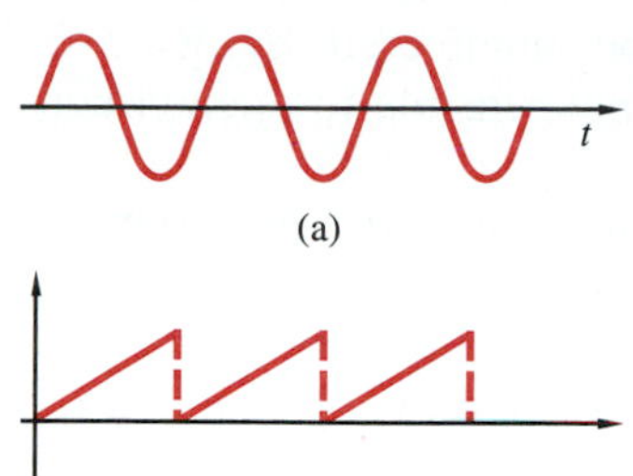

Figure 13
Examples of time dependence of voltage source.

The model of the battery consists of two interconnected branches as shown in Figure 12b: a voltage source and a resistor. R_0 is called the *internal resistance* of the battery.

The Current Source

Without any further consideration of real devices (like the battery) from which models (like the voltage source) can be created, but simply by analogy with a voltage source, we can conceive of another hypothetical device, as follows:

> A *current source* is an ideal two-terminal component having the property that the current through the device is independent of the voltage across its terminals; that is, no matter what is connected to the terminals of a current source, causing any amount of voltage, the current is unaffected.

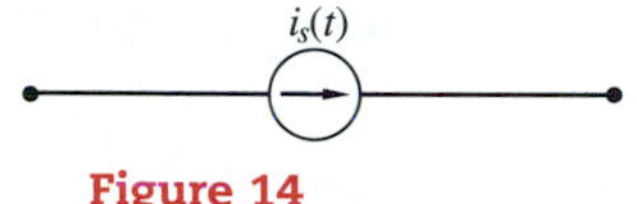

Figure 14
Current source.

The symbol for a current source is a circle inside which appears the current reference arrow, as shown in Figure 14. In general, the current of such a source is a function of time, such as the functions in Figure 13.

The two circuit components introduced in this section are ideal components. They approximate real devices. Although their properties are said to be independent of terminal conditions, peculiar things can happen under extreme conditions. To clarify, consider Figure 15, which shows each source with a variable load (a resistor) connected to the terminals. (The arrow through the resistor implies that the value can be varied.) Nothing significant happens for moderate values of resistance. But in Figure 15a suppose the resistance is gradually reduced to very low values. The current, v_s/R, will therefore become very large. In the limit as R approaches zero, the current becomes infinite. For a real generator, anything approaching an infinite current is likely to have severe consequences for its longevity.[3] Thus, although the definition of an ideal voltage source includes connecting anything to the terminals, it is not sensible to short-circuit a voltage source. The voltage across a short circuit is zero, while the definition of a voltage source requires the voltage to remain unchanged. The problem is that we are forcing two incompatible idealizations on each other; don't do it.

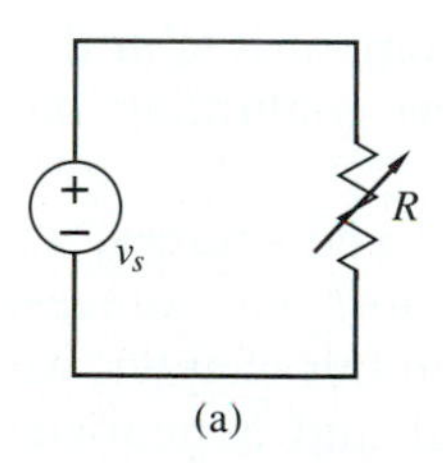

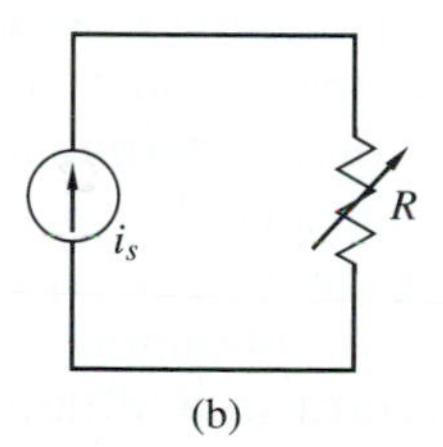

Figure 15
Voltage source and current source with loads.

[3] "Have severe consequences for its longevity" is a long-winded euphemism for "burn it up." You ought to experience, at least once, use of the typical language of politics in an engineering setting!

Similarly, suppose the resistance in Figure 15b is gradually increased to very high values. Again there is no difficulty unless the resistance goes to infinity—an open circuit. Then there will be a contradiction, since there can't be a current through an open circuit, on the one hand, while the definition of a current source requires the source current to be unaffected by what is connected to the terminals, on the other hand. It is not sensible to open-circuit the terminals of a current source.

So far, three components have been introduced: a resistor and two sources. Someone with a creative bent decided to construct a circuit by connecting one each of these three things, as shown in Figure 16. They have the numerical values given. Her goal was to calculate the power absorbed by, or supplied to, each of the three components.

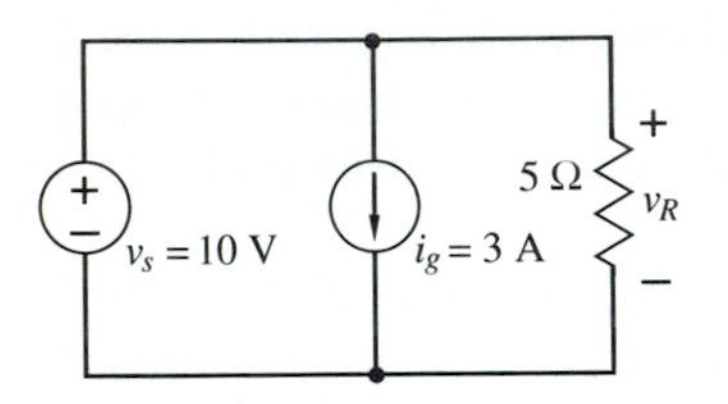

Figure 16
Example circuit.

1. She started with the resistor. Since the voltage of a voltage source does not depend on what is connected to its terminals and since, by KVL (if you need to review this, go back to Chapter 1), the resistor voltage is the same as that of the voltage source, $v_R = 10$ V. Assuming standard references for R, the resistor current is $v_R/R = 10/5 = 2$ A. Thus, the power *into* the resistor and absorbed there is $p_R = v_R i_R = 20$ W.

2. Next, the current source. By the same reasoning, its voltage is also 10 V. Hence, the power into the current source is 30 W. Thus, the current source, too, is absorbing power.

3. Finally, the voltage source. Choosing its current reference to be *up* (nonstandard) and applying KCL at the upper node, its current will be $3 + 2 = 5$ A. Hence, the power *supplied by* the voltage source is 50 W.

Isn't it reassuring that the power *supplied by* one of the components in this circuit equals the total power *absorbed by* the others, thus confirming the conservation of energy?

With the introduction of Kirchhoff's two laws, Ohm's law, and sources, we are in a position to begin studying circuits consisting of resistors and sources. At this point, it might make sense to do just that. However, in this chapter, we have followed an approach that combines an experimental and a heuristic outlook to arrive at ideal models—so far, of resistors, voltage sources, and current sources, all two-terminal components. It would also make sense to pursue this same approach to arrive at other ideal models of two-terminal circuit components, even though the study of circuits that include such components will be postponed a bit. That's what we will do. Some of you may wish to go through the rest of this chapter rapidly, without obtaining a thorough understanding, with the expectation that you will return to it with a vengeance later.

3 THE CAPACITOR

There exist many physical devices which cannot be adequately described by an interconnection of resistors alone. We shall consider one set of such devices in this section.

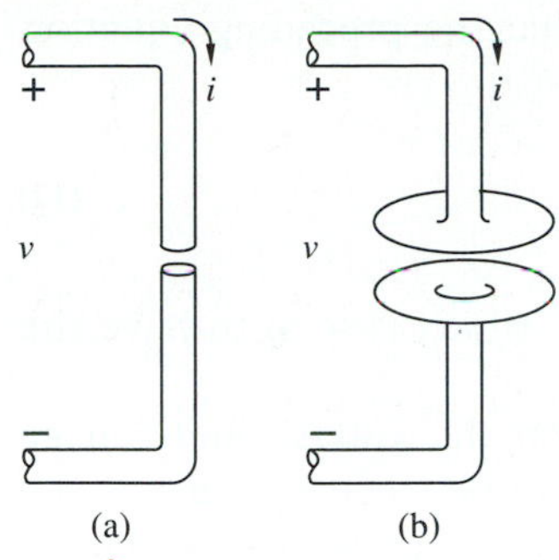

Figure 17
Idea of a capacitor.

Consider a conductor that has been cut in two as shown in Figure 17a. (The effect can be emphasized by supplying to the ends so formed a set of conducting plates as in Figure 17b.) The conductor forms part of a physical circuit. We assume that there is a current in this device and a voltage across it.[4] The current is, by definition, the charge q that flows past a given cross section of the conductor per unit time; that is:

$$i = \frac{dq}{dt} \tag{9}$$

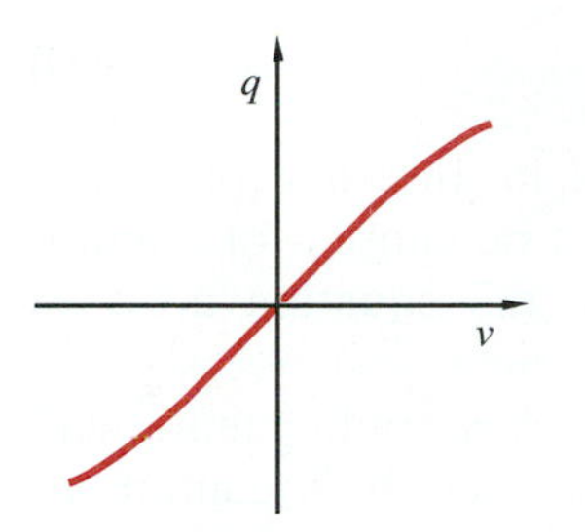

Figure 18
Charge-voltage relationship of a capacitor.

The question now arises how the voltage in Figure 17 is related to the current. To answer the question, we fall back on observation. It is reasonable to expect that the amount of charge on the plates will depend on the geometry (size and shape of the plates and their distance apart) as well as on the properties of the material, called a *dielectric,* between the plates. If charge is plotted as a function of voltage using a wide variety of geometrical shapes, plate spacings, and materials, curves such as the one shown in Figure 18 are obtained.

A physical device with two terminals whose q-v relationship is a curve such as the one in Figure 18 is called a *capacitor.* If the same curve is obtained for the same device at all other times, then time does not enter into the relationship and the capacitor is said to be *time-invariant.*[5]

Current-Voltage Relationship

As a function of voltage, the charge can be expressed generally as $q = f(v)$. When this is inserted into the definition of current in (9) and the chain rule for differentiation is invoked, we get:

$$i = \frac{dq}{dt} = \left(\frac{dq}{dv}\right)\frac{dv}{dt} \tag{10}$$

Now $dq/dv = f'(v)$ is the slope of the q-v curve. If the q-v relationship is a straight line, the slope is a constant; otherwise the slope varies with voltage. In

[4] You may question the assertion that there exists a current in a conductor that is cut in two. It is, indeed, perplexing. How can the charge jump across the gap? Actually, charge is transferred in and out of the terminals but none is conducted across the gap. Of course, if the voltage becomes too high, charge *will* leap across the gap, thus causing a spark. The air or other material between the plates is said to break down in that case. Short of such a spark, the continuity of current is maintained by what James Clerk Maxwell (1831–1879), the creator of "Maxwell's equations," called the *displacement current.* As far as circuit theory is concerned, only the terminals of a branch are available for external observation. What transpires within the branch, between the terminals, is not dealt with; it is a matter for field theory. If charge is transferred in and out of the terminals, a current will indeed be observed there.
[5] Suppose a little child played the parallel plates like an accordion, moving them closer and spreading them apart periodically. Then different q-v curves would be obtained at different times. In that case, the capacitor would be *time-dependent.* We will not consider time-dependent capacitors in this book.

either case, we can give dq/dv a label, say, C, so that the preceding equation becomes:

$$i = C\frac{dv}{dt} \tag{11}$$

C is called the *capacitance.* Its unit is the *farad* (F), which is seen to have the dimensions of *coulombs per volt.*[6]

In the nonlinear case, the value of C depends on the voltage and can be expressed as $C(v)$. Although a nonlinear relationship between C and v is the norm in the physical world (for example, the capacitance across a semiconductor junction) and nonlinear capacitors find many useful applications, in this book we will limit ourselves to the ideal model of *linear* capacitors; hence, we will take the capacitance C to be constant. For the linear case, then:

$$q = Cv \tag{12}$$

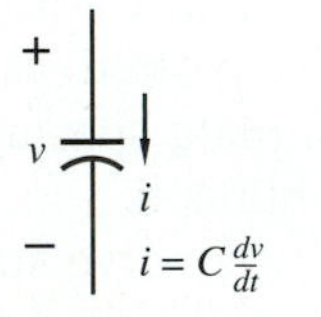

Figure 19
Circuit symbol for C.

The circuit symbol for a capacitor is shown in Figure 19. In order for (12) to be valid (with a positive sign), standard references must be chosen for v and i, as shown. If either reference is reversed, the sign of the equation must be changed.

In Chapter 1, the principle of conservation of charge was briefly mentioned. In accordance with this principle, the charge at a node must be a continuous function of time. Let's apply this principle to a capacitor which may be connected to other components in a circuit. Each terminal of the capacitor will form a node with the other components connected there. Thus, the charge flowing into the capacitor through the other branches must be continuous. Since charge and voltage are proportional, the voltage of a capacitor also must be a continuous function of time. This is an extremely significant conclusion and will find widespread application as we go along.

If, indeed, a capacitor voltage *could* be discontinuous at some point, its derivative at that point, which is proportional to the current, from (11), would become infinite, hardly something to be experienced in the real world.

EXERCISE 1

The voltage across a certain 10-μF capacitor has the following waveform. It rises linearly from 0 to 100 V in the time interval from 0 to 1 s. Then it falls linearly from 100 to -100 V in the interval from 1 to 3 s. Finally, it rises linearly back to 0 in the interval from 3 to 5 s. (a) Sketch the voltage

[6] The farad is named after Michael Faraday (1791–1867), the English scientist. Note that the farad is an inconveniently large unit. To appreciate just how large, a 1-farad parallel-plate capacitor with an air dielectric and a spacing of 1 mm between plates must have an area of the order of 100 km^2! Faraday's formal education never got beyond simple reading, writing, and arithmetic. He educated himself, while he was apprenticed to a bookbinder from the age of 14, by reading everything he could lay his hands on. By chance he read an article on electricity in an encyclopedia, and he was hooked. To Faraday we owe the idea of electromagnetic induction, the dynamo, electrolysis, the basic concept of field theory, and much more.

waveform. (b) Assuming standard references, sketch the waveform of the capacitor current, giving appropriate units. □

EXERCISE 2

The voltage across the terminals of a certain 0.5-μF capacitor is $500 \cos 1000\, t$. Find the peak value of the current waveform.

ANSWER: 250 mA. □

Voltage-Current Relationship

Equation (11) gives the capacitor current in terms of its voltage. This expression can be inverted to express the capacitor voltage in terms of its current by integrating it, say, between a fixed time t_0 and an arbitrary time t:

$$\int_{t_0}^{t} i(x)\, dx = \int_{t_0}^{t} C \frac{dv}{dx}\, dx = C \int_{v(t_0)}^{v(t)} dv \tag{13}$$

The upper and lower limits of the integral with respect to v correspond to the values of time in the limits of the preceding integral. Completing the integration and rewriting, we get:

$$v(t) = \frac{1}{C} \int_{t_0}^{t} i(x)\, dx + v(t_0) \tag{14}$$

where $v(t_0)$ is the *initial value* of the voltage at time t_0. This characteristic of the capacitor is something new. For the resistor, the current at any instant of time depends on the voltage precisely at that same time. But for the capacitor, we find that the voltage at time t depends on what the historical record of the current was in the past, up to time t.[7]

The full capacitor voltage is made up of two terms. One of them is a constant, representing the voltage of the capacitor at the initial time t_0, the *initial voltage*. The other term represents the voltage of a capacitor with no initial charge. Thus, an initially charged capacitor is indistinguishable at its terminals from a combination of an initially uncharged capacitor and constant-voltage source (a battery), as illustrated in Figure 20. By KVL, $v_C = v_C' + V$, where v_C' is the voltage on an initially uncharged capacitor and V is the voltage of a battery.

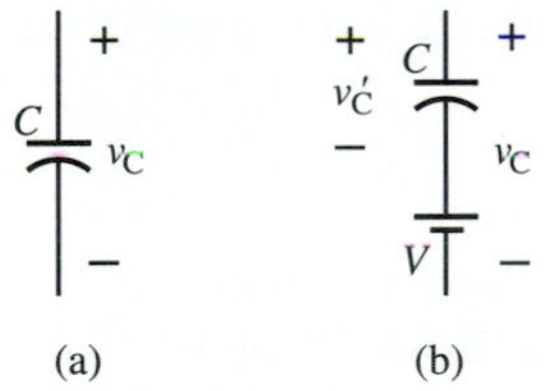

(a) (b)

Figure 20 Initially uncharged capacitor with battery equivalent to initially charged capacitor.

If the capacitor voltage is specified as a function of time, for example $v_C(t) = 50 \sin 400t$, the current is easily obtained by differentiation in accordance with (11). If, instead, the current is given as a function of time, together with the initial voltage, then integration will give the voltage, in accordance with (14). However, sometimes there is no single function

[7] Unlike earlier times when humans ascribed some of their own characteristics to the natural objects they encountered in daily life (like trees and animals), the anthropomorphism of present-day people in technological societies tends to ascribe human qualities to our fabricated technological objects. Thus, in view of (14), we say that a capacitor "remembers," or has "memory."

specifying the voltage or current over all time. Different relationships might be specified over different time intervals, perhaps even graphically.

EXAMPLE 3 Suppose the current in a 0.5-μF capacitor is given by the waveform in Figure 21a and the initial capacitor voltage at $t = 0$ is specified as 20 V. The voltage is determined from (14) by integration. The integrals of the two linear portions of the current curve will be quadratics. The integral of the constant portion will be linear. Furthermore, the capacitor voltage must be continuous everywhere—in particular, where the current changes slope. The result is given in Figure 21b. (Please work out the details.)

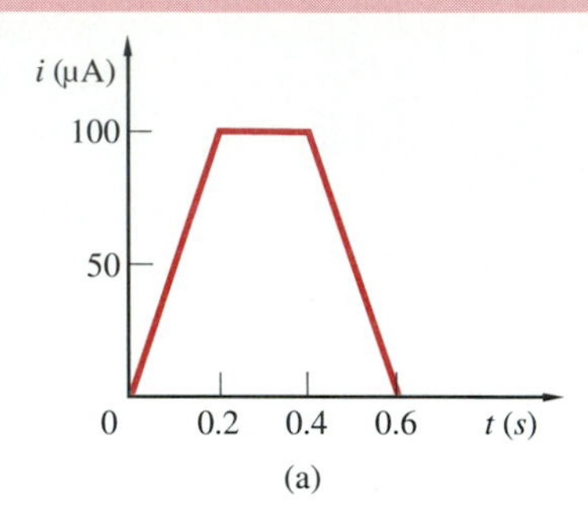

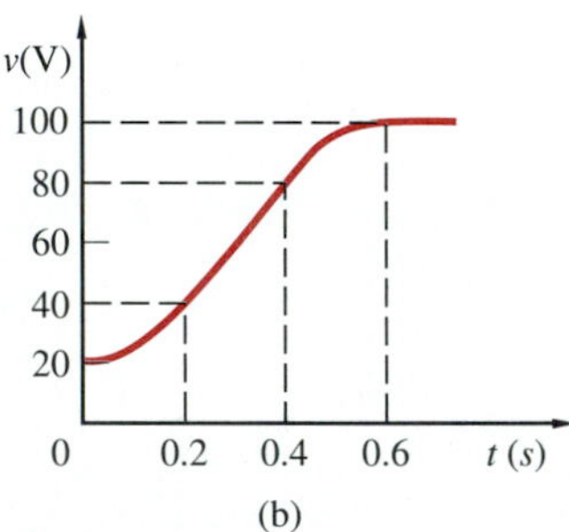

Figure 21
Capacitor voltage from current.

EXERCISE 3

Construct a problem which specifies (a) a value of capacitance of a capacitor and (b) a function of time as the voltage across the capacitor, and which then asks to find the capacitor current as a function of time. Include appropriate units for all quantities. Then solve the problem. □

EXERCISE 4

The current through a 2-μF capacitor is 10 mA from $t = 0$ to 20 ms; at that time it drops discontinuously to -20 mA and stays there from 20 ms to 40 ms; finally, at the latter time, it rises discontinuously to 0, where it remains. At $t = 0$ the voltage across the capacitor is 50 V. Assuming standard references, sketch the voltage across the capacitor. □

Voltage Rating of a Physical Capacitor

In the real world, physical capacitors are manufactured in a wide variety of different sizes, shapes, and configurations. In all cases, however, they consist of two conducting objects separated by an insulating dielectric. Whatever value the voltage across the terminals of a capacitor may have, it is experienced by the dielectric. If the voltage increases above a certain limit, the insulating material breaks down, producing a short circuit between the two conductors.

This limit depends on two things: the *dielectric strength* of the insulating material and the thickness of the insulation. The dielectric strength, S_d, is defined as the maximum voltage per unit thickness that can be applied across a slab of the material before it breaks down. This is not an exactly fixed quantity, because it depends on such things as imperfections in the material, possibly humidity, and other factors. Thus, it is necessary to use the "worst case," representing the smallest value of dielectric strength for all possible conditions. The dielectric strength of dry air is about 3 kV/mm, while that of fused quartz is 600 kV/mm. Others lie between 10 and 40 kV/mm. The *voltage rating* of the capacitor is the dielectric strength times the thickness of the insulation. For a given material, the thinner the insulation, the smaller the voltage rating.

When a capacitor appears in a physical circuit, we need to know not only its capacitance value but also its voltage rating. If it turns out in the operation of the circuit that the voltage across the capacitor will exceed its voltage rating, then the capacitor should be replaced by another one with a higher voltage rating.

Capacitor Energy

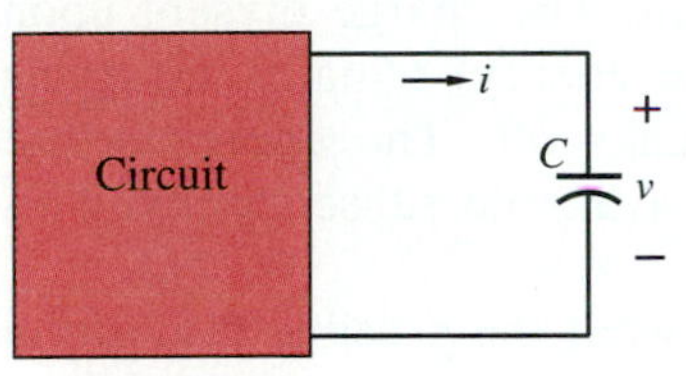

Figure 22
Capacitor in a circuit.

Let us now establish an expression for the energy delivered to a capacitor when it is connected in a circuit, as shown in Figure 22. The references for the capacitor voltage and current are standard, so the power *into* the capacitor is $p = vi$. Using the i-v relationship from (11) results in:

$$p = vi = Cv\frac{dv}{dt} \tag{15}$$

Hence, the energy delivered to the capacitor over an interval of time from 0 to t will be:

$$w_C = \int_0^t Cv\frac{dv}{dx}\,dx = C\int_{v(0)}^{v(t)} v\,dv = \tfrac{1}{2}C(v^2(t) - v^2(0))$$

or:

$$w_C(t) - w_C(0) = \tfrac{1}{2}Cv^2(t) \tag{16}$$

That is, the *net* energy delivered to a capacitor at any time t starting from a time chosen as the time origin is proportional to the square of the voltage at that time. Since the charge on a capacitor is proportional to its voltage, it is equally true that the energy is proportional to the square of the charge. (Using $q = Cv$, determine the expression for energy in terms of q.)

An interesting observation results from a comparison of this expression for the energy delivered to a capacitor and the similar expression in (4) for the energy delivered to a resistor. We note that, for the capacitor, this energy does not depend on the variation of capacitor voltage over the interval, but on the values of the voltage at the endpoints of the interval. This is different for the case of the resistance, as (4) shows. There, the voltage or (current) is under the integral sign; the energy will depend on the voltage *waveform*—on its variation over time—as well as on its value at the endpoints of the interval.

EXERCISE 5

A 2-μF capacitor is connected to a circuit. The two functions that follow represent the voltage across the capacitor on two different occasions:

$$v_1(t) = t(t - 2)$$
$$v_2(t) = 6 - t$$

The initial energy in the capacitor is 1 mJ in the first case and half that amount in the seond. Find the net energy (include units) delivered to the capacitor from $t = 0$ to $t = 3$ s in each of the two cases. Comment on the results.

ANSWER: 9 μJ, the same in both cases. Only the final value of v has an effect, not the initial energy or the variation of v with t. □

Energy Storage in a Capacitor

The question arises as to what happens to the energy delivered to the capacitor. Is it dissipated as in the case of the resistor? Suppose that, at some time, the capacitor is disconnected from the circuit. The charge present upon disconnecting (and hence also the voltage and the energy) remains, assuming that the capacitor is ideal so that no charge leaks off. The energy is not dissipated but *stored* in the capacitor. It will be available subsequently. This matter will be examined further in Chapter 6.

Indeed, thinking more generally, we might ask what the possible disposition of energy can be in the process of transmitting energy from where it is produced (converted from some other form) to where it is used.

1. We have just seen that energy can be *stored* (as in a capacitor) and thus be available for subsequent use.

2. We have also seen that, if a resistor is in the transmission path, some energy will be *dissipated* there—converted into heat. Of course, if that resistor is just incidental to the specific purpose of the energy in that case (unlike the case of a toaster, whose very purpose is to dissipate heat), this dissipation of energy is a waste. (The energy lost in power lines used to transmit energy from power plants to users constitutes such a waste, and great efforts are made to minimize this energy loss.) Another form of dissipation is energy which is radiated in the form of electromagnetic waves that propagate out into space. In some instances, as in the case of an antenna, radiation is the desired outcome. But if the radiation is incidental to the operation of the circuit, it is again a waste.

3. Conversion to heat is only one kind of conversion. Energy can also be converted to other forms: for example, to mechanical motion in the case of electric motors; indirectly to sound in a loudspeaker, where the electrical energy is first converted to diaphragm motion, which then produces sound; or to radiated energy in the form of electromagnetic waves, as in an antenna. But these forms of conversion occur in systems which are not strictly circuits.

In electric circuits, then, energy is either dissipated (including in the form of radiation) or stored.

4 THE INDUCTOR

We now continue with the development of circuit models of two-terminal physical devices. This time the physical device is a coil of wire. The coil may take one of a number of different shapes, as shown in Figure 23. It may even consist of a single turn of wire. The core may consist of some magnetic or nonmagnetic solid material or just air.

Observations of the phenomena that result when an electric current passes through one or more coils of wire were made at approximately the same time (1831, some five years after the publication of Ohm's law) by both Michael Faraday in England and Joseph Henry in the United States. Suppose the current in the coil results from applying a voltage source to the coil. If the voltage is suddently increased, it is observed that the current does not follow this increase immediately but builds up gradually. Similarly, if the voltage is suddenly reduced, the current again does not immediately fall but decreases slowly. This property is similar to the property of inertia in a moving body. The rate at which the current follows a sudden change in voltage is found to depend on the geometry (size and shape), the number of turns of the coil, and the material of the core.

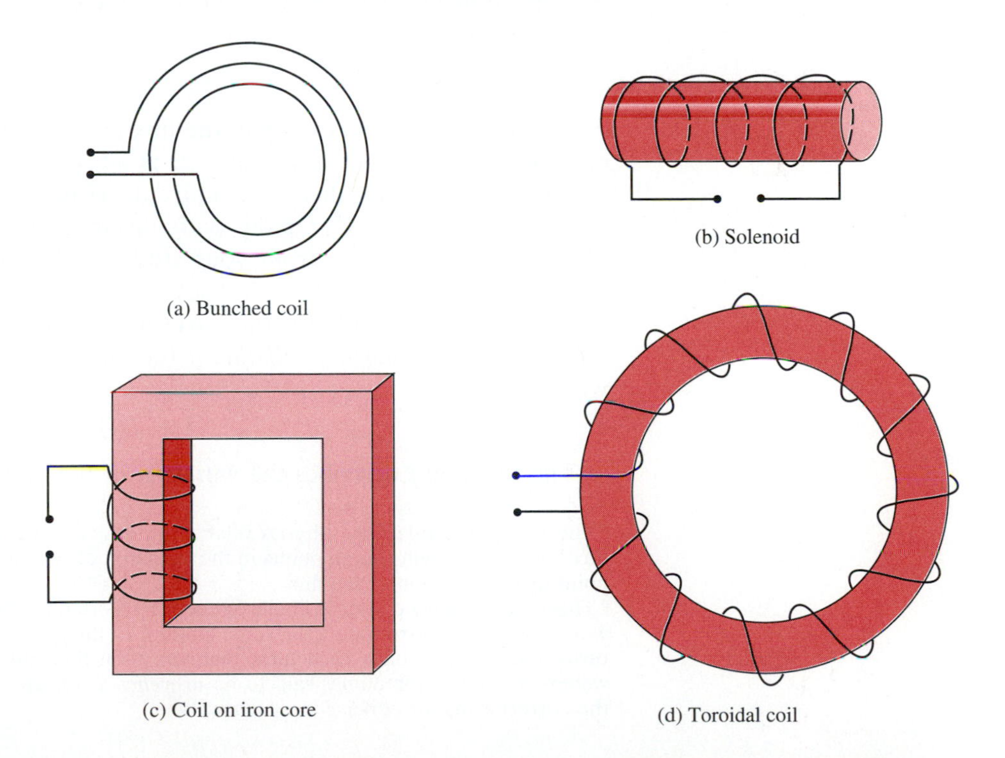

Figure 23
Coils of wire.

Magnetic Flux

Faraday advanced the idea of *magnetic flux* in an effort to explain his observations. It isn't easy to give a definition of magnetic flux that is both satisfactory and accurate without invoking other field concepts. This much is clear: whenever there is an electric current, there exists a magnetic field in the space surrounding the current. (In the case of a permanent magnet, a field is produced by the spin of electrons in the magnetic material. This spin of electrons can also be viewed as a movement of charge and, thus, a current.) A measure, or property, of the field is the concept of magnetic flux.

An elementary way of visualizing the flux is to imagine field lines filling up all of space. Their direction at any point represents the direction of the magnetic field at that point, as determined by the orientation of a compass needle under the influence of the field. The density of the lines passing through an area is a measure of the strength of the magnetic field there. The field is strongest in the vicinity of the current producing the field. The lines of magnetic flux are closed upon themselves, as seen by observing the distribution of iron filings sprinkled in the vicinity.

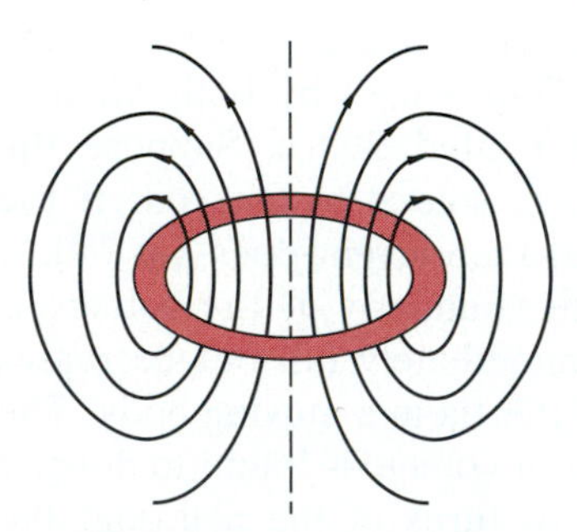

Figure 24
Magnetic flux linking a coil.

Figure 24 shows some turns of wire (say, N turns) tightly wound into a coil carrying a current i. (For simplicity the source of current is not shown.) Some of the flux lines are shown piercing the plane of the coil. The flux is said to *link* the coil.[8]

The flux linking a turn is labeled ϕ. If we assume that the flux linking each turn is the same, then the total flux linking the coil, which is called the *flux linkage* and is designated by the symbol λ, is $\lambda = N\phi$. That is, the flux linkage is proportional to the number of turns of the coil. The unit of flux linkage is the *weber* (Wb).[9]

Flux Linkage

Leaving this specific example, the flux linking a coil can be generated by the current in the coil itself or by currents in adjacent circuits. In general, then, λ is dependent on more than one current. However, since the strength of a magnetic field falls off rapidly with distance (as the inverse square of the distance, you will learn from your study of fields), unless special efforts are made to place a coil near another source of flux, the flux linkage of a particular coil will depend mainly on the current in that coil itself. We will make the assumption here that this *self-flux* is the only one of significance. Thus, we can write:

$$\lambda = f(i) \tag{17}$$

The function f describes the variation of flux linkage with current. It can be

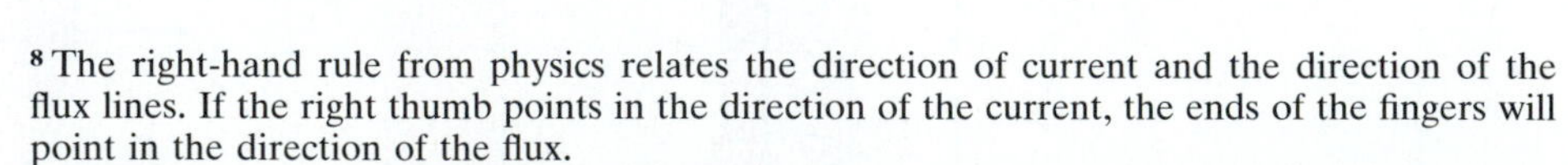

[8] The right-hand rule from physics relates the direction of current and the direction of the flux lines. If the right thumb points in the direction of the current, the ends of the fingers will point in the direction of the flux.

[9] This unit was named after Wilhelm Weber (1804–1891), a German physicist. Coils are often wound so that approximately the same amount of flux links each of the turns. If there are N turns and a flux ϕ links each turn, then the total flux linkage will be $\lambda = N\phi$. If ϕ is in webers, then λ is sometimes said to be in *weber-turns.* However, *turns* is dimensionless, so the correct units for either ϕ or λ are webers.

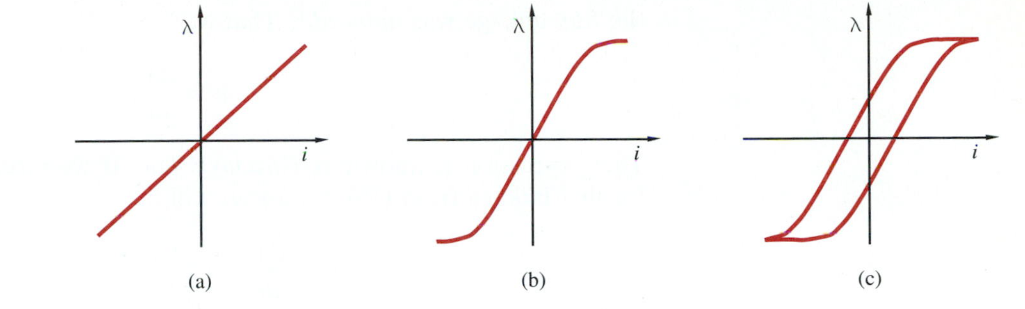

Figure 25
Variation of λ with current: (a) linear; (b) nonlinear; (c) multivalued, with hysteresis.

expected to depend on the geometry of the core and the magnetic properties of the core material.

Figure 25 shows several curves relating λ and i. In Figure 25a, the core consists of a nonmagnetic material (air or waxed paper), and in Figure 25b it is iron or some ferromagnetic material (ferrite or permalloy). The λ-i curve for a ferromagnetic core is highly nonlinear, whereas an air core yields a linear relationship between λ and i. The curve in Figure 25c is given only for completeness. If the current in the coil is varying periodically, the corresponding variation in the flux linkage lags behind. Thus, the flux linkage for a given value of current depends on whether the current is increasing or decreasing. The width of the "hysteresis" curve depends on the material of the core. We will neglect this aspect of magnetic cores in this book and limit ourselves to the linear case. In the linear case, the expression for λ can be written as:

$$\lambda = Li \tag{18}$$

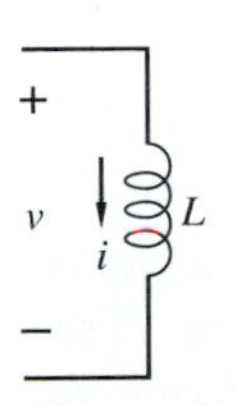

Figure 26
Circuit model for L.

The parameter L linking flux linkage to current is constant. This parameter is called the *inductance.* Its circuit symbol, shown in Figure 26, resembles a coil, but it is a purely hypothetical model.

The unit of inductance is the *henry* (H), which, from (18), has the dimensions of *webers per ampere.*[10]

For any given coil of wire, the inductance depends on a number of factors: the properties (permeability) of the medium surrounded by the coil, the geometry of the coil (for example, toroidal with circular or rectangular cross section, solenoidal, or pancake-shaped), and the number of turns. Over a period of time, formulas have been developed giving L in terms of each of these factors. For our purposes, the significant factor is the number of turns N. It turns out that the inductance is proportional to the square of the number of turns. Thus:

$$L = KN^2 \tag{19}$$

where K is dependent on the permeability of the medium and the geometry of the coil. We will make use of this idea in Chapter 10.

Voltage-Current Relationship of an Inductor

One of the most significant discoveries of Faraday was what he called electromagnetic *induction.* He found that whenever the flux linking a coil varies with time, a current appears in the coil. It was as if a voltage had been applied to the coil. He said that a voltage equal to the time rate of change of the flux linkage was *induced.*[11] That is:

$$v = \frac{d\lambda}{dt} \tag{20}$$

This expression is known as *Faraday's law.* If we insert into it the expression for flux linkage from (18), the result will be:

$$v = \frac{d\lambda}{dt} = \left(\frac{d\lambda}{di}\right)\frac{di}{dt} = L\frac{di}{dt} \tag{21}$$

On the right-hand side, (18) has been used for λ and the linear case has been assumed. (For the nonlinear case, there would be an added term dL/di, which goes to zero in the linear case.) Compare this v-i relationship of the inductor with the i-v relationship of the capacitor in (11); note the amazing symmetry between the two.

Since the voltage across an inductor is proportional to the derivative of the current, it is possible to solve for the current in terms of the voltage by integrating equation (21). Thus:

$$i(t) = \frac{1}{L}\int_{t_0}^{t} v(x)\,dx + i(t_0) \tag{22}$$

where $i(t_0)$ is the *initial value* of the current at time t_0.

Compare this with (14), which is the v-i relationship of a capacitor. Again we find that the inductor "remembers" its history. The value of inductor current at any time depends on all its past values up to that time. Again we find that the inductor current has contributions from two places. One is a constant—its initial value—which can be represented by a constant-current source. The other is from an inductor that has no initial current. This is illustrated in Figure 27. The equivalence of these two follows from KCL: $i_L = i_L' + I = i_L' + i_L(0)$. The current i_L' is the current in an inductor; its value at the initial time t_0 is zero.[12]

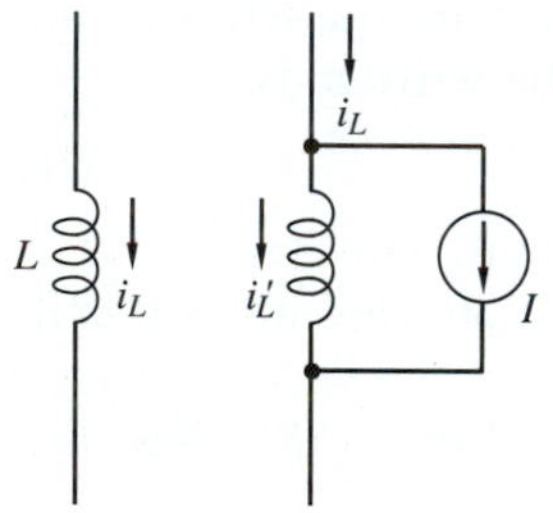

Figure 27 Inductor with initial current equivalent to initially relaxed inductor with a constant-current source.

EXERCISE 6

Construct a problem that specifies (a) the value of inductance of an inductor, (b) a function of time as the voltage across the inductor, and (c) an initial value

[10] The unit of inductance was named after the American Joseph Henry (1797–1878), who carried out experiments on electromagnetic induction at about the same time as Faraday.

[11] In your study of physics you probably carried out an experiment in which you plunged a magnet rapidly through a coil and noted the movement of the needle of a galvanometer connected to the coil, indicating a variation of current. The current results from the voltage induced in the coil.

[12] It is incomplete to take inductance as the only parameter in the model of a coil of wire. The coil will inevitably display resistance as well. Indeed, under some circumstances, it is necessary to include not only resistance in the model but also capacitance, since the charges in one turn of the coil are in proximity to those in adjacent turns. In most applications, however, taking into account only the major effect represented by the inductance is adequate.

of the inductor current with a specified orientation, and then asks to find the inductor current as a function of time. Appropriate units should be included for all quantities. Then solve the problem. □

EXERCISE 7

Construct a problem that should have the following features. It should specify the inductance value of an inductor and a function of time that is to be the current through it, all with appropriate units. The problem should ask for the inductor voltage (including units) as a function of time. Then solve the problem. □

Inductor Energy

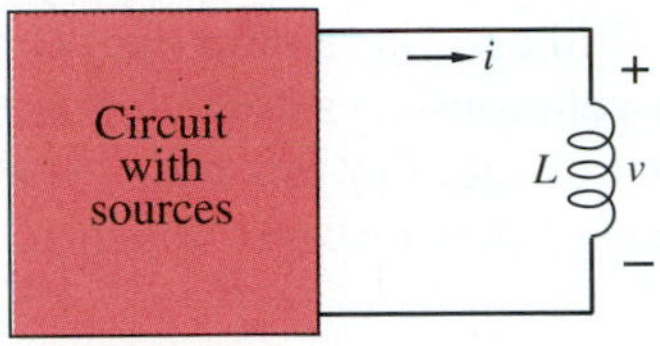

Circuit with sources — i → , L, v, + , −

Figure 28
Inductor in a circuit.

Suppose now that an inductor is connected to a circuit containing energy sources, as shown in Figure 28. As in the case of a capacitor shown in Figure 22, the question is, How much energy is delivered to the inductor? The references for the inductor current and voltage are standard, so the power *into* the inductor is $p = +vi$. Using the v-i relationship from (21) leads to:

$$p = vi = Li\frac{di}{dt} \tag{23}$$

The energy delivered to the inductor over time is obtained by integrating. Thus:

$$\begin{aligned} w_L &= \int_0^t Li\frac{di}{dx}\,dx = L\int_{i(0)}^{i(t)} i\,di \\ &= \tfrac{1}{2}L(i^2(t) - i^2(0)) \end{aligned} \tag{24}$$

or:

$$w_L(t) - w_L(0) = \tfrac{1}{2}Li^2(t) \tag{25}$$

That is, the net energy delivered to an inductor at time t starting from a time chosen as the time origin is proportional to the square of the current at that time. Since the flux linkage is proportional to the inductor current, it is equally true that the energy is proportional to the square of the flux linkage. (Determine the expression for the energy in terms of the flux linkage of the inductor.)

As described earlier for the capacitor, it is true for the inductor also that the energy delivered to L over a time interval depends on the value of current at the ends of the interval, not on the current waveform over the entire interval. Again, this is unlike the case for the energy delivered to a resistor.

An illustration of this conclusion is obtained by considering Exercise 5, where two different functions of time were specified as the capacitor voltage. Suppose, instead, that those functions are the currents in a given inductor on two different occasions, and that the problem is to find the net energy delivered to the inductor over the specified time interval. Confirm that the same calculations will lead to the same conclusion.

As in the case of the capacitor, we again ask, What happens to the energy delivered to the inductor? Again, the energy is not dissipated but *stored* in the inductor. We will again examine this matter in Chapter 6.

Conservation of Flux Linkage

It was noted in the preceding discussion that a magnetic field displays a phenomenon similar to mechanical inertia. If the voltage across a coil of wire (or its model, an inductor) changes suddenly, the current responds only gradually; it does not suddenly change its value. Thus, inductance in an electrical system is analogous to mass in a mechanical system. Indeed, the state of motion of a massive body cannot be changed suddenly. A fundamental postulate of mechanics is the *law of conservation of momentum,* which states that the total momentum of a system of particles cannot be changed instantly—except by an infinite force, something which is in very short supply. If inductance is analogous to mass, then flux linkage is analogous to momentum. Thus, it is reasonable to assume that a relationship similar to the law of conservation of momentum will apply to flux linkage. On this basis, we postulate the existence of a *principle of conservation of flux linkage,* which is stated as follows:

> The sum of all flux linkages around any closed path will be continuous; it cannot be changed instantly.

As is true of all other fundamental principles, we apply such principles and derivative consequences. If at any time we should find that these consequences do not agree with reality (measurements), then the principle would be overturned. This has not yet happened with the principle of conservation of flux. Note its analogy with the principle of conservation of charge.

In case there is just a single inductor in a particular circuit, then, since $\lambda = Li$, the conservation of flux linkage reduces to the statement that *the current in an inductor must be continuous*; it cannot be changed instantly.

The "Bilateral" Concept: A Digression

During your academic career, you will no doubt consult other references on circuits. You may encounter the term *bilateral* used to describe a two-terminal linear element which is said to have the same properties for one polarity of voltage as for the other, so that, "when the polarity of the voltage reverses, the current reverses, and vice versa." Although it isn't evident from this description, the implication is that the v-i characteristic is symmetrical around the origin. On this basis, linear resistors, capacitors, and inductors are said to be bilateral. The linear v-i curve of the resistor in Figure 3, for example, demonstrates this property. The resistor couldn't fail to have this property and still be linear; if a device is linear, it is *automatically* bilateral in the sense specified. For *linear* two-terminal components, therefore, the bilateral concept is redundant and of no value. Linearity *demands* that the component be bilateral in the sense given. But, you might ask, what about *nonlinear* two-terminal devices? Let's bring up the subject when we discuss the nonlinear device called a diode in the following section. And what about multiterminal devices? Again, let's consider that when we take up those devices in Chapter 4.

5 THE DIODE

The component to be introduced in this section is inherently nonlinear. Its presence in a circuit would require approaches different from those that can be used for linear circuits. There is reason, therefore, for omitting the consideration of such a device here. However, if you are not exposed to nonlinear devices early, you might get the impression that *everything* is linear or that all circuits can be treated by methods appropriate to linear circuits. In many places in this book we will make statements like "Such-and-such is true for linear circuits." If nothing is treated *but* linear circuits, however, you might acquire the impression that this "such-and-such" works all the time, for all kinds of circuits. On the other hand, if you are exposed to both linear and nonlinear devices and circuits, you will not have a tendency to apply linear procedures to all circuits you meet. This would be a decided advantage to you.

That said, you may wish to omit this section in a first reading. None of the material to be developed subsequently for linear circuits will be thereby affected. You will, however, need to omit problems containing the device discussed here.

A *diode* is a two-terminal electronic device whose conducting properties are different in the two possible directions of current flow. The most common physical diode consists of a *semiconductor* junction between two types of semiconductor materials, referred to as p type and n type. (You don't need to know anything about this to go on, just as you did not need to know anything about the physics of magnetic materials in order to learn about inductors.) An empirical current-voltage relationship of a typical diode is shown in Figure 29a. Note that the curve is not drawn to scale. In order to show the significant variations in the curve, a unit on the negative voltage axis is about 100 times a unit on the positive voltage axis; and a unit on the positive current axis is more than 10,000 times a unit on the negative current axis.

Clearly, the i-v characteristic of the diode is highly nonlinear. As the voltage increases in the positive (forward) direction, the current is at first quite small. After the voltage V_f (the *forward offset* voltage) is reached, the current rises rapidly. In the negative (reverse) direction the current remains almost constant

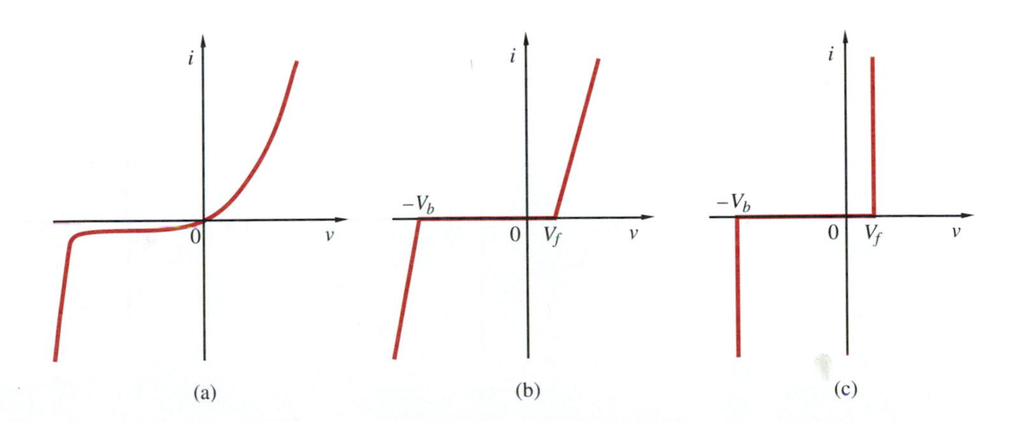

Figure 29 Current-voltage relationship of a typical diode and its approximations.

at a small negative value until voltage $-V_b$ (the *reverse breakdown* voltage) is reached, after which the negative current increases precipitously with almost no further change in the voltage. Typical values of the two significant voltages are 0.6 V for V_f and 30 to 100 V for V_b.

This empirical curve can be approximated fairly well by the curve shown in Figure 29b. In the central portion (the *nonconducting* region), the current is zero from $-V_b$ to V_f. On either side, the curve is approximated by straight lines with large but finite slopes. A further idealization can be made, as shown in Figure 29(c), where the slopes of the two lines are infinite. A device having this characteristic curve will be in any one of three possible conditions, or *states*: (1) when there is any nonzero forward current, the voltage is V_f (this is the *forward-biased* region); (2) when there is any nonzero reverse current, the voltage is $-V_b$ (this is the *reverse breakdown* region); and (3) when the voltage lies between $-V_b$ and $+V_f$, the current is zero (the *nonconducting* region).

Ideal Diode

In many applications of diodes in circuits, by design, it is not anticipated that the diode voltage will fall below $-V_b$. Hence the breakdown region will not come into play. For such applications, the model for a diode reduces to the one shown in Figure 30a. Remember that (at least for semiconductor diodes) V_f is a few tenths of a volt. If other voltages in the circuit are appreciably higher than that, then V_f can be neglected. Ultimately, then, the diode curve reduces to the one shown in Figure 30b. The circuit symbol for a diode having this ideal characteristic is shown in Figure 30c. This is called an *ideal diode.*

The ideal diode will be in one of two possible states:

Conducting, forward-biased, ON: $v = 0, i > 0$
Nonconducting, reverse-biased, OFF: $i = 0, v < 0$

Thus, the ideal diode is a two-state device like a switch; it represents either an open circuit or a short circuit. In a given case, it is necessary only to determine the state of the diode and to replace it by the appropriate short circuit or open circuit. The state of the diode will depend on conditions in the rest of the circuit, external to the diode. We will study procedures for determining the state in Chapter 3.

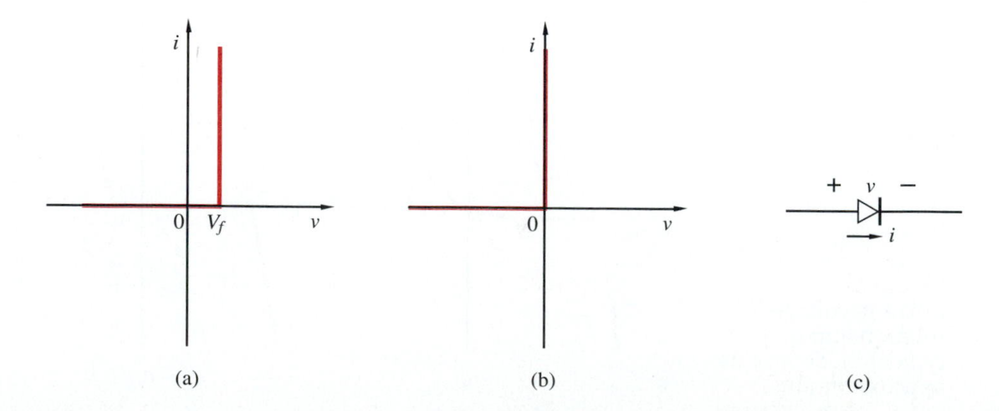

Figure 30 Ideal-diode characteristics and circuit symbol.

Note that the power delivered to the diode will be $p = vi$. Whenever the current has a nonzero value ($i > 0$), the voltage is zero. And whenever the voltage has a nonzero value ($v < 0$), the current is zero. Hence, the product of v and i is *always* zero. *No power is dissipated in an ideal diode*; it is a lossless device.

Diode Models

If we compare the two curves in Figure 30, we see that the only difference is that the vertical part of the characteristic occurs at V_f in part (a) as compared with 0 in part (b). That is, the ON state of the diode, when $i > 0$, will occur when $v = V_f$ in this case rather than $v = 0$. Figure 31a shows a circuit model of a diode whose characteristic is the one in Figure 30a. When the diode is OFF, the circuit representation is still an open circuit. But, now, when the diode is ON, the circuit reduces to a battery of V_f volts. Actually, the model in Figure 31a can serve equally well for the ideal diode; it is necessary in that case only to set the battery voltage V_f to 0.

Now let us reexamine Figure 29b but still assume that the breakdown region will not be encountered. The part of the characteristic to the right of $-V_b$ differs from Figure 29c only in that the slope of the curve in the conducting region is finite. That is, in this region, the current is proportional to the voltage. What is it whose current is proportional to its voltage? A resistor, of course. Thus, an appropriate model would be the one shown in Figure 31b: a series connection of an ideal diode, a battery, and a resistance R_f whose value is equal to the reciprocal of the slope of the i-v curve in the conducting region.

Finally, it is clear that the reverse direction constitutes the mirror image of the forward direction except that (1) the battery is reversed, (2) the battery has a numerical voltage V_b, and (3) the resistance has a different value because the slope of the curve in the breakdown region is not the same as that in the forward region. Thus, a more general model will be the one shown in Figure 31c. Note that both diodes in this model cannot be conducting at the same time; the forward diode D_f will be conducting only if v exceeds V_f, and the reverse diode D_r will be conducting only if v is negative and its magnitude exceeds V_b. (Confirm this by applying *KVL*.)

As noted, the diode characteristic is nonlinear. However, in Figure 29b, portions of the curve were approximated by straight lines. Although the overall characteristic in that diagram is still nonlinear, the curve is made up of segments, or pieces, which are linear. Thus, it is referred to as *piecewise linear*. As noted earlier in this section, to carry out an analysis with a piecewise linear device, it is necessary to determine (from the rest of the circuit) which "piece"

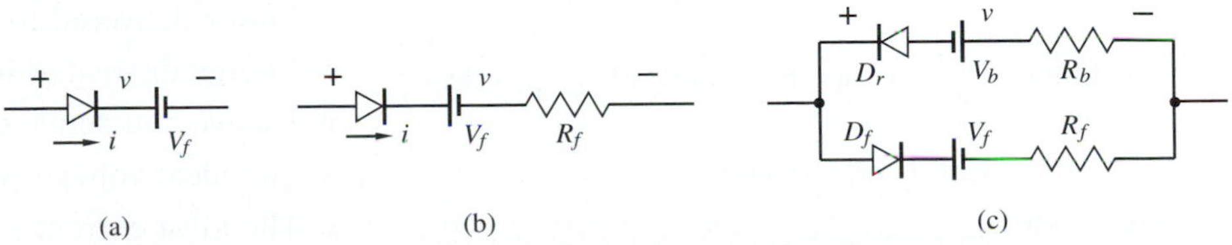

+ v −
i V_f
(a)

+ v −
i V_f R_f
(b)

+ v −
D_r V_b R_b
D_f V_f R_f
(c)

Figure 31
Increasingly nonideal diode models.

is operative and to replace the device with the linear portion appropriate to that piece. More of this later.

Nonideal Diodes

The circuit symbol in Figure 30c represents an ideal diode. A question might arise how we would represent a diode which is not ideal. Should we have different symbols for the different models shown in Figure 31? That would be excessive. However, we will use the symbol shown in Figure 32a to represent a physical diode. This differs from the symbol of an ideal diode only in that the triangle is not white but filled in. Depending on how accurate an analysis is required, any one of the models in Figure 31 could be used to represent the physical diode. We will assume that the diodes are ideal in most of the applications of diodes in this book.

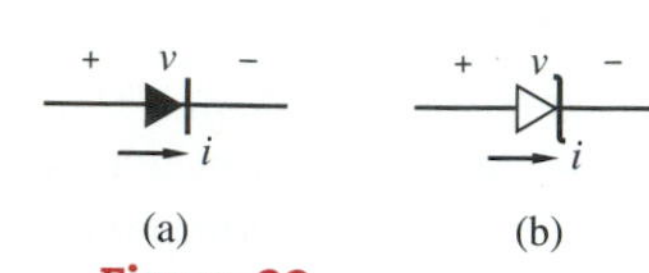

Figure 32
Real-diode symbols.

It was previously noted that, in many applications, circuit conditions are such that reverse voltages as large as V_b are seldom encountered. The converse is also true; that is, there are other applications where it is *required* for the voltage across two points to stay approximately constant. This is achieved by connecting a diode with the appropriate polarity across those two points so that it remains in its breakdown region. Diodes made for this purpose are called *Zener diodes.* The circuit symbol for a Zener diode is shown in Figure 32b.

Bilateralness Again: Another Digression

How about the bilateral concept when it comes to two-terminal *nonlinear* components? Recall that a two-terminal component is said by some to be bilateral if its v-i relationship is symmetrical about the origin. We saw in an earlier section that saying "linear component" automatically implies that the component has the bilateral property. The class "two-terminal linear" subsumes the class "two-terminal bilateral." On the other hand, if we say "nonlinear," does that automatically imply the class "nonbilateral"? Certainly the diode is not bilateral; its behavior for negative voltage does not match its behavior for positive voltage. However, it is possible to conceive a nonlinear curve which is symmetrical about the origin, as shown in Figure 33. Whatever the device may be with the characteristic shown here, it is nonlinear but bilateral. The class "nonlinear" includes subclasses that are "bilateral" and "nonbilateral." We will not study such devices in this book.

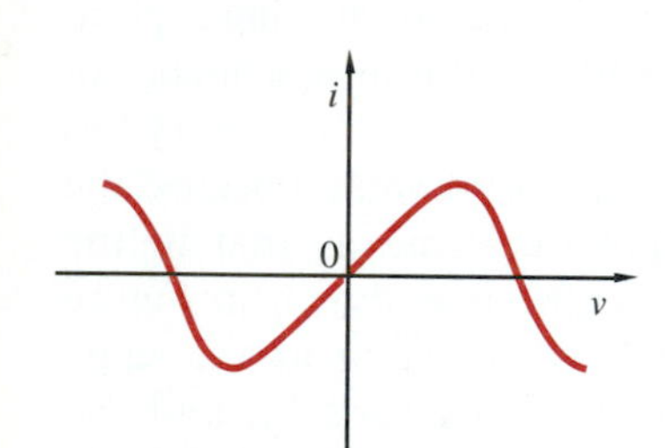

Figure 33
A characteristic curve of a nonlinear two-terminal device that is bilateral.

SUMMARY

In this chapter, the following topics were introduced:

- The linear resistor as a model of a physical resistor
- Short circuits and open circuits
- Ohm's law
- Power delivered to a resistor
- Energy dissipated in a resistor (Joule's law)
- Passive and active components
- The ideal voltage source
- The ideal current source

- The i-v and v-i relationships of a linear capacitor
- Energy stored in a capacitor
- Principle of conservation of charge; continuity of capacitor charge
- Magnetic flux and flux linkage
- The v-i and i-v relationships of a linear inductor
- Energy stored in an inductor
- Principle of conservation of flux linkages; continuity of inductor current
- Units: ohm (Ω), farad (F), henry (H), weber (Wb)
- The diode and its models
- Forward and reverse bias

PROBLEMS

1. A 45-V battery of dry cells with internal resistance R_0 is providing a current of 2 A to an external resistor R across which the voltage is 35 V. Find the values of R and R_0. (A diagram would be helpful.)

2. Two voltage measurements are made on an automobile storage battery.

a. With the battery terminals open, the voltage is found to be 12 V.

b. With a 1-Ω resistor connected to the terminals of the battery, the voltage is measured to be 11.6 V.

Find the voltage of the battery and its internal resistance. (Draw appropriate diagrams.)

3. The resistor in the circuit of Figure P3 has a resistance of 100 kΩ and a power rating of 5 W. The source current has a value of 14 mA and the load current is 4 mA, with the references shown.

a. Find the current through and the voltage across the resistor.

b. Decide whether the resistor should be replaced by one having a higher power rating.

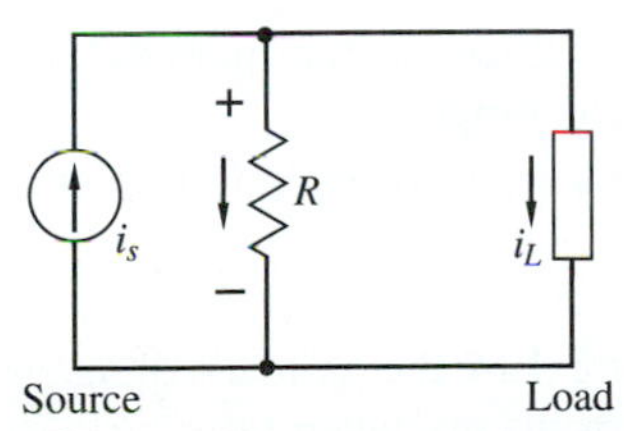

Figure P3

4. A resistor is connected across a 90-V source, resulting in a 180-mA current. What is the value of R, and what is the lowest power rating the resistor can have in order to safely connect it to this source?

5. A resistor dissipates 500 W of power when the voltage across it is 100 V DC. Find the value of (a) the resistance, (b) the current, and (c) the energy consumed in 20 s (include units).

6. The voltage and current of a two-terminal component with standard references are given as $i(t) = 4t$ A and $v(t) = e^{-2t}$ V between $t = 0$ and 10 s, and $i(t) = 0$ otherwise. Find out whether this component is active or passive, at least in this interval.

7. Repeat Problem 6 if $v(t) = 10 \cos 40t$ V and $i(t) = 2 \sin 40t$ A in the same interval of t.

8. A real source, with the model shown in the dashed box in Figure P8, has a variable resistor connected to its terminals. At a particular setting of this resistor, the current is found to be $i_s = 5$ mA. The resistor setting is changed so as to increase the voltage across it. Specify whether the value of i_s increases, decreases, or stays the same, giving your reasons.

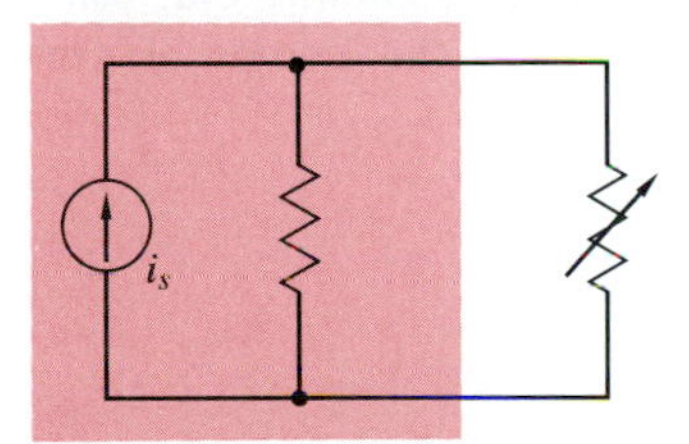

Figure P8

9. Not to be outdone by the creative student who suggested the circuit of Figure 16 in the text, you have proposed the circuit shown in Figure P9, which is also a connection of three ideal components introduced in this chapter. Now you need to find the power *delivered to* each of the three components. Go to it! Is the law of conservation of energy satisfied? If one or more of the values of power you calculate turn out to be negative, what would this imply?

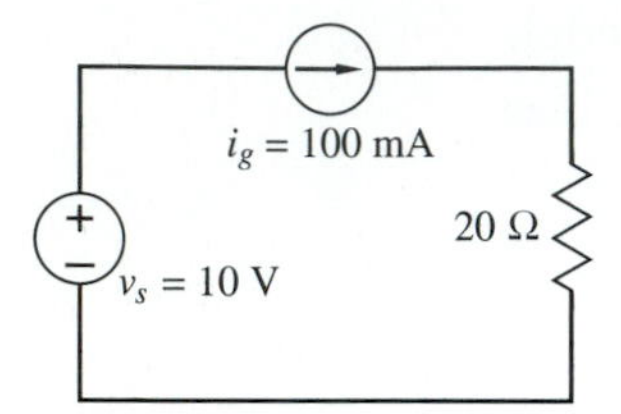

Figure P9

10. The creative student strikes again!

a. This time, select a value of R in Figure P10 so that both sources deliver power and the power delivered by the voltage source is half as much as that delivered by the current source.

b. Suppose, instead, the power delivered by the voltage source is to be twice as much as the power delivered by the current source. Now how much is R?

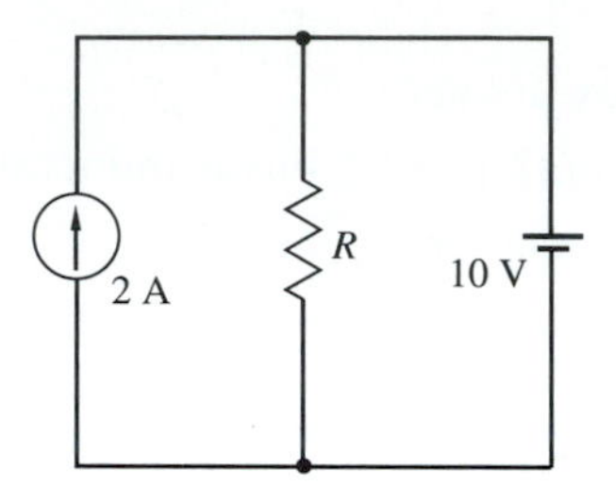

Figure P10

11. a. Using fundamental laws in the diagram of Figure P11a, derive an expression relating the voltage to the current at the terminals and plot the result.

b. Repeat for the diagram in Figure P11b.

c. Repeat for the diagram in Figure P11c.

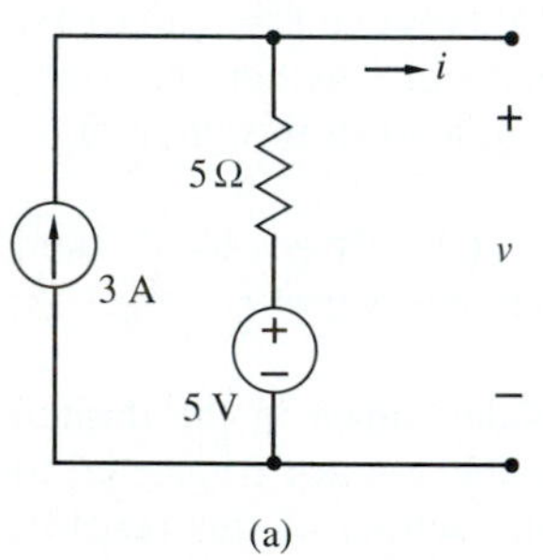

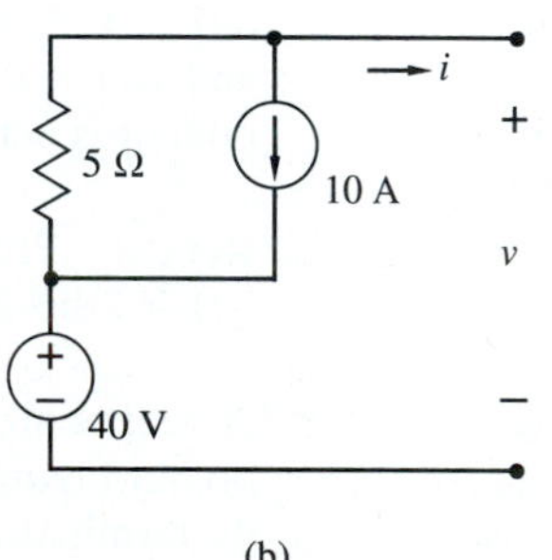

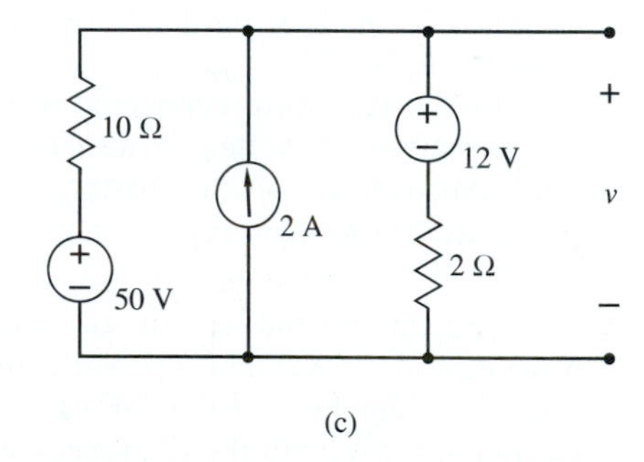

Figure P11

12. The diagram in Figure P12 shows the *external characteristic* (that is, the terminal v-i relationship) of a DC generator. Determine approximate values of the parameters of two equivalent circuits that can represent the generator over the range of current $0 < i < 40$ A.

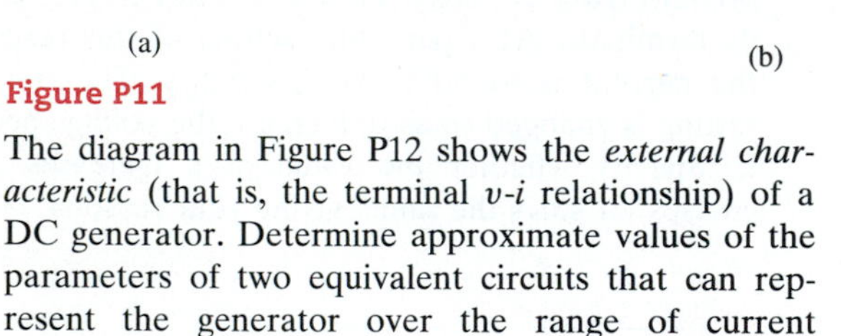

Figure P12

13. The curve in Figure P13 gives the voltage-current relationship of a physical source.

a. Find the approximate values of the parameters, valid for the interval $0 \leqslant i \leqslant 10$ A, of one model of the source.

b. Find approximate values of the parameters of another model in this range.

c. Repeat if the models are to be valid in the interval $0 \leqslant i \leqslant 25$ A.

14. In the circuit of Figure P14 the source voltage and one of the resistor currents are given.

a. Find values of all other resistor currents.

b. Assume standard references and find all the resistor voltages.

c. Find the power supplied by the source.

d. Think of a way to confirm part (c), and do so.

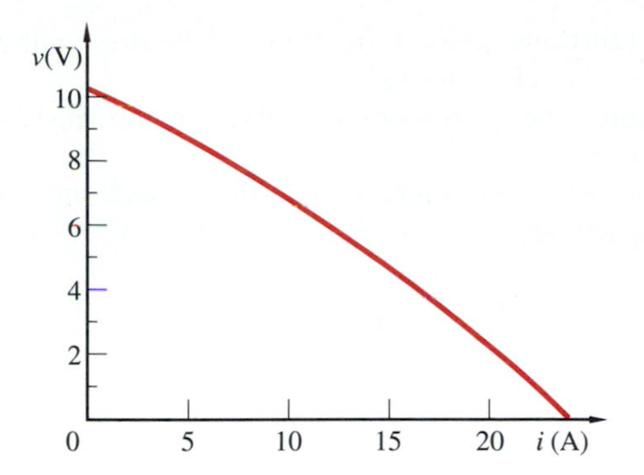

Figure P13

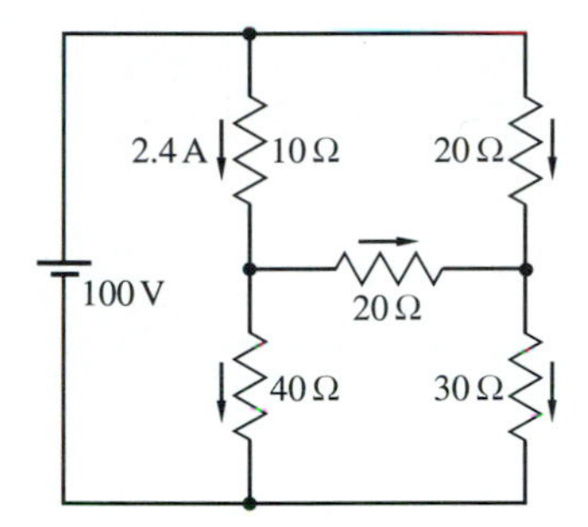

Figure P14

15. In the circuit of Figure P15, the source voltage and current i_R are given as follows: $v_s = 100 \cos 400t$ V and $i_R = 0.06 \cos 400t - 0.02 \sin 400t$ A.
    a. Find the current i_C in the capacitor.
    b. Find an expression for the power into the capacitor.
    c. Also find the current in R_1.

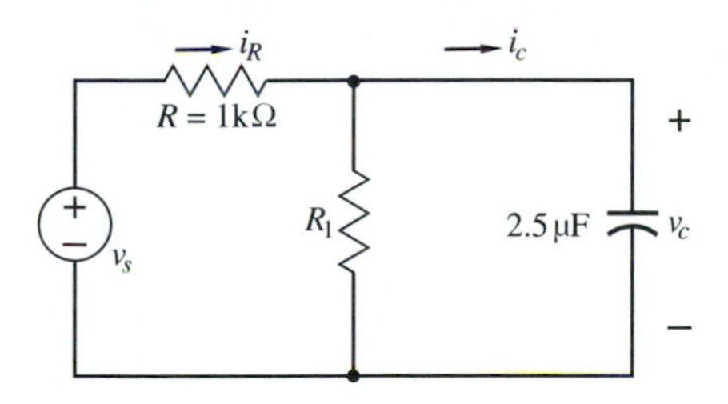

Figure P15

16. In the circuit of Figure P16 the two resistances are equal. The known quantities are the source current $i_s(t)$ and the voltage across the resistor on the left, $v_1(t)$. Applying the fundamental circuit laws and the appropriate v-i relations of components,
    a. Find an expression for the inductor voltage v_L in terms of component values and the known quantities $i_s(t)$ and $v_1(t)$.
    b. Find also an expression for the current i in the unknown branch.

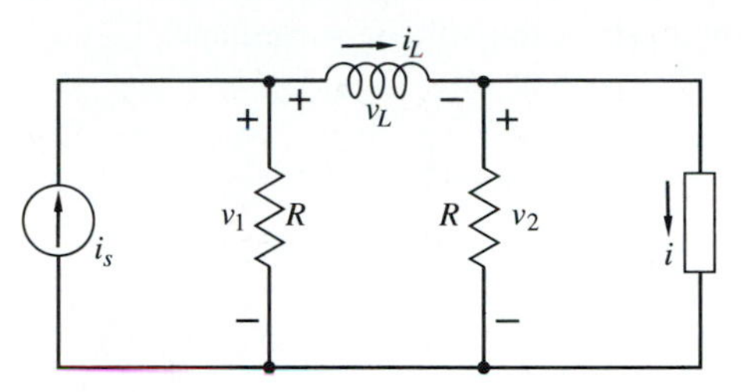

Figure P16

17. In the circuit shown in Figure P17, find the source current $i_s(t)$ that will result in $v_R(t) = -20e^{-2t}$ V.

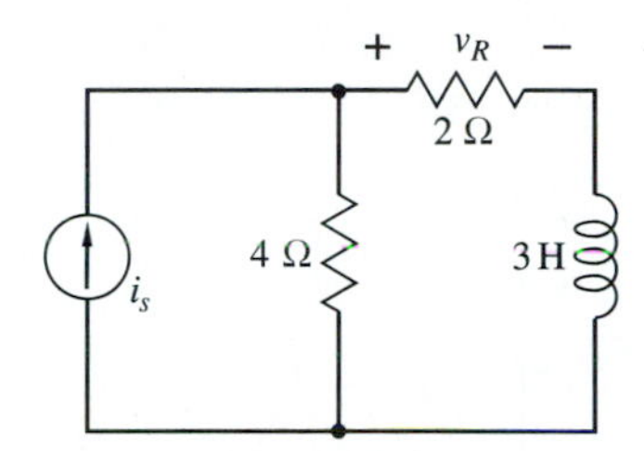

Figure P17

18. A voltage source v_s is connected to a 0.5-μF capacitor. The charge delivered to the top terminal is shown by the waveform in Figure P18.

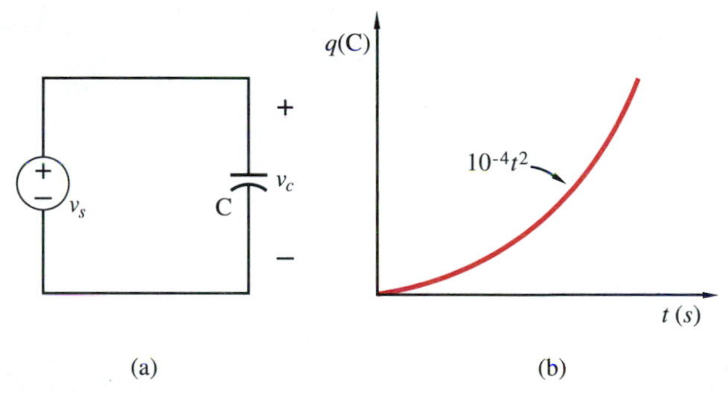

Figure P18

a. Find the power delivered to the capacitor as a function of time.
b. Find the energy stored in the capacitor at $t = 5$ s.
c. Find the voltage across the capacitor at $t = 2$ s.
(Include units in all cases.)

19. The voltage v across a 2-H inductor has the waveform shown in Figure P19.

a. Sketch the inductor current i (assuming standard references); include units.

b. Interpret the relationship between $i(0)$ and $i(40)$ on the basis of the voltage waveshape.

20. The functions plotted in Figure P20 are voltages across a 20-μF capacitor.

a. Sketch the corresponding curves of the capacitor current.

b. On the same axes, sketch the power into the capacitor.

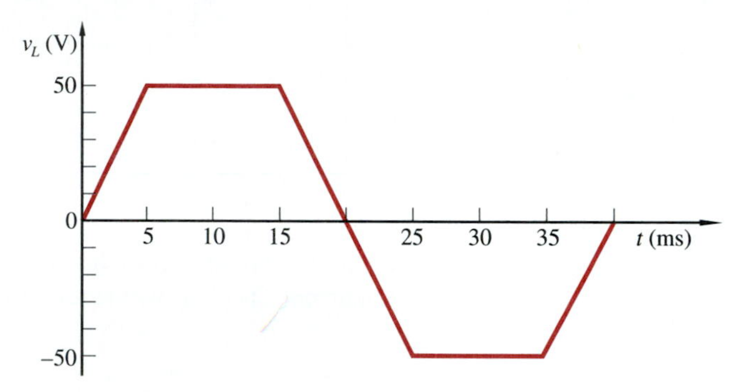

Figure P19

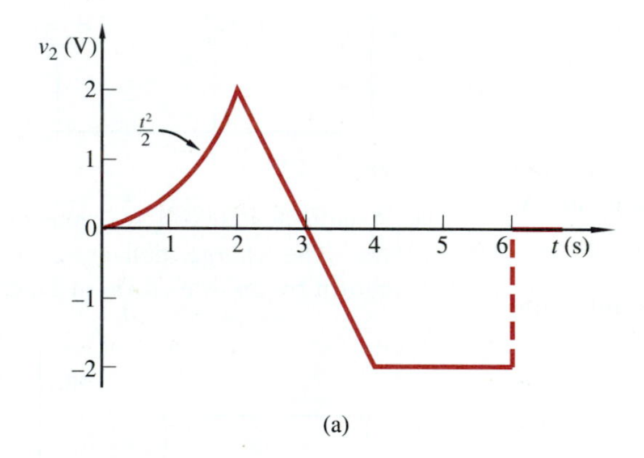

(a)

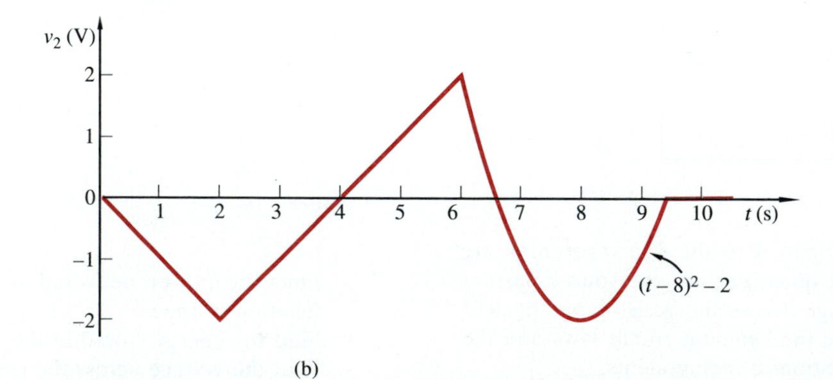

(b)

Figure P20

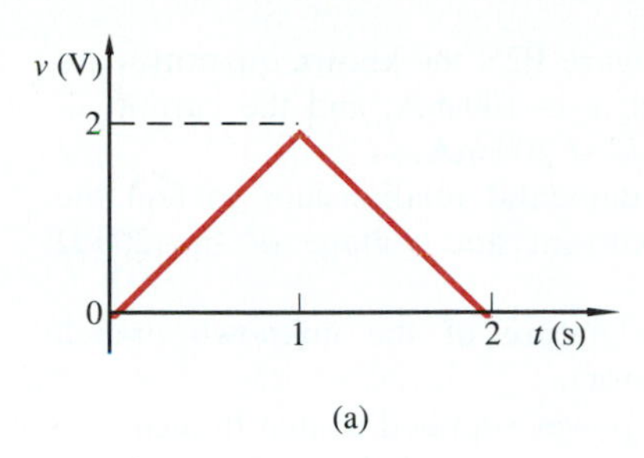

(a)

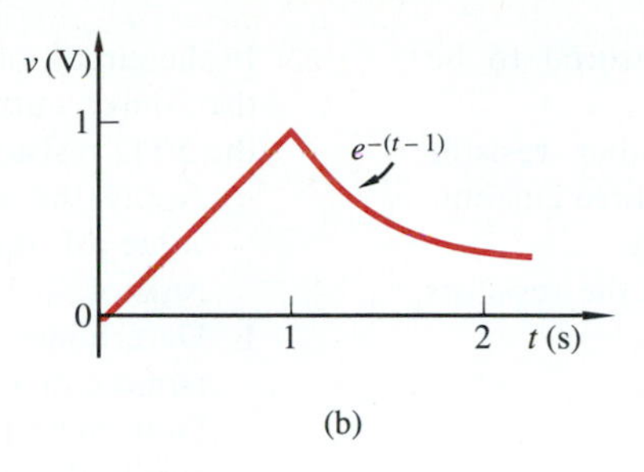

(b)

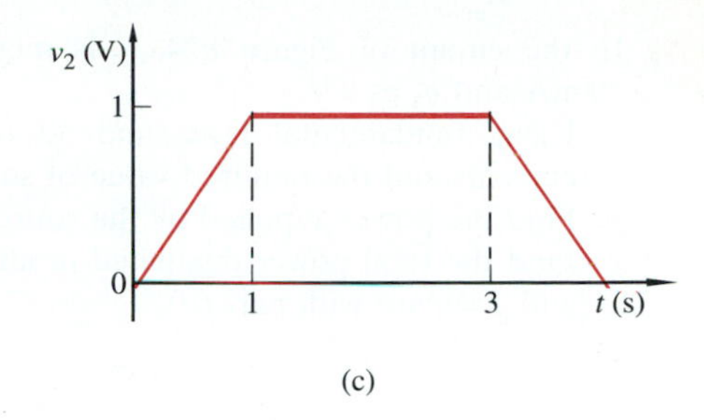

(c)

Figure P21

c. Also sketch the energy stored in the capacitor at each instant of time.

21. Each of the functions in Figure P21 is the voltage across a 2-μF capacitor.

a. Sketch the capacitor current on the same axes.

b. Sketch also the energy stored in the capacitor.

c. Repeat part (a) if each function is the voltage across a 10-mH inductor.

d. Repeat part (b) for the energy stored in the inductor.

22. The waveform in Figure P22a is the current in a 2-H inductor.

a. Sketch to scale the voltage v across the inductor, the power p into the inductor, and the energy w stored in the inductor. Label each curve, showing the equation used to derive it.

b. Suppose the energy stored in the inductor can all be converted into mechanical energy. How high could a 165-lb (75-kg) person be raised with the maximum energy stored?

c. You want to generate the same current waveform through the inductor, but only a nonideal source is available, as shown in the box in Figure P22b. Plot the shape v_s is required to have in order to produce the desired current.

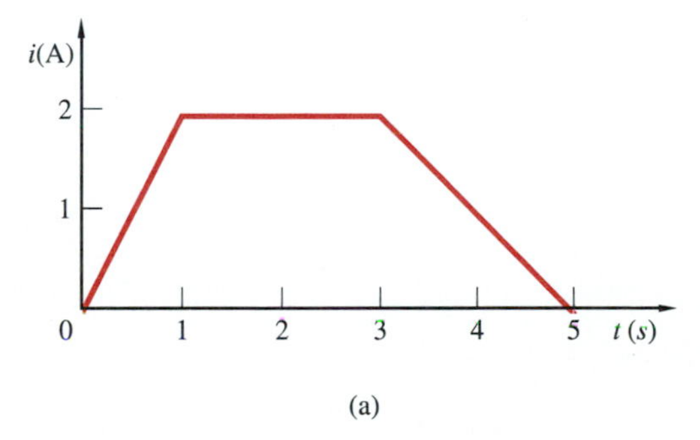

(a)

3 Ω
+
−
v_s
2 H
i
source

(b)

Figure P22

23. Voltage v in Figure P23 is 40 V at a particular instant of time soon after the circuit is established, but before all variable changes are completed.

a. Applying all the fundamental laws at your disposal, find the value of the current i_L at that instant.

b. Find also the derivative di_L/dt at that instant.

c. Find the values of i_L and v_L after the passage of a long time when no further changes are taking place in any variables.

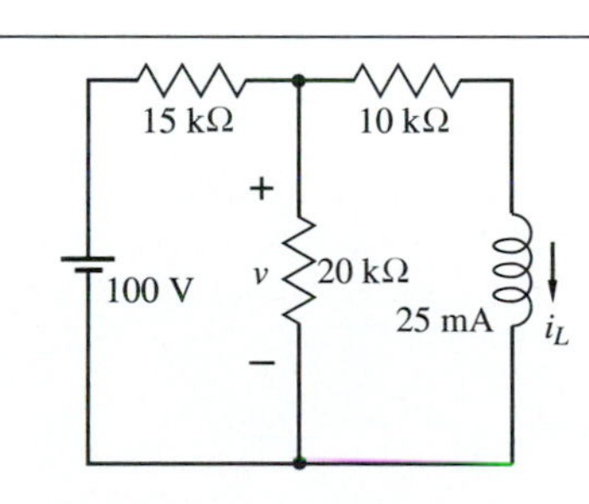

Figure P23

24. In the circuit of Figure P24, i_1 is measured to be 40 mA and v_2 as 9 V.

a. Using fundamental laws, find all other resistor currents and the required value of source current.

b. Find the power supplied by the source.

c. Find the total power dissipated in all the resistors and compare with part (b).

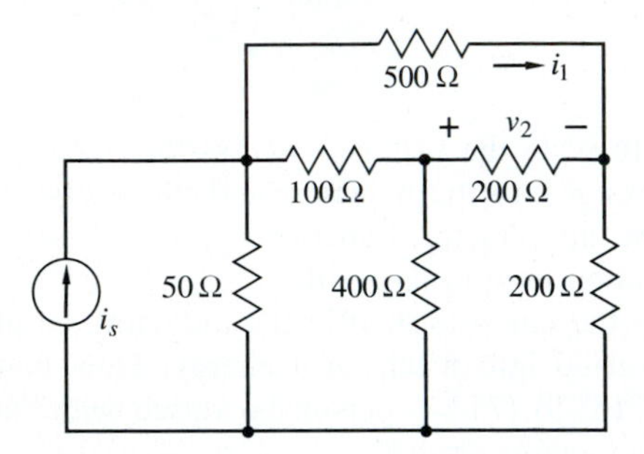

Figure P24

25. In the circuit of Figure P25, the known quantities are the source current, $i_s = 100$ mA, and the current in the 50-Ω resistor, $i_R = 200$ mA.

a. Apply the fundamental relationships to find the value of the current and voltage of the 200-Ω resistor.

b. Determine the nature of the unknown branch (source or resistor).

c. Determine the power supplied to that branch.

d. Verify that there is a power balance.

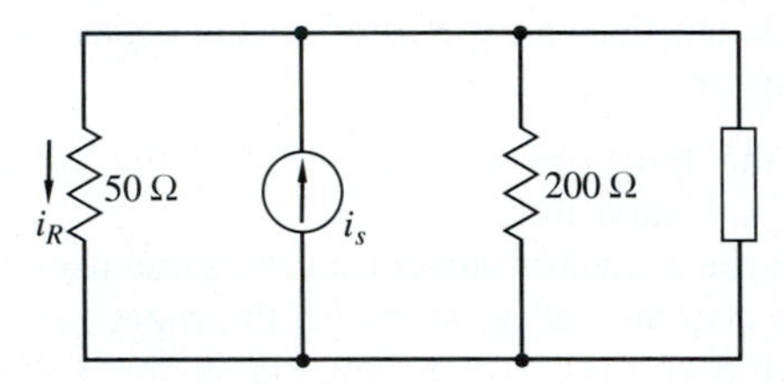

Figure P25

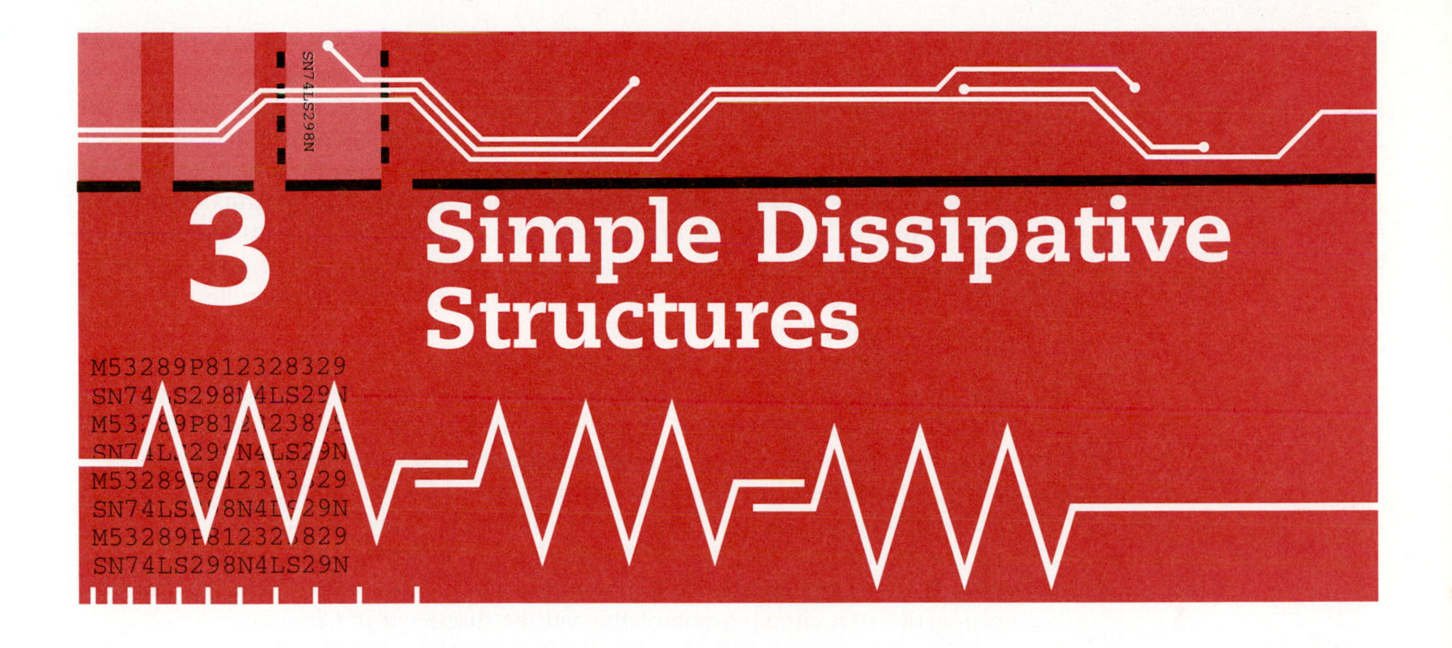

3 Simple Dissipative Structures

Generally speaking, engineers perform at least three related and overlapping functions: (1) analysis, (2) synthesis, and (3) design. *Analysis* consists of an examination and study of a whole structure in order to determine something about its parts. *Synthesis* consists of combining separate parts to construct a whole system whose characteristics, presumed to be desirable, have been specified beforehand. *Design* is a process wherein the results of synthesis are subjected to additional criteria (such as economics, aesthetics, safety, resource conservation, environmental factors, and the like) to arrive at an optimal end product. Throughout their education, engineers must study and engage in each of these activities.

Many of the problems an electrical engineer is called upon to solve involve finding the voltage across, the current through, or the power delivered to some electrical device when that device is connected in a circuit of other devices. This is a problem of analysis. Conversely, an engineer may be asked to specify the characteristics of some electrical device or circuit connected in a more extensive circuit with other electrical devices so that the voltage, current, or power absorbed somewhere else in the circuit will be equal to some specified value. This is a problem of synthesis.

In the preceding two chapters we described the circuit laws governing the relationships among the voltages (KVL) and among the currents (KCL) of devices that are connected in a circuit. These circuit laws are independent of the nature of the specific components making up the circuit. We also examined the voltage-current relationships of a number of two-terminal electrical devices. The next step is to complete the process of circuit analysis by combining these two parts—the circuit laws and the device characteristics.

In the preceding two chapters we also developed models of certain physical electrical devices. It is our expectation that the electrical behavior of interconnections of these models (and others to be introduced subsequently) will approximate, to any desired degree, the actual behavior of physical devices and their interconnections. In this book we shall be concerned mainly with the model, and, most of the time, problems will be formulated in terms of models. From time to time, however, we will formulate some problems in terms of physical devices, just to remind you that this aspect of the overall function of an engineer—namely, to construct an appropriate model to describe a physical situation—should not be forgotten.

An electrical network has two distinct aspects. One of these has to do with the connectedness of components. Thus, the components which form the network are interconnected at their terminals. The junctions formed at the common terminals of two or more components are the nodes in the network. Similarly, one or more closed loops are formed by the interconnections. Thus, no matter what the constituents of the circuit may be, there is a definite geometrical aspect to a circuit. A considerable amount of knowledge exists in a branch of mathematics called *topology* that deals with these geometrical properties of a circuit. Some of this will be discussed in Chapter 5.

The second aspect deals with the modifications of signals by the circuit. According to Ohm's law, the only way in which resistors modify a signal is to multiply it by a constant. On the other hand, inductors and capacitors perform the operations of differentiation and integration. Thus, when a signal is transmitted through a circuit consisting of R, L, and C components, it is modified by different combinations of differentiation, integration, and multiplication by a constant. If you will now pause and think of other operations which might conceivably be performed on signals—such as addition, multiplications, amplification, delaying their transmission, modulation, limiting the amplitude (called *clipping*), changing the average value (called *clamping*)—you will appreciate that a model consisting only of R, L, and C components is quite limited in the ways that it can modify a signal. (Some further operations are possible when diodes are included.) We shall later introduce other devices, but as long as we are restricted to a linear model, some desirable operations will always remain beyond our reach.

This line of thought leads to another consideration. At the beginning of the book we considered observations in the physical world and postulated a model which would serve to explain the observed phenomena. We now see a whole other vista. When confronted with the task of synthesizing a circuit to perform a particular function, we need not limit our thinking to those operations that can be performed on signals by presently available devices. We might ask for a device to perform some operation which has not heretofore been performed. This, in turn, might spur investigation, and eventually invention, of one or more devices to perform that operation. This is an exciting thought. Rather than develop a model to account for the behavior of existing physical devices, we might first *postulate* a model of a desired physical device; then we would seek to invent a device whose behavior would be approximately the same as that of the model.

In this book we shall not reach a level of expertise which will permit exploring such ideas. Nevertheless, these possibilities should be kept in mind as one of the goals toward which you should strive in your professional education.

In this chapter we will take only a small bite. We will limit ourselves in two ways: (1) by considering only simple circuit structures and (2) by including, in addition to voltage and current sources, only devices that do not store energy: resistors and diodes. In subsequent chapters we will introduce the two-terminal energy-storing devices described in Chapter 2, as well as three- and four-terminal devices, and we will study methods of analysis appropriate to network structures of any size.

A word about terminology. To describe the nature of the circuits to be treated here, we might have used the term *resistive* to imply the lack of energy storage. But this might be misleading, since diodes are included. Another possibility is the term *algebraic,* which calls attention to the nature of the equations (no derivatives or integrals), rather than to the nature of the elements. The final choice, both here and in Chapter 4, is *dissipative,* which conveys the same meaning as *resistive* but without the danger of possible misunderstanding.

Another word about units. Whenever a mathematical expression relating some variables and parameters in a circuit is written, the equation must be *dimensionally homogeneous*; that is, each term in the equation must have the same dimensions. Thus, in the equation

$$v_1 + v_2 - v_3 = 0$$

each term is dimensionally *voltage* and each of the voltages must be measured in the same units: normally volts, but possibly millivolts or another unit. Furthermore, it must be stated what the units are. In a given case, if some of the values are specified in one unit and some in another, they must all be converted to the same units before insertion in the equation. In an equation like:

$$v = Ri$$

if all quantities are measured in their SI units, then the equation will be dimensionally homogeneous. However, the equation can still be dimensionally homogeneous even if some of the units are not SI but one unit compensates for another. Thus, R in kilohms and i in milliamperes will still give v in volts. If we agree that all units are SI units, then it is not essential to show each unit in an equation explicitly. In this book we will sometimes follow this practice. Whenever values are given without designating a unit, it will be assumed that SI units are implied. Whenever a value is in non-SI units, it must be shown explicitly.[1]

[1] As someone who presumably knows what he is doing, an author can take such liberties and shortcuts. However, early in the process of learning, you should make a habit of specifying all units. This has the virtue, at least, of compelling you to think about what the units in a given case should be. This way, you can earn the right to take shortcuts later on.

1 SERIES AND PARALLEL CIRCUITS

We will begin by studying two very simple, but very useful, circuit structures.

Series Circuits

Figure 1a shows a physical resistor connected to a real battery with some connecting wires. A circuit model of this interconnection is shown in Figure 1b. The real battery is represented by a voltage source and the internal resistance of the battery. The resistance of real wires is proportional to their length; for the length involved here, the wire resistance will be negligible. Nevertheless, we have shown it for completeness. Note that, in a circuit model, the lines joining together the components of the model are not wires; they have no resistance but simply show the interconnections in a convenient manner.

The components in the circuit model in Figure 1b form a loop. By KCL applied at each of the nodes, the actual current in each of the components is found to be the same.

We say that branches of a circuit that are connected end to end in such a way that they carry the same current (if there were to be a current) are connected *in series.*

The relative placements of components have nothing to do with their being connected in series or not. Figure 2 shows six different arrangements of components; in which of these arrangements are all the components connected in series?[2] The placement of components in a circuit diagram has to do with clarity and convenience. It is clear that all the components in Figure 1, including the source, are connected in series. Let V_{R_0}, V_{R_w}, and V_R be the voltages across the resistors, assuming standard references. An expression relating these voltages can be obtained by applying Kirchhoff's voltage law around the single loop comprised by the circuit. (Do this before you go on reading.) The result will be:

$$\begin{aligned} v_{R_0} + v_{R_w} + v_R &= v_s \\ R_0 i + R_w i + Ri &= v_s \\ (R_0 + R_w + R)i &= v_s \end{aligned} \tag{1}$$

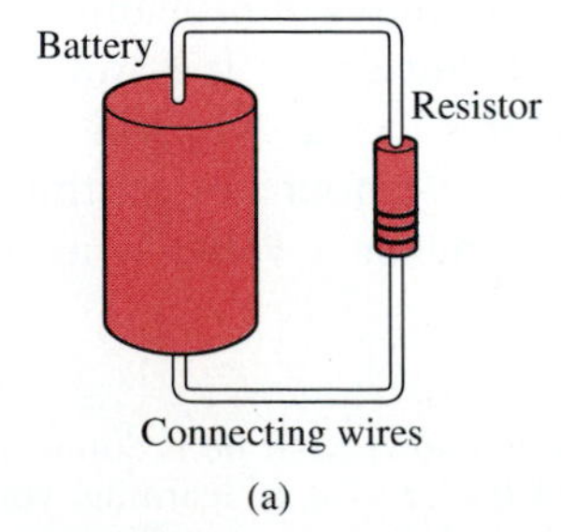

(a)

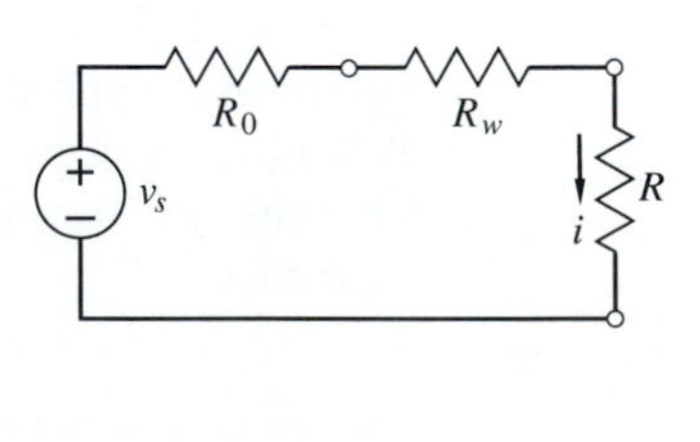

(b)

Figure 1 Resistor connected to a battery.

[2] All but (b) and (d).

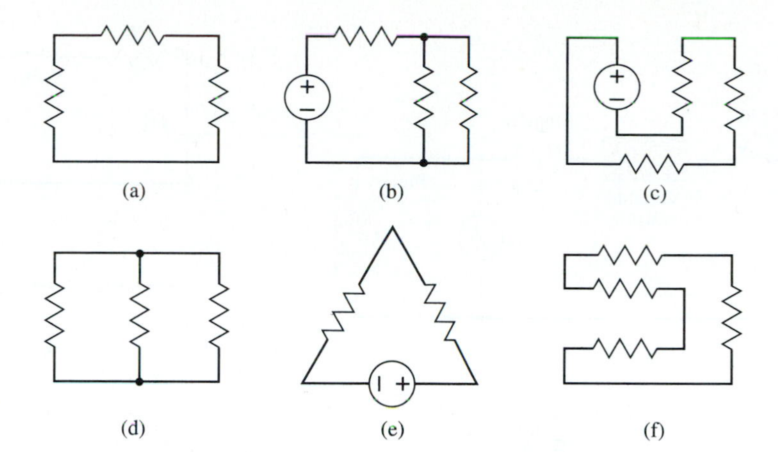

Figure 2
Which interconnections are in series?

In the second line, Ohm's law was used to replace all the resistor voltages, and in the last line i was factored from each term. The form of the last line is equivalent to the statement that some constant times the current equals the source voltage. We conclude that

> A series connection of resistors can be replaced by a single *equivalent resistor* whose resistance is equal to the sum of the series resistances.

This result permits simplifications in the analysis of many circuits. What we have done for three resistors in series can be done for any number, by mathematical induction.

EXERCISE 1

By using KLV show that a series connection of voltage sources can be replaced by a single voltage source whose voltage is the algebraic sum of the series-connected voltages, with an appropriate reference. □

EXAMPLE 1

The diagram in Figure 3a shows an arrangement for charging a battery by a battery charger through a variable resistor. The battery and charger voltages are measured by voltmeters and the current by an ammeter. A model of the arrangement is shown in Figure 3b. Initially we have no knowledge of the voltages and internal resistances of the battery or the charger, but a series of measurements are made, as follows:

a. *With the switch open:*

Voltmeter V_c indicates 18 V.
Voltmeter V_b indicates 12 V.

(What does the ammeter indicate?) Since no current is flowing, there is no

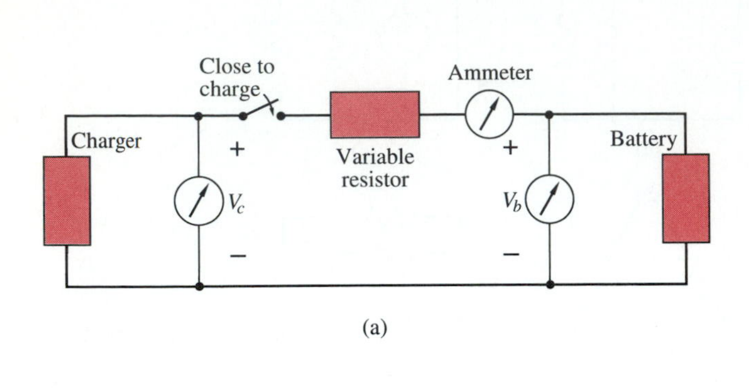

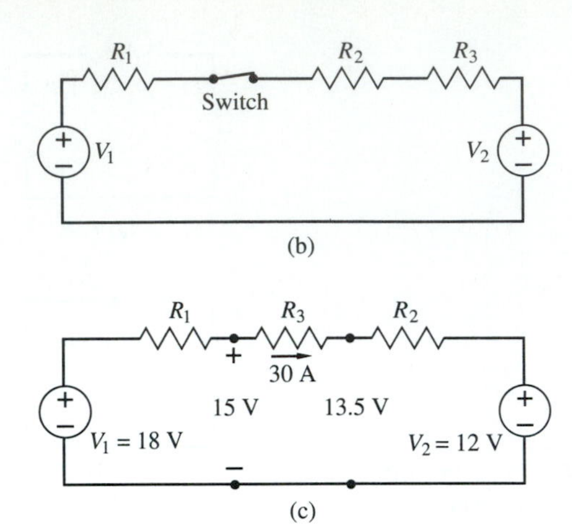

Figure 3
Charging a battery.

voltage across any of the resistances, so the voltmeters read the correrspondingsource values. Because of the given voltmeter connections, the source voltages are $v_1 = 18$ V and $v_2 = 12$ V, with the references as shown in Figure 3b.

b. *With the switch closed:*

Voltmeter V_c indicates 15 V.
Voltmeter V_b indicates 13.5 V.
Ammeter indicates 30 A.

These values are shown in Figure 3c. We can write an expression for the voltage across each resistor in terms of the current by Ohm's law. We can write another expression that includes each of these voltages by KVL. The results will be:

$$v_{R_1} = R_1 i = 18 - 15 = 3$$

$$v_{R_2} = R_2 i = 13.5 - 12 = 1.5$$

0.15 Ω
i = 30A
6 V
R_3

Figure 4
Equivalent circuit for battery-charging circuit.

Since i is known, we can now solve for R_1 and R_2. The results are $R_1 = 0.1\ \Omega$ and $R_2 = 0.05\ \Omega$. All components in Figure 3c are connected in series. R_1 and R_2 can be replaced by a single equivalent resistor. Similarly, the two sources can be replaced by a single source. The result of doing this is shown in Figure 4. Adjusting the value of R_3 changes the value of the current. For the setting that gives the measured value $i = 30$ A and the values of v_{R_1} and v_{R_2} previously determined, the voltage across the 0.15-Ω resistor, with standard references, is 4.5 V. Hence, by KVL, the voltage across R_3 (again with standard references) is $6 - 4.5 = 1.5$ V. Hence, the required value of R_3 is $1.5/30 = 0.05\ \Omega$.

VOLTAGE DIVIDER A simple circuit structure of some importance will be considered in this section. A part of the circuit in Figure 5a consists of two resistors connected in series. The part of the circuit in the rectangle can be anything so long as the voltage across it is 100 V; for simplicity, you can think of it as a voltage source. The two resistors in series can be replaced by a single resistor having a resistance of 200 Ω. Thus the current will be $i = 100/200 = 0.5$ A. The voltage across each resistance now follows from Ohm's law. Thus, $v_1 = 25$ V and $v_2 = 75$ V. We see that the total voltage across the two series resistors is distributed between them in proportion to the resistances (150 is 3 times as big as 50, so the voltage across the 150-Ω resistor is 3 times as much as that across the 50-Ω resistor). Hence, although it would more appropriately be called a voltage distributor, this structure is called a *voltage divider*.

The voltage divider is important enough to require a general expression for the distributed voltages. The general structure is shown in Figure 5b. From Ohm's law and Kirchhoff's voltage law:

$$v_1 = R_1 i \qquad v_2 = R_2 i \tag{2a}$$

$$v = v_1 + v_2 = (R_1 + R_2)i \tag{2b}$$

Now solve for i in the last equation and substitute into the first. The result will be:

$$v_1 = \frac{R_1}{R_1 + R_2} v \tag{3a}$$

$$v_2 = \frac{R_2}{R_1 + R_2} v \tag{3b}$$

$$\frac{v_1}{v_2} = \frac{R_1}{R_2} \tag{3c}$$

The fraction of the voltage that appears across R_1 is in the ratio of R_1 to the total series resistance $R_1 + R_2$, and similarly for the voltage across R_2.

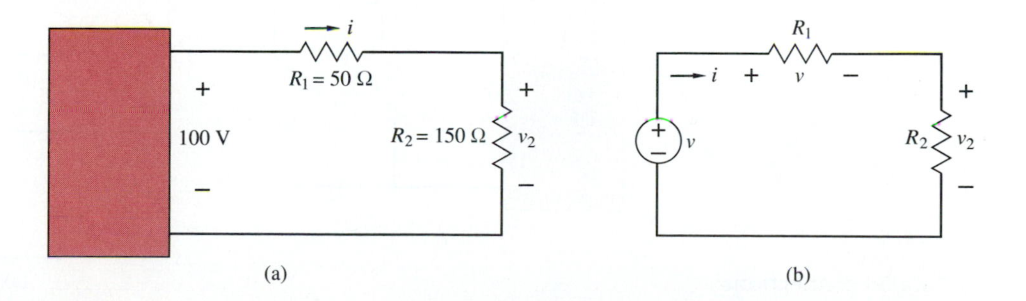

Figure 5
Voltage-divider circuit.

EXAMPLE 2 A 10-Ω resistor is connected across the terminals of a 12-V battery having an internal resistance of 0.2 Ω. The problem is to find the voltage across the resistor.

First draw the circuit, showing the internal resistance of the battery explicitly. (Do it.) The structure is that of a voltage divider, so the voltage across the 10-Ω resistor will be 10/(10 + 0.2) times the battery voltage, or 11.76 V.

EXERCISE 2

Two resistors connected in series are fed by a voltage source. The resistance of one of them is 3 kΩ and the voltage across the other one is found to be 4 V; the current is found to be 2 mA. Use Ohm's law and KVL appropriately to find (a) the voltage across the 3-kΩ resistor, (b) the resistance of the second resistor, and (c) the voltage of the source. Then use the voltage-divider relationship to confirm your answers.

ANSWER: 6 V, 2 kΩ, 10 V. □

Parallel Circuits

Another simple structure of similar importance is shown in Figure 6a; two resistors are connected in a circuit in such a way that the voltage across each of them is the same, by KVL.

We say that the branches of a circuit that are connected end to end in such a way that they have the same voltage (when there is a voltage) are connected *in parallel.*

Again, the placement of the components relative to each other is not relevant, only the manner of their connection. For example, two components may be at

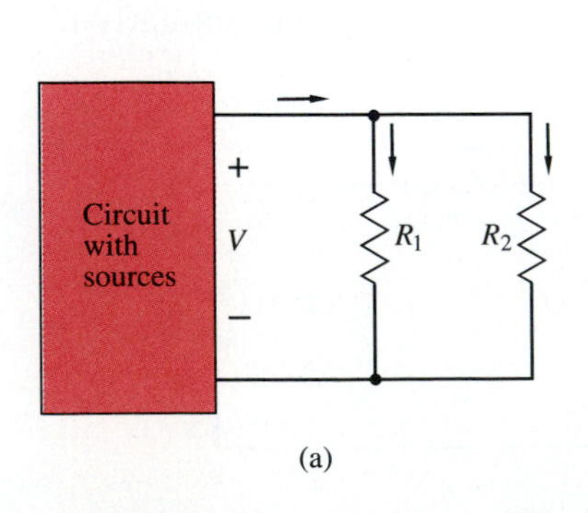

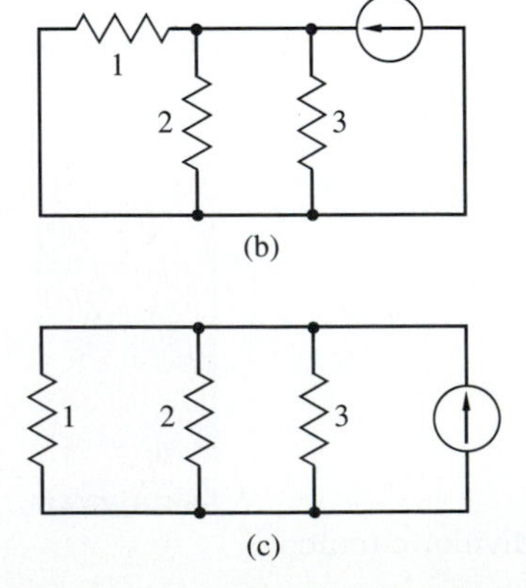

Figure 6
Parallel connections.

right angles geometrically but still be connected in parallel electrically. Electrically, the arrangement in Figure 6b behaves in the same way as the one in Figure 6c.

We previously saw that resistors connected in series could be replaced by a single equivalent resistor. Now we would like to explore a similar possibility for resistors connected in parallel. An expression relating the individual resistor currents to the total current i can be obtained by applying Kirchhoff's current law at the upper junction. Then we can apply Ohm's law in the form $i = Gv$. The result will be:

$$\begin{aligned} i_1 + i_2 &= i \\ G_1 v + G_2 v &= i \\ (G_1 + G_2)v &= i \end{aligned} \tag{4}$$

The last line is in the form of Ohm's law: some constant times the voltage equals the current. We conclude that

> A parallel connection of resistors can be replaced by a single *equivalent resistor* whose conductance is equal to the sum of the conductances of the parallel resistors.

This result permits some simplification in the analysis of circuits. The proof was carried out for two parallel resistors. The proof for more resistors connected in parallel can be obtained by mathematical induction. That is, a third resistor would be in parallel with the resistor equivalent to the first two, so the result would apply to this combination, etc. Sometimes it is more convenient to deal with the equivalent *resistance,* rather than the conductance, of two resistors in parallel. This is easily done, since $R = 1/G$. Thus, for two resistors we can write:

$$\begin{aligned} G &= G_1 + G_2 \\ \frac{1}{R} &= \frac{1}{R_1} + \frac{1}{R_2} \\ R &= \frac{R_1 R_2}{R_1 + R_2} \end{aligned} \tag{5}$$

The last line is obtained by solving for R from the second line. A nice mnemonic way to remember the expression for the equivalent resistance of two resistors in parallel is "*product over sum.*" (Warning: This is not valid for any number of resistors in parallel except two.)

Another interesting observation relates to the relative value of the equivalent resistance of two resistors connected in parallel: the value is less than either of the two parallel resistances; when the two resistances are the same, the equivalent resistance is equal to half of each resistance.

EXERCISE 3

By using KCL, show that current sources connected in parallel can be replaced by a single current source, with an appropriate reference, whose current is the algebraic sum of the parallel-connected current sources. □

CURRENT DIVIDER In a series-connected circuit fed by a voltage source, the voltage is distributed (divided) among the resistors in proportion to each R value compared with the total resistance. A similar relationship exists for the distribution of currents, this time among a set of parallel-connected resistors fed by a current source.

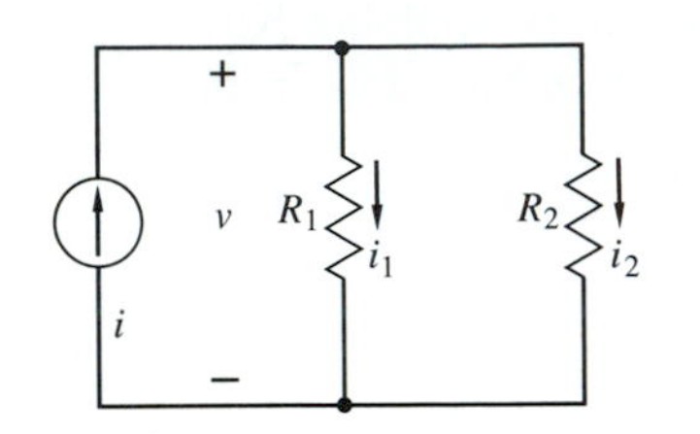

Figure 7
Current-divider circuit.

Figure 7 shows two resistors connected in parallel and fed by a current source. Applying Ohm's law and KCL results in:

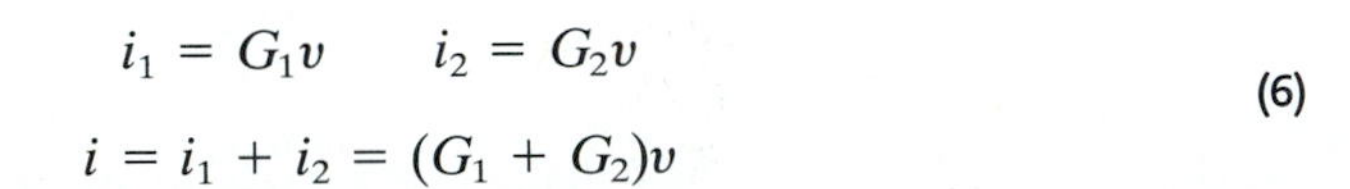

$$i_1 = G_1 v \qquad i_2 = G_2 v$$
$$i = i_1 + i_2 = (G_1 + G_2)v \tag{6}$$

Solving for v in the last line and substituting into the first line permits solving for each of the two branch currents, as follows:

$$i_1 = \frac{v}{R_1} = \frac{1}{R_1}\left(\frac{i}{G_1 + G_2}\right) = \frac{R_2}{R_1 + R_2}i \tag{7a}$$

$$i_2 = \frac{v}{R_2} = \frac{1}{R_2}\left(\frac{i}{G_1 + G_2}\right) = \frac{R_1}{R_1 + R_2}i \tag{7b}$$

$$\frac{i_1}{i_2} = \frac{R_2}{R_1} \tag{7c}$$

Compare these expressions with (3). We see that the total current is distributed (divided) between the two parallel resistors in *inverse* ratio to the values of the resistances. That is, the larger of the two parallel resistors carries the smaller current.

EXAMPLE 3 Three resistors are connected in parallel and fed by a 20-mA current source. The values of the resistances are 2, 4, and 6 kΩ, respectively. Use the current-divider relationship to find the currents in each of the resistors.

SOLUTION Let the three resistor currents be, respectively, i_2, i_4, and i_6. The conceptually simplest approach when finding the current in one resistor is to consider replacing the other two resistors by an equivalent. These equivalent values, respectively, are determined as follows:

To find i_2: $4 \times 6/(4 + 6) = 2.4\ \text{k}\Omega$
To find i_4: $2 \times 6/(2 + 6) = 1.5\ \text{k}\Omega$
To find i_6: $2 \times 4/(2 + 4) = 1.33\ \text{k}\Omega$

Now we apply the current-divider principle three times:

$$i_2 = \frac{2.4}{2.4 + 2}(20) = 10.91\ \text{mA}$$

$$i_4 = \frac{1.5}{1.5 + 4}(20) = 5.45\ \text{mA}$$

$$i_6 = \frac{1.33}{1.33 + 6}(20) = 3.64\ \text{mA}$$

(As a check, the three currents add to 20 mA, as they should.)

EXERCISE 4

Two resistors connected in parallel are excited by a current source. The voltage across the combination is found to be 50 V. One resistance is 5 kΩ, while the current through the other one is 5 mA. Use Ohm's law and KCL to find (a) the current through the 5-kΩ resistor, (b) the resistance of the second resistor, and (c) the current of the source. Confirm your answers by using the current-divider relationship.

ANSWER: 10 mA, 10 kΩ, 15 mA. □

Equivalent Resistance

It is often desired to find the resistance of a circuit of resistors that are connected neither all in series nor all in parallel. In some structures, successive application of the series and parallel rules can accomplish the result.

EXAMPLE 4 The resistors in Figure 8a are not all connected in series, nor are they all connected in parallel. (Remember that when no units are specified, SI units are assumed.) Nevertheless, the 10-Ω and 20-Ω resistors *are* in series, as are the

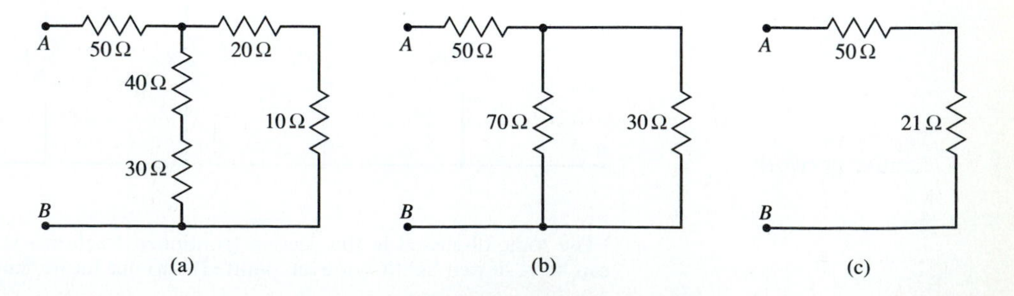

Figure 8 Consecutive applications of series and parallel equivalences.

30-Ω and 40-Ω ones. These two pairs can be replaced by equivalent resistances, as shown in Figure 8b. Now the 30-Ω and 70-Ω resistors are in parallel and can be replaced by their parallel equivalent, as shown in Figure 8c. Now the remaining two resistors are in series. Thus, the entire circuit of Figure 8a can be replaced by a single 71-Ω equivalent resistance at terminals *AB*.

EXERCISE 5

By successive application of series equivalents and parallel equivalents, find the value of a single resistance which is equivalent to the entire circuit of Figure 9 at terminals *AB*.

ANSWER: 50 Ω. □

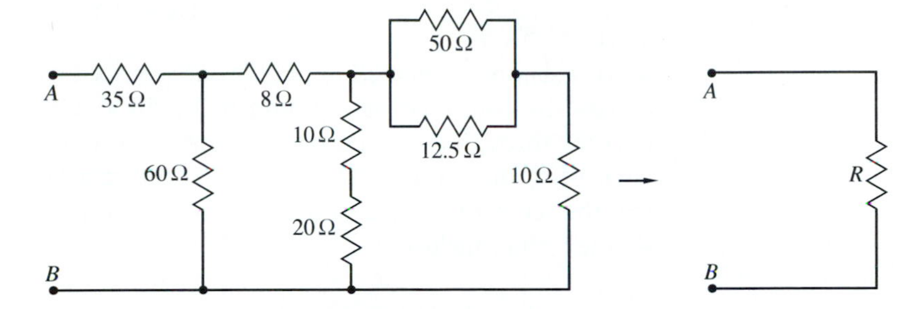

Figure 9 Find the equivalent resistance at *AB*.

Ladder Network[3]

The circuits of Figure 8a and Figure 9 are examples of a class of networks called *ladders.* A more general form of the ladder network is shown in Figure 10. (The circuit terminates somewhere, of course.) Any series or parallel combinations that might have appeared in the branches have been replaced by their single-resistor equivalents. Different approaches can be taken, depending on what unknown it is desired to find. If the equivalent resistance as seen from

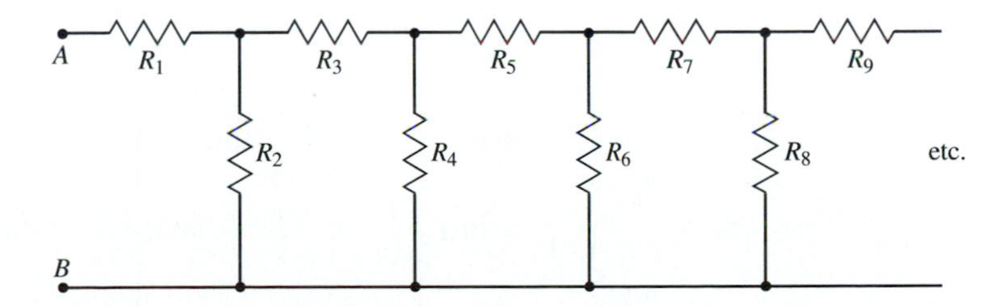

Figure 10 Ladder network.

[3] The topic discussed in this section (continued fractions) is not of primary significance and can be reviewed lightly (or even omitted) without future handicap.

the input terminals is required, then the same process as used in Figure 9 can be used for any ladder. That is, work from right to left and successively apply series and parallel equivalences.

Even in this case, however, the following approach which proceeds from left to right can be tried. The values R_a, R_b, R_c, etc., are the values of equivalent resistance looking to the right as we remove one resistor at a time, starting on the left. Thus, R_a is obtained by removing R_1; it is the parallel equivalent of R_2 and R_b, where R_b is the resistance of what is left looking to the right after R_2 is also removed. Now R_b is the series equivalent of R_3 and R_c, which is what is left after the removal of R_3; etc.

Using R for the series resistors and conductance G for the parallel resistors, we can write the equivalent resistance mathematically as:

$$R = R_1 + R_a$$

$$G_a = G_2 + G_b$$

$$R_b = R_3 + R_c$$

$$G_c = G_4 + G_d$$

etc.

When these are substituted one into the other, appropriately inverting the R's and G's, the result becomes:

$$R = R_1 + R_a = R_1 + \frac{1}{G_a} = R_1 + \frac{1}{G_2 + G_b}$$

$$= R_1 + \cfrac{1}{G_2 + \cfrac{1}{R_b}} = R_1 + \cfrac{1}{G_2 + \cfrac{1}{R_3 + R_c}}$$

$$= R_1 + \cfrac{1}{G_2 + \cfrac{1}{R_3 + \cfrac{1}{G_4 + \cfrac{1}{R_5 + \cdots}}}}$$

Notice how, for the odd-numbered resistors (those in the *series arms* of the ladder), it is the resistance that appears and for the even-numbered ones (those in what are called the *shunt arms* of the ladder) it is the conductance.

The mathematical form of this expression is called a *continued fraction.* It continues until it is terminated. Once it is terminated, finding its value is a matter of arithmetic, starting from the end.

Thus, to find the resistance of the ladder in the example of Figure 9, we would write:

$$R = 35 + \cfrac{1}{\cfrac{1}{60} + \cfrac{1}{8 + \cfrac{1}{\cfrac{1}{30} + \cfrac{1}{10 + 10}}}}$$

$$= 35 + \cfrac{1}{\cfrac{1}{60} + \cfrac{1}{8 + \cfrac{1}{\frac{1}{12}}}} = 35 + \cfrac{1}{\cfrac{1}{60} + \cfrac{1}{20}}$$

$$= 35 + \frac{1}{\frac{1}{15}} = 35 + 15 = 50$$

(Isn't it fortunate that the answer is the same as obtained before!)

2 SOURCE TRANSFORMATIONS

In the preceding chapter we saw that a circuit model of a physical battery consists of a constant-voltage source in series with a resistor whose resistance is the "internal resistance" of the battery. Indeed, power supplies and signal generators of all types can be represented by a model such as the circuit shown in Figure 11a. The voltage of the source can be a constant, a sinusoid, a square wave, or some other function of time. Assuming that some circuit is connected across terminals AB, a current i will flow and there will be a voltage v across the terminals. By KVL and Ohm's law, the expression relating v and i will be:

$$v = v_0 - R_0 i \tag{8}$$

This expression can now be solved for i, with the result:

$$i = \frac{v_0}{R_0} - \frac{1}{R_0}v = i_0 - \frac{1}{R_0}v \tag{9}$$

Since v_0/R_0 is dimensionally current, it has been replaced by the quantity i_0. This equation can be viewed as an expression of KCL at point A. The current directed out of terminal A is apparently the difference of two other currents.

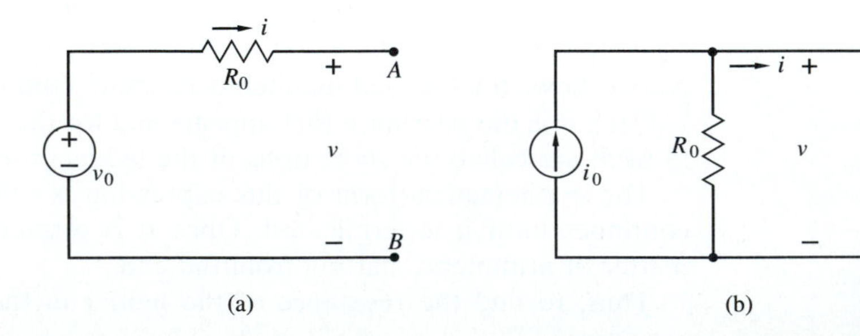

Figure 11
Generator equivalent models.

On this basis we construct the circuit of Figure 11b, consisting of a current source in parallel with a resistance R_0. (Apply KCL at terminal A and verify that (9) results.)

Equivalent Circuits

The two circuits in Figure 11 viewed from terminals AB have the same relationship between voltage and current. As far as anything external connected to terminals AB is concerned, there would be no difference whether one or the other of these two circuits is connected there. We make this the basis of a definition:

Two circuits, each with two terminals, are said to be *equivalent* at these terminals if they have the same voltage-current relationships there.

In Section 1, when we discussed replacing a series connection or a parallel connection of resistors with a single *equivalent resistance,* it was precisely because of their having the same v-i relationship. See (1) and (5).

It is common to refer to the two circuits in Figure 11 as the *voltage-source equivalent* and the *current-source equivalent.* The voltage source is in series with a resistor and the current source is in parallel with the same resistor. Note how the three quantities v_0, i_0, and R_0 are related:

$$v_0 = R_0 i_0 \tag{10}$$

(*This is not an expression of Ohm's law,* although it may appear to be so. The voltage and the current are not even in the same circuit.)

ACCOMPANIED SOURCES Because the combinations of elements shown in Figure 11 occur very often, we need some terminology to refer to them simply, instead of saying "a voltage source in series with a resistor," or the like. We use the following terminology:

An *accompanied voltage source* is a voltage source in series with a linear two-terminal component, such as a resistor.
An *accompanied current source* is a current source in parallel with a linear two-terminal component, such as a resistor.

EXAMPLE 5

In the circuit of Figure 12a it is desired to find the voltage across the 50-Ω resistor. The approach we will take is to replace the accompanied current

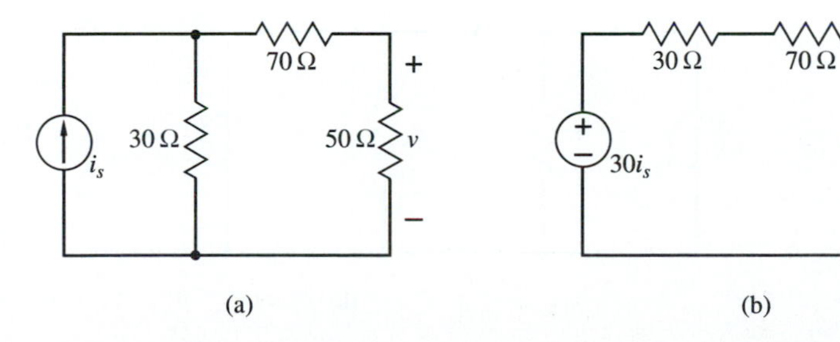

Figure 12 Replacing an accompanied current source by an accompanied voltage source.

source by its voltage-source equivalent. The result is shown in Figure 12b. The structure of this circuit is that of a voltage divider; applying the voltage-divider rule leads to:

$$v = \frac{50}{50 + 70 + 30} 30i_s = 10i_s$$

Other approaches can be used to solve the same problem. For example, the 50- and 70-Ω resistors are in series and can (mentally) be replaced by a single 120-Ω equivalent. The current in this equivalent can be found by the current-divider rule. Ohm's law then gives the desired voltage. (Do this and verify the previous result.)

Simplifying Circuit Structures

"Solving" a circuit means finding the voltages and currents in the components of the circuit when the source values are given. So far in this chapter we have discussed a number of simple concepts: series and parallel combinations, voltage dividers, current dividers, and voltage-source and current-source equivalents. Successive applications of these ideas often permit solving a circuit. (Of course, there are also Kirchhoff's two laws and Ohm's law.) We will now illustrate several approaches to doing this.

Consider the circuit in Figure 13a. We will initially solve for the voltage across R_4; then, using this information in the original circuit, we will find all other voltages and currents. As a first step, we replace the accompanied voltage source by its current-source equivalent. The result is shown in Figure 13b. Here we find two current sources in parallel; they can be replaced by a

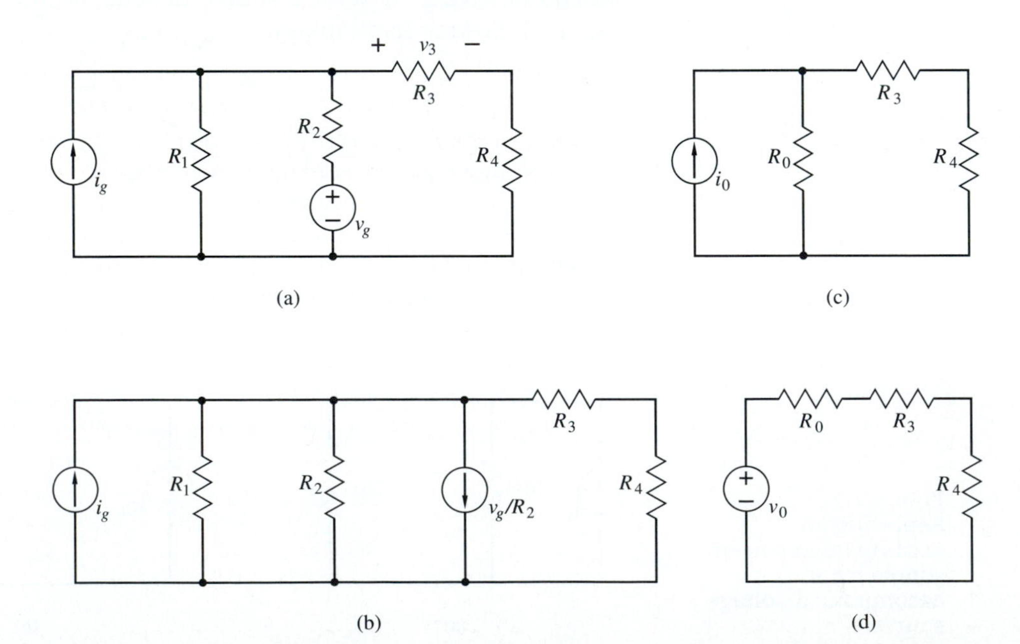

Figure 13
Circuit to be solved.

single equivalent current source, say, i_0; but note carefully the reference of current source v_g/R_2. Similarly, R_1 and R_2 can be replaced by their parallel equivalent, say, R_0 (Figure 13c). These two quantities are:

$$i_0 = i_g - \frac{v_g}{R_2}$$
$$R_0 = \frac{R_1 R_2}{R_1 + R_2} \tag{11}$$

The accompanied current source i_0 with resistance R_0 can now be converted to a voltage-source equivalent; the resulting structure is shown in Figure 13d. The voltage of the source is:

$$v_0 = R_0 i_0 = \frac{R_1 R_2}{R_1 + R_2}\left(i_g - \frac{v_g}{R_2}\right) \tag{12}$$

This structure is a voltage divider, so, finally, the desired voltage can now be found; it is:

$$v_4 = \frac{R_4}{R_0 + R_3 + R_4} v_0 \tag{13}$$

The solution just carried out proceeds step by step by gradually converting parts of the circuit to simpler equivalent forms. The reduction of structure is achieved by using (1) source equivalents and (2) series and parallel combinations of branches repeatedly. A word of caution here: This approach does not work for all possible circuit structures.

The result so far is just one voltage. However, now that v_4 is known, we can return to the original circuit and successively solve for all other variables. (Keep examining the original circuit while following this discussion.) When v_4 is known, for example, Ohm's law gives the current through R_4, which is the same as the current through R_3; now Ohm's law again gives us the voltage across R_3. Then KVL gives the voltage across the combination of R_3 and R_4. But that voltage is the same as the voltage of all other branches to the left. From a knowledge of this voltage, the current through R_1 follows by Ohm's law, and from this voltage and the source voltage, an application of KVL gives us the voltage across R_2. Ohm's law then gives the current through R_2. (Go through the details, please.)

Let's review this process. To "solve a circuit" (for those circuits having a series-parallel, or ladder, structure) we concentrate on one branch. By successive conversions from voltage-source equivalent to current-source equivalent, and vice versa, and by series and parallel combinations of branches, the rest of the circuit is converted to a simple form which permits solving for the desired voltage or current. We next use this value in the original circuit and successively apply Ohm's law, KVL, and KCL to solve for all other variables one at a time.

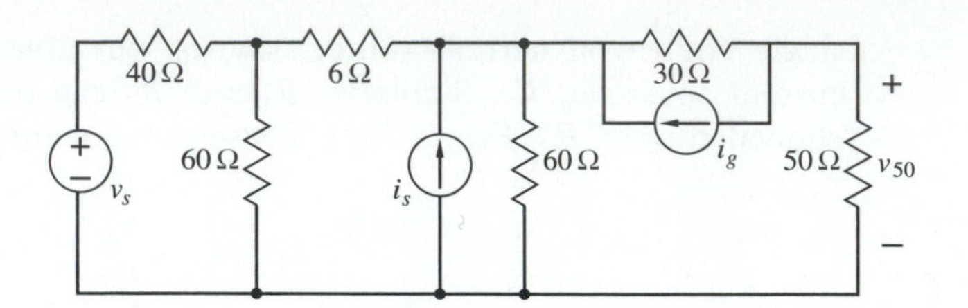

Figure 14
Example circuit.

EXERCISE 6

Carry out the procedure just outlined to solve for all unknown voltages and currents in Figure 14.

ANSWER: $v_{50} = \frac{1}{2}v_s + 10i_s - 15i_g$. □

Working Back from the Output

In the preceding section we discussed an approach to solving a circuit which involves (1) gradually transforming the structure of the circuit to a simpler form, (2) solving for the current in the resulting simple series circuit—which is one of the unknowns in the original circuit—(or the voltage in the resulting parallel circuit), and (3) using this information back in the original circuit to solve successively for all other variables.[4]

Using the simple circuit of Figure 15 as an example, we will now describe another approach which can be used in some cases. To avoid cluttering the diagram, not all variables will be shown and standard references will be assumed for all resistor voltages and currents. In this circuit it is desired to find voltage v_2, given the source voltage. First a verbal description will be given, followed by a mathematical development. The reasoning goes like this.

Suppose v_2 were known. Then the current through R_2 would follow from Ohm's law: $i_2 = v_2/R_2$. By KCL, this is the same as the current through R_4: $i_4 = i_2$. Now the voltage across R_4 will be known, again by Ohm's law: $v_4 = R_4 i_4$. By KVL, the voltage across R_3 is the sum of the voltages across R_4 and R_2, both of which are known: $v_3 = v_4 + v_2$. From this voltage, the current through R_3 is found by Ohm's law: $i_3 = v_3/R_3$. The current through R_1 is now found by KCL: $i_1 = i_2 + i_3$. Finally, the voltage across R_1 follows by Ohm's law.

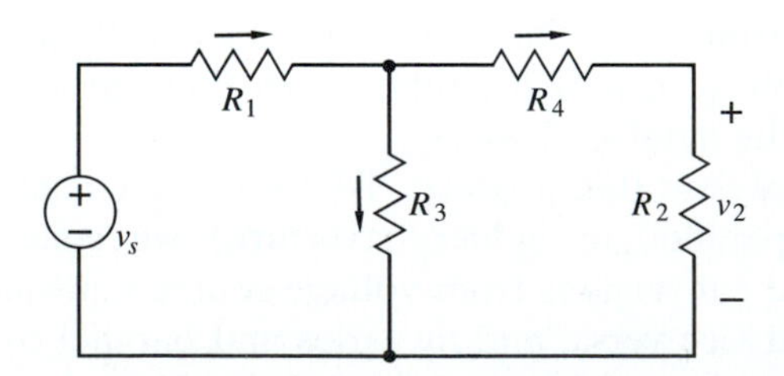

Figure 15
Back-to-front solution.

[4] You will greatly improve the depth of your learning if, after reading such a review, you do more than simply go on. It would be immensely helpful to you if you interact with what you are reading, by returning to the original description of this process as you read each step and confirming to yourself that you understand what is being done.

Notice that, starting from the right end of the circuit, each successive current or voltage has been expressed in terms of the unknown voltage v_2. Finally, KVL around the left loop brings in the source voltage. The steps carried out in this verbal development will now be carried out mathematically:

$$i_2 = \frac{v_2}{R_2} = i_4 \qquad v_4 = R_4 i_4 = \frac{R_4}{R_2} v_2$$

$$v_3 = v_2 + v_4 = \left(1 + \frac{R_4}{R_2}\right) v_2 \qquad i_3 = \frac{v_3}{R_3} = \frac{1}{R_3}\left(1 + \frac{R_4}{R_2}\right) v_2$$

$$i_1 = i_2 + i_3$$

$$\begin{aligned} v_1 &= R_1 i_1 + v_3 = R_1(i_2 + i_3) + v_3 \\ &= R_1\left[\frac{1}{R_3}\left(1 + \frac{R_4}{R_2}\right) + \frac{1}{R_2}\right] v_2 + \left(1 + \frac{R_4}{R_2}\right) v_2 \\ &= K v_2 \end{aligned}$$

where K is the constant multiplying v_2 in the preceding line. Thus:

$$v_2 = \frac{v_1}{K}$$

Note that all variables in the circuit were expressed in terms of v_2 in the previous expressions. Now that v_2 has been determined in terms of the source voltage, all other variables can be solved in terms of the source voltage. Thus, for example:

$$v_3 = \left(1 + \frac{R_4}{R_2}\right)\left(\frac{v_1}{K}\right)$$

It is clear from this example that the same procedure can be used for a circuit of any size so long as the structure permits the successive application of Kirchhoff's laws and Ohm's law, thereby expressing all circuit variables in terms of just one of them. The procedure can always be applied to a ladder structure, for example. Circuits to which this approach cannot be applied require other methods which will be taken up in Chapter 5.

3 SOME BASIC PRINCIPLES

A number of ideas of a fundamental nature will be introduced in this section. They not only provide useful procedures for finding the solution to circuit problems, but they give important insights into the general behavior of circuits. Such insights are often helpful in visualizing the "forest"—seeing the broad picture—instead of merely performing the mechanics of specific procedures, such as finding the series or parallel equivalent of two resistors.

Linearity and Superposition

Each of the electrical components introduced in Chapter 2 is characterized by a relationship between two variables, their voltage and current. Using general notation, such a relationship can be expressed as $y = f(x)$, where x and y are independent and dependent variables, respectively, and f is a function relating the two variables. Procedures which are appropriate in analyzing a circuit depend on the nature of this relationship. A typical example is Ohm's law ($v = Ri$), which has the form $y = kx$, where k is a constant. This is a *linear* relationship.

The linear relationship $y = kx$ has two significant properties. Suppose the independent variable is doubled; what happens to the dependent variable? If x is replaced by $2x$, we see that y becomes $k(2x) = 2(kx) = 2y$. That is, if x is doubled, y is doubled; or if x is tripled, y is tripled; etc. That's one property.

Let's view x as an excitation (input) and y as the response (output). What would happen if there were two different excitations acting simultaneously? If the excitation is x_1, let the response kx_1 be labeled y_1, and if the excitation is x_2, let the response kx_2 be y_2. Now suppose the excitation is $x_1 + x_2$; then $y = k(x_1 + x_2) = kx_1 + kx_2 = y_1 + y_2$. That is, the new output is the sum of the individual outputs.

Finally, suppose each excitation, acting simultaneously, is multiplied by an arbitrary constant; what happens to the response? By going through the details, you can show that the new response will equal the sum of the individual responses, each multiplied by the corresponding constant. That is, if:

$$\text{Input} = k_1x_1 + k_2x_2$$

then:

$$\text{Output} = k_1y_1 + k_2y_2$$

This result, which is a consequence of linearity, is referred to as the *principle of superposition.*[5]

Glance back at all the circuits considered so far in this chapter. They all consist of resistors and sources. Solving for any voltage and current involves the application of Kirchhoff's laws, Ohm's law, and specialized techniques derived from them, such as source conversions and divider rules. KCL and KVL involve only the addition of variables; Ohm's law is just a linear relationship between two variables. Solving for the unknown variables in a circuit involves such operations as addition, multiplication by a constant, and substituting from one equation into another. Such operations maintain linearity. That is, linear relationships remain linear if we add them, multiply them by constants, and substitute them into other linear equations. Review the results in Figures 12 through 14, for example, and observe that every solution is a linear combination of all the source variables.

For the linear dissipative circuits under consideration in this chapter, the

[5] For mathematicians, indeed, the last set of relationships actually constitutes a *definition* of linearity. Thus, if $y = f(x)$, then the function is said to be linear if and only if $f(k_1x_1 + k_2x_2) = k_1f(x_1) + k_2f(x_2)$.

principle of superposition can sometimes be used to determine a circuit response when there are two or more sources.

EXAMPLE 6 An illustration is provided by the circuit in Figure 16a. Our goal is to find the "output" voltage labeled v. We will do so by first finding the value of v when each source acts alone, with the other source removed, and then *superposing* the two results. "Removing" a voltage source means making it go to zero, which is to say replacing it by a short circuit. (Similarly, removing a current source—making it go to 0—means replacing it with an open circuit.) In the present case, the results of removing one source at a time are shown in Figure 16b and c.

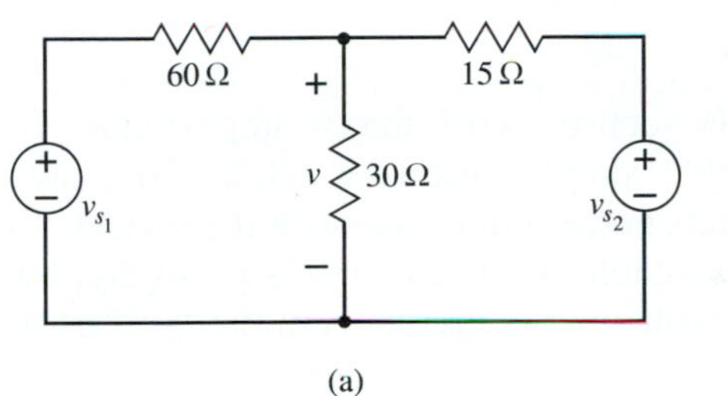

(a)

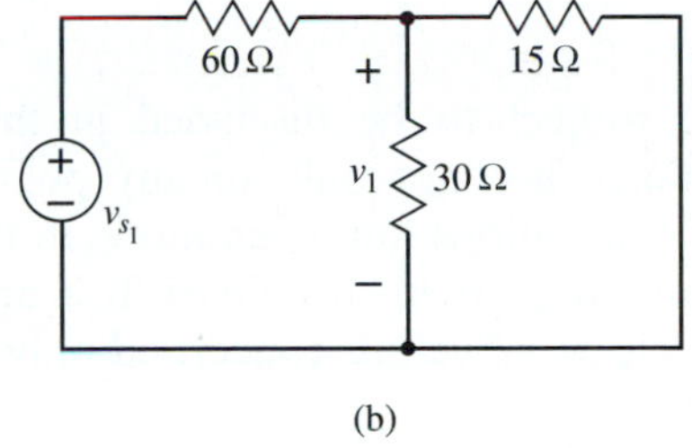

(b)

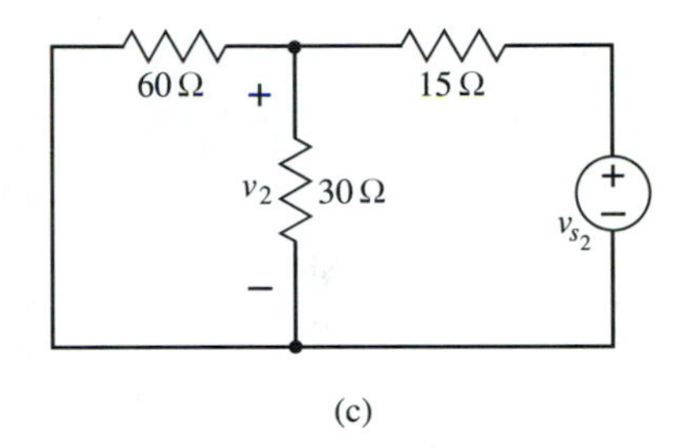

(c)

Figure 16 Application of the principle of superposition; $v_{s_1} = 21 \sin 400t$; $v_{s_2} = 14 \cos 400t$.

In Figure 16b, the desired voltage is labeled v_1; the 30-Ω and 15-Ω resistors are in parallel, their parallel-equivalent resistance being 10 Ω. The resulting structure is a voltage divider, so the desired voltage can be found easily. In Figure 16c, the desired voltage is labeled v_2; now the 30-Ω and 60-Ω resistors are in parallel, their equivalent resistance being 20 Ω. Again the structure is a voltage divider. The results of these calculations will be:

$$v_1 = \left(\frac{10}{70}\right)v_{s1} = 3 \sin 400t$$

$$v_2 = \left(\frac{20}{35}\right)v_{s2} = 8 \cos 400t$$

$$v = v_1 + v_2 = 3 \sin 400t + 8 \cos 400t$$

Although the principle of superposition applies generally to any linear circuit, its use does not always lead to a simplification in the calculations required to solve a circuit problem. In more complex circuits than this example, even leaving only one source active at a time may require a substantial analysis procedure to arrive at each partial result. When all the sources are acting concurrently, only one analysis is needed. Often this is not much more complicated than the case of only one source, even if one is solving the problem by hand.

EXERCISE 7

Two students had a heated argument about the subject of superposition. In a circuit such as the one in Figure 16 with two sources, it was required to find the power delivered to one of the resistors, say, the 30-Ω resistor in that figure. One of the students wanted to find the power delivered by each source acting alone and then add the two, by superposition. The other student argued that this would not work. Decide who is right and why. Find the power delivered to the 30-Ω resistor from the voltage found in Example 6. Then find it by the method suggested by the first student. Do you get the same value? □

The Thévenin Theorem

The subject to be discussed in this section is of major importance. Its importance lies not only in any possible simplifications it might bring into circuit calculations but, even more, in the conceptual framework it provides for contemplating circuit problems. It is applicable to linear circuits generally, but in this chapter, we are concerned only with simple circuits containing resistors and sources.

MOTIVATION FOR THÉVENIN'S THEOREM The topic will be introduced with the circuit in Figure 17a. Our objective is to find the voltage v_2 across what can be considered a load resistance. In the preceding section, we proceeded with a sequence of transformations, as shown in Figure 17b and c. In Figure 17b, the first step, we have replaced the accompanied voltage source by an equivalent accompanied current source. Resistors 1 and 4 are then in parallel, and that combination accompanies the current source. The current source accompanied by the two parallel resistors is now replaced by an equivalent accompanied voltage source, as shown in Figure 17c. What has been done overall is to replace the entire structure within the dashed rectangle in

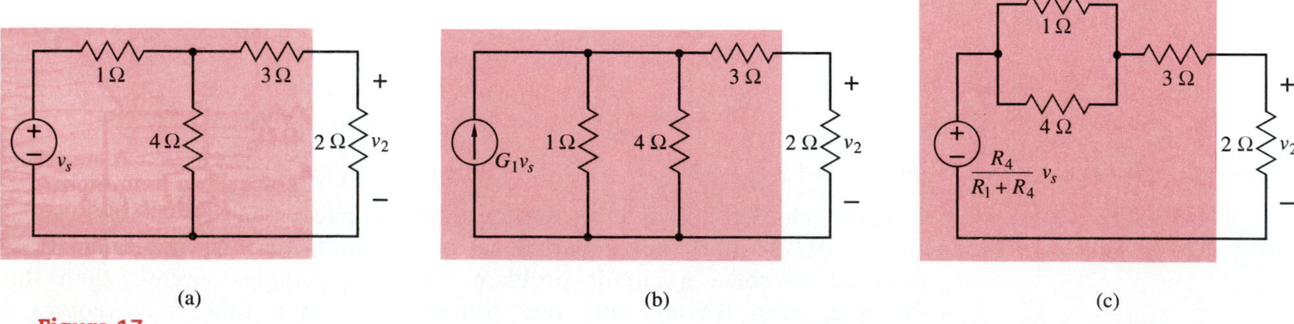

Figure 17
Motivation for Thévenin's theorem.

Figure 17a by an accompanied voltage source whose voltage, v_T, is found to be:

$$v_T = \left(\frac{R_4}{R_1 + R_4}\right)v_s \tag{14}$$

We now make an interesting observation. Suppose the load resistance is removed from the original circuit; what would the voltage be across the open terminals left there? We note that no current would flow in branch 3 in such a case. The rest of the circuit to the left has the structure of a voltage divider; the voltage across R_4, then, is exactly what is given in the preceding expression for v_T. Since there is no current in R_3, this value is the voltage across the open-circuited output also. At least in this example, then, instead of going through the sequence of transformations shown in Figure 17 to find the equivalent source in part (c), we can open-circuit the output and calculate the voltage there.

From an examination of Figure 17, we make another observation. Notice the entire resistance accompanying the source in part (c). The series combination of R_3 with the equivalent of R_1 and R_4 in parallel is exactly what would be seen looking into the terminals of the dashed box in part (a) if v_s were removed (that is, replaced by a short circuit). Again, to find the equivalent resistance in part (c), it isn't necessary to go through the intermediate transformations; the result can be found from the original circuit.

It would be interesting to speculate whether a similar conclusion applies to more general circuits. Such a result would be very useful. We will now devote some time to proving this result.

PROOF OF THÉVENIN'S THEOREM[6] Figure 18 illustrates a linear, dissipative

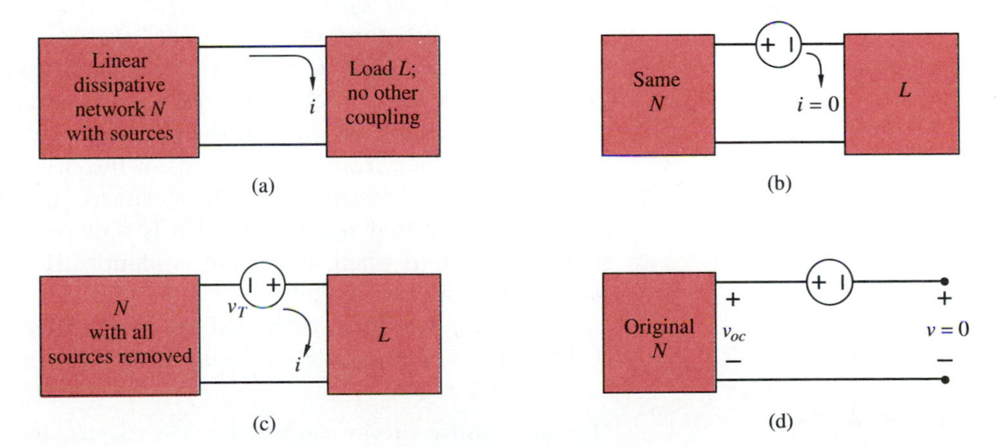

Figure 18
Deriving Thévenin's theorem.

[6] Some of you may wish to skip the proof and just accept the theorem. This is definitely not recommended. The proof is not onerous and will give you some insight into network behavior. However, if you insist, skip about two pages and go to the paragraph that starts, "This result was first advanced by M. L. Thévenin. . . . "

network that has been separated into two parts. The part on the right is to be considered as a load connected to the part on the left. The only stipulation is that there should be no coupling between the two parts, except through the direct connection shown.[7] We will be concentrating on the left part, and the result we achieve will depend on that part's being a linear, resistive circuit. However, nothing we do will involve the properties of the load, the part on the right. Therefore there is no restriction on this part, even as to its being linear.

As a consequence of the sources, there is a current i between the two parts. Now let's insert a voltage source in series with the two parts and adjust its value as a function of time so that the current between the two parts is reduced to zero. According to the principle of superposition, the total current in a branch is the sum of the currents owing to all sources in the circuit, including the source we added. The current resulting from all the sources originally in this circuit is i. Since the *total* current is zero, the current resulting from the source we added must be $-i$. Suppose the reference for the added source is reversed; then the current it produces will also be reversed. The conclusion is that if we (1) reverse the reference of the added source *and* (2) *remove all the original sources,* the same current will result in the load as existed in the original circuit. Hence, the circuit shown in Figure 18c is equivalent to the original circuit, as far as the load is concerned. The only remaining task is to find what the value of v_T should be.

To carry out this task, examine Figure 18b. Under those conditions, there is no current at the terminals of the load and so no voltage across its terminals either. Hence, open-circuiting the terminals, as shown in Figure 18d, would change nothing; the voltage across these terminals would still be zero. By KVL, we find that:

$$v_T = v_{oc} \tag{15}$$

where v_{oc} is the voltage across the original network when its terminals are open-circuited—the *open-circuit voltage.*

From the structure of Figure 18c, we see that, as far as the load is concerned, the original network N is equivalent to the series connection of two things: one is a voltage source, whose voltage equals the open-circuit voltage at the terminals of the original network; the second is a resistance equivalent to the resistance seen from the terminals of the original network when all sources are removed, or *inactive.* A voltage source is inactive when its voltage is identically zero; that will happen if it is replaced by a short circuit. A current source is inactive when its current is identically zero; that will happen if it is replaced by an open circuit.[8]

This result was first advanced by M. L. Thévenin in France in 1885; it is known as *Thévenin's theorem.* The two-terminal circuit consisting of

[7] For the resistive circuits considered in this chapter, there can't be any coupling except through direct connection. However, in circuits to be taken up subsequently, magnetic and other types of coupling are possible.

[8] Note that we don't short-circuit a voltage source to make it inactive; nor do we open-circuit a current source to make it inactive. Those operations are no-nos. Instead, we say that the two sources are *replaced by* short circuits or open circuits, respectively.

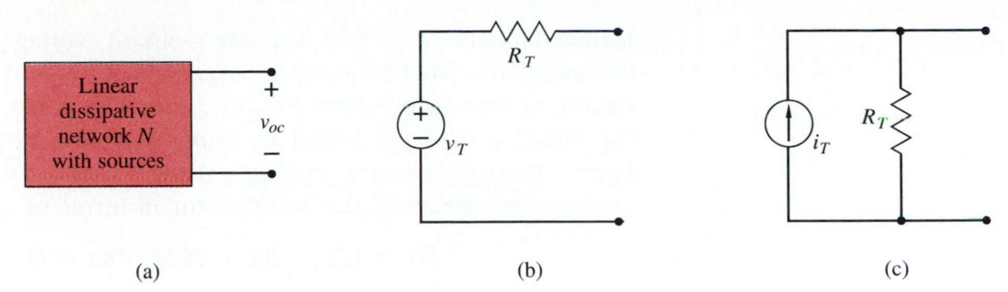

Figure 19
The Thévenin equivalent circuit.

the accompanied source in Figure 19b is called the *Thévenin equivalent* of the linear, dissipative circuit shown in Figure 19a, the equivalence holding at the pair of terminals of the circuit. The Thévenin equivalent voltage v_T is the open-circuit voltage at the terminals of the original network. The Thévenin equivalent resistance R_T is the resistance seen looking back from the terminals of the network when all its sources are removed.

Recall that the concept of equivalence of two circuits means that the voltage-current relationships at the terminals of the two circuits are identical. Thus, as far as a load connected at the terminals is concerned, the same voltage and current would appear at the load regardless of which of the two is connected to the load. Obviously, it is much simpler to calculate the load current if the circuit is replaced by its Thévenin equivalent.

EXAMPLE 7 The preceding rather extensive proof of Thévenin's theorem may have masked the simplicity of applying it. Let's therefore, as an example, find the Thévenin equivalent of the circuit shown in Figure 20a. To find v_T we leave the output

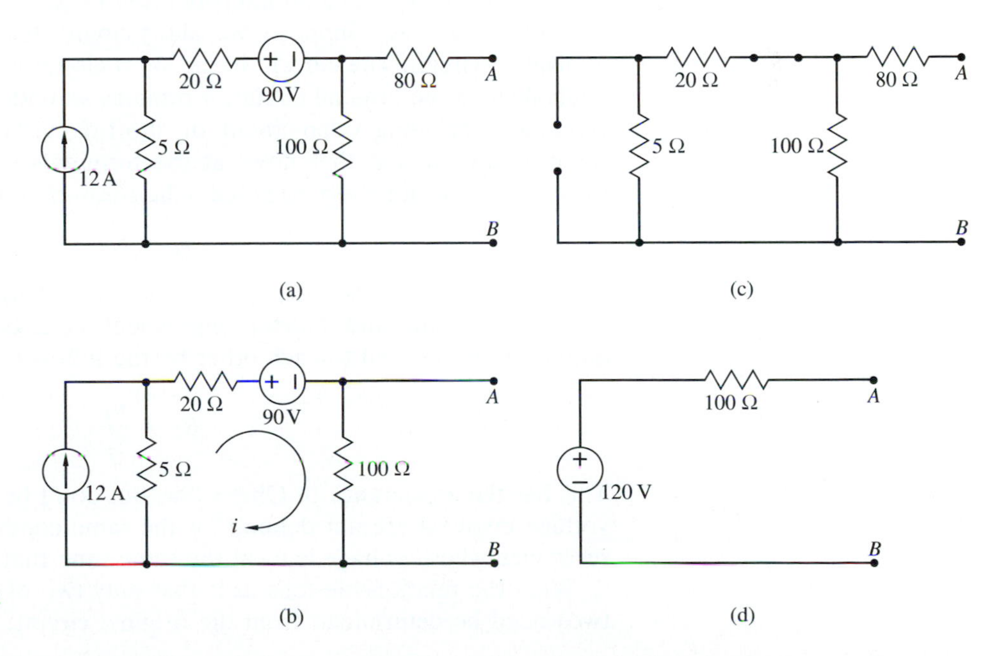

Figure 20
Example of the Thévenin equivalent circuit.

terminals open and find the open-circuit voltage there. Since there is no current in the 80-Ω resistor, its voltage is zero, so it can be replaced by a short circuit, as shown in Figure 20b. The output voltage, which is the voltage across the 100-Ω resistor, is found by applying KVL around the closed path in the figure, then eliminating voltages using Ohm's law, and then using KCL to express the current in the 5-Ω resistor in terms of i, resulting in:

$$5(i - 12) + 20i + 90 + 100i = 0$$

$$i = 0.24 \text{ A}$$

$$v_T = 100i = -24 \text{ V}$$

Next, to find the Thévenin equivalent resistance, both sources are removed by replacing the current source by an open circuit and the voltage source by a short circuit, as shown in Figure 20c. The series combination of the 5-Ω and the 20-Ω resistors is in parallel with the 100-Ω resistor, and this combination is in series with the 80-Ω resistor. Hence, R_T is found to be $80 + 25(100)/(25 + 100) = 100\ \Omega$.

THE NORTON EQUIVALENT It was earlier established that an accompanied current source can be made equivalent to an accompanied voltage source by properly selecting the current of the source. Hence, not only can a given linear dissipative circuit be replaced by a Thévenin equivalent, but it can also be replaced by an accompanied current source whose accompanying resistance is the same as the Thévenin resistance and whose source current is $i_T = v_T/R_T$, as shown in Figure 19c. This possibility was first revealed by E. L. Norton in 1935, and the circuit is known as the *Norton equivalent.*

It is possible to obtain an interpretation for the Norton equivalent current in the following way. Suppose we short-circuit the terminals of the Norton equivalent circuit. The current in the short circuit will be i_T. Since this circuit is equivalent to the original circuit, it remains so under any load condition at the terminals, including open-circuit or short-circuit conditions. Thus, i_T is the same as the current that flows at the terminals of the original circuit when those terminals are short-circuited—the *short-circuit current*:

$$i_T = i_{sc} \tag{16}$$

RELATIONSHIPS AMONG v_T, I_T, AND R_T Three quantities are involved in the Thévenin and Norton equivalent circuits: v_T, i_T, and R_T. These quantities are related to each other by the following relationship:

$$R_T = \frac{v_T}{i_T} \tag{17}$$

This has the appearance of Ohm's law, but don't be distracted; the current and voltage involved are not defined for the same condition of circuit. There is no resistance whose voltage is v_T at the same time that its current is i_T.

What the relationship tells us is that only two of these three quantities (any two) need be determined from the original circuit; the third follows from this relationship. Sometimes a specific pair of these quantities can more easily be calculated than the third. In such cases, those are the two to calculate from the original network; then you find the third from (17).

Maximum Power Transfer

We will continue examining some fundamental circuit relationships. A common situation in many areas of electrical engineering is the occurrence of weak signals which must be processed. For example, in the detection and measurement of signals of medical origin, in the telemetry of astronomical data, or in signals transmitted for communications purposes, every effort must be made to maintain as much of the signal as possible after its reception. The signal source is not under our control in such cases, so we can't do anything about enhancing the incoming signal before it arrives.

A diagram is shown in Figure 21a. The signal originates somewhere and is to be received or measured. A simplified model is shown in Figure 21b. The signal source is represented by its Thévenin equivalent (an accompanied voltage source) and the receiver as a load resistance. The source parameters (v_T and R_T) are not under our control but the load *is*. The question is, Is it possible to vary R_L so that the maximum amount of power coming from the signal source is transmitted to the receiving system represented by R_L? That's the question being considered here.

The power delivered to R_L is found from the current:

$$p = R_L i^2 = R_L\left(\frac{v_T}{R_T + R_L}\right)^2 = v_T^2 \frac{R_L}{(R_T + R_L)^2} \tag{18}$$

In this expression, R_T and v_T are fixed; the only variable under our control (as designers of the receiving system) is the load resistance. Thus, p is a function of R_L.

The maximum (or minimum) of a function $p(x)$ is found by differentiating p with respect to x, setting the result equal to zero, and solving for the value of x. In this case, x is R_L. The result of this process is

$$\begin{aligned} \frac{dp}{dR_L} &= v_T^2\left(\frac{(R_T + R_L)^2 - 2R_L(R_T + R_L)}{(R_T + R_L)^4}\right) = 0 \\ R_L &= R_T \end{aligned} \tag{19}$$

(Carry out the details.) We know that this value of R_L gives an extreme value, but we don't yet know if this value is a maximum or a minimum. The test is to

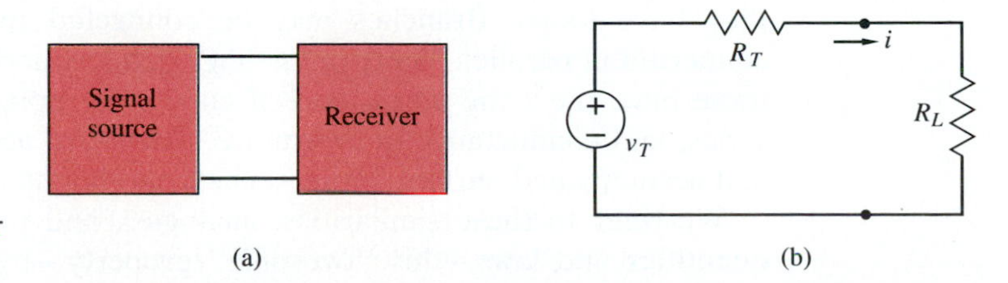

Figure 21
Finding maximum power transfer.

take the second derivative. If its value is negative for the given critical point, then the extreme is a maximum; otherwise it is a minimum. Carry out the details and confirm that the result of (19) is a maximum.

Another approach might be to examine what happens at the extreme values of R_L: 0 and ∞, corresponding to a short circuit and an open circuit. At both extremes, no power will be delivered to R_L. Thus the power delivered rises from zero and then falls to zero again. In between, it must reach a maximum.

The actual value of the maximum power is obtained by inserting the critical value of R_L from (19) into the expression for power in (18). You carry out the details; the result will be:

$$p = \frac{v_T^2}{4R_T} \tag{20}$$

Another interesting question is, How much power is lost in the resistance accompanying the source when the maximum power is transmitted? Since the two resistances are the same and they carry the same current, the answer is easy. As much power is lost in R_T as is delivered to R_L when the latter is a maximum. This may appear to be a big waste; but, while that may be true, it is not the determining factor. The amount of power involved in the cases under consideration is measured in microwatts or milliwatts, not kilowatts or more, and so the cost of the power is not a concern. Getting the most you can into your receiving system is the goal.

Suppose, however, that the conditions of the problem are different, that the amount of power involved is in kilowatts or more, and that the resistance accompanying the source—within limits—*is* within our control, while the load is fixed. If the previously determined condition for maximizing the power transfer is used, half of those kilowatts will be wasted—not a happy result. Decide what to do in this case to maximize the power transfer.

4 THE PRINCIPLE OF DUALITY

We will now introduce a phenomenon which may have intruded itself upon your consciousness to some extent already. The laws and procedures being discussed, the physical quantities, the structure of circuits, all seem to occur in pairs. First there are Kirchhoff's two laws which refer to current and voltage, respectively. Then there is Ohm's law, which can be written either as $v = Ri$ or as $i = Gv$. The v-i relationships of capacitors and inductors have the same form except that current and voltage are interchanged. Circuits have nodes and they have loops. Branches may be connected in series and they may be connected in parallel. There are voltage dividers and there are current dividers; these have the same appearance if current is replaced by voltage, parallel by series, and conductance by resistance. There are accompanied voltage sources and accompanied current sources which can be made equivalent to each other.

We refer to these similarities, analogies, and parallelisms between certain quantities and laws—this "twosome" property—by the general term *duality.*

This is a somewhat vague term, and we must specify more precisely what we mean by it if the concept is to be useful. To be a little more precise, but still somewhat vague, we'll say that two entities are *dual* if they play similar roles in the formulation of procedures in circuit theory that fall under the general heading of node analysis and loop analysis, subjects which we have not even explored yet. The difficulty is that if we wait to introduce the concept until everything has been discussed, then we will not benefit from any insights which the idea could provide in understanding the subjects being discussed. For this reason, it's better to suffer from some vagueness for a while.

Some of the aspects of duality are based on the geometrical (what we will later call the *topological*) structure of circuits—on the interconnection of components at nodes, thereby forming closed paths—but other aspects are physical. KVL refers to a loop and KCL refers to a node, so a loop and a node are topological duals. Voltage and current are themselves dual physical quantities. From the Ohm's law expressions $v = Ri$ and $i = Gv$, we see that resistance and conductance are dual quantities, since one of these expressions can be obtained from the other by interchanging v and i, and R and G. Similarly, from $q = Cv$ and $\lambda = Li$, we see that L and C are duals, as are charge and flux linkage. The latter two are not topological but physical duals. The accompanying table gives a partial tabulation of duals. In the chapters that follow, as we introduce new topics, we will further extend the table.

Dual Quantities and Laws

Voltage	Current
KVL	KCL
Loop	Node
Number of independent loops	Number of independent nodes
Charge	Flux linkage
Resistance	Conductance
Inductance	Capacitance
Series	Parallel
Short circuit	Open circuit
Voltage-source equivalent	Current-source equivalent

Let's now try to formulate the concept of duality a little more precisely:

Two circuit branches are said to be *dual* if the expression for the voltage in terms of the current of the first branch has the same mathematical form as the expression for the current in terms of the voltage of the second branch.

Two networks, N_1 and N_2, are *dual networks* if (1) the current equations of N_1 are the same as the voltage equations of N_2, with current replaced by voltage; (2) the voltage equations of N_1 are the same as the current equations of N_2, with voltage replaced by current; and (3) the branches of N_1 are the duals of the branches of N_2.

From the definition, two dual networks must have the same number of branches; the number of independent current equations of one of them must equal the number of independent voltage equations of the dual.

To see how the concept of duality can be of use in circuit analysis, let's look back at the current divider in Figure 7 and the expressions for the distribution of current given in equation (7). The current divider, which consists of the parallel connection of two branches to which a total current i_g is supplied, is the dual of the voltage divider, which consists of the series connection of two branches to which a total voltage v_g is supplied. Hence, we should expect that the equations expressing the distribution of current can be obtained from the equations expressing the distribution of voltage of a voltage divider, with all quantities replaced by their duals. This is indeed the case, as you can verify by comparing (3) and (7) and noting the duality between voltage and current, and resistance and conductance. On this basis, having obtained the expressions for voltage division in (3), we could have immediately written the expression for current division in (7), without going through any further analysis.

In later work, we'll use this concept of duality to obtain results about one circuit when the corresponding results about the dual circuit are at hand.

5 APPLICATIONS IN SYNTHESIS

In the preceding parts of this chapter, we have discussed procedures for "solving a circuit," that is, computing voltages and currents of all branches when component values and source values are given. Often, instead, it may be required to find the value of one or more components so that one or more conditions can be satisfied in the circuit—for example, in order that the voltage or current of some branches have prespecified values.

To solve such problems, no new principles are needed. The same basic laws and procedures apply, except that now the unknowns are not specific voltages or currents but component values.

EXAMPLE 8

The circuit in Figure 22 is a *Wheatstone bridge.* What is required is to find the value of unknown resistance R for which voltage $v_{AB} = 0$, the other resistance values being given.

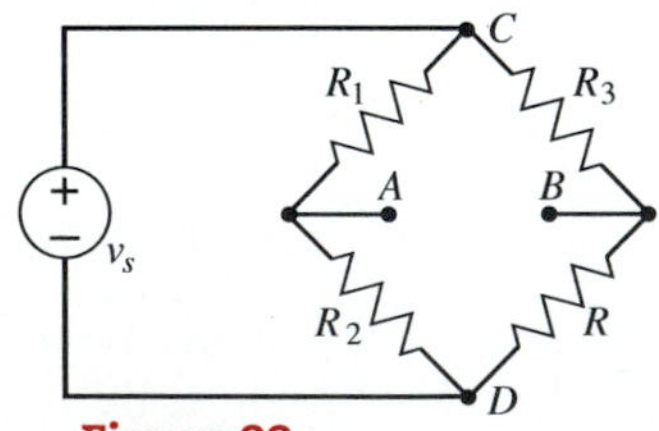

Figure 22
Wheatstone bridge circuit.

The problem can be solved by finding an expression for v_{AB} in terms of the circuit components, then setting it to zero and solving for R. Since nothing is connected between A and B, each of the branches going from C to D consists of two series-connected resistors. The voltages from A to D and from B to D can each be obtained by the voltage divider formula. Thus:

$$v_{AD} = \frac{R_2}{R_1 + R_2} v_s \qquad v_{BD} = \frac{R}{R + R_3} v_s$$

$$v_{AB} = v_{AD} - v_{BD} = \left(\frac{R_2}{R_1 + R_2} - \frac{R}{R + R_3}\right) v_s = 0$$

The last step follows from an application of KVL around the path ABD. Solving for R is now a simple matter of algebra:

$$R = \frac{R_2 R_3}{R_1}$$

This is often referred to as the *condition of balance* of the bridge. The Wheatstone bridge is a very sensitive instrument for measuring values not only of resistance, but also of inductance and capacitance. (In the latter cases, of course, some of the branches besides the unknown branch must be inductances or capacitances.)

EXERCISE 8

Another circuit that requires application of what we have learned so far to the determination of an unknown resistance is shown in the diagram of Figure 23. The two unknown resistors have the same value of resistance. Find the value of R for which the value of v_2 has its maximum possible value. (Hint: Express v_2 as a function of R.)

ANSWER: R = 7.07 Ω. □

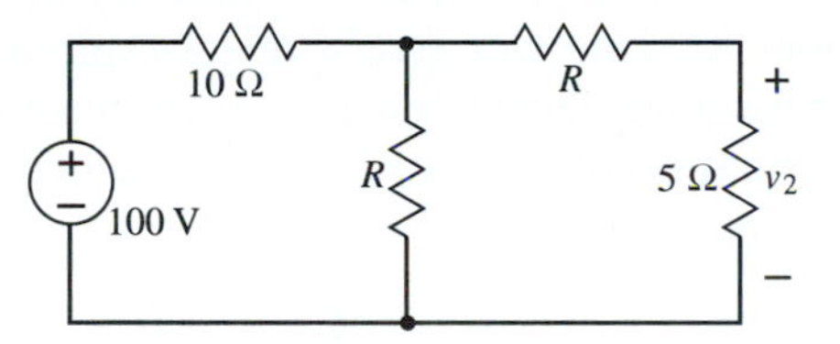

Figure 23
Find R for maximum v_2.

Attenuator Design

In Example 8 and again in Exercise 8, a single resistance value was to be found in order to satisfy a single condition. In other problems, it may be required to satisfy more than one condition. In such cases, it will be possible to find the value of more than one unknown component to satisfy the given conditions.

EXAMPLE 9 In some cases where a physical generator is to supply current to a load, the resulting current may be so high as to damage the load. It is desired to reduce the amount of current supplied to the load without changing what is supplied by the generator. This can be accomplished by inserting an *attenuator* between the generator and the load, as shown in Figure 24. The attenuator is a combination of resistors which, in this example, is the two-resistor upside-down ell structure shown in Figure 24c. If the load were to be connected directly to the generator, the resulting current would be i_g. This is presumably too high, so we wish to attenuate it by an *attenuation factor* A. That is, in Figure 24c the current i in R_L is i_g/A. On the other hand, the current supplied by the generator is to be the same whether the load is connected directly to the

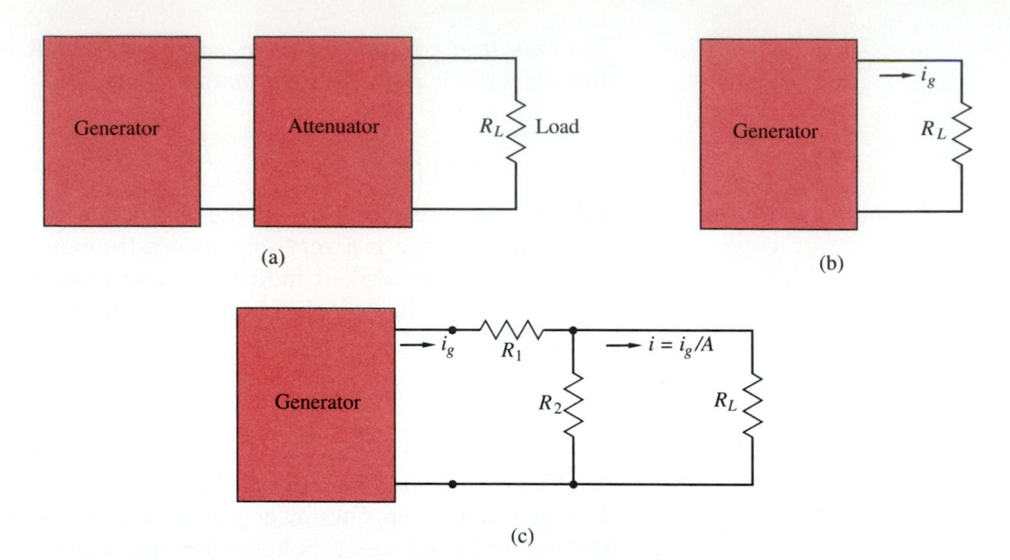

Figure 24
Designing an attenuator.

generator or through the attenuator. That is to say, the equivalent resistance "seen by" the generator looking out from its terminals is to be the same as R_L.

The solution is obtained by writing mathematical expressions for these two conditions. First, in Figure 24c, R_L and R_2 are connected in parallel. The current-divider formula will give a relationship between i and i_g. Second, the equivalent resistance facing the generator terminals is R_1 in series with the parallel combination of R_L and R_2. This is to equal R_L. These two conditions can be expressed as follows:

$$i = \frac{i_g}{A} = \frac{R_2}{R_2 + R_L} i_g$$

$$R_L = R_1 + \frac{R_2 R_L}{R_2 + R_L}$$

The first of these equations can be solved for R_2, since that is the only unknown. Then, after substituting this value for R_2, the second equation can be solved for R_1. The results are:

$$R_2 = \frac{R_L}{A - 1}$$

$$R_1 = \left(\frac{A - 1}{A}\right) R_L$$

(Carry out the algebraic details to confirm these results.)

Digital-to-Analog (D/A) Converter

Electrical signals can be described generically as either in *analog* (also *continuous*) or in *digital* (also *discrete*) form. Analog signals are those of everyday experience—continuous streams of sound or light, for example, or

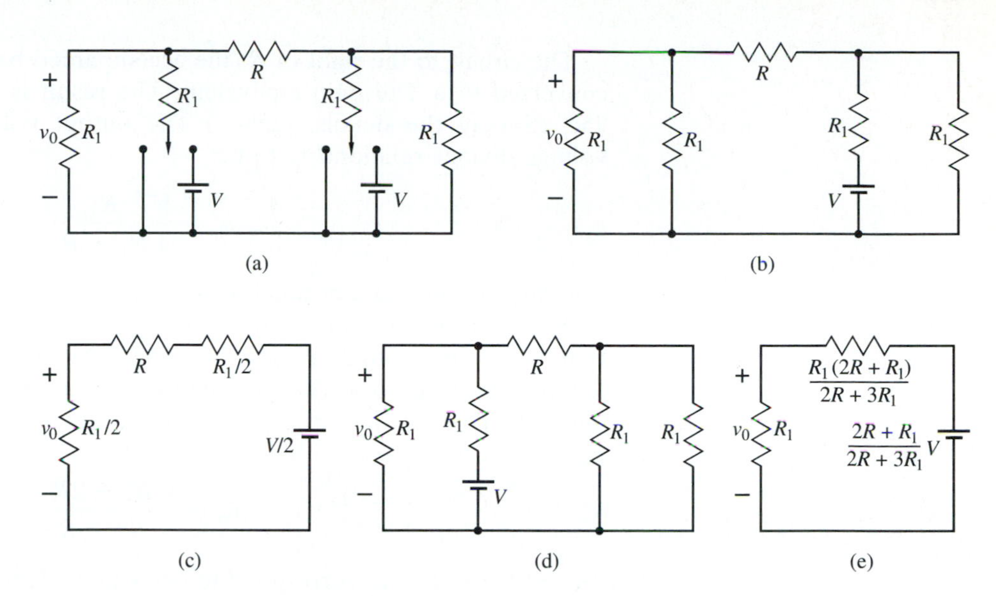

Figure 25
D/A converter.

the signals in radio music. On the other hand, digital signals are those represented by binary numbers. The binary digits 1 and 0 can represent the presence or absence of something. Thus, the three-position binary number 101 is interpreted as having a signal present in the third and first positions, and absent in the second position. The decimal equivalent of this number is $2^2 + 0 \times 2^1 + 2^0 = 5$.

An important problem in electrical engineering is to convert from one type of signal to the other. In this section we will take up a circuit that converts signals from digital to analog form. A possible circuit for converting a 2-bit digital signal is shown in Figure 25a.[9]

Each of the bits is represented by the battery-switch combination. When the value of a bit is 1, the corresponding switch is connected to the battery; when it is 0, the switch is connected to ground. (The switch is controlled by the signal.) All of the resistors except one have the same value. Our objective is to determine the resistor values so that the output voltage v_o gives the decimal equivalent of the binary inputs.

When both switches are connected to ground, both binary bits are 0. Sure enough, in this case there will be no output, corresponding to decimal 0. When the switches are connected to give a binary 01 (left switch to ground, right switch to battery), the circuit reduces to the one shown in Figure 25b. The two R_1 resistors on the left are in parallel and can be replaced by a resistance $R_1/2$. Everything else in the circuit can be converted to an equivalent accompanied voltage source by Thévenin's theorem.

[9] Don't get me wrong! The designation "2-bit" is not intended as a judgment about its worth, only as a description of its length. Another concern you may have is to question how the structure of this circuit was chosen. This is where engineering creativity and ingenuity come in. This structure was no doubt arrived at after long study and immersion in the subject. Don't feel discouraged that you couldn't have immediately created this structure.

The circuit to the right of R (the accompanied battery in parallel with R_1) is converted to a Thévenin equivalent. The result is the series circuit in Figure 25c. (Supply the details, please.) The output voltage is now found by the voltage-divider relationship to be:

$$v_o = \frac{V}{4}\frac{R_1}{R_1 + R}$$

This corresponds to a decimal digit of 1.

Next, the switches are connected to give a decimal 2: left switch to battery, right switch to ground, as in Figure 25d. This time, everything but R_1 on the left is converted to a Thévenin equivalent, as in Figure 25e. (Carry out the details to confirm this circuit.) The output voltage is found from this circuit to be:

$$v_o = \frac{V}{4}\frac{R_1 + 2R}{R_1 + R}$$

This expression is to correspond to decimal 2. Hence, we set this value of v_o to twice the value of the previous one, yielding:

$$\frac{R_1 + 2R}{R_1 + R} = 2\left(\frac{R_1}{R_1 + R}\right)$$

$$R_1 = 2R$$

With these resistor values, the output voltage in the two preceding cases will be $V/6$ and $V/3$; the latter is twice the former, as it should be. Note that the value of R is arbitrary and can be chosen for convenience, the only important criterion being $R_1 = 2R$. The circuit is often referred to as an R-$2R$ D/A converter.

As a final step, both switches are connected to the batteries, corresponding to binary 11, or decimal 3. Go through the details to confirm that the output voltage this time is $V/2$ (independent of the values of R and R_1); this is 3 times the value corresponding to binary 01, as it should be.

The design of a D/A converter with more bits should be fairly obvious; add one more R branch and one more battery/switch-resistor branch for each added bit. This is left as a problem for you to carry out.

6 DIODE CIRCUITS

All of the methods discussed in this chapter so far apply to linear circuits of resistors and sources. We shall now consider circuits containing one or more diodes in addition to these.[10]

[10] This section is obviously dependent on the section on diodes in the preceding chapter. If you chose not to study that material there, then either you have to go back to it now or you omit this section also.

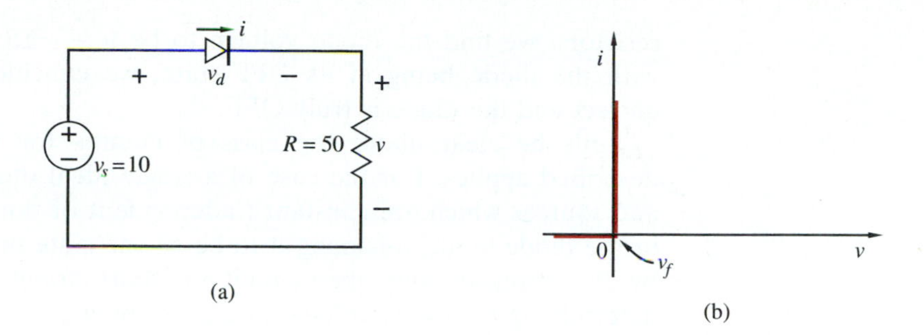

Figure 26
(a) Diode circuit and (b) ideal characteristic; $v_f = 0$.

First consider the simple circuit shown in Figure 26 in which a diode is connected in series with a source and a resistor. The first question to be answered is, What model should be used for the diode? We shall initially assume the diode is ideal, with an *i-v* characteristic as shown in Figure 26b, but with the offset voltage V_f equal to zero. (Remember that, in practical semiconductor diodes, the offset voltage is a few tenths of a volt.) The ideal diode acts as a switch; it is either ON, in which case the diode acts as a short circuit, so that its *voltage* is zero, or OFF, in which case the diode acts as an open circuit, so that its *current* is zero.

The State of an Ideal Diode

When an ideal diode is in a circuit containing resistors and sources, the first question that must be answered is, *What state is the diode in*? When the question is answered, the diode is replaced by the appropriate equivalent: either a short circuit or an open circuit. The result is a circuit of resistors and sources only.

In order to *determine* the state of the diode in Figure 26, let us *assume* a state and then determine if this state is consistent with the resulting conditions in the circuit. Suppose that the diode is in the OFF condition in Figure 26. In this case, there will be no current and, hence, no voltage across the resistor. By KVL, the diode voltage will be $v_d = 10$. Is that consistent with the diode being OFF? Well, if the diode is OFF, then its voltage must be negative, so there is an inconsistency. The assumption that the diode is OFF is not confirmed by the behavior of the circuit; hence, the diode must be ON.

Our original assumption was chosen arbitrarily. Suppose, instead, we had assumed that the diode is ON. Since it then acts as a short circuit, its voltage will be zero. Therefore, the entire source voltage is across the resistor; by Ohm's law, the current is found to be $i = 0.2$. Since this is a positive current through the diode, it *is* consistent with the assumption that the diode is ON.

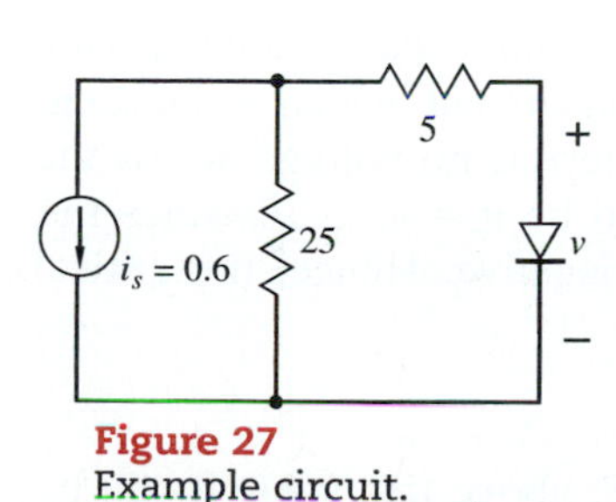

Figure 27
Example circuit.

As a second example, consider the circuit in Figure 27. To determine the state of the diode, let's try assuming it is OFF. It can then be replaced by an open circuit, so there will be no current in the diode. (Remember that, if no units are shown, then SI units are assumed; therefore the resistor values are in ohms and the value of the source current is in amperes.) The current in the 5-Ω resistor is also zero, by KCL. Thus the entire source current flows through the 25-Ω resistor. By KVL around the path consisting of the diode and the two resistors, we find the diode voltage to be $v = -15$ V. Since this is consistent with the diode being in its OFF state, we conclude that our assumption is correct and the diode is truly OFF.

Let's be clear about the class of circuits for which the procedure just described applies. For the case of a single ideal diode in a circuit of resistors and sources which are constant (independent of time), we determine the state of the diode by (1) assuming it to be in one state or the other, (2) replacing it by its corresponding open-circuit or short-circuit equivalent, and then (3) determining the diode voltage or current by analyzing the rest of the circuit. If this value is consistent with the value corresponding to the assumed state of the diode, then the diode must be in the state assumed. If not, the diode is in the opposite state.

EXERCISE 9

Suppose that a circuit of linear resistors and constant sources includes an ideal diode. The procedure just described is summarized in the tables that follow. Complete the tables.

ASSUMED STATE: OFF

Sign of v_d from trial solution	Negative	Positive
Assumed state correct?	______	______
Actual state of diode	______	______

ASSUMED STATE: ON

Sign of i_d from trial solution	Negative	Positive
Assumed state correct?	______	______
Actual state of diode	______	______

□

TIME-VARYING SOURCES When an ideal diode is in a circuit with constant sources, the diode will take on a specific state (ON or OFF) and remain in that state. However, if a source voltage or current is varying with time, the diode may be in one state for a certain range of source values and switch to the other state for other values. What is important in this case is to determine the critical value of source voltage or current at which the diode switches from one state to the other.

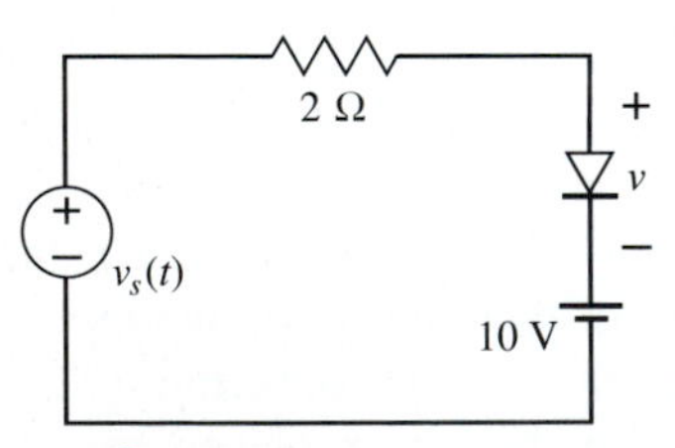

Figure 28
Diode circuit with time-varying source.

A simple example is shown in Figure 28. The source has a time-varying voltage $v_s(t)$. Let's assume that the diode is OFF; its current, which is the same as the current in the rest of the circuit, is zero. There is no voltage across the resistor, so, by KVL, the diode voltage is found to be $v = v_s - 10$. Since the diode is to be OFF, the diode voltage must be negative. Hence, the critical value of v_s is found to be:

$$v_s \leq 10$$

Below the value $v_s = 10$ V, the diode will be OFF; above this value it will be ON.

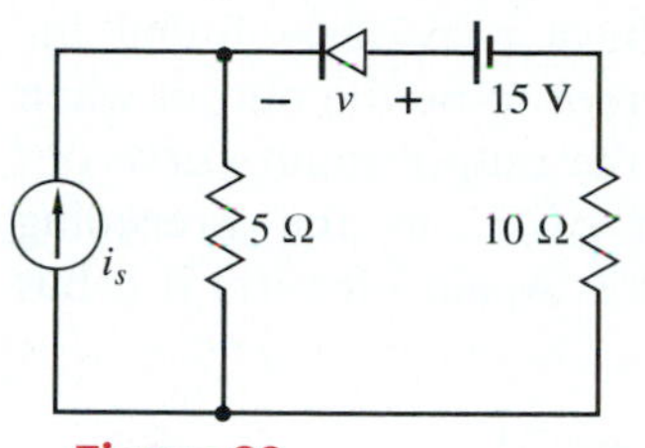

Figure 29
Exercise circuit.

An equally satisfactory approach is to assume that the diode is ON. Replacing the diode by a short circuit results in a circuit with two sources in series with a resistor. The current is easily found to be:

$$i = \frac{v_s - 10}{2} \geqslant 0$$

Since, if the diode is ON, its current must not be negative, we find the same critical value of source voltage at which the diode will switch from one state to the other.[11]

EXERCISE 10

In the circuit shown in Figure 29, the source is a time-varying current $i_s(t)$. Determine the critical value of i_s by assuming the diode is OFF, and confirm it by assuming the diode is ON.

ANSWER: 3 A. □

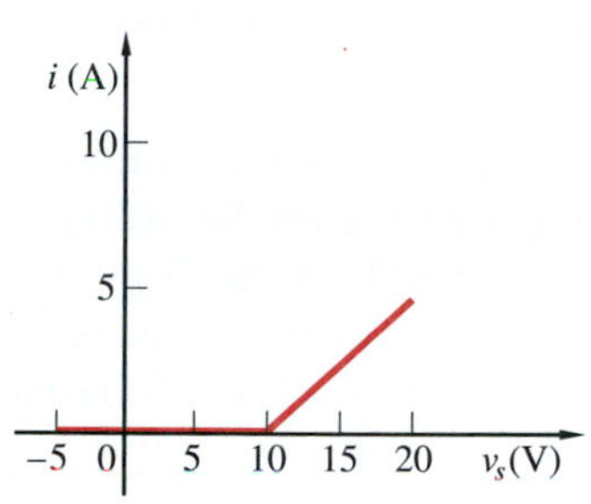

Figure 30
Example output-input graph.

PLOTTING OUTPUT CURVES In the circuits discussed in the preceding section, the time-varying source voltage or current can be considered an "input" signal, or an *excitation.* Any one of the resulting currents or voltages in the circuit can be looked upon as a *response* to this excitation, or an "output." We concentrated on obtaining the critical switching value of a time-varying source. Now we shall discuss two ways of displaying various current and voltage responses graphically.

Look back at Figure 28, where we found:

$$i = 0 \qquad \text{for } v_s \leqslant 10$$
$$i = 0.5v_s - 5 \qquad \text{for } v_s \geqslant 10$$

Each of these expressions describes a straight line, with i as dependent variable and v_s as independent variable. They are plotted in Figure 30. This type of graph, where an output variable is plotted against an input, is called an *output-input* graph.

The output-input graph does not provide information about the actual variation of the output as a function of time. To obtain this information it is necessary to know the variation of the input with time. Suppose the source voltage varies with time as shown in Figure 31a. Using the same time scale, a set of axes is drawn under the voltage plot. Graphically, we find the time at which the source reaches 10. The current output will be zero at all times for which v_s is less than this value, 10. When $v_s > 10$, then $i = (v_s - 10)/2$. Thus, at each instant of time for which $v_s > 10$, we find the value of v_s from the graph, subtract 10, and divide by 2. This value for i is then plotted against the corresponding time, with the result shown in Figure 31b.

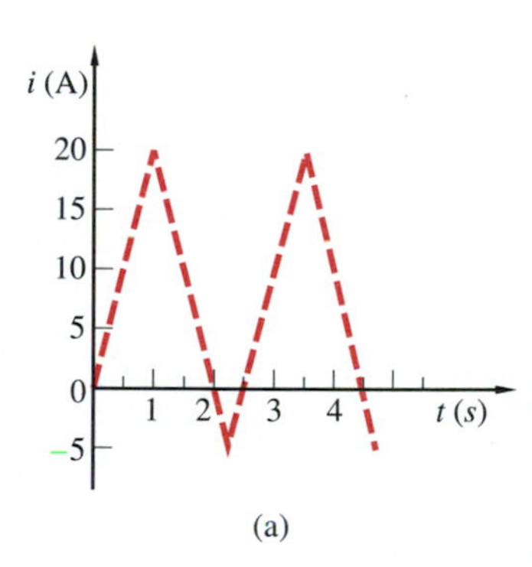

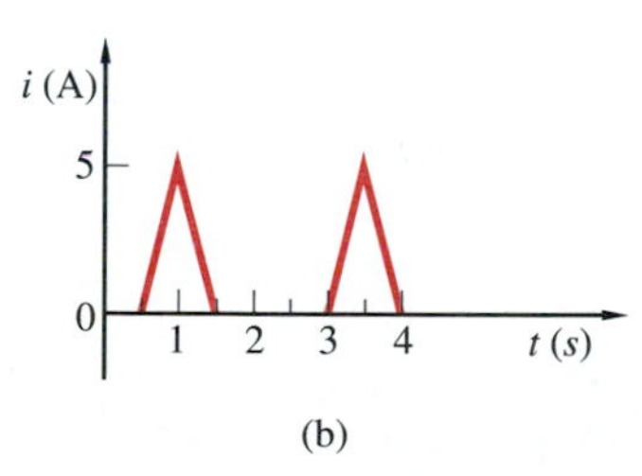

Figure 31
Output versus time; from input and output-input curve.

[11] Although, in problems like this, it can be assumed that the diode is either in its ON state or in its OFF state, there may be a difference in complexity of the circuit that results after the assumption is made. Sometimes it is possible to tell ahead of time which assumption will lead to the simpler circuit. Obviously, that's the assumption to make.

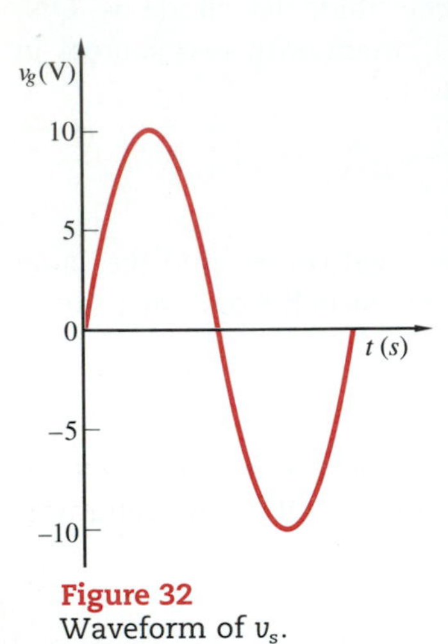

Figure 32
Waveform of v_s.

EXERCISE 11

In the exercise circuit of Figure 29, you should have previously found the critical value of i_s to be 3 A. Now take the diode current to be the output when the diode is ON and find an expression for it. Plot the output-input curve of i against i_s. Let i_s have the triangular waveform of v_s in the preceding paragraphs (same scale) and plot the waveform for i. Again plot $i(t)$ if i_s has the sinusoidal waveform shown in Figure 32.

ANSWER: $i = (3 - i_s)/3$. The graphs are given in Figure 33. □

Applications

Diodes find many applications in electronic circuits. This section will describe a few of these.

RECTIFIERS An important application of diodes is to convert voltages that alternate periodically between positive and negative values to voltages that are undirectional, never negative. In common (and loose) terminology this process is said to be a conversion from alternating current to direct current. (The terminology is "loose" because DC usually implies constant, not varying with time, which, as we shall see, is not the case here.)

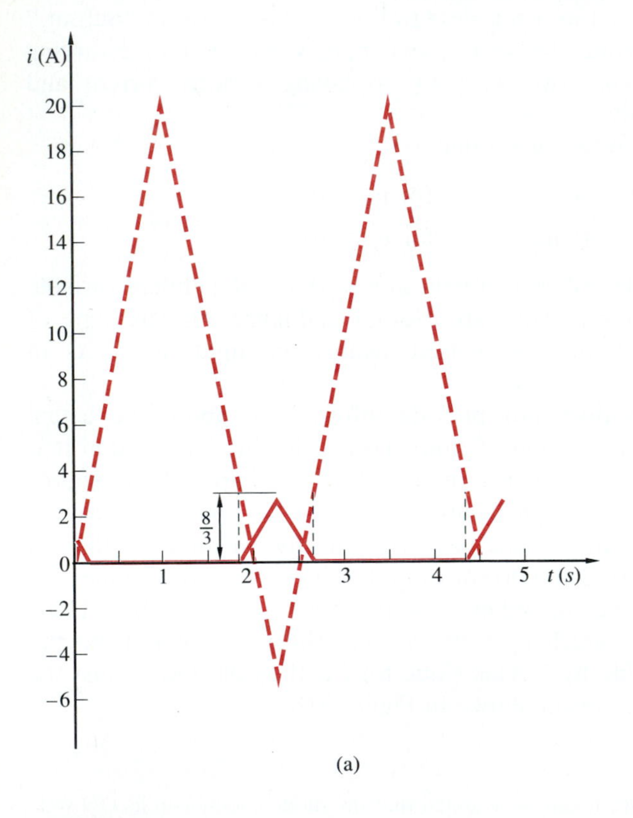

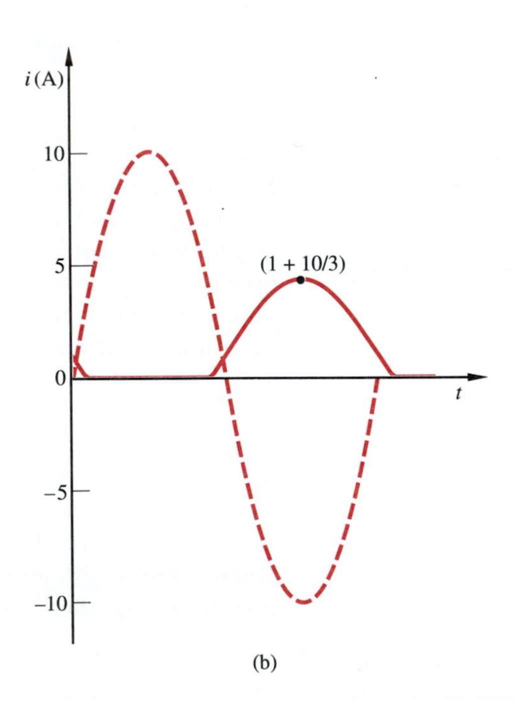

Figure 33
Output waveforms for exercise circuit.

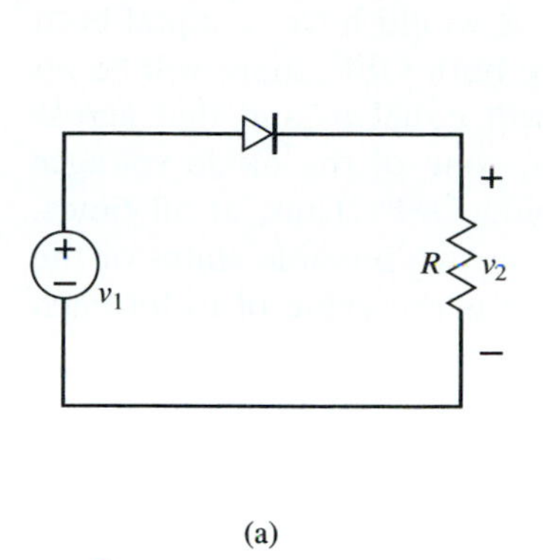

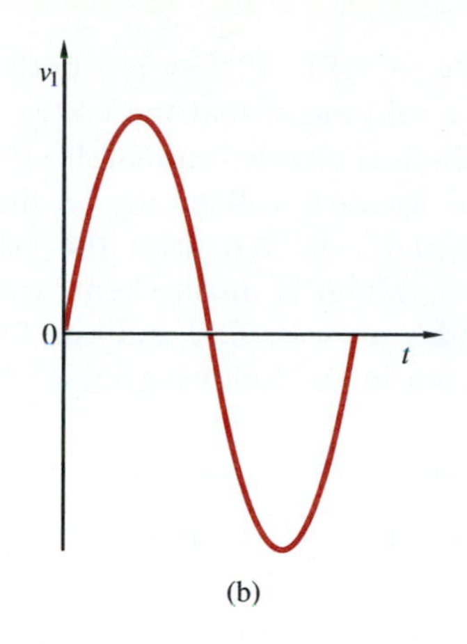

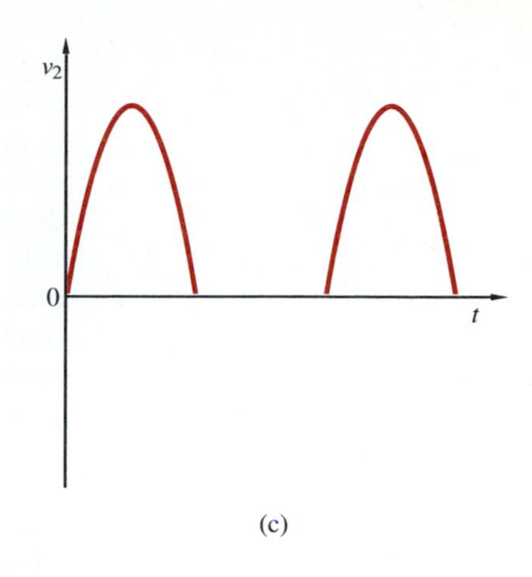

Figure 34
Half-wave rectifier circuit.

A simple circuit that accomplishes this function is shown in Figure 34. By using the method for determining the critical switching value previously discussed, we find that the diode will be ON for $v_1 > 0$. For this diode state, the entire source voltage appears across the resistor, so $v_2 = v_1$. When v_1 is negative, the diode is OFF and the current is zero, so $v_2 = 0$. Suppose the input voltage is the sinusoid shown in Figure 34b. Then the output will have the waveform shown in Figure 34c. The output is the same as the input whenever the input is positive.

The output is said to be *rectified,* and so the circuit is a *rectifier.* The negative part of the input is suppressed and not utilized in the output; the circuit is, thus, a *half-wave* rectifier. The loss of half the input wave can be a disadvantage in some applications. A possible remedy is shown in Figure 35. The two sources are identical.[12]

To find out the states of the diodes, let's first assume that both have the

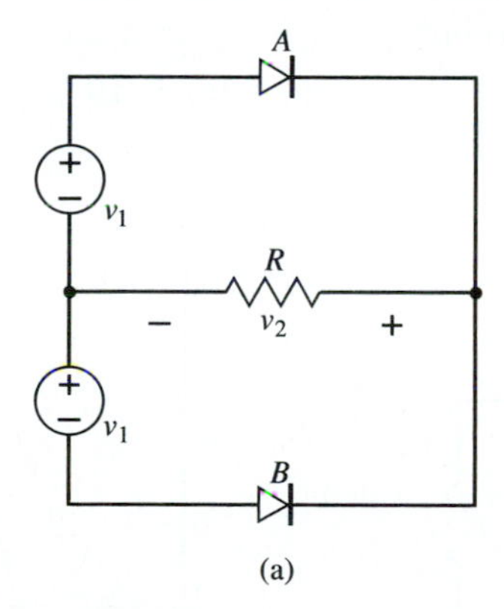

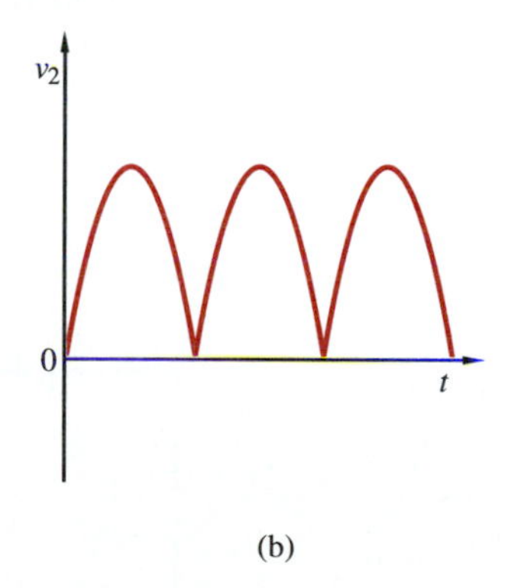

Figure 35
Full-wave rectifier circuit.

[12] This is a simplified diagram. In practice, the effect of two sources is obtained by using a device called a *transformer* which has two identical secondary windings. Transformers will be introduced in Chapter 10, although an ideal version will be described earlier, in Chapter 4.

same state: ON or OFF. If they are both ON, both can be replaced by short circuits. That would mean that the voltage across R would have to equal both v_1 and $-v_1$, which is clearly impossible. If they are both OFF, there will be no current, so the forward voltage across diode A will equal v_1 and that across diode B will equal $-v_1$. Whatever the value of v_1, one of the diode voltages will be positive, which is inconsistent with its being OFF. Thus, at all times, one of the diodes must be ON and one OFF. These two possible states of the diodes are shown in the following table. Also given is the value of v_2 for each range of v_1.

	$v_1 > 0$	$v_1 < 0$
Diode A	ON	OFF
Diode B	OFF	ON
v_2	v_1	$-v_1$

Suppose v_1 has the sinusoidal waveform shown in Figure 34b. The output voltage will be the same as it was in the half-wave rectifier case when v_1 is positive. But when v_1 is negative, $v_2 = -v_1$, which is a positive quantity. During this time, the output will be a reflection of v_1 around the time axis. A graph of the output is shown in Figure 35b. Because both the positive and the negative parts of the input are utilized, this circuit is called a *full-wave* rectifier.

COMPARATORS In some applications, it is useful to know if an input voltage which varies with time will ever exceed a specified value. Reviewing the half-wave rectifier circuit, we see that it can be considered to be a circuit which detects whenever the input voltage has a value greater than zero.

Suppose a battery is added in series with the diode, as shown in Figure 36a. In this circuit, when the diode is OFF, there will be no current, so the output voltage will equal zero. In this state, from KVL, the forward diode voltage will be $v_1 - 10$. The critical switching voltage is, therefore, 10 V. When v_1 exceeds this value, the diode turns ON. When the diode is ON, the output voltage can be found from KVL to be:

$$v_2 = v_1 - 10$$

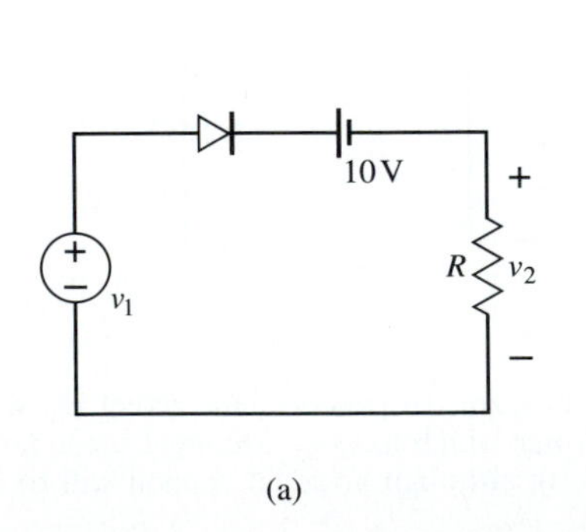

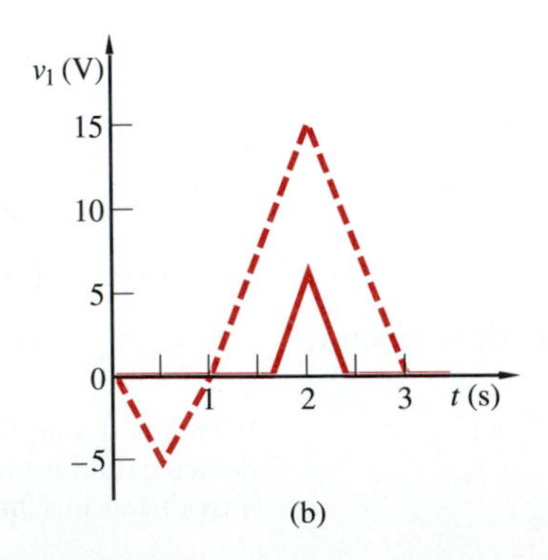

Figure 36
Comparator circuit.

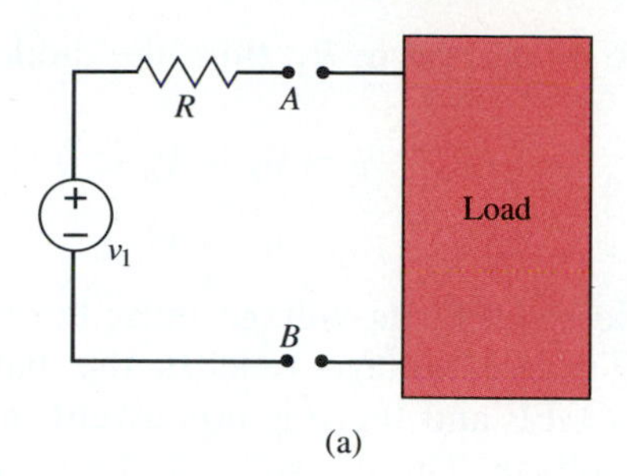

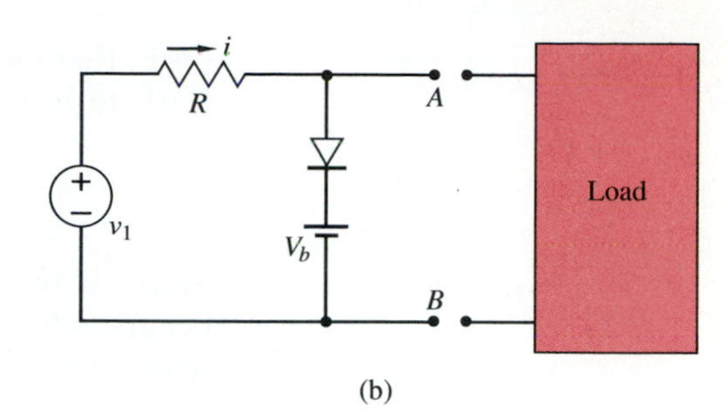

Figure 37
Load-voltage protection.

To summarize: There will be no output voltage until the input exceeds 10 V. After that, the output equals the input reduced by 10. That is, the circuit compares the input with the battery voltage (10 V in this case) and provides an output only if the input exceeds this value. It can, for this reason, be called a *comparator*.

Suppose the input is the dashed curve in Figure 36b. The output will have the waveform shown in the solid curve. It has a nonzero value only for input values that exceed 10.

VOLTAGE LIMITERS In the preceding application, we were able to *detect* when an input voltage exceeded a predetermined level. But, in the operation of some electronic systems, it is important to *prevent* the voltage across two points from ever exceeding a predetermined level, to *limit* the voltage. The operation of a diode as a switch permits this capability.

Consider the circuit in Figure 37. A source (with internal resistance R in ohms) is to be applied to a circuit which we will designate as a load. But if the source voltage becomes too high, some devices in the load might be damaged. Therefore we connect a diode in series with a battery across terminals AB. When the diode is OFF, this branch is effectively disconnected from the circuit and so it doesn't influence its operation. But when the diode is ON (a short circuit), then the voltage across the terminals AB will equal the battery voltage, no matter what the value of the input voltage v_1 may become. All that remains is to determine the diode switching voltage.

The details of this depend on whether the load is connected or not. We will consider both cases. Figure 38a shows the case without the load. If the diode is

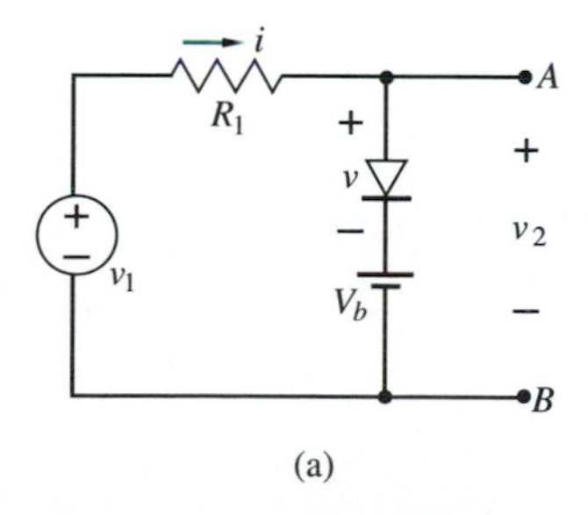

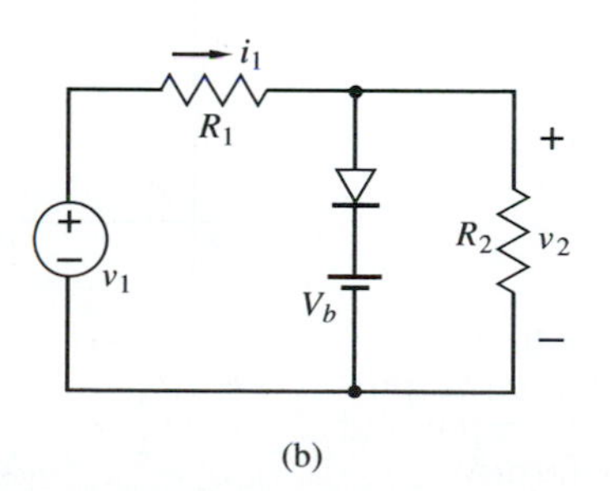

Figure 38
Voltage-limiter circuit: (a) without load, (b) with load.

OFF, there will be no current in R_1; thus, the diode voltage is determined from KVL to be:

$$v = v_1 - V_b < 0$$
$$v_1 < V_b$$

Since, if the diode is OFF, its voltage must be negative, the switching value occurs when the source voltage reaches the battery voltage. Furthermore, when the diode is OFF and there is no current in R_1, the output voltage will equal the source voltage. These results can be summarized as follows:

$$v = v_1 \quad \text{for} \quad v_1 < V_b$$
$$v_2 = V_b \quad \text{for} \quad v_1 > V_b$$

The same information can be given graphically in the output-input curve, as shown in Figure 39a, where the battery voltage is taken to be 10 V. Suppose the input voltage is sinusoidal with a peak value of 20 V, as shown in Figure 39b. For all negative values and for positive values up to 10 V, the output voltage equals the input. But at all times when v_1 exceeds 10 V, the output remains at 10. This is also shown in Figure 39b.

Let us now consider the case when the load is connected at terminals AB. Again, the details will depend on what the load circuit consists of. For simplicity, we will consider only a resistance load, as shown in Figure 40. Again our purpose is to find the output voltage v_2 as a function of v_1. A useful first step is to redraw the circuit twice, once for each state of the diode, replacing the diode with the appropriate equivalent. This is shown in Figure 40b and c.

When the diode is ON, the battery is directly across the output, just as in the previous circuit. When the diode is OFF, the circuit is like a voltage divider, and so the output voltage can be obtained in terms of the input from the voltage-divider expression. The only remaining problem is to determine the critical value of the input voltage for which the diode switches state. We will

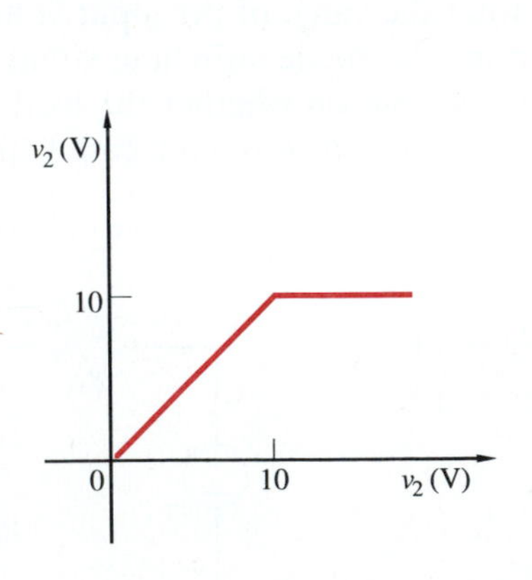

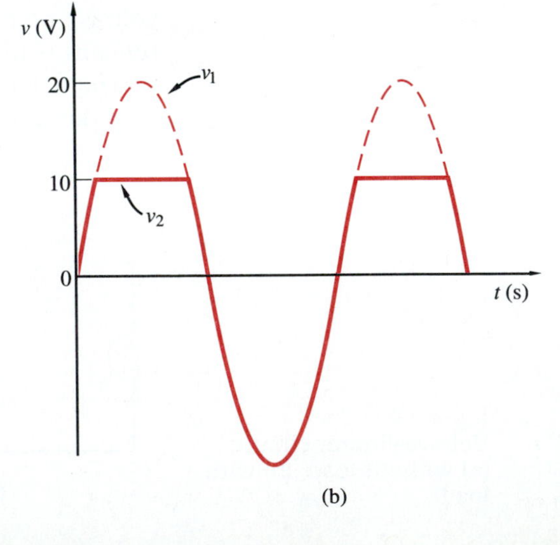

Figure 39
Graphical results for voltage limiter: (a) output-input curve, (b) input and output waveforms.

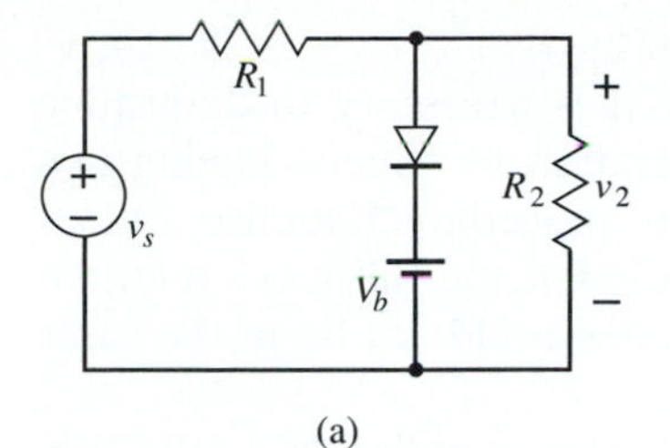

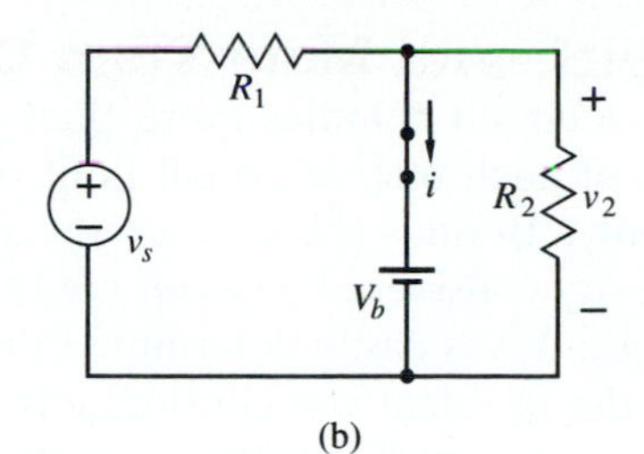

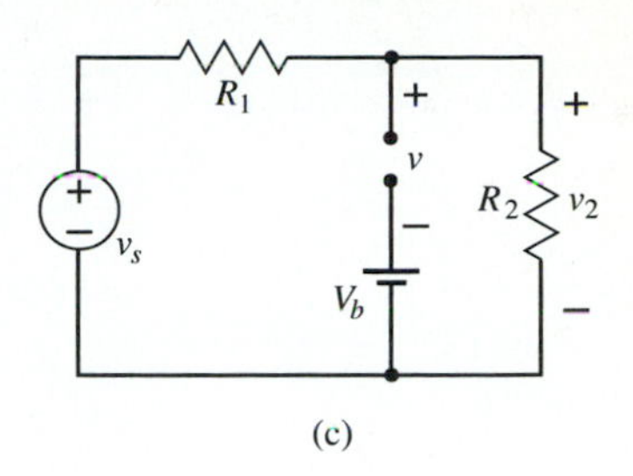

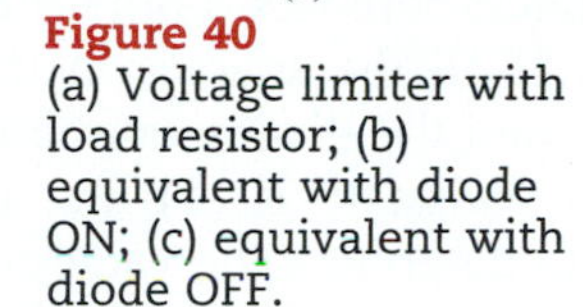

Figure 40
(a) Voltage limiter with load resistor; (b) equivalent with diode ON; (c) equivalent with diode OFF.

do this by (1) assuming that the diode is ON, (2) finding an expression for the diode current, and then (3) taking this expression to be positive. From Figure 40b, the diode current can be obtained by KCL in terms of the currents through the two resistors; thus:

$$i = \frac{v_s - V_b}{R_1} - \frac{V_b}{R_2} > 0$$

$$v_s > \left(1 + \frac{R_1}{R_2}\right)V_b$$

The critical value of v_s is found by setting the diode current >0.

EXERCISE 12

Let $R_1 = R_2$ and let $V_b = 10$. Draw the output-input curve for the circuit in Figure 40. Let v_s be a sinusoid of peak value 30 V. Draw this function in dashed lines and superimpose on it the graph of the output voltage. □

EXERCISE 13

Instead of limiting the positive excursions of a voltage, suppose we would like to limit the negative excursions. Suggest how to modify the circuit of Figure 40 to achieve this goal. Then carry out an analysis and obtain the critical switching value for v_s.

ANSWER: Reverse both the diode and the battery. □

EXERCISE 14

Suggest a circuit that will limit the positive excursions of the voltage at its terminals to 6 V and the negative excursions to 10 V. This might be called a double-limiting circuit. Determine the critical switching values of the input voltage. □

Circuits with More Than One Diode

When a circuit contains more than one diode, it is necessary to determine the state of each one. We shall limit our consideration to circuits having two diodes only. Besides the case of the immediately preceding Exercise 14, we have already considered a circuit containing two diodes: the full-wave rectifier. In that case it was easily determined that both diodes could not be in the same state, so the problem was considerably simplified.

Generally we must examine each of the possible states: both ON, both OFF, or the two cases of each diode in a different state. If all the sources in the circuit are constant, then no switching will take place and the diodes can be in only one state. The procedure for determining this one state is (1) first to assume each one of the four possible states, (2) then to determine the values of diode currents or voltages (as appropriate), and, finally, (3) to verify whether these values are consistent with the assumed conditions. (Of course, once consistent values have been confirmed, that's the correct state and the remaining possibilities don't have to be checked.)

However, when there is at least one time-varying source, then the state of each diode may switch as time goes on and we will have to determine the critical values at which the diode states switch. Then we will have to determine the output in each of the intervals between switching events.

As an illustration, consider the circuit of Figure 41a. Suppose we assume that both diodes are OFF. The circuit is redrawn for this state in Figure 41b. From KVL, the voltage across diode *B* is found to be $8 - 2 = 6 > 0$. This is inconsistent with diode *B* being OFF; hence, this state is impossible. The situation for the remaining states is summarized in the table that follows.

Diode A	Diode B	Input voltage	Output voltage
OFF	ON	$v_1 < 4$	$v_2 = 4$
ON	ON	$4 < v_1 < 8$	$v_2 = v_1$
ON	OFF	$v_1 > 8$	$v_2 = 8$

You should confirm these results for each of the states by redrawing the circuit, with each diode replaced by its equivalent, determining the switching values of v_1, and then finding the output voltage.

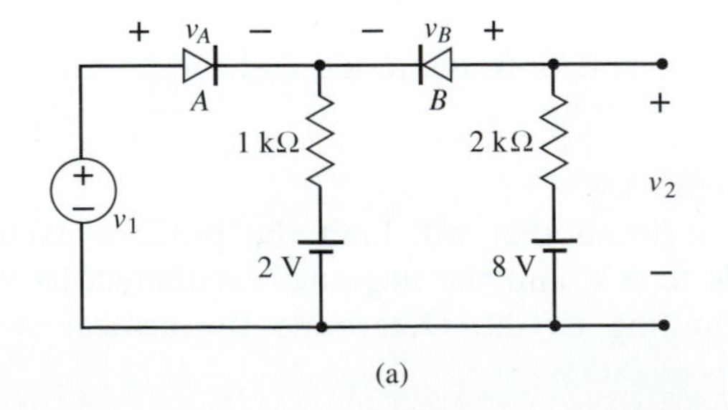

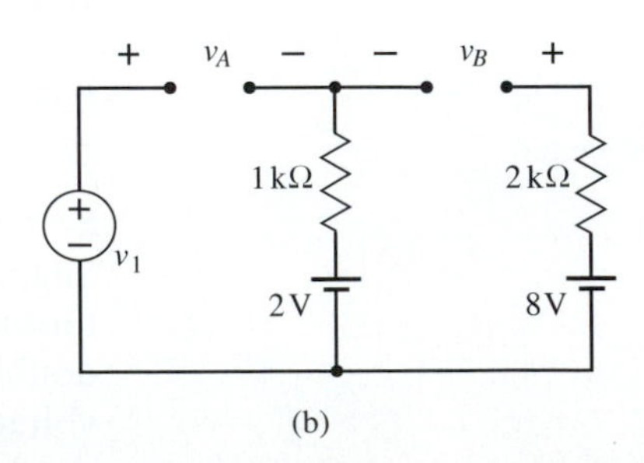

Figure 41
Circuit with two diodes.

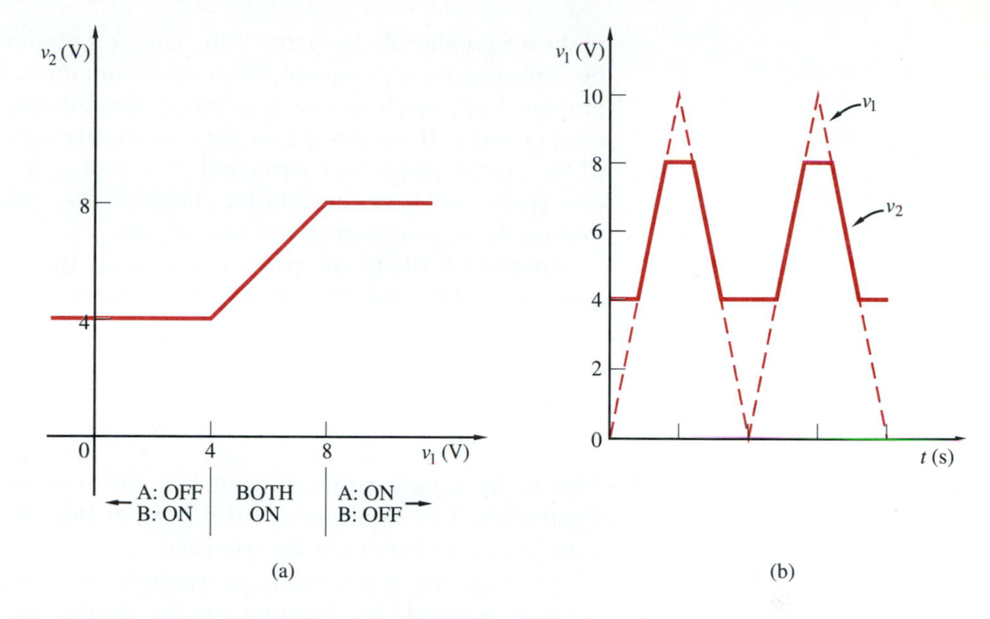

Figure 42
Graphical results:
(a) output-input graph,
(b) output waveform.

The output-input graph is obtained by plotting the information in the table; it is shown in Figure 42a. Figure 42b shows an input waveform and the corresponding output waveform. The output is the same as the input for values between 4 and 8 V. For inputs less than 4 V, the output is constant at 4. For input values greater than 8 V, the output is limited to 8 V.

Diode Models with Graphical Characteristics

So far we have dealt exclusively with the ideal model of a diode. Not much would change if we were to include also the offset voltage V_f in the model. The only change would be that when the diode is ON, its voltage would have to equal V_f rather than zero. You should repeat some of the preceding examples with this change and see if you encounter any difficulties; you shouldn't.

In Figure 43a assume that the diode is a real diode whose i-v characteristic

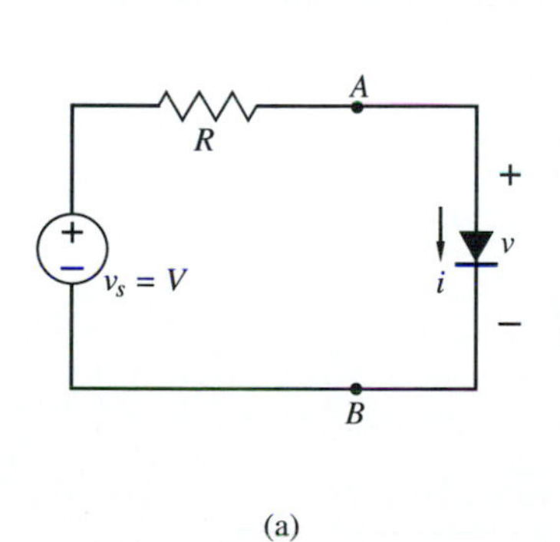

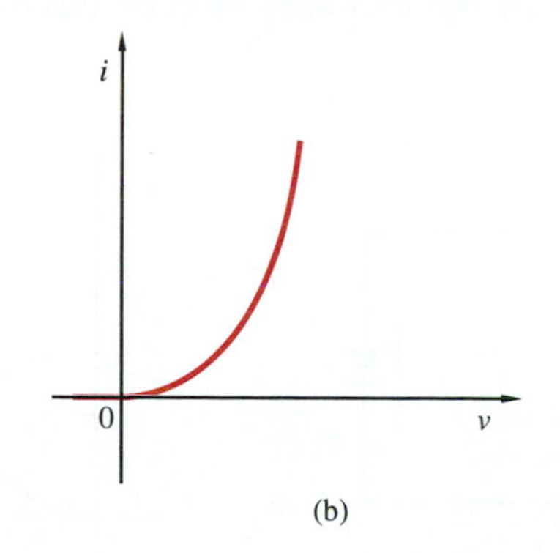

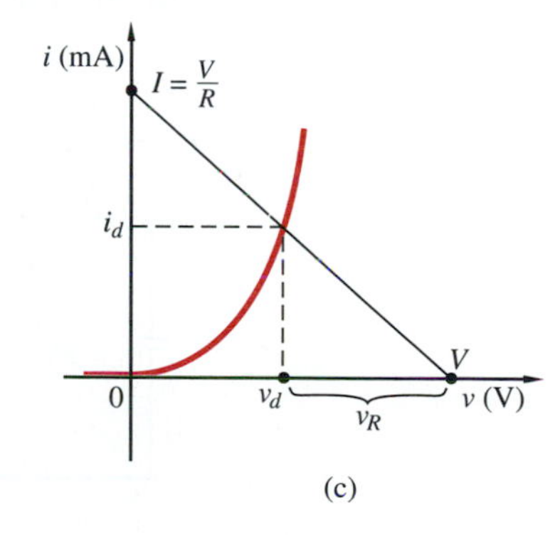

Figure 43
Diode circuit with real diode.

is shown graphically in Figure 43b. This is a nonlinear curve which we will not be replacing by a piecewise linear approximation. So how can we proceed? A graphical approach is one possibility. We already have a graph relating the diode i and v. If we can get an expression relating the same i and v for the rest of the circuit (the linear part) and then plot it, a simultaneous solution of the two graphs will give the solution. Graphically, "simultaneous solution" means finding the intersection of the two graphs.

Applying KVL to the part of the circuit to the left of terminals A and B, and using Ohm's law to eliminate the resistor voltage, leads to:

$$v = V - Ri$$
$$i = \frac{V}{R} - \frac{1}{R}v \tag{21}$$

This is the equation of a straight line with i as the ordinate shown plotted in Figure 43c. The values of i and v lying at the intersection of the straight line with the diode curve are the solution.

Although the preceding is an example, it demonstrates a general nonlinear solution method. By focusing on the diode, the rest of the circuit can be considered to be a *load* on the diode. For this reason, the straight line representing the i-v relationship of the load is called a *load line,* and the method itself is referred to as *load-line analysis.*

EXERCISE 15

The i-v characteristic of the diode in Figure 44a is drawn to scale in Figure 44b. Find the value of the diode voltage and current graphically by drawing the load line. Find the resistor voltage by Ohm's law and confirm its value by applying KVL.

ANSWER: $v = 0.5\text{ V}$, $I = 15\text{ mA}$. □

In the preceding example, the source voltage was a constant. What happens if the voltage is varying with time but everything else is the same, as in Figure 45a? The slope of the line is $1/R$, so that remains fixed. The intersections of the line with the two axes at each instant of time are proportional to the

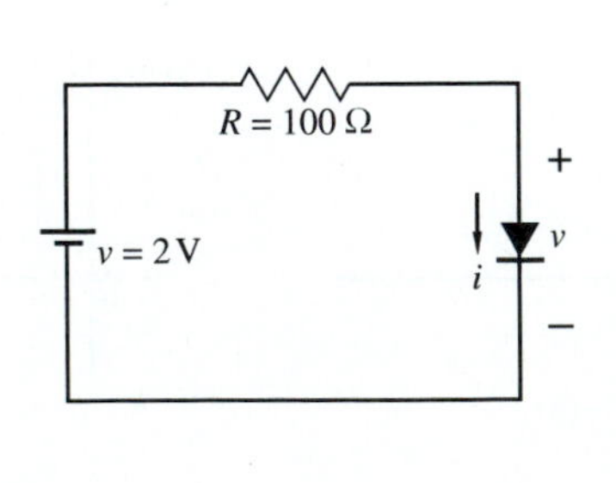

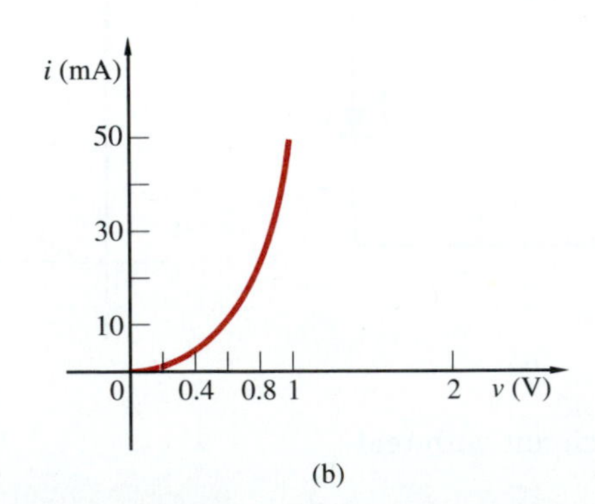

Figure 44
Circuit for Exercise 15.

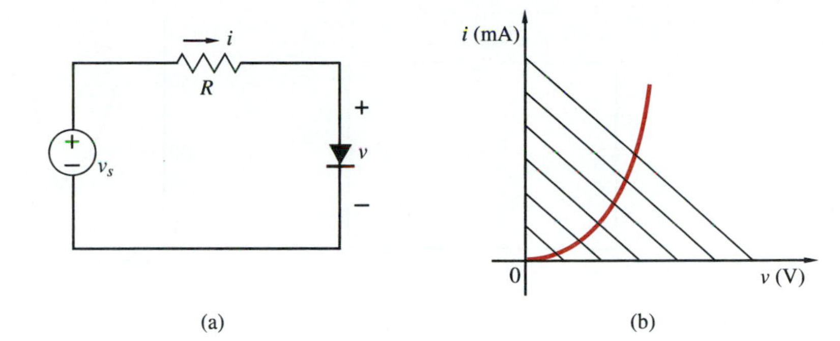

Figure 45
Real diode circuit with time-varying source.

source voltage. (The horizontal intercept is the source voltage itself, while the vertical intercept is this voltage divided by R.) As the source voltage increases, the intercepts move away from the origin; and as this voltage decreases, the intercepts move toward the origin, with the load line at each voltage remaining parallel to itself. This is illustrated in Figure 45b. If the waveform for v_s were known, we could obtain the value of v_s for each instant of time, draw the load line for that value of v_s, read off the values of v and i from the intersection of this load line with the diode curve, and then plot these values against time for each instant of time. This is a somewhat tedious, mechanical procedure—although simple and easily carried out—which amounts to repeating the same steps over and over. We shall not pursue the matter any further here.

Power Dissipation in a Diode

Up to this point a very important and practical feature of a diode has not been considered; this is its power dissipation capability. The dissipation of power in any electrical device results in the generation of heat and, thus, a rise in temperature. Any given electrical device will operate properly only when its temperature does not exceed a certain limit. This makes it necessary to place an upper limit on the power dissipated in the device. This limit is called the *maximum power rating*. Real diodes have such ratings.

Looking back at Figure 44, where the references for v and i are standard references, the power dissipated in the diode is $p = vi$. Placing an upper limit on the product of v and i is equivalent to restricting the range of the diode curve over which the diode is permitted to operate. The nature of this restriction is conveniently displayed by plotting the relationship between i and v when the power is at the maximum rated value. If the maximum power rating of the diode is P_m watts, then the dissipated power must satisfy $p = vi \leqslant P_m$.

The equals sign gives the boundary of the region defined by this limitation in the i-v plane. This boundary is given by:

$$i = \frac{P_m}{v} \tag{22}$$

This is the equation of a hyperbola in the i-v plane.

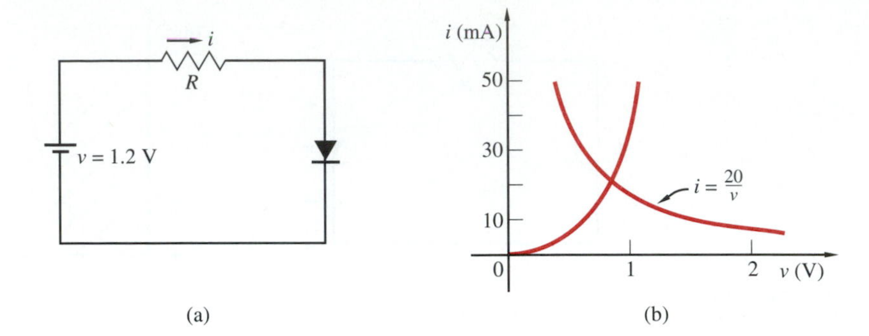

Figure 46
Permissible power dissipation in a diode.

The diagram in Figure 46 shows a diode characteristic on which has been superimposed the maximum power dissipation hyperbola, for the case $P_m = 20$ mW. The shaded region is out of bounds; the only part of the diode characteristic it is safe to be on is the portion outside the shaded region. We cannot allow the load line to intersect the diode characteristic in the shaded region. This means that, for any given battery voltage, the slope of the load line will have an upper limit. Since the slope of the load line is the reciprocal of the resistance, this will impose a lower bound on the resistance. Conversely, if the resistance (and hence the slope) is fixed, there will be an upper limit on the battery voltage.

SUMMARY

In this chapter we have initiated the analysis of electric circuits, after having established the basic laws (Kirchhoff's laws and Ohm's law, power and energy) in the preceding chapters. Topics of importance include:

- Series connections of like components: resistors and v-sources
- Parallel connections of like components: resistors and i-sources
- Equivalent resistance of series-connected resistors
- Equivalent G (and R) of parallel-connected resistors
- The voltage-divider relationship
- The current-divider relationship
- Equivalent resistance of a series-parallel circuit by successive application of series and parallel equivalences
- Continued-fraction expansion
- Determination of branch v and i in a ladder network
- Accompanied voltage- and current-source equivalents
- Solving for branch variables in a ladder network by sequential application of basic laws
- Simplifying a circuit structure by source equivalences
- Linearity and superposition
- Thévenin and Norton equivalent circuits
- Maximum power transfer
- The principle of duality
- Finding resistance values to satisfy specified conditions
- Digital-to-analog converters
- The state of diodes in a circuit with constant sources
- Diode switching conditions in circuits with time-varying sources

- Output-input curves in ideal-diode circuits.
- Diode circuits for specific purposes: peak-detecting, limiting
- Load-line analysis with diode graphical characteristic
- Maximum power dissipation of real diodes

PROBLEMS

1. Find the equivalent resistance looking to the right from the terminals of the circuits in Figure P1:
   a. By using a sequence of series and parallel combinations of resistances, starting from the far end.
   b. By using continued fractions.

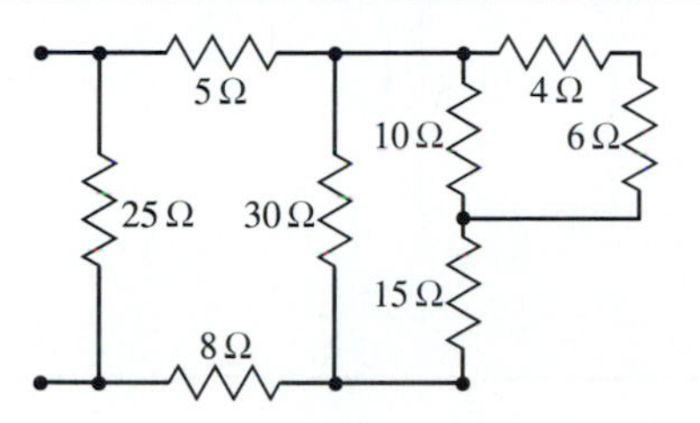

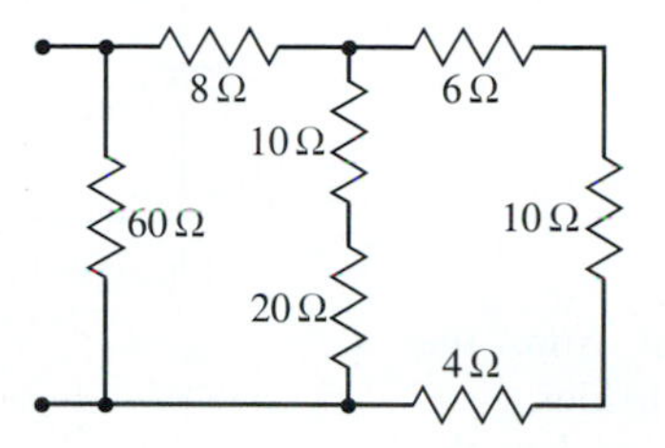

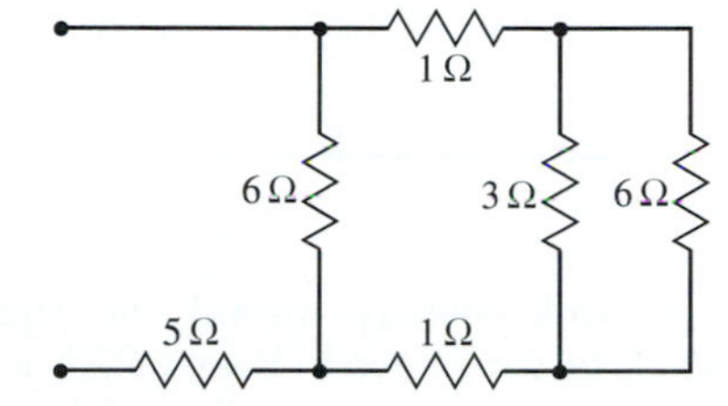

Figure P1

2. Find the parameters of an accompanied source which, at its terminals AB, will be equivalent to the circuit shown in Figure P2.

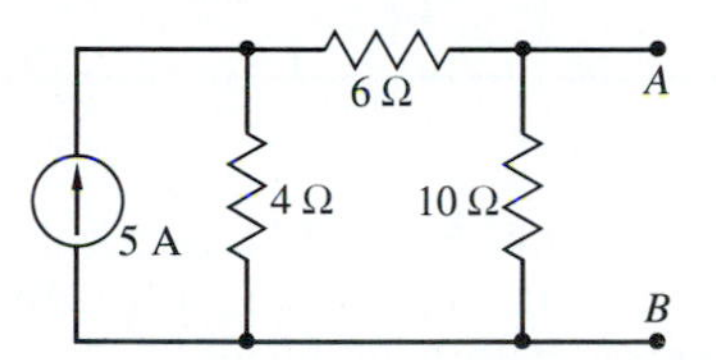

Figure P2

3. A laboratory power supply generally has an accompanied source as an equivalent circuit. When a 10-Ω resistor is connected to a certain power supply, the measured voltages across the terminals is 10.42 V. Then, when a 2-Ω resistor is connected, the voltage is measured to be 6.85 V.
   a. Find the parameters of the accompanied-source equivalent of the power supply.
   b. When a 6-Ω resistor is connected across the terminals, find the power dissipated in it and the power supplied by the source.

4. Find the value of the source voltage in Figure P4 in order for the output voltage to equal 20 V.

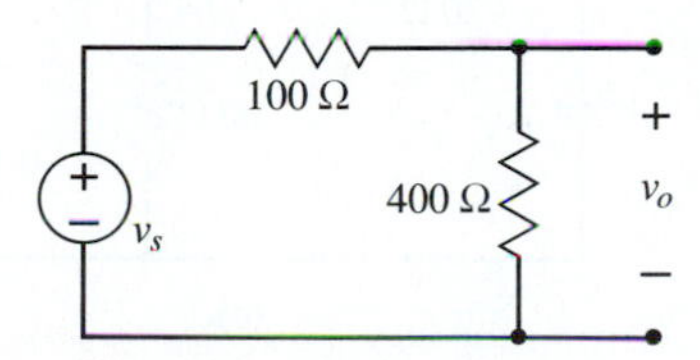

Figure P4

5. a. Find the power dissipated in the 120-Ω resistor in Figure P5.
   b. Find also the voltage across the source.

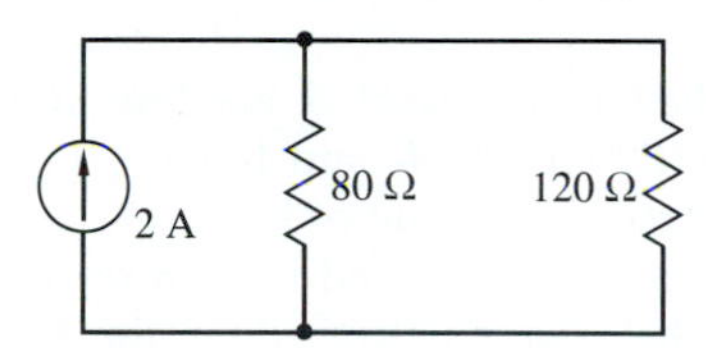

Figure P5

6. In Figure P6, the power dissipated in the 16-Ω resistor is found to be 9 W.
   a. Find the required source current i_s.
   b. Find the power supplied by the source.
   c. Confirm that this power equals the sum of the power dissipated by all the resistors.

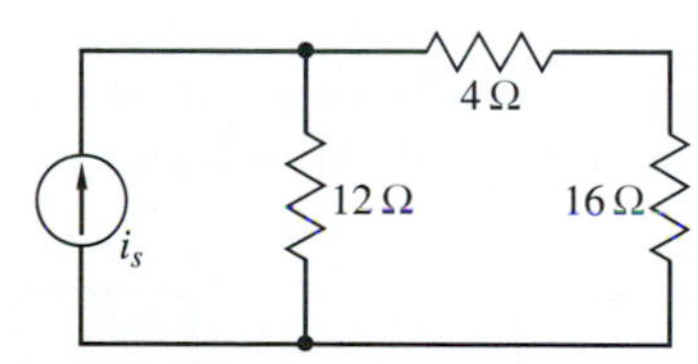

Figure P6

7. Assume that the voltage across the 5-Ω load in Figure P7 is 1 V.
   a. On this basis, starting from the load and using only fundamental laws successively, determine the

value of source current needed to give this value of output.

b. Since the value of the source current is given, find the actual value of v_5.

c. Also find the voltages and currents of all other resistors in the network.

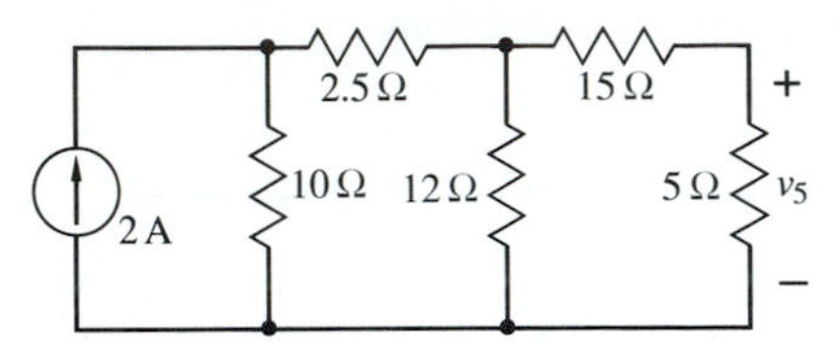

Figure P7

8. The network looking toward the right from the terminals of the source in Figure P7 is a ladder.

a. Find the equivalent resistance of this ladder by the process of continued fractions.

b. Confirm your answer by using a sequence of series and parallel combinations, starting from the right.

c. Find the voltage across the source and compare with the value found in Problem 7.

9. The power supplied to the load R and the voltage across it in Figure P9 are 500 W and 100 V.

a. Determine the required value of v_s.

b. Determine the power dissipated in each resistor.

c. Confirm that the power delivered by the source equals the total power dissipated elsewhere.

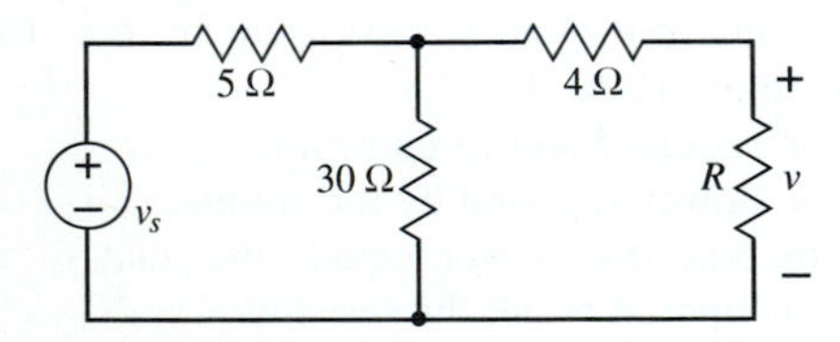

Figure P9

10. The diodes in the circuits of Figure P10 are ideal. Determine the state of the diode in each circuit.

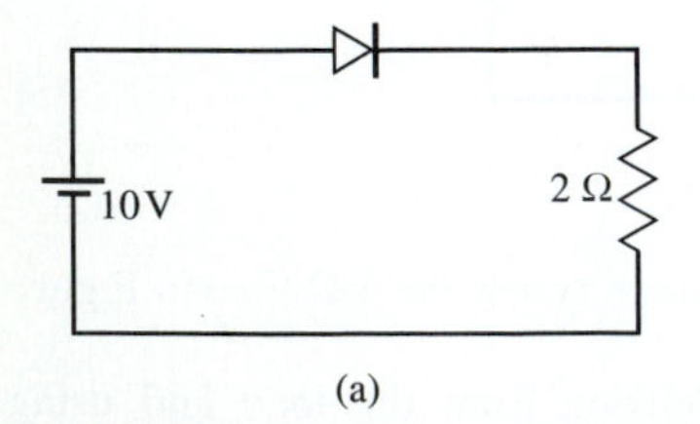

(a)

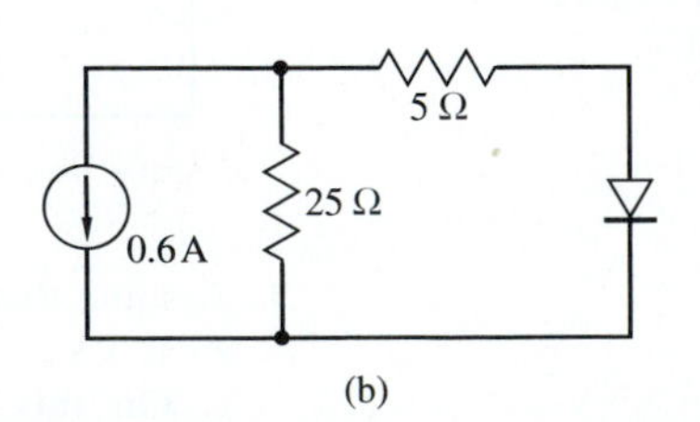

(b)

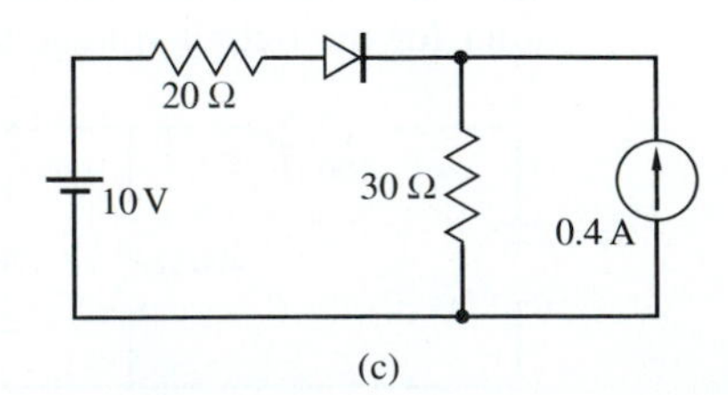

(c)

Figure P10

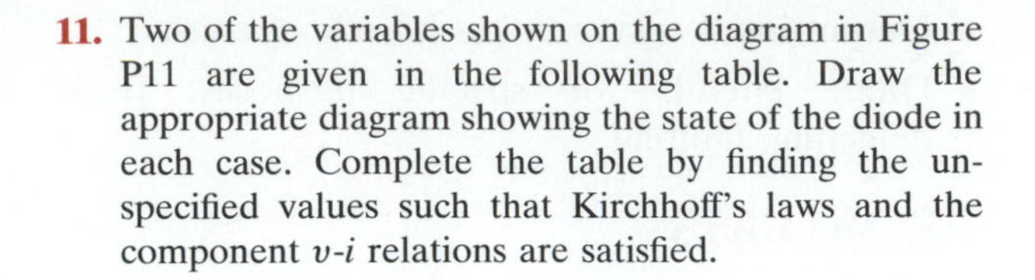

11. Two of the variables shown on the diagram in Figure P11 are given in the following table. Draw the appropriate diagram showing the state of the diode in each case. Complete the table by finding the unspecified values such that Kirchhoff's laws and the component v-i relations are satisfied.

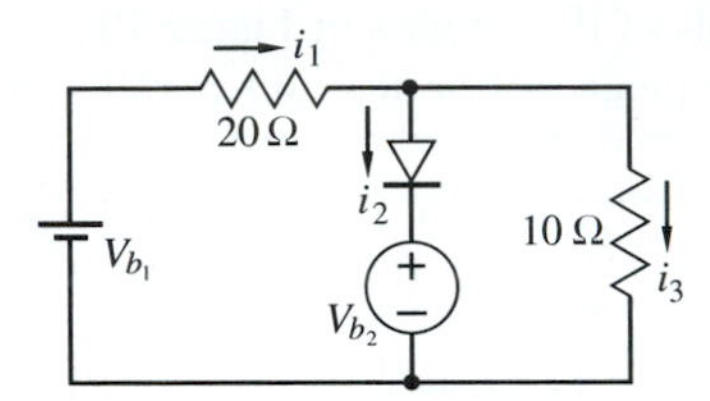

Figure P11

V_{b1} (V)	V_{b2} (V)	i_1 (A)	i_2 (A)	i_3 (A)
30	0			
30		1		
		2	4	
	−20		1.5	

12. The diodes in the circuit of Figure P12 are ideal.

a. Determine the current (units too) in each resistor.

b. Repeat for reversed battery terminals.

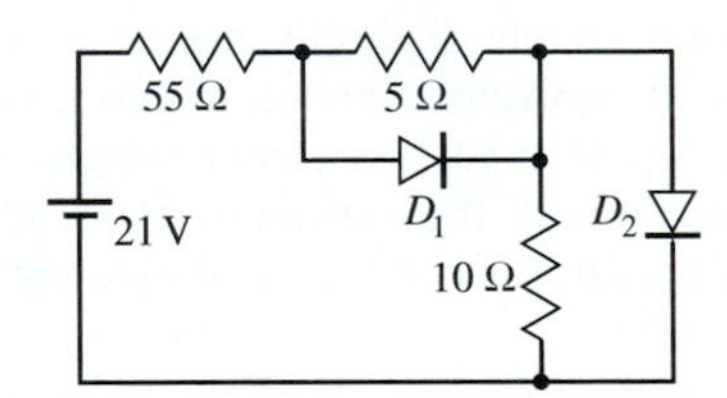

Figure P12

13. The diodes in the circuit of Figure P13 are ideal. After determining the states of the diodes, find values of the output voltage and the three currents shown.

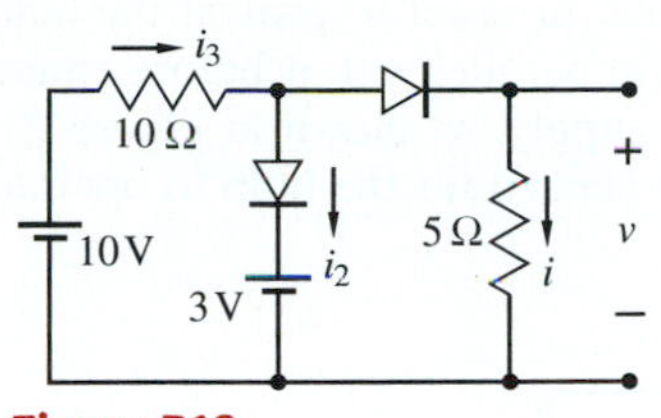

Figure P13

14. The diode in Figure P14 is ideal.
    a. Determine the critical value of v_s for which the diode changes state.
    b. Draw the curve relating v_{AB} to v_s.
    c. For the input voltage shown, draw the curve of v_{AB} against time from $t = 0$ to 10 ms.

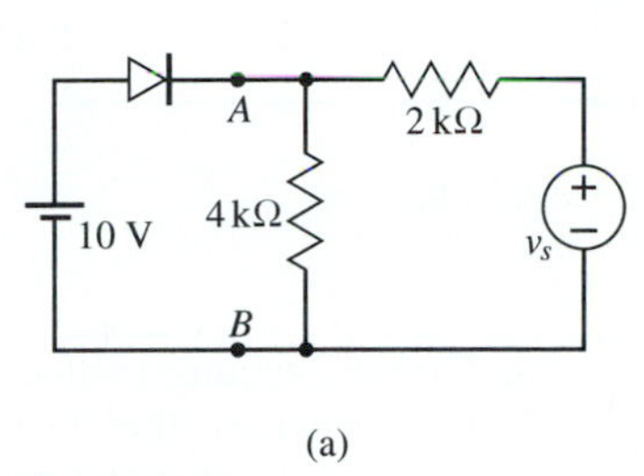

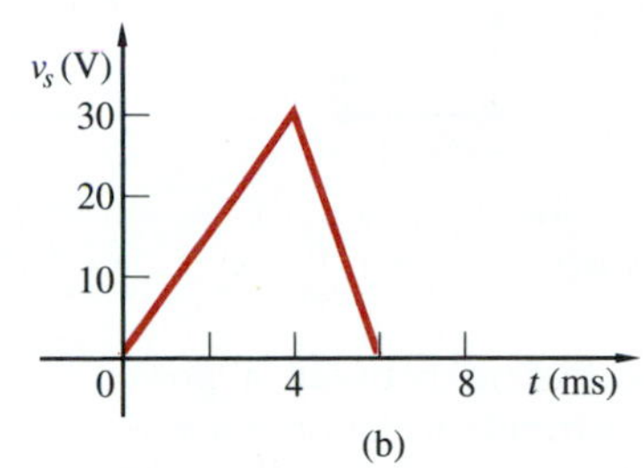

Figure P14

15. The circuit in Figure P15 is an early version of an AND gate. The two input sources can take on one of two possible voltages: either 0 or 1 V. There are four possible combinations of these voltages.
    a. Determine the state of the diodes for each of the four combinations of source voltages.
    b. From the results of part (a), determine the output voltage v_3 for each combination of input voltages.
    c. Make a table listing the input and the output voltages and justify the name of the circuit.

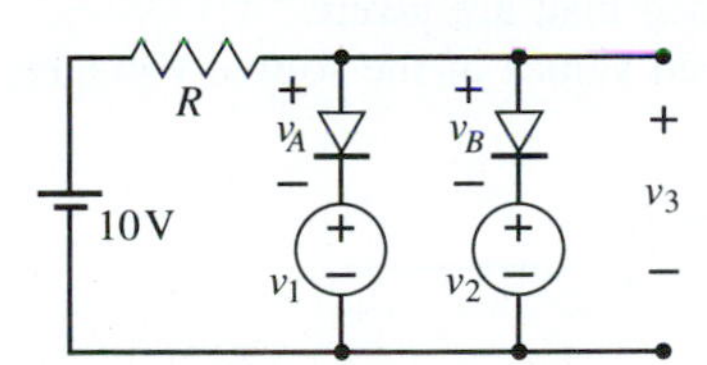

Figure P15

16. The circuit in Figure P16 is an early version of an OR gate. Both the battery and the two diodes have been reversed from the polarities in Figure P15. The input sources again can take on only the values 0 and 1 V.
    a. Determine the state of the diodes for each combination of source voltages.
    b. From the results of part (a), determine the output voltage v_3 for each combination of input voltages.
    c. Make a table listing the input and output voltages and justify the name of the circuit.

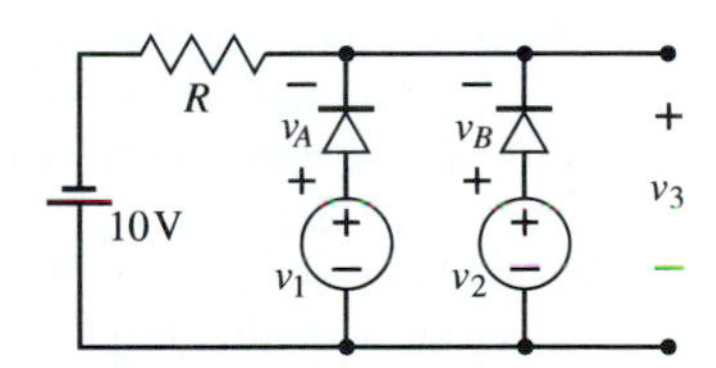

Figure P16

17. You are a consulting engineer in a small shop using electric ovens for baking enamel. A model of the setup is shown in Figure P17. The devices labeled F are fuses. The shop has far exceeded its fuse budget for the year because the fuses keep burning out.

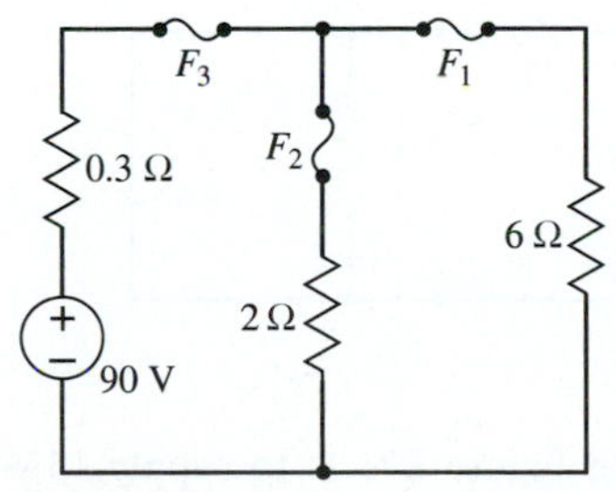

Figure P17

Fuses come in multiples of 5 A. Rhe shop has been using fuses having the following capacities: F_1, 15 A, and F_2 and F_3, 50 A. Your job is to recommend the size of each fuse in order to permit a 50% approximate margin above the expected currents before the fuse burns out. Your consulting fee is on a "per job" basis, so it's to your advantage to move fast.

18. A 20-W DC light bulb is designed to operate at 8 V. An available power supply has the equivalent circuit shown in Figure P18a. In order to protect the bulb, a resistor R is placed in parallel with it before connecting it to the power supply, as shown in Figure P18b. Find the value of R needed for the bulb to operate at its rated values.

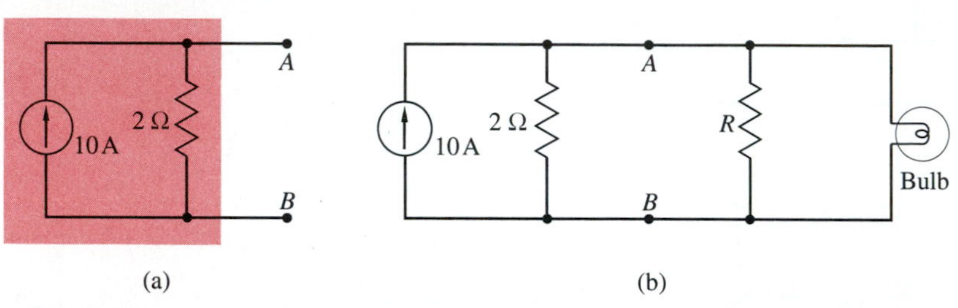

Figure P18

19. A two-wire transmission system between a power-generating and a power distribution station has $n - 1$ support towers at equal intervals between the two stations. The line resistance (both lines together) of each of the n lengths of line between the stations is represented by a resistance R, the total resistance of the lines being nR. There is some current leakage between the two lines. This is represented by a lumped conductance G_L between the lines located at the position of each tower. The voltage of the line at the mth tower is labeled V_m. Using only fundamental laws, find an expression for V_m in terms of V_{m-1} and V_{m+1}, the voltages at the towers just before and just after the mth tower. (A diagram would help.)

20. The 10-A current source in Figure P20 is to supply 1500 W of power.

a. Determine the required value of R.

b. Specify whether the 4-A source is delivering or absorbing power, and determine how much.

c. Find the voltage across each source.

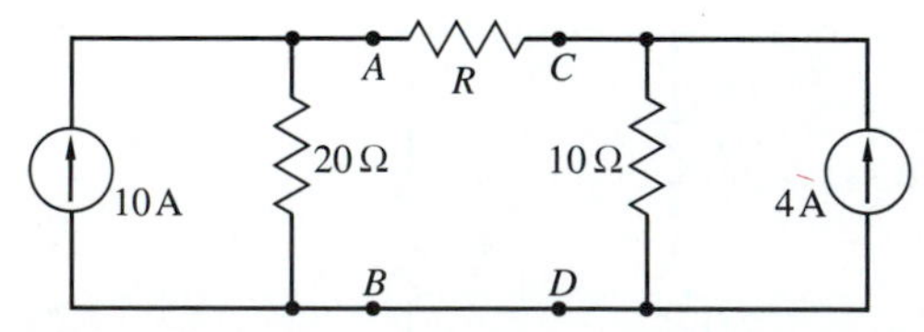

Figure P20

21. The 30-V source in Figure P21 is to supply 15 W of power.

a. Find the required value of R.

b. Find whether the 25-V source is delivering or absorbing power and determine how much.

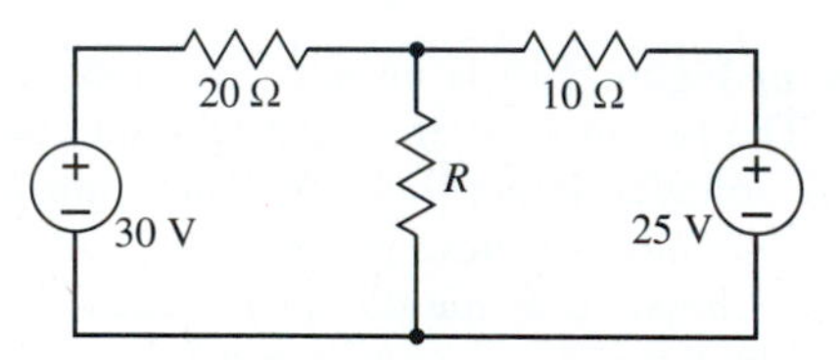

Figure P21

22. Figure P22 represents a three-wire power distribution system. The resistors represent the line resistances of the connecting wires. The voltage and the power requirements of each load are given.

a. Find the required values of the source voltages.

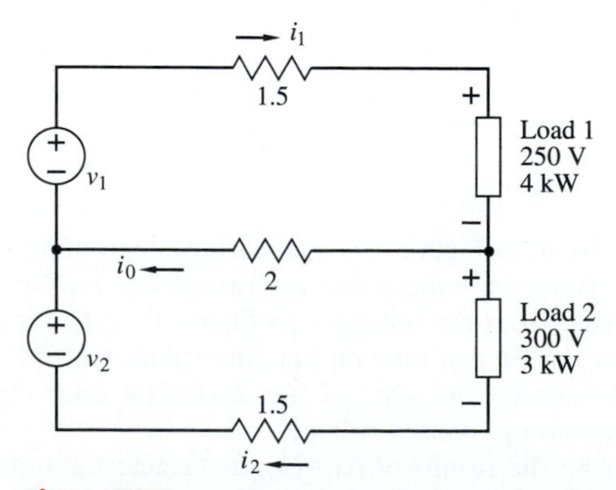

Figure P22

b. Find the efficiency of this distribution system; that is, find the fraction of power going to the loads compared with the total power supplied by the generators.

23. A power supply is represented by a voltage source v_s accompanied by a resistor R. (These are presumed fixed and not under our control to change.) A voltage reduction (division) d is obtained by the voltage divider arrangement in Figure P23a; that is, $v_{AB} = dv_s$. When nothing is connected to terminals AB, the circuit is equivalent to an accompanied voltage source.

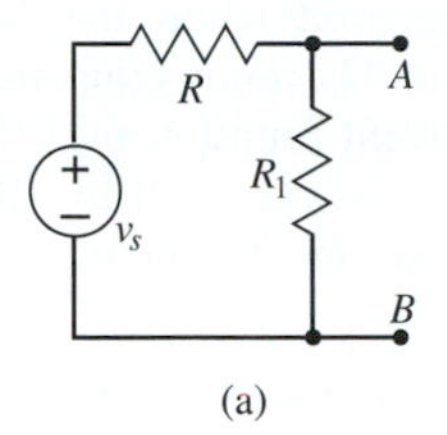

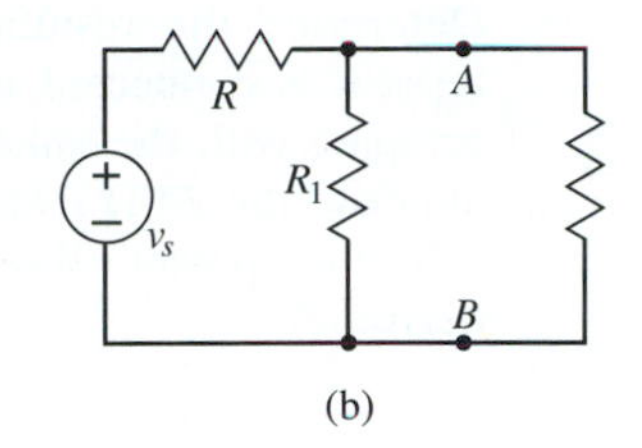

Figure P23

a. Find the values of the equivalent source voltage and the accompanying resistor in terms of v_s, R, and d; draw the diagram.

b. Sometimes, however, a load R_L will be connected to the terminals, as in Figure P23b. When this happens, the terminal voltage will be reduced. A small reduction can be tolerated, but not a large one. Let p be the smallest percent of the no-load voltage that can be tolerated when the divider terminals are loaded by R_L. Find the smallest permissible value of R_L in terms of p, d, and R.

c. To get a numerical "feel," take $R = 20\ \text{k}\Omega$, $d = \frac{1}{2}$, and $p = 90\%$; then take $d = \frac{1}{10}$ and repeat.

24. An available accompanied source has a voltage v_s. Smaller values of voltage are needed in two different applications, namely, $v_s/4$ and $v_s/10$. The voltage divider arrangement in Figure P24a is proposed.

a. Determine values of R_1 and R_2 in terms of R to provide the desired voltages.

b. The voltages at terminals A and B are $v_s/4$ and $v_s/10$ when no load is connected at those terminals. Under operation, however, this will not be the case. The equivalent circuit looking back from each terminal of the divider is an accompanied source, as shown in Figure P24b. Determine R_A and R_B in terms of R to provide the desired voltages.

c. Specify the smallest possible load resistance in terms of R at each terminal if the voltage there is not to decrease under load conditions by more than 10% from its no-load value.

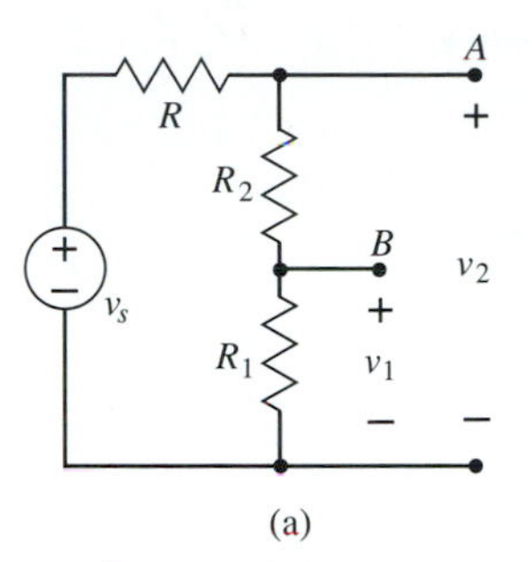

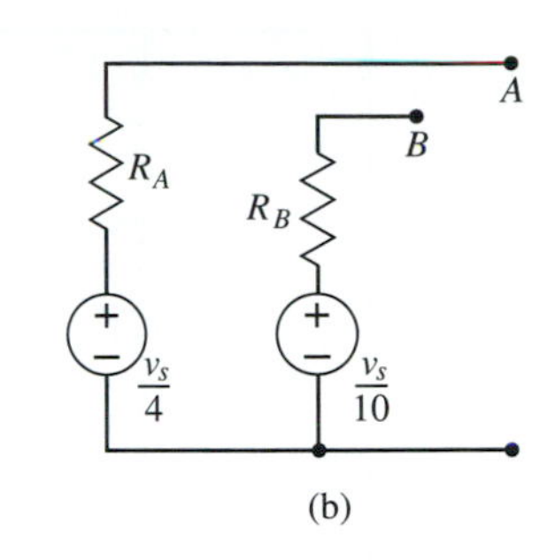

Figure P24

25. The circuit in Figure P25 is similar to the circuit of Figure P24a. The taps are to provide voltages of $v_1 = 0.3v_s$ and $v_2 = 0.6v_s$. (The upper tap, of course, gives *all* of v_s.) When nothing is connected to the taps, the power supplied by the source must be no more than 200 mW.

a. Determine the required values of the three resistors if the source voltage is $v_s = 100$ V.

b. When connections are made to one or both of the taps, the taps will be loaded down by the external loads. The voltages will, therefore, change. Suppose the changes are not to be more than 5%. Determine the smallest permissible values of the load resistances.

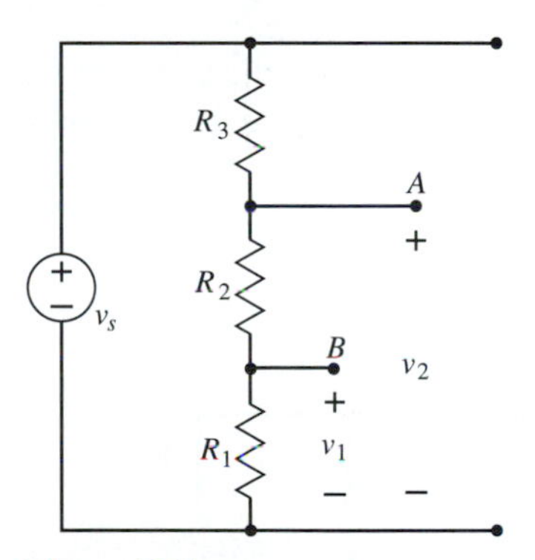

Figure P25

26. The circuit shown in Figure P26 is similar to that of Figure P25 except that it has a total of four taps instead of three. The source again is a 100-V source. The taps are to provide voltages of $v_1 = 0.2v_s$,

$v_2 = 0.5v_s$, and $v_3 = 0.8v_s$. When nothing is connected to the taps, the power supplied by the source must equal 100 mW. Find the required values of resistance.

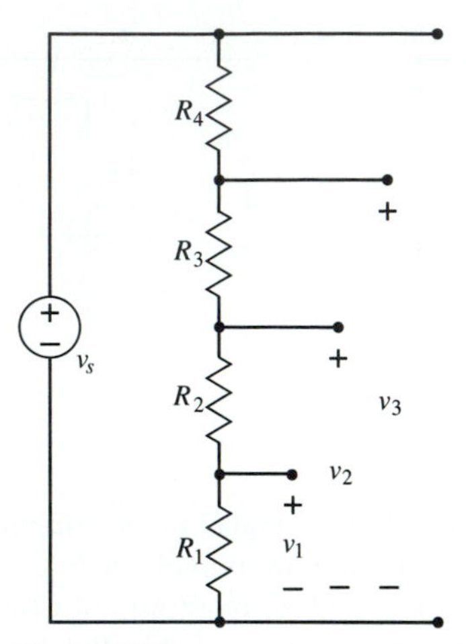

Figure P26

27. In the circuit of Figure P27:
   a. Find the value of R in order to maximize the power delivered to it.
   b. For that value of R, find the value of this maximum power.

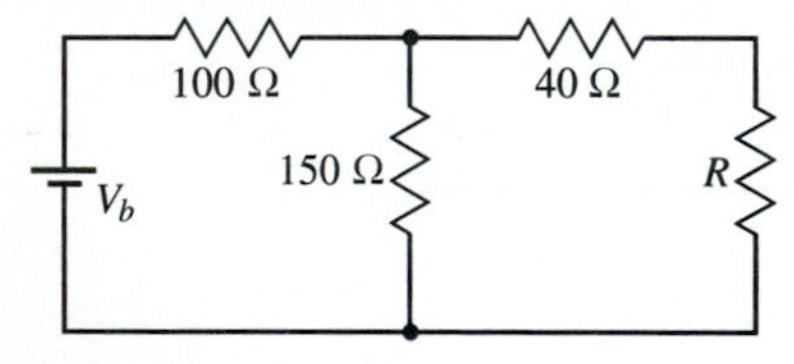

Figure P27

   c. Sketch against R the variation of the power delivered to R.

28. a. Use the principle of superposition in Figure P28 to find *any one* of the resistor currents (you choose).
   b. Then, by successive applications of KCL, KVL, and Ohm's law, find all other resistor currents.
   c. Determine the power supplied by each source (it can be positive or negative).
   d. Assume that the 24-Ω resistor is a load. Find the Thévenin equivalent of the remainder of the circuit, first determining the two that are the easiest to determine from the set of three parameters v_T, i_T, and R_T.
   e. Determine the resulting current when the 24-Ω resistor is connected to the Thévenin equivalent; compare with the same current found in part (b).
   f. Suppose the 24-Ω resistance is increased by 10%; will the power dissipated in it increase or decrease?

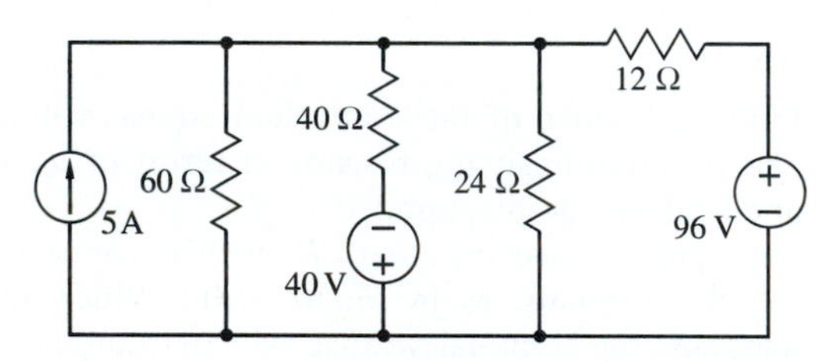

Figure P28

29. In each of the circuits in Figure P29, determine the indicated variable (a) using the principle of superposition and (b) by other methods, without using superposition. (c) Evaluate the amount of effort expended and make a judgment in each case whether the use of superposition is "cost-effective."

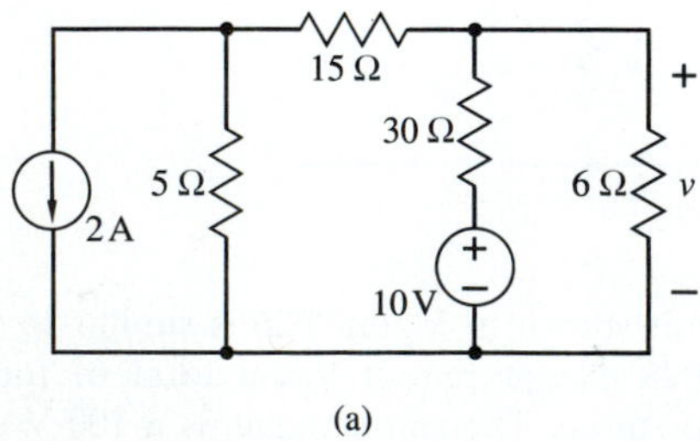
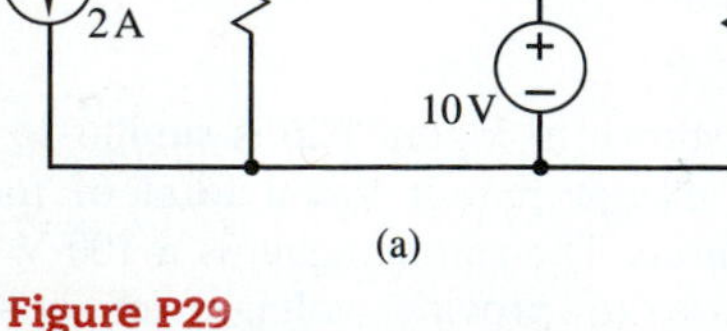

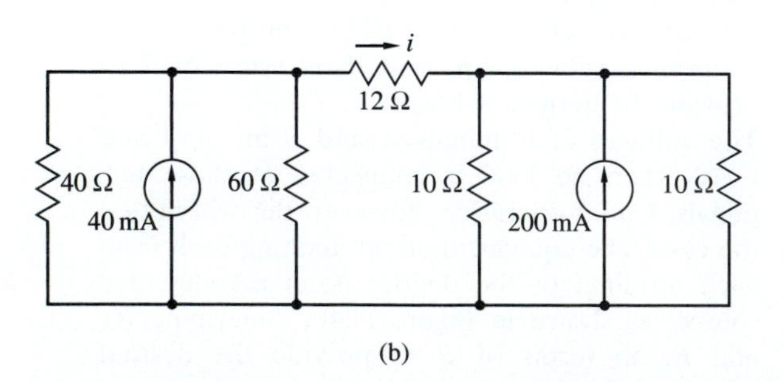

Figure P29

30. a. Find the Thévenin equivalent circuit looking to the left from the terminals of the 6-Ω resistor in Figure P29a.

b. Find the voltage v and compare with the value found in Problem 29.

31. a. Find the Thévenin equivalent voltage and resistance at terminals AB in Figure P31.

b. Independently find the Norton equivalent current and confirm the relationship among those three quantities. (It might help to redraw the circuit.)

c. Find the maximum power that would be delivered to an appropriate resistor to be connected at AB.

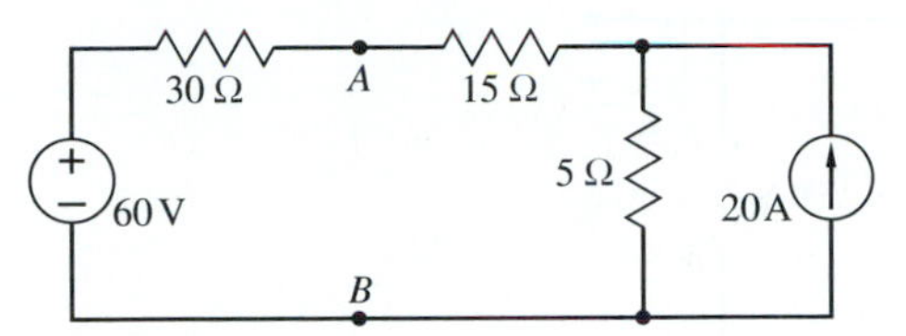

Figure P31

32. The power in watts delivered to resistor R in Figure P32 is labeled P for some particular values of the components. The voltage of the source is then increased by 10%. Find the new value of the power delivered to R in terms of P.

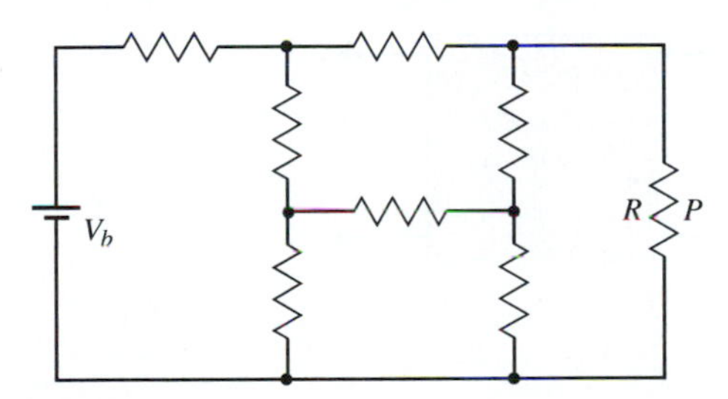

Figure P32

33. a. Find the Thévenin equivalent voltage and resistance at terminals AB in the circuit of Figure P33.

b. Independently find the Norton equivalent current and confirm the relationship among the three quantities.

c. Determine the value of R_L so that as much power as possible is delivered to it from the circuit to the left of AB, and find the value of this power.

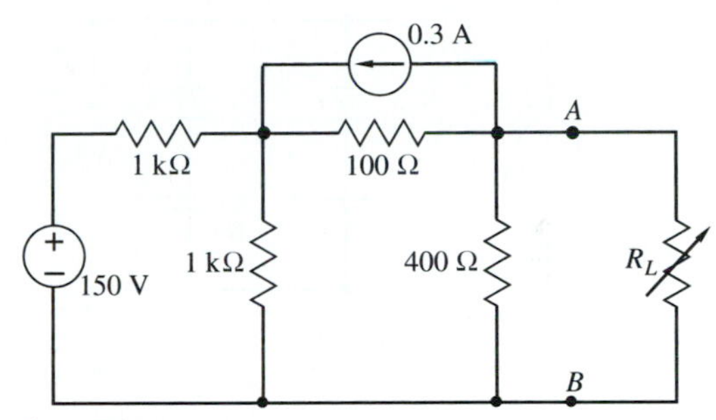

Figure P33

34. The diode in the circuit of Figure P34 is ideal. The input voltage has the waveform shown.

a. Determine the value of input voltage v_1 at which the diode switches state.

b. Draw an output-input curve.

c. Sketch the output waveform on the same axes as the input waveform, showing the points at which the diode switches.

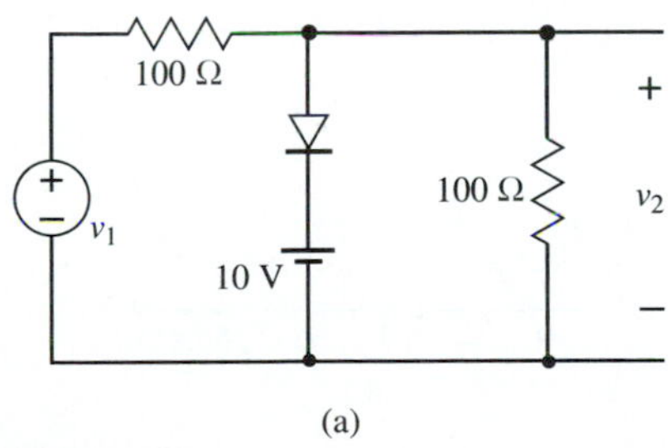

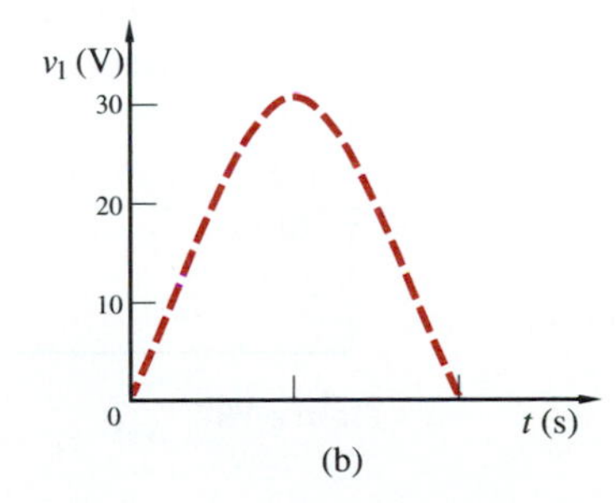

Figure P34

35. The diodes in the circuits in Figure P35 are assumed to be ideal. Voltage v_1 is a time-varying input and v_2 is the output. For each circuit:

a. Find an expression for the output in terms of the input for each state of the diode.

b. Plot the output-input curve.

c. Plot the output as a function of time for the input waveform in Figure P34.

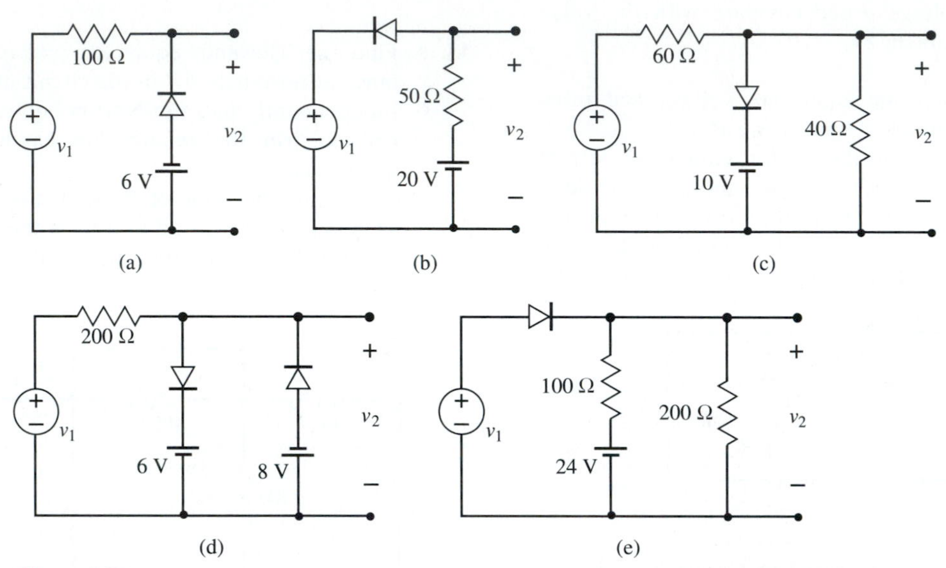

Figure P35

36. Design a comparator to go in the box in Figure P36 (that is, draw a diagram and find the numerical values of components) that will produce a nonzero output voltage only when the input satisfies the condition $v_1 < -20$ V.

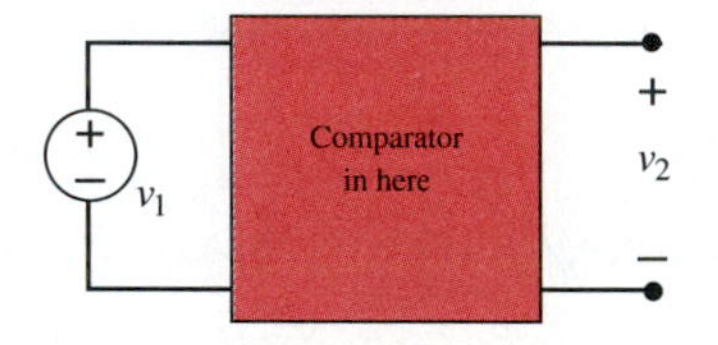

Figure P36

37. The diode in Figure P37 has the i-v characteristic shown. Find the value of power supplied by the voltage source.

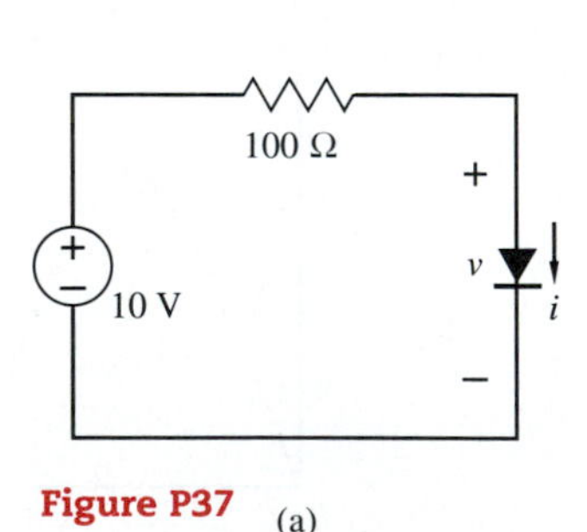

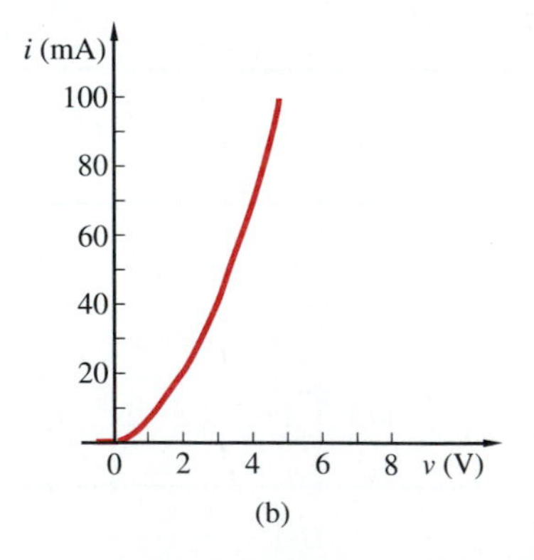

Figure P37 (a) (b)

38. The diode in Figure P38 has the i-v characteristic shown. The maximum permissible power dissipation is given as $P_m = 1$ W.

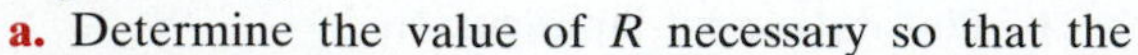

a. Determine the value of R necessary so that the diode voltage v will equal 4 V.

b. For this value of R, determine the diode power. Does it exceed P_m?

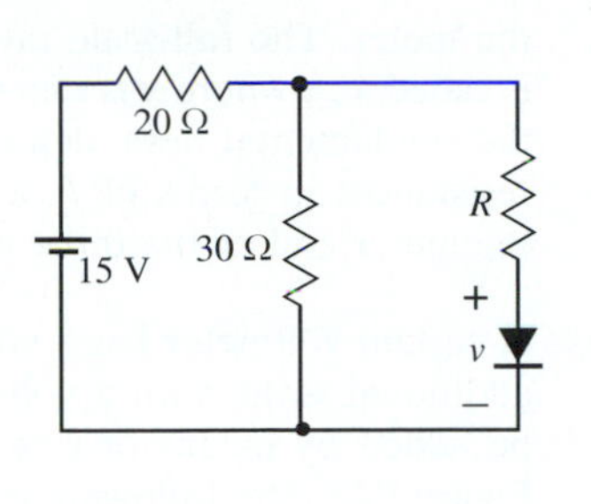

(a)

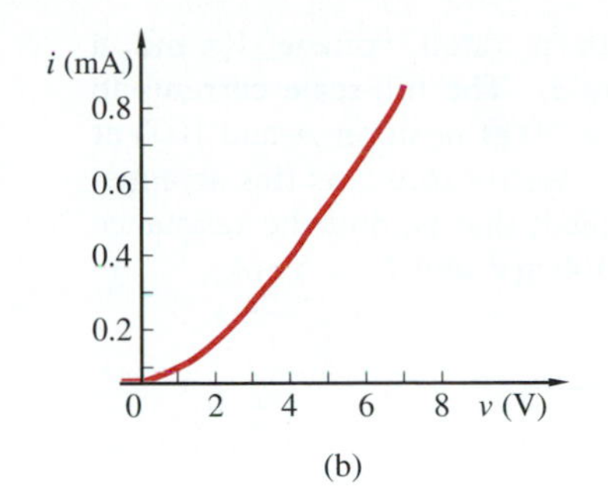

(b)

Figure P38
Designing an ammeter.

ELECTRICAL INSTRUMENTS

The following development is needed for understanding the next seven problems, the first of which should be completed before tackling any of the others.

One of the most useful electrical instruments of the analog variety is based on the interaction of two magnetic fields, one from a permanent magnet, the other caused by a current which is to be measured. The current is passed through a coil of wire, with a pointer attached to it, pivoted between the poles of the permanent magnet. Such an arrangement is called a *d'Arsonval meter* movement. The force causing the coil to rotate and, hence, also causing the deflection of the pointer is proportional to the current. An instrument for measuring the current is created once a scale is placed under the pointer so that readings can be made. The maximum current that can be measured (called the *full-scale current,* I_{fs}) occurs when the pointer reaches the limit of its excursion, which is typically between 90° and 180°. The voltage across the coil when the current has its full-scale value is the *rated* voltage, V_r.

For purposes of circuit analysis, the meter can be represented by a small resistance R_m, the resistance of the coil of wire, which will equal the rated voltage divided by the full-scale current: $R_m = V_r/I_{fs}$. For typical meters, I_{fs} varies from about 10 μA to about 10 mA. What can be done, then, if larger currents are to be measured? The solution is to place a resistor across the meter and to fix its value based on the largest expected current to be measured, as shown in the accompanying diagram. From the current-divider relationship:

$$i_m = \frac{R}{R + R_m} i$$

All but a small part of the current to be measured is to go through the resistor R; hence, R must be small compared with R_m.

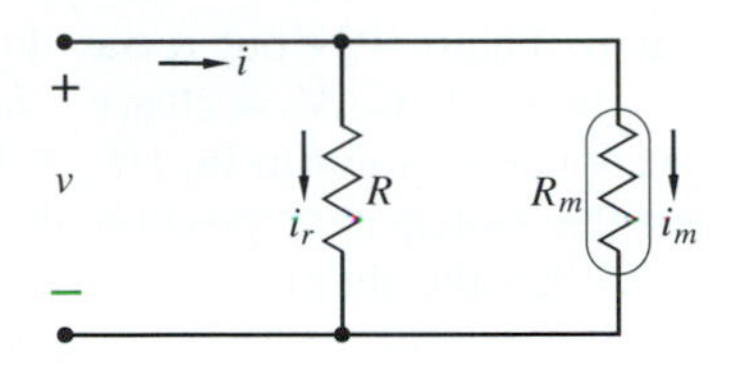

Let (capital) I be the maximum value of current i to be measured; it corresponds to I_{fs} in the meter movement. Then, solving the preceding expression for R and using the value of the meter resistance in terms of the rated voltages and the full-scale meter current yields:

$$R = \frac{V_r}{I - I_{fs}}$$

(Verify!) Thus, for a meter with a voltage rating of 50 mV and a full-scale current of 1 mA (thus, a meter resistance of 50 Ω) and a full-scale external current $I = 1$ A, the value of shunt (parallel) resistance required is 5/99.9. For accuracy, this must be a high-precision resistor.

Ammeters normally have more than one scale, with a switch to bring each of the scales into operation.

The same meter movement can be used with other combinations of external resistors and batteries in order to form voltmeters and ohmmeters. The following sequence of problems relates to such instruments. All that is required to design the instruments is an application of the fundamental electrical laws together with the parameters of the meter movement just discussed.

39. Figure P39 shows an ammeter circuit with two scales and a switch to select the appropriate scale. It uses a meter movement with a rated voltage V_r and a full-scale meter current I_{fs}. The full-scale currents to be measured are $I/I_{fs} = 50$ at position A and 1000 at position B. Design the *Ayrton shunt,* as this arrangement of resistors is called; that is, find the resistance values. Assume $V_r = 100$ mV and $I_{fs} = 5$ mA.

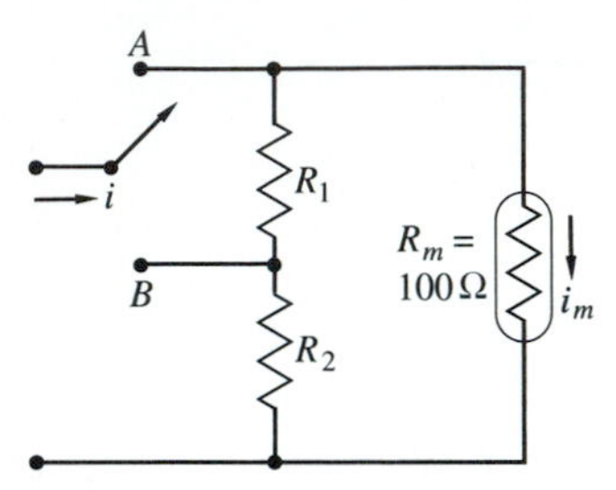

Figure P39

40. Figure P40 is similar to Figure P39 but it has three scales instead of two. Let $V_r = 20$ mV and $I_{fs} = 0.1$ mA. Full-scale currents are to be $I/I_{fs} = 10$, 100, and 1000 when the switch is at position A, B, and C, respectively. Design the shunt.

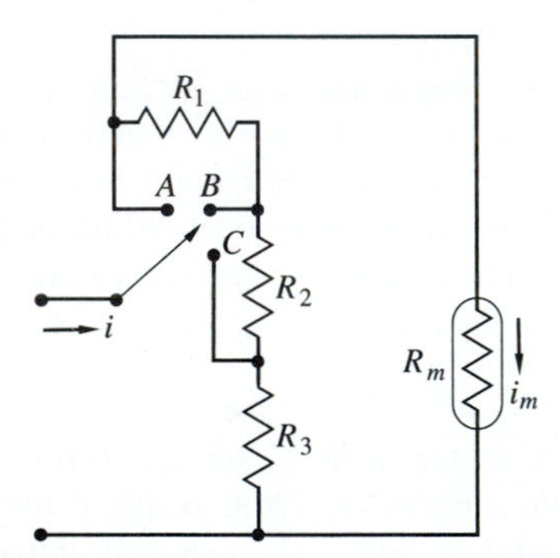

Figure P40

41. The circuit in Figures P41 is a voltmeter with three scales. The external resistors are now *in series* with the meter. The full-scale rated voltage for each scale is called V_{rx}, where x is one of the scales. By applying the fundamental laws, determine expressions for the resistances in terms of I_{fs} and the voltage ratings of the meter and of the three voltage scales.

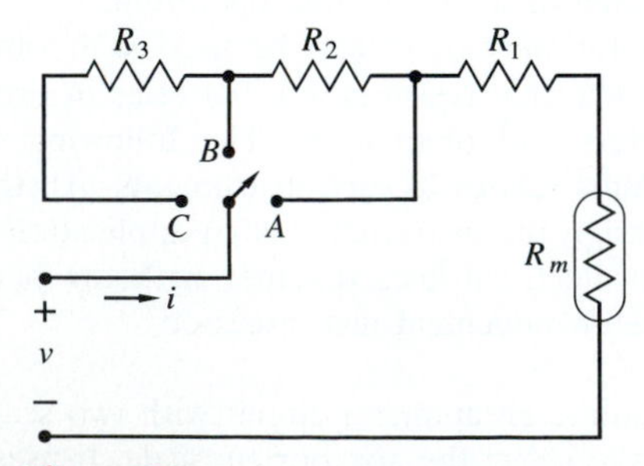

Figure P41

42. A certain voltmeter has a voltage rating of 100 V. An additional scale, with a voltage rating of 250 V, is to be added by means of a series resistor, as shown in Figure P42. The following measurements are made at terminals BC:

a. when the voltmeter is connected to an electronically regulated power supply whose voltage is kept constant, the voltmeter reads 75 V;

b. when a precision 100-kΩ resistor is connected in series with the same supply, the voltmeter reads 50 V. Find the required value of R to make the voltage rating 250 V at terminals AC.

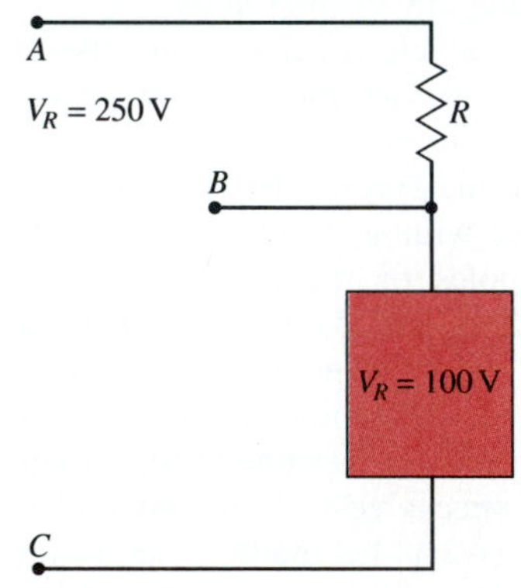

Figure P42

43. A series ohmmeter circuit is shown in Figure P43. Resistor R_1 is adjustable. (The internal resistance of the battery can be lumped with it.) To use the instrument, terminals AB are first short-circuited and R_1 is adjusted so that full-scale deflection results. When any other resistance is connected to the terminals, the current, and hence the deflection of the meter, will be reduced. Let R_h represent the value of resistance R which results in half-scale deflection, or a current $I_{fs}/2$.

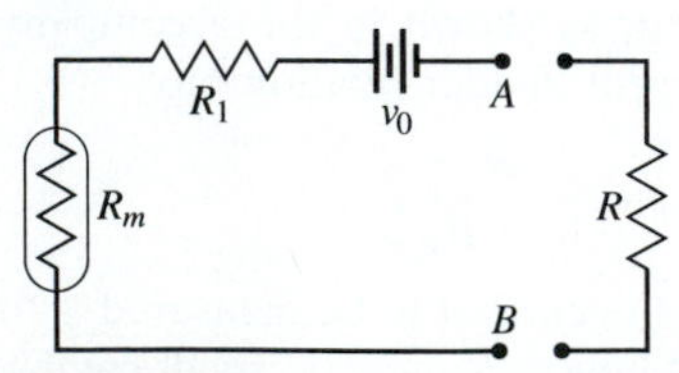

Figure P43

a. Determine the value of R_h in terms of V_b and I_{fs}.
b. Let k be a fraction (<1) of full-scale current corresponding to a general value of R. Determine the value of R in terms of k and R_h. Is the deflection a linear function of the resistance to be measured?
c. A series ohmmeter uses a meter movement with a voltage rating of 25 mV and a full-scale current of 1 mA. Assume that the internal resistance of the battery can be neglected. Determine R_1 and V_b in terms of R_h.

44. The diagram of a shunt ohmmeter is shown in Figure P44. It differs from the series ohmmeter in that maximum deflection occurs when the terminals are left open. The notation in both cases is identical.
a. Determine the value of R_h in terms of R_1 and of V_r and I_{fs} of the meter.
b. Let the meter current corresponding to a general value of R be kI_{fs}, with $k < 1$. Determine the value of R in terms of k and R_h.
c. Let the meter parameters be $I_{fs} = 0.5$ mA and $V_r = 50$ mV, and assume that the battery is ideal. Determine values of V_b and R_1 in terms of R_h.

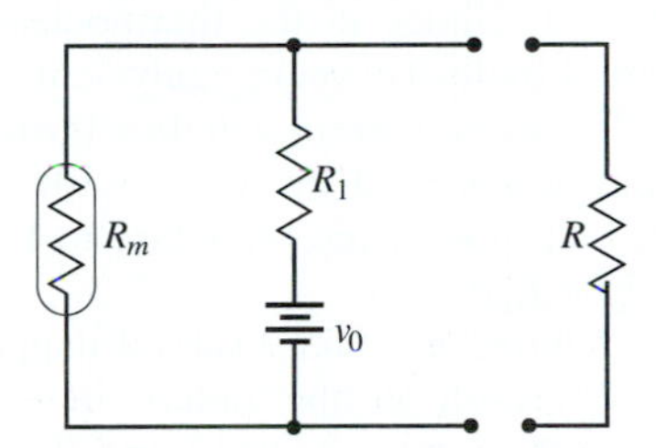

Figure P44

45. The circuit of a voltmeter-ammeter is shown in Figure P45. Terminals VC are the voltmeter terminals and AC the ammeter terminals (C stands for *common*). The voltage scale has a maximum value of V_{max} and the current scale a maximum value of I_{max}. When the ammeter current is at its maximum, the voltage across terminals AC should not exceed V_{max}. The voltage rating and full-scale current of the meter movement have their usual designations. Determine the three resistor values in terms of the ratings described.

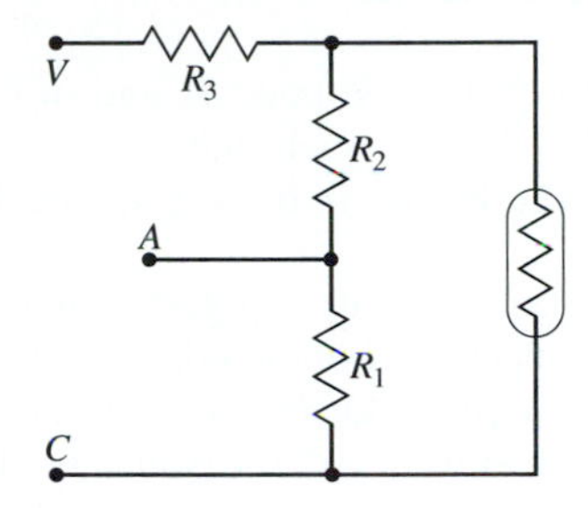

Figure P45

DESIGN PROBLEMS

1. The unknown resistor in Figure DP1 is to absorb a power of P watts.
a. Determine the highest value of P possible.
b. Choose a value of P slightly less than the maximum (say, eight-ninths of the maximum) and find one or more values of R which will satisfy this condition.
c. For each value of R, find the power supplied by, or absorbed by, each of the sources and specify which it is.
d. Choose the value of R such that the least net power is supplied by the two sources.

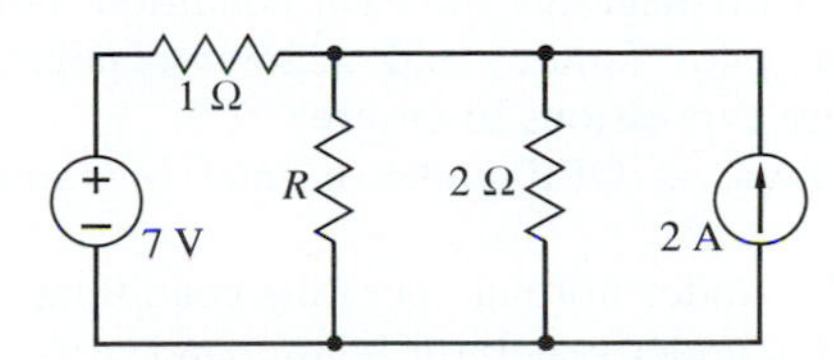

Figure DP1

2. The circuit in Figure DP2 is proposed as the design for a 3-bit D/A converter. Carry out an analysis similar to the one on the 2-bit A/D converter in the text and determine the values of R_a and R_b.

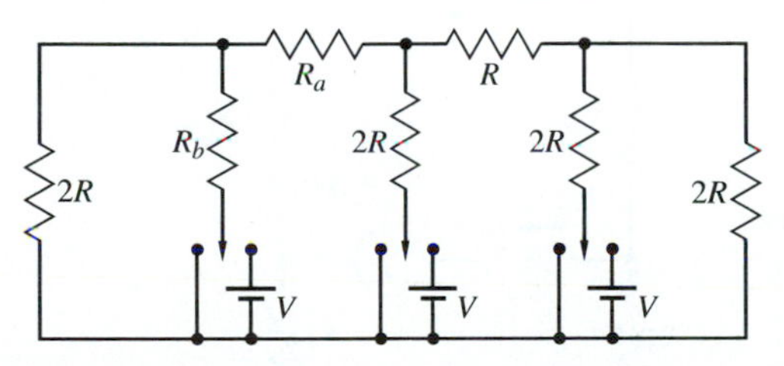

Figure DP2

3. A mechanical engineer is designing a system in which there is danger of excessive oil pressure. A circuit is needed to give a warning signal whenever the oil pressure in the system exceeds 30 psi (pounds per square inch). A pressure-sensing device is available which gives a DC output voltage V_0 proportional to pressure in psi: $V_0 = 0.33p$ V. The internal resistance of the sensor is 0.5 Ω. You have been assigned to design the circuit. Available to you are several diodes that can be assumed to be ideal, as well as a range of batteries and resistors. Design a circuit that will give a "trigger signal" whenever the oil pressure exceeds the maximum allowable.

4. The bridge circuit in Figure DP4 is to be used to measure the resistance $R_x = R \pm \Delta R$ of a semiconductor device which can carry a maximum current I; that is, $i \leq I$. The value of R around which R_x varies is fixed and known and the change in resistance is small; that is, $\Delta R \ll R$. The microammeter to be used in the bridge is given so that its resistance R_m is known and fixed.

a. Choose values of the battery voltage V_b and of the three resistors R_1, R_2, and R_3 such that:
1. When $\Delta R = 0$, the bridge is balanced; that is, $i_m = 0$.
2. For each value of ΔR, the largest possible microammeter deflection is obtained.

b. The value of ΔR is *approximately* proportional to the microammeter reading. Find the value of the proportionality "constant."

c. For the following numerical values, specify the maximum percentage error made in calibrating the microammeter dial to read resistance in accordance with this proportionality constant:

$$R = 1\ \text{k}\Omega, \quad R_m = 10\ \Omega,$$
$$\Delta R_{\max} = 50\ \Omega, \quad I = 1\ \text{mA}$$

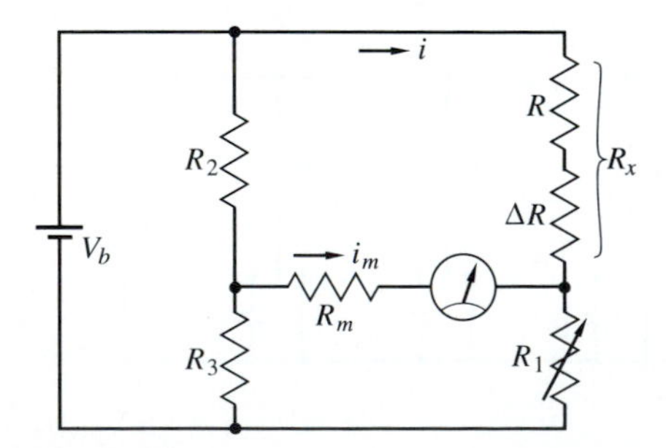

Figure DP4

5. This problem is concerned with the design of a gas-flame monitoring system. A pilot is used to ignite the main flame in a heating unit under the control of a thermostat. If there is a failure in the pilot and the pilot flame goes off for some reason, the main flame will not be ignited and the escaping gas may cause an explosion. A monitoring system is needed to detect whenever there is a failure.

A proposed monitoring system is shown in Figure DP5. The dashed boxes represent the following devices:
1. A relay coil with resistance R_c operating the main gas valve
2. A thermoelectric generator energized by the pilot flame
3. A failure detector consisting of a battery, resistance, and diode
4. A pilot-flame controller at a remote console

Simplified electrical equivalents of these devices are shown inside the dashed boxes.

Under normal operation, (a) the heat from the pilot flame generates a voltage in the thermoelectric generator, represented by its Thévenin equivalent; (b) the diode will be OFF, so no current will flow through the failure detector, represented by R_d ($i_d = 0$); and (c) enough current will flow in the relay coil to keep the gas line open ($i_c > I_{\text{crit}}$).

When there is a failure, two things should happen: (a) a current should result in the failure detector ($i_d > 0$), thus alerting the remote console, and (b) the current in the relay coil should fall below a critical value ($i_c < I_{\text{crit}}$), thus causing the relay to pull open and close the main gas valve. If there is no gas in the main gas line, none can escape.

There can be two types of failure:

a. *Pilot failure.* When the pilot flame fails, no thermoelectric voltage will be generated, so $V_g = 0$; but R_g is still there. In this case, let the current in the failure detector be $i_d = I_{pf}$.

b. *Generator failure.* In this case, the generator branch opens. Let the current in the failure detector be $i_d = I_{gf}$.

The problem is to design this system. As an aid, draw appropriate circuit diagrams for each condition: normal operation, pilot failure, and generator failure. Write inequality expressions to ensure:
1. That the diode is OFF under normal operating conditions;
2. That $i_c < I_{\text{crit}}$ under normal operating conditions;
3. That $I_{gf} < I_{\text{crit}}$ under generator failure; and
4. That $I_{pf} < I_{gf}$ under pilot failure.

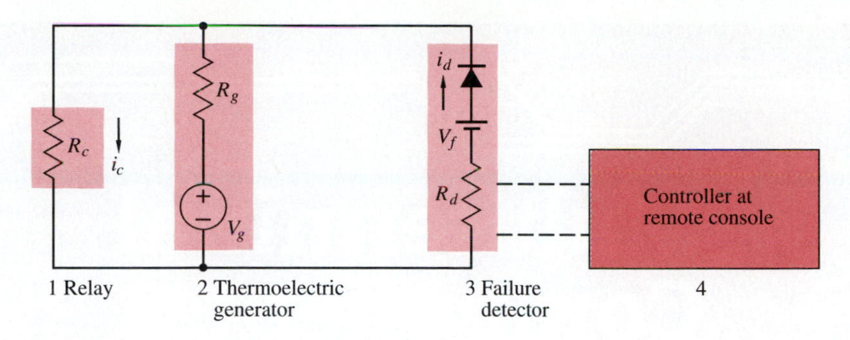

Figure DP5

The following design values are given:

$$V_f = 1.5\,\text{V}, \qquad V_g = 3\,\text{V}, \qquad I_{\text{crit}} = 1\,\text{mA}$$

Assume that the values of circuit resistances can change by ±25% under extreme operating conditions. Determine nominal values for R_c, R_d, and R_g that will satisfy the design requirements, introducing large enough factors of safety into the preceding inequalities so that the conditions are met even for the extreme variations of the resistance values. Using the nominal values of the resistances, determine the values of I_{gf} and I_{pf} and the corresponding values of i_c for the two types of failure. (You should not consider the job complete until you check to see if all conditions are satisfied with your design values.)

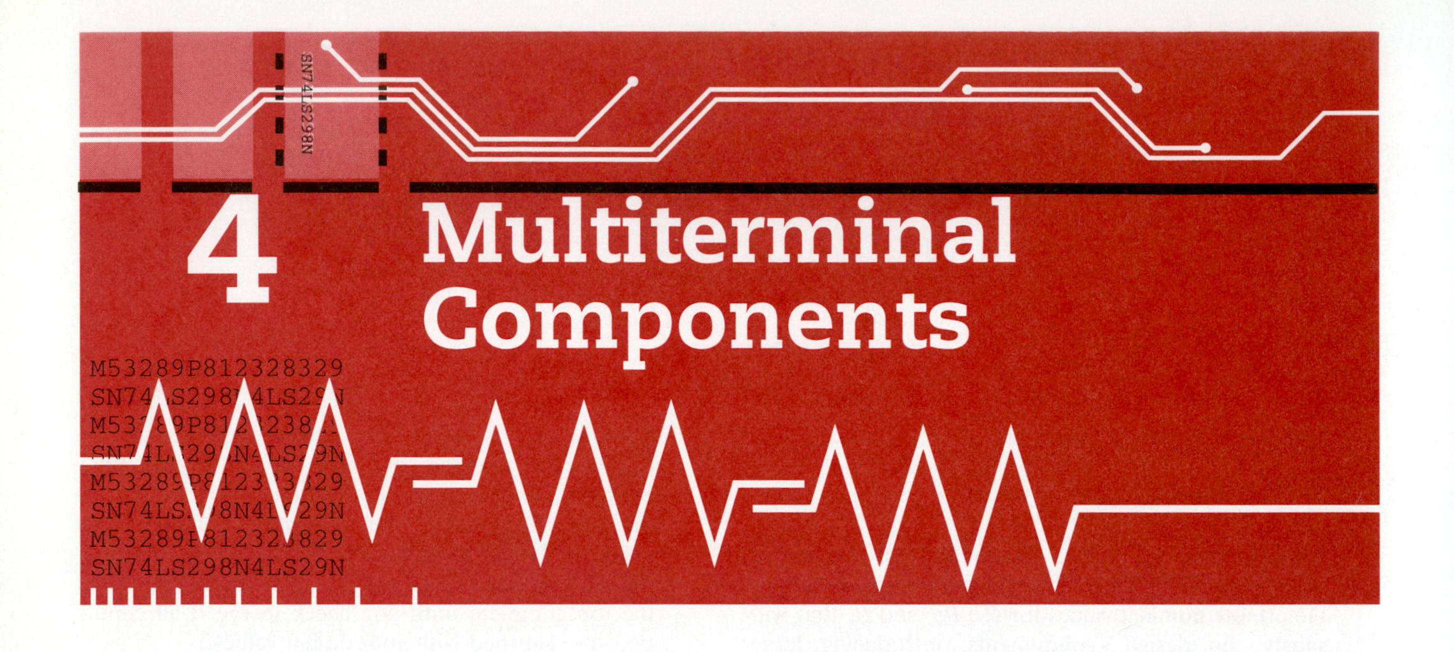

4 Multiterminal Components

So far, the only electrical components we have introduced are two-terminal components. Of those, in addition to sources, only resistors and diodes have been included in the circuits studied so far. Resistors can carry out only multiplication of a voltage or current signal by a constant, and diodes act as switches. We *did* introduce capacitors and inductors, but we have not yet studied circuits containing such components. Since capacitors and inductors perform the mathematical operations of differentiation and integration, such study promises to be very interesting. However, we will defer this study for yet a while.

In this chapter, we will introduce electrical components having more than two terminals. Circuits containing such components perform a variety of operations not otherwise possible on voltage and current signals.

1 THE IDEAL TRANSFORMER

In Chapter 2 we discussed the observations first made by Faraday concerning the appearance of a voltage across the terminals of a coil of wire when it is in the presence of a time-varying magnetic field. There we gave explicit consideration to the case in which the magnetic field depends only on the current in the coil itself. (Refer to that discussion to refresh your memory.)

Now consider the situation illustrated in Figure 1. It consists of two coils of wire wound on the same core. The coils are said to be *magnetically* or *inductively coupled.* Each of the coils is connected to a circuit; at least one

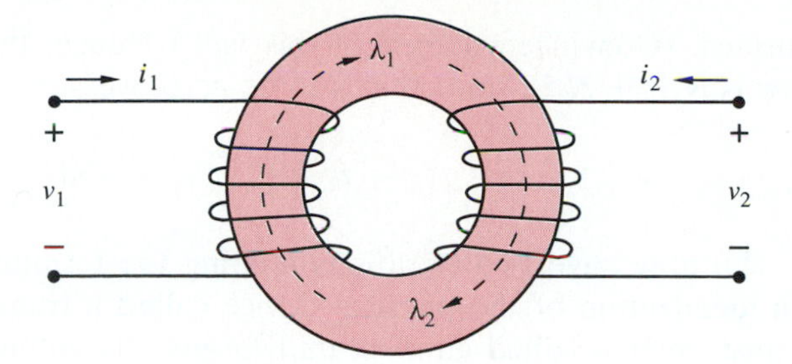

Figure 1
A physical transformer.

circuit contains an electrical source whose voltage or current varies with time. Such a device is called a *transformer,* for reasons which will become clear as we discuss its properties. The device has four terminals; two pairs of terminals, really. In normal use, one side of the transformer, called the *primary,* is energized, and a load is connected to the other side, called the *secondary.*

Magnetic flux is generated by the currents in each coil. The flux linking each coil comes partially from each current. Depending on how the coils are wound on the core and on the permeability property of the core, some fraction of the flux generated by the current in one coil will not link the second coil and a possibly different fraction of the flux generated by the second current will not link the first coil. We say some of the flux *leaks.*

We now make the idealization that *all* the flux generated by *each* current links *both* coils—that is, that the flux linking each coil is the same, so that there is no *leakage* flux. But, from Faraday's law, the voltage induced in a coil is $d\lambda/dt = N\,d\phi/dt$. Hence, with this idealization, the only thing that distinguishes the two voltages is the corresponding number of turns, since $d\phi/dt$ is the same for both coils. If the turns are labeled N_1 and N_2, respectively, then the ratio of the two voltages will be:

$$\frac{v_1}{v_2} = \frac{N_1}{N_2} = n \qquad v_1 = nv_2 \tag{1}$$

The quantity $n = N_1/N_2$ is called the *turns ratio.*

This relationship constitutes part of the idealized model of a transformer. Since there are two pairs of terminals, there ought to be two relationships among the variables. The second relationship is obtained by invoking *Ampère's law,* which is usually stated in terms of magnetic field concepts.[1]

If we assume that the core of the transformer is ideal in the sense that very little (ideally, zero) magnetic field intensity is required to establish the flux in the core, then, by Ampère's law, the net current going through the surface bounded by the inner diameter of the toroid in Figure 1 is almost zero—ideally, zero. Each of the N_1 turns of the primary, carrying current i_1, and each of the N_2 turns of the secondary, carrying current i_2, passes through that

[1] You have probably encountered Ampère's law in a physics course. It states that the line integral of the magnetic field intensity H around a closed path is equal to the net current piercing a surface bounded by the closed path.

surface. (Convince yourself of this fact.) Hence, the net current in Ampère's law is $N_1 i_1 + N_2 i_2$; since this is to be zero, we get:

$$i_2 = -\frac{N_1}{N_2} i_1 = -n i_1 \tag{2}$$

We now have two equations relating the terminal currents and voltages of an idealization of the physical device called a transformer. This model of the transformer is called an *ideal transformer.* Its voltage-current relationships are collected here:

$$\begin{aligned} v_1 &= n v_2 \\ i_2 &= -n i_1 \end{aligned} \qquad \begin{bmatrix} v_1 \\ i_2 \end{bmatrix} = \begin{bmatrix} 0 & n \\ -n & 0 \end{bmatrix} \begin{bmatrix} i_1 \\ v_2 \end{bmatrix} \tag{3}$$

where n is the turns ratio N_1/N_2. Note that these relationships are algebraic. The matrix form is also given. (Expand the matrix form and confirm that the same scalar equations result. See Appendix A.)

The circuit symbol for an ideal transformer is shown in Figure 2. This symbol might suggest two physical coils of wire in close proximity, but it doesn't represent a physical component, merely a circuit model.[2]

In Figure 1 it is possible to show the sense in which each current encircles the core. Starting from the upper terminals—toward which both current references are directed—both coils encircle the core in the same sense. Hence they both produce flux having the same sense through the core.

However, it isn't possible to show how the coils encircle the core in the circuit symbol in Figure 2. We need a method for showing this; the dots shown in the symbol (one on each side of the transformer) comprise that method. We will not provide a complete explanation of this scheme, postponing it until Chapter 10 devoted to transformers. Here it is enough to note that the ideal transformer equations in equation (3) require that both voltage reference plus marks be located at the terminals carrying the dots and that the $v - i$ references be standard so that, with the arrows located as shown, they are both directed *toward* the dots. Variations on the equations when this is not the case will be worked out in Chapter 10. For now, we will always use the references

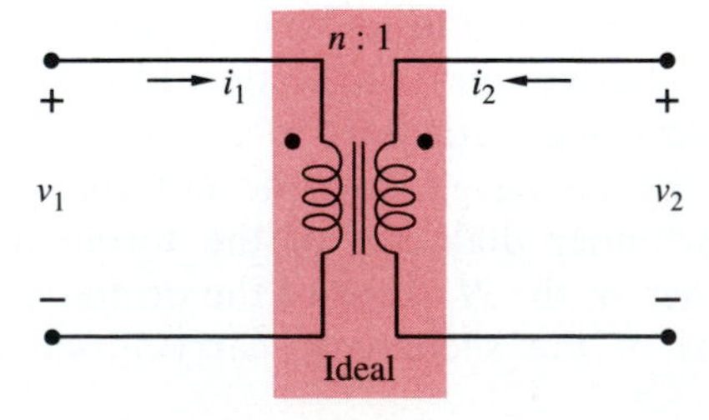

Figure 2 Circuit symbol of an ideal transformer.

[2] The set of two vertical lines between the two sides of the ideal transformer is a symbol for a ferromagnetic core. Such a core presents an easy path for the magnetic flux, greatly reducing the leakage flux.

relative to the dots as shown. If one of the dots is reversed, then the corresponding voltage and current references must also be reversed.

Ideal-Transformer Properties

An ideal transformer has two significant properties which we will now explore.

First let's find the energy that is supplied to an ideal transformer when its terminals are excited. The power supplied from its two pairs of terminals to the ideal transformer in Figure 2 is easily found to be:

$$\begin{aligned} p_i &= v_1 i_1 + v_2 i_2 \\ &= (nv_2)i_1 + v_2(-ni_1) \equiv 0 \end{aligned} \tag{4}$$

That was easy! The power—and hence the energy input—is identically zero. The ideal transformer is a *lossless* device.

Note from the ideal-transformer v-i relations that if $n < 1$, the secondary voltage will be larger than the primary voltage. In this case, the transformer is said to be a *step-up* transformer. Similarly, if $n > 1$, the transformer is said to be a *step-down* transformer. But because the device is lossless, the vi product on the secondary side has the same magnitude as the vi product on the primary side. Thus, if the voltage is stepped *up*, then the *current* must be stepped *down*, and vice versa.

Next, let's find out what happens when a resistor is connected to one side of an ideal transformer, as shown in Figure 3. We need to combine the v-i relations of the ideal transformer with the circuit tools at our disposal. The following sequence of steps is carried out in succession: first the transformer voltage equation, then Ohm's law for R, and then the transformer current equation.

$$v_1 = nv_2 = n(-Ri_2) = -nR(-ni_1) = n^2Ri_1 \tag{5}$$

Thus, we find that when the secondary side of an ideal transformer is loaded with a resistance R, there appears to be a resistance n^2R at the primary side. The resistance has been *transformed* by the square of the turns ratio; hence the name transformer.

This property of an ideal transformer is very useful for *matching* purposes, as illustrated in Figure 4. A load with resistance R_L is to be connected to a source circuit represented by a voltage source and an internal resistance R_T. What is called the source circuit could be part of a signal-transmission system

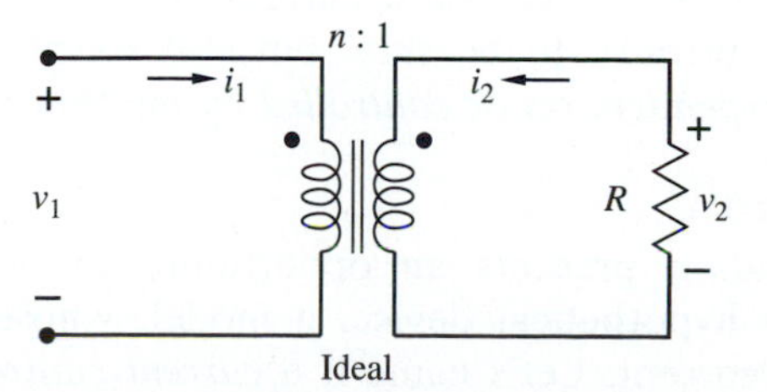

Figure 3
Ideal transformer loaded with a resistor.

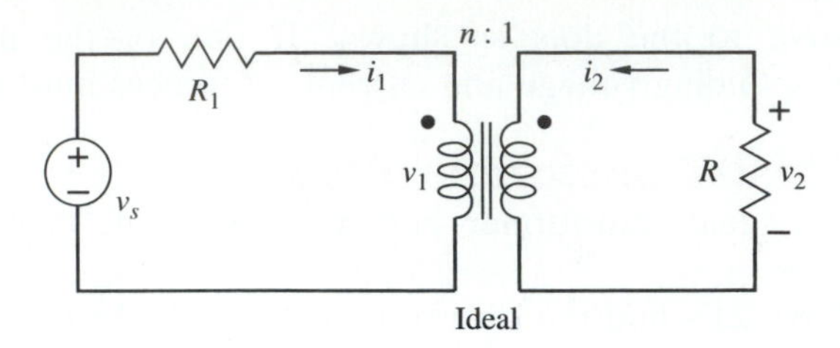

Figure 4
Matching a load to a source with an ideal transformer.

which is to deliver a signal to the load, assumed to be resistive. The parameters of the source circuit are fixed, as are those of the load. However, to extract maximum power from the source, the load presented to the source circuit should be matched to its Thévenin equivalent resistance. To achieve this goal, an ideal transformer is inserted between the source and the load. The resistance presented to the source circuit through the ideal transformer is n^2R_L, which should equal R_T for maximum power. Hence, the required value of the turns ratio is $n = \sqrt{R_T/R_L}$. Such a value is easily achieved.

EXERCISE 1

A transmission circuit with a Thévenin equivalent resistance of 400 Ω is to work into a 100-Ω load. Find the turns ratio of an ideal transformer that will match the load to the source. Compare the power transmitted to the load resistance under the matched condition with the power delivered to the load if it is connected directly to the circuit without the transformer.

ANSWER: $n = 2$; $p(\text{matched})/p(\text{direct}) = 25/16$. □

2 CONTROLLED SOURCES

In its own right, the ideal transformer is a useful model of a real electrical device. However, it has an added virtue which we will now explore. Glance back at the relationships among the transformer voltages and currents in (3). Unlike a resistor, for example, where the voltage is proportional to the current, here one voltage is proportional to the other voltage and one current is proportional to the other current.

This device is also different from a source, where a voltage (or a current) is described by a specified function of time. On the other hand, it is similar to a source in that a voltage (or a current) is *specified*; proportional to another voltage (or current), to be sure, but still specified. We say one voltage (or current) is *dependent* on or *controlled* by another voltage (or current).

Definition

This situation presents an opportunity to be creative. Let's *create* (by postulate) a hypothetical device, a model, whose current is proportional to some other current. Let's name it a *current-controlled current source* (*CCCS*,

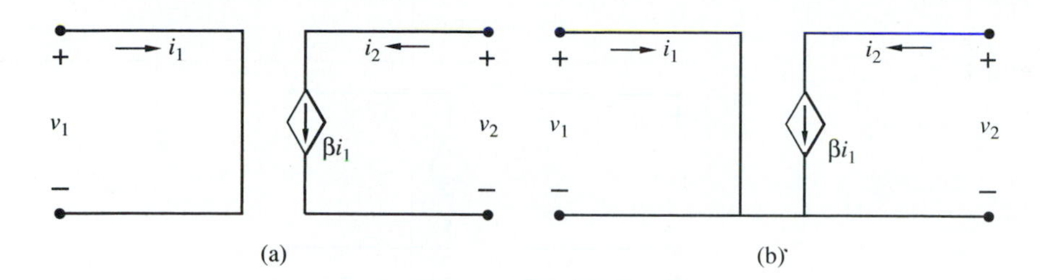

Figure 5
A current-controlled current source.

for short). A circuit model is shown in Figure 5a. The diamond-shaped symbol for this "source" distinguishes it from the circular symbol for an honest-to-goodness independent source. It is a device with two pairs of terminals, which can be reduced to three, as in Figure 5b, when two terminals, one from each pair, are connected together. The *controlling* current, the current in a short circuit, is i_1.

The designation "source" for such a component is unfortunate, because this component is *not* a source as previously defined. It is a dissipative component that has some of the features of a source, as will become clear shortly. To distinguish it from an actual source, which is "independent," a controlled source is sometimes called a "dependent" source. The diamond-shaped symbol helps distinguish a controlled source from an independent source.

A controlled source has two variables: a controlling variable and a controlled variable. In Figure 5, both of these variables are current, but either or both of the variables could be voltage. Hence, it is possible to create *four* different controlled sources. A complete set is shown in Table 1. (Examine each of them carefully and become familiar with their properties.) Observe that each of these has two pairs of terminals, called *ports.* The input port is either an open circuit or a short circuit. The parameter relating the output variable to the input variable is different in all cases. One is a resistance, one a conductance, one a voltage ratio, and one a current ratio.

Having defined controlled sources as ideal circuit elements, let's return to the equations for the ideal transformer in (3). Each scalar equation can be implemented by means of a controlled source. The combination takes the form of Figure 6. (Confirm that the equations obtained from the circuit in the box are the same as the ones for the ideal transformer in (3).) Two circuits represented by the same equations are said to be equivalent, just like the equivalent resistance representing two resistors in series or parallel. Thus, the circuit of controlled sources in Figure 6 is equivalent to the ideal transformer.[3]

In the preceding development, controlled sources were introduced to account for the relationships among the voltages and currents of an ideal transformer. Actually, the ideal transformer itself is already an abstract model of a real device; there was no real need to introduce yet other abstract models

[3] The controlled sources considered in this chapter are *linear.* That is, a plot of the controlled variable against the controlling variable will be a straight line. It is possible to conceive of other cases where this relationship is not linear, but we will not deal with such cases in this book.

Table 1

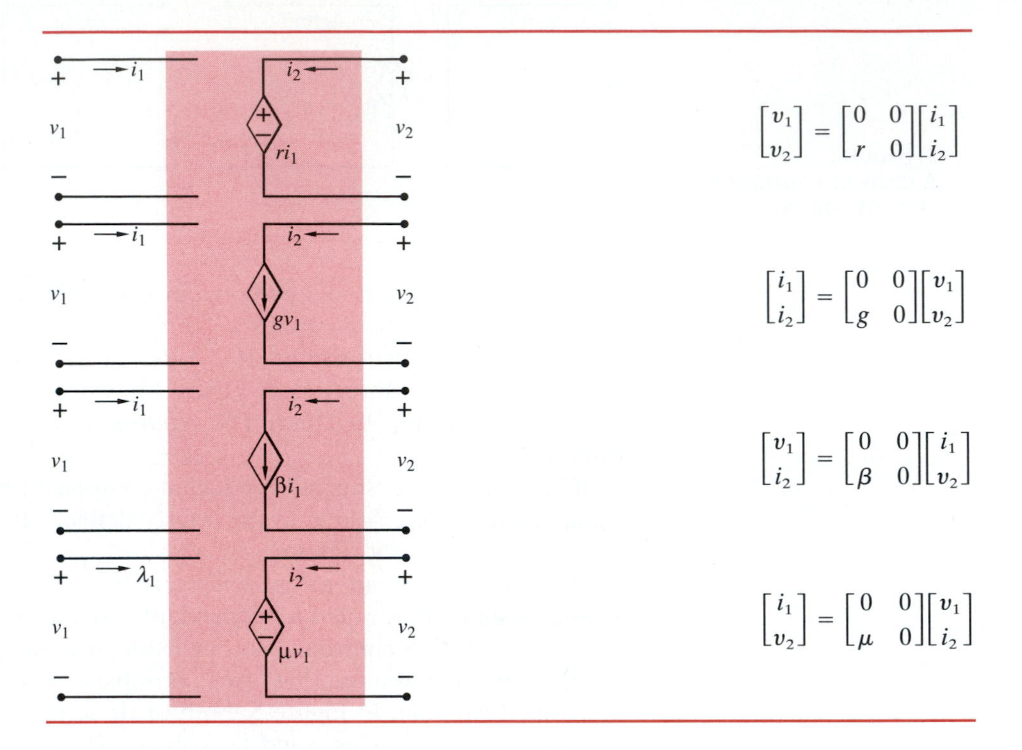

to represent it. However, would it be satisfactory just to state that there are such things as the controlled sources in Table 1, and that you should learn about them?

It turns out that there are other real devices (transistors) whose behavior under certain operating conditions more naturally leads to the introduction of controlled sources. The introduction of controlled sources might seem better motivated from an analysis of a transistor circuit. However, familiarity with the characteristics of transistors would be necessary for pursuing this line. Since this is not assumed to be in your background, a prohibitive digression would be necessary to supply sufficient background on transistors to make the controlled sources appear realistic. For those of you who would want to see such a development, an optional section is provided at the end of this chapter.

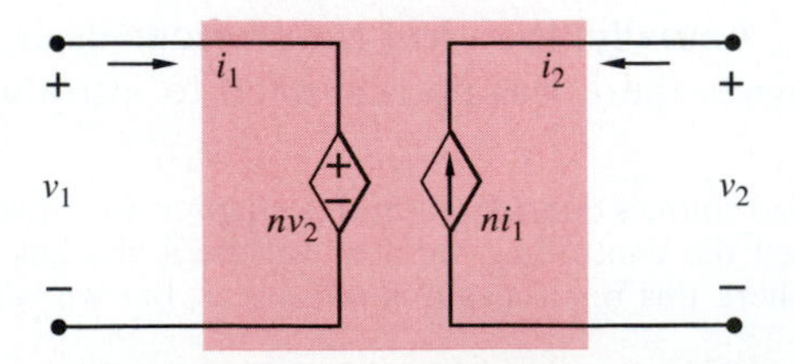

Figure 6 Controlled-source equivalent of an ideal transformer.

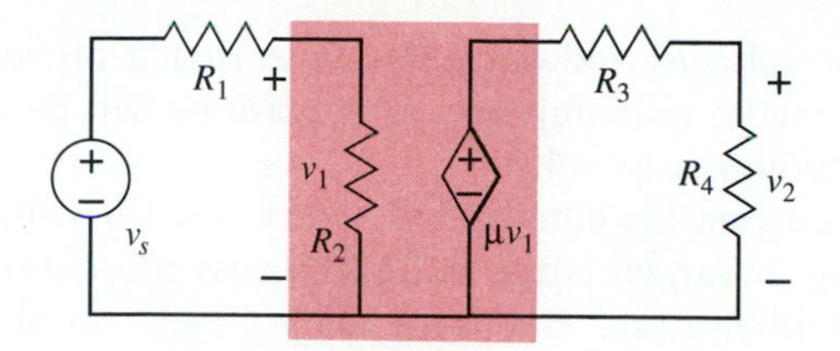

Figure 7
Model of an amplifier containing a controlled source.

Circuits with Controlled Sources

Controlled sources take their place as another set of circuit elements that, together with other elements, provide models for circuits of real components. We will examine a number of circuits containing controlled sources in this section.

An initial illustration is shown in Figure 7. This is, incidentally, a simplified model of an amplifier circuit. The objective is to find the output voltage in terms of the excitation voltage v_s. It is clear that the output voltage can be found in terms of the controlled source voltage by the voltage-divider relationship. Since the controlling varible, v_1, can also be obtained by a voltage-divider relationship from the input voltage, the result immediately follows:

$$v_1 = \frac{R_2}{R_1 + R_2} v_s \qquad v_2 = \frac{R_4}{R_3 + R_4} \mu v_1$$

$$v_2 = \frac{\mu R_2 R_4}{(R_1 + R_2)(R_3 + R_4)} v_s$$

Without the controlled source, each voltage-divider relationship is a fraction less than 1. The output would then be smaller than the input voltage. However, since μ can be large, the output can be made a larger replica of the input in the controlled-source circuit.

Another amplifier circuit is shown in Figure 8, this time containing a current-controlled current source. The amplifying device is inside the box, the rest of the components being connected to it externally. Assume that the value of R is large enough that this resistor can be neglected (considered an open circuit) compared with the 5-kΩ resistor. The objective is to determine all

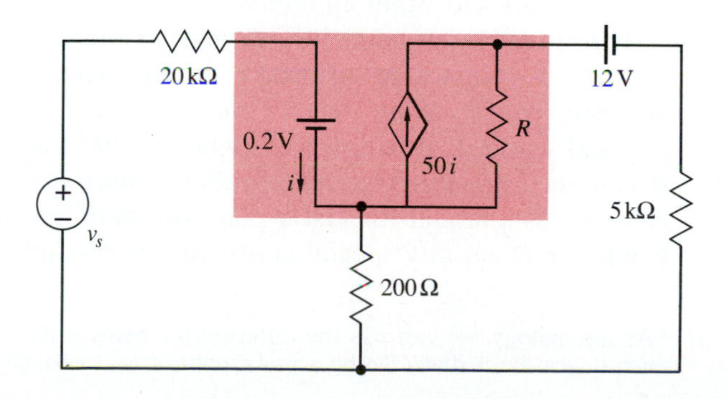

Figure 8
Amplifier circuit.

unknown voltages and currents. Note that if all resistor values are in kilohms and currents in milliamperes, all equations will be dimensionally homogeneous with the voltages in volts.

Once current i is determined, every resistor voltage will be known. The only remaining unknown is the voltage across the controlled source; call it v_{cs} with standard references. But with all voltages on that closed path known, the voltage across the controlled source is determined by KVL. By KCL, the current through the 200-Ω resistor is $i + 50i$. Hence, by KVL around the left loop (and Ohm's law), we find:

$$20i + 0.2 + 0.2 \times 51i = v_s$$

$$i = \frac{v_s - 0.2}{30.2} \text{ mA}$$

$$v_{cs} = 12 - 5i - 0.2 \times 51i$$

$$= 12 - 15.2i \text{ V}$$

Once v_s is given, everything else will become known.

EMBARRASSING MOMENTS WITH CONTROLLED SOURCES Once hypothetical elements of a model have been introduced, they can be arranged in arbitrary configurations to create circuits. It is conceivable that the arbitrary arrangements of ideal elements may not correspond to real circuits, because such circuits may require *specific* combinations of elements from the model, not arbitrary connections, to represent them adequately. Thus, a real battery will inevitably have an internal resistance. Nevertheless, most of the time, representing a real battery with a constant ideal voltage source will provide enough accuracy.

Controlled sources, on the other hand, are a different matter. They are like other components in some ways and unlike them in other ways. For *physical* resistors, independent sources, capacitors, inductors, and transformers there are idealizations we call *ideal* resistors, sources, capacitors, inductors, and transformers, respectively. The latter are *models* of the former. But a controlled source is itself an ideal model; there is no such thing as a *physical* controlled source. There *are* electronic devices (such as transistors, which will be discussed in the last section of this chapter) which, under certain conditions of operation, can be represented by controlled sources, *in specific connections with* resistors and other elements.

It is possible to conceive of, and to draw on paper, a circuit containing controlled sources in arbitrary configurations with other elements. This arbitrary configuration may or may not be a model for a real circuit in the physical world. If it is not, we might say that the circuit is *contrived,* not real.[4] It is indeed possible that a controlled source in some configuration with other elements is a model for just a plain ordinary resistor or for a physical battery. Could we claim any special credit for successfully analyzing such a circuit? A

[4] As an analogy, we can use the criterion for being a duck: it must look like a duck *and* it must quack like a duck. To be a real circuit, it isn't enough just to *look* like a circuit.

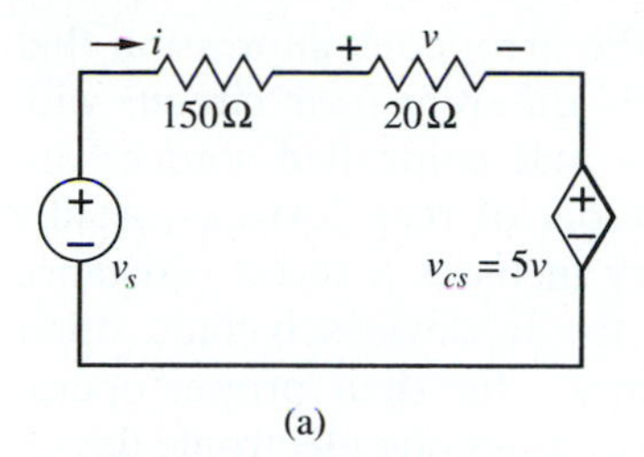

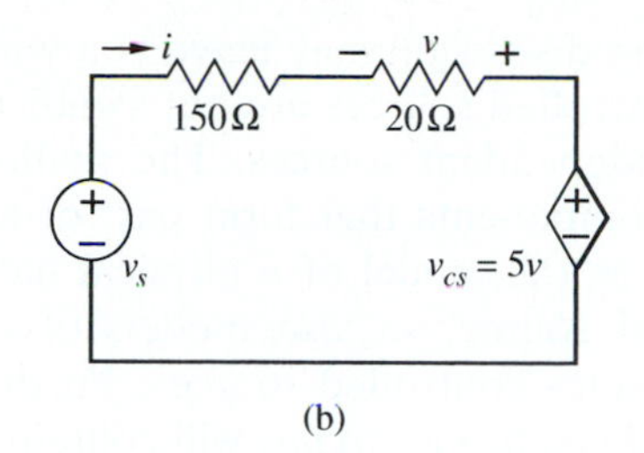

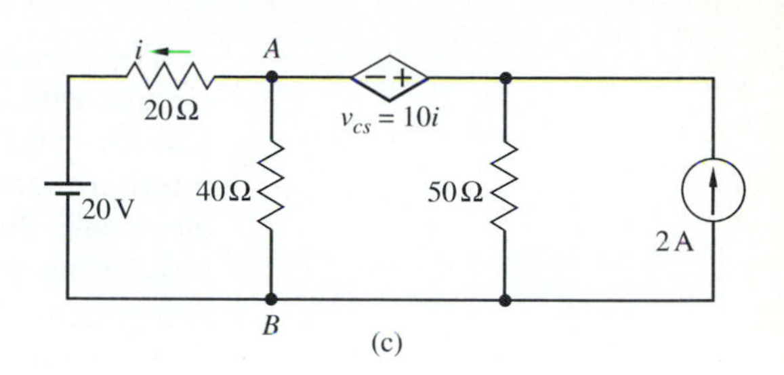

Figure 9
Trivial examples of circuits with controlled sources.

few examples are shown in Figure 9. (You will find many other examples in other books you might consult.)

The circuit in Figure 9a contains a voltage-controlled voltage source. Since the controlling voltage v is proportional to current i, however, the controlled voltage v_{cs} can be written as $v_{cs} = 5v = 100i$. What device is it whose voltage is proportional to its own current?[5] Thus, the controlled source here acts just like a resistor; the circuit is contrived. Solving such circuits would in no way demonstrate proficiency in dealing with circuits with controlled sources in configurations that are truly models of real devices.

Figure 9b is almost the same circuit, except that the polarity of the controlling voltage has been reversed. Now $v = -20i$, so the voltage of the controlled source becomes $v_{cs} = -100i$. This is the equation of a straight line going through the origin with a negative slope. A device having such a v-i relationship would be a resistor with a negative resistance. This is something new; the controlled source here is acting like a negative resistance. We could have simply admitted two-terminal negative resistances as elements of circuits. Your ability to solve this circuit says little about your ability to deal with circuits containing controlled sources in realistic, authentic configurations.[6]

Other circuits of the type illustrated in the preceding examples are possible; for example, their duals. Figure 9c shows a slightly less trivial example. By KCL the current i_{cs} in the controlled source (with standard references) is the sum of i and $v_{AB}/40$. But v_{AB} can be found in terms of i by KVL and Ohm's law as $v_{AB} = 20i + 20$. Hence, the controlled voltage in terms of i can be converted to $v_{cs} = 20i_{cs}/3 - 10/3$. (Work out the details.) A branch exhibiting this v-i relationship is a $\frac{20}{3}$-Ω resistor in series with a $\frac{10}{3}$-V battery, with appropriate polarity. In this example we again find that the controlled source is contrived; it can be replaced by a battery in series with a resistor.

[5] In contemporary language, you don't have to be a rocket scientist to answer that.
[6] The power *into* a negative resistance $-R$ is $-Ri^2$. This means the device does not absorb power but supplies it; it is active, not passive. Indeed, electronic devices, of which controlled sources are models, have the possibility of displaying a negative resistance under certain conditions. Later in the chapter, a real circuit (called a *negative converter*) will be introduced that has exactly the property that it displays at one pair of terminals the negative of the element connected at a second pair of terminals.

The preceding discussion may leave you with the erroneous impression that circuits with controlled sources are not significantly different from circuits with resistors and independent sources. The reality is that controlled sources are multiterminal components that form part of a model of real devices—usually electronic. Just as the model of a physical battery includes a series resistance besides an ideal source, so also models of electronic devices include other components besides controlled sources. Furthermore, for their proper operation, meaningful electronic *circuits* will contain more than one electronic device and still other components besides these; hence, their models will be rather extensive circuits. As such, they might require analysis procedures more general than those we have considered so far. Since such procedures will not be taken up until the following chapter, further consideration of circuits with controlled sources will be deferred until then. Nevertheless, a few such circuits will be provided in the problems at the end of this chapter. See how many trivial, contrived circuits like the preceding ones in Figue 9 you can detect among these.

3 ONE-PORTS AND TWO-PORTS[7]

In a network the pairs of terminals where current or voltage signals can enter or leave are called *ports*. Metaphorically speaking (with thoughts of ships sailing in or out), we call a device with just two terminals—that is, with only one place where a voltage can be applied or extracted, and only one place to send a current as an input or take a current as an output—a *one-port network*, or simply a *one-port*. But if there are two pairs of terminals into which to send an excitation and out of which to extract a response, the circuit or component is called a *two-port network*, or *two-port*. This digression from the topic at hand will discuss representations of one-ports and two-ports.

One-Ports

Figure 10a shows an arbitrary source-free linear circuit with one pair of terminals: a *one-port*. The port is characterized by a voltage and a current.

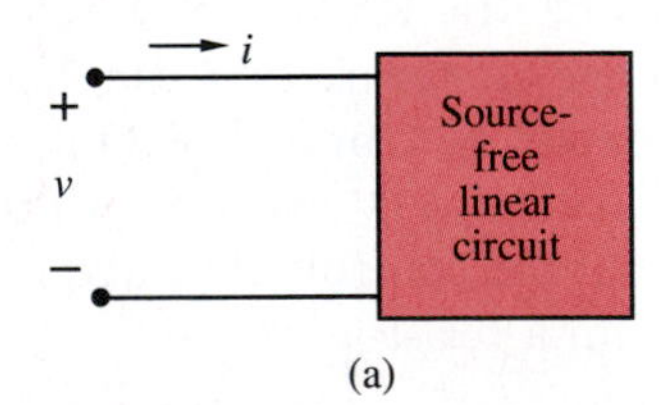

(a)

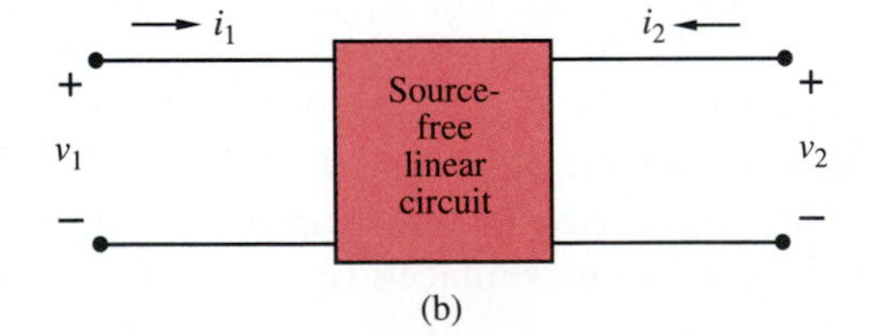

(b)

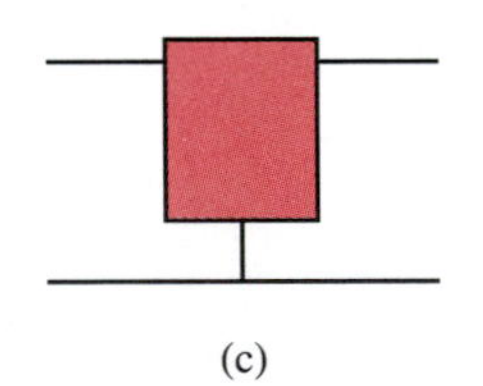
(c)

Figure 10
One-port and two-port circuits.

[7] The purpose of this section is to introduce you to some ideas that are fundamental in network analysis. This very brief introduction, however, is inadequate to give you mastery or make you feel at home in the subject. That will come later, in Chapter 15. Right now, we need a passing acquaintance with the basic ideas and the terminology.

Assuming only dissipative components in the circuits, the one-port is described by the relationship between its voltage and current, namely:

$$v = Ri \quad \text{(current-controlled)}$$

or

$$i = Gv \quad \text{(voltage-controlled)} \tag{6}$$

where R is the equivalent resistance of the circuit seen from its terminals and G is the equivalent conductance. The first relationship is said to be current-controlled because the voltage is controlled by, or is dependent on, the *current.* In this case, we say that the circuit has a *resistance representation.* In the second case, the current is controlled by the *voltage* and the circuit is said to have a *conductance representation.*

Two-Ports

Figure 10b shows a general linear circuit with two pairs of terminals. It is called a *two-port* and is described by two voltages and two currents. Note that the references at each port are standard references; from the outside the currents are oriented into that terminal of each port which carries the voltage + reference.

In Figure 10b the two ports are made up of four distinct terminals. On the other hand, in Figure 10c, each port has a terminal in common with the other port. Such two-ports are said to be *common-terminal* two-ports. Sometimes this common terminal is grounded; then such two-ports are called *grounded two-ports.*

Two of the port variables can be taken as independent, the other two being dependent on, or controlled by, the first two. Thus with either the two currents or the two voltages as independent variables, the following pairs of equations can be written:

$$\begin{aligned} v_1 &= r_{11}i_1 + r_{12}i_2 \\ v_2 &= r_{21}i_1 + r_{22}i_2 \end{aligned} \quad \text{(current-controlled)} \tag{7}$$

$$\begin{aligned} i_1 &= c_{11}v_1 + c_{12}v_2 \\ i_2 &= c_{21}v_1 + c_{22}v_2 \end{aligned} \quad \text{(voltage-controlled)} \tag{8}$$

If, for a given two-port, it is possible to write current-controlled equations, we say the two-port has a *resistance representation.* The r_{ij} coefficients in the resistance representation in (7) are called the *resistance parameters.* If it is possible to write the voltage-controlled equations, we say the two-port has a *conductance representation.* The c_{ij} coefficients in the conductance representation in (8) are called the *conductance parameters.*[8]

[8] The notation in these equations presents a dilemma. The usual symbol for conductance is g, so the conductance parameters in these equations would normally be labeled g_{ij}. However, what will be called the hybrid g parameters in (10) are universally labeled g_{ij}, not only for the dissipative case now under discussion but in the transformed case to be treated in later chapters. If we were to use g_{ij} for the conductance parameters, then we would need a nonstandard notation for these hybrid parameters. Here, we have chosen the lesser of two evils, and have opted to use the c_{ij} notation for the conductance parameters. (Note that c is the first letter of conductance just as r is the first letter of resistance.)

If the circuit contains only two-terminal linear resistors, then both a resistance and a conductance representation will exist. However, we have seen devices where only one of these representations exists. The current-controlled voltage source, for example, has a resistance, but not a conductance, representation. (Confirm this.) The converse is true for the voltage-controlled current source.[9]

It's even worse than that! *The VCVS and the CCCS have neither resistance nor conductance representations.* Such components are said to have a *hybrid* representation. For linear, dissipative two-ports, the following hybrid representations can be written:

$$\begin{aligned} v_1 &= h_{11} i_1 + h_{12} v_2 \\ i_2 &= h_{21} i_1 + h_{22} v_2 \end{aligned} \qquad \text{(hybrid } h\text{)} \tag{9}$$

$$\begin{aligned} i_1 &= g_{11} v_1 + g_{12} i_2 \\ v_2 &= g_{21} v_1 + g_{22} i_2 \end{aligned} \qquad \text{(hybrid } g\text{)} \tag{10}$$

Remember that this was a digression. You are not expected to learn and remember all details of the preceding equations; we'll get to that later. For now, you should become familiar with (1) the port terminology, (2) the possible representations of the port variables in a general way, and (3) the fact that circuits and their representations can be voltage-controlled, current-controlled, or neither, in which case they are said to be *hybrid.*

Reciprocity

A question that might pique your curiosity arises from an examination of either of the equation sets (7) and (8). The parameter r_{12} relates the voltage resulting in port 1 from a current excitation in port 2. How is this coefficient related to r_{21}, which involves the voltage resulting in port 2 from a current excitation in port 1? A similar question arises about the relationship between c_{12} and c_{21} in (8).

To be specific, let's deal with the latter question. Figure 11a shows a linear two-port excited at each port by voltage sources except that the voltage source at port 1 has the value 0, a short circuit. In Figure 11b it's the excitation at

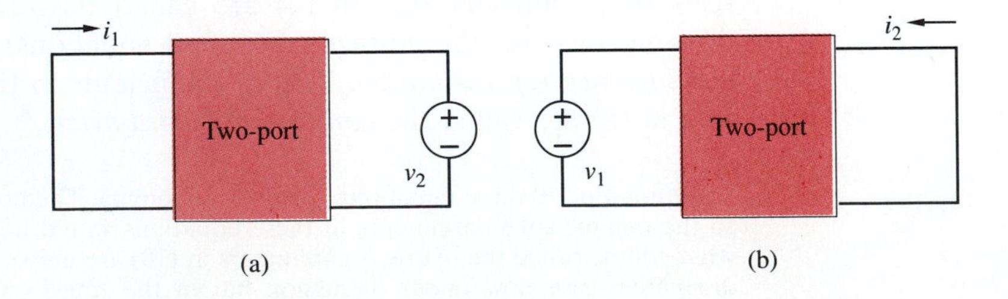

Figure 11 The concept of reciprocity.

[9] Indeed, as you will see in the next chapter, that's why it isn't possible to construct loop equations for a circuit containing a VCCS or node equations for a circuit containing a CCVS, without carrying out some preliminary procedures.

port 2 which is zero. We find the following results in these two cases:

$$\begin{aligned} i_1 &= c_{12} v_2 \quad \text{Figure 11a} \\ i_2 &= c_{21} v_1 \quad \text{Figure 11b} \end{aligned} \tag{11}$$

Now let's make the sources v_2 in Figure 11a and v_1 in Figure 11b the same. Then, either the resulting currents i_1 and i_2 are the same or they are not. If they are the same, we say the two-port is *reciprocal.* If they are not the same, the two-port is *nonreciprocal.* From (11) we see that a two-port is reciprocal if $c_{12} = c_{21}$.

We will examine one example of each kind. The first is the second entry in Table 1, where we find that $c_{12} = 0$ and $c_{21} = G$. Since these are not the same, this two-port is nonreciprocal. A second example is shown in Figure 12a. We excite this two-port with the same voltage v, once at port 2 in Figure 12b and once at port 1 in Figure 12c, the opposite port being short-circuited in each case. In one case we find $c_{12} = -1/R$, and in the other case $c_{21} = -1/R$ also. Hence this two-port is reciprocal.

It is evident from these examples that linearity *does not* imply reciprocity in a two-port. That is, a linear two-port circuit may be reciprocal or it may not. Reciprocity is a property of an entire linear two-port circuit, not of the components that make up the circuit.[10]

Suppose a two-port consists of two-terminal linear resistors only, as in the example in Figure 12; will the two-port be reciprocal in general? We found it to be true in the example. It is true in general also, but we will defer further consideration to Chapter 15. We also found by example that a certain two-port containing a controlled source is nonreciprocal; it is tempting to consider that a general truth. While it may be so in almost all cases, it isn't necessarily true that two-ports containing controlled sources *must be* nonreciprocal. Several controlled sources might be interconnected with resistors in such a way that the resulting two-port is reciprocal.

Another question might occur to you. Reciprocity was defined in terms of conductance parameters $c_{12} = c_{21}$. Is there a similar relationship between the

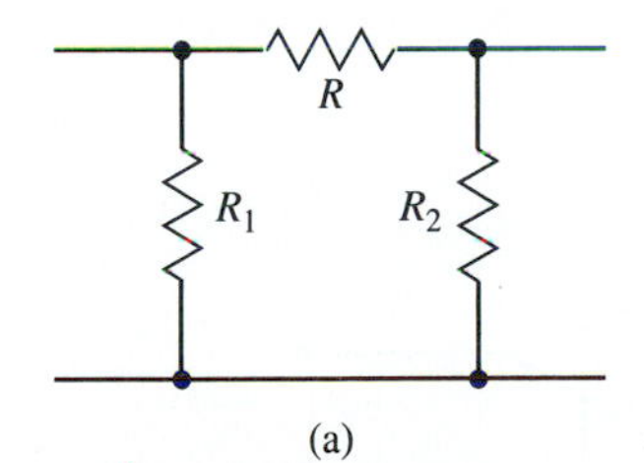

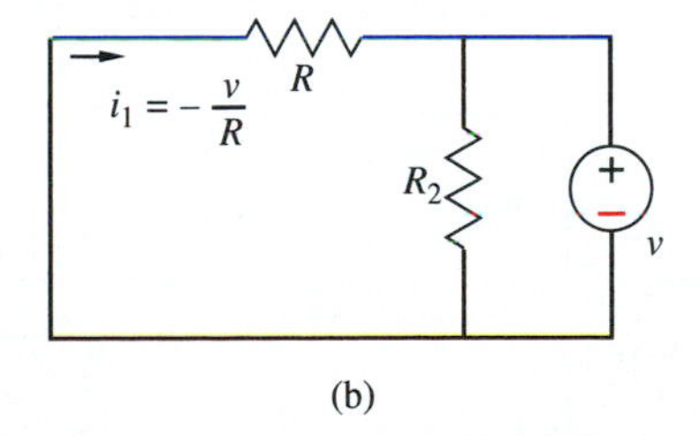

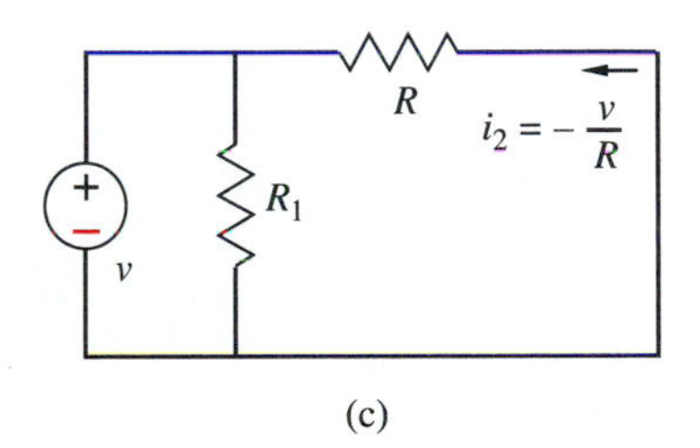

Figure 12
Example of reciprocal two-port.

[10] One might use the term *bilateral* in place of reciprocal, as some authors do, but reciprocal is the term almost universally used not only in circuits but in electromagnetic fields, where it describes the properties of a medium between a transmitting antenna and a receiving antenna, for example.

resistance parameters r_{12} and r_{21}? If we were to go through the necessary algebra we would find that:

$$r_{21} = \frac{c_{21}}{c_{12}} r_{12} \tag{12}$$

(Confirm this result.) Hence for a reciprocal two-port with $c_{21} = c_{12}$, it is also true that $r_{21} = r_{12}$.[11]

4 OPERATIONAL AMPLIFIERS

An operational amplifier (*op amp,* for short) is a multiterminal electronic device whose use has become ubiquitous for applications in instrumentation, communications, and controls. It comes in standard packages of various kinds; a common package is the *dual in-line* package (DIP) shown in Figure 13. The actual design of an op amp circuit utilizes tens of transistors and associated components. Two of the seven terminals of an op amp (the offset null terminals) are for the purpose of improving its performance and will not be included in our subsequent considerations. The symbol for an op amp is shown in Figure 14a. The two balanced supply voltages (typically 15 V) are connected to an external ground. These voltages are required for the proper operation of the op amp but are not involved in the operation of any other circuit to which

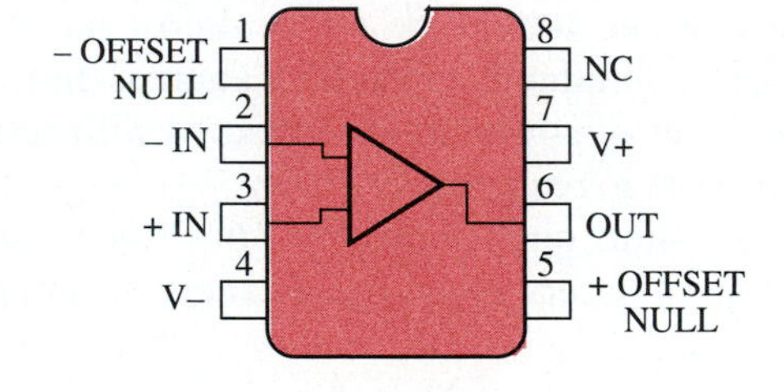

Figure 13
Op-amp dual in-line package.

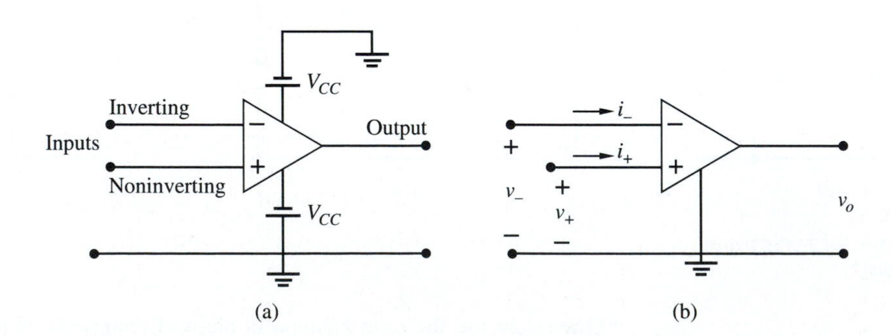

Figure 14
Op-amp symbol.

[11] This result will be established for arbitrary reciprocal two-ports, not just dissipative ones, in Chapter 15.

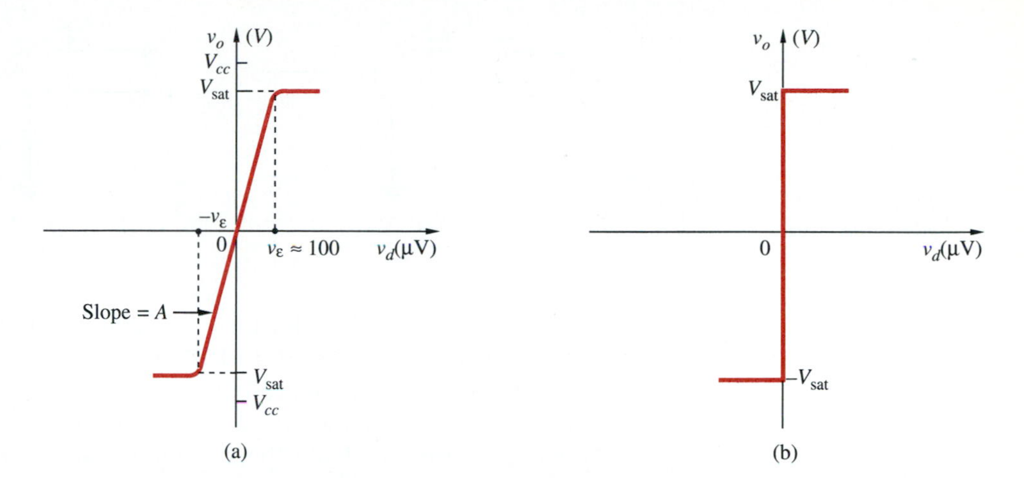

Figure 15
Op-amp output-input characteristic.

the op amp is connected. They are, therefore, usually omitted in schematic diagrams, as shown in Figure 14b. (There is a danger in doing this which is avoided if we show a ground connection, even though there is no pin for a ground connection in the actual device. We will clarify this matter after we have had a chance to discuss the operation of the device.)

The two input terminals are called *inverting* (designated by a minus sign) and *noninverting* (designated by a plus sign) for reasons which will become clear as we proceed. By measurements made at the terminals, the following is determined experimentally:

1. The two input currents (called the *input bias currents*) have very small values (microamperes or less).

2. The output voltage is a function of the *differential input voltage* $v_d = v_+ - v_-$ shown sketched in Figure 15a. Over a small range of inputs around the origin ($-v_\varepsilon$ to $+v_\varepsilon$), the output is an approximately linear function of the differential input: $v_o = Av_d$. Typical values of the slope A, called the *open-loop gain,* range from 10^5 to 10^6. The scales on the voltage axes are different, the output voltage scale being about A times the input voltage scale. For differential inputs more than a few tens of microvolts, the output voltage saturates at V_{sat} ($-V_{sat}$ for negative inputs).

Note that, in circuits, the op amp is a four-terminal device. This count does not include terminals required to connect the supply voltages and terminals used to connect external circuits for the purpose of reducing even further the input bias currents, labeled *offset null* in the diagram of Figure 13.

Ideal Op Amp

We now make an approximation. Because v_ε and the input bias currents are so small, we might as well assume they are zero; and because the gain is so large, we might as well assume it is infinite. With these assumptions, the

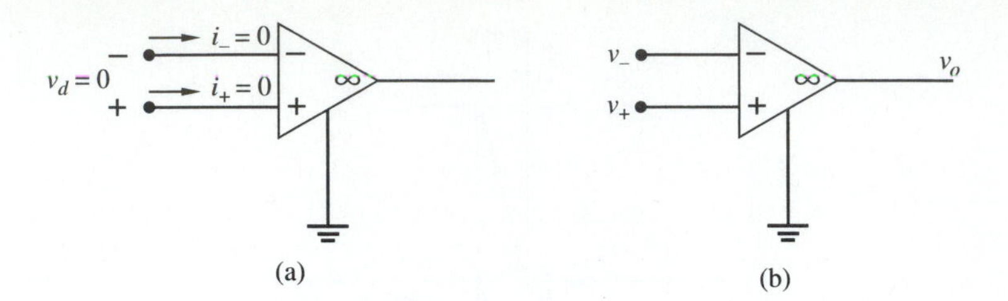

Figure 16
Ideal op amp.

output-input curve takes the form of Figure 15b. The saturated value of the output voltage, V_{sat}, is just 10 to 15% less than the supply voltage V_{CC}; for this reason it is sometimes replaced by V_{CC} for simplicity.

The resulting model is called an *ideal* op amp. The corresponding symbol is shown in Figure 16a. Note the symbol ∞ inside the triangle, representing the fact that A is infinite. In the future, we will not show explicitly that v_d, i_-, and i_+ are zero on the diagram; it may be more useful to label the two (inverting and noninverting) voltages, as in Figure 16b. Although voltage references have not been shown here, it is implied that the reference + is located at the terminal where the voltage symbol is shown and that the reference − is at the position of ground. Gradually, we will wean ourselves away even from this practice of labeling these voltages. Since the inverting and noninverting terminals are already labeled with − and + signs, the locations of v_- and v_+ voltages and their references are implied.

There seems to be something peculiar about the ideal op amp. No currents enter its two input terminals, so there seems to be an open circuit across these terminals; and because the differential input voltage is zero, there seems to be a short circuit there also! But there isn't really a direct short circuit across the inputs; it's only that the two input voltages are always equal. Therefore we call it a *virtual* short circuit. (If *virtual reality* is possible, why not a virtual short circuit?)

Now it should also be clear why we can't completely neglect the power-supply terminals, to the extent of omitting even the ground terminal. Without the ground terminal, KCL applied to the op amp as a whole would require the output current always to be zero, since the two input currents are zero. This would not make sense.

5 LINEAR OP-AMP CIRCUITS

When the op amp is operating on the vertical part of its output-input characteristic, it is operating in its linear region. This region is characterized by

$$v_+ - v_- = 0$$

$$-V_{sat} < v_o < V_{sat}$$

When analyzing a circuit containing an op amp, we *assume* that the op amp is operating in the linear region and then obtain a solution. If we find that the condition just shown for linear operation is not satisfied, then the solution does not apply. That is, just obtaining a solution assuming linear operation is not enough; we must also confirm the linearity condition. Examples will be shown for the specific op amp circuits that follow. (This has some similarity with the approach taken in Chapter 3 with diode circuits, where a diode state is assumed and then confirmed.)

We will now study a number of important op-amp circuits which perform specific operations on an input signal.

The Inverting Amplifier

The circuit for what is called an *inverting amplifier* is shown in Figure 17. It includes two resistors in addition to an op amp, assumed to be ideal. The usual circuit tools (KCL, KVL, and Ohm's law) and the op-amp conditions ($i_+ = 0 = i_-$ and $v_d = v_+ - v_- = 0$) must be applied to solve the circuit. But the op-amp conditions are so simple that they can be applied mentally without the need for formally invoking them. A voltage is applied at the input of this circuit, and the requirement is to find the output voltage.

Since no current is entering the inverting terminal ($i_- = 0$), KCL at the terminal gives $i_1 = i_2$. Since there is no voltage across the input terminals, these terminals are both at ground voltage (zero). Hence, by KVL, the input voltage is all across R_1. Thus, $i_1 = v_i/R_1 = i_2$. For the same reason, applying KVL around the path including the output, R_2, and the inputs, we find $v_o = -R_2 i_2$. Combining the last two expressions leads to:

$$v_o = -\left(\frac{R_2}{R_1}\right)v_i = Kv_i \tag{13}$$

where $K = -R_2/R_1$ is called the *voltage gain.* It is the *closed-loop* gain as opposed to the *open-loop* gain A. The negative sign means that the output voltage is inverted relative to the input; that's the origin of the name *inverting amplifier.*

There is only one step left: to confirm the linearity condition, namely, that $|v_o| < V_{\text{sat}}$. Using the preceding equation, this leads to the following condition

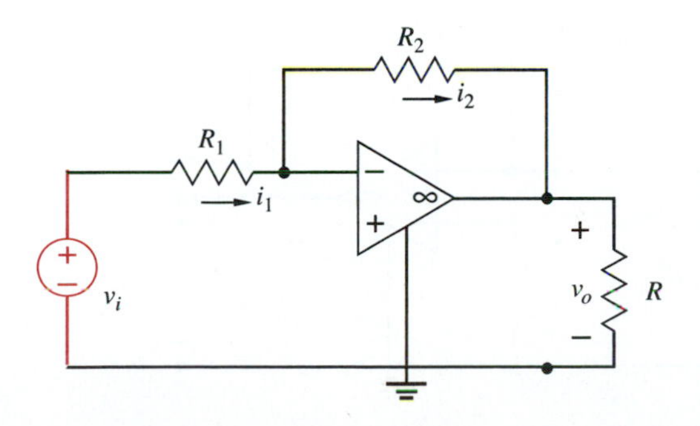

Figure 17
Inverting amplifier.

on the input voltage:

$$|v_i| < \frac{R_1}{R_2} V_{\text{sat}} \tag{14}$$

Thus, the circuit in Figure 17 acts as a linear inverting amplifier with a voltage gain $K = -R_2/R_1$ so long as the input voltage satisfies the condition $|v_i| < (R_1/R_2)V_{\text{sat}}$.

Note the manner in which R_2 is connected from the output back to the inverting input. R_2 is said to be a *feedback resistor,* since it "feeds back" the output voltage to the input.

EXERCISE 2

Suppose $V_{\text{sat}} = 13.5$ V (10% less than V_{CC}) and $R_2 = R_1$. Find the value of the voltage gain and the limitation on the input voltage for linear operation. Why do you think this circuit is called an *inverter,* or a *phase inverter*?

ANSWER: $v_O/v_i = -1$; $v_i < 13.5$ V. □

The Noninverting Amplifier

Suppose we want an amplifier but without the sign inversion. A possible circuit is shown in Figure 18. Here, unlike the previous case, the input voltage is applied to the noninverting terminal. However, the feedback resistor is still returned from the output to the inverting input.

To determine the output voltage, we again use all of the relevant tools. Thus, because of the virtual short circuit, all the input voltage is applied across R_1. Hence, $i_1 = v_i/R_1$. Also, since no current enters the inverting input terminal, $i_2 = i_1$, by KCL. Finally, by KVL around the path that includes the output, resistor R_2, the virtual short circuit across the input terminals, and the input voltage, we find that $v_o = R_2 i_2 + 0 + v_i$. When these are all combined, the result is

$$v_o = \left(1 + \frac{R_2}{R_1}\right)v_i$$

$$= Kv_i \qquad \text{where} \qquad K = \left(1 + \frac{R_2}{R_1}\right) \tag{15}$$

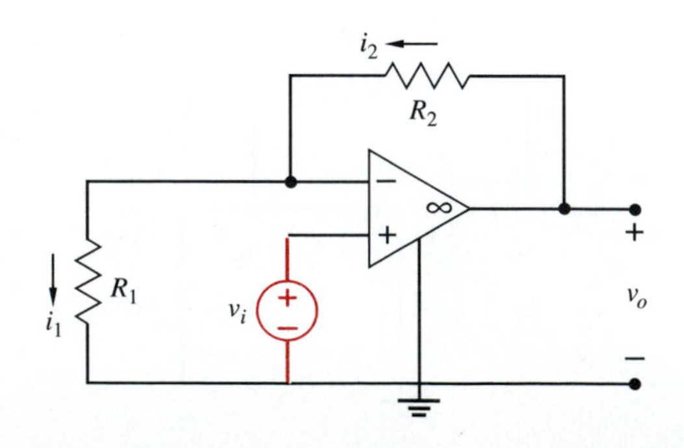

Figure 18
Noninverting amplifier.

In this case the voltage gain is a positive constant (hence the name *noninverting*) greater than 1, no matter what the resistance values may be.

EXERCISE 3

Find the limitation on the input voltage imposed by the linearity condition.

ANSWER: $v_i < \dfrac{R_1}{R_1 + R_2} V_{sat}$. □

The Voltage Follower

Consider what would happen to the voltage gain of the preceding circuit if we set the feedback resistance to zero, that is, if we replace it by a short circuit. The gain becomes 1, independent of the value of R_1. Since R_1 has no influence on the gain, we might as well remove it. The result becomes the circuit in Figure 19a. This is an op amp, excited at its noninverting input, with a direct feedback connection from the output to the inverting input.

Since the gain is 1, $v_o = v_i$. Thus, the output voltage is equal to the input voltage; this is why the circuit is called a *voltage follower.* A little thought will show that, for linear operation, this whole circuit can be replaced by a voltage-controlled voltage source, as shown in Figure 19b.

Why would anybody go to the trouble of designing a circuit which seems to do nothing but multiply an input voltage by 1? In Figure 19a the input is shown as a source. In practice, this voltage v_i might be the output of an extensive circuit. If a load were connected directly to this circuit, it could interact with the circuit, maybe causing the voltage applied to the load to be different from v_i. However, with the voltage follower between v_i and the load, no such interaction will occur. (Since no current enters the op-amp input, no current is supplied by the circuit whose output is v_i.) Thus, the voltage follower performs the function of buffering and is, consequently, also called a *buffer.*

EXERCISE 4

In the circuit in Figure 20a, a box is inserted between a source circuit and a load. Inside the box there is only a conducting path. In Figure 20b, the box

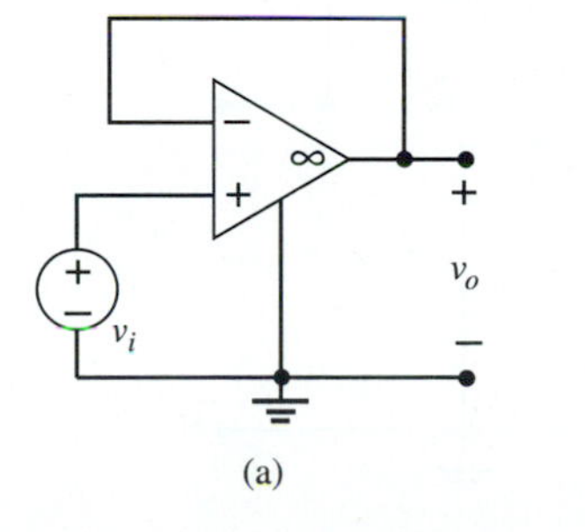

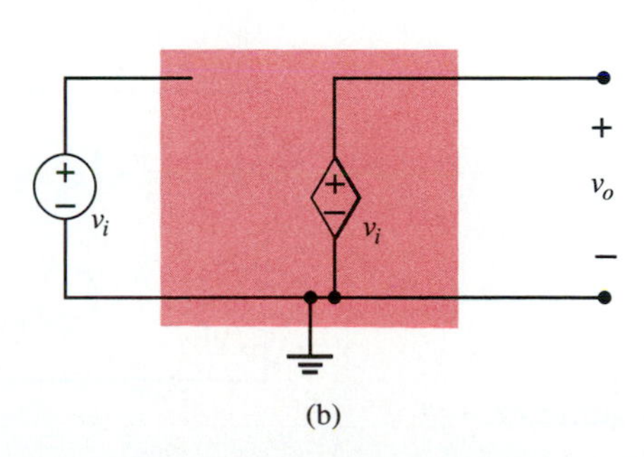

Figure 19
Voltage-follower circuit.

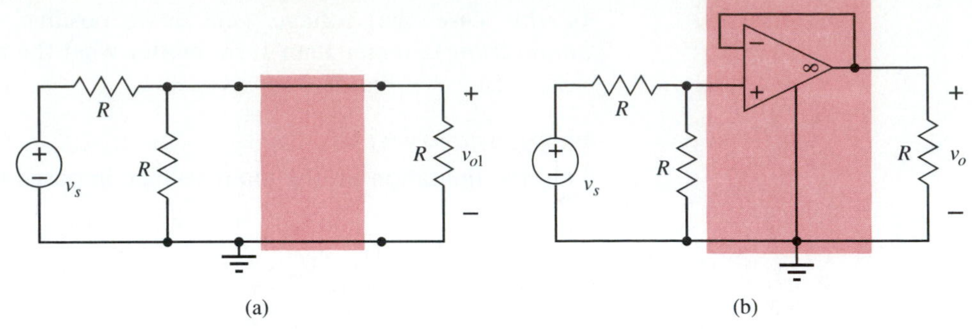

Figure 20
Output voltage with and without a buffer.

contains a buffer. Find the output voltage in each case and compare. Do you see how the presence of a buffer makes a difference?

ANSWER: $v_{o1} = \frac{1}{3}$; $v_{o2} = \frac{1}{2}$. □

The Voltage Summer

The circuit shown in Figure 21 is an inverting amplifier with more than one source providing an input at the inverting terminal. If you were asked to be creative and imagine what the output voltage would be, you might reason as follows. If each source were acting alone, we would have an inverting amplifier, giving $v_o = Kv_i$. With both sources acting together, maybe the output would be $v_o = K_1v_{i1} + K_2v_{i2}$. Let's see if that conjecture can be confirmed.

Because of the virtual short circuit, KVL applied to the closed paths through each of the sources, their resistors, and the virtual short circuit leads to two equations. KVL applied to the path consisting of the output, the feedback resistor, and the virtual short circuit leads to a third equation. Finally, KCL applied to the node at which the three resistors are connected leads to a fourth equation. (Remember that no current enters the inverting input terminal.) These equations are as follows:

$$i_1 = \frac{v_{i1}}{R_1} \qquad i_2 = \frac{v_{i2}}{R_2} \qquad i_3 = i_1 + i_2$$
$$v_o = -R_3i_3 = -R_3(i_1 + i_2) = -\frac{R_3}{R_1}v_{i1} - \frac{R_3}{R_2}v_{i2} \tag{16}$$

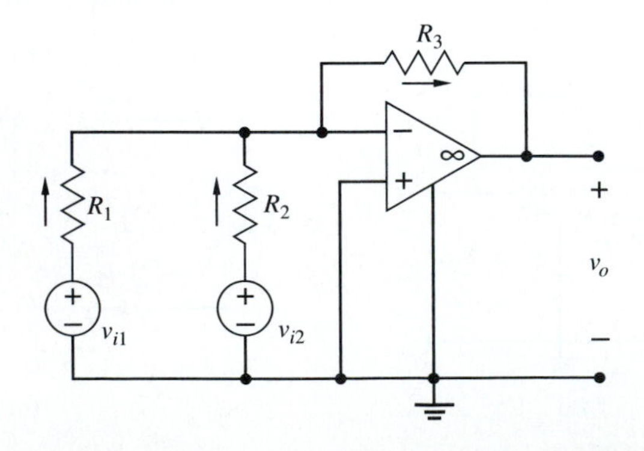

Figure 21
Voltage summer.

The expression for i_3 was substituted into the expression for v_o, after which the two individual currents were inserted. Sure enough, the output equals the sum of the inverted and amplified inputs, as you conjectured. (Determine the restriction which linearity imposes on the inputs.)

Because this circuit gives an output which is a linear "sum" of inputs, it is called a *summer*. (It's a cheerful circuit!)

6 NONIDEAL OP-AMP MODELS

Several assumptions were made when we introduced the ideal op amp in an earlier section. One assumption was that the currents entering the input terminals were zero. From another perspective, that means that the equivalent resistance R_i looking into the op amp from its input terminals is infinite. Another assumption was that of infinite open-loop gain A. In that case, since the output is finite, it must mean that the difference voltage v_d across the two inputs is zero; otherwise, $v_o = Av_d$ would become infinite. Finally, since the voltage produced across the op-amp output terminals is not changed if a load connected there is changed, the op-amp output must be like an ideal voltage source, with no accompanying resistance.

Suppose now that these assumptions are relaxed so that:

1. The input resistance R_i, though large, has a finite value;
2. The open-loop gain A, though large, is not infinite; and
3. The accompanying output resistance, though small, is not zero.

Under these circumstances, a model of the op amp would take the form shown in Figure 22. Typical values of the parameters for existing op amps are $A = 10^5$, $R_i = 1\ \text{M}\Omega$, and $R_o = 50\ \Omega$.

Obviously, using this model for an op amp, the previously carried-out analysis of various circuits now becomes somewhat more complex. We will here initiate more refined analyses for some of the circuits and leave for you the job of supplying the details.

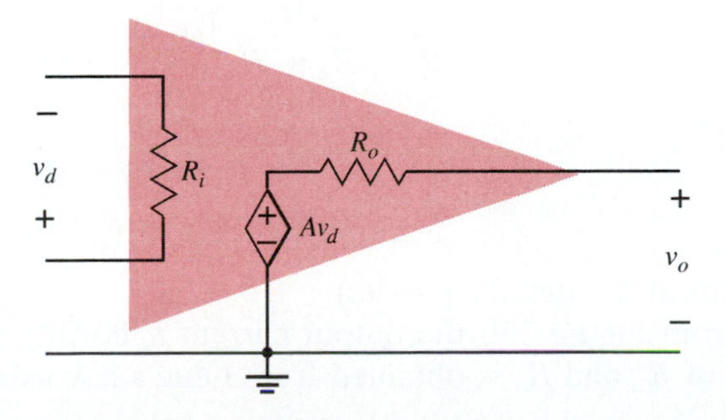

Figure 22
Nonideal op-amp model.

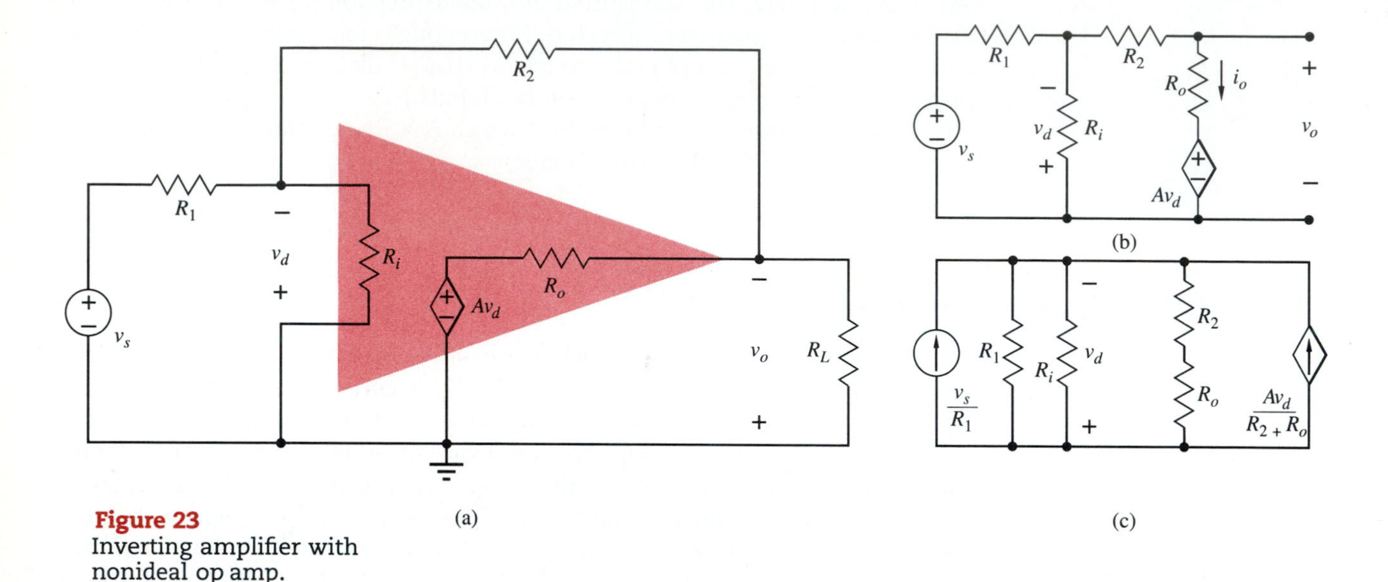

Figure 23
Inverting amplifier with nonideal op amp.

First, consider the circuit of the inverting amplifier shown in Figure 23a. Whereas v_d was previously 0, it is nonzero now; finding it is part of the analysis. Let us initially assume that the load R_L is not connected. For convenience, the circuit is redrawn in Figure 23b. In the next chapter we will discuss methods of analysis to deal with circuits such as this. For the present, we will rely on the simpler methods of the preceding chapter. Note that resistors R_2 and R_o are in series.

One possibility you might think of for simplifying the circuit is to convert the part of the circuit consisting of v_s, R_1, and R_i into a Thévenin equivalent. This procedure would fail because the controlling voltage v_d would be lost in the process; but we can't do without it, because the controlled source depends on it. An alternative is to replace the accompanied sources (including the controlled source) by current-source equivalents, as shown in Figure 23c. In that circuit, all branches are in parallel; hence the voltage v_d is the product of the equivalent current source and the equivalent resistance, which we will label R_p. Thus:

$$v_d = -\left(\frac{v_s}{R_1} + \frac{Av_d}{R_2 + R_o}\right)R_p = -\frac{R_p(R_2 + R_o)/R_1}{AR_p + R_2 + R_o}v_s \tag{17}$$

where

$$R_p = \frac{R_1R_i(R_2 + R_o)}{R_1R_i + (R_2 + R_o)(R_i + R_1)}$$

(Confirm the details, please.)

From Figure 23b the output current i_o flowing through the series combination of R_2 and R_o is obtained from Ohm's law using the branch voltage as the difference in voltages at the two ends of the branch. The output voltage is now obtained as follows:

$$i_o = \frac{-v_d - Av_d}{R_2 + R_o}$$
$$v_o = R_o i_o + Av_d = \frac{AR_2 - R_o}{R_2 + R_o} v_d \tag{18}$$

(Again it's up to you to supply the details.)

Finally, we insert v_d from (17) into (18) to obtain the final expression for the output voltage. If you carry out the details you will obtain:

$$v_o = \frac{-\dfrac{R_2}{R_1}\left(1 - \dfrac{R_o}{AR_2}\right)}{1 + \dfrac{1}{A} + \dfrac{1}{A}(R_2 + R_o)\left(\dfrac{1}{R_i} + \dfrac{1}{R_1}\right)} v_s \tag{19}$$

What happens to this expression for the gain if A goes to infinity? It reduces to $-R_2/R_1$, which is exactly the gain for an inverting amplifier with an ideal op amp.

You can get a feel for the degree of accuracy of the gain for the ideal case by using typical values for the various parameters in the preceding equation. The second term within parentheses in the numerator, for example, should be compared with 1. Using $R_o = 50\,\Omega$, $A = 10^5$, and realistic values for R_2 in the kilohm range, we will be comparing something of the order of 10^{-7} with 1, a completely negligible value. Similar conclusions result by comparing the remaining terms in the denominator with 1.

The general conclusion from this exercise is that very little error is made if the op amp is taken as ideal, at least in the case of the inverting amplifier. You will find similar conclusions if you carry out an analysis similar to the preceding one for other linear op-amp circuits. That will be left for you to do.

One other question should be posed before leaving the subject. Suppose there is a resistive load R_L at the output terminals of the op amp, as suggested in Figure 23a; will it have a large influence on the outcome? Without going through a complete analysis, note that R_L and the branch containing the controlled source can be combined into an accompanied controlled source by source transformations and parallel combination of R_L and R_o. Thus, in the preceding analysis, everywhere R_o appears it should be replaced by the parallel equivalent of R_L and R_o. Also, Av_d has to be replaced by $Av_dR_L/(R_L + R_o)$. If R_L is no smaller than kilohms, then very little error is made by taking it to be infinite. In any case, you should work out the details.

In the case of an ideal op amp, the output voltage is independent of R_L, so R_L has no effect on the proper operation of the circuit. Hence, it need not be shown in the circuit. However, op-amp circuits are connected to *something* at their output. We say they *work into* or they *drive* that something. That something will constitute a physical load on the op amp, so its presence ought to be acknowledged with an R_L, even though, in the case of an ideal op amp, it has no influence on the output.

7 GYRATORS AND NEGATIVE CONVERTERS[12]

The preceding section described a number of very useful circuits that can be designed with an op amp and a few resistors. These are not the only possibilities. In this section we shall study two more classes of devices which can be realized by means of op amps.

The Negative Converter

Consider the circuit in Figure 24. Although the feedback resistor is familiar, the presence of a resistor in series with the op-amp output lead and the connection of the noninverting input terminal to the circuit output terminal (not the op-amp output) is strange. Nevertheless, we can proceed, as before, to apply the tools at our disposal to determine relationships among the output and input variables.

By applying KVL to the path consisting of the circuit output, the input, and the virtual short circuit, we find that $v_2 = v_1$. Because no current flows into the op-amp terminals, the current in R_1 is i_1 and the current in R_2 is i_2. Finally, by KVL applied around the path consisting of the two resistors and the virtual short circuit, we find that $R_1 i_1 = R_2 i_2$.

Now suppose that a resistor R is connected at the output of this circuit. Ohm's law will give $v_2 = -Ri_2$. With this load connected, let us find the relationship between the voltage and current at the circuit input. Using the results of this and the preceding paragraph, we get:

$$v_1 = v_2 = -Ri_2 = -R\left(\frac{R_1}{R_2} i_1\right) = -\frac{R_1}{R_2} Ri_1 \tag{20}$$

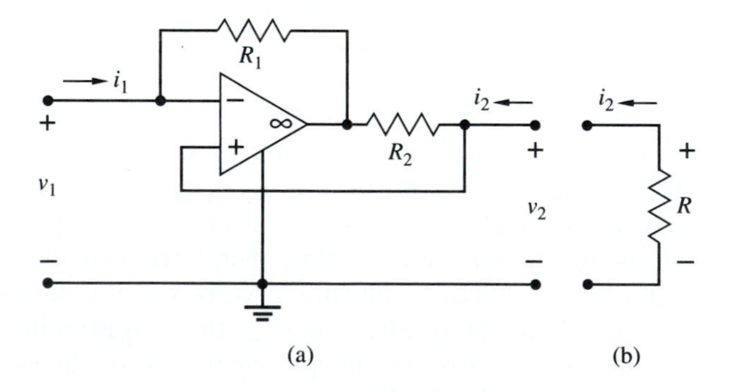

Figure 24
Negative-converter circuit.

[12] Although the circuits described here are very interesting, this section may be omitted without causing you a future handicap.

This is an interesting result. When the original circuit is terminated in a resistor R, the voltage-current relationship at the input terminals is that of a resistance which is proportional to the negative of R. For this reason, this circuit is called a *negative converter* (NC). The quantity R_1/R_2 is called the *conversion ratio.*

Thus, although linear resistors are known to have positive values, we have been able to design a two-port circuit which has the properties of a negative resistance at one port when it is loaded by a positive resistance at the second port.

EXERCISE 5

In the preceding circuit, apply the linearity condition to determine the limitation on voltage v_1 for linear operation.

ANSWER: $|v_1| \leq RV_{sat}/(R + R_2)$. □

EXERCISE 6

Using the op-amp v-i relationships and other circuit tools at your disposal, show that the circuit in Figure 25 is a negative converter. Find its conversion ratio. Determine the limitation on the input imposed by the linearity condition.

ANSWER: $k^2 = R_1/R_2$; $|v_1| \leq RV_{sat}/(R + R_2)$. □

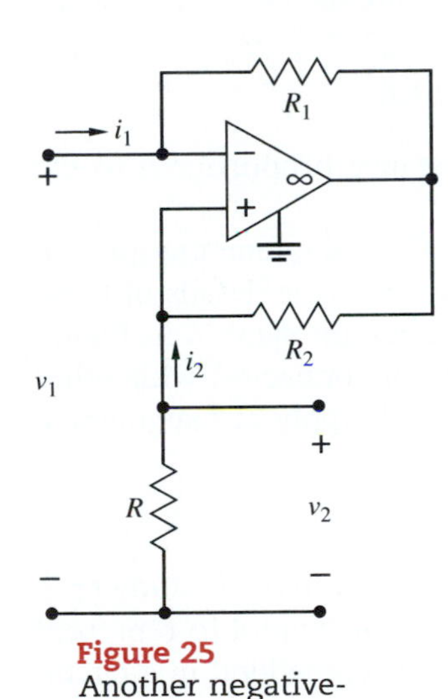

Figure 25
Another negative-converter circuit.

Observe that, without the load R, the negative converter in Figure 25 is a two-port. Sometimes focusing on the properties of the two-port itself, without regard to the details of its construction, is desirable. Indeed, besides the circuit shown here, there are other op-amp circuit designs for a negative converter. Hence, it is useful to create a circuit symbol for the NC. Such a symbol is shown in Figure 26a.

In the preceding development we found the v-i equations of the NC to be $v_1 = v_2$ and $i_2 = (R_1/R_2)i_1$. That is, the input and output voltages are the same; it is the current which is converted. Hence this case is referred to as a *current* negative converter. The port equations take the form on the left:

Current NC	Voltage NC
$v_1 = v_2$	$v_1 = -k^2 v_2$
$i_2 = k^2 i_1$	$i_2 = -i_1$
$\begin{bmatrix} v_1 \\ i_2 \end{bmatrix} = \begin{bmatrix} 0 & 1 \\ k^2 & 0 \end{bmatrix} \begin{bmatrix} i_1 \\ v_2 \end{bmatrix}$	$\begin{bmatrix} v_1 \\ i_2 \end{bmatrix} = \begin{bmatrix} 0 & -k^2 \\ -1 & 0 \end{bmatrix} \begin{bmatrix} i_1 \\ v_2 \end{bmatrix}$

(21)

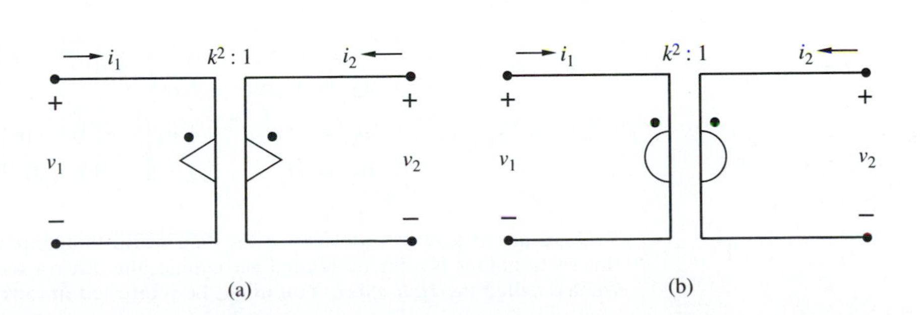

Figure 26
Circuit symbols for negative converters. (a) Current converter: $i_2 = -k^2 i_1$; (b) voltage converter: $v_1 = -k^2 v_2$.

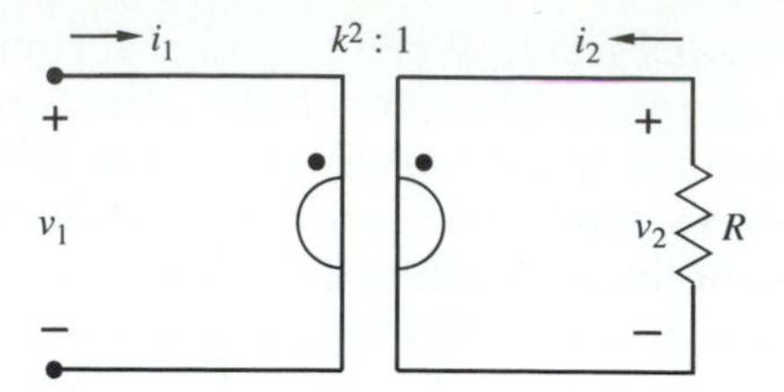

Figure 27
A voltage NC terminated in R.

where k^2 is a positive constant. The relationships have been written in matrix form also.

Both Figure 26b and the right column of (21) refer to another variety of negative converter, the *voltage* NC, although we have not seen a circuit which embodies such a device. Nevertheless, let's consider a hypothetical voltage NC terminated in a resistance R, as shown in Figure 27. Using the NC relationships and $v_2 = -Ri_2$ by Ohm's law leads to:

$$v_1 = -k^2 v_2 = -k^2(-Ri_2) = -k^2 R i_1$$

This is again the v-i relationship for a negative resistance proportional to the load resistance R.

To summarize: There exist a number of different implementations of negative converters. When we are not interested in the specific details of these circuits, we can represent the NC by the appropriate circuit symbol in Figure 26, either current NC or voltage NC. When an NC is connected with other components in a circuit, previous methods of analysis still apply but augmented by the v-i relations of the NC.

The Gyrator

In the preceding section on negative converters, we started by studying a number of op amp circuits. We introduced a special circuit symbol to represent one type of negative converter. When this component is embedded in a circuit including other components, we don't need to consider the insides of the NC, only the v-i relationships at the ports.

Now we will take a different approach. Instead of studying a specific op-amp circuit and determining its v-i relationships, we will *postulate* a device having a prescribed set of v-i relations, *without even knowing if such a thing exists.* (Is this being creative, or what?)[13]

We define a *gyrator* as a two-port model having one of the following v-i relationships:

$$\begin{aligned} i_1 &= gv_2 \\ i_2 &= -gv_1 \end{aligned} \qquad \begin{bmatrix} i_1 \\ i_2 \end{bmatrix} = \begin{bmatrix} 0 & g \\ -g & 0 \end{bmatrix} \begin{bmatrix} v_1 \\ v_2 \end{bmatrix} \tag{22a}$$

$$\begin{aligned} v_1 &= -ri_2 \\ v_2 &= ri_1 \end{aligned} \qquad \begin{bmatrix} v_1 \\ v_2 \end{bmatrix} = \begin{bmatrix} 0 & -r \\ r & 0 \end{bmatrix} \begin{bmatrix} i_1 \\ i_2 \end{bmatrix} \tag{22b}$$

[13] There are at least two devices in the real world that display the effect being described in this section. One is a ferrite-loaded waveguide, the other a semiconductor device that utilizes what is called the *Hall effect.* You might be interested in looking into these someday.

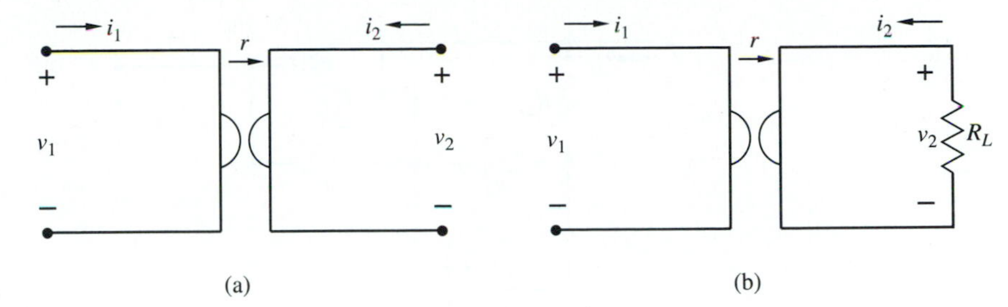

Figure 28
Circuit symbol of gyrator.

where $r = 1/g$ is the *gyration resistance,* in ohms (g is the *gyration conductance*). The circuit symbol for a gyrator is shown in Figure 28a. Notice how the two "bumps" face each other for a gyrator instead of, as in the case of the negative converter, being directed away from each other.

Now suppose the output side of the gyrator is loaded with a resistance R_L (Figure 28b). The relationship between the input voltage and current can be determined by the following sequence of steps:

$$v_1 = -ri_2 = -r\left(-\frac{v_2}{R_L}\right) = \left(\frac{r}{R_L}\right)ri_1 = \frac{r^2}{R_L}i_1 \tag{23}$$

Thus, when the output side of a gyrator is loaded with a resistance R_L, it appears that the equivalent *resistance* seen from the input side is proportional to the load *conductance.* The output resistance seems to be an inverted, twisted, or *gyrated* version of the input. (That's why it's called a gyrator.)

Let's do one more thing: calculate the energy delivered from its ports to the gyrator. The total power input from the ports is $p = v_1 i_1 + v_2 i_2$. Assuming the circuit was turned on in the dim past ($-\infty$) and goes on until time t, the energy delivered is obtained by integrating the power between those two points. The result will be:

$$w(t) = \int_{-\infty}^{t} (v_1 i_1 + v_2 i_2)\, dx = \int_{-\infty}^{t} [(-ri_2)i_1 + (ri_1)i_2]\, dx \equiv 0 \tag{24}$$

No energy is dissipated in the gyrator! It is a *lossless* two-port.

Finally, consider the circuit shown in Figure 29. It is claimed that this circuit behaves as a gyrator, a claim we want to confirm. This circuit, with two op amps, is more complex than all the preceding ones in this chapter. Nevertheless, its analysis is not complicated if you remember the two conditions for an ideal op amp: zero voltage across the op-amp input terminals and zero current into those terminals.

Thus, because of zero op-amp terminal currents, the following results hold:

The current through R_1 is the same as i_1.
The current through R_2 is the same as the current through R_4.
The current oriented up through R_3 is the same as i_2.

Because of the virtual short circuit across the op-amp terminals:

The input voltage v_1 appears entirely across R_3.

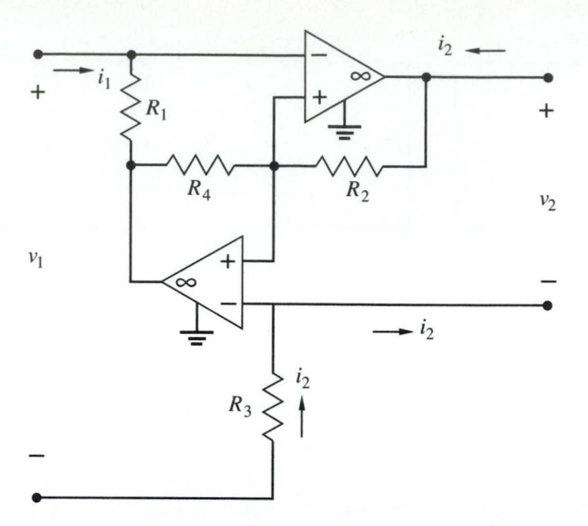

Figure 29
Gyrator circuit.

The output voltage v_2 is the voltage across R_2.
R_1 and R_4 form a closed path around which KVL can be applied.

Applying these results leads to the following equations:

$$\begin{aligned} v_1 &= -R_3 i_2 \\ R_4 i_4 &= R_1 i_1 \\ v_2 &= R_2 i_4 = \frac{R_1 R_2}{R_4} i_1 \end{aligned} \tag{25}$$

The first and last of these are in the form of the gyrator relations in (22b). Hence:

$$\begin{aligned} r = R_3 &= \frac{R_1 R_2}{R_4} \\ \frac{R_1}{R_4} &= \frac{R_3}{R_2} \end{aligned} \tag{26}$$

The last equation is an easily achieved condition on the resistors in order for the circuit in Figure 29 to act as a gyrator. Note that, in this particular design, the value of r can range up to very high values—many thousands. This is true for other designs (not described here) also.

There is only one difficulty with the circuit in Figure 29: there is no common (ground) terminal between the input and the output. In some cases, a common ground may be a requirement; in such cases this circuit would not be appropriate.

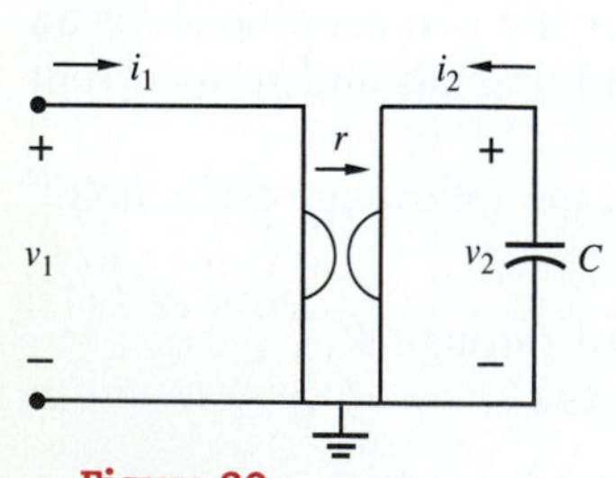

Figure 30
Inductor simulation: capacitor-loaded gyrator.

INDUCTOR SIMULATION Although components were introduced in Chapter 2 whose v-i relations involve the mathematical operation of differentiation (the capacitor and the inductor), everything else we have done so far has involved only algebraic equations. We will consider what happens if this peculiar component called a gyrator is terminated at one pair of terminals by a capacitor, as shown in Figure 30.

Recall from Chapter 2 that the capacitor *i-v* relationship is $i = C\,dv/dt$. This expression (appropriately modified for the references in the figure) and the gyrator equations constitute a set of three equations. Substituting one into the other consecutively leads to:

$$v_1 = -ri_2 \quad i_2 = -C\frac{dv_2}{dt} \qquad v_2 = ri_1$$

$$v_1 = -r\left(-C\frac{dv_2}{dt}\right) = rC\frac{d}{dt}(ri_1) = (r^2C)\frac{di_1}{dt}$$

$$= L\frac{di_1}{dt} \qquad \text{where} \qquad L = r^2C \tag{27}$$

The end result is a relationship between the input voltage and current which is exactly that of an inductor (see Chapter 2). This is an extraordinary result. Among all components in electronic circuits, the inductor is the most difficult to fabricate, the heaviest, and the bulkiest. By means of a gyrator, it is possible to *simulate* an inductor, avoiding all those disadvantages. You should be aware, however, that this simulated inductor cannot be used in those applications (such as in power systems) where heavy currents are involved. Nevertheless, this simulated inductance finds many uses. (If you were a supplier, what would you ship to a customer who wanted a 1000-H inductor?)

8 TRANSISTORS AND THEIR MODELS[14]

A history of technology would undoubtedly award to the transistor the distinction of being one of a handful of basic devices without which life as lived in the last few years of the twentieth century would be impossible.[15] A

[14] This is the promised optional section that develops controlled sources as models of transistors under certain operating conditions. If you would like a more extensive development of how controlled sources arise out of an analysis of transistor amplifiers, this is for you. If, on the other hand, you are satisfied with the manner in which controlled sources were introduced via ideal transformers earlier, you may wish to skip this section. Nothing in the rest of the book will depend on this material.

[15] The transistor was invented through the collaboration of three individuals and news of it was announced by Bell Telephone Laboratories in 1953. The inventors were William Shockley, John Bardeen, and James Brattain. Fabrication methods, secondary principles of operation, and new structures have evolved over time, but the fundamental principles are those described at that time. One characteristic of many creative scientists and engineers is to visualize new directions of inquiry and new combinations of materials and ideas that lead to radically new conceptions. Of course, many such creative thoughts and endeavors lead to dead ends; the creatively conceptualized devices don't work and are discarded. Thomas Edison's brilliant concept (no pun intended) of the incandescent light bulb, for example, was followed by many failures in attempting to find the right material for the filament. It is tragic that in the last two decades of his life, William Shockley's creative mind turned to the proposition that race is a major factor in determining human intelligence. Who knows what other benefits to the world could have flowed from his creative mind had he not become so obsessed.

transistor is a three-terminal electronic device whose basic structure consists of a piece of one type of semiconductor material sandwiched between two pieces of a second type of semiconductor.

The goal in this section is to develop models of a transistor that are valid over certain conditions of operation. Enough of a study of transistors will be undertaken so that you will appreciate the nature of the models that are introduced and will have some sense of their limitations. However, this brief study will in no way substitute for a study of electronics and electronic circuits.[16]

Transistor Operation

From the earliest days of electricity, materials have been classified into two categories on the basis of how well they conduct electricity. On one side are the metals (copper, silver, aluminum) that conduct electricity very well; they are known as *conductors.* On the other side are those materials that conduct electricity poorly (glass, ceramics, polystyrene); they are known as *insulators* or *dielectrics.* But there is another class of crystalline materials (including silicon, gallium arsenide, and germanium) whose conduction properties lie somewhere between good conductors and good insulators. They are known as *semiconductors.*

Pure semiconductors are not of great interest; but when a semiconductor is *doped* by introducing minuscule amounts of certain other substances into the semiconductor crystal, an entirely new behavior results. One of the resulting materials is called *n-type*; a second is called *p-type.* In n-type materials there is an excess of free electrons, which are negative charge carriers, and in p-type materials there is an excess of positive charge carriers, called *holes.*

A semiconductor *junction* is formed when a piece of n-type material abruptly, over a very short distance, is made to change to p-type material. The region in which the material changes from n-type to p-type is called the *transition region.* A difference of potential (voltage) is established across the transition region by virtue of a preponderance of negative charges on one side of the junction and positive charges on the other. Charges tend to move across the junction (by diffusion); this makes the potential of the n-type material more positive relative to the p-type material and creates a potential *barrier*, inhibiting the further diffusion of charges. If an external voltage is applied across the junction so as to reduce this potential barrier, current flow is facilitated. This is the case if the polarity of an external voltage tends to make the potential of the p-type material more positive relative to that of the n-type material, as shown in Figure 31a. The junction here is *forward-biased.* The opposite is shown in Figure 31b, where the junction is *reverse-biased.*

One class of transistor is the *junction transistor,* made by a "back-to-back" arrangement of two semiconductor junctions, as illustrated schematically in

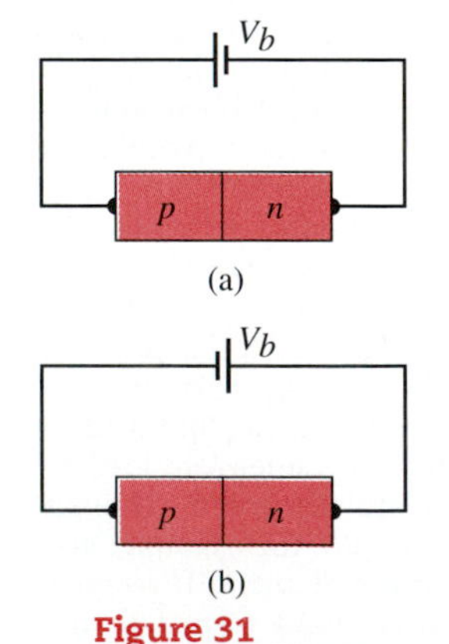

Figure 31
(a) Forward-biased and (b) reverse-biased semiconductor junction.

[16] Another reason for going beyond the controlled-source model and bringing into your consciousness some knowledge of the structure of a transistor is to make it possible for you to use the circuit analysis software named SPICE more intelligently. SPICE requires listing the nodes of a transistor and providing enough of its characteristics so that the software can carry out an analysis.

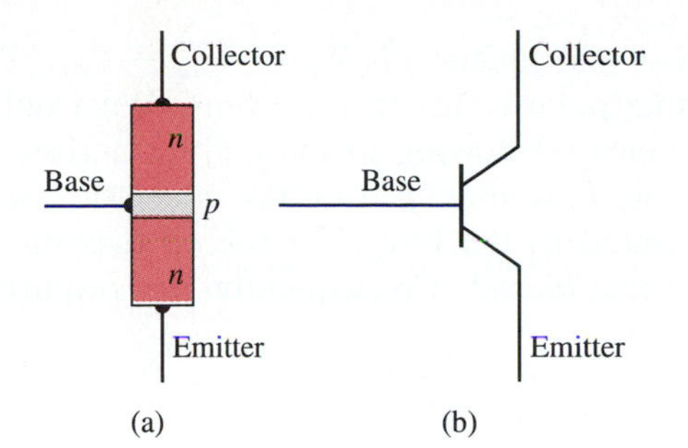

Figure 32
(a) Schematic and (b) circuit symbols of *npn* junction transistor.

Figure 32a. What is shown is an *npn* transistor; it is a three-terminal device whose terminals are given the names shown. The names *emitter* and *collector* are suggestive; charge carriers are "emitted" at the emitter and "collected" at the collector. But the name of the third terminal, *base,* is not suggestive of its function, which is to *control* the flow of the charge carriers. For the transistor to operate properly, the transistor junctions must be properly biased. The arrow on the emitter is part of the standard symbol for the transistor; it designates the orientation of the *forward current* in the base-emitter junction, as being *from* the base *to* the emitter. To be significant, however, the base should exercise its function of controlling the current flowing between the emitter and collector without requiring a large amount of external current supplied to the base.

By this time you are probably wondering whether the "sandwich" in Figure 32a of a single *p*-type region between two *n*-type regions couldn't be reversed, thereby creating a *pnp* transistor. The answer is affirmative. The *pnp* is an equally valid transistor. The only difference would be that the required bias voltages would be reversed. The difference is also manifested by reversing the direction of the arrow on the emitter, indicating the orientation for the forward current of the *pnp* transistor. Our purposes here will be satisfied adequately if we stick with the *npn* transistor.

An *npn* transistor with biasing batteries is shown in Figure 33. Note that the base-emitter junction is forward-biased with V_{BB} connected as shown (with the positive terminal connected to the *p* region). This is the proper bias for the normal operation of this junction. Now consider the base-collector junction. There is no battery across that junction directly, but, by KVL, the

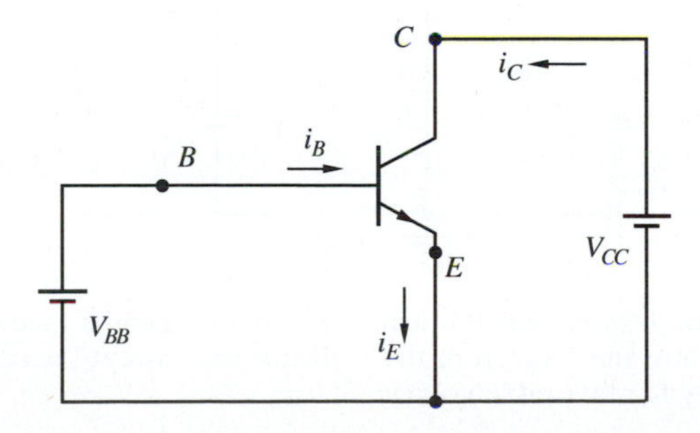

Figure 33
Transistor with biasing batteries.

voltage across that junction is $V_{BC} = V_{BB} - V_{CC}$. The question is, What is the proper biasing polarity for this junction? We would like to arrange it so that the forward current flowing to the emitter across the base-emitter junction is supplied to the base mainly from the collector, rather than from any external circuit connected to the base. For this to happen, the base-collector junction must be reverse-biased. Consequently, we require that V_{CC} be greater than V_{BB}.[17]

Transistor Amplifier

We now turn our attention to an important application: a transistor amplifier. Such a basic circuit is shown in Figure 34. Aside from the biasing batteries, there are two resistors, one connected to the collector and one to the base. There is also a signal source, v_s. The currents in the three terminals of the transistor are not independent but are related by KCL. Similarly, the three voltages across terminal pairs are related by KVL, as follows:

$$\begin{aligned} i_E &= i_B + i_C \\ v_{BC} &= v_{BE} - v_{CE} \end{aligned} \tag{28}$$

Two other equations can be obtained by writing KVL equations around the two closed paths in the circuit and using Ohm's law to eliminate the resistor voltages. When these two are solved for the base and collector currents, the result is:

$$\begin{aligned} i_B &= -\frac{1}{R_1} v_{BE} + \frac{V_{BB}}{R_1} \\ i_C &= -\frac{1}{R_L} v_{CE} + \frac{V_{CC}}{R_L} \end{aligned} \tag{29}$$

Both these are equations of straight lines. We now have four relationships among six variables, all obtained by applying KVL, KCL, and Ohm's law. To obtain a solution for these variables, we need two more relationships; they

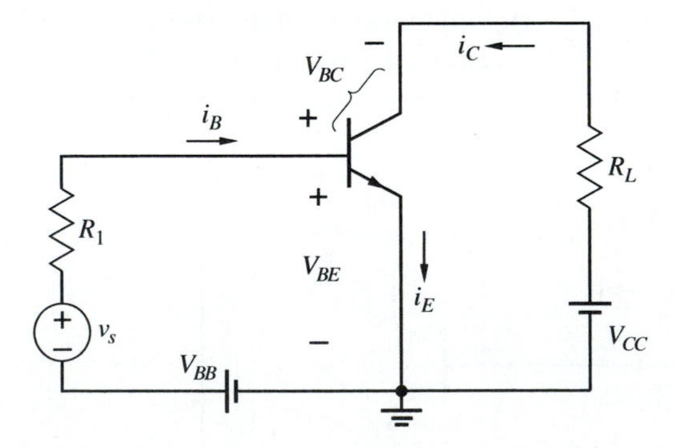

Figure 34
Transistor amplifier circuit.

[17] When the transistor is fabricated, the base region is made very thin. Then charge carriers that drift into the base from the collector are "swept" across the base and the base-emitter junction by the favorable voltage.

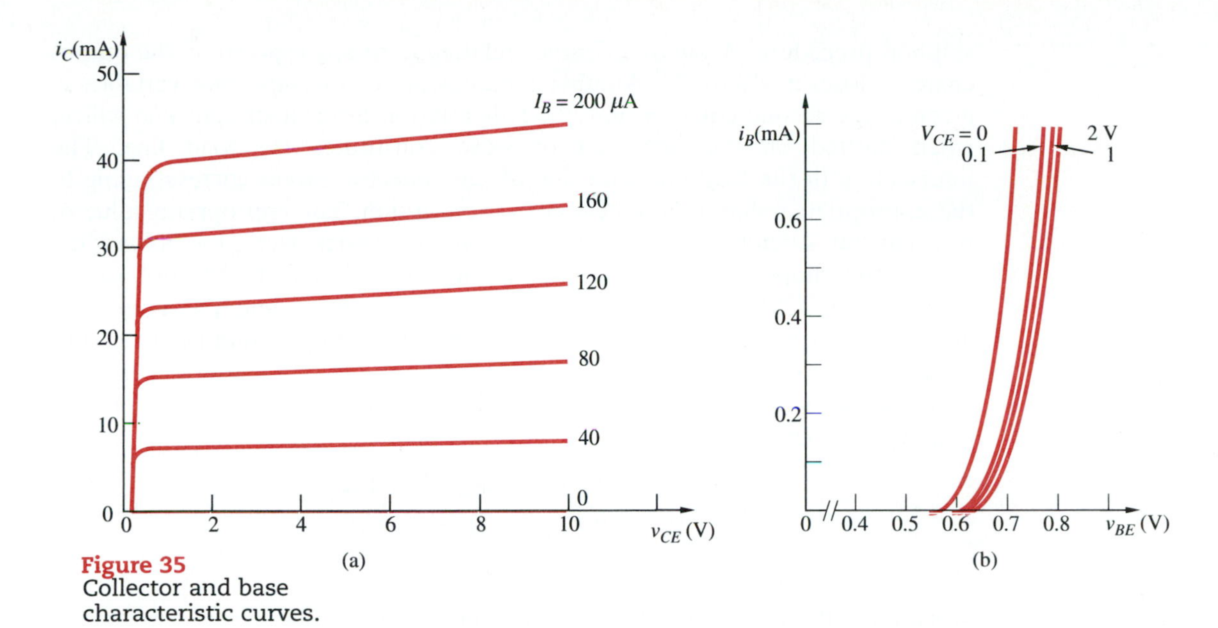

Figure 35
Collector and base characteristic curves.

must come from the properties of the transistor itself, since all other circuit equations have been used.

CHARACTERISTIC CURVES OF A TRANSISTOR The transistor is a three-terminal device with two independent currents and two independent voltages. Relationships among these variables are obtained by making measurements in the amplifier circuit. The results of such measurements for a particular transistor are shown in Figure 35, where there are two sets of curves. The *collector* (or *output*) characteristics are plots of i_C against v_{CE}, with i_B as a parameter. They are obtained by holding the base current constant at a selected value (by adjusting R_1, for example) and making a run of i_C against v_{CE}. This is repeated for other values of base current. The curves are seen to be highly nonlinear. The *base* (or *input*) characteristics are plots of v_{BE} against i_B (or vice versa), holding v_{CE} constant. These curves are also nonlinear, but the curves for almost all values of v_{CE} are bunched. That is, for a given value of i_B, the value of v_{BE} is approximately the same for all values of v_{CE} that are greater than, say, 1 V.

Look at the input characteristic. As the input current ranges from low to high values, the base-emitter voltage moves over a range of about 0.5 V. Whether this has a large or a small influence in the base-emitter loop depends on the value of V_{BB}, as evident from (29). Under usual operating conditions, V_{BB} is 10 times larger than the largest value of v_{BE}. Thus, we might neglect the value of V_{BE} entirely, or take its value to remain constant at 0.2 or 0.3 V. The results obtained would be correct to a high degree of accuracy.

LOAD-LINE ANALYSIS When dealing with a nonlinear diode curve in Chapter 3, we introduced the concept of a load line and described a graphical solution procedure. A family of curves relating i_C to v_{CE} is given by the output characteristics in Figure 35. Another relationship for the same two variables is given in the second equation in (29). This relationship is a straight line which, when plotted on the same set of axes, constitutes the load line. The intersection of the load line with one of the collector curves corresponding to the appropriate value of base current is the solution. The appropriate value of base current is determined from the input characteristics. Here, too, a load line can be superimposed on the curves, this time represented by the first of the equations in (29). However, as discussed in the preceding paragraph, an approximate value of the base current can be obtained by setting v_{BE} in (29) to be a few tenths of a volt.

To illustrate this process, we will choose the following values in the amplifier of Figure 34:

$$V_{CC} = 15\text{ V} \qquad R_L = 0.5\text{ k}\Omega$$
$$V_{BB} = 2.2\text{ V} \qquad R_1 = 20\text{ k}\Omega$$
$$v_{BE} \approx 0.2\text{ V}$$

Assuming there is no signal (the static case), the base current is $i_B = (2.2 - 0.2)/20 = 0.1$ mA. The load line equation is:

$$i_C = 30 - 2v_{CE} \qquad \text{mA} \tag{30}$$

This line is shown plotted on a set of collector characteristics in Figure 36. The intersection of the load line with the curve corresponding to $i_B = 0.1$ mA is called the *quiescent* (or *operating*) *point,* quiescent because it applies to the case of no signal (no excitation). The coordinates of this point give the quiescent values of i_C and v_{CE}.

INCREMENTAL VARIABLES Continue to consider the amplifier circuit in Figure 34. When there is no signal, the voltages and currents are at their static,

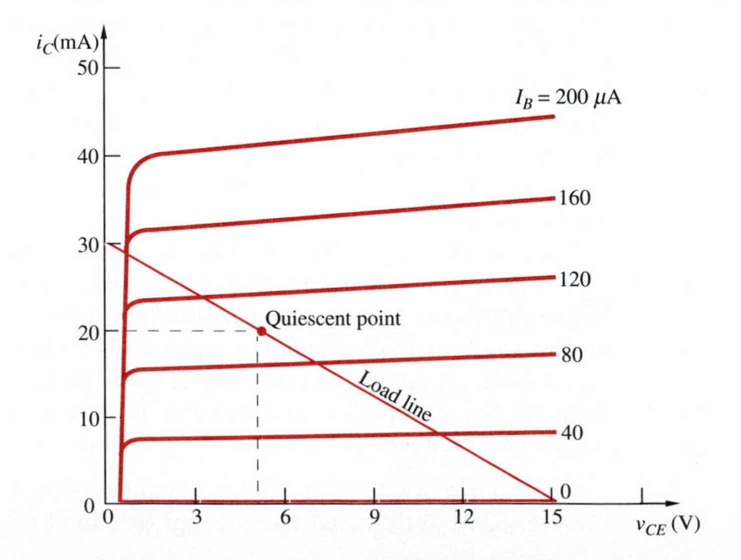

Figure 36
Determination of quiescent (operating) point.

operating values. When the signal source is turned on, each of the variables takes on an instantaneous value which varies with time. For many purposes, it is convenient and useful to deal not with the total instantaneous values but with the *changes* from the operating values, the so-called *incremental values,* or the *signal components* of the variables.

It should be clear from the preceding discussion that a notational quagmire is coming up. A number of categories of variables are under discussion: the total instantaneous variables, the constant operating variables, and the incremental variables. We would be in notational difficulties if we used the same current or voltage symbols to represent all three of these classes. It is customary to adopt the following convention:

1. Constant, static (DC) values are represented by capital letters with capital subscripts: I_C, V_{CE}, I_B.

2. Total instantaneous values are represented by lowercase letters with capital subscripts: i_C, v_{CE}, i_B.

3. Incremental values, *changes* from DC values, are represented by lowercase letters with lowercase subscripts: i_c, v_{ce}, i_b. Incremental values can also be written mathematically. Thus, $i_c = \Delta i_C$, $v_{ce} = \Delta v_{CE}$.

The graphical input and output characteristic relationships of the transistor can be expressed mathematically as:

$$\begin{aligned} v_{BE} &= f_1(i_B, v_{CE}) \\ i_C &= f_2(i_B, v_{CE}) \end{aligned} \tag{31}$$

where the f's are mathematical functions which are expressed graphically by the input and output characteristics. In the notational convention just described, the variables are total instantaneous values. These functions can be expanded in power series.[18] At the same time, the total values on the left are expressed as the sum of the static value plus the incremental value. The result is:

$$\begin{aligned} v_{BE} &= V_{BE} + v_{be} = f_1(I_B, V_{CE}) + r_\pi i_b + \mu v_{ce} + \text{higher-order terms} \\ i_C &= I_C + i_c = f_2(I_B, V_{CE}) + \beta i_b + g_o v_{ce} + \text{higher-order terms} \end{aligned} \tag{32}$$

The arguments of the functions (first terms) on the right are the DC values of the "independent" variables, so these terms are constants. The constant terms on both sides of the equations must be equal, and so must the time-varying terms. When the incremental values are small, the higher-order terms will be negligible compared with the first-order terms. Hence, the higher-order terms can be neglected. In this case, for the incremental terms we

[18] You are no doubt familiar with power series of functions of a single variable: $f(x) = f(x_0) + a_1(x - x_0) + a_2(x - x_0)^2 + \cdots$. Here x_0 is the point around which the function is being expanded, for example, the operating point. The form for two variables is similar to this except that there is more than one term for each power on the right.

get:

$$\begin{aligned} v_{be} &= r_\pi i_b + \mu v_{ce} \\ i_c &= \beta i_b + g_o v_{ce} \end{aligned} \tag{33}$$

The smaller the incremental variables, the better is the approximation obtained by neglecting the higher-order terms. On this basis, the linear equations just obtained are called the *small-signal* equations, and the analysis of the circuit carried out with them is the small-signal analysis, valid only when variations around the operating point are small.

The coefficients in these equations are constants; the notation used is fairly standard. Dimensionally, r_π is resistance and g_o is conductance (that's why they were given those designations), but μ and β are dimensionless. Consider the second equation; it appears to be a statement of KCL. Similarly, the first one seems to be an expression of KVL.

If we had not previously known of controlled sources, attempting to generate a circuit to represent these equations would lead to their creation. Since controlled sources have already been proposed, however, we recognize that the first equation in (33) is represented by a controlling voltage v_{ce} in series with a resistance r_π carrying a current i_b, the voltage of the series connection being v_{be}. Similarly, the second equation is seen to express the incremental collector current as the sum of the current in two branches, one of which is a CCCS whose current is controlled by the incremental base current, and the other is a resistor whose conductance is g_o. The complete circuit representing the small-circuit equations of a transistor is shown in Figure 37a.

Some confusion is possible if this diagram is compared with the diagrams of controlled sources in Table 1. The controlling current of the CCCS, i_b in Figure 37, seems to be the current in a resistor, not in a short circuit, as expected from Table 1. Yes, but the resistor is in series with the short circuit, so their currents are the same. Similarly, the controlling voltage of the VCVS seems to be the voltage across a resistor, not across an open circuit. Again, yes, but . . . (you complete the rest).

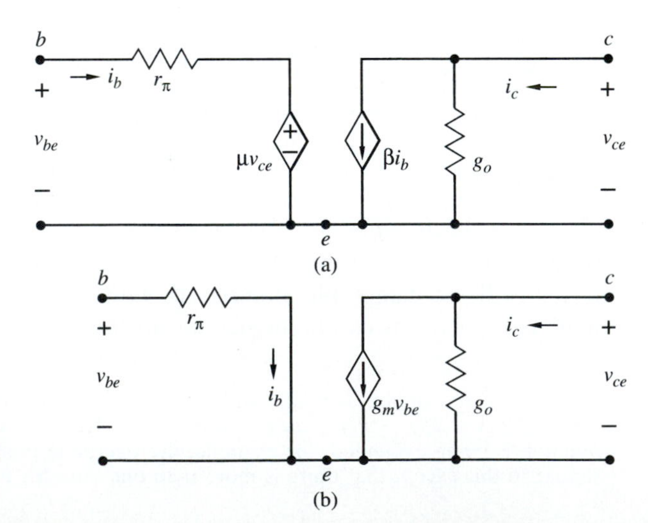

Figure 37
Equivalent circuit created from equations.

There are still several questions we have not considered. First, not all the four parameters in (33) produce equally important effects. The effect of μ, for example, is relatively small. Sometimes it is neglected, thus removing the controlled source from the input side of Figure 37. Second, the choice of independent and dependent variables is not always made as we have made it. Thus, the roles of i_b (as independent) and v_{be} (as dependent) can be interchanged.

If both these things are done, then $i_c = g_m v_{be} + g_o v_{ce}$, where $g_m = \beta / r_\pi$. A model of the device (an equivalent circuit) is shown in Figure 37b; it is similar to the one in Figure 37a, except that the controlled source on the input side is not present and the controlled current source on the output side is voltage-controlled. (Confirm the equations from the model.)

This has been a *very* quick but perhaps more satisfactory and natural approach for controlled sources to make their appearance. You are probably not totally secure in your understanding of all the details of this development, but that's alright. The purpose of this optional section is to show that the appearance of controlled sources among our ideal circuit elements is not arbitrary, but a natural development arising out of the need to account for the operation of something real, the transistor. In your later studies of electronic devices and circuits you will acquire a much deeper understanding of the details.

SUMMARY

This chapter has introduced multiterminal dissipative components. The major topics include:

- The ideal transformer and its v-i equations
- Matching with an ideal transformer
- Controlled sources: CCCS, CCVS, VCCS, VCVS, and their v-i relations
- Current-controlled, voltage-controlled, and hybrid equations
- One-ports and two-ports
- Reciprocal and nonreciprocal two-ports
- Operational amplifiers and their characteristics
- Ideal op amps and the conditions for linear operation
- The inverting-amplifier circuit
- The noninverting amplifier and the voltage follower
- The summing amplifier
- Equivalent circuit of a nonideal op amp
- The negative converter: the current and voltage NC
- The gyrator and its properties: inductor simulation
- Junction transistors
- A transistor amplifier circuit
- Empirical characteristic curves of transistors

PROBLEMS

1. A model of the signal source in Figure P1 is an accompanied voltage source with voltage $v_1(t)$ and internal resistance R_1. The op amp is assumed to be ideal.

   a. How much power is delivered by the signal source?

   b. How much power is delivered to the load resistance?

c. From where is this power supplied?

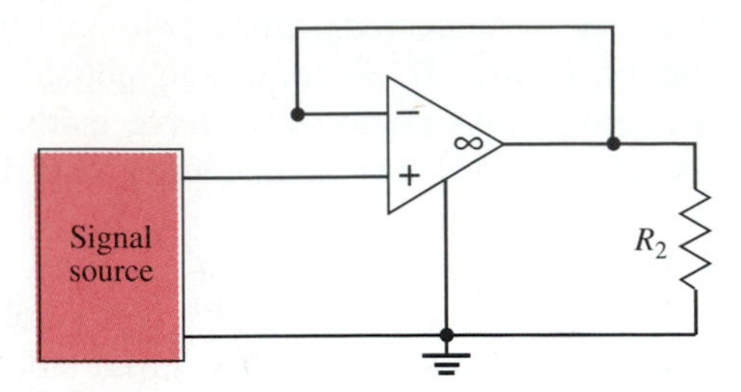

Figure P1

2. a. In the circuit of Figure P2, find the value of v and the power dissipated in the 12-kΩ resistor.
   b. A buffer amplifier (voltage follower) is to be inserted between the points marked A and B; draw the resulting circuit. Determine the new value of v and the new value of power dissipated.
   c. None of the parameter values in Figure P2 can be changed. Suggest how the value of v found in part (b) can be doubled.

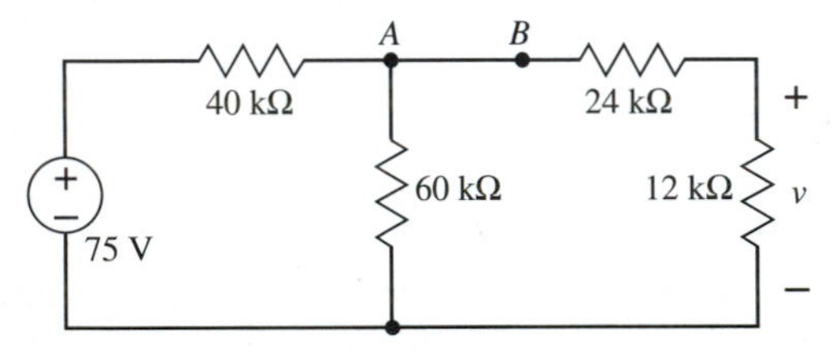

Figure P2

3. From its terminals AB, the circuit in Figure P3 is to constitute a load on another circuit.
   a. Find an expression for the equivalent resistance at terminals AB.
   b. Specify the values of R_2/R_1 for which this is a passive circuit.
   c. How is the result modified if a load resistance R_3 is connected to the output terminals?

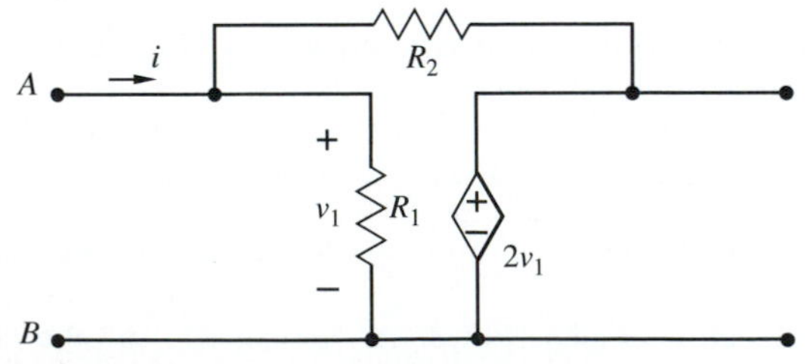

Figure P3

4. Find the equivalent resistance seen from the terminals of the circuit in Figure P4.

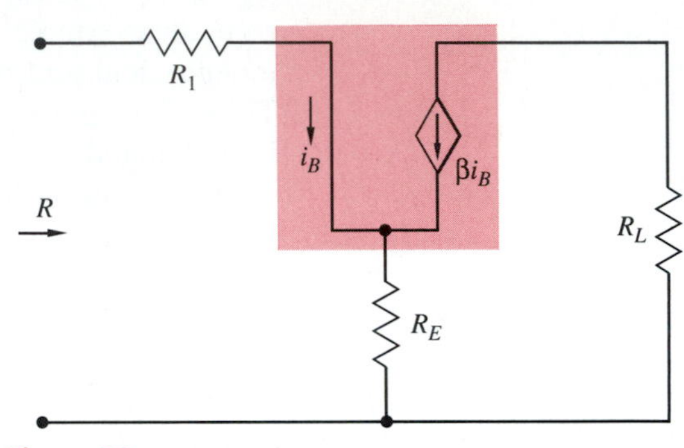

Figure P4

5. Two measurements are made at the terminals of a power supply. With the terminals open, the voltage is measured to be $50 \cos 100t$; with the terminals short-circuited, the current through the short circuit is found to be $2 \cos 100t$.
   a. Find the parameters of a model of the power supply as an accompanied voltage source.
   b. This power supply is to be connected to a load resistance $R_L = 400\ \Omega$ through an ideal transformer, as shown in Figure P5. Determine the transformer turns ratio n so that the load resistance is matched to the internal resistance of the power supply.
   c. Find the power supplied to the load as a function of time.

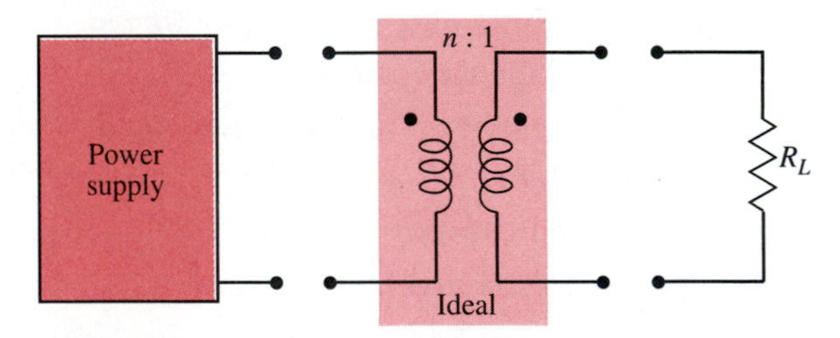

Figure P5

6. The following equations represent a two-port having standard voltage and current references. Construct a model (an equivalent circuit) representing each set of equations.

a. $i_1 = 0.5v_1$; $i_2 = 2v_1 + 3v_2$

b. $v_1 = 100i_1 - 10i_2$; $v_2 = 20i_2$

c. $i_1 = 0.05v_1 - 0.01v_2$; $i_2 = -0.01v_1 + 0.03v_2$

d. $i_1 = 0.2v_1$; $v_2 = 30v_1 + 10i_2$

e. $v_1 = 25i_1 + 2v_2$
$i_2 = -4i_1 + 0.1v_2$

f. $v_1 = 600i_1 + 200i_2$
$v_2 = 200i_1 + 500i_2$

7. a. By applying the properties of ideal op amps and the basic circuit laws, obtain an expression for the output voltage in terms of input voltage in the circuit of Figure P7.

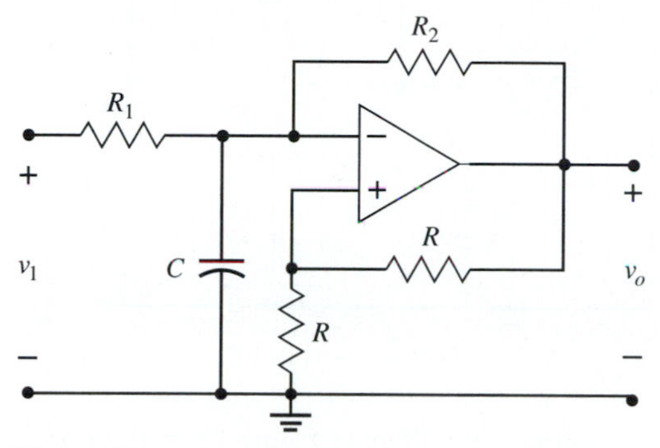

Figure P7

b. Let $R_1 = R_2$. What would be an appropriate name for the circuit?

c. Suppose $R_1 = R_2 = 10\,\text{k}\Omega$, $C = 10\,\text{nF}$, and $v_1 = 10\cos 1000t$. Determine an expression for $v_o(t)$.

8. In the circuit of Figure P7 the capacitor is initially uncharged when a unit step at $t = 0$ is applied at the input. Assume $R_1 = R_2 = 10\,\text{k}\Omega$ and $C = 0.1\,\mu\text{F}$.

a. Find an expression for $v_o(t)$ for $t \geqslant 0$.

b. The saturation voltage of the op amp is 10 V. Will the op amp saturate? If so, how long will it operate linearly after the step is applied?

9. a. In the circuit of Figure P9, by using two sets of current excitations at the ports, one of the currents being zero in each set, determine the resistance parameters r_{ij}. Is the two-port reciprocal or nonreciprocal? (It would help to draw appropriate diagrams.)

b. Reverse the polarity of the controlled source and repeat part (a).

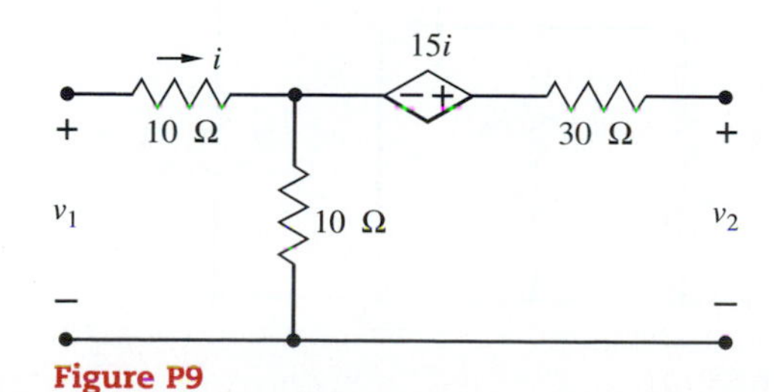

Figure P9

10. a. By using two sets of voltage excitations in the circuit of Figure P10, one voltage of which is zero in each case, determine the conductance parameters c_{ij}. Is the two-port reciprocal or nonreciprocal?

b. Reverse the polarity of the controlled source and repeat part (a).

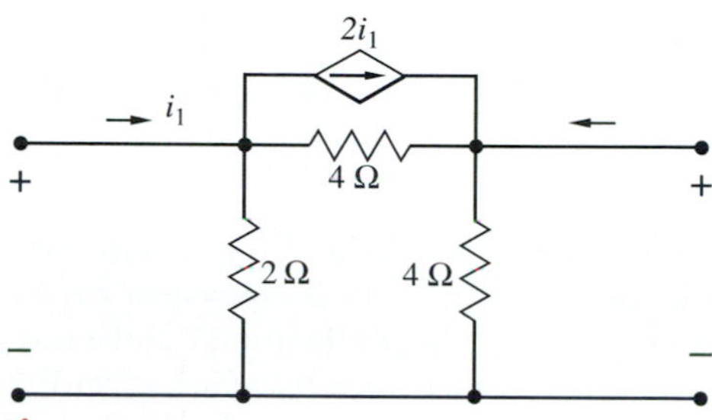

Figure P10

11. The network shown in Figure P11 is to be represented by the following set of equations. Find the values of these hybrid parameters, assuming $R = 0$. (Hint: What do the equations become if $v_2 = 0$, or if $i_1 = 0$?).

$$v_1 = h_{11}i_1 + h_{12}v_2$$
$$i_2 = h_{21}i_1 + h_{22}v_2$$

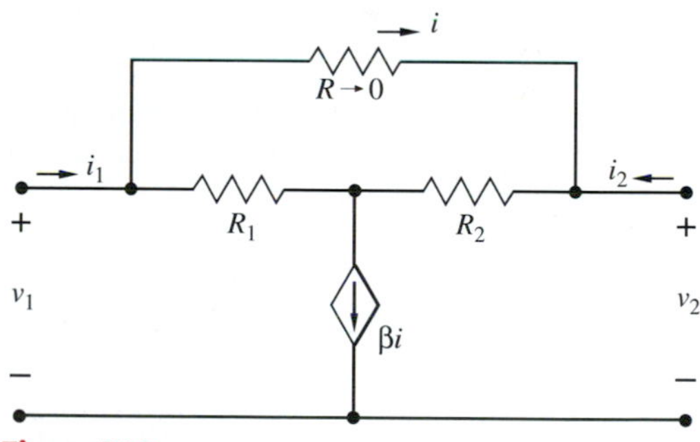

Figure P11

12. Assume the op amp in Figure P12 is ideal and is operating in its linear region.

a. Find an expression for the output voltage.

b. Assume that the supply voltage is $V_{CC} = 20\,\text{V}$. Find the condition satisfied by the source voltages and resistors for linear operation.

c. Let $V_{\text{sat}} = V_{CC}$, $R_1 = 1\,\text{k}\Omega$, $R_2 = 5\,\text{k}\Omega$, and $v_1 = 2\,\text{V}$. Find the range of values of v_2 for which the op amp remains in its linear range (does not saturate).

d. Repeat part (c) for v_1 if $v_2 = 2$ V.

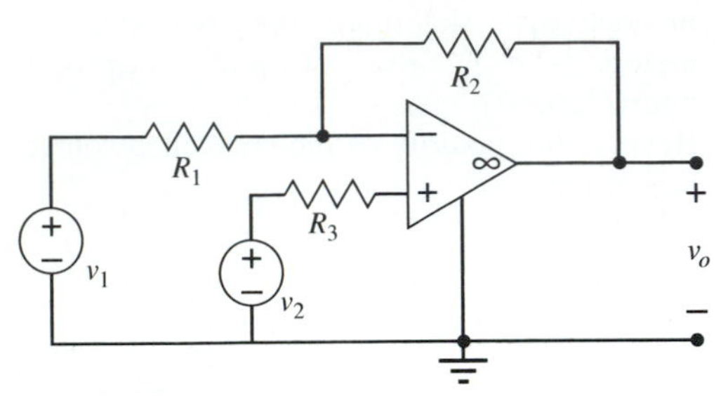

Figure P12

13. The two-port circuit in Figure P13 is a more complete (what is called low-frequency) equivalent circuit of a transistor. (Compare it with Figure 37 in the text.) By applying current sources (one of which is zero) at the ports, determine the resistance parameters r_{ij}.

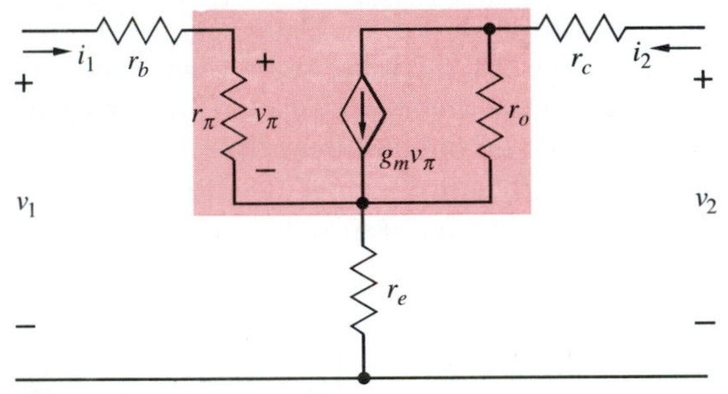

Figure P13

14. In the circuit of Figure P14, find the current i in the 5-Ω load. (Hint: Use a property of the ideal transformer to simplify the circuit.)

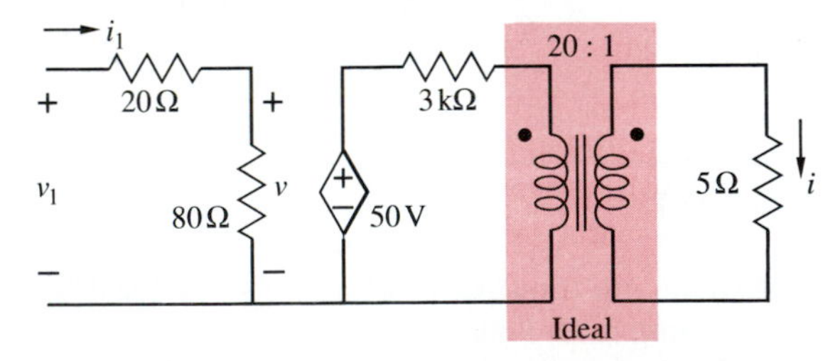

Figure P14

15. The noninverting amplifier circuit in Figure 18 in the text is redrawn in Figure P15. It is claimed that the circuit inside the box acts as an ideal current source. That is, although R_2 (considered as a load) can be varied, its current will remain unchanged over some range of values of R_2.

a. Find an expression for i and verify that it is independent of R_2.

b. Find the range of values of R_2 for which this result holds.

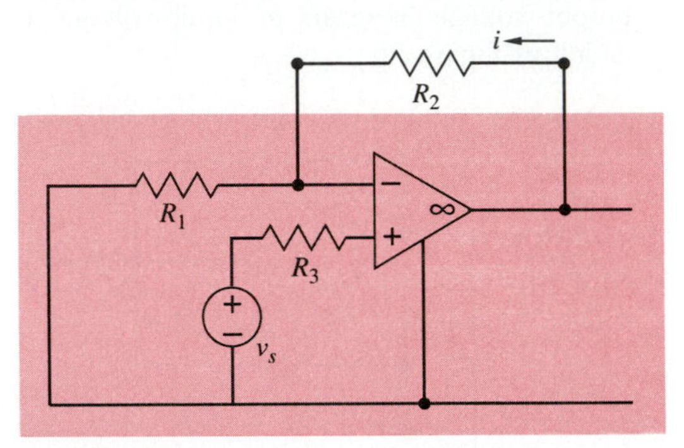

Figure P15

16. The summing amplifier in Figure 21 in the text yields an output which is a weighted sum of the inputs—but the signs are inverted. (After all, it is an inverting amplifier.) It would be useful to have a summer which does not invert signs. One possibility is to make the output of an inverting summer into the input to an inverting amplifier, thus changing the sign. Another suggestion is the circuit shown in Figure P16.

a. Obtain an expression for the output voltage and confirm that it is a positive weighted sum of the inputs.

b. Let the weights in the weighted sum be labeled w_a and w_b; determine an expression for R_a if $w_a = 4w_b$.

c. Repeat part (b) if $w_a = w_b/2$.

d. Would the circuit operate in the same way if more accompanied voltage sources were connected to the noninverting terminal of the op amp?

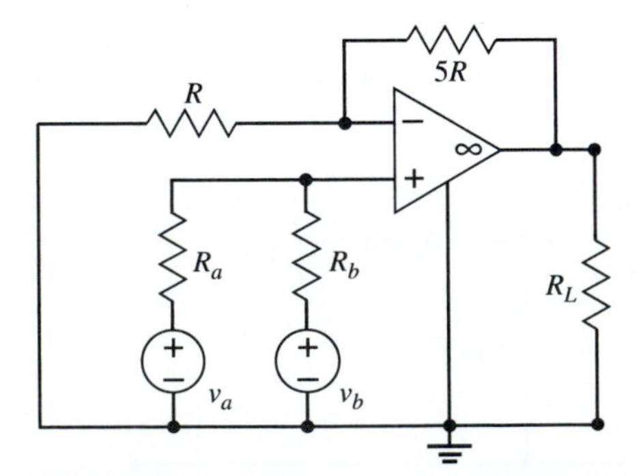

Figure P16

17. The op amp in the inverting amplifier circuit in Figure P17 is not ideal; it has finite input resistance R_i and open-loop gain A, although its output resistance R_o is 0.

a. Determine the equivalent resistance seen by the accompanied source looking into the op amp input terminals.

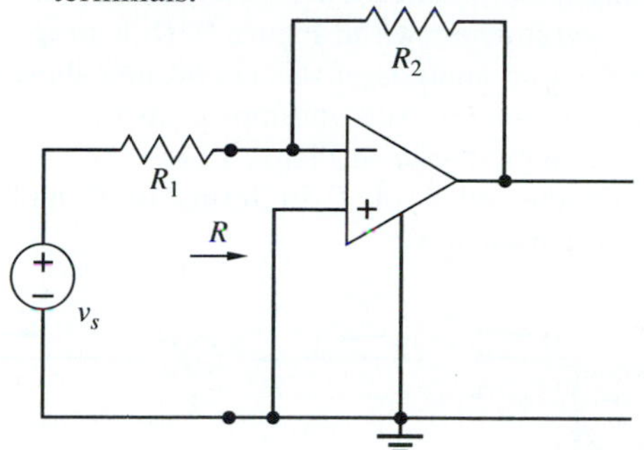

Figure P17

b. Specify the value of R_2/R_1 so that the input resistance found in part (a) is just 1% less than its value for infinite R_i.

c. Make a judgment about how good (or poor) an approximation it is to take R_i to be infinite for normal values of R_2 and R_1 (that is, of closed-loop gain).

18. The op amp in the noninverting amplifier circuit in Figure P18 is nonideal, although R_o is assumed to be zero. The circuit is redrawn in Figure P18b with the op amp flipped over 180°.

a. Determine an expression for the gain v_o/v_s that includes the internal parameters A and R_i.

b. Determine an expression for the resistance R seen by the accompanied voltage source. Show that this resistance is even greater than the already high R_i by a factor equal to the open-loop gain A over the closed-loop gain of the amplifier, $(R_1 + R_2)/R_1$.

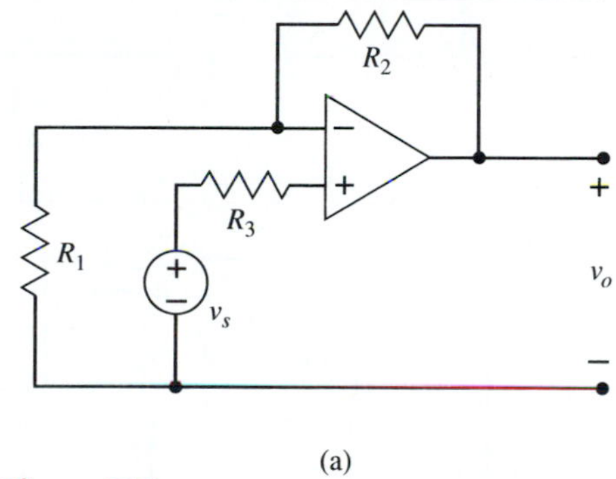

(a)

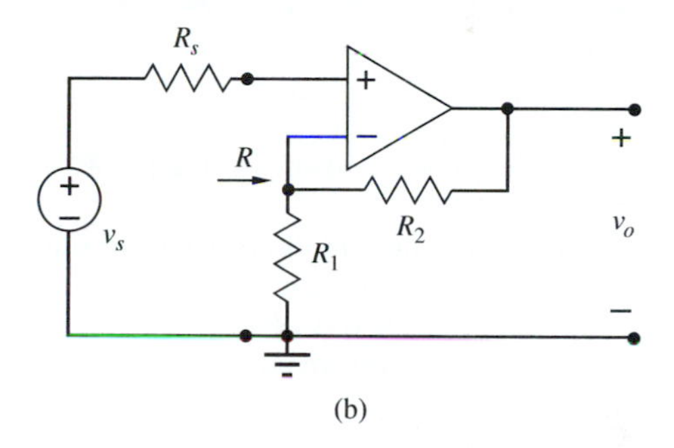

(b)

Figure P18

19. The circuit of Figure P18b, in which the op amp is now assumed to be ideal, is redrawn in Figure P19a without R_s since it has no effect. It is to be considered a two-port with ports AB and CD, as shown in Figure P19b.

a. Derive a pair of equations for the port currents i_1 and i_2 in terms of the port voltages v_1 and v_2.

b. Provide a name for what these equations represent.

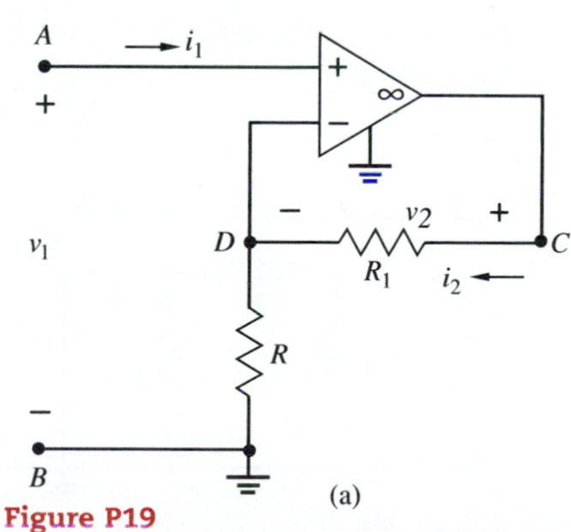

(a)

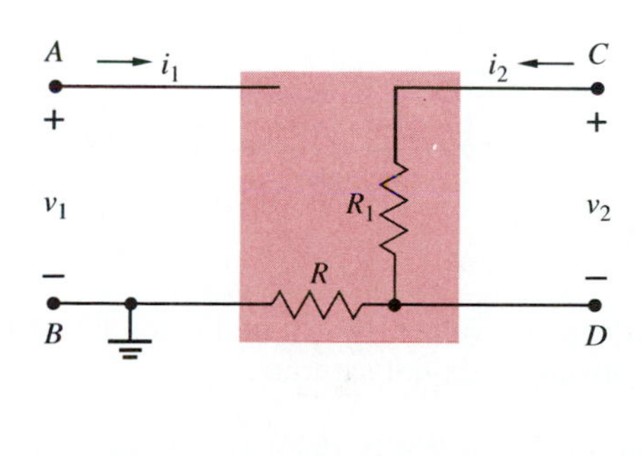

(b)

Figure P19

20. Determine whether the circuit in Figure P20 is active or passive. Does the answer depend on the component values?

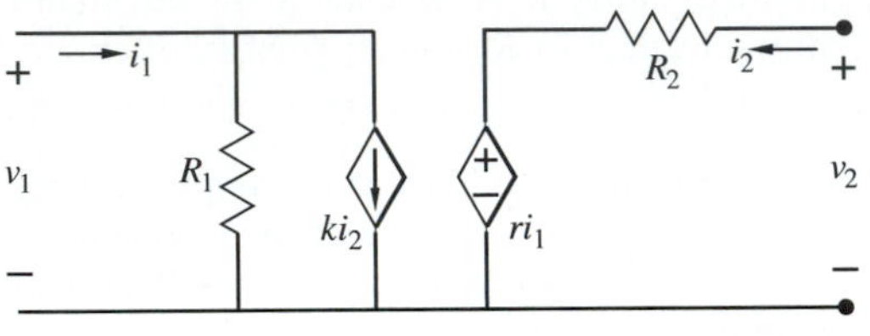

Figure P20

21. In the simulated inductor circuit in Figure 30 in the text, one end of the inductor is at ground voltage; that is, the inductor is a grounded component. In many circuits, however, an inductor is "floating"; that is, as in the filter circuit in Figure P21a, neither terminal of the inductor is at ground. A different simulating circuit is needed in such cases. The one with two gyrators shown in Figure P21b is proposed.

a. Carry out an analysis of this circuit and show that it has the same v-i relationships at its terminals as the floating inductor in Figure P21c.

b. Specify the value of L in terms of C and the gyrator parameters.

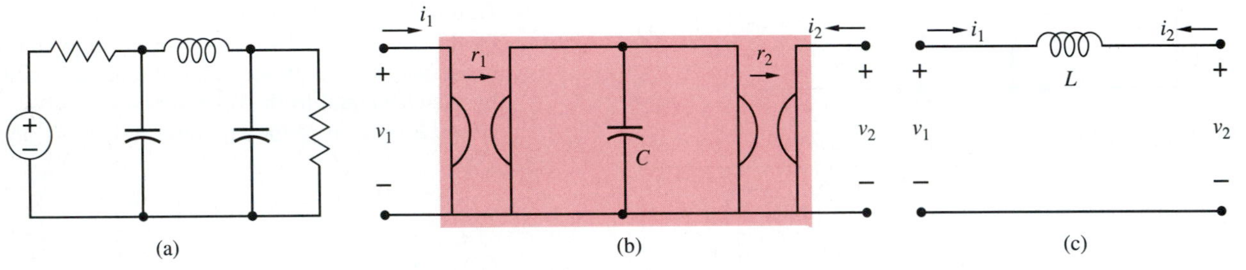

Figure P21

22. In the summing amplifier shown in Figure P22, suppose that $v_1 = v_2 + v_x$.

a. Determine a condition among the resistors so that the output voltage is independent of v_2.

b. Find the amplifier gain for this case.

c. Determine the condition for linear operation.

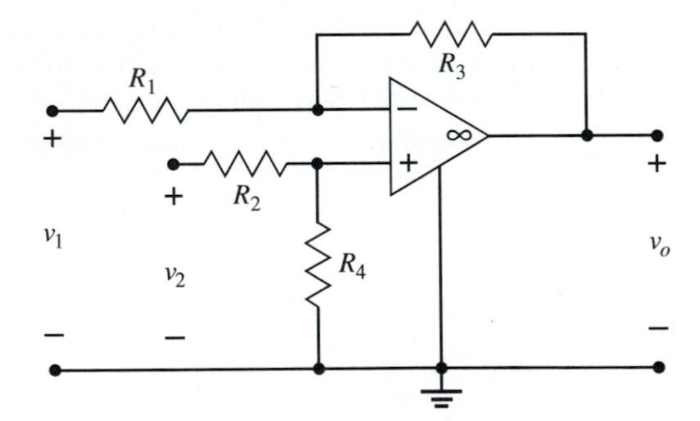

Figure P22

23. Find the range of values of v_{s1} in Figure P23 for which the op amp does not saturate.

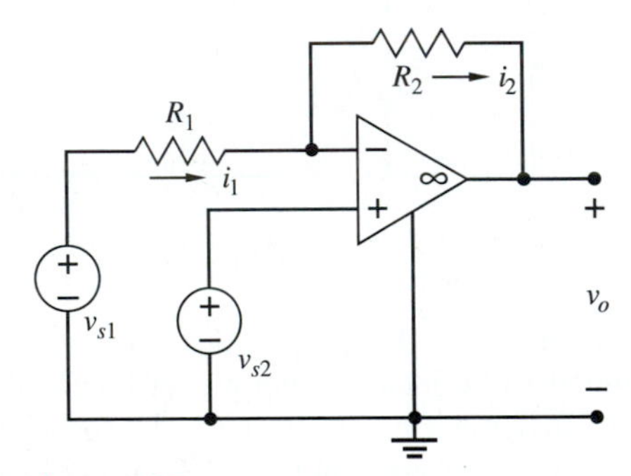

Figure P23

24. The op amp in Figure P24 is ideal. The circuit is that of a noninverting amplifier with the input voltage being provided through a voltage divider. Find the limitation on the ratio R_b/R_a for linear operation of the op amp.

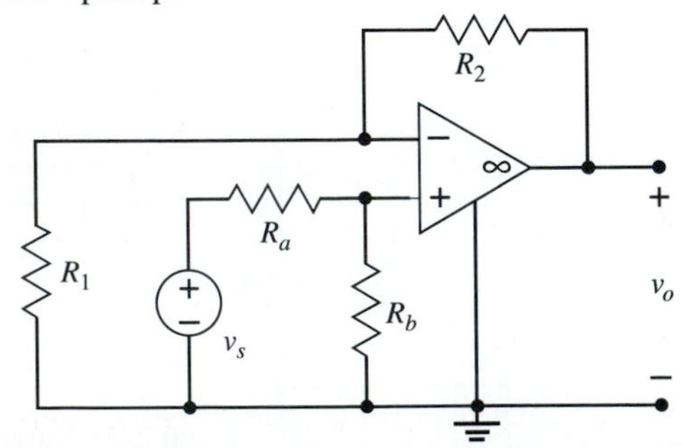

Figure P24

25. Suppose a source is added in Figure P24 in series with R_1 at the inverting terminal, as shown in Figure P25.

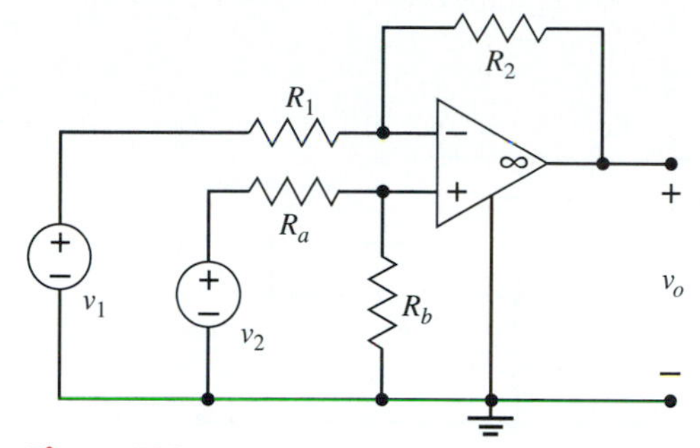

Figure P25

a. Find an expression for the output voltage.

b. Find a relationship among all the resistors such that the output is proportional to the difference of the two input voltages. This circuit might be called a *difference amplifier.*

26. The structure of the difference-amplifier circuit in Figure P25 can be generalized by adding more inputs to both the inverting and noninverting op-amp terminals. A circuit with two inputs at each terminal is shown in Figure P26.

a. By extrapolating from previous results, make an educated guess what an expression for the output voltage might be.

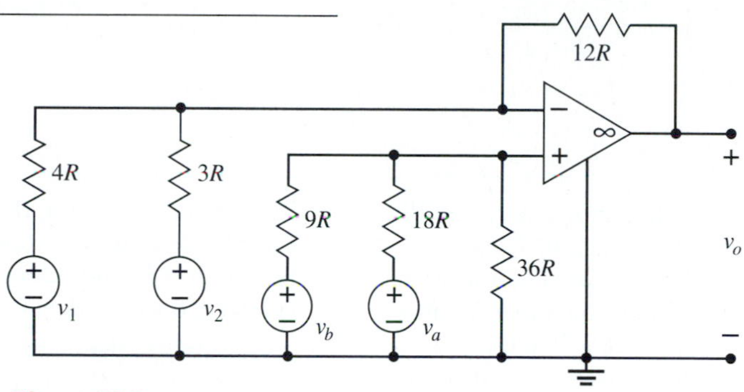

Figure P26

b. Confirm (or reject) your guess by determining an expression for the output voltage.

c. Suppose the input source voltages are all the same: v_s. Determine the condition on v_s that will maintain linear operation of the op amp.

27. The op amp in the inverting amplifier circuit in Figure P27a is not ideal only in that it has a finite open-loop gain A. (Assume it is ideal in all other respects.)

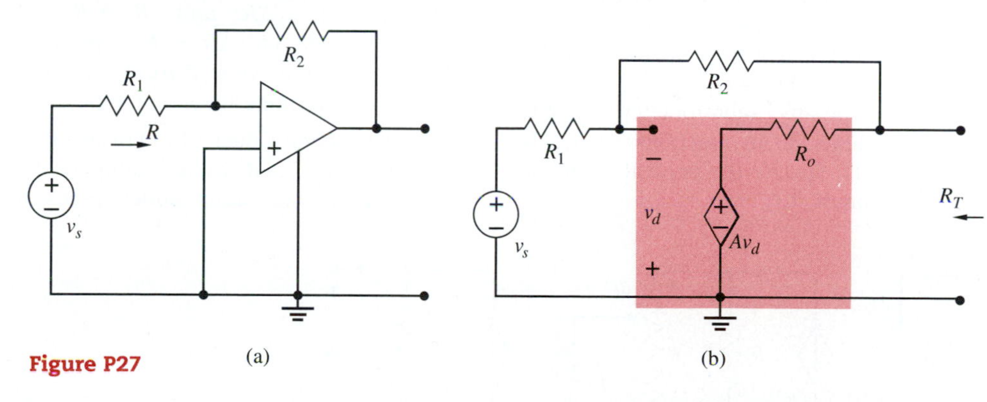

Figure P27

a. Show that the input resistance R looking in at the op-amp terminals is approximately R_2/A. Compare with the case of an ideal op amp. (Draw an appropriate equivalent circuit.)

b. Now suppose that the op-amp input resistance R_i is also not infinite. Again find the input resistance of the circuit at the op-amp terminals. What conclusion can you draw about the need to take the op-amp R_i into account?

c. Finally, suppose that R_o is not zero but R_i is again infinite. (A is still finite.) Determine the Thévenin resistance looking in from the output terminals of the circuit whose model is shown in Figure P27b. How valid is it to neglect R_o in this case?

28. The open-loop gain A and the input resistance R_i of the op amp in the noninverting amplifier of the circuit in Figure P28 are finite. Find the input resistance R in this circuit. Show that it is approximately:

$$R = R_i \frac{A}{(R_1 + R_2)/R_2} = R_i \frac{\text{open-loop gain}}{\text{closed-loop gain}}$$

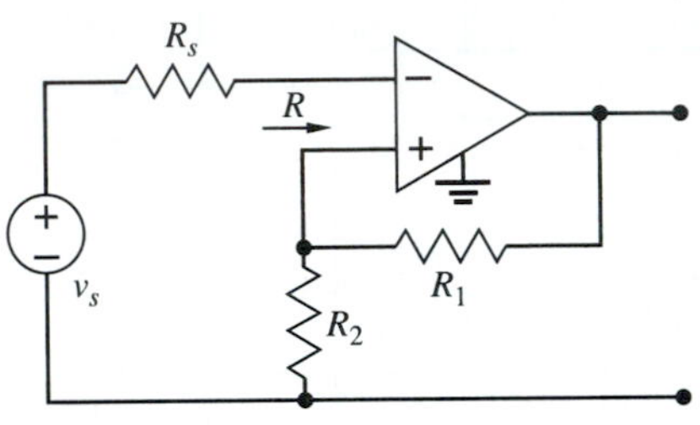

Figure P28

DESIGN PROBLEMS

1. A potentiometer is a device with a movable wiper W in contact with a resistor R. The fractional displacement of the wiper from end 2 of the resistor is x ($0 \leq x \leq 1$); the resistance between the wiper W and end 2 is xR and the resistance between W and end 1 is $(1 - x)R$. A potentiometer is to be used to control the overall gain of a two-stage amplifier, as illustrated in Figure DP1.

a. Let the gain be $G(x) = v_2/v_s$. For the following numerical values, find the expression for $G(x)$ and put it in the form $G(x) = k(x)x$: $R_s = 50\ \Omega$, $R_1 = R_2 = 50\ \text{k}\Omega$, $R_L = 500\ \Omega$, $g_m = 0.1$ S.

b. If k were a constant, then the gain would be proportional to x. But k is a function of x. Determine a value of R so that:

1. At full potentiometer displacement ($x = 1$), the gain is no less than 1500; and
2. The percent error at any potentiometer setting is as low as possible. (The fractional error is $|k(1) - k(x)|/k(1)$. Specify the percent error for your chosen value of R.)

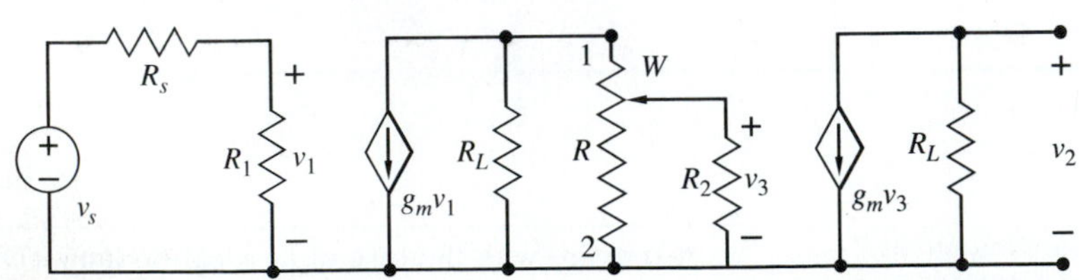

Figure DP1

2. The circuit containing the inverting amplifier shown in Figure DP2 is to represent a current source at terminals AB connected to a resistor R_L. The op amp is assumed to be ideal.

a. Assuming the circuit is operating linearly and taking $R_1 = R_2$ and $R_3 = R_4$, show that the current delivered to R_L is independent of R_L. (This is the definition of a current source.)

b. For the same resistance values as in part (a), find the limits on R_L for which this circuit will act as an ideal current source.

c. Now suppose the conditions among the resistors in part (a) hold only approximately; that is, $R_2 = R_1 + \Delta R$ and $R_4 = R_3 + \Delta R$, in which $\Delta R \ll R_1$ and $\Delta R \ll R_3$. Specify what the circuit becomes, as viewed from the terminals AB of R_L, and determine its parameters.

d. The actual values of physical resistors deviate from their nominal values. Suppose that the resistors used in the circuit under discussion have a tolerance of 5%. The circuit is to operate as a current source with a nominal value of 1 mA at terminals AB, and not deviate from this by more than 5%. Determine the required nominal values of all other resistors as R_L varies from 0 to 5 kΩ.

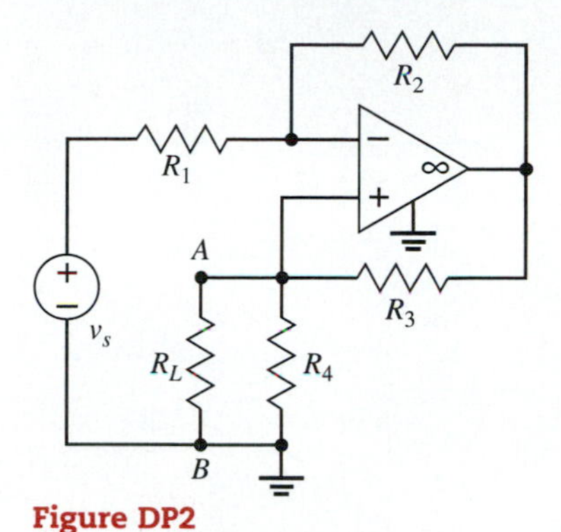

Figure DP2

3. The small-signal equivalent circuit of a transistor amplifier with feedback is shown in Figure DP3. (You don't need to know anything about transistors to solve this problem.)
   a. The following parameter values are given: $R_s = 8$, $R_1 = 1$, $R_f = 30$, and $R_2 = 5$, all in kΩ, and $\beta = 50$. Determine the voltage gain $G = v_2/v_s$.
   b. Repeat part (a) for $\beta = 40$ and for $\beta = 60$. Describe how the gain varies with β.
   c. With the values of R_s, R_1, R_2, and β given in part (a), determine the feedback resistance R_f so that the voltage gain will equal 5.

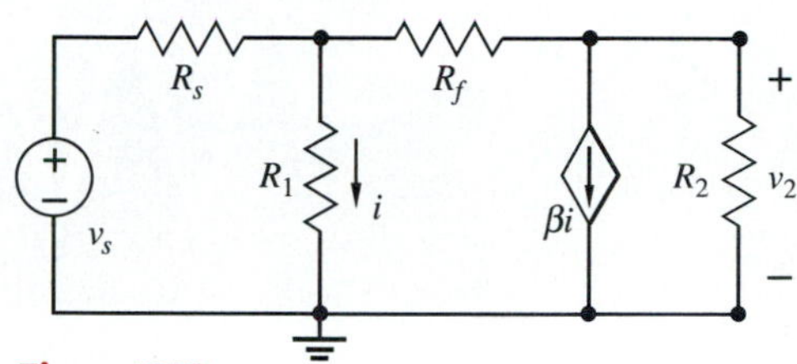

Figure DP3

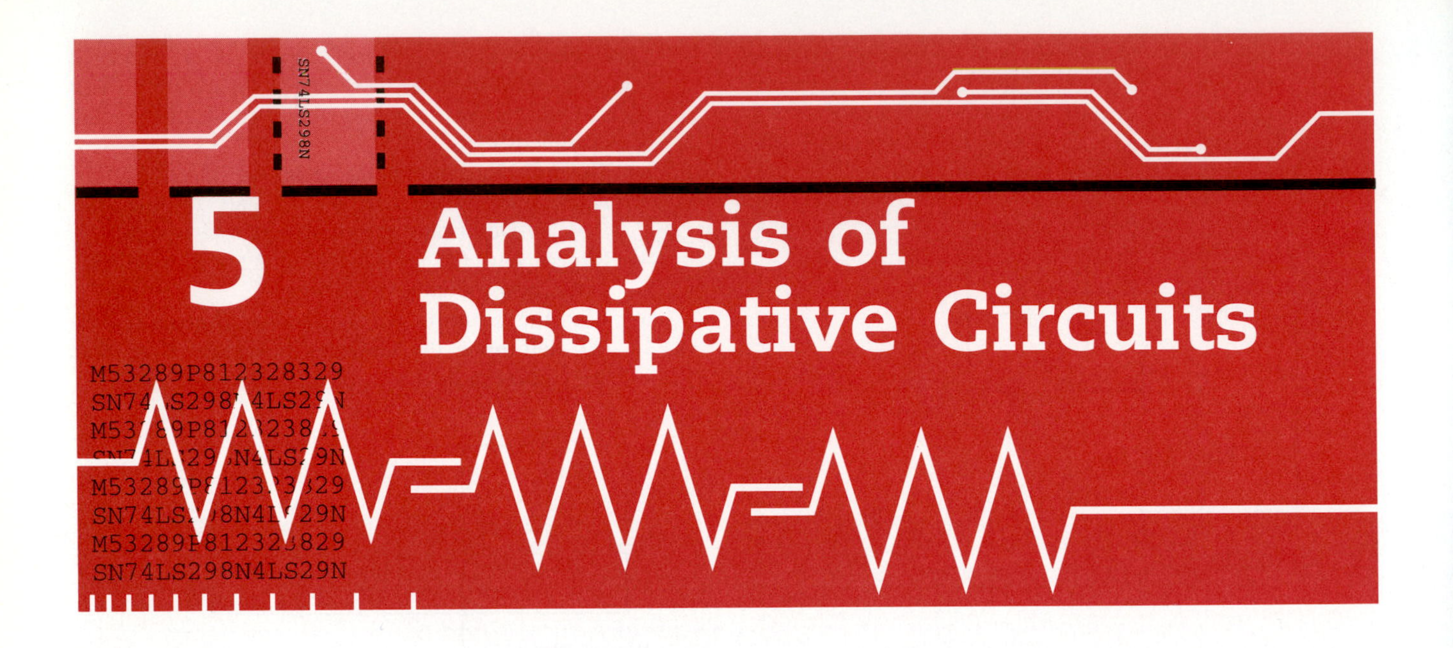

5 Analysis of Dissipative Circuits

The stage has now been set for a major attack on the analysis of circuits. Preceding chapters have introduced two-terminal and multiterminal components as models of real electrical devices and specified their voltage-current relationships. Also presented were the laws governing the equilibrium of voltages and currents in an electrical circuit—the two laws of Kirchhoff. Finally, in some relatively simple circuit structures, such as a ladder network, we carried out procedures which replaced certain subcircuits (series or parallel resistors or accompanied sources) by simpler equivalents to arrive at a simple series or parallel circuit for which a single equation with a single unknown could be written. With this unknown determined, all other unknowns can be found in a sequence of steps applied to the original circuit.

This chapter is devoted to the development of methods of analysis applicable to linear networks of arbitrary structure and complexity containing all varieties of dissipative components. At times, you might become overwhelmed by what looks like a mass of details that must be remembered. As you proceed, though, keep in mind that there are several aspects to the problem being addressed.

First, there is stage setting. This includes the development of different approaches, various points of view, alternative plans of attack, special cases to be considered, even making conjectures which may or may not lead to fruitful results. Second, there are the details of specific procedures or algorithms: "Do this, then do that, then do the other thing." These analytic procedures require a great deal of attention to detail. Finally, there are conditions: in what cases, under what conditions, for what classes of circuits can this or that procedure be carried out?

In Chapter 3, certain procedures were developed for a limited class of circuits, limited both in terms of structure (series-parallel) and in terms of contents (resistors and independent sources). What perhaps might be considered most desirable are analytic procedures that are universal, that is, procedures applicable to any and all classes of circuits. This would require the least amount of judgment and decision making. Given a circuit, just follow the universal algorithm and turn the crank; out will come the answer. (The use of computer software such as SPICE has some of this flavor.) Fortunately, that is not the case. There exist several general approaches, each widely applicable, though not universal without modifications. Thus, judgment, thought, and decision making among alternatives are needed.

In order to be in a position to use judgment, and to decide among alternatives, you have to become familiar with the details of each of the general approaches, the pros and cons of using one method over another based on (1) circuit structure and (2) circuit contents.

The plan of attack in this chapter is to apply the basic laws (Kirchhoff's two laws and the component v-i relations) to the nodes and closed paths of a circuit to arrive at a set of *simultaneous equations* in a number of unknowns. Solving these equations simultaneously produces solutions for all the unknowns at once. A number of different sets of equations are possible by employing different sets of variables. The process of arriving at these equations is simplified if, at the outset, we agree to choose standard references for the component voltages and currents. Therefore this is the convention we will use.

1 SYSTEMATIC FORMULATION OF CIRCUIT EQUATIONS

Let us initially limit ourselves to circuits containing only resistors, in addition to sources. (The restriction will be lifted later in the chapter.) Each resistor has a voltage and a current which are both initially unknown. Each source connected in the circuit also has a voltage and a current, but one of these is initially known: the voltage is known for a voltage source and the current is known for a current source. That's a complication because it prevents obtaining a simple count of the number of unknowns on the basis of a count of the number of branches; we would have to say there are so many unknowns for the resistors and so many for the sources.

Treatment of Sources

This complication can be overcome if we assume that sources are *always* accompanied by resistors—a voltage source is always in series with a resistor and a current source in parallel with one, as shown in Figure 1. The v and i shown in the figure pertain to the entire branch, including the source and the resistor. We previously defined a source to be *accompanied* when, if it is a voltage source, it is in series with a resistor and, if it is a current source, it is in parallel with a resistor.

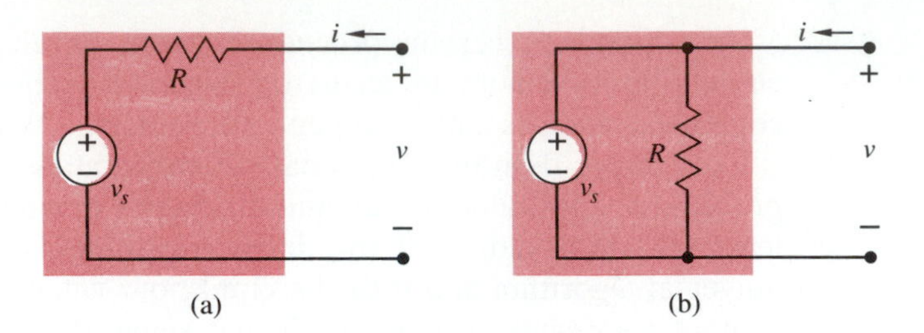

Figure 1
Accompanied sources.

This isn't as arbitrary as it looks. Remember that sources are ideal devices, models of real devices. A *real* source, such as a battery, always has some "internal" resistance, which shows up in series with the voltage-source representation of the real source and in parallel with the current-source representation. That said, it is also true that ideal models of devices can be pushed around on paper and so a circuit of model components can end up with unaccompanied sources. We shall subsequently discuss the possibility of converting these to accompanied sources. For the present, with the goal of getting a count on the number of variables, let's assume that all sources come accompanied.[1]

The v-i equations of the composite branches in Figure 1 can be written as follows:

$$v = Ri + v_s \quad \text{(1a)}$$

$$i = Gv + i_s \quad \text{(1b)}$$

where $G = 1/R$ for the circuit of Figure 1b. These are dual equations pertaining to the dual circuits in Figure 1. Since the source variables, voltage and current for the two cases, are known, each of these branches has two initially unknown variables, just the same number of unknowns as a lone resistor. Indeed, the preceding equations could apply equally well to solitary resistors; in this case the source values would be zero.

That gives us the answer: assuming all sources are accompanied and hence not counted as separate branches, the number of branch unknowns in a circuit is just twice the number of branches. Therefore, to solve for that many unknowns, there must be that many *independent* equations.

Each Ohm's law equation is independent of all others, since it includes a

[1] There is a "downside" to everything! Assuming all sources to be accompanied is useful for the purpose at hand. However, in the circuit-analysis software called PSPICE, counting branches and unknowns is unimportant. (PSPICE is the MicroSim Corporation version of SPICE, acronym for *Simulation Program with Integrated-Circuit Emphasis,* originally developed for computer-aided design at the University of California at Berkeley.) A circuit to be analyzed by PSPICE is specified by identifying each separate component by the nodes between which it lies, and giving the value of the branch parameter. To use PSPICE, therefore, sources are normally identified as separate components. However, PSPICE also allows the introduction of "models" having any number of terminals. Such models (e.g., for transistors, op amps, or diodes) are built into the program; other models can be specified by a user. Accompanied sources, thus, could be introduced into PSPICE as models. You are unlikely to experience any difficulty in using PSPICE on account of these different possible ways of treating sources.

resistance not included in any other Ohm's law equation. That leaves a need for as many more independent equations as there are branches, and these must come from the Kirchhoff equations.

Branch Variables

The first step in the analysis of circuits is to select appropriate variables. We have already discussed the branch *v-i* (*i-v*) equations. We shall initially proceed with the branch voltages and currents as the variables to be used in the analysis. Subsequently, we'll find more convenient ones.

We will introduce the general procedure to be developed in this section by means of a relatively simple example. Once you have acquired some facility with the approach, we will consider more complex circuits.

Let's start with the circuit shown in Figure 2. The objective is to find voltage v_2 and the value of the power delivered by the current source. If that were all we wanted to do, we could use the procedures developed in Chapter 3. We could replace the current source in parallel with its accompanying resistor by its voltage-source equivalent. The result would have the structure of a voltage divider, as shown in Figure 2b. The output voltage is then easily obtained by inspection to be 0.4 sin 400t V. (Carry out the details and confirm this.)

The power delivered by the current source cannot be found in this equivalent circuit, however, because the current source does not even exist here. We must return to the original circuit and, using our newly found knowledge of v_2, proceed working from the right side to the left side of the circuit, finding all the branch currents and voltages one at a time, including the voltage across the current source. The power can then be obtained. (Carry out the details to confirm that the power is 16.8 mW.)

But our interest in this circuit is to use it as a vehicle for introducing more powerful methods of solution which can be used for more extensive circuits. We observe that there are three branches in this circuit, including the branch consisting of the accompanied source; thus there are six unknowns: three voltages and three currents. Three equations relating these six unknowns come from Ohm's law. There is just one loop, and so there is only one KVL equation relating the voltages. There are three nodes, but we will write KCL equations at only two of them. (The reason for this will be discussed subsequently.)

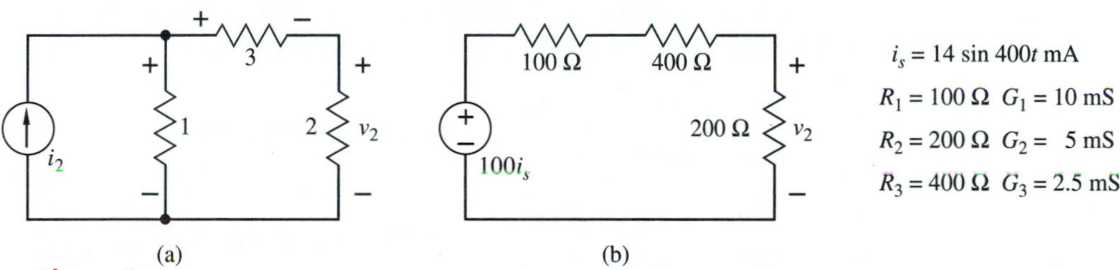

Figure 2
Find output voltage and power supplied by source.

These six equations are as follows:

$$
\begin{array}{lll}
\text{Ohm's law:} & v_1 = R_1 i_1 & i_1 = G_1 v_1 \\
& v_2 = R_2 i_2 & i_2 = G_2 v_2 \\
& v_3 = R_3 i_3 & i_3 = G_3 v_3 \\
\text{KVL:} & \multicolumn{2}{c}{v_3 + v_2 - v_1 = 0} \\
\text{KCL:} & \multicolumn{2}{c}{i_1 + i_3 = i_s} \\
& \multicolumn{2}{c}{i_2 - i_3 = 0}
\end{array}
\tag{2}
$$

These six equations in six unknowns will need to be solved simultaneously. All three voltages appear in the KVL equation and can all be eliminated at once by using the Ohm's law equations in the current-controlled form $v = Ri$. This will leave an equation involving all three of the currents. Two of the currents can be eliminated by using the two KCL equations, leaving an equation with a single unknown. These steps can be carried out as follows:

$$
\begin{aligned}
400i_3 + 200i_2 - 100i_1 &= 0 \\
400i_3 + 200i_3 - 100(i_s - i_3) &= 0 \\
700i_3 &= 100i_s \\
i_3 &= 2 \sin 400t \text{ mA} \\
v_2 &= 200i_3 = 0.4 \sin 400t \text{ V}
\end{aligned}
$$

(Fortunately, this answer for the output voltage agrees with the one previously obtained! Did you doubt it?)

To review: We wrote three sets of equations: Ohm's law, KVL, and KCL. Into the KVL equation (there was only one in this example) we substituted Ohm's law, thus eliminating all the voltages and leaving only the currents as variables. Into this equation we substituted KCL, eliminating two of the currents and leaving a single equation in a single unknown. (In more complex circuits, there will be more than one KVL equation, so in the final step more than one equation will be left to solve simultaneously, in an equal number of variables.)

But—you might ask—what prompted you to start with KVL and substitute the others into it? Why not start with another set? It's good that you should ask. Let's , instead, start with the KCL equations, substitute Ohm's law (in the $i = Gv$ form), and then substitute KVL. The result of these steps is as follows:

$$
\begin{aligned}
10v_1 + 2.5v_3 &= 1000i_s \\
5v_2 - 2.5v_3 &= 0
\end{aligned}
\qquad \text{(Ohm's law into KCL)}
$$

$$
\begin{aligned}
10v_1 + 2.5(v_1 - v_2) &= 1000i_s \\
5v_2 - 2.5(v_1 - v_2) &= 0
\end{aligned}
\qquad \text{(KVL into result)}
$$

$$
\begin{aligned}
12.5v_1 - 2.5v_2 &= 1000i_s \\
-2.5v_1 + 7.5v_2 &= 0
\end{aligned}
\qquad \text{(collect terms)}
$$

(After numerical values were inserted for the conductances in the first pair of equations above, each of the equations was multiplied by 1000 to avoid decimals.) The final result after collecting terms is two equations in two unknowns. The second of these two equations gives $v_1 = 3v_2$. When this is inserted into the first equation of the two, the result for v_2 is found to be $0.4 \sin 400t$, in agreement with the previous solution.

For this particular circuit, the first approach we used (KVL, then Ohm's law, then KCL) seemed to be simpler, since it resulted in fewer equations (only one) to solve at the end. The second approach (KCL, then Ohm's law, then KVL) resulted in more equations. Nevertheless, in what follows, we will concentrate on the second approach.[2]

MORE ON BRANCH VARIABLES The simple example just treated was inadequate to clarify the general principles of the formulation of circuit equations we are in the process of developing. We will now consider a somewhat more complex example to get some clues to what the general case might be. The circuit is shown in Figure 3. The numbers on the branches are not resistance values but branch numbers. Branch 1 is the current source with its accompanying resistor, for example.

The nodes have also been identified numerically. Confusion is possible among the different sets of numerical designations, at least for a while. Perhaps a different set of identifying symbols (possibly alphabetical, possibly color-coded) should be used for the nodes. We will stick with the numerical designation, at the possible expense of some early confusion. A voltage reference is shown for each branch but without showing a voltage label. Current labels and references are not shown explicitly either, but we assume that the references of all branches are standard.

The variables we will use are *branch voltages* and *branch currents.* To be proper and unambiguous, whenever there is a chance for confusion, we ought to use an appropriate subscript (say, b) to indicate that the variables are branch variables. We will do so here temporarily, even though it is cumbersome; later on, we'll earn the right to become a little sloppy, when there is less chance for confusion. As a first step let's write the equations resulting from

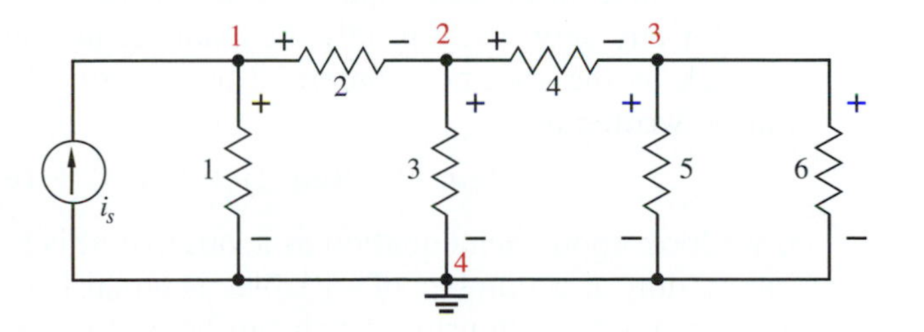

Figure 3
Example circuit to be analyzed.

[2] A significant reason is that PSPICE is based on this approach.

Ohm's law; using the form $i = Gv$, they are:

$$\begin{aligned} i_{b1} &= G_1 v_{b1} - i_s \\ i_{b2} &= G_2 v_{b2} \\ i_{b3} &= G_3 v_{b3} \\ i_{b4} &= G_4 v_{b4} \\ i_{b5} &= G_5 v_{b5} \\ i_{b6} &= G_6 v_{b6} \end{aligned} \tag{3}$$

There are 6 equations in 12 unknowns. They all look alike, of course, except for the one for the composite branch containing the source. (You should confirm the sign in front of the current source by drawing a separate diagram for this branch and showing reference arrows for each of the currents involved.)

The next step is to apply Kirchhoff's laws. There are four nodes; let's write KCL at each of them, using "away from" as the node reference:

$$\begin{aligned} i_{b1} + i_{b2} \qquad\qquad\qquad\qquad &= 0 \\ i_{b2} + i_{b3} + i_{b4} \qquad\qquad &= 0 \\ -i_{b4} + i_{b5} + i_{b6} &= 0 \\ -i_{b1} \qquad - i_{b3} \qquad - i_{b5} - i_{b6} &= 0 \end{aligned} \tag{4}$$

Linear Independence of Equations

Suppose we now add all of these equations. The right-hand sides are all zero anyway, so their sum is also zero. But the interesting thing is that the sum of the left sides also gives zero. This is easiest to see if you look upon these equations as an array; each equation constitutes a row of the array. But now concentrate on the columns, each one of which corresponds to one current. That current appears exactly twice in each column, once with a plus sign and once with a minus sign. It is not surprising then that each column adds up to zero.

Indeed, the same reasoning applies in a network of *any* complexity. Each branch is connected to exactly two nodes and the current reference is necessarily toward one node and away from the other. Hence, each branch current *must* appear in exactly two KCL equations written at all the nodes, necessarily once with a plus sign and once with a minus sign. Thus, that the columns in a set of KCL equations that includes all the nodes add up to zero is true for any network, not only the example in Figure 3.

Back to the specific example. The relationship among the KCL equations can be written as:

$$(\text{eq. } 1) + (\text{eq. } 2) + (\text{eq. } 3) + (\text{eq. } 4) = 0$$

If we look upon each equation as a variable, this looks like a statement that the summation of a number of variables is equal to zero. This is an example of a more general relationship which can be written as:

$$\sum_{j=1}^{n} k_j x_j = 0 \tag{5}$$

where each k_j is a constant. If such a linear relationship exists among the n x_j variables, then we say that these quantities are *linearly dependent,* not *linearly independent.* The existence of such a relationship means that the equation can be solved for any one of the variables as a linear combination of all the others.

In the case of the KCL equations (each of which is one of the variables in (5)), they are clearly linearly dependent. Thus any one of them can be obtained as a linear combination (in this case, the specific "linear combination" is the "negative sum") of the others. The upshot of all this is that we don't need to write KCL equations at all the nodes of a circuit; we can omit any one. But after we omit one equation, how can we be sure that the remaining equations are not linearly dependent? The answer is that we can't be sure and we'll have to prove it eventually. At this point let's continue with the example and come back to this question later in this chapter.

So far, equations have been written applying Ohm's law and KCL. The final step is to write the KVL equations for this circuit (the circuit in Figure 3). Here we meet a bigger problem. While it is easy to count the number of nodes, *identifying* all the loops (closed paths) is not easy for extensive circuits. (Look back at Figure 12 in Chapter 1; it has the same number of branches as our circuit here, and we found it to have seven loops.)

In the present case, there are six loops; they consist of the following sets of branches: $\{1,2,4,5\}$, $\{1,2,4,6\}$, $\{1,2,3\}$, $\{3,4,5\}$, $\{3,4,6\}$, and $\{5,6\}$. Using a clockwise traversal, the KVL equations written for these loops are:

$$\begin{aligned}
-v_{b1} + v_{b2} \qquad\quad + v_{b4} + v_{b5} \qquad\quad &= 0 \\
-v_{b1} + v_{b2} \qquad\quad + v_{b4} \qquad\quad + v_{b6} &= 0 \\
-v_{b1} + v_{b2} + v_{b3} \qquad\qquad\qquad\qquad &= 0 \\
-v_{b3} + v_{b4} + v_{b5} \qquad\quad &= 0 \\
-v_{b3} + v_{b4} \qquad\quad + v_{b6} &= 0 \\
-v_{b5} + v_{b6} &= 0
\end{aligned} \tag{6}$$

That's a lot of equations, but how many of them are independent? A little bit of manipulation, adding, and subtracting show the following relationships to hold:

$$\begin{aligned}
(\text{eq. } 3) + (\text{eq. } 4) &= (\text{eq. } 1) \\
(\text{eq. } 3) + (\text{eq. } 5) &= (\text{eq. } 2) \\
(\text{eq. } 5) - (\text{eq. } 4) &= (\text{eq. } 6)
\end{aligned} \tag{7}$$

That is, the preceding set of KVL equations is linearly dependent; the first two and the last one can be obtained from the other three, by (7). The question again is, Are those the only linear combinations among the equations? Can fewer than three equations be independent? Again, we will defer answering the question in a general way until later in this chapter.

For the present, let's ask the question, How many independent equations do we *need*? Let's use the following general notation for a circuit:

$$b = \text{number of branches}$$

$$n = \text{number of nodes}$$

In the circuit of Figure 3, $b = 6$ and $n = 4$. We found that there are no more than three independent KCL equations, one less than the number of nodes, or $n - 1$. And we promised to show later that there are no fewer than $n - 1$ either, so there must be exactly $n - 1$. To solve for 12 ($2b$) unknowns, we need 12 independent equations: $b = 6$ of them come from Ohm's law and $n - 1 = 3$ come from KCL. That leaves $2b - b - (n - 1) = b - (n - 1) = 3$ independent equations required from KVL. And that's exactly the number we obtain from the six KVL equations, using the linear relationships in (7).

Summary of Branch-Variable Analysis

In summary: Given a linear circuit consisting of resistors and sources in which all sources are accompanied, we write a set of b Ohm's law equations. Next we write KCL equations at all the nodes except one, $n - 1$ of them. This omitted node is chosen arbitrarily, for convenience. Often a particular node is "ground," so this is a convenient one to omit. (In Figure 3 this is node 4.) Finally, we write $b - (n - 1)$ KVL equations; a system is needed here to choose loops that will lead to independent equations. In a subsequent section we will describe such a system. For now, it is clear that the final equations must, necessarily, include all the branches in the circuit, although this necessary condition is not sufficient. (See Problem 20.)

For convenience, all the equations for the preceding example are collected here:

Ohm's law	**KCL**	**KVL**
$i_{b1} = G_1 v_{b1} - i_s$	$i_{b1} + i_{b2} = 0$	$-v_{b1} + v_{b2} + v_{b3} = 0$
$i_{b2} = G_2 v_{b2}$	$-i_{b2} + i_{b3} + i_{b4} = 0$	$-v_{b3} + v_{b4} + v_{b5} = 0$
$i_{b3} = G_3 v_{b3}$	$-i_{b4} + i_{b5} + i_{b6} = 0$	$-v_{b3} + v_{b4} + v_{b6} = 0$
$i_{b4} = G_4 v_{b4}$		
$i_{b5} = G_5 v_{b5}$		
$i_{b6} = G_6 v_{b6}$		

(8)

Another thought comes to mind when contemplating all these equations together: Why not solve all 12 of these equations simultaneously in one big operation? The first step in such an approach would be to immediately eliminate half the unknowns by substituting the i-v (voltage-controlled) equations from Ohm's law into the KCL equations, leaving six equations with six voltage unknowns, and then proceed from there. Yet another possibility is to write Ohm's law in current-controlled form $v = Ri$ instead of $i = Gv$, and then to substitute the result into KVL, this time eliminating all the voltage variables and leaving six equations in six current unknowns. None of this seems to be greatly appealing, so let's try something different.

2 GRAPHS AND CIRCUIT TOPOLOGY

The approach to circuit analysis in the preceding section, which concentrates on the branch voltage and current variables, leaves something to be desired. What we need is an approach that involves different sets of varibles, fewer in number, that will permit an organized attack on the problem. We will initiate that search now.

What we are going to discuss may appear like a diversion to you, some abstractions whose relevance you may fail to see initially but will gradually appreciate.[3]

Recall that Kirchhoff's laws pertain to the manner in which components in a circuit are interconnected; they do not at all depend on the specific nature of the components. Indeed, we introduced these laws in Chapter 1 before introducing any of the circuit components. Hence, in order to concentrate on these laws and their consequences, it would be valuable to unclutter the circuit and to suppress other information contained in a circuit diagram, such as whether a particular component is a resistor or controlled source, even whether it is linear or nonlinear. This is done by representing each component simply as a line segment. The abstraction that results is called a *linear graph,* sometimes also a *circuit graph* to emphasize what it is representing.[4] This section will be devoted to a brief study of linear graphs. It will be necessary at the outset to introduce a lot of new terminology whose merits may not be entirely clear to you when first encountered. They will become clear as we proceed; patience will be rewarded.

Definition of a Graph and Its Parts

Suppose each component of an electric circuit is represented by a line segment to be called (surprise!) a *branch.* The endpoints of the branches are connected at *nodes.* Branches whose terminals fall on a node are said to be *incident* at that node. A *graph* is defined as a collection of branches and nodes in which each branch connects two nodes. A sequence of branches traversed in going from one node to another is called a *path.* If there exists at least one

[3] In earlier times, when people were stricken with what we now call malaria, they had chills and fevers, and their bodies were thought to be filled with "bad humors." The remedy was to "bleed" them; one method of bleeding was to attach to their bodies suction cups from which some of the air was evacuated by lighting a match inside just before attaching. But then someone came along who suggested draining the nearby swamps to prevent the fever. This person was regarded with some skepticism and pity. "Draining a swamp to prevent fevers? You must have lost your marbles somewhere! Do you know how much work that takes? What's the relationship anyway?" Well, this person had learned of the new germ theory of disease. Swamps breed mosquitoes. Mosquitoes carry germs; they bite people, transmit the germs, and make people sick. Get rid of the mosquitoes and you prevent the fever. . . . Hang in there while we drain the swamp. You'll see the relationship later.

[4] *Linear* in this case is an adjective referring to "line." A *linear graph* is a graph of lines, *not* a graph which is, or represents, a straight line. This usage is different from the usage in the expression *linear equation,* where the equation is that of a line and describes and represents a particular straight line.

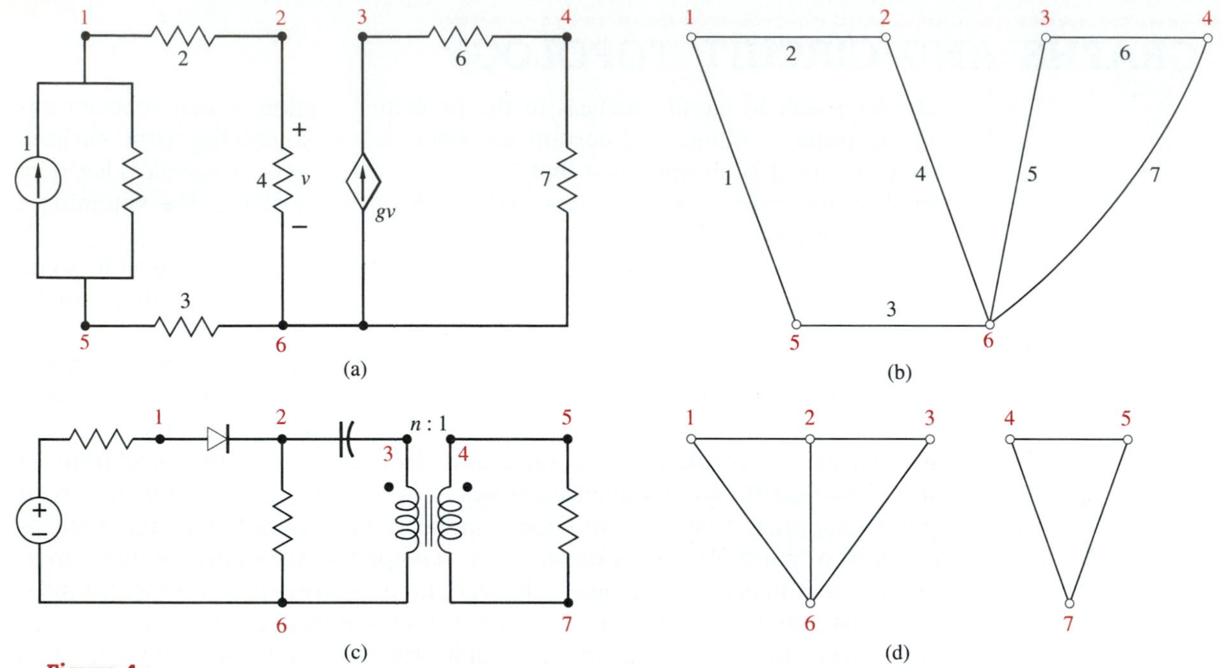

Figure 4
Two circuits and their graphs.

path from each node in a graph to every other node, we say the graph is *connected.* Figure 4 shows two circuits and their graphs, one of which is connected and the other of which is not. These circuits have components that are not resistors or sources, that are not linear, that don't have only two terminals, and are included to show the wide applicability of the ideas in this section.

A *subgraph* G_1 of a graph G is a collection of branches and nodes every one of which that is contained in G_1 is also contained in G. A subgraph may be connected or not; it can be its entire parent graph G or only one branch from G, or just a node, or anything in between.

In a circuit, the reference for each branch voltage and current can be arbitrarily chosen. We will here assume, however, that *all branch references are standard.* It is, therefore, not necessary to show both voltage and current references, since one of them implies the other. In the graph, this single branch reference will be indicated by giving the branch an *orientation*; this is done by placing an arrow on the branch, as shown in Figure 5. Because the branches are oriented or directed, such a graph is called an *oriented* or a *directed* graph, or a *digraph* for short.

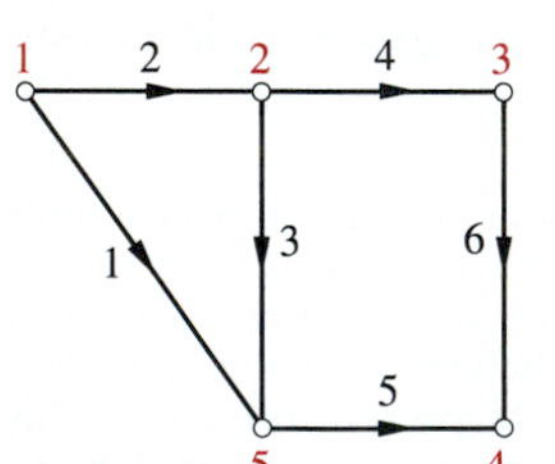

Figure 5
An oriented or directed graph (a digraph).

Structure of a Graph

In a connected graph, there may be more than one path from one node to another. In such a case, it will be possible to start at a node and return to the same node after traversing a *closed path* of branches, without traversing any

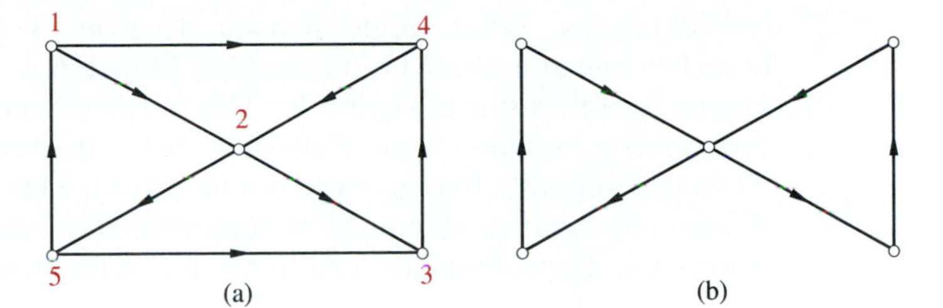

Figure 6
Intuitive idea of a closed path.

branch more than once. We all have an intuitive idea of what a closed path, or loop, is, but the idea must be made more precise if it is to be useful.

LOOPS To illustrate the concept of a loop, consider the graph in Figure 6a. Starting at node 1, follow the sequence determined by the branch orientations through nodes 2, 3, and 4, back to 2, and then through 5 and back to 1. For clarity, the graph is redrawn in Figure 6b, after the branches not in this sequence are removed. (Removing a branch means to delete it from the graph.) The sequence of branches starting at node 1 and returning to node 1 along the oriented branches shown certainly constitutes a path. (Verify this by checking the definition of a path.) This path is certainly closed, but it doesn't fit our intuitive notion of a loop. By observing the subgraph in Figure 6b, we note that every node—except node 2—has two branches connected to it.

This observation (and a little thought) leads us to the following definition:

A *loop* is a *connected subgraph* of a graph that has exactly two branches of the subgraph incident at each of its nodes.

Note that it is two branches of the *subgraph* that are to be incident at each node, not two branches of the parent graph. There may be more branches of the *graph* incident at one or more of the nodes of the subgraph, without negating the existence of a loop.

In the graph in Figure 7a, for example, the subgraph consisting of the set of branches $\{1, 4, 6\}$ constitutes a loop. (To confirm this, you might redraw the graph with all other branches removed and check the definition.) Note that there are three branches of the graph connected to node 3, but one of them does not belong to the subgraph, so it doesn't count.

TREES Another important concept can be clarified from the following

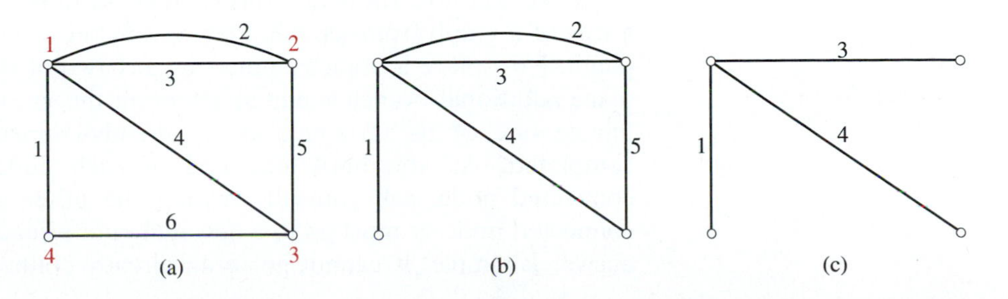

Figure 7
Example graph.

considerations. What would happen if, given a graph, we start removing branches one at a time? For example, if branch 6 is removed from the graph in Figure 7a, the result is Figure 7b. This is still a connected graph. Suppose we now remove branches 2 and 5 also. The result is shown in Figure 7c. This is still a connected graph; that is, there is a path from each node to every other node. If any one remaining branch is removed, however, the graph is no longer connected. These thoughts lead to the following definition:

A *tree* of a connected graph is a connected subgraph consisting of a set of branches which, together, connect all the nodes of the graph without forming any loops.

Thus, the subgraph in Figure 7c constitutes a tree. Generally, a graph has more than one tree. If you were asked to do so, what would *you* call the branches of a tree? Why, *twigs,* of course!

EXERCISE 1

List the twigs of each tree in Figure 7a. (There are a total of 10 trees.) □

Remember (1) that a tree is connected and (2) that it has no loops. Now suppose we have identified a tree of a graph, as in Figure 7c, and we add another branch of the graph to the tree. Will the subgraph remain connected? Of course. Will it now contain a loop? Every node of the graph is on the tree and there is a path in the tree from each node to every other node (because the tree is connected), so the added branch (which lies between two nodes) must form a loop with the twigs that were already on a path between those two nodes. The answer is, If another branch of a graph is added to one of the trees, the result is no longer a tree, because it contains a loop.

COTREES AND LINKS Once a tree has been selected in a graph, the remaining branches not on the tree are called *links.* Note that, given a graph, its branches cannot uniquely be identified as twigs or links. Since there are several trees, a given branch may be a twig on one tree, but a link on another tree. Only *after* a tree has been identified can the branches of a graph be classified as twigs or links. Collectively, the subgraph consisting of all the links is called a *cotree.* A graph is made up of a tree and its cotree; the set of all its branches is made up of a set of twigs and a set of links.

Look back at Figure 7c showing a tree of the graph in Figure 7a. The graph has four nodes and the tree has three twigs, one fewer than the number of nodes, that is, $n - 1$, where n is the number of nodes. With you serving as an artist, we will now show that this is a general result by attempting to construct a tree of a graph from scratch. Place a number of nodes, say, six, on a sheet of paper. Now place a branch connecting any two of the six nodes. The plan is to place additional branches one at a time in the graph, with each added branch having one of its terminals at an already-connected node, until a tree is completed. As you hook one end of each added branch to an already-connected node, ask yourself whether the other end can go to an already-connected node or must go to a new node not already on the growing tree. The answer is simple: it cannot go to an already connected node, because then a loop would be formed and the result could not be a tree; thus, the branch must go to a new node not already connected.

Notice that the first branch placed necessarily connects two nodes. Each added branch necessarily brings only one additional node into the tree, since one end of the added branch is anchored to the partial tree. Hence, it takes exactly $n - 1$ twigs to complete the tree. Assuming that the total number of branches in the graph is b, that leaves $b - (n - 1)$ branches as links. To summarize:

$$b = \text{number of branches in a graph}$$

$$n = \text{number of nodes in a graph}$$

$$n - 1 = \text{number of twigs in a tree}$$

$$b - (n - 1) = \text{number of links in a cotree}$$

EXERCISE 2

List the set of links for each tree of the graph in Figure 7 you determined in Exercise 1. □

AN ASIDE: THE KÖNIGSBERG BRIDGE PROBLEM The theory of linear graphs is useful in many applications besides electric circuits. A problem of historical interest, solved by Leonhard Euler when he was 29 years old in 1736, was the Königsberg bridge problem. It seemed that a river flowed through the German city of Königsberg. Two islands in the river were interconnected with each other and with the two banks of the river by seven bridges, as shown in Figure 8a. The citizens of Königsberg liked to take a Sunday stroll along the river and on the islands.[5] They wanted to know if they could take their stroll across all seven of the bridges without crossing any one of them more than once. Euler solved their problem by inventing graph theory and showing that it couldn't be done. If the bridges are the branches of a graph, with the nodes corresponding to the four pieces of land, the graph representing Königsberg, its islands, and its bridges takes the form shown in Figure 8b. Euler's proof

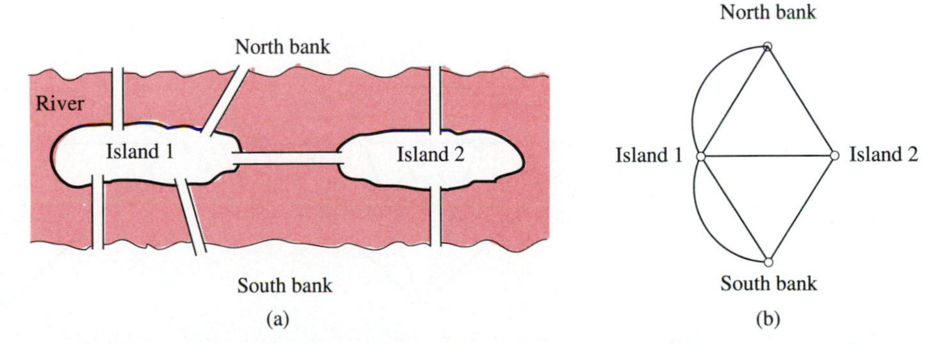

Figure 8
The Königsberg circuit.

[5] Nowadays, they are probably jogging.

consisted of showing that it is impossible to have a simple path consisting of all the branches of a graph (that is, without retracing any branches) unless exactly two of the nodes have an odd number of branches connected to them. Unfortunately, in the Königsberg bridge graph, there are four nodes having an odd number of branches. (Who said graph theory doesn't have *practical* applications?)

In depriving the citizens of Königsberg of part of the fun in taking a Sunday stroll, Euler generated basic knowledge in the field of mathematics which later became significant in the field of electric circuits—which didn't even exist when Euler did his work.

Independent Voltage Equations

By this time you may be wondering why so much time is being spent on what appears to be an abstract topic. Besides solving the old Königsberg bridge problem, graph theory answers some questions about the number of independent KVL and KCL equations which we had previously left illustrated or guessed at but unanswered. We'll start answering those questions now.

The graph in Figure 9a represents a circuit: it has $b = 9$ branches and $n = 5$ nodes. The number of twigs in a tree must be 4. One of the trees, consisting of twigs $\{1, 2, 4, 7\}$, is shown in Figure 9b. Suppose link 6 is added to the tree, as shown in the dashed line in the figure. This link forms a loop with twigs $\{1, 2, 7\}$. If the voltages of the twigs are known, the voltage of this link will be determined by KVL around this loop; thus, the link voltage is dependent on twig voltages. Now remove link 6 and add another link, say, 9. (Draw the diagram yourself.) Again, by KVL around the loop so formed, we find that the voltage of link 9 is not independent, but is dependent on the voltages of twigs. On the other hand, in the tree alone, there are no loops, so none of the twig voltages can be obtained from other independent voltages via KVL. The conclusion is:

The number of independent *voltages* in a graph equals the number of twigs, namely, one less than the number of nodes: $(n - 1)$.

Review the procedure just followed. Given a graph, a tree is selected. To the subgraph consisting of this tree we add one link at a time. The loop so formed by each link (consisting of that link and some twigs) can be uniquely

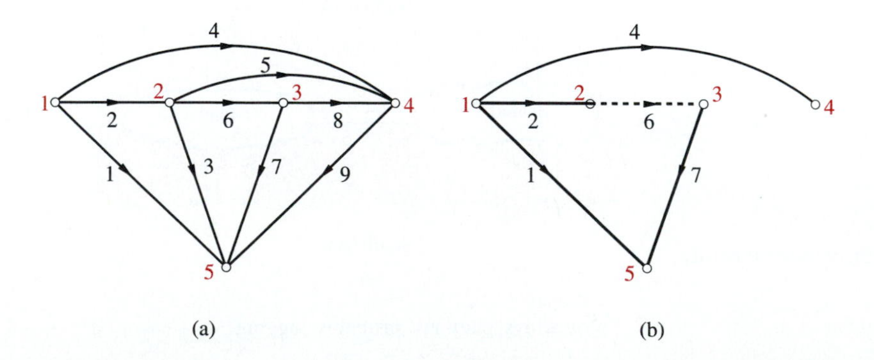

Figure 9
Number of independent voltages.

identified with the link. We call such a loop a *fundamental loop*. More precisely,

> A *fundamental loop* is a loop formed by adding any one link to a tree. The branches of the loop will consist of twigs and this one link.

In any graph, there are as many fundamental loops as there are links, namely, $b - (n - 1)$. A KVL equation can be written around each fundamental loop. Since each loop contains a branch voltage (that of the link) that does not appear in any other of these loops, this equation must be independent. We conclude that:

> In a graph having b branches and n nodes, the number of independent *KVL equations* is exactly $b - (n - 1)$.

In the graph in Figure 9a, $b - (n - 1) = 5$.

EXERCISE 3

(a) For the tree shown in Figure 9b, find the other four fundamental loops. (b) Choose another tree of the graph in Figure 9a and find all the fundamental loops.

ANSWER: (a) {1, 2, 3}, {1, 4, 9}, {1, 4, 8, 7}, {2, 4, 5}. □

In an earlier discussion of the orientation of a loop for the application of KVL, whether clockwise or counterclockwise, the answer was that the orientation is arbitrary. Reversing the orientation of a loop amounts to a multiplication of the KVL equation by -1. In the case of a fundamental loop in a graph, we identify its orientation with that of its defining link. Once the orientation of a link is specified, there is no longer any ambiguity about the orientation of the corresponding fundamental loop.

Node Voltages as a Basis Set

It was illustrated in the previous section that the set of all b branch voltages in a circuit represented by a graph can be determined by the voltages of a smaller set of branches: $(n - 1)$. This smaller set is a *basis set* in terms of which the larger set is determined. The particular basis set of voltages discussed was the set of twig voltages for some tree.

We shall now introduce a more useful basis set of voltages, a set on which a major analytic procedure is founded. This set of voltages is chosen as follows. First, one of the nodes in a graph (circuit) is chosen as a *datum* node, with respect to which the voltages of all other nodes will be measured. The node selected as a datum (sometimes called reference) node is arbitrary; it is chosen for convenience. The voltage *from* any other node, say, node *j*, *to* this datum node is labeled v_{nj} and is called the *node voltage* of node *j*. (The subscript *n*, for node, will be used here temporarily; we will discard it later for simplicity.)

In Figure 5, for example, suppose node 5 is chosen as the datum node. Then v_{n1} is the node voltage of node 1, the voltage from node 1 to node 5. Similar meanings attach to v_{n2}, v_{n3}, and v_{n4}, but what could v_{n5} mean? It means the voltage of node 5 relative to itself, which is obviously zero.

The voltage of any *branch* in a graph (circuit) is the difference of two node voltages—the nodes at the two ends of the branch. Thus, in Figure 5, the voltage of branch 4 equals $v_{n2} - v_{n3}$. The voltage of branch 3, on the other hand, appears to violate the general statement, since it equals just node voltage v_{n2}. But this is really the node-voltage difference $v_{n2} - v_{n5}$, which equals v_{n2}, since v_{n5} is zero.

We shall now establish that the set of node voltages in a graph (circuit) is a basis set from which *all* branch voltages can be obtained.

To establish the desired result, note that all nodes of a graph lie on a tree. Since, by definition of connected graph, a tree connects all the nodes, there must be a path *in the tree* from each node to every other node, especially from each nondatum node to the datum node. Thus, by KVL, the voltage of each nondatum node relative to that of the datum node—namely, the node voltage of that node—must be a linear combination of twig voltages. Thus node voltages can be expressed in terms of twig voltages for some tree.

The converse is also true. Thus, each twig voltage is the difference in voltage of the nodes at its two ends—and "difference" is a particular linear combination. If the node voltages are known, the twig voltages can be obtained and, hence, also the link voltages. The conclusion is:

The $n - 1$ node voltages in a graph constitute a basis set in terms of which all b branch voltages of the graph can be determined.

Cut Sets

Now that the truth of the number of independent voltage equations has been established, how about the number of independent current equations?

To set the stage, consider the graph in Figure 10. Focus on one part of the graph which includes several nodes; it is identified in color in the diagram. This part of the graph is connected to the rest of the graph by a set of branches marked with an $\times$. The set of all the branches connected to nodes in this part of the graph falls into two subsets: (1) those branches that run between two nodes *within* the group and (2) those branches that connect a node within the group and a node external to the group. The $\times$-marked branches constitute this second subset.

Now consider writing KCL equations at each node within the colored set and then adding the equations. This set of equations will include both the first and the second categories of branches. The orientation of the current in

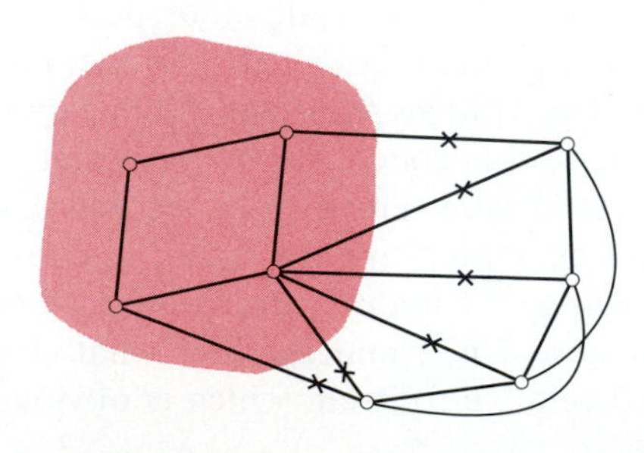

Figure 10
A supernode and cut sets.

a branch connecting two nodes within this set will be toward one node and away from the other, so when the equations are added the terms corresponding to this current will cancel each other. The only currents remaining in the summation will correspond to the ×-marked branches. We might conceive of this group of nodes as a *super*node, the ×-marked branches being connected to the supernode. The result of this paragraph can be stated as a generalization of KCL: *the sum of currents of all branches connected to a supernode is zero.*

But this example illustrates a more significant concept. The ×-marked branches have the distinctive property that if they were removed from the initially connected circuit, the circuit would become disconnected, or cut, into two parts. We make this observation the basis for a definition:

> A cut set is a set of the smallest number of branches of a graph which, when cut, divide the graph into two parts.

Make note of the three conditions: (1) this set of branches *cuts the graph,* (2) it cuts it into *two parts,* and (3) *no smaller number* of branches will do the same thing. The six ×-marked branches in Figure 10 constitute a cut set. Besides these branches, it is possible to cut one or more additional branches in Figure 10 without increasing the number of separate parts. (Find some.) The set of branches consisting of the previous six *plus* these new ones *will not* be a cut set. On the other hand, if any one of the six is not cut, the graph will remain connected, so the remaining branches will not constitute a cut set.

A cut set divides a graph into two groups of nodes—two supernodes. As seen in the example, KCL is satisfied at a supernode. A KCL equation written at a supernode is called a *cut-set equation.* Each cut-set equation is a linear combination (simply a sum) of KCL equations at nodes. Indeed, the set of branches connected to a single node is also a cut set; one part of the divided graph is simply an isolated node in this case. If we can determine the number of independent cut-set equations, this will also be the number of independent KCL equations at nodes. This is the task we will now undertake.

The number of twigs in any tree of any graph is $n - 1$. If any twig is cut, the tree will be cut into two parts (one of which may be a solitary node). A particular case is shown in Figure 11a, where the tree is in solid lines and the

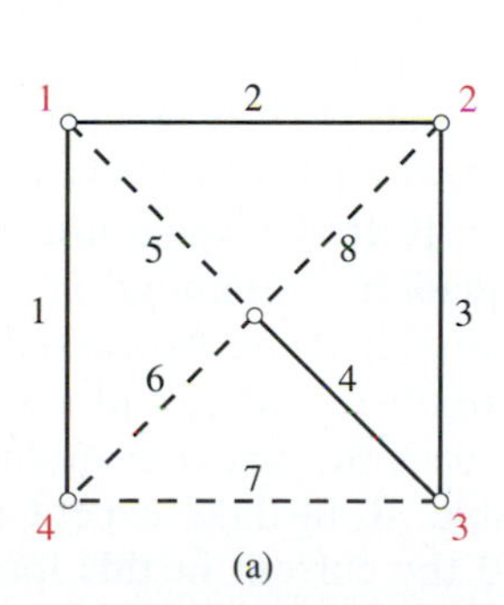

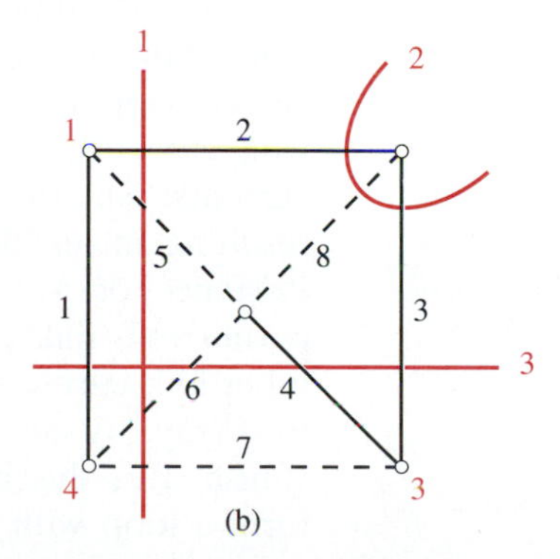

Figure 11
A tree (solid) and its cotree (dashed).

links are in dashed lines. If twig 2 is cut, the tree will be in two parts, with nodes 1 and 4 in one part and nodes 2, 3, and 5 in the other. Now look at the links that are connected between these two sets of nodes; they are {5, 6, 7}. If we assume that they, in addition to twig 2, are cut, then, not just the tree, but the whole graph will be in two parts. Thus, this set of branches—the twig and the three links—constitutes a specific kind of cut set which we will make the basis of a definition:

Given a tree of a graph, a *fundamental* cut set is defined as a cut set which includes a single twig of this tree, all its remaining branches being links.

Since a tree contains $n - 1$ twigs, there are exactly $n - 1$ fundamental cut sets. The set of KCL equations at all fundamental cut sets in a graph will be independent, since each equation includes one current (that of the twig) that is included in no other equation and so is unobtainable by a linear combination of the other equations. The conclusion is that *at least* $n - 1$ cut-set equations are independent. It still remains to be shown that a larger set of equations is not independent, so let's do that in the next section.

EXERCISE 4

For the tree composed of the set of branches {4, 5, 6, 8} in Figure 11, list the branches of each fundamental cut set. (Redraw the graph with solid twigs and dashed links.) Is there anything special about the KCL equations of these cut sets? □

Independent Current Equations

How many twigs can a cut set include? Fundamental cut sets contain exactly one twig. Is it possible for a cut set to include no twigs? The answer is no; there cannot be a cut set consisting of links alone, without at least one twig. This is so because, even if all links are cut, a tree will still remain, and a tree connects all the nodes and so keeps the graph connected. The conclusion is that any cut set which is not a fundamental cut set must contain more than one twig. As illustration, assuming the tree in Figure 11a, look at the cut sets in Figure 11b; each is the set of branches intersected by one of the colored lines. Cut set 1 contains one twig (fundamental), cut set 2 contains two twigs, and cut set 3, three twigs.

Now contemplate writing a cut-set equation for a cut set having two or more twigs; call the equation E. From the fundamental cut-set equations, each twig current can be expressed in terms of link currents. Using these expressions for twig currents, we can eliminate the twig currents in E, leaving only link currents. But the coefficient of each link current in this equation must be 0, resulting in an identity: $0 = 0$. If this were not the case, it would mean the existence of a current equation containing link currents only. This would permit one link current (say, i_k) to be expressed as a linear combination of other link currents only. Can this be? If all link currents except i_k were reduced to 0 (say, by cutting the links), the linear combination would force i_k also to vanish. But this is impossible; if all links except link k are cut, this link will form a loop with twigs and the current in this loop cannot be affected by the vanishing of all other link currents. (Draw a graph, choose a tree, and remove all links but one to demonstrate this argument to yourself.) The conclusion is: *any other cut-set equation in a graph can be obtained as a linear combination of the fundamental cut-set equations*; *no other cut-set equations are independent.* The upshot of all of this is that:

In a graph having n nodes, the number of independent KCL equations is exactly $n - 1$.

This ends the necessary digression into topology. A substantial number of concepts have been introduced and terms defined. Some of these (loop, node, branch) were already in your vocabulary. Others (cut set) may not have been so familiar. But we are now ready to use them all to establish procedures to analyze the most complex of circuits.

3 NODE EQUATIONS

One of the results established in the preceding section on topology is that all the branch voltages in a circuit can be expressed in terms of a smaller number of voltages which constitute a basis set. In particular, the set of node voltages, which are $n - 1$ in number, constitutes such a basis set. We will now introduce a circuit analysis procedure which is based on the node voltages.

Defining the Variables

Recall that the first step is to choose one of the nodes of a circuit as a *datum* node to which the voltages of all other nodes are referred. Most conveniently, this is the node which is omitted when KCL equations are written. Then, the voltage *from* each node *to* this datum node is a *node-to-datum* voltage, or simply a *node* voltage. In the preceding section, we identified a node voltage by a subscript n followed by the number of the node. However, this notation is cumbersome; henceforth, we will omit the n. For example, in Figure 3, v_3 is the voltage of node 3 relative to that of node 4, the ground node.

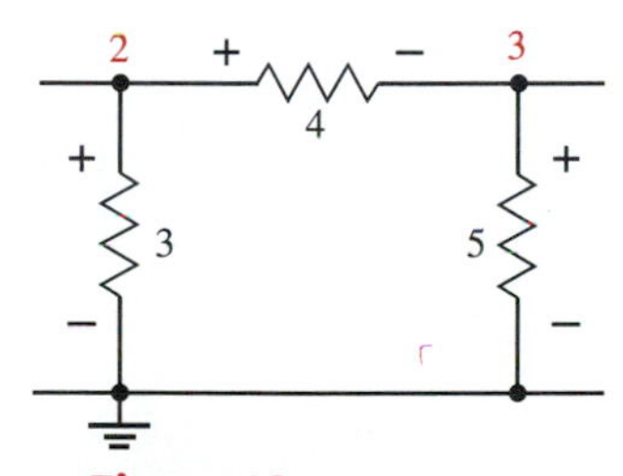

Figure 12
Branch voltages relative to node voltages.

Each branch is connected between two nodes, one of which can be datum (ground). If one of the nodes of a branch is the datum node, then the node voltage of the nonground node is the same as that branch voltage. In Figure 12 we have isolated a section of the circuit from Figure 3 that includes branches 3, 4, and 5. Branch 5 lies between node 3 and datum node 4. Hence, v of branch 5 = v of node 4. On the other hand, branch 4 lies between nodes 2 and 3. By applying KVL around the closed path formed by branches 3, 4, and 5, you can determine that

$$v_{b4} = v_2 - v_3$$

Indeed, we can say that *any* branch voltage is *always* the difference of the voltages of its two end nodes. Of course, if one of the two end nodes is the datum node, its voltage relative to the datum is 0, so, in this case, the branch voltage equals a node voltage. Defining the node voltages as done here automatically satisfies the KVL equations of the circuit.

Node-Voltage Analysis Procedure

The analysis procedure has now become clear:

1. Write the KCL equations at all the nondatum nodes in the circuit ($n - 1$ equations in b unknowns).
2. Eliminate the currents by substituting the branch i-v relations (Ohm's law) in the voltage-controlled form $i = Gv$ (still leaving $n - 1$ equations in b unknowns).
3. Eliminate the b branch voltages by expressing them in terms of the $n - 1$ node voltages ($n - 1$ equations in $n - 1$ unknowns).

BRANCH i-v EQUATIONS INTO KCL We will now illustrate this approach for the earlier example in Figure 3, for which all the equations are given in (8). The process of substituting the branch equations from the first column into KCL in the second column results in the following:

$$\begin{aligned} G_1 v_{b1} - i_s + G_2 v_{b2} &= 0 \\ -G_2 v_{b2} + G_3 v_{b3} + G_4 v_{b4} &= 0 \\ -G_4 v_{b4} + G_5 v_{b5} + G_6 v_{b6} &= 0 \end{aligned} \tag{9}$$

These are three equations in six branch-voltage variables.

ELIMINATING BRANCH VOLTAGES The next step is to express branch voltages in terms of node voltages. This can be done by manipulating the KVL equations in the right column of (8) or simply by inspection of the circuit diagram, which is much simpler. In either case, the result is

$$\begin{aligned} v_{b1} &= v_1 \\ v_{b2} &= v_1 - v_2 \\ v_{b3} &= v_2 \\ v_{b4} &= v_2 - v_3 \\ v_{b5} &= v_3 \\ v_{b6} &= v_3 \end{aligned} \tag{10}$$

Next we insert these equations into (9), resulting in

$$\begin{aligned} G_1 v_1 - i_s + G_2(v_1 - v_2) &= 0 \\ -G_2(v_1 - v_2) + G_3 v_2 + G_4(v_2 - v_3) &= 0 \\ -G_4(v_2 - v_3) + G_5 v_3 + G_6 v_3 &= 0 \end{aligned} \tag{11}$$

Success! Three equations in three unknowns.

REARRANGING TERMS The final step is to rearrange and collect terms so that each unknown appears only once in each equation:

$$\begin{aligned} (G_1 + G_2)v_1 - G_2v_2 &= i_s \\ -G_2v_1 + (G_2 + G_3 + G_4)v_2 - G_4v_3 &= 0 \\ -G_4v_2 + (G_4 + G_5 + G_6)v_3 &= 0 \end{aligned} \tag{12}$$

These three algebraic equations in three unknowns can be solved by standard mathematical methods. Cramer's rule and Gaussian elimination are discussed in Appendix A. Useful software for this purpose is MATLAB. We shall concentrate here on methods of *setting up* the equations.

MATRIX FORM OF ALGEBRAIC EQUATIONS A compact way to organize the writing of a set of linear algebraic equations is in *matrix form.* The preceding set can be written as follows:

$$\begin{bmatrix} G_1 + G_2 & -G_2 & 0 \\ -G_2 & G_2 + G_3 + G_4 & -G_4 \\ 0 & -G_4 & G_4 + G_5 + G_6 \end{bmatrix} \begin{bmatrix} v_1 \\ v_2 \\ v_3 \end{bmatrix} = \begin{bmatrix} i_s \\ 0 \\ 0 \end{bmatrix} \tag{13}$$

There are three *arrays* in this equation. Two of them are *vectors* representing the node voltages and the source(s). The third is the coefficient *matrix* of conductances. By identifying each array by a single letter, the equation can be written compactly as:

$$\mathbf{Gv} = \mathbf{i}_s \tag{14}$$

Boldface is used to designate an array (matrix or vector). If you don't have a background in matrix algebra, a brief introduction is provided in Appendix A.

The coefficient matrix in this example has some properties which are general for the class of circuits being treated in this section, namely, a circuit of linear resistors and independent sources. First we note that the matrix is *symmetrical,* meaning that an element on one side of the diagonal has the same value as the similarly located element on the other side. This means $G_{jk} = G_{kj}$. For example, $G_{13} = 0 = G_{31}$. Second, we see that each element on the main diagonal is the sum of conductances connected to the node corresponding to its row of the matrix. Finally, note that branch G_2 lies between nodes 1 and 2 and the matrix element in position 12 (which is the same as the element in position 21) is $-G_2$. The generalization of this is that the off-diagonal element in position jk is the negative sum of all conductances lying between nodes j and k.

These properties permit writing the node equations for a circuit of linear resistors directly by inspection of the circuit, without going through the steps we have previously outlined. However, this approach does not apply when the circuit includes two-port dissipative components, as we shall discover in a subsequent section. Learning specialized procedures for each class of circuits may be of limited utility, so we will not emphasize it here, although you may wish to use such specialized procedures whenever you are sure they apply to the case at hand.

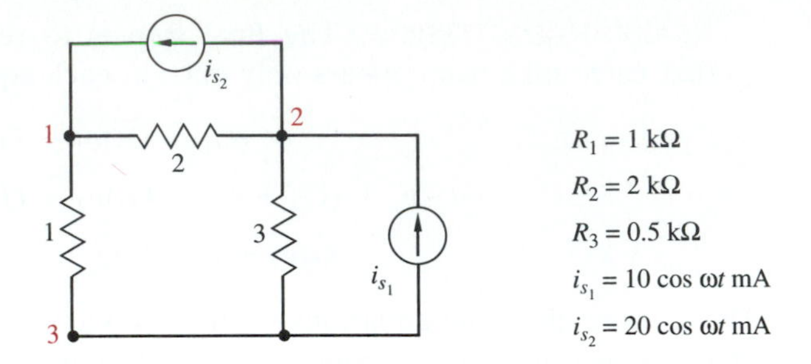

Figure 13
Exercise circuit.

EXERCISE 5

Use the step-by-step approach described in this section to arrive at a set of node equations for the circuit in Figure 13. What will be the maximum voltage from node 1 to ground and from node 2 to ground? Use the approach based on an inspection of the circuit and confirm your previous results.

ANSWER: $(v_1)_{\text{max}} = 90/7$ V, $(v_2)_{\text{max}} = 10/7$ V. □

IRRELEVANCE OF BRANCH REFERENCES Look back at (11), which resulted from substitutions of Ohm's law and KVL into KCL in (8). Before any parentheses are removed, some terms are preceded by a minus sign; for example, the first term in the second equation. That happens because fixed references are given for the currents; for example, the current in branch 2. If a branch lies between two nondatum nodes, its reference will be directed away from one node and toward the other (for example, current 2 is directed away from node 1 but toward node 2), so that current will enter into one of the equations preceded by a plus sign and the other equation preceded by a minus sign; for example, current 2 will enter the equation for node 2 with a minus sign.

Suppose we change the reference of the current in branch 2 *when we write KCL at node* 2. (What happens to the voltage reference of branch 2?) Since we are assuming standard references at all resistor branches, the voltage reference will also change. But if both the current and voltage references are reversed, there will be no change of sign in Ohm's law. Thus, there will be *two* sign reversals in the equations pertinent to node 2: once in the sign preceding the current i_{b2} in KCL, and once when the voltage of branch 2, with its now opposite reference, is expressed in terms of the node voltages. The end result is that *the node equations will be the same.*

EXERCISE 6

Repeat the process described in the last paragraph of assuming that the references of all currents are leaving node 3 in Figure 3 when KCL is written for node 3. Verify that the same result given in the last equation of (11) is obtained. □

The conclusion from this discussion is that when we are planning to write node equations for a circuit, *no voltage or current references of branches need be shown.* Of course, there are some assumptions (or conventions) that must

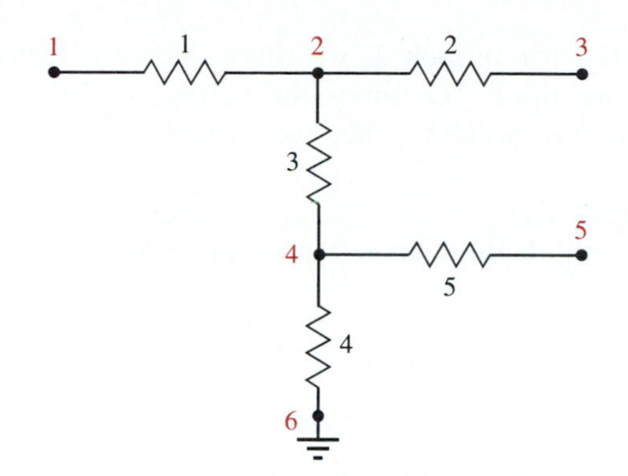

Figure 14
Segment of a circuit.

be kept in mind: (1) all branch-variable references are assumed to be standard, and (2) each reference + of a node voltage is assumed to be at the corresponding node, with the reference minus at the datum node.

Combining Steps in the Process

Since the branch *i-v* equations (Ohm's law in this case) are so simple, and since each branch voltage equals either a node voltage or the difference between two node voltages, is there really any necessity to formally write the equations down, or can this be done mentally, just by inspection? Consider the segment of a circuit shown in Figure 14. The numbering of nodes and branches follows the system used in the preceding section. No fixed branch references have been shown, because, when each node is under consideration, we are planning on (1) taking each relevant branch current reference to be oriented away from that node and (2) choosing standard references for all resistor branch variables.

Let's concentrate on node 2 and write the three steps needed to arrive at the node equation for this node: KCL, then Ohm's law, then branch voltages in terms of node voltages:

$$
\begin{gathered}
i_1 + i_2 + i_3 = 0 \qquad \text{(KCL)} \\
\begin{aligned}
i_1 &= G_1 v_{b1} & & & v_{b1} &= v_2 - v_1 & & \\
i_2 &= G_2 v_{b2} & & (i\text{-}v) & v_{b2} &= v_2 - v_3 & & \text{(KVL)} \\
i_3 &= G_3 v_{b3} & & & v_{b3} &= v_2 - v_4 & &
\end{aligned}
\end{gathered} \tag{15}
$$

We plan to substitute $i_1 = G_1 v_{b1}$ into the KCL equation and then replace the branch voltage by $v_2 - v_1$. Both of these are trivial steps. Instead of writing these equations explicitly, we can "think" them and just write the final answer.

Here's how the process might go. We are planning to write a node equation at node 2. Instead of writing the sum of *all* currents leaving that node, we concentrate on one branch at a time, say, branch 1.

Think: "Current leaving node 2 through branch 1"; then
Think: "$G_1 v_{b1}$"; then
Write: "$G_1(v_2 - v_1)$."

Similarly, for branch 2, we think "current leaving node 2 through branch 2"; then we think "G_2 times the voltage of branch 2, with voltage reference *from* node 2 *to* node 3"; then we write $G_2(v_2 - v_3)$.

EXERCISE 7

Use this "think-think-write" process to write a node equation at node 4 in Figure 14.

ANSWER: $G_3(v_4 - v_2) + G_4v_4 + G_5(v_4 - v_5) = 0.$ □

Treating Voltage Sources in Node Equations

The only sources in circuits encountered so far for which we wrote a set of node equations were (accompanied) current sources. What happens if a circuit contains a voltage source? Assuming it is accompanied, we have no problem.

Figure 15 shows a circuit containing an accompanied voltage source. The circuit has four nodes. Can we choose any node we please as the datum node? Yes; the datum node is chosen for convenience, so we choose node 4.

Branch 1 consists of the resistor and source in series. The v-i equation of this branch is easily written, from which we solve for i in terms of v; this constitutes the branch i-v equation:

$$v_{b1} = R_1 i_1 + v_s$$
$$i_1 = G_1(v_{b1} - v_s)$$

We now proceed, as before, with the "think-think-write" approach. At node 1, we assume that all current references are directed away from the node, all branches have standard references, and each node-voltage reference + is located at the corresponding node. Here is the resulting node equation for node 1:

$$G_1(v_1 - v_s) + G_2v_1 + G_3(v_1 - v_2) + G_4(v_1 - v_3) = 0$$

(You should confirm each term.) The node equations at the other two nodes will not be influenced by the voltage source, so there is nothing new in obtaining those equations. They will be left for you to complete as an exercise.

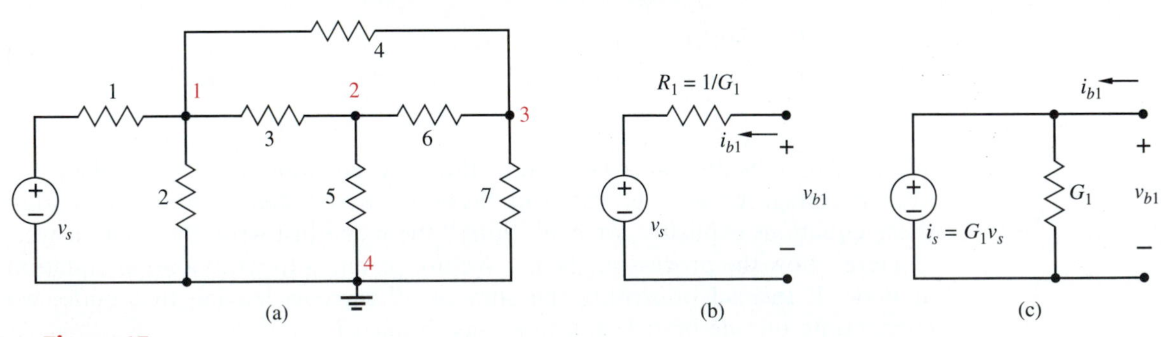

Figure 15
Node equations for a circuit containing an accompanied voltage source.

Glance back at Figure 15. Suppose the accompanied voltage source is replaced by its current-source equivalent, as described in Chapter 3. This equivalent is shown in Figure 15c. The *i-v* equation of this new branch is exactly the same as that of the branch with the voltage source. (That's not surprising, since equality of the *i-v* equations is exactly what equivalence means.) Thus, the node equation will come out the same; there really is no need to make a source conversion.

UNACCOMPANIED SOURCES[6] In the preceding discussion of node equations, it was assumed that all sources are accompanied—a voltage source by a series resistor and a current source by a parallel resistor. We will now discuss what to do if a source is not accompanied.

First, let's consider the trivial situations shown in Figure 16. The voltage across the resistor in Figure 16a is known (v_s), so its current is determined by Ohm's law. The presence of R does not change v_s; hence it does not change the voltage appearing across the rest of the circuit. As far as that remaining circuit is concerned, R might just as well be removed. When this is done, the source may then become accompanied, in which case we proceed as in the previous section. Or removing R may create a situation like the one to be described in "The Voltage-Source Shift," below, where we will discuss a remedy.[7] The presence of R *will* add v_s/R to the current in v_s, however.

In Figure 16b, i_s appears in series with a resistor. The resistor current is i_s and, hence, known; so is its voltage, by Ohm's law. The presence of R changes neither the source current nor the current flowing into the circuit to the right of R. As far as that circuit is concerned, R might just as well be replaced by a short circuit. If that is done, then the current source may become accompanied, in which case the problem is solved. If not, it may have two or more branches connected to its terminals, as in the cases to be described now. The only thing the presence of R *does* influence is the voltage across i_s.

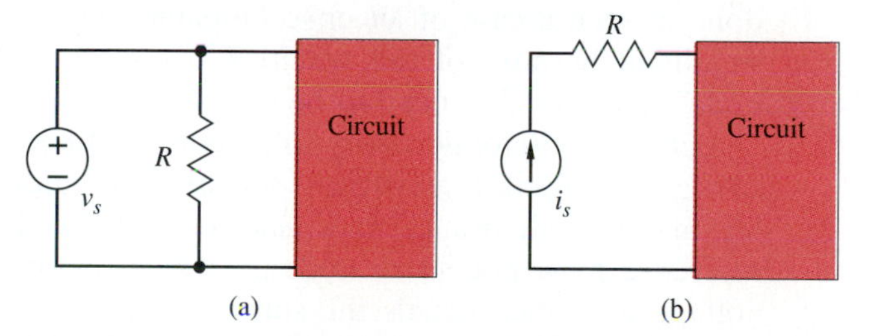

Figure 16
(a) Resistor in parallel with voltage source; (b) resistor in series with current source.

[6] This section might be viewed as specialized and not of primary importance. You may wish to go over it lightly or omit it altogether. However, the procedures described here permit claiming that node analysis is possible to carry out for *all* linear circuits; it is not limited to only those circuits whose sources are accompanied.

[7] A possible exception might conceivably occur if the current in R is the controlling variable of a controlled source inside the circuit. Removing R can then conceivably influence the remaining circuit. Note, however, that the current in R is proportional to v_s. Hence, the controlling variable can be changed to v_s, so the statement in the text is valid after all. A similar discussion will apply for the situation in Figure 16b.

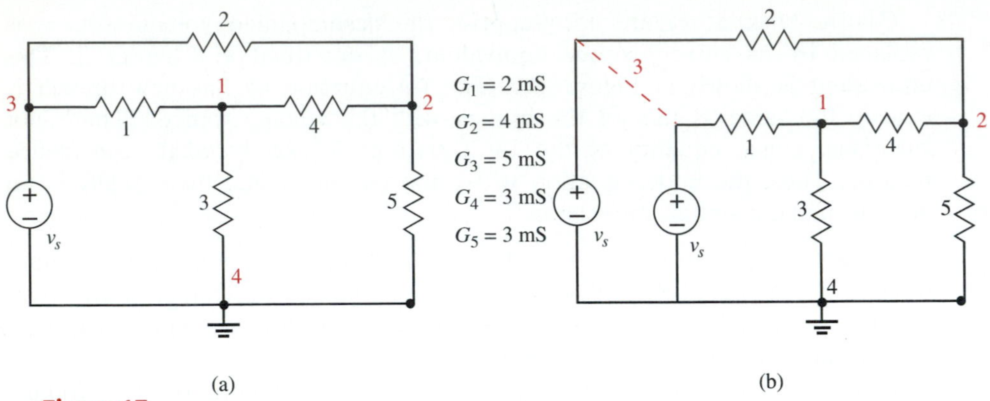

Figure 17
Unaccompanied voltage source in circuit.

VOLTAGE SOURCE AS SUPERNODE The two preceding situations (a resistor in parallel with a voltage source and a resistor in series with a current source) came about because of circuit idealizations. It is unlikely that such configurations will occur in practical circuits. Nevertheless, it would be impossible to understand the real world without using some kind of circuit models, so such idealized circuits can make their appearance from time to time. Let's now consider some less thoroughly abstract situations.

The circuit in Figure 17a includes an unaccompanied voltage source. There are a total of four nodes. One of these, say, 4, is selected as the datum. But the voltage of node 3 is known; it is equal to the source voltage. There is no need to write a node equation at this node, since its voltage is not an unknown, so there will be just two node equations. There *is* one unknown related to the voltage source, and that is its current. Once the two node equations are solved, all the branch voltages and currents will be determined. Then, the source current can be obtained from KCL at node 3. Thus, nothing special needs to be done in such a case of an unaccompanied voltage source; indeed, the work is simplified because of the resulting decrease in the number of simultaneous node equations that need to be solved.

Let's take the opportunity to practice writing node equations for this circuit. Using the "think-think-write" sequence at node 1, we would think: current leaving through branch 1, which is G_1 times the voltage of branch 1 (with reference *from* node 1 *to* node 3), which, in turn, is node voltage v_1 − node voltage v_3 (which equals the source voltage). Doing the same at node 2 and using the given numerical values leads to the following equations:

$$2(v_1 - v_s) + 5v_1 + 3(v_1 - v_2) = 0$$
$$4(v_2 - v_s) + 3(v_2 - v_1) + 3v_2 = 0$$

Note that the conductances are given in millisiemens, so each coefficient in these equations should have a factor 10^{-3}. However, those factors are removed by multiplying each equation by 10^3. Indeed, each equation can be multiplied through by *any* constant without changing the result.

The solution of these equations can be obtained by methods described in Appendix A or by use of appropriate computer software, such as MATLAB. The result is:

$$v_1 = 0.505v_s \qquad \text{and} \qquad v_2 = 0.352v_s$$

Now suppose we also want to find the power delivered by the source to the rest of the circuit. For this purpose we need the source current; it can be obtained by KCL applied at node 3. Assuming nonstandard references for the source current, we find:

$$\begin{aligned} i_s &= G_1(v_s - v_1) + G_2(v_s - v_2) \\ &= 2(1 - 0.505)v_s + 4(1 - 0.352)v_s \text{ mA} \\ &= 3.58v_s \text{ mA} \end{aligned}$$

Hence, the power delivered by the source will be:

$$p_s = v_s i_s = 3.58v_s^2 \text{ mW}$$

Another way of looking at this circuit is fruitful. Consider a cut set consisting of branches 1, 2, 3, and 5. This cut set separates the circuit into two parts, one of which consists of just the voltage source. (Draw the diagram.) Thus, the voltage source can be considered as a supernode. (See Figure 10.) Each branch of the cut set connects this supernode to the rest of the circuit. KCL can be written for this cut set and the currents then eliminated by Ohm's law. The result will be an equation which is not independent of the previous ones. Show that it is the same as the sum of the two previously obtained equations at nodes 1 and 2.

The supernode formed by a voltage source merges the two nodes at its terminals. In the present case, nodes 3 and 4 become one node. Since node 4 is the datum, it isn't surprising that KCL at this supernode gives the sum of the equations at the other nodes.

THE VOLTAGE-SOURCE SHIFT Yet another approach can be taken in the preceding circuit, as shown in Figure 17b. Node 3 is split into two and a source v_s is connected to each node. Note that the voltage across the two halves of node 3 is found to be zero by KVL around the path containing the two sources. Hence no current would flow in the short circuit created by joining the upper ends of the sources. That would put the two sources in parallel, so one of them would be redundant and could be removed, leaving us with the original circuit. Thus, as far as the rest of the circuit besides the sources is concerned, the two situations in Figure 17a and b are completely equivalent.

In Figure 17b, however, both sources are accompanied. Each source and its accompanying resistor constitute a single branch. In this circuit there are three nodes, so only two node equations result. By writing them, you should verify that the same equations result from this circuit as from the original one.

A generalization of the process carried out in Figure 17b can be described as follows. An unaccompanied voltage source has more than one resistor connected to each of its terminals. (That's what makes it unaccompanied.) Imagine sliding or shifting the source through one of its terminals into each of the branches connected there. The resulting circuit is equivalent to the original one in all respects except one: the current in the original source cannot be found from the new circuit, since the identity of the source has vanished. But once all other voltages and currents are found, then the source current is determined by KCL from the original circuit. The process of shifting a source as described is called the *voltage-shift* (or *v-shift*). Confirm for yourself that what was done in the preceding example fits the general process described here.

EXAMPLE 1 As a practical application of this approach, consider the circuit of Figure 18.[8] The three-terminal component lying between nodes B (base), C (collector), and E (emitter) constitutes a model of a transistor under quiescent operating conditions. The base-emitter voltage is assumed to be a constant V_B. The collector supply voltage, V_{CC}, is not in series with a single resistor but it has two resistors joined at each of its terminals. In accordance with the voltage-shift theorem, we shift V_{CC} through its upper terminal into each of the branches (R_1 and R_C) connected there, and then we split the node. The result (after redrawing for convenience) is shown in Figure 18b. R is assumed to be very large and is replaced by an open circuit.

As an exercise, replace the V_{CC}-R_1-R_2 combination by a Thévenin equivalent. Then, from the left-hand loop, find i_B in terms of the given parameters. Finally, from the right-hand loop, find the voltage v_{CE} across the controlled

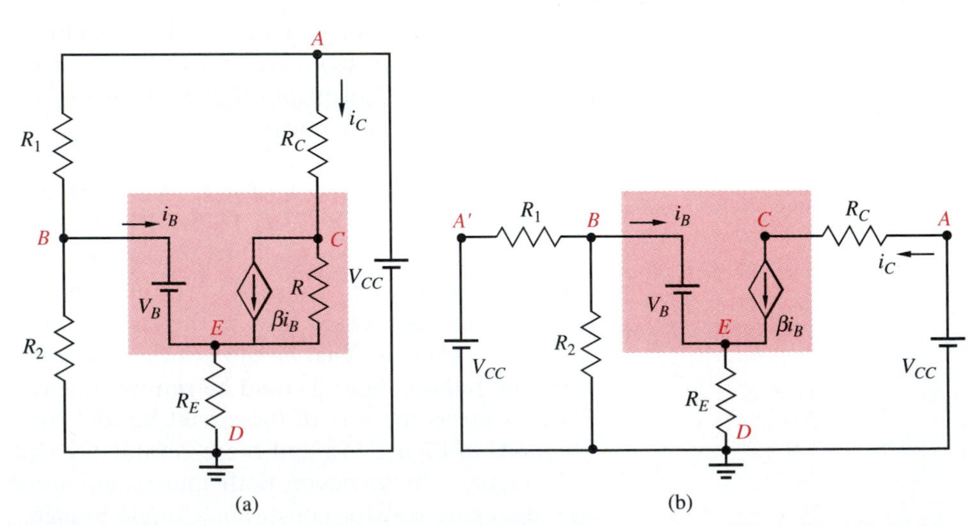

Figure 18 Application of the voltage-source shift to a transistor circuit.

[8] If you skipped the optional section on transistors in Chapter 4, some of the terminology related to transistors in the example may be unfamiliar to you. Nevertheless, there should be nothing strange about the circuit itself; you will not be handicapped in following the discussion.

source, noting that v_{CE} without R_C would be augmented by $-R_C\beta i_B$:

$$i_B = \frac{V_{CC}\dfrac{R_2}{R_1 + R_2} - V_B}{\dfrac{R_1R_2}{R_1 + R_2} + (\beta + 1)R_E}$$

$$v_{CE} = V_{CC} - [\beta R_C + (\beta + 1)R_E]i_B$$

Note that, because the circuit in this example contains a controlled source, it is more general than the class of circuits (resistive) to which we have been limited so far. But neither did we use any of the methods discussed in this chapter, except for the v-shift.

EXERCISE 8

Construct a circuit containing an unaccompanied voltage source which has three branches connected to each of its terminals. (No cheating; they can't all be in parallel.) Redraw the diagram after applying the v-shift. □

4 LOOP AND MESH EQUATIONS

Before starting the subject of this section, let's review what we have done so far in the treatment of dissipative circuits. Given a circuit, we count the number of branches b and the number of nodes n. (For this purpose, sources can be removed; replace voltage sources by short circuits and current sources by open circuits.) We know that $n - 1$ KCL equations are independent and $b - (n - 1)$ KVL equations are independent. It's easy to drop one node—chosen as a datum node—and write KCL equations at all the other nodes. Next—not as easy—we have to select $b - (n - 1)$ independent loops around which to write KVL equations. One surefire set of loops is the set of fundamental loops for some choice of tree. And finally, there are the voltage-curent relationships of the branches: Ohm's law, when the branches are simple resistors.

One approach to "solving the circuit" is to solve all $2b$ of these equations simultaneously. We entertained this idea briefly but gave it up as too complicated.[9] Rather, starting with the $n - 1$ KCL equations, we eliminated

[9] But we did that without taking the power of computers into account. It turns out that this approach isn't out of the question at all, even for circuits containing not only dissipative components, but inductors and capacitors as well. Indeed, all the branch v-i equations can be combined with *both* the independent node equations *and* loop equations to arrive at a superset of equations called the *tableau equations.* It turns out that there are some advantages to the numerical solution of such equations, even though there are many more of them. We will not advance far enough in this book to treat this subject. But these ideas should convince you that an exploratory approach will vastly enhance your engineering education; that, although knowledge of the steps of various algorithms (such as the method of node equations) may be useful, asking why *this* step is taken instead of *that,* or *what if* we were to do the other thing instead, is much more important.

all the branch currents and replaced them by the branch voltages by using Ohm's law. The final step was to replace the b branch voltages by $n - 1$ (fewer) node voltages using KVL. The result was $n - 1$ *node equations* in $n - 1$ unknowns. If $n - 1$ is a low single-digit number, "hand" solutions are "doable"; otherwise, the appropriate computer software is advisable.

In this section we will take up another sequence for applying the fundamental equations. Instead of first KCL, then Ohm's law, then KVL, we reverse the order. First we write the $b - (n - 1)$ KVL equations in the b branch voltage variables around an appropriate set of loops. We then eliminate these voltages, using Ohm's law to replace them by the branch currents. The final step is to replace the b branch currents by a set of $b - (n - 1)$ appropriate currents. (This number is the same as the number of equations.) The selection of these current variables will be an important part of this approach.

Defining Loop Variables

Let's begin the development with a rather modest example, as shown in Figure 19a. This circuit is a model for two generators (with their internal resistances) feeding a load R_3 in parallel. Counting each accompanied source as a branch, the number of branches is $b = 3$. There are two nodes, so $n = 2$. Thus, the number of independent KVL equations is 2. Parenthetically, if we wanted to solve this problem in the simplest way, we would use node equations, since only one equation would be needed. (Do you agree?)

The digraph representing the circuit is shown in Figure 19b. Since $n = 2$, the number of twigs in a tree is 1. Branch 3 (the load) is chosen as a twig and shown in the solid line; the other two branches (the source branches) are the links. Each link, together with the single twig in this case, identifies a fundamental loop. The loops consist of branches $\{1, 3\}$ and $\{2, 3\}$. The three sets of equations for this network can now be written, as follows:

$$\begin{aligned} v_{b1} + v_{b3} &= 0 \\ v_{b2} - v_{b3} &= 0 \end{aligned} \qquad \text{(KVL)} \qquad \begin{aligned} v_{b1} &= -v_s + R_1 i_1 \\ v_{b2} &= v_g + R_2 i_2 \\ v_{b3} &= R_3 i_3 \end{aligned} \qquad (v\text{-}i)$$

$$-i_1 + i_2 + i_3 = 0 \qquad \text{(KCL)}$$

(Since there can't be any confusion in this simple case, we did not use the b subscript for the branch currents.) The next step is to eliminate the branch

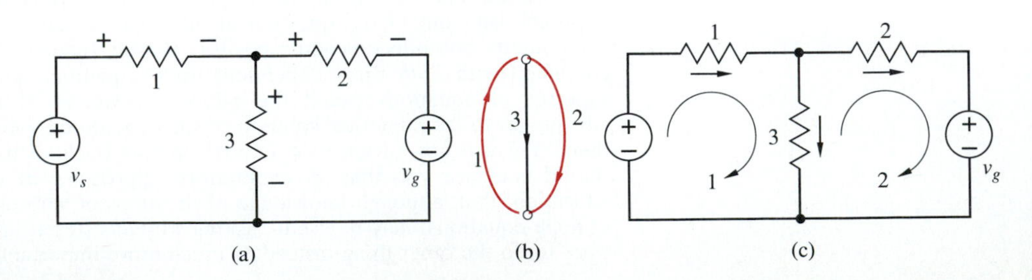

Figure 19
Two generators feeding a load in parallel.

voltages from the KVL equations using the v-i equations. Then we eliminate the twig current (i_3) using KCL. These two steps, sequentially, are as follows:

$$-v_s + R_1 i_1 + R_3 i_3 = [-v_s + R_1 i_1] + [R_3(i_1 - i_2)] = 0$$
$$v_g + R_2 i_2 - R_3 i_3 = [v_g + R_2 i_2] - [R_3(i_1 - i_2)] = 0$$

Notice the terms on the right sides in brackets. Each term represents the voltage of one of the branches expressed in terms of the currents of the links defining the fundamental loops.

Combining Steps in the Process

The "think-think-write" procedure utilized in the case of node equations can also be used in the loop-equation case. We concentrate on one branch at a time (say, branch 3) and we:

Think: "voltage of branch 3"; then
Think: "R_3 times the current in branch 3"; then
Write: "$R_3(i_1 - i_2)$."

This is exactly one of the terms in brackets on the right in the preceding equations. All other branches (in general, not just in this example) would be just like this except for two possibilities. One is the case of a branch that is a link in a fundamental loop; in the final "write" step, then, only that link current would appear, not the difference of two currents. The second possibility is a branch that includes a voltage source, like branches 1 and 2 in this example. Then, in the second "think" step the source voltage, with appropriate sign, would be an additional term.

Finally, to continue with the example, the terms are collected and rearranged; the result is:

$$(R_1 + R_3)i_1 - R_3 i_2 = v_s$$
$$-R_3 i_1 + (R_2 + R_3)i_2 = -v_g$$

These two simultaneous equations in two unknowns are easily solved by standard methods. They are known as *loop equations,* the loops being the fundamental loops for the chosen tree. As is possible with node equations, they, too, can be written in matrix form, as follows:

$$\begin{bmatrix} R_1 + R_3 & -R_3 \\ -R_3 & R_2 + R_3 \end{bmatrix} \begin{bmatrix} i_1 \\ i_2 \end{bmatrix} = \begin{bmatrix} v_s \\ -v_g \end{bmatrix}$$
$$\mathbf{Ri} = \mathbf{v}_s$$

Defining Meshes and Mesh Currents

The procedure followed for this rather simple example can be carried out for circuits of much greater complexity. However, for one class of circuits of modest complexity, we can follow a simplified procedure to arrive at similar results more expeditiously.

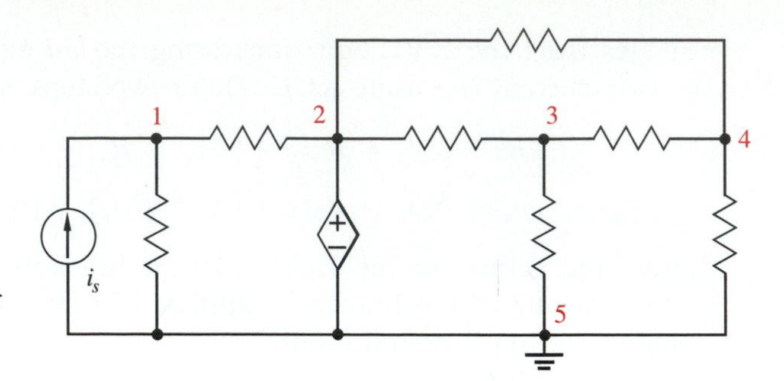

Figure 20
A planar circuit which would become nonplanar by adding a branch between nodes 1 and 3.

A network (or graph) is said to be a *planar* network (or graph) if it can be drawn on a plane without any two of its branches crossing each other. An example of a planar network is shown in Figure 20. Now try to connect a resistor between nodes 1 and 3. It can't be done on the plane without causing branches to cross each other; hence, that network would be nonplanar.

The simple circuit in Figure 19a, which is planar, is redrawn in Figure 19c. In this circuit—in *any* planar circuit—the branches separate the plane into nonoverlapping areas that resemble "windows." A *mesh* is defined as a set of branches that forms a loop but that encloses no other branch of the circuit. Thus, a mesh surrounds a "window." It's easy to see that, in Figure 19c, there are two such meshes; they are identified by the branch sets {1, 3} and {2, 3}. But these were exactly the fundamental loops identified in Figure 19b.

Even though, in this circuit, we don't generate any new equations by this approach, we now have a new viewpoint. With each mesh we identify a circulating *mesh current,* as indicated in Figure 19c. In this example, the mesh currents are identical with the link currents defining the fundamental loops. (Check back in Figure 19b to confirm this.) Then every branch current equals either a mesh current or the difference between two mesh currents—just like the relationship between branch voltages and node voltages in a node-voltage analysis. Temporarily using m as a subscript to represent meshes, the branch currents in terms of mesh currents are:

$$i_1 = i_{m1} \qquad i_2 = i_{m2} \qquad i_3 = i_{m1} - i_{m2}$$

ALGORITHM FOR MESH EQUATIONS The procedure for writing mesh equations is now clear. Given a planar circuit:

1. Identify the meshes and assign a mesh current to each, showing the reference on the diagram.

2. Write (or think of writing) a KVL equation for each mesh in terms of the branch voltages, traversing the mesh in the same orientation as the mesh current.

3. Eliminate (or think of eliminating) the branch voltages by Ohm's law, replacing them with the branch currents.

4. Replace each branch current that is not itself a mesh current by the difference or sum (depending on their orientation) of the currents of the two meshes between which the branch lies.

In the example we have been considering, with the branch current orientations shown, branch current 1 is the same as mesh current 1, branch current 2 is the same as mesh current 2, and branch current 3 is the difference of mesh currents 1 and 2, since branch 3 is common to meshes 1 and 2. Hence, for KVL around mesh 1, we would think: the voltage of branch 1, which is minus the source voltage v_s plus resistor voltage 1, which itself is R_1 times branch current 1, which is the same as mesh current 1. After all that "thinking," we write the term $-v_s + R_1 i_{m1}$. Continuing with the KVL equation for mesh 1, we "think" of adding the voltage of branch 3, which is R_3 times branch current 3, which itself is mesh current 1 minus mesh current 2; we end up "writing" $R_3(i_{m1} - i_{m2})$. Continuing in this way for mesh 2 also results in the two equations:

$$-v_s + R_1 i_{m1} + R_3(i_{m1} - i_{m2}) = 0$$

and

$$R_2 i_{m2} + v_g - R_3(i_{m1} - i_{m2}) = 0$$

Rearranging and collecting terms results in the following pair:

$$(R_1 + R_3)i_{m1} - R_3 i_{m2} = v_s$$
$$-R_3 i_{m1} + (R_2 + R_3)i_{m2} = -v_g$$

It should be clear that a mesh is a loop; it fits the definition of a loop as being a subgraph which has exactly two of its branches incident at each of its nodes. Hence, whatever conclusions we reach about loops in general must apply to the special category of loops called meshes. (Can you make the same statement about the converse?)

Because we had to introduce new ideas, make definitions, and provide explanations as we went along in the preceding development, you may have lost track of the simplicity of the procedure, so let's contemplate a somewhat more extensive circuit.

EXAMPLE 2 The circuit shown in Figure 21 is given. Our objective is to carry out a mesh

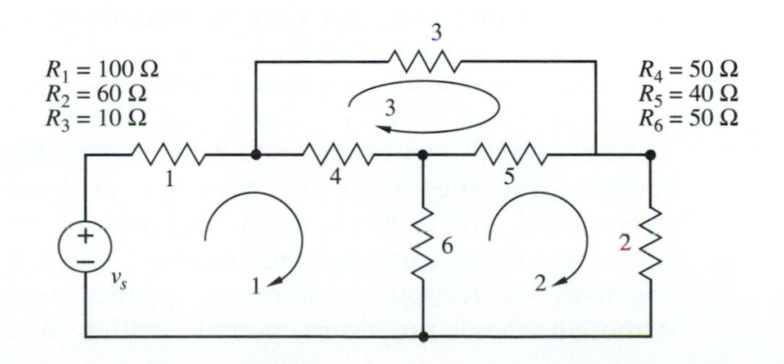

Figure 21
Mesh analysis of a bridged-tee circuit.

analysis. The first step is to establish the number of independent mesh equations. Counting the accompanied source as a single branch, this circuit has $b = 6$ branches and $n = 4$ nodes. The number of independent loop equations, therefore, is 3.

Although the mesh orientations are shown in this circuit, no branch references are shown. The reason is that, as in the case of node analysis, we will not use fixed references for the branch variables. Instead, when writing the equation for a particular mesh, we will assume that branch current references coincide with the orientation of that mesh current. With this convention, branch current references change from mesh to mesh. Just as we did in the case of node equations, we also assume that all branches have standard references. Hence, branch voltage references will also change from mesh to mesh. Consequently, *the sign in Ohm's law will not change* and the ultimate equations will be the same whether we use fixed branch references or not.

In the four-step procedure outlined earlier in this section, the step of identifying the mesh currents appears on the circuit diagram. The next step is (1) to write KVL around each mesh, then (2) to insert the branch v-i equations, and then (3) to express the branch currents in terms of mesh currents, ending with equations whose unknowns are mesh currents. For simplicity, we will not use the m subscript for mesh currents. Carrying out the "think-think-write" process, we will show only the final result here:

$$-v_s + 100i_1 + 50(i_1 - i_3) + 50(i_1 - i_2) = 0$$

$$60i_2 + 50(i_2 - i_1) + 40(i_2 - i_3) = 0$$

$$10i_3 + 40(i_3 - i_2) + 50(i_3 - i_1) = 0$$

(Carry out each step and confirm the result.) Rearranging and collecting terms is then a simple step. In matrix form, it is:

$$\begin{bmatrix} 200 & -50 & -50 \\ -50 & 150 & -40 \\ -50 & -40 & 100 \end{bmatrix} \begin{bmatrix} i_1 \\ i_2 \\ i_3 \end{bmatrix} = \begin{bmatrix} v_s \\ 0 \\ 0 \end{bmatrix}$$

$$\mathbf{Ri} = \mathbf{v}$$

In the last line, each array has been given a letter designation; **R** is the resistance matrix and **i** and **v** are the vectors of source currents and voltages.

Although this expression came from an example, again we can identify some general properties valid for a circuit of linear resistors. The matrix is symmetrical. The elements on the diagonal are the sums of resistances on the contours of the corresponding meshes. The off-diagonal elements are the negatives of resistances common to the corresponding pair of meshes. Although these properties permit writing a set of mesh equations by inspection of a circuit diagram, they do not apply to circuits that include multiterminal dissipative components, such as controlled sources. Hence, memorizing the structure of the matrix in the preceding example is of limited utility. Nevertheless, you may wish to use such simplifications in those cases to which they apply.

What would happen in the analysis of Figure 21 if the voltage source were unaccompanied, that is, if $R_1 = 0$ instead of $100\,\Omega$? Nothing dramatic, only that the coefficient of i_1 in the first mesh equation would be reduced by 100. That is, nothing special needs to be done in loop (or mesh) analysis if a circuit contains an unaccompanied voltage source.

One more observation about the preceding circuit. Try using some of the methods in Chapter 3 to arrive at a solution. Series and parallel equivalences, source transformations, voltage and current dividers, and other methods introduced there all fail to yield a solution. Without branch 3, we would be successful in this attempt. Branch 3 is a "bridge" over branches 4 and 5. Branches 4, 5, and 6 constitute a "tee" structure; with branch 3, the result is a "bridged tee." The entire circuit in the diagram is not in series-parallel form, so the methods of Chapter 3 are not applicable.

Treating Current Sources in Loop Equations

The only sources appearing in the preceding development were voltage sources. No big complication results from the introduction of an *accompanied* current source.

Such a circuit, having five branches and four nodes, is shown in Figure 22a. The source and its accompanying resistor are counted as a single branch. Hence the number of independent loop equations is 2. The circuit diagram identifies two meshes. The *i-v* equation of the branch involving the current source (taking the current reference to coincide with that of mesh current 2) is:

$$i_5 = G_5 v_5 - i_s$$
$$v_5 = R_5 i_5 + R_5 i_s$$

(Note that no branch subscripts are used, because there is no danger of confusion here.) The second line is obtained by solving the first line for the branch voltage. This *v-i* relationship is what would be used when replacing the branch voltages in the KVL equation for mesh 2. As an exercise, carry out the "think-think-write" process and write the two mesh equations, using the preceding *v-i* relationship for the accompanied source (do it before looking at

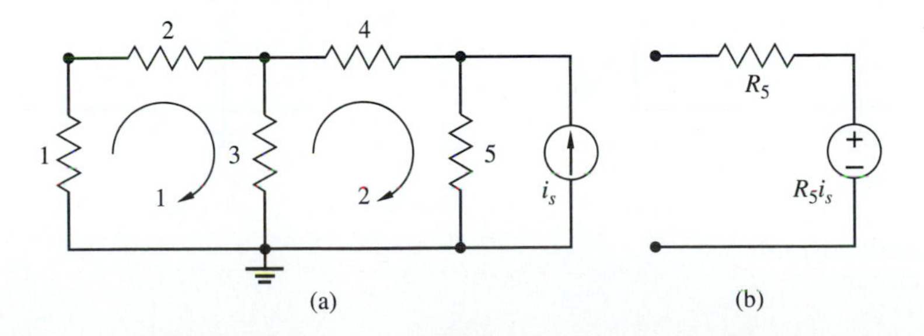

Figure 22
Current source in mesh analysis.

the answer that follows):

$$R_1 i_1 + R_2 i_1 + R_3(i_1 - i_2) = 0$$
$$R_4 i_2 + R_5 i_2 + R_5 i_s + R_3(i_2 - i_1) = 0$$

Another approach that can be taken, when there is an accompanied current source in a circuit which is a candidate for mesh analysis, is to convert to a voltage-source equivalent. For the case at hand, the equivalent accompanied voltage source is shown in Figure 22b. If this equivalent replaces the accompanied current source in Figure 22a, the resulting mesh equation for mesh 2 will be identical with the equation found already. Hence, there is no real need to make the source conversion.

UNACCOMPANIED CURRENT SOURCES As just noted, no big complication results from the presence of an accompanied current source in carrying out a loop analysis, but what about an *unaccompanied* current source? An example is shown in Figure 23a. The circuit is the same as the one in Figure 17a except that the voltage source is replaced by a current source. This circuit has $b = 5$ nonsource branches and $n = 4$ nodes. Hence, there will be $n - 1 = 3$ independent KCL equations. But now there will be only $b - (n - 1) = 2$ independent KVL equations. Thus, there will be just two loop (or mesh) equations. For the purpose of identifying the meshes to be used for writing mesh equations, *and only for this purpose,* the unaccompanied current source can be replaced by an open circuit. As an exercise, write a pair of mesh equations around the two meshes that do not include the current source in the original circuit. Once these equations are solved, all currents and voltages in the circuit will become known, except for the voltage across the current source. This can now be determined by writing a KVL equation around any loop containing the current source.

The presence of unaccompanied current sources, if anything, reduces the amount of effort when carrying out a loop analysis, since one fewer loop

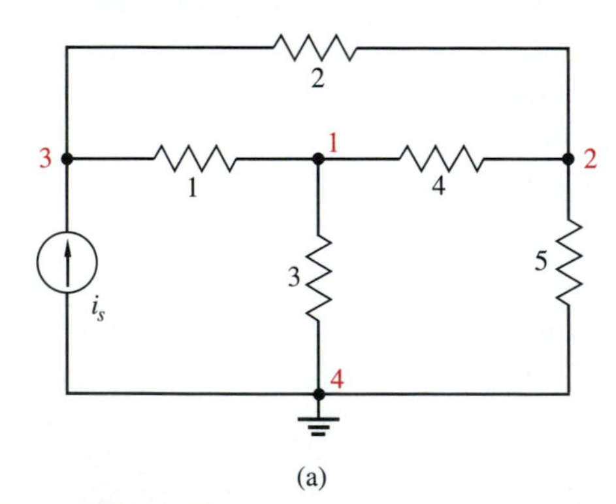

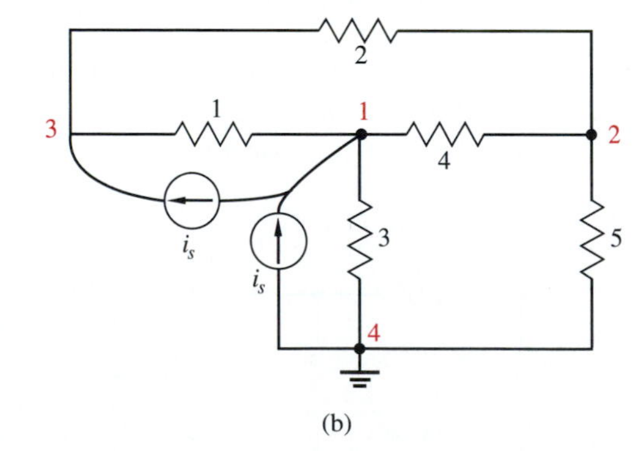

Figure 23
Unaccompanied current source in a circuit.

equation (around a loop containing that current source) will be required for each unacompanied source. But then one additional equation must be written for each current source to determine its voltage.

THE CURRENT-SOURCE SHIFT When confronted with the task of performing a node analysis in the presence of an unaccompanied voltage source, one of the approaches we took was the voltage shift in Figure 17. A similar (dual) approach can be taken in the presence of an unaccompanied current source when performing a loop analysis. Again we do something peculiar, as shown in Figure 23b. We again introduce a second source, but this time connected to node 1, as shown. Note that no current flows to node 1 from the two equal sources connected there. Hence, the connection of the sources to node 1 could be opened. That would put the two sources, carrying the same current, in series. Hence one of them would be superfluous and could be removed, while still maintaining the same current. That would return us to the original circuit.

In the new circuit, a current source i_s is directed away from node 4 and toward node 3; there is no net current source at node 1. This is exactly the situation in the original circuit. Thus the two circuits are equivalent in every respect except that the voltage across the original current source cannot be found in the new circuit because the identity of the original source is lost there. Once the appropriate node equations are written and solved for the node voltages, the resistive branch voltages in the original circuit are easily obtained. Then, in the original circuit, the voltage across the source is obtained by applying KVL around any loop containing the current source. This step is the only one that cannot be carried out in the new circuit. Observe that carrying out an i-shift results in an open circuit at the position of the original current source. Thus, the number of loops (or meshes) has been reduced by 1, just as we observed before. The process of moving the current source as described is called the *current-shift* (or *i-shift*).

A generalization of this procedure can be described as follows. Given a circuit with an unaccompanied current source having more than one resistor connected at each of its terminals (but none of them connected across its two terminals), shift the current source around any loop containing it so that it is placed in parallel with each resistor on that loop. The resulting circuit will be equivalent to the original circuit in every respect except one, namely, that the voltage across the current source cannot be found in the new circuit. To find this voltage requires returning to the original circuit after all other unknowns have been found in the new circuit. Confirm that what was done in the preceding example fits this description of the general process.

EXERCISE 9

Construct a planar circuit with an unaccompanied current source having at least three resistors around each mesh on which the current source lies. Apply the i-shift and redraw the diagram. □

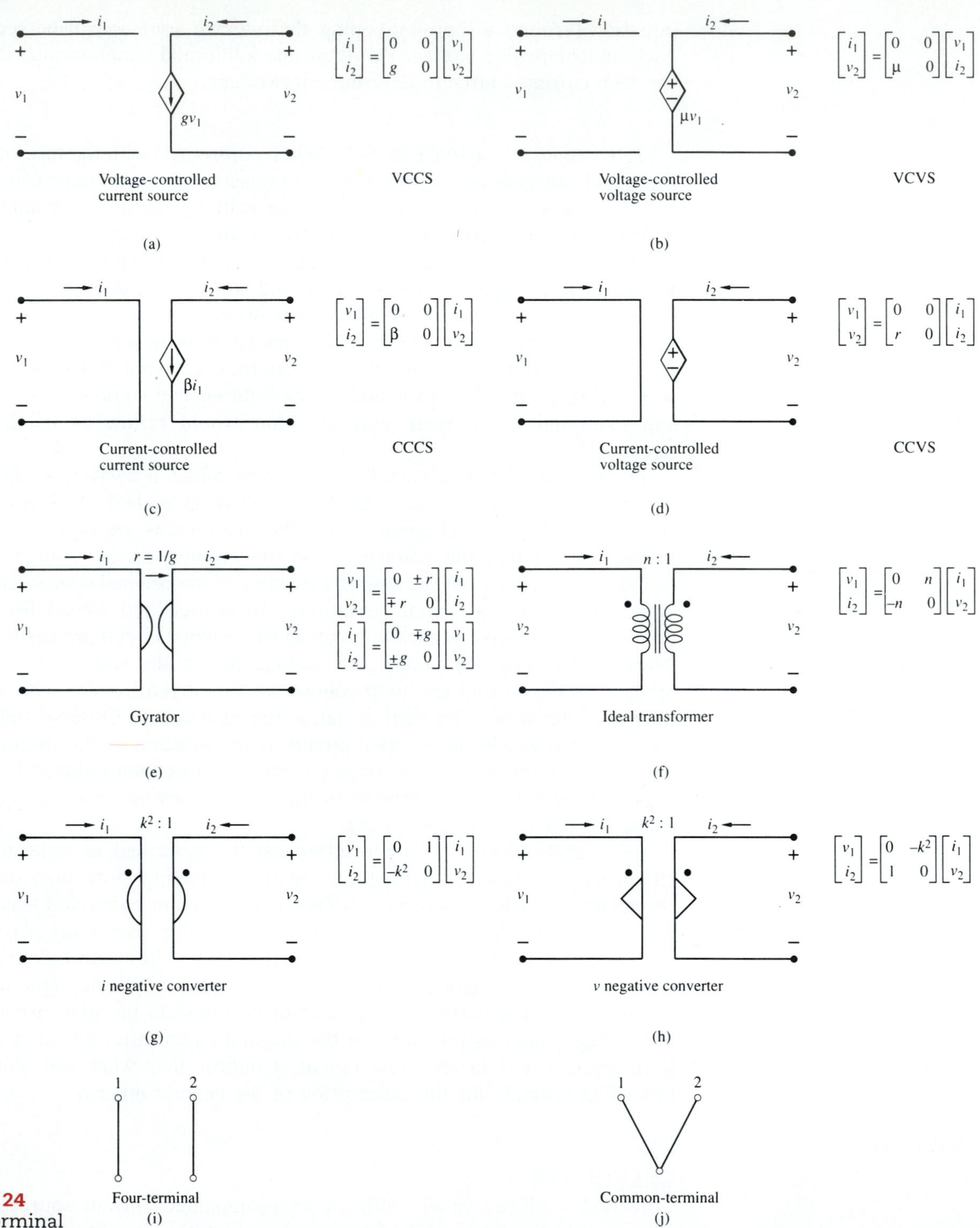

Figure 24 Four terminal components and their graphs

5 ANALYSIS OF GENERAL DISSIPATIVE CIRCUITS

Until now, except for an example, the only components, in addition to independent sources, that were included in the circuits we treated were resistors. We will now generalize to include also the dissipative multiterminal components introduced in the preceding chapter. These primarily include controlled sources and ideal transformers, but also gyrators and negative converters. The ideal op amp is not included in this group, although a nonideal op amp with a model consisting of a controlled source, together with input and output resistances, is included.

Models and Graphs of Multiterminal Components

The components under consideration here have four terminals, two belonging to an input branch and two to an output branch. Each of the components and their defining equations are shown in Figure 24, together with a two-branch graph representing each one. The graph is not connected; however, on many occasions, two terminals, one from each side, of these four-terminal components are connected together, often as a common ground, thus converting them to three-terminal components and turning their graphs into connected graphs. This possibility is also shown.

While one of the two branch equations for the controlled sources is degenerate (i_1 or $v_1 = 0$), the equations for the ideal transformer, the gyrator, and the negative converter are full; that is, neither of them is degenerate. This makes it possible to represent each of these components by an equivalent four-terminal circuit consisting of two controlled sources. The equivalent circuits are shown in Figure 25. Thus, if we were to augment the class of

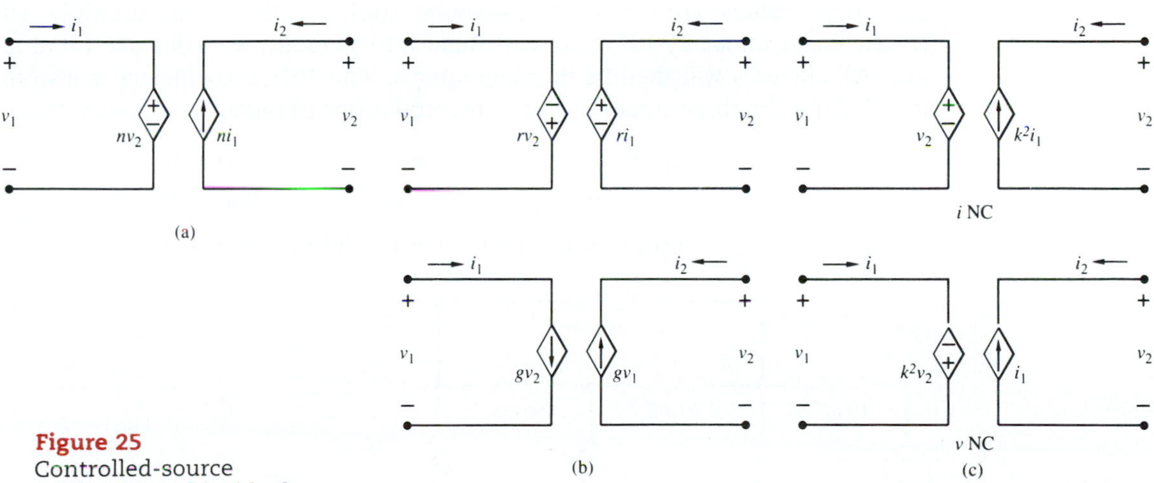

Figure 25
Controlled-source equivalents of (a) ideal transformer, (b) gyrator, and (c) negative converter.

circuits we have been studying by adding controlled sources to resistors and independent sources, we would be treating a relatively broad class of circuits.

Permissible Components in Node Analysis

Review the procedure for conducting a node analysis of a circuit. Into the KCL equations we insert the branch i-v relations in the voltage-controlled form $i = Gv$. This representation is always possible for a resistor, but what about controlled sources? Looking at Figure 24, you will see that only a voltage-controlled current source (VCCS) and a gyrator have equations that can be expressed in this form. Hence, the class of circuits for which a node analysis can be carried out directly, without modification or special procedures, includes only VCCSs and gyrators besides independent sources and two-terminal resistors. If any other multiterminal devices appear in a circuit, something else must be done before a node analysis is possible. These other cases will be treated shortly; for now let's examine circuits containing permissible components.

CIRCUITS CONTAINING A VOLTAGE-CONTROLLED CURRENT SOURCE

One of the multiterminal components that permits a node analysis is a VCCS.

EXAMPLE 3 A circuit that includes a voltage-controlled current source is shown in Figure 26. Since this controlled source has a conductance representation, a node analysis is possible. We choose node 4 as a datum node. There are three node-voltage unknowns, so a set of three node equations can be written. For nodes 1 and 2, there is nothing that is different from a resistive circuit. One of the branches connected to node 3, however, is the controlled source. When we carry out the "think-think-write" process for this branch, the current i_{cs} is eliminated not by Ohm's law in terms of a difference of node voltages but by the controlling equation of the VCCS in terms of the controlling voltage. The numerical values are all in millisiemens (mS). Rather than multiply all conductance values by 10^3 to convert them into siemens, we will leave them in mS. All currents will then be in milliamperes. The VCCS coefficient is also in mS. Taking the three nodes in order, the equations obtained in this way are:

$$10(v_1 - v_s) + 20v_1 + 30(v_1 - v_3) + 40(v_1 - v_2) = 0$$

$$0(v_2 - v_1) + 50(v_2 - v_3) + 20v_2 = 0$$

$$30(v_3 - v_1) + 10(v_1 - v_3) + 50(v_3 - v_2) = 0$$

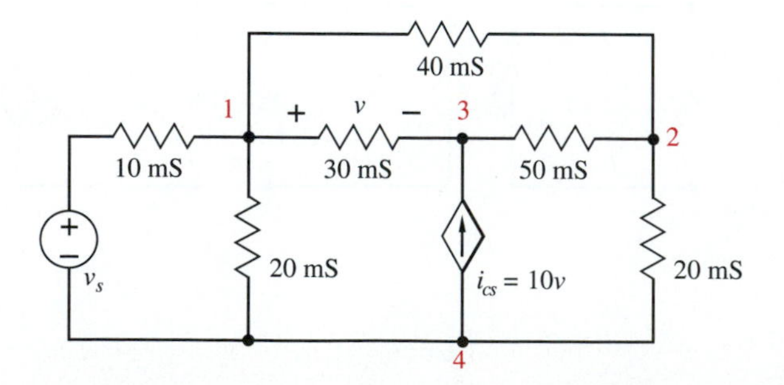

Figure 26
Node equations in a circuit with a VCCS.

In the last equation, the controlling voltage of the controlled source has been replaced by $v_1 - v_3$. (Make sure you confirm these equations.) The presence of the VCCS causes no special difficulties. Dimensionally, each term is a current in mA. Further numerical simplification is possible by dividing each equation through by 10. The equations are still dimensionally homogeneous with the voltages in volts. Collecting terms, the new equations can be rewritten in both scalar and matrix forms as:

$$\begin{aligned} 10v_1 - 4v_2 - 3v_3 &= v_s \\ -4v_1 + 11v_2 - 5v_3 &= 0 \\ -2v_1 - 5v_2 + 7v_3 &= 0 \end{aligned} \qquad \begin{bmatrix} 10 & -4 & -3 \\ -4 & 11 & -5 \\ -2 & -5 & 7 \end{bmatrix} \begin{bmatrix} v_1 \\ v_2 \\ v_3 \end{bmatrix} = \begin{bmatrix} v_s \\ 0 \\ 0 \end{bmatrix}$$

Because of the controlled source, we would not expect the matrix to be symmetrical—and it isn't.

Each node voltage is obtained by solving these equations. The solution is straightforward, and you are urged to carry it out. (Use MATLAB if it is available to you.) With the node voltages determined, every branch current will follow from Ohm's law, except for the current in the controlled source. This can be obtained from the controlling relationship or from KCL at node 3. Do it both ways and confirm that the same result is obtained.

CIRCUITS CONTAINING A GYRATOR[10] The only other component that has a conductance representation and permits a node analysis without modification is a gyrator.

EXAMPLE 4 The circuit in Figure 27a contains a gyrator, with a common terminal. In Figure 27b the gyrator has been replaced by a pair of VCCSs whose controlling port

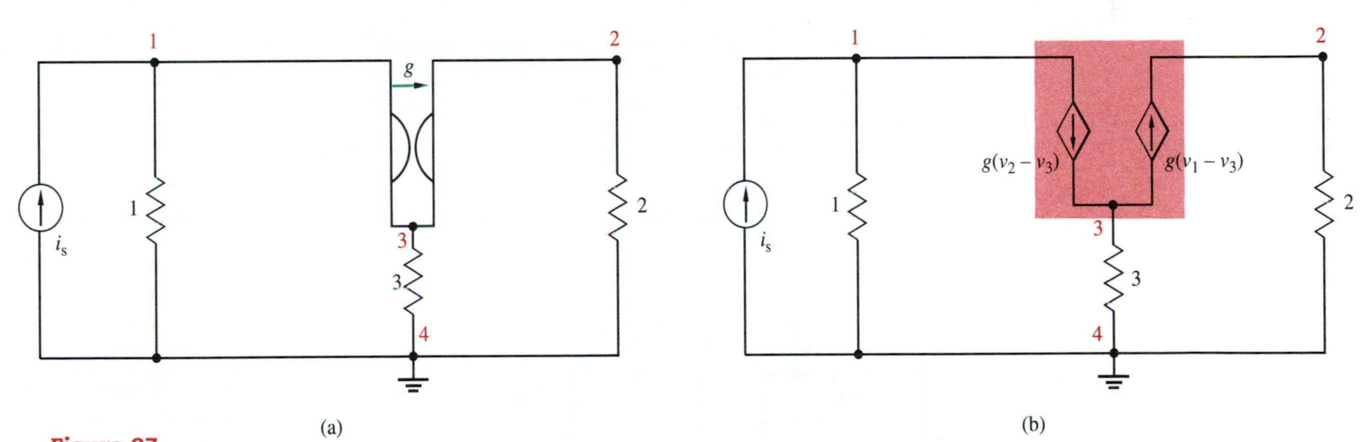

Figure 27
Node equations for a gyrator circuit.

[10] If you skipped the section on gyrators in Chapter 4, you can skip this section also.

voltages have been expressed in terms of the node voltages. (You should confirm these experessions.) The goal is to compose a set of node equations (there will be 3) describing this circuit. We arbitrarily select node 4 as the datum.

To write the node equations at node 1, we "think": the current leaving node 1 through the composite branch that includes the current source (which is $G_1v_1 - i_s$) plus the current leaving node 1 through the input port of the gyrator equals zero. In a similar manner, we "think" KCL at the other two nodes and we "think" of replacing the currents by the appropriate i-v equations. Then we express all branch voltages in terms of the corresponding node voltages and write the result as follows:

$$\begin{aligned} (G_1v_1 - i_s) + g(v_2 - v_3) &= 0 \qquad & G_1v_1 + gv_2 - gv_3 &= i_s \\ -g(v_1 - v_3) + G_2v_2 &= 0 \qquad & -gv_1 + G_2v_2 + gv_3 &= 0 \\ g(v_1 - v_3) - g(v_2 - v_3) + G_3v_3 &= 0 \qquad & gv_1 - gv_2 + G_3v_3 &= 0 \end{aligned}$$

To arrive at the set on the right, terms in the left-hand set of equations were collected and rearranged. In matrix form, this can be rewritten as

$$\begin{bmatrix} G_1 & g & -g \\ -g & G_2 & g \\ g & -g & G_3 \end{bmatrix} \begin{bmatrix} v_1 \\ v_2 \\ v_3 \end{bmatrix} = \begin{bmatrix} i_s \\ 0 \\ 0 \end{bmatrix}$$

$$\mathbf{Gv} = \mathbf{i}_s$$

The node conductance matrix in this case, still called **G**, no longer has the properties that are valid for a linear resistive circuit. We see that it is not symmetrical, and we would not be able to construct it by inspection of the circuit. Not only is it not symmetrical, it is *antimetrical*; that is, $a_{ij} = -a_{ji}$, just as in the case of the gyrator.

EXERCISE 10

Write a set of node equations for the circuit of Figure 28. From this set, construct the node conductance matrix **G**. Is it symmetrical? (Note that the numbers beside the resistors are not numerical values in Ω but identifications—

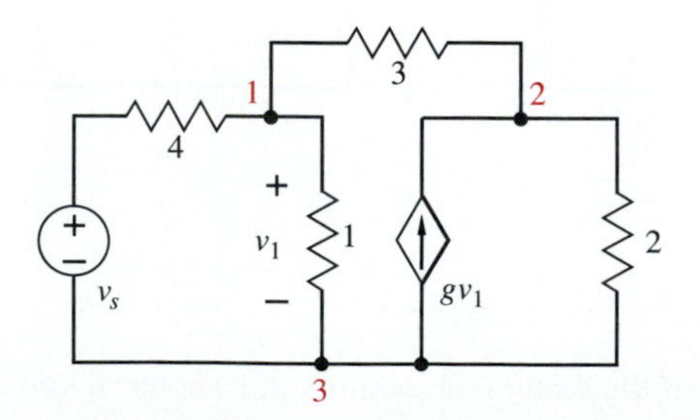

Figure 28
Exercise 10 circuit.

e.g., R_1, G_2.

ANSWER: $\begin{bmatrix} (G_1 + G_3 + G_4) & -G_3 \\ -(G_3 + g) & (G_2 + G_3) \end{bmatrix}$ □

CIRCUITS CONTAINING A CURRENT-CONTROLLED CURRENT SOURCE The equation describing a CCCS is in the form $i_{cs} = \alpha i_{\text{in}}$. What prevents a node analysis in a circuit with a CCCS is the *controlling* current i_{in}. We would be able to proceed if this could be converted to a voltage, but how is that possible? (1) It's easily done if there is a resistor in series with the controlling branch, since the current in this branch would be proportional to the resistor voltage. Although, in any practical circuit, a resistor is bound to exist in any current path, we can always concoct a circuit in which this is not the case. (2) Even if the controlling current is not the current in a resistor, KCL might permit it to be written in terms of resistor currents.

Figure 29a shows a case in point. Here the controlling current is the current in the voltage source. By KCL it can be written as $i_1 + i_2$, each term of which is the current in a resistor. If node 3 is chosen as the datum, only two node voltages are unknown, v_1 and v_2. (The voltage source can be taken as a supernode.) At node 2, we "think" KCL; then we "think" of eliminating the currents in favor of the branch voltages. However, an added step is needed: replacing $i_{cs} = \alpha i$ by $i_{cs} = \alpha(i_1 + i_2)$, and then eliminating i_1 and i_2 in favor of the resistor voltages. In the final step, the branch voltages are expressed in terms of the node voltages. The result obtained by this process is the following:

$$G_3(v_1 - v_2) + G_4 v_1 + G_2(v_1 - v_s) = 0$$
$$G_1(v_2 - v_s) + G_3(v_2 - v_1) + \alpha[G_1(v_s - v_2) + G_2(v_s - v_1)] = 0$$

The terms are now collected to obtain:

$$(G_2 + G_3 + G_4)v_1 - G_3 v_2 = G_2 v_s$$
$$-(G_3 + \alpha G_2)v_1 + [G_1(1 - \alpha) + G_3]v_2 = [G_1(1 - \alpha) - \alpha G_2]v_s$$

(Confirm these equations.) As expected, the coefficient matrix isn't symmetrical.

These equations are standard node equations whose variables are node

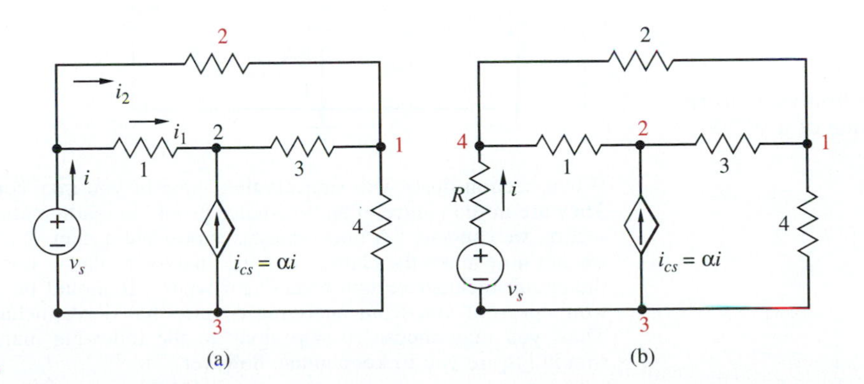

Figure 29 Node equations in a circuit containing a CCCS.

voltages and whose coefficients are dimensionally conductance. After the equations are solved for the node voltages, the only remaining unknown in the circuit is the controlling current in the voltage source. It is found by applying KCL. Thus, even though a CCCS does not have a conductance representation, a node analysis is possible in a circuit with such a component, leading to standard node equations, but at the expense of some preliminary manipulation of the circuit.

An alternative approach is to *augment* the circuit by *adding* a resistor in series with the controlling current. Then the controlling variable can again be made a voltage. But, you might object, it won't be the same circuit; how can we find the solution of *one* circuit by creating and solving a different circuit? Yes, but once you have a solution that includes the extraneous resistance, set that resistance equal to zero; the circuit will revert to the original form and the solution will reduce to what it should be for the original circuit.

EXERCISE 11

Carry out for the circuit in Figure 29b the process just described, and confirm the previous set of equations. □

Modified Node Equations with Controlled Voltage Sources[11]

As noted before, controlled voltage sources of either type do not have a conductance representation. Their presence in a circuit, thus, prevents a node analysis. We will now describe a variation on the node analysis procedure that leads to a modified set of equations that can be called *quasi-node equations.*

MODIFIED EQUATIONS WITH A VOLTAGE-CONTROLLED VOLTAGE SOURCE The circuit in Figure 30 includes a VCVS whose presence

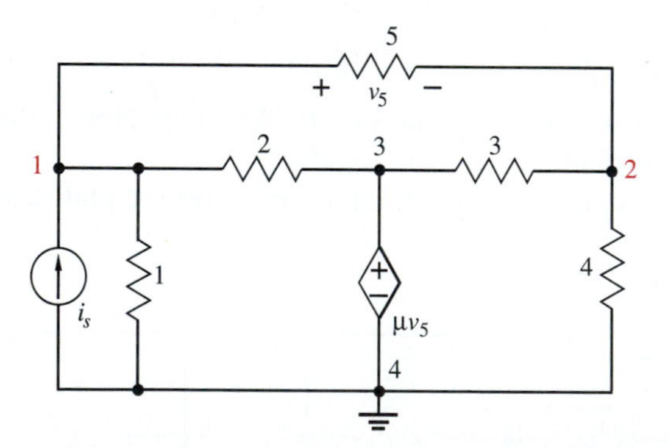

Figure 30
"Node analysis" in the presence of a VCVS.

[11] This section deals with subjects that some of you may consider of secondary importance. They are in the nature of an "existence proof" in mathematics. Without the discussion of this section we can say, "A node analysis is possible except. . . . " With the added material here, we can now make the claim, "A node analysis is *always* possible if you do this or that when the circuit includes certain component types." It should be included for completeness but it won't prevent you from analyzing circuits that don't include these offending components. Thus, you may choose to skip along to the following major section. Intellectual curiosity should inspire you to keep going, however.

prevents the writing of a complete set of node equations. Suppose node 4 is chosen as the datum. How should the other terminal of this controlled source be treated? If the controlled source had been an *independent* voltage source, the voltage of node 3 would have been known, leaving only two unknown node voltages. Node equations would be written at those two nodes only. The independent source would constitute a supernode.

In the present case, the voltage of node 3 is *not* known. Nevertheless, the controlled source still constitutes a supernode (thus joining node 3 with datum node 4), and we could still write node equations at nodes 1 and 2 in the usual way. The result of this process will be:

$$G_1 v_1 + G_2(v_1 - v_3) + G_5(v_1 - v_2) = i_s$$
$$G_3(v_2 - v_3) + G_4 v_2 + G_5(v_2 - v_1) = 0$$

These are two equations in three unknowns. However, the controlling relationship of the controlled source gives one more equation in these same three variables:

$$v_3 = \mu(v_1 - v_2)$$

This equation *is not a node equation.* Each term in a node equation is dimensionally a current; not so in this equation, where each term is a voltage. Nevertheless, it constitutes an independent equation whose variables are node voltages. Together with the previous two node equations, it permits solving for all node-voltage unknowns.

Once the node voltages are known, all branch currents, *except the one in the controlled source,* are determined from Ohm's law. Finally, the current in the controlled source is obtained in terms of other currents using KCL at node 3.

The equations obtained in this example constitute a set of modified node equations we have called *quasi-node equations.* The variables are all node voltages, although not all of the equations are in the usual form of node equations, obtained by summing currents. The terms in one of the equations are dimensionally voltage, not current.

How general is this example? Is it possible that other circuits containing VCVSs can differ so greatly from this example that similar results cannot be obtained? The answer is no. The only thing special in this circuit is that one terminal of the VCVS was chosen as the datum. But choosing one terminal of a source as the datum *always* facilitates the analysis, not just for controlled sources but for independent sources as well.[12] The same general ideas will apply if neither terminal of the controlled source is taken as the datum.

[12] As a general goal in your professional life, whenever there are choices to be made which are otherwise arbitrary, always make the choice that leads to simplification and convenience. In some situations, however, it may not always be clear which choice does this. In the case of choosing datum nodes, however, it *is* clear: choose the datum node as one terminal of as many sources as possible. In the example here, the independent and controlled sources have one terminal in common and this one has been chosen as the datum. One caveat, however. Sometimes (in fact, most of the time) engineering choices are not arbitrary, and each has different consequences—for safety, for cost, for environmental impact, for ethics. Obviously, the preceding discussion does not apply to these cases; simplification and convenience are not the criteria in such cases.

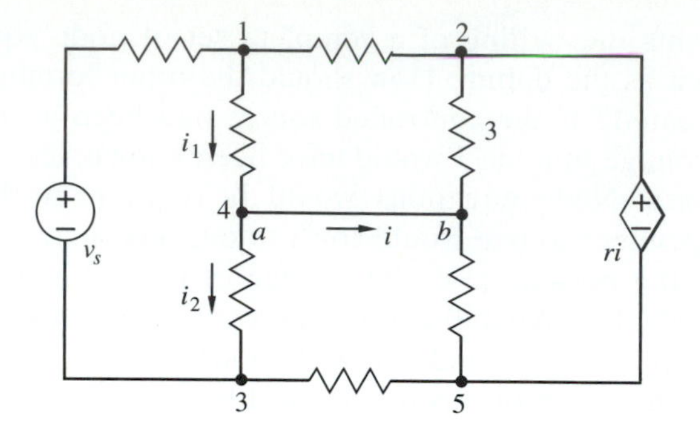

Figure 31
Node equations in circuit with a CCVS.

As an exercise, choose node 2 as datum and attempt to write a set of node equations. Again the controlled source can be taken as a supernode. A node equation at this supernode and one at node 1 still constitute two equations in three unknowns. As before, the controlled-source relationship yields yet another equation; in this case, it will be $v_3 - v_4 = \mu v_1$. There are, thus, three equations in three node-voltage unknowns. Again, after these are solved, all branch currents, except the one in the controlled source, are obtained from Ohm's law. The current in the controlled source follows from KCL applied at one of the controlled-source terminals. This set of equations also can be considered as a modified set of node equations. Go through the details to obtain these equations.

MODIFIED EQUATIONS WITH CURRENT-CONTROLLED VOLTAGE SOURCES Now let's consider the other kind of controlled voltage source, a CCVS. Now it's the *controlling* branch that will cause the trouble, just as it was for the CCCS. An example is shown in the circuit of Figure 31. What is needed again is somehow to convert the controlling variable to a voltage. This is easily done if there is a resistor in the controlling branch, as already described, since the current is proportional to the resistor voltage. Hence, the controlling variable becomes the resistor voltage and we would be back to the case of the preceding section. This is not the case in Figure 31. However, as with the CCCS in Figure 29a, KCL is invoked to express the controlling current in terms of other currents (in this case $i = i_1 - i_2$). It is possible that these other currents are the currents in resistors, as is the case here. Then the controlling variables can again be made resistor voltages, and we can proceed as with the VCVS. As an exercise, write the expression for the controlled voltage of the source in terms of node voltages in this circuit.

The alternative approach of (1) augmenting the circuit with a resistor, R, in the controlling branch, (2) writing and solving the resulting quasi-node equations, and (3) then letting R go to zero is also valid. An example of this alternative follows.

EXAMPLE 5 In Figure 32a there is already a resistor in series with the controlling branch of the controlled source, but, to illustrate the procedure just described, we will add another series resistor, as in Figure 32b. Node 5 is chosen as datum. A

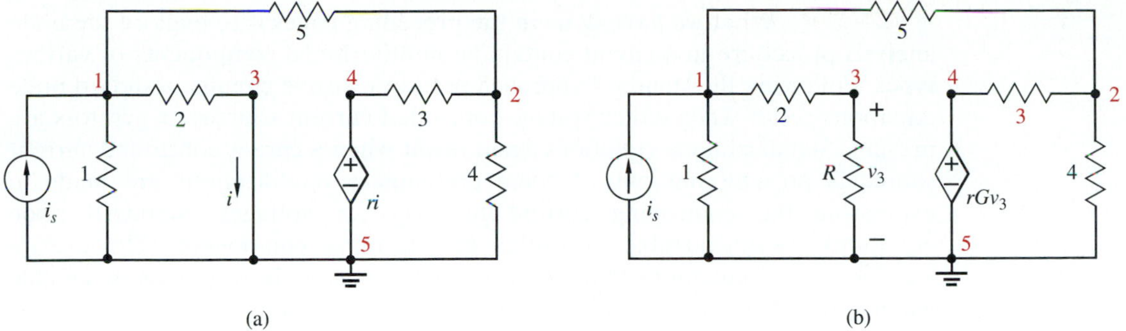

Figure 32
Augmenting a circuit containing a VCVS, then letting $R \to 0$.

node equation at node 4 is unnecessary, since its voltage is proportional to another node voltage, v_3. The remaining node equations are:

$$G_1 v_1 + G_2(v_1 - v_3) + G_5(v_1 - v_2) = i_s$$

$$G_4 v_2 + G_3(v_2 - v_4) + G_5(v_2 - v_1) = 0$$

$$G_2(v_3 - v_1) + G v_3 = 0$$

There are four unknowns in these three equations; a fourth equation comes from the contolled source: $v_4 = rGv_3$. The solution of this modified set of node equations gives all the node voltages. The solutions for v_2 and v_3, for example, are:

$$v_3 = \frac{G_2 R i_s}{(G_1 + G_2 + G_3)(1 + RG_2) - RG_2^2 - \dfrac{G_5[rG_2G_3 + G_5(1 + RG_2)]}{G_3 + G_4 + G_5}}$$

$$\to 0 \quad \text{as} \quad R \to 0$$

$$v_2 = \frac{[rG_2G_3 + G_5(1 + RG_2)]i_s}{(G_3 + G_4 + G_5)[(G_1 + G_2 + G_5)(1 + RG_2) - RG_2^2] - G_5[rG_2G_3 + G_5(1 + RG_2)]}$$

$$\to \frac{(rG_2G_3 + G_5)i_s}{(G_1 + G_2 + G_5)(G_3 + G_4 + G_5) - G_5(rG_2G_3 + G_5)} \quad \text{as} \quad R \to 0$$

(You should confirm these expressions by solving the preceding equations.) When $R = 0$ is substituted into these solutions, v_3 goes to zero, as it should. If you want to convince yourself that the expression obtained for v_2 is the correct one, you can repeat the process, this time using the existing R_2 to change the dependence of the controlled source from i to the voltage across R_2. Now there can be no worry about having changed the circuit.

REVIEW What we have done in the preceding pages is to explore the node analysis procedure in a circuit containing multiterminal components of various types. Following the standard approach used in resistive circuits, standard node equations result when either voltage-controlled current sources or gyrators are present. Standard node equations again result when a curent-controlled current source is present, but only if some preliminary modifications are made in expressing the controlling current in terms of voltages. Standard node equations are *not* possible with other multiterminal components. However, a modified set of equations that we have called quasi-node equations is possible when controlled voltage sources of either type are present.

Permissible Components in Loop Analysis

Let's now turn our attention to a loop analysis of circuits containing multiterminal components. The loops involved can be meshes or any other appropriate set of loops, such as the fundamental loops. Recall that the procedure consists of first writing KVL equations and then eliminating the branch voltages using the branch v-i equations in the current-controlled form $v = Ri$. It must be possible for the branch relationships to take on this form if the procedure is to work.

Looking back at Figure 24, you can see that the only multiterminal components whose v-i relationships can be written in the required form are current-controlled voltage sources (CCVSs) or gyrators. In circuits including such devices, a loop analysis can proceed directly.

To illustrate the procedure, consider the circuit in Figure 33, which contains a CCVS. This is a planar circuit with four nodes and six branches, including the controlled source.[13] Thus there are three independent KVL equations. The three meshes are identified by the circulating mesh currents. In terms of the mesh currents, the controlling current is $i_1 - i_2$. Using the "think-think-write" procedure, the mesh equations are written as:

$$-v_s + R_1 i_1 + R_5(i_1 - i_2) = 0$$

$$R_5(i_2 - i_1) + R_2 i_2 + R_4(i_2 - i_3) + r(i_1 - i_2) = 0$$

$$-r(i_1 - i_2) + R_4(i_3 - i_2) + R_3 i_3 = 0$$

[13] Strictly speaking, the controlling branch of the source is the short circuit through which the controlling current is flowing. But the short circuit is in series with resistor R_5 and the two currents are the same. Therefore we eliminate the "short-circuit" branch and also the "node" between it and R_5. All of this seems to be trivial, but two points can be made as to why it would be important at least to pass the idea through your mind. One is that the same thinking applies to two resistors in series when carrying out a loop analysis; we might combine the two series resistors and eliminate the node between them from the count of branches and nodes. The second is that if we were to write an algorithm and develop some software for carrying out this analysis, it would be simplest not to have the program do something special for special types of branches but to treat all branches alike—even short-circuit or open-circuit controlling branches of controlled sources. It might be, however, that other considerations outweigh this one in special cases.

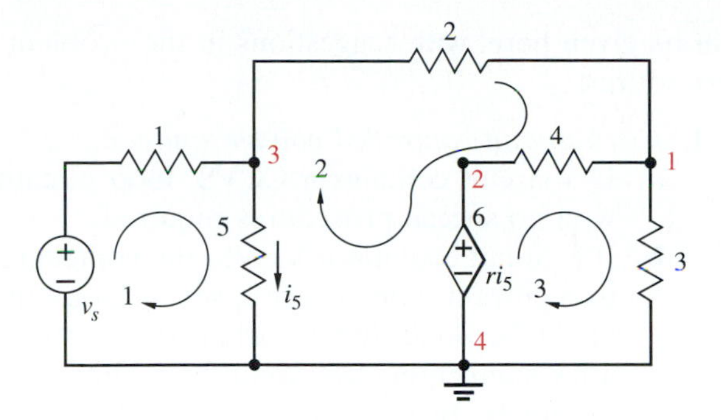

Figure 33
Circuit with controlled source.

The rearranged result can be written in matrix form as follows:

$$\begin{bmatrix} R_1 + R_5 & -R_5 & 0 \\ r - R_5 & R_2 + R_4 + R_5 - r & -R_4 \\ -r & r - R_4 & R_3 + R_4 \end{bmatrix} \begin{bmatrix} i_1 \\ i_2 \\ i_3 \end{bmatrix} = \begin{bmatrix} v_s \\ 0 \\ 0 \end{bmatrix}$$

$$\mathbf{Ri} = \mathbf{v}_s$$

The coefficient matrix here is called the *loop resistance matrix.* Notice again that the matrix is not symmetrical and it can't be written by inspection directly from the circuit diagram.

The second equation in this set seems to have fuller coefficients than the other two equations. By examining the circuit in Figure 33 we discover that mesh 2 traverses both the controlling branch and the controlled branch of the controlled source. What would happen if we made a selection of loops different from the set of all meshes? For example, suppose we add the last two equations in the preceding set. The result will be:

$$R_2 i_2 + R_3 i_3 + R_5(i_2 - i_1) = 0$$

A glance at the circuit reveals that this is the loop equation for the loop consisting of resistors 2, 3, and 5. This equation, together with the first and last of the mesh equations, constitutes a set of independent loop equations which are simpler than the set of mesh equations. Thus, when a circuit contains controlled sources, the mesh equations are not necessarily the simplest set of loop equations. When presented with the problem of constructing a set of loop equations for a circuit, we could shop around for an appropriate set that may not be the set of mesh equations. The only caveat is that the set of loop equations chosen must be independent.

Modified Loop Equations[14]

The presence of the three other kinds of controlled sources besides the CCVS prevents a standard loop analysis. The counterpart of what was done in the corresponding situation for a node analysis is the way to handle these cases, with appropriate interchanges of controlled-source types. A summary

[14] This section is the counterpart for loop equations of the optional section for modified node equations that you may have skipped. You may want to skip this section also.

will be given here, with suggestions in the problem sets for you to validate the procedures.

1. *Circuits with controlled voltage sources:*
   - **a.** If a circuit contains a CCVS, loop equations can be written directly with no special procedures required.
   - **b.** If a circuit contains a VCVS, the controlling variable must be changed to a current, after which a set of loop equations can be written. This might be done directly when the controlling branch is a resistor, or after augmenting the circuit by adding a resistor across the controlling terminals. In the latter case, we let the resistor value go to *infinity* after a solution is obtained.

2. *Circuits with controlled current sources:*
   - **a.** If a circuit contains a CCCS, the controlling variable is appropriate (a current) but the controlled variable is not. If the controlled source is accompanied, the controlled variable can be converted to the voltage of the combined branch. This amounts to a source transformation; the situation reverts to category 1(a). If the source branch is not accompanied, an i-shift can be performed to make it so.
   - **b.** If a circuit contains a VCVS, both controlling and controlled variables are inappropriate. Hence both procedures of 1(b) and 2(a) can be carried out.

What was just completed is an exhaustive (and exhausting!) development that includes all possible controlled sources. It's clear that a loop or node analysis—appropriately modified—is always possible. Rather than provide you with an unerring algorithm to follow toward the goal of obtaining the appropriate equations, we have described several different viewpoints: the forming of composite branches when sources (independent or controlled) are accompanied; v-shifts and i-shifts so that unaccompanied sources (independent or controlled) can be made accompanied; cut sets or supernodes for handling unaccompanied voltage sources (independent or controlled); initially augmenting a circuit to convert unaccompanied controlled sources to accompanied ones, with a subsequent removal of the augmenting resistor ($R \to 0$ if in series, $R \to \infty$ if in parallel).

The purpose of all this is to expand your powers of imagination—to induce you to contemplate the question "what if?" when presented with a situation not exactly covered by routine procedures.

6 DUALITY REVISITED

The concept of duality was introduced in Chapter 3. We promised to extend the concept as we proceeded. In this chapter a considerable number of new ideas was introduced, and many are each others' dual. In the discussion of graphs, for example, a tree was defined, as well as its complement, a cotree. The branches on a tree (the twigs) are the topological duals of those not on the tree (the links). The following is an extension of the table and duals first given in Chapter 3:

Extended Table of Duals

Node equations	Loop (mesh) equations
Tree	Cotree
Twig	Link
Cut set	Loop
Voltage-source shift	Current-source shift
VCCS	CCVS
VCVS	CCCS

It was already noted in Chapter 3 that if two networks, N_1 and N_2, are duals, the number of independent KCL equations of one equals the number of independent KVL equations of the other and they both have the same number of branches, b. If n_1 and n_2 are the number of nodes of the two networks, the following relationship must be satisfied by the two dual networks:

$$n_2 - 1 = b - (n_1 - 1)$$

An example is provided in Figure 34. Although we have not yet carried out analyses of circuits containing capacitors or inductors, one of each is included in the circuit in Figure 34a. The dual of this circuit is shown in Figure 34b. It is obtained by replacing each branch connected at a node in Figure 34a by a dual branch connected around a loop and replacing each branch found around a loop in Figure 34a with its dual branch connected to a node. This includes replacing the short circuit in series with R_4 whose current controls the CCVS in Figure 34a by the open circuit in parallel with the dual of R_4 whose voltage controls the VCCS in Figure 34b.

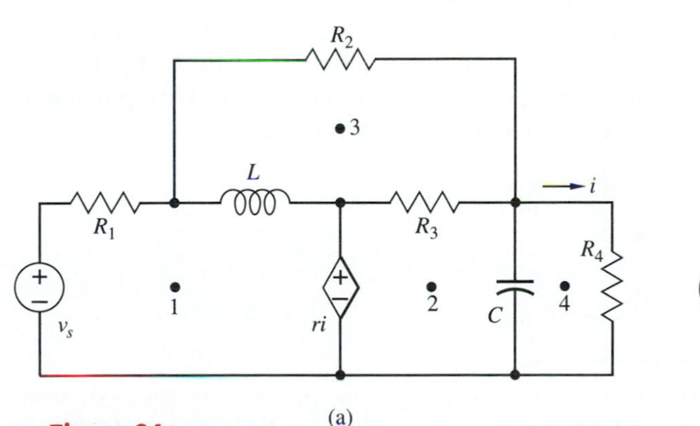

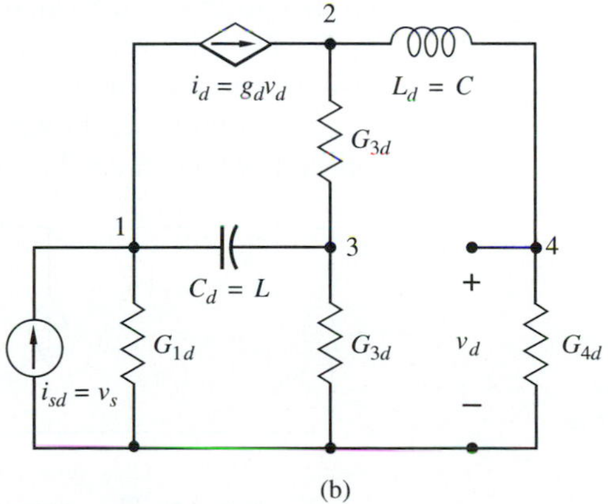

Figure 34
Dual circuits.

7 NETWORK THEOREMS REVISITED

We first undertook the study of methods of circuit analysis in Chapter 3, where the linear circuits were restricted to resistors and sources. There, in particular, we studied the principle of superposition and Thévenin's theorem and found that their use facilitated the solution of circuit problems. Now that we have introduced multiterminal dissipative devices, it would be appropriate to reexamine these two topics to see whether they apply in these more general circuits and, if they do, whether any modifications in procedures might be necessary.

Superposition Again

The principle of superposition is an immediate consequence of linearity. So long as the circuits under consideration are linear, superposition will apply, even though the circuits may contain controlled sources, ideal transformers, gyrators, or other multiterminal elements. The only caveat is to watch out for controlled sources; they are *not* (independent) sources but dissipative components. When applying superposition, only *independent* sources are removed.

EXAMPLE 6 The example shown in Figure 35 is a simplified transistor amplifier circuit. The objective is to find the output voltage, using superposition. The two *independent* sources are removed one at a time and the resulting partial output voltages are calculated (draw the diagram for each case); then the partial results are added, as follows:

$$V_{CC} = 0: \quad i_B = \frac{v_s}{R_s + R_B} \qquad v_{o1} = -\beta R_L i_B = -\frac{\beta R_L}{R_s + R_B} v_s$$

$$v_s = 0: \qquad i_B = 0 = \beta i_B \qquad v_{o2} = V_{CC}$$

$$v_o = v_{o1} + v_{o2} = V_{CC} - \frac{\beta R_L}{R_s + R_L} v_s$$

Confirm this result by solving for the output voltage without using superposition. Observe that when $v_s = 0$, i_B will be 0, so the controlled source current will also be zero. That is, the controlled source seems to be removed (open-circuited in this case) when v_s is removed. It wasn't done by us, the circuit analyzers, however; it was done by the structure of this particular circuit, which caused i_B to go to zero when v_s was removed.

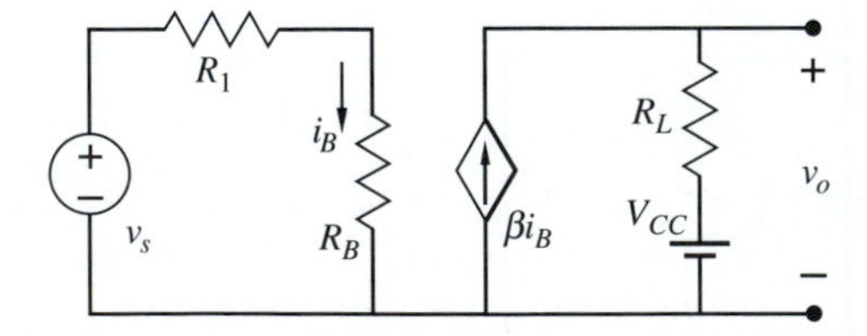

Figure 35 Example for superposition.

To repeat: The principle of superposition applies to linear circuits, those containing multiterminal dissipative components as well. The only potential

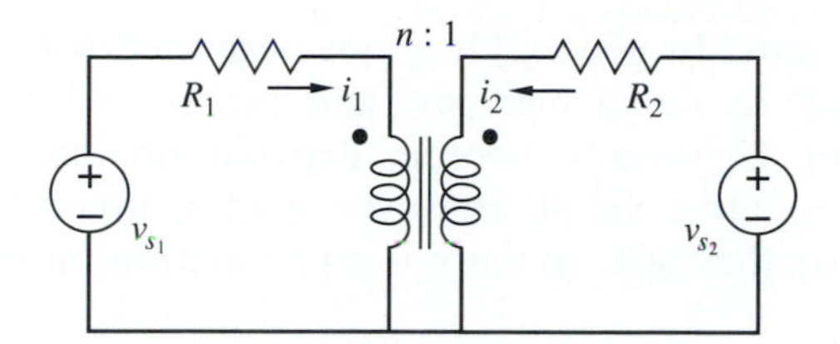

Figure 36
Exercise circuit.

source of difficulty is a controlled source; just keep in mind that this component *is not a source*.

EXERCISE 12

Apply the principle of superposition to find current i_1 in the circuit of Figure 36. Then, replace the ideal transformer by an equivalent circuit consisting of controlled sources and repeat the process. Does this help in confirming that controlled sources are not sources? □

The Thévenin Theorem Again

Carrying out a node or loop analysis results in a set of equations in the basis variables: node voltages or loop currents. Solving these equations means finding the solutions for *all* of the variables. Very often, however, we are interested in the voltage or current of just one branch. In such cases, the power and generality of these methods are wasted. Different methods to deal with such cases would be useful. Some such approaches were already considered in Chapter 3; we have just reexamined one of them for the more general linear circuits being considered now. We shall now reexamine Thévenin's theorem (and its Norton corollary) when a circuit includes multiterminal linear dissipative components.

Recall that the circuit structure to which Thévenin's theorem applies is the one shown in Figure 37a, one part of which is a linear, dissipative network, the other part (on the right) being considered a load on the linear part on the left. The only stipulation is that there should be no coupling between the two parts, except through the direct connection shown. For a circuit of two-terminal resistors and sources, that's the only possible coupling. However, with multiterminal components, other coupling mechanisms exist.

For Thévenin's theorem to be valid, for example, it is not possible for one winding of an ideal transformer to lie in the left part and the second winding in

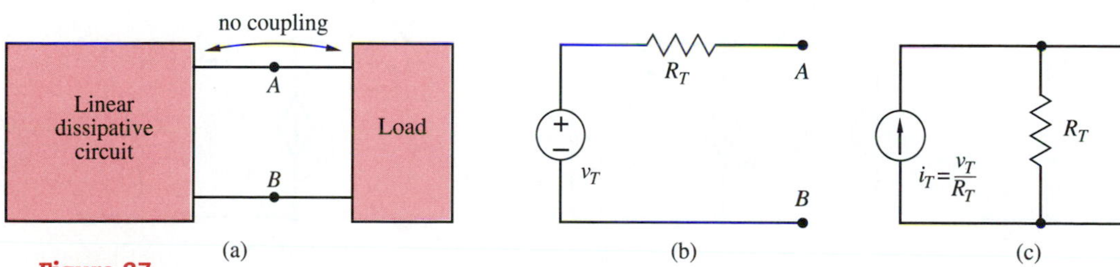

Figure 37
Structure for Thévenin's theorem.

the right part. In general, it is not permitted for one port of a multiterminal component to lie in one part and the second port in the other part. The validity of Thévenin's theorem depends on one part of the circuit being a linear, dissipative circuit. However, nothing involves the properties of the load, the part on the right, so there is no restriction on this part, even as to its being linear.

By way of review, recall that the Thévenin equivalent has the form shown in Figure 37b, where the Thévenin voltage v_T is the voltage at terminals AB of the original network when the load is removed and the terminals are left open: the open-circuit voltage. Also, the Thévenin equivalent resistance R_T is the resistance seen from the terminals of the circuit when all sources are removed. The only other caveat is the one emphasized for the principle of superposition. The Thévenin equivalent resistance is to be found by removing all *sources* in the circuit. But "sources" *do not* include controlled sources, only the *independent* sources.

Finally, the Norton equivalent shown in Figure 37c is a corollary of the Thévenin equivalent. The current source i_T is the current at the terminals of the original circuit when those terminals are short-circuited: the short-circuit current. The three quantities are related by $v_T = R_T i_T$.

EXAMPLE 7 The circuit in Figure 38a contains a controlled source. Our objective is to find the Thévenin equivalent. By KCL, the current in R_1 is found to be $(\alpha + 1)i$. A loop equation around the loop consisting of the voltage source and the three resistors contains only i as an unknown. The open-circuit voltage is seen to be $R_3 i$. Hence:

$$R_1(\alpha + 1)i + R_2 i + R_3 i = v_s$$

$$v_T = v_{oc} = R_3 i = \frac{R_3}{(\alpha + 1)R_1 + R_2 + R_3} v_s$$

To find the Thévenin resistance, let us first find the short-circuit current; the appropriate circuit is shown in Figure 38b. The desired current is seen to be identical with i. The current in R_1 is still $(\alpha + 1)i$. Again a loop equation around the outer loop will yield i. Thus:

$$i_{sc} = i = \frac{v_s}{(\alpha + 1)R_1 + R_2}$$

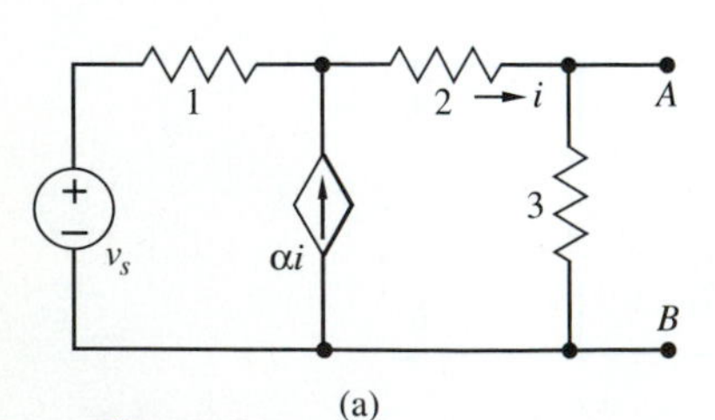

(a)

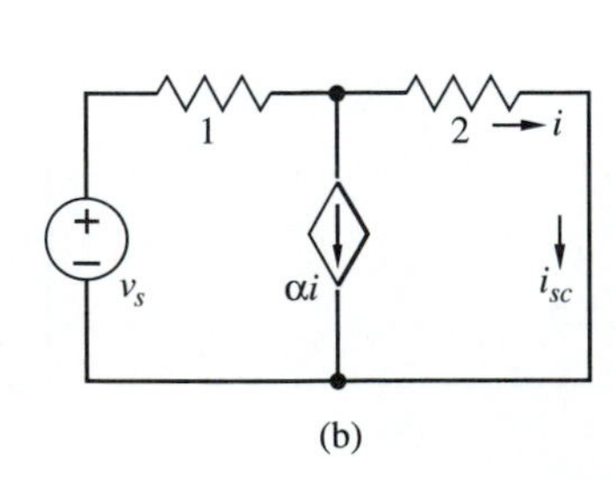

(b)

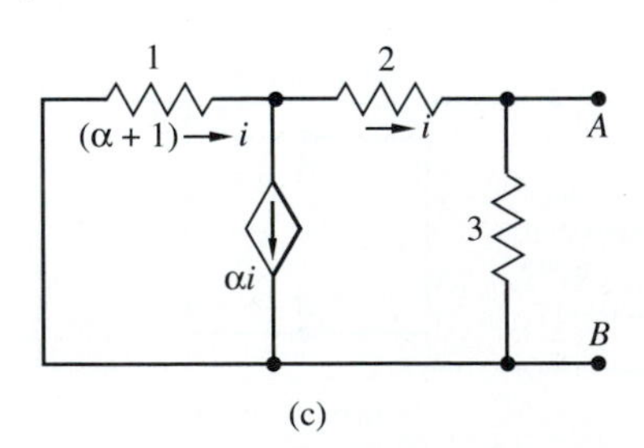

(c)

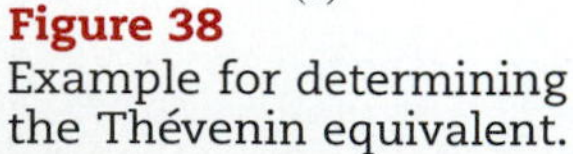
Figure 38
Example for determining the Thévenin equivalent.

The Thévenin resistance is now found by taking the ratio of the last two equations. Thus:

$$R_T = \frac{v_T}{i_T} = \frac{\dfrac{R_3}{R_1(\alpha + 1) + R_2 + R_3} v_s}{\dfrac{1}{R_1(\alpha + 1) + R_2} v_s}$$

$$= \frac{R_3[R_1(\alpha + 1) + R_2]}{R_1(\alpha + 1) + R_2 + R_3}$$

This looks like a fairly complicated expression, but it has a simple interpretation. (Look it over and state your interpretation.)

Finally, let's confirm this value of R_T by finding it directly, that is, by removing the voltage source and finding the resistance at terminals AB. The appropriate circuit is shown in Figure 38c. The controlled source is a complication; its presence prevents us from performing such maneuvers as combining resistances in series or parallel. We are forced to fall back on fundamental concepts!

How *was* the resistance of a two-terminal component defined? Why, as the ratio of the voltage across it to the current through it, with standard references. If a resistor sits unexcited in a circuit, there will be no current or voltage. In order for such things to exist, the resistor must be excited. If it is excited by a voltage source, we calculate (or measure) the current and then take the appropriate ratio. If it is excited by a current source, we calculate the voltage and again take the appropriate ratio. Since there are two choices of excitation, which should we choose? Whichever is most convenient, of course.

Actually, there is nothing novel about this, even without controlled sources. For all circuits except series-parallel ones, the approach just described *has to be* used, since the circuit cannot be reduced to a very simple form by series and parallel combinations of branches and by source transformations.[15]

In the present case, a voltage source applied at terminals AB is simpler. (Draw the relevant diagram.) R_3 will be directly across it, so its presence will not influence the current in the rest of the circuit. Thus, if v_g is applied (with reference + at A), the current going in at A is $v_g/R_3 - i$. The current i can again be found from a loop equation around the outer loop. Hence:

$$-(\alpha + 1)R_1 i - R_2 i = v_g$$

$$i_g = \frac{v_g}{R_3} + \frac{v_g}{(\alpha + 1)R_1 + R_2}$$

From the last expression, it appears that the Thévenin resistance equals the parallel combination of R_3 and the series connection of R_2 and $(\alpha + 1)R_1$. This confirms the value of R_T and is precisely the interpretation which you were asked to provide when R_T was previously determined.

[15] In some circuits it may be possible to use a circuit transformation to simplify their structure, after which series and parallel simplifications may be possible. Such transformations, however, may involve more effort than using the definition of equivalent resistance.

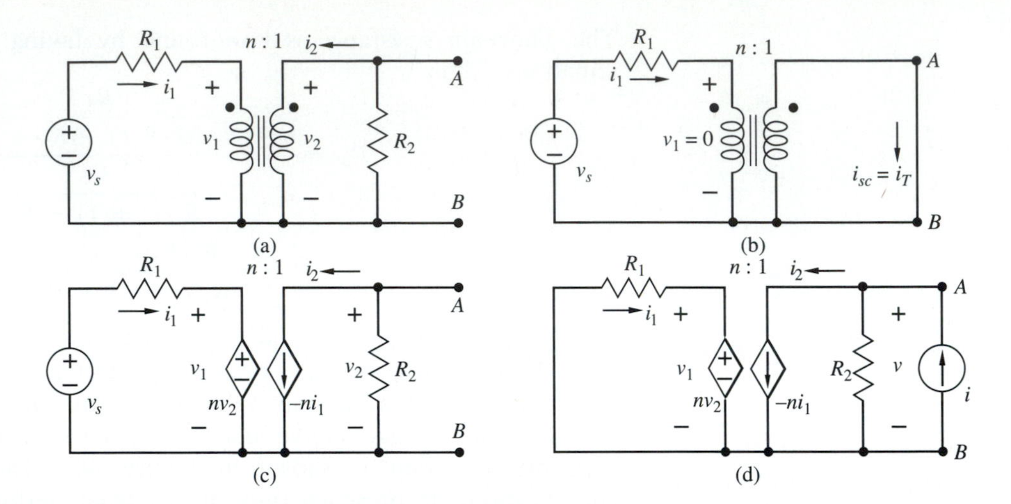

Figure 39
Thévenin equivalent of a circuit with an ideal transformer.

EXAMPLE 8 As another example, consider the circuit in Figure 39a, which includes an ideal transformer characterized by $v_1 = nv_2$ and $i_2 = -ni_1$. The open-circuit voltage at the output equals the transformer secondary voltage. Two circuit equations, one for each side of the transformer, can be written and then combined with the two transformer equations. The four equations are:

$$v_T = -R_2 i_2 \qquad i_1 = \frac{v_s - v_1}{R_1}$$

$$i_2 = -ni_1 \qquad v_1 = nv_2 = nv_T$$

The solution for v_T is obtained by successive substitution of variables:

$$v_T = \frac{nR_2}{R_1 + n^2 R_2} v_s$$

Next let us find the short-circuit current. The relevant circuit is shown in Figure 39b. With a short circuit across its terminals, R_2 plays no role, so it is removed. The transformer secondary voltage is, thus, 0; so also is the primary voltage, since $v_1 = nv_2$. The short-circuit current is now easily found:

$$i_T = -i_2 = -(-ni_1) = \frac{nv_s}{R_1}$$

Then taking the ratio of v_T to i_T $(= i_{sc})$ yields R_T, as follows:

$$R_T = \frac{v_T}{I_T} = \frac{R_1 R_2}{R_1 + n^2 R_2}$$

EXERCISE 13

Determine the Thévenin equivalent resistance at the output of the circuit in Figure 39a after the voltage source is replaced by a short circuit, and confirm the previous result. (Hint: What do you see from the output of an ideal transformer when the input terminals are loaded with a resistance?) □

Pursuing the preceding example a little further, suppose the ideal transformer is replaced by an equivalent involving two controlled sources, as in Figure 39c. The goal in this circuit is to determine the Thévenin equivalent resistance. Let's emphasize again: *controlled sources are not sources.* To find R_T from the output terminals, only the independent source v_s is replaced by a short circuit. (Check to see what kind of weird result you would obtain if you removed the controlled sources also.)

Removing the independent source, let's this time apply a current source at terminals *AB*, as shown in Figure 39d. Successively applying Ohm's law for R_2, KCL at the upper terminal of the current source, the transformer current relationship, and Ohm's law for R_1 leads to the following results:

$$v = R_2(i - i_2) = R_2(i + ni_1) = R_2\left(i - \frac{nv_1}{R_1}\right) = R_2\left(i - \frac{n^2v}{R_1}\right)$$

$$R_T = \frac{v}{i} = \frac{R_1R_2}{R_1 + n^2R_2} = \frac{R_2(R_1/n^2)}{R_2 + R_1/n^2}$$

We interpret the last form as R_2 in parallel with R_1/n^2.

EXERCISE 14

Use a voltage source as excitation in Figure 39d instead of the current source shown there and redetermine the Thévenin resistance. Which calculation is easier? □

The Compensation Theorem

In the preceding section we observed that replacing a subnetwork by something equivalent to it can lead to a simplification that permits easier determination of some voltage or current variable. (On a smaller scale, replacing two or more series resistors or parallel resistors by a single equivalent does the same thing.) In this section, we will consider another such situation where a subnetwork is replaced by something very simple—a source.

A general situation is depicted in Figure 40, where a two-terminal subnetwork *B* is shown isolated from the rest of the network. Suppose that, by some means or other, the current i and the voltage v of subnetwork *B* have been determined. Then, it would be possible to place a voltage source $v_g = v$ in parallel with subnetwork *B* without affecting any other voltage or current, since the voltage across those terminals is already v. But now subnetwork *B* can be removed entirely, since it is in parallel with a voltage source, without affecting subnetwork *A*. The result, shown in Figure 40b, is to replace a two-terminal subnetwork by a voltage source whose voltage is equal to the already calculated voltage of the subnetwork.

A similar discussion applies that involves the already determined current in subnetwork *B*. Suppose we place a current source with $i_g = i$ in series with

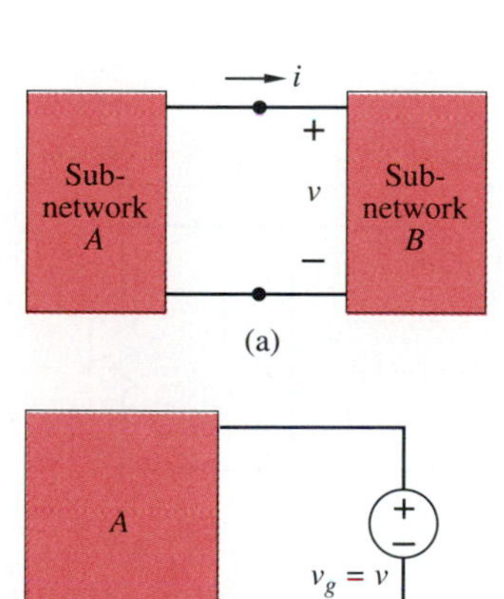

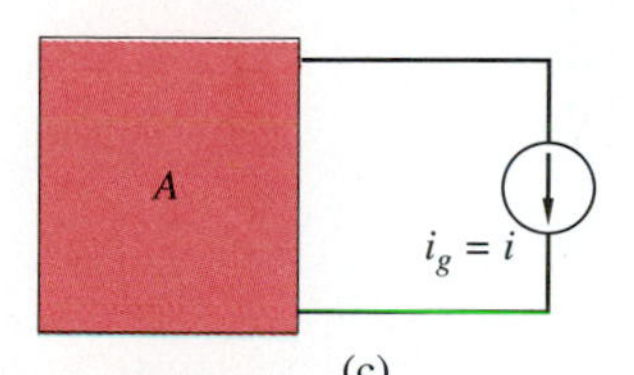

Figure 40
The compensation theorem.

subnetwork *B*. Such a source would have no influence on the remaining currents and voltages, since the current into subnetwork *B* is already *i*. But now, since subnetwork *B* is in series with a current source, its presence will have no effect on the remaining voltages and currents and can be safely removed. The result, shown in Figure 40c, is to replace a two-terminal subnetwork by a current source whose current is the previously calculated value of the current in the subnetwork.

The results just described have been dignified by the name *compensation theorem* or (by some authors) *substitution theorem.* Note that, in order to apply the theorem, it is first necessary to know the voltage (or the current) of a two-terminal subnetwork. We shall consider several applications in the next two chapters. For now, let's consider Figure 41 as an example. The numbers

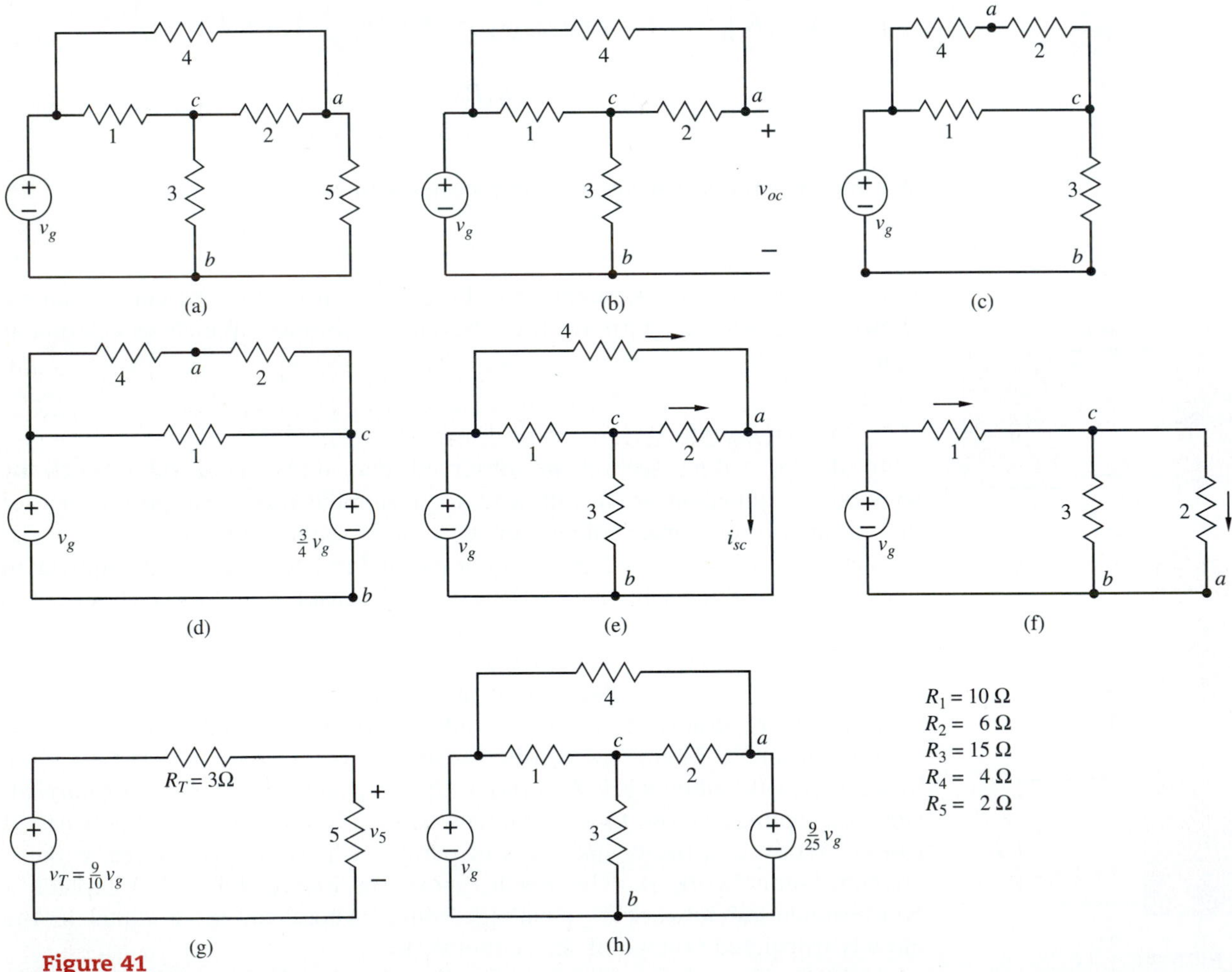

Figure 41
Applying the compensation theorem.

on the branches are not resistance values but branch numbers. The ultimate objective is to find the current in branch 1. One way would be to construct and solve a set of mesh equations. But suppose we proceed as follows:

1. First find the Thévenin equivalent of the circuit to the left of terminals *ab*. Then use this to find the voltage across branch 5 from the voltage-divider relationship.

2. Then, using the compensation theorem, replace branch 5 with a voltage source. The resulting circuit is much simpler for finding the desired current i_1.

In pursuit of this strategy, the circuit for finding the open-circuit voltage is shown in Figure 41c. Since no current is leaving the open-circuited terminals, branches 2 and 4 are in series—they carry the same current—and their combination is in parallel with branch 1, as shown in Figure 41c. Note that, so far, no work has been done aside from redrawing diagrams. Now, by series and parallel combination of resistances and the voltage-divider relationship, the voltage across R_3 is found:

$$v_3 = \frac{15}{15 + \dfrac{10(4+6)}{10+4+6}} v_g = \frac{3}{4} v_g$$

Although it was not part of the initial strategy, we can now use the compensation theorem to complete finding the open-circuit voltage across terminals *ab*. Thus, returning to Figure 41c, we replace branch 3 by a voltage source whose voltage, determined from the voltage-divider expression in Figure 41c, is $3v_g/4$, as shown in Figure 41d. In this circuit, branch 1 is directly in parallel with the series connection of the two voltage sources. Hence, its presence can have no influence on the rest of the circuit, so it can be removed. The resulting circuit has the structure of a voltage divider, so the voltage across branch 2 can be found easily. The desired open-circuit voltage is then the sum of this voltage and the previously found v_3. Thus:

$$v_2 = v_{ac} = \left(\frac{6}{6+4}\right)\left(v_g - \frac{3}{4} v_g\right) = \frac{3}{20} v_g$$

$$v_{oc} = v_2 + v_3 = \tfrac{3}{4} v_g + \tfrac{3}{20} v_g = \tfrac{9}{10} v_g$$

As the next step in finding the Thévenin equivalent of the original circuit, let's now find the short-circuit current at terminals *ab*. From KCL in Figure 41e, this is the sum of the currents in branches 2 and 4. But branch 4 is now directly in parallel with the voltage source. (Trace the path from the upper terminal of the source, through branch 4, to the lower terminal of the source to confirm this.) Two conclusions result from that: (1) its current i_4 is $v_g/4$ and (2) this branch will have no influence on the voltages and currents in other branches of the circuit. With branch 4 removed, the result is shown in Figure 41f, where branches 2 and 3 are in parallel. One can proceed in one of two ways:

1. Find the current supplied by the source as v_g divided by the total resistance connected to it and then use the current-divider relationship to find the current in branch 2; or
2. Find the voltage across the parallel combination of branches 2 and 3 by the voltage-divider relationship and then use Ohm's law to find the current in branch 2.

You do the first of these in your spare time; we will do the second one here. After finding i_2, we add it to the previously found i_4 to get i_{sc}. With v_{oc} and i_{sc} both determined, the Thévenin equivalent impedance is determined by taking their ratio. Thus:

$$i_2 = \frac{1}{6}v_2 = \frac{1}{6}\frac{\frac{6(15)}{6+15}}{\frac{6(15)}{6+15}+10}v_g = \frac{1}{20}v_g$$

$$i_{sc} = i_2 + i_4 = \tfrac{1}{20}v_g + \tfrac{1}{4}v_g = \tfrac{3}{10}v_g$$

The Thévenin equivalent resistance R_T is now found:

$$R_T = \frac{\frac{9}{10}v_g}{\frac{3}{10}v_g} = 3\,\Omega$$

The resulting Thévenin equivalent circuit, loaded with branch 5, is shown in Figure 41g. The voltage across this branch is next determined by the voltage-divider relationship to be:

$$v_5 = \frac{2}{2+3}\frac{9}{10}v_g = \frac{9}{25}v_g$$

Now, using the compensation theorem, branch 5 in the original circuit can be replaced by a voltage source having this voltage, as shown in Figure 41h. In this circuit there are only two nodes, and hence only one unknown node voltage. A node equation written as node c is all that is now needed. Thus:

$$\tfrac{1}{10}(v_c - v_g) + \tfrac{1}{15}v_c + \tfrac{1}{6}(v_c - \tfrac{9}{25}v_g) = 0$$

$$v_c = \tfrac{23}{100}v_g$$

With v_c determined, it is a simple matter to find any other branch voltage or current, including the current in branch 1.

By this time you might be thinking that, to avoid constructing and solving a few mesh or node equations, a great deal of work was needed to reach a solution in this example. If you review the preceding process, you will observe that, while there seem to be many steps in arriving at the final result, each is a very simple step—e.g., redrawing simplified circuit diagrams, combining resistors in parallel or series, applying the voltage-divider or current-divider relationships, writing a single node equation, or replacing a branch by a voltage source. Even though there are many steps, the chances of making errors when carrying out simple steps are much smaller than the chances of error when carrying out more complicated steps, even though there are fewer of them.

SUMMARY

This is a major chapter of the book and includes a lot of the "meat" of network analysis. Major topics in the chapter include the following:

- Treatment of sources in analysis: accompanied sources.
- Analysis using branch variables.
- Idea of the linear independence of equations.
- Circuit topology and linear graphs.
- Definitions of loop, tree, cotree, twig, link, cut set.
- Number of independent KCL equations in a circuit: $n - 1$.
- Number of independent KVL equations in a circuit: $b - (n - 1)$.
- Carrying out a node analysis for a circuit—scalar and matrix.
- Multiterminal components not permitted for node analysis.
- Modified node equations to accommodate such components.
- Carrying out a loop (mesh) analysis—scalar and matrix form.
- Multiterminal components not permitted for loop analysis.
- Modified loop equations to accommodate such components.
- Extension of duality to topology and multiterminal components.
- The voltage-source shift and the current-source shift.
- Extension of the superposition principle to circuits with multiterminal components; how to treat controlled sources.
- Extension of Thévenin's and Norton's theorems to circuits with multiterminal components.
- The compensation theorem.

PROBLEMS

1. A certain linear graph has n nodes and each node is connected to every other node. Is the graph planar or not: (a) If $n = 4$? (b) If $n = 5$? If the graph is nonplanar in either case, what is the minimum number of branches whose removal will render the graph planar?

2. On the basis of circuit structure and composition:
   a. Specify the number of independent voltage variables and the number of independent current variables in Figure P2a.
   b. Repeat for the circuit in Figure P2b.
   c. Repeat part (b) but with R_1 short-circuited.
   d. Repeat part (b) but with R_2 open-circuited.

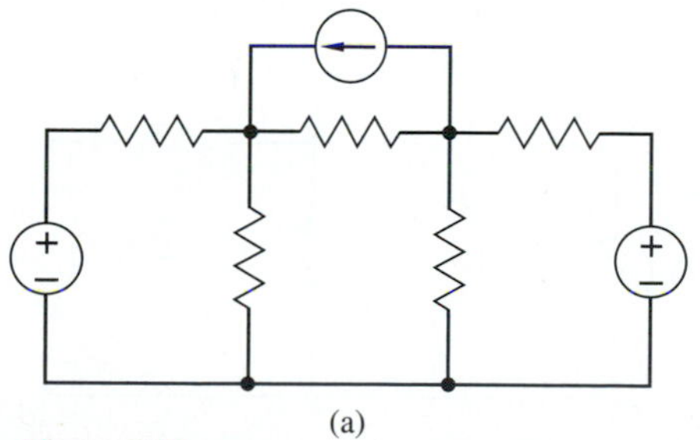
(a)

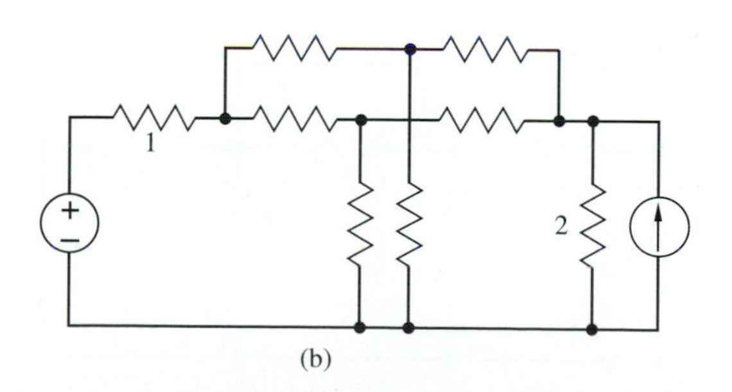

(b)

Figure P2

3. The numbers shown on the branches in the circuit of Figure P3 are identifying numbers, not component values.
   a. Draw a directed linear graph representing the circuit.
   b. List the branches constituting a tree for six trees in the graph. (There are 21 trees.)
   c. Write a set of KCL equations at the nodes of the circuit in terms of the branch currents.
   d. Identify 10 loops in the circuit by listing the branches they contain. (Be systematic; there are 12 loops.)
   e. Write KVL equations around each of these loops.
   f. From among the KVL equations, choose a set that is independent. Show how each of two of the remaining KVL equations can be obtained from the independent set.

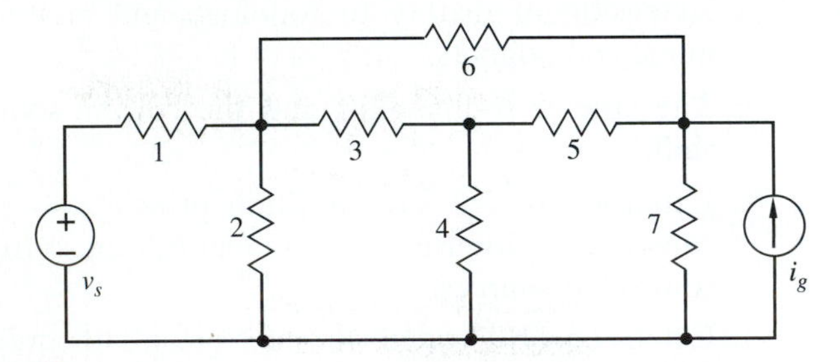

Figure P3

4. The numbers on the branches in the circuit in Figure P4 are identifying numbers, not component values. Assume standard references for all branches and place current references on all branches.
   a. Draw a directed graph representing the circuit.
   b. Specify 10 trees (there are 41) by listing their branches.
   c. Write a set of KCL equations at the nodes of the circuit in terms of the branch currents.

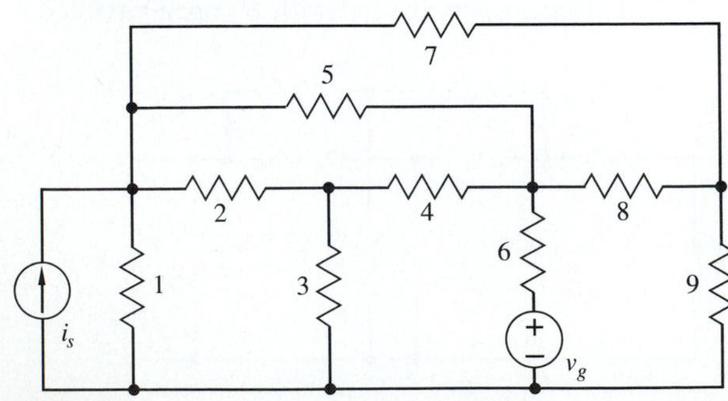

Figure P4

   d. Choose a tree and write a set of fundamental cut-set equations. One or more of these may be identical with a KCL equation at a node. Show how those that are not so can be obtained by linear combinations of KCL equations at nodes.
   e. Specify any five loops that are not meshes by listing their branches. Write KVL equations for these loops.
   f. Show how each of them can be obtained as a linear combination of KVL equations around the meshes.

5. A convention described in the text is the counting of an accompanied source (v-source in series with a resistor or i-source in parallel with one) as a single branch. A more general configuration is shown in Figure P5. This can be taken as the form of *any* branch in a resistive circuit. In specific cases one or two of the three ideal components included here could be missing.
   a. Write the relationship between the overall v and i for this branch.
   b. Draw and label a diagram of a branch which is the dual of the one shown.
   c. Write the v-i relationship of the new branch and compare with the one in part (a). Can this configuration also serve as the general form of every branch?

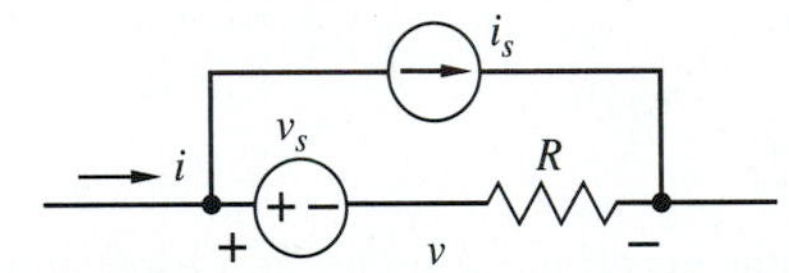

Figure P5

6. A node analysis is to be carried out in the circuit of Figure P6.
   a. Take node 3 as datum and write a pair of node equations at nodes 1 and 2.
   b. Again take node 3 as datum for defining node voltages but this time use node 3 as one of the

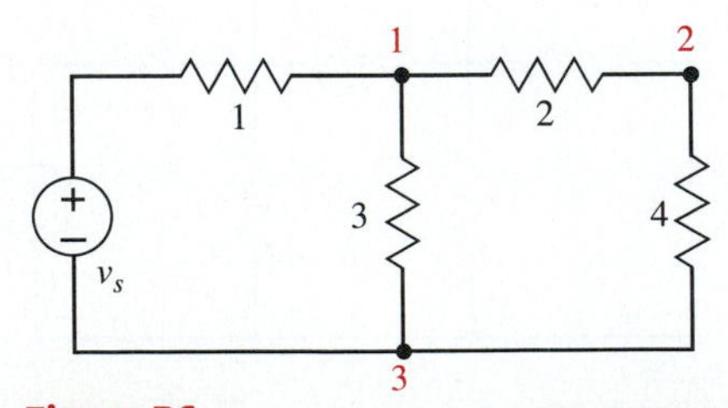

Figure P6

nodes at which node equations are written, taking either node 1 or 2 as the second node.

c. Finally, choose node 2 as datum node and write node equations at nodes 1 and 2.

d. Compare the three sets of equations. How are their determinants related?

e. What conclusions can you draw?

7. Although it may not be evident from the procedure followed in arriving at a set of mesh equations for a given circuit, two different choices for sets of loops are made in the process. Once, a set of meshes is selected for defining mesh currents. Without explicitly saying so, the same set is then used for applying KVL.

a. Write standard mesh equations for the circuit in Figure P6.

b. Choose mesh currents as in part (a), but this time let one of the loops for applying KVL and writing loop equations be the loop around the outer contour of the circuit.

c. Finally, let one of the loop currents be the current around the outer contour, the other being one of the mesh currents. But apply KVL to the meshes, arriving at a set of loop equations.

d. Compare the resulting equations; are they the same or different? Determine how the determinants of the three sets of equations are related. Any conclusions?

8. Some sophomores studying circuits had enlisted the help of some seniors and had just completed a marathon problem-solving session. The seniors had been telling them how easy students had it these days, not like the old days when the seniors were really given tough circuit problems to solve! One of the problems they suggested, taken from one of their old books, was the circuit in Figure P8, which had as many independent sources as resistors. "Now *that's* tough to solve," they said. The sophomores were at first impressed with this complicated circuit with apparently many nodes, loops, and sources.

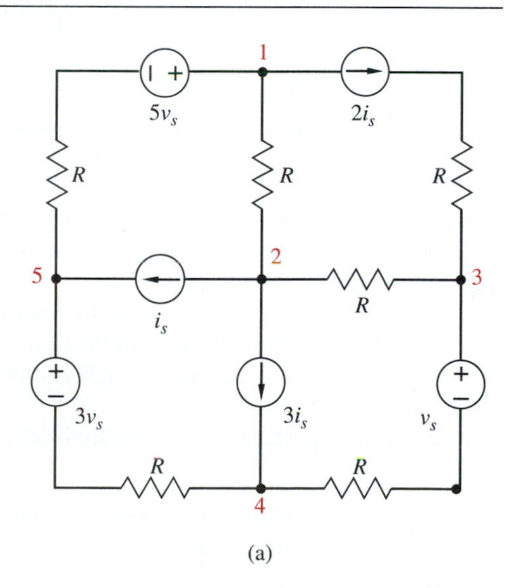

(a)

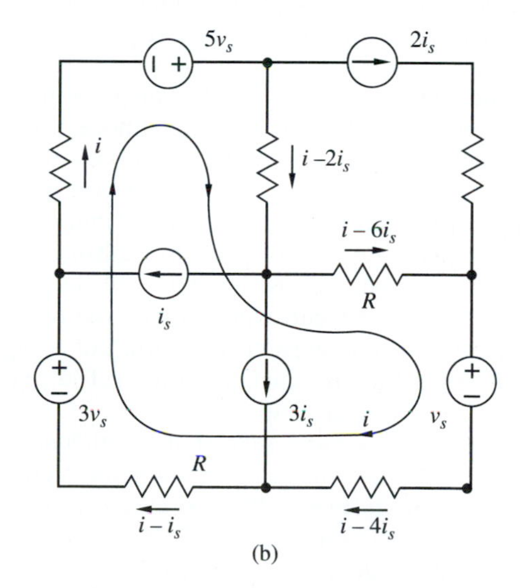

(b)

Figure P8

After a while, though, one of the students (student A) made some observations. First of all, he noted, this looks like a contrived circuit, not one that's a model of a real circuit. Where would you ever find such a collection of sources in a real circuit, he wondered. But even if you accept it at face value, what's the big deal? Not that anybody would want to do it, but consider using the principle of superposition and solving the circuit six times, each time with only one of the sources operating, the others being replaced by short circuits (voltage sources) or open circuits (current sources). The result each time will be a simple series or parallel circuit with one source; just one equation to solve each time. It may be tedious but hardly tough. Or think of replacing everything but one of the resistors, say, the vertical one in the middle, by a Thévenin equivalent. When all the sources are removed, the Thévenin equivalent resistance will be just the series connection of just a few of the resistors; hardly tough to calculate. The open-circuit voltage is obtained trivially also; it turns out to be $v_T = 9v_s$. The voltage across and the current through the vertical resistance are obtained trivially from the Thévenin equivalent. All other resistor currents and voltages follow, one by one, from KCL and KVL. No simultaneous equations need to be solved. Again, tedious but simple.

"Yeah, but what about mesh equations? Gotcha!" a senior gloated. It became clear; the circuit had been contrived just for the purpose of practicing what were called "mesh equations." The sophomores put their heads together and student A made some more observations. The resistor on the upper right in series with the $2i_s$ current source is apparently there just to provide confusion. Its current is known ($2i_s$), so its voltage is also known, by Ohm's law. Its presence would have absolutely no effect whatsoever on the rest of the circuit, so it could just as well be replaced by a short circuit. To count the number of loop equations needed, we can replace all unaccompanied curent sources by open circuits. The result is a single loop.

That's the loop that contains all the voltage sources and all the other five resistors, starting at node 1 and going through nodes 2, 3, 4, and 5, then back to 1, without traversing any current sources. A single KVL equation is written around this loop; then Ohm's law is used to replace resistor voltages by resistor currents, which are then expressed in terms of the one loop current and source currents by KCL. The result is one equation in one current unknown; not a big deal. And not a "mesh equation" either! The diagram is redrawn in Figure P8b.

a. Carry out the procedure suggested by student A and solve for i, the single current unknown. Verify the expressions in Figure P8b for all the resistor currents and determine the value of all the currents in terms of R and the source values.

b. Determine also the voltage across each current source.

9. Another example of a tough problem suggested by the seniors advising the sophomores is shown in Figure P9. While it doesn't have as many sources as the preceding example, two of the sources are controlled sources and *all* are unaccompanied. By now, student B, a friend of student A, had become alerted to distrust problems with a lot of unaccompanied sources. She felt that this circuit, too, was contrived. After examining it a bit, she came to the conclusion that the VCVS and the CCCS were just decoys for resistors. By Ohm's law she found that the two controlling variables were proportional to each other. By some further simple applications of KVL, KCL, and Ohm's law she found the current in the VCVS to be proportional to its voltage and the voltage across the CCCS to be proportional to its current. Hence, each could be replaced by an appropriate resistor and the circuit rendered trivial.

a. Carry out numerically the plan of student B and confirm her conjectures.

b. Determine also the power supplied by the 5-A current source.

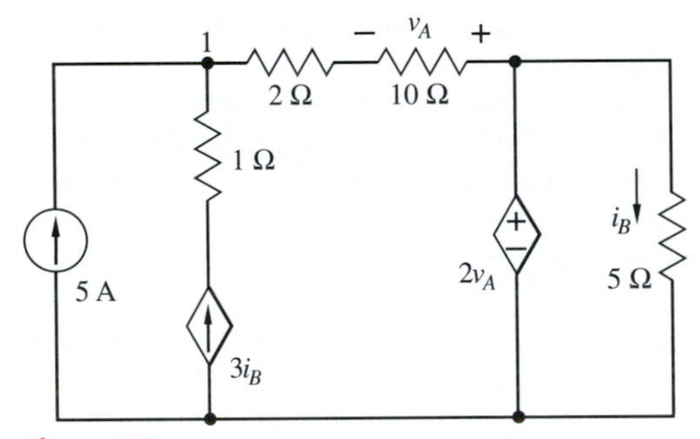

Figure P9

10. The objective in the circuit of Figure P10 is to find the output voltage v_2 and the input current i_1 in terms of source voltage v_s.

a. Describe several strategies for finding v_2.

b. Carry out each of them to find v_2.

c. Compare the amount of effort involved in each case.

d. Repeat parts (a) to (c) for finding i_1.

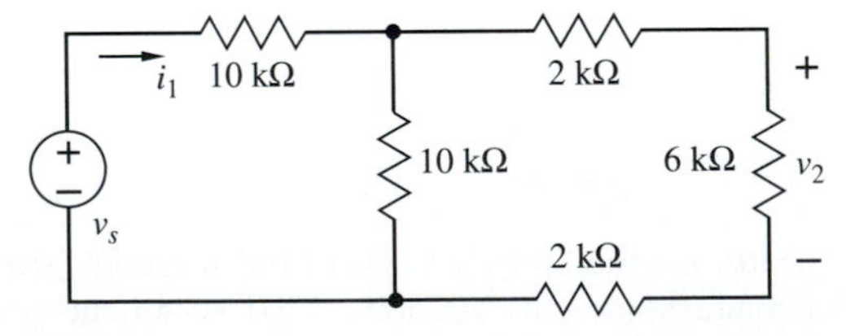

Figure P10

11. The current through the switch in the diagram of Figure P11 is found to be 100 mA when the switch is closed. When the switch is opened, the voltage across it is found to be 50 V. If the switch is replaced by a small resistance (say, 10 Ω or so), will the power delivered by the network to the 200-Ω load increase or decrease? Explain.

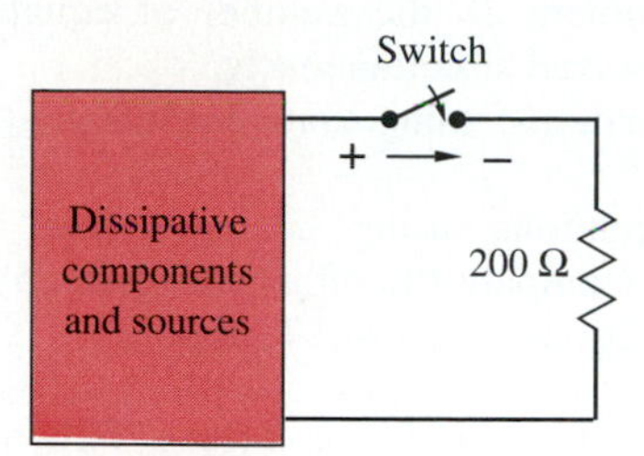

Figure P11

12. Find the Thévenin equivalent at terminals AB in the circuit of Figure P12.

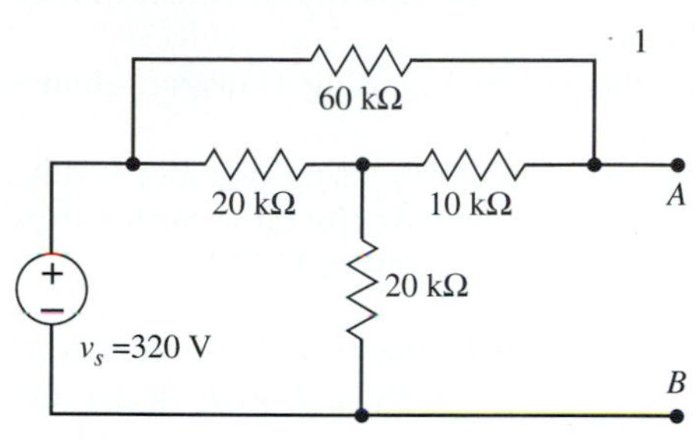

Figure P12

13. **a.** Find the Thévenin equivalent at the output terminals in the circuit of Figure P13.
b. Using a different method of your choice, find v_2.
c. Use yet another method and find v_2. Which approach do you prefer, and why?

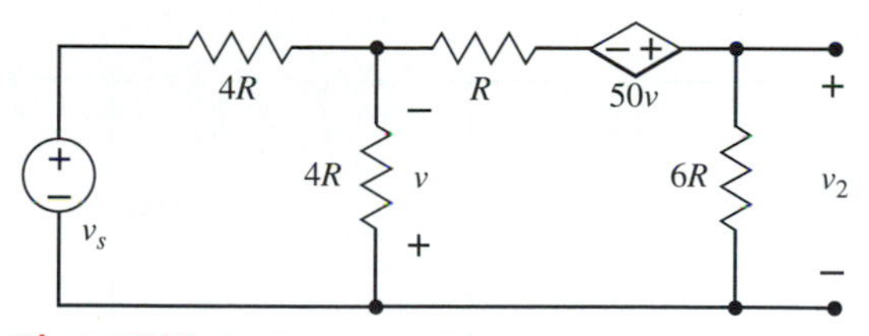

Figure P13

14. Because of its structure, the circuit in Figure P14 is called a *lattice*.
a. From the count of the number of nodes and branches, specify the number of independent voltage variables.
b. Specify also the number of independent current variables.
c. Carry out a node analysis using the bottom terminal of the source as datum. Determine the output voltage v_2 and the power being supplied by the source.

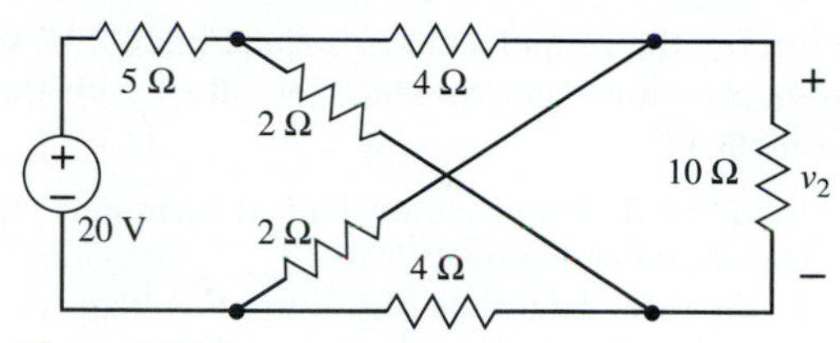

Figure P14

15. **a.** Redraw the circuit in Figure P14 so that it is planar.
b. Use a mesh analysis to find the output voltage v_2 and the power supplied by the source. (They had better be the same as you obtained in Problem 14.)

16. **a.** Find the Thévenin equivalent at terminals AB in Figure P16.
b. Confirm the value of the Thévenin resistance R_T by evaluating it in some other way.

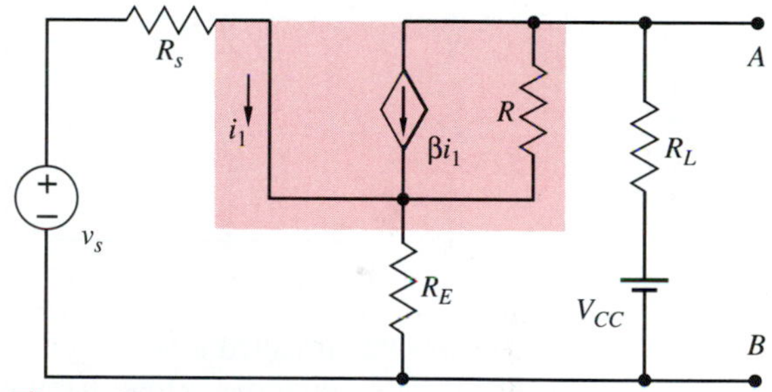

Figure P16

17. In Figure P17:
a. Solve for the output voltage v_2 using node equations.
b. Determine the power supplied by both the independent source and the controlled source. Any surprises?

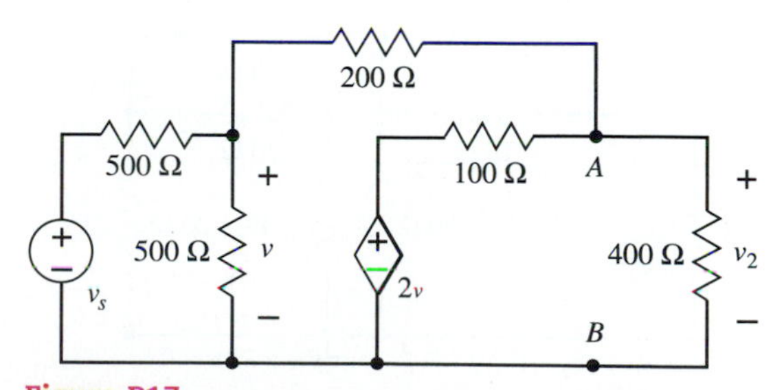

Figure P17

18. Solve for the output voltage v_2 in Figure P17 using mesh equations, and compare the effort with that in Problem 17.

19. **a.** Find the Thévenin equivalent at terminals AB in the circuit of Figure P17.
b. From this, determine the output voltage v_2 and compare the effort with that of Problems 17 and 18.

20. This problem is concerned with the number of independent KVL equations needed to solve for all the variables in a circuit. The branches in the circuit of Figure P20 are numbered. The two loops identified by branches $\{1,3,4\}$ and $\{2,3,5,6\}$ include *all* branches of the circuit. By attempting a solution, show that the KVL equations for these two loops alone, together with KCL equations and Ohm's law, are not sufficient to solve for the variables in the circuit. Thus, inclusion of all of a circuit's branches in a loop analysis, though necessary, is not sufficient to obtain a solution.

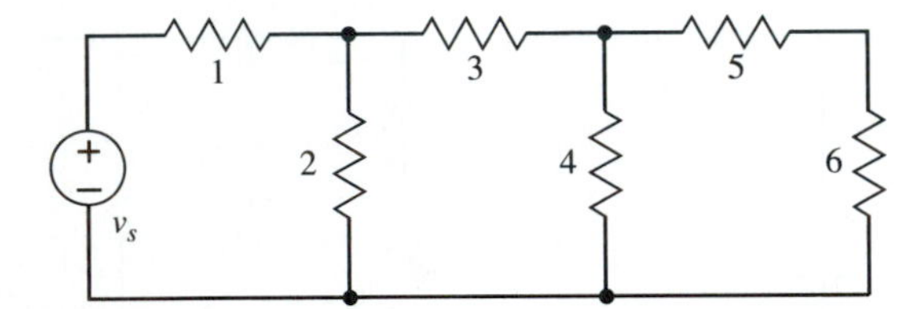

Figure P20

21. The circuit in Figure P21 is called a *twin tee.*
a. Write a set of node equations. How many equations need to be solved simultaneously?
b. Solve this set of node equations for v_2 using Gaussian elimination.
c. Solve the equations again using Cramer's rule and compare the effort.
d. Solve the equations using MATLAB.
e. Determine the power supplied by the source.
f. Use PSPICE to determine the circuit variables.

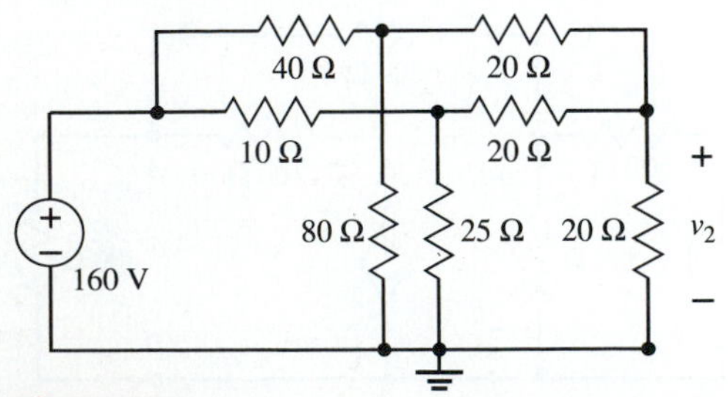

Figure P21

22. **a.** Redraw the circuit in Figure P21 to make it planar, and write a set of mesh equations. Compare with Problem 21 the number of equations that must be solved simultaneously.
b. Solve the equations, using Gaussian elimination, and find v_2.
c. Solve the equations, using Cramer's rule, and again find v_2. Compare the effort with part (b).
d. Solve the equations, using MATLAB.

23. **a.** With the twin-tee circuit of Figure P21 as it stands, number the branches and specify a tree.
b. On the basis of this tree, list the branches in a set of fundamental loops.
c. Write a set of loop equations for these fundamental loops.
d. Solve the equations, using Gaussian elimination, and find v_2.
e. Solve the equations, using Cramer's rule, and again find v_2. Compare the effort with part (d).
f. Solve the equations, using MATLAB.

24. **a.** In the circuit of Figure P24, decide how to handle the controlled source, and write a set of node equations.
b. Solve the node equations for v_2, using Gaussian elimination.
c. Solve the node equations for v_2, using Cramer's rule. How does the relative effort in this case compare with that in the preceding two problems?
d. Find the Thévenin equivalent resistance at the output.

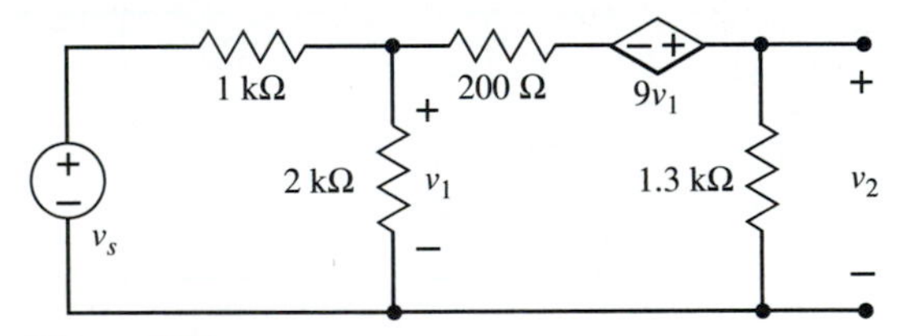

Figure P24

25. Find the output voltage in the circuit of Figure P24 again, this time using mesh equations. Do this using both Gaussian elimination and Cramer's rule. Obviously, the result must be the same as in Problem 24.

26. In the circuit of Figure P26, let $R_L = R/2$, $g = \frac{1}{10}$ mS, and $R = 1$ kΩ.
a. Carry out a node analysis and solve for voltage v_2 using Gaussian elimination.
b. Repeat using Cramer's rule.

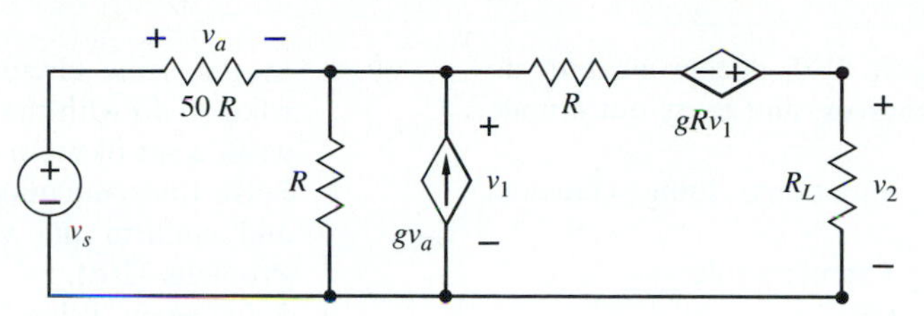

Figure P26

c. Now take a skeptical outlook and consider whether the controlled sources can be replaced by resistors (either positive or negative) and possibly independent sources. Do this by examining the controlling variables to determine if a controlled current can be expressed in terms of its own voltage or a controlled voltage can be expressed in terms of its own current. In this way reduce the circuit to a single loop and determine v_2.

27. In the bridge network of Figure P27, the load resistance R_L is to be adjusted so that the maximum power is delivered to it.
 a. Find the required value of R_L.
 b. For this value of R_L, find the power supplied by the source.
 c. How much of this power is delivered to R_L?

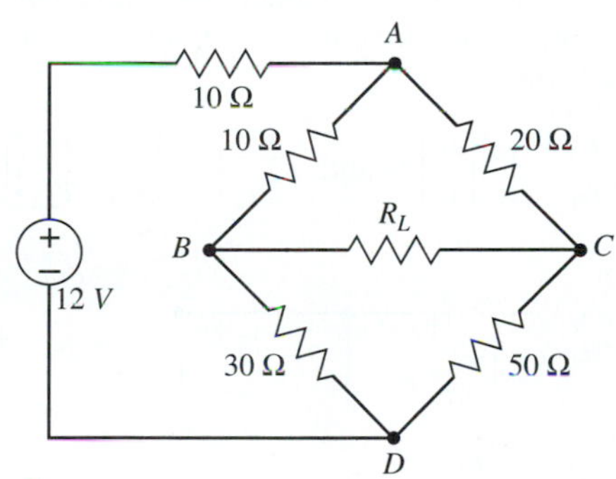

Figure P27

28. The diagram in Figure P28 represents a three-wire distribution system. The source voltages are to be adjusted so that the loads will operate at the rated value given on the diagram.
 a. Find these values.
 b. Each circuit breaker is set to open whenever there is a greater than 100% increase in the corresponding current from its rated value (the value corresponding to the rated power and voltage). During operation, load 1 becomes short-circuited but the resistance of load 2 remains unchanged. Determine whether each circuit breaker will open or stay closed.

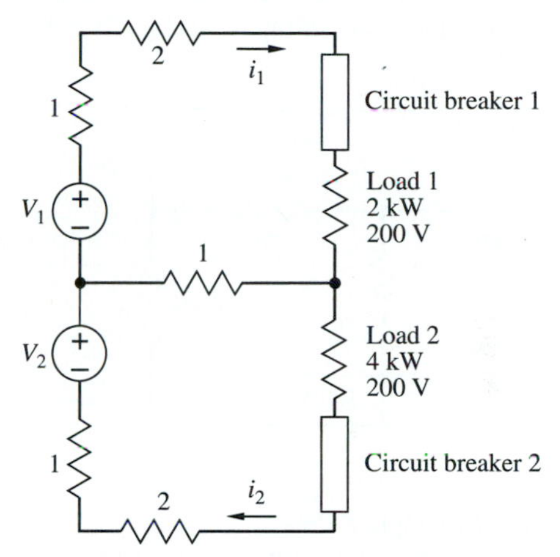

Figure P28

29. a. In the circuit of Figure P29, decide what to do with the controlled sources, and write a set of mesh equations.
 b. Solve for the mesh currents, using Gaussian elimination.
 c. Repeat using Cramer's rule.
 d. Determine the power supplied by the 10-V source and the two controlled sources.
 e. Use PSPICE to determine the branch currents and confirm the values of the mesh currents previously found, by KCL.

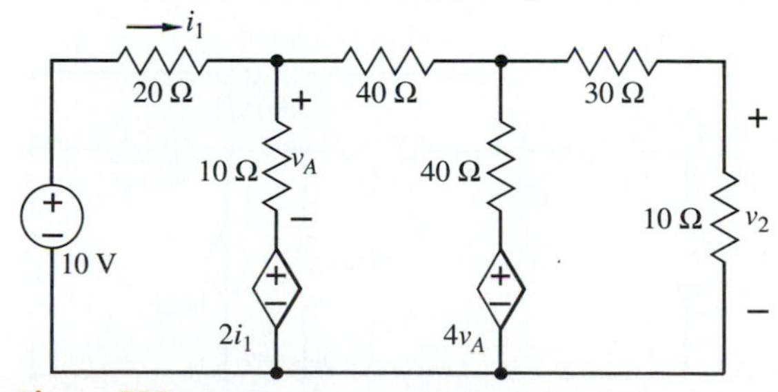

Figure P29

30. **a.** In the circuit of Figure P30, decide what to do with the controlled sources, and carry out a node analysis.
b. Solve the resulting equations, using Gaussian elimination.
c. Solve them also using Cramer's rule.
d. Repeat, using MATLAB.
e. Determine whether each of the sources is actually supplying power to the circuit or absorbing power.
f. Determine the node voltages, using PSPICE.

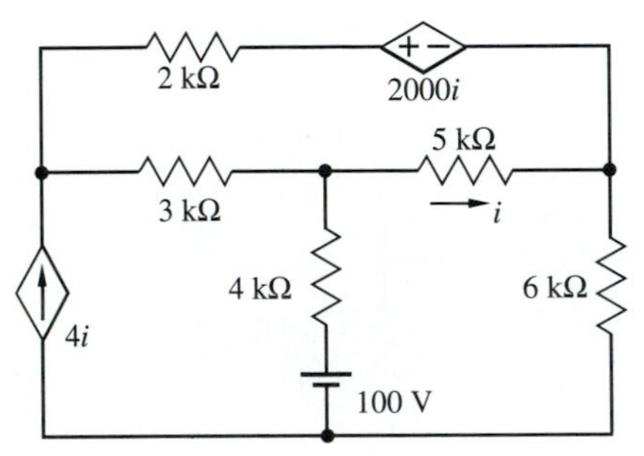

Figure P30

31. In the circuit of Figure P30:
a. Decide how to deal with the controlled sources, and write a set of mesh equations. How many simultaneous equations is it necessary to solve?
b. Solve these equations, using Gaussian elimination.
c. Repeat, using Cramer's rule.
d. Use the current shift on the CCCS and write the resulting mesh equations. How do they compare with the ones in part (a)?

32. **a.** Write a set of node equations in the circuit of Figure P32.
b. Solve for v_1 and v_2 using Gaussian elimination.
c. Solve again for v_1 and v_2 using Cramer's rule.
d. Use PSPICE to solve for v_1 and v_2. Compare the effort by each method.
e. Find the power supplied by the independent source and by the controlled source.

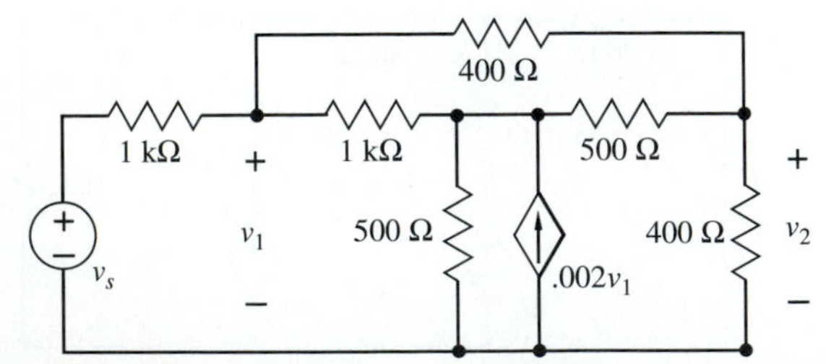

Figure P32

33. **a.** In the same circuit, that of Figure P32, decide what to do with the controlled current source, and write a set of mesh equations.
b. Solve these equations using Gaussian elimination, and confirm the values of v_1 and v_2 found in Problem 32(b).
c. Solve again, using Cramer's rule.
d. Write a set of loop equations for a different set of loops.
e. By appropriate mathematical operations, confirm that the equations in this set are linear combinations of the mesh equations.

34. An ideal transformer has neither a resistance ($v = Ri$) nor a conductance ($i = Gv$) representation. Hence, in a circuit containing an ideal transformer, neither node equations nor loop equations can be written without some preliminary steps. However, it is possible that an ideal transformer is connected in such a way in a circuit that it is *augmented,* as shown in Figure P34, where it is accompanied by a resistor in series with the input and another resistor in parallel with the output. The entire structure can be taken as a single multiterminal component. The only remaining problem is to determine the v-i relationship of the entire component. Determine both a resistance and a conductance representation for the entire component in Figure P34 in the following forms:

$$\begin{bmatrix} v_1 \\ v_2 \end{bmatrix} = \begin{bmatrix} r_{11} & r_{12} \\ r_{21} & r_{22} \end{bmatrix} \begin{bmatrix} i_1 \\ i_2 \end{bmatrix} \quad \text{and} \quad \begin{bmatrix} i_1 \\ i_2 \end{bmatrix} = \begin{bmatrix} c_{11} & c_{12} \\ c_{21} & c_{22} \end{bmatrix} \begin{bmatrix} v_1 \\ v_2 \end{bmatrix}$$

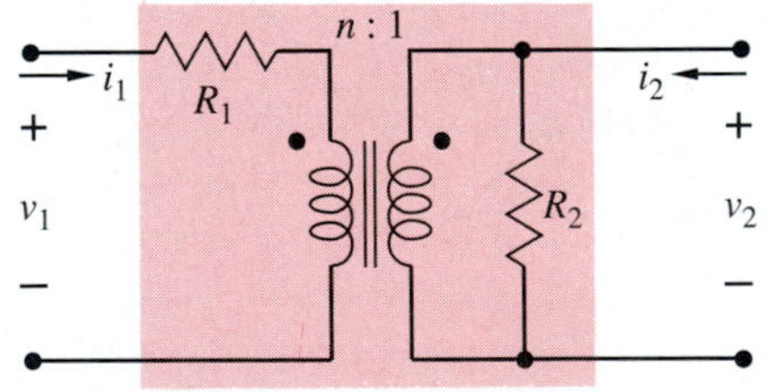

Figure P34

35. An augmented transformer as described in Problem 34 forms part of a circuit, as shown in Figure P35.
a. Write a set of mesh equations and determine the output voltage v using either Gaussian elimination or Cramer's rule.
b. Find also the power supplied by the source.
c. Determine the Thévenin equivalent at terminals AB and confirm that the same output voltage is obtained.

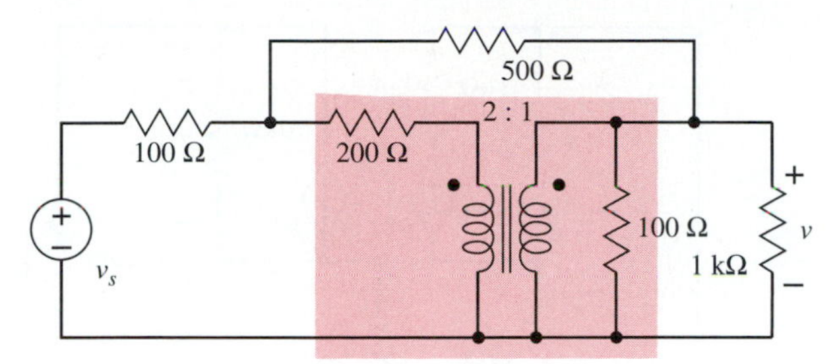

Figure P35

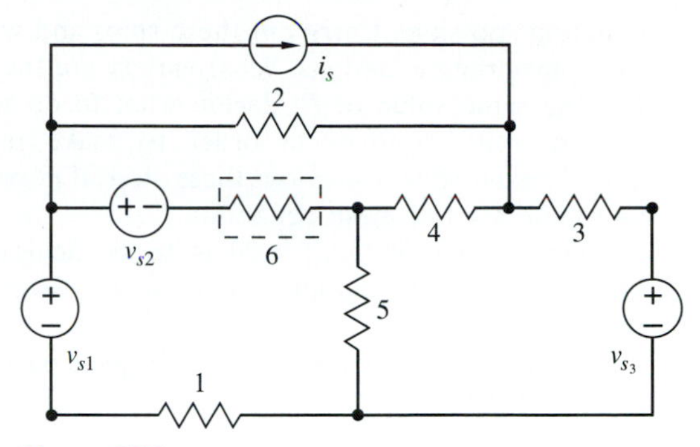

Figure P36

36. Two cases are to be considered in the circuit of Figure P36. In the first case, we assume that R_6 is small enough that it can be neglected (assumed 0).

a. Specify the number of independent KVL and KCL equations on the basis of the values of b and n.

b. How many independent equations will be required for carrying out a node analysis? How many for loop analysis?

c. Write whichever set in part (b) is fewer; take your pick if the numbers are the same.

d. Now suppose that R_6 cannot be neglected. Repeat parts (a) and (b).

37. a. Decide how to handle the controlled source in Figure P37 so as to permit the writing of standard node equations.

b. Choose the bottom terminal of the voltage source as datum and write a set of node equations, using your plan in part (a). Assume that the branch-identifying numbers are values of conductance in millisiemens, and let $\alpha = 2$.

c. Solve for the node voltages, using Gaussian elimination.

d. Solve again, using Cramer's rule.

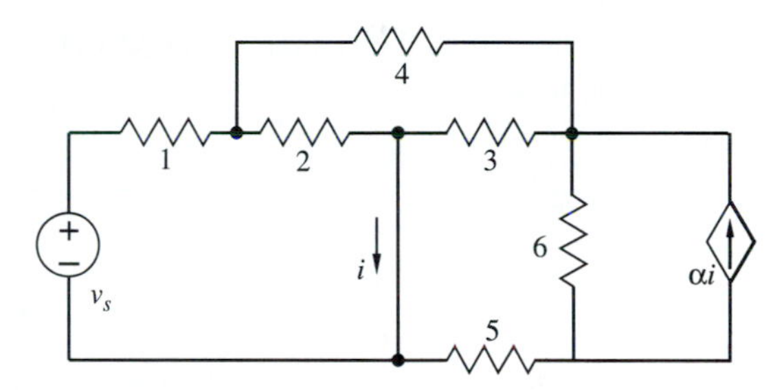

Figure P37

DESIGN PROBLEMS

1. The objective of this problem is not the design of a circuit or the determination of component values in a circuit to satisfy some specified conditions. Nevertheless, the problem is one of "design" in the sense that its objective is to create something: a certain theorem concerning a class of circuits.

Given a tree of a circuit, one set of loops in this circuit is the category of *fundamental loops*. For planar circuits, another category is the set of *meshes*. The problem is to discover the conditions under which all the meshes of a planar circuit are fundamental loops for some tree.

In your search for such a theorem, construct a few graphs and use the definition of a fundamental loop to specify whether certain specific branches related to meshes must be twigs or links if the meshes are to be fundamental loops. Along the way, determine the answer to the following question: Is it possible that the manner in which the branches of a circuit are laid out will determine whether or not a set of branches in a planar graph constitutes a mesh? Give your reasoning, and illustrate with simple graphs having four or five nodes.

2. The circuit in Figure DP2 includes a model of a transistor amplifier. (You don't need knowledge of transistor amplifiers to solve this problem.) The controlled-source parameters are assumed to be $\beta = 100$ and $\mu = 2.5 \times 10^{-4}$.

a. Suppose the only unspecified resistor has the value $R = 5\,\text{k}\Omega$. Decide how to treat the controlled sources in order to make node or quasi-node equations possible. Carry out these steps and write the appropriate equations. What variety are they?

b. For the same value of R, decide what to do with the controlled sources in order to make mesh equations possible. Carry out these steps if possible and write a set of mesh equations.

c. A current gain of $|i_o/i_s| = 80$ is to be achieved. Determine an appropriate value of R, assuming that $\mu = 0$.

d. Assuming the original value of μ, repeat part (c) using PSPICE.

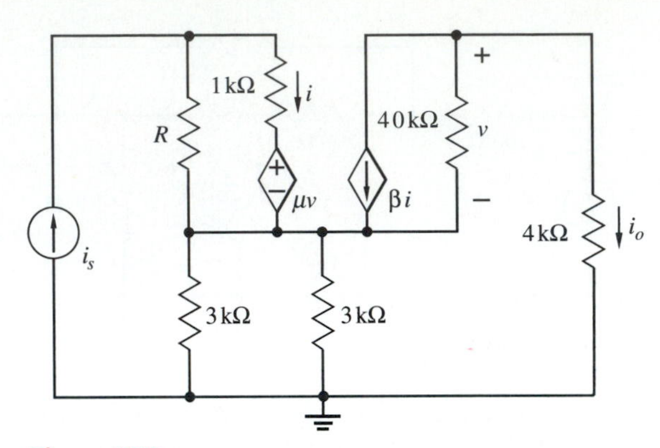

Figure DP2

3. The circuit shown in Figure DP3 is the small-signal equivalent of a two-transistor amplifier.

a. Assuming that the value of the unspecified resistor is $R = 10\text{ k}\Omega$, make appropriate modifications and write a set of node equations.

b. Solve for the current gain i_o/i_s using some numerical method.

c. Repeat part (b) but using a different numerical method.

d. Repeat part (b) but this time using MATLAB if available to you.

e. The value of R is no longer given but it is required that the current gain be $|i_o/i_s| = 20$. Determine the necessary value of R.

f. Repeat part (e) but this time using PSPICE if available to you.

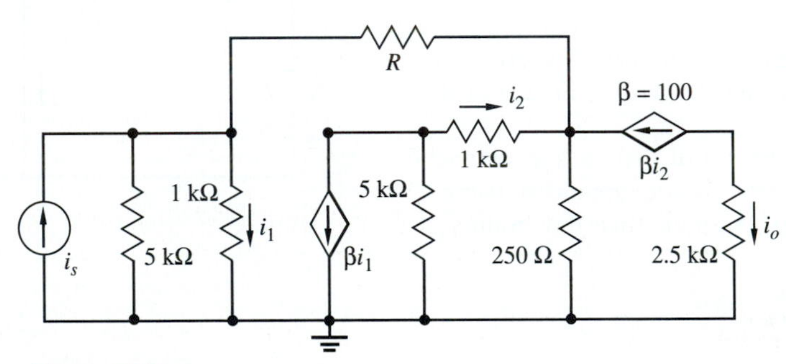

Figure DP3

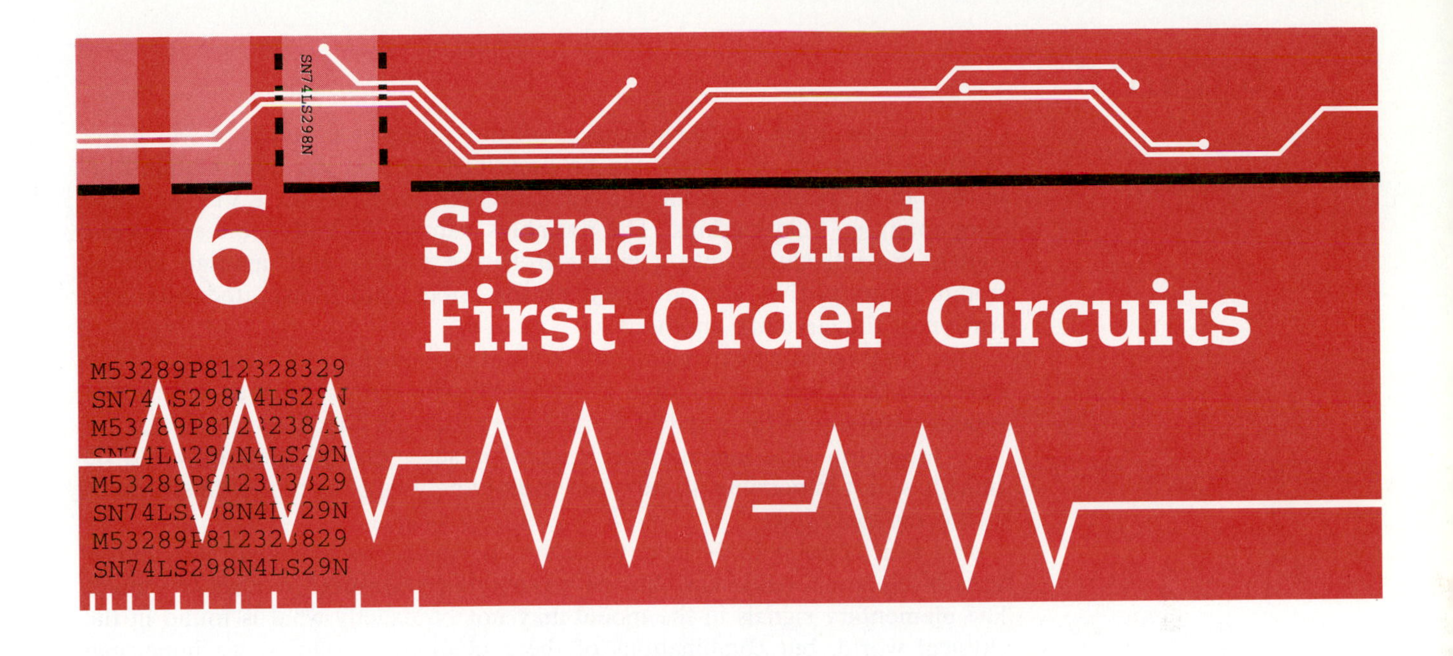

6 Signals and First-Order Circuits

All the preceding linear circuits we have studied have consisted of dissipative components only. In such circuits, the response to an excitation signal is simply proportional to that signal. If more than one input signal is simultaneously active, the response is just a linear combination of the responses to each exciting input. Thus, the specific nature of the signal waveforms does not come into play.

Now consider an expanded class of circuits that includes capacitors and inductors, in addition to any previously studied components. Since the voltage and current in these elements are not simply proportional to each other but one is the derivative of the other, we would not expect the waveform of an output signal to be the same as the waveform of an input signal.

We are now ready to take up the study of arbitrary circuits made up of any combination of all the elements of our model. But we won't challenge more than one complexity at a time, not only because it might be easier to learn that way, but because some of the simpler circuits are very important in their own right. In this chapter, we will take up the study of circuits that include a single energy-storing element—one capacitor or one inductor—in addition to dissipative components. Such circuits might be designated *single-energy* or *first-order* circuits.

Since the specific nature of signal waveforms will now be significant, as a prelude to the study of such circuits we will first undertake a study of some common signals which are important in engineering.

1 SIGNALS

Until now attention has been focused on the variables in terms of which the electrical behavior of circuits can be described and on the interrelationships among them resulting from the fundamental laws: KCL, KVL, and the component v-i relationships. The variables can be looked upon as *signals*. Signals are introduced into a circuit from the outside and undergo various modifications as they are processed by the circuit and transmitted from one portion of the circuit to another.

One method of classifying signals is by their *waveform* or *waveshape*. The waveform describes the variation of the signal with time, its shape. Signals that appear naturally in the physical world have relatively complicated waveforms. (The waveform of a human speech signal is such an example.) It would be difficult, if not impossible, to carry out useful analyses of circuits with signals of such complexity only.[1]

Just as it is possible to make models to represent complicated interconnections of electrical devices, so also is it possible to construct models of signals. The elementary signals in the model may not be exactly what is found in the physical world, but combinations of these elementary signals, we hope, can represent any arbitrary waveform. Then, once it is found how a given circuit acts on a simple signal, the action of the circuit on a more complicated waveform will be known, because the complicated waveform can be decomposed into simpler ones. (As we shall discuss later, this procedure is valid only for linear circuits.)

Periodic Functions

Many natural phenomena are of a repetitive or cyclic nature: the swinging of pendulums, the rise and fall of tides, and the yearly change of seasons are cyclic. A variable that describes the instantaneous state of a cyclic phenomenon, such as the displacement of a pendulum, is said to be a *periodic* function or a periodic wave.

Periodic functions occur frequently in electrical phenomena. The electric power system operates with periodic waves; the oscillator in a radio transmitter produces periodic waves; and the sweep signal which constructs the picture on the face of a television tube is periodic. You need to be able to recognize, classify, describe mathematically, and manipulate the mathematical expressions of periodic functions. This section will be devoted to that task.

There is a wide variety of simple waveforms that can be used as elementary signals. In a general way, they can be classified as periodic or not periodic, i.e.,

[1] Some signals, such as the noise generated in a physical resistor due to the Brownian motion of the charge carriers, are completely irregular and are called *random signals*. For random signals, it is possible to carry out an analysis based on the very randomness itself. The properties of such signals can be described in terms of probability theory. We shall not discuss random signals in this book.

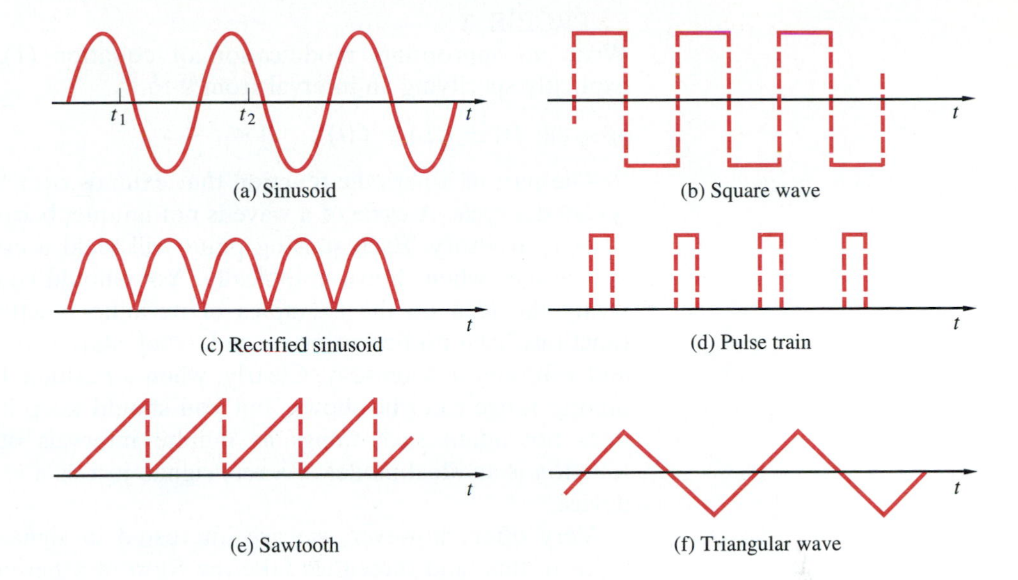

Figure 1
Periodic waveforms.

aperiodic. Some examples of periodic waveforms are shown in Figure 1 and of aperiodic waveforms in Figure 2.

A periodic function assumes a sequence of values over a certain interval of time and continually repeats these values over each succeeding interval. The shortest time interval over which the function repeats itself is called the *period*, designated by T. Suppose a function repeats its sequence of values every 5 seconds; will it repeat its sequence of values every 10 seconds also? Every $5n$ seconds, where n is an integer? Yes, it will; that's why we defined the period as the shortest of these intervals. If the function is $f(t)$ and its period is T, then the mathematical definition of a periodic function is:

$$f(t + nT) = f(t) \qquad -\infty < t < \infty \tag{1}$$

Whatever value the function has at some time, it will have the same value T (or any multiple of T) units of time later. The interval of time between t_1 and t_2 in Figure 1a, for example, is one period.

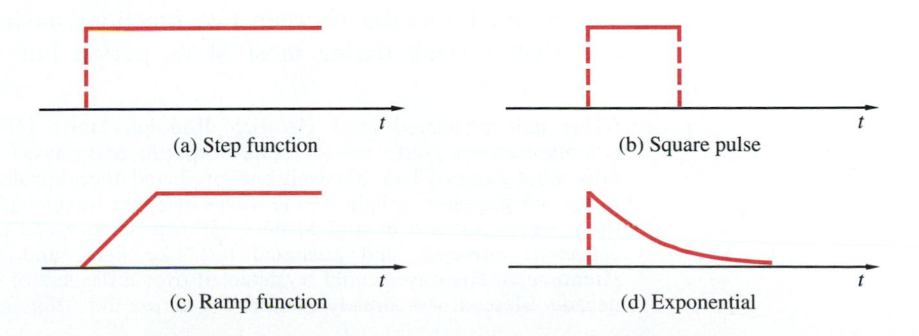

Figure 2
Aperiodic functions.

EXERCISE 1

Write an appropriate modification of equation (1), valid for all time, but explicitly specifying an interval from 0 to ∞.

ANSWER: $f(t \pm nT) = f(t) \qquad 0 < t < \infty.$ □

The part of a periodic function that extends over the interval of one period is called a *cycle.* A cycle of a wave is not unique, because the starting point of a cycle is arbitrary. Each starting point will yield a cycle which has a different appearance when shown graphically. You should convince yourself of this by trying the idea on the examples of periodic functions in Figure 1. Periodic functions have no beginning or end. They started in the dim past of antiquity and will outlast doomsday. Clearly, when a periodic function is portrayed, this infinite range can't be shown, but you should keep it in mind. The real world does not admit such things as infinite intervals of time. Thus, a periodic function is an ideal model of a real signal, just as a resistor is a model of a real device.

Very often, however, we are interested in signals that started at a finite point in time and thereafter take the form of a periodic function. Don't make the mistake of calling such functions periodic. The effect of such a signal on a circuit a long time after it has been switched on, however, is practically indistinguishable from the effects of a truly periodic function—of the same waveform, of course.

The period gives a measure of how long a cycle lasts. The converse of this is the question, How many cycles are there in a given interval, say, in one unit of time? The number of cycles a periodic function completes in one second is called its *frequency,* labeled f. If the period of a wave is 0.02 s, for example, the number of intervals of that length in 1 second is $1/0.02 = 50$, so the frequency is 50 cycles per second. The international unit for frequency is the *hertz,* abbreviated Hz. A hertz is, thus, one cycle per second.[2] Hence, the relationship between the period and the frequency of a periodic function is:

$$f = \frac{1}{T} \tag{2}$$

Let's now turn to a consideration of some quantitative aspects of the contents of periodic signals. Probably the first one that comes to mind is how high the signal goes—its *amplitude* at the highest point, its *peak* value. This is certainly a useful number to describe a function, but such a number alone would not distinguish between two functions having the same amplitude: (1) one that is small during most of its period but shoots up to a high peak

[2] This unit is named after Heinrich Rudolph Hertz (1857–1894), a pioneer in radio communications. Hertz was a German experimental physicist who corroborated experimentally what James Clerk Maxwell had predicted theoretically in 1873, that electromagnetic waves are the same as light. Using waves of 24 cm wavelength, he found experimentally that these waves traveled in straight lines with the same speed as light waves; that they can be reflected, refracted, and polarized just like light; and that they exhibit interference phenomena. His waves could be detected over a disance of only about 20 m, but in a short decade, Marconi was already transmitting across the Atlantic.

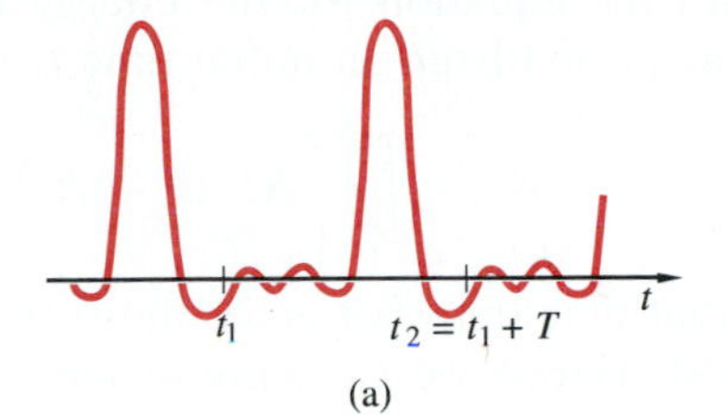

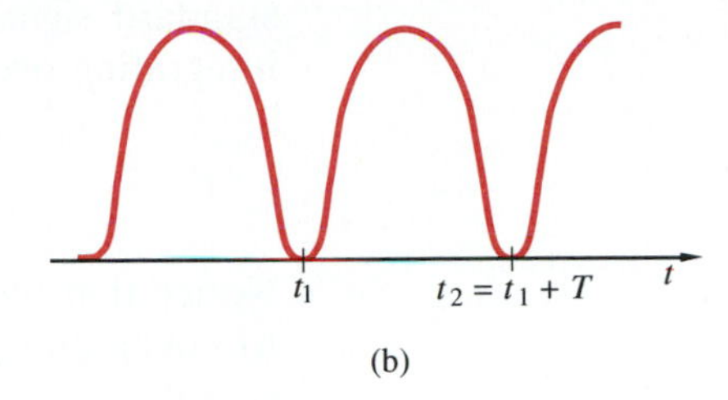

Figure 3
Two functions with the same peak.

momentarily and (2) another that remains near the same peak over a substantial part of the cycle. Two such functions are shown in Figure 3.

One way to distinguish between such functions is to find their *average* value. To find the average of a set of numbers, we add them and divide by their number. To find the average value of a function of time, we integrate the function (find the area under the curve) over an interval of time and then divide by that time interval. The value obtained will depend on the averaging interval. A unique value will result only if we agree to average over one or more complete cycles of a periodic function. Thus, in terms of the two periodic signals in Figure 3, we integrate from t_1 to t_2 and divide by the period $T = t_2 - t_1$. Although both functions have the same period, their *full-cycle* average value will be different.

EXAMPLE 1

Suppose the sawtooth waveform in Figure 1 represents a voltage. Let's find its average value, assuming its amplitude is 10 V and its period is 2 s. Figure 4 shows one cycle of the wave.

The slope of line will be 10/2 = 5 V/s, so the equation of the line is $v = 5t$. The average is found by integrating this expression over one period and dividing by the period. Thus:

$$\text{Average} = \int_0^2 5t\,dt = \left.\frac{5t^2}{2}\right|_0^2 = 10\ \text{V}\cdot\text{s}$$

Figure 4
One cycle of sawtooth.

In cases like this, where the geometry of the function permits, it is easier to find the area under the curve geometrically. The shape of the sawtooth wave is triangular; hence, the "area under the curve" is half the base times the height. Does this yield the same result?

ROOT-MEAN-SQUARE (RMS) VALUE Another important quantitative measure of a periodic signal can be derived in the following way. For purposes of discussion, let's assume that the signal represents a current. If a current i goes through a resistor R for an interval of time, some energy will be dissipated as heat. A quantitative measure of the signal is obtained if we compare this heat with the amount that would be dissipated if some standard signal were to go through the same resistor. Again, to obtain a unique, meaningful value, the time interval should be one period of the periodic signal. As a standard signal, we choose what is called DC, a constant, say, I. Inserting this constant as a standard signal into the equation for the energy dissipated in a resistor and integrating over one period from an initial time t_1 to $t_1 + T$, we get:

$$w_R = \int_{t_1}^{t_1+T} RI^2\, dt = RI^2T$$

Notice that the initial time does not enter the result at all; all that is needed is to cover one period. Hence we can choose any convenient initial time, say, $t_1 = 0$.

The next step is to repeat this process using the general signal $i(t)$ instead of the standard signal. In this case, the energy dissipated will be:

$$w_R = \int_{t_1}^{t_1+T} Ri^2(t)\, dt$$

We want the energy dissipated by the actual signal over one period to be the same as the energy dissipated by the standard signal over the same interval. Therefore we equate the preceding two equations and then solve for I. Note that R cancels from both sides of the equation. This result will be:

$$I = \sqrt{\frac{1}{T}\int_{t_1}^{t_1+T} i^2(t)\, dt} \equiv I_{\text{rms}} \tag{3}$$

This measure of a signal has been defined as its *rms value,* the letters being the initials of *root-mean-square.* The operations carried out on a signal to obtain its rms value are done backward from the name: first we *square* the signal; then we take its average, the *mean*; and finally we take the *square root.* The initial time t_1 can be chosen as any convenient time.

Another name used for the rms value for signals which are currents (or voltages) originates from asking how effective a given current or voltage is in producing heat as compared with a DC current or voltage; hence, the term *effective value.*

The fact that here we developed the concept of rms value in terms of the dissipation of heat is not evident when we look at the expression in the preceding equation. The concept of the rms value of a signal also arises in other applications where the signals are not currents or voltages and heat dissipation is not the question. So, generally speaking, for any periodic function $f(t)$, the rms value is defined as:

$$F_{\text{rms}} = \sqrt{\frac{1}{T}\int_0^T f^2(t)\, dt} \tag{4}$$

Let's now use this concept of rms value to determine the average energy stored in a capacitor when the capacitor is subjected to a periodic waveform of voltage. In Chapter 2 we found the energy stored in a capacitor at any time to be $w_C = Cv^2(t)/2$. Our goal, then, is to find the average value of w_c over one period T, assuming that the voltage is periodic. The result is:

$$W_C = \frac{1}{2}C\left[\frac{1}{T}\int_0^T v^2(t)\, dt\right] = \frac{1}{2}CV_{\text{rms}}^2 \tag{5}$$

By comparing the square root of the quantity within brackets on the right with the expression in (3), we note that it is the rms value of the voltage. We have used a capital W to represent the average energy stored.

EXERCISE 2

Determine the rms value of the sawtooth waveform one cycle of which was shown in Figure 4.

ANSWER: 5.77. □

EXERCISE 3

Find the rms value of the triangular wave in Figure 1(f), assuming that it represents a voltage with a peak value of 5 V and a period of 80 ms.

ANSWER: 50 V. (Did you take advantage of symmetries?) □

THE SINUSOIDAL FUNCTION The sinusoidal function of time is one of the most important periodic functions in all of engineering. We will now discuss in some detail the quantitative features of a sinusoidal function. For purposes of discussion, assume that the signal is a voltage. A general sinusoidal voltage function can be written:

$$v(t) = V_m \cos(\omega t + \phi) \tag{6}$$

A plot of this function is shown in Figure 5. Two scales are shown for the abscissa, one for time t and one for the *phase variable* ωt. As time goes on, the function alternately takes on positive and negative values: hence, it is commonly referred to as an *alternating* function. (This is where the designation *AC*, or *alternating current,* comes from.) To determine the number of cycles that the function goes through per unit time, note that the change in ωt over one period is 2π. Hence, the time elapsed during one cycle (the period) is $T = 2\pi/\omega$. The frequency, being the number of cycles per unit time, is simply the reciprocal of this, namely, $f = 1/T = \omega/2\pi$. ω is referred to as the *angular frequency,* sometimes also the *radian frequency*; $\omega = 2\pi f = 2\pi/T$. As the time variable traverses one period, the phase variable traverses an angle of 2π radians.

Aside from the variable t, there are three parameters in the sinusoidal function in (6): (1) The peak (or maximum) value, or amplitude, is V_m. (2)

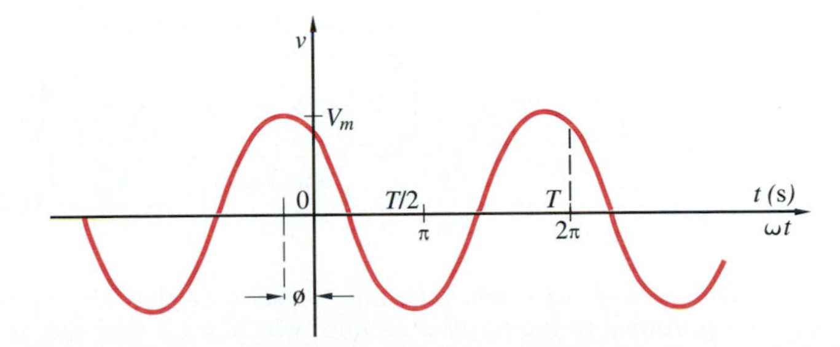

Figure 5
Sinusoidal function.

The symbol ϕ represents the entire argument of the cosine function at $t = 0$, so it is called the *initial angle.* It is measured from the positive peak; that is, $\phi = 0$ means that the vertical axis has been chosen to go through the positive peak. The term *phase,* or *phase angle,* is also used to describe the initial angle. (3) The third parameter is the angular frequency.

In the diagram, ϕ is a positive number, but it can also take on negative values. Observe that $\cos(\omega t + \phi \pm 2\pi n)$ is an expression for the same sinusoidal function as $\cos(\omega t + \phi)$. Thus, the initial angle is not unique; any multiple of 2π can be added to or subtracted from it. In order to avoid this ambiguity, we specify that ϕ will lie in the range between $-\pi$ and π. An angle lying in this range is said to be the *principal value* of the angle.

Observe that the sine function $V_m \sin \omega t$ and the cosine function $V_m \cos \omega t$ are special cases of the sinusoidal function given in the preceding equation, with the initial angle being $\phi = -\pi/2$ for the sine and $\phi = 0$ for the cosine. (Confirm this.) The general sinusoid can be written either as a sine function or as a cosine function with arguments that differ by π radians. This follows from the trigonometric identities:

$$\cos x = \sin\left(x + \frac{\pi}{2}\right)$$
$$\sin x = \cos\left(x - \frac{\pi}{2}\right) \qquad (7)$$

Average Value: Let's now calculate for the sinusoid some of the quantitative measures discussed in an earlier part of this section, starting with the average value. Since the areas under the negative half and the positive half of the cycle are the same, the full-cycle average value of the sinusoid is zero. Sometimes, however, another average—the *half-cycle* average value—is useful. This is defined as the average value of the positive half cycle.[3] The positive half cycle will be covered if we choose the time origin so that $\phi = -\pi/2$ and integrate from 0 to $T/2$. In view of the preceding trigonometric identity, the general sinusoid then reduces to $\sin \omega t$. Thus, the half-cycle average becomes:

$$V_{\text{avg}} = \frac{1}{T/2}\int_0^{T/2} V_m \sin \omega t \, dt = -2\frac{V_m}{T}\frac{\cos \omega t}{\omega}\Big|_0^{T/2}$$
$$= \frac{2}{\pi}V_m \approx 0.637V_m$$

RMS Value: The next measure is the rms value of a sinusoid. For convenience, choose the origin of time to make $\phi = 0$:

$$V_{\text{rms}} = \sqrt{\frac{1}{T}\int_0^T V_m^2 \cos^2 \omega t \, dt} = \sqrt{\frac{V_m^2}{2T}\int_0^T (1 + \cos 2\omega t)\, dt}$$
$$= \sqrt{\frac{V_m^2}{2T}\left(t + \frac{\sin 2\omega t}{2\omega}\right)\Big|_0^T} = \frac{V_m}{\sqrt{2}} \approx 0.707V_m \qquad (8)$$

[3] One reason why this measure is useful is that the readings of some instruments are proportional to the *rectified average,* which, for a sinusoid, is the half-cycle average.

In the second step we used the trigonometric identity $2\cos^2 x = 1 + \cos 2x$. You should confirm all the details of this calculation, *after which* you will have earned the right to memorize this very important expression for the rms value of a sinusoid in terms of its peak value.

The sinusoid is important not only in its own right as a signal, but also because, as Chapter 13 will describe, almost any other periodic signal can be decomposed into a linear combination of sinusoids having different frequencies. At this time we'll show one simple example in which a relatively complex periodic signal can be resolved into a sum of sinusoids. (You can skip the example if you are prepared to accept tentatively the claim made in the first sentence of this paragraph; you won't thereby miss anything fundamental, but you'll miss something interesting.)

EXAMPLE 2 **AMPLITUDE-MODULATED WAVE** Consider the waveform shown in Figure 6. This is a so-called *amplitude-modulated* wave. (The AM radio band carries such signals.) A general expression for such a wave can be written:

$$g(t) = f(t) \sin \omega_0 t$$

The *modulating signal* $f(t)$ is any continuous function of time, and the sinusoid is called the *carrier*. The function $f(t)$ plays the role of the amplitude of the sinusoid, although a time-varying amplitude. In the case of the signal in Figure 6:

$$f(t) = 1 + K \sin \omega_1 t$$

The entire waveform becomes:

$$\begin{aligned} g(t) &= (1 + K \sin \omega_1 t) \sin \omega_0 t \\ &= \sin \omega_0 t + K \sin \omega_1 t \sin \omega_0 t \end{aligned}$$

The second term in this expression can be modified by using the following trigonometric identity:

$$2 \sin x \sin y = \cos (x - y) - \cos (x + y)$$

Substituting this into the previous expression leads to:

$$g(t) = \sin \omega_0 t + \frac{K}{2} \cos (\omega_0 - \omega_1)t - \frac{K}{2} \cos (\omega_0 + \omega_1)t$$

The relatively complex waveform shown in Figure 6 is expressed here as a sum of three sinusoids. The frequency of one of them is the carrier frequency. The

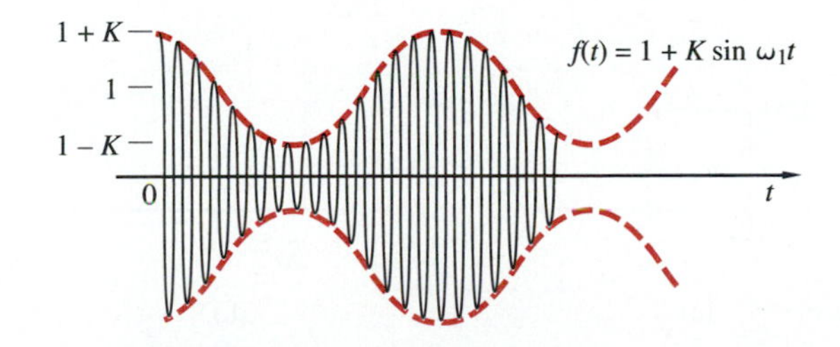

Figure 6. Amplitude-modulated wave.

other two have frequencies less than and greater than the carrier frequency by the frequency of the modulating wave.

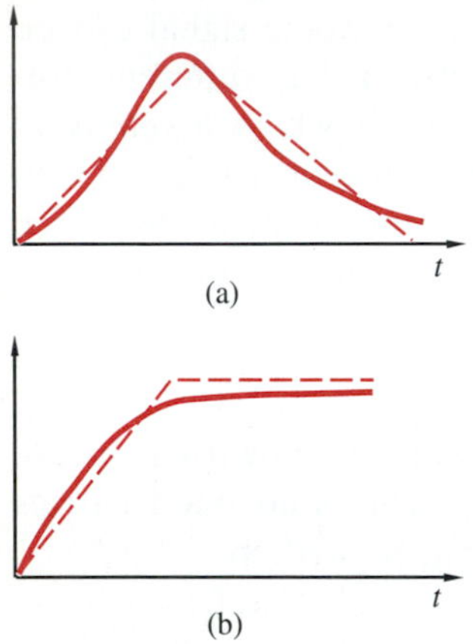

Figure 7
Approximations of real aperiodic signals.

Some Elementary Aperiodic Functions

Let us now turn our attention to some simple aperiodic waveforms, a few of which were illustrated in Figure 2. You might be concerned with these choices, since signals containing straight-line segments or discontinuities are hardly to be expected in the physical world.

Consider, however, the waveform given by the solid curve in Figure 7a. This is a smooth curve that might represent a voltage or current in a physical circuit. The triangle shown by the dashed curve is an approximation of the solid curve. Would you expect the electrical behavior of a circuit in response to the triangle to differ greatly from its behavior in response to the solid curve? Our reason for dealing with the triangle rather than the solid curve is the greater ease of handling it mathematically without greatly jeopardizing accuracy. The same type of discussion would apply to the curves in Figure 7b.

RAMP FUNCTION AND PULSES We turn now to a consideration of two very basic waveforms that play the same sort of role for aperiodic signals that a sinusoid plays for periodic signals. Consider the ones shown in Figure 8a. The curve above is called a *ramp* function. The curve rises linearly from 0 for a time interval Δt, then remains constant. The slope of the curve is zero everywhere except in the interval Δt, where it is a constant. Thus, the pulse waveform shown below the ramp is the derivative of the ramp. Conversely, the ramp is the integral of the pulse. The derivative-integral relationship is easily visualized from the graphs. At the ends of the interval Δt, the derivative is not defined. The area under the pulse is the height A of the ramp.

STEP FUNCTIONS AND IMPULSES Now suppose that the interval Δt is made smaller while the height A of the ramp is held fixed. This means that the slope of the ramp will increase. Thus, the pulse will become narrower and taller, as shown in Figure 8b, but the area under it will stay constant at the

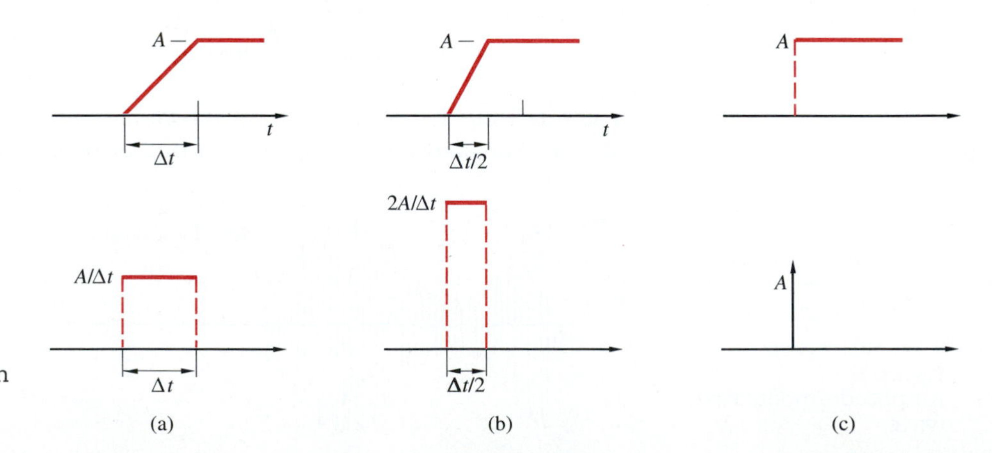

Figure 8
(a) and (b) Ramp function and pulse; (c) step function and impulse.

value A. If we keep on reducing the interval Δt until it reaches 0 in the limit, the ramp will take the form of a discontinuous jump, or step, of height A, as shown in Figure 8c. This waveform is called a *step function.*

In this process, the pulse will become narrower and taller until, in the limit, it attains an infinite height for a zero time interval. This waveform is called an *impulse.* At each stage as Δt is reduced, the area under the pulse remains constant. This is assumed to be true even in the limit as $\Delta t \to 0$. The area A is called the *strength* of the impulse. An impulse is represented graphically by a vertical arrow, as shown in Figure 8c, its strength being written alongside the arrow. For each value of Δt, the pulse is the derivative of the ramp and the ramp is the integral of the pulse, except at the ends of the interval, where the derivative is not defined.

Note that in the limit as $\Delta t \to 0$, the endpoints of the interval merge. Thus, strictly speaking, the impulse and the step do not have the derivative-integral relationship of the ramp and the pulse. It is conceptually useful, however, to interpret the derivative of a function at a discontinuity as an impulse.[4]

If the discontinuity of the step function is of unit height, we say that the function is a *unit step*; its companion is a *unit impulse.* If the step occurs at $t = 0$, we designate the unit step as $u(t)$. Thus:

$$\begin{aligned} u(t) &= 0 \qquad t < 0 \\ u(t) &= 1 \qquad t \geqslant 0 \end{aligned} \tag{9}$$

What if the discontinuity occurs at some time other than $t = 0$? Note that $u(t)$ denotes the function which is 0 whenever its argument t is negative and which is 1 whenever its argument is positive. If we wish to designate a step which occurs at, say, $t = 5$ s, we should write u with an argument that is negative for $t < 5$ and positive for $t > 5$. The desired argument is, obviously, $t - 5$. Thus, in general, a step function which occurs at a specific time $t = t_1$ can be written as $u(t - t_1)$:

$$\begin{aligned} u(t - t_1) &= 0 \qquad t < t_1 \\ u(t - t_1) &= 1 \qquad t \geqslant t_1 \end{aligned} \tag{10}$$

This notation agrees with the previous expression for the unit step when $t_1 = 0$.

EXERCISE 4

Write an expression for a function having a constant value K for all past time until $t = t_1$, when it drops to 0. Draw this function.

ANSWER: $f(t) = Ku(-(t - t_1)) = Ku(t_1 - t)$. □

[4] A mathematical theory of what are called *distributions* was introduced in 1950 by the French mathematician Laurent Schwarz. It places the impulse on a sound mathematical footing. The impulse, then, is not a function but a distribution. The derivative-integral relationship discussed in the text is mathematically rigorous so long as we are treating mathematical entities called distributions, but not for the ordinary point functions of calculus we are used to. For our purposes in this book, we will continue to use the concept of an impulse, while trying to be careful to avoid saying impulse "function." (See how many times you can catch lapses in this attempt.)

EXERCISE 5

Using step functions, write an expression for a voltage pulse $v(t)$ of 10 V amplitude and extending from $t = 2$ to 6 s.

ANSWER: $v(t) = 10[u(t - 2) - u(t - 6)]$ V. □

EXERCISE 6

Write an expression for a current step-stool function which rises in 2-mA steps from 0 to 4 mA and falls in the same increments to 0 in 20 s. The top of the function is twice as wide as the duration of each step. Draw the function also.

ANSWER: $i(t) = 2[u(t) + u(t - 5) - u(t - 15) - u(t - 20)]$ mA. □

Functions such as the ones in the preceding exercises are called *piecewise constant* functions. Such functions maintain a constant value over an interval of time, then jump discontinuously to another constant value which they maintain for another interval, and continue to jump to other constant values. Such functions can be easily specified as a linear combination of step functions, each occurring at some specific time.

Now, let's turn to the impulse. An impulse whose strength is unity is called a *unit impulse* and is designated $\delta(t)$, assuming that the impulse occurs at $t = 0$. If the strength is K, the impulse is designated $K\delta(t)$. Since the step is the integral of the impulse, we write:

$$\int_{-\infty}^{t} K\delta(x)\, dx = Ku(t) \tag{11}$$

The only contribution to the integral comes from the point $t = 0$. Thus, if the integration were performed up to some negative value of time, the integrand would be zero over the entire range of integration, so the integral would yield 0. If the upper limit of the integral is positive, the integral will equal the strength of the impulse.

EXERCISE 7

Write an expression similar to (11) except that the impulse occurs at $t = t_1$. □

The impulse has some peculiarities, only one of which will be considered here. Suppose the impulse in (11) represents a current and is to have a strength A. You might think that A has the dimensions of current, just as it would in the expression $i(t) = A \cos \omega t$. Such is not the case. In Figure 8, A is the area under the pulse. If the pulse represents a current, then the area, which is the strength of the impulse, dimensionally, is charge: current times time. Similarly, the strength of a voltage impulse dimensionally is flux linkage. Consideration of other properties of an impulse will be deferred until such time as they will be used.

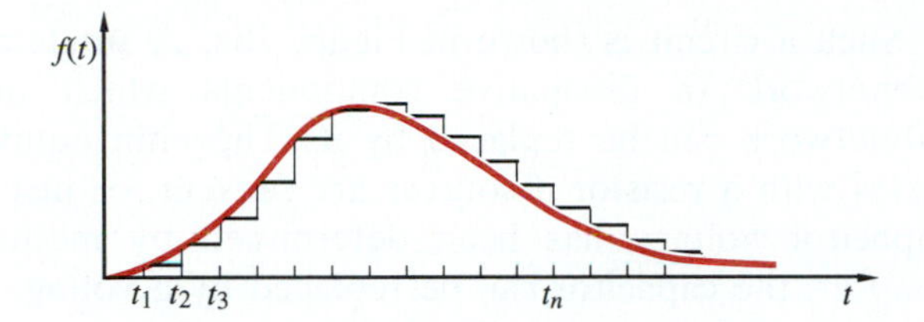

Figure 9
Approximation by step functions.

APPROXIMATION OF AN APERIODIC FUNCTION BY STEPS[5] Just as we claimed that an arbitrary *periodic* function can be represented as a linear combination of sinusoids of different frequencies, so we also claim that an arbitrary *aperiodic* function can be represented by a linear combination of step functions. The following development discusses this idea briefly.

Figure 9 shows a smooth curve which is approximated by the "staircase" function shown. This staircase is a piecewise constant function; it is a combination of step functions, each one having a different strength and each displaced from the preceding one by a fixed time interval. The approximation can be improved by taking smaller and smaller intervals between steps.

Using the notation in (10) to represent a displaced step, the piecewise constant function in Figure 9 can be approximated by a linear combination of displaced steps:

$$f(t) = K_1u(t - t_1) + K_2u(t - t_2) + \cdots + K_nu(t - t_n) + \cdots$$

where the K's are the jumps at the discontinuities. If the intervals between steps are allowed to approach zero, the summation will become an integral and the approximate equality will become exact.

We shall not pursue this topic any further here. It is not our purpose to be exhaustive, but to give an indication of how the elementary signals we have introduced can be used to represent any arbitrary signal. These ideas are the basis for powerful methods of analysis which we shall discuss in Chapter 14.

2 RESISTANCE-CAPACITANCE CIRCUITS[6]

After this brief discussion of periodic and aperiodic signals, we now return to circuits. Those that include capacitors and dissipative components but no inductors are referred to as *resistance-capacitance circuits,* or *RC* circuits, for short, even though multiterminal dissipative components might be included. In this chapter, we will consider *RC* circuits that include only one capacitor.[7]

[5] This section isn't essential to the material that follows, so it can be omitted from a first reading without resulting in a future handicap for you.
[6] If you skipped over the study of capacitors and inductors in Chapter 2, or if you feel the need for review, you should return to Chapter 2 for a read.
[7] This description of "only one capacitor" includes the possibility of a capacitor-only two-terminal subcircuit equivalent to a single capacitor. The simplest such subcircuit might be two capacitors in series or in parallel.

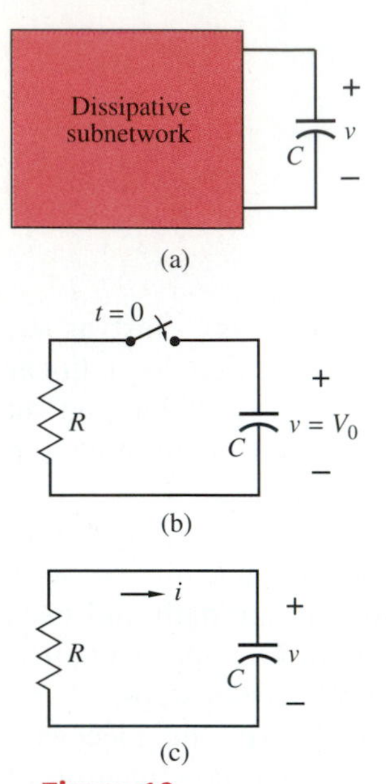

Figure 10
RC circuits: (a) general case; (b) initially charged capacitor; (c) discharging through a resistor.

Such a circuit is shown in Figure 10a. A single capacitor is connected to a subnetwork of dissipative components which may include sources. The subnetwork can be replaced by its Thévenin equivalent: a voltage source in series with a resistor if sources are present, or just a resistor if not. Once the capacitor voltage has been determined by methods to be studied in this chapter, the capacitor can be replaced by a voltage source on the basis of the compensation theorem. The voltage or current anywhere in the dissipative subnetwork can now be solved by algebraic methods described in the preceding chapters. Thus, the most general circuit under study here can be reduced to a series connection of a single capacitor, a single resistor, and a single voltage source. This section is limited to the case where the voltage source is absent.

Resistance-Capacitance Circuits; Zero-Input RC Circuits

Since a capacitor can store energy, it is possible for a response to take place in an *RC* circuit *even without* an exciting signal, other than the signal that provided the stored energy in the past.

A case in point is shown in Figure 10b. A capacitor which has been initially charged to a voltage V_0 (by some mechanism or other) is suddenly connected to a resistor by some switching event, as shown. The switch, which was initially open, is closed at time $t = 0$. (We ought to be more general and say that the switching instant is some arbitrary time t_0. But there is no absolute origin for time, so, for convenience, we will choose the initial time to be $t = 0$.) What is of interest is finding the capacitor voltage and current after the switching.

A Differential Equation

The circuit in Figure 10c applies for the interval of time after the switch is closed. This is a simple circuit with two nodes; hence, there is only one node equation. As we are by now accustomed, we "think" of adding the two currents oriented away from the top node; then we "think" of the branch relationships to eliminate currents in favor of voltages, remembering that standard references are assumed. In this case, KVL (trivially) requires the two branch voltages to be equal, both equal to $v(t)$. The final result is:

$$C\frac{dv}{dt} + \frac{1}{R}v = 0$$

or, dividing through by C:

$$\frac{dv}{dt} + \frac{1}{RC}v = 0 \tag{12}$$

This is something new. Because a derivative of the variable appears in addition to the variable itself, this equation is a member of a class called *differential equations.* Such equations can be described in terms of their properties: the highest derivatives they contain (their *order*), whether or not they are linear (that is, whether the variable or its derivatives appear raised to any power higher than 1), whether the derivatives are *ordinary* or *partial,* and the nature of their coefficients (constant or time-varying). Our equation here is an *ordinary, linear differential equation of the first order, with constant coefficients.*

Another significant property of equations is clarified by asking what would happen if the dependent variable (in this case v) is multiplied by some constant. Suppose v in the preceding equation is replaced by kv, where k is a constant. Then each term in the equation is multiplied by k. But this will really have no influence, since this constant can be canceled out, leaving the original equation. A linear equation which remains unchanged when the dependent variable is multiplied everywhere by a nonzero constant is said to be *homogeneous.* Thus, in addition to all the other descriptive things it is, the differential equation in (12) is homogeneous.

EXERCISE 8

Using the circuit laws, confirm that the differential equation satisfied by the current i in Figure 10 is identical with the one satisfied by v. □

To find the capacitor voltage, we will have to solve (12). By *solving* a differential equation, we mean finding a function of the independent variable which, when substituted into the differential equation, will cause it to be satisfied. When we say that an equation is *satisfied* by a function, we mean that insertion of the function into the equation will result in an identity, such as $0 = 0$. Solving differential equations is one of the things in which an engineer engages as an adjunct to the process of analysis and design. Many methods have been devised over the centuries for solving such equations, one that sounds logical being to "undifferentiate," or to integrate directly.

Let's carry out this process in (12) after dividing through by v. Since we are interested in the solution for any time after $t = 0$, we'll integrate from 0 to t:

$$\begin{aligned} \int_0^t \frac{1}{v}\frac{dv}{dx}\,dx &= -\frac{1}{RC}\int_0^t dx \\ \int_{v(0)}^{v(t)} \frac{dv}{v} &= -\frac{t}{RC} \end{aligned} \tag{13}$$

INITIAL VALUE OF CAPACITOR VOLTAGE On the left side of the first line of this equation, because $(dv/dx)\,dx = dv$, the integration with respect to x is changed to an integration with respect to v, with the limits changed appropriately. The lower limit will be the value of v at $t = 0$. Here we meet a problem. Just before the switching event, the capacitor voltage is V_0. The differential equation applies to the time *after* the switching, however, and the lower limit of the integral must be the value of the voltage immediately after the switching. The distinction is made by designating these two values—before and after switching—as $v(0-)$ and $v(0+)$, respectively. The question becomes, Can we say anything about the zero-plus value of the capacitor voltage, knowing the zero-minus value?

The answer is given by the *law of conservation of charge,* which was discussed in Chapter 1. (Review that discussion if you feel the need.) As applied to a circuit, this principle states that electric charge cannot be transferred into or out of a node instantaneously. In terms of Figure 10, that means charges cannot be suddenly moved into or out of the capacitor terminals. Since the capacitor voltage is proportional to its charge, that means the capacitor voltage cannot change instantly either; that is, the *capacitor voltage is a continuous function of time.* Whatever the voltage was just prior to the switching, that's what it must be just afterward. Thus, the lower limit of the integral must be V_0.

In the present simple case, the following reasoning can also be used. Suppose the capacitor voltage does change instantaneously at some instant of time. That means the voltage will have a step discontinuity. The capacitor current, being the derivative of the charge, which is proportional to the voltage, will then become infinite at the discontinuity; it will contain an impulse. But the capacitor current and the resistor current are the same, and because the resistor voltage is proportional to its current, the resistor voltage will also become infinite. But that's a contradiction, since the resistor voltage equals the capacitor voltage—which is not infinite. The conclusion from this reasoning is that the capacitor voltage *in this circuit* cannot undergo a discontinuous change.

THE EXPONENTIAL Armed with this information, we carry out the integration in the previous equation, getting the result:

$$\int_{V_0}^{v} \frac{du}{u} = \ln u \Big|_{V_0}^{v} = -\frac{t}{RC}$$

The symbol for the dummy variable of integration has again been changed to avoid confusion with the upper limit.[8] The integral of du/u is the natural logarithm function. After inserting the limits and then inverting the logarithm, we get:

$$\ln \frac{v}{V_0} = -\frac{t}{RC} \qquad \textbf{(14)}$$

$$v(t) = V_0 e^{-t/RC}$$

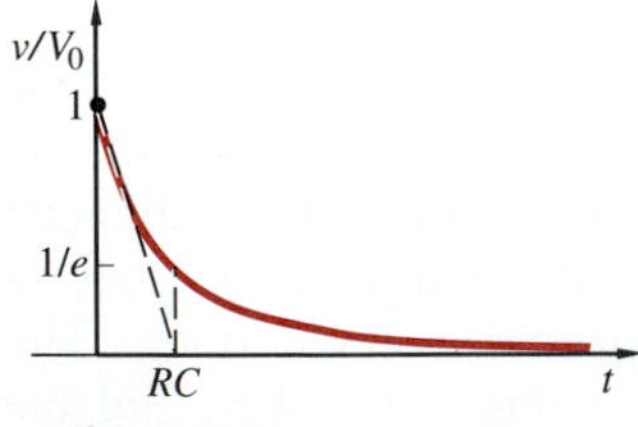

Figure 11
Exponential decay.

(Remember that you should carry out the suggested operations in every case to confirm the claimed result.)

A plot of the exponential is shown in Figure 11. Examine this curve carefully. The capacitor voltage is seen to start at a value V_0 at time $t = 0$ and to decay gradually (exponentially, we say) as time goes on.

THE TIME CONSTANT The rate at which the curve falls is determined by the product of R and C, which appears in the exponent. By differentiating the exponential and setting $t = 0$, we find that $-1/RC$ is the initial slope of the exponential, for $t = 0$. That is to say, if the curve were to fall linearly with

[8] In the future, we will not continually call attention to this maneuver every time we do it.

the same slope it has at $t = 0$, it would reach zero at $t = RC$, as illustrated by the dashed line in the figure. The RC product is called the *time constant* of the circuit. (Check to confirm that its dimensions are indeed those of time.)

However, when $t = RC$ (one time constant), (14) shows that the voltage has reached $1/e \approx 37\%$ of its initial value. (For practice, and to satisfy your curiosity, work out the corresponding percentages for 2 and 3 time constants.) When $t = 4$ time constants, the voltage will have reached $1/e^4 \approx 2\%$ of its initial value. The exponential function does not become zero until t approaches infinity. But it becomes so small after a relatively short time that, for practical purposes, we can then assume it to be zero. The accuracy of electrical measurements using everyday instruments is often of the order of 1 or 2%. Therefore, once the exponential has gotten below 2% of its initial value in 4 time constants, we couldn't even measure it without precision instruments. In engineering, it is commonly taken that an exponential has completed its total excursion in about 4 time constants.

The exponential function rivals the sinusoid as the single most important mathematical function you will meet in engineering, and you should become intimately acquainted with all its aspects. (The sinusoid itself can be represented by exponentials, as we shall see in Chapter 8.)

THE NATURAL RESPONSE Let us turn back to the differential equation in (12) and examine it in some detail with a view toward finding another method of solution. Such a method might be useful in other cases, besides the one under discussion, when the preceding approach may be impossible. The equation states that the rate of change of the function $v(t)$ we are seeking is proportional to the function itself.[9] What mathematical function is there that has the property that its derivative has the same form as the function itself? The only one is the exponential function. Therefore, let's take a guess that the solution we are seeking might be an exponential, and tentatively write:

$$v(t) = Ke^{st} \tag{15}$$

where K and s are constants. We are not yet committed to the values of either. If this is to be a solution, we should be able to insert it into the differential equation and see if the equation is satisfied. Doing the substitution results in the following:

$$Kse^{st} + \frac{Ke^{st}}{RC} = Ke^{st}\left(s + \frac{1}{RC}\right) = 0 \tag{16}$$

There are three factors on the right side of the first equals sign: the constant K, the exponential e^{st}, and the quantity $(s + 1/RC)$. If $K = 0$, then, from (15), the voltage will be identically zero. This is indeed a possible solution but a trivial one. If the capacitor has no initial charge when the switch is closed,

[9] This situation is encountered over and over again in many branches of engineering and science. To name a few, it arises in the radioactive decay of matter and in the rate of recombination of ions in a solution. If you obtain a fundamental grasp of the problem and its solution here, you will, at the same time, obtain some insights into other problems whose formulation leads to a similar equation.

nothing would happen in the circuit and the voltage would indeed remain zero. This trivial solution is not of great interest. The exponential—the second factor in the expression—is not zero for any finite value of the exponent. That leaves $s + 1/RC$; it will be zero only if the value of s is fixed at $-1/RC$.

With this value of s, the assumed solution in (15) will indeed satisfy the equation, and it will do so *no matter what the value of K might be.* To determine the specific value of K for our problem, we need one independent condition that the solution must satisfy; and, sure enough, there *is* one such condition: the value of capacitor voltage just after switching at $t = 0$. By the law of conservation of charge, the capacitor voltage at this time is V_0, the voltage just before switching. Putting $t = 0$ and $v = V_0$ in (15) leads to $K = V_0$.

With these values of K and s, the tentatively assumed solution becomes identical with the one we found before by direct integration. The advantage of the present method is that it can be applied in other, more difficult cases when direct integration might be difficult or impossible.

As for the current, since $i = v/R$, the current in the resistor (which is the same as that in the capacitor) will have the same waveform as the voltage but will differ by a constant multiplier.

Note that there is no external source in this network. The signal which arises is a consequence of the energy initially stored in the capacitor. For this reason, this response is referred to as the *natural response,* or the *free* (*unforced*) *response,* or the *zero-input response.* The natural behavior of the circuit is an exponential decay. As we go along, there will be much more to say about this subject.

ENERGY DISSIPATED IN THE RESISTOR What happens to the energy that was initially stored in the capacitor? Check back in Chapter 2 to find the following expressions for the energy stored in a capacitor and the energy dissipated in a resistor, from the time of switching up to time t:

$$w_C(t) = \tfrac{1}{2}Cv^2(t) \tag{17a}$$

$$w_R(t) = R\int_0^t i^2(x)\,dx \tag{17b}$$

These expressions are valid at any time after $t = 0$. Since the initial capacitor voltage is V_0, the initial energy stored in the capacitor is $w_C(0) = CV_0^2/2$.

The energy dissipated in the resistor from the time of switching up to time t can be calculated by using v from (15) in $i = -v/R$ and inserting it into (17). The result will be:

$$\begin{aligned} w_R(t) &= R\int_0^t \left(\frac{V_0}{R}e^{-x/RC}\right)^2 dx = \frac{V_0^2}{R}\int_0^t e^{-2x/RC}\,dx \\ &= \frac{V_0^2}{R}\left(\frac{e^{-2x/RC}}{-2/RC}\right)\Bigg|_0^t = \frac{1}{2}CV_0^2 - \frac{1}{2}CV_0^2e^{-2t/RC} \end{aligned} \tag{18}$$

Two observations can be made from the right side. Note first that the last term on the right is just $Cv^2(t)/2$, which is the energy stored in the capacitor at the instant t. Thus:

$$w_R(t) = \tfrac{1}{2}CV_0^2 - w_C(t) \tag{19}$$

This expression tells us that the energy dissipated in the resistor at time t (starting from $t = 0$) is the initial energy stored in the capacitor less the energy still stored in the capacitor at time t.

The second observation to be made from (19) is that, as t approaches infinity, the last term (the energy stored in the capacitor) will approach zero. Thus, a long time after the switch is closed, all the energy that was initially stored in the capacitor will be dissipated in the resistor and none will be left stored in the capacitor. It inspires great confidence that the law of conservation of energy is upheld by our solution!

EXAMPLE 3 The circuit in Figure 12a had been in existence for a long time when the switch opened at $t = 0$. Solve for the capacitor voltage and the current through R_1 thereafter. (You will be given a summary of the solution and numerical answers, but you will have to work out details.)

Before the switching, the capacitor voltage and all other variables had reached constant equilibrium values, so $dv_C/dt = 0$, meaning there was no current in C. Just before the switch is opened, then, $v_C(0-)$ is determined from a purely resistive circuit, with C removed. (Draw this circuit. Then, either convert the entire circuit seen from the terminals of C to a Thévenin equivalent so that the Thévenin voltage v_T becomes the initial capacitor voltage; or replace the source and the 200-Ω and 50-Ω resistors by a Thévenin equivalent, and then use the voltage-divider relationship to obtain the voltage across the 120-Ω resistor, which is $v_C(0-)$. The latter might be easier, but do both procedures anyway to provide a check.) The result is $v_C(0-) = 7.2$ V.

After $t = 0$, the circuit is shown in Figure 12c; it can be reduced to a single 80-Ω resistor in series with the capacitor, as in Figure 12d. Hence, the time constant is $80 \times 12.5\ \mu s = 1$ ms. The capacitor voltage will fall exponentially from its initial value of 7.2 V to 0: $v_C(t) = 7.2e^{-1000t}$ (t in seconds) or $= 7.2e^{-t}$ (t in milliseconds).

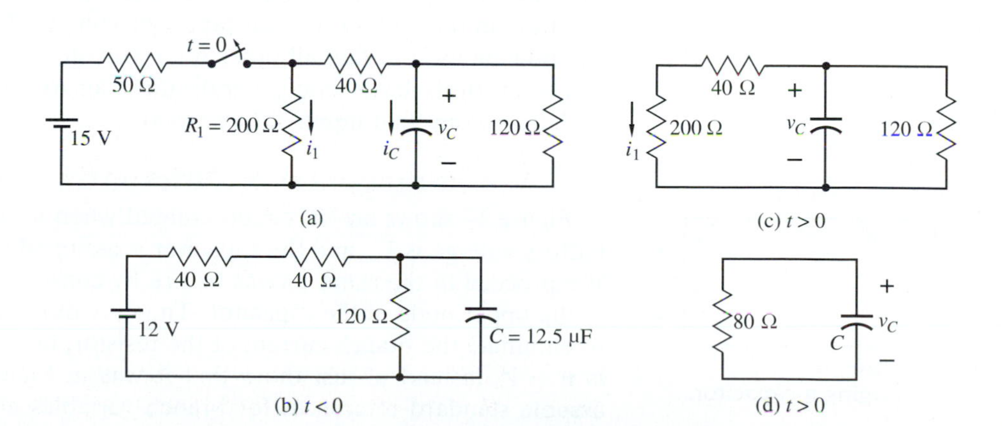

Figure 12
Circuit of Example 3.

Once $v_C(t)$ is determined, all other voltages and currents in the circuit can be obtained algebraically, by any of the methods of the preceding chapters. Thus the current in R_1 (the 200-Ω resistor) is found from Figure 12c. Applying the compensation theorem, replace C by a voltage source having a voltage given by the solution for $v_c(t)$. Being in parallel with a voltage source, the 240-Ω resistor will have no influence on the rest of the circuit. The 200-Ω and 40-Ω resistors form a voltage divider whose terminal voltage is $v_c(t)$. Therefore the voltage across R_1 will be 200/240 times $v_c(t)$, or $6e^{-t}$. Hence, current i_1 is $30e^{-t}$ mA (t in ms).

3 RESISTANCE-CAPACITANCE CIRCUITS WITH INPUT

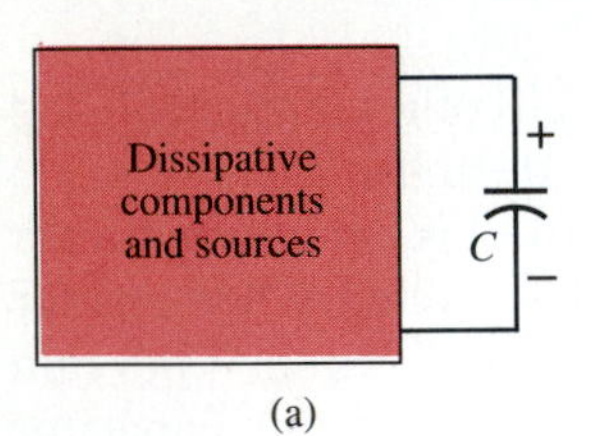

(a)

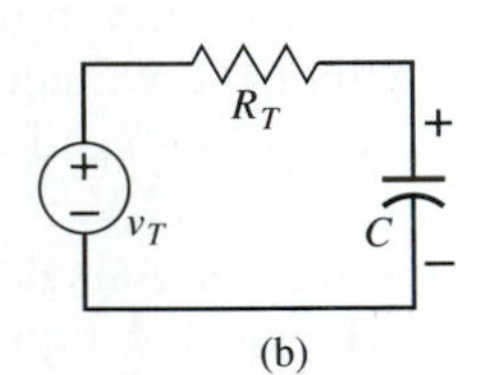

(b)

Figure 13 Single capacitor in a circuit with dissipative components and, possibly, sources.

The preceding sections have dealt with a circuit with a single capacitor and resistor whose activity results only from the initial value of the capacitor voltage. We can generalize this situation in two directions: (1) more than one resistor or other dissipative component in the circuit, everything else being the same, and (2) voltage and/or current source inputs, everything else being the same.

The first of these generalizations is easily handled; an example was just given. Looking out from the terminals of the initially charged capacitor, everything else is a two-terminal circuit of dissipative components and sources, as shown in Figure 13. Hence, by Thévenin's theorem, it can be replaced by an equivalent consisting of a resistor in series with a source. Hence, multiple dissipative components introduce no new complication. Once the capacitor voltage is found, using its value in the original circuit will lead to the solution for every other current and voltage by algebraic methods discussed in preceding chapters.

An alternative approach is to write a set of node equations with one terminal of the lone capacitor taken as the datum node. Then, only one of the node equations will be a differential equation. All other variables except the capacitor voltage can be eliminated algebraically, leaving a single differential equation in one unknown—the capacitor voltage. The solution of this equation can now be used to find all other node voltages and, hence also, any current.

Let's turn to the second generalization: an RC circuit with an input. We will initially assume that inputs are constant.

A Nonhomogeneous Differential Equation

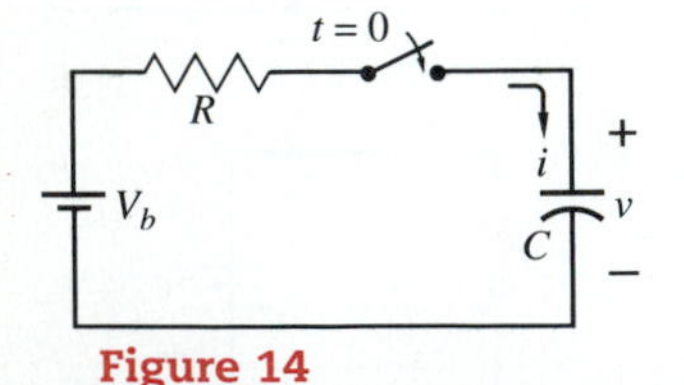

Figure 14 Charging a capacitor.

Figure 14 shows an RC circuit created when a switch is closed at $t = 0$. The battery voltage is V_b, and the capacitor is assumed to have an initial voltage V_0. We proceed in the same way as before by considering writing a node equation at the upper node of the capacitor. The only difference now occurs when we go to eliminate the branch current of the resistor, because now the resistor voltage is $v - V_b$ instead of just the v that it was in Figure 10. (Remember that we assume standard references for branch variables and we assume that currents are oriented away from the node at which a node equation is being written.) For $t > 0$, the equation satisfied by the capacitor voltage is:

$$C\frac{dv}{dt} + \frac{v - V_b}{R} = 0$$

$$RC\frac{dv}{dt} + v = V_b \tag{20}$$

The equation obtained is a linear differential equation of the first order with constant coefficients—but it is *nonhomogeneous,* as you can discover by testing it. The only difference from the previously studied differential equation is the input on the right. This function is said to *excite, drive,* or *force* the circuit. One approach to solving this equation is the one previously undertaken, namely, to integrate directly, after isolating the derivative on one side of the equation and dividing both sides by $v - V_b$. Carry this out as an exercise while we take another approach here.

Suppose that, by some method (even by revelation in a dream), we have found some solution of the nonhomogeneous equation which we label v_f. We refer to this as the *forced response* (or the forced solution), because it originates from the fact that the circuit is forced. What would be the consequence of adding to this the general solution of the *homogeneous* equation obtained by setting the battery voltage to zero? Say this is v_n, the *natural response.* Let's substitute the sum of these two into the nonhomogeneous differential equation and see what happens:

$$RC\frac{d}{dt}(v_f + v_n) + (v_f + v_n) = V_b$$

$$\left(RC\frac{dv_n}{dt} + v_n\right) + \left[\left(RC\frac{dv_f}{dt} + v_f\right) = V_b\right] \tag{21}$$

The second line is obtained by removing the parentheses in the first line and collecting terms. In the second line some terms are within brackets and the remainder between parentheses. Within brackets we find the nonhomogeneous differential equation; since v_f was chosen to satisfy this equation, the quantity in brackets disappears, leaving only the quantity in parentheses. But v_n is supposed to satisfy the homogeneous differential equation, so this quantity in parentheses is also zero. The conclusion is that if there is a particular solution of the nonhomogeneous differential equation and we add to it the general solution of the homogeneous differential equation, the result will still be a solution, a general solution, of the nonhomogeneous equation.

The preceding conclusion provides an approach to solving nonhomogeneous differential equations. By some means or other (which we will shortly discuss), a particular solution to the nonhomogeneous differential equation is found. The general solution of the homogeneous equation is also found, separately. Their sum provides us with the solution we are seeking.

FORCED RESPONSE: SOLUTION OF A NONHOMOGENEOUS EQUATION The equation whose solution we are seeking is (20). Since the input is a constant, we might conjecture that the forced solution might also be a constant, say, A. There is no harm in trying it. Since the derivative of a constant is zero, when $v_f = A$ is substituted for v in (20), the result becomes:

$$v_f = A = V_b$$

(To confirm that this is a solution, substitute V_b for v in the equation.) We had previously found the general solution of the homogeneous equation to be $Ke^{-t/RC}$. Hence, the general solution of the nonhomogeneous equation is:

$$v(t) = V_b + Ke^{-t/RC} \tag{22}$$

To determine the value of K, we must impose an independent condition regarding the state of the capacitor voltage at the switching time. By the principle of conservation of charge, the capacitor voltage immediately after switching is the same as it was just before switching, namely, V_0. We can solve for K in (22) by setting $t = 0$, recalling that $e^0 = 1$, and using $v(0) = V_0$ for the initial capacitor voltage. Then, putting this value of K in the immediately preceding solution gives:

$$v(t) = V_b + (V_0 - V_b)e^{-t/RC} \tag{23}$$

(You can confirm that this is a solution by verifying that it satisfies the equation, including the initial value and the value after the passage of a long time.) Compare this with the solution obtained by direct integration which you were asked to do.

Let's study this solution a little further. There are two terms: the exponential, which would be present with or without an input, and the constant term equal to the input-voltage value. The exponential is *transient*—it gradually disappears after the passage of enough time constants of time. After the transient has died down and the circuit has reached equilibrium, there remains a voltage across the capacitor which is steady and equal to the battery voltage. We refer to this as the *equilibrium state,* which, since it is steady, is the *steady state.* By KVL, we conclude that in the steady state there is no voltage across the resistor, and hence no current through it.

Let's confirm this about the current by seeking the equation satisfied by the current after the circuit has been established. By KVL applied to the circuit of Figure 14 and Ohm's law to eliminate the resistor voltage, we get $Ri + v = V_b$. Multiplying through by C and differentiating, we obtain the following equation:

$$RC\frac{di}{dt} + i = 0 \qquad \rightarrow \qquad i(t) = Ke^{-t/RC}$$

This is a homogeneous differential equation, with exactly the same time constant as the one applicable for the voltage. Hence the solution will be the exponential shown. The constant K represents $i(0+)$, the value of the current just after the switching. Unlike the capacitor voltage, there is no principle that requires the current to be continuous. Prior to the switching, there was no current. Since the capacitor voltage cannot suddenly change from its initially relaxed value, the entire battery voltage is initially across the resistor. (*Initially relaxed* is a metaphorical term about a capacitor that carries no initial voltage.) Hence, the initial current is V_b/R.

We can obtain the same result from the circuit and the solution for the capacitor voltage in two ways: from $i = (v - V_b)/R$ algebraically and from the capacitor i-v relationship by differentiation. Carry out the details to confirm the expression for the current.

RESPONSE OF AN RC CIRCUIT TO A STEP INPUT Look back at Figure 14. The circuit is established by the closing of a switch which suddenly applies a constant source to the circuit. Now refer back to the discussion of step functions earlier in this chapter. A step function is a discontinuous jump from 0 to a constant value—which is somewhat like the closing of the switch in the circuit.

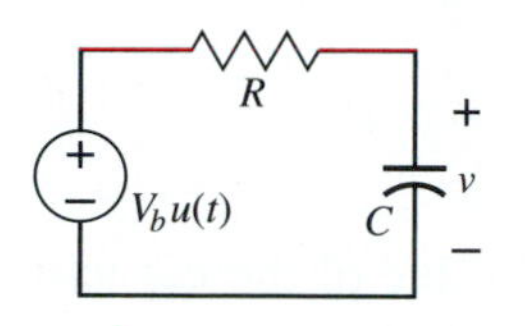

Figure 15
RC circuit excited by a step function.

We can remove the switch and battery and replace them by a source whose voltage is a step function, as shown in Figure 15. This circuit is completely equivalent to the one in Figure 14 after the closing of the switch—but not before. The distinction is that, before the closing of the switch, the circuit in Figure 14 is open, so that a charge could be maintained on the capacitor. However, in Figure 15, there is a path for current, so the capacitor is discharged before $t = 0$. Whatever response results is a consequence of the step, without any contribution from an initial capacitor voltage. Thus, the solution given in the preceding equation will be applicable in this case, but with $V_0 = 0$.

RESPONSE OF AN RC CIRCUIT TO A RAMP FUNCTION Next, let's consider the same circuit, but this time with a ramp function as the excitation, as shown in Figure 16. The ramp function is conveniently written as $tu(t)$. The differential equation, valid for all time, is:

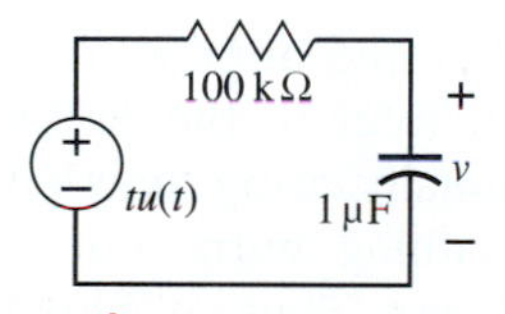

Figure 16
RC circuit excited by a ramp.

$$RC\frac{dv}{dt} + v = tu(t) \tag{24}$$

We will follow the plan of adding the general solution of the homogeneous equation to some solution of the nonhomogeneous equation, obtained in any way we can. Our approach will be to argue as follows.

For $t > 0$, the right side is a linear function of t, a straight line with a specific slope and a specific intercept. The left side of the equation is a linear combination of the solution and its derivative. What kind of function can the solution be so that it and its derivative add up to a linear function? When stated this way, the solution is obvious: it must be a linear function. We don't know yet whether it will have the same slope or the same intercept. Hence, to cover all contingencies, we write:

$$v_f(t) = (A + Bt)u(t)$$

where we are not yet committing ourselves as to the values of A and B. Substituting this into the previous differential equation, for $t > 0$ we get:

$$(RC)B + (A + Bt) = t$$

This should be an identity, true for all values of time, not just some specific values. This requires the constant terms to balance on both sides of the equation and the coefficients of t likewise. From this, we find:

$$B = 1$$

$$A = -RCB = -RC$$

Hence, the forced solution is:

$$v_f(t) = (-RC + t)u(t)$$

Thus, the slope of the forced response is the same as that of the excitation, but the intercept is not. The natural response had already been found to be an exponential. So the complete response is:

$$v(t) = (-RC + t + Ke^{-r/RC})u(t)$$

As before, the constant K is found by invoking the initial value of the capacitor voltage, which, in this case, is 0. Thus:

$$v(0) = 0 = -RC + K$$

$$K = RC$$

The final solution, therefore, becomes:

$$v(t) = [t - RC]u(t) + RCe^{-t/RC}u(t) \tag{25}$$

Here again we find two terms: the transient exponential and another term that results from the specific driving function. We cannot refer to this as the steady state, since there is nothing "steady" about a continually rising function. But the circuit certainly is in equilibrium (assuming nothing burns out or becomes nonlinear with the continually increasing voltage). Thus, although what remains after the transient has died down cannot be called the steady state here, it can still be called the equilibrium state. Of course, whenever the equilibrium state is a "steady" function, it is also the steady state.

EXERCISE 9

Assume that the excitation in Figure 16 is $t^2u(t)$ instead of the ramp. This is a special case of a general quadratic. Assume that the forced response is also a quadratic, a general one. Find the coefficients in this assumed response so that the differential equation is satisfied. Then write the complete solution and apply the initially relaxed condition to determine the unspecified constant.

ANSWER: $v_f = t^2 - 2RCt$. □

RESPONSE OF AN RC CIRCUIT TO AN EXPONENTIAL So far, we have examined the forced response of a first-order circuit to a step, a ramp, and—through an exercise—a quadratic in time. The process followed for these waveforms has been to assume that the forced response has the same form as the excitation, namely, a polynomial in t whose highest power is the same as the highest power in the excitation. The coefficients in the assumed response are determined by inserting the solution into the differential equation and arguing that the result must be valid for all time.

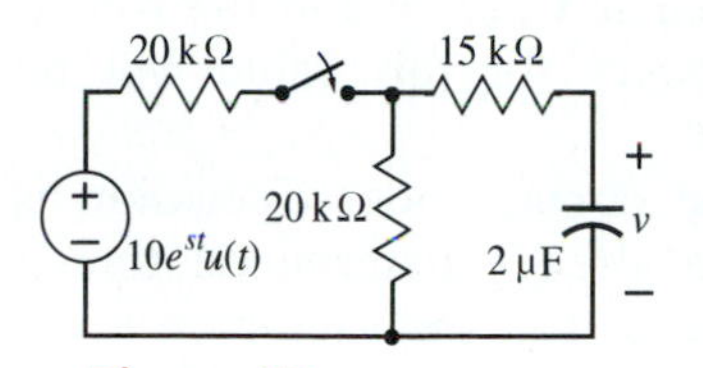

Figure 17
RC circuit excited by an exponential.

Now let's study one final excitation function, an exponential. The circuit in Figure 17 is excited by an exponential e^{st}, starting at $t = 0$, when the switch is closed. Alternatively, the excitation can be taken as $e^{st}u(t)$ without the need to show a switch. We are interested in just the forced response.

The differential equation for the capacitor voltage can be obtained in a number of ways: node equations, loop equations, or otherwise. Simplest, perhaps, is to replace the rest of the circuit to the left of the capacitor by a Thévenin equivalent. Carry out the details to find an exponential voltage source $5e^{st}u(t)$ in series with a 25-kΩ resistor. The time constant will be $\frac{1}{20}$ s = 50 ms. The differential equation will be:

$$\frac{dv}{dt} + 20v = 200e^{st}u(t)$$

In seeking the forced response, we argue as before that the derivative of the desired function, when added to a multiple of itself, must yield the exponential on the right. The only possible way to satisfy this is for the forced response to be an exponential also. Two quantities are associated with the exponential: the exponent and the coefficient. It is not possible for the exponent of the forced response to be different from that of the excitation. (Try it and see if the differential equation can be satisfied.) The coefficient, on the other hand, has to be adjusted in order for the differential equation to be satisfied.

Thus, putting $v_f = Ae^{st}$ into the preceding differential equation results in $A = 200/(s + 20)$, so the forced response is $v_f = [200/(s + 20)]e^{st}u(t)$. (Go through the details.)

Nothing has been said about the value of s; it can be either positive or negative. The only value excluded is $s = -20$, since the coefficient of the exponential in the solution would become infinite. But note that the time constant of the circuit is $\frac{1}{20}$ s, so its natural frequency is -20. That is, the only value of s for which the assumed forced response *is not* a solution is s equal to the natural frequency of the circuit.[10]

The solution in this unique case is not difficult to find by standard methods of solving differential equations. However, you aren't expected to be familiar with these methods. It is also possible to arrive at a solution by assuming its form and testing the assumption on the differential equation. But the

[10] You may be a little uneasy about an exponential with a positive real exponent—a function that increases without limit. "Where," you might ask, "would one find such a function?" Suppose we charge a capacitor to some reasonable voltage level and then discharge it through a resistor, with a reasonable time constant. We take the precaution of making a tape recording of the decaying capacitor voltage—for as long as our interest holds up. Now suppose that we replay the tape—only in reverse. What was recorded as a decaying exponential becomes a growing exponential—at least until the end of the tape! (This suggestion is the brainchild of William H. Huggins, Professor Emeritus, Johns Hopkins University, an inspiring teacher and esteemed colleague.)

justifications usually used to arrive at the desired form are tortured and farfetched, and therefore unsatisfying. The answer seems to be pulled out of a hat. Hence, we shall simply ignore this one particular value of s in the forced response to an exponential excitation. The mystery for this value will be resolved in Chapter 14.

The complete response for the voltage in the circuit under discussion is obtained by adding the natural response to the already determined forced response. For concreteness, assume that the value of s in the excitation (and, hence, in the forced response) is $s = -10$. Hence, the complete response will be:

$$v(t) = (20e^{-10t} + Ke^{-20t})u(t)$$

The second term is the natural response. All that remains is to apply the initial value of the capacitor voltage to determine the constant K. (Do it.)

EXERCISE 10

In the same circuit, Figure 17, suppose the source is a sinusoid $v_g(t) = A\cos(\omega t + \theta)$. Your goal is to find the forced response by assuming its form and testing it on the differential equation to determine the parameters. (The excitation has three parameters: ω, A, and θ. Which of these would you expect to be the same in the forced response, and which possibly different?) □

Response of an RC Circuit to a Pulse

A very important circuit problem is the determination of the response of an RC circuit to a pulse waveform or to other pulselike waveforms. Refer to Figure 18, which shows an RC circuit to which a voltage pulse is applied.

The pulse might be created by the double switching arrangement shown, or it might be the result of a source whose waveform is given by:

$$v_s(t) = 15[u(t) - u(t-4)]$$

Just before $t = 0$, the source voltage is 0; just after, it is 15 V. Right at $t = 0$ there is a discontinuity. A similar description applies for the time $t = 4$ ms, except that the discontinuity goes the other way. Assume that the circuit is initially relaxed, meaning that the initial capacitor voltage is zero.

Note first that the time constant is 2 ms; hence, the pulse width is 2 time constants. Two intervals of time are of interest: while the pulse is ON and after it goes OFF. To avoid powers of 10, we will use milliseconds for the unit of time. The differential equations satisfied during these two intervals are as

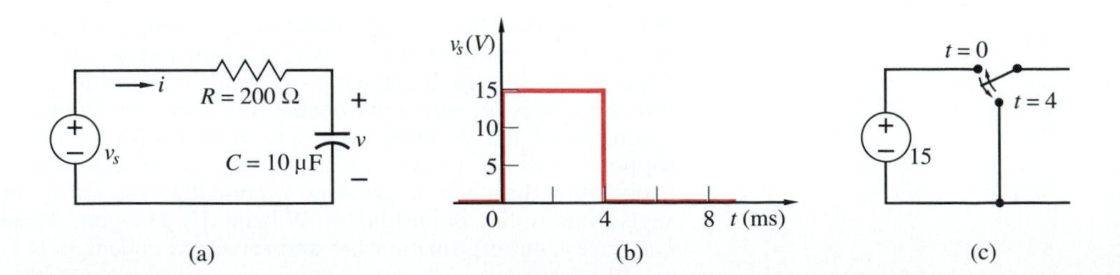

Figure 18
Voltage pulse applied to an RC circuit.

follows:

$$2\frac{dv}{dt} + v = 15 \qquad 0 \leq t \leq 4$$
$$(t \text{ in ms})$$
$$2\frac{dv}{dt} + v = 0 \qquad t \geq 4$$

In the first interval (pulse ON), the general solution will be:

$$v(t) = 15 + Ke^{-t/2} \qquad (t \text{ in ms})$$

This expression is valid between 0 and 4 ms, including the ends of the interval, since the capacitor voltage is continuous. Setting $v(0) = 0$ in this expression, we determine that $K = -15$. Thus, during the ON time of the pulse, the capacitor voltage is:

$$v(t) = 15(1 - e^{-t/2}) \qquad (t \text{ in ms})$$

At the end of the pulse, the capacitor voltage becomes $v(4) = 15(1 - e^2) \approx 13$ V.

In the second interval, $t \geq 4$ ms and the differential equation is homogeneous. But now we have a problem. In previous problems the initial time was taken as $t = 0$; now it is $t = 4$. What do we do? There are two possible ways of proceeding. First, for this part of the problem, redefine the time origin, letting the end of the pulse start at $t' = 0$, where $t' = t - 4$, and use the previous value of $v = 13$ at $t' = 0$. Thus:

$$v(t') = Ke^{-t'/2} \quad \text{for} \quad t' \geq 0 \quad \rightarrow \quad v(0) = 13 = K$$
$$v(t) = 13e^{-(t-4)/2} = 13e^2e^{-t/2} = 96e^{-t/2}, \qquad t \geq 4$$

After finding the value of K, we replace t' by $t - 4$.

The second approach is to stick with time t, write the solution as $v(t) = Ke^{-t/2}$, set $v(4) = 13$ (found from the solution in the first interval), and determine K. You should carry out the details and confirm that the same answer is obtained.

Now let's determine the current in the same circuit. One possibility is to obtain the differential equations for i in each interval and then proceed to solve them. Another possibility is to use the capacitor i-v relationship. You do those two; here we'll find the current by first finding the resistor voltage by KVL and then using Ohm's law. Thus, there is no necessity to solve a differential equation for the current; once a differential equation has been solved for the capacitor voltage, the current is obtained strictly by algebraic means. From the circuit:

$$i(t) = \frac{1}{R}[v_s(t) - v(t)]$$
$$i(t) = \tfrac{1}{200}[15 - 15(1 - e^{-t/2})]$$
$$= 75e^{-t/2} \text{ mA} \qquad 0 < t < 4 \text{ ms}$$
$$i(t) = \tfrac{1}{200}[0 - 96e^{-t/2}]$$
$$= -480e^{-t/2} \text{ mA} \qquad t > 4 \text{ ms}$$

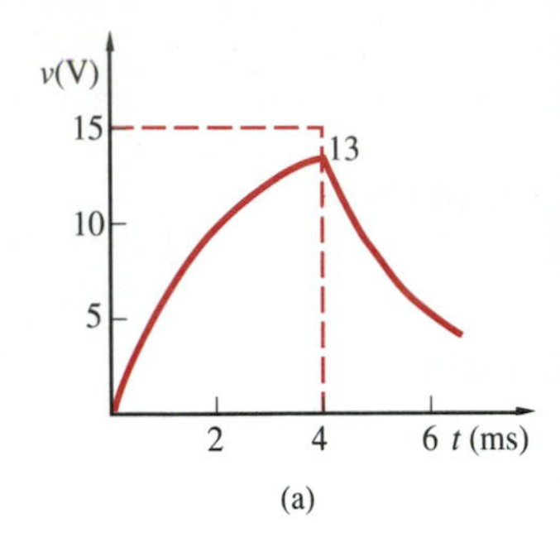

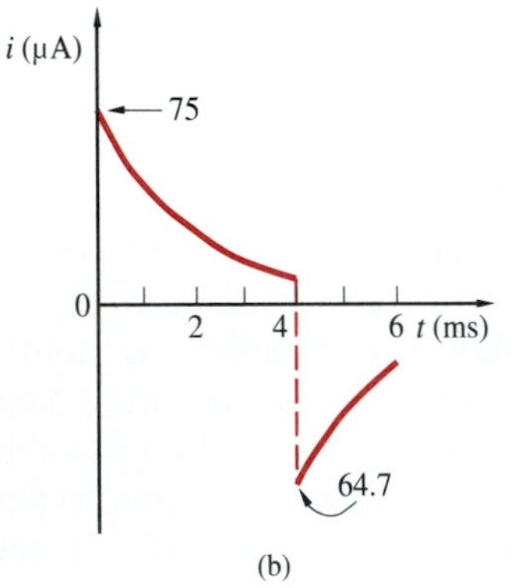

Figure 19
Pulse response of capacitor voltage and current.

Plots of the pulse response of the capacitor voltage and current are given in Figure 19. Examine the general characteristics of these curves. First, note that the capacitor voltage is continuous. It was 0 before the beginning of the pulse and rises continuously toward the height of the pulse. It never reaches there, because the pulse ends too soon. After the end of the pulse, the capacitor voltage falls exponentially toward 0 and will eventually reach there if no other pulse comes along.

The current, on the other hand, was 0 before the beginning of the pulse and jumps discontinuously to its maximum value. Thereafter, it falls exponentially toward 0 but never reaches it before the pulse ends. At that point, it drops discontinuously, reversing direction, and then rises toward its steady-state value of 0.[11]

The shapes of neither the voltage nor the current resemble the shape of the input pulse. The reason has to do with the size of the circuit time constant relative to the pulse width. In the present case, the RC time constant is found to be 2 ms, which is just half the pulse width; or, looking at it in reverse, the pulse width is 2 time constants. We know that a capacitor voltage completes 98% of the excursion between its initial and final values in 4 time constants. Thus, if we reduce the time constant by a factor of 2 or more, the capacitor voltage will rise to a value much closer to the height of the pulse before the pulse ends. The waveform of the capacitor voltage will, then, be much closer to the shape of the pulse. Why don't you do the following? Change the resistance in the circuit to 100 Ω and make a plot of the capacitor voltage; then reduce the resistance again to 50 Ω and repeat.

What happens to the current when we reduce the time constant? The current will also complete more of its excursion before the end of the pulse. Indeed, if the time constant is made small enough compared with the pulse width, the current will take the shape of two spikes at the beginning and end of the pulse. There will be a discontinuity at each end followed by a rapid exponential change toward the steady-state value of 0. Using the values of time constant you were asked to try in the preceding paragraph, make a plot of the current and confirm what has just been described.

EXERCISE 11

In the circuit of Figure 18, the capacitance is the same but the value of the resistance is 200 kΩ. The source waveform is $v_s(t) = 10[u(t) - u(t - 0.2)]$.

[11] Notice the detail in which these two paragraphs describe the curves. Whenever Sherlock Holmes looks over the scene of a crime, he takes note of every detail and can describe everything precisely to Watson when he is explaining how he solved the crime. He also relates a specific detail to something he already knows. (The room was a mess, with everything strewn about. There was only one shoe under the bed, a size 13 Birkenstock. So the murderer is a 250-lb 6-ft 2-in. blond Swede; you see, Englishmen don't wear Birkenstocks, so the shoe belonged to the murderer; it fell off during the struggle, which, we know from the condition of the the scene of the crime, must have taken place. Swedes wear Birkenstocks—and almost all Swedes are blond. QED.) When you look at a diagram, a graph, an equation, you should get in the habit of describing to yourself all the details, relating the details to each other and to information you already know. This will help greatly, not only in the learning process now, but in your future engineering practice.

This is a 10-V pulse whose width is 0.2 s. Compare the pulse width with the time constant; find the solutions for the capacitor voltage and current and plot these on the same time scale as the input pulse.

ANSWER: The pulse width is one tenth of a time constant. Very little change in either the capacitor voltage or the current will occur over the width of the pulse. The current waveform will approximate a pulse and the voltage will rise only to about a tenth of the input voltage by the end of the pulse. □

Let's briefly review what you should know from your study of this chapter so far. Given a circuit consisting of a single capacitor and any number of dissipative components and sources, you should be able to convert all but the capacitor into an equivalent consisting of a single resistor in series with a voltage source. In the resulting circuit, you should be able to write a differential equation for the capacitor voltage. You should be able to find a solution for this equation (1) when there is no source and the circuit is responding to just the initial voltage of the capacitor and (2) when the source voltage is a step, a ramp, a quadratic, an exponential, a sinusoid, and a pulse—or combinations of them. Indeed, not much more in new principles is needed to solve the problem for other source-voltage waveforms also. We haven't considered any others because the solution process becomes a little tedious and also because we will later introduce other methods of solution which are simpler and more general, that can handle any waveform you are likely to meet.

STEADY-STATE RESPONSE OF AN RC CIRCUIT TO A PERIODIC WAVE A pulse is a "piecewise constant" function, as previously described. The solution process would be the same for any piecewise constant input function, no matter how many intervals there were. At each discontinuity, the value of the capacitor voltage from the preceding interval would be the initial voltage for the next interval.

One such piecewise constant function is shown in Figure 20. Here we have a rectangular wave which has been going on for some time. It is the source

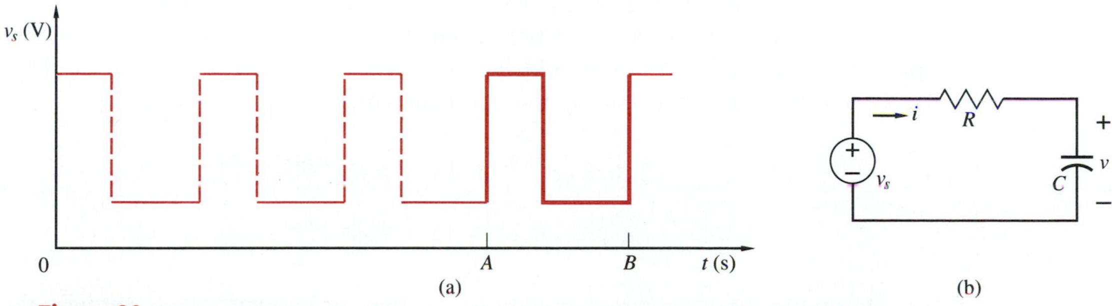

Figure 20
Rectangular wave in an RC circuit.

voltage in the RC circuit in the figure. What we would like to do is to find the capacitor voltage (and, knowing that, the current) over the interval of one period, say, the period from A to B. If that interval occurs many periods after the input wave is initiated, then the response voltage or current is very nearly periodic by then.

If we assume periodicity, then a simplified solution is possible. The alternative is to find the solution for each interval, starting from the first one, and matching up capacitor voltage values at the end of one interval and at the beginning of the next. This is repeated until we reach the specific interval for which a solution is to be found.

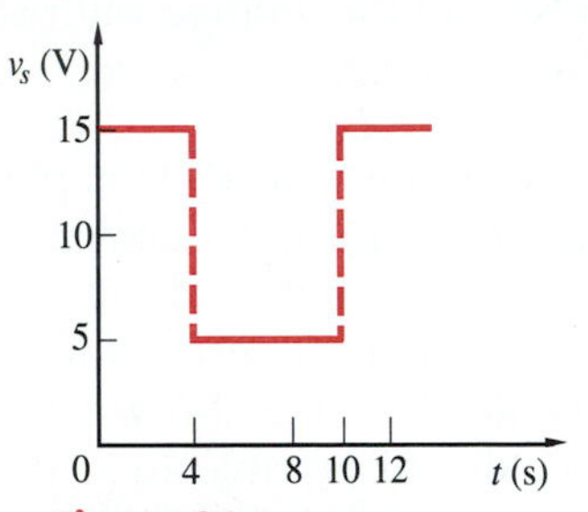

Figure 21
One period of a rectangular wave.

The interval from A to B is shown in Figure 21. We'll measure time from point A. We should really use general notation, but, for simplicity, numerical values will be used for the time axis and the voltage values in the two intervals. The differential equations satisfied by the capacitor voltage in the two intervals and their solutions are as follows:

Time Interval	Differential Equation	Form of Solution
$0 \leqslant t \leqslant 4$	$RC\frac{dv}{dt} + v = 5$	$v(t) = 5 + K_1 e^{-t/RC}$
$4 \leqslant t \leqslant 10$	$RC\frac{dv}{dt} + v = 15$	$v(t) = 15 + K_2 e^{-t/RC}$

(Don't forget: don't just *read* these results; work them out yourself and confirm them.)

There is one unknown constant in the solution in each interval. The source voltage is discontinuous at $t = 0$, 4, 10, and 14 s, etc. But at those times the capacitor voltage is not discontinuous. Therefore the value of v at $t = 4$ in the solution for the first interval equals the value of v at that same time in the solution for the second interval. That is, $v_1(4) = v_2(4)$. This will give one equation relating K_1 and K_2. Another equation will result by comparing the solution at the beginning of the first interval at $t = 0$ with the solution at the end of the second interval one period later at $t = 10$ s. Since the response is periodic, these two values should be the same. That is, $v_1(0) = v_2(10)$. That gives another equation relating the two constants. These equations are:

$$5 + K_1 e^{-4/RC} = 15 + K_2 e^{-4/RC}$$

$$5 + K_1 e^{0} = 15 + K_2 e^{-10/RC}$$

These are two *algebraic* equations in two unknowns, even though the coefficients include exponentials. Those exponentials are not variables, but constants. We won't go though the details of solving for the two constants (you should do that); the solution will be:

$$K_1 = 10\left(\frac{1 - e^{-6/RC}}{1 - e^{-10/RC}}\right)$$

$$K_2 = 10\left(\frac{1 - e^{4/RC}}{1 - e^{-10/RC}}\right)$$

You may think these expressions appear formidable, but they are just constants whose numerical values can be easily calculated once the RC time constant is specified. Using the values of these two constants, the solution for the capacitor voltage is completely determined.

Multiple Switching

So far the circuits considered have not been modified after they were first constituted, so that the time constants have stayed the same. We shall now briefly consider the possibility of switching events that cause the time constant of a circuit to change.

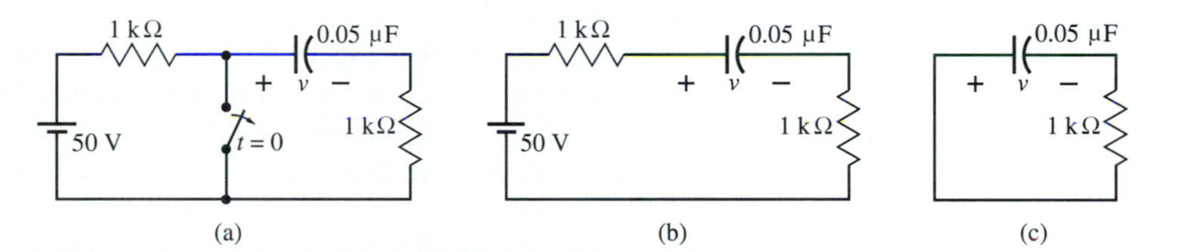

Figure 22
(a) Circuit with multiple switching events; (b) switch open; (c) switch closed.

One such case is illustrated in Figure 22. The switch has been in the closed position for a long time, long enough for the capacitor voltage to reach a steady-state value. The switch opens at $t = 0$ and remains open for 100 μs, at which instant it closes again and remains closed thereafter. The circuits corresponding to these intervals are shown in Figure 22b and c. The differential equations and their solutions in these two intervals are as follows:

Interval	Time Constant	Equation	Solution
$0 \leqslant t \leqslant 100\ \mu$s	100 μs	$100\dfrac{dv}{dt} + v = 50$	$50(1 - e^{-t/100})$ (t in μs)
$t \geqslant 100\ \mu$s	50 μs	$50\dfrac{dv}{dt} + v = 0$	$31.6e^{(t-100)/50}$ (t in μs)

(As you go about confirming these results, the only thing you might consider nonroutine is the second interval. To find the solution for the second interval, we need to know the capacitor voltage at the beginning of that interval, which is also the end of the first interval. Putting $t = 100\ \mu s$ in the solution for the first interval gives 31.6 V for the initial capacitor voltage in the second interval.)

SWITCHING WITH A DIODE[12] Recall that an ideal diode behaves like a switch. When we studied diodes in resistive circuits in Chapter 3, the only reason that diodes could change state was the variation with time of a source variable. But a diode in an *RC* circuit might change state even with DC sources because capacitor and other voltages and currents are varying with time after some initial switching event.

In the circuit of Figure 23a, the capacitor is initially relaxed (no pep!) when the switch is closed at $t = 0$. (We can think of the 20-V battery and the switch as a step voltage source.) The first miniproblem to be solved is to determine the state of the diode right after the switch is closed. Recall that this is done by assuming a state and then, from the behavior of the circuit, confirming whether or not the assumption is correct. Let's assume that the diode is ON. Since the initial capacitor voltage is 0, KVL around the diode-capacitor loop will require a current through the 3-MΩ resistor oriented upward—in the diode reverse direction. Hence, the diode cannot be ON; it must be OFF. The circuit for the period immediately after the closing of the switch is shown in Figure 23b. The time constant is 10 s.

In this circuit, the capacitor voltage will rise exponentially toward the battery voltage, 20 V. But as soon as it reaches 15 V, the diode will turn ON. The time of this switching must be determined after a solution for the capacitor voltage is obtained in this interval; $v = 15$ is inserted for the voltage and the expression solved for t; the details will be carried out in what follows. The circuit that results after the diode changes state is the one shown in Figure 23c. Now you can have a chance to practice on your Thévenin equivalent. You carry out the details; the result is an equivalent voltage source of 18 V and an equivalent resistance of 1.2 MΩ. For this circuit, the time constant is 6 s. The

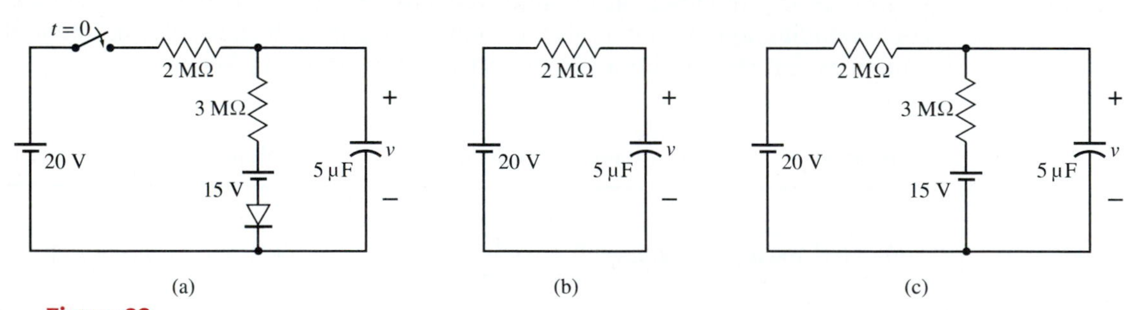

Figure 23
(a) RC circuit with diode; (b) diode OFF; (c) diode ON.

[12] If you skipped the study of diodes in Chapter 2 and diode circuits in Chapter 3, you aren't prepared for this section. Either skip it or go back to study the previously omitted material.

differential equations and their solutions for the two circuits shown in Figure 23b and c corresponding to the two intervals are:

Intervals	**Differential Equations**	**Solutions**
$0 \leqslant t \leqslant 13.9\text{ s}$	$10\dfrac{dv}{dt} + v = 20$	$v(t) = 20 + K_1 e^{-t/10}$
$t \geqslant 13.9\text{ s}$	$6\dfrac{dv}{dt} + v = 18$	$v(t) = 18 + K_2 e^{-t/6}$

(The switching time, $t = 13.9$ s, is determined in what follows.)

By now, evaluation of the constants in the solution should be routine for you: set t equal to the initial time and v equal to the initial voltage. In the first interval, $v = 0$ when $t = 0$; hence, $K_1 = -20$. At the end of the interval, $v = 15$. This requires that $1 - e^{-t/10} = \frac{3}{4}$, so $e^{-t/10} = \frac{1}{4}$ and $t = 10 \ln 4 = 13.9$ s. (This is how the value in the preceding table was obtained.)

In the second interval, $v = 15$ when $t = 13.9$. When these values are inserted into the solution for v in this interval, the value of K_2 is found as follows: $18 + K_2 e^{-(10 \ln 4)/6} = 15$, so $K_2 = -3e^{(10 \ln 4)/6} = 30.24$. This can be inserted in the solution for v in both forms: either as the constant 30.24, or as the exponential raised to the indicated power. The final form of the solution will be:

$$v(t) = 20(1 - e^{-t/10}) \qquad 0 \leqslant t \leqslant 10 \ln 4 = 13.9\text{ s}$$

$$v(t) = 18 - 3e^{-(t-10 \ln 4)/6}$$

$$= 18 - 30.24e^{-t/6} \qquad t \geqslant 10 \ln 4 = 13.9\text{ s}$$

These equations are shown plotted in Figure 24. Although the curve is continuous, there is a change in slope when going from one interval to the other. This would have been more evident if the time constants had been radically different in the two intervals.

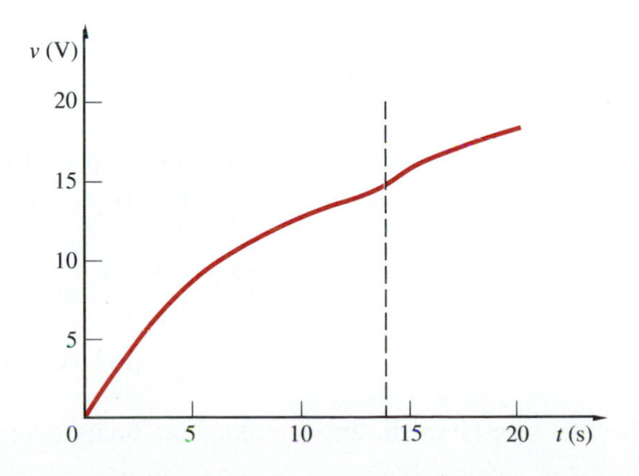

Figure 24
Graph of capacitor voltage in Figure 23.

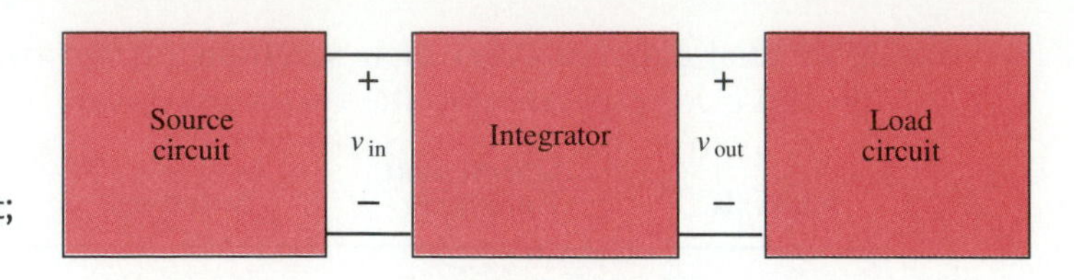

Figure 25
Desired integrator circuit; $v_{out} = K \int v_{in}(t)\,dt$.

Op-Amp RC Integrator

In many applications it would be very desirable if there existed a circuit which accepted a signal at its input and produced another signal at its output proportional to the integral of its input signal. Figure 25 illustrates this situation. We know that a capacitor current is proportional to the derivative of its voltage; conversely, a capacitor voltage is proportional to the integral of its current. However, a simple capacitor cannot fulfill the desired function of the circuit in Figure 25. If a capacitor replaced the integrator in the diagram, the load and source circuits would interact and help to determine the current that the capacitor would be integrating. In other words, in the desired circuit of Figure 25, the integrator should also serve to isolate the load circuit from the source circuit.

The requirement immediately brings to mind an op amp; but an op amp alone does not integrate, so a capacitor must be involved somehow. Look back at the circuit of the simple inverting amplifier in Figure 17, Chapter 4. Suppose one of the resistors is replaced by a capacitor. The two possibilities are shown in Figure 26. The first of these will be left as a problem for you to analyze and to determine the mathematical operation it performs on the input.

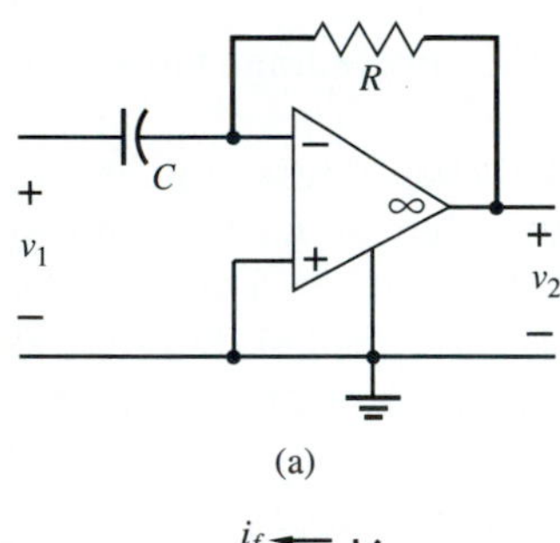

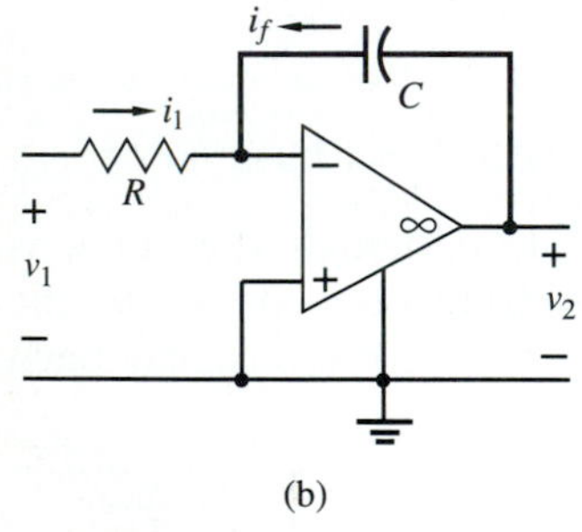

Figure 26
(a) Differentiator and (b) integrator circuits.

In Figure 26b, we assume the op amp is ideal. (That means: no input current into either of the input terminals, and a virtual short circuit across those terminals.) We apply KVL to the loops formed (1) by the input, the resistor, and the virtual short circuit, and (2) by the output, the capacitor, and the virtual short circuit. Into the results we insert the *i*-*v* relations of the resistor and capacitor, yielding:

$$i_1 = \frac{1}{R} v_1$$
$$i_f = C\frac{dv_2}{dt} \tag{26}$$

Next we apply KCL at the node formed at the inverting input of the op amp. Because no current enters the op amp terminal, one of the two currents identified in (26) is the negative of the other. Thus:

$$\frac{1}{R} v_1(t) = -C\frac{dv_2}{dt}$$
$$v_2(t) = -\frac{1}{RC}\int_0^t v_1(x)\,dx + v_2(0) \tag{27}$$

The first line gives the input voltage as being proportional to the derivative of the output. Let's call the time at which the circuit is established $t = 0$. Then, integrating the first line from 0 up to an arbitrary time t gives the second line. The initial value of the capacitor voltage may or may not be zero.

This is the desired integrator. Any circuit connected at the output of the integrator will not load the circuit producing the input voltage v_1.

Recall that the output voltage cannot exceed V_{sat}. For the initially relaxed case, (27) will impose the following restrictions:

$$\left|\int_0^t v_1(x)\,dx\right| \leqslant RCV_{sat} \tag{28}$$

There are two quantities that are affected here: the functional variation of the input voltage (the values it takes on over time) and the value of the upper limit—the time over which the circuit acts as an integrator before the op amp saturates. For any given value of the upper limit, there will be a limitation on the input voltage, and for any given input voltage there will be a limit on the time over which the circuit operates as an integrator.

EXERCISE 12

(a) Suppose in the preceding integrator circuit that the input voltage is a 5-V step function and that the op amp saturation voltage is 15 V. Determine for how long after the initiation of the step the circuit will continue to act linearly as an integrator. (b) Now suppose it is required that the circuit act as an integrator for at least 10 time constants after the onset of the step function. Find the maximum step voltage permitted.

ANSWER: (a) 3 time constants; (b) 1.5 V. □

Circuits with Controlled Sources

Although *RC* circuits with op amps were treated in this chapter, only ideal op amps were considered. The controlled source, which forms part of the model of an op amp, is not explicitly evident in the ideal case. What kinds of complications would be introduced by a controlled source in an *RC* circuit? (Review the section on circuits with controlled sources in Chapter 4 before proceeding.)

A simple amplifier circuit is shown in Figure 27a. The input voltage $f(t)$ is applied at $t = 0$ by some switching arrangement. This can be handled analytically by setting $v_s(t) = f(t)u(t)$. Voltage v_1 is easily obtained in terms

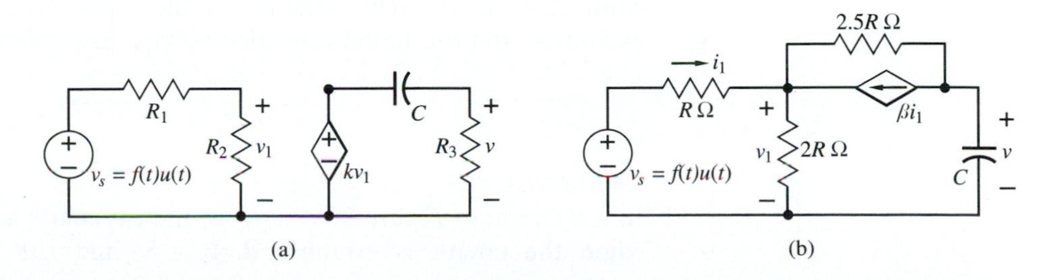

Figure 27
Amplifier circuits.

of v_s by the voltage-divider relationship. The controlled-source voltage, therefore, is proportional to the independent-source voltage v_s. In essence, then, the right-hand loop is a simple series RC circuit with an independent source. Since there are no complications beyond those you know how to handle, we won't pursue this case any further. You are urged to solve for the complete output-voltage response.

EXAMPLE 4 A highly simplified amplifier circuit is shown in Figure 27b. A source v_s is applied at $t = 0$. The objective is to find the output voltage $v(t)$ for $t > 0$.

It may be tempting to convert to a single-loop circuit by replacing the source and the two resistors by a Thévenin equivalent, but that would not work. Doing this would eliminate the controlling current i_1. This would make the controlled voltage indeterminable. We need another approach. As something to do in your spare time, consider writing a set of loop equations and see if that can be successful.

Here we'll try node equations; taking the bottom node as datum, the node voltages will be v_1 and v. The usual "think-think-write" approach can be carried out. The only problem might arise when we "think" current leaving node 1 through the controlled source, which is $-\beta i_1$. But $i_1 = (v_s - v_1)/R$, so there is no problem. The node equations become:

$$\frac{v_1 - v_s}{R} + \frac{v_1}{2R} + \beta\frac{v_1 - v_s}{R} + \frac{v_1 - v}{2.5R} = 0$$

$$C\frac{dv}{dt} + \beta\frac{v_s - v_1}{R} + \frac{v - v_1}{2.5R} = 0$$

(Confirm these equations.) The first equation is algebraic; it can be solved for v_1 and inserted into the second. Carry out the details and find that:

$$v_1 = \frac{4}{10\beta + 19}v + \frac{10(\beta + 1)}{10\beta + 19}v_s$$

$$\frac{dv}{dt} + \frac{6}{(10\beta + 19)RC}v = \frac{4 - 5\beta}{(10\beta + 19)RC}v_s$$

As expected, the capacitor voltage satisfies a first-order differential equation. Its time constant can be increased by increasing the β (current gain) of the controlled source. The solution of the equation can now proceed once the excitation and the initial capacitor voltage are specified.

EXERCISE 13

In the circuit of Figure 27b, suppose the capacitor has an initial voltage of 10 V when the circuit is established, $\beta = 5$, and the RC product is 10 ms. (a)

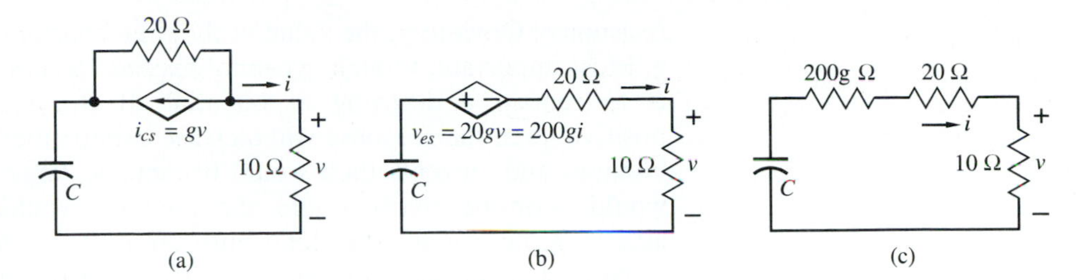

Figure 28
A controlled source in an RC circuit whose effect is merely that of a resistor.

Suppose the excitation is a unit step $u(t)$. Find the complete response for the capacitor voltage and for the input current i_1. (b) Repeat if the excitation is $2e^{-t/6}$. Sketch the response in each case. □

PARENTHETICAL ASIDE As described when controlled sources were first introduced in Chapter 4, treating a controlled source as an abstract component, quite independent of the physical devices of which it is a model, can lead to circuits which are unrealistic. Circuits can be contrived just so as to introduce a controlled source. In such cases it can easily turn out that the controlled source displays no special characteristics but behaves just like an ordinary resistor or a resistor accompanying an independent source.

Such a circuit is shown in Figure 28a. The circuit was established by some switching operation which left the capacitor with an initial voltage. The controlled source and its accompanying resistance can be converted to a voltage-source equivalent, as shown in Figure 28b. The controlling variable v can be replaced by $10i$ using Ohm's law applied to the 10-Ω resistor. Hence, the voltage of the controlled source here is proportional to its own current. We all know that something with that characteristic is an ordinary resistor. Thus, this circuit consists of a capacitor in series with an equivalent resistor—not basically different from the first circuit treated in this chapter. The controlled source here simply has the effect of a resistor, as shown in Figure 28c.

EXERCISE 14

It isn't necessary first to convert to a controlled voltage source to demonstrate the nature of this device. Keeping the circuit of Figure 28a as is, by appropriate application of Kirchhoff's laws and Ohm's law, find a relationship between the voltage and current of the voltage-controlled current source. Your answer, with standard references, should be $v_{cs} = -30i_{cs}$. Find the equivalent resistance of this branch in parallel with the accompanying 20-Ω resistor. Is the result the same as the previous one? □

Suppose now that the orientation of the source current in Figure 28a is reversed. All of the preceding analysis still applies, including the result that the voltage of the source is proportional to its current in Figure 28b. However, because of the voltage-source polarity with respect to the current, the proportionality constant is negative. Hence, the "resistance" replacing the source is $-200g$ Ω, a negative resistance. This can cause a major difference of behavior, depending on the size of $200g$ compared with the rest of the series resistance. Obviously, the value of the time constant can be changed by varying g. Most important, though, g cannot exceed 150 mS if the exponential response is to remain a decaying exponential. If the characteristic value becomes positive, then the response will *increase* exponentially. In a physical circuit, the voltages and currents then would become so large that the limits of linearity would soon be reached and the analysis would no longer apply. That's assuming the components don't burn up in the meantime!

Note that the same properties that result from the preceding circuit can be obtained by inserting negative resistance in place of the controlled source.[13]

EXERCISE 15

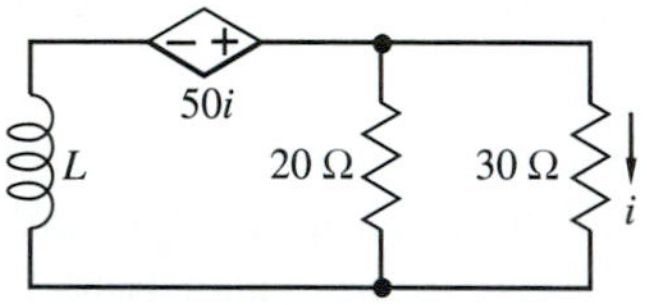

The circuit in Figure 29 is set up by some switching event. Express the current on which the controlled source depends in terms of the current through the source itself. (a) From the resulting v-i relationship of the controlled source, specify the nature and parameter value of a device equivalent to the controlled source. (b) What happens if the polarity of the controlled source is reversed?

Figure 29
Exercise circuit.

ANSWER: (a) −20-Ω resistor; (b) 20-Ω resistor. □

In summary, controlled sources are models abstracted from certain electronic devices. As such, they can be introduced in circuits with other models, such as resistors, capacitors, and inductors. No special difficulties will be encountered in the analysis of such circuits. For example, in a first-order circuit being treated here, the dissipative circuit viewed from the terminals of the energy-storing component (including any controlled sources) can be replaced by a Thévenin equivalent. However, if the placement of a controlled source in a circuit is made without regard for its origins as part of a real device, it is likely that the resulting circuit will seem artificial, with the controlled source playing the role of a resistor (positive or negative) or a combination (such as an accompanied source) of the usual components.

4 RESISTANCE-INDUCTANCE CIRCUITS

We are now ready to turn our attention to circuits containing a single inductor (including interconnections of inductors equivalent to a single one) together with dissipative components and sources. In any such circuit, the inductor can be isolated and everything else included in a two-terminal circuit as shown in Figure 30a. The circuit connected to the inductor can be replaced by a Thévenin equivalent, as shown in Figure 30b (or by a Norton equivalent). The

[13] In an optional section in Chapter 4, a physical two-port device called a *negative converter* was discussed. If you did not skip that section, you know that this device exhibits a negative resistance at one of its ports if the other port is loaded with a (positive) resistor. A negative resistance is thus achievable.

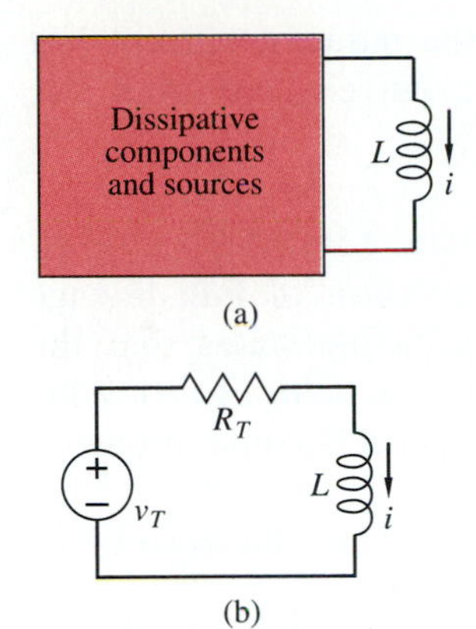

(a)

(b)

Figure 30
Single inductor in a circuit with dissipative components and sources.

next step would be to apply the circuit laws to arrive at the appropriate equation satisfied by the variables in this circuit.

Because much of the analysis here has obvious parallels with what has already been done with *RC* circuits, it is not necessary to spend as much time on the development.

Differential Equation and Time Constant

Rather than deal with this general case, we will start with the specific circuit shown in Figure 31a. The switch has been in position *A* for a long time—long enough that a steady state has been reached and the current in the inductor is no longer changing. Thus, there will be no voltage across the inductor, since the voltage is proportional to di/dt. The switch is suddenly moved to position *B* at a time we take as $t = 0$.[14] The resulting circuit is the one shown in Figure 31b.

A trivial loop equation written around this loop leads to:

$$L\frac{di}{dt} + R_1 i = 0 \quad \rightarrow \quad \frac{di}{dt} + \frac{R_1}{L}i = 0 \tag{29}$$

This homogeneous equation is to be compared with the one in (12) corresponding to the *RC* circuit in Figure 10. They have the same form. L/R_1 plays the same role as the *RC* time constant does; it is the time constant of the *RL* circuit. Hence, the solution will take the form of the following exponential:

$$i(t) = Ke^{-R_1 t/L} \tag{30}$$

As in the case of the *RC* circuit, the constant *K* must be determined by setting $t = 0$ and invoking the initial current. Here we encounter a problem similar to

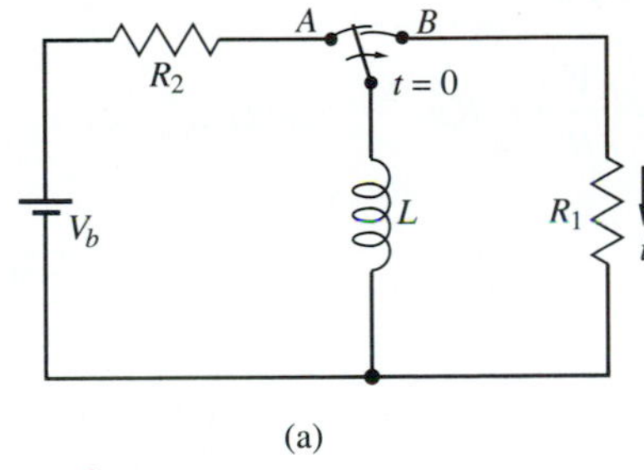

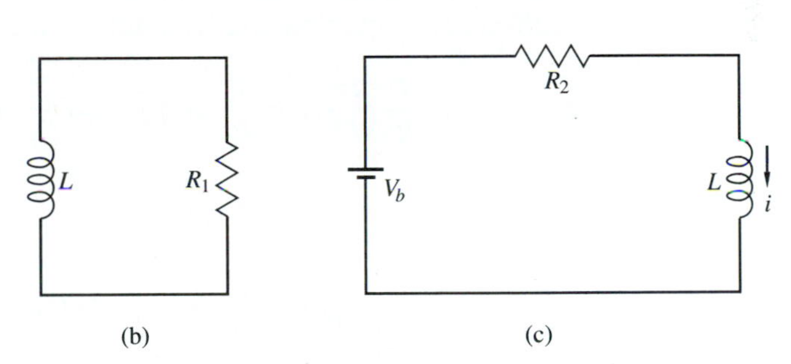

(a) (b) (c)

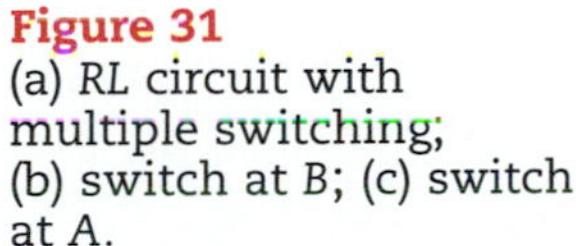

Figure 31
(a) *RL* circuit with multiple switching;
(b) switch at B; (c) switch at A.

[14] Note that two switching events are involved: opening the contact at *A* and closing the contact at *B*. If the events are carried out in that order in the physical world, an electrical arc is likely to occur at *A* as the current that exists is interrupted. To avoid arcing, the contact at *B* must be made first, before the one at *A* is broken, even though the two events are almost simultaneous. This type of switch is called a *make-before-break* switch.

the one we met for the capacitor voltage. We know the inductor current just before the switching event; how is it related to the current just after the switching?

Initial Current and Conservation of Flux Linkage

The answer is provided by the principle of conservation of flux linkage described in Chapter 2. For a single inductor, this principle states that the current in the inductor cannot change instantly. Since $v = L\,di/dt$, if i (or λ) is discontinuous (changes its value instantaneously), then v will become infinite—surely an uncommon occurrence.

Setting $t = 0$ in (30) leads to $i(0) = V_b/R_2 = Ke^0 = K$. Thus, the zero-input solution for the inductor current is:

$$i(t) = \frac{V_b}{R_2} e^{-R_1 t/L} \tag{31}$$

EXERCISE 16

The circuit of Figure 32 has been operating for a long time when the switch is closed (say, at $t = 0$). Find the value of the inductor voltage both at $t = 0-$ and $t = 0+$.

ANSWER: $v(0-) = 0$, $v(0+) = 40$ V. □

Now let's return to Figure 31. A long time after the switch was moved from position A to position B (long enough so that there is no longer any current), it is moved back to position A. The relevant circuit is shown in Figure 31c. For convenience, we again take the switching instant to be $t = 0$. The only difference from the previous circuit is that now the circuit is excited by a step function. There is also a difference in the initial current. The appropriate differential equation and its solution are as follows:

$$\frac{L}{R_2}\frac{di}{dt} + i = \frac{V_b}{R_2} u(t) \quad \rightarrow \quad i(t) = \left(\frac{V_b}{R_2} + Ke^{-R_2 t/L}\right) u(t)$$

$$i(0) = 0 = \frac{V_b}{R_2} + K \quad \rightarrow \quad K = -\frac{V_b}{R_2} \tag{32}$$

$$i(t) = \frac{V_b}{R_2}(1 - e^{-R_2 t/L}) u(t)$$

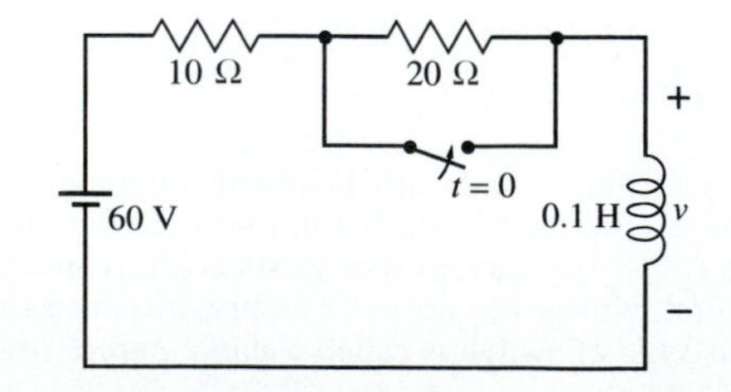

Figure 32
Problem: Find $v(0+)$, the initial inductor voltage (SI units).

Note that the time constant is not the same as it was in the previous part of the problem.

No new principles would be introduced if we proceeded with RL circuits having more extensive dissipative subcircuits attached to a single inductor and excited by other inputs. Hence, we will terminate the discussion here and leave further details to the Problems section.

5 REVIEW

In this chapter we have been studying the behavior of circuits having the following structure: a single energy-storing element (capacitor or inductor) together with any number of dissipative components and sources. By applying the circuit laws, a single first-order differential equation can be obtained whose variable is the capacitor voltage or the inductor current. This can be done in one of two ways: (1) everything but the energy-storing element can be replaced by a Thévenin (or Norton) equivalent; or (2) a node or loop analysis can be carried out in which the capacitor or inductor appears in only one loop equation or one node equation; all other equations will be algebraic and so all other variables can be eliminated algebraically, leaving as a variable only the capacitor voltage or inductor current in a single differential equation.

The solution of this differential equation will require the determination of just three things: the time constant, the equilibrium solution, and the constant multiplier of the exponential in the natural response. The time constant is determined by the circuit parameters; the equilibrium solution by the nature of the input; and the constant by the initial condition on the capacitor voltage or the inductor current applied to the complete response, not just to the zero-input response.

EXERCISE 17

The switch in Figure 33 has been open for a long time; it is closed at $t = 0$. Find the inductor current and the voltage v across the 50-Ω resistor for all time thereafter.

ANSWER: Time constant = 0.8 ms; steady-state current: $i(\infty) = 64$ mA; initial current: $i(0) = 0.1$ A.

□

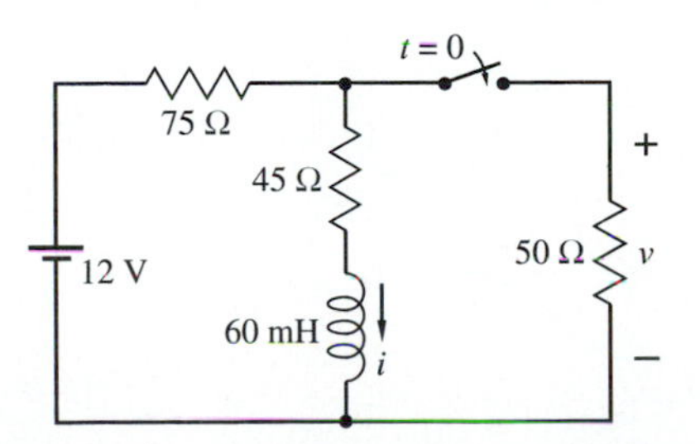

Figure 33 Exercise problem (SI units except as indicated).

SUMMARY

- Periodic signals: sinusoids, square waves, saw-tooth waves
- Root-mean-square (rms) value
- Aperiodic signals: ramp, pulse, step, impulse
- First-order *RC* circuit
- Linear differential equation with constant coefficients
- Natural and forced responses
- Continuity of capacitor voltage
- The exponential function
- Charging and discharging a capacitor
- Time constant of *RC* and *RL* circuits
- Response of an *RC* circuit to a step, a ramp, a quadratic, a sinusoid, an exponential, a pulse
- Steady-state response of an *RC* circuit to a rectangular wave
- Switching an *RC* circuit with a diode
- Op amp integrator circuit
- *RC* circuits with controlled sources
- First-order *RL* circuit
- Continuity of inductor current
- Response of an *RL* circuit to step and other excitations

PROBLEMS

1. For each of the waveforms in Figure P1 find the average value and the rms value.

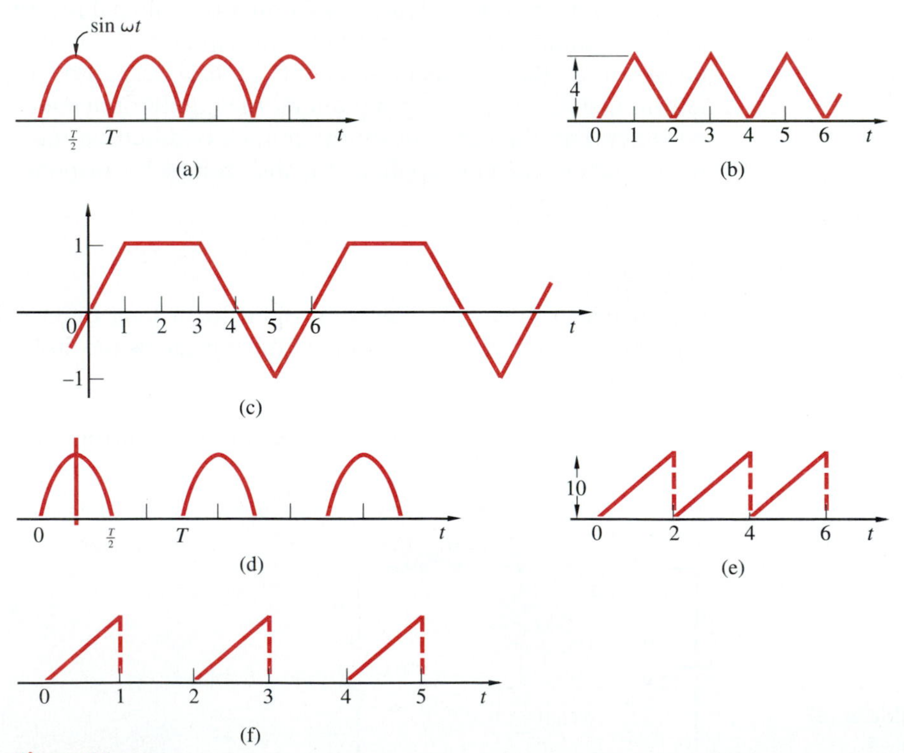

Figure P1

2. Write an expression for each of the following functions:

a. An exponential with a 0.5-s time constant which starts at $t = 3$ s and equals zero before that.

b. A cosine function with angular frequency 200 rad/s which is zero until $t = -5$ s.

3. In Exercise 7 in the text you extended (11) in the text to an impulse occurring at $t = t_1$.

a. Suppose the integrand in (11) is multiplied by 5 before integrating. Write the new expression that results.

b. Next, suppose that the impulse is multiplied by $\cos x$ before integrating. Now find the resulting expression.

c. Finally, suppose the impulse is multiplied by a general function $f(x)$ before integrating. Determine the resulting expression. This can be viewed as a general theorem; it is called the *sampling property* of an impulse.

4. Just as series-connected and parallel-connected resistors can be replaced by an equivalent resistor whose resistance is determined by an appropriate combination rule, so also for series-connected and parallel-connected capacitors and inductors. The purpose of this problem is to determine these combination rules.

The branch relation $i = C\,dv/dt$ for capacitors is voltage-controlled, and $v = L\,di/dt$ for inductors is current-controlled. Simply by analogy with the combination rules for resistors that are based on the corresponding voltage-controlled and current-controlled branch relations, specify a combination rule for:

a. Two inductors L_1 and L_2 in series

b. Two capacitors C_1 and C_2 in parallel

c. Two capacitors C_1 and C_2 in series

d. Two inductors L_1 and L_2 in parallel

(Drawing diagrams will help.)

e. Confirm these, using Kirchhoff's laws and the branch relations.

5. A relaxed 0.01-μF capacitor in series with an 80-Ω resistor is suddently connected to a 20-V battery by a switch.

a. Determine the initial value of the capacitor current.

b. Determine the value of the capacitor charge after a long time.

c. How many time constants will it take for the capacitor voltage to reach within 10% of its final value?

d. Determine the value of the energy dissipated in the resistor up to the time in part (c).

6. At $t = 0$, the switch in Figure P6 is moved from A to B.

a. Find the value of i just after the switching.

b. Find the value of dv/dt just after the switching.

c. Write the differential equation satisfied by v for $t \geqslant 0$.

d. Specify the value of the time constant.

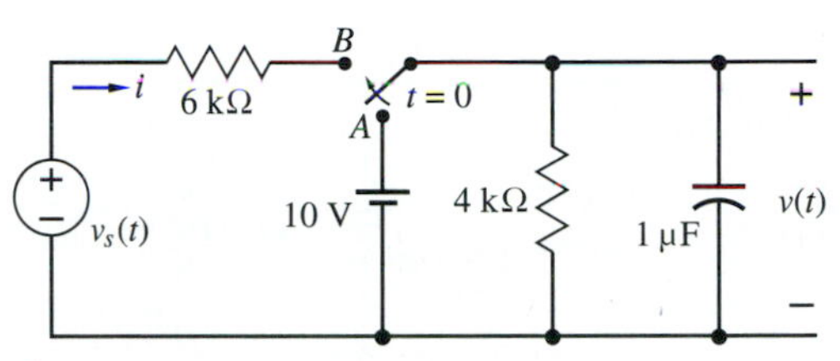

Figure P6

7. Just before the switch is closed at $t = 0$ in Figure P7, the voltage across the capacitor is $v = -10$ V.

a. Find the value of capacitor voltage immediately after the switching.

b. Find the values of the currents indicated by the arrows right after the switching.

c. Find the value of the time constant valid for $t > 0$.

d. Find the differential equation satisfied by the capacitor current for $t > 0$.

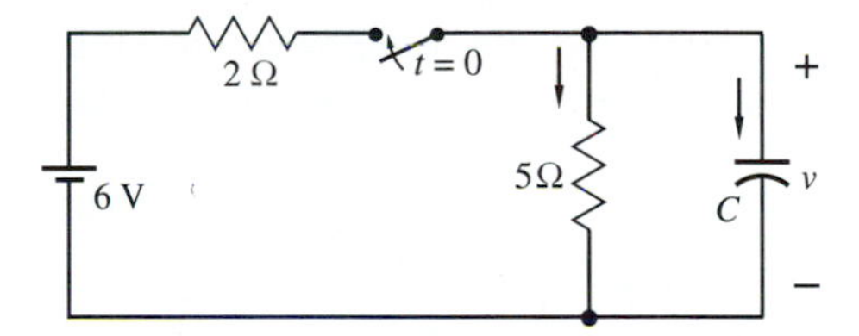

Figure P7

8. The switch in Figure P8 has been open for a long time; it is closed at $t = 0$.

a. Determine the value of capacitor current (standard references) immediately after the switch is closed.

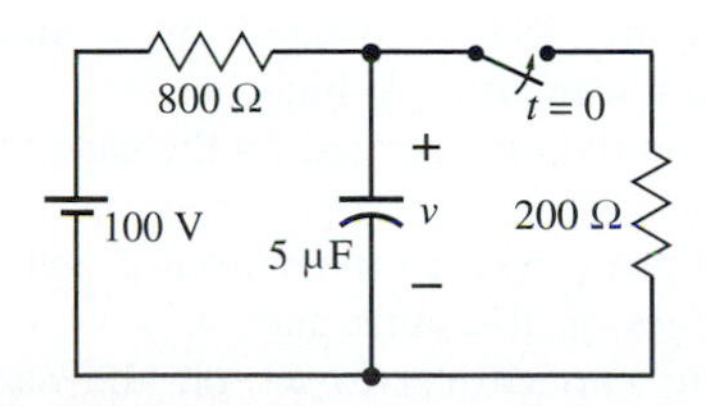

Figure P8

b. Write the differential equation satisfied by the current in the 200-Ω resistor.
c. Specify the time constant.
d. Determine the energy dissipated in the 800-Ω resistor from 0 to 2 ms.
e. Find the value of the capacitor voltage at $t = 10$ s.

9. The switch in the circuit of Figure P9 has been open since before breakfast when it is closed at $t = 0$. Find $i(t)$ for $t \geqslant 0$.

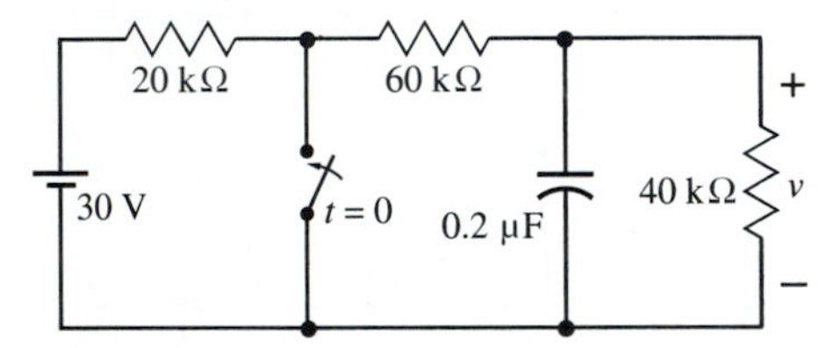

Figure P9

10. The switch in Figure P10 is moved from A to B at $t = 0$.
a. Find the energy stored in the capacitor just before $t = 0$.
b. Determine an expression for the capacitor voltage for $t \geqslant 0$.
c. Determine an expression for $i(t)$ for $t \geqslant 0$.
d. Sketch each of these variables in the interval $0 \leqslant t \leqslant 600\ \mu$s.
e. One second later, the switch is moved back from B to A. Determine expressions for $v(t)$ and $i(t)$ for $t \geqslant 1$ s.

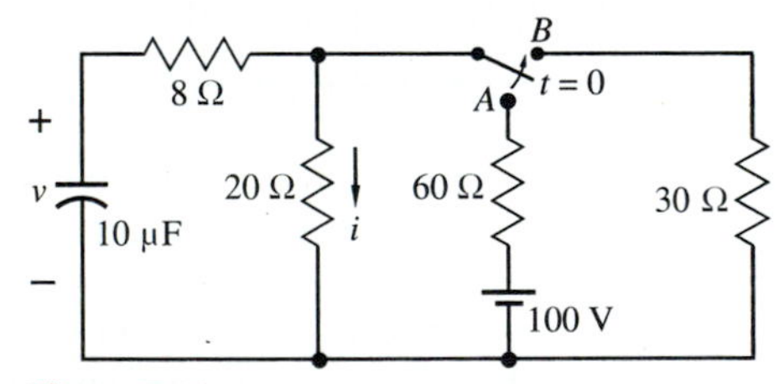

Figure P10

11. The circuit in Figure P11 is excited by a single voltage pulse 0.1 μs long and 1 V high.
a. Find the maximum value reached by the capacitor voltage.
b. Draw the input pulse. Sketch the capacitor voltage for $0 < t < 0.2\ \mu$s on the same axes.
c. Sketch also the capacitor current on the same axes.

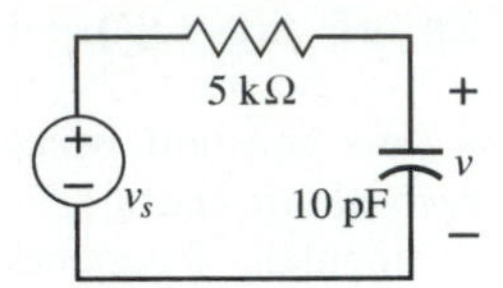

Figure P11

12. The switch in Figure P12 has been in position A for a minute or more when at $t = 0$ it is switched to position B. It stays at that position for 2 ms and is then moved back to A.
a. Find the solution for the capacitor voltage and sketch it as a function of time.
b. On the same axes, sketch the voltage v_{BO} across the switch.

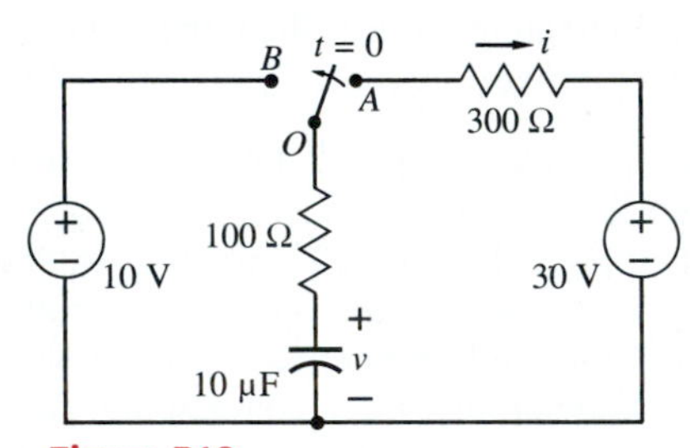

Figure P12

13. The switch in Figure P13 is thrown from position A to B at $t = 0$.
a. Find the initial value ($t = 0+$) of the current through R_3.
b. Find the complete solution for the capacitor voltage v as a function of time.
c. Determine the energy stored in the capacitor after the passage of a long time.

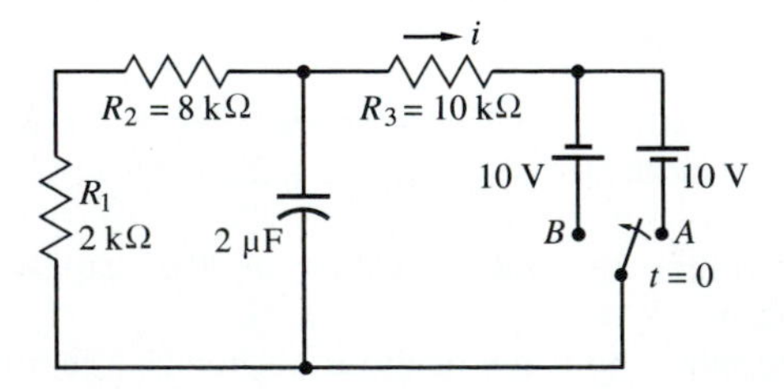

Figure P13

14. After the switch in Figure P13 has been in position B

for a whole second, it is returned to position A.

a. Again find the initial value (after its return to A) of current i. (A diagram would be useful.)

b. Find the complete response of the capacitor voltage following the return of the switch to A.

c. Find also the complete response for the current $i(t)$.

15. A coil of wire (whose model is an inductor L in series with a resistor R) is connected in series with a 20-V battery and a 10-Ω resistor. When conditions have reached equilibrium, the voltage across the 10-Ω resistor is found to be 16 V. Some time later, the terminals of the coil are short-circuited together. During this time it is found that the current in the coil decays with a time constant of 200 ms.

a. Find the value of the inductance. (A diagram would be helpful.)

b. Sketch the current in the coil from the time it is short-circuited.

16. The switch in Figure P16 has been open for a long time when it is closed at $t = 0$.

a. Find the current i just before the switch is closed.

b. Find the voltage v just after the switching.

c. Find the currents i and i_1 a long time after the switch closes.

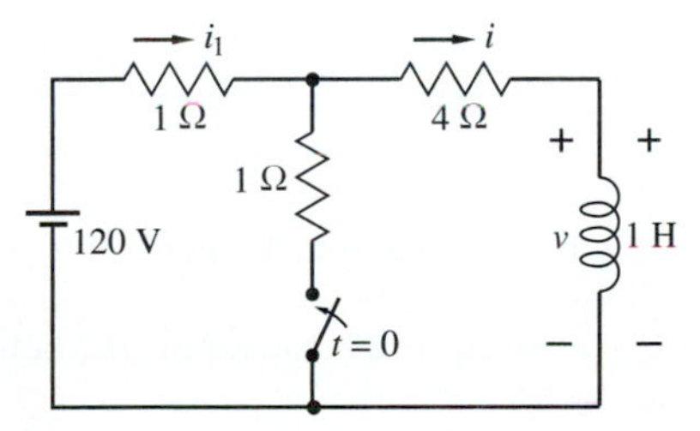

Figure P16

17. From a previous connection, the current, i, in the circuit of Figure P17 with the switch closed has a value of 2 A at $t = 0$. The switch is opened at $t = 10$ ms.

a. Determine $i(t)$ for $0 < t < 10$ ms and $t \geqslant 10$ ms.

b. Sketch $i(t)$ for the period $0 \leqslant t \leqslant 20$ ms.

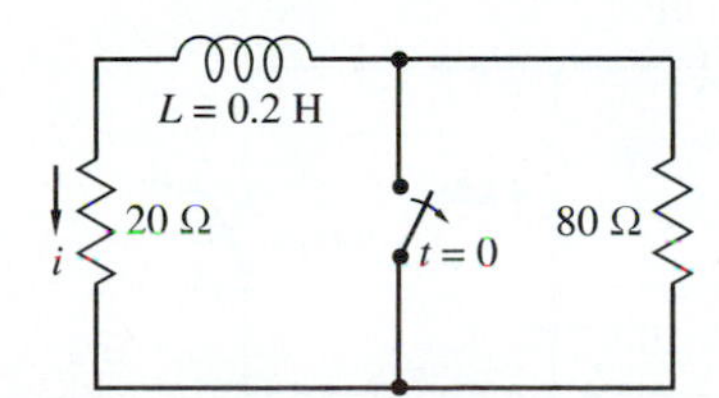

Figure P17

18. The switch in Figure P18 has been closed for a long time when it is opened at $t = 0$.

a. Determine the current in the inductor and in each of the two resistors just before the switch is opened.

b. Repeat for the same currents just after the switch is opened.

c. Find the time constant for $t > 0$.

d. Find the solution for the inductor current for $t \geqslant 0$.

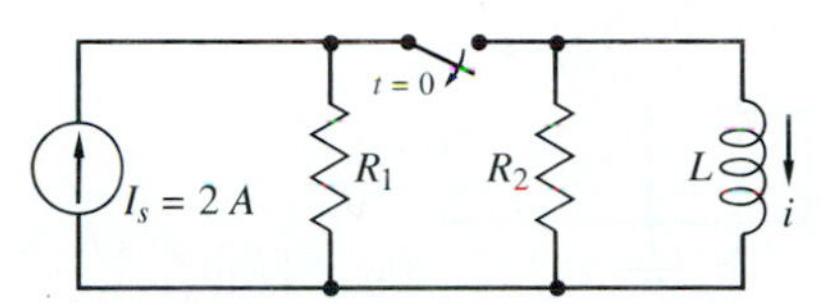

Figure P18

19. The switch in Figure P19 has been closed for a long time; it opens at $t = 0$.

a. Find the energy stored in the inductor just before the switching.

b. Find an expression for the inductor current for $t \geqslant 0$.

c. Calculate the total energy dissipated in the resistors as $t \to \infty$. Compare with the energy found in part (a).

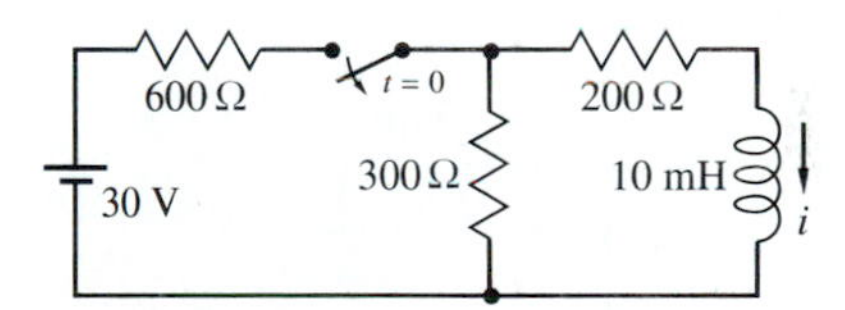

Figure P19

20. After being open for a long time, the switch in Figure P20 is closed at $t = 0$.

a. Find $v(t)$ for $t > -1$ s and sketch it to scale.

b. Draw a tangent to the curve at $t = 0+$ and use it to find the time constant.

c. Compare with the value obtained in part (a).

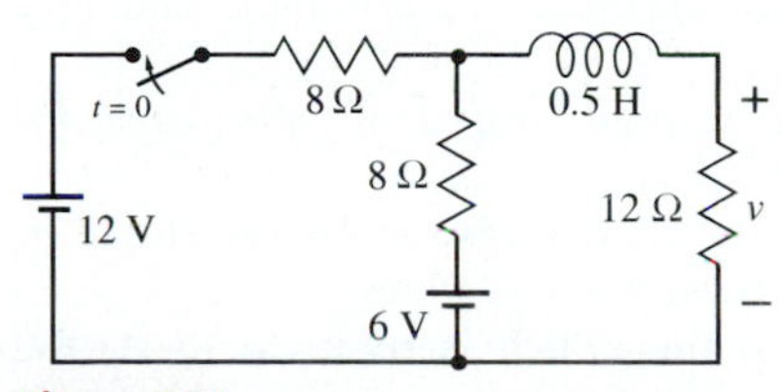

Figure P20

21. a. By applying the basic circuit laws and the properties of an op amp, determine the equation satisfied by the output voltage v_2 in Figure P21.

b. Let V_{sat} be the saturation voltage of an op amp. Determine a condition on the circuit parameters and the input voltage for the op amp to continue operating in its linear region.

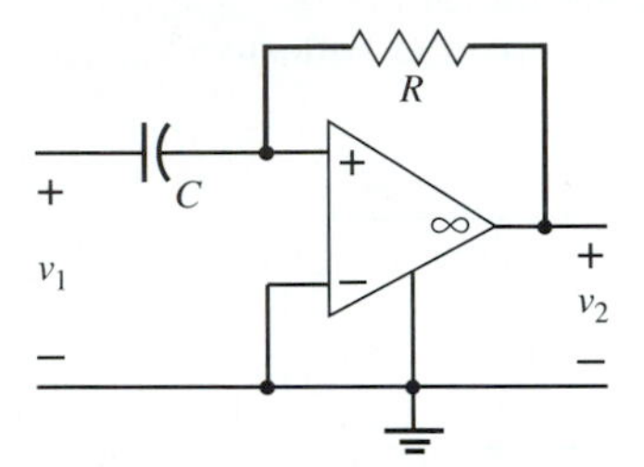

Figure P21

22. Both capacitors in Figure P22 are uncharged when the switch is closed at $t = 0$.

a. What is the state of the diode just after switching?

b. Find the time at which the diode switches.

c. Find a relationship between the two capacitor currents when the diode is ON.

d. Write two differential equations, each satisfied by one of the capacitor voltages when the diode is ON.

e. Find the values of the capacitor voltages a long time after switching.

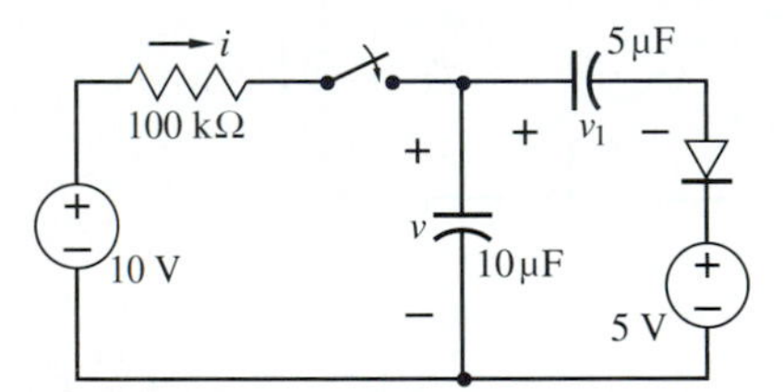

Figure P22

23. The switch in Figure P23 has been in position A for a long time when it is moved to position B instantaneously at $t = 0$.

a. Find $v_C(t)$ for $t \geqslant 0$.

b. Using only algebraic relationships, find $v(t)$ for $t \geqslant 0$.

c. Find the energy stored in the capacitor at $t = 75$ ms.

d. Account for the decrease in energy stored in the capacitor from $0 \leqslant t \leqslant 75$ ms.

e. At $t = 1$ s, the switch is returned to A. Repeat part (a) for $t \geqslant 1$ s.

f. Repeat part (b) for $t \geqslant 1$ s.

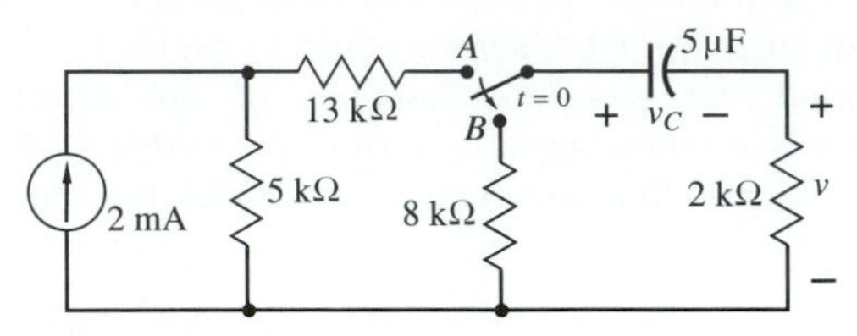

Figure P23

24. The capacitor in Figure P24 is charged to 27.2 V by a mechanism that is not shown. Switch S_1 is closed at $t = 0$.

a. Find the capacitor voltage for $t \geqslant 0$.

b. What fraction of the energy initially stored remains in the capacitor at $t = 1$ ms?

c. Switch S_2 is closed at $t = 2$ ms. Find the capacitor voltage for $t \geqslant 2$ ms.

d. Repeat part (c) if the battery is replaced by a voltage source with $v_s(t) = 5(t - 2)u(t - 2)$, t in ms.

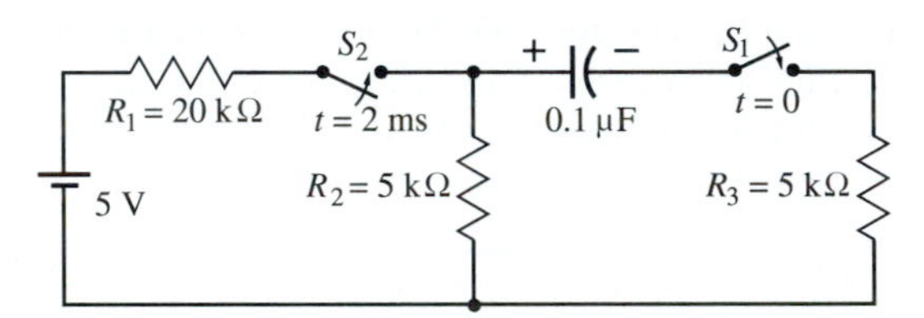

Figure P24

25. The switch in Figure P25 has been open for a long time when it is closed at $t = 0$.

a. Find the energy stored in the capacitor just before the switch is closed.

b. Find the current in the capacitor just after the switch is closed.

c. Find the capacitor voltage for $t \geqslant 0$.

d. The switch is opened again at $t = 100\ \mu$s. Find the capacitor voltage for $t \geqslant 100\ \mu$s.

e. Sketch the capacitor voltage for $-20 \leqslant t \leqslant 300\ \mu$s.

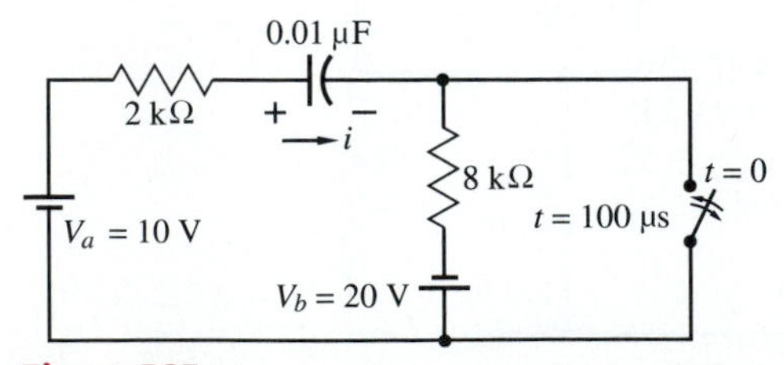

Figure P25

26. The switch in Figure P26 has been at position A overnight when in the morning it is moved at t 5 0 from A to B.

a. How much energy is stored in the capacitor at t 5 0?

b. Find the capacitor voltage for $t > 0$.

c. Find the current i in the capacitor for $t > 0$.

d. At t 5 2 ms the switch is returned to position A. Find the capacitor voltage and current for $t > 2$ ms.

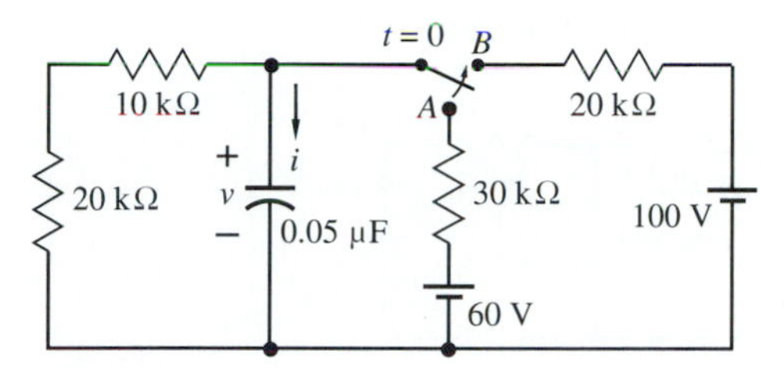

Figure P26

27. The capacitor voltage in Figure P27 equals 10 V when the switch is closed at t 5 0.

a. Is the diode ON or OFF just after the switch closes?

b. Find the time at which the diode switches.

c. Specify the time constants both before and after the diode switches.

d. Sketch the capacitor voltage to scale for t . 2 10 ms.

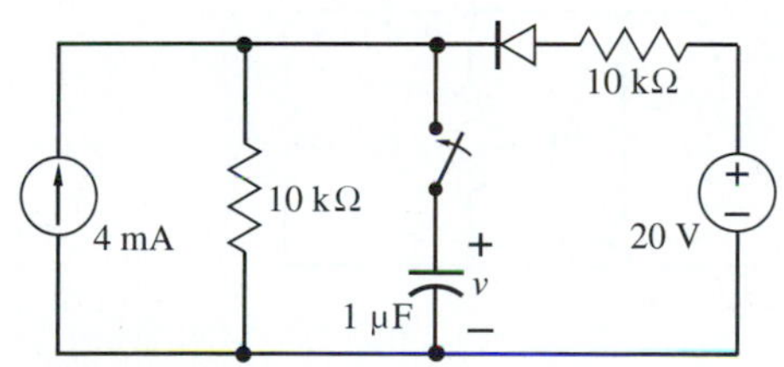

Figure P27

28. The circuit in Figure P28 is in equilibrium with the switch in position A. At t 5 0 the switch is moved from A to B.

a. Find the initial energy stored in the capacitor.

b. Find $y(t)$ for $t > 0$.

c. Find the capacitor current using only algebra. Confirm it from the capacitor i-y relation.

d. At 0.3 ms the switch is returned to A. Find $y(t)$ for $t > 0$.

e. Sketch $y(t)$ for $0 < t < 2$ ms.

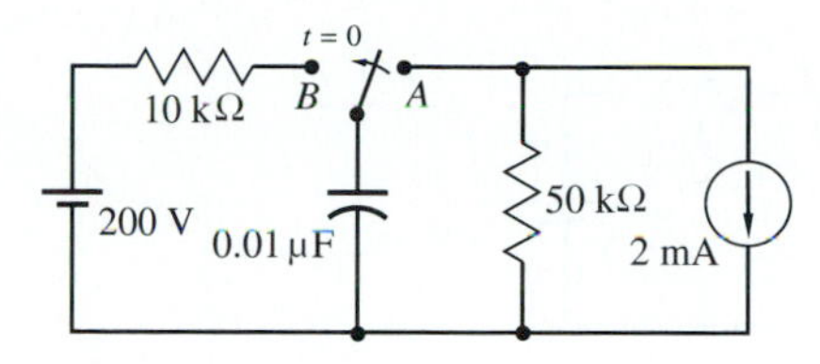

Figure P28

29. The feedback path in Figure P29 contains a "leaky" capacitor whose model is a resistor in parallel with an ideal capacitor.

a. Find the differential equation satisfied by output voltage $y_2(t)$.

b. Suppose the input voltage is a unit step occurring at t 5 0 and the initial capacitor voltage is zero. Find the solution of the equation in part (a) for $t > 0$.

c. Suppose that V_{sat} 5 ±10 V. Determine the condition on the parameters if linear operation is to be maintained over all time.

d. Suppose the leaky-capacitor time constant is 15 ms, C 5 100 nF, and R_1 5 10 kΩ. Determine how long after the step input the op amp will saturate.

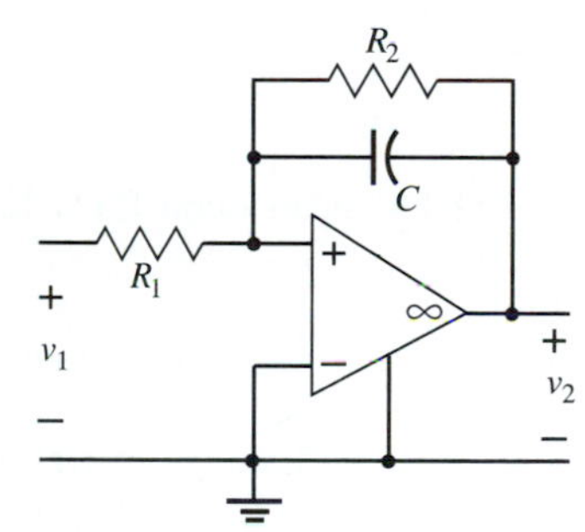

Figure P29

30. The capacitor in Figure P30 is uncharged when the switch closes at t 5 0.

a. Determine the power supplied by the battery as a function of time and the total energy it supplies to the circuit.

b. Determine the energy stored in the capacitor when it is completely charged.

c. Determine the energy dissipated in the resistor during the entire time the capacitor is charging.

d. How are the energies in the last three parts related?

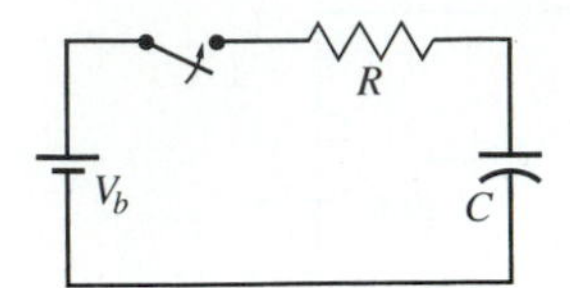

Figure P30

31. The switch in Figure P31 has been open for a long time; it is closed at $t = 0$.

a. Find the energy stored in the inductor at that time.

b. Find $i(t)$ for $t \geqslant 0$.

c. Find the energy stored in the inductor at $t = 10$ ms.

d. The switch is reopened at $t = 30$ ms. Find $i(t)$ for $t \geqslant 30$ ms.

e. Sketch $i(t)$ for $-5 \leqslant t \leqslant 40$ ms.

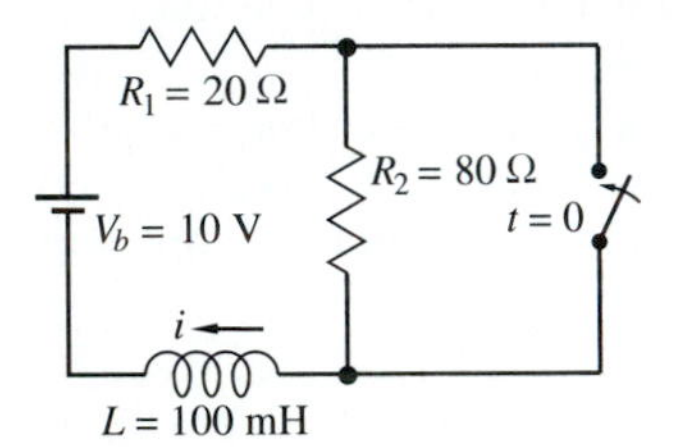

Figure P31

32. The switch in Figure P32 has been open for a long time; it closes at $t = 0$.

a. Find $i(t)$ for $t > 0$.

b. Using only algebra, find the inductor voltage at $t = 0$.

c. The switch reopens at $t = 25\ \mu$s. Find $i(t)$ for $t \geqslant 25\ \mu$s.

d. Find the energy stored in the inductor at $t = 50\ \mu$s.

e. Sketch $i(t)$ for $-10 \leqslant t \leqslant 50\ \mu$s.

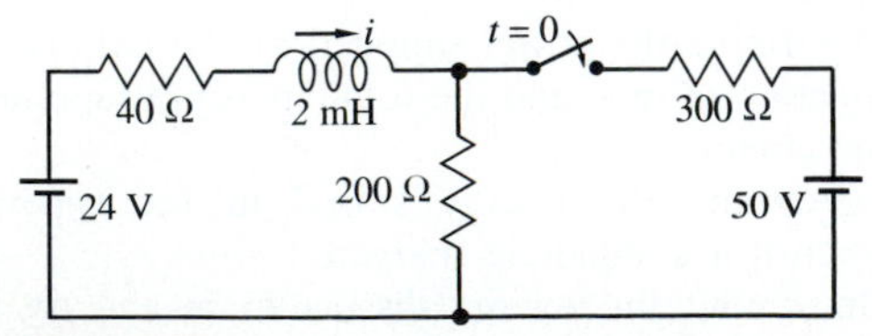

Figure P32

33. The capacitor in Figure P33 is initially relaxed. The voltage source is a pulse generator whose voltage is:

$$v_s(t) = 10[u(t) - u(t - T)]\ \text{V}$$

The pulse width T is adjustable.

a. Find $v(t)$ for $0 \leqslant t \leqslant T$; $T = 100$ ms.

b. Find $i(t)$ for $0 \leqslant t \leqslant T$; $T = 100$ ms.

c. Repeat part (a) for $T = 200$ ms.

d. Repeat part (a) for $T = 500$ ms.

e. Sketch $v(t)$ from parts (a), (c), and (d) on the same axes as v_s. Describe what you learn by a comparison of these sketches.

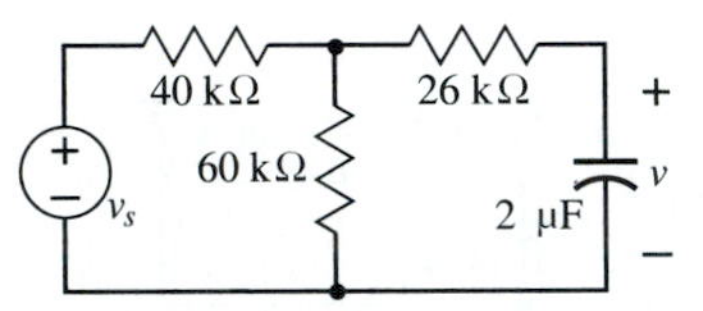

Figure P33

34. The circuit in Figure P34 is in equilibrium with the switch in position A when the switch is moved to position B at $t = 0$.

a. Find $v(t)$ for $t \geqslant 0$.

b. Find $i(t)$ for $t \geqslant 0$.

(It may be helpful to draw separate diagrams.)

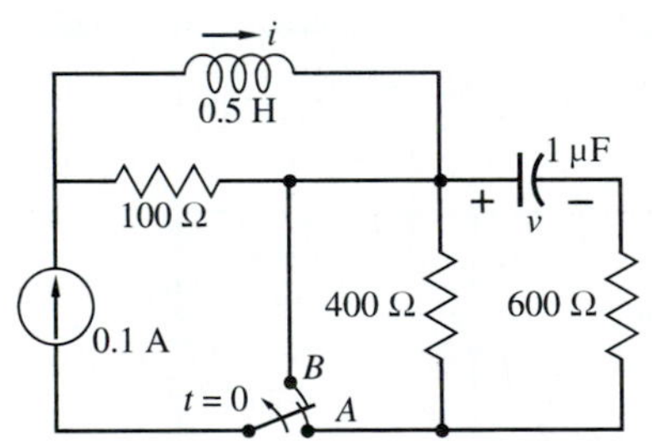

Figure P34

35. The circuit in Figure P35 is in equilibrium with the switch at position A. At $t = 0$ the switch is moved to position B.

a. Find the inductor current $i(t)$ for $t \geqslant 0$.

b. Using only algebra, find the inductor voltage for $t \geqslant 0$.

c. Confirm your answer to part (b) using the inductor v-i relationship.

d. At $t = 100$ ms, the switch is moved back to position A. Find the new expression for $i(t)$ for $t \geqslant 100$ ms.

e. Sketch $i(t)$ for $0 \leqslant t \leqslant 200$ ms.

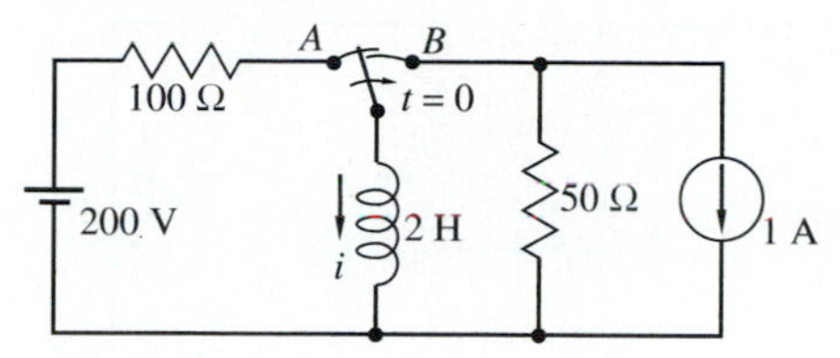

Figure P35

36. The capacitor in Figure P36 is initially uncharged when a step function is applied at $t = 0$.

a. Write a differential equation for the capacitor current i in response to the step.

b. Solve the differential equation for the capacitor current for $t > 0$, leaving no unspecified constants.

c. Write and solve the differential equation for the capacitor voltage for $t > 0$, leaving no unspecified constants.

d. From the capacitor voltage and using strictly algebraic means, find the capacitor current for $t > 0$. Compare with the result in part (b).

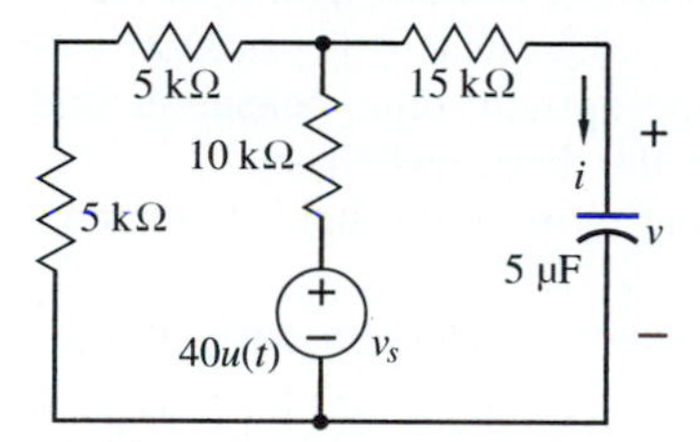

Figure P36

37. Switch 2 in Figure P37 has been closed with switch 1 in position A for at least 1 s when switch 1 is moved to position B at $t = 0$. One millisecond later, switch 2 is opened. Ten milliseconds after that, switch 1 is returned to position A.

a. Find the solutions for the capacitor voltage and current, and sketch them to scale in the interval $0 \leqslant t \leqslant 13$ ms.

b. Determine the energy stored in the capacitor at $t = 0$, 1, 10, and 13 ms.

c. Confirm the last three values by integrating the power into the capacitor over the appropriate intervals.

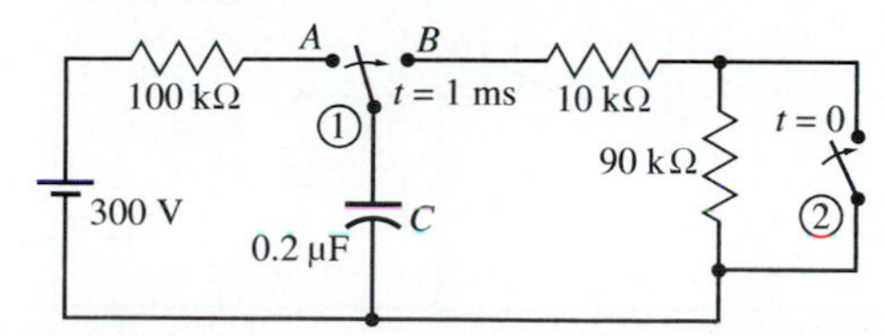

Figure P37

38. A 1-V pulse is applied to the initially relaxed circuit in Figure P38 at $t = 0$.

a. The pulse duration is 4 ms. Determine $v(t)$ for $0 \leqslant t \leqslant 8$ ms.

b. Suppose that the pulse duration is 20 ms instead. Determine $v(t)$ for $0 \leqslant t \leqslant 40$ ms.

c. Sketch $v(t)$ to scale using the input pulse as the basis for the scale in each case.

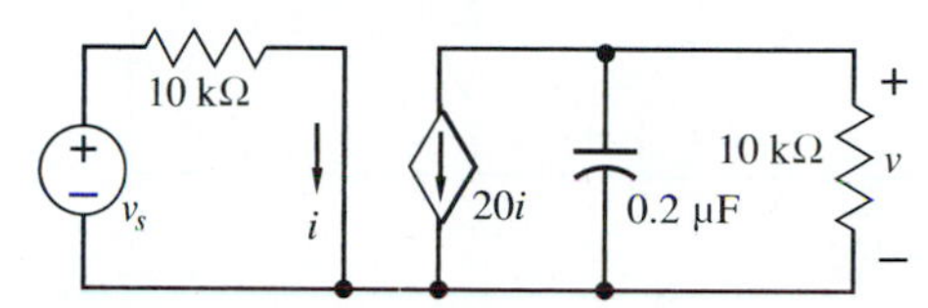

Figure P38

39. At $t = 0$, a 10-mA current pulse (rectangular) is applied to the initially relaxed circuit of Figure P39.

a. Describe two possible approaches for determining expressions for $i(t)$ during the pulse and for an equal interval after the end of the pulse.

b. Find $i(t)$ in those two intervals by one of your approaches.

c. Sketch the result to scale on the graph of the pulse.

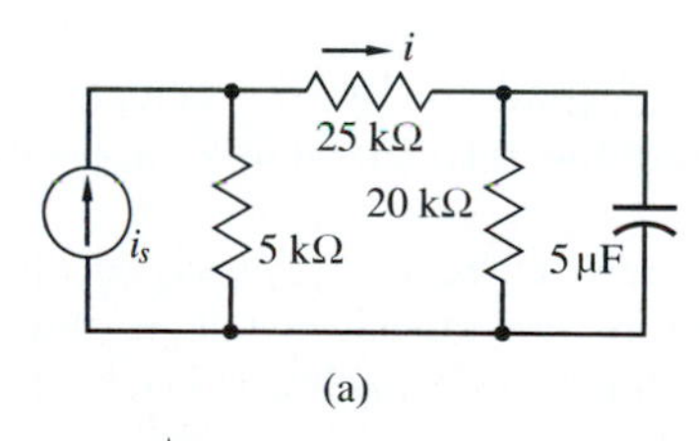

(a)

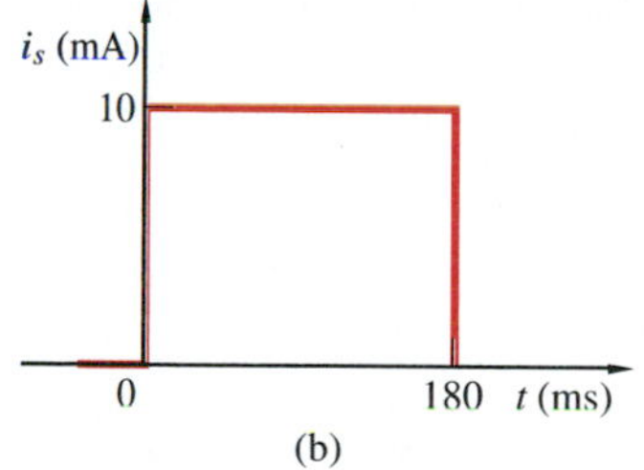

(b)

Figure P39

40. The initially relaxed circuit in Figure P40 represents a feedback amplifier. It is excited with a step function $v_s = u(t)$.

a. Using any approach you know, find the differential equation satisfied by the capacitor voltage v.

b. Determine the complete capacitor-voltage response, leaving no unspecified constants.
c. Determine all resistor currents without any further calculus.

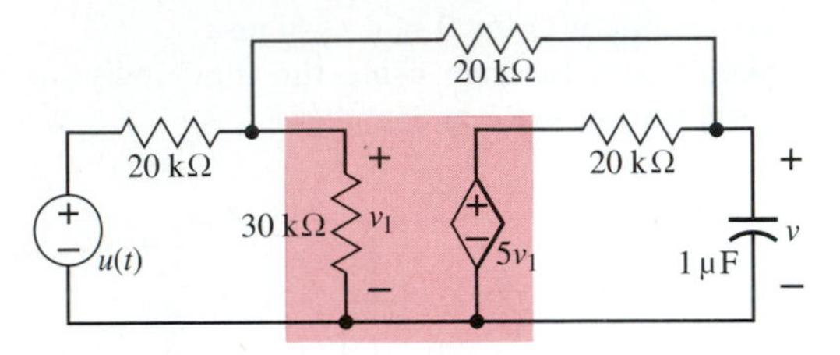

Figure P40

41. The circuit in Figure P41 is initially relaxed when an exponential $v_s(t) = 5e^{2t}$ is applied at $t = 0$. The overall objective is to find the capacitor voltage and all resistor currents for $t > 0$.
   a. Find the differential equation satisfied by the capacitor voltage v by converting the rest of the circuit to its Thévenin equivalent.
   b. Find this equation again by applying KVL around the two meshes and eliminating all resistor voltages in terms of mesh currents, leaving as voltage only the capacitor voltage.
   c. Find the natural response.
   d. Determine the forced response.
   e. Apply the initial condition to the complete response to determine any previously unspecified constants.
   f. Find the two mesh currents by purely algebraic means in terms of the capacitor voltage and the source voltage. Since only algebra is involved (no further differentiations), the currents will have the same functional dependence on time as the capacitor voltage except possibly for the values of the coefficients. Confirm this.

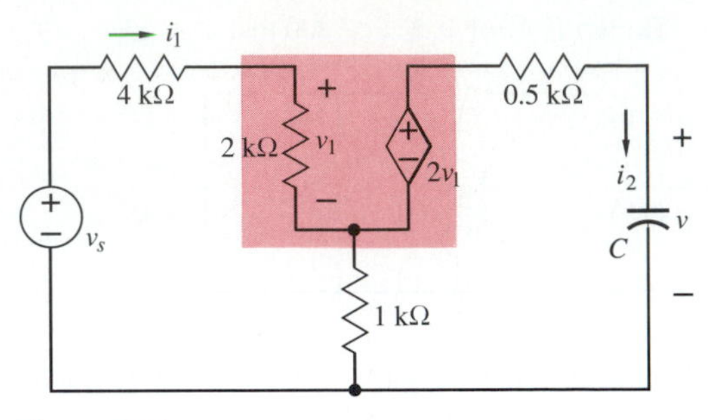

Figure P41

42. The capacitor in Figure P42 is initially uncharged when the switch is closed at $t = 0$.
   a. Determine the state of the diode before $t = 0$.
   b. Determine the state of the diode *immediately after* the switch is closed.
   c. Determine the time after the switch is closed when the diode changes state.
   d. Sketch the capacitor voltage v_C for several time constants after the switch is closed, showing clearly the appropriate time constants and the point at which the diode switches.
   e. Also sketch the current through R_2 in the same interval.

(*Note*: Draw equivalent circuits for each condition.)

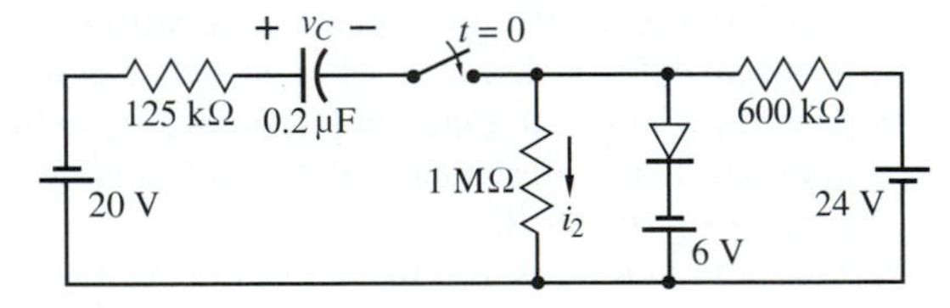

Figure P42

DESIGN PROBLEMS

1. The circuit in this problem has the same structure as that in Problem 42. The capacitor in Figure DP1 is initially uncharged when the switch is closed at $t = 0$. The design objective is to determine values of the battery voltage and resistors R_1 and R_2 in order to satisfy certain conditions to be specified. It would be helpful to draw equivalent circuits for each condition of the circuit.

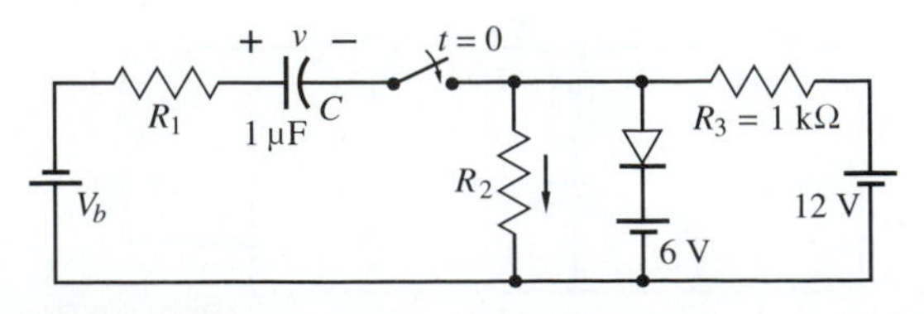

Figure DP1

a. Determine a range of values of R_2 for which the diode will be ON for $t < 0$. Then choose a value of R_2 in this range so that the current in the diode is at least 2 mA.

b. Determine ranges of values of R_1 and V_b such that the diode will turn OFF as soon as the switch is closed.

c. Choose values of V_b and R_1 in this range so that the diode will turn ON again exactly 1 ms after it turned OFF.

d. Sketch the capacitor voltage v over a sufficient interval of time to show all significant variations in the voltage; show all appropriate time constants and the points on the curve where the diode switches.

e. Sketch the current through R_2 in the same interval.

2. The device across the capacitor in Figure DP2 is nonlinear. Let's call it a *glow tube.* At low voltages it is normally an open circuit. When the voltage across the device increases and reaches a relatively high *threshold voltage* V_T, a physical process is activated (such as the breakdown of a conducting gas) which causes current conduction through the device, accompanied with a glow. During the conduction phase, the device operates like a linear resistor R. At some point, controlled by the voltage across the tube, conduction stops. That is, conduction, and hence the glow, stops when the voltage reaches a low *sustaining* value, V_S. During the time that the glow is extinguished, the capacitor charges toward the battery voltage. When the capacitor voltage reaches the threshold value V_T, the glow lamp turns on and the capacitor starts discharging (mainly) through the glow lamp until the lamp is extinguished. Then the cycle is repeated. By properly selecting component values in the circuit, the entire system can constitute a self-contained, periodically flashing lighting system. The *duty cycle* is the time during which the lamp is ON.

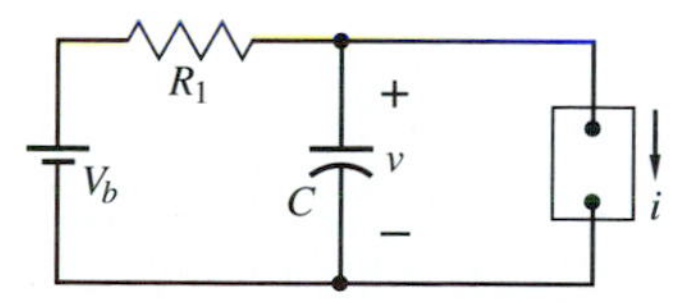

Figure DP2

a. For a device with a threshold voltage of $V_T = 20$ V, sustaining voltage $V_S = 8$ V, and $R = 5$ kΩ, let $V_b = 30$ V, $R_1 = 1$ MΩ, and $C = 10\,\mu$F. Determine expressions for $v(t)$ and $i(t)$ for both the ON portion and the OFF portion of one period in the steady state. Confirm that the capacitor voltage at the end of a period equals its value at the beginning of a period. Sketch $v(t)$ and $i(t)$. Specify the duty cycle. How often will the lamp flash each minute?

b. The problem now is one of design. The following parameters of a glow lamp are given: threshold voltage $V_T = 100$ V, $R = 10$ kΩ, and sustaining voltage $V_S = 20$ V; also given are the period of the output $T = 10$ s and a duty cycle of $T_d = 3$ s. Determine appropriate values of the circuit parameters C, R_1, and V_b.

3. The circuit for this problem is similar to the one in Figure DP2 except that the glow tube is current-controlled. That is, after the tube turns on, conduction will be extinguished when the *current* in the tube reaches a low *sustaining current*, I_S. All other characteristics of the tube are the same.

a. For a device with a threshold voltage $V_T = 60$ V, sustaining current $I_S = 5$ mA, and $R = 30$ kΩ, let $V_b = 100$ V, $R_1 = 2$ MΩ, and $C = 5\,\mu$F. Determine expressions for $v(t)$ and $i(t)$ for both the ON portion and the OFF portion of one period in the steady state. Confirm that the capacitor voltage at the end of a period equals its value at the beginning of a period. Sketch $v(t)$ and $i(t)$. Specify the duty cycle. How often will the lamp flash each minute?

b. The problem now is one of design. The following parameters of the glow tube are given: $V_T = 80$ V, $I_S = 8$ mA, and $R = 40$ kΩ; also, the desired period of the output is specified as $T = 10$ s and the duty cycle is $T_d = 4$ s. Determine the appropriate values of circuit parameters C, R_1, and V_b.

4. The circuit in Figure DP4a is similar to the one in the preceding design problems. The neon tube in the diagram consists of a pair of electrodes enclosed in a glass envelope containing neon at low pressure. It behaves approximately like a switch. Its characteristics are similar to those of the glow tube in the preceding project, with some differences. When the tube voltage v_t is below a certain breakdown voltage V_B (similar to the threshold voltage in the preceding projects), there is little ionization and the tube does not conduct; it is approximately an open circuit. When the tube voltage reaches the breakdown voltage V_B, the gas ionizes rapidly; the tube voltage drops rapidly (say, instantaneously) to a low value V_m—called the *maintaining* voltage—and stays constant at that value (like a battery) until the tube current has been reduced to a low value called the *sustaining* current, I_S. When the tube current reaches I_S, ionization ceases and the tube again becomes an open circuit.

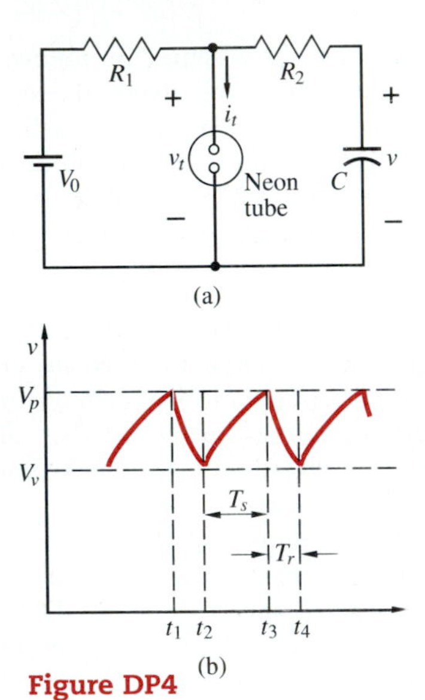

Figure DP4

The desired output voltage across the capacitor is shown in Figure DP4b. When the tube stops conducting, the capacitor charges from the battery. The tube voltage also rises until it reaches its breakdown value. The capacitor then discharges (mainly) through the tube, dropping from its peak value V_p, when the tube voltage equals its breakdown value, to its valley, V_V, when the tube voltage reaches its maintaining value, at which time the cycle starts again. The frequency of oscillation is the reciprocal of the time required for one complete cycle in the steady state. Ideally the return time T_r should be zero; more practically, it should be small compared with the sweep time T_s.

The goal of the project is to design a *relaxation oscillator*. The specifications are as follows:

$$V_p = 80\text{ V}, \qquad V_V = 60\text{ V}, \qquad f = 10^4\text{ Hz}$$

The characteristics of an available tube are:

$$V_B = 90\text{ V}, \qquad V_m = 10\text{ V}, \qquad I_S = 10\text{ mA}$$

The source voltage is $V_b = 120$ V.

The design consists of the following:

a. Determine a range of resistor values that permit the circuit to act as a relaxation oscillator.

b. Find appropriate values of R_1, R_2, and C within the range of permissible values.

c. Specify the values of sweep time and return time achieved by your design.

d. Suppose that your return time is too high. Specify values of R_1 and R_2 within the permissible range which will reduce the T_r/T_s ratio. (What does that do to C?)

e. After all this work, the chief engineer comes into your lab and tells you that new specs require a larger ratio of peak-to-valley voltage, say, 80/50. Specify new values of the circuit parameters. (How does that influence the peak-to-valley voltages?)

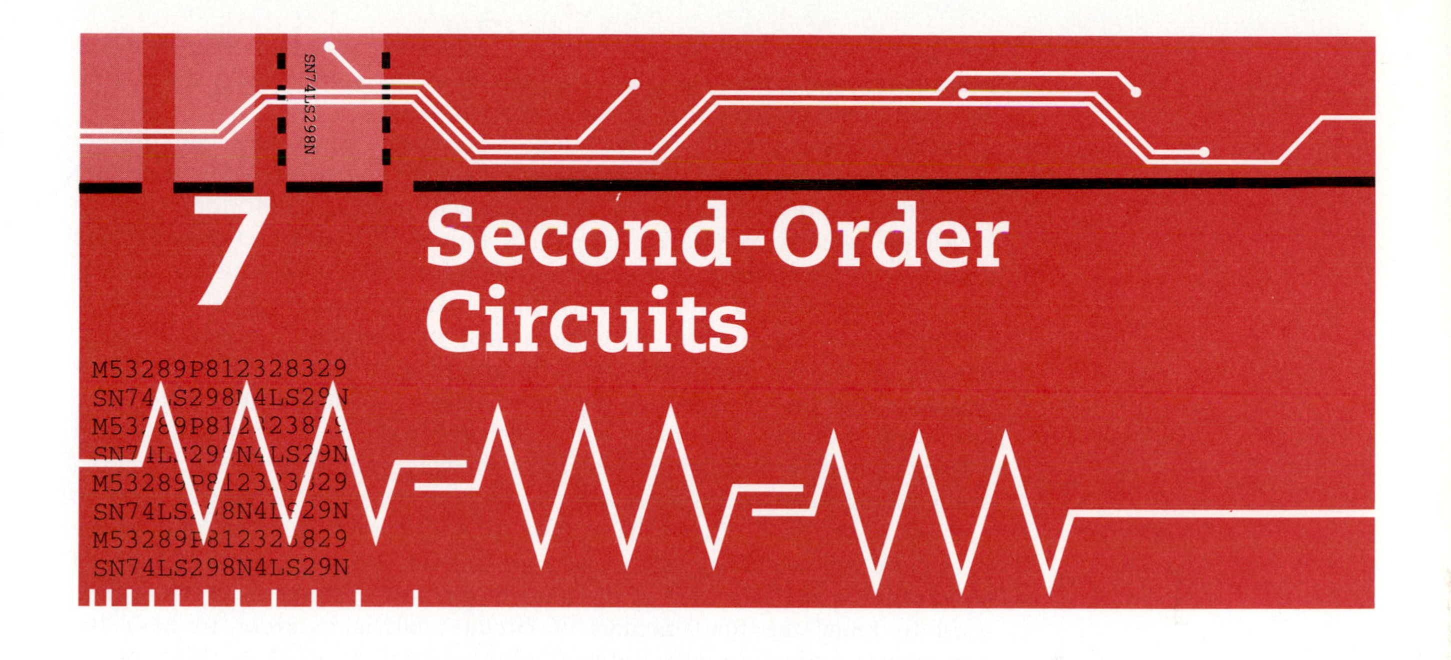

7 Second-Order Circuits

The preceding chapter described the behavior of linear, time-invariant circuits containing only one energy-storing component (together with dissipative components, of course). The energy-storing component may or may not have some initially stored energy when the circuit is excited. The current and voltage responses in such a circuit consist of two components. One, the forced response, consists of a function having the same form as the excitation. The other, the natural, or transient, response, consists of an exponential $e^{st} = e^{-t/T}$ which decays after a few time constants T. Only the natural response, a transient, will result, of course, if the circuit is established (by closing a switch, for example) at some time when the capacitor has an initial voltage or the inductor has an initial current but there is no other excitation.

We should now consider circuits of *any* complexity, with *any* number of energy-storing components. However, we will delay the general case a little longer and in this chapter treat only those circuits which contain exactly two energy-storing devices. There are two major reasons for doing this: (1) many circuits, important in their own right, contain just two energy-storing components, so by studying them we will be dealing with an important set of circuits; and (2) some important concepts which are applicable to circuits of any arbitrary complexity can be introduced without the tedious algebraic and other complications of dealing with that complexity.

For each circuit containing a single energy-storing component in the preceding chapter we found a first-order differential equation. The behavior of *any* voltage and current variable is described by such an equation. It might be conjectured that circuits containing more than one energy-storing component, say, n of them, will be described by an nth-order differential equation; in particular, if n is 2, the describing equations should be second-order differential equations.[1] We shall see that this conjecture is valid.

In this chapter, then, we will take up circuits containing two distinct energy-storing components: in addition to dissipative components, two distinct capacitors, two distinct inductors, or one capacitor and one inductor.

All the circuits considered here (and, indeed, throughout the book) are models (equivalent circuits) of dozens of different practical systems and devices used in hundreds of applications. Such systems might contain thermal, mechanical, or other types of components, in addition to electrical ones. Knowledge of the properties of such components would be needed for understanding the operation of the entire system. As stated in Chapter 1, in this book we can deal only with electrical models. Unfortunately, that eliminates some of the glamor. Instead of showing some space-age application, using the language of satellites, robots, or laser systems, all we can do is to present . . . an *RLC* circuit! You have to be aware that, even though the circuits we will be dealing with might appear prosaic to you, they are models of significant applications used in the most modern of engineering systems. You need to know the fundamentals of circuit analysis in order to have the competence in the future to deal with all those applications that excite you.

We, could, of course, use the language of a space-age application, and then say that our simple circuit is a model of some particular satellite system or high-tech medical instrumentation system that has been discussed in the popular press recently. Of course, you would have to take the author's word for it, since the background necessary for you to *derive* the model from a knowledge of the physical system would not be provided. Doing this would serve no educational purpose but would be strictly for PR.

1 RESISTOR-CAPACITOR CIRCUITS

As the first type of circuit to be considered, we will start with circuits containing two distinct capacitors.

Obtaining the Differential Equations

A circuit in this class is shown in Figure 1. Since a derivative appears only when a capacitor current is expressed in terms of its voltage (and not the other way around), the first equations to write ought to include capacitor currents. That suggests KCL, so let's think of writing KCL at the two nondatum nodes in Figure 1. Into these equations we substitute the i-v relationships of the

[1] No cheating! Two capacitors in parallel, for example, should be counted as one. That's because the two voltages are not independent; they are equal. Initial values of these voltages cannot be specified independently. The same is true for two inductors in series. In more complex circuits, it is possible for other constraints among capacitor voltages or inductor currents to reduce the *effective* count of these components below their *actual* count.

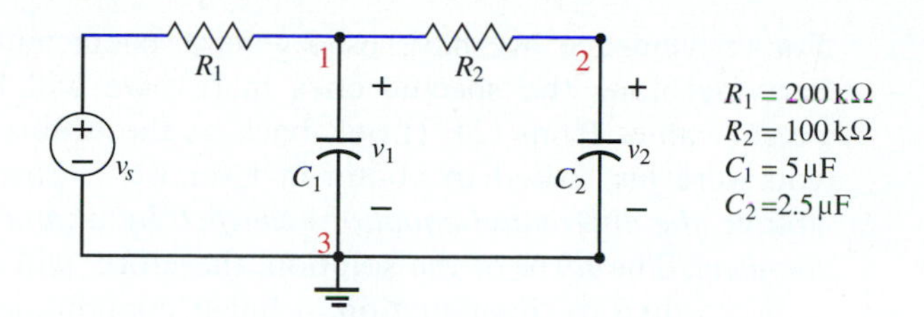

Figure 1
Circuit with two capacitors (low-pass filter).

branches. The result of these steps is:

$$C_1\frac{dv_1}{dt} + G_1(v_1 - v_s) + G_2(v_1 - v_2) = 0$$

$$C_2\frac{dv_2}{dt} + G_2(v_2 - v_1) = 0 \tag{1}$$

This is a set of two first-order differential equations in two unknowns, v_1 and v_2. They can be combined in more than one way to arrive at one equation in a single unknown. One possibility is to solve the second equation for v_1 and then insert it into the first. After some algebraic manipulation, the result will be:

$$\frac{d^2v_2}{dt^2} + \left(\frac{G_2}{C_2} + \frac{G_1 + G_2}{C_1}\right)\frac{dv_2}{dt} + \frac{G_1G_2}{C_1C_2}v_2 = \frac{G_1G_2}{C_1C_2}v_s \tag{2}$$

As expected, this is a linear, nonhomogeneous second-order differential equation with constant coefficients.

EXERCISE 1

Carry out the other possibility, of solving the first equation for v_2 and substituting into the second. The result should be a differential equation with v_1 as the variable.

ANSWER: Except for the forcing function it is identical with the equation for v_2. □

Solution for the Natural Response

From our experience with first-order circuits, we would expect that the solution of (2) would consist of two parts. One part—the natural response—would exist even in the absence of a source, simply from the excitation provided by the initial energy possibly stored in the capacitors when the circuit is established. This part is a solution of the homogeneous equation. The other part—the forced response—is in response to the driving or forcing function and takes on the shape of that function. It is a solution of the nonhomogeneous equation. We will postpone consideration of the forced response temporarily.

Let us assume that the circuit in Figure 1, without the source, is established by some switching operations such that each of the capacitor voltages has an initial value: $v_1(0)$ and $v_2(0)$. With $v_s = 0$ in (2), the differential equation to be solved has the form:

$$\frac{d^2v_2}{dt^2} + a_1\frac{dv_2}{dt} + a_0v_2 = 0 \tag{3}$$

For convenience, we have used general coefficients a_1 and a_0 here instead of carrying along the specific ones in (2); we will later replace these by their actual values from (2). (Look back at the differential equation for v_1 which you were just asked to obtain in Exercise 1. Since, in the present case, v_s is absent, *the differential equations satisfied by v_1 and v_2, being homogeneous, are identical.* The form of the solution, therefore, will also be identical.)

According to this equation, a linear combination of the function v_2 and its first and second derivatives is to add to zero. Since this must be identically true—true for all time—that means the function and its derivatives must have the same functional form. The only function for which this is true is an exponential. Thus, as we did in the first-order case in Chapter 6, we *assume* that the solution is an exponential:

$$v_2(t) = Ke^{st} \tag{4}$$

(There is no danger of going wrong here. If the assumption turns out to be false, the equation will not be satisfied and we will have to go back to the drawing board.) Since an exponent should be dimensionless, s has the dimensions of reciprocal time.

When this expression is substituted into the differential equation and the indicated differentiations are carried out and the terms collected, the result becomes:

$$(s^2 + a_1 s + a_0)Ke^{st} = 0 \tag{5}$$

The product of three factors is to be zero; that means at least one factor should vanish. What about K? If K is zero, then, from (4), the entire solution will be zero. Is that acceptable? It certainly is; it is indeed the appropriate solution when the circuit is initially relaxed—that is, when the initial values of the capacitor voltages are zero. But that's an uninteresting nonproblem; it is the trivial case of no excitation, no initial values, and so no response.

As for the exponential, it does not vanish for any finite value of t. Hence, the only nontrivial possibility is that the quantity in parentheses is zero:

$$s^2 + a_1 s + a_0 = 0$$

or, using the specific values of a_1 and a_0 from (2):

$$s^2 + \left(\frac{G_2}{C_2} + \frac{G_1 + G_2}{C_1}\right)s + \frac{G_1 G_2}{C_1 C_2} = 0 \tag{6}$$

CHARACTERISTIC EQUATION AND CHARACTERISTIC VALUES The process of assuming an exponential solution for the differential equation satisfied by the variables in a second-order circuit has led to a polynomial in the variable s, which was the assumed coefficient of t in the exponent. The equation resulting from this process is called the *characteristic equation*; it is a quadratic in this case and so has two roots, as follows:

$$\begin{aligned} s &= -\frac{1}{2}\left(\frac{G_2}{C_2} + \frac{G_1 + G_2}{C_1}\right) \pm \sqrt{\frac{1}{4}\left(\frac{G_2}{C_2} + \frac{G_1 + G_2}{C_1}\right)^2 - \frac{G_1 G_2}{C_1 C_2}} \\ &= -\frac{1}{2}\left(\frac{G_2}{C_2} + \frac{G_1 + G_2}{C_1}\right) \pm \sqrt{\frac{1}{4}\left(\frac{G_2}{C_2} + \frac{G_2 - G_1}{C_1}\right)^2 + \frac{G_1 G_2}{C_1^2}} \\ &= -A \pm B \end{aligned} \tag{7}$$

$$s_1 = -A + B \qquad s_2 = -A - B$$

The roots of the characteristic equation are called the *characteristic values.* As you know, the nature of the roots depends on the *discriminant*—the quantity under the square root. The question is, Can this discriminant be negative?

After a considerable amount of manipulation, the discriminant can be written in the form of the second line. (You can expand the squares under the radicals in both lines and verify that both lines are the same.) It is seen to be the sum of two positive quantities and, hence, *it cannot be negative.* Furthermore, by comparing B with A from the first line, it is clear that, since B is less than A, both characteristic values must be distinct negative real numbers, independent of the element values. *This is a major result.* Although we have found it to be true for a particular RC circuit, it can be verified by other methods to be true for all linear, passive RC circuits. Thus:

The characteristic values of linear passive RC circuits are distinct, real, and negative.

The proof of this result requires concepts more advanced than we will treat in this book; we will proceed as if it were proved.

Let us now go on with the solution. We have found two distinct values of s in the exponential which will satisfy the differential equation. The general solution should then have two exponentials, as follows:

$$v_2(t) = K_1 e^{s_1 t} + K_2 e^{s_2 t} \tag{8}$$

This takes care of capacitor voltage v_2. To find the other capacitor voltage, we can repeat the entire process starting from the differential equation for v_1 that you obtained in Exercise 1. The solution of this equation will have the same form as (8) but with constants different from K_1 and K_2. What is significant is that the characteristic values are the same. As an alternative, we can return to the second equation in (1)—but with $v_s = 0$—and substitute into it the solution for $v_2(t)$. The result is:

$$\begin{aligned} v_1(t) &= R_2 C_2 \frac{dv_2}{dt} + v_2 \\ &= (R_2 C_2 s_1 + 1) K_1 e^{s_1 t} + (R_2 C_2 s_2 + 1) K_2 e^{s_2 t} \end{aligned} \tag{9}$$

The magnitude of the reciprocal of each characteristic value plays the role of a time constant. The solution for each voltage consists of two decaying exponentials, each decaying at a different rate. To pursue this matter quantitatively, let's use the numerical values of the elements given in Figure 1. Using (7), the characteristic values are found to be:

$$s_1 = -1 \qquad s_2 = -6$$

With these values inserted into (8) and (9), the solutions become:

$$v_1 = 0.75K_1e^{-t} - 0.5K_2e^{-6t}$$
$$v_2 = K_1e^{-t} + K_2e^{-6t}$$

We now have general solutions for both capacitor voltages. These solutions are "general" because they are valid independent of the initial conditions, which, in this case, are the initial values of the two capacitor voltages. It is essential that the general solution contain two arbitrary constants so that, in any particular case, the constants can be adjusted to ensure that any given initial conditions are satisfied.

Suppose that the capacitors were initially charged to $v_1(0) = 5\text{ V}$ and $v_2(0) = 2\text{ V}$ when the circuit was established by means of some switching operation. When these are inserted into the preceding equations and appropriate algebraic manipulations are carried out, the solutions for the constants are found to be $K_1 = 3.6\text{ V}$ and $K_2 = 1.4\text{ V}$. Hence, the specific solutions satisfying both the differential equations and these initial conditions are:

$$v_1(t) = 2.7e^{-t} - 0.7e^{-6t}$$
$$v_2(t) = 3.6e^{-t} + 1.4e^{-6t}$$

(You should confirm these expressions by supplying the details.)

Plots of these curves are shown in Figure 2. The plot of v_2 behaves as our intuition would suggest: that is, we would expect the sum of two decaying exponentials to be like a single decaying exponential. But $v_1(t)$ behaves somewhat counterintuitively. We see that there is a slight upswing in the curve, reaching a slight peak, before it decays. The maximum in this case occurs at a time equal to approximately half the smaller time constant. After the passage of a time equal to three or four of the smaller time constants, that exponential has almost disappeared and the curve is approximated by the exponential having the larger time constant.

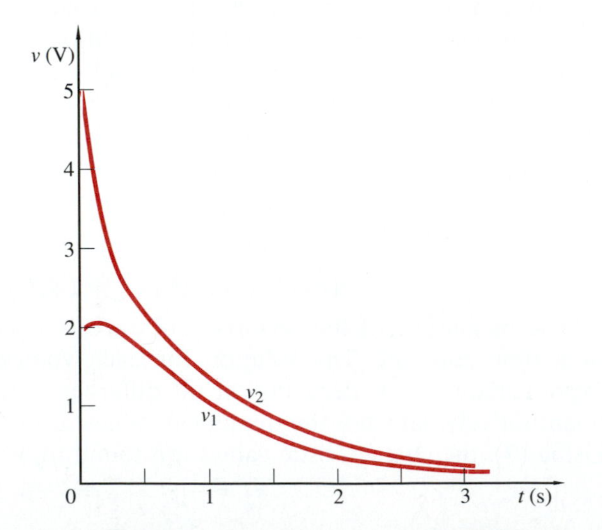

Figure 2
Plots of $v_1(t)$ and $v_2(t)$ for the example of Figure 1.

EXERCISE 2

In the example of the preceding paragraphs, confirm that there is indeed a maximum in the curve of $v_1(t)$ by differentiating v_1 and setting the derivative equal to zero. For the linear combination of two exponentials in general, show that a maximum occurs only when the coefficients of the two exponentials have opposite signs. □

2 STATE EQUATIONS

Given a circuit with two capacitors, the steps we have taken so far have been (1) to apply the circuit laws and obtain a second-order differential equation for each capacitor voltage; and (2) to solve the two equations. Up to this point we have dealt only with the natural response, assuming that there is no external excitation and that the natural response is a result of initial values of capacitor voltages (or inductor currents, when there are inductors, as we shall see shortly).

Along the way to reaching a second-order differential equation, for the circuit treated so far we first arrived at the pair of first-order differential equations in (1). These are repeated here, slightly rewritten, assuming that the circuit is source-free:

$$\begin{aligned} \frac{dv_1}{dt} + \left(\frac{G_1 + G_2}{C_1}\right)v_1 - \frac{G_2}{C_1}v_2 &= 0 \\ \frac{dv_2}{dt} - \frac{G_2}{C_2}v_1 + \frac{G_2}{C_2}v_2 &= 0 \end{aligned} \tag{10}$$

This is a pair of first-order differential equations whose variables are the capacitor voltages. Collectively, these variables are called *state variables* (*state* meaning "condition"). The interesting thing about the state variables is that, once a solution for them has been obtained, all other variables can be determined algebraically, that is, without the need to solve differential equations. In Figure 1, for example, the voltage across each resistor follows by KVL; then the resistor currents follow by Ohm's law. The capacitor currents are then obtained by KCL from the resistor currents; there is no need to obtain them from $i = C\,dv/dt$. What is true for this example is true in general.

Solution of State Equations

We already know that the solutions for the capacitor voltages are exponentials. Nevertheless, let's proceed as if this is not established and assume that $v_1(t) = K_1e^{st}$ and $v_2(t) = K_2e^{st}$, where K_1 and K_2 are to be determined. When these are inserted into the preceding equations, the result is:

$$\left[\left(s + \frac{G_1 + G_2}{C_1}\right)K_1 - \frac{G_2}{C_1}K_2\right]e^{st} = 0$$

$$\left[-\frac{G_2}{C_2}K_1 + \left(s + \frac{G_2}{C_2}\right)K_2\right]e^{st} = 0$$

As before, we argue that the exponential is not zero for any finite time; hence, each quantity in brackets multiplying an exponential must be zero. The result is two algebraic equations in two unknowns, as follows:

$$\left(s + \frac{G_1 + G_2}{C_1}\right)K_1 - \frac{G_2}{C_1}K_2 = 0$$
$$-\frac{G_2}{C_2}K_1 + \left(s + \frac{G_2}{C_2}\right)K_2 = 0 \tag{11}$$

Solve for K_1 from the first and substitute into the second:

$$\left[s^2 + \left(\frac{G_1 + G_2}{C_1} + \frac{G_2}{C_2}\right)s + \frac{G_1 G_2}{C_1 C_2}\right]K_2 = 0 \tag{12}$$

For a nontrivial solution, K_2 cannot be zero. (If $K_2 = 0$, then, according to (11), $K_1 = 0$ also.) Hence, the quantity in brackets must be zero. But this is exactly the characteristic equation we first met in (6)! From here on, everything that was done previously now applies. Indeed, how could you be surprised at this result? Would you expect the variables in a circuit to behave in two different ways just because we went about finding them in two slightly different ways?

EXERCISE 3

As an alternative approach, solve for the ratio K_2/K_1 from each of the equations in (11). Should they be equal? Verify that the same characteristic equation results. □

STATE EQUATIONS IN MATRIX FORM Let's now return to the pair of first-order differential equations in [10] with a view toward further processing them to achieve useful results. After transposing terms so that only derivatives are on one side, this pair of equations can be written in matrix form (see Appendix A) as follows:

$$\frac{d}{dt}\begin{bmatrix} v_1 \\ v_2 \end{bmatrix} = \begin{bmatrix} -\dfrac{(G_1 + G_2)}{C_1} & \dfrac{G_2}{C_1} \\ \dfrac{G_2}{C_2} & -\dfrac{G_2}{C_2} \end{bmatrix}\begin{bmatrix} v_1 \\ v_2 \end{bmatrix} \tag{13}$$

$$\frac{d}{dt}\mathbf{x} = \mathbf{A}\mathbf{x} \tag{14}$$

In (14), the vector of state variables has been labeled $\mathbf{x}$ and the matrix of coefficients has been labeled $\mathbf{A}$. (Remember, vectors and matrices are in **boldface**.) Such first-order equations whose variables are the state variables are called *state equations*. More specifically, the equations are written so that the first derivatives are on one side and everything else is on the other side.

State equations are first-order differential equations whose variables (in the present example) are capacitor voltages. (As we shall see, in general the variables include inductor currents.) In this chapter we are dealing with circuits that have only two state variables. However, the procedures we are following are applicable generally to circuits with any number of energy-storing components. Indeed, the matrix form in (14) applies to state equations of *any* order.

CHARACTERISTIC EQUATION FROM THE MATRIX FORM Let's think about the state equations in matrix form and repeat the steps in the solution of the scalar form that led from (3) to (8). We assume exponential solutions for v_1 and v_2 and substitute them into (13). As before, we argue that e^{st} can be canceled from the result, leaving equations involving the constants K_1 and K_2. Without performing all the intermediate steps here (although you are urged to do so), the result will be:

$$s\begin{bmatrix} K_1 \\ K_2 \end{bmatrix} - \mathbf{A}\begin{bmatrix} K_1 \\ K_2 \end{bmatrix} = \mathbf{0} \tag{15}$$

By expanding out this matrix equation, you will find that this is simply a rewriting of (11). Both terms in the equation contain the vector of variables K_1 and K_2; why not combine the terms? To do so means combining the coefficients of the vectors. However, $\mathbf{A}$ is a matrix, while s is a scalar. Remember that addition of two matrices requires them to be of the same order; we seem to be at an impasse. But, wait; all is not lost. Multiplying a scalar by unity does not change its value. Applying this idea to the present situation means we should multiply the scalar s by a unit matrix ($\mathbf{U}$) *of second order.* The resulting matrix can now be added to $\mathbf{A}$. The result of all this will be:

$$(s\mathbf{U} - \mathbf{A})\begin{bmatrix} K_1 \\ K_2 \end{bmatrix} = \mathbf{0} \tag{16}$$

Note that this is just a rewriting of the preceding equation; thus its expanded, or scalar, form is the same as (11).

EXERCISE 4

Find the determinant of matrix $s\mathbf{U} - \mathbf{A}$ in the preceding equation and confirm that det $(s\mathbf{U} - \mathbf{A}) = 0$ is identical with the characteristic equation in (7). □

We can demonstrate the conclusion of Exercise 4 generally by solving for the unknowns in (11)—which is the scalar form of (16)—using Cramer's rule given in (18) in Appendix A. In the present case, $\mathbf{Y}$ on the right side in the appendix is zero, so we find that:

$$K_1 = \frac{0}{\det(s\mathbf{U} - \mathbf{A})}$$

$$K_2 = \frac{0}{\det(s\mathbf{U} - \mathbf{A})}$$

Since zero over something is zero, this leads to the trivial conclusion that K_1 and K_2 are both zero, . . . *unless* det $(s\mathbf{U} - \mathbf{A}) = 0$. If that's true, the right-hand sides of the preceding equations are in the indeterminate form $\frac{0}{0}$. This still does not give a solution for the constants, but at least it allows them to be nonzero. (Remember that, to find these constants, the initial values must be invoked.)

There it is: a nontrivial solution of the state equations requires that $\det(s\mathbf{U} - \mathbf{A}) = 0$, which is the characteristic equation. Note: Although we are dealing here with the second-order case, the general form of the preceding equations applies to any order, and $\det(s\mathbf{U} - \mathbf{A}) = 0$ is the characteristic equation for a state equation of any order in (14).

Another RC Circuit

Everything done so far has concerned a single second-order *RC* circuit. It's time to take another example. As another *RC* circuit, let's consider the second-order circuit shown in Figure 3, obtained from the circuit in Figure 1 by interchanging the capacitors with the resistors.[2] Again the state variables are the capacitor voltages, but now there is no way to choose the datum node so that the capacitor voltages are the same as the node voltages. The closest we can come is to choose the datum node in the peculiar way shown. The nondatum nodes are labeled *A* and *B*, and the capacitor voltages are still labeled v_1 and v_2. (Remember that for a three-node circuit, the sum of KCL equations at any two of the nodes gives the KCL equation at the third node. Thus, no matter how the datum is chosen, the KCL equation at the datum can be obtained by adding the other two equations. Hence, there is no need for this peculiar choice of datum—except to add spice to life.)

The state equations still follow from applying KCL at the two nondatum nodes and replacing the branch currents by the *i*-*v* relations of the branches. But now care is needed in eliminating the branch voltages in favor of the desired variables, which are now the capacitor voltages. Thus, for example, the current in R_2 is $G_2(-v_s + v_1 + v_2)$. With that caution, writing node equations leads to:

$$C_1\frac{dv_1}{dt} + G_1(v_1 - v_s) + G_2(v_1 + v_2 - v_s) = 0$$

$$C_2\frac{dv_2}{dt} + G_2(v_1 + v_2 - v_s) = 0$$

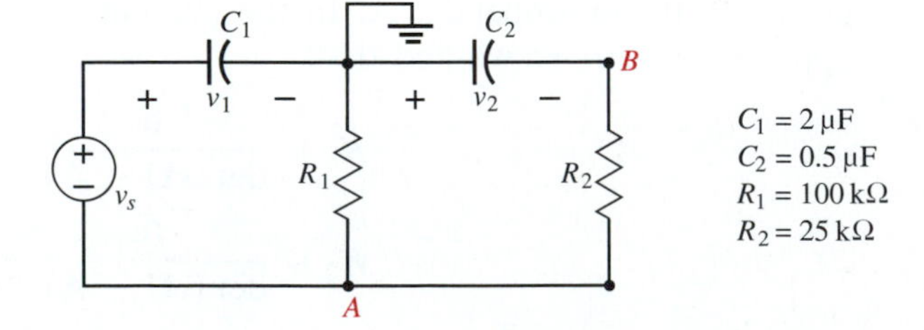

Figure 3
Circuit with two capacitors (high-pass filter).

[2] The present circuit is a *high-pass filter,* as opposed to the one in Figure 1, which is a *low-pass filter.* These filtering operations will be clarified in Chapter 13.

(You should confirm this result.) Placing this in standard matrix form and using the numerical values given in the figure yields:

$$\frac{d}{dt}\begin{bmatrix} v_1 \\ v_2 \end{bmatrix} = \underbrace{\begin{bmatrix} -25 & -20 \\ -80 & -80 \end{bmatrix}}_{\mathbf{A}}\begin{bmatrix} v_1 \\ v_2 \end{bmatrix} + \begin{bmatrix} 25 \\ 80 \end{bmatrix} v_s$$

With the coefficient matrix $\mathbf{A}$ determined, the characteristic equation is found from $\det(s\mathbf{U} - \mathbf{A}) = 0$ to be:

$$\det(s\mathbf{U} - \mathbf{A}) = \det\begin{bmatrix} s+25 & 20 \\ 80 & s+80 \end{bmatrix}$$
$$= s^2 + 105s + 400 = 0$$

Then, the approximate roots of the characteristic equation are $s_1 = -101$ and $s_2 = -4$; this again exemplifies the general principle that the characteristic values of *RC* circuits are negative real.

Reviewing the last few paragraphs, we find that the general form—including a single excitation—in which the second-order state equations can be written is:

$$\frac{d}{dt}\begin{bmatrix} x_1(t) \\ x_2(t) \end{bmatrix} = \begin{bmatrix} a_{11} & a_{12} \\ a_{21} & a_{22} \end{bmatrix}\begin{bmatrix} x_1(t) \\ x_2(t) \end{bmatrix} + \begin{bmatrix} b_{11} \\ b_{21} \end{bmatrix} v_s(t)$$
$$\frac{d}{dt}\mathbf{x} = \mathbf{A}\mathbf{x} + \mathbf{B}\mathbf{v} \tag{17}$$

We used $\mathbf{v}$ to designate the source vector, although this is not the symbol that is used generally.

Low-Pass Op-Amp Circuit

As a final example of a second-order *RC* circuit, consider the one shown in Figure 4. The op amp in the circuit is assumed to be ideal. (You may wonder at the lack of units on the components and at their unrealistic values—if they are

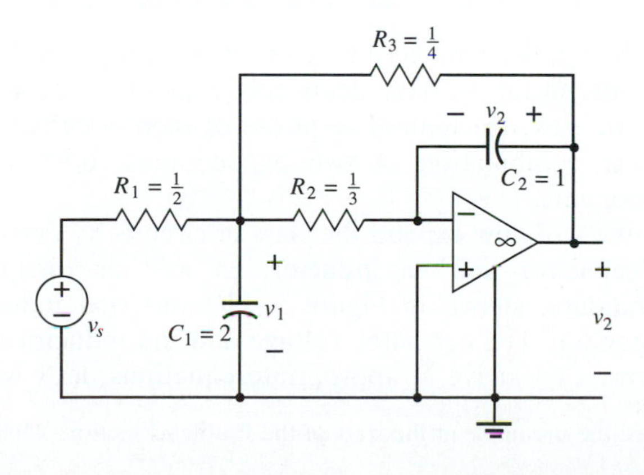

Figure 4
Low-pass op-amp circuit.

SI. They are *not* SI units; they are normalized values and, hence, dimensionless.[3])

Using the properties of an ideal op amp (zero input currents and a virtual short circuit across the input terminals), the voltage across R_2 is the same as the voltage v_1 of capacitor C_1; the current in R_2 is the negative of the current in C_2, which is $C_2\, dv_2/dt$. Using the latter result and the result of applying KCL at the node where the three resistors and C_1 are connected (and then inserting the i-v relationships), we get:

$$2\frac{dv_1}{dt} + 2(v_1 - v_s) + v_1 + 4(v_1 - v_2) = 0$$

$$\frac{dv_2}{dt} + v_1 = 0$$

(Carry out the details.) Collecting terms and rewriting, we get:

$$\frac{dv_1}{dt} = -4.5v_1 + 2v_2 + v_s$$

$$\frac{dv_2}{dt} = -3v_1$$

This can now be placed in matrix form, from which the characteristic equation can be easily obtained:

$$\frac{d}{dt}\begin{bmatrix} v_1 \\ v_2 \end{bmatrix} = \begin{bmatrix} -4.5 & 2 \\ -3 & 0 \end{bmatrix}\begin{bmatrix} v_1 \\ v_2 \end{bmatrix} + \begin{bmatrix} 1 \\ 0 \end{bmatrix} v_s$$

$$\det(s\mathbf{U} - \mathbf{A}) = \begin{vmatrix} s + 4.5 & -2 \\ 3 & s \end{vmatrix} = s^2 + 4.5s + 6$$

$$s = -2.25 \pm j0.97$$

We find something very interesting here: although the circuit is an RC circuit, the presence of the op amp *causes the characteristic values to be complex, not real.* We will not pursue the ramifications of this observation here but defer it to the next section.

3 RESISTOR-CAPACITOR-INDUCTOR CIRCUITS

Although we claimed a certain amount of generality in some of the preceding results, so far we have dealt only with some specific second-order RC circuits. As we saw, the natural response of such circuits that are passive consists of a linear combination of two exponentials, both with distinct, negative, real exponents.

We will now expand the class of circuits by examining those containing both a capacitor and an inductor, as well as dissipative components. A first candidate, shown in Figure 5, contains one inductor, one resistor, and one capacitor. The capacitor voltage and the inductor current have been explicitly shown. To arrive at appropriate equations, let's write KVL around the single

[3] See the preamble at the start of the Problems section, which explains the parameter values.

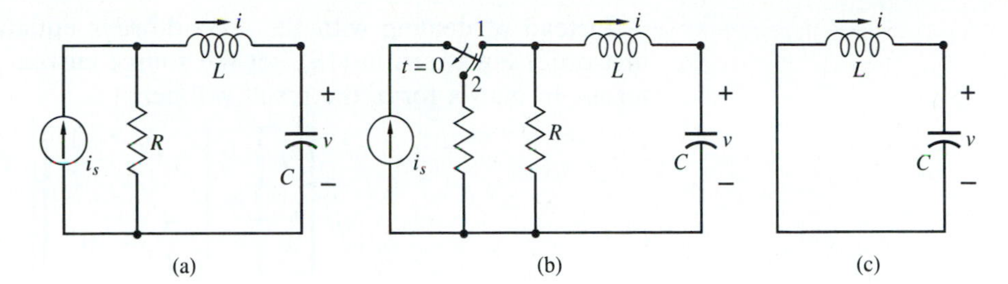

Figure 5
Candidate *RLC* circuit for analysis.

loop, at the same time inserting the v-i relations of the resistor and inductor but not the capacitor. Then let's write KCL at the junction between C and L and insert the i-v relation of the capacitor. The results will be:

$$L\frac{di}{dt} + Ri + v = Ri_s$$
$$C\frac{dv}{dt} - i = 0 \tag{18}$$

This is a set of two first-order differential equations in two unknowns, i and v.[4] They can easily be rewritten to place derivatives only on one side of the equation, in the standard form of state equations. For completeness, however, let's first convert them into second-order equations. This is easily done by solving the second equation for i and inserting into the first, or solving the first equation for v and inserting into the second. The result of these operations (which you should confirm) is the following:

$$\frac{d^2v}{dt^2} + \frac{R}{L}\frac{dv}{dt} + \frac{1}{LC}v = \frac{R}{LC}i_s$$
$$\frac{d^2i}{dt^2} + \frac{R}{L}\frac{di}{dt} + \frac{1}{LC}i = \frac{R}{L}\frac{di_s}{dt} \tag{19}$$

As expected, these are linear, nonhomogeneous second-order differential equations with constant coefficients. They differ from each other only in the excitation function on the right-hand side. Hence, we would expect that the forms of the natural solutions would be identical.

State Equations

We shall now assume an initial current in the inductor and an initial voltage across the capacitor, but no source. Such a situation can easily be achieved by the switching arrangement in Figure 5b, for example. The switch, which has been in position 1 for some time, is switched to position 2 at $t = 0$. The resulting circuit has no source but does have a nonzero initial inductor current and capacitor voltage.

[4] These variables—the capacitor voltage and the inductor current—are the only variables in a circuit whose initial values can be specified independently, the *state variables.* It makes sense, therefore, to write the circuit equations in such a way that other variables are eliminated, leaving only state variables in the differential equations.

Instead of dealing with the second-order equations, we will return to the first-order equations in (18), set the source current to zero, and rearrange the terms. In matrix form, the result will be:

$$\frac{d}{dt}\begin{bmatrix} i \\ v \end{bmatrix} = \begin{bmatrix} -\frac{R}{L} & -\frac{1}{L} \\ \frac{1}{C} & 0 \end{bmatrix}\begin{bmatrix} i \\ v \end{bmatrix} \tag{20}$$

$$\frac{d\mathbf{x}}{dt} = \mathbf{Ax}$$

This is the matrix form of the state equations. The coefficient matrix $\mathbf{A}$ is readily identified. The characteristic equation is obtained from det $(s\mathbf{U} - \mathbf{A}) = \mathbf{0}$. Thus:

$$\det(s\mathbf{U} - \mathbf{A}) = \det\begin{bmatrix} s + \frac{R}{L} & \frac{1}{L} \\ -\frac{1}{C} & s \end{bmatrix} = s^2 + \frac{R}{L}s + \frac{1}{LC} = 0 \tag{21}$$

Characteristic Values

As expected, the characteristic equation is a quadratic, so it has two roots; they can be obtained by the quadratic formula:

$$\begin{aligned} s_{1,2} &= -\frac{R}{2L} \pm \sqrt{\left(\frac{R}{2L}\right)^2 - \frac{1}{LC}} \\ &= -\frac{R}{2L}\left(1 \pm \sqrt{1 - \frac{4L}{R^2C}}\right) \end{aligned} \tag{22}$$

As you already know, the nature of the roots depends on the discriminant. When the discriminant is positive or negative, the two roots are distinct—real when the discriminant is positive, and complex when the discriminant is negative. In the special case when the discriminant is 0, the two roots are equal. This case requires some special treatment which is more easily carried out after the material in Chapter 14 is taken up. Hence, we will temporarily postpone consideration of this special, critical case.[5]

As we found in the case of the second-order *RC* circuit, there are two characteristic values, so, when the characteristic values are distinct, the natural solutions for v and i will contain two exponentials. However, when the solution for either v or i alone is obtained, that of the other will follow from one of the equations of (18). Thus, suppose we write the natural solution for v as a linear

[5] It might be argued that this case is not worth pursuing, anyway. For the discriminant to be zero, we would require that $L = R^2C/4$ exactly. Since practical component values are not usually accurate to better than 5 or 10%, even if values in the model are selected to satisfy this condition, it is highly improbable that the values needed for the critical case will be achieved when the circuit is actually built. Nevertheless, the critical case constitutes a boundary between distinct real characteristic values and imaginary ones.

combination of two exponentials. Then the solution for i is obtained from the second equation in (18). The two solutions will be:

$$\begin{aligned} v(t) &= K_1 e^{s_1 t} + K_2 e^{s_2 t} \\ i(t) &= C s_1 K_1 e^{s_1 t} + C s_2 K_2 e^{s_2 t} \end{aligned} \tag{23}$$

We now have formal solutions for both the capacitor voltage and the inductor current.

The expressions for the characteristic values in (22) are valid whether the values are real or complex. However, the shapes of the function $v(t)$ will be different for the two cases, so we will take them up in turn.

REAL CHARACTERISTIC VALUES The characteristic values in (22) are real when:

$$\begin{aligned} \left(\frac{R}{2L}\right)^2 &> \frac{1}{LC} \\ R &> 2\sqrt{\frac{L}{C}} \end{aligned} \tag{24}$$

For these values of R, the quantity under the square root in the last line of (22) is less than 1. Hence, *when the characteristic values are real, they will always be negative.*

When the characteristic values are distinct and real, the natural response for the present case will not be different from the case of a second-order RC circuit, so there isn't much new that you can learn. However, for practice let's choose the following numerical values: $L = 1$ mH, $C = 2.5\ \mu$F, and $R = 50\ \Omega$. The inequality in (24) is satisfied and the characteristic values are found from (22) to be:

$$\begin{aligned} s_1 &= -40{,}000\ \text{s}^{-1} \\ s_2 &= -10{,}000\ \text{s}^{-1} \end{aligned}$$

With these values inserted, the solutions in (23) become:

$$\begin{aligned} v(t) &= K_1 e^{-40{,}000t} + K_2 e^{-10{,}000t} && (t \text{ in s}) \\ &= K_1 e^{-t/25} + K_2 e^{-t/100}\ \text{V} && (t \text{ in } \mu\text{s}) \\ i(t) &= -0.1 K_1 e^{-t/25} - 0.025 K_2 e^{-t/100}\ \text{A} && (t \text{ in } \mu\text{s}) \end{aligned}$$

These expressions are in the familiar $e^{-t/T}$ form of the first-order case. (Pick some values of t in seconds and verify that the first line in the expression for $v(t)$ gives the same values of exponents as the second line does for the corresponding number of microseconds.)

The conclusion is that when the characteristic values are distinct and real, the natural solution of the second-order differential equation (linear, homogeneous, time-invariant) is like that of the first-order equation—except that there are two time constants and two exponentials. Recall the discussion, in Section 1, on RC circuits, of the monotonic decay as opposed to the decay after an initial rise. In the present case, in order for the latter to apply, the signs of K_1 and K_2 must be different. Of course, the values of these two constants will depend on the initial values.

COMPLEX CHARACTERISTIC VALUES When the inequality in (24) is reversed, the characteristic values are distinct and complex. In this case, from (20), they can be written as:

$$\begin{aligned} s_{1,2} &= -\frac{R}{2L} \pm j\sqrt{\frac{1}{LC} - \left(\frac{R}{2L}\right)^2} \\ &= \alpha \pm j\omega_d \end{aligned} \tag{25}$$

Two new parameters have been defined here: α is called the *damping coefficient* and ω_d the *damped frequency,* for reasons which will become clear shortly. For convenience, we will define one other parameter also: $\omega_0 = 1/(LC)^{1/2}$. For future reference, these parameters are collected here:

$$\begin{aligned} \omega_0 &= \frac{1}{\sqrt{LC}} \\ \alpha &= -\frac{R}{2L} \\ \omega_d &= \sqrt{\frac{1}{LC} - \left(\frac{R}{2L}\right)^2} \\ &= \sqrt{\omega_0^2 - \alpha^2} \end{aligned} \tag{26}$$

With the values in (25) for s_1 and s_2 inserted into (23), the solution for $v(t)$ becomes:

$$\begin{aligned} v(t) &= K_1 e^{(\alpha + j\omega_d)t} + K_2 e^{(\alpha - j\omega_d)t} \\ &= e^{\alpha t}(K_1 e^{j\omega_d t} + K_2 e^{-j\omega_d t}) \\ &= e^{\alpha t}(A \cos \omega_d t + B \sin \omega_d t) \\ &= V e^{\alpha t} \cos(\omega_d t + \theta) \end{aligned} \tag{27}$$

There will be similar expressions for $i(t)$. A great deal of mathematical manipulation went into all this. The first line is clear; the complex characteristic values from (25) were substituted into the solution in (23). The second step should also be clear; all we did was use the law of exponents $e^{(x+y)} = e^x e^y$ and then factor the common exponential from the two terms. The third step uses Euler's theorem ($e^{jx} = \cos x + j \sin x$) to express each exponential with an imaginary exponent in terms of cosines and sines; then the cosine terms are collected together and so are the sine terms. This is a bit more work. The multipliers of these terms have been named A and B. (You can find expressions for A and B in terms of K_1 and K_2 if you want.)

Finally, a linear combination of a cosine and a sine having the same argument can be written as a cosine function at a nonzero initial angle. (It can also be written as a sine function at a different initial angle.) The best way to determine the equivalences among A, B, V, and θ is to expand the cosine function in the last step using the trigonometric identity relating to the cosine of the sum of two angles. (See inside front cover.) You can carry out this step if you want.

What is significant is that, in each of the lines in (27), there are two unknown constants whose values are yet to be determined (K_1 and K_2, A and

B, or V and θ). You might think that there are two other constants: α and ω_d. You would be right that these are constants for a given circuit, but these two are not in the same category as the others; their values are known in terms of the component values in the circuit.

At this point, let's again consider a numerical example. This time let L and C have the same values as before, but reduce the value of R to 20 Ω. Then, from (25), the characteristic values are found to be:

$$s_{1,2} = -10^4 \pm j10^4\sqrt{3}$$

With these values, the last line in the solution for $v(t)$ in (27) becomes:

$$v(t) = Ve^{-10,000t}\cos(10,000\sqrt{3}\,t + \theta)$$

For simplicity in all of the preceding development, we did not carry along the solution for $i(t)$ in all its various forms. As a final step now, we will obtain an expression for the inductor current corresponding to the last form of the solution for $v(t)$. From $i = C\,dv/dt$:

$$\begin{aligned} i(t) &= C\frac{dv}{dt} \\ &= -0.025Ve^{-10,000t}[\cos(10^4\sqrt{3}t + \theta) + \sqrt{3}\sin(10^4\sqrt{3}t + \theta)] \end{aligned}$$

Again we find that, in the expressions for v and i, a total of two constants remains to be determined from the initial energy storage in the circuit: V and θ.

What happens if the circuit we are dealing with is lossless, that is, if the resistance is zero, as shown in Figure 5c? Then, $\alpha = 0$ and $\omega_0 = \omega_d$. In this case, the characteristic values are purely imaginary: $s_{1,2} = \pm j\omega_0$. The exponential $e^{-\alpha t}$ now disappears from (27), and a natural response is a free, undamped oscillation with an angular frequency ω_0. What could be more natural than to refer to ω_0 as the *undamped frequency*? Looking back at (26), we see that the damped frequency is always less than the undamped frequency.

In any physical circuit, resistance will inevitably be present, although it can be made small by proper design. Suppose it were possible to buy a negative resistance. Then one could place it in series with R in Figure 5, or elsewhere in the circuit, and thus cancel the effect of R. We have seen in earlier chapters that the effect of a negative resistance can be achieved with a negative converter or by an appropriate controlled source. The only caveat is that the value of negative resistance is critical. A little too much of it, and the circuit will become unstable, resulting in exponentially increasing voltages and currents—at least long enough to destroy the circuit.

Satisfying Initial Conditions

At the instant when the circuit is established, there might be a nonzero capacitor voltage and/or a nonzero inductor current. Let's say these are $v(0) = V_0$ and $i(0) = I_0$.[6] These are the values that (23) reduces to when we set

[6] Recall that the initial capacitor voltage and inductor current are presumably known just before the circuit is established at $t = 0$, namely, $v(0-)$ and $i(0-)$. By the principles of conservation of charge and conservation of flux linkage, these are equal to the respective 0+ values, labeled $v(0)$ and $i(0)$.

$t = 0$. Thus:

$$\begin{aligned} v(0) &= K_1 + K_2 = V_0 \\ i(0) &= Cs_1K_1 + Cs_2K_2 = I_0 \end{aligned} \tag{28}$$

These are two linear equations in the two constants K_1 and K_2 and can readily be solved in terms of the initial values V_0 and I_0. (You should carry out the solution.)

The preceding equations apply whether the characteristic values are real or complex. However, for complex characteristic values, one or another of the other forms in (27) may be more convenient. Thus, if we use the last form for $v(t)$ (and also in $i(t) = C\,dv/dt$), we will get:

$$\begin{aligned} v(0) &= V \cos\theta = V_0 \\ i(0) &= CV[-\omega_d e^{\alpha t} \sin(\omega_d t + \theta) + \alpha e^{-\alpha t} \cos(\omega t + \theta)]_{t=0} \\ &= CV[\alpha \cos\theta - \omega_d \sin\theta] = I_0 \end{aligned} \tag{29}$$

Again, this is a pair of equations, this time in the two unknowns V and θ. However, since θ appears in terms of its sine and cosine, now the solution is more complicated. As an exercise, you should carry out the details of the solution and confirm the following:

$$V^2 = V_0^2 + \frac{1}{\omega_d^2}\left(\frac{I_0}{C} - \alpha V_0\right)^2 \tag{30a}$$

$$\cos\theta = \frac{V_0}{V} \tag{30b}$$

$$\sin\theta = \frac{1}{V\omega_d}\left(\alpha V_0 - \frac{I_0}{C}\right) \tag{30c}$$

There seems to be a confused mass of different quantities in these expressions. To cut through the confusion, note that there are, besides C, three different pairs of quantities: (1) α and ω_d are combinations of the circuit parameters in accordance with (26); (2) V_0 and I_0 are initial values; and (3) V and θ are the constants that characterize the general solution—their values are dependent on initial conditions and circuit parameters, as evident in (30).

Care must be exercised in using these equations. Equation (30a) is used to determine V without problem. Once V is determined, the cosine follows easily from (30b). The cosine is an even function, however, so the angle is $\theta = \pm\cos^{-1}(V_0/V)$. Which angle should be used? The answer is found by checking for the sign from the expression for $\sin\theta$. With α negative and all other quantities positive, the sign of the angle must be negative. In other cases, the signs of V_0 and I_0 will influence the answer. You might ask, Why not forget about the expression for the cosine and use only the one for the sine? The answer is pure laziness! The cosine is simpler; and all we have to do in the expression for the sine is just to check on the sign of the angle.

Graphical Portrayal of Natural Response

What has been done so far can be summarized as follows. Given a circuit with two energy-storing components, we have constructed the state equations.

(In the preceding case, their variables are the capacitor voltage and inductor current.) For completeness we have also constructed a second-order differential equation satisfied by each state variable. From the state equations we have constructed the characteristic equation whose roots are the characteristic values. These values may be real and negative or a pair of complex conjugates. General solutions have been obtained for each of the state variables. When the characteristic values are real (and distinct), the solution consists of two decaying exponentials. When the characteristic values are complex, the solution consists of a sinusoid multiplied by a decaying exponential. In both cases, these solutions contain two unknown constants whose values are to be determined from the initial condition—the initial capacitor voltage and inductor current.

A plot of the solution against time would be instructive. Since we have already seen such plots for the case of real characteristic values, we will here limit ourselves to the complex case. The equations for the constants in the solution in terms of the initial values were given in (30) for complex characteristic values. Suppose the initial values are as follows:

$$V_0 = 3\text{ V}$$
$$I_0 = 150\text{ mA}$$

Using numerical values of the parameters previously given and these initial values, (30) gives the following values for the constants:

$$V = 6 \qquad \theta = -60° \qquad (\cos\theta = 0.5, \sin\theta = -0.866)$$

With these values of V and θ, the expressions for $v(t)$ and $i(t)$ become:

$$v(t) = 6e^{-10{,}000t}\cos(10{,}000\sqrt{3}t - 60°)$$
$$i(t) = -0.15e^{-10{,}000t}[\cos(10^4\sqrt{3}t - 60°) + \sqrt{3}\sin(10^4\sqrt{3}t - 60°)]$$

(Confirm that these expressions yield the correct initial values: $v(0) = 3$ V and $i(0) = 0.15$ A.)

Curves of $v(t)$ and $i(t)$ for complex characteristic values can easily be plotted. (Do it!) In contrast with the curves for distinct real characteristic values, which decay exponentially, these curves are oscillatory, but they still decay as time goes on. The term *damping* is used to describe this decaying feature: the case of complex characteristic values is called the *underdamped* case, while the case of distinct real characteristic values is called the *overdamped* case. (The case of coincident real characteristic values, which we have postponed considering, is called the *critically damped* case, lying as it does between the other two cases.)

4 THE NATURAL FREQUENCIES OF LINEAR NETWORKS

Let us now review the development in this chapter and pick out the salient ideas, separated from the mass of algebraic detail. The natural response of a

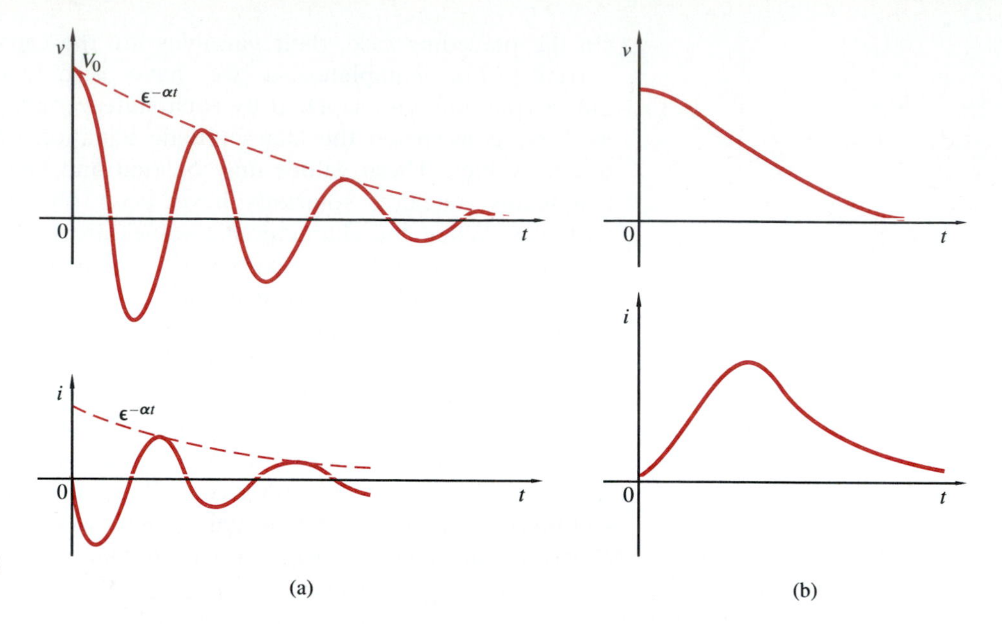

Figure 6
Natural response: (a) damped oscillatory case; (b) nonoscillatory case.

circuit with two energy-storing elements, a capacitor and an inductor, consists of a sum of two exponential functions of time. Depending on the values of parameters, the exponents (characteristic values) may be either real or complex.

When the characteristic values are real, the response is the sum of two decaying exponentials (nonoscillatory case). When the characteristic values are complex, they are a pair of conjugates; the exponentials can then be expressed in terms of sinusoidal functions multiplied by a real decaying exponential. The response oscillates, in this case while decaying exponentially. In the special case when $R = 0$, there is no damping and the response is purely oscillatory. (In this case, does it make sense to refer to the natural response as "transient"?)

Sketches of possible voltage and current waveforms are shown in Figure 6. In the damped oscillatory case, the exponential $e^{-\alpha t}$ forms an "envelope" which (what else?) envelops the oscillations.

Suppose that the resistance in the circuit in Figure 5 is varied from 0 up to a high value. The response at first will be purely oscillatory, the angular frequency being ω_0. As R is increased, the frequency of oscillation, ω_d, will change, as seen in (26). As the resistance goes through the critical value for which $\alpha = \omega_0$, the response will change to a nonoscillatory character. Instead of changing our terminology, we can still speak of a "frequency," even though this term with its usual meaning is no longer descriptive of the response behavior.

Nevertheless, if we do think of any value whatsoever of the quantity s as being a "frequency," a door presenting a whole new vista of exciting possibilities will be opened. This may not be apparent to you immediately, but as we proceed, the implications of this process of thought and the power of the techniques which can be brought to bear on circuit problems with this point of view will be amply demonstrated.

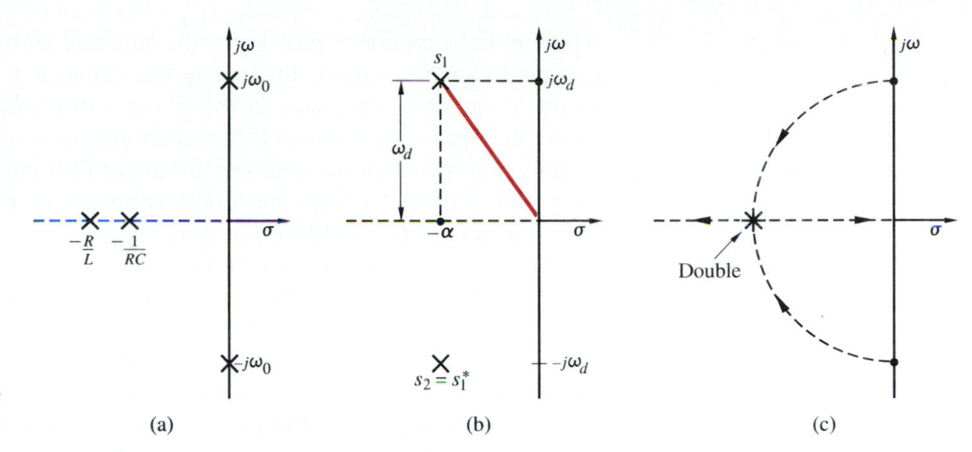

Figure 7
Natural frequencies in the complex-frequencyplane.

Thus, let's use the name *natural frequencies* for the characteristic values obtained as the roots of the characteristic equation of a circuit. Since these values are generally complex, we further call them the *complex natural frequencies.* It seems strange, if not a downright fantasy, to imagine such a thing as a "complex frequency." What can it possibly mean?, you might ask. What it means is that we are giving the name *complex natural frequency* to the roots of a certain polynomial which is characteristic of circuits containing energy-storing elements; right now, that's all.

The variable in terms of which this polynomial is written is s; hence, it is natural to refer to s as the *complex-frequency variable.* For the preceding circuit, we found two specific values of this variable, s_1 and s_2, which we called the *natural frequencies* of the circuit. In the preceding chapter we found circuits with a single energy-storing element to be characterized by a first-order polynomial which, of course, has but a single root. In the first-order case the natural frequency is a real number. But real numbers are special cases of complex numbers, so here also there is a "complex" frequency whose imaginary part happens to be zero.

These ideas can be further clarified by means of diagrams of the complex plane shown in Figure 7. Any point in the plane represents a value of s. Hence, it is called the *complex-frequency plane,* or *s-plane.* The variable s can be written in terms of its real and imaginary parts as:

$$s = \sigma + j\omega \qquad \sigma = \text{Re}(s) \qquad \omega = \text{Im}(s)$$

The axes in the diagram are labeled in terms of these quantities. Real numbers are represented by points on the σ axis and imaginary numbers are represented by points on the $j\omega$ axis. Thus, real numbers $s = -1/RC$ and $s = -R/L$, representing the natural frequencies of first-order RC and RL circuits, are illustrated by the crosses on the negative real axis in Figure 7a. Similarly, the natural frequencies $s = \pm j\omega_0$ in the undamped oscillatory case are shown by crosses on the imaginary axis.

In the damped oscillatory case, the natural frequencies have both real and imaginary parts. The two complex conjugate points $s_{1,2} = -\alpha \pm j\omega_0$ are marked by crosses in Figure 7b. The real part shows how fast the response dies out, whereas the imaginary part gives the angular frequency of oscillation. As the damping changes, say, by increasing the value of R from 0 on up, the locations of the complex frequencies in the plane will change, as shown by the dashed curve in Figure 7c, from purely imaginary, to complex, and then to two real values, one less than ω_0 and one greater. (This curve is referred to as the *root locus,* an idea which finds important applications in control systems. We won't give it any further consideration here.)

One special situation—that of two equal characteristic values—has not been considered in the preceding development. This is the boundary between the oscillatory (said to be *underdamped*) case and the nonoscillatory (said to be *overdamped*) case; hence, it is referred to as the *critically damped* case as already identified. It is represented by the double crosses on the negative real axis of the root locus in Figure 7c. The analytical form of the solution in (23) does not apply in this case. Nevertheless, the *qualitative* behavior of the response will be much the same as that of the nonoscillatory case. We shall postpone examining the details of the critically damped case, since it would not add significantly to our understanding of the natural behavior of circuits at this time. We will return to it in Chapter 14.

5 OTHER SECOND-ORDER *RLC* CIRCUITS

So far, we have treated only one particular *RLC* circuit that resulted in complex characteristic values. To gain some further experience and to strengthen what you have learned so far, we will introduce some other second-order circuits, find their state equations, and examine the properties of their characteristic values.

Parallel Resonant Circuit

As a first candidate, we will examine what is called the *parallel resonant* circuit shown in Figure 8a. It has three components, one of each kind, all in parallel. An excitation is shown, although we will consider only the natural response at this time. (This is to be compared with the previous circuit, in which—in the source-free case—the three components were all connected in series: a *series resonant* circuit.)

To arrive at the state equations, we use all available tools: KCL, KVL, and

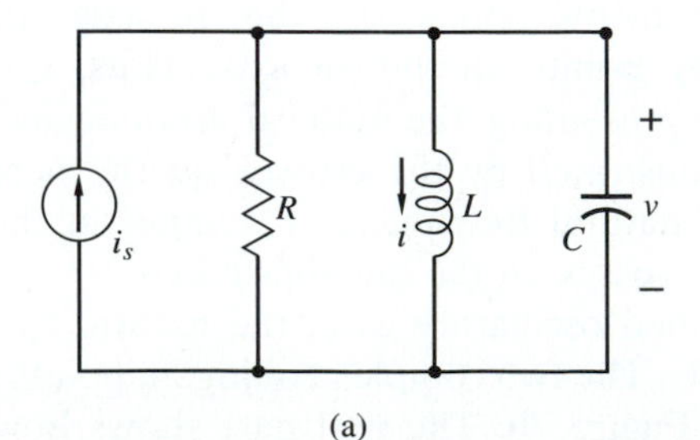

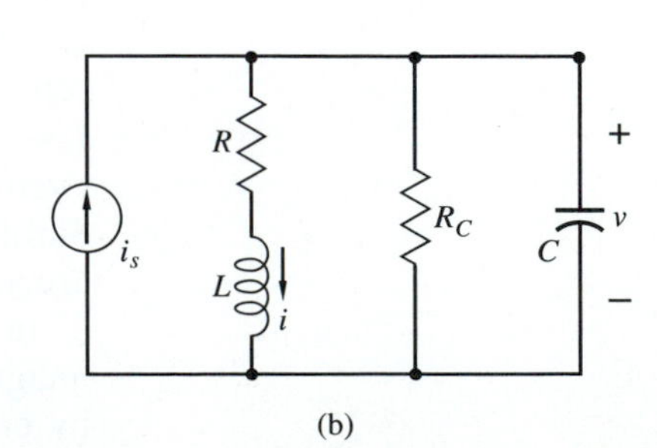

Figure 8
Parallel resonant circuit.

the component v-i relationships in a specific order. Thus, let's write KCL at the nondatum node, inserting the i-v equations of R and C but not L. Then we apply KVL around the loop consisting of L and C while inserting the v-i relationship for L. You should confirm that, with the source missing, the result of all this is the following:

$$C\frac{dv}{dt} + Gv + i = 0$$
$$L\frac{di}{dt} - v = 0 \tag{31}$$

These equations are exactly the duals of (18), as we should expect, since the two circuits are each other's dual. In matrix form they become:

$$\frac{d}{dt}\begin{bmatrix} v \\ i \end{bmatrix} = \begin{bmatrix} -\frac{G}{C} & -\frac{1}{C} \\ \frac{1}{L} & 0 \end{bmatrix}\begin{bmatrix} v \\ i \end{bmatrix} \tag{32}$$
$$\frac{d\mathbf{x}}{dt} = \mathbf{A}\mathbf{x}$$

There is no need to repeat all the steps of (1) assuming an exponential solution, (2) substituting into the equations, (3) eliminating the exponential, and (4) ending with a set of equations in which the unknowns are the constants multiplying the exponentials. Indeed, once the differential equations are obtained in the form of (32) with the matrix $\mathbf{A}$ determined, the characteristic equation is obtained by setting det $(s\mathbf{U} - \mathbf{A})$ to zero:

$$\det(s\mathbf{U} - \mathbf{A}) = \det\left(\begin{bmatrix} s & 0 \\ 0 & s \end{bmatrix} - \begin{bmatrix} -\frac{G}{C} & -\frac{1}{C} \\ \frac{1}{L} & 0 \end{bmatrix}\right)$$
$$= \det\begin{bmatrix} s + \frac{G}{C} & \frac{1}{C} \\ -\frac{1}{L} & s \end{bmatrix}$$
$$= s^2 + \frac{G}{C}s + \frac{1}{LC} = 0$$

Notice that this characteristic equation can be obtained from the one for Figure 5c by interchanging the dual quantities: L with C, C with L, and R with G.

All other considerations pertaining to the solution that were examined for Figure 4c would follow: the nature of the characteristic values (real or complex), the nature of the response (exponentially decaying or oscillatory), etc.

Parallel Circuit with Leaky Capacitor

As we shall see in Chapter 11, resonant circuits are immensely useful in many applications, particularly in communications. What makes these circuits useful is the presence of L and C. We would be better off in some applications if the resistor weren't there at all. However, a coil of wire, of which the inductor is a model, inherently includes some resistance in series with the inductance. Even a capacitor will have some "leakage," which can be represented by a resistor in parallel with the capacitor, although a real capacitor is more nearly ideal than a real inductor. Thus, if a real inductor and a real capacitor were connected in parallel, the model would include a resistor in series with L and another in parallel with C.

Such a model is shown in Figure 8b. The approach to arriving at state equations should be clear by now. KCL is applied at one node of the capacitor. All currents, except the inductor current, are replaced using the component i-v equations. Then KVL is applied around a loop containing C and L. All voltages, except for the capacitor voltage, are eliminated from the result by component v-i relations.

The two resulting equations are the state equations. The details are left for you to carry out; the result in matrix form is:

$$\frac{d}{dt}\begin{bmatrix} v \\ i \end{bmatrix} = \begin{bmatrix} -\dfrac{1}{R_C C} & -\dfrac{1}{C} \\ \dfrac{1}{L} & -\dfrac{R}{L} \end{bmatrix}\begin{bmatrix} v \\ i \end{bmatrix} + \begin{bmatrix} \dfrac{1}{C} i_s \\ 0 \end{bmatrix}$$

Notice that a vector representing the source is also included. This would be needed to find the complete response and is included here only for completeness, since so far we have not dealt with the forced response. Once the coefficient matrix $\mathbf{A}$ is determined, the characteristic equation can be found from det $(s\mathbf{U} - \mathbf{A}) = \mathbf{0}$. This will be left for you to carry out as an exercise.

Integrodifferential Equations: An Afterthought

We took a certain approach in arriving at the equations characterizing the voltage and current in an RLC circuit. We selected the order of applying KCL, KVL, and the v-i relations of the components in such a way that the variables in the final differential equations are only the state variables—the capacitor voltage and the inductor current. That is, we did not strictly follow the procedure of loop equations or node equations in which all the final variables are loop currents or node voltages only. What happens to the analysis if we *do* write loop equations or node equations instead?

Look back at Figure 5b with the source removed; it contains a single loop whose loop current can be identified with the inductor current i. If we write a loop equation around this loop, the result will be:

$$L\frac{di}{dt} + Ri + \frac{1}{C}\int i(t)\,dt = 0 \tag{33}$$

This is not a differential equation but an *integrodifferential* equation. On the other hand, it is just a single equation instead of two first-order differential equations.

Note that, if this equation is differentiated, it becomes identical with the second equation in (19) with the source absent.

To solve (33), we assume an exponential solution $i(t) = Ke^{st}$, as before, and substitute into this equation. The result will be:

$$\left(Ls + R + \frac{1}{Cs}\right)Ke^{st} = 0$$

By the same kind of reasoning used before, we conclude that the quantity within parentheses must be zero. The rewritten result is:

$$s^2 + \frac{R}{L}s + \frac{1}{LC} = 0$$

This is exactly the characteristic equation in (21)! Hence, the same characteristic values as before will result. Can you really be surprised at that?

EXERCISE 5

Follow the procedure just outlined using the circuit in Figure 8a. However, this time use node equations. Again, a single integrodifferential equation, this time in a voltage variable, should result. □

It may sometimes appear more convenient to follow the approach just outlined in which the resulting equations are not differential equations but integrodifferential equations. No new information is provided by this approach, however, and no particular simplifications are introduced by it. It also lacks the generalizability that is a property of state equations. When state equations are written in matrix form, for example, they apply for *all* orders, not just first or second order. Similarly, the characteristic equation $\det(s\mathbf{U} - \mathbf{A}) = \mathbf{0}$ applies for all orders.

In later chapters, when we are interested in the steady-state response, the integrodifferential equations discussed in this section might be useful. It is, therefore, a good idea to include them in your repertoire.

6 THE COMPLETE RESPONSE

In all the circuits considered so far, we assumed that there was no external excitation. This was done in order to concentrate on the natural response without being distracted by the details of calculating the forced response. We are now ready to tackle the overall problem.

We will start with a variation of the circuit in Figure 5. As shown in Figure 9, the accompanied current source in Figure 5 has been replaced by its

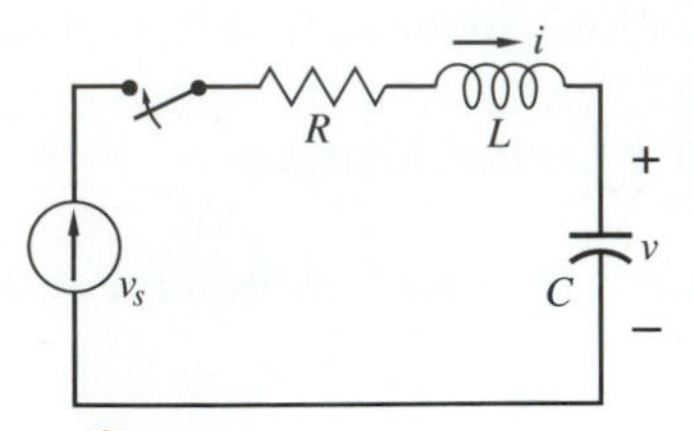

Figure 9
A series *RLC* circuit excited by a source.

Thévenin equivalent. We assume that the circuit is established by closing a switch at $t = 0$. To concentrate on the effects of the excitation, let's assume that the initial values are zero. Nothing of importance will be different if that is not the case.

The approach to determining the complete response to the excitation is the same as that used in first-order circuits:

1. Find the general solution to the homogeneous differential equations, which, in the second-order case, will contain two unspecified constants; this is the natural response.

2. Find a particular solution to the nonhomogeneous equations. This is also called an *equilibrium response.*

3. Add the two to obtain the general overall response, which will still include the two unspecified constants.

4. Apply the initial conditions in order to determine the values of the constants.

This chapter has so far been devoted to the first step. Our present task is step 2. As it was in the first-order case, we expect the forced response—a particular solution of the nonhomogeneous equations—to resemble the excitation. For example, if the excitation is a constant, we expect the forced response to be a constant; if the excitation is a sinusoid, we expect the forced response to be a sinusoid.

Step Response

To start, let's assume that the source in Figure 9 is a battery having a voltage V_b. To describe the source over all time we can take v_s to be $v_s = V_b u(t)$, where $u(t)$ is a step function. The appropriate differential equations and their natural solution were given in (18) through (23). The state equations, including the excitation, are repeated here:

$$\frac{d}{dt}\begin{bmatrix} i \\ v \end{bmatrix} = \begin{bmatrix} -\dfrac{R}{L} & -\dfrac{1}{L} \\ \dfrac{1}{C} & 0 \end{bmatrix}\begin{bmatrix} i \\ v \end{bmatrix} + \begin{bmatrix} \dfrac{1}{L} \\ 0 \end{bmatrix} u(t) \tag{34}$$

$$\frac{d\mathbf{x}(t)}{dt} = \mathbf{A}\mathbf{x}(t) + \mathbf{f}(t)$$

The vector $\mathbf{f}(t)$ in the last line is the one that includes the step function in the preceding line.

The natural solution of these equations was given in the general case in (23). To this must be added the forced solution. Since the excitation after $t = 0$ is a constant, we expect the equilibrium responses of v and i to be constants also. Their values, say, V_e and I_e, are established by arguing as follows.

If the equilibrium solution of the capacitor voltage is constant, its current (being $C\,dv/dt$) is zero. This is the same as the equilibrium inductor current, so the inductor voltage (being $L\,di/dt$) is zero. By KVL, then, the equilibrium capacitor voltage equals the battery voltage. Thus, $V_e = V_b$ and $I_e = 0$.

When these equilibrium solutions are added to the natural solutions in (23), the resulting general solution will be:

$$\begin{aligned} v(t) &= K_1 e^{s_1 t} + K_2 e^{s_2 t} + V_b \\ i(t) &= C s_1 K_1 e^{s_1 t} + C s_2 K_2 e^{s_2 t} \end{aligned} \tag{35}$$

The next step is to evaluate the unspecified constants using the initial conditions. Let's designate the initial values as $i(0) = I_0$ and $v(0) = V_0$. Setting $t = 0$ in the preceding equations and using these initial values, we get:

$$\begin{aligned} v(0) &= K_1 + K_2 + V_b = V_0 \\ i(0) &= C s_1 K_1 + C s_2 K_2 = I_0 \end{aligned} \tag{36}$$

After some algebraic manipulations (which you should carry out), the solution for the unspecified constants is obtained:

$$\begin{aligned} K_1 &= \frac{I_0 - C s_2 (V_0 - V_b)}{C(s_1 - s_2)} \\ K_2 &= \frac{-I_0 + C s_1 (V_0 - V_b)}{C(s_1 - s_2)} \end{aligned} \tag{37}$$

These results apply to both real and complex natural frequencies.

EXAMPLE 1

REAL NATURAL FREQUENCIES Let's use the numerical values previously given that lead to real natural frequencies: $L = 1\,\text{mH}$, $C = 2.5\,\mu\text{F}$, and $R = 50\,\Omega$. From (22), the natural frequencies were found to be:

$$\begin{aligned} s_1 &= -40{,}000\,\text{s}^{-1} \\ s_2 &= -10{,}000\,\text{s}^{-1} \end{aligned}$$

Suppose the battery voltage is $V_b = 10\,\text{V}$ and the initial values are $V_0 = 5\,\text{V}$ and $I_0 = 0$. Then, from (37) the solutions for the constants become $K_1 = 1.67$ and $K_2 = -6.67$. Substituting all these values into (35) leads to the complete solution:

$$\begin{aligned} v(t) &= 1.67e^{-4\times 10^4 t} - 6.67e^{-10^4 t} + 10 \\ i(t) &= 0.167(-e^{-4\times 10^4 t} + e^{-10^4 t}) \end{aligned}$$

You can easily show that these equations do, indeed, satisfy the initial conditions.

EXAMPLE 2 **COMPLEX NATURAL FREQUENCIES** With the same L and C but R reduced to 20 Ω, the natural frequencies were previously found to be:

$$s_{1,2} = -10{,}000 \pm j10{,}000\sqrt{3}$$

Let the initial conditions this time be $V_0 = 5$ V and $I_0 = 0.1$ A. Then, from (37), the constants have the following values:

$$K_1 = -2.5 + j0.289 = 2.517\angle 173.4^\circ$$
$$K_2 = -2.5 - j0.289 = 2.517\angle -173.4^\circ$$

These are each other's conjugates, as they should be. (See Appendix B for a review of complex numbers.) Note that the complex natural frequencies are also each other's conjugates. Hence, s_2K_2 in (35) is the conjugate of s_1K_1, and so each term in the expression for $i(t)$ is the conjugate of the other. Thus the first term can be written:

$$Cs_1K_1e^{s_1t} = Me^{j\theta}e^{\alpha t}e^{j\omega t} = Me^{\alpha t}e^{j(\omega t+\theta)}$$

where M stands for magnitude. Since the second term is just the conjugate of this, the sum of the two terms will equal twice the real part of this expression. Thus, the only real work is to find M and θ. The same considerations apply to the expression for the voltage.

With these calculations, the solutions for the voltage and current in (35) become:

$$v(t) = 5.034e^{-10{,}000t}\cos(10^4\sqrt{3}t + 173.4^\circ) + 10$$
$$i(t) = 0.2517e^{-10{,}000t}\cos(10^4\sqrt{3}t - 66.6^\circ)$$

As an exercise, show that these expressions do satisfy the initial conditions and the differential equations.

Two-Stage Op-Amp Integrator

Problem 29 in Chapter 6 described an op-amp integrator with a leaky capacitor in the feedback path. What happens if the output from one such amplifier is the input to a second one, as shown in Figure 10? Although you

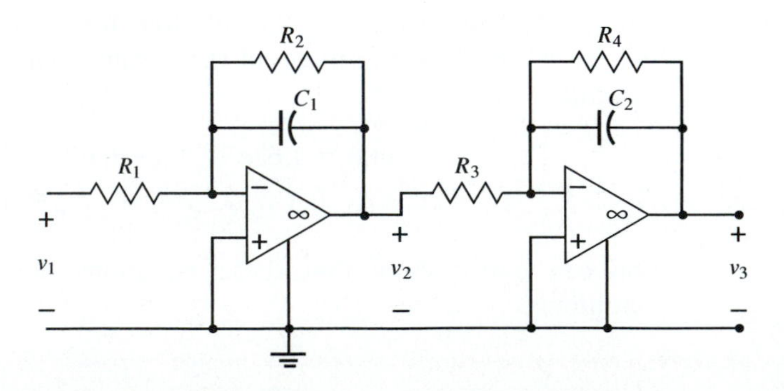

Figure 10
Two-stage op-amp integrator.

should already have found the differential equation satisfied by the output in that problem, we will repeat the process here.

Assuming an ideal op amp with a virtual short circuit across its terminals and no current into them, the voltage across R_1 is v_1 and the voltage across the feedback branch is v_2. Hence, the current in R_1 equals the sum of the currents in R_2 and C_1, with proper attention to the references. Thus:

$$C_1 \frac{dv_2}{dt} + \frac{1}{R_2} v_2 = -\frac{1}{R_1} v_1 \tag{38}$$

An identical expression applies to the second "stage" (as it is called) except for the names of variables and parameters. In this stage, v_3 replaces v_2 in the first stage, v_2 replaces v_1, C_2 replaces C_1, etc. Hence, changing these names in the last equation gives:

$$C_2 \frac{dv_3}{dt} + \frac{1}{R_4} v_3 = -\frac{1}{R_3} v_2 \tag{39}$$

These two equations, suitably rewritten, are the state equations. In matrix form, they become:

$$\frac{d}{dt}\begin{bmatrix} v_2 \\ v_3 \end{bmatrix} = \underbrace{\begin{bmatrix} -\dfrac{1}{R_2C_1} & 0 \\ -\dfrac{1}{R_3C_2} & -\dfrac{1}{R_4C_2} \end{bmatrix}}_{\mathbf{A}} \begin{bmatrix} v_2 \\ v_3 \end{bmatrix} + \begin{bmatrix} -\dfrac{1}{R_1C_1} \\ 0 \end{bmatrix} v_1 \tag{40}$$

The characteristic equation comes from the determinant of $(s\mathbf{U} - \mathbf{A})$:

$$\begin{aligned} \det(s\mathbf{U} - \mathbf{A}) &= \det\begin{bmatrix} s + \dfrac{1}{R_2C_1} & 0 \\ \dfrac{1}{R_3C_2} & s + \dfrac{1}{R_4C_2} \end{bmatrix} \\ &= \left(s + \frac{1}{R_2C_1}\right)\left(s + \frac{1}{R_4C_2}\right) \end{aligned} \tag{41}$$

It's already (conveniently) in factored form, so we see that the characteristic values are real and negative. Note that R_2C_1 and R_4C_2 are the time constants of the leaky capacitors.

We learn something else from this example in combination with the previous op-amp circuit in Figure 4: the mere presence of an op amp (or of controlled sources in general) in a circuit does not guarantee the nature of the characteristic values as real or complex.

Suppose the input voltage in Figure 10 is a unit step. We would expect the equilibrium solution for both v_2 and v_3 to be constants starting at $t = 0$. Hence, the derivatives in (40) would contribute nothing. The equilibrium (forced) response, obtained from (40), would be:

$$v_{2f} = -\frac{R_2}{R_1}v_1 = -\frac{R_2}{R_1}u(t)$$
$$v_{3f} = -\frac{R_4}{R_3}v_{2f} = \frac{R_2R_4}{R_1R_3}u(t) \tag{42}$$

To pursue this development without excessive algebraic manipulations, let's revert to numerical values. Let $R_1 = 25$, $R_2 = 100$, $R_3 = 50$, and $R_4 = 200$, all in kΩ; and $C_1 = 0.5$ and $C_2 = 0.1$, both in μF. Hence, the preceding forced response becomes:

$$v_{3f} = 16u(t)$$

The overall response is the combination of the forced response and the natural response, consisting of two exponentials:

$$v_2(t) = (-4 + K_1e^{-20t} + K_2e^{-50t})u(t)$$
$$v_3(t) = (16 + K_3e^{-20t} + K_4e^{-50t})u(t)$$

The four constants in these equations cannot all be independent; they are constrained by (38) and (39)—or (40), in matrix form. Substituting these solutions into (38) and (39) leads to $K_2 = 0$ and $K_1 = -0.15K_3$, leaving only two independent constants. Thus, the solutions become:

$$v_2(t) = -(4 + 0.15K_3e^{-20t})u(t)$$
$$v(t) = (16 + K_3e^{-20t} + K_4e^{-50t})u(t)$$

(Remember, you should be supplying all the details.)

Recall that the time constant of the second stage is $\frac{1}{50}$ s. The corresponding exponential is missing from the solution for the output of the first stage. Is this reasonable? It certainly is! You should be surprised if it *isn't* missing, since the parameters of the second stage should have no influence on anything that happens in the first stage.

The remaining two constants are determined from initial capacitor voltages. Suppose both capacitors were initially relaxed. From Figure 10 it is clear that the capacitor voltages are v_2 and v_3, respectively. Hence, setting $v_2(0) = 0 = v_3(0)$ in the solutions leads to $K_3 = -80/3$ and $K_4 = 32/3$. The final solution becomes:

$$v_2(t) = -4(1 - e^{-20t})u(t)$$
$$v_3(t) = 16(1 - \tfrac{5}{3}e^{-20t} + \tfrac{2}{3}e^{-50t})u(t)$$

Summary Observation

- The *form* of the natural response of a circuit results from the appropriate differential equations, which, in turn, depend on the circuit structure and the parameter values. Given the circuit, you apply the circuit laws to arrive at the differential equations.

- The *form* of the forced, or equilibrium, response depends on the excitation. Together with the differential equations, the excitation determines not only the form, but the entirety of the equilibrium response, including its coefficients.

- In general form, there will be some unspecified coefficients in the natural response, as many as there are distinct energy-storing components. The *complete* response—both natural and equilibrium—will include these unspecified coefficients; they are to be determined from the initial conditions: initial capacitor voltages and/or inductor currents.
- These initial values must be available *somehow*. Either they must be specified, much as the circuit parameter values are specified; or they must be determinable from some previously existing circuit, prior to an event—such as some switching event or the application of an excitation—that establishes the circuit whose response is to be determined.

7 RESISTANCE-INDUCTANCE CIRCUITS

The first type of circuit considered in this chapter was a second-order *RC* circuit with two capacitors. We then went on to circuits containing one capacitor and one inductor. For completeness, we will now briefly examine circuits which are second-order by virtue of having two distinct inductors. The study will be brief because no new principles are introduced.

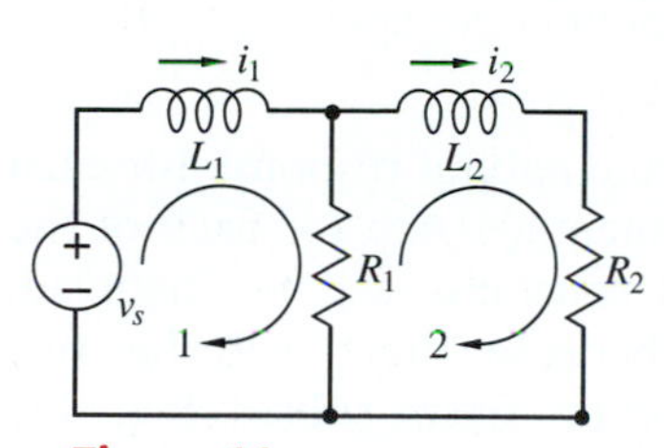

Figure 11
Low-pass filter circuit with two inductors.

Consider the circuit in Figure 11. It is similar to the one in Figure 3 except that the capacitors are replaced by inductors. The circuit is set up at $t = 0$ through some switching mechanism (not shown) resulting in possibly nonzero initial inductor currents. Noting that the inductor currents are identical with the mesh currents, we find that the mesh equations for this circuit are:

$$
\begin{aligned}
L_1 \frac{di_1}{dt} + R_1(i_1 - i_2) &= v_s(t) \\
L_2 \frac{di_2}{dt} + R_1(i_2 - i_1) + R_2 i_2 &= 0
\end{aligned}
\tag{43}
$$

Each of these equations is a first-order linear differential equation. The set can be rewritten in the form of state equations and then put in matrix form, as follows:

$$
\begin{aligned}
L_1 \frac{di_1}{dt} &= -R_1 i_1 + R_1 i_2 + v_s \\
L_2 \frac{di_2}{dt} &= R_1 i_1 - (R_1 + R_2) i_2
\end{aligned}
\tag{44}
$$

$$
\frac{d}{dt}\begin{bmatrix} i_1 \\ i_2 \end{bmatrix} = \begin{bmatrix} -\dfrac{R_1}{L_1} & \dfrac{R_1}{L_1} \\ \dfrac{R_1}{L_2} & -\dfrac{R_1 + R_2}{L_2} \end{bmatrix} \begin{bmatrix} i_1 \\ i_2 \end{bmatrix} + \begin{bmatrix} \dfrac{v_s(t)}{L_1} \\ 0 \end{bmatrix}
\tag{45}
$$

The characteristic equation is easily obtained from det ($s\mathbf{U} - \mathbf{A}$):

$$\det(s\mathbf{U} - \mathbf{A}) = \det\left(\begin{bmatrix} s & 0 \\ 0 & s \end{bmatrix} - \begin{bmatrix} -\dfrac{R_1}{L_1} & \dfrac{R_1}{L_1} \\ \dfrac{R_1}{L_2} & -\dfrac{R_1 + R_2}{L_2} \end{bmatrix}\right)$$

$$= \det\begin{bmatrix} s + \dfrac{R_1}{L_1} & -\dfrac{R_1}{L_1} \\ -\dfrac{R_1}{L_2} & s + \dfrac{R_1 + R_2}{L_2} \end{bmatrix} \tag{46}$$

$$= s^2 + \left(\frac{R_1}{L_1} + \frac{R_1 + R_2}{L_2}\right)s + \frac{R_1 R_2}{L_1 L_2} = 0$$

This should be compared with (6), which gives the corresponding equation for the second-order *RC* circuit. It is clear that the present equation can be obtained from (6) by duality. (Go through the details.) The conclusions reached about *RC* circuits, therefore, apply to *RL* circuits also. That is:

> The characteristic values of linear passive *RL* circuits are distinct, real, and negative.

The next step in the solution would be to write the natural response for each mesh current as a linear combination of two exponentials, one for each of the natural frequencies. A total of four unspecified constants will be involved. However, they will not all be independent, because there will be two relationships among these four constants resulting from substituting the solutions into the original differential equations in (40) or (41) but with $v_s = 0$. Two other relationships are obtained when the complete solutions for the currents for $t = 0$ are set equal to their initial values. Since the process is straightforward and differs little from what has been done for *RC* and *RLC* circuits, we will not pursue the details any further.

EXAMPLE 3 The switch in the circuit of Figure 12 has been closed for a long time when it is opened at $t = 0$. A solution for the inductor currents is required for all time following the opening of the switch.

Although the circuit contains a capacitor, which must be accounted for

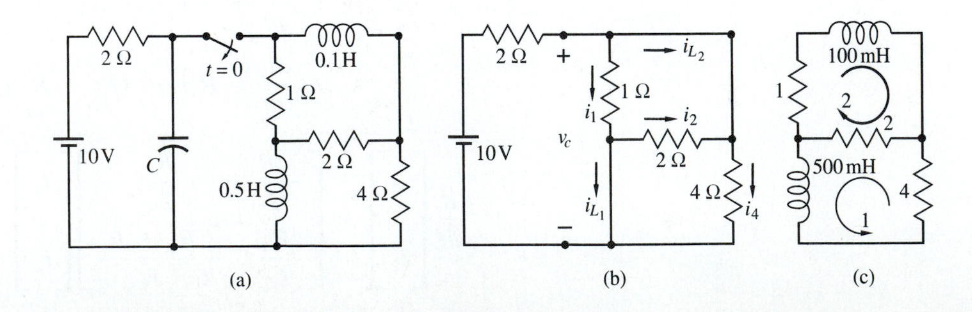

Figure 12
Example circuit.

when determining the initial values, it is not part of the circuit after the switch is opened. The circuit is complex enough that even determining the initial values requires some work.

Before the switch is opened, the circuit is in equilibrium and no further changes are taking place; neither the capacitor voltage nor the inductor currents are changing. Hence, the capacitor is like an open circuit and the inductors are like short circuits during that period. The relevant circuit is shown in Figure 12b. The three resistors to the right of the capacitor are all in parallel across the position of the capacitor. Their equivalent resistance is $\frac{4}{7}\,\Omega$. (Confirm this.) This resistance forms a voltage divider with the accompanied battery. Hence, $v_C(0-)$, which is the same as the parallel resistor voltages, can be found from the voltage-divider relationship. The resistor currents follow from Ohm's law. Finally, by KCL, the two inductor currents can be found in terms of the resistor currents. You should fill in the details; the results are as follows:

$$v_C(0) = \frac{10(\frac{4}{7})}{2 + \frac{4}{7}} = \frac{20}{9} = 2.222 \text{ V}$$

$$i_{L_1}(0) = i_1(0) - i_2(0) = 2.222 + 1.111 = 3.333 \text{ A}$$

$$i_{L_2}(0) = i_4(0) - i_2(0) = 0.556 + 1.111 = 1.667 \text{ A}$$

Since v_C and i_L cannot change suddenly, these $0-$ values apply also for $0+$.

After the switch is opened, the circuit takes the form shown in Figure 12c. With the currents chosen as shown, the two mesh equations are:

$$0.5\frac{di_{L_1}}{dt} + 4i_{L_1} + 2(i_{L_1} + i_{L_2}) = 0$$

$$0.1\frac{di_{L_2}}{dt} + 2(i_{L_2} + i_{L_1}) + i_{L_2} = 0$$

or, in matrix form:

$$\frac{d}{dt}\begin{bmatrix} i_{L_1} \\ i_{L_2} \end{bmatrix} = \begin{bmatrix} -12 & -4 \\ -20 & -30 \end{bmatrix}\begin{bmatrix} i_{L_1} \\ i_{L_2} \end{bmatrix}$$

The characteristic values are found from the characteristic equation:

$$\det(s\mathbf{U} - \mathbf{A}) = \det\begin{bmatrix} s + 12 & 4 \\ 20 & s + 30 \end{bmatrix}$$

$$= s^2 + 42s + 280 = 0$$

$$s_1 = -8.3 \qquad s_2 = -33.7$$

Since the equilibrium solution is zero, the complete response is just the natural response. Thus:

$$i_{L_1}(t) = K_1e^{-8.3t} + K_2e^{-33.7t}$$

$$i_{L_2}(t) = K_3e^{-8.3t} + K_4e^{-33.7t}$$

There are four unspecified constants in these two expressions. However, two relationships among them come from the original differential equations. Two other relationships come by imposing the previously determined initial values of the inductor currents. The details will be left for you to work out. The results will be:

$$i_{L_1}(t) = 2.58e^{-8.3t} + 0.75e^{-33.7t}$$

$$i_{L_2}(t) = -2.38e^{-8.3t} + 4.05e^{-33.7t}$$

(Verify that these reduce to the correct initial values.)

8 A THIRD-ORDER CIRCUIT: STRETCHING THE IMAGINATION

Wait a minute! Isn't this chapter titled "Second-Order Circuits"? Yes, but circuits march on. The purpose of this very brief section is to show you that you have all the tools necessary to handle higher-order circuits. If you had to, you *could* find the appropriate differential equations for such circuits.

The circuit in Figure 13 has two capacitors and an inductor, in addition to resistors. To arrive at the state equations, we write KVL around loops on which inductors appear and KCL at nodes to which capacitors are connected. Thus, a KVL equation is written around the indicated loop. From this equation we eliminate the voltages of all branches except the capacitors. The result is:

$$L\frac{di}{dt} + R_3 i + v_2 - v_1 = 0$$

Then we write KCL at the two numbered nodes; from these we eliminate all currents except the inductor current. The result is:

$$C_1\frac{dv_1}{dt} + i + \frac{1}{R_1}(v_1 - v_s) = 0$$

$$C_2\frac{dv_2}{dt} + \frac{1}{R_2}v_2 - i = 0$$

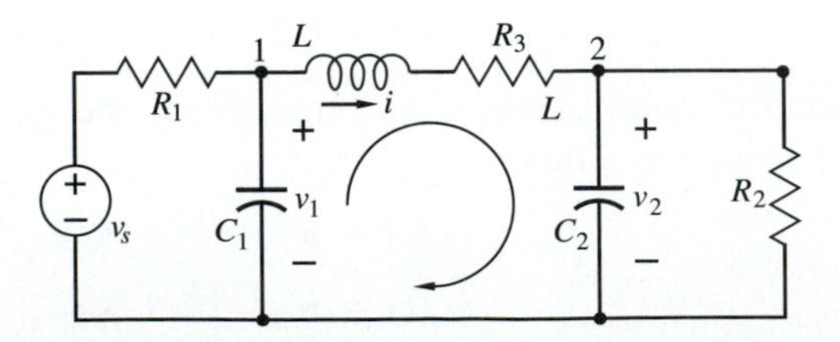

Figure 13
A third-order circuit.

These are the state equations: three first-order differential equations in three unknowns. In matrix form, they become:

$$\frac{d}{dt}\begin{bmatrix} i \\ v_1 \\ v_2 \end{bmatrix} = \begin{bmatrix} -\frac{R_3}{L} & \frac{1}{L} & -\frac{1}{L} \\ -\frac{1}{C_1} & -\frac{1}{R_1C_1} & 0 \\ \frac{1}{C_2} & 0 & -\frac{1}{R_2C_2} \end{bmatrix}\begin{bmatrix} i \\ \frac{v_s}{R_1} \\ 0 \end{bmatrix} + \begin{bmatrix} 0 \\ \frac{1}{R_1C_1} \\ 0 \end{bmatrix} v_s$$

(Confirm these equations.)

The characteristic equation is found as before by setting the determinant of $s\mathbf{U} - \mathbf{A}$ equal to zero. The roots of that equation are the characteristic values. We will not pursue the matter here. However, the ease of obtaining these equations should convince you that you are fully capable of constructing the state equations of circuits whose order is higher than the second.

SUMMARY

- Differential equations of second-order *RC* circuits
- Negative real characteristic values of second-order *RC* circuits
- State equations of second-order *RC* circuits in matrix form
- Characteristic equation from matrix form of the state equations
- Differential equations of series *RLC* circuit
- State equations of series *RLC* circuit
- Characteristic values of *RLC* circuit: real and complex
- Damping coefficient and damped and undamped frequency
- Underdamped, overdamped, and critically damped characteristic values
- Natural frequencies of linear networks; complex-frequency variable
- Parallel *RLC* circuit: state equations and natural frequencies
- Parallel *RLC* circuit with a leaky capacitor
- Step response of *RLC* circuits: real and complex natural frequencies
- Two-state op-amp integrator: state equations, step response
- Second-order *RL* circuits
- Characteristic values of linear, passive *RL* circuits
- Complete solution of *RL* circuits
- Third-order circuits

PROBLEMS

In some of the problems starting in this chapter, the values of some circuit parameters may seem strange to you (e.g., $C = 1$). Using such convenient numerical values permits concentrating on the important ideas in the chapter, rather than obscuring the issues with a lot of numerical calculation. What makes this intellectually acceptable is the concept of *normalization,* or scaling, which won't be considered until Chapter 11. You will find there that by proper scale changes, convenient sizes of parameter values that facilitate numerical work can be obtained.

Thus, the numerical values that appear in some of the

problems here will be dimensionless, normalized values; they are not farads, for example. What can be said about the numerical values is that they are consistent; thus, RC in normalized units is normalized time and LC in normalized units has the dimensions of the square of normalized time.

This discussion may not be entirely clear to you at this point. One alternative would be to label the numerical values as farads, henries, and ohms (as some authors do) and not mention the idea of normalization at all. Many of you would probably rebel at that, since those would be outlandish, unnatural values. The other alternative is to stick with typical practical values always and carry the burden of a lot of extra numerical calculation that does nothing for the understanding. The alternative undertaken here seems more satisfying.

1. The objective of this problem is to establish that the characteristic equation resulting from second-order differential equations in each of the two variables in Figure 3 in the text is the same as the characteristic equation obtained in the text for the pair of first-order state equations.
   a. Start with the state equations given in the text for the circuit of Figure 3. Derive a second-order differential equation for v_2 by solving the second state equation for v_1 and substituting into the first.
   b. Confirm that the resulting characteristic equation is the same as the one in the text.
   c. Derive a second-order differential equation for v_1 also and confirm that it has the same characteristic equation.

2. Suppose there is no external excitation in a series RLC circuit.
   a. By performing mathematical operations on the state equations, establish the result that the time derivative of the total energy stored in the circuit is negative. That is:

$$\frac{d}{dt}(w_C + w_L) < 0$$

   b. Repeat part (a) for a parallel RLC circuit.

3. The circuit of Figure P3 has been in equilibrium (that is, no further changes are occurring as time goes on) when the switch is closed at $t = 0$.
   a. Find the initial ($t = 0+$) value of the capacitor voltage and of the currents in L and C.
   b. Write loop equations after the switch is closed, using the loops shown in the figure. In addition to a derivative, there will be an integral.
   c. Using the capacitor v-i relationship, change one of the variables in the preceding equations to the capacitor voltage, thereby eliminating the integral. The two variables are now i_1 and v.

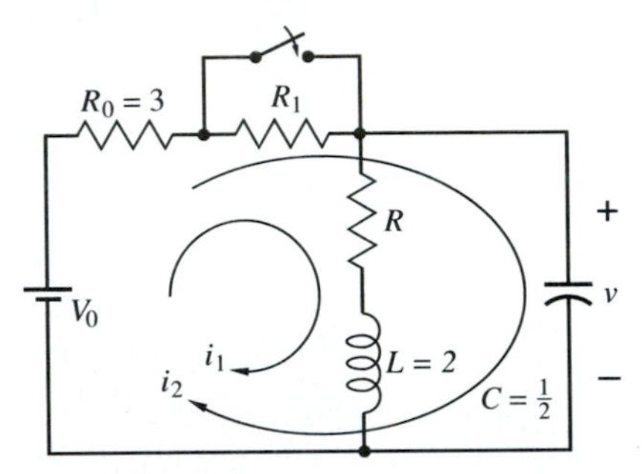

Figure P3

   d. By elimination, reduce this pair of equations to a single second-order differential equation. Find the natural frequencies corresponding to each of the following values of R: 0, 2, 3, 10.

4. The circuit in Figure P4 is in equilibrium with the switch closed when it is opened at $t = 0$.
   a. Determine the initial values of the capacitor voltage v and the inductor current i.
   b. Write a pair of state equations whose variables are v and i.
   c. From these, determine the characteristic values.
   d. From the state equations derive a second-order differential equation whose variable is v.
   e. Confirm that this equation leads to the same characteristic values.
   f. Repeat parts (d) and (e) but for variable i.

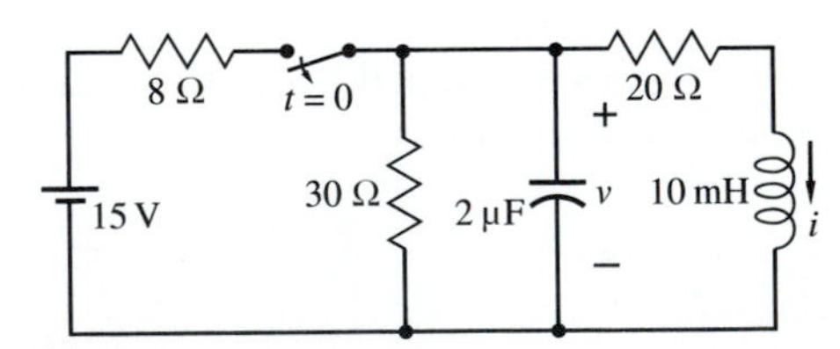

Figure P4

5. The switch in Figure P5 has been closed for a long time when it is opened at $t = 0$.
   a. Specify the $t = 0+$ values of v and i (shown on the diagram).

b. Write a pair of first-order differential equations (the state equations) in which the variables are v and i.

c. Convert these two equations into a single second-order differential equation in which the variable is i.

d. The general solution of the equation in part (c) is:

$$i(t) = K_1 e^{s_1 t} + K_2 e^{s_2 t}$$

Find the characteristic values s_1 and s_2 from the single equation in part (c).

e. Determine the characteristic values from the pair of first-order equations in part (b); are there any surprises?

f. Repeat part (c) for the variable v.

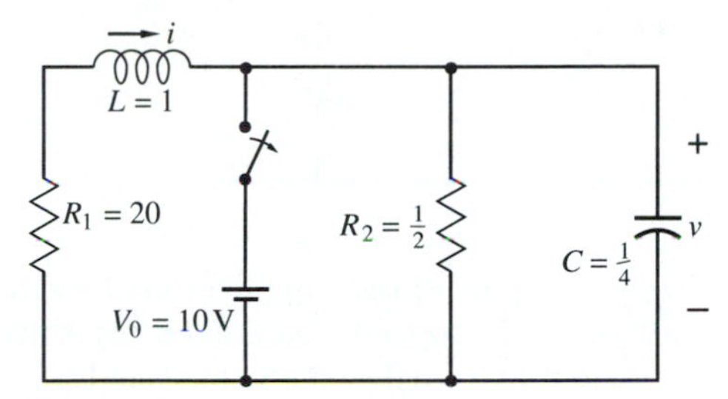

Figure P5

6. Determine the natural response for each of the following differential equations (or sets of equations):

a. $$\frac{dv}{dt} + 100v - 100i = t + 2$$

$$\frac{di}{dt} + 50v + 20i = 0$$

b. $$\frac{d}{dt}\begin{bmatrix} v \\ i \end{bmatrix} = \begin{bmatrix} -10 & -5 \\ -20 & -40 \end{bmatrix}\begin{bmatrix} v \\ i \end{bmatrix} = \begin{bmatrix} 0 \\ 2e^{-10t} \end{bmatrix}$$

c. $$\frac{d^2v}{dt^2} + 2\frac{dv}{dt} + 5v = t^2 + 2t + 3$$

d. $$\frac{d^2i}{dt^2} + 6\frac{di}{dt} + 25 = 10\cos 10t$$

7. Find the equilibrium response for each case in Problem 6.

8. The equivalent of switching on a battery V_b in a circuit is obtained by replacing the battery and switch combination by means of a step function $V_b u(t)$. In the same way, the equivalent of removing a battery V_b from a circuit (leaving a short circuit) is to replace it with the step function $V_b u(-t)$, a step that decreases from V_b to 0 at $t = 0$. Rather than simply specify initial capacitor voltages or inductor currents, the scheme described in the preceding sentence might be used to make you struggle to determine them. This scheme will be inflicted on you only in this problem and the next.

In the circuit of Figure P8:

a. Derive the state equations.

b. Find the complete solution for $v(t)$ and $i(t)$ for $t > 0$, leaving no unspecified constants.

c. Using only algebra, find the initial value of the capacitor current.

d. At what time does the current in the resistor again reach its initial value?

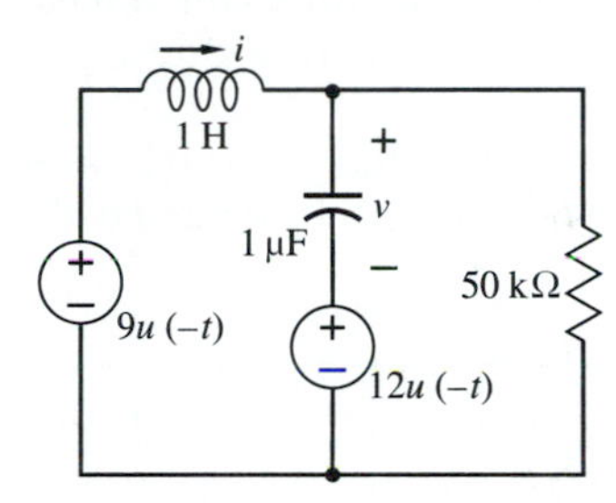

Figure P8

9. The initial capacitor voltage in the circuit of Figure P9 is 10 V and the initial inductor current is 20 mA. A source voltage $K_1 u(-t)$ is to be placed in series with the resistor, and another, $K_2 u(-t)$, in series with the inductor so that the resulting initial values are the ones given. Find the values of K_1 and K_2, and draw the resulting circuit.

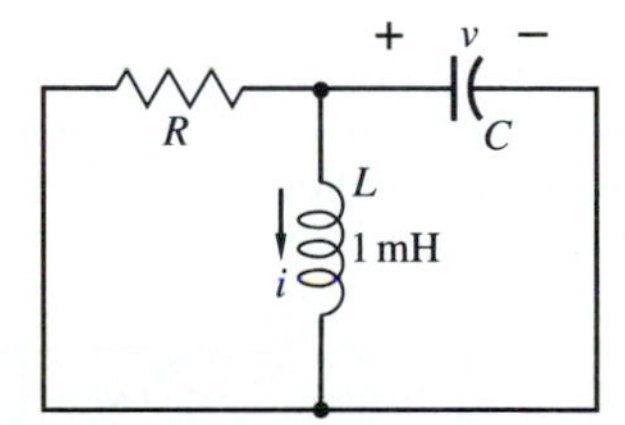

Figure P9

10. Just before the switch in Figure P10 is moved from A to B, the capacitor voltage is 20 V and the inductor current −200 mA. The ultimate goal is to find the voltage $v(t)$ for $t > 0$. Carry out a strategy session with yourself, thinking through the options, and decide on the simplest approach. Then carry it out and find $v(t)$.

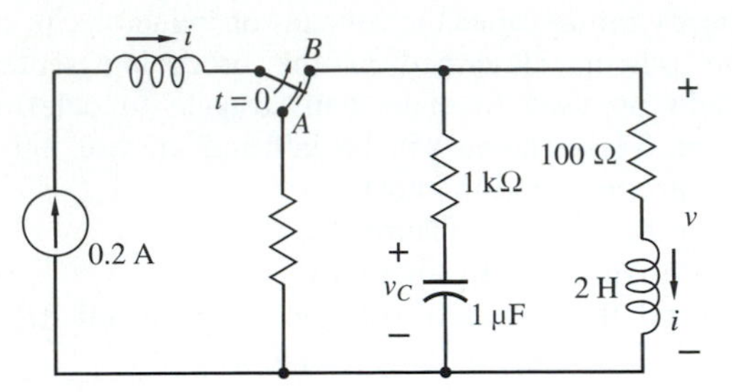

Figure P10

11. The circuit in Figure P11 is in equilibrium when the switch is opened at $t = 0$.
  a. Find the maximum value of R for which the capacitor voltage is not oscillatory after the opening of the switch.
  b. Find the maximum value attained by the capacitor voltage and the time at which it occurs for this R.

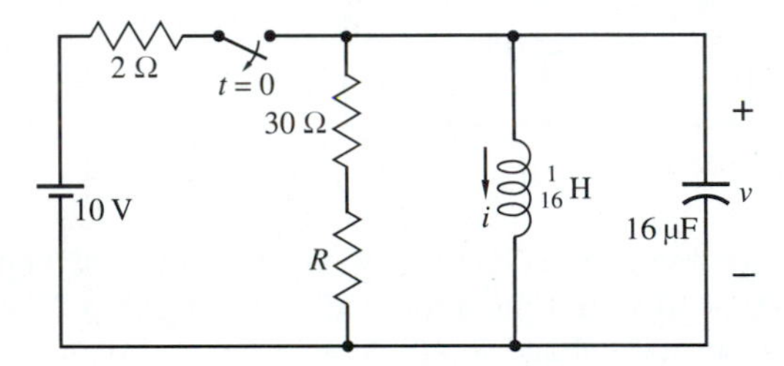

Figure P11

12. The circuit in Figure P12 has been in equilibrium with switch S_2 closed and switch S_1 in position A when, at $t = 0$, S_1 is moved to position B. At $t = 1$ (normalized), S_2 is opened.
  a. Determine the inductor current for all $t > 0$.
  b. Show the locations of the pertinent sets of natural frequencies in the complex plane.

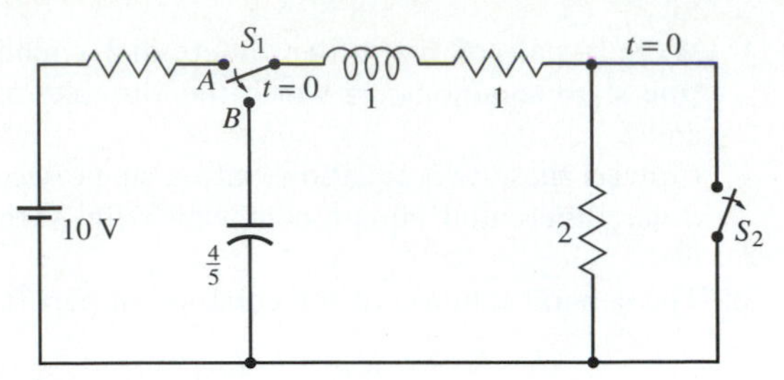

Figure P12

13. In Figure P13, find the voltage across the 2-unit capacitor following the step current, leaving no unspecified constants. Take the initial values to be zero.

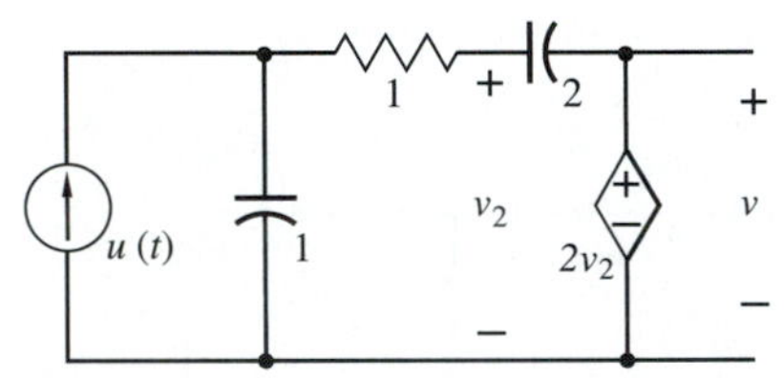

Figure P13

14. The circuit in Figure P14 is initially relaxed when the pulse voltage v_s is applied. Determine the solution for $v(t)$; write it as a single expression valid for $t > 0$. $v_s = 10[u(t) - u(t - 5)]$.

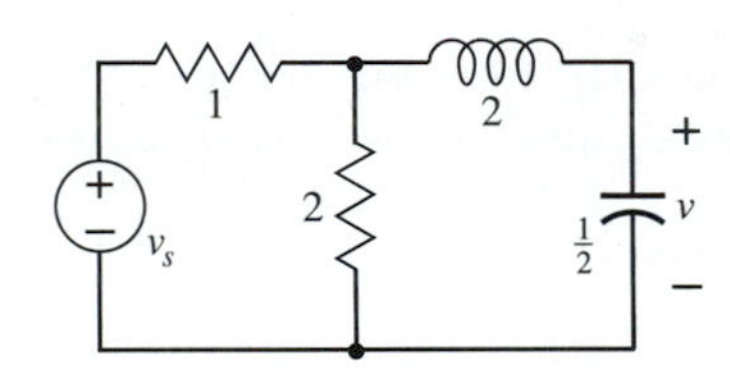

Figure P14

15. The circuit in Figure P15 is in equilibrium when switch S_1 is opened at $t = 0$. Switch S_2 is opened 2 units of time later. Determine the capacitor voltage and the inductor current in the two intervals $0 \le t \le 2$ and $t > 2$.

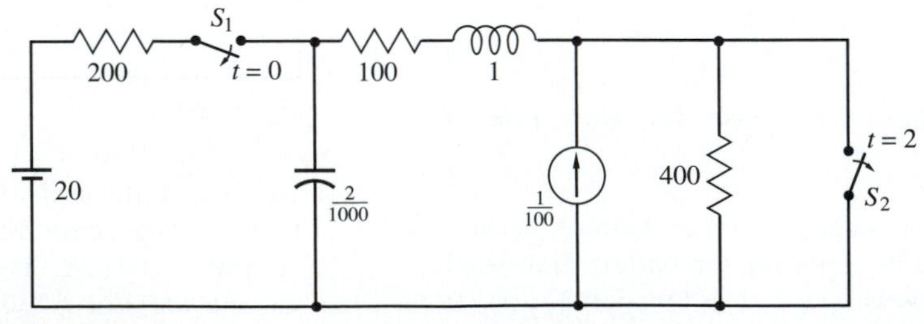

Figure P15

16. The circuit in Figure P16 is initially relaxed when a step input is applied.

a. Write the state equations for the circuit.

b. Determine the characteristic values; take $g = 1\ \mu$S.

c. Find the complete solution for the capacitor voltages for $t > 0$.

d. Using only algebra, find the resistor and capacitor currents.

e. Find an expression for the total energy stored in the circuit for $t > 0$.

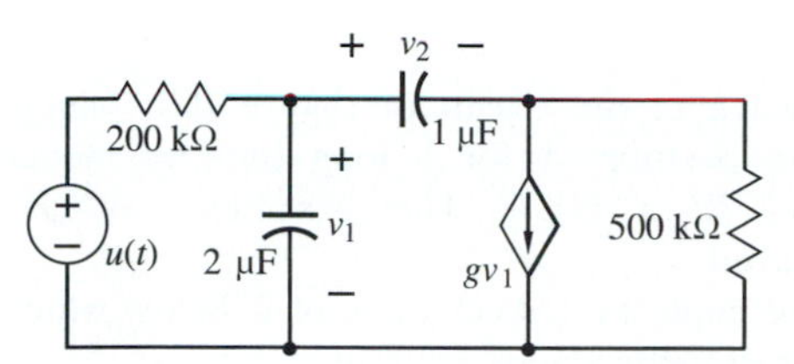

Figure P16

17. The circuit in Figure P17 is in equilibrium when the switch is moved from A to B at $t = 0$.

a. Write a pair of state equations satisfied by v and i for $t > 0$.

b. Determine the characteristic equation and the characteristic values.

c. Write a second-order differential equation satisfied by v for $t > 0$. Confirm that the same characteristic values result.

d. Repeat part (c) for the inductor current i.

e. Find $v(t)$ and $i(t)$ for $t > 0$.

f. Determine an expression for the total energy stored in the circuit for $t > 0$.

g. Show that its rate of increase with time is negative, and interpret what that means.

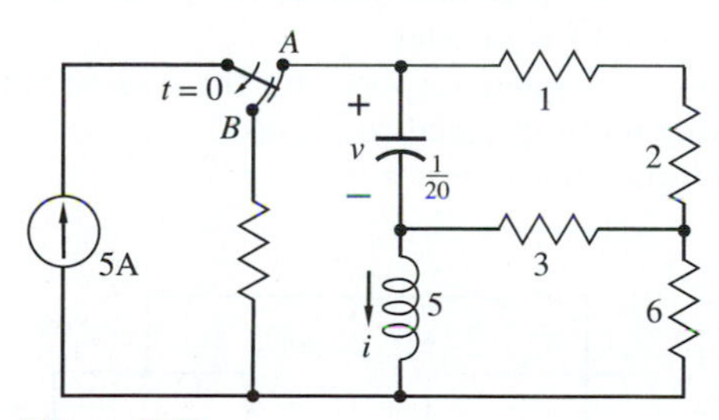

Figure P17

18. This problem illustrates that "unexpected" things can happen in a circuit. The circuit in Figure P18 is in equilibrium when the switch is moved from A to B at $t = 0$. In equilibrium, the inductor acts as a short circuit and the capacitor as an open circuit, resulting in a short-circuit load across the accompanied source. Thus, the initial capacitor voltage is zero and the initial inductor current is $v_s/R_s = 20/2$. For a fixed source voltage, the value of the initial inductor current is inversely proportional to the resistance R_s. If that resistance is very small, then . . .?

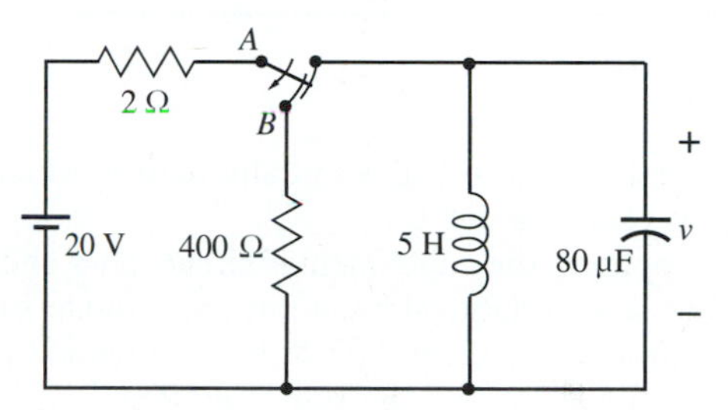

Figure P18

What can be expected from a parallel RLC circuit left alone with an initial inductor current? In the long run, both the inductor current and the capacitor voltage will vanish. Therefore the capacitor voltage starts out from 0 and ends up at 0. In between, it must rise and then fall. If the natural frequencies are complex, the capacitor voltage will oscillate on its trip from zero to zero. The peak that it reaches in the first cycle of its oscillation will be the largest, since the voltage will be damped as time goes on. If the natural frequencies are real, on the other hand, the capacitor voltage will rise from zero to a maximum only once on its way to its ultimate zero. With a source voltage that isn't very high to begin with (20 V), how high can those maxima be?

a. Make a thoughtful estimate; then determine the solution for the capacitor voltage for $t > 0$ and determine its maximum value.

b. Suppose the 2-Ω resistance is changed to 100 Ω. Find the solution for the capacitor voltage for $t > 0$ and again find the maximum. How close was your original estimate?

19. The circuit in Figure P19 is initially relaxed when a unit step is applied at $t = 0$. The parameter values are normalized. (This is a symmetrical low-pass filter with resistive terminations at both its input and output.)

a. Write a set of three first-order differential equations which constitute the state equations.

b. Make a stab at determining the characteristic values. (Use MATLAB if accessible to you for confirmation.)

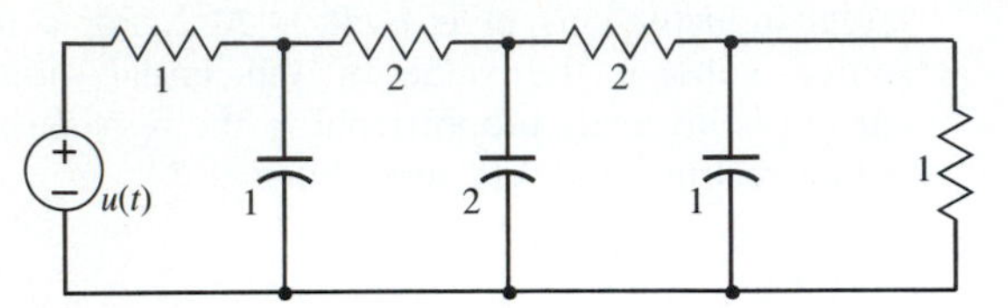

Figure P19

20. The circuit in Figure P20 is initially relaxed when the switch is closed at $t = 0$.
   a. By applying the fundamental circuit laws and the component relationships in an appropriate order, obtain a set of three first-order differential equations that constitute the state equations.
   b. Find the characteristic values and place them on the complex-frequency plane. (Use MATLAB for confirmation.)
   c. Determine the ultimate values (as $t \to \infty$) of the state variables.
   (An appropriate circuit diagram for this condition would be useful.)
   d. Write an expression for the general solution for each state variable, using the characteristic values.
   e. If you feel like doing it, use the state equations to reduce down to 3 the number of unknown constants to be determined. Then apply the initial conditions to determine these constants.

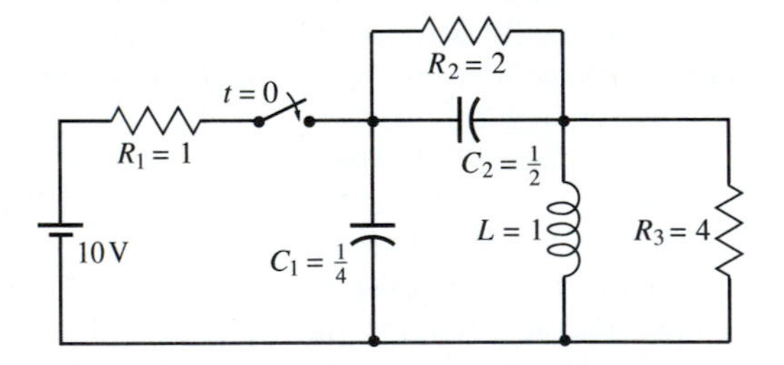

Figure P20

21. From a previous excitation, the capacitor voltage in Figure P21 at the time just before the switch is closed ($t = 0$) is 2 V and its initial derivative is 5000 V/s.
   a. Find the initial value of the inductor current.
   b. Derive the state equations valid for $t > 0$.
   c. Find the complete response for $v(t)$ and $i(t)$. Which will die out sooner, the transient (natural response) or the equilibrium response? Does that bother you?

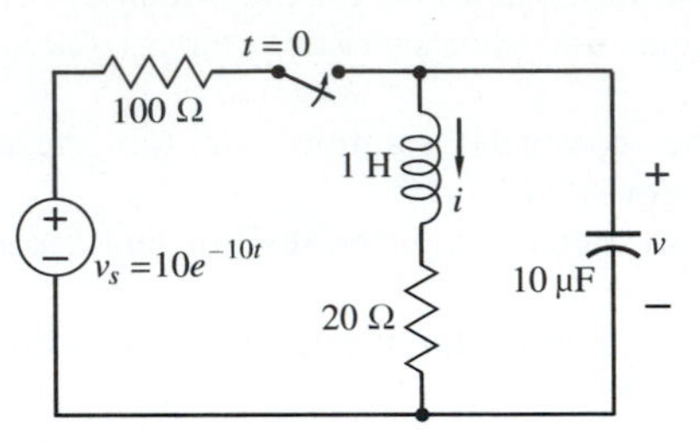

Figure P21

22. The switch in the circuit of Figure P22, which has been in position A for a long time, is moved to position B at $t = 0$. The parameter values are normalized.
   a. Determine the critical value of C below which the characteristic values are real.
   b. With $C = 2$ (normalized), find $v(t)$ and $i(t)$ (complete response) for $t > 0$.
   c. Sketch these functions for 20 units of time.

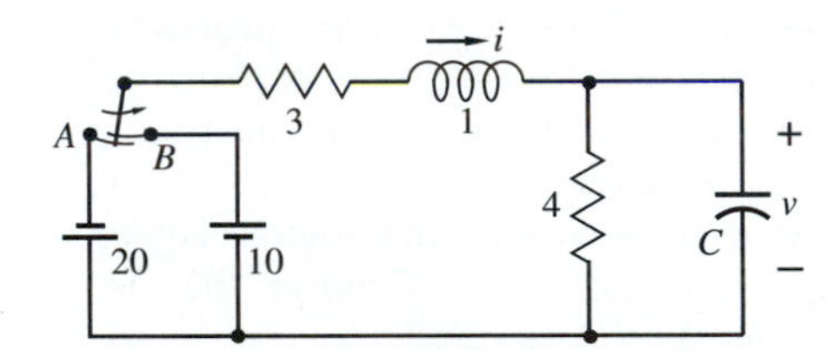

Figure P22

23. The capacitors in the circuit of Figure P23 are initially relaxed when the circuit is excited by a step. The given values are normalized.
   a. Find the state equations.
   b. From these, find the characteristic equation and the natural frequencies.
   c. Find the complete response for the state variables, leaving no unspecified constants.
   d. Determine $v_o(t)$ for $t > 0$.

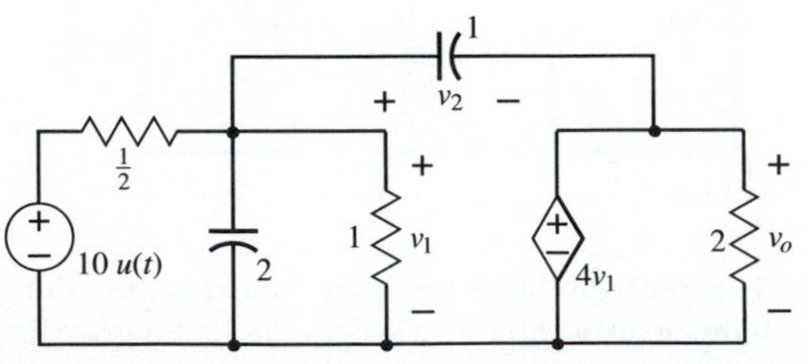

Figure P23

24. The values in the circuit of Figure P24 are normalized. The circuit has been operating for a long time when the switch is closed at $t = 0$, connecting the initially relaxed capacitor C_1 to the circuit.

a. Determine the state equations and, from these, the second-order differential equations satisfied by the capacitor voltages.

b. Find the complete response for the capacitor voltages, with all constants specified, confirming that they satisfy the initial conditions.

c. Without any further calculus, determine the two capacitor currents (standard references) as functions of time. Using the initial values of the capacitor voltages in the circuit, confirm the initial values of the capacitor currents obtained from these equations.

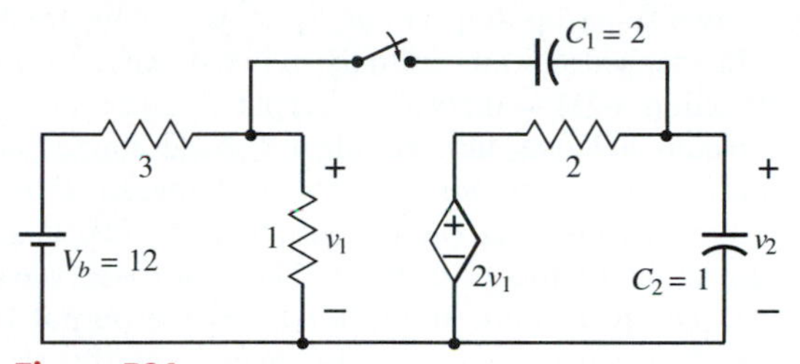

Figure P24

25. The circuit in Figure P25 is initially relaxed when the following pulse is applied: $v_s(t) = 10t[(u(t) - u(t - 0.01)]$ V. Determine the complete output-voltage response (leaving no unspecified constants) and sketch it on the same axes as the input pulse.

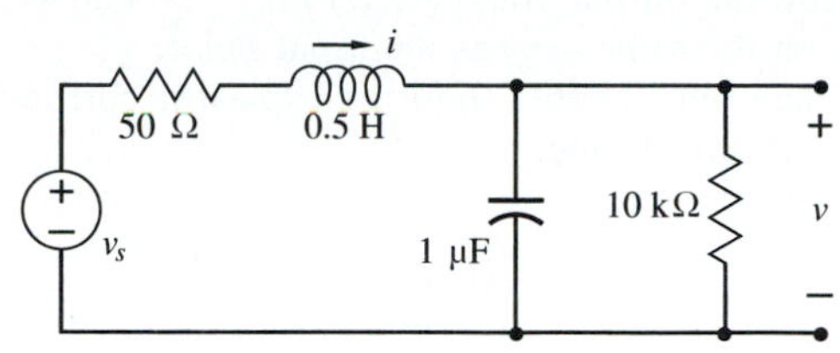

Figure P25

26. Both switches in the circuit of Figure P26 have been open for a long time. At $t = 0$, switch S_1 is closed. Switch S_2 is closed 100 μs later.

a. Find the characteristic values in the two intervals of time by whatever means you can. (If you want practice, find them by some other means also.)

b. Determine the complete response for the capacitor voltage $v(t)$ and the inductor current $i(t)$ for all $t > 0$ and sketch them over an interval up to about 400 μs.

c. Using only algebra, determine the derivatives of the capacitor voltage and the inductor current as functions of time after $t = 100\ \mu$s.

d. Suppose the value of L is changed to 5 mH. Repeat parts (a) and (b).

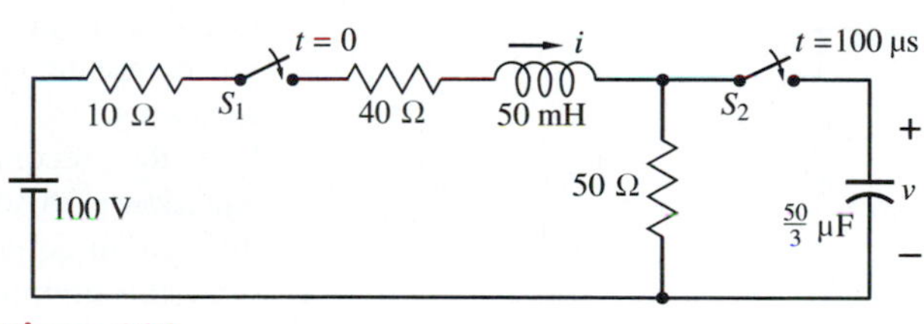

Figure P26

27. Both switches in Figure P27 have been open for an hour. At $t = 0$, switch S_1 closes; S_2 closes 1 ms later.

a. Find the value of the capacitor voltage at $t = 1$ ms.

b. Find the characteristic values of the circuit after $t = 1$ ms.

c. Determine the complete response for $v(t)$ and $i(t)$ for $t > 0$.

d. Sketch the curves for an interval of several milliseconds.

e. Without calculus, determine expressions for the inductor voltage and the capacitor current for $t > 1$ ms.

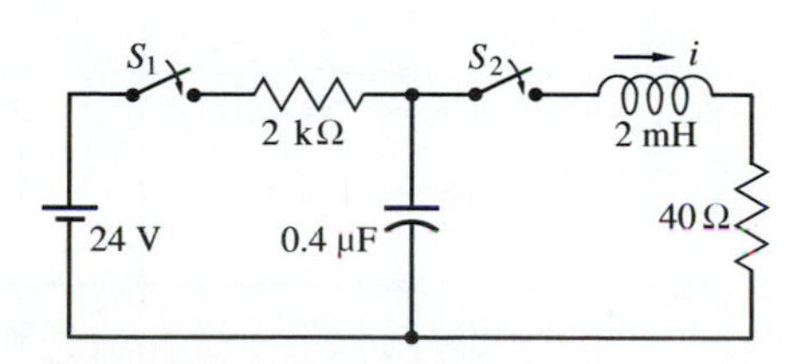

Figure P27

28. The amplifier circuit in Figure P28 is initially relaxed. The pulse voltage $v_s(t) = 10[u(t) - u(t - 10^{-4})]$ is applied.
 a. Find the characteristic values of the circuit.
 b. Find the output voltage $v_2(t)$ for $t > 0$ and sketch it on the same axes as the input pulse.
 c. Using only algebra, find the capacitor currents as functions of time.

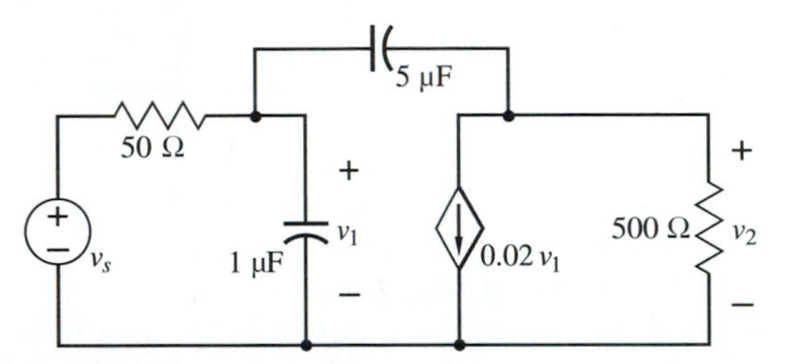

Figure P28

29. The switch in the circuit of Figure P29 has been open for a long time. It is closed at $t = 0$.
 a. Determine the values of the capacitor voltages a long time after the switch is closed.
 b. Determine the characteristic values of the circuit for $t > 0$.

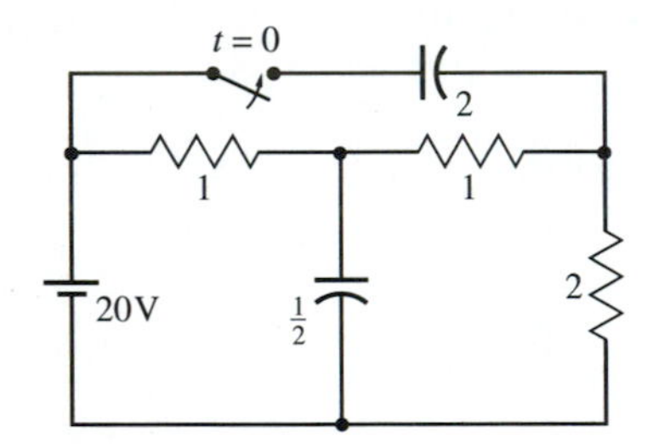

Figure P29

 c. Find the complete solution for the capacitor voltages, leaving no unspecified constants. Confirm the values found in part (a).
 d. Find expressions for the capacitor currents as functions of time using only algebra.
 e. Derive the differential equations satisfied by the capacitor currents.
 f. Confirm that the expressions for the currents found in part (d) satisfy the equations in part (e).

In the op-amp circuits that follow, the op amps are assumed to be operating in their linear region. After you have obtained a solution, you must confirm that the conditions for linearity hold. If, instead, an op amp saturates, you must determine for how long the circuit operates linearly.

30. A two-stage op-amp circuit is shown in Figure P30. The capacitors are initially relaxed when a ramp function $v_s(t) = 0.1tu(t)$ is applied. Since an ideal op amp behaves like an ideal voltage source at its output, there is no interaction between the two stages, except that the output from the first stage is the input to the next stage. There are two ways to proceed to a solution: (1) First find the output from the first stage alone, without regard for the second stage; then, using this as its input, find the output of the second stage. (2) Treat the entire circuit as a unit.
 a. Using the first approach, find the complete $v_1(t)$, leaving no unspecified constants.
 b. Use this as the input to the second stage and find the complete $v_2(t)$, again leaving no unspecified constants.
 c. Use the second approach and find the state equations. Also derive a single second-order differential equation for the output voltage v_2.
 d. Find the complete solutions for the state equations and compare with parts (a) and (b).

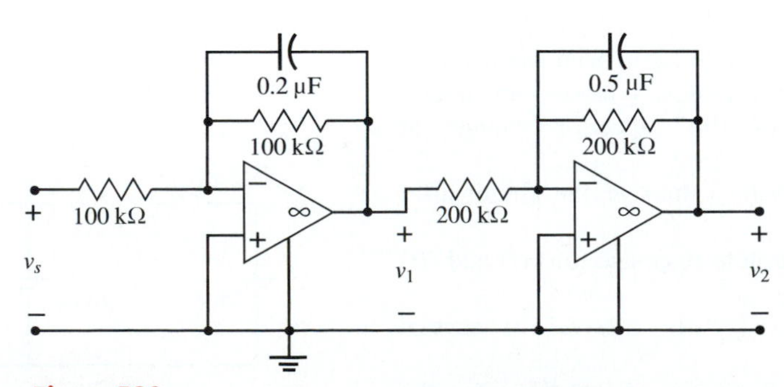

Figure P30

31. The capacitors in Figure P31 are initially uncharged when a step voltage is applied at $t = 0$.

a. Find the state equations.

b. Determine the characteristic values, assuming $G_2 = \frac{7}{8}G_1$.

c. Determine the complete solution, using $a = G_1/C$ for short.

d. Repeat parts (b) and (c), assuming $G_2 = 3G_1$.

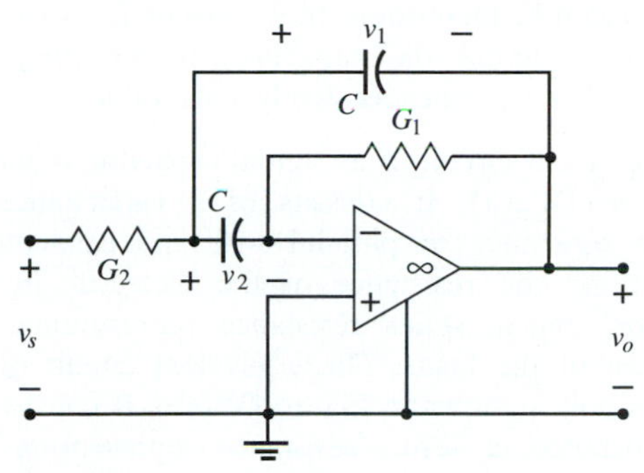

Figure P31

32. Let the numerical values of the parameters in the circuit in Figure P31 be as follows: $C_1 = C_2 = 0.1\ \mu\text{F}$, $R_1 = 125\ \text{k}\Omega$, and $R_2 = \frac{1}{7}\ \text{M}\Omega$.

a. Find the equilibrium response if the excitation applied at $t = 0$ is $v_s = 2 \cos 160t$.

b. Repeat if the frequency of the excitation is cut in half.

33. The same excitation as in Problem 32 is to be applied to the circuit in Figure P33.

a. Derive the state equations for this circuit and determine the natural frequencies.

b. Find the complete response, leaving no unspecified constants, for each of the two state variables.

c. Compare the amplitude of the equilibrium response in this circuit with that in Problem 32 for each of the two variables. What conclusions can you draw about the nature of the two circuits?

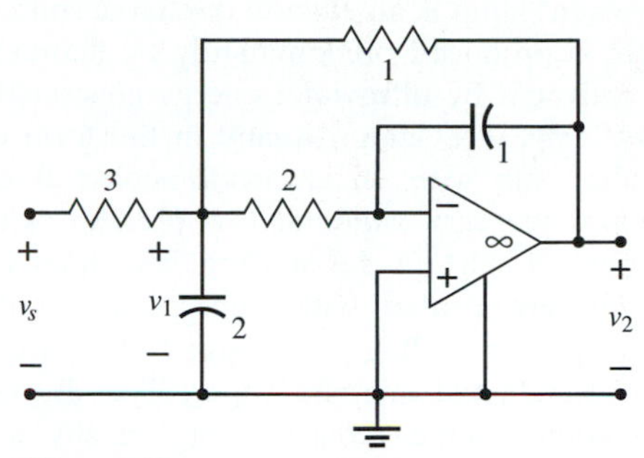

Figure P33

34. The capacitors in the circuit of Figure P34 are initially relaxed when a step voltage is applied. Determine the complete response for the output voltage.

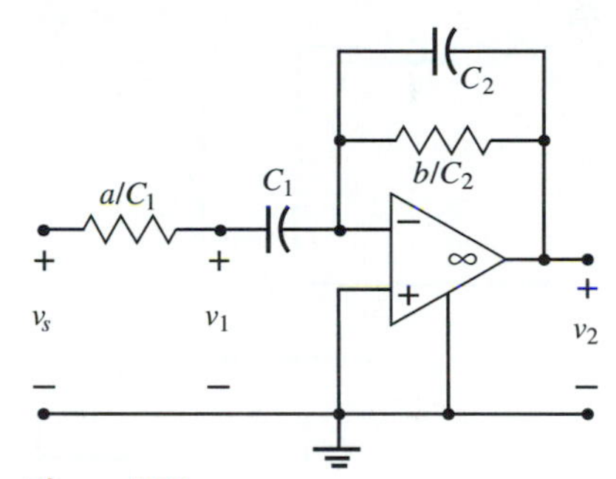

Figure P34

35. The capacitors in Figure P35 are uncharged at $t = 0$ at the time the switch is closed.

a. Derive the state equations for this circuit and find the characteristic values.

b. Find the second-order differential equation satisfied by the output voltage.

c. Determine the complete solution for the output voltage, leaving no unspecified constants.

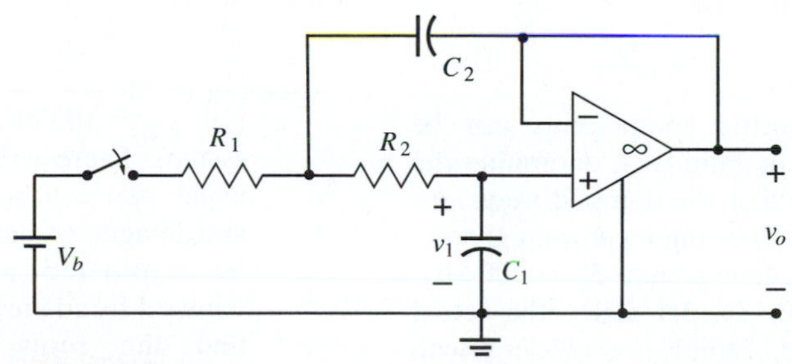

Figure P35

DESIGN PROBLEMS

1. The fluorescent lamp is an electric discharge source in which light is produced predominantly by fluorescent powders activated by ultraviolet energy generated by a mercury lamp. The lamp, usually in the form of a long tubular bulb with an electrode sealed at each end, contains mercury vapor at low pressure with a small amount of inert gas for starting. The inner walls of the bulb are coated with fluorescent powders. When the proper voltage is applied an "arc" is produced through the mercury vapor. This discharge generates some visible radiation, but mostly ultraviolet radiation. The ultraviolet in turn excites the phosphors to emit light. Like most electric discharge lamps, fluorescent lamps require a high voltage to start the discharge and a lower voltage to sustain the discharge once it is started. Figure DP1 represents an idealized model of a fluorescent lamp with its starting circuitry. Switch $S2$ is normally closed. To turn on the lamp, switch $S1$ is closed first, then, after a delay of 5 seconds, switch $S2$ opens.

   In Figure DP1, assume $v_s = 10\,\text{V}$, $R = 1.25\,\Omega$, and $C = 15.625\,\text{mF}$. Determine the value of L for which the voltage across the capacitor, representing the lamp, reaches 1.5 times its steady state value.

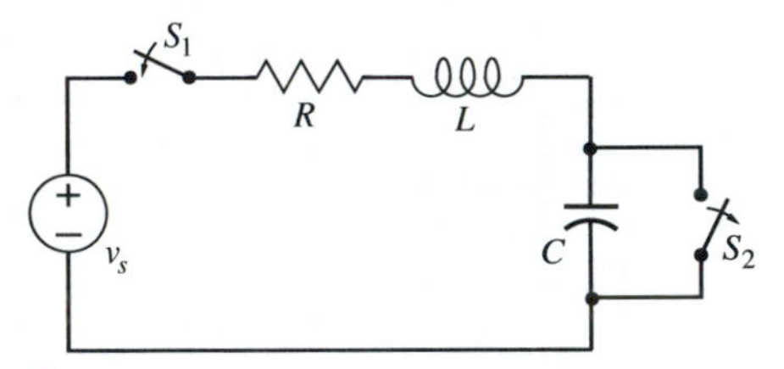

Figure DP1

2. The equivalent circuit of an actual capacitor is shown in Figure DP2(a). It consists of a capacitance, a leakage resistance in parallel with the capacitance representing the resistance of the dielectric in the capacitor, and a series resistance representing the resistance of the leads. The equivalent circuit of an inductor coil is shown in Figure DP2(b). It consists of an inductance, a series resistance representing the resistance of the wire used for the coil, and a parallel capacitance representing the interturn capacitance of the coil. In a physical parallel RLC circuit, these "parasitic components" should be taken into account, as shown in Figure DP2(c). Let $L = 1\,\text{mH}$.

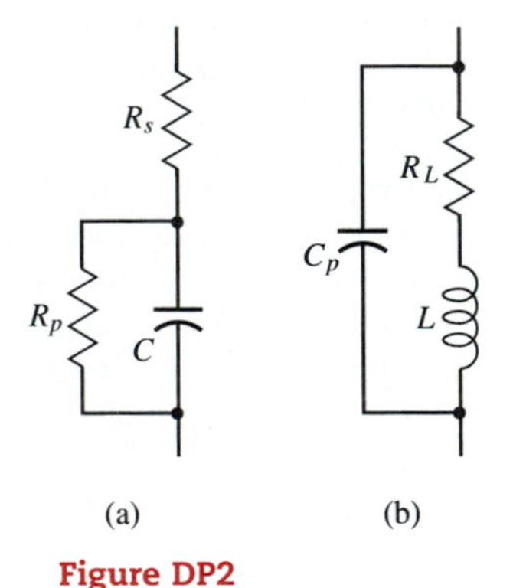

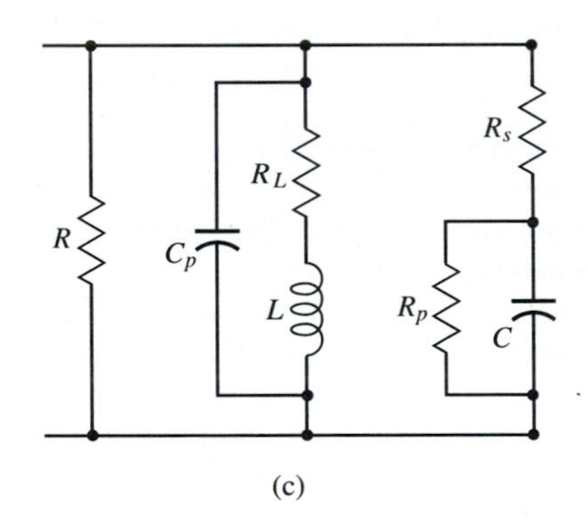

Figure DP2

   a. Assuming all the parasitic components can be ignored (are they zero or infinity?), determine the values of R and C such that the damped frequency is $\approx 10^5$ rad/sec and the damping coefficient is 500.
   b. Neglecting C_p, but assuming that $R_p = 100\,\text{M}\Omega$, $R_s = 0.1\,\Omega$, and $R_L = 0.5\,\Omega$, determine the actual damped frequency and damping coefficient with the values of R and C determined in part (a).
   c. Let $C_p = 100\,\text{pF}$. Now you have a third-order system. There will be three natural frequencies, of which two will be complex and one real. The real and imaginary parts of the complex roots can still be considered as the damping coefficient and damped frequency of the system. Using MATLAB, find the roots and determine the damping coefficient and the damped frequency.

d. Determine the value of R required to have the specified damping coefficient with all the parasitic components taken into consideration.

3. A DC motor connected to a load is shown in Figure DP3a. Under normal operating conditions the switch is in position 1. The equations governing the operation of a DC motor (simplified) are as follows: $v_g = K\omega$ and $T = Ki$, where v_g is the back voltage (often called *emf*) indicated at the terminals of the motor, ω is the angular velocity of the shaft of the motor in rad/s, T is the torque generated by the motor in N-m, i is the motor current, and K is a constant (more or less) characteristic of a given motor under given conditions. For the motor under consideration we will assume that $K = 1.22$ V-s.

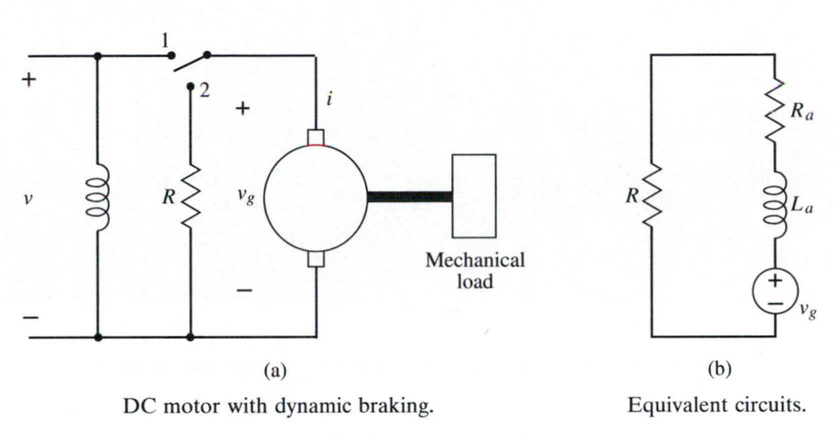

(a) DC motor with dynamic braking.

(b) Equivalent circuits.

Figure DP3

The resistance R is used to stop the motor rapidly when it is turned off; hence the term *dynamic braking.* When the switch is thrown to position 2, the DC machine acts as a generator until it stops completely. Under this condition, the DC machine can be represented as a voltage source in series with a resistance $R_a = 0.2\ \Omega$ and an inductance $L_a = 0.5$ H feeding the resistance R, as shown in Figure DP3b.

The mechanical portion of the system is governed by the following equation:

$$J\frac{d\omega}{dt} + B\omega = T$$

where $J = 6.25$ kg-m^2 is the moment of inertia of the load and $B = 0.5$ kg-m^2/s is the damping factor, and ω and T are as defined above.

Choose ω and i as your state variables and write the differential equations of the system after the switch is thrown into position 2. Assuming that the angular velocity is 188.5 rad/s and that the current i is 60 A just before the switch is moved, determine the value of R such that the angular velocity is at most 1/10 of its original value 15 seconds after the switch is moved to position 2 and that the magnitude of the current never exceeds 70 A.

8 The Sinusoidal Steady State

In the introductory paragraphs of Chapter 6, the following projection appeared: "We are now ready to take up the study of arbitrary circuits made up of any combination of all the elements of our model." So far, we have barely broken the ice. The limitations have been significant: only one or two energy-storing elements and only some simple sources.

We found that the complete response of such circuits consists of two parts: the zero-input (input-free, or natural) response, which is transient; and the forced, or equilibrium, response, which predominates when the transient dies down. In proceeding to generalize what has been done so far, we can take one of several directions. As one possibility, we can gradually take up circuits containing more and more energy-storing elements and find the complete response to more and more complex input waveforms. Another approach is to forget temporarily about the natural response and concentrate on the equilibrium response of circuits to a particular, but highly significant, waveform. This is what we will undertake in this chapter; we will return later to the other task.

The waveform to be considered here is the sinusoid. Its importance stems from several considerations. First, the power system (generation, transmission, distribution, and utilization in hundreds of applications) utilizes the sinusoidal waveform. Except for surges and transients, which are generally undesirable and often cause damage to the system and to appliances connected to it, the system operates in the steady state. Second, as mentioned in Chapter 6, sinusoids are basic components from which other signals, both periodic and aperiodic, can be constructed. Thus, if the responses of linear circuits to sinusoids are well understood, then understanding their responses to other signals will be facilitated.

The goal here is to find the steady-state response of arbitrary circuits to inputs that are sinusoidal. The circuits may contain any number of all the kinds of components introduced in preceding chapters. In order to avoid becoming enmeshed in extraneous details, we'll start with a simple circuit so as to clarify the answers to a number of questions that spring up. You should review the material on sinusoids in Chapter 6 before proceeding.

1 STEADY-STATE RESPONSE OF A SIMPLE CIRCUIT

A simple *RL* circuit excited by a sinusoidal source is shown in Figure 1. In the preceding chapter we would have been interested in the complete solution, say, for the inductor current. Now we want just the forced response. By applying the circuit laws, we find the equation satisfied by the inductor current to be:

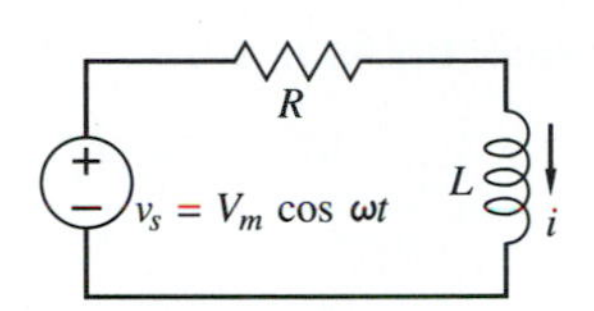

Figure 1
Circuit excited by a sinusoidal source.

$$L\frac{di}{dt} + Ri = v_s(t) = V_m \cos \omega t \tag{1}$$

Since the origin of time is arbitrary, we have chosen it to give the initial angle of the source sinusoid a value of zero.

On the basis of the experience in the preceding two chapters, although admittedly a small sample, we would expect the forced response to be similar to the excitation. That is, we would expect it to be a sinusoid having the same frequency, but not necessarily the same amplitude or initial angle. After all, a linear combination of the desired response and its derivative on the left is to add up to the source voltage. This could not happen if the frequency of the response sinusoid were different from that of the source.

Euler's Identity: Sinusoids as Exponentials

But there is one more concept to introduce before making our play. An important expression that relates sinusoids to exponentials is *Euler's identity*; it states:

$$e^{\pm jx} = \cos x \pm j \sin x \tag{2}$$

$$\begin{aligned} \cos x &= \text{Re}\,(e^{jx}) \\ \sin x &= \text{Im}\,(e^{jx}) \end{aligned} \tag{3}$$

Thus, the cosine in (1) can be expressed as the real part of an exponential with an imaginary exponent: $v_s(t) = V_m \text{Re}\,(e^{j\omega t})$. Not only that, but the expected response can also be expressed in that form. Thus, the anticipated steady-state response can be written:

$$\begin{aligned} i(t) &= I_m \cos(\omega t + \phi) = I_m \text{Re}\,(e^{j(\omega t + \phi)}) \\ &= \text{Re}\,[(I_m e^{j\phi})e^{j\omega t}] \end{aligned} \tag{4}$$

Note that the operations of taking the real part and multiplication by a *real* constant are commutative; they can be interchanged, and that's what we did in the last step. (Consult Appendix B if you feel the need to review complex algebra.)

When these are inserted into (1) and the indicated operations carried out, we get:

$$\text{Re}\,(j\omega L I_m e^{j(\omega t+\phi)}) + \text{Re}\,(R I_m e^{j(\omega t+\phi)}) = \text{Re}\,(V_m e^{j\omega t})$$
$$\text{Re}\,\{[(R + j\omega L)I_m e^{j\phi} - V_m]e^{j\omega t}\} = 0$$

Again, some manipulations were performed that were not shown. The operations of differentiating and taking the real part are commutative. So also are addition and taking the real part. The latter means that the sum of real parts of complex quantities is the same as the real part of the sum. These lead to the last line of the preceding expression, which is of the form:

$$\text{Re}\,(\text{constant} \cdot e^{j\omega t}) = 0$$

This is to be identically true *for all time.* Since the exponential is not 0 for all time, the constant must be zero. Thus:

$$(R + j\omega L)I_m e^{j\phi} - V_m = 0$$
$$I_m e^{j\phi} = \frac{V_m}{R + j\omega L} \tag{5}$$

(A less convincing argument would examine the case for $t = 0$, a particular value of time; whatever is true for all time must be true for *any* particular time, the argument would go. Since $e^{j0} = 1$, we end up with the same result. However, the argument is weak. What is true for one point in time is not necessarily true for all time.)

This expression is extremely significant. Note that the sinusoid representing the response current in (4) has three parameters: the frequency, the amplitude I_m, and the phase angle ϕ. The frequency is known to be the same as the frequency of the source, so the only unknowns are the amplitude and initial angle. The right-hand side of (5) is a complex quantity whose magnitude and angle are exactly the amplitude and initial angle of the desired steady-state inductor current. Once this complex quantity is determined, we multiply it by $e^{j\omega t}$ and then take its real part to obtain the sinusoid.

There's one other convention we need to make, for various reasons. The normal power-system AC household voltage in the United States is labeled 110 volts. (You will find that number on the nameplates of all kinds of electrical devices and appliances, from light bulbs to toasters and radios.) This number does not refer to the *amplitude* of the sinusoidal wave but to its rms value. As you recall, the relationship between the two is $X_m = \sqrt{2}X_{\text{rms}}$, where X can stand for any variable, including voltage and current. Furthermore, because of their principles of operation, many voltmeters give readings that correspond to the rms values of the signals they are measuring. Hence, it would be useful in expressions like (5) if rms values were used instead of amplitudes. If each of the quantities V_m and I_m in that equation were written as $\sqrt{2}$ times the corresponding rms value, the $\sqrt{2}$ would be canceled and there would be exactly the same relationship between the rms values as between the amplitudes.[1]

Phasors

The kind of situation we are addressing in this chapter is this. In a circuit of arbitrary complexity, all sources are assumed to be sinusoidal, all having the same frequency. The steady-state current or voltage response anywhere in the circuit is expected to be sinusoidal at the same frequency. The only things unknown about the response sinusoids are their amplitudes (or rms values) and their initial angles.

We define a *phasor* as follows:

> A phasor representing a sinusoid is a complex quantity whose magnitude is the rms value of the sinusoid and whose angle is the initial angle of the sinusoid.

Thus, if the variable is $x(t)$, the phasor X is related as:

$$x(t) = \sqrt{2}X_{\text{rms}} \cos(\omega t + \phi) \quad \leftrightarrow \quad X = X_{\text{rms}} e^{j\phi} \tag{6}$$

Note that it is the *cosine* form of the sinusoid that is involved in the definition. As you become familiar with the idea that the magnitude of the phasor is the rms value of the sinusoid, we will gradually discard the use of the subscript rms, for the sake of simplicity and convenience.

REPRESENTATION OF A SINUSOID BY A PHASOR The double-headed arrow symbol $\leftrightarrow$ is not an equality; it shows a mutual transformation from a sinusoidal function of time to a complex quantity (a phasor), and vice versa. The idea is illustrated by the following sinusoids and their phasors:

$$i(t) = 20\cos\left(\omega t + \frac{\pi}{3}\right) \quad \leftrightarrow \quad I = \left(\frac{20}{\sqrt{2}}\right)e^{j\pi/3}$$

$$v(t) = 7.07\sin\left(\omega t + \frac{\pi}{8}\right) \quad \leftrightarrow \quad V = 5e^{-j3\pi/8}$$

You might disagree with the angle of the phasor in the second case, but remember that the sine function must be expressed as a cosine before the phasor angle can be specified. This can be done using a trigonometric identity which expresses a sine function as a cosine at an initial angle of $-\pi/2$.

The transformation from the sinusoidal function of time to the phasor is easy: one simply copies the amplitude and initial angle from the time function. The only complication is the need (1) to divide the amplitude by $\sqrt{2}$ and (2) to subtract $\pi/2$ from the initial angle if the sinusoid happens to be given as a sine function. In the opposite direction, if anything, the transformation is even easier. One multiplies the magnitude of the phasor by $\sqrt{2}$ to get the amplitude,

[1] Remember, you are not expected to simply *read* such a discussion, but you are expected to carry out the discussed operations on the equations mentioned and to *confirm* the conclusion.

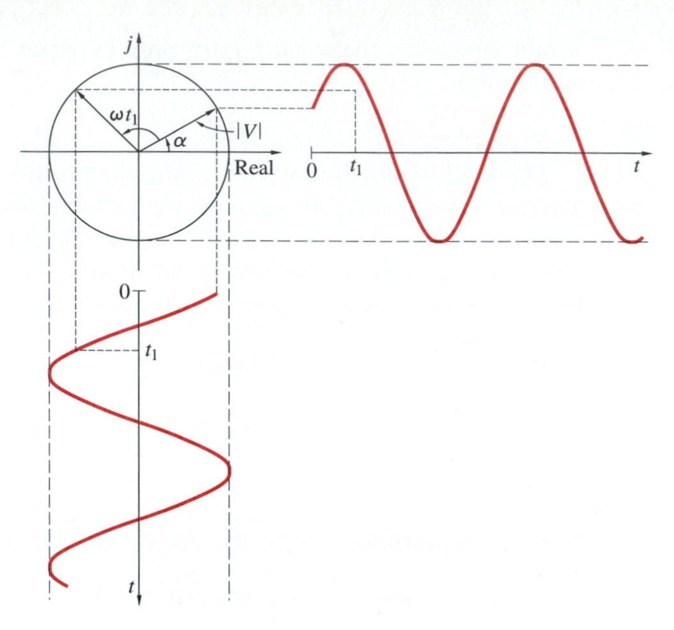

Figure 2
Rotating phasor.

and one just copies the angle as the initial angle of the cosine function. More formally and conceptually, of course, one is supposed to multiply the phasor by $e^{j\omega t}$ and then take the real part.

EXERCISE 1

Find the phasors corresponding to the following sinusoids:

a. $v(t) = 155.6 \sin(377t - 30°)$

b. $i(t) = 353.5 \cos(6283t + 67.5°)$ □

EXERCISE 2

The following complex numbers are phasors; write the corresponding sinusoids both as cosine functions and as sine functions:

a. $V = 10\underline{/-45°}$, **b.** $I = 353.5\underline{/135°}$, **c.** $50e^{j\pi/6}$ □

GRAPHICAL RELATIONSHIP BETWEEN PHASOR AND SINUSOID A simple graphical interpretation can be given for the relationship between a sinusoidal function and its phasor. Since a phasor, say, $V = |V|\, e^{j\alpha}$, is a complex number, it can be represented in the complex plane by a directed line making an angle α with the real axis as shown in Figure 2. When the phasor is multiplied by $e^{j\omega t}$, the entire angle becomes a time-varying quantity $(\alpha + \omega t)$. As time goes on, the angle increases at a uniform rate. The directed line representing the phasor appears to rotate in the counterclockwise direction at an angular velocity ω. As the line rotates, it will have a projection on the real axis, as follows:

$$\text{Real-axis projection} = \text{Re}\,(Ve^{j\omega t}) = |V| \cos(\omega t + \alpha) \tag{7}$$

Because the phasor is defined in terms of the rms value, this projection differs from the sinusoid by the factor $\sqrt{2}$, which can be easily supplied. The projection is shown in Figure 2. The sinusoid is obtained by drawing a time axis vertically downward and choosing a convenient scale for t. For each value of t, the top of the rotating line is projected downward until it intersects a horizontal line drawn perpendicular to the t-axis at the same value of t. One such point is labeled t_1 in the figure.

To recover a sinusoid—conceptually—from its phasor, we would: lay out the phasor in the complex plane as a line drawn at its initial angle; stretch it by a factor of $\sqrt{2}$; cause it to rotate counterclockwise at an angular velocity ω; and, finally, take its projection on the real axis. Of course, none in their right mind would want to perform this series of steps graphically; it's much simpler to do it algebraically. The graphical view, however, gives just one more way of looking at things, and as a wise old person once said, "A major purpose of eduction is to provide numerous viewpoints."[2] As a corollary, you can see that a projection on the imaginary axis is also possible and it also generates a sinusoid:

$$\text{Imaginary-axis projection} = \text{Im}\,(Ve^{j\omega t}) = |V|\sin(\omega t + \alpha) \tag{8}$$

The sinusoid so obtained is also shown in the figure.

In a given problem, if two sinusoids of the same frequency are involved, we can consider each to be generated by a rotating line. Since the two rotating lines have the same angular velocity, their *relative* position does not change, but depends solely on the difference of their initial angles, or their *phase difference.* Thus if:

$$\begin{aligned} v_1 &= \sqrt{2}\,|V_1|\cos(\omega t + \alpha) \\ v_2 &= \sqrt{2}\,|V_2|\cos(\omega t + \beta) \end{aligned} \tag{9}$$

then the phasors will be $|V_1|\,e^{j\alpha}$ and $|V_2|\,e^{j\beta}$. They can be represented in the complex plane by the directed lines shown in Figure 3a. These are the positions of the rotating lines at $t = 0$. The angle between the two lines, which is $\alpha - \beta$, is labeled θ. At some later time, the lines will have rotated to a new position, as shown in Figure 3b, but their lengths and the angle between them will still be the same. A diagram such as the one in Figure 3a is called a *phasor diagram.*

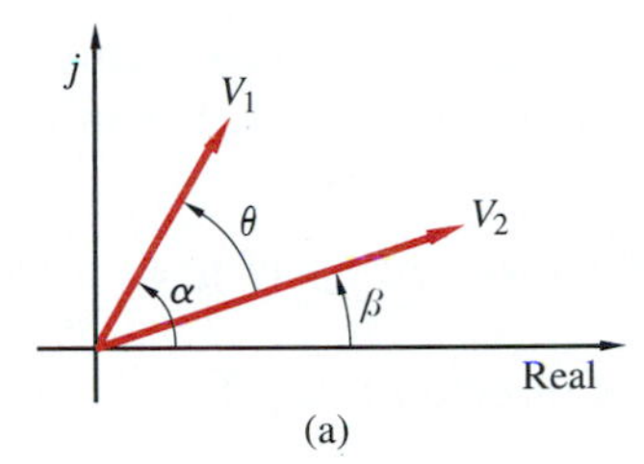

(a)

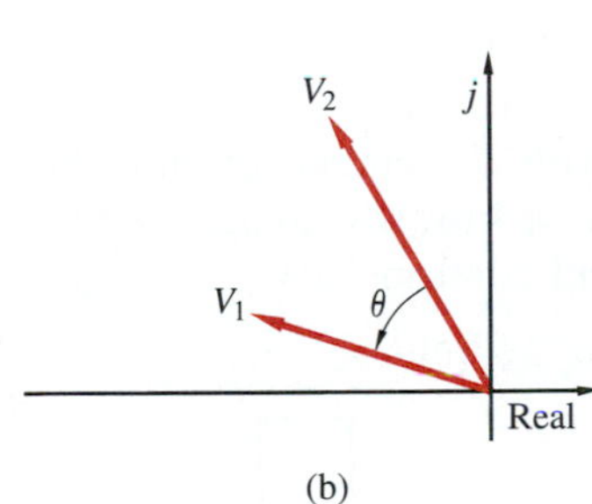

(b)

Figure 3
Phasor diagram.

Clearly, if there are other sinusoids involved in the same problem, the phasors can all be shown in the same diagram, with the appropriate lengths and angle relative to each other. Since the origin of time is arbitrary and the absolute locations of the directed lines are not important, but only their relative locations, it may be convenient to choose one of the phasors in a horizontal position, along the real axis. This phasor is said to be taken as the *reference for phase.*[3]

[2] Anyway, some wise old person *should* have said it!

[3] A notational problem arises when dealing with phasors, since reference is to be made both to a phasor and to its magnitude. If a phasor represents a current, what does I refer to: the phasor or its magnitude? The same difficulty would arise with other complex quantities, such as the complex frequency or the impedance, to be defined in the following section. This problem is sometimes resolved by using special symbols to represent phasors, such as boldface. Since boldface is almost universally used to designate matrices, we shouldn't use the same thing to represent phasors also. In this book we resolve the problem by using no special notation for phasors but indicating their magnitude by magnitude bars, at least when there is the possibility of confusion.

2 FUNDAMENTAL LAWS IN TERMS OF PHASORS

So far in this chapter, we have laid the groundwork and the justification for finding the steady-state response of linear electrical networks to sinusoidal excitations in terms of exponential functions and phasors. Not much will be gained in the way of computational effort, however, if, for each problem, we must go through the entire process of writing node equations or loop equations, which will be differential equations; substituting the exponentials for the sinusoids; and then solving the equations. What would happen if, instead, we went back to the fundamental laws themselves—Kirchhoff's laws and the v-i relationships of components—and examined them under sinusoidal steady-state conditions? That's what we will now do.

A linear circuit of arbitrary complexity, containing any number of each type of component, is given. The sources are all sinusoidal of the same frequency. Let's consider KCL at node m; it can be written as:

$$\sum_{k=1}^{b} a_{mk} i_k(t) = a_{m1} i_1 + a_{m2} i_2 + \cdots + a_{mb} i_b \tag{10}$$

The summation extends over *all* branches of the circuit, even though many of them are not incident at node m. As usual, b is the number of branches in the circuit; the coefficients a_{mk} are as follows:

a_{mk} = +1 if the current reference for branch k is directed away from node m.
a_{mk} = −1 if the current reference for branch k is directed toward node m.
a_{mk} = 0 if branch k is not connected to node m.

EXERCISE 3

To get a feel for this, draw a circuit diagram with a number of branches connected at a node. Number the branches and arbitrarily assign current references to them. Now write KCL at the node and confirm (10). □

Equation (10) can also be written in matrix form, as follows:

$$\mathbf{A}_m \mathbf{i}(t) = 0 \qquad \mathbf{A}_m = [a_{m1} \quad a_{m2} \cdots a_{mb}] \qquad \mathbf{i}(t) = \begin{bmatrix} i_1 \\ i_2 \\ \vdots \\ i_b \end{bmatrix}$$

In the steady state, all the currents will be sinusoids of the same frequency. Hence, for the kth current we can write:

$$i_{mk} = \text{Re}\,(\sqrt{2} I_{mk} e^{j\omega t}) = \sqrt{2}\,\text{Re}\,(I_{mk} e^{j\omega t})$$

where I_{mk} is a phasor. If this is inserted into (10) and it is noted that $\sqrt{2}$ appears in every term of the equation so that it can be canceled out, the result will be:

$$\sum_{k=1}^{b} a_{mk}\,\text{Re}\,(I_k e^{j\omega t}) = \text{Re}\left[\left(\sum_{k+1}^{b} a_{mk} I_k\right) e^{j\omega t}\right] = 0$$

(Do you remember why the real part and the summation operations can be interchanged?) The only way this equation can be satisfied for all t is for the sum multiplying the exponential to be zero. Thus:

$$\sum_{k=1}^{b} a_{mk} I_k = a_{m1} I_1 + a_{m2} I_2 + \cdots + a_{mb} I_b = 0 \qquad (11)$$

$$\mathbf{A}_m \mathbf{I} = \mathbf{0}$$

This is in exactly the same form as KCL written in (10), except that the variables here are phasors instead of functions of time. The conclusion is that in the sinusoidal steady state, *KCL is satisfied by the phasors representing the branch currents that themselves satisfy KCL.*

EXERCISE 4

Repeat the process just completed and show that, in the sinusoidal steady state, the phasors representing the branch voltages around any closed path satisfy KVL. □

Now there remains the problem of finding out how the branch v-i relationships can be expressed in terms of phasors. There are two categories to distinguish: the algebraic relations (for resistors, ideal transformers, controlled sources, and other dissipative components), and the differential ones (for capacitors and inductors). In the case of the algebraic relationships, one sinusoid is multiplied by a constant to get a second sinusoid. This means that the corresponding phasors are related by the same constant. Thus, if $v(t) = Ri(t)$ and if these variables are sinusoids with phasors V and I, respectively, then $V = RI$ also, and *Ohm's law is satisfied by phasors in the steady state.* The same is true for all other algebraic relationships. For ease of referring to the sinusoidal steady state where the variables are represented by the corresponding phasors, we say that this is the *phasor domain.* Thus, we have already established that KVL, KCL, Ohm's law, and all other algebraic voltage-current relationships are satisfied in the phasor domain.

For the capacitor, $i(t) = C\,dv(t)/dt$. In the sinusoidal steady state, both i and v are sinusoids, represented by phasors I and V. Representing each sinusoid by the real part of the corresponding phasor multiplied by $e^{j\omega t}$, and utilizing the commutative properties of some of the mathematical operations involved, we obtain the following sequence of results:

$$\begin{gathered} \operatorname{Re}(Ie^{j\omega t}) = C \operatorname{Re}\left[\frac{d}{dt}(Ve^{j\omega t})\right] \\ \operatorname{Re}[(I - j\omega CV)e^{j\omega t}] = 0 \\ I = j\omega CV \end{gathered} \qquad (12)$$

This is the I-V relationship for a capacitor in the phasor domain. It is strictly algebraic in nature and consitutes a tremendous simplification over the derivative relationship in the time domain.

EXERCISE 5

Carry out a development similar to that of the preceding pargraphs to determine a relationship between the phasor voltage and current of an inductor.

ANSWER: $V = j\omega LI.$ □

The implications that have been shaping up in this section have great significance. All the methods of analysis carried out earlier for dissipative networks were based on Kirchhoff's laws and the v-i relationships of the components. If these equations have now been successfully expressed in the phasor domain, where all the relationships are algebraic, it will surely be possible to extend the methods of analysis already discussed for dissipative circuits to more general networks containing any number of L's and C's together with dissipative components in the steady state. That's the task to which we will now turn our attention.

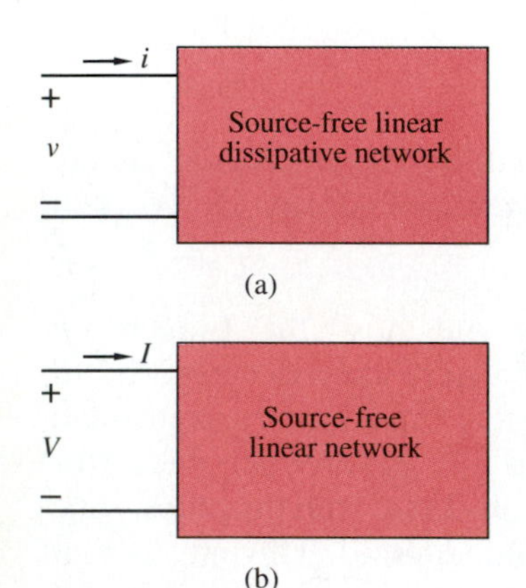

Figure 4 Definition of impedance and admittance.

Impedance and Admittance

Along the way to exploiting the ideas introduced in the preceding section, we need to define some new concepts and terminology. Look at Figure 4a, which shows an arbitrary linear dissipative source-free circuit. Assuming standard references, as shown, the equivalent resistance at the terminals of this circuit is simply the ratio of the terminal voltage, taken as a response, to the exciting current, independent of their waveforms. If the current is the response to an exciting voltage, then the equivalent conductance is given as the ratio of the current to the voltage. The conductance, of course, is the reciprocal of the resistance.

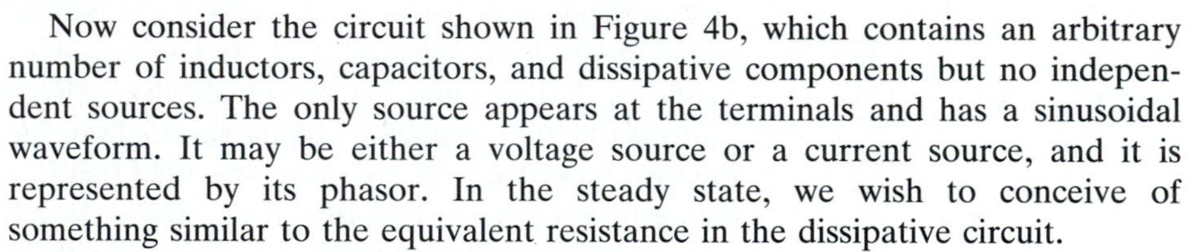

Now consider the circuit shown in Figure 4b, which contains an arbitrary number of inductors, capacitors, and dissipative components but no independent sources. The only source appears at the terminals and has a sinusoidal waveform. It may be either a voltage source or a current source, and it is represented by its phasor. In the steady state, we wish to conceive of something similar to the equivalent resistance in the dissipative circuit.

For the circuit in Figure 4b we define the *impedance,* designated by Z, and its reciprocal, the *admittance,* designated by Y, as follows:

$$Z \equiv \frac{V}{I}$$
$$Y \equiv \frac{I}{V} = \frac{1}{Z} \tag{13}$$

Note very carefully that impedance and admittance are defined as ratios of *phasors,* not of time functions.

Suppose the excitation is the voltage, the current being the response. Once the admittance is known, the response current phasor is determined by the product of Y and the excitation phasor V. Similarly, if the current is the excitation, then the response-voltage phasor is determined by the product of Z and I. Thus:

The sinusoidal steady-state response of a source-free linear circuit with one pair of terminals is completely determined from a knowledge of the impedance Z or the admittance Y.

The impedance and the admittance are defined at a single pair of terminals. The circuit is excited, or *driven,* at these terminals and the response is taken at the same terminals. Hence, these quantities are called the *driving-point* impedance and admittance.[4]

Since voltage and current phasors are generally complex, their ratio—the impedance or admittance—will also be complex. Hence, in rectangular form, we can write:

$$Z = R + jX \tag{14a}$$

$$Y = G + jB \tag{14b}$$

These forms show the real and imaginary parts explicitly. Over time, they have acquired certain names:

R = real part of Z: resistance component
X = imaginary part of Z: reactance component
G = real part of Y: conductance component
B = imaginary part of Y: susceptance component

The names *resistance* and *conductance*—and the symbols R and G—are unfortunate choices, because each of these names and symbols already represents a circuit parameter. For specific circuits or specific frequencies, each of the four components just defined may be constant, but generally, they are functions of frequency. Thus, the "resistance" component of an impedance is not generally constant, much less a single circuit parameter. Since these names and symbols are almost universally used, however, we will also continue to use them, however reluctantly.

IMPEDANCE AND ADMITTANCE OF SINGLE COMPONENTS In preparation for considering more complex circuits, let's examine the impedance and admittance of single circuit components. The resistor needs little comment; its voltage and current phasors have the same angle, and we say they are *in phase.* The same result would apply to multiterminal dissipative components.[5]

For the inductor:

$$Z_L = \frac{V_L}{I_L} = j\omega L \qquad X_L = \omega L \tag{15a}$$

$$Y_L = \frac{1}{Z_L} = \frac{1}{j\omega L} = -j\frac{1}{\omega L} \qquad B_L = -\frac{1}{\omega L} \tag{15b}$$

[4] The reason is to distinguish them from similar quantities later to be defined when the excitation is at one pair of terminals and the response at another pair, called the *transfer* impedance and admittance.

[5] For some of the controlled sources and other multiterminal components, concepts other than impedance and admittance are needed, since the branch relations are not between a current and a voltage but between two currents or two voltages. At least these branch relations are algebraic, i.e., proportionalities. Hence, the phasor relationships will involve the same proportionalities. In this respect, they are similar to resistors.

Both the impedance and the admittance of the inductor are purely imaginary. The *inductive reactance* is $X_L = \omega L$, which is positive for positive frequency. The *inductive susceptance* is $B_L = -1/\omega L$, which is negative for positive frequency. The angle of the voltage phasor is 90° greater than that of the current phasor. We say that the voltage and current are 90° *out of phase,* or they are *in quadrature,* with the voltage *leading* the current.[6]

For a capacitor:

$$Z_C = \frac{V_C}{I_C} = \frac{1}{j\omega C} = -j\frac{1}{\omega C} \qquad X_C = -\frac{1}{\omega C} \tag{16a}$$

$$Y_C = \frac{I_C}{V_C} = j\omega C \qquad B_C = \omega C \tag{16b}$$

Again, both the impedance and the admittance are imaginary. The *capacitive reactance* is $X_C = -1/\omega C$, which is negative for positive frequency. The *capacitive susceptance* is $B_C = \omega C$, which is positive for positive frequency. The angle of the voltage phasor is 90° less than that of the current phasor. Again, the voltage and current are 90° *out of phase,* but this time the voltage *lags* the current.

Perhaps a better understanding of leading and lagging angles can be obtained from a consideration of the sinusoids shown in Figure 5. Suppose the dashed curve represents the current in an inductor. The voltage, being proportional to the derivative of the current, will be represented by the solid curve. A positive peak of the voltage curve occurs 90° before the closest positive peak of the current curve; hence, the voltage leads the current by 90°.

It is also true that a positive peak of the voltage curve occurs 270° after the previous positive peak of the current curve. Hence, you might be tempted to

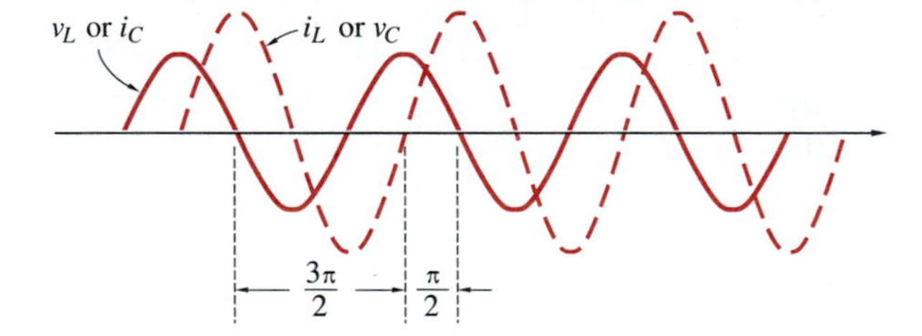

Figure 5
Steady-state voltage and current in an inductor or a capacitor.

[6] I can see some of you bristling: "What does he mean, *positive* frequency," you might be asking; "isn't frequency—the number of cycles completed by a periodic function per unit of time—*inherently* positive?" It *is,* as first defined; but concepts are elastic: they can be stretched in new dimensions that weren't imagined when they were first introduced. Thus, $\cos \omega t = \text{Re}\, e^{j\omega t}$; but what is $\text{Re}\, e^{-j\omega t}$? Why, $\cos(-\omega t)$, of course. If ω is positive (angular) frequency, then surely $-\omega$ is *negative* (angular) frequency! Indeed, $j\omega$ is *imaginary* frequency, and, before you know it, we have stretched the concept to include *complex* frequency. All of this does not mean that we can act like the Mad Hatter (in *Alice in Wonderland*) and take words (and ideas) to mean anything we want them to mean. What it signifies is that the meaning of concepts can sometimes be extended and generalized to new situations, provided that, when the new situation is reduced to the old one, then the concept reduces to the original one. Look for other examples of this in your studies.

say that the voltage lags the current by 270°. Although this would not be incorrect, 270° is not within the range we have agreed upon for the angle, namely, the principal value between −180° and +180°.

The same description would apply for a capacitor, but with the words *voltage* and *current* interchanged.

EXERCISE 6

Write a statement for a capacitor like the preceding statement about an inductor voltage and current but interchanging the words *voltage* and *current*. Review the statement in connection with Figure 5 to verify that it makes sense. □

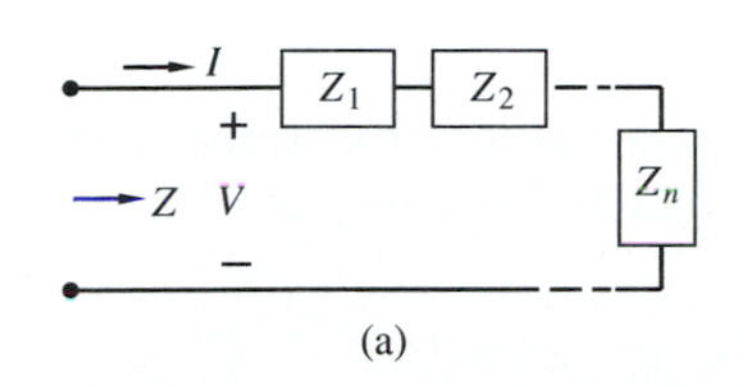

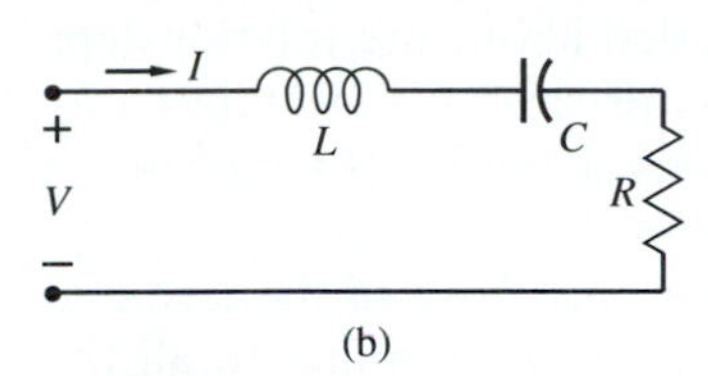

Figure 6 Series-connected branches.

SERIES CONNECTIONS AND VOLTAGE DIVIDERS As the next order of complexity, let's consider a series connection of branches, as shown in Figure 6a. Each branch is labeled in terms of its impedance, and the terminal voltage and current have been specified by their phasors. If we write KVL in terms of phasor voltages and use the V-I relationships of the branches, the result will be:

$$\begin{aligned} V &= Z_1 I + Z_2 I + \cdots + Z_n I \\ &= (Z_1 + Z_2 + \cdots + Z_n)I = ZI \\ Z &= Z_1 + Z_2 + \cdots + Z_n \end{aligned} \tag{17}$$

Thus, the impedance of a set of two-terminal branches connected in series equals the sum of the impedances of the individual branches. (Does this result depend on the choice of branch voltage and current references? Try different references and confirm.)

EXERCISE 7

Find an expression for the impedance of the circuit in Figure 6b in terms of ω and element values.

ANSWER: $Z = V/I = R + j\omega L + 1/j\omega C = R + j(\omega L - 1/\omega C)$. □

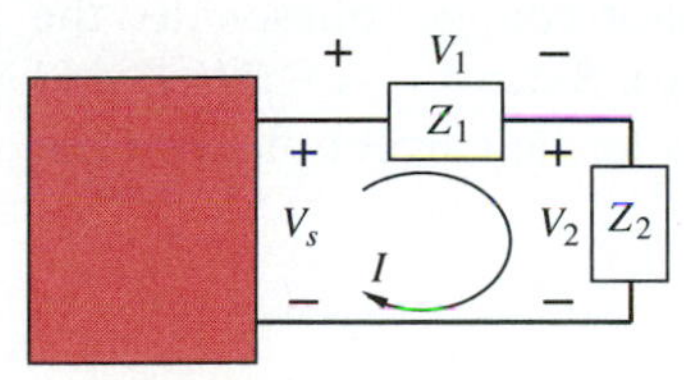

Figure 7 Voltage divider.

As a simple case of a series connection, consider Figure 7, which consists of two branches in series, excited by a sinusoidal voltage source. The objective is to determine the steady-state voltage across one of the two branches.

The structure is that of a voltage divider, first discussed in Chapter 3 for resistive circuits. The phasor voltage V_2 is obtained in the same way it was in the resistive case. The phasor current supplied by the source to the equivalent impedance seen from its terminals is the voltage phasor multiplied by the equivalent admittance, or divided by the equivalent impedance, since $Y = 1/Z$. The desired voltage phasor then is Z_2 times this current phasor. The details are as follows:

$$\begin{aligned} I &= \frac{V_s}{Z_1 + Z_2} \\ V_2 &= Z_2 I = \frac{Z_2}{Z_1 + Z_2} V_s \end{aligned} \tag{18}$$

The form is the same as the voltage-divider relationship for resistive circuits, but in terms of impedance rather than resistance. A similar result is true for the phasor voltage across Z_1, except that Z_1, not Z_2 would appear in the numerator. In words, the phasor voltage across the series combination of two branches is distributed between the two branches in the ratio of the impedance of each branch to the total impedance. Clearly, this result can be generalized to more than two branches in series.

EXERCISE 8

Using the voltage-divider concept, express the voltage across the resistor in Figure 6b in terms of the element values and the angular frequency ω.

ANSWER: $V_R/V = R/[R + j(\omega L - 1/\omega C)]$. □

It is often the case that only a rough idea is needed about the relationships among various voltages and currents in a circuit; for example, whether one voltage is greater than another or whether a certain voltage leads a current or lags it. In such cases, a phasor diagram may be useful.

In Figure 6b, for example, we may be interested in the phase of the terminal voltage relative to that of the current. Since the current is common to all the components, we use the angle of the current as a reference by taking it to be zero. We lay out a horizontal line to represent the current phasor. A scale (say, in amperes per centimeter) is established so that the length of this line is the magnitude of the phasor current—its rms value. The voltage of the resistor is in phase with the current. Hence, the line representing the resistor-voltage phasor lies along the line representing the current. The length of this line will depend on the voltage scale chosen—say, in volts per centimeter. These two lines are shown in Figure 8a.

The terminal voltage is the sum of the three element voltages. We intend to add the capacitor-voltage phasor and the inductor-voltage phasor to the resistor-voltage phasor, assuming standard references. Since $V_C = -jI/\omega C$ and $V_L = j\omega LI$, both of these phasors are at right angles to the current phasor (and

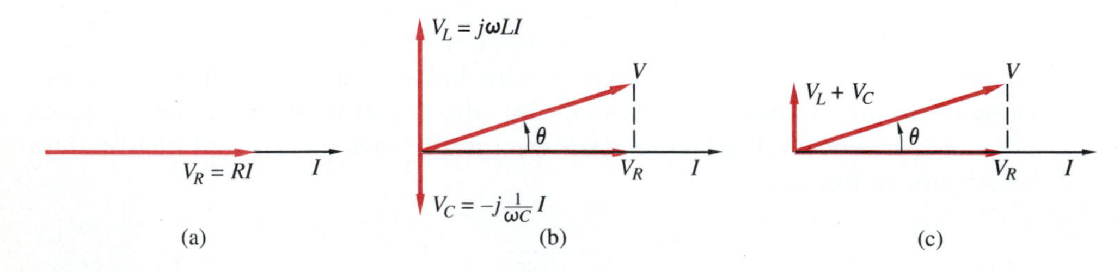

Figure 8
Phasor diagram for Figure 6b.

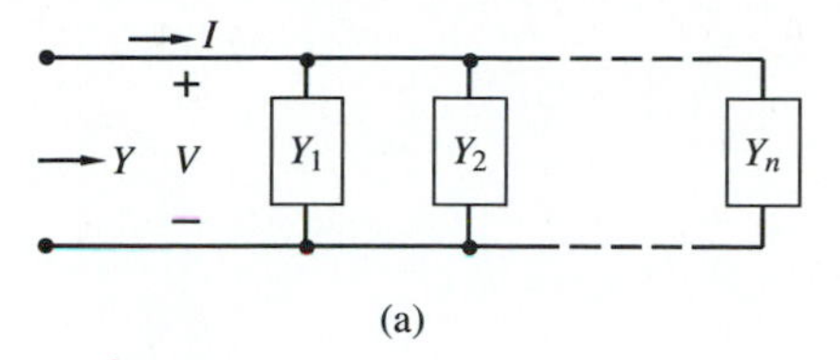

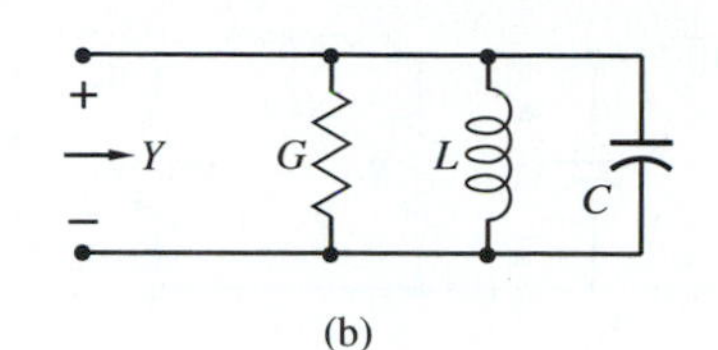

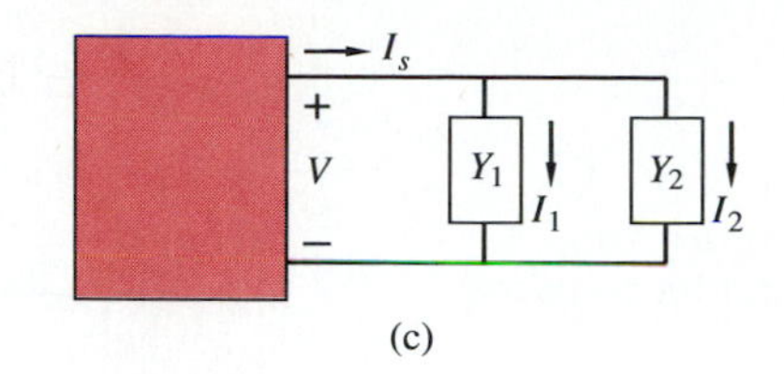

Figure 9
Parallel-connected branches and current dividers.

the resistor-voltage phasor), one leading and one lagging. Depending on their magnitudes, one or the other might dominate. One possibility is shown in Figure 8b, where V_L is assumed to have a larger magnitude. The final result, shown in Figure 8c, is obtained by graphical addition. In this case, where the inductive reactance dominates, the entire circuit is said to be *inductive.* The voltage leads the current by an angle θ. In the opposite case, the circuit would be said to be *capacitive* and the voltage would lag the current.

PARALLEL CONNECTIONS AND CURRENT DIVIDERS The duals of the preceding circuits are illustrated in Figure 9. In Figure 9a, a number of branches are shown connected in parallel and excited by a sinusoidal current source represented by its phasor. We can now invoke the principle of duality to obtain:

$$Y_{\text{eq}} = Y_1 + Y_2 + \cdots + Y_n \tag{19}$$

Thus, the admittance of a parallel connection of two-terminal branches is the sum of the admittances of the individual branches.

The case shown in Figure 9c has the structure of a current divider. Our objective is to find the fraction of the input current phasor that is supplied to one of the two branches. The result can be written by duality with the voltage divider:

$$I_2 = \frac{Y_2}{Y_1 + Y_2} I_s \tag{20}$$

In words, this says that the phasor current supplied to the parallel combination of two branches is distributed to each branch as the ratio of the admittance of that branch to the total admittance.

EXERCISE 9

Confirm (19) and (20) by invoking the fundamental laws. □

LADDER NETWORKS A number of successive steps were carried out for resistive ladder networks in Chapter 3 to arrive at solutions. These included series and parallel connections of branches, voltage-divider and current-divider relationships, source transformations (later generalized to Thévenin's

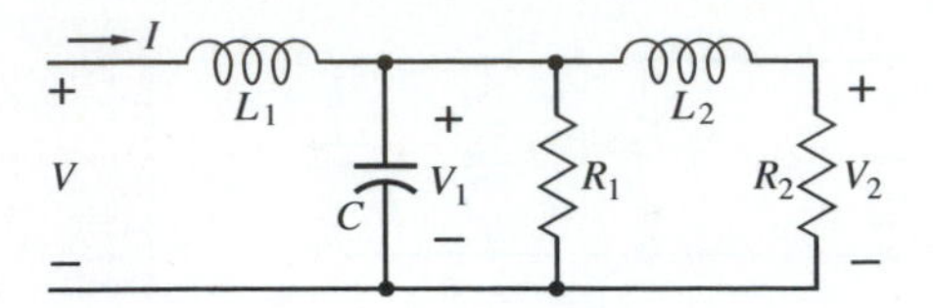

Figure 10
Ladder network.

theorem), and continued-fraction expansions. All of these apply equally well in the phasor domain. (Thévenin's theorem will be taken up in a later section.) A single example should suffice to illustrate the idea.

Such an example is given in Figure 10. The equivalent impedance at the terminals of the ladder is easily found by continued fractions as follows:

$$Z = \frac{V}{I} = j\omega L_1 + \cfrac{1}{j\omega C + G_1 + \cfrac{1}{j\omega L_2 + R_2}}$$

Only algebra is needed to place this in more compact, collected form. You should do that in your spare time.

The output voltage can be obtained in terms of V in small steps. Thus, it can be written in terms of V_1 from the voltage-divider relationship. The input current can then be obtained by KCL applied at the node formed at the junction of the two inductors. Then the input voltage will be $V = j\omega L_1 I + V_1$. These steps are as follows:

$$V_2 = \frac{R_2}{j\omega L_2 + R_2} V_1 \qquad V_1 = \frac{j\omega L_2 + R_2}{R_2} V_2$$

$$\begin{aligned} I &= \left(j\omega C + G_1 + \frac{1}{j\omega L_2 + R_2}\right) V_1 \\ &= [(j\omega C + G_1)(j\omega L_2 + R_2) + 1]\frac{V_2}{R_2} \\ V &= j\omega L_1 I + V_1 \\ &= j\omega L_1[(j\omega C + G_1)(j\omega L_2 + R_2) + 1]\frac{V_2}{R_2} + \frac{j\omega L_2 + R_2}{R_2} V_2 \\ \frac{V_2}{V} &= \frac{R_2}{j\omega L_1[(j\omega C + G_1)(j\omega L_2 + R_2) + 1] + j\omega L_2 + R_2} \end{aligned}$$

(The final form for I is obtained by inserting the expression for V_1 in terms of V_2 from the first line.) Writing the denominator of the last line as a complex number in rectangular form is just a matter of algebra.

EXERCISE 10

Take the series connection of L_2 and R_2 in Figure 10 as a single branch in parallel with R_1 and C. Taking that entire parallel combination as a single

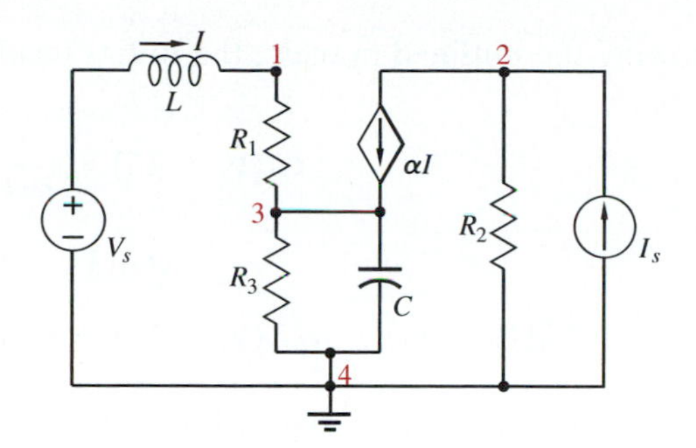

Figure 11
Candidate circuit for node analysis.

branch, use the voltage-divider idea to write V_1 in terms of V. Then use it again to determine V_2. Confirm the previous result. □

3 CIRCUIT EQUATIONS

In the study of dissipative circuits carried out in Chapter 5, a number of general methods of analysis (node equations, and loop and mesh equations) were described. These procedures rested on different ways of combining the fundamental circuit laws, resulting in sets of simultaneous equations to be solved. Since, in the sinusoidal steady state, the fundamental laws are satisfied by phasors, these same sets of equations can be written in terms of phasors. Recall, however, that the presence of certain multiterminal dissipative components requires some modifications in the procedures.

An example is provided in Figure 11. This circuit contains a CCCS, which has neither an impedance nor an admittance representation. However, the controlling current is in series with a resistor, so by taking this accompanying resistor as part of the multiterminal component, it can be converted to a voltage-controlled current source. Hence, a node analysis is possible.

Node Equations

In the preceding circuit, there are a total of four nodes; hence, three node equations will be needed. Remember the process:

1. First choose one node as a datum, say, node 4.
2. Then, concentrating on each node, assume that the current reference of each branch connected there is directed away from the node; standard references are assumed for all branches.
3. Next, "think" KCL in terms of branch-current phasors at each node, and then "think" of eliminating branch-current phasors in favor of branch-voltage phasors using the branch I-V relationship.
4. Finally, replace each branch-voltage phasor by the difference between two node-voltage phasors and "write" the result.

Following the outlined process, the initial result is:

$$G_1(V_1 - V_3) + \frac{1}{j\omega L}(V_1 - V_s) = 0$$

$$\alpha G_1(V_1 - V_3) + G_2V_2 = I_s$$

$$G_1(V_3 - V_1) - \alpha G_1(V_1 - V_3) + (G_3 + j\omega C)V_3 = 0$$

Terms are now collected; in matrix form, the result becomes:

$$\begin{bmatrix} G_1 + \dfrac{1}{j\omega L} & 0 & -G_1 \\ \alpha G_1 & G_2 & -\alpha G_1 \\ -(1+\alpha)G_1 & 0 & (1+\alpha)G_1 + G_3 + j\omega C \end{bmatrix} \begin{bmatrix} V_1 \\ V_2 \\ V_3 \end{bmatrix} = \begin{bmatrix} \dfrac{1}{j\omega L}V_s \\ I_s \\ 0 \end{bmatrix}$$

(Observe the matrix of coefficients; is it symmetrical?)

The solution for the node voltage phasors is now obtained by inverting the coefficient matrix. However, not all the cofactors are needed; since one of the elements of the input vector is 0, all the cofactors that would have multiplied this element are not needed. Symbolically, by Cramer's rule the solution is:

$$V_1 = \frac{\Delta_{11}}{\Delta}\frac{V_s}{j\omega L} + \frac{\Delta_{21}}{\Delta}I_s$$

$$V_2 = \frac{\Delta_{12}}{\Delta}\frac{V_s}{j\omega L} + \frac{\Delta_{22}}{\Delta}I_s$$

$$V_3 = \frac{\Delta_{13}}{\Delta}\frac{V_s}{j\omega L} + \frac{\Delta_{23}}{\Delta}I_s$$

There is still a considerable amount of work to do to evaluate the determinant and the cofactors in this expression, but it is mainly routine.[7]

Loop Equations

The process for carrying out a loop analysis (or mesh analysis for planar circuits) was developed in Chapter 5. Recall that it includes the following steps:

1. Select a set of independent loops. (A surefire way is to construct a tree

[7] As pointed out in Appendix A, Cramer's rule is not the most computationally efficient method of solution, although it is symbolically convenient. Numerically, Gaussian elimination would be most efficient, although the gain is not great for low-order matrices. Of course, software packages, such as MATLAB, are available for carrying out the solution of linear algebraic equations like this set. These would be best to use if all you wanted was a numerical answer in a particular circuit with given parameter values, without concern for understanding.

and choose the fundamental loops for that tree. When the circuit is planar, the meshes will serve adequately.)

2. Taking one loop at a time, we assume that the references of all branches are standard references. Branch-voltage references are so chosen that, when traversing that loop, each branch reference plus is enountered first. Another way of saying this is that each branch voltage is taken to be a potential decrease (a drop) in the direction of transversal of the loop.

3. "Think" KVL in terms of branch-voltage phasors; then "think" of eliminating branch-voltage phasors in favor of branch-current phasors using the V-I relationships.

4. Then replace each branch-current phasor in terms of the loop-current phasors by an appropriate relationship and "write" the result. (For general loop equations, this relationship is not always the difference between two loop currents.) In a circuit with b branches and n nodes, there will be $b - (n - 1)$ independent loop equations.

As a mundane example, consider the circuit in Figure 12. The controlled source here is a current-controlled voltage source, the kind that is exactly suited for a loop analysis.

In the present example, we will write mesh equations; there will be two of them. (Confirm that. The accompanied current source is a single branch.) Note that nothing special needs to be done about the controlled source; in the last step of the procedure outlined above, where the branch currents are expressed in terms of loop currents, the controlling current of the controlled source is also so expressed. The result of applying this procedure is as follows:

$$(R_1 + j\omega L_1)I_1 + j\omega L_2(I_1 - I_2) + rI_2 + R_3(I_1 - I_s) = 0$$

$$\frac{1}{j\omega C}I_2 + R_2 I_2 - rI_2 + j\omega L_2(I_2 - I_1) = 0$$

$$\begin{bmatrix} (R_1 + R_3) + j(\omega L_1 + \omega L_2) & r - j\omega L_2 \\ -j\omega L_2 & R_2 - r + j\omega L_2 - j\dfrac{1}{\omega C} \end{bmatrix} \begin{bmatrix} I_1 \\ I_2 \end{bmatrix} = \begin{bmatrix} R_3 I_s \\ 0 \end{bmatrix}$$

The last step is obtained by rearranging terms and putting the result in matrix form. To further pursue the solution, we will utilize the given numerical

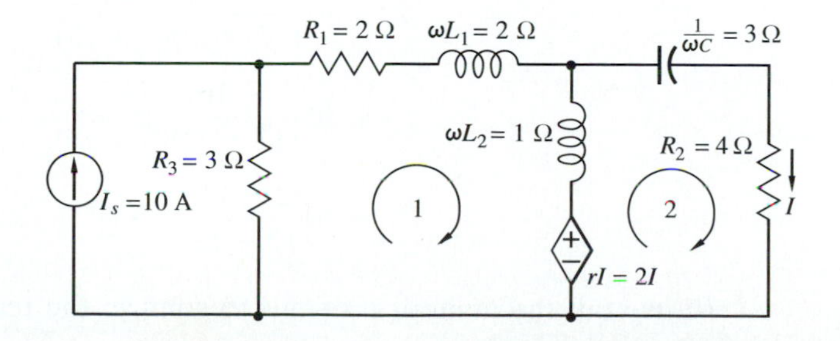

Figure 12
Candidate circuit for loop analysis.

values. When these are inserted into the last form, the result is:

$$\begin{bmatrix} 5 + j3 & 2 - j1 \\ -j1 & 2 - j2 \end{bmatrix} \begin{bmatrix} I_1 \\ I_2 \end{bmatrix} = \begin{bmatrix} 30 \\ 0 \end{bmatrix}$$

The determinant of the matix is:

$$\Delta = (5 + j3)(2 - j2) + j(2 - j1) = 17 - j2$$

Hence, from Cramer's rule, the solutions for the mesh-current phasors will be:

$$I_1 = 30\frac{\Delta_{11}}{\Delta} = 30\frac{2 - j2}{17 - j2} = 3.89 - j3.07$$

$$I_2 = 30\frac{\Delta_{12}}{\Delta} = 30\frac{j1}{17 - j2} = -0.20 + j1.74$$

Once the mesh-current phasors are known, all the branch-current phasors become known; from them, all the voltage phasors are determined.

Although only two algebraic equations are involved in this example and there is little difference in computational efficiency between Cramer's rule and Gaussian elimination, we will apply the latter to this problem for practice and comparison.

Look back at the matrix form of the two equations. For the forward elimination step, we multiply the first of the algebraic equations by $j/(5 + j3)$ and add to the second. The result is:

$$\left(2 - j2 + \frac{1 + j2}{5 + j3}\right)I_2 = \frac{j30}{5 + j3}$$

Solving this for I_2 is just a matter of complex arithmetic:

$$I_2 = \frac{j30}{(2 + j2)(5 + j3) + (1 + j2)} = \frac{j30}{17 - j2} = -0.20 + j1.74$$

In the back substitution step, the solution for I_2 is to be inserted into the first algebraic equation. The result will be:

$$(5 + j3)I_1 = 30 - (2 - j1)I_2 = \left[30 - \frac{j30(2 - j1)}{17 - j2}\right]$$

$$I_1 = 30\frac{17 - j2 - j2 - 1}{(17 - j2)(5 + j3)}$$

$$= 30\frac{16 - j4}{(17 - j2)(5 + j3)}$$

$$= 30\frac{2 - j2}{17 - j2} = 3.89 - j3.07$$

(Carry out the numerical details to confirm the result.) As they should be, the

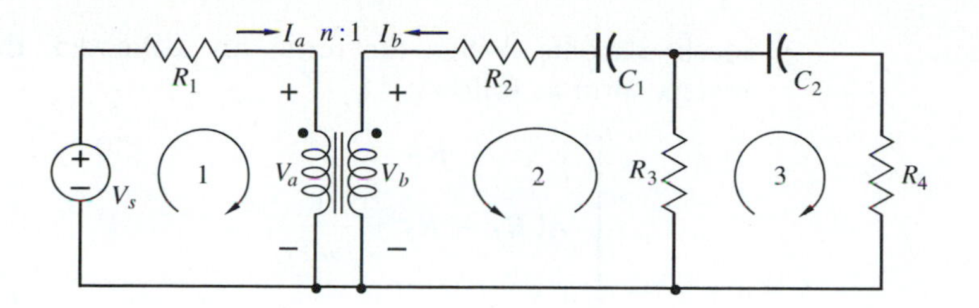

Figure 13
Candidate circuit for loop analysis.

values for the currents obtained by Gaussian elimination and by Cramer's rule are the same.

EXERCISE 11

Carry out a mesh analysis of the circuit in Figure 11, for which node equations were obtained earlier. How many mesh equations are there? □

As a final example, we will take the circuit shown in Figure 13, which includes an ideal transformer. There are three meshes. (Verify that there are three independent loop equations.)

With the two sides of the transformer designated *a* and *b*, as in Figure 13, its equations are:

$$V_a = nV_b \qquad \text{and} \qquad I_b = -nI_a$$

The first step in mesh analysis is to "think" of KVL in terms of branch-voltage phasors. But the next step doesn't work in this case; we can't eliminate the transformer voltage phasors in favor of the transformer current phasors, because the ideal transformer doesn't have an impedance representation. So now what?

Although mesh equations cannot be constructed, we can still proceed with the basic plan, but without eliminating the transformer voltage phasors. Here's the result of pursuing the first two steps in the modified process:

$$R_1 I_1 + V_a = V_s$$

$$R_3(I_2 + I_3) + \frac{1}{j\omega C_1} I_2 + R_2 I_2 + V_b = 0$$

$$\frac{1}{j\omega C_2} I_3 + R_4 I_3 + R_3(I_2 + I_3) = 0$$

These are three equations in five unknowns, but when we include the preceding transformer equations, they constitute five equations in five unknowns. (Note that $I_b = I_2$ and $I_a = I_1$.) Now, using the transformer equations, $V_a\,(=nV_b)$ and $I_2\,(=-nI_1)$ can be eliminated from the preceding three equations, leaving three equations in three unknowns; but not all of them are mesh currents. When the terms are collected, the results can be written in matrix form as follows:

$$\begin{bmatrix} R_1 & 0 & n \\ -n\left(R_2 + R_3 + \dfrac{1}{j\omega C_1}\right) & R_3 & 1 \\ -nR_3 & R_3 + R_4 + \dfrac{1}{j\omega C_2} & 0 \end{bmatrix} \begin{bmatrix} I_1 \\ I_3 \\ V_b \end{bmatrix} = \begin{bmatrix} V_s \\ 0 \\ 0 \end{bmatrix}$$

The final step of writing the solution by Cramer's rule is left for you to carry out.

Reconsider the circuit in Figure 13 and assume that the objective is to determine I_1, I_3, and, V_b—the same three variables as in the preceding matrix equation—by whatever means. Recall also that, for a resistive (dissipative) load R on the secondary side of an ideal transformer, the equivalent resistance seen from the primary side is n^2R. In the sinusoidal steady state, the corresponding result would be that a load impedance Z on the secondary side is transformed to n^2Z on the primary side. (Derive this result by expressing Z in terms of the secondary voltage and current phasors, and then by using the transformer relationships.)

One method of solution would therefore proceed as follows. Let Z be the impedance of the circuit looking to the right from the secondary terminals of the transformer. Then

$$Z = R_2 + \frac{1}{j\omega C_1} + \cfrac{1}{\cfrac{1}{R_3} + \cfrac{1}{\cfrac{1}{j\omega C_2} + R_4}}$$

$$V_a = \frac{n^2 Z V_s}{R_1 + n^2 Z}$$

$$I_1 = \frac{V_s}{R_1 + n^2 Z}$$

Although the expression for Z looks formidable, it is a simple matter of algebra to simplify it. If numerical values were known, Z would reduce simply to a complex number. With the expression for I_1, one-third of the objective has now been reached.

The secondary voltage and current phasors follow from the transformer V-I relationships. Thus

$$V_b = \frac{1}{n} V_a = \frac{1}{n} \frac{n^2 Z V_s}{R_1 + n^2 Z}$$

$$I_b = -nI_a = -nI_1 = \frac{-nV_s}{R_1 + n^2 Z}$$

(What do you think $V_b/(-I_b)$ should be? Do these equations confirm that?)

The second of the desired unknowns is found by inserting the previously determined expression for Z into the expression for V_b. Finally, I_3 can be obtained from I_b by the current-divider relationship. Thus

$$I_3 = \frac{R_3}{R_3 + R_4 + \dfrac{1}{j\omega C_2}}(-I_b) = \frac{nR_3V_s}{(R_1 + n^2Z)\left(R_3 + R_4 + \dfrac{1}{j\omega C_2}\right)}$$

If you review the steps in this process, you will find that the three unknowns were determined one at a time by a sequence of rather simple steps. This is to be compared with the approach which attempted to use loop equations, where all three unknowns are determined simultaneously by solving a set of linear algebraic equations.

As an exercise, complete both solutions and compare both the results and the effort required in the two cases.

Summary of Circuit Equations

To summarize: When all source variables in a circuit are sinusoids of the same frequency, all circuit variables in the steady state will be sinusoids of that same frequency. Since KVL, KCL, and the voltage-current relationships of the components are satisfied by the phasors representing these sinusoids, all previously used approaches to solve for voltages and currents can be applied, taking all variables to be phasors. In particular, constructing and solving node equations and loop equations (or mesh equations, in the planar case) can be carried out with phasors.

When certain multiterminal components are present in the circuit, node equations or loop equations may not be possible to construct (without some modifications, at least).[8] In such cases, all is not lost; a number of different actions can be taken:

1. Branches accompanying the offending components can sometimes be combined with them to permit the previously forbidden equations to be constructed.

2. If there are no accompanying branches, a voltage-shift or current-shift maneuver can be carried out (as described in Chapter 5), causing sources to become accompanied.

3. Accompanying resistive branches can be added to the circuit; after a solution of the resulting equations has been obtained, the added resistance is allowed to go to zero (if in series with a voltage source) or to infinity (if in parallel with a current source).

4. At other times, the process of constructing the appropriate equations can proceed up to the point where only node voltages or only loop currents

[8] This is where ingenuity comes in, rather than mechanically following the steps of an algorithm; wouldn't a computer do that better, anyway?

are to be left. The offending components prevent this from happening. The analysis can be continued, however, by eliminating some other variables instead, and ending with a set of mixed-variable equations. This is what happened with the ideal transformer in the preceding example.

4 CIRCUIT THEOREMS

When studying dissipative circuits, you were introduced to a number of useful theorems that facilitated calculations of circuit responses to external excitations. We shall now examine these theorems for their applicability in the phasor domain. Because many of the concepts to be discussed are intuitive and routine, we will not provide examples to illustrate them all. Besides, in many cases the results are useful largely for conceptual reasons, and not necessarily for calculational purposes.

Superposition

In Chapter 5, we found the validity of the superposition principle to be based on the linearity of the circuits under discussion, not on the nature of the sources. The conclusion that the response to several inputs acting simultaneously is the same as the sum of the responses to the individual inputs acting alone is not changed if what we are seeking is the sinusoidal *steady-state* response. Thus, the superposition theorem would apply in the phasor domain also.

We shall not provide a detailed example to illustrate this theorem. For computational purposes it does not greatly assist in finding the solution of a circuit problem. Thus, if a set of equations (loop, node, or mixed) is constructed for a given problem and a solution is effected by inverting a matrix, the same determinant and cofactors will come into play whether the source vector is chosen with all sources acting simultaneously or the set of equations is solved several times, once for each source acting alone. The latter may indeed entail more work. A similar conclusion follows if the solution is to be obtained by Gaussian elimination.

Thus, the concept of superposition in linear circuits is a powerful conceptual tool, but it has limited utility for numerical work. The idea that the overall solution of a network problem can be obtained by adding component solutions is of major significance, and it leads to important theoretical conclusions. We shall deal with one of the consequences in Chapter 13.

The preceding computational argument does not apply in the case where the sinusoidal sources in a circuit have different frequencies. A special case under this category occurs when, in addition to signal sources, DC sources may be present to provide appropriate *bias* voltages. In such cases, a solution would *require* a superposition approach, finding the DC response and the AC, or signal, response separately.

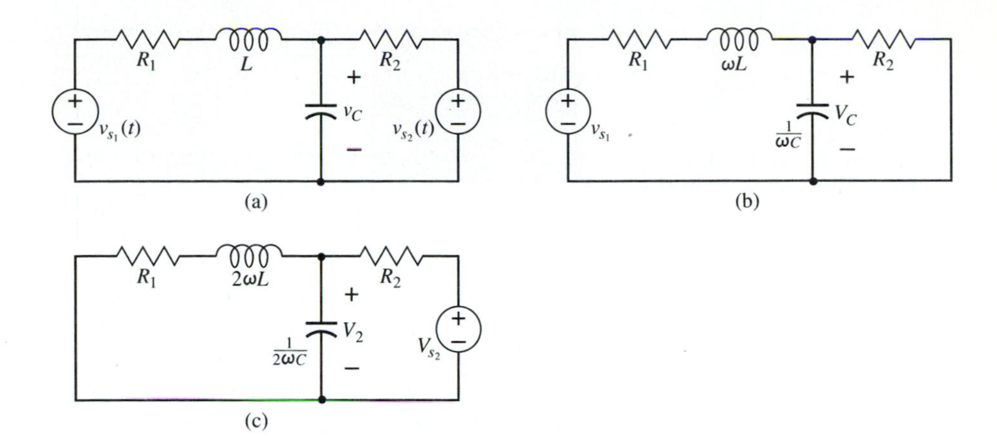

Figure 14
Circuit with sources of different frequencies.

EXAMPLE 1 The sources in Figure 14a are $v_{s1} = 20\sqrt{2}\cos\omega t$ and $v_{s2} = 10\sqrt{2}\cos(2\omega t + \pi/4)$. They obviously have different frequencies. The impedances of L and C at these frequencies will be different. The objective is to find the capacitor voltage $v_C(t)$. The element values, in ohms are $R_1 = 50$, $R_2 = 100$, $\omega L = 50$, and $1/\omega C = 100$.

The principle of superposition will be used. The relevant circuits, first with v_{s1} acting alone and then with v_{s2} acting alone, are shown in Figure 14b and c, respectively. The sources are represented by their phasors at the relevant frequency. The impedances of L and C are also shown at the corresponding frequencies. For finding the capacitor voltage, each of the last two circuits has the structure of a voltage divider. In the first case, C is in parallel with R_2, and in the second case, it is in parallel with R_1 in series with L. The phasors V_1 and V_2, which correspond to two different frequencies, are easily determined as:

$$V_1 = \frac{V_{s1}}{1 + j1} = 14.14\underline{/-45^\circ}$$

$$V_2 = \frac{1 + j2}{-1 + j4} V_{s2} = 5.42\underline{/4.4^\circ}$$

(Carry out the details.)

This is as far as we can go with phasors. The corresponding sinusoids are now written at their appropriate frequencies. The final result is obtained by superposition:

$$V_C(t) = 20\cos(\omega t - 45^\circ) + 7.6\cos(2\omega t + 4.4^\circ)$$

Thévenin and Norton Equivalents

Recall that, for a dissipative circuit, the Thévenin equivalent at a port is a voltage source in series with a resistance, and that the Norton equivalent is a current source in parallel with the same resistance, the current source and

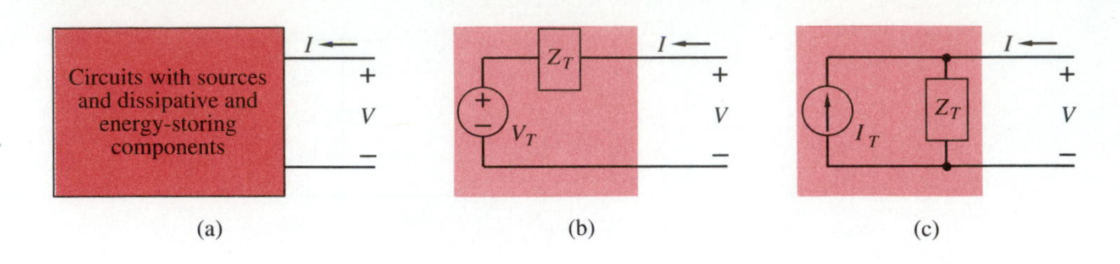

Figure 15
Thévenin and Norton equivalent circuits in the sinusoidal steady state.

voltage source being appropriately related. (If you don't remember the relationship, refresh your memory in Chapter 5 before proceeding.)

Nothing that was used to establish these theorems for dissipative circuits is untrue for general linear circuits in the phasor domain, that is, in the sinusoidal steady state when phasors are used to represent the variables. Thus, the circuits in Figure 15b and c constitute the Thévenin and Norton equivalents of the one-port circuit shown in Figure 15a.

An application is illustrated in Figure 16. It is required to find the output voltage in the phasor domain. If all but the capacitor were replaced by a Thévenin equivalent, finding V_2 would then be a simple matter. Finding the Thévenin impedance requires removing the voltage source (replacing it by a short circuit). The result of doing this is shown in Figure 16b. The impedance is that of L and R_1 in parallel, the combination being in series with R_2.

To find the Thévenin voltage phasor, the capacitor is removed and the open-circuit voltage phasor $V_{oc} = V_T$ evaluated, as illustrated in Figure 16c.

Since there is no current in R_2, this voltage will be the same as V_1, which is found by the voltage-divider relationship to be:

$$V_{oc} = \frac{R_1}{R_1 + j\omega L} V$$

The complete Thévenin equivalent is shown in Figure 16d. Now the output voltage phasor can be found by the voltage-divider relationship from Figure

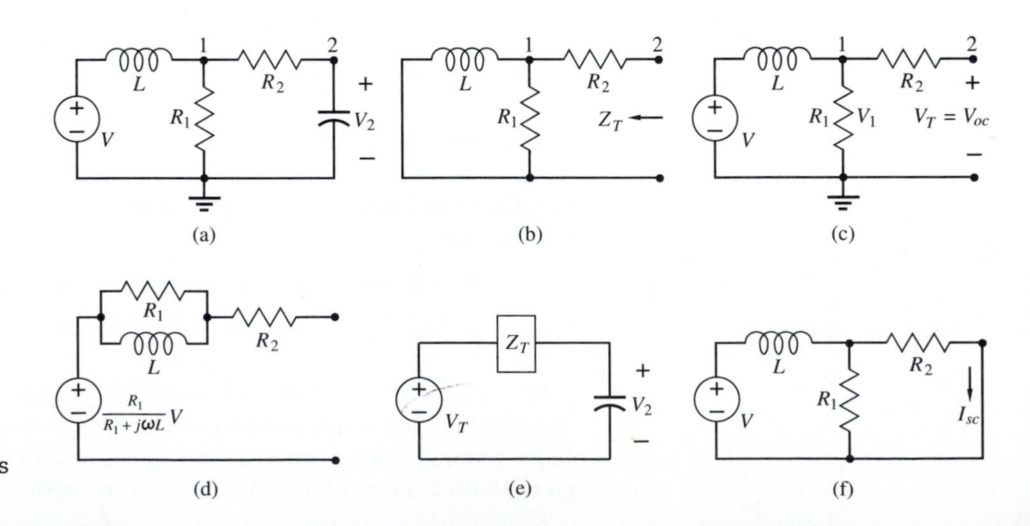

Figure 16
Application of Thévenin's theorem.

16e; it is:

$$V_2 = \frac{1/j\omega C_2}{1/j\omega C_2 + Z_T} V_T = \frac{V_T}{1 + j\omega C_2 Z_T}$$

with
$$Z_T = R_2 + \frac{j\omega L R_1}{R_1 + j\omega L}$$

$$V_T = \frac{R_1}{R_1 + j\omega L}$$

An alternative approach after finding the Thévenin impedance would be to determine the short-circuit current phasor, as depicted in Figure 16f. We see that R_1 and R_2 are in parallel, the short-circuit current being that through R_2. The total phasor current delivered by the voltage source is found as the phasor voltage divided by the total impedance connected to the source terminals. After this is determined, the desired short-circuit phasor current is obtained by the current-divider relationship. The details are:

$$I_{\text{source}} = \frac{V}{j\omega L + R_1 R_2/(R_1 + R_2)}$$

$$I_{sc} = \frac{R_1}{R_1 + R_2} I_{\text{source}} = \frac{R_1 V}{j\omega L(R_1 + R_2) + R_1 R_2}$$

Now the Thévenin voltage phasor will equal the short-circuit current phasor times the Thévenin impedance. (Carry out the details and confirm the result previously obtained.)

For some circuits it will not be possible to find the Thévenin equivalent impedance simply by combining two-terminal impedances in series and parallel, as done in the example in the preceding paragraphs. This will occur in non-series-parallel circuits and when multiterminal components are present. An example is shown in Figure 17. The Thévenin equivalent impedance of everything to the left of the terminals of capacitor C_2 is required (with the source V_s inactive, of course). The impedance at a pair of terminals is defined as the ratio of the voltage phasor, taken as the output, to the current phasor at the terminals, taken as the input.

Thus, in Figure 17b, with phasor I assumed to be a source, we can apply KVL around the two loops in the circuit, then, using the V-I relationships, eliminate the branch voltages in favor of the branch currents, and then express

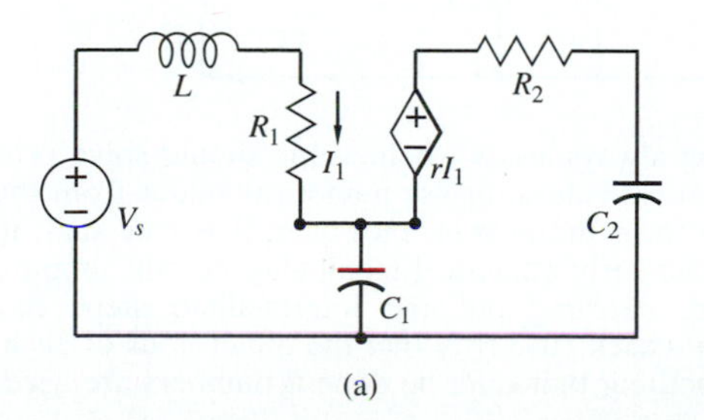

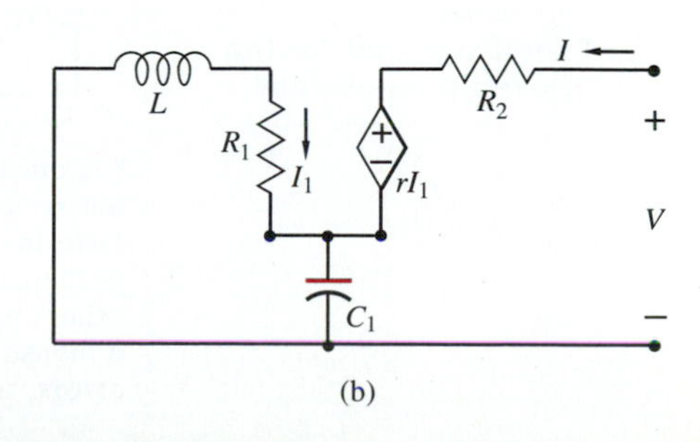

Figure 17 Thévenin impedance found from definition: $Z_T = V/I$.

all the branch currents in terms of the two currents I_1 and I. In this case, the only unknown appearing in the equation around the left loop is I_1, so this equation can be solved for I_1 and substituted into the second equation, where V also appears as an unknown. The details are as follows, keeping all parameters in literal terms first and then inserting the numerical values $\omega C_1 = 1$, $\omega L = 3$, $r = 1$, $R_1 = 2$, and $R_2 = 3$, all normalized:

$$j\omega L I_1 + R_1 I_1 + \frac{1}{j\omega C_1}(I_1 + I) = 0 \quad \Rightarrow \quad I_1 = \frac{-1/j\omega C_1}{R_1 + j\omega L + 1/j\omega C_1} \cdot I$$

$$V = R_2 I + r I_1 + \frac{1}{j\omega C_1}(I_1 + I) = \left(R_2 + \frac{1}{j\omega C_1}\right)I + \left(r + \frac{1}{j\omega C_1}\right)I_1$$

$$Z = \frac{V}{I} = R_2 + \frac{1}{j\omega C_1} - \left(r + \frac{1}{j\omega C_1}\right)\frac{1/j\omega C_1}{R_1 + j\omega L + 1/j\omega C_1}$$

$$Z = (3 - j1) - \frac{(1 - j1)(-j1)}{2 + j3 - j1}$$

$$= 3 - j1 + \frac{1 + j1}{2 + j2} = 3.5 - j1$$

To complete finding the Thévenin equivalent, the open-circuit voltage or the short-circuit current is also needed; this task will be left for you to carry out.[9]

EXERCISE 12

Find the Thévenin equivalent of the circuit in Figure 18 looking back from the terminals of R_L. Find the Thévenin impedance both from the relationship between open-circuit voltage and short-circuit current and from rendering inactive the sources internal to the circuit.

ANSWER: $V_T = 0.983\angle 59.6°$; $Z_T = 0.734\angle 23.8°$ □

VOLTAGE REGULATION Thévenin's theorem is a powerful tool that finds

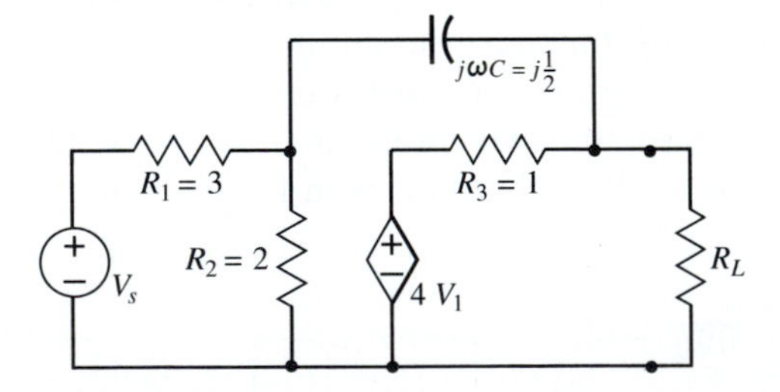

Figure 18
Exercise circuit for the Thévenin equivalent.

[9] A question always arises whether one should solve problems in literal terms first and then insert numerical values, or use numerical values from the first. There is a certain amount of taste involved in the answer. However, it is true that, if numerical values are used earlier, perhaps some early numerical manipulations will simplify the resulting relationships. On the other hand, carrying out the intermediate steps in literal form permits a continual dimensional check (to verify that the dimensions of each term are proper) in order to avoid errors, something that can't be done if numbers are used too early.

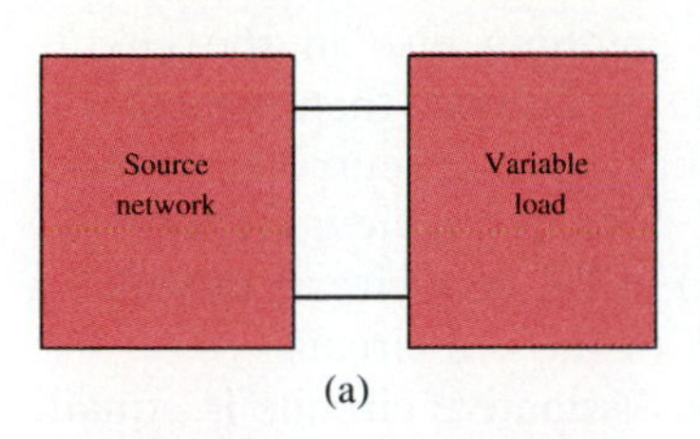

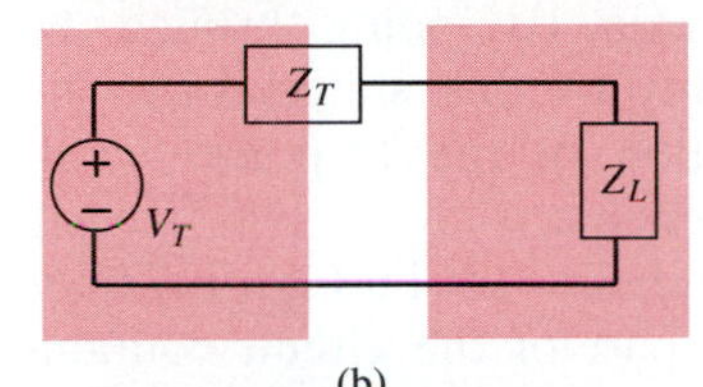

Figure 19
Circuit for defining voltage regulation.

wide application in network analysis. It is useful in the determination of another common concept, which is the subject of this section. The concept has to do with how closely a physical generator (or a two-terminal network containing sources) approaches an ideal voltage source.

To introduce the concept, consider the diagram in Figure 19a. A load having impedance Z is connected to a two-terminal network containing sinusoidal sources. A desired goal is to determine how much the magnitude of the voltage across the load varies as the load is varied.

A possible basis of comparison is the open-circuit voltage, often called the *no-load* voltage. A measure of this variation of load voltage is called the *voltage regulation* (VR), defined as follows:

$$\text{VR} = \frac{(\text{no-load voltage}) - (\text{loaded voltage})}{\text{no-load voltage}} \tag{21}$$

Although not explicitly stated, voltage magnitudes are implied here. If the voltage does not change from the no-load condition, then the network behaves like a voltage source and the regulation is zero. (Low values of regulation are best.)

To show something more quantitative, let's replace the network by its Thévenin equivalent, as shown in Figure 19b. The no-load voltage magnitude is simply $|V_T|$. The regulation can now be found. Thus:

$$\begin{aligned} \text{Voltage regulation} &= \frac{|V_T| - \left|\dfrac{Z}{Z_T + Z} V_T\right|}{|V_T|} \\ &= 1 - \left|\frac{Z}{Z_T + Z}\right| = 1 - \frac{|Z/Z_T|}{|1 - Z/Z_T|} \end{aligned} \tag{22}$$

The voltage regulation reduces to zero for an open circuit (infinite Z), as it should. For other loads, the regulation depends on the relative magnitudes of the load and the Thévenin impedance. (In the denominator of the fraction, the relative angle of the two impedances is also a factor.) When the Thévenin impedance is small relative to the load impedance, then the regulation is small, which is said to be *good* regulation. The regulation is said to be *poor* when it is large. The closer a network with sources approximates a voltage source, the better will be the regulation.

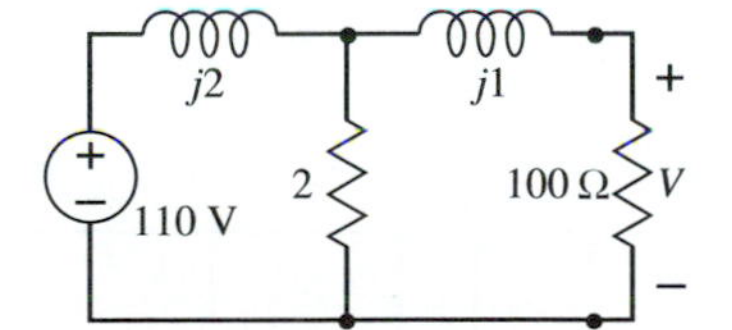

Figure 20
Exercise circuit for determining voltage regulation.

EXERCISE 13

A circuit whose regulation is to be calculated is shown in Figure 20. A voltage regulation of 1 or 2% is considered good. Determine the voltage regulation of this circuit and specify whether it would be considered good.

ANSWER: VR ≈ 1%; good. □

The Compensation Theorem

As described in Chapter 5 for dissipative circuits, once the solution for the voltage across a pair of nodes is known, the *compensation theorem* permits replacing a part of the circuit lying between those two nodes by a voltage source having that voltage, without influencing anything else in the circuit. Similarly, once the current in a part of a circuit is known, that part of the circuit can be replaced by a current source having that current, without influencing anything else in the circuit. (If you need to, review Chapter 5.) Then, any other voltage and current can be found in the new circuit in which a voltage source (or a current source) replaces part of the old circuit.

The procedure used to establish that result for dissipative circuits is equally valid for general linear circuits in the phasor domain. (By going through the details, you should satisfy yourself of the validity of this statement.) The theorem can be used to simplify the computation of the response in an extensive network by breaking it down into components.

Thus, suppose everything to the right of terminals AB in Figure 21a is replaced by its equivalent impedance Z. (If this part of the circuit contains sources, then it is replaced by its Thévenin equivalent.) The result is shown in Figure 21b. This is a simpler circuit than the original one. In this simpler circuit we proceed to solve for voltage V across terminals AB, besides any other variable we are interested in. Once that voltage is determined, we replace everything to the left of AB by a voltage source whose voltage is V, as shown in Figure 21c. This is again a simpler circuit than the original. Thus, by applying the compensation theorem, we are able to reduce a large problem to two smaller ones. The solutions of these two smaller problems will presumably require less effort than the original problem.

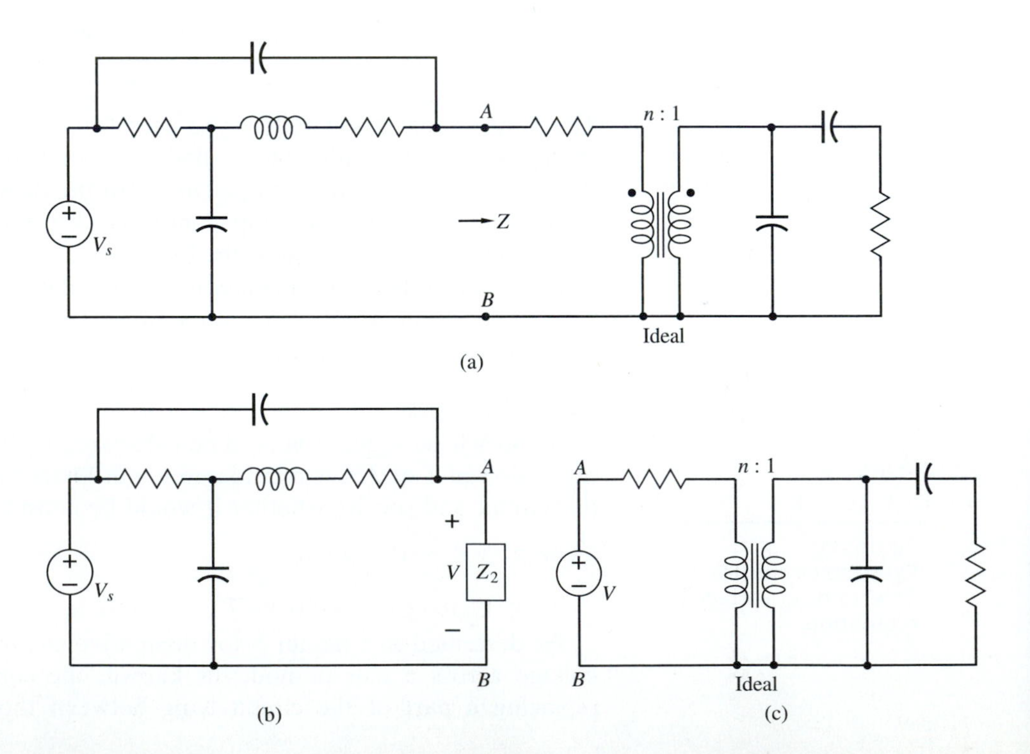

Figure 21
An application of the compensation theorem.

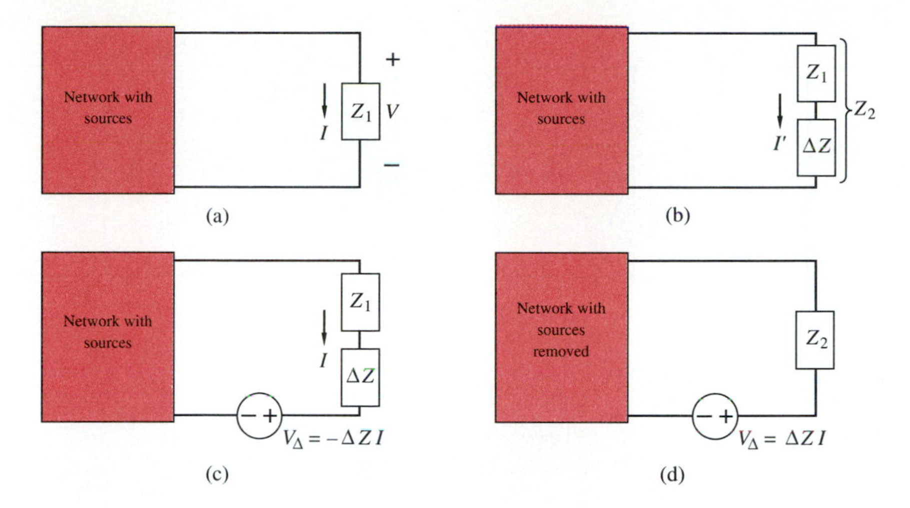

Figure 22
Finding the effect of an impedance variation by the compensation theorem.

EFFECT OF AN IMPEDANCE VARIATION Another useful application of the compensation theorem will now be described. In Figure 22a one branch of an extensive network, whose impedance is Z_1, is explicitly shown. By any method available, the phasor-domain solution for the voltage V and current I of the impedance has been determined. Now suppose the impedance of the branch changes from Z_1 to $Z_2 = Z_1 + \Delta Z$, as shown in Figure 22b. The current in this branch will change to $I' = I + \Delta I$, and all other voltages and currents in the network will also change. What we would like to do is to find these *changes* throughout the network.

It is always possible, of course, to treat the network with the changed value of Z_1 as a new network and to proceed to solve for all the variables by any method. There is nothing wrong with that, but why not simplify life if at all possible?

Suppose that, as illustrated in Figure 22c, a voltage source V_Δ is inserted in the branch and its value adjusted until it is just equal in magnitude to the voltage across ΔZ but opposite in sign. That is, the voltage across the combination of ΔZ and the new source is zero. Hence, this new network is equivalent to the original one in Figure 22a before the change in the impedance. This means that the current in the branch is the original current I; hence, the voltage of the added source to make this happen must be $V_\Delta = -\Delta Z\, I$.

The net effect of the added voltage source is to cancel the effect of the change in the impedance. If we now change the sign of the voltage of the source, making it $V_\Delta = +\Delta Z\, I$, the effect will be identical to that of changing the impedance. (Note that the voltage source and current I have standard references.)

On the basis of the preceding discussion, in order to compute the change in any voltage and current phasor in a network after a change ΔZ in some impedance, (1) we first find the phasor current I in that impedance *before* the change. Then (2) we insert, in series with the impedance, a voltage source with phasor voltage $V_\Delta = \Delta Z\, I$, forming standard references with I. Then (3) we

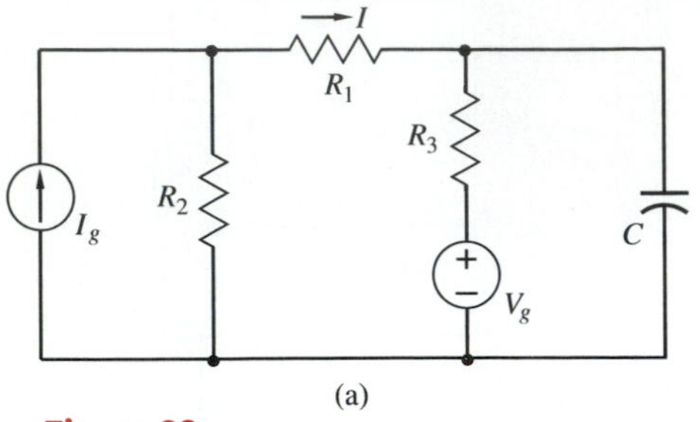

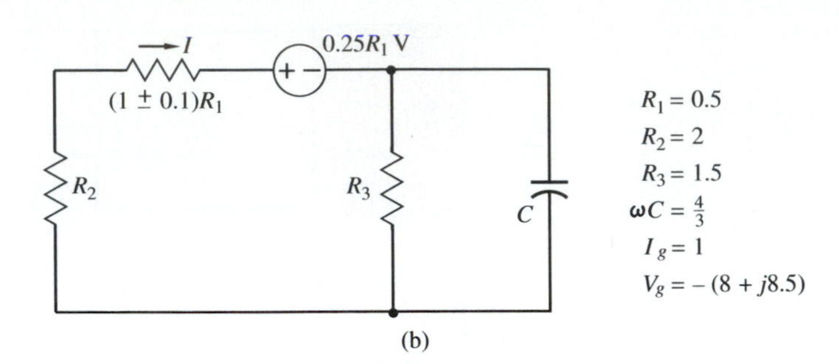

Figure 23
Change in response due to a parameter variation, using the compensation theorem.

remove all the sources in the original network. The phasor currents and voltages anywhere in the resulting network are the *changes* caused by the change in the one impedance.

The process is illustrated in the circuit of Figure 23a. The value of R_1 can vary by 10% in either direction. We want to know how this variation influences the voltages across the other branches. With the given numerical values, the value of I is found to be $I = 2.5$ A. (You should confirm this.) The change in impedance is $\Delta R_1 = 0.1R_1$. Hence, the compensating source voltage is $\Delta R_1 I = 0.1R_1(2.5) = 0.25R_1$ V.

To find the changes in the other branch voltages, we insert this voltage source and remove the original sources; the resulting circuit is shown in Figure 23b. With the parallel combination of R_3 and C taken as a single branch, the circuit has a voltage-divider structure. Hence:

$$\Delta V_{R_2} = \frac{R_2(\pm 0.25R_1)}{R_2 + (1 \pm 0.1)R_1 + R_3/(j\omega CR_3 + 1)}$$
$$= 0.086e^{j11.9°} \quad \text{or} \quad 0.089e^{-j167.4°}$$

$$\Delta V_C = \Delta V_{R_3} = \frac{R_3/(1 + j\omega CR_3)}{R_2}\Delta V_{R_2}$$
$$= 0.029e^{-j52.6°} \quad \text{or} \quad 0.03e^{j129.2°}$$

It is evident that there isn't much change in the magnitude of the various voltages as R_1 changes over a 20% range. However, the angle of each voltage changes by about 180° over this range.

In the process just described, there is no limitation on the magnitude of the impedance variation. This is not an approximate relationship that holds for small variations, but an exact one that holds for an impedance variation of any size.

Figure 24
Compensation theorem exercise.

EXERCISE 14

The load resistance R_2 in the circuit of Figure 24 can vary by 20% in either direction from its nominal value of 10 Ω. Use the compensation theorem to

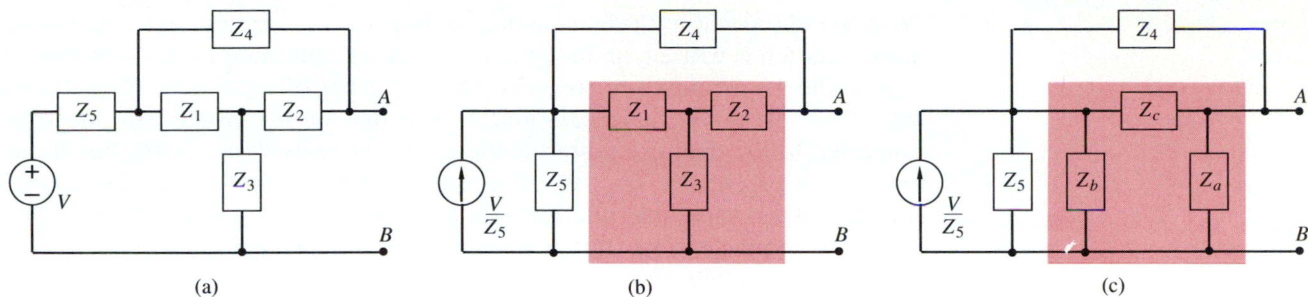

Figure 25
Circuit simplification using equivalent circuits.

determine the variation of the capacitor voltage ΔV_C as a result of the variation in R_2. The nominal values of the parameters, in ohms, are: $R_1 = 2$, $R_2 = 10$, $\omega L = 1.5$, and $1/\omega C = 10$. The source current phasor is 700 mA.

ANSWER: $\Delta V_C = 0.10\underline{/51.6^\circ}$ to $0.018\underline{/121.9^\circ}$; nominal value $V_C = 0.013\underline{/-18.4^\circ}$. □

Tee-Pi or Wye-Delta Transformations

It was observed earlier in this chapter that simple analytical procedures (such as addition of impedances in series or admittances in parallel, voltage-divider and current-divider calculations, and continued-fraction expansions) can be applied in series-parallel, or ladder, circuits. However, such simplifications are not available in such circuits as the one in Figure 25a, for example. On many occasions in circuit analysis, a simplification is achieved when a portion of a circuit is replaced by something *equivalent* to it. Such is the case when several impedances in series are replaced by a single impedance, or a subcircuit with two terminals is replaced by a Thévenin equivalent.

Suppose it were possible in Figure 25b to replace the tee structure (so called because it resembles the letter T) shown in color by the structure shown in Figure 25c (called a pi structure because it is shaped like the letter Π). Now we find that Z_5 and Z_b are in parallel, as are Z_4 and Z_c. When they are replaced by their parallel equivalents, the result is a ladder structure, which permits simple analysis.

In all previous equivalent circuits, the equivalence was valid at a pair of terminals. In the case of the colored circuits in Figure 25, however, four terminals are involved. The bottom two terminals are common, however, so the circuits within the boxes can be looked upon as two-ports. (Review the material on two-ports in Chapter 4 if you need a refresher.)

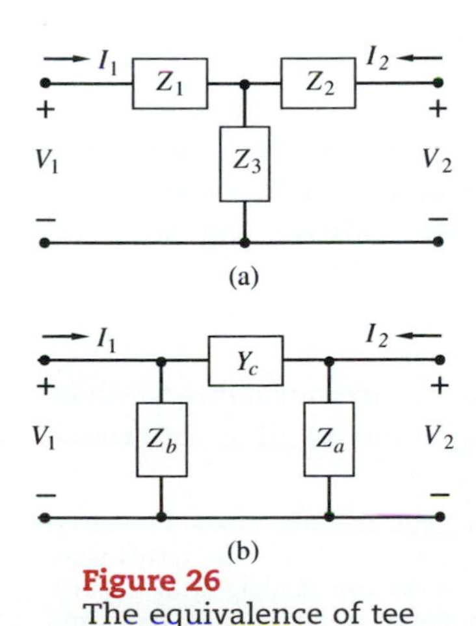

Figure 26
The equivalence of tee and pi circuits.

These two structures are shown isolated in Figure 26. The equations relating the phasor port variables are as follows:

$$\begin{aligned} V_1 &= (Z_1 + Z_3)I_1 + Z_3 I_2 \qquad & I_1 &= (Y_b + Y_c)V_1 - Y_c V_2 \\ V_2 &= Z_3 I_1 + (Z_2 + Z_3)I_2 \qquad & I_2 &= -Y_c V_1 + (Y_a + Y_c)V_2 \end{aligned} \tag{23}$$

(You should confirm these equations.) For the tee circuit, an impedance representation is written, and for the pi circuit, an admittance representation.[10]

In order to permit comparison of the two pairs of equations, they should both have the same representation. Either one or the other pair must be converted to the opposite representation. This is easily done. In the equations for the tee, solve the first equation for I_2 and substitute into the second equation. Then solve that one for I_1 in terms of the two voltage phasors. The change is complete when this is inserted into the preceding equation for I_2. Carry out the details; the result will be:

$$\begin{aligned} I_1 &= \frac{Z_2 + Z_3}{Z_1Z_2 + Z_1Z_3 + Z_2Z_3} V_1 - \frac{Z_3}{Z_1Z_2 + Z_1Z_3 + Z_2Z_3} V_2 \\ I_2 &= -\frac{Z_3}{Z_1Z_2 + Z_1Z_3 + Z_2Z_3} V_1 + \frac{Z_1 + Z_3}{Z_1Z_2 + Z_1Z_3 + Z_2Z_3} V_2 \end{aligned} \tag{24}$$

By comparing these equations with the equations for the pi circuit, we see that they can be made identical if the following relationships exist between the impedances of the tee and the admittances of the pi:

$$Y_a = \frac{Z_1}{D} \qquad Y_b = \frac{Z_2}{D} \qquad Y_c = \frac{Z_3}{D} \tag{25}$$

where $$D = Z_1Z_2 + Z_1Z_3 + Z_2Z_3$$

EXERCISE 15

Carry out a similar derivation, this time starting with the admittance representation of the pi, solving for the voltage phasors in terms of the currents, and then comparing with the impedance representation of the tee. The result should be:

$$Z_1 = \frac{Y_a}{D'} \qquad Z_2 = \frac{Y_b}{D'} \qquad Z_3 = \frac{Y_c}{D'} \tag{26}$$

where $$D' = Y_aY_b + Y_aY_c + Y_bY_c$$

□

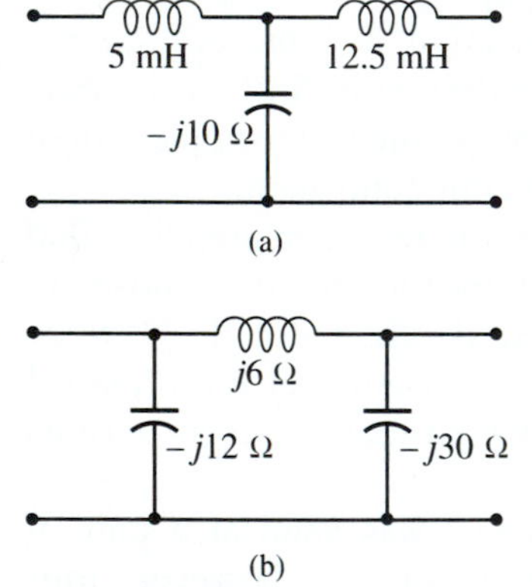

Figure 27 Circuits whose two-port equivalents are to be found.

EXERCISE 16

A tee circuit is shown in Figure 27a. The frequency is $\omega = 400$ rad/s. Find the numerical values of the impedances of the branches of a pi which is equivalent to the tee. Draw the branches of the equivalent pi and label them with their numerical values. The pi is equivalent to the tee only at the specified frequency. □

EXERCISE 17

A pi circuit is shown in Figure 27b. The given values are normalized values of impedance. Determine the admittances of the branches of a tee which

[10] This terminology is new but should be understandable. In Chapter 4, the circuits involved were dissipative only and did not include energy-storing components. There the terms were *resistance representation* and *conductance representation.* Now we are dealing with general linear circuits in the steady state. Impedance and admittance replace resistance and conductance, respectively.

is equivalent to the pi. Draw the equivalent tee circuit showing the values of the parameters. Is this tee equivalent to the pi at all frequencies or at only one frequency?

ANSWER: $Y_1 = -j\frac{1}{2}\,S$, $Y_2 = -j\frac{1}{5}\,S$, $Y_3 = j\frac{1}{10}\,S$. □

EXERCISE 18

Suppose all three impedances in the tee circuit shown in Figure 26a are equal, say, Z_{tee}. Find the impedances Z_{pi} in an equivalent pi circuit.

ANSWER: $Z_{\text{pi}} = 3Z_{\text{tee}}$, all equal. □

SUMMARY

- Representing a sinusoid by a phasor
- Graphical relationship between a phasor and a sinusoid
- Kirchhoff's laws in terms of phasors
- Phasor diagrams; reference for phase
- Voltage-current relationship of components in terms of phasors
- Concepts of impedance Z and admittance Y
- Single-component impedances and admittances
- Reactance X and susceptance B
- Evaluation of impedance and admittance of interconnections of components
- Voltage dividers and current dividers in terms of Z or Y
- Circuit analysis in terms of phasors: node and loop equations
- Thévenin and Norton equivalents in the phasor domain
- Voltage regulation
- The compensation theorem in the phasor domain
- Effect of impedance variation on steady-state response
- Tee-pi and pi-tee transformations

PROBLEMS

The subject of AC power has been deferred to the next chapter. In many of the problems to follow, it would be normal to ask questions about power dissipated, power supplied, or energy stored. We can't do it yet but will do so in the next chapter, so save your problem solutions in this chapter to use again later for further work.

1. In the circuit of Figure P1, the source voltage and the resulting AC steady-state current are as follows: $v_s(t) = 100\cos 400t$ V and $i(t) = 0.707\sin(400t + 45°)$ A. Find the value of R and a condition that L and C must satisfy. Specify the maximum possible value of L.

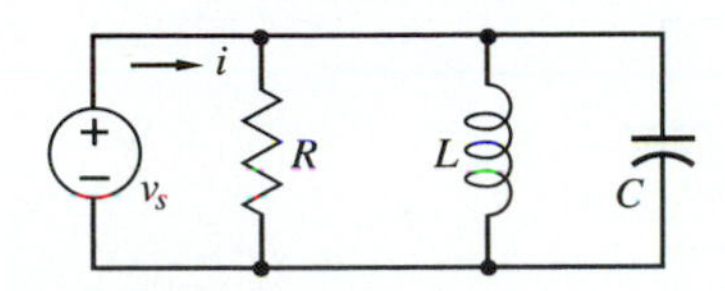

Figure P1

2. The steady-state voltage and current at the terminals of the circuit of Figure P2a are:

$$v(t) = 100\cos\left(400\pi t + \frac{\pi}{3}\right)\text{ V}$$

$$i(t) = 5\cos\left(400\pi t + \frac{2\pi}{3}\right)\text{ A}$$

   a. Determine the impedance of the circuit in both polar and rectangular form. Someone claims that the box in Figure P2a contains the circuit in Figure P2b. Someone else disagrees and claims that it contains the circuit in Figure P2c. You are to settle the controversy.
   b. Check each possibility; if you agree with the champion of either case, specify the component values.

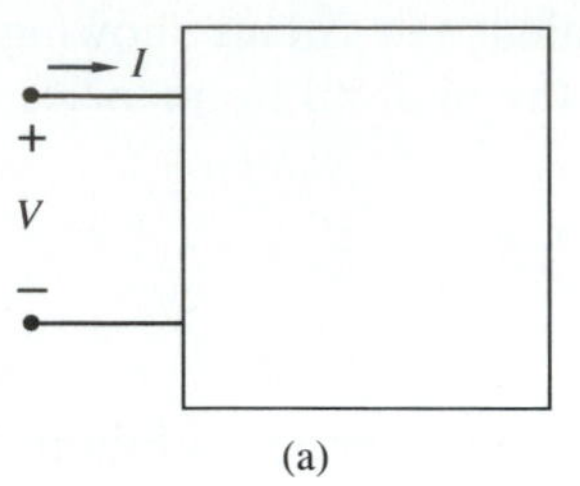

(a)

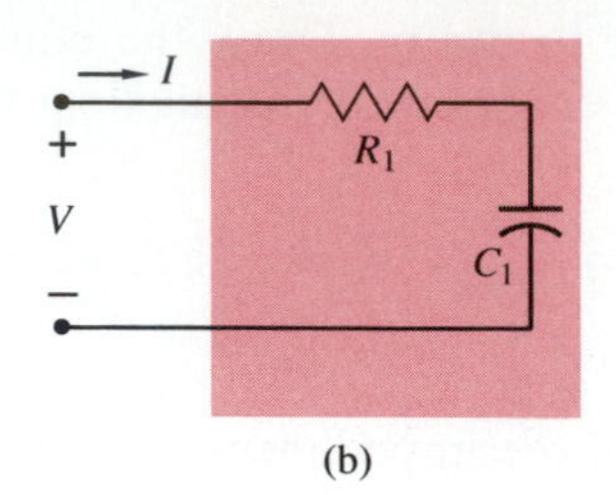

(b)

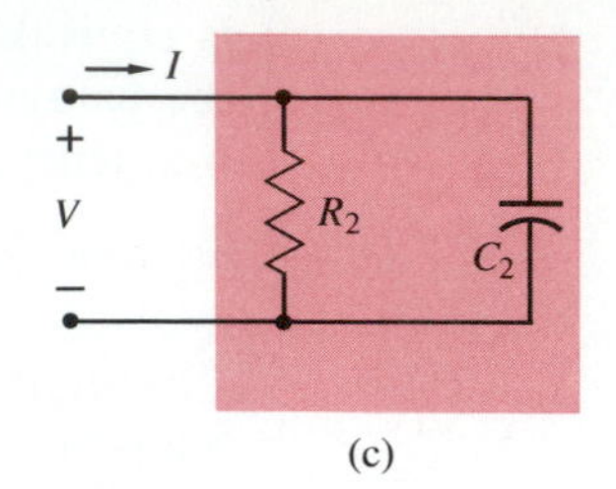

(c)

Figure P2

3. In the circuit of Figure P3:
   a. Find the phasors V_L, V_C, and I_C.
   b. Construct a phasor diagram to scale, showing the phasors of the two sources and all the preceding quantities. Take V_s as the reference for phase.

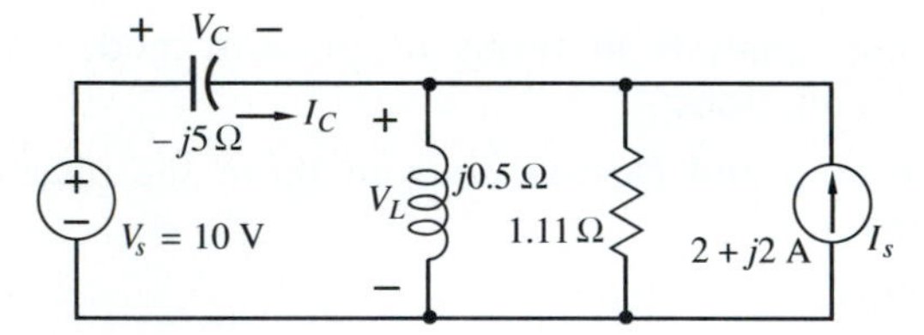

Figure P3

4. Two two-terminal circuits are equivalent at their terminals if they have the same voltage-current relationship. In the AC steady state, equivalence implies equality of the two impedances (or admittances). Two circuits could be equivalent in the steady state at *all* frequencies or at only one frequency.

   Determine the impedances of the two circuits in Figure P4 as a function of frequency and confirm that they are identical. In that case, the two circuits are equivalent at all frequencies.

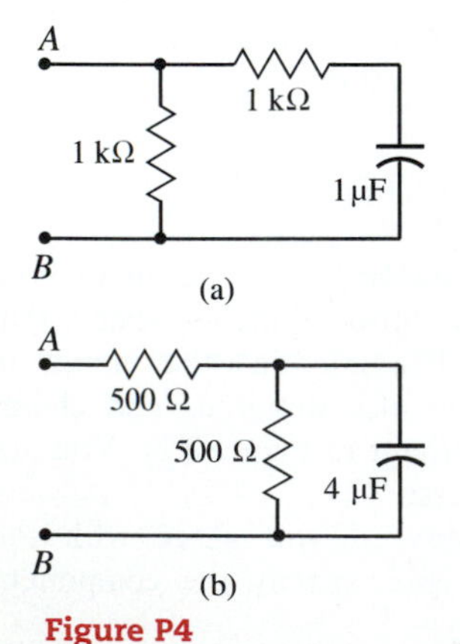

Figure P4

5. Two two-terminal circuits are shown in Figure P5. Assume they are to be equivalent at a frequency ω.
   a. Determine the parameters R_1 and L_1 in terms of the frequency ω and R_2 and L_2.
   b. Conversely, find the parameters R_2 and L_2 in terms of R_1, L_1, and ω.

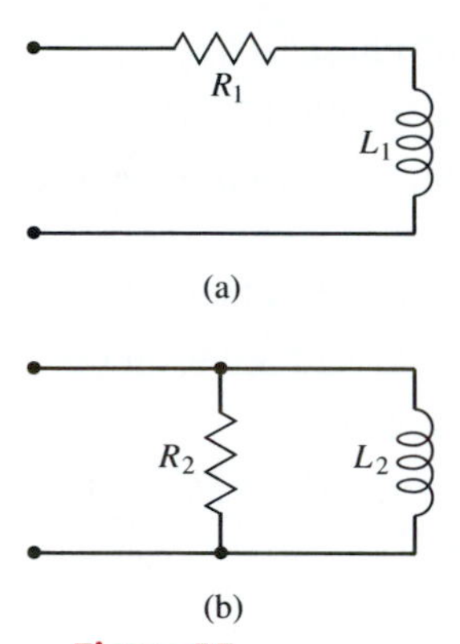

Figure P5

6. The two-terminal circuits in Figure P6 are to be equivalent at a frequency ω. Find the parameters of one in terms of the frequency and the parameters of the other, and vice versa.

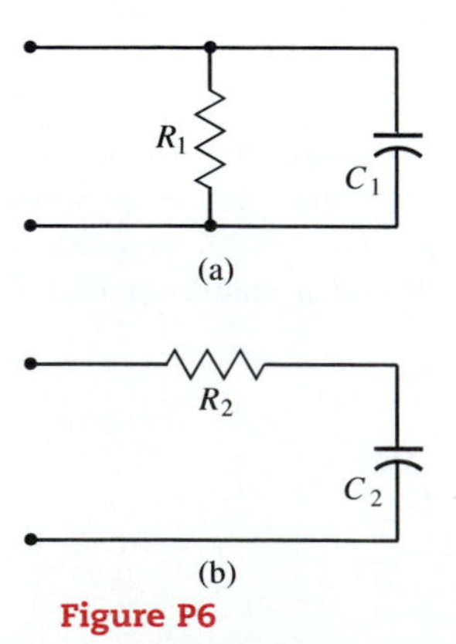

Figure P6

7. The values shown in the two-terminal circuit in Figure P7 are impedances in Ω.

a. Find the impedance at the terminals of the circuit.

b. Suppose $I_3 = 4 + j1$. Successively use Ohm's law for phasors, the voltage-divider and current-divider relationships, and Kirchhoff's laws to determine I_4 and V_0.

c. Repeat part (b) but using a different order and a different combination of the steps.

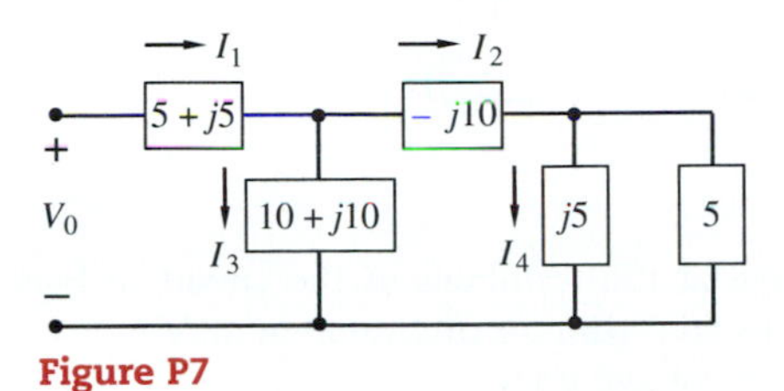

Figure P7

8. The capacitor in the circuit of Figure P8 is variable.

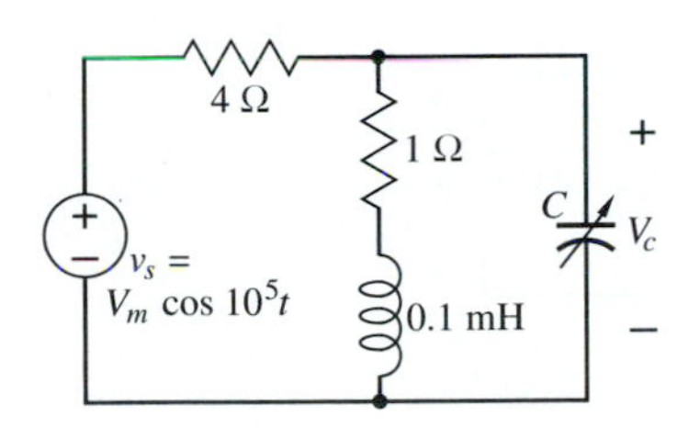

Figure P8

a. Find the value of C for which the steady-state voltage across it is a maximum.

b. For this value of C, find the magnitude of the phasor voltage across C compared with the phasor source voltage.

9. The value of C in Figure P9 is to be varied until the resulting voltage across the terminals is in phase with I_s. Find the required value of C.

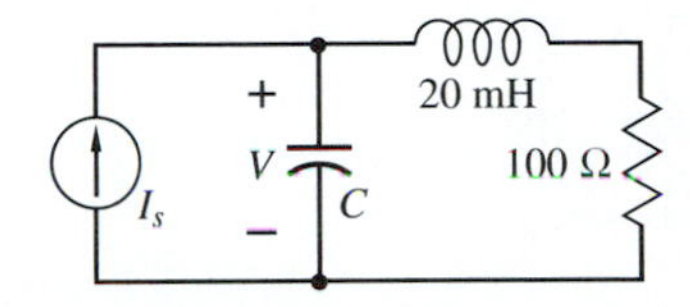

Figure P9

10. On many occasions, repeated steady-state calculations are to be made on a given network with various conditions at the terminals. The calculations are to be made at some specific frequency, say, the power frequency. It is often useful to simulate two-terminal parts of the network with simple equivalent circuits. Some two-terminal circuits are shown in Figure P10.

a. Find the value of the input impedance or admittance of each circuit.

b. Assume that each of these circuits is to be simulated by each of the two-element circuits in Figures P5 and P6. Find the element values of the simulating circuits at the given frequency 377 rad/s in each case.

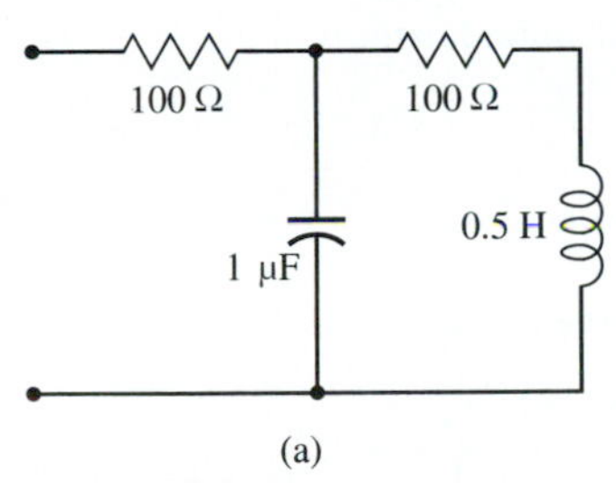

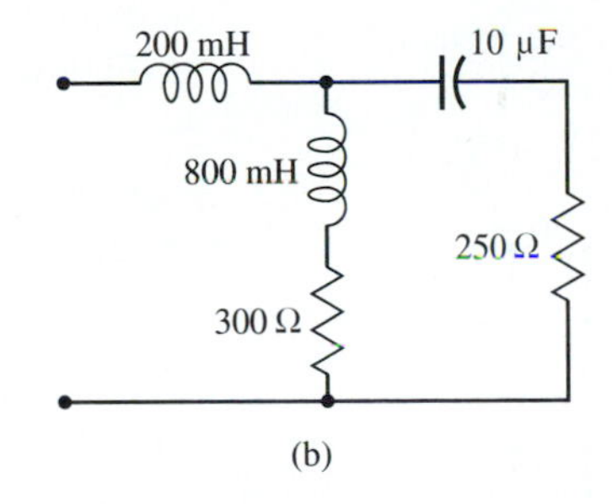

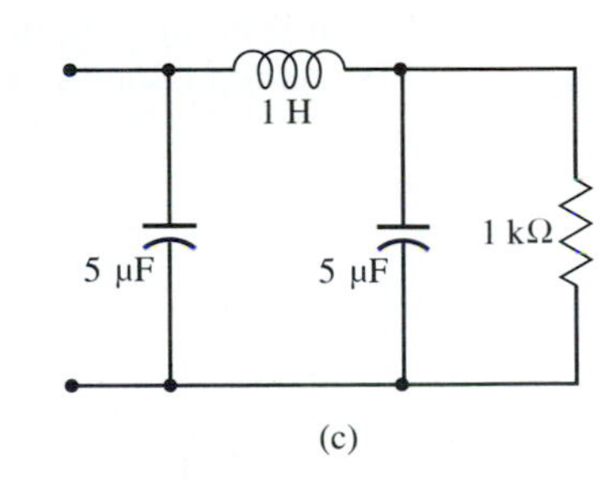

Figure P10

11. Laboratory measurements are made on a coil of wire in series with a resistor at a frequency of 60 Hz, as illustrated in Figure P11. The following values are obtained: $|V_1| = 50$, $|V_2| = 60$, and $|V| = 90$, all in volts. Assume that the coil of wire can be represented by a series connection of L and R.

a. Find the numerical values of L and R analytically.

b. Confirm these values graphically, by means of phasor diagrams.

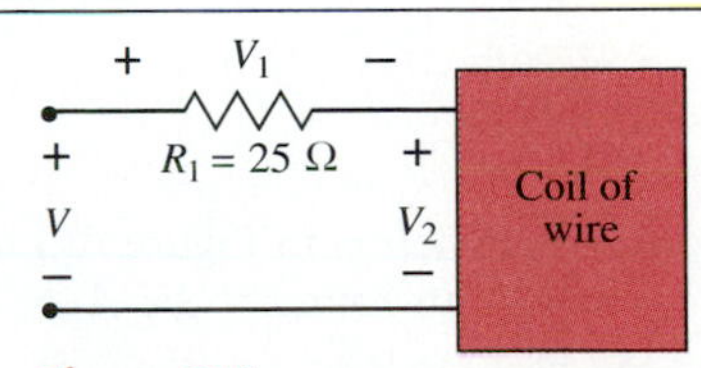

Figure P11

12. The device inside the box in Figure P12 is to be represented by a model consisting of the parallel connection of a resistor and a capacitor, as shown. Laboratory measurements using a 400-Hz sinusoidal source give the following data: $|I_1| = 100$, $|I_2| = 200$, and $|I| = 250$, all in mA.

a. Find the numerical values of R and C by analytical means.

b. Confirm these results graphically.

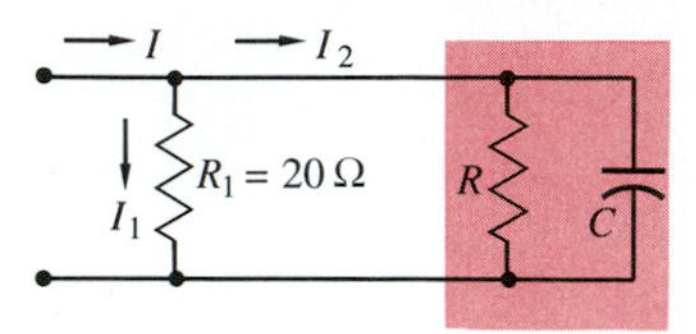

Figure P12

13. In the preceding two exercises, the nature of the branches is known. In the present exercise, the nature of the branches forming a node in Figure P13 is not known. The AC steady-state currents are measured by ammeters whose readings are A_1, A_2, and A_3, in amperes. Two of the ammeter readings are $A_1 = 6$ A and $A_2 = 2$ A. The problem is to find the possible values of the third current I_3 both analytically and graphically. Choose the angle of I_1 as the reference for phase and let $I_2 = A_2 e^{j\alpha}$ and $I_3 = A_3 e^{j\beta}$.

a. Find algebraic expressions for A_3 and β in terms of A_2 and α and the given measured values.

b. Find the desired values from a phasor diagram drawn to a convenient scale by laying out I_1 horizontally and drawing from its tip the locus of all possible points representing the tip of I_2.

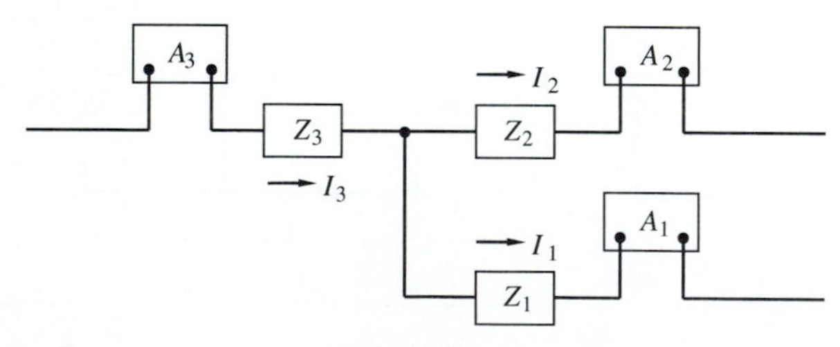

Figure P13

14. The readings of the three ammeters in Figure P13 for a particular set of branches are found to be $A_1 = 8$, $A_2 = 5$, and $A_3 = 6$, all in amperes.

a. Draw two possible phasor diagrams, and find two possible values of α and β.

b. Confirm these values analytically.

15. The three voltages in Figure P15 are measured to be $|V_1| = 100$, $|V_2| = 150$, and $|V| = 200$, all in volts. Choose one of the voltages as the reference for phase and determine the possible angles of the other two, both graphically and analytically.

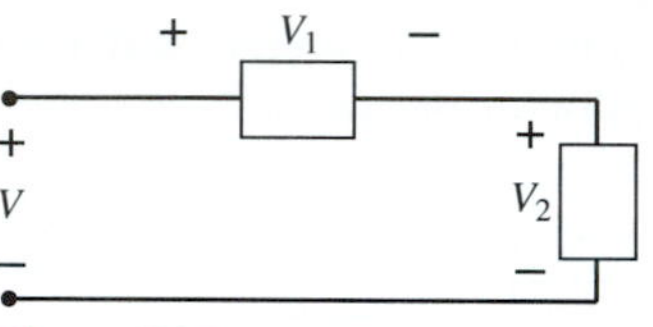

Figure P15

16. The voltage at the terminals of the circuit in Figure P16 is sinusoidal with an rms value of 10 V.

a. Find the voltage $v_2(t)$.

b. Find the input current $i(t)$.

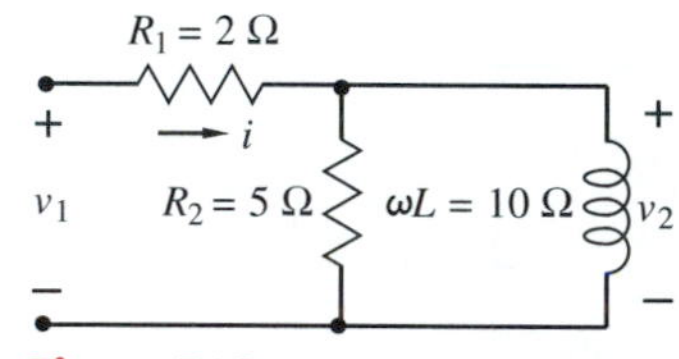

Figure P16

17. Phasor current I_1 in Figure P17 is given as $I_1 = j10$ A.

a. Find the phasor input current I.

b. Find the output-voltage phasor V_2.

c. Find the input impedance.

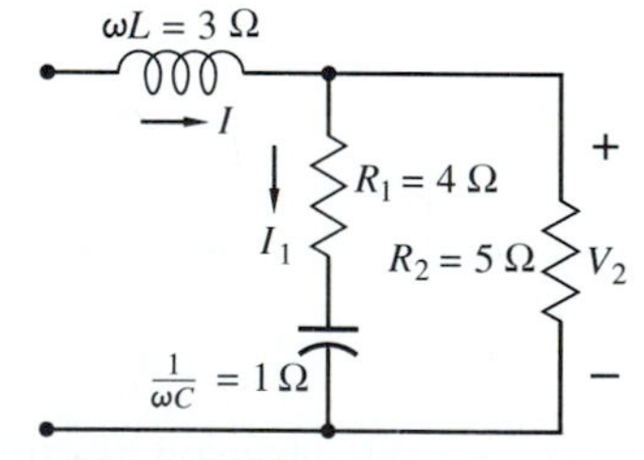

Figure P17

18. The component values in Figure P18 are $L = 0.1$ H, $C = 1\ \mu$F, $R_1 = 100\ \Omega$, and $R_2 = 1$ kΩ; $v_s(t) = 141.4 \cos 400t$ V. Find the phasor current I in Figure P18:

a. By using Thévenin's theorem

b. By using the ladder-network approach

Find the phasor current in the source:

c. By the ladder-network approach

d. By first finding the admittance of the network looking from the terminals of the source

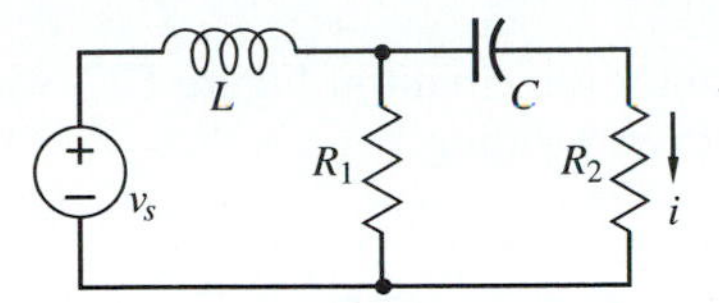

Figure P18

19. Repeat Problem 18 for the network of Figure P19. The component values are $R_1 = 2\ \text{k}\Omega$, $R_2 = 1\ \text{k}\Omega$, $L = 1$ H, and $C = 0.5\ \mu$F; $v_g(t) = 282.8 \cos 377t$ V.

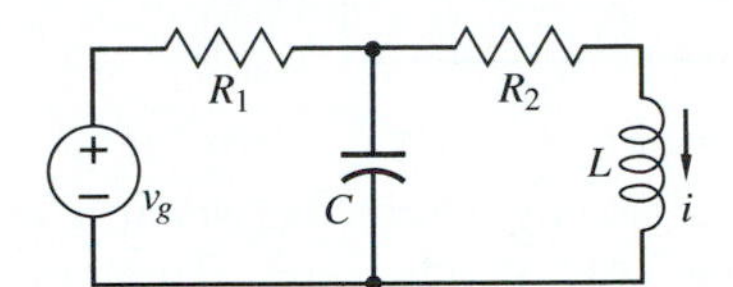

Figure P19

20. In the circuit of Figure P20:

a. Find the input impedance from the terminals of the source.

b. Find the output voltage using Thévenin's theorem.

c. Corroborate these values by finding them again using the back-to-front approach.

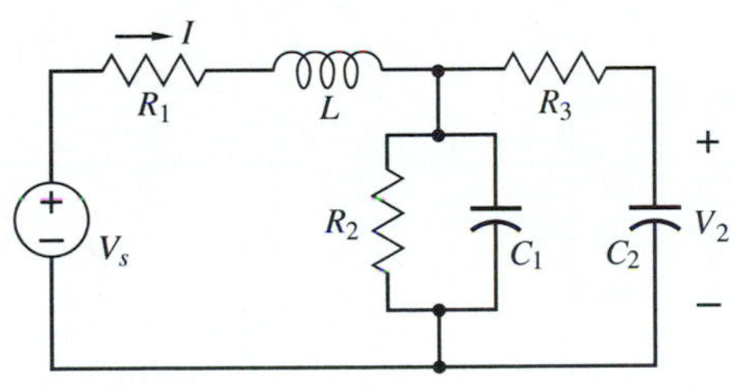

Figure P20

21. The driving-point impedance of the circuit in Figure P21 at a frequency of 400 rad/s is to be $Z = 5\ \text{k}\Omega$. Find the required values of C and L.

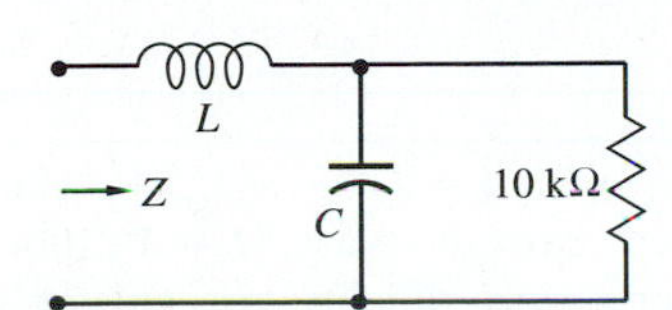

Figure P21

22. At an angular frequency of ω rad/s, the impedance at the terminals of the circuit in Figure P22 is to be $2R$. Find expressions for the required values of L and C.

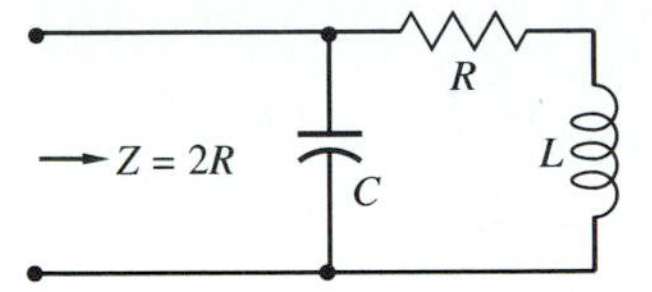

Figure P22

23. Find the impedance of the circuit in Figure P23 at a frequency of 400 rad/s:

a. With the terminals AB open, as shown

b. With the terminals short-circuited

c. With a 100-Ω resistor connected between A and B

Express the impedance in both polar form and in rectangular form.

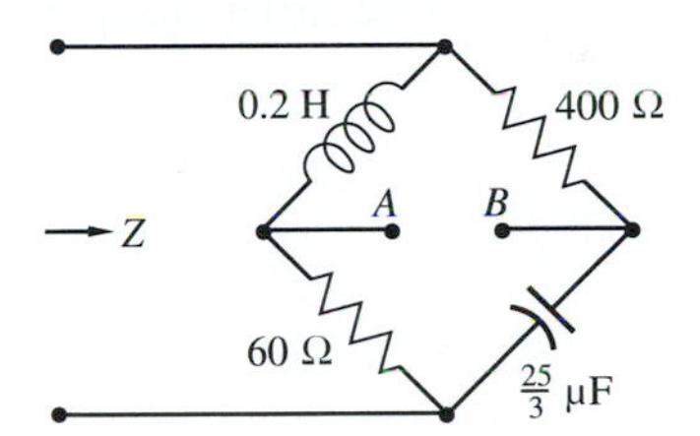

Figure P23

24. Use I in Figure P24 as the reference for phase, and draw a phasor diagram that includes all the voltages. Since the value of $|V|$ is given, this establishes the scale value. Give numerical values of all the voltages, including the phase.

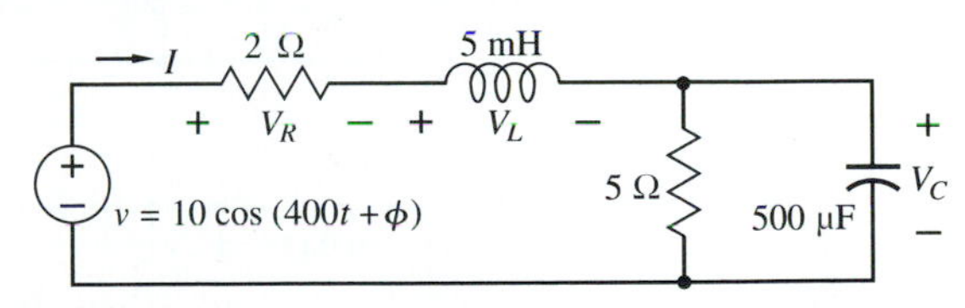

Figure P24

25. In the circuit of Figure P25, $R = 200\ \Omega$. At $\omega = 6000$ rad/s, $Z_L = j100\ \Omega$ and $Z_C = -j100\ \Omega$.

a. Find the input admittance at this frequency.

b. Take the angle of I_R as the reference for phase and draw a phasor diagram that includes the voltage and current phasors of each element, as well as the terminal voltage and current phasors.

c. Find the angular frequency at which the impedance will be purely real.

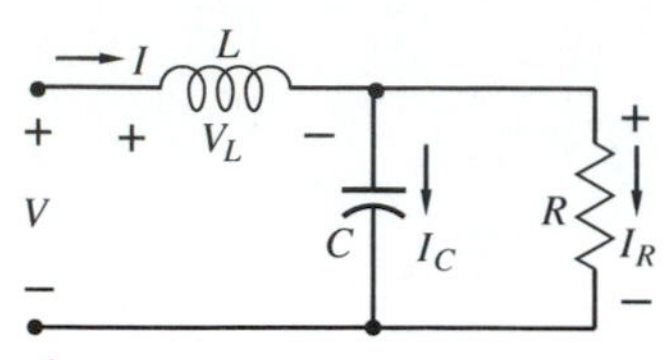

Figure P25

26. The object in this problem is to find the output voltage V_2 by several methods. (Save the early solutions for comparisons with the later ones if you don't do all parts at the same time.)

a. Find the input impedance in the circuit of Figure P26; from this, find the input current. Then use the current-divider idea to find the current in the 10-Ω resistor; then find V_2.

b. Find V_2 by working backward from the output side.

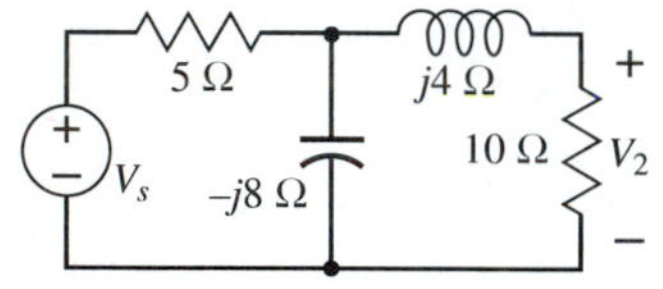

Figure P26

c. Write node equations and solve them to find V_2.

d. Write loop equations and solve them to find V_2.

e. Apply Thévenin's theorem looking back from the terminals of the right-hand resistor and then use the voltage-divider concept to find V_2.

27. Repeat Problem 26 for the circuit in Figure P27, with appropriate changes in wording.

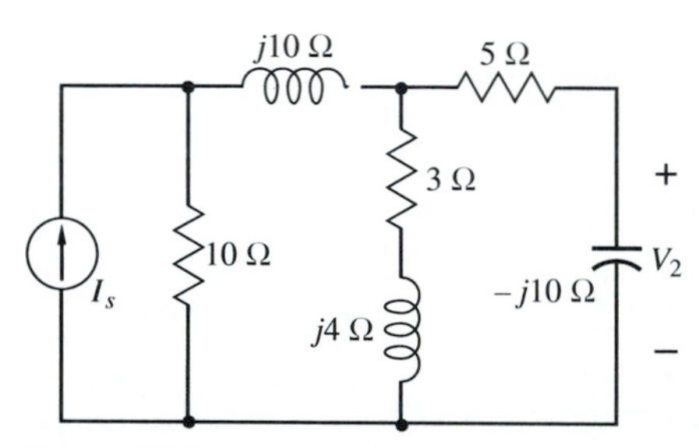

Figure P27

28. Some students in a circuits course had spent a few weeks studying the AC steady state. They were complaining that there were no "modern" components in any of the circuits. They sought out a senior, who suggested the circuit in Figure P28a, which consists of an accompanied voltage across which are connected one each of all the kinds of components he could think of, including a controlled source (a CCCS). The two independent sources were specified to be sinusoids of the same frequency. The objective was to find phasor V.

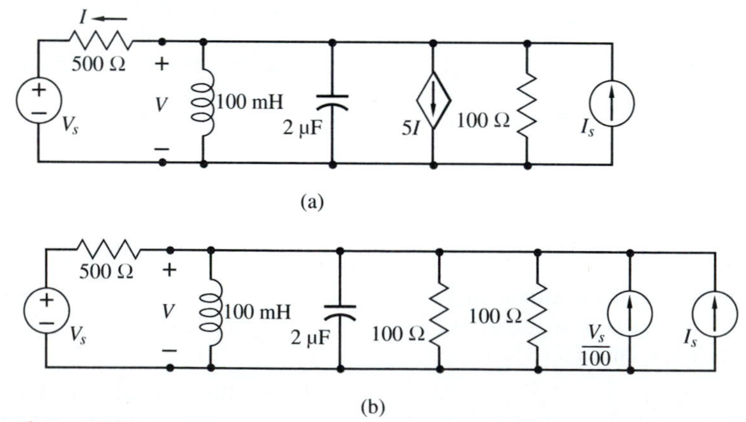

Figure P28

The students were impressed with the controlled source but were having trouble dealing with it, until someone suggested solving for the controlling variable I. From the terminals of the accompanied source they quickly came up with $I = (V - V_s)/500$. Thus, the controlled-source current was $5I = V/100 - V_s/100$. The first term was quickly recognized as an expression of Ohm's law and the second term as an independent current source with reference opposite to that of the CCCS. The circuit was then redrawn as in Figure P28b, but the luster was gone: there was no "modern" component in the equivalent. One node equation was all that was needed to solve for V, hardly challenging. Nevertheless, supply the details and solve for phasor V in terms of V_s and I_s and the frequency. (The moral of the story is: Beware of contrived circuits in which arbitrarily introduced controlled sources do not represent models of actual electronic devices in a bona fide circuit, but simply give the appearance of being up to date.)

29. The function of the circuit in Figure P29 is to act as a *phase shifter*, that is, to produce an output-voltage phasor V_3 which has the same magnitude as that of V_1 and an angle that leads the angle of V_1 by 90°. Find V_3 and determine how closely (within how many percent or degrees) the circuit meets these objectives.

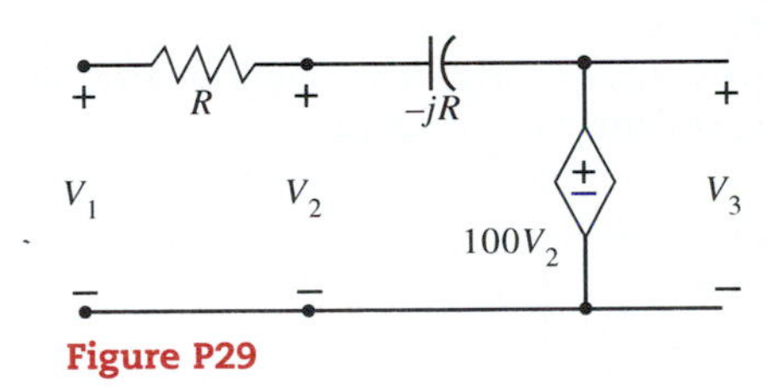

Figure P29

30. The op amp in the circuit of Figure P30 is ideal. The circuit is operating in the AC steady state.

a. Determine the voltage gain V_o/V_1 in terms of Z_1 and Z_2.

b. Suppose $Z_1 = R_1$ and Z_2 is the parallel combination of R_2 and C. Determine a condition that the two resistors must satisfy if the circuit is to continue operating linearly.

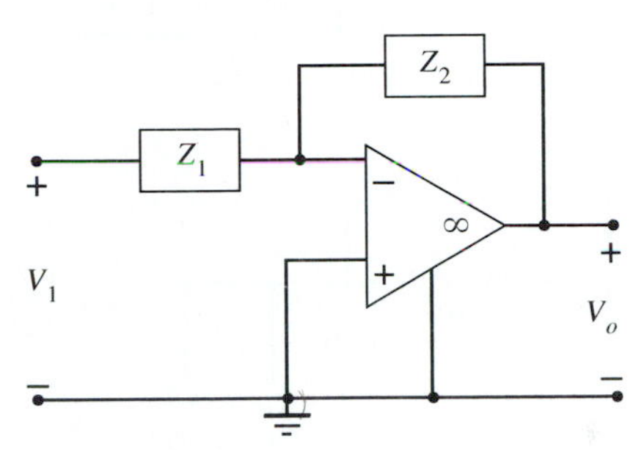

Figure P30

31. By appropriately manipulating the circuit of Figure P31, reduce it to the form of Figure P30. The frequency of the source is 10^4 rad/s.

a. Use the result of Problem 30 to determine the voltage gain V_o/V_s as a function of frequency.

b. Assuming that the saturation voltage is 15 V, is it possible for the op amp to saturate? If so, determine the permitted values of R for the op amp to maintain linear operation.

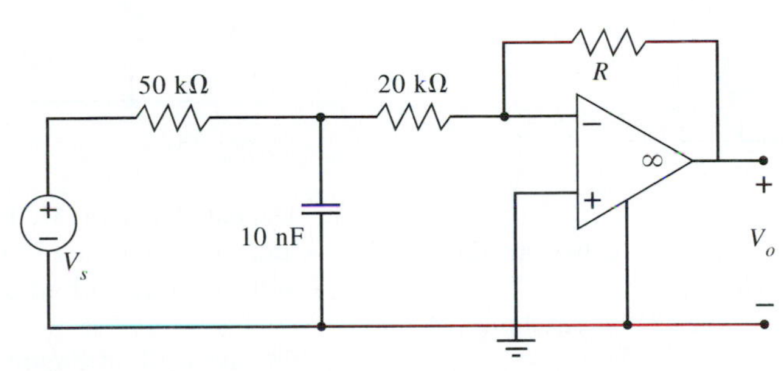

Figure P31

32. The objective of this problem is to carry out a loop analysis of the circuit in Figure 11 in the text. Add a resistor in parallel with the controlled source, making it an accompanied source. Carry out a loop analysis and solve for the variables. Then let the added resistance $\rightarrow \infty$ and compare the results with those previously found by node equations in the text. (Or compare with the results of Exercise 11.)

33. Use loop equations in the circuit of Figure P33 to find the voltage across the 4-Ω resistor and the voltage across the current source.

34. First convert the controlled source in Figure P34 into a VCCS; then write and solve a set of node equations. From the result determine the currents in the inductor and the capacitor, and in the controlled source.

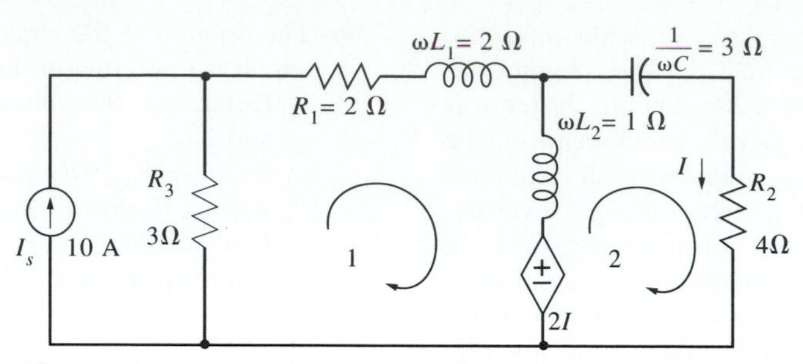

Figure P33

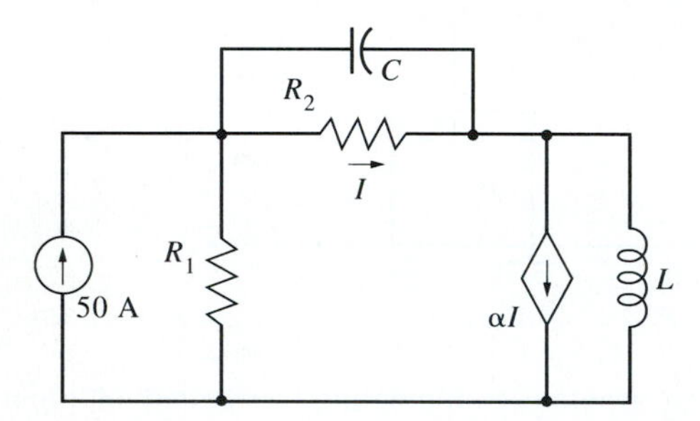

Figure P34

35. Determine the Thévenin equivalent from terminals AB in the circuit of Figure P35.

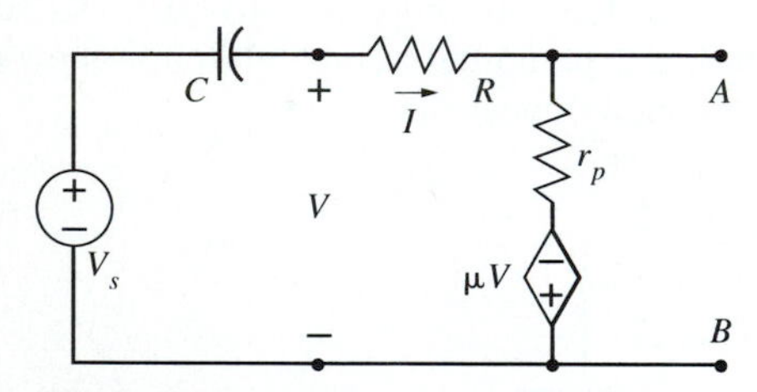

Figure P35

36. The circuit in Figure P36 is to behave like an ideal sinusoidal source at terminals AB.
   a. Determine the required value of the reactance of the capacitor.
   b. Determine the equivalent source-voltage phasor.

37. The network in Figure P37 is a loaded bridged-tee network.
   a. Write a set of node equations.
   b. Solve these equations for the node voltages, using Gaussian elimination.
   c. Repeat, using Cramer's rule.
   d. Repeat, using MATLAB.
   e. Find the node voltages, using PSPICE.
   f. Determine the phasors of the currents in the capacitors.
   g. Repeat parts (a)–(f) after inserting a 2-Ω resistor to accompany the voltage source.

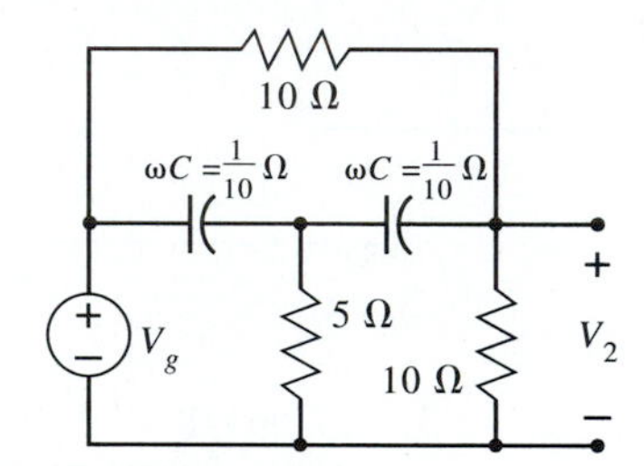

Figure P37

38. Repeat Problem 37, this time using loop equations and including the 2-Ω resistor accompanying the voltage source. Omit part (e).

39. The network in Figure P39 is a twin-tee network.
   a. Write a set of node equations.

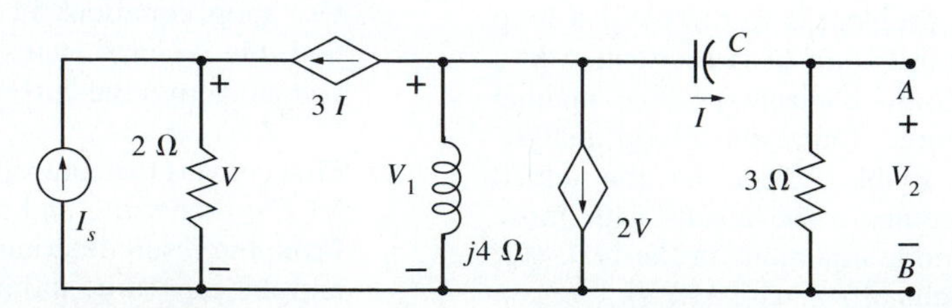

Figure P36

b. Solve the equations for the node-voltage phasors, using Gaussian elimination.
c. Repeat, using Cramer's rule.
d. Repeat, using MATLAB.
e. Find also the phasors of the inductor currents.
f. Use PSPICE to solve for all node voltages and compare.

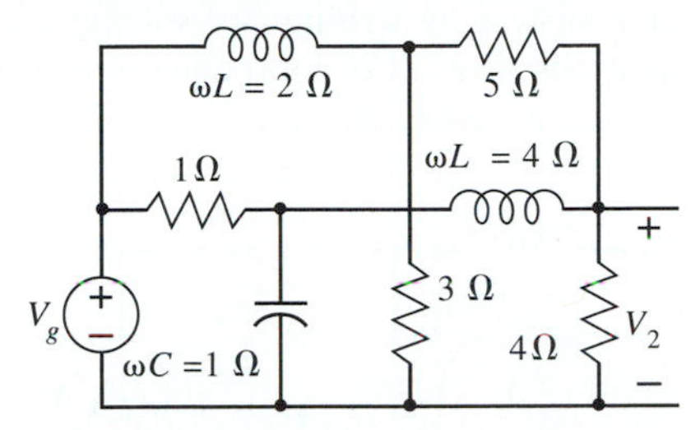

Figure P39

40. Redraw Figure P39 so that it is planar. Repeat each of parts (a) through (d) in Problem 39, but this time writing and solving mesh equations.
e. Confirm that the same inductor-current phasors are obtained.
f. Find also all the node-voltage phasors, and confirm that the same values are obtained.

41. The circuit in Figure P41 is a bridge used to measure capacitance. The resistance R_x is varied until the bridge is *balanced.* Balance is said to occur when $|V_A| = |V_B|$.
a. Assuming a fixed (and known) frequency ω, determine the value of the capacitance C in terms of ω and the value of R_x at balance, say, R_{xb}.
b. Let $R_x = 2R_{xb}$. Find the value of V_{AB}.
c. Now let $R_x = R_{xb}/2$. Find the value of V_{AB}. Do these results show anything?

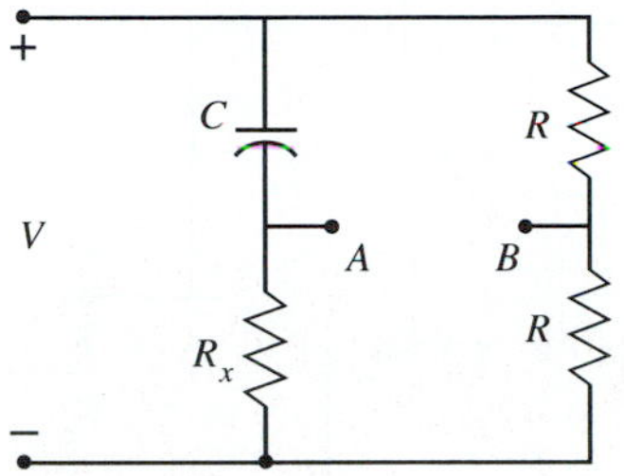

Figure P41

42. The circuit in Figure P42 is a bandpass filter. (Confirm that there is no output at very small and very large frequencies by replacing the capacitors by appropriate equivalents.) Find the phasor-voltage gain V_2/V_1 in terms of ω and the parameter values, assuming that the op amp remains in its linear region of operation.

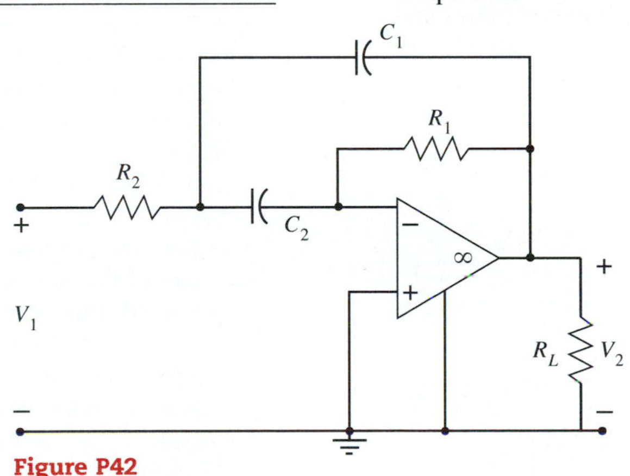

Figure P42

43. The circuit in Figure P43 is a low-pass circuit. (Again, on the basis of the behavior of the capacitors, confirm that there will be no output at very high frequencies but there will be some output at very low frequencies.) Find the numerical value of the phasor-voltage gain, assuming the angular frequency is 100,000 rad/s.

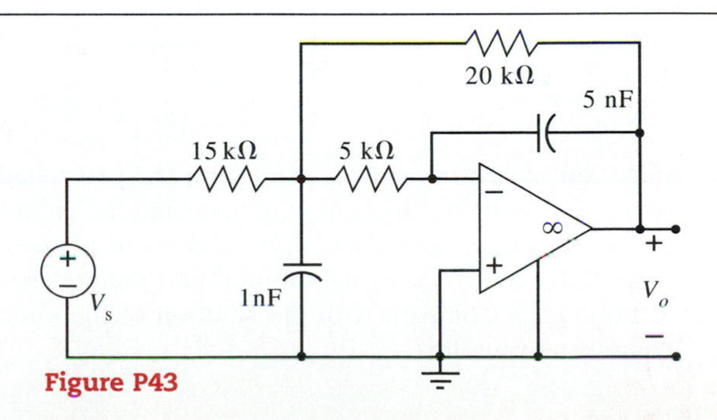

Figure P43

44. The circuit in Figure P44 is an all-pass circuit. (That is, the magnitude of the voltage gain equals 1 for all frequencies.) Find an expression for the phasor-voltage gain V_2/V_1 in terms of the parameter values and the angular frequency ω. (Observe the circuit and describe its features to yourself. For example, which inputs of the op amp are excited? To which terminals is the feedback returned?) The value of K $(= R_1/4R_2)$ is critical. Suppose K is 1% more than $R_1/4R_2$; determine the approximate effect that this will have on the gain.

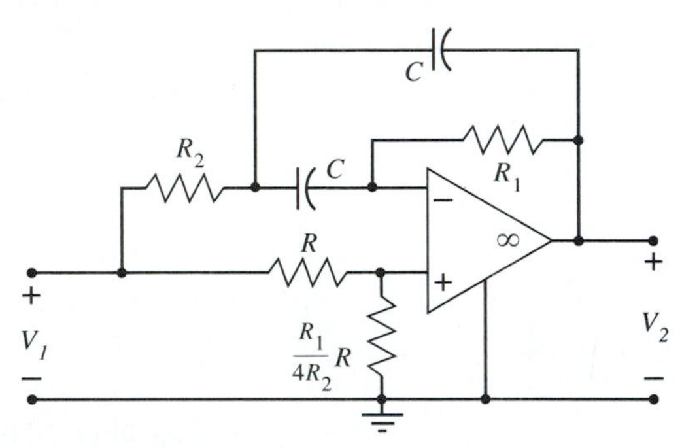

Figure P44

45. The circuit in Figure P45 is another bandpass circuit. (Again, observe its features and describe them to yourself. Also, confirm the behavior of the circuit for very low and very high frequencies.) Find the phasor-voltage gain in terms of the parameter values and the frequency.

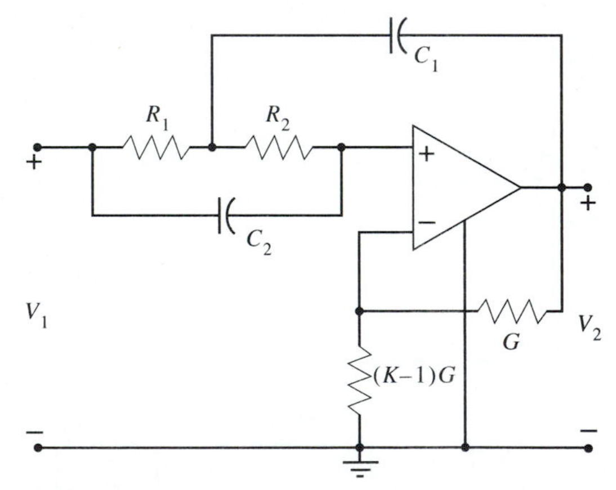

Figure P45

46. Carry out a current-shift operation on the controlled source in Figure 11 in the text, converting the circuit to two separate parts. One part consists of a single loop. Solve the resulting circuit, and find the voltages at the nodes. Compare with the solution of the node equations in the text.

47. Find the solution of the set of loop equations resulting from Figure 13 in the text.

48. The objective in this problem is to find the voltage labeled V in the circuit of Figure P48.

a. Do so by replacing everything to the right of the terminals that define V by a Thévenin equivalent.

b. Confirm your answer by writing and solving a set of node equations or a set of loop equations. (Do both if you want a double check.)

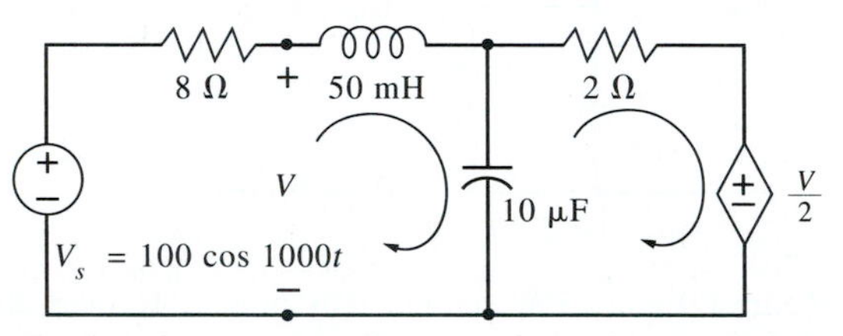

Figure P48

49. As noted in Problem 7 in Chapter 5, two distinct choices of loops are made when a loop analysis is carried out: (1) loops for defining loop currents and (2) loops around which KVL is applied. These two sets of loops can be chosen to be the same—and most of the time they *are* so chosen, without a second thought. That is, KVL is applied around the same loops used to define the loop currents. However, nothing prevents these two sets of loops from being different.

In Figure P49, the circulating arrows define the mesh currents.

a. For the purpose of applying KVL, choose only one of the meshes, the second loop being the one around the outer periphery of the circuit, and write a set of loop equations.

b. Show that the two equations so obtained are linearly independent.

c. Obtain the second mesh equation as a linear combination of the two equations in part (a).

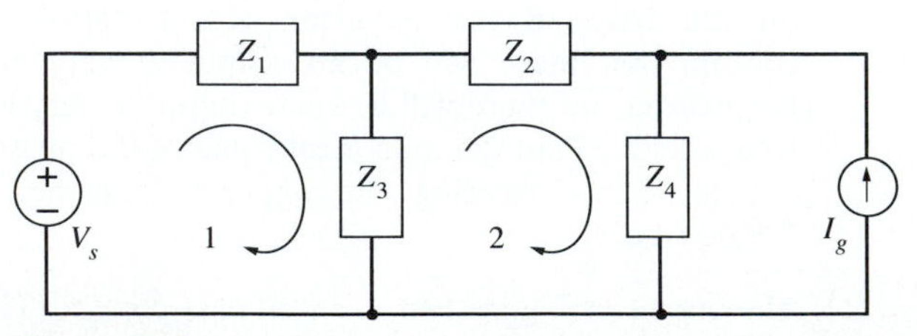

Figure P49

50. The objective of this problem is to take an analytical approach using loop equations to prove the compensation theorem. Suppose that in Figure 22 in the text the branch whose impedance is to be changed from Z to $Z + \Delta Z$ appears in only one loop and this loop is taken as loop 1. (Can this always be done?) The loop currents after the change of impedance can be identified with a prime, those before the change being unprimed. Remember that it is the changes in currents and voltages generally that are to be determined. If the changes in loop currents are found, the change in any other current and voltage can be obtained in terms of these.

Construct a proof of the compensation theorem based on loop equations. (Hint: How would you get the *change* in loop currents?)

DESIGN PROBLEMS

1. This problem concerns some practical aspects of wiring a large area to provide electrical lighting without excessive voltage loss in the wiring. Electrical loads are normally steady, requiring a fixed current at a rated voltage. However, sometimes there is a demand for more power, thus a distinction is made between *continuous* load and *maximum* load.

Assume that a storage area similar to the one shown in Figure DP1 is to be provided with electrical power. The area is subdivided into four equal subareas, and each subarea must be fed through a separate circuit from the panel board at which electrical power is provided. Each circle in the figure indicates the location of an outlet capable of delivering a continuous steady-state current of 5 A. The power source is 230 V, and copper conductors are to be used.

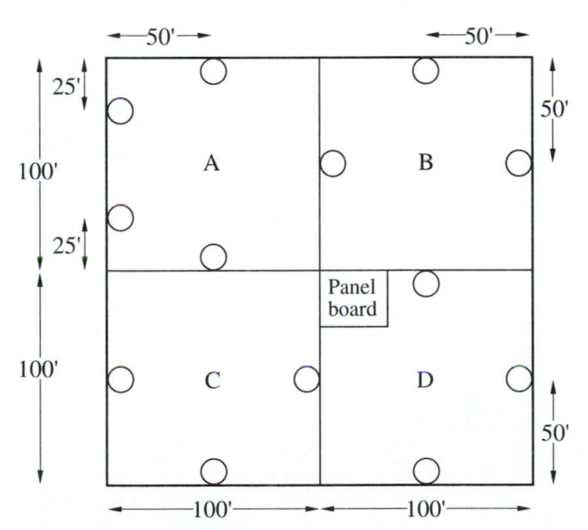

Figure DP1

Table DP1a gives the National Electrical Code branch circuit current-carrying capacities. Table DP1b gives the resistance of copper conductors of various sizes. Calculate the required currents in each circuit from the panel board to each outlet on that circuit. Estimate the length of each circuit and make sure that the voltage drop in the wiring is less than 3% of the source voltage by choosing the appropriate wire size.

Table DP1a

Branch circuit rating	15 A	20 A	30 A	40 A	50 A
Minimum size (AWG) copper conductors	14	12	10	8	6
Maximum load	15 A	20 A	30 A	40 A	50 A
Continuous load	12 A	16 A	24 A	32 A	40 A

Table DP1b

Size AWG	DC resistance Ω/1000 ft
14	2.57
12	1.62
10	1.018
8	0.6404
6	0.410

2. The single-phase induction motor is used widely in home appliances. The starting torque in the single-phase induction motor is produced by means of two windings whose currents are out of phase with each other. The starting torque is proportional to the magnitude of these currents as well as to the sine of the angle between them. This problem involves a capacitor-start single-phase motor; in which one of the windings, called the *main winding*, is always connected to the driving source. The second winding, called the *auxiliary winding*, is connected only when the motor is started. A capacitor and a centrifugal switch are inserted in series with the auxiliary winding. The capacitor accentuates the phase difference of the currents, while the centrifugal switch opens when the motor speed reaches a certain value. A simplified circuit model of such a motor, valid when the motor is just starting, is shown in Figure DP2, where $r_a = 12\ \Omega$ and $L_a = 17.25$ mH are the resistance and inductance of the auxiliary winding; $r_m = 1.9\ \Omega$ and $L_m = 7$ mH are the resistance and inductance of the main winding; $Z_f = 3.28 + j2.69\ \Omega$ is the effective impedance of the rotor as seen from the main winding; $a = 1.6$ is the turns ratio of the two windings and a^2Z_f is the effective impedance of the rotor as seen from the auxiliary winding. The design requirements are that the phase angle between I_f and I_a be at least 60° and that $|I_a| < 6$ A. Choose a value of C that satisfies these requirements.

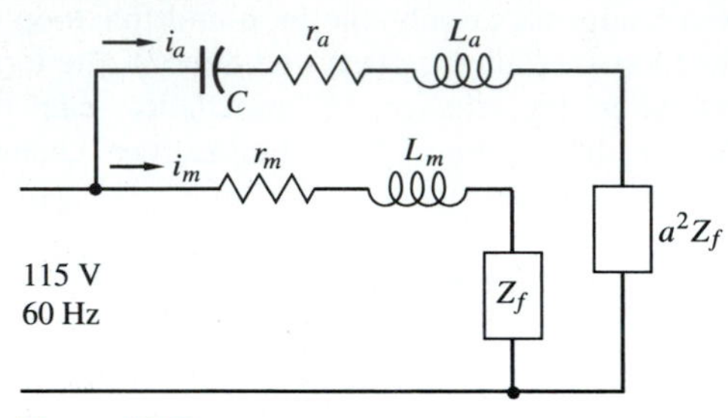

Figure DP2

9 Power and Energy in the Sinusoidal Steady State

Electric power systems throughout the world produce electricity whose waveshape is sinusoidal.[1] Electrical appliances and other devices and equipment are designed to operate most efficiently when the input electricity has a sinusoidal waveform. Typically, an electrical appliance operates over a period of time that covers thousands of cycles of the input voltage. (The number of 60-Hz cycles in one minute is 3600, and even a toaster operates for around that length of time after it has been turned on.) At radio frequencies (around 1 MHz and higher), there are millions of cycles in one second. Even if the time constants involved in the equipment are a small fraction of a second, this corresponds to many RF cycles. Hence, any transient that occurs after a device is turned on, or after some interference, has long since disappeared and we can safely assume that the sinusoidal steady state obtains for most of the time that such devices are operating.

Common electrical devices convert to other forms the electrical energy that is used to operate them: to light (lamp); to sound (radio); to heat (toaster, broiler, iron, water heater); to motion (mixer, fan, clock, sewing machine); or to combinations of forms (clothes dryer, hair dryer, television, air conditioner). As just argued, these devices are all operating in the steady state. It is essential, therefore, to understand the nature of energy—and its time derivative, power—in the sinusoidal steady state.

[1] The frequency is 60 Hz in the United States and most other countries, but in some countries it is 50 Hz.

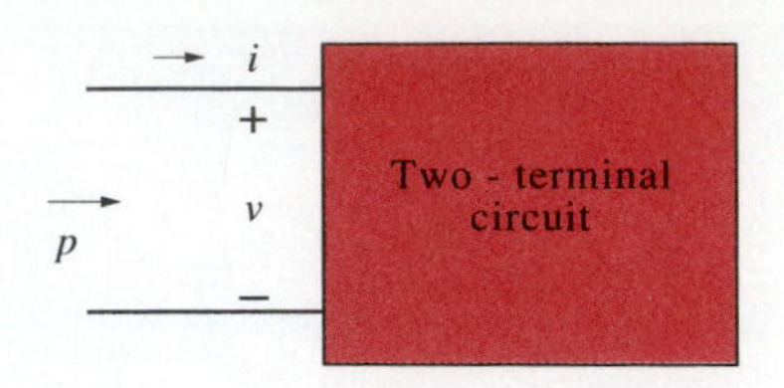

Figure 1
Power input at a pair of terminals.

1 AVERAGE POWER

The basic concepts of power and energy for arbitrary waveforms were introduced in Chapter 2, where the energy-dissipating and energy-storing properties of the circuit elements were also discussed. You should review those sections if you need a refresher. Here we shall turn to a consideration of power and energy in the sinusoidal steady state.

A two-terminal circuit is shown in Figure 1. It is presumed to be excited at the terminals by a sinusoidal source, either voltage or current. Standard references are chosen for v and i, so the input power is $p = vi$. The voltage and current at the terminals are sinusoidal; the sinusoids and their phasors are given by:

$$v(t) = \sqrt{2}\,|V| \cos(\omega t + \alpha) \quad \leftrightarrow \quad V = |V|\,e^{j\alpha}$$
$$i(t) = \sqrt{2}\,|I| \cos(\omega t + \beta) \quad \leftrightarrow \quad I = |I|\,e^{j\beta}$$

With standard references for the voltage and current, the reference for power corresponding to $p = +vi$ is directed into the terminals, as shown. The power at any instant of time is given by:

$$\begin{aligned} p(t) = v(t)i(t) &= 2\,|V|\,|I| \cos(\omega t + \alpha) \cos(\omega t + \beta) \\ &= |V|\,|I| \cos(\alpha - \beta) + |V|\,|I| \cos(2\omega t + \alpha + \beta) \end{aligned} \tag{1}$$

Assuming that the voltage and current are in volts (V) and amperes (A), respectively, the power is in watts (W).

The last form of (1) is obtained by using trigonometric identities for the sum and difference of two angles. (Confirm its validity.) Whereas both the voltage and the current are sinusoidal, the instantaneous power contains—in addition to a sinusoidal term—a constant term. Furthermore, the frequency of the sinusoidal term is twice that of the voltage or current. Plots of v, i, and p are shown in Figure 2 for specific values of α and β. The power is alternately

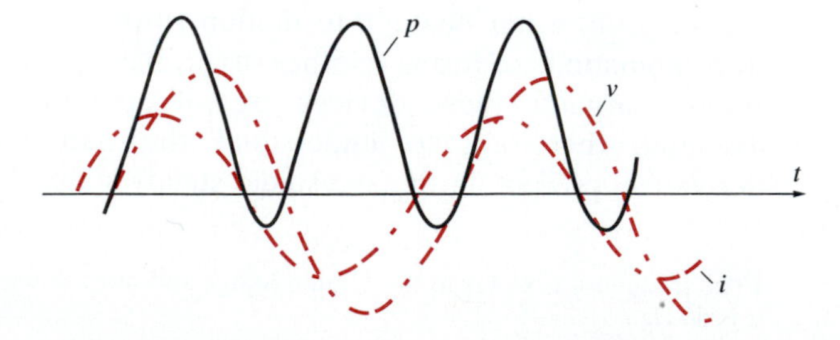

Figure 2
Instantaneous voltage, current, and power.

positive and negative. This means that power actually flows into the terminals sometimes and out of them some other times.

The energy which is transmitted into the network over some interval of time is found by integrating the power over this interval. If the area under the positive part of the power curve were the same as the area under the negative part, the net energy transmitted over one cycle would be zero. For the values of α and β used in the figure, however, the positive area is greater, so that there is a net transmission of energy toward the right. The energy flows back from the network to the source over part of the cycle, but, on the average, more energy flows toward the network than away from it.

EXERCISE 1

Assuming standard references, the phasor voltage and current at the terminals of a one-port are $V = 120\underline{/75^\circ}$ and $I = 4\underline{/45^\circ}$ in V and A, respectively. (a) Write an expression for the instantaneous power at the terminals. (b) Determine the constant term and the double-frequency term as in (1). (c) Sketch the waveforms of $v(t)$, $i(t)$, and $p(t)$. □

Power in Terms of RMS Values and Phase Difference

Consider the question from another point of view. Equation (1) shows the power to consist of a constant term and a sinusoid. The average value of a sinusoid is zero, so this term will contribute nothing to the net energy transmitted. Only the constant term will contribute. This constant term is the average value of the power. This is evident either from Figure 2 or by integrating (1) over one cycle. Denoting the average power by capital P and letting $\theta = \alpha - \beta$, the average power becomes:

$$P = |V|\,|I| \cos\theta \tag{2}$$

This is a very important result; it states that the average power supplied to the input terminals of a circuit is the product of the rms value of the voltage, the rms value of the current, and the cosine of their phase difference. (Since the cosine is an even function, it does not matter whether the difference is taken as $\alpha - \beta$ or $\beta - \alpha$.)[2]

EXERCISE 2

To get a feel for the numerical values involved, let the rms values of voltage and current be 100 V and 2 A. Plot the average power against angle θ from 0 to 90°. We might accept a drop in the average power down to 85% of the

[2] In this chapter a number of new terms, with their symbols and units and interrelationships, will be introduced. Just reading these and going on is *not* a good approach to your education. Notice how an interpretation of (2) was given following the equation. You should examine each equation in detail, confirming its derivation, formulating verbal descriptions of its meaning, and noting how the variables are related to each other. Typically, in a book, the derivation of an equation includes intermediate steps. You may want a record of the beginning point and the final result. Keep a notebook where you can record these, as well as your verbal descriptions. Such an effort will pay off.

maximum as being close to the maximum. How much of the interval between 0 and 90° is traversed before the average power is reduced to that much?

ANSWER: Almost 32°. □

Power in Terms of Voltage and Current Phasors

Sometimes the expression for average power in (2) is not the most convenient to use either for theoretical or for numerical purposes. We shall, therefore, determine alternative ways of expressing the average power. With the voltage and current phasors previously given, the impedance of the circuit in Figure 1 will be:

$$Z = \frac{V}{I} = \frac{|V|\,e^{j\alpha}}{|I|\,e^{j\beta}} = \frac{|V|}{|I|}e^{(\alpha-\beta)} = |Z|\,e^{j\theta} \tag{3}$$

The angle of the impedance is seen to be the phase difference between the voltage and the current. It is the cosine of this angle which appears in the expression for average power.

Remembering that the cosine can be expressed as the real part of an exponential, we can rewrite the previous expression for P in (2) as:

$$\begin{aligned} P &= |V|\,|I|\,\mathrm{Re}\,(e^{j\theta}) = \mathrm{Re}\,[|V|\,|I|\,e^{j(\alpha-\beta)}] \\ &= \mathrm{Re}\,[(|V|\,e^{j\alpha})(|I|\,e^{-j\beta})] \\ &= \mathrm{Re}\,(VI^*) \end{aligned} \tag{4}$$

The second line is obtained by breaking up the exponential in the previous line by the law of exponents. The first factor between square brackets in this line is identified as the phasor voltage, and the second factor as the conjugate of the phasor current. The last line then follows. It expresses the average power in terms of the voltage and current phasors and is sometimes more convenient to use, especially in theoretical developments.

Power in Terms of Z and I or of Y and V

Two additional forms can be obtained for P by substituting for either V or I^* its equivalent in terms of impedance or admittance. Thus, the last expression in (4) becomes:

$$P = \mathrm{Re}\,(ZII^*) = |I|^2\,\mathrm{Re}\,(Z) = R(\omega)\,|I|^2 \tag{5a}$$

$$P = \mathrm{Re}\,(VV^*Y^*) = |V|^2\,\mathrm{Re}\,(Y^*) = |V|^2\,\mathrm{Re}\,(Y) = G(\omega)\,|V|^2 \tag{5b}$$

In the first step of (5b) we have used the fact that the conjugate of a product, $(VY)^*$, equals the product of conjugates, V^*Y^*. The second step in each equation of (5) is obtained by noting that a complex quantity multiplied by its conjugate equals the square of its magnitude. The next step in the second equation follows from the fact that the real part of the conjugate of a complex quantity is the same as the real part of the quantity itself. Finally, the real parts of Z and Y have been expressed in the usual notation as R and G, respectively. But remember that these do not refer to the resistance or conductance of a single resistor; to emphasize this fact the explicit dependence on frequency is shown.

Several different expressions, all equivalent, for the average power supplied to a two-terminal circuit have now been obtained. In any given situation, one might be easier to apply than the others.

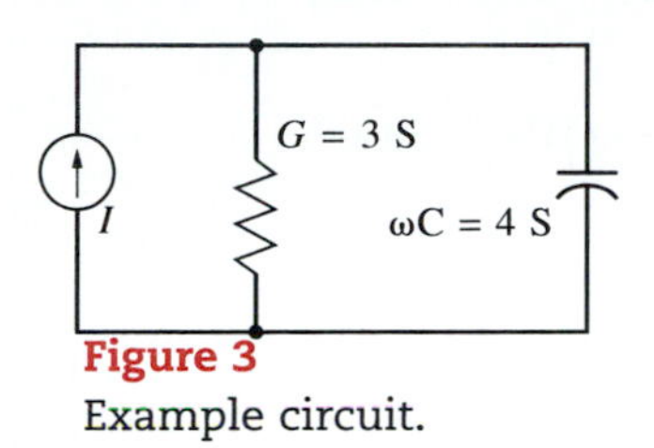

Figure 3
Example circuit.

An illustrative example is shown in Figure 3. A sinusoidal current source with an rms value of $|I| = 10$ A feeds a resistor and a capacitor in parallel. The objective is to find the average power delivered to the circuit in the steady state. Because the input is a current, the most convenient form to use for P is $|I|^2 \operatorname{Re}(Z)$:

$$P = |I|^2 \operatorname{Re} Z = 100 \operatorname{Re} \frac{1}{Y} = 100 \operatorname{Re} \frac{1}{G + j\omega C}$$

$$= 100 \operatorname{Re} \frac{3 - j4}{3^2 + 4^2} = 100\left(\frac{3}{25}\right) = 12 \text{ W}$$

Because the two branches are in parallel and the admittance is easily obtained, Z has been replaced by $1/Y$ in an intermediate step. The simplest way to determine the real part of the reciprocal of a complex quantity in rectangular form is to rationalize the denominator, as shown.

An alternative form for the average power would be to multiply the rms values of V and I and the cosine of the phase difference. In this form, since I is given, we would need to calculate V and $\cos\theta$. Now $V = ZI$; finding V requires first finding Z:

$$Z = \frac{1}{Y} = \frac{1}{3 + j4} = \frac{1}{5e^{j53.2^\circ}} = \frac{1}{5}e^{-j53.2^\circ}$$

The magnitude of Z serves to determine the magnitude (rms value) of V, and the angle of Z is the phase difference between V and I. Thus:

$$|V| = |Z|\,|I| = \tfrac{10}{5} = 2$$

$$\cos\theta = \cos(-53.2^\circ) = \tfrac{3}{5}$$

$$P = |V|\,|I| \cos\theta = 2 \times 10 \times \tfrac{3}{5} = 12 \text{ W}$$

We are in good luck; the answers by the two approaches agree.

EXERCISE 3

A lossy inductor whose impedance is $1 + j1$ is placed in parallel with the circuit components in Figure 3. Determine the average power delivered by the current source, using two different expressions.

ANSWER: $P = 14.3$ W. □

2 COMPLEX, REACTIVE, AND APPARENT POWER

The average power is expressed in (4) as the real part of a complex quantity VI^* which also has an imaginary part. Let S designate this complex quantity, and write it in rectangular form:

$$S = VI^* = |V|\,|I|\,e^{j\theta} = |V|\,|I| \cos\theta + j\,|V|\,|I| \sin\theta$$
$$= P + jQ \tag{6}$$

where

$$P = |V|\,|I| \cos\theta \tag{7a}$$
$$Q = |V|\,|I| \sin\theta \tag{7b}$$
$$|S| = |V|\,|I| \tag{7c}$$

We already know that P is the average power. Since it is the real part of some complex quantity, it would be reasonable to call it the *real power.* The complex quantity S, whose real part is P, is therefore called the *complex power.* The magnitude of the complex power, $|S| = |V|\,|I|$, is the product of the rms values of voltage and current. It is called the *apparent power,* and its unit is the *volt-ampere* (VA). To be consistent, then, we should call Q the *imaginary power.* This is not usually done, however; instead, Q is called the *reactive power.* It has the same dimensions as S, but its unit is called a *VAR* (volt-ampere reactive).[3]

EXERCISE 4

A given circuit with two terminals is operating in the sinusoidal steady state. The reactive power supplied to the circuit is half the apparent power. Find the average power in terms of the apparent power.

ANSWER: $P = \sqrt{3}Q = \sqrt{3}\,|S|/2.$ □

Phasor and Power Diagrams

It is possible to give a graphical interpretation useful for clarifying and understanding the preceding relationships and for the calculation of power. Figure 4 is a phasor diagram of V and I in a particular case. The phasor voltage

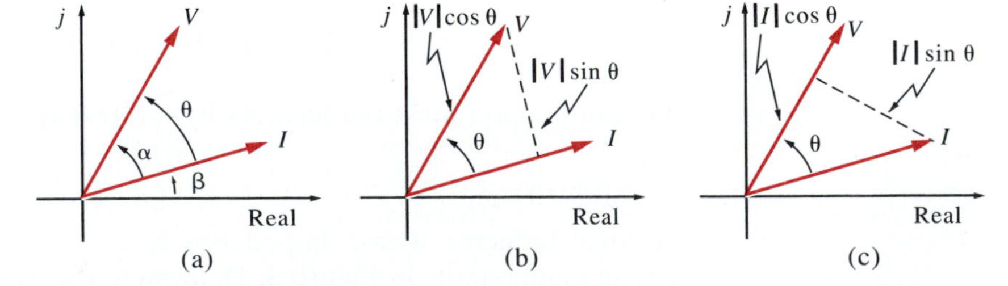

Figure 4
In-phase and quadrature components of V and I used in calculation of P and Q.

[3] Don't throw up your hands because of this shower of names. There is a certain logical relationship in these designations; they are not simply arbitrary. Familiarity will soon come and you will learn to associate the names with the "faces."

can be resolved into two components, one parallel to the phasor current (or *in phase with I*) and another perpendicular to the current (or *in quadrature with* it). This is illustrated in Figure 4b. Hence, the average power P is the magnitude of phasor I multiplied by the in-phase component of V; the reactive power Q is the magnitude of I multiplied by the quadrature component of V.

Alternatively, one can imagine resolving phasor I into two components, one in phase with V and one in quadrature with it, as illustrated in Figure 4c. Then P is the product of the magnitude of V with the in-phase component of I; and Q is the product of the magnitude of V with the quadrature component of I. Real power is produced only by the in-phase components of V and I. The quadrature components contribute only to the reactive power.

The in-phase or quadrature components of V and I do not depend on the specific angles of V and I, but only on their phase *difference.* One can imagine the two phasors in Figure 4 to be rigidly held together and rotated around the origin. So long as the angle θ is held fixed, all the discussion of this section will still apply. It is common to use the current phasor as the reference for angle, that is, to choose $\beta = 0$ so that phasor I lies along the real axis. Then $\theta = \alpha$.

This is illustrated in Figure 5a, which shows the same V and I phasors, but with I taken along the real axis.[4] An interesting point is illustrated in Figure 5b. A graphical display of the impedance and its real and imaginary parts is a

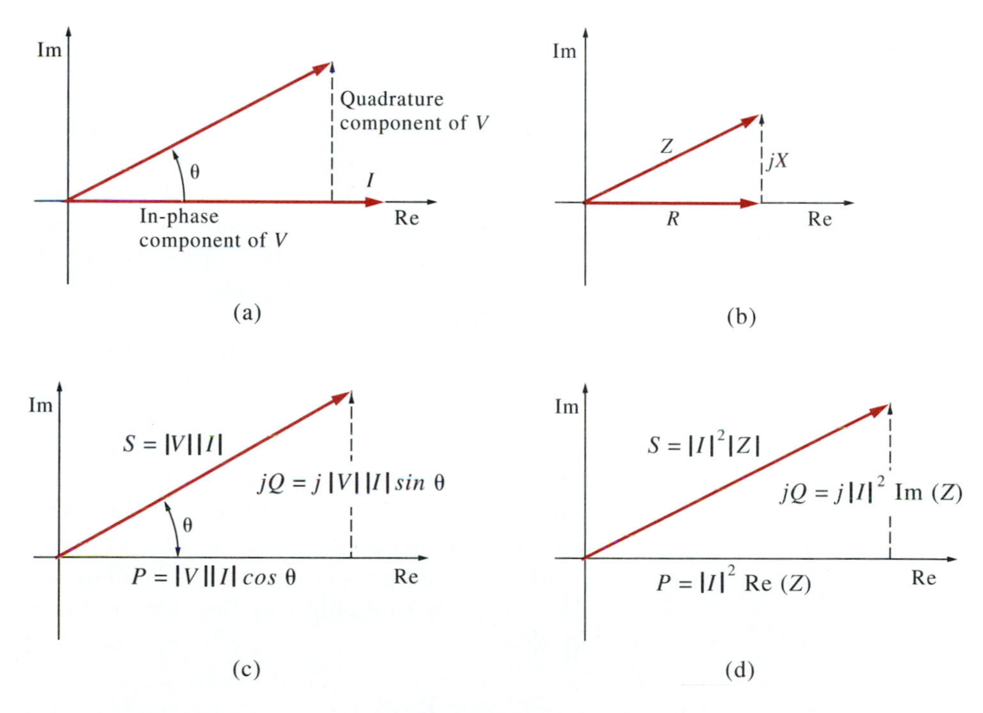

Figure 5
Impedance and power triangles.

[4] Note that two different quantities (voltage and current) with different units are placed on the same set of axes. The scales for the two quantities are obviously different, have no relationship to each other, and can be chosen strictly for convenience: so many volts per centimeter or amperes per centimeter.

triangle congruent with the triangle formed by the phasor voltage and its in-phase and quadrature components. Furthermore, since the angle of the complex power is the same as that of the impedance, the impedance triangle can be converted to the one shown in Figure 5c, called the *power triangle.* The horizontal leg is the real (average) power, and the vertical leg is the reactive power; the hypotenuse represents the complex power, its length being the apparent power.

EXERCISE 5

The voltage and current phasors of a one-port with standard references, are $85 + j85$ and $5 \angle 30°$, in V and A, respectively. (a) Find the components of I that are in phase and in quadrature with V. (b) Conversely, find the components of V that are in phase and in quadrature with I. (c) Find the average power, the reactive power, and the apparent power; include units.

ANSWER: (a) In phase: 4.83 A; quadrature: 1.29 A; (b) in phase: 116.1 V; quadrature: 31.8 V; (c) $P = 580.6$ W; Q 155.6 VAR; $|S| = 601$ VA. □

Reactive and Complex Power in Terms of Z or Y

Just as alternative expressions were found for the average power in (5), it is also possible to find similar expressions for the reactive power, starting from its definition. Thus:

$$Q = \text{Im}\,(VI^*) = \text{Im}\,(ZII^*) = |I|^2\,\text{Im}(Z) = |I|^2\,X(\omega) \tag{8a}$$

$$Q = \text{Im}\,(VI^*) = \text{Im}\,(VV^*Y^*) = |V|^2\,\text{Im}\,(Y^*) = -|V|^2\,B(\omega) \tag{8b}$$

in (8b) we used the fact that the imaginary part of a complex number is the negative of the imaginary part of its conjugate.

Since, in (5) and (8), the real and imaginary parts of the complex power are related to the real and imaginary parts of the impedance and admittance, we should be able to express the complex power S in terms of Z or Y. Thus, if we combine the preceding expressions with the corresponding ones in (5), the result will be:

$$S = |I|^2 R + j\,|I|^2 X = |I|^2 Z \tag{9a}$$

$$S = |V|^2 G - j\,|V|^2 B = |V|^2\,Y^* \tag{9b}$$

Thus, if the excitation of a two-terminal circuit is a current (of constant amplitude), the complex power is directly proportional to the impedance of the circuit. Similarly, if the excitation is a constant-amplitude voltage, the complex power is directly proportional to the conjugate of the admittance of the circuit.

The preceding results permit a relabeling of the power triangle in Figure 5d. This amounts to multiplying the sides of the impedance triangle in Figure 5b by $|I|^2$.

Power Factor

For any given circuit it is useful to know what part of the total complex power is real (average) power and what part is reactive power. This is usually expressed in terms of the *power factor* F_p, defined as the ratio of real power to apparent power:

$$\text{Power factor } F_p \equiv \frac{P}{|S|} = \frac{P}{|V|\,|I|} = \cos\theta \tag{10}$$

The general relationship based on the definition reduces to just $\cos\theta$ when the expression $P = |V|\,|I|\cos\theta$ is used. Because the power factor is simply $\cos\theta$, θ itself is called the *power-factor angle.*

Since the cosine is an even function, specifying the power factor does not reveal the sign of θ. Remember that θ is the angle of the impedance. If θ is positive, this means that the current lags the voltage; we say that the power factor is a *lagging power factor.* On the other hand, if θ is negative, the current leads the voltage and we say this represents a *leading power factor*.

The power factor will reach its maximum value, unity, when the voltage and current are in phase. This will happen in a purely resistive circuit, of course. It will also happen in more general circuits for specific element values and specific frequencies.

We can now obtain a physical interpretation for the reactive power. When the power factor is unity, the voltage and current are in phase and $\sin\theta = 0$. Hence, the reactive power is zero. In this case, the instantaneous power is never negative. This case is illustrated by the power waveform in Figure 6a; the power curve never goes negative, and there is no exchange of power between the source and the circuit.

At the other extreme, when the power factor is zero, the voltage and

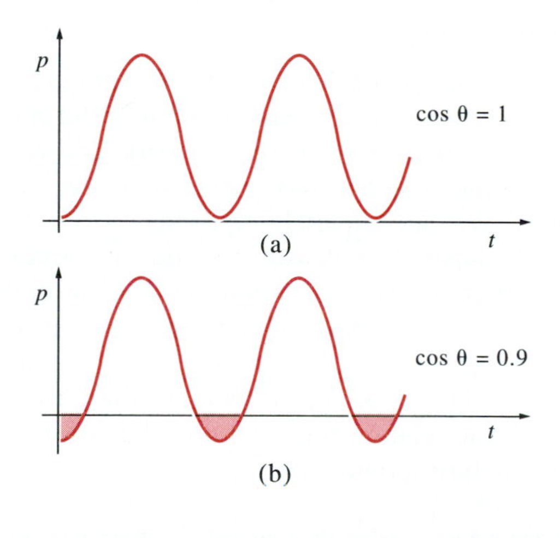

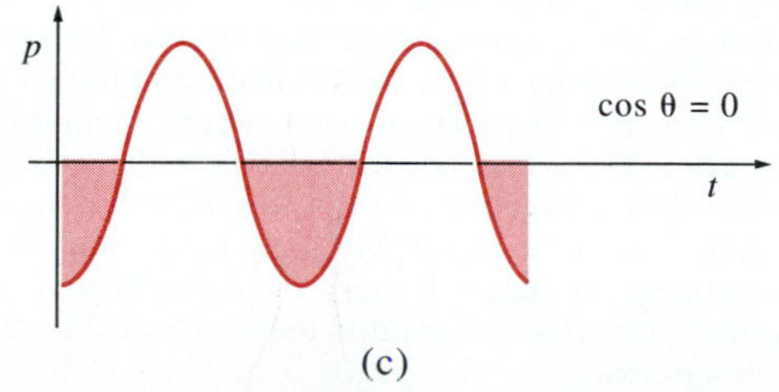

Figure 6
Power waveform for different power factors.

current are 90° out of phase and $\sin \theta = 1$. Now the reactive power is a maximum and the average power is zero. In this case, the instantaneous power is positive over half a cycle (of the voltage) and negative over the other half. All the power delivered by the source over half a cycle is returned to the source by the circuit over the other half. This is shown in Figure 6c.

For intermediate values of the power factor, as in Figure 6b, some of the power supplied by the source over part of the cycle is returned by the circuit to the source, but not all of it. There is a nonzero average power delivered by the source.

It is clear, then, that the reactive power is a measure of the energy exchanged between the source and the circuit without being used by the circuit. Although none of this exchanged energy is dissipated by or stored in the circuit and it is returned unused to the source, nevertheless it is temporarily made available to the circuit by the source.[5]

Review

By the nature of the subject under discussion, a large number of quantities come into play:

1. Voltage and current: as functions of time and as phasors (magnitudes and angles)
2. Impedance and admittance: real and imaginary parts, magnitudes and angles
3. Power: average (real), reactive, complex, apparent; power factor (lagging and leading)

The normal approach in the analysis of networks is, given a network, its parameter values, and the excitation, to determine the voltage, current, and/or power at various places in the network. However, in the area of AC power networks, this is not always the sequence. Sometimes, desired values of outcomes are incompletely specified (for example, voltage and current magnitudes—possibly measured—and/or average power and power factor at various places); what is wanted, for example, is details of the circuit (impedances, component values), reactive power, and voltage and current angles. In such cases, an algorithmic approach (for example, carrying out a node analysis) is less appropriate. Rather, for any given problem, you must draw on basic relationships and all relevant tools at your command and apply them in an appropriate order.

[5] Power companies charge their industrial customers not only for the average power they consume but for the reactive power they return. There is a reason for this. Suppose a given power system is to deliver to a load a fixed amount of average power at a constant voltage amplitude. Since $P = |V|\,|I| \cos \theta$, the current magnitude will be inversely proportional to the power factor. If the reactive power is high, the power factor will be low, so a high current will be required to deliver the given power. The conductors carrying a relatively large current to the customer must be correspondingly larger and better insulated, which means a larger capital investment in plant. It may be cost-effective for customers to try to reduce the reactive power they require, even if they have to buy additional equipment to do so, as we shall shortly discuss.

EXAMPLE 1 An electrical heating element is rated at 500 W and 120 V. However, the only 60-Hz source available is at 220 V. It has been suggested that something be placed in series with the heater so as to reduce the voltage to the desired value. A resistor is out of the question, since it would waste energy. Somebody suggested an inductor, but the only coils available include some resistance which equals one-fifth the inductive reactance. The problem is to find: (a) the required value of L, (b) the power dissipated in the physical inductor, (c) the voltage across it, and (d) the power factor at the source.

A circuit diagram is shown in Figure 7. Note that the voltages 120 and 220 V are rms values, or phasor magnitudes.

A possible plan of attack is as follows. The magnitude of the impedance presented to the source can be obtained from the magnitudes of the source voltage and its current. The latter is the same as the magnitude of the heater current, which follows from the heater ratings. (With the heater assumed to be purely dissipative, the rms heater current follows immediately from the expression for power; $|I| = 500/120 = 4.17$ A.) The magnitude of the impedance can also be written in terms of the parameter values, of which only L is unknown. This unknown is determined by equating these two expressions. Thus:

a. $$R_h = \frac{120}{4.17} = 28.8\,\Omega \qquad R = \frac{\omega L}{5} = 0.2\omega L$$

$$|Z| = \frac{220}{|I|} = \frac{220}{4.17} = 52.8\,\Omega$$
$$= \sqrt{(R + R_h)^2 + (\omega L)^2} = \sqrt{(0.2\omega L + 28.8)^2 + (\omega L)^2}$$

Squaring both sides and simplifying leads to the quadratic:

$$1.04(\omega L)^2 + 11.52\omega L - 1958.4 = 0$$

whose solution is $\omega L = 38.21$, so that $L = 38.21/120\pi = 101.4$ mH.

b. The power dissipated in the real inductor is:

$$P_{\text{ind}} = R\,|I|^2 = \frac{4.17^2\omega L}{5} = 17.39(7.64) = 132.9\text{ W}$$

c. Let the reference for phase be the angle of the current. Then $I = |I|\angle 0 = 4.17$ A. Since the impedance of the inductor has been bound,

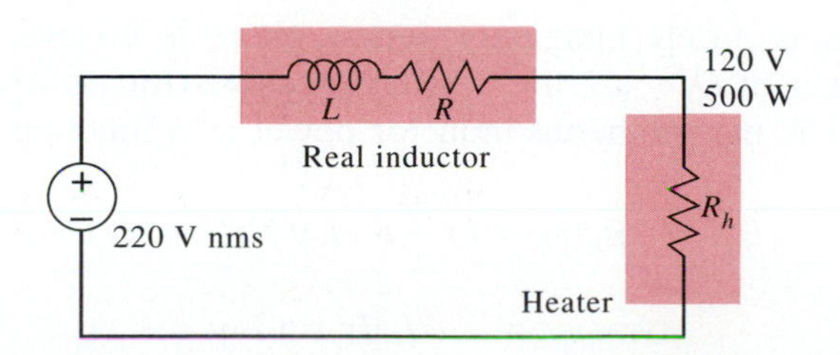

Figure 7
Heater problem.

its voltage is:

$$V_{\text{ind}} = 4.17(7.64 + j38.21) = 162.5 \underline{/78.7°}$$

d. More than one approach is possible to determine the power factor at the source. Thus:

$$F_p = \frac{R + R_h}{|Z|} = \frac{7.6 + 28.8}{52.8} = 0.69$$

$$F_p = \frac{P}{|S|} = \frac{500 + 132.9}{(220)(4.17)} = 0.69$$

3 POWER AND ENERGY RELATIONS OF INDIVIDUAL COMPONENTS

In the preceding sections we have dealt with general two-terminal circuits described by their impedance or admittance. It would be of great interest to examine individual circuit components and determine their power and energy properties.

Dissipative Case

Suppose the general circuit in Figure 1 is strictly dissipative, with no energy-storing components; the equivalent impedance at the terminals will be resistance R. In this case, the complex power in (6) becomes:

$$S = VI^* = RII^* = R\,|I|^2 = P \tag{11}$$

This is purely real; there is no reactive power in this case. As a matter of fact, the instantaneous power, $p = vi = Ri^2$, is always positive and at no time flows away from the circuit. The waveform in Figure 6a represents this case.

Inductive Case

Next assume that the general circuit in Figure 1 is an inductor. (It could be a circuit of inductors only, replaced by a single equivalent inductor.) In this case, the complex power will be:

$$S = VI^* = j\omega L\,|I|^2 = jQ \tag{12}$$

This is purely imaginary; hence, there is no average power, only a reactive power of $Q = \omega L\,|I|^2$, which is positive (for positive ω).

The instantaneous inductor power as a function of time is:

$$\begin{aligned} p_L(t) &= Li\frac{di}{dt} = L\sqrt{2}\,|I| \cos \omega t\,(-\omega\sqrt{2}\,|I| \sin \omega t) \\ &= -\omega L\,|I|^2 \sin 2\omega t \end{aligned}$$

(The angle of the current was taken to be 0, for convenience.) The second line follows from the trigonometric identity $2 \sin x \cos x = \sin 2x$. The power is a sinusoidal function at twice the frequency of the current. As time goes on, the power is alternately positive and negative. No net power flows into the inductor over time, in agreement with the previous result. This corresponds to Figure 6c.

As power flows in and out of the inductor, what about the energy stored there? You will recall from Chapter 2 that the energy stored in an inductor as a function of time is $Li^2/2$. Again we assume that the angle of the current sinusoid is 0. Thus:

$$w_L = \tfrac{1}{2}Li^2 = L\,|I|^2 \cos^2 \omega t = \tfrac{1}{2}L\,|I|^2(1 + \cos 2\omega t)$$

The right side follows from the trigonometric identity $2\cos^2 x = 1 + \cos 2x$. Since the cosine varies between -1 and $+1$, the right-hand side will vary between 0 and $L\,|I|^2$. Since $\cos 2\omega t$ has an average value of zero, the average value of the stored energy, which we shall label W_L, will be:

$$W_L = \tfrac{1}{2}L\,|I|^2 \tag{13}$$

This entity is never negative; it is 0 only when the current is identically 0.

If we compare the expression for the average energy stored in an inductor in the sinusoidal steady state with the instantaneous stored energy for any current $i(t)$, we see that they have exactly the same form. In one case (the average stored energy), however, the current is the rms value of a sinusoid and in the other it is a function of time.

Plots of the current, power, and stored energy against time are given in Figure 8. The power and energy curves have twice the frequency of the current. It is clear that, as power flows into and out of the inductor, the stored energy rises to a peak value and then falls to zero. Although the power is negative over half the cycle, the stored energy is never negative. (A negative

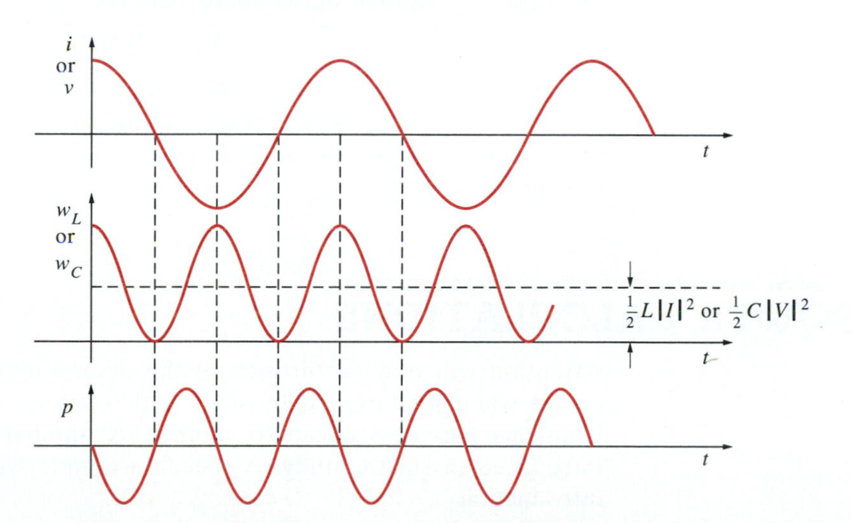

Figure 8
Power and stored energy in inductor or capacitor.

stored energy would mean that net energy was actually being supplied by the inductor, a physical impossibility.)

Capacitive Case

Finally, assume that the two-terminal circuit is a capacitor, or a circuit of capacitors represented by a single equivalent capacitor. In this case, the complex power is:

$$S = VI^* = \frac{1}{j\omega C} II^* = -j\frac{1}{\omega C}|I|^2 = -j\omega C\,|V|^2 = jQ$$

Again there is no average power, only reactive power, which, in this case, is negative.

Again let's look at the instantaneous values and this time choose $v = \sqrt{2}\,|V| \cos \omega t$. The instantaneous power flowing into the capacitor will be:

$$\begin{aligned} p_C &= Cv\frac{dv}{dt} = C\sqrt{2}\,|V| \cos \omega t\,(-\omega\sqrt{2}\,|V| \sin \omega t) \\ &= -\omega C\,|V|^2 \sin 2\omega t \end{aligned}$$

(The right side follows from the trigonometric identity $2 \sin x \cos x = \sin 2x$.) This expression is strikingly similar to the corresponding one previously obtained for the inductor power. It is necessary only to replace L by C and current by voltage.

Finally, let's again consider the energy stored in the capacitor; it will be:

$$w_C = \tfrac{1}{2}Cv^2 = C\,|V|^2 \cos^2 \omega t = \tfrac{1}{2}C\,|V|^2\,(1 + \cos 2\omega t)$$

Again there is great similarity between this equation and the corresponding one for inductor energy, except for a change of symbols. Thus, the curves in Figure 8 can equally well represent the power and stored energy of a capacitor, if the current curve is now labeled voltage. The entire discussion regarding the in-and-out flow of power and the storage of energy need not be repeated; it applies in the present case as well—with appropriate changes in the words. In this case, the average stored energy will be:

$$W_C = \tfrac{1}{2}C\,|V|^2 \tag{14}$$

Again we note that this expression has the same form as that of the instantaneous energy stored in a capacitor for any voltage $v(t)$, except that, in the present case, the voltage is the rms value of a sinusoidal function instead of a function of time.

4 POWER CALCULATIONS

Attention will now be directed to the determination of power supplied to a circuit when it is excited by sinusoidal sources, or to the power transferred from one part of a circuit to another. A number of important concepts are introduced in such a study. A specific example will be considered by way of introduction.

EXAMPLE 2 The two-terminal circuit shown in Figure 9 is supplied by a sinusoidal voltage source having an effective (rms) value of 10 V. Find the average power and the reactive power supplied by the source.

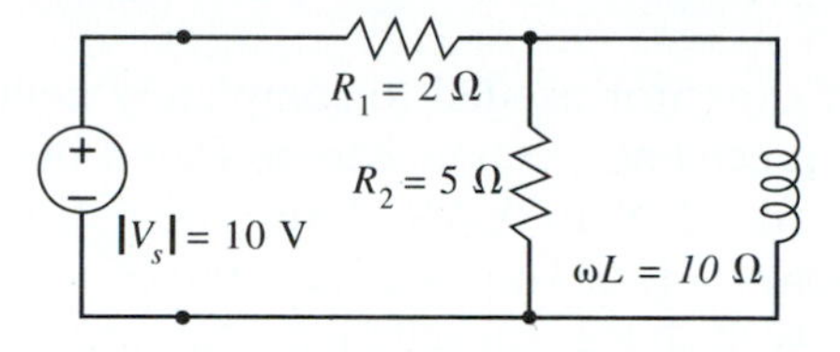

Figure 9
Illustrative example.

One approach would be to compute the current supplied by the source, thus determining the power factor. Then, $|V|\,|I| \cos\theta$ and $|V|\,|I| \sin\theta$ determine P and Q, respectively. An alternative approach is to use $|V_s|^2 Y^*$ to find the complex power. Since the source voltage is already given, it remains only to find the admittance of the two-terminal circuit connected to the source. This can be written as follows:

$$Y = \frac{1}{Z} = \frac{1}{R_1 + \dfrac{1}{1/R_2 + 1/j\omega L}} = \frac{1}{2 + \dfrac{1}{\frac{1}{5} - j\frac{1}{10}}} = \frac{1}{2 + \dfrac{10}{2 - j1}}$$

$$= \frac{2 - j1}{14 - j2} = \frac{2.24e^{-j26.5^\circ}}{14.2e^{-j8.1^\circ}} = 0.158e^{-j18.4^\circ}$$

The complex, average, and reactive power can now be written:

$$S = |V_s|^2\, Y^* = 100(0.158e^{j18.4^\circ}) = 15 + j5\ \text{VA}$$

$$P = 15\ \text{W}$$

$$Q = 5\ \text{VAR}$$

As you might expect, the reactive power is positive, corresponding to an inductive circuit. (Carry out the first approach previously suggested and compare.)

Power-Factor Correction

Suppose the circuit in the preceding example corresponds to a manufacturing plant supplied by the local utility. Although the reactive power in this example is not excessive, the rate charged by the utility will be larger than the rate charged to those customers having close to zero reactive power. What can be done so that the plant can reduce its reactive power? (In a manufacturing plant, obviously the power level would be much higher. To be more realistic, let's increase the scale by 10^4.)

The answer is not difficult to find. The reactive power now happens to be positive, resulting from the fact that Y^* has a positive imaginary part (or Y has a negative imaginary part). If we placed in parallel with the terminals of the circuit a component whose admittance has a positive imaginary part, we could, wholly or in part, cancel the original negative imaginary part. That would result in what is called a *power-factor correction.* The admittance of a capacitor is $j\omega C$, so a capacitor will definitely satisfy the requirement for a positive susceptance.

The size of capacitor needed to completely neutralize the inductive reactive power in the preceding example can be found by setting the reactive power of the capacitor to -5×10^4 VAR. Using $\omega = 377$ rad/s (corresponding to the power frequency of 60 Hz), we get $100\omega C = 5 \times 10^4$ or $C = 500/377 = 1.33$ F. This is a huge capacitance! Nevertheless, a manufacturing plant may have to make a capital investment in such a capacitor in order to avoid large operating costs. A trade-off is involved. Perhaps a smaller (and cheaper) capacitor can be used to correct for only a part of the power factor, thus reducing the capital cost but not reducing the operating cost as much as wanted.

Maximum Power Transfer

In the chapter on dissipative circuits (Chapter 5), we found that the power delivered by a two-terminal circuit to a load resistor connected to its terminals could be maximized if the load resistance could be varied and made equal to the Thévenin equivalent resistance of the circuit. This result has a counterpart for general linear circuits in the sinusoidal steady state.

The diagram in Figure 10 illustrates a two-terminal linear circuit at whose terminals an impedance Z_L is connected. The circuit is assumed to be operating in the sinusoidal steady state. The problem to be addressed is this: given the two-terminal circuit, how can the load impedance be adjusted so that the maximum possible average power is transferred from the circuit to the impedance?

The first step is to replace the circuit by its Thévenin equivalent, as shown in Figure 10b. The current phasor in this circuit is $I = V_T/(Z_T + Z_L)$. The average power transferred by the circuit to the impedance is:

$$P = |I|^2 \operatorname{Re}(Z_L) = \frac{|V_T|^2 \operatorname{Re}(Z_L)}{|Z_T + Z_L|^2}$$

$$= \frac{|V_T|^2 R_L}{(R_T + R_L)^2 + (X_T + X_L)^2} \tag{15}$$

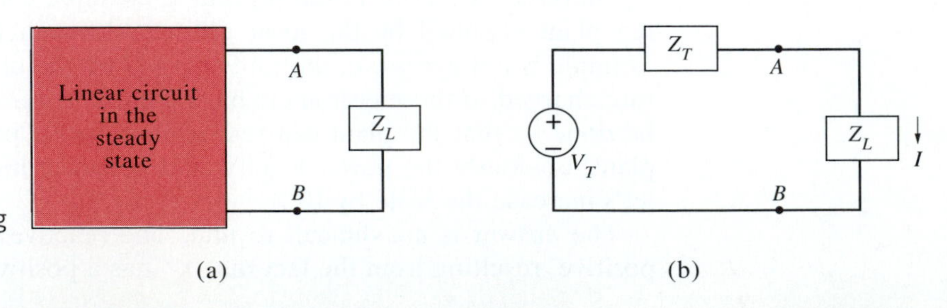

Figure 10
A linear circuit delivering power to a load in the steady state.

There are a lot of different symbols in this expression; the question is, Which of them are fixed (constant) and which can vary? Since the two-terminal circuit is fixed, the Thévenin voltage and impedance are constants; only the load (that is, R_L and X_L) can be varied. The preceding equation, then, expresses a dependent variable (P) in terms of two independent ones (R_L and X_L).

What is required is to maximize P? For a function of a single variable, this is done by setting the derivative of the function equal to zero. When there is more than one independent variable, then the *partial* derivatives with respect to each of the independent variables must be set equal to zero; that is, $\partial P/\partial R_L = 0$ and $\partial P/\partial X_L = 0$. Carrying out these differentiations (with the denominator of the last line in the preceding expression labeled D for convenience) leads to:

$$\frac{\partial P}{\partial R_L} = \frac{|V_T|^2 [(R_L + R_T)^2 + (X_L + X_T)^2 - 2R_L(R_L + R_T)]}{D^2} \tag{16a}$$

$$\frac{\partial P}{\partial X} = \frac{-2\,|V_T|^2 R_L(X_L + X_T)}{D^2} \tag{16b}$$

Equating each of these expressions to zero leads to:

$$X_L = -X_T \tag{17a}$$

$$R_L = \sqrt{R_T^2 + (X_L + X_T)^2} \tag{17b}$$

When the first of these is inserted into the second, the result is $R_L = R_T$. Hence, the maximum average power will be transferred to the load when:

$$Z_L = Z_T^* \tag{18}$$

That is, for maximum power transfer, the load impedance must equal the conjugate of the Thévenin impedance of the circuit.[6]

We have been a little rash in claiming that the average power has been *maximized* for the stated condition. All we can really say is that this is the condition for an extreme value of P—either a maximum or a minimum. To be on safe ground, we would have to check. Setting $Z_L = Z_T^*$ in the expression for P in (15), we find that:

$$P_{\max} = \frac{|V_T|^2}{4R_T} \tag{19}$$

That this is indeed a maximum value can be established by noting that the denominator of the expression for P contains a term $X_L + X_T$. The contribu-

[6] If the power-delivering circuit in Figure 10 contains no reactive components, then the Thévenin equivalent parameters will be constant. If it does contain reactive components, however, then both V_T and Z_T will be frequency-dependent. Nevertheless, both will be constant if the frequency is fixed, as assumed in the preceding development. But if the frequency is not fixed, what has just been done will not be valid. It may still be approximately correct if the range of frequency change is so small that V_T and Z_T are appreciably constant.

tion of this term to the denominator will be the smallest when $X_L = -X_T$; and small denominator means large ratio.

In the preceding discussion of maximum power transfer, it was assumed that the Thévenin impedance of the source circuit was fixed but that the load impedance could be varied in any arbitrary way. Suppose, however, that there are limitations on how the load impedance can vary; then what? Obviously, that will depend on the nature of the restriction. There are several possibilities:

1. It may be possible to vary the real and imaginary parts of the load impedance over a limited range of values only. In this case, because of the term $X_L + X_T$ in the denominator, the smallest value of the denominator will occur when this term is made as small as possible, within the restriction on the variation of X_L. Then R_L must be chosen as close as possible to its value in (17b).

2. It may be that the magnitude of the load impedance can be varied but its angle is fixed. This case will not be discussed but will be left as a problem for you.

3. It may be that the angle of the load impedance can be varied but the magnitude is fixed. In that case, the angle of the load impedance can be set equal to the negative of that of the Thévenin impedance. Hence, $Z_L = KZ_T^*$, where K is a positive constant. It was observed in Chapter 8 that an ideal transformer converts an impedance Z_L connected to one pair of terminals to an impedance n^2Z_L as seen from the other pair of terminals. Thus, the present problem is solved by inserting an ideal transformer with an appropriate turns ratio between the source circuit and the load impedance.

EXERCISE 6

The load in the circuit of Figure 11 is to be adjusted so that the maximum power will be delivered to it at a fixed frequency of $\omega = 1000$ rad/s. (a) Determine the required values of R and L. (b) Now assume that the frequency is not fixed but varying, but that, nevertheless, the parameters of the Thévenin equivalent of the source circuit are approximately constant. With R and L fixed at their maximum-power values, determine the frequency range over which the power delivered to the load does not deviate by more than 10% from the maximum as the frequency varies. (c) Confirm that the assumption of

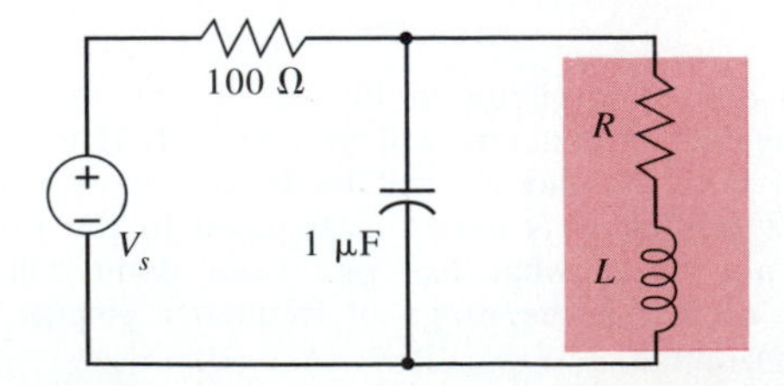

Figure 11
Maximum-power exercise.

approximately constant V_T and Z_T is satisfied over the range of frequency in question.

ANSWER: (a) $R = 99\,\Omega$, $L = 9.9$ mH; (b) $334 < \omega < 1667$ rad/s; (c) $98.64 < |Z_T| < 100.05\,\Omega$, compared with $99.5\,\Omega$, <1% change. □

5 THREE-PHASE CIRCUITS

Figure 12a represents the basic circuit for considering the flow of power from a single sinusoidal source to a load. The power can be thought to cross an imaginary boundary surface (represented by the dotted line in the figure) separating the source from the load. Suppose that:

$$v(t) = \sqrt{2}\,|V| \cos(\omega t + \alpha)$$
$$i(t) = \sqrt{2}\,|I| \cos(\omega t + \beta)$$

Then, as determined in (1), the power to the load at any instant of time is:

$$p(t) = |V|\,|I|\,[\cos(\alpha - \beta) + \cos(2\omega t + \alpha + \beta)] \tag{20}$$

The instantaneous power has a constant term and a sinusoidal term at twice the frequency. The quantity in brackets fluctuates between a minimum value of $\cos(\alpha - \beta) - 1$ and a maximum value of $\cos(\alpha - \beta) + 1$. This fluctuation of power delivered to the load has certain disadvantages in some situations where the transmission of power is the purpose of a system. An electric motor, for example, operates by receiving electric power and transmitting mechanical

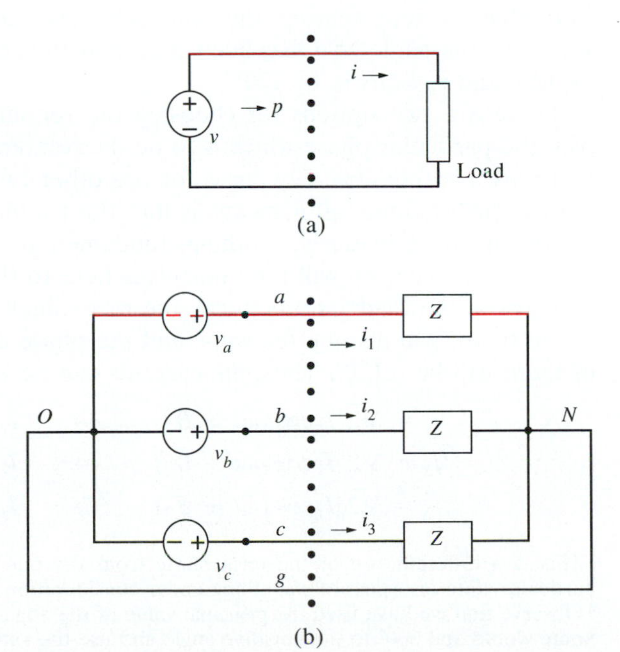

Figure 12
Flow of power from source to load.

(rotational) power at its shaft. If the electric power is delivered to the motor in spurts, the motor is likely to vibrate. For satisfactory operation in such a case, a physically larger motor, with a larger shaft and flywheel, will be needed to provide more inertia for smoothing out the fluctuations than would be the case if the delivered power were constant.

This problem is overcome in practice by the use of what is called a *three-phase* system. This section will provide a brief discussion of three-phase power systems.

Consider the circuit in Figure 12b. This arrangement is a combination of three simple circuits like the one in Figure 12a connected in such a way that each one shares the return connection between O and N. The three sources can be viewed collectively as a single source, and the three loads—which are assumed to be identical—can be viewed collectively as a single load. Then, as before, the dotted line represents an imaginary surface, separating the source from the load, across which the power is transmitted. Each of the individual sources and loads is referred to as one *phase* of the three-phase system.[7]

Relationships among Voltages and Currents

The three sources are assumed to have the same frequency; for this reason, they are said to be *synchronized.* It is also assumed (1) that the three voltages have the same rms values and (2) that the phase difference between each pair of voltages is ±120° ($2\pi/3$ rad). Thus, the voltages can be written:

$$\begin{aligned} v_a &= \sqrt{2}\,|V|\cos(\omega t + \alpha_1) &&\leftrightarrow && V_a = |V|\,e^{j0^\circ} \\ v_b &= \sqrt{2}\,|V|\cos(\omega t + \alpha_2) &&\leftrightarrow && V_b = |V|\,e^{-j120^\circ} \\ v_c &= \sqrt{2}\,|V|\cos(\omega t + \alpha_3) &&\leftrightarrow && V_c = |V|\,e^{j120^\circ} \end{aligned} \tag{21}$$

The phasors representing the sinusoids have also been shown. For convenience, the angle of v_a has been chosen as the reference for angles; v_b *lags* v_a by 120° and v_c *leads* v_a by 120°.[8]

There are two options for choosing the sequence of the phases. Once we pick the particular phase which is to be the reference for angles and name it *a,* there are two possible sequences for the other two: either *abc* or *acb.* This is hardly earthshaking; all it means is that the leading and lagging angles can be interchanged. Obviously, nothing fundamental is different in the second sequence. Hence, we will limit ourselves here to the *abc phase sequence.*

Because the loads are identical, the rms values of the three currents shown in the figure will also be the same and the phase difference between each pair of them will be ±120°. Thus, the currents can be written:

$$\begin{aligned} i_1 &= \sqrt{2}\,|I|\cos(\omega t + \beta_1) &&\leftrightarrow && I_1 = |I|\,e^{j\beta_1} \\ i_2 &= \sqrt{2}\,|I|\cos(\omega t + \beta_2) &&\leftrightarrow && I_2 = |I|\,e^{j(\beta_1 - 120^\circ)} \\ i_3 &= \sqrt{2}\,|I|\cos(\omega t + \beta_3) &&\leftrightarrow && I_3 = |I|\,e^{j(\beta_1 + 120^\circ)} \end{aligned} \tag{22}$$

[7] This is a different use of the term *phase* from the one meaning "angle"; it refers to a particular state or aspect of something under consideration, like the phase of the moon.

[8] Observe that we have used the principal value of the angle lying between −180° and +180°. Some would add 360° to the negative angle and use the value 240° instead.

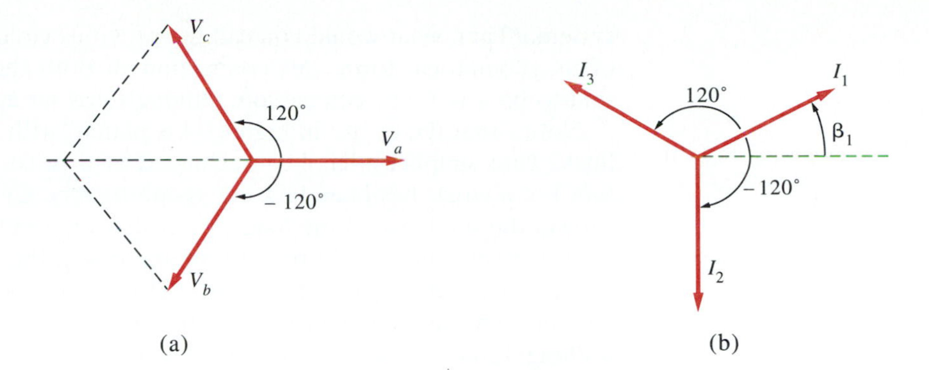

Figure 13
Voltage and current phasor diagrams.

Perhaps a better form for visualizing the voltages and currents is a graphical one. Phasor diagrams for the voltages and the currents separately are shown in Figure 13. The value of angle β_1 will depend on the load. Something of the utmost significance is clear from these diagrams. First, V_b and V_c are each other's conjugate. Thus, if we add them, the imaginary parts cancel and the sum will be real, as illustrated by the construction in the voltage diagram. Furthermore, the construction shows this sum to be negative and equal in magnitude to V_a. Hence, *the sum of the three voltages is zero.* The same is true of the sum of the three currents, as you can establish graphically by a similar construction.

EXERCISE 7

Confirm analytically that the sum of the voltages is 0 and that the sum of the currents is 0 in (21) and (22) by converting the phasor voltages and currents into rectangular form. □

By Kirchhoff's current law applied at node N in Figure 12b, we find that the current in the return line is the sum of the three currents in (22). But since this sum was found to be zero, the return line carries no current. Hence, it can be removed entirely without affecting the operation of the system. The resulting circuit is redrawn in Figure 14. It can be called a *three-wire* three-phase

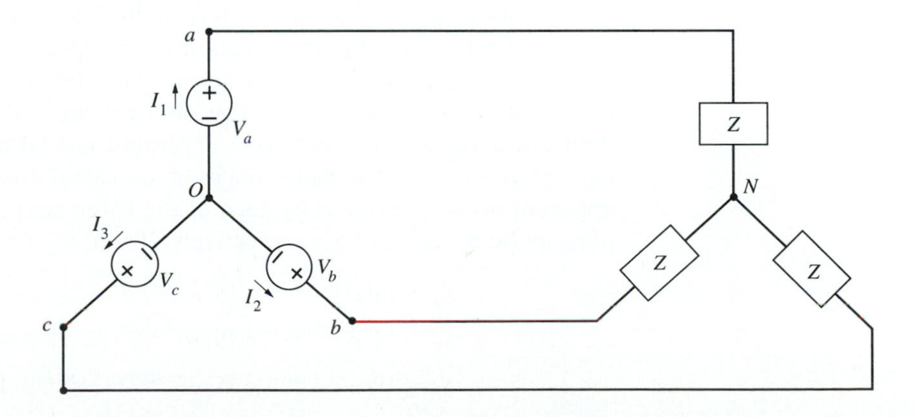

Figure 14
Wye-connected three-phase system.

system. (Then what would you call the previous circuit in Figure 12b?) Because of its geometrical form, this connection of both the sources and the loads is said to be a wye (Y) connection, although it is an upside-down Y.

Notice that the circuit in Figure 14 is planar, with no lines crossing any other lines. That simplicity has been achieved at a price. Notice how the sequence (*abc*) of sources has been laid out geometrically. Clearly, with the connections shown, the sequence of the loads is not the same as that of the sources. Having the same sequence would require interchanging the connections of the b and c sources with the bottom two loads. Doing that would result in one branch crossing another. (Confirm that by drawing the connections.) However, nothing fundamental would change with either connection, assuming equal loads as we have done.

Line Voltages

Look again at the three-wire three-phase system in Figure 14. The neutral point O is not accessible (as will be explained shortly), so phase voltages cannot be measured. The voltages that are available for measurement are the *line-to-line,* or simply the *line,* voltages: V_{ab}, V_{bc}, and V_{ca}. By KVL:

$$\begin{aligned} V_{ab} &= V_a - V_b = |V| - |V|\,e^{-j120^\circ} = \sqrt{3}\,|V|\,e^{j30^\circ} \\ V_{bc} &= V_b - V_c = |V|\,e^{-j120^\circ} - |V|\,e^{j120^\circ} = \sqrt{3}\,|V|\,e^{-j90^\circ} \\ V_{ca} &= V_c - V_a = |V|\,e^{j120^\circ} - |V| = \sqrt{3}\,|V|\,e^{j150^\circ} \end{aligned} \tag{23}$$

The interesting result is that all the line-voltage magnitudes are equal at $\sqrt{3}$ times the phase-voltage magnitude. Thus, a 220-V line voltage corresponds to a phase voltage of 127 V. The line-voltage angles have the same mutual relationships that the phase-voltage angles have; they are separated by $\pm 120^\circ$.

EXERCISE 8

Starting with the phasor diagram of phase voltages in Figure 13, determine the line-voltage phasors graphically, thus confirming (23). □

Power Relationships

The instantaneous power delivered by each of the sources has the form given in (20), consisting of a constant term representing the average power and a double-frequency sinusoidal term. The latter, being sinusoidal, can be represented by a phasor also. The only caveat is that a different frequency is involved here, so this power phasor should not be mixed with the voltage and current phasors in the same diagram or calculations. Let $|S| = |V|\,|I|$ be the apparent power delivered by each of the three sources, and let the three power phasors be S_a, S_b, and S_c, respectively. Then:

$$\begin{aligned} S_a &= |S|\,e^{j(\alpha_1+\beta_1)} = |S|\,e^{j\beta_1} \\ S_b &= |S|\,e^{j(\alpha_2+\beta_2)} = |S|\,e^{j(-120^\circ+\beta_1-120^\circ)} = |S|\,e^{j(\beta_1+120^\circ)} \\ S_c &= |S|\,e^{j(\alpha_3+\beta_3)} = |S|\,e^{j(120^\circ+\beta_1+120^\circ)} = |S|\,e^{j(\beta_1-120^\circ)} \end{aligned} \tag{24}$$

It is evident that the phase relationships among these three phasors are the same as the ones among the voltages and the currents. That is, the second leads the first by 120° and the third lags the first by 120°. Hence, just as in the case of the voltages and the currents, the sum of these three power phasors will also be zero. This is a very significant result. It constitutes the motivation for using three-phase power over the pulsating power of a single-phase system. Although the instantaneous power delivered by each load has a constant component and a sinusoidal component, when the three powers are added, the sinusoidal components add to zero, leaving only the constants. Thus, the total power delivered to the three loads is constant.

To determine the value of this constant power, let's use (20) as a model. The contribution of the kth source to the total (constant) power is $S \cos(\alpha_k - \beta_k)$. You can easily verify that $\alpha_k - \beta_k = \alpha_1 - \beta_1 = -\beta_1$. The first equality follows from the relationships among the α's from (21) and among the β's from (22). The choice of $\alpha_1 = 0$ leads to the last equality. Hence, each phase contributes an equal amount to the total average power. If P is the total average power, then:

$$P = P_a + P_b + P_c = 3P_a = 3|V||I|\cos(\alpha_1 - \beta_1) \tag{25}$$

Although the angle α_1 has been set equal to zero, for the sake of generality we have shown it explicitly in this equation.

A similar result can be obtained for the reactive power. The reactive power of the kth phase is $|S|\sin(\alpha_k - \beta_k) = |S|\sin(\alpha_1 - \beta_1)$. If Q is the total reactive power, then:

$$Q = 3|S|\sin(\alpha_1 - \beta_1)$$

EXERCISE 9

Find the relationship among the power factor of a three-phase power system and the power factors of the individual phases. □

Balanced Source and Balanced Load

What has just been described is a *balanced* three-phase three-wire power system. The three sources in practice are not three independent sources but consist of three different parts of the same generator. The same is true of the loads.[9] What has been described is ideal in a number of ways. First, the circuit

[9] An AC power generator consists of (1) a *rotor* which is rotated by a *prime mover* (say, a turbine) and produces a magnetic field, which also rotates; and (2) a *stator* on which are wound one or more coils of wire. In three-phase systems, the number of coils is 3. The rotating magnetic field induces a voltage in each of the coils. The frequency of the induced voltage depends on the number of magnetic poles created on the rotor and the speed of rotation. These are fixed so as to "synchronize" with the 60-Hz frequency of the power system. The 120° leading and lagging phase relationships among these voltages are obtained by distributing the conductors of the coils around the circumference of the stator so that they are separated geometrically by 120°. Thus, the three sources described in the text are in reality a single physical device, a single generator. This is the reason why the common point O in Figure 14 is inaccessible. Similarly, the three loads might be the three windings on a three-phase motor, again a single physical device. Or they might be the windings of a three-phase transformer.

can be *unbalanced*—for example, because the loads are somewhat unequal. Second, since the real devices whose ideal model is a voltage source are coils of wire, each source should be accompanied by a branch consisting of the coil inductance and resistance. Third, since the power station (or the distribution transformer at some intermediate point) may be at some distance from the load, the parameters of the physical line carrying the power (the line inductance and resistance) must also be inserted in series between the source and the load.

The analysis of this section does not apply to an unbalanced system. An entirely new analytical technique is required to do full justice to such a system.[10] In this book, we shall not pursue this matter any further. An understanding of balanced circuits, however, is a prerequisite for tackling the unbalanced case.

The last two of the conditions that make the circuit less than ideal (winding and line impedances) introduce algebraic complications but change nothing fundamental in the preceding theory. If these two conditions are taken into account, the appropriate circuit takes the form shown in Figure 15. Here the internal impedance of a source (the winding impedance labeled Z_w) and the line impedance Z_l connecting that source to its load are both connected in series with the corresponding load. Thus, instead of being Z, the impedance in each phase is $Z + Z_w + Z_l$. Hence, the rms value of each current is:

$$|I| = \frac{|V|}{|Z + Z_w + Z_l|}$$

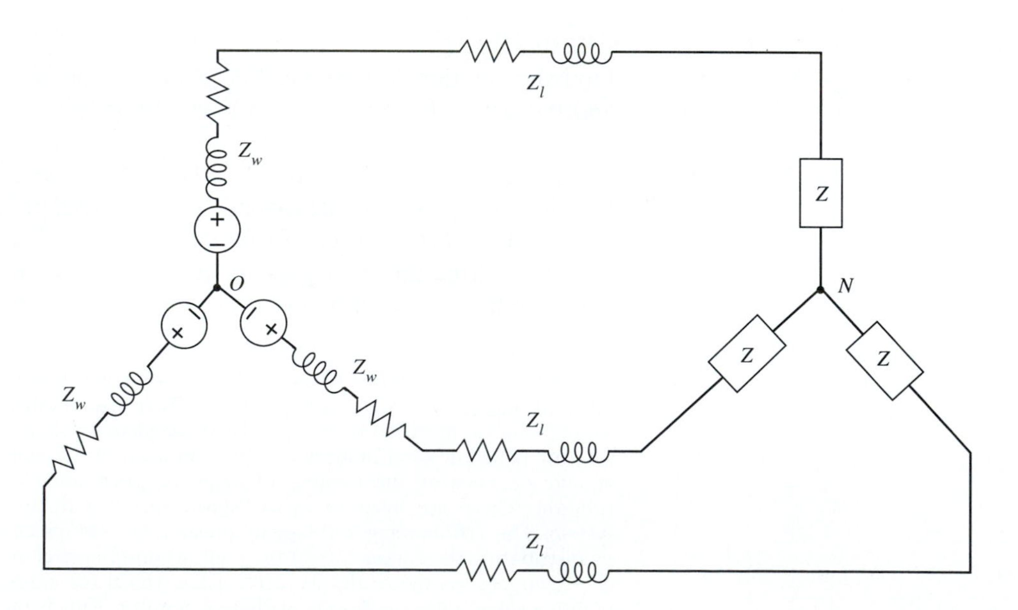

Figure 15
Three-phase circuit with nonzero winding and line impedances.

[10] The technique for analyzing unbalanced circuits utilizes what are called *symmetrical components*.

instead of $|V|/|Z|$. All other results we had arrived at remain unchanged, namely, that the sum of the phase currents is zero and that the sum of the phase powers is a constant. The detailed calculations become a little more complicated, that's all.

Other Types of Interconnections

All of the preceding development was based on a wye connection for both the sources and the loads. Although the upside-down Y structure looks geometrically a little different from an upside-down tee circuit, electrically the two are exactly the same. The wye is not, however, the only possible way to interconnect the phases of a three-phase system. Another possibility, the *delta* connection, so named because it looks like the Greek letter, is shown in Figure 16b. (In this figure the boxes can represent either sources or impedances.)

In the preceding chapter, we found that, by proper choice of the branch parameters, a tee could be made equivalent to a pi at the terminals. We note that the delta is just an upside-down pi. As a pi, the junction between branches A and B is usually extended as the common terminal of a two-port. The tee-pi equivalence given in (25) in Chapter 8 applies also to the equivalence between a wye and a delta of impedances.

If the structures in Figure 16 are to be equivalent, the voltages V_{ab}, V_{bc}, and V_{ca} (the line voltages) should be the same in both circuits. Similarly, the currents into the terminals (the line currents should be the same in both). Note that, in the delta connection, the phase voltages are not in evidence; the only voltages available are the line voltages. Thus the voltages in the delta are the line voltages given in (23). In the wye, the phase currents are also the currents in the lines. For the delta, however, the line currents are the difference between two phase currents, as noted in Figure 16b. For the line currents, a set of equations similar to (23) can be written in terms of the phase currents. Since the same 120° difference of angle exists between the phase currents as between the phase voltages, we would expect that the result for currents would be similar to the result for voltages in (23), namely, that the line-current magnitudes in a delta connection would be $\sqrt{3}$ times the phase-current magnitudes.

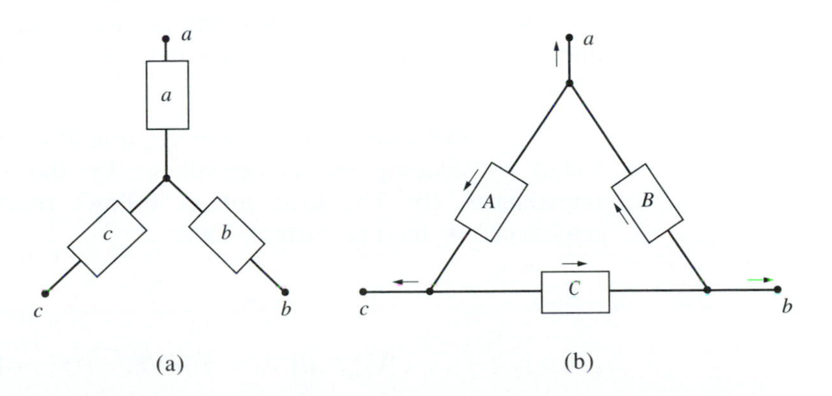

Figure 16
(a) Wye and (b) delta connections.

EXERCISE 10

(a) Write a set of equations similar to (23) relating the line-current phasors in a delta connection to the phase-current phasors. (b) Using the phase-current phasors from (22), determine the line-current phasors using complex arithmetic. (c) Repeat part (b) using a phasor diagram. □

EXERCISE 11

In a particular case, the boxes in Figure 16 represent the load impedances of a three-phase circuit. Assuming that the three impedances are the same in each case, determine a relationship between the impedances of the wye, Z_Y, and those of the delta, Z_Δ, if the circuits are to be equivalent.

ANSWER: $Z_\Delta = 3Z_Y$. □

In a three-phase circuit either the sources or the loads, or both, can be replaced by a delta equivalent, so we can conceive of four different three-phase circuits: wye-wye, wye-delta, delta-wye, and delta-delta. Not only can we conceive of them, they are actually utilized in practice.

Carrying out detailed calculations for these four cases would not be very educational. There is nothing fundamentally different in analyzing each of these four circuits. Once you understand the basic properties described here, you should be able to make the calculations for each case if you have to. Some of the problems will give you the opportunity to do so. Indeed, because of the wye-delta equivalence, any one of the three other cases can always be converted to the wye-wye connection anyway, without any influence on the line voltages and currents, or consequently on the power either. Here we will end with an example of a wye-wye connection.

EXAMPLE 3

A balanced 120-V three-wire three-phase transmission system in a wye-wye connection is represented by the circuit in Figure 15. We will assume that the winding impedances are negligible but that the line impedances are given by $Z_l = 0.1 + j0.2\,\Omega$. Each load impedance is $Z = 20 + j5\,\Omega$. The following quantities are to be determined: (a) the line-current magnitude; (b) the magnitude of the voltage across each load; (c) the average power, reactive power, and apparent power delivered to the load by each phase; (d) the average power, reactive power, and apparent power delivered by each source; and (e) the fraction of the power delivered by the system that is lost in the lines.

SOLUTION The solution is completely straightforward. (a) The line current is found by dividing the phase voltage by the sum of the load and line impedances. (b) The load voltage follows from the product of the load impedance by the line current. Thus:

$$|I| = \frac{120}{\sqrt{(20 + 0.1)^2 + (5 + 0.2)^2}} = 5.78\text{ A}$$

$$|V_L| = |I|\,|Z| = 5.78\sqrt{20^2 + 5^2} = 119.16\text{ V}$$

(c) The power calculations then follow:

$$|S_L| = |V_L|\,|I| = 119.16(5.78) = 688.7 \text{ VA}$$
$$|S_L| = |I|^2\,|Z_L| = 5.78^2\sqrt{20^2 + 5^2} = 688.7 \text{ VA}$$
$$P_L = R_L\,|I|^2 = 20(5.78)^2 = 668.2 \text{ W}$$
$$Q_L = X_L\,|I|^2 = 5(5.78)^2 = 167.0 \text{ VAR}$$
$$= \sqrt{|S_L|^2 - P_L^2} = \sqrt{688.7^2 - 668.2^2} = 166.8 \text{ VAR}$$

(There is a slight difference in the values of Q_L found by the two methods. The first value is likely to be more accurate, since accuracy is lost whenever two large numbers of the same order are subtracted.)
(d) Perhaps the best way to find the power delivered by the sources is to determine the power lost in the line and then add this to the load power. (Another way is to add the load and line impedances. Carry out this approach and confirm the result that follows.) Thus:

$$P_l = 0.1\,|I|^2 = 3.34 \text{ W} \qquad P_s = 3.34 + 668.2 = 671.5 \text{ W}$$
$$Q_l = 0.2\,|I|^2 = 6.68 \text{ VAR} \qquad Q_s = 6.68 + 167.0 = 173.7 \text{ VAR}$$

(e) Finally, the fraction of the source power that is lost in the line is 3.34/671.5 = 0.005, or 0.5%.

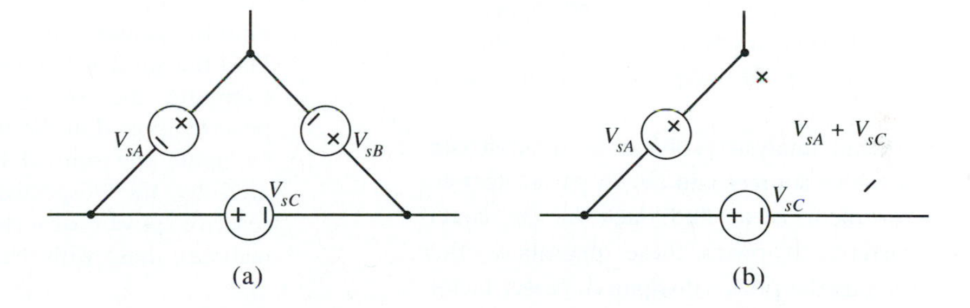

Figure 17
A delta connection of sources.

As a final observation, note that something interesting happens when voltage sources are connected in a delta, as illustrated in Figure 17. The delta constitutes a loop whose branches are all voltage sources. Since KVL must be satisfied, the source voltages must satisfy the following:

$$V_{sA} + V_{sB} + V_{sC} = 0$$

Hence, the three voltages cannot all be independent; any two of them will determine the third. Three sources are not really needed in the delta connection; one of them can be removed, as shown in Figure 17b. The only caveat would be that the two physical generators of which the ideal sources are the models must have the capacity to deliver the requisite amount of power.

SUMMARY

- Average power expressed in terms of rms values and phase
- Average power expressed in terms of voltage and current phasors
- Average power expressed in terms of impedance and current
- Average power expressed in terms of admittance and voltage
- Reactive, complex, and apparent power
- Reactive and complex power in terms of impedance or admittance
- Phasor diagrams and power diagrams
- Power factor and power-factor angle
- R, L, and C power and energy relationships
- Power-factor correction
- Maximum power transfer from a circuit to a load
- Balanced three-phase circuits
- Relationships among phase voltages and phase currents
- Line voltages and relationships with phase voltages
- Average and reactive power in three-phase circuits
- Phase sequence
- Wye and delta connections

PROBLEMS

The source of power in power systems typically supplies multiple loads, all connected in parallel. This is true for the overall power system, which serves multiple clients, or for a single plant, where multiple electrical loads (including motors, induction heaters, and lights) are powered in parallel. Such loads are generally inductive; power-factor correction is sometimes economically viable.

The typical circuit analysis problem most often considered is one in which sources and circuit parameters are specified; the problem is to find each of the circuit voltages and currents. Knowing these quantities, other information—such as the power dissipated, power factor, reactive power, energy stored here or there—can be calculated. However, in power-system problems, what are often specified are the characteristics of each load (in terms of its normal power rating and power factor, for example) and the properties of the transmission lines carrying the power from the source to the loads. It might be that finding the parameters of the transmission lines is part of the problem when some properties of the sources (such as the average or reactive power supplied) are specified. Some of the questions here will be phrased in this way.

Single-phase three-wire power distribution systems fall in this category also. Such systems permit servicing two loads at one voltage level and a third load at a double-voltage level. A few problems related to such systems will be included here.

1. The sinusoidal voltage v_1 in Figure P1 has an rms value of 10 V.
   a. Find the steady-state voltage $v_2(t)$.
   b. Find the steady-state current $i_1(t)$.
   c. Compute the average power and the reactive power supplied at the terminals.
   d. Compute the sum of the average powers dissipated by the respective resistors; also find the reactive power absorbed by the inductor, and compare these with the results in part (c).

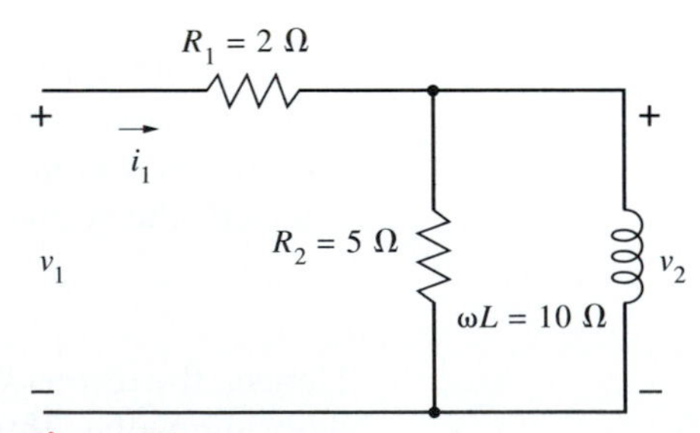

Figure P1

2. The parameters in Figure P2 have the following values: $R_1 = R_2 = 10\ \Omega$, $\omega L_1 = \omega L_2 = 15\ \Omega$, $R_3 = 40\ \Omega$. The sinusoidal sources are adjusted so

that $I_2 = j0.75I_1$. The average power dissipated in R_3 is to be 25 kW.

a. Determine phasors I_1 and I_2 and the source-voltage phasors V_1 and V_2.
b. Find the average power supplied by each source.
c. Confirm the balance of reactive power.

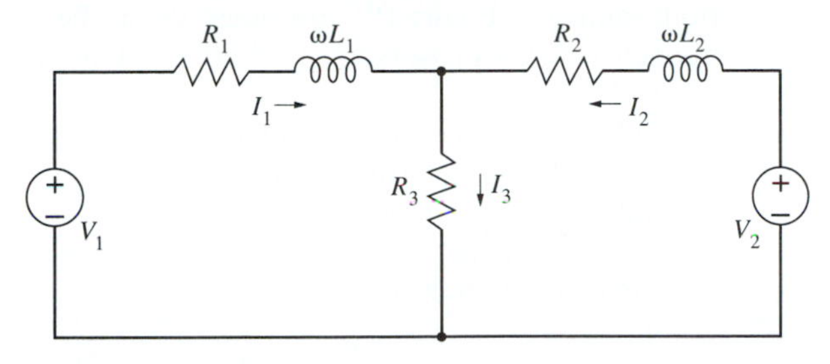

Figure P2

3. For the same circuit (Figure P2), the parameter values have been changed as follows: $R_1 = 10$, $R_2 = 20$, $R_3 = 80$, $\omega L_1 = 30$, and $\omega L_2 = 10$, all in ohms. This time the sources are to be adjusted so that $I_1 = (1 + j1)I_2$.

a. Find the current phasors I_1 and I_2.
b. Find the average power and the reactive power supplied by each source.

4. The complex power supplied to a two-terminal circuit is found to be $S = 2.7 + j4.5$ (kW and kVAR) when the steady-state current directed into the terminals is $i(t) = 30\sqrt{2}\cos(100t + 20°)$.

a. Find the impedance Z of the circuit.
b. Find the voltage $v(t)$ across the terminals of the circuit.
c. Find the value of the capacitor or inductor (as the case may be) which must be placed in series with Z in order to make the reactive power vanish. How is the real power affected thereby?

5. The two-terminal circuit in Figure P5 absorbs an average power of $P = 500$ W in the AC steady state. The current and voltage phasor magnitudes are 6 A and 120 V, respectively. The angular frequency is 400 rad/s.

a. Find the power factor.
b. Find the complex power and the reactive power delivered to the network.
c. Find the angle of the current relative to that of the voltage.

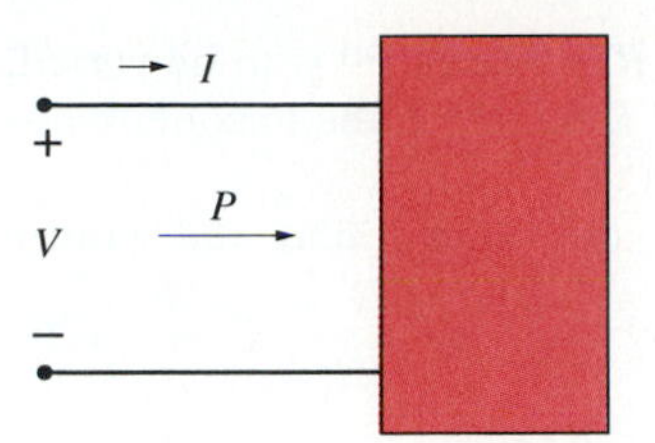

Figure P5

6. The two-terminal circuit within the box in Figure P6 represents an induction motor. It is absorbing an average power $P = 2$ kW at a lagging power factor of 0.8 from a 220-V, 377-rad/s voltage source. A capacitor is to be connected across the terminals such that the reactive power of the combination is 500 VAR. Find the required value of capacitance.

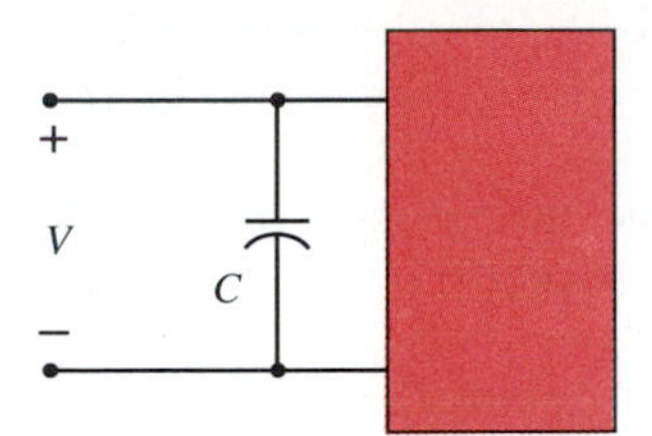

Figure P6

7. Now the circuit in Figure P6 represents an induction furnace. It draws an apparent power of 40 kVA at a lagging power factor of 0.5 from the same power line as in Problem 6. The added capacitor is to correct the power factor to 0.9 lagging.

a. Find the value of C.
b. Find the magnitude of the phasor current in the furnace.

8. A load Z_L is supplied from a 220-V sinusoidal source through a line with parameters as shown in Figure P8. The load dissipates an average power $P_L = 1500$ W and a reactive power $Q_L = -995$ VAR. Find the average power and the reactive power supplied by the source.

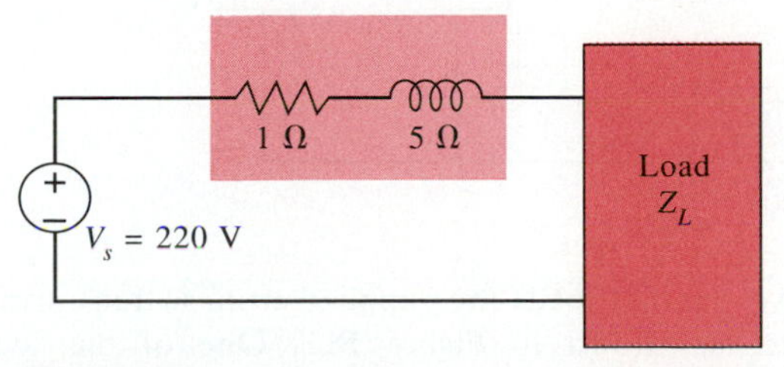

Figure P8

9. The load resistance R in Figure P9 is to be varied.

a. Find the value of R for which the maximum power is dissipated in it.

b. If R has twice this value, find the power it dissipates.

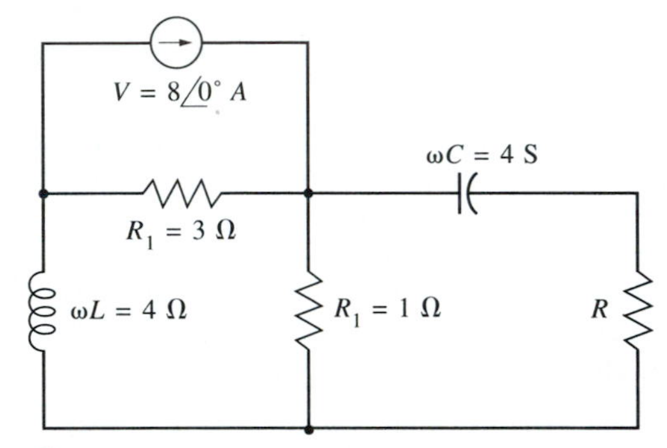

Figure P9

c. If C is also variable, find the new values of R and ωC such that the maximum power is dissipated in R.

d. Find this maximum power.

10. The average power absorbed by a two-terminal circuit is P at a power-factor angle θ. Determine the reactive power Q in terms of P and θ.

11. The complex power absorbed by each of the two parallel two-terminal circuits in Figure P11 is known at the given voltage. The complex power is specified either in terms of the real and reactive power or in terms of the real power and the power factor. Find the impedances of the loads in terms of the known quantities.

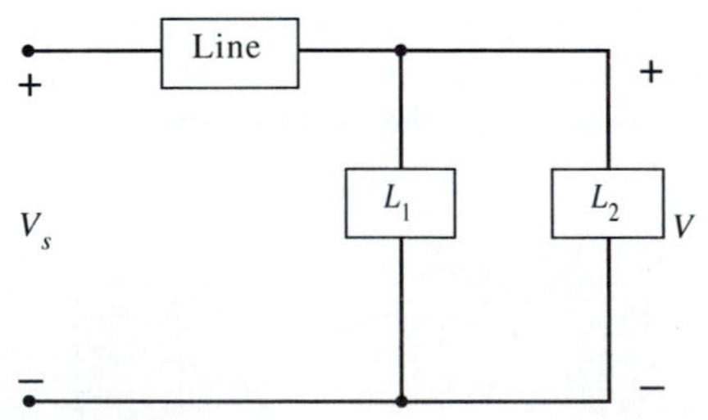

Figure P11

12. Two parallel loads are supplied from a transmission line, as shown in Figure P11. One of the loads absorbs 1 kW at a leading power factor of 0.6; the other load absorbs 1.5 kW at a lagging power factor of 0.8. The voltage phasor across the loads has a magnitude of 100 V. The line impedance is $0.2 + j0.4\ \Omega$.

a. Find the impedance of the parallel combination.

b. Find the source voltage and the power factor at which the source is operating.

13. Both sources in Figure P13 are sinusoidal at the same frequency. Their phasors are $V_s = 20 + j24$ and $I_s = -3 + j1$.

a. Find the average power and the reactive power delivered by the current source.

b. Find the average power supplied to the 5-Ω load.

c. Find the average power delivered by the voltage source and its power factor.

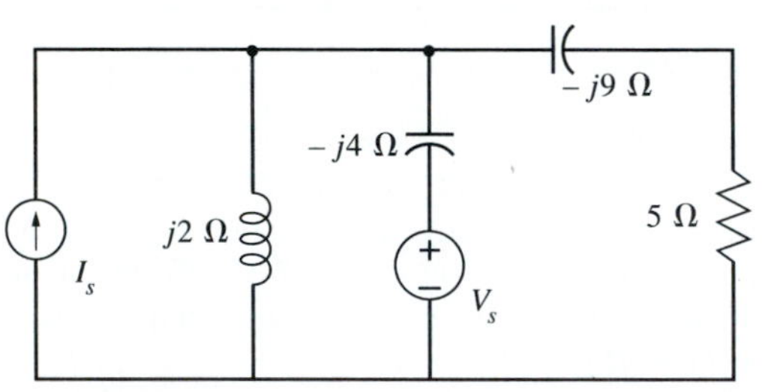

Figure P13

14. The total power absorbed by two parallel loads is P at a power factor F_p and a voltage $|V|$. One of the loads absorbs a power P_1 at a lagging power factor F_{p1}. Find an expression for the admittance of the second load in terms of these quantities. Use the general expression to find the admittance for the following numerical values:

a. $P = 10$ kW at 0.8 power factor, lagging, and $P_1 = 8$ kW at 0.6 power factor, lagging

b. $P = 20$ kW at unity power factor and $P_1 = 20$ kW at 0.8 power factor, lagging

15. The power dissipated in R_2 in Figure P15 is 500 W. Find the source voltage V_g and the complex power supplied by the source.

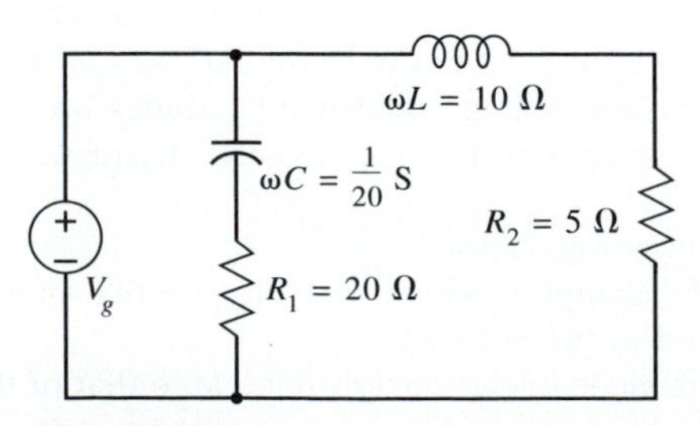

Figure P15

16. The source voltage in Figure P16 is $v_s(t) = 20\sqrt{2}\cos 1000t$. It powers a load through a line, as shown.

a. Determine the average steady-state power supplied to the load.

b. Determine the power supplied by the source.

c. Confirm that the difference is dissipated in the line.

d. Confirm the balance of reactive power.

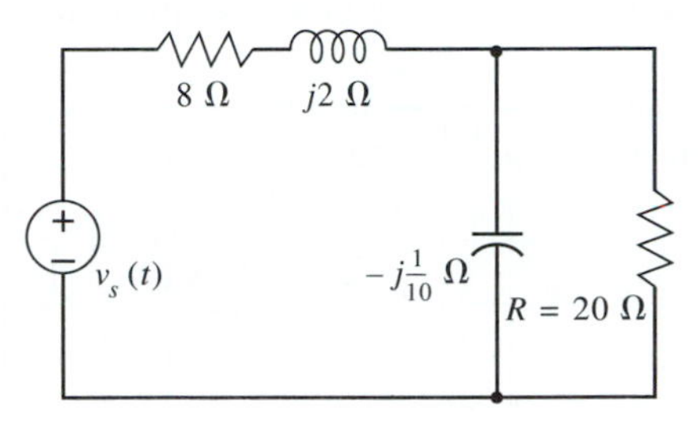

Figure P16

Chap/9 BalaP16

17. The Thévenin equivalent of a two-terminal circuit in the AC steady state is shown in Figure P17. The values of V_T and Z_T are to be determined by measurements at the terminals. With the terminals open, the phasor voltage V_{AB} is found to be V_1. With a known load Z_L connected to the terminals, the phasor voltage is found to be V_2. In terms of these measurements, find the impedance Z which should be connected to terminals AB so that the average AC power delivered to it is a maximum.

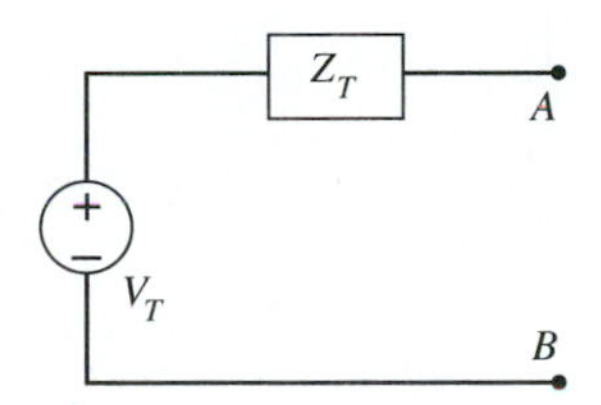

Figure P17

18. A 377-rad/s power system supplying an industrial load through a transmission line is represented by the circuit in Figure P18. The impedance of the line is $Z_T = 0.5 + j2\,\Omega$, and the load has an impedance of $Z_L = R + jX = 5 + j6\,\Omega$. The load-voltage phasor V has a magnitude of 250 V.

a. Find the average power, reactive power, and power factor of the load.

b. Find the transmission line power loss and the required value of source-voltage phasor V_s.

c. A capacitor is to be connected across the load in order to correct the load power factor to unity. Find the required value of capacitive reactance.

d. Recalculate the transmission loss and the new value of $|V_s|$. By what percentage is each of these two quantities reduced from its earlier values?

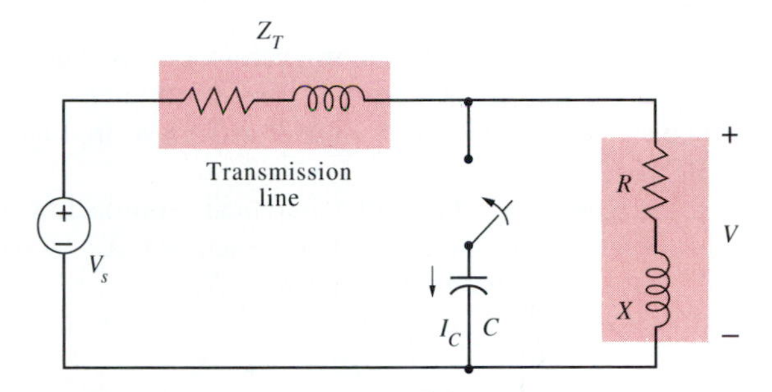

Figure P18

19. A certain factory draws 100 kW at 0.8 power factor (lagging) from a power line rated at 2200 V, 60 Hz. The power distribution transmission line has an equivalent impedance of $5 + j12\,\Omega$. Assume that the load voltage is 2200 V, and take this voltage as the phase reference.

a. Determine the load-current phasor.

b. Determine the required voltage at the power plant.

c. Determine the value of a capacitor to be used to correct the power factor to 0.95.

d. Construct a phasor diagram, after power-factor correction, showing the load voltage and current, the capacitor current, the voltage across the line impedance, and the voltage at the power plant.

20. The electrical characteristics of a small factory can be represented by the circuit in Figure P20. The average power absorbed by the heaters is 20 kW, and the apparent power of the motors is 100 kVA at a lagging power factor of 0.75. The line voltage V, taken as the phase reference, is 2000 V.

a. Determine the power factor at the plant and the current drawn, I.

b. Specify a device to be connected across the line to correct the power factor to unity. If a choice must be made, choose the cheapest.

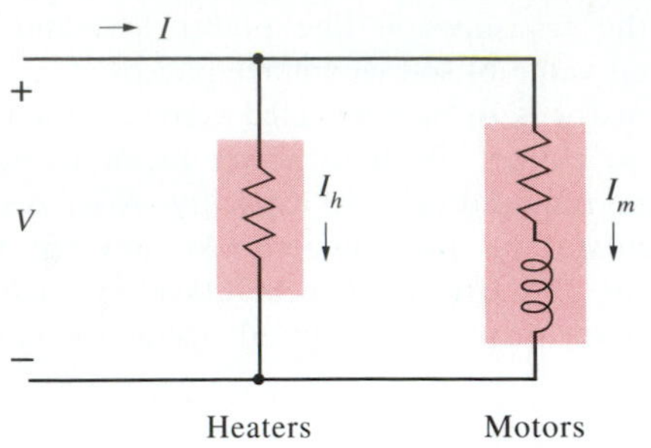

Figure P20

21. The magnitude of the input-current phasor and the input power are measured (the wattmeter is not shown) to be $|I_1| = 6$ A and $P = 1.5$ kW in Figure P21.
   a. Suggest two different analytical approaches to determining the capacitive reactance X_C. Carry out one of these and determine X_C.

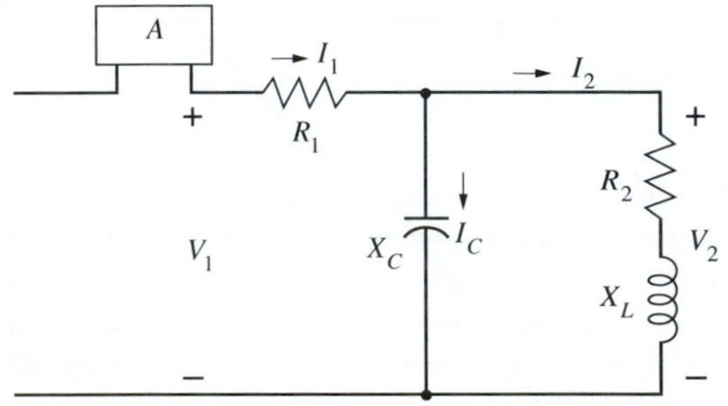

Figure P21

   b. Determine the magnitude of phasor V_2 and the phase difference between V_2 and I_2.

22. The 1000-rad/s, 100-V AC source in Figure P22 supplies power to a load Z through an intervening network.
   a. Determine the value of Z for which its average power is a maximum.
   b. Determine this value of maximum power.
   c. Under this condition, determine whether the controlled source is supplying power or absorbing it.

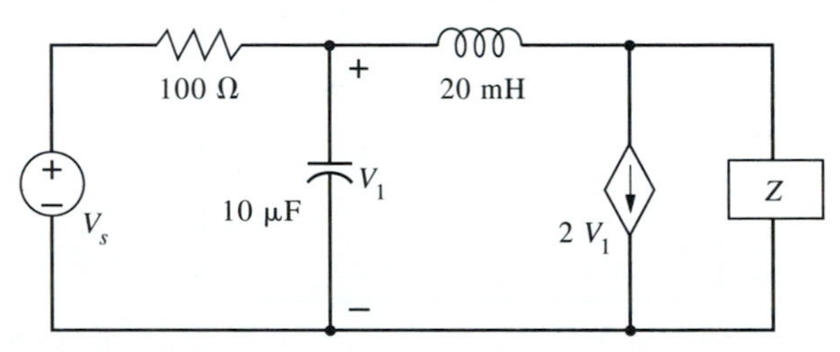

Figure P22

23. A 10-Ω load is supplied with power through an amplifier in the circuit of Figure P23.
   a. Find the average power consumed by the load.
   b. Determine the power supplied by the source. Comment on the utility of the amplifier.
   c. Determine the value to which the load resistance should be changed to increase the power dissipated in the load to a maximum.

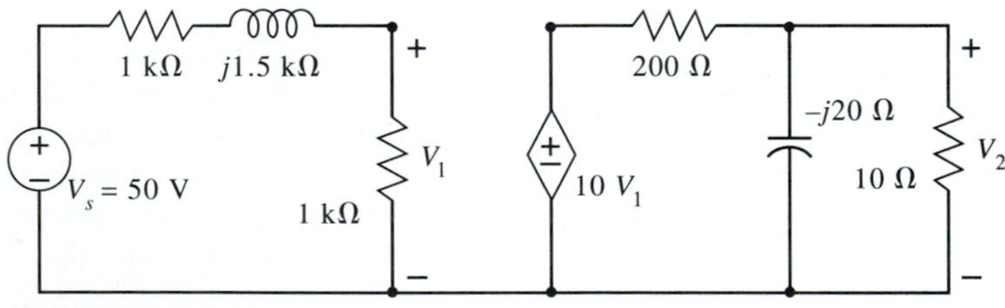

Figure P23

Chap/9 BalaP23

24. A 440-V, 400-rad/s sinusoidal source supplies two loads in parallel. One load absorbs 16 kW at a lagging power factor of 0.9; the other load draws 16 kVA of apparent power at a lagging power factor of 0.8.
   a. Neglect the line impedance and determine the power supplied by the source and the power factor under which the source is operating.
   b. Assume that the line impedance is $2 + j1\ \Omega$. Again determine the power supplied by the source and its power factor. Also find the source-voltage magnitude.
   c. What is the percentage increase in the power demand on the source in part (b)?

25. The 400-rad/s, 120-V sinusoidal source in Figure P25 supplies an apparent power of 500 VA to the two loads. The average power supplied to the inductive load is 200 W, and its reactive power is 80 VAR.

Fax & Office Services in Vallejo

CA Check Cashing
240 Tennessee St.
cross street is Sacramento
648-2825
F

Digital Copy
2124 Sonoma Blvd
cross street is Kentucky
554-2679
F S C N

FedEx Office Print & Ship
742 Admiral Callaghan
Redwood St. exit off fwy 80
644-4990
F S C N

Minuteman Press
812 Tennessee St
cross street is Broadway
647-1000
F C

Office Max
117 Plaza Dr.
near Cinema 14, Costco
643-8857
F S C

Postal Annex
1830 Springs Rd.
Near O'Reilly Auto Parts
648-0605
F C N

UPS Store (3 locations)
164 Robles Dr.
in Glen Cove
642-1915
F C N

UPS Store (3 locations)
3505 Sonoma Blvd
cross St. is Redwood; is in Vallejo Plaza
648-1161
F C N

UPS Store (3 locations)
55 Springstowne Ctr.
corner of Springs Rd. & Oakwood
554-2628
F C N

Solano County Library
John F. Kennedy Library
505 Santa Clara St
Vallejo, CA 94590-5922
07/15

F - Fax Send & Receive **S** - Scanner **C** - Color Copies **N** - Notary

a. Find the equivalent impedance of the two parallel loads.
b. Find the average power and the reactive power absorbed by the capacitive load.
c. Determine the power factor at which the source is operating.

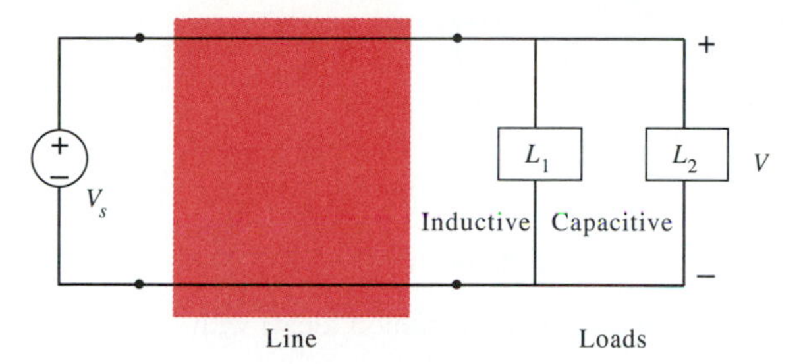

Figure P25

26. Two loads are connected to a 660-V AC source through a transmission line, as shown in Figure P25. The loads have the following properties. The average power of load L_1 is 15 kW at a lagging power factor of 0.8; the average power of load L_2 is 25 kW, and its reactive power is 10 kVAR.
a. Neglect the parameters of the transmission line and determine the phasor current from the source.
b. Determine the impedance presented to the source by the combined load.
c. Now assume that the line parameters are not zero and that the loads still operate at 660 V. Under this condition the complex power supplied by the source is found to increase by 0.5% from its previous value. Find the parameters of the line and the required value of V_s.

27. Three loads in parallel are supplied from a 440-V, 400-rad/s sinusoidal source through a line whose parameters are assumed to be negligible. The apparent power drawn by one of the loads (say, Z_1) is 4 kVA at a lagging power factor of 0.6. The second load (Z_2) dissipates an average power of 8 kW with a reactive power of 6 kVAR. The resulting phasor current from the source is found to be $I = 30e^{j\pi/6}$.
a. Find the impedance of the third load (Z_3).
b. Find also the average and reactive power it absorbs.

28. The phase voltage of a balanced three-phase wye-connected source has a magnitude of 120 V. It is connected to a balanced, wye-connected load of impedance $Z_L = 15 + j10\,\Omega$ by a three-wire line whose impedance per phase is $Z_l = 0.2 + j0.8\,\Omega$.
a. Determine the line currents. How are they related to the phase currents?
b. Determine the line voltages at the source end of the line. How are they related to the phase voltages there?
c. Determine the line voltages at the load end of the line.
d. Determine the total real and reactive power delivered to the load.
e. Determine the total real and reactive power delivered by the source.
f. What percent of the power supplied by the source is dissipated in the line?

29. Two balanced three-phase loads are connected in parallel. One is a wye with branch impedances of $8 + j6\,\Omega$, the other, a delta with branch impedances of $10\,\Omega$. The line voltage at the load is 208 V. Each of the lines connecting the loads to a balanced three-phase source has an impedance Z_l. Two cases are to be considered: (1) $Z_l = 0$ and (2) $Z_l = 0.05 + j0.2$.
a. Draw and label a diagram representing the circuit.
b. Determine the line currents for each of the cases.
c. Determine the line voltages at the source in the two cases.
d. Determine the total real and reactive power supplied by the three-phase source in each case.

DESIGN PROBLEMS

1. In power transmission systems one problem is the losses in the transmission lines themselves. This problem illustrates why transmission of electrical energy is done at a much higher voltage than the generated voltage or the voltage at which electrical energy is normally used. Figure DP1 represents such a system. The load is rated 480 kW at 120 V with a lagging power factor of 0.8. The impedance of the transmission line is $Z_l = 1.2 + j3.0\,\Omega$. The voltage of the generator is $V_g = 4$ kV. Assume that the transformers are ideal.

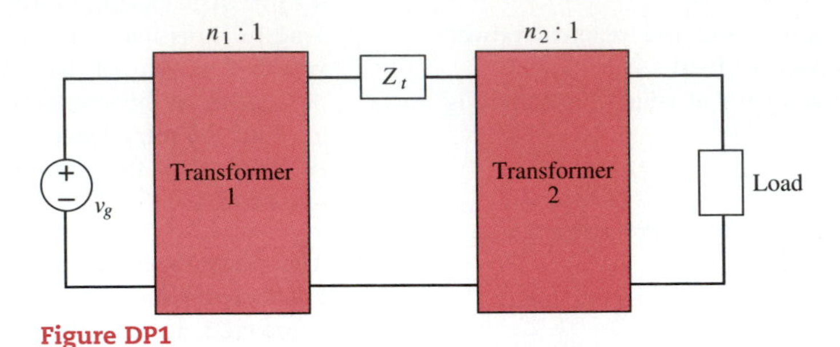

Figure DP1

a. Specify the turns ratios n_1 and n_2 such that 120 volts is supplied to the load and the losses in the transmission line are 0.5% of the total generated power.
b. For the same turns ratios, specify the reactance that should be placed in parallel with the load to reduce the losses to 0.4% of the total generated power.

2. A 100-kHz sinusoidal source which can be represented as in Figure DP2a is to be connected to a load

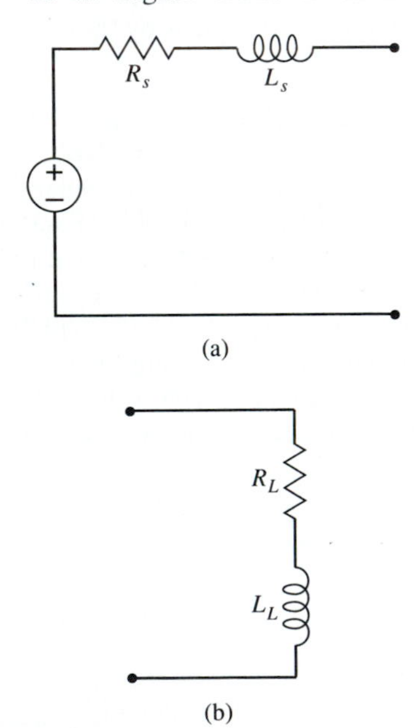

Figure DP2

represented as in Figure DP2b. The components have the following values: $R_s = 0.5\,\Omega$, $L_s = 1\,\mu$H, $R_L = 8\,\Omega$, $L_L = 10\,\mu$H. It is desired to have maximum power transfer to the load. The following devices are available:

transformers (assumed ideal) with turns ratio of 6:1, 5:1, 4:1, and 3:1

capacitors of 5 μF, 4 μF, 3 μF, and 2 μF

Design a circuit connection to have maximum power transfer to the load, subject to the constraint of using only one or more of the available components.

3. The *wattmeter* is an instrument that measures power. The connection of a wattmeter is shown in Figure DP3a. This wattmeter provides a reading of $|V|\,|I| \cos\theta$, where $|V|$ is the magnitude of the voltage across the voltage coil, $|I|$ is the magnitude of the current through the current coil, and θ is the angle between the two. It is assumed that the current through the voltage coil is negligible. The total power measurement in a 3-phase system can be performed by using only two wattmeters, as shown in Figure DP3b, where both the source voltages and the load are balanced.
   a. Determine the relationship between the readings of the two wattmeters and the total power supplied to the load.
   b. Given that the readings of wattmeter 1 and wattmeter 2 are 1150 W and 575 W respectively, determine the angle of the load impedance Z.

4. The diagram in Figure DP4 is not a circuit diagram; it is a schematic showing a power distribution system. Plants 1 and 2 represent the location where electrical energy is generated and Loads 1 and 2 represent the location where this energy is consumed. The parameters on the diagram are normalized. Normalization is referred to in power distribution systems as *per unit* (p.u.). Thus the impedances of the transmission line

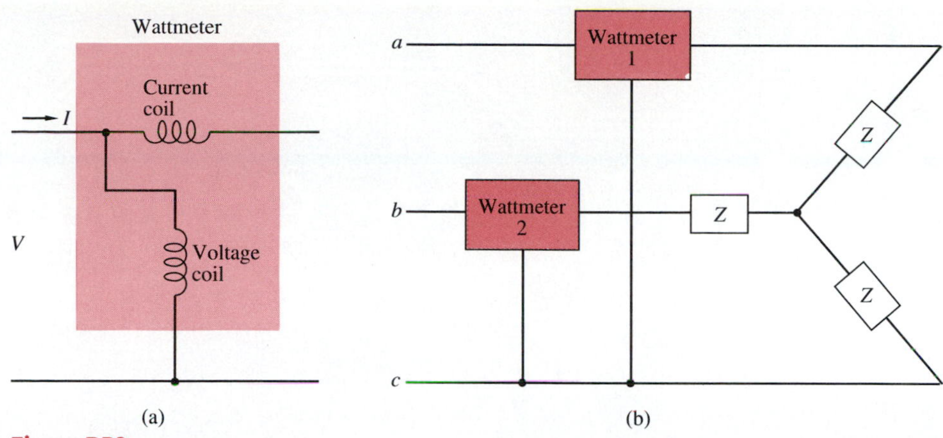

Figure DP3

sections are:

$Z_a = 0.025 + j0.075$ p.u. $\qquad Z_b = 0.01 + j0.06$ p.u.

$Z_c = 0.025 + j0.075$ p.u. $\qquad Z_d = 0.09 + j0.25$ p.u.

$Z_e = 0.025 + j0.075$ p.u.

(These include both forward and return line impedances.)

The load requirements are as follows. Load 1 requires a current of $0.85 - j0.28$ p.u. and Load 2 requires a current of $1.3 - j0.33$ p.u. The voltage at the point marked as 1 p.u. is taken to have 0 phase angle. Draw a lumped circuit diagram for the power distribution system shown. It is possible to vary the magnitude and the angle of the voltage at each plant and change the power generation at each plant. The design requirement is to minimize the losses in the transmission lines. Design the system; i.e., determine the voltages at plants 1 and 2.

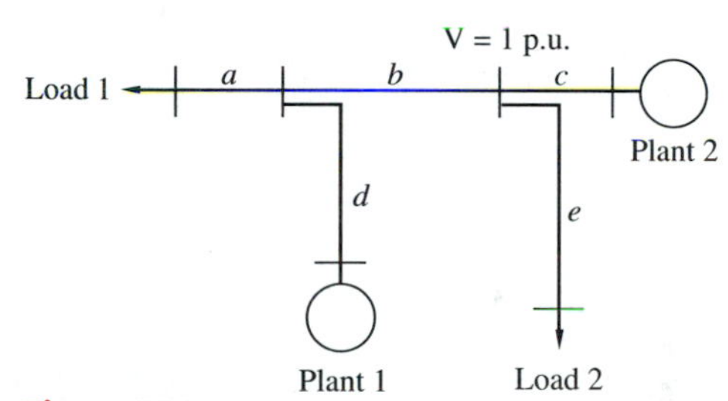

Figure DP4

10 Inductive Coupling and Transformers

All of the multiterminal circuit components introduced in Chapter 4 have v-i relationships which are algebraic. (What seems to be an exception—the inductor-simulation circuit using a gyrator—really is not. The derivative showing up in that circuit results from the capacitor, not the gyrator.) In this chapter we will be concerned with a multiterminal component whose v-i relationship is not algebraic. It finds application in many areas; power systems could not operate without it, and it is also useful in communications systems and other areas. An idealized model of the subject of this chapter, called an *ideal transformer,* was introduced in Chapter 4.

In Chapter 2 we discussed the observations first made by Faraday concerning the appearance of a voltage across the terminals of a coil of wire when it is in the presence of a time-varying magnetic field. There we explicitly considered the case in which the magnetic field depends only on the current in the coil itself. (If you skipped that discussion on your first pass, study it now. In any case, refer to it to refresh your memory.)

We will now consider the case where a time-varying magnetic field produced by a current in one coil links a second coil and thus induces a voltage there. If the second coil is in a circuit, it, too, will carry a current. This current will also contribute to the magnetic field, so that the total field is created by contributions from both currents.

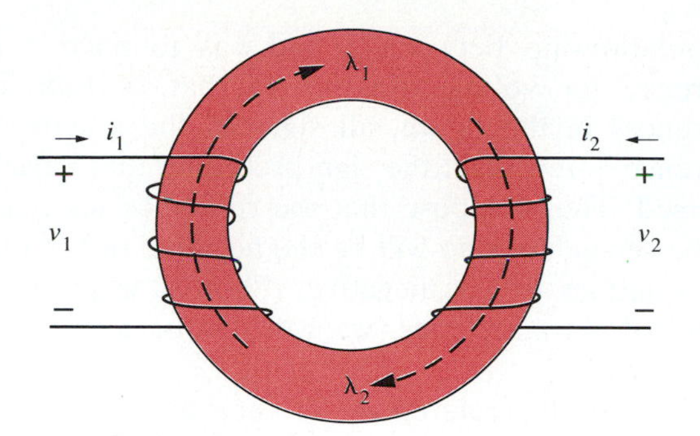

Figure 1
A physical transformer.

1 THE TWO-WINDING TRANSFORMER

Consider the situation illustrated in Figure 1. It consists of two coils of wire wound on the same core. Each of the coils is connected to a circuit, at least one of which contains a time-varying electrical source. The coils are said to be *magnetically* or *inductively coupled.* Such a device is called a *transformer,* for reasons which will become clear as we discuss its properties. The device has four terminals (really, two pairs of terminals), so it is a two-port. In normal use, one side of the transformer, called the *primary,* is energized and a load is connected to the other side, called the *secondary*—although both sides can be energized.

Transformer Models

An idealized model of a transformer was introduced in Chapter 4. We shall now develop one or more less idealized models to represent this physical device, dealing initially with the general time domain and then moving on to the phasor and complex-frequency domains. Recall from the discussion of inductors in Chapter 2 that an electric current is always accompanied by a magnetic field. This field is enhanced and concentrated if the wire in which the current flows is wound into a coil. In Chapter 2, we dealt with the flux produced by a single current—the self-flux. In the linear case, the flux linkage was given by $\lambda = Li$, where L is a constant. In the present case, the flux linking each of the two coils (λ_1 and λ_2) has contributions from both currents. That is, $\lambda_1 = f_1(i_1, i_2)$ and $\lambda_2 = f_2(i_1, i_2)$. Generally, the functions f_1 and f_2 are nonlinear. However, this book is limited to linear relationships between flux linkages and currents. Hence, the flux linkages of the two coils can be written:

$$\begin{aligned} \lambda_1 &= L_{11}i_1 + L_{12}i_2 \\ \lambda_2 &= L_{21}i_1 + L_{22}i_2 \end{aligned} \tag{1}$$

The L_{ij} coefficients in these equations are dimensionally inductance; they are the inductance parameters of the transformer.

References for the flux linkages are shown in the figure. From your study of physics recall the right-hand rule relating current and flux-linkage references if the relationship between λ and i is to have a plus sign.[1] Also recall the reference for voltage relative to that of flux linkage. For the choice of references in the figure, all signs in the equations are positive. If any one reference is reversed, the sign preceding that variable in the equations will be reversed. Thus, suppose that the reference for i_2 is reversed; the new current on the secondary side will be the negative of the old i_2, so the sign before i_2 in the equations will be negative. (Shortly, when we have developed the circuit model of a transformer, we will discuss a more systematic approach to the references.)

Note that the references for the flux linkages in the figure have the same orientation around the core; we say they are *aiding*. If one of them had been reversed, we would then have said that they were *opposing*.

The double subscripts on the inductance parameters in the preceding equations have the following significance. The first subscript corresponds to the coil whose flux linkage is in question; the second corresponds to the current producing the flux linkage. Thus, L_{21} is the coefficient relating the flux linkage produced in coil 2 (the first subscript) to current 1 (the second subscript).

According to Faraday's law (look back at Chapter 2), the voltages induced in the coils will equal the rate of change with time of the corresponding flux linkages. That is:

$$v_1 = \frac{d\lambda_1}{dt} \qquad v_2 = \frac{d\lambda_2}{dt} \tag{2a}$$

$$v_1 = L_{11}\frac{di_1}{dt} + L_{12}\frac{di_2}{dt} \qquad v_2 = L_{21}\frac{di_1}{dt} + L_{22}\frac{di_2}{dt} \tag{2b}$$

Recall from Chapter 2 that the sign relating v to λ is positive for the voltage and flux references in the figure. If any one reference for v or λ is reversed, then in (2a) the sign in the equations preceding that variable will become negative.

EXERCISE 1

Write equations relating flux linkages to currents, and voltages to flux linkages, for each of the coupled coils in Figure 2, paying particular attention to the sign of each term. Specify whether the fluxes are aiding or opposing.

ANSWER: (a) opposing, (b) aiding. □

SELF AND MUTUAL INDUCTANCES An immediate question is, How is L_{21} related to L_{12}? That is, is the flux linkage of coil 2 produced by current 1 the

[1] If you grasp the coil in the right hand, with the fingers oriented in the same way around the coil as the current, then the thumb will point in the reference direction of the flux linkage.

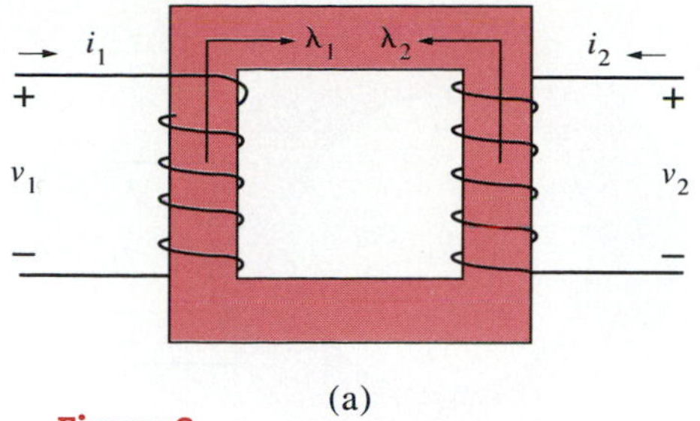

(b)

Figure 2
Determine equations for flux linkages and voltages.

same as, or different from, the flux linkage of coil 1 produced by current 2? Empirical measurements show these quantities to be approximately the same. Therefore, we will *postulate* that, in the model we are seeking, $L_{12} = L_{21}$; and we will let the label of the common value be M. (The same result can be obtained from an argument based on energy, but since the argument is lengthy and detailed, we will not provide it here.)[2]

Looking back at (1) and (2), without the need for subscripts in the quantities $L_{12} = L_{21}$, there is no longer any need for double subscripts on L_{11} and L_{22}, so we will set $L_{11} = L_1$ and $L_{22} = L_2$. Hence, (1) can be rewritten more simply as:

$$\lambda_1 = L_1 i_1 + M i_2$$
$$\lambda_2 = M i_1 + L_2 i_2 \tag{3}$$

Note that L_1 and L_2 are coefficients relating the flux linkage of a coil to its respective current; they are, thus, *self-inductances,* the same as the self-inductance of a single coil discussed in Chapter 2. The coefficient M relates the contribution of the current in one coil to the flux linkage of the other coil; it is called the *mutual inductance.* All of these parameters obviously have the same unit, the henry (H).

Equation (2) can also be rewritten, replacing L_{12} and L_{21} by M, L_{11} by L_1, and L_{22} by L_2. The result will be:

$$v_1 = \frac{d\lambda_1}{dt}$$
$$v_2 = \frac{d\lambda_2}{dt} \tag{4a}$$

$$v_1 = L_1 \frac{di_1}{dt} + M \frac{di_2}{dt}$$
$$v_2 = M \frac{di_1}{dt} + L_2 \frac{di_2}{dt} \tag{4b}$$

[2] The condition $L_{12} = L_{21}$ is an example of a class of two-ports that was briefly introduced in Chapter 4 and that will be studied at greater length in Chapter 15. This property is referred to as *reciprocity,* and such two-ports are said to be *reciprocal.*

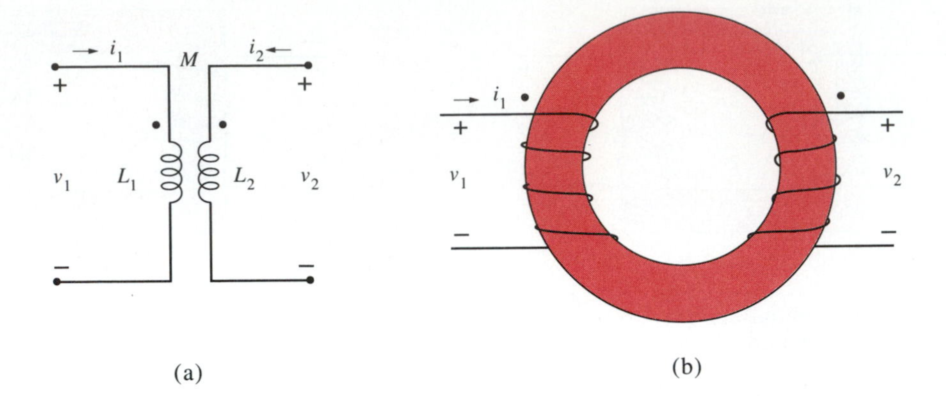

Figure 3
Transformer model.

Circuit Symbol for a Transformer

Let's recapitulate what we have done so far. Starting from a physical device having two pairs of terminals, we have arrived at a mathematical model represented by (3) and (4). Since the device is reciprocal, this model is characterized by three parameters: two self-inductances, L_1 and L_2; and a mutual inductance, M.

The circuit symbol for the model of the transformer of Figure 1 is shown in Figure 3a. (The diagram of the physical transformer is also repeated in Figure 3b.) The symbol might suggest two physical coils of wire in close proximity, but this is not a physical component, simply a circuit model. It is clear that there are three sets of variables—voltage, current, and flux linkage—and so there must be three sets of references. In the physical device, the reference for flux linkage can be readily shown, because we can see in what sense the current encircles the core, but not so in the model. Hence, some convention must be adopted in the model to show the references for flux linkages relative to those for currents and voltages.

DOT CONVENTION FOR FLUX REFERENCES The convention commonly adopted is to place a dot at one terminal of each pair, as in Figure 3, to show the orientation of each winding. The dots are placed so that currents oriented the same way relative to the dots (both entering or both leaving) produce flux linkages that are aiding. The signs in each of the equations in (3) and (4) depend on the locations of the references for v and i relative to the dots. If both current references are oriented toward the dots, then the flux linkages are aiding and the signs in both equations in (3) are all positive. If both current references are oriented away from the dots, then the flux linkages are again aiding but this time the signs preceding both flux linkages are negative. Finally, if the current references are oppositely oriented relative to the dots (one toward and one away), the flux linkages are opposing and the signs of the terms involving the current that is directed away from the dot must be negative.

Similarly, if both voltage-reference plus signs are at the dots, then all signs

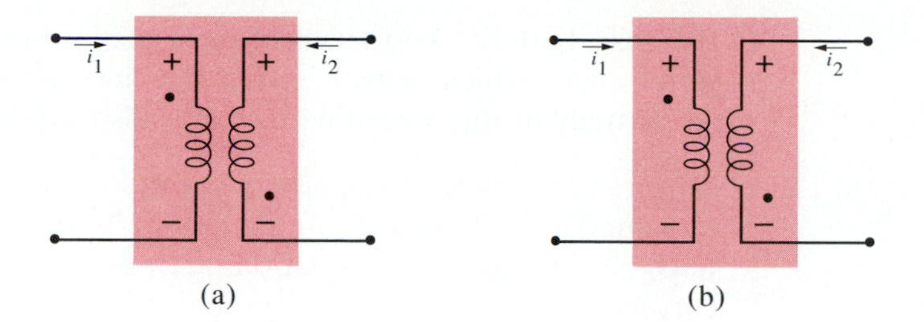

Figure 4
Models, including dots, of the transformers in Figure 2.

in (4a) will be positive. If neither reference is at the dots, then both equations in (4a) will carry a negative sign. Finally, if the voltage references have opposite relationships relative to the dots, then the equation in (4a) corresponding to the voltage whose reference is not at the dot will have a negative sign.

The preceding discussion is illustrated in Figure 4, which shows models—including the dot positions—for the two transformers in Figure 2. In the first case, i_1 is directed toward the dot and i_2 is directed away. Also, the reference for v_2 is located at the dot, while that for v_1 is not.[3] Hence, according to the preceding discussion, the signs in the equations will be as follows:

$$\begin{aligned} \lambda_1 &= L_1 i_1 - M i_2 \\ \lambda_2 &= M i_1 - L_2 i_2 \end{aligned} \tag{5a}$$

$$\begin{aligned} v_1 &= \frac{d\lambda_1}{dt} = L_1 \frac{di_1}{dt} - M \frac{di_2}{dt} \\ v_2 &= -\frac{d\lambda_2}{dt} = -M \frac{di_1}{dt} + L_2 \frac{di_2}{dt} \end{aligned} \tag{5b}$$

The first pair was substituted into the second to arrive at the right-hand side. Note how the negative signs in the (5a) for flux linkage in terms of current and in (5b) for voltage in terms of flux linkage have yielded v-i equations in which the mutual-inductance terms have negative signs.

For the second model in Figure 4, using the discussion about references and dots, the following equations result:

$$\begin{aligned} -\lambda_1 &= L_1 i_1 + M i_2 \\ -\lambda_2 &= M i_1 + L_2 i_2 \end{aligned} \tag{6}$$

$$\begin{aligned} v_1 &= -\frac{d\lambda_1}{dt} = L_1 \frac{di_1}{dt} + M \frac{di_2}{dt} \\ v_2 &= -\frac{d\lambda_2}{dt} = M \frac{di_1}{dt} + L_2 \frac{di_2}{dt} \end{aligned} \tag{7}$$

[3] The description of a current reference relative to the dot depends on the placement of the current arrow. If the arrow is placed directed in from the terminals, the description here is proper. However, if the arrow is placed alongside the model for the winding, the descriptions "away from the dot" and "toward the dot" would be reversed. The almost universal practice is the former one.

In this case, both the λ-in-terms-of-i and the v-in-terms-of-λ equations all have negative signs, which cancel, giving the final v-i equations without negative signs. A study of the preceding results leads to the following conclusion:

> In a transformer model, when the references of the currents and voltages (assumed to be standard references) are oriented similarly relative to the dots, all signs in the v-i equations are positive; when one set of voltage and current references is oriented in the opposite way, the signs of the mutual terms in the v-i equations will be negative.

EXERCISE 2

There is one remaining configuration of the references that was not examined, namely, the case in which the dots in Figure 4a are reversed. Carry out the details for this case, and verify that the same v-i equations result as those in (5). □

Note that the dots on the circuit symbol for a transformer describe something about the way the physical coils are wound on the core of the physical transformer from which the model has been abstracted. Therefore, they are not something which we choose arbitrarily, but something imposed by the transformer manufacturer. On the other hand, the choice of voltage and current references is up to us; we can choose them arbitrarily.

In order to simplify life, we will *always* choose *standard* references for transformer voltages and currents. Hence, as outlined in the preceding paragraphs, there will be only two possible combinations of all the references:

1. Both current references are directed toward (or both away from) the dots, in which case both voltage references will be located at (or both away from) the dots, so that all signs in the v-i equations will be positive.

2. The two currents are oppositely oriented relative to the dots (one toward and one away from), in which case the two voltage references will also be oppositely oriented relative to the dots; in this case, the signs of the mutual terms in the v-i equations will be negative.

It is possible in the preceding development to take a different point of view, and some textbook authors do. Since the only negative signs appear in the mutual *terms,* it is conceivable to associate the sign with the mutual *inductance* itself. Thus, so long as standard references are chosen for the voltages and currents, *all* signs in the v-i equations can be made positive. But now the burden is on M. In case 2 just described, the sign of M will be negative. This is a perfectly valid approach; however, it requires a circuit parameter in a passive circuit to be negative. Although no rule of nature is thereby violated, it gives a somewhat uneasy feeling. In this book, we will assume that the mutual inductance, like the self-inductances, is positive. Then the signs of the mutual terms in the equations will be positive, except when the current references are oppositely oriented with respect to the two dots, in which case those signs will be negative.

2 COUPLING OF TRANSFORMER COILS

The two coils of a transformer are said to be *coupled*; not conductively coupled, but *inductively*. Not all of the flux resulting from one of the currents links the other coil. The relative value of flux linkage generated in each of the coils by a current in one of them is expressed in terms of the *closeness*, or *tightness*, of the coupling. A measure of the tightness of coupling can be obtained by taking the ratio of λ_1 and λ_2 in each of two special cases. In one case, the left-hand side in Figure 3a is left open, causing $i_1 = 0$; in the second case, the right-hand side in Figure 3a is left open, causing $i_2 = 0$. From (3) we get:

$$\left.\frac{\lambda_1}{\lambda_2}\right|_{i_1=0} = \frac{M}{L_2} \tag{8a}$$

$$\left.\frac{\lambda_1}{\lambda_2}\right|_{i_2=0} = \frac{L_1}{M} \tag{8b}$$

$$\frac{\left.\dfrac{\lambda_1}{\lambda_2}\right|_{i_1=0}}{\left.\dfrac{\lambda_1}{\lambda_2}\right|_{i_2=0}} = \frac{M^2}{L_1 L_2} = k^2 \tag{8c}$$

The magnitude of the right side of either (8a) or (8b) may be greater than unity or less than unity, without violating any law of nature in either case. That is, M can be greater than or less than either L_1 or L_2.

If we take the ratio of these two expressions, we get the one in (8c). This ratio of the flux-linkage ratios is a positive constant we have called k^2. (It is positive even when M is permitted to take on negative values.) The constant k is called the *coupling coefficient*. What are its permissible values? That's a very important question whose answer we will now seek.

Size of Coupling Coefficient

Our approach in answering the preceding question will be to argue that the total energy supplied to a transformer from its terminals can never be negative. *Negative* energy supplied *to* the transformer means that the transformer itself is supplying energy to the circuits connected to its terminals.[4]

To obtain the total power into the transformer in Figure 3a, we multiply the

[4] This would be in the nature of a perpetual-motion machine, something that many unsophisticates have tried to invent over the years. Just imagine what a blessing it would be if you could pick up some wire at the hardware store, wind two coils with it, and then sit back and enjoy the energy which your device would deliver. Unfortunately, you couldn't sell it, because other people could go out and do the same thing; they wouldn't have to pay you for it.

first equation in (4b) by i_1 and the second by i_2 and then add them. The result will be:

$$v_1 i_1 = i_1 L_1 \frac{di_1}{dt} + i_1 M \frac{di_2}{dt}$$

$$v_2 i_2 = i_2 M \frac{di_1}{dt} + i_2 L_2 \frac{di_2}{dt}$$

$$\begin{aligned} v_1 i_1 + v_2 i_2 &= L_1 i_1 \frac{di_1}{dt} + M\left(i_1 \frac{di_2}{dt} + i_2 \frac{di_1}{dt}\right) + L_2 i_2 \frac{di_2}{dt} \\ &= \frac{d}{dt}\left[\frac{1}{2}(L_1 i_1^2 + 2M i_1 i_2 + L_2 i_2^2)\right] \end{aligned} \tag{9}$$

(You can verify that the second line of the last equation is equivalent to the first line by performing the indicated differentiation.)

The left side of the last equation is the total power input to the transformer. Since power is the time derivative of energy, the quantity inside the brackets in the last line represents the energy supplied to the transformer up to any instant of time from the time when the currents were zero. In Chapter 2 we found that the energy stored in a simple inductor is $w_L = Li^2/2$. Let's extend the use of the symbol w_L to include the energy stored in a transformer. Thus, for a transformer, the energy stored is:

$$w_L = \tfrac{1}{2}(L_1 i_1^2 + 2M i_1 i_2 + L_2 i_2^2) \tag{10}$$

This is an example of a nonlinear function; specifically, it is a *quadratic* function of the transformer currents. The energy stored in the transformer can never be negative at any time. (If it could be, it would mean that the transformer could be a net supplier of energy—giving more than was supplied to it externally.) Suppose that, at the time in question, $i_2 = 0$. Then, $w_L = L_1 i_1^2/2 \geqslant 0$; hence, L_1 cannot be negative. (Now you check the case when $i_1 = 0$ and confirm that L_2 cannot be negative.)

There is still more information in (10). The first and last terms are always positive for all values of the currents. The only way for the whole expression to be negative is for the middle term to be negative. Since M is positive, the onlly way *that* can happen is for the two currents to have opposite signs. Therefore let's write $i_1 = -x i_2$, where x is a positive number, and substitute it into (10). Since w_L is to be nonnegative, we end up with the requirement:

$$x^2 - 2\frac{M}{L_1}x + \frac{L_2}{L_1} \geqslant 0 \tag{11}$$

(You should supply the details of the calculation.) The left side of this expression is a quadratic in x whose discriminant is proportional to $M^2 - L_1 L_2$. If the discriminant is not negative, then the quadratic will have

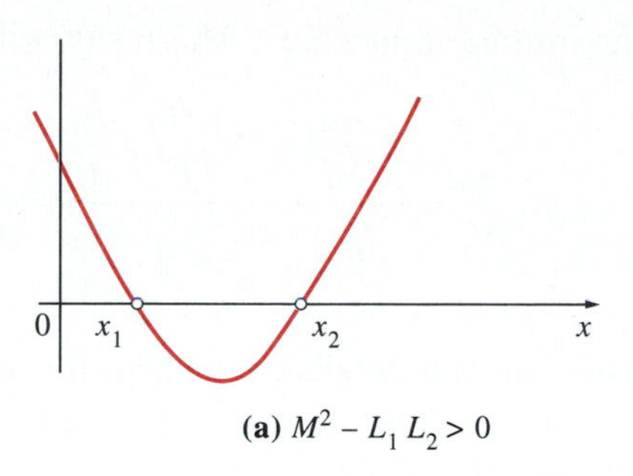

(a) $M^2 - L_1 L_2 > 0$

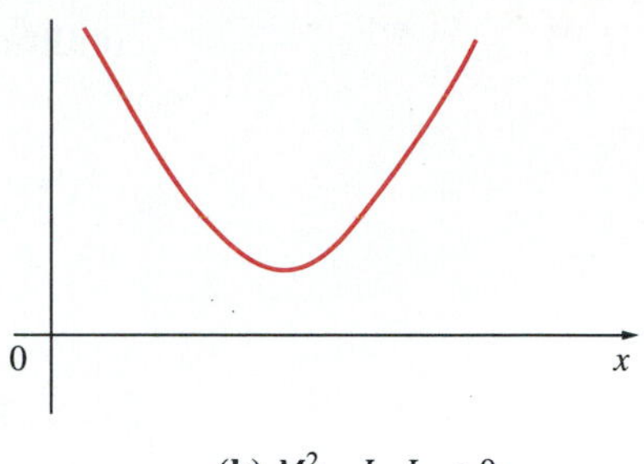

(b) $M^2 - L_1 L_2 < 0$

Figure 5
Candidates for the variation of energy stored in a transformer:
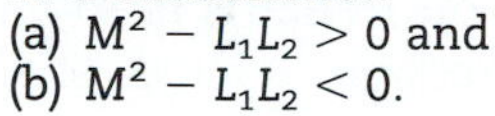
(a) $M^2 - L_1L_2 > 0$ and
(b) $M^2 - L_1L_2 < 0$.

two real roots. If not, the roots will be complex. A plot of the quadratic is shown in Figure 5 for the two cases.

If the roots of the quadratic are real, then there will be a range of values of x (which is related to the currents; remember?) over which the quadratic (and so the energy stored in the transformer) is negative. Can't be! Thus, the discriminant can't be positive. Therefore, the roots of the quadratic are complex and the plot takes the form of Figure 5b. (In the limit, if the discriminant is zero, the parabola will rest on the x-axis but it will not go negative.) The upshot of all of this is:

$$M^2 - L_1L_2 = L_1L_2(k^2 - 1) \leqslant 0$$

and so

$$k^2 \leqslant 1 \tag{12}$$

That's the answer we were seeking: *the coupling coefficient cannot exceed 1.* If the coupling coefficient is 0, that means the mutual inductance is zero and so there is no coupling between the two coils.

The Perfect Transformer

If the coupling coefficient of a transformer takes on its limiting value $k = 1$, the transformer is said to be *perfect,* or *perfectly coupled,* or *unity-coupled.*

Let's now rewrite the definition of k from (8c) and impose the requirement that k cannot exceed 1. We get:

$$M = k\sqrt{L_1L_2} \leqslant \sqrt{L_1L_2} \tag{13}$$

That is, the mutual inductance generally is no greater than the geometric mean of L_1 and L_2; for a perfect transformer ($k = 1$), *M equals* this geometric mean: $M = \sqrt{L_1L_2}$. It is certainly possible for M to be greater than either of the two self-inductances, but not greater than both of them simultaneously.

MODEL OF A PERFECT TRANSFORMER Let's now return to the *v-i* equations of a transformer in (4b), using the condition $M^2 = L_1L_2$ for a perfect transformer, and manipulating the equations to see if something useful results. Let's take the ratio of v_1 and v_2 in (4b), substituting the preceding condition on the mutual inductance. The result will be:

$$\frac{v_1}{v_2} = \frac{L_1 \dfrac{di_1}{dt} + \sqrt{L_1 L_2}\dfrac{di_2}{dt}}{\sqrt{L_1 L_2}\dfrac{di_1}{dt} + L_2 \dfrac{di_2}{dt}} = \sqrt{\frac{L_1}{L_2}} \tag{14}$$

(You should work out the details to confirm this equation.) That is to say, the primary-to-secondary voltage ratio is a simple expression related to the primary-to-secondary inductance ratio.

Let's define a parameter n as:

$$n = \sqrt{\frac{L_1}{L_2}} \tag{15}$$

Recall from Chapter 2 that the inductance of a coil is proportional to the *square* of the number of turns in the coil. Hence, the parameter n just defined is simply the *turns ratio,* the ratio of the number of turns of the primary to the number of turns of the secondary—inherently a positive number.

With $M^2 = L_1 L_2$ and $n^2 = L_1/L_2$, both L_1 and L_2 can be expressed in terms of M and n, as follows:

$$L_1 = nM$$

$$L_2 = \frac{M}{n} \tag{16}$$

We turn next to the transformer currents. Since their derivatives appear in (4b), it will be necessary to integrate these equations from 0 up to some time t. Let's also substitute the equations in (16) for the self-inductances:

$$i_1(t) = -\frac{1}{n} i_2(t) + \frac{1}{L_1}\int_0^t v_1(x)\,dx + \left[i_1(0) + \frac{1}{n} i_2(0)\right] \tag{17}$$

(You should again carry out the details to confirm this expression.) The form of all but the first term on the right suggests an inductor.

EXERCISE 3

A transformer has the following parameters: $n = 4$, $k = 0.9$, and $L_1 = 160$ mH. Find the secondary and mutual inductances.

ANSWER: $L_2 = 10$ mH, $M = 36$ mH. □

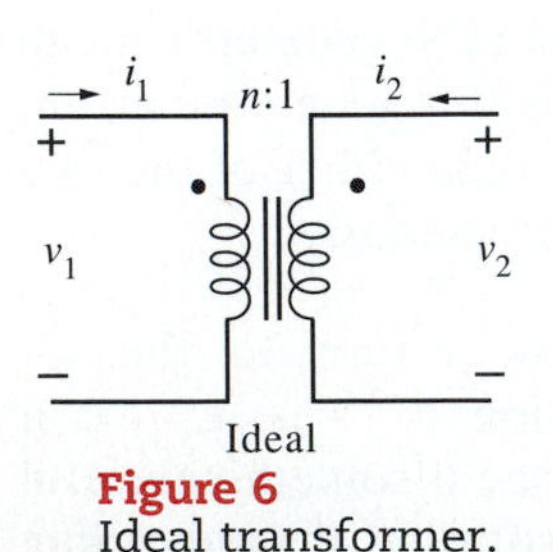

Figure 6
Ideal transformer.

The Ideal Transformer

Now let's turn our attention to the first term on the right in (17). You should be getting a glimmer of recognition, especially if you take into account the voltage relationship of (14). Recall from Chapter 4 the model of an *ideal* transformer, redrawn in Figure 6.

Recall also from Chapter 4 that the v-i relations of the ideal transformer are:

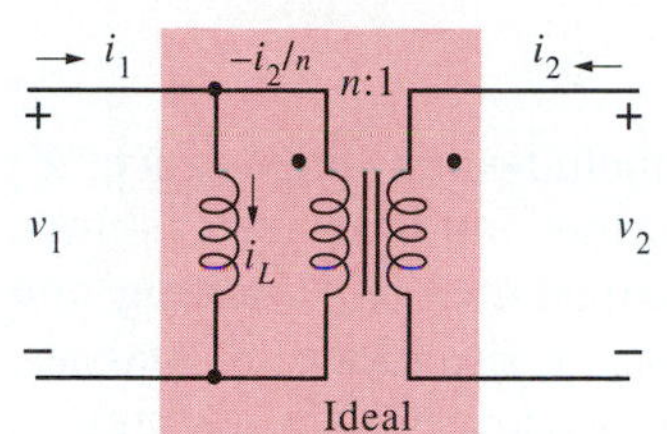

Figure 7
Perfect-transformer model: inductor and ideal transformer.

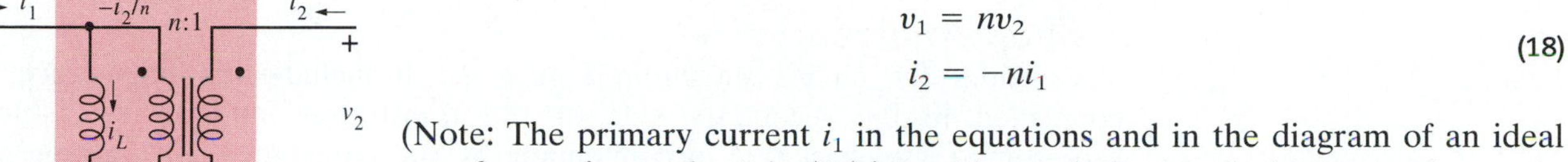

$$
\begin{aligned}
v_1 &= n v_2 \\
i_2 &= -n i_1
\end{aligned}
\tag{18}
$$

(Note: The primary current i_1 in the equations and in the diagram of an ideal transformer is not the i_1 in (17).) Looking at (17) with the ideal transformer in mind, we can conceive of the diagram in Figure 7 as equivalent to the perfect transformer. You can easily verify that both the voltage relationship in (14) and the current relationship in (17), which apply to the perfect transformer, are satisfied.

IDEAL-TRANSFORMER PROPERTIES In Chapter 4 we already introduced two important properties of ideal transformers:

1. The ideal transformer is lossless. This is determined by evaluating the total power input from its ports, which turns out to be 0.
2. The equivalent resistance seen from the input port is n^2R when the output port is terminated in a resistance R.

From the preceding development, we can make another observation. First, note that the first model of a transformer we introduced in this chapter is characterized by three parameters: the self and mutual inductances. For the special case of a perfectly coupled transformer, only two parameters, n and an inductance, are needed. Finally, with a further level of abstraction, we arrive at the ideal transformer, characterized by a single parameter, n.

Looking over Figure 7, you will note that, if the inductance L_1 across the input were removed, the perfect transformer would reduce to an ideal transformer. If the current through that inductance were identically zero, that would be tantamount to removing it. Since the current is inversely proportional to L_1 (see (17)), the current would go to zero if the inductance were large, approaching ∞. But then, since $n^2 = L_1/L_2$, the turns ratio would also become infinite—unless L_2 also approached infinity!

In summary: An ideal transformer is a perfectly coupled transformer whose self-inductances both approach infinity but in such a way that their ratio stays constant.

3 TRANSFORMER NETWORKS IN THE AC STEADY STATE

So far in this chapter, we have been dealing with the time domain and have placed no restriction on the time dependence of voltages and currents. Let us now limit ourselves to sinusoidal signals and consider only the steady-state response. In this case, the voltages and currents are represented by their phasors. We shall deal with AC circuits that include transformers.

Loop Analysis

Consider the circuit shown in Figure 8a. It includes an impedance Z_2 connected to the secondary side of the transformer and an AC source represented by phasor V_g accompanied by an impedance Z_1 exciting the primary side. The first order of business is to decide on the mode of analysis; let's use loop analysis, choosing the loops shown in Figure 8b. To distinguish the transformer primary and secondary currents from the loop currents, the phasors of the former will be labeled I_p and I_s (and V_p and V_s for the voltage phasors).

In order to carry out a loop analysis, the transformer must have an impedance representation. For the references shown, the steady-state transformer equations are:

$$\begin{aligned} V_p &= j\omega L_1 I_p + j\omega M I_s \\ V_s &= j\omega M I_p + j\omega L_2 I_s \end{aligned} \tag{19}$$

These equations are in the form $V = ZI$, an impedance representation; hence, a loop analysis is possible.

We next write KVL around the two loops, using Ohm's law for phasors to replace the voltages of the two impedances:

$$\begin{aligned} Z_1 I_1 + V_p &= V_g \\ Z_2 I_2 - V_s &= 0 \end{aligned} \tag{20}$$

Next we observe that $I_p = I_1$ and $I_s = -I_2$; when we combine the preceding two sets of equations, they yield the following loop equations:

$$\begin{aligned} (Z_1 + j\omega L_1)I_1 - j\omega M I_2 &= V_g \\ -j\omega M I_1 + (Z_2 + j\omega L_2)I_2 &= 0 \end{aligned} \tag{21}$$

The negative signs in the mutual terms are a consequence of the fact that the reference of I_2 is the opposite of that of the transformer secondary current; this causes the loop current references to be oppositely oriented relative to the dots.

Suppose we had reversed the reference for loop current I_2. Then it would

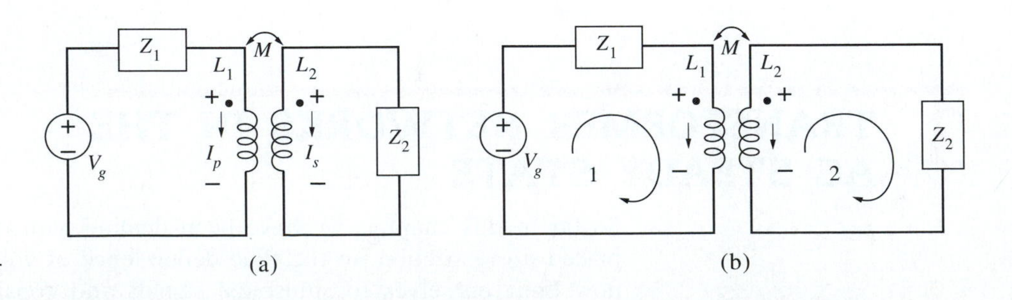

Figure 8
Circuit with transformer in the steady state.

have been identical with secondary current I_s, and all signs in the final equations would have been positive.

EXERCISE 4

Suppose that both loop-current references are chosen counterclockwise in Figure 8. Make appropriate changes in sign in (19) and (20), and confirm that the loop equations are identical with those in (21) except that the source voltage now appears with a negative sign. □

We come to the following conclusion. Given a transformer in a circuit, always choose standard references for the *transformer* voltages and currents, with the references on both sides similarly placed relative to the dots. Choose the *loop*-current references for convenience. If the loop-current references coincide with (or are both opposite to) the corresponding transformer currents, the signs of the mutual terms in the equations will be positive. If one of the loop currents is reversed from this condition, then the sign of the mutual terms will be negative. On the right-hand side of the equation, the sign of a source voltage that lies on a loop will be negative if it constitutes a drop in the direction of the loop current.

TRANSFORMER RESPONSE WHEN LOADED BY AN IMPEDANCE Let us now return to the original problem for which the loop equations were given in (21) and take the case where $Z_1 = 0$, that is, the case where a transformer is inserted between a sinusoidal source and a load impedance. The presence of Z_1 only complicates the expressions, adding nothing of importance. The currents are:

$$I_1 = \frac{Z_2 + j\omega L_2}{\Delta} V_g \qquad I_2 = \frac{j\omega M}{\Delta} V_g \tag{22}$$

$$\Delta = -\omega^2(L_1 L_2 - M^2) + j\omega L_1 Z_2 = j\omega L_1(j\omega L_2 + Z_2) + \omega^2 M^2$$

where Δ is the determinant of the equations in (21) with $Z_1 = 0$. From the currents it is possible to find any of the network functions: input impedance, voltage gain, and current gain. They are:

$$\begin{aligned}
Z &= \frac{V_g}{I_1} = \frac{\Delta}{Z_2 + j\omega L_2} = j\omega L_1 + \frac{\omega^2 M^2}{Z_2 + j\omega L_2} \\
G_{21} &= \frac{V_2}{V_1} = \frac{j\omega M Z_2}{\Delta} = \frac{j\omega M Z_2}{j\omega L_1(Z_2 + j\omega L_2) + \omega^2 M^2} \\
H_{21} &= \frac{I_2}{I_1} = \frac{j\omega M}{Z_2 + j\omega L_2}
\end{aligned} \tag{23}$$

What happens to these expressions if the transformer is perfectly coupled—or even ideal? Using $L_1 L_2 - M^2 = 0$ for a perfect transformer and a little manipulation, the preceding network functions become:

$$\begin{aligned} Z &= \frac{j\omega L_1 Z_2}{j\omega L_2 + Z_2} = n^2 Z_2 \frac{1}{1 + Z_2/j\omega L_2} \\ G_{21} &= \frac{M}{L_1} = \frac{1}{n} \\ H_{21} &= \frac{j\omega \sqrt{L_1 L_2}}{j\omega L_2 + Z_2} = \frac{n}{1 + Z_2/j\omega L_2} \end{aligned} \tag{24}$$

In the first and third of these, we used $L_1 = n^2 L_2$ to help arrive at the final result. Aside from the factor n^2, the impedance is that of the parallel combination of $j\omega L_2$ and Z_2.

Observe that the voltage-gain function does not depend on the load impedance; it is the same as that of an ideal transformer. Recall that, for an ideal transformer, the inductances approach infinity. In that case, the expression for the impedance approaches $n^2 Z_2$ and that for the current gain approaches n. (Recall that I_2 is the opposite of the secondary current; that accounts for the absence of the minus sign that appears in the definition of the ideal transformer.) Are these the expressions you would expect for an ideal transformer loaded by an impedance Z_2?

VOLTAGE LEVELS OF THE TWO SIDES An interesting observation can be made by considering the circuit in Figure 8—redrawn in Figure 9—and the loop equations in (21) obtained from the circuit. In Figure 9a one side of the transformer has been connected to the other side through a branch. This branch can consist of any components, including a source. Clearly, this branch will have no effect on the transformer equations in (19) or on KVL or on the relation between the loop currents and the transformer primary and secondary currents; hence, it will have no effect on the loop equations.

Now, suppose the branch is a DC voltage source. That means that the voltage levels on the two sides of the transformer can be different without affecting the operation of the circuit in which the transformer is connected. In particular, the voltage can be zero, as in Figure 9b. In this case one end of one side of the transformer is connected to one end of the other side and the voltage levels are the same. In this case, while still a two-port, the transformer is a three-terminal device. The common terminal is usually taken as "ground."

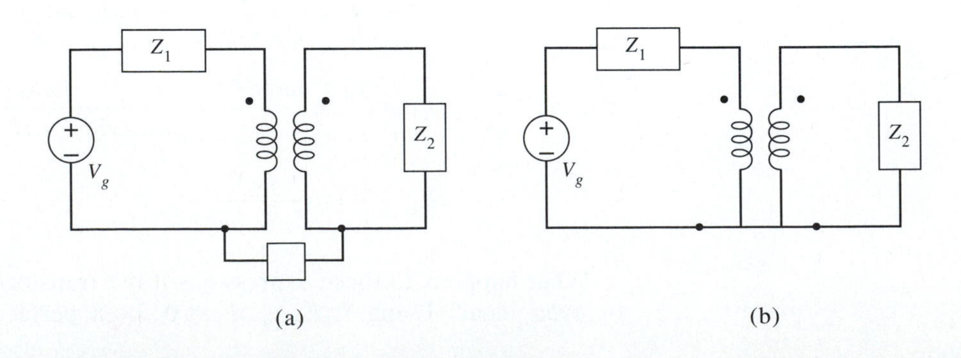

Figure 9
Transformer circuit of Figure 8 with a connection from one part to the other.

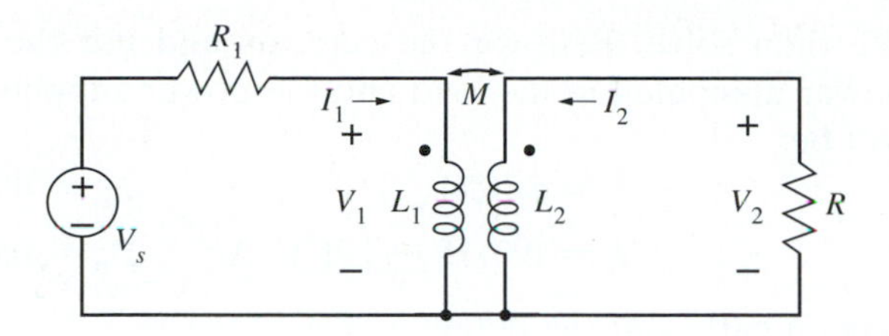

Figure 10
Example transformer circuit.

EXERCISE 5

The parameter values in the transformer circuit in Figure 10 are as follows: $L_1 = 20\ \text{mH}$, $L_2 = 500\ \text{mH}$, $k = 0.8$, $R_1 = 20\ \Omega$, and $R = 120\ \Omega$. The source is a 100-V, 400-rad/s sinusoid. (a) Find the turns ratio and the mutual inductance of the transformer. (b) Determine the steady-state phasor voltage across the output R. (c) Find the average power dissipated in the load resistance. (d) Find the average and reactive power supplied by the source. □

DEGREE OF APPROXIMATION IN ASSUMING AN IDEAL TRANSFORMER When faced with analyzing a given circuit containing a transformer, we might be interested in getting an approximate idea of the effect of the transformer by assuming that the transformer is ideal. How bad an error would we make in replacing the actual transformer with an ideal one? It isn't possible to answer this question in general, but to gain some idea of the size of the error, let's consider a specific case.

Look back at the transformer circuit shown in Figure 10. V_s is the phasor of a 400-rad/s sinusoidal source. Suppose $R_1 = 10\ \Omega$ and $R = 100\ \Omega$; also that the transformer is ideal, with $n^2 = \frac{1}{4}$. The four voltage and current variables are related by the two transformer relationships and two loop equations; they are:

$$10I_1 + V_1 = V_s \qquad V_1 = \frac{V_2}{2}$$

$$V_2 = -100I_2 \qquad I_2 = -\frac{I_1}{2}$$

These equations can now be solved for the two currents; from these, we can also calculate the power dissipated in the load and the power delivered by the source. The results are:

$$I_1 = 0.0286V_s\ \text{A} \qquad P_L = 0.0204\,|V_s|^2\ \text{W}$$

$$I_2 = -0.0143V_s\ \text{A} \qquad P_{\text{in}} = 0.0286\,|V_s|^2\ \text{W}$$

(Carry out the details.)

Next, let's assume that the parameter values of the transformer are as follows: $L_1 = 500\ \text{mH}$, $L_2 = 2\ \text{H}$, and $k = 0.95$. Choosing loop-current references to coincide with those of the transformer currents, we now write a pair of loop equations, which you should confirm:

$$(10 + j\omega L_1)I_1 + j\omega MI_2 = V_s$$

$$j\omega MI_1 + (100 + j\omega L_2)I_2 = 0$$

We then solve these for the currents and use the solutions to find again the power dissipated in the load and the power supplied by the source. The results will be:

$$I_1 = 0.0251\,\underline{/22^\circ}V_s\ \text{A} \qquad P_L = 0.014\,|V_s|^2\ \text{W}$$
$$I_2 = 0.0118\,\underline{/-151^\circ}V_s\ \text{A} \qquad P_{\text{in}} = 0.0233\,|V_s|^2\ \text{W}$$

(Again carry out the details.)

A comparison of the two calculations shows that assuming an ideal transformer leads to values of power that are too high. (By what percentage?) Nevertheless, if only a rough idea is needed, taking the transformer to be ideal reduces the effort. Perhaps larger inductances—closer to the infinite value that corresponds to an ideal transformer—would lead to a better approximation.

EXERCISE 6

Increase the inductances in the example in the preceding paragraphs by a factor of 20, and repeat the calculations for the power dissipated in the load and the power supplied by the source. Has the approximation improved? Now increase the coupling coefficient to $k = 1$ and again check the approximation, using the new inductance values. □

Node Analysis

Let us now contemplate carrying out a node analysis of the circuit in Figure 11. For a node analysis to be possible, the transformer must have an admittance representation, in the form $I = YV$. To get this, we return to the transformer equations in (19) and solve for the currents in terms of the voltages:

$$I_1 = \frac{L_2}{j\omega\Delta}V_1 - \frac{M}{j\omega\Delta}V_2 \qquad \Delta = L_1L_2 - M^2$$
$$I_2 = -\frac{M}{j\omega\Delta}V_1 + \frac{L_1}{j\omega\Delta}V_2 \tag{25}$$

where Δ is the determinant of the equations in (19) from which $-\omega^2$ has been removed. We find that an admittance representation exists (and so a node analysis is possible) unless $L_1L_2 - M^2 = 0$. But this is the condition for a perfectly coupled transformer. The conclusion is that a node analysis is possible for a transformer, but not if the transformer is perfect.

The next thing is to write KCL equations at the nodes. There are four nodes, as shown in the figure. But the current equation at node 3 will be satisfied if it is satisfied at node 1. Similarly, the current equation at node 4

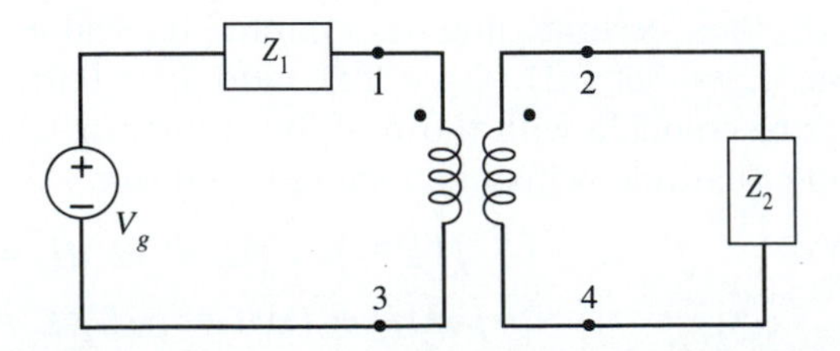

Figure 11
The circuit of Figure 8, redrawn.

will be satisfied if it is satisfied at node 2. Hence, there are only two independent KCL equations. Here we meet a difficulty: this conclusion disagrees with the previous result, first discussed in Chapter 5, which claims that, with n nodes, there should be $n - 1$ independent KCL equations. Indeed, from Chapter 5 we would have to conclude that the number of independent KVL equations is $b - (n - 1) = 1$, which disagrees with the loop analysis just carried out.

This difficulty is easily overcome. Until now, the circuits we have considered have all been *connected,* in one part, with a conductive connection from any one node to any other node. This is not the case for the transformer circuit in Figure 11, which is in two parts, with an *inductive* connection between the two parts. In the preceding paragraph we established that there are two independent KCL equations. Now 2 happens to equal the number of nodes less the number of separate parts of the circuit.

If we let p stand for the number of separate parts of a circuit, then, for this example, the number of independent KCL equations is $n - p$. This is indeed a general result, true for all networks, although we will not prove it for the general case here. On this basis, the number of independent KVL equations will be $b - (n - p)$. For the present example, this will be $4 - (4 - 2) = 2$, which agrees with the number of loop equations previously obtained. When the circuit is connected (the number of parts is $p = 1$), this expression agrees with our usual one from Chapter 5.

The remaining steps in obtaining the node equations are straightforward: writing KCL at one node from each part; using Ohm's law and the transformer current equations from (25); and relating the transformer primary and secondary voltages to the node voltages. The details will not be provided here, but you are urged to complete the process.

Transformer Not Connected as a Two-Port

Looking back at the preceding development, you will notice that the transformer is inherently a two-port component. It serves such purposes as (1) isolating two subcircuits from each other, (2) impedance matching, and (3) raising or lowering voltage levels. It is possible, however, to make connections to the two sides of a transformer in such a way that its two-port character is changed. The two sides might be connected in series, for example, and other external connections made. Not to worry. The basic laws still apply; loop or node equations can still be written; and such connections introduce no difficulty in obtaining a solution.

The network in Figure 12 is an example. A reasonable plan is to write a set

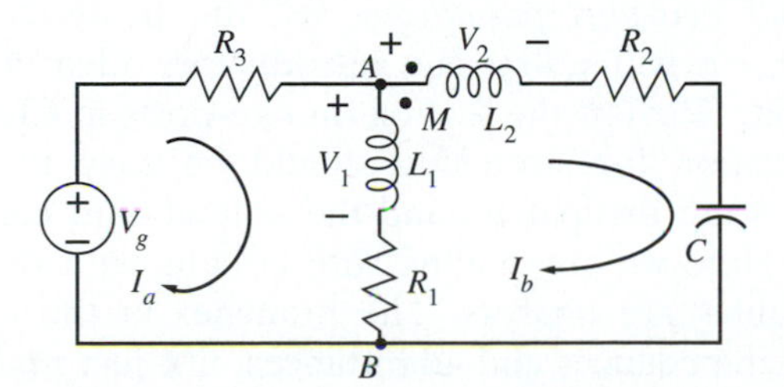

Figure 12
Transformer in a network.

of mesh equations. The mesh currents are labeled I_a and I_b. In the step where the voltages in the expression for KVL are to be eliminated in favor of the currents using the V-I relations, we perform that step for all branches except the transformer branches:

$$R_3 I_a + V_1 + R_1(I_a - I_b) = V_g$$
$$V_2 + \left(R_2 + \frac{1}{j\omega C}\right)I_b + R_1(I_b - I_a) - V_1 = 0 \tag{26}$$

The next step is to eliminate the transformer voltages. Always assuming standard references, the proper expressions for the primary and secondary currents in terms of the mesh currents are $I_1 = I_a - I_b$ and $I_2 = I_b$. Hence, the transformer equations are:

$$V_1 = j\omega L_1(I_a - I_b) + j\omega M I_b$$
$$V_2 = j\omega M(I_a - I_b) + j\omega L_2 I_b \tag{27}$$

When these are substituted into the preceding equations and terms are collected, the result in matrix form becomes:

$$\begin{bmatrix} R_1 + R_3 + j\omega L_1 & -(R_1 + j\omega L_1 - j\omega M) \\ -(R_1 + j\omega L_1 - j\omega M) & R_1 + R_2 + j\omega(L_1 + L_2 - 2M) + \dfrac{1}{j\omega C} \end{bmatrix} \begin{bmatrix} I_a \\ I_b \end{bmatrix} = \begin{bmatrix} V_g \\ 0 \end{bmatrix}$$

The solution is straightforward; carry it out.

4 TRANSFORMER EQUIVALENT CIRCUITS

Notwithstanding the circuit in the preceding example, most of the time transformers are connected as two-ports. The behavior at the terminals of the transformer, therefore, can be described in terms of the sets of two-port parameters introduced briefly in Chapter 4, in connection with the introduction of controlled sources.

Further development of two-ports and their description by means of what are called *two-port parameters* will be postponed until Chapter 15. Here, however, we will utilize the introductory ideas from Chapter 4 and stretch them a bit. (Review the section on two-ports in Chapter 4 if you feel the need.) One direction for stretching should be easy to appreciate. The circuits in Chapter 4 are dissipative, and the voltage and current variables are functions of time. Here we are dealing with circuits with reactive components also, and the variables are phasors. The branches in the circuits are characterized by complex impedances and admittances, not just real numbers.

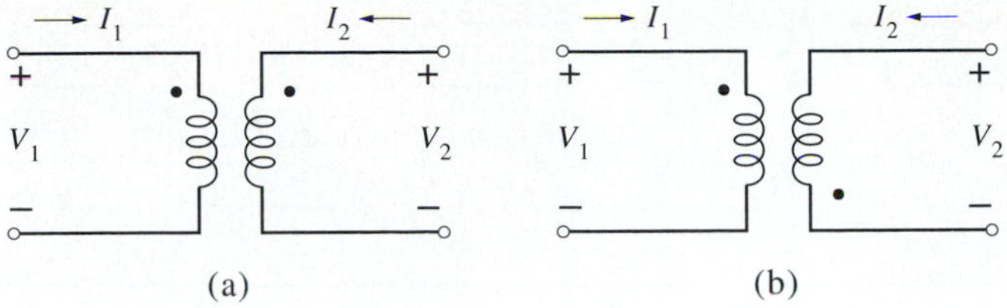

Figure 13 Transformer as two-port: possible combinations of dots and voltage-current references.

Our objective now is to examine the equations of a transformer in the phasor domain and to seek to find other circuits that have the same equations. The outside world connected to the terminals would not be able to tell whether there is a transformer or one of these other circuits between the ports, because, at the ports, the equations are exactly the same.

Consider the transformers in Figure 13 as two-ports. The two-port voltage and current references, on the one hand, and the standard references always chosen for the primary and secondary voltages and currents, on the other hand, coincide in both diagrams, which cover both possible arrangements of the dots. Because the two currents in Figure 13b are oppositely oriented relative to the dots, the sign in front of the mutual term in the corresponding equations will be negative. The transformer equations for the two cases are:

$$\begin{array}{ll} \textbf{(a)} & \textbf{(b)} \\ V_1 = j\omega L_1 I_1 + j\omega M I_2 & V_1 = j\omega L_1 I_1 - j\omega M I_2 \\ V_2 = j\omega M I_1 + j\omega L_2 I_2 & V_2 = -j\omega M I_1 + j\omega L_2 I_2 \end{array} \tag{28}$$

Look back at (7) in Chapter 4 and imagine that you have a pair of goggles that changes all variables to phasors and all real coefficients to complex impedances. Instead of (8) in Chapter 4, the goggles would show the following pair of equations:

$$\begin{aligned} V_1 &= z_{11} I_1 + z_{12} I_2 \\ V_2 &= z_{21} I_1 + z_{22} I_2 \end{aligned} \tag{29}$$

The z_{ij} coefficients are impedances rather than resistances. A comparison of this pair of equations with the ones in (28) shows the latter to have exactly this form. Hence, the transformer parameters in (28) can be identified with these impedance parameters as follows:

$$\begin{array}{rcc} & \textbf{(a)} & \textbf{(b)} \\ z_{11} & j\omega L_1 & j\omega L_1 \\ z_{22} & j\omega L_2 & j\omega L_2 \\ z_{21} = z_{12} & j\omega M & -j\omega M \end{array} \tag{30}$$

The only difference is the negative sign of the mutual term when the two currents are oppositely oriented with respect to the dots.

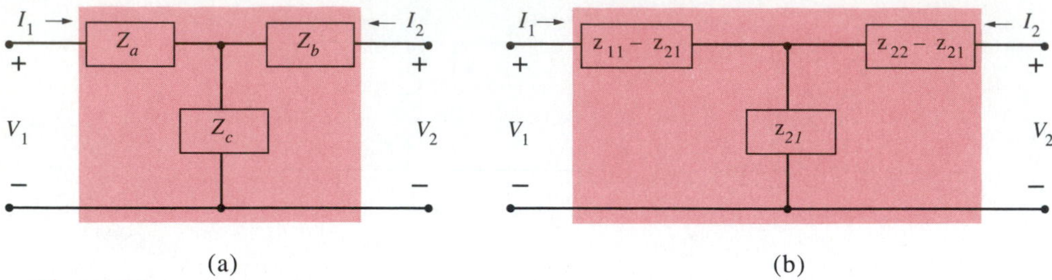

Figure 14
Tee-network equivalent of a general two-port.

Tee Equivalents

The objective now is to seek one or more two-port circuits that are equivalent to the transformer. Probably the most common two-port structure is the tee circuit shown in Figure 14a. The impedances of the branches are presumed to be known. It would be immensely useful if the equations relating the voltages and currents of this two-port could be made to look like those in (29).

Suppose the output port is left open (so that $I_2 = 0$). It is a simple matter to determine from the circuit that the impedance at the input port is $V_1/I_1 = Z_a + Z_c$ and that the output voltage is $V_2 = Z_c I_1$. With $I_2 = 0$, the relationships in (29) simplify to $V_1 = z_{11}I_1$ and $V_2 = z_{21}I_1$. By comparing these two sets of relationships, we find that $z_{11} = Z_a + Z_c$ and $z_{21} = Z_c$. Inverting these relationships leads to $Z_c = z_{21}$ and $Z_a = z_{11} - z_{21}$. (Remember: Don't just read this, but write it all out and derive the results yourself.)

Now we interchange the ports and carry out a similar development. Leave port 1 open; get two V-I relationships from the circuit and see what (29) reduces to in this case. Go through the details to find that $Z_c = z_{12} = z_{21}$ and $Z_b = z_{22} - z_{21}$. The impedances of the resulting tee are specified in terms of the z_{ij} parameters in Figure 14b.

By using the open-circuit impedance parameters of the transformer from (30), the branch impedances of the tee are found to be those shown in Figure 15 for the two cases. The series branch inductances in Figure 15a involve a difference. Since M can take on values greater than either of the two self-inductances (but not both), one or the other of these two inductances may

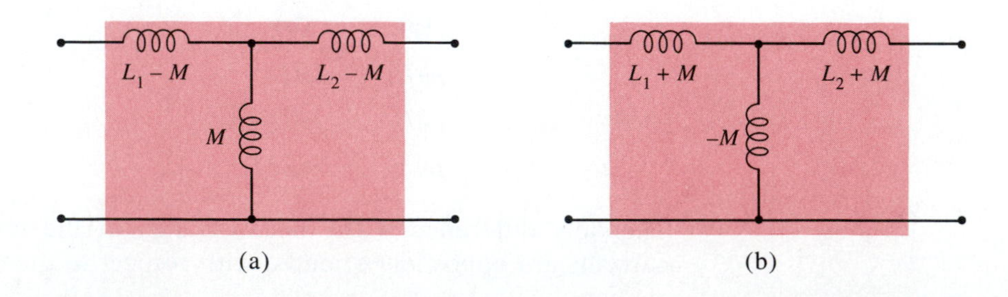

Figure 15
Tee equivalent of a transformer (partial).

be negative. In the second case, in Figure 15b, the shunt inductance is always negative. In either of these cases, the tee equivalent of the transformer will not be physically realizable (it can't be fabricated); nevertheless, it can still be used for the purpose of calculating the external behavior. Furthermore, a modified equivalent without negative elements will be described shortly.

TEE EQUIVALENT WITH IDEAL TRANSFORMER There is still something lacking in the tee "equivalents" in Figure 15 which makes the equivalence only partial. A transformer provides *isolation* between the primary and secondary sides, whereas the tee network does not. In the tee, the bottom terminals on the two sides are conductively connected; the equivalent condition for the transformer is that its two sides be at the same voltage level. Hence, the tee alone is not a complete equivalent of the transformer. But that problem can be easily remedied. A device that provides isolation and does nothing else is a 1:1 ideal transformer. Let's generalize this idea and explore it by connecting an n:1 ideal transformer, say, on the output side, as in Figure 16a. To be specific, let's assume that the dots are located as in Figure 13a.

The output side of the tee is to be connected to the input side of the ideal transformer. We need to exercise care in labeling the variables at these ports to make sure that the normal port references are maintained. As labeled, $V_1' = V_2'$ and $I_1' = -I_2'$. The plan of attack is first to write the z-parameter equations for the tee alone; then to express the output variables of the tee in terms of the input variables of the ideal transformer, as just identified; and finally to use the ideal-transformer V-I equations: $V_1' = nV_2$ and $I_2 = -nI_1'$. The resulting z-parameters in terms of the unknown inductances are equated to the original z-parameters of the transformer, from which L_a, L_b, and L_c can be determined. Thus:

$$V_1 = j\omega(L_a + L_c)I_1 + j\omega L_c I_2' \qquad V_1 = j\omega(L_a + L_c)I_1 + j\omega L_c \frac{I_2}{n}$$

$$V_2' = j\omega L_c I_1 + j\omega(L_b + L_c)I_2' \qquad nV_2 = j\omega L_c I_1 + j\omega(L_b + L_c)\frac{I_2}{n}$$

$$L_c = nM \qquad L_a = L_1 - nM \qquad L_b = n^2 L_2 - nM \tag{31}$$

The completed equivalent circuit is shown in Figure 16b.

You are no doubt curious as to what values n can take on. A look back shows that no particular restrictions were placed on n; it can take on *any* real value. Thus a transformer has an infinite number of copies of its equivalent

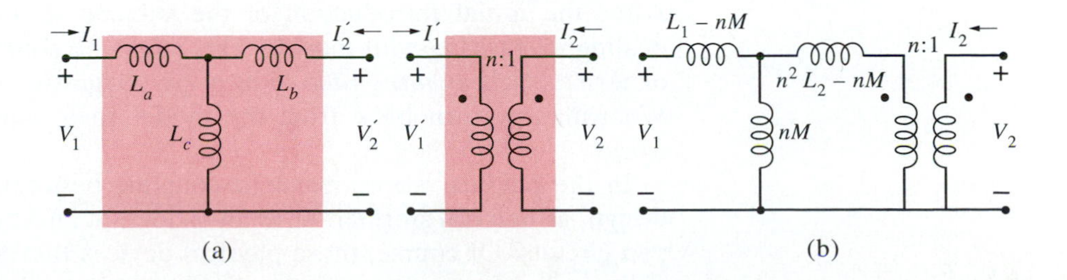

Figure 16 More complete equivalent circuit of transformer.

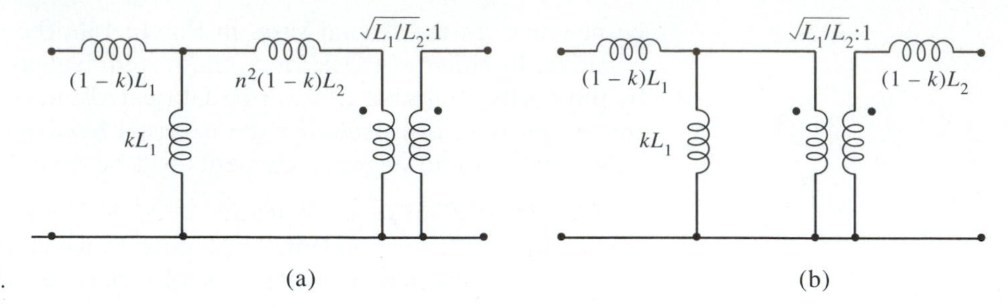

Figure 17
Transformer equivalent circuit with $n = (L_1/L_2)^{1/2}$.

circuit, all having the same structure but with different numerical values. In particular, when $n = 1$, the equivalent circuit reduces to the tee we started with, followed by a 1:1 ideal transformer.

Another interesting value of n is $n = (L_1/L_2)^{1/2} = kL_1/M = M/kL_2$. The resulting equivalent circuit is shown in Figure 17a. This circuit can be further modified by remembering the impedance-transforming properties of an ideal transformer: the impedance reflected into the primary side is n^2 times the impedance on the secondary side. The modification amounts to shifting the output series inductance to the secondary side after reducing its value by a factor of n^2, as shown in Figure 17b. (Confirm that the correct impedance equations result.)

Observe that, for a perfectly coupled transformer, the coupling coefficient k is unity. Hence, both series inductances in Figure 17b vanish. The resulting circuit is an ideal transformer with an inductance L_1 across the primary side. This confirms what we know about a perfect transformer. The larger the value of L_1 (and also L_2, to maintain a constant n), the closer a perfect transformer approaches an ideal one.

EXERCISE 7

In Figure 16b each of the two series inductances is a difference of two terms. Select a value of n so that one of these two inductances vanishes. Draw and label the resulting equivalent circuit of the transformer. Repeat for another value of n. □

5 PHYSICAL TRANSFORMERS

After the initial introduction of the subject of this chapter, we have been dealing exclusively with models. This might be somewhat obscured by the use of terms, such as *turns ratio,* which are suggestive of actual physical devices. We will now turn back from the model to a consideration of the physical world.

In the physical world, magnetic coupling between two circuits may exist by design, as in a transformer, or as an incidental effect caused by the proximity of two circuits. Of course, those physical devices that are designed to produce in one circuit a response to an excitation in another circuit through the agency of magnetic coupling are called transformers. It is unfortunate that the same name is used for a model of the physical device. To distinguish the two, we could call the physical device a *physical transformer.*

A physical transformer consists of two or more coils of wire wound on a common core. It may take one of many forms, such as a toroid or a three-legged rectangular frame. Very often, whenever large values of inductance are required, for example, ferromagnetic materials are used for the core. The inductances depend on the number of turns in the windings, the material of the core, and the geometry of the windings. If N_1 and N_2 are the numbers of turns of the two windings, the turns ratio n is N_1/N_2. The inductance of a winding is proportional to the square of the number of turns. In a physical transformer, then, $L_1/L_2 = N_1^2/N_2^2 = n^2$.

The size of the coupling coefficient k depends on how well the flux paths linking the two windings coincide. The core materials and the method of winding the coils have an influence on k. For high-permeability cores, the coupling coefficient can be made to approach very close to unity. If the coils are formed by winding two wires simultaneously, the result is called a *bifilar* winding. Such a winding would lead to a k very close to 1. Thus, actual physical transformers can approximate perfect transformers quite well.

In the physical world, however, windings will inevitably have some resistance. Thus, a model of a transformer should include winding resistances, as illustrated in Figure 18. Furthermore, there will be an inevitable capacitive effect among the turns of the coils, although windings are designed to minimize the capacitance. The effect of these capacitances may be negligible at low frequencies of operation. For high enough frequencies, however, the effect of these interturn and interwinding capacitances can become quite significant. An adequate model in such cases might resemble the one shown in Figure 18b.

Very often, magnetic coupling occurs in physical systems when it is not planned for or desired. A common example is the inductive interference caused by power lines on communications lines. In such cases, measures are taken to reduce the inadvertent coupling. The effects of undesirable magnetic fields can be reduced by *shielding,* the placing of iron materials around the system being influenced. When analyzing a circuit that experiences external incidental magnetic coupling, this coupling must be accounted for in the model representing the circuit.

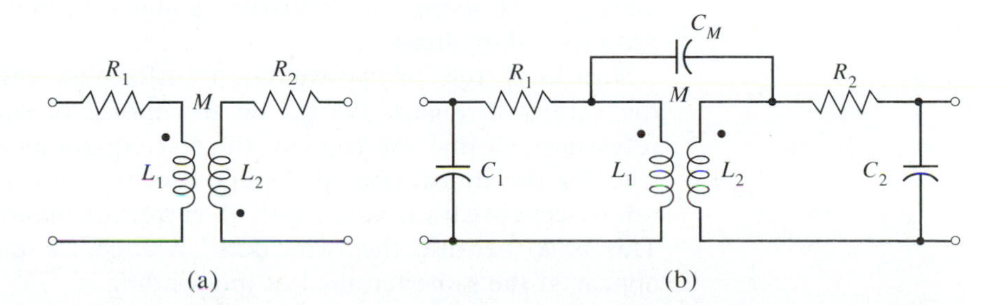

Figure 18
Models of a physical transformer.

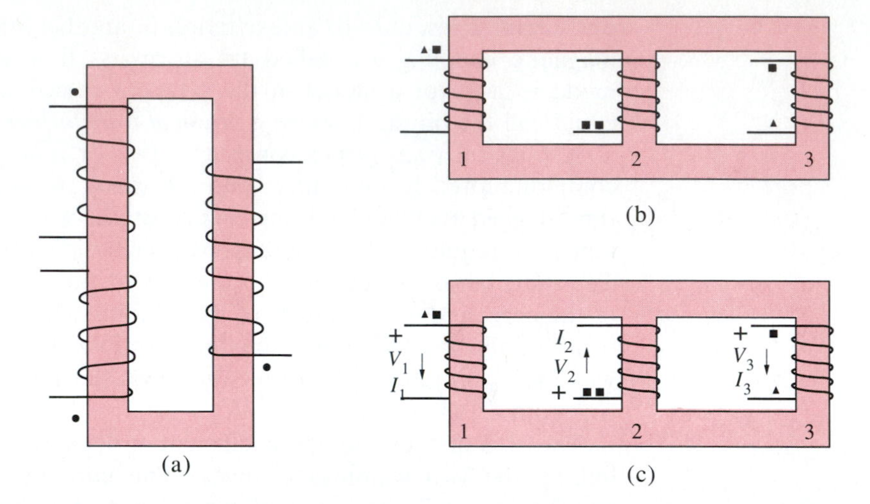

Figure 19
Three-winding transformers.

Multiterminal Transformers

So far in this book, only two coupled coils have been considered. Similar ideas can be applied if more than two windings are involved. Figure 19 shows two different arrangements of three coils wound on the same core. Consider first the one in Figure 19a. Recall that a dot is placed on one terminal of a pair to indicate the polarities of the flux linkages. If each current reference is oriented similarly relative to the dot (toward it or away from it), the fluxes are aiding. (Confirm that the dots are properly placed in Figure 19a.)

When there are only two windings, there is no ambiguity: the flux produced by one current will be either aiding or opposing the flux produced by the other. For the three windings in Figure 19a there is again no ambiguity: the fluxes produced by all three currents will be aiding if they are all oriented the same way relative to the dots. However, with three (or more) windings, this may not always be the case. That is, with just one dot per winding, there may be no way of locating the dots so that similarly oriented currents produce aiding fluxes.

This is the case for the three-winding transformer shown in Figure 19b. The remedy is to assign a different set of polarity markings for each pair of windings. Different-shaped symbols can be used for this purpose—dots, triangles, or squares, for example. The markings for each pair are placed according to the same principles as for a two-winding transformer. In Figure 19c there is the need for three sets of "dots." Confirm these flux-polarity markings, verifying that currents similarly oriented for each pair of "dots" produce aiding fluxes.

Just as is true for two-winding transformers, three-winding transformers of the variety in Figure 19a permit the choice of standard voltage and current references so that the signs in the *V-I* equations are all positive. This is not true for the transformer in Figure 19b. It is not possible to choose standard references for all variables, with all currents similarly oriented toward the dots. This is so because the two "dots" needed for each winding do not always appear at the same terminal of the winding.

The diagram in Figure 19b is redrawn in Figure 19c; the voltage and current references are chosen as shown. With the windings numbered from 1 to 3, let's label the mutual inductances with double subscripts: M_{12}, M_{13}, M_{23}. Then the transformer V-I relationships will be:

$$\begin{aligned} V_1 &= j\omega L_1 I_1 + j\omega M_{12} I_2 - j\omega M_{13} I_3 \\ V_2 &= j\omega M_{12} I_1 + j\omega L_2 I_2 + j\omega M_{23} I_3 \\ V_3 &= -j\omega M_{13} I_1 + j\omega M_{23} I_2 + j\omega L_3 I_3 \end{aligned} \tag{32}$$

Notice that the sign associated with the terms involving M_{13} is negative because I_1 and I_3 are oppositely oriented relative to the triangular "dots." The signs before the other mutual terms are positive because the other currents, in pairs, are similarly oriented relative to their "dots."

If the three-winding transformer in Figure 19c forms part of a circuit, a loop analysis can be carried out in the usual way, with (32) used to eliminate the transformer voltages in favor of currents.

The purpose of this section is to bring to your attention some practical considerations that arise with inductive coupling and transformers. It is useful to be aware of these whenever you are dealing with physical devices. In such cases, steps that may be appropriate to take, such as shielding, are not relevant to the models we are dealing with. Hence, no exercises are suggested for you to work out in this section.

SUMMARY

- Inductive, or magnetic, coupling
- Flux linkages, aiding and opposing
- Mutual inductance and transformers
- Faraday's law for coupled coils
- Circuit model of a transformer
- Dot convention for flux references
- Signs of mutual terms in transformer equations
- Coupling coefficient and the limit on its size
- The ideal transformer and its properties
- A perfectly coupled transformer and its model
- Transformers in the steady state
- Input impedance, voltage gain, and current gain in a transformer circuit
- Loop and node analysis in transformer circuits
- Equivalent circuits of transformers
- Physical transformers and more complete models
- Multiterminal transformers

PROBLEMS

1. The circuit in Figure P1 is excited by a 400-rad/s sinusoidal source whose phasor is V_s. The current phasors are specified to be $I_1 = 4e^{j0}$ A and $I_2 = 2e^{j\pi/4}$ A. Determine the source-voltage and load-voltage phasors.

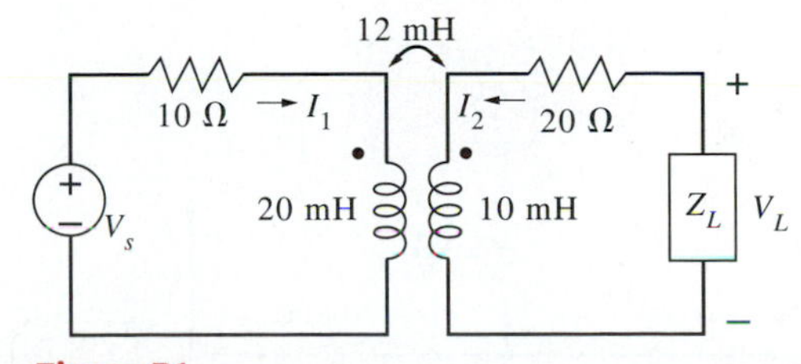

Figure P1

2. No energy is stored in the transformer in Figure P2 when the source voltage is turned on at $t = 0$. The objective is to find $v_2(t)$ across the open-circuit output terminals. The parameter values are $R = 100\ \Omega$, $L_1 = 400$ mH, $L_2 = 100$ mH, $k = 0.75$, and $v_s(t) = \sqrt{2}\ 100 \cos 1000tu(t)$.

a. Find the steady-state response, expressing it eventually as a function of time.

b. Find the natural response. Determine the complete response for $t > 0$, with no unspecified constants.

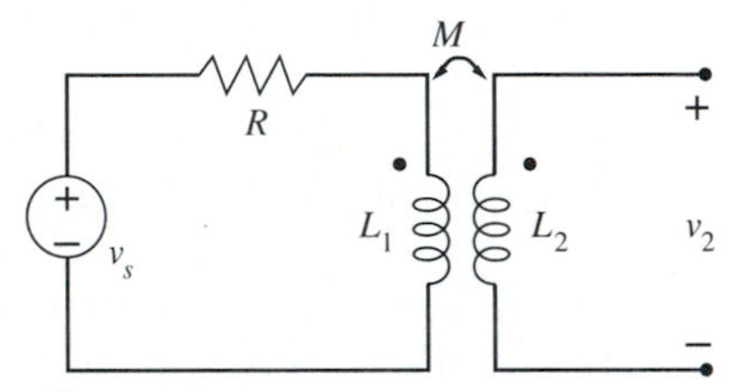

Figure P2

3. The transformer in Figure P3 has the following parameters: $L_1 = 40$ mH, $L_2 = 250$ mH, and $M = 100$ mH. It is excited by a 50-V, 1000-rad/s sinusoidal voltage source.

a. Find the coupling coefficient of the transformer.

b. Determine the numerical value of the impedance at the terminals.

c. Find the numerical value of the voltage gain V_2/V_1.

d. Determine the power dissipated in the load.

e. Determine how each of the preceding quantities would change if one of the dots were reversed.

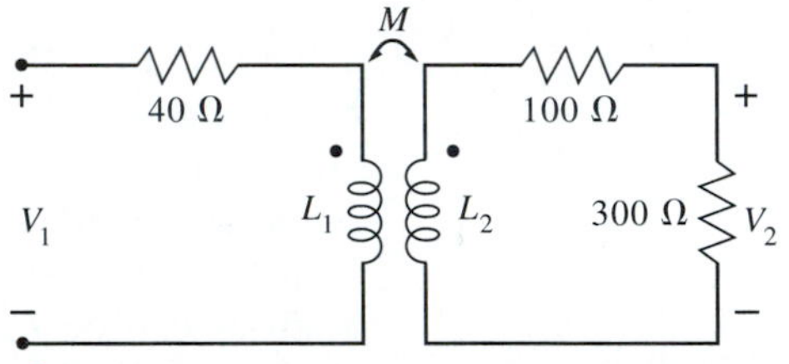

Figure P3

4. The transformers in Figure P4 are ideal. The load resistor is to be chosen to maximize the power delivered to it from the source whose phasor is V_s.

a. Determine the required value of R.

b. For this value of R, determine the amount of power dissipated in it.

c. Determine the current in each of the other resistors also.

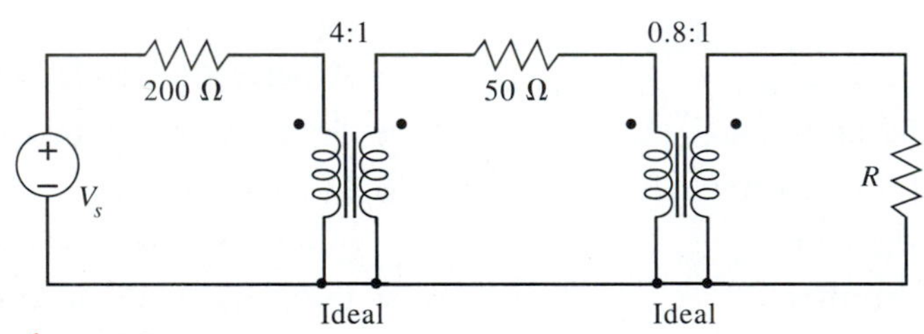

Figure P4

5. The source in Figure P5 is a 120-V, 400-rad/s sinusoidal source. The coupling coefficient of the transformer is 0.5.

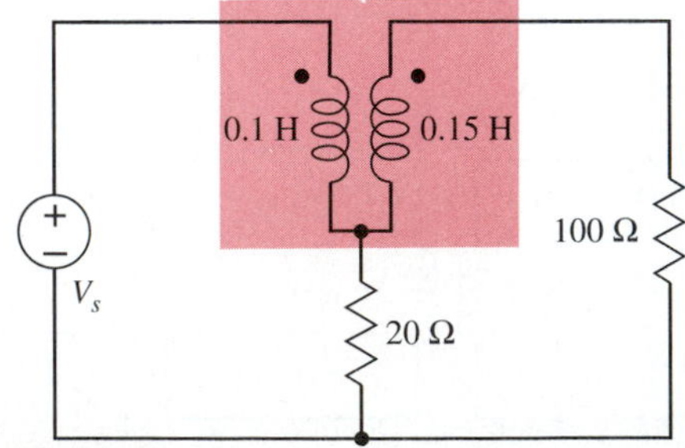

Figure P5

a. Determine the power supplied to the 100-Ω load.

b. Compare this with the power supplied to the network by the source.

c. If the power supplied to the load is not the largest possible amount, specify to what the load should be changed in order for the power to be maximized.

6. a. From the transformer equations in the AC steady state, construct a model of a transformer that includes two current-controlled voltage sources.

b. Using this model, find the input impedance of the circuit in Figure P6, in which the two sides of the transformer are connected in series.

c. How does the impedance change if one of the dot locations is reversed?

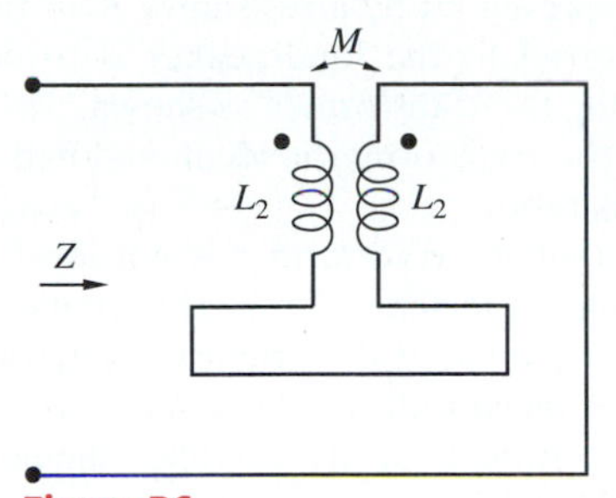

Figure P6

7. a. From the equations of a transformer given in (25) in the text, construct a transformer model that includes two voltage-controlled current sources.

b. Using this model, find the input admittance of the circuit in Figure P7, in which the two sides of the transformer are connected in parallel.

c. Suppose one of the dot locations is reversed. How does the admittance change?

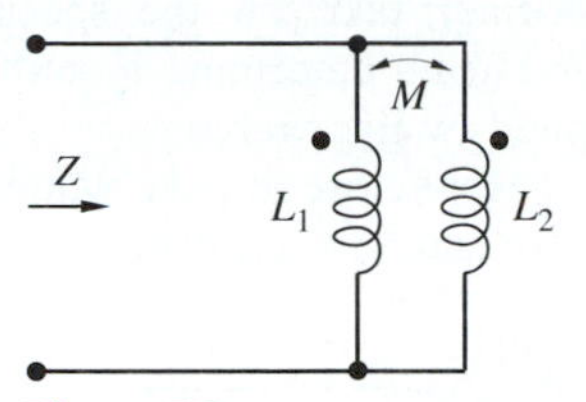

Figure P7

8. The networks in Figure P8 are excited by a sinusoidal source of 1 unit voltage and 1 unit frequency. (All values are normalized.) In each case determine the numerical value of the phasor input current I_1 and output voltage V_2.

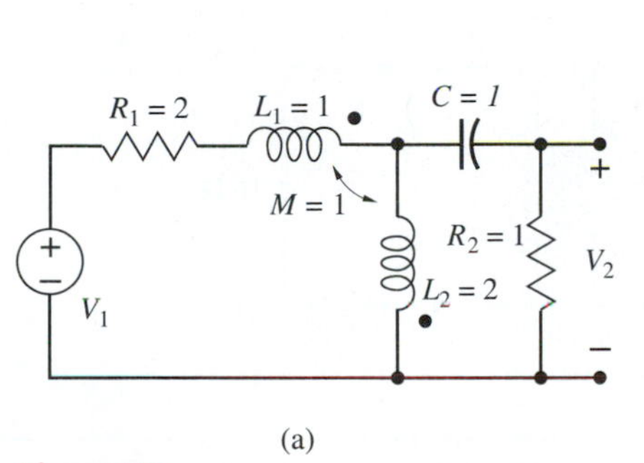

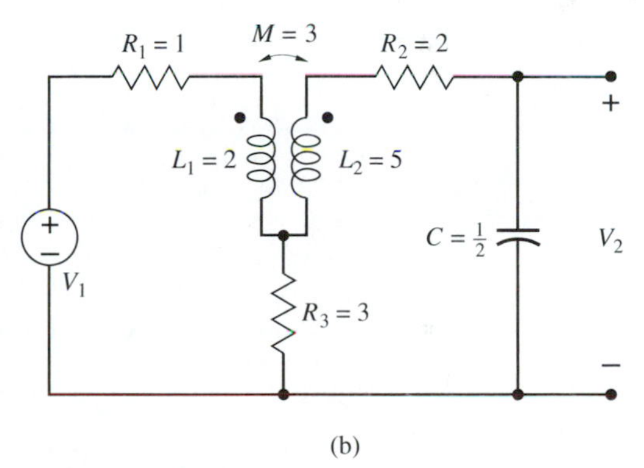

Figure P8

9. The following parameters are specified for the transformer in Figure P9: $n = 2$, $k = 0.95$, and $M = 3.8$ mH.

a. Find the self-inductances.

b. As a first approximation, the transformer is to be considered ideal. Assuming the circuit is excited by a 100-V, 1000-rad/s sinusoidal voltage source, find the impedance at the terminals of the circuit.

c. Find the power dissipated in the 20-Ω load.

d. Repeat part (b) with the full parameters of the transformer, dropping the assumption that it is ideal.

e. Repeat part (c) under these conditions also.

f. Compare the results for the two cases and describe the differences; e.g., what happens to the impedance magnitude, the load power, or the power factor under which the source is operating?

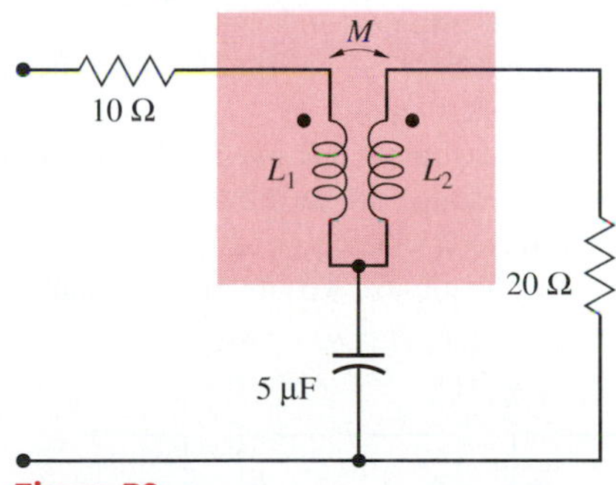

Figure P9

10. The circuit in Figure P10 is excited by a 70-V, 200-rad/s sinusoidal source. The parameters of the transformer are $L_1 = 960$ mH, $L_2 = 60$ mH, and $k = 0.99$.

a. Initially assume that the transformer is ideal (what would n be?), and determine phasor V_2.
b. Determine the average power supplied by the source.
c. Now let the transformer take on the specified parameter values, and again determine V_2 and the average power supplied by the source.
d. Compare the results. How close are the values in the ideal case to these values, in percent?

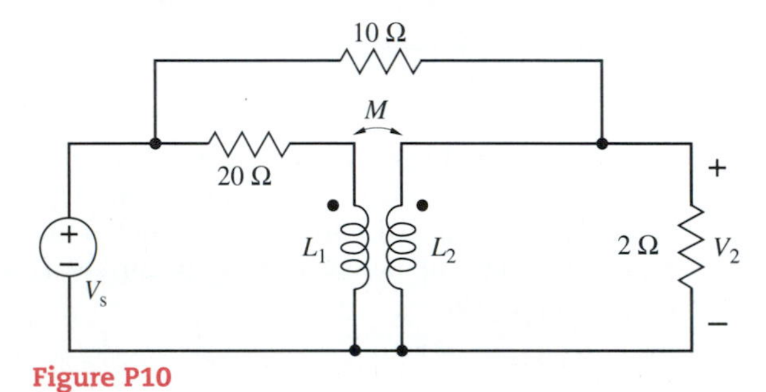

Figure P10

11. Figure P11 shows the Thévenin equivalent of an amplifier which is to drive a loudspeaker. The loudspeaker is represented by a resistance $R = 16\ \Omega$. The power delivered to the loudspeaker is to be a maximum by using the transformer as shown.
a. Assume that the transformer is ideal, and find the required turns ratio.
b. Next assume that the transformer is not ideal but has a coupling coefficient $k = 1$. Determine the smallest possible value of the primary reactance in order that the magnitude of the impedance presented at the terminals of the amplifier should be at least 90% of 900 Ω.
c. If the amplifier is to operate over a frequency range of 200 to 15,000 rad/s, what is the smallest permissible value of the primary inductance?
d. With the value of the inductance found in part (c), find the power delivered to the load at 200 rad/s and compare it with the power delivered if the transformer were ideal.

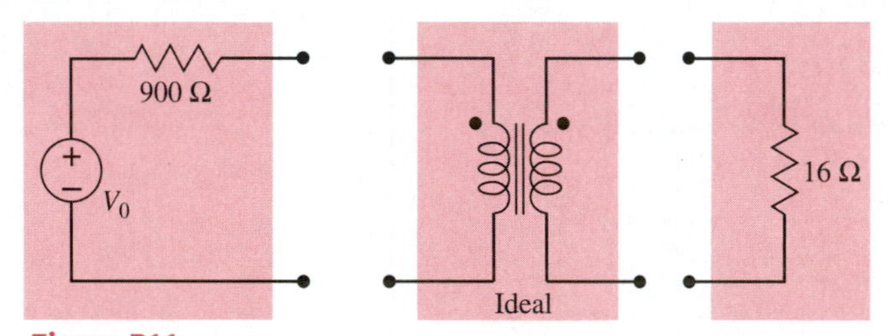

Figure P11

12. Two identical transformers are connected as shown in Figure P12. (There is no other coupling from one transformer to the other.)
a. Find the impedance at terminals AB when terminals CD are left open.
b. Repeat when terminals CD are short-circuited.
c. Find the impedance at terminals CD when terminals AB are left open.
d. Repeat when terminals AB are short-circuited.

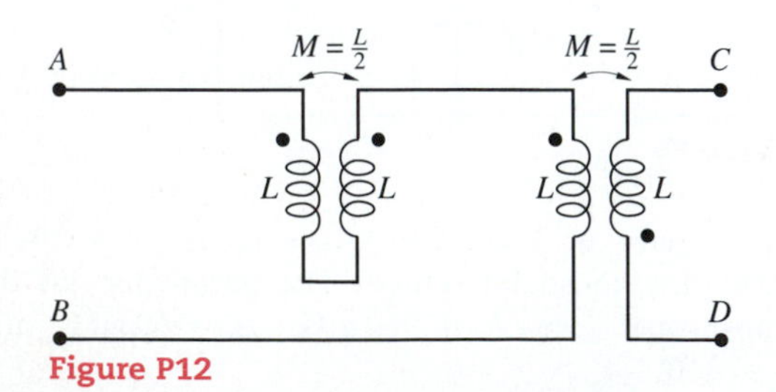

Figure P12

13. The falling step function $100u(-t)$ is a neat way of saying that the circuit in Figure P13 has been excited by a 100-V voltage for a long time when the voltage drops to zero at $t = 0$.
a. Find the energy stored in the transformer at $t = 0$.
b. It takes an infinite length of time for this energy to be completely dissipated in the 20-Ω resistor. How long does it taken for 90% of the energy to be dissipated?

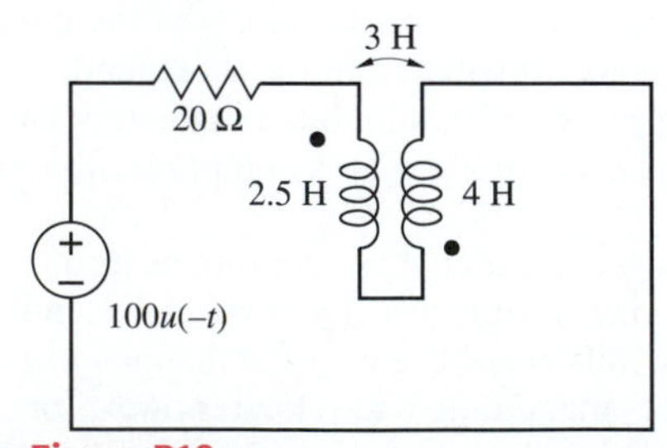

Figure P13

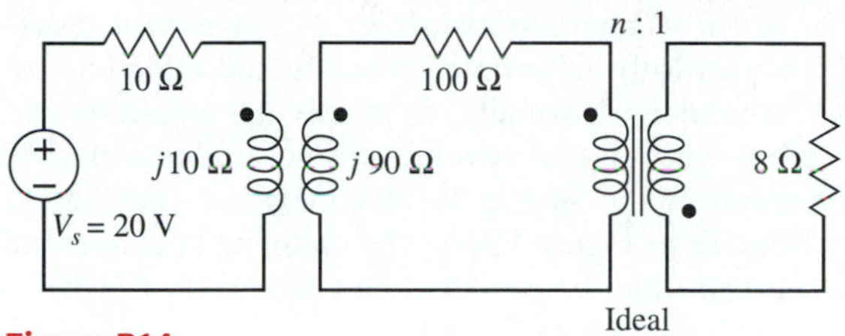

Figure P14

14. a. Find the turns ratio of the ideal transformer in Figure P14 so that the maximum power will be dissipated in the 8-Ω load assuming the other transformer is perfect.
    b. For this value of n find this maximum power.
    c. For the same value of n, find the power dissipated in the other two resistors.

15. Write loop equations for each of the circuits shown in Figure P15, with particular emphasis on the signs of the mutual terms.

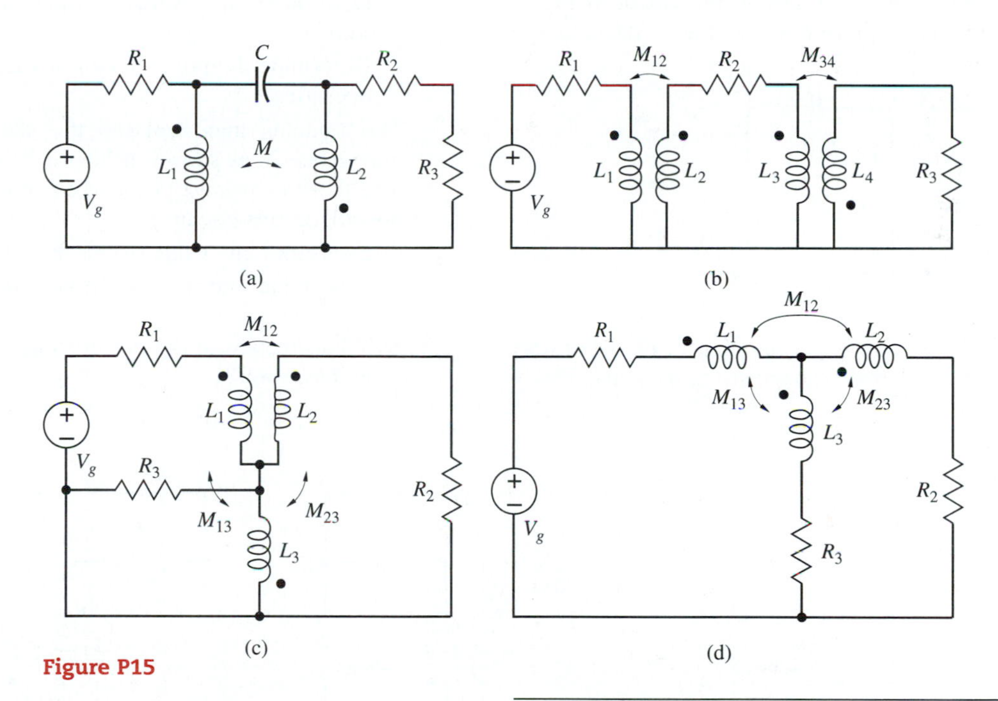

Figure P15

16. A transformer is sometimes constructed from a single coil of wire, taken as the primary, by *tapping* at some point on the coil to form the secondary. Such an *autotransformer* is shown in the circuit of Figure P16. The primary winding of 1000 turns has an inductance of $L_1 = 1.25$ H. The tap is taken 40% of the way up from the bottom terminal. The coupling coefficient is 0.96.
    a. The circuit is excited by a 120-V, 400-rad/s sinusoidal source. Determine the steady-state output voltage V.
    b. Determine the average power supplied by the source.

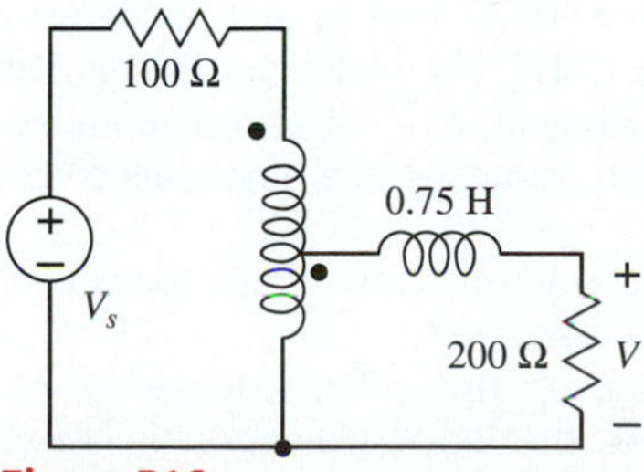

Figure P16

17. Letting I_1 and I_2 be the phasors of the primary and secondary currents of a transformer, prove that the average energy stored in the transformer is:

$$W_L = L_1 I_1 I_1^* + M I_1 I_2^* + M I_1^* I_2 + L_2 I_2 I_2^*$$

18. The secondary of the transformer in Figure P18 is short-circuited. By computing the average energy stored, prove that the coupling coefficient is no greater than 1.

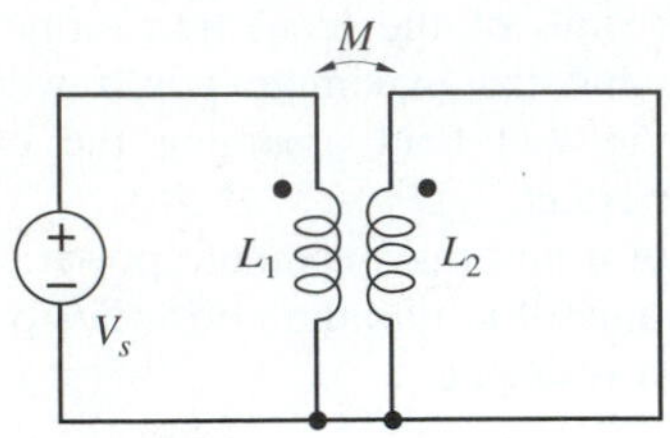

Figure P18

19. Find the value of M so that the steady-state voltage and current at the terminals of the circuit in Figure P19 are in phase at a frequency of $\omega = 1000$ rad/s.

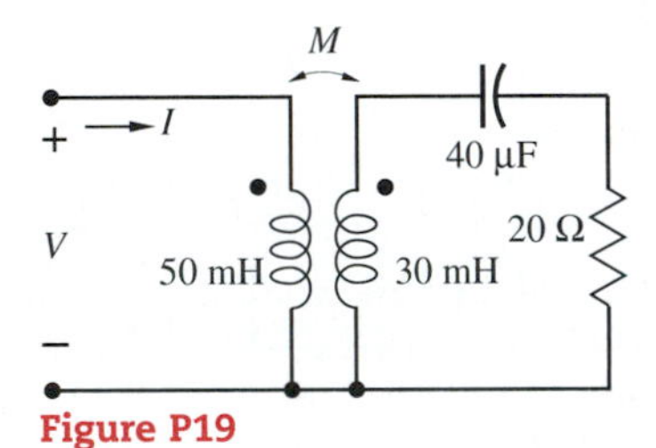

Figure P19

20. Some students were experimenting in the laboratory and set up the circuit shown in Figure P20a. This is just a standard transformer circuit with a resistor in series with the primary winding and a load across the secondary. Normally, terminals AB would be excited, but one of the adventuresome students decided to connect the source to terminals AC, leading to the circuit in Figure P20b. The coupling coefficient of the transformer is $k = 0.8$; the source is a 1-kV, 400-rad/s sinusoidal source.

a. Determine the average power dissipated in the 400-Ω load.

b. Determine the average power supplied by the source.

c. Determine the power factor at which the source is operating.

The students then replaced the 400-Ω load by an impedance Z, as shown in Figure P20c, and adjusted it so that the average power it dissipates is the largest possible in this circuit.

d. Determine the value of this impedance.

e. Determine the average power delivered to the load.

f. What fraction of the power supplied by the source is this power?

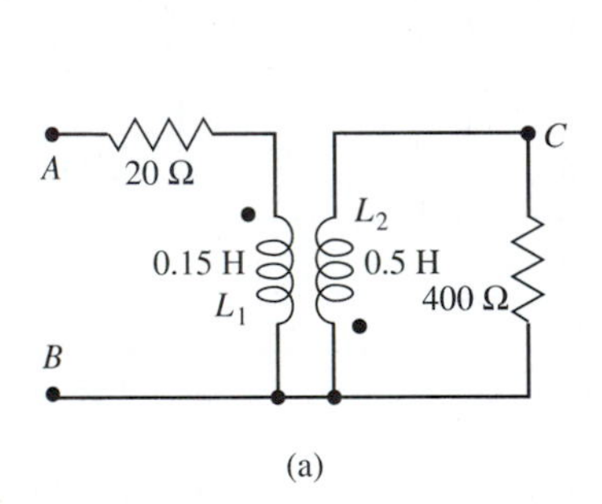

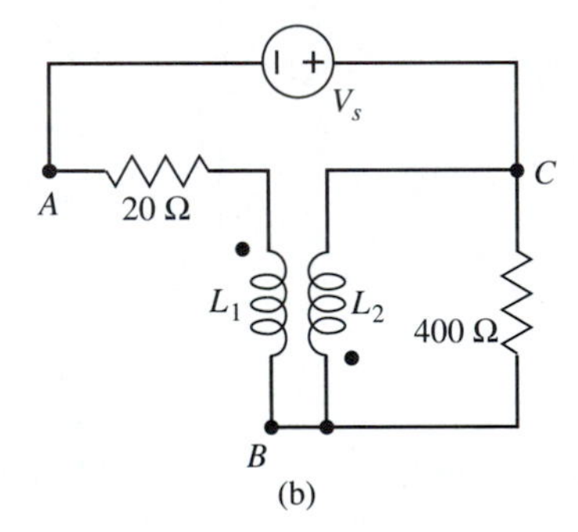

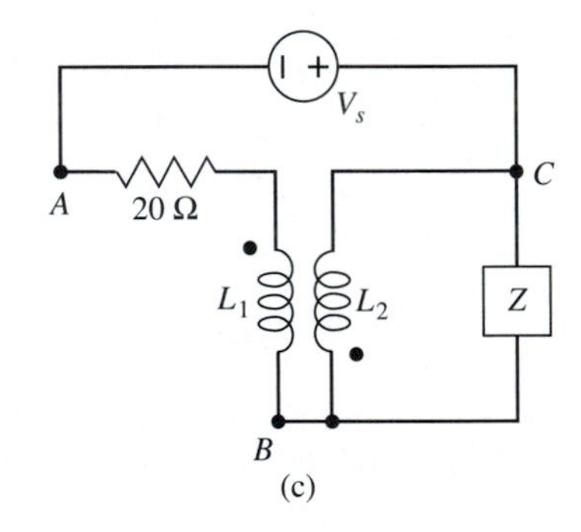

Figure P20

21. Continuing the experimentation they started in Problem 20, the adventuresome students now changed the load back to a 400-Ω resistor, and they reconnected the source to terminals AB, as illustrated in Figure P21a.

a. Find the average power dissipated in the load.

b. Also determine the average power delivered by the source.

c. Which of the two connections of the source provides to the 400-Ω load a larger fraction of the power generated by the source?

In the continuing spirit of experimentation, they again replaced the 400-Ω load by an impedance Z, as shown in Figure P21b. They adjusted this impedance until the power dissipated in it was a maximum.

d. Determine the value of the impedance for this condition.

e. Under this condition, how much power is delivered to the load?

f. What fraction of the power delivered by the source reaches the load? Compare the results in this problem with the results in Problem 20.

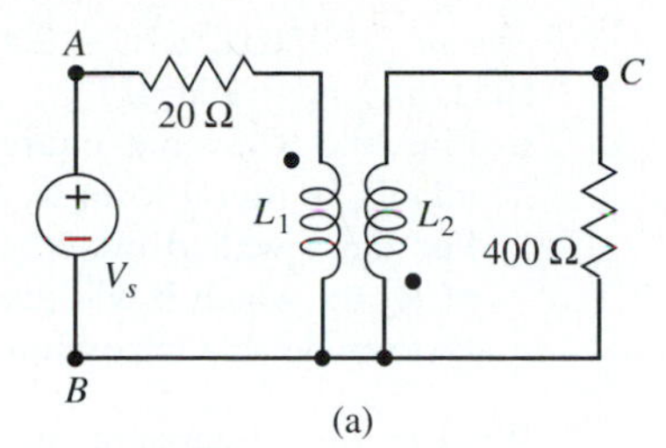

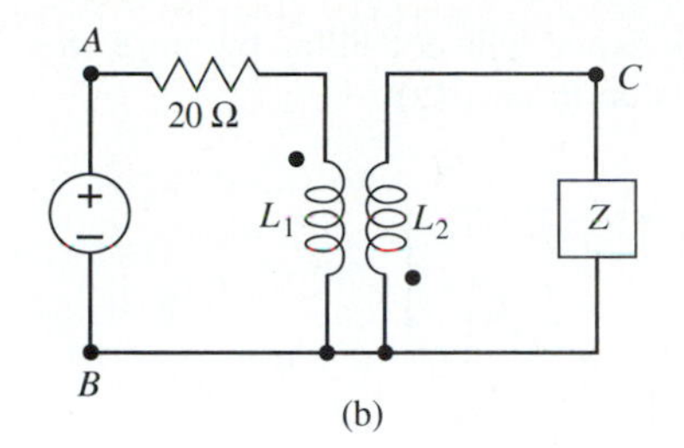

Figure P21

22. Still continuing the experimentation, the students changed the load to a 2-μF capacitor, and they changed the transformer to one having a turns ratio $n = 1.5$, a coupling coefficient $k = 0.6$, and a mutual inductance $M = 0.1$ H. This time they changed to a 120-V, 800-rad/s source and connected it in the same unorthodox manner, as shown in Figure P22. Determine the average and reactive power supplied by the source.

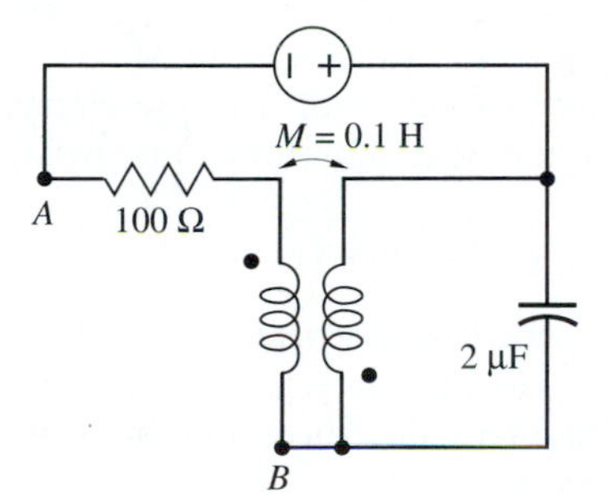

Figure P22

23. The enterprising students strike again! This time they went a little further. Not only did they connect the source as they had in Figures P20b and P22, but they also changed the location of the load, as shown in Figure P23. The source is a 200-V, 500-rad/s voltage.

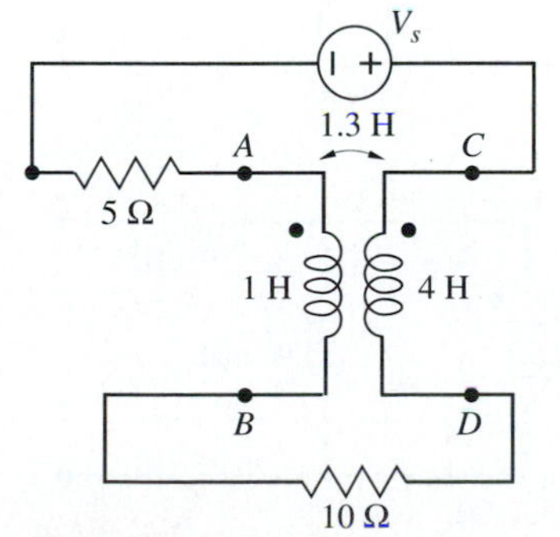

Figure P23

a. Find the impedance of the circuit from the terminals of the source.

b. Find the average power and the reactive power supplied by the source. Show that the average power dissipated in the two resistors equals the average power supplied by the source.

24. Before ending their experimentation, the students tried one other connection. For the same transformer and source, they changed the resistors and added a 5-μF capacitor between points A and D, as shown in Figure P24.

a. Again find the impedance from the source terminals.

b. What has the power factor become?

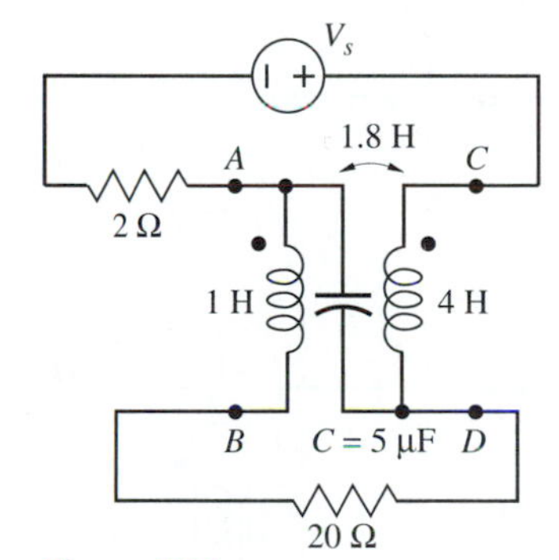

Figure P24

25. Not to be outdone by her adventuresome classmates, a creative student tried an unorthodox connection of a load to a transformer, as in Figure P25. The transformer has the following parameters: $L_1 = 4$ H, $L_2 = 1$ H, and $k = 1$.

a. Assume the transformer is ideal and determine the impedance at the terminals of the circuit.

b. Use the given transformer parameters, instead, and again find the input impedance. (There's a hard way and an easy way.)

c. Find a limitation on R so that the magnitude of

the input impedance will not differ by more than 5% from its value in part (a).

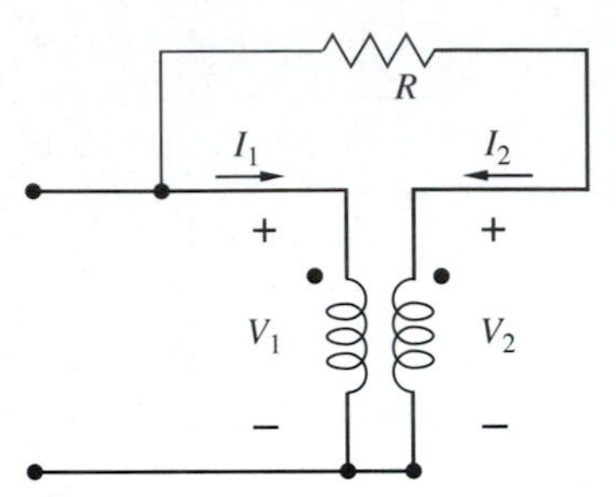

Figure P25

26. The parameter values in the circuit of Figure P26 are $\omega L_1 = 150\,\Omega$, $\omega L_2 = 200\,\Omega$, $\omega M = 100\,\Omega$, $R = 50\,\Omega$, and $V_g = 100\angle 0°$.

a. Find the Thévenin equivalent at the output terminals, in literal form as a function of frequency.

b. For the specified numerical values, find the value of R_L for which it will dissipate maximum power.

c. Determine this maximum power.

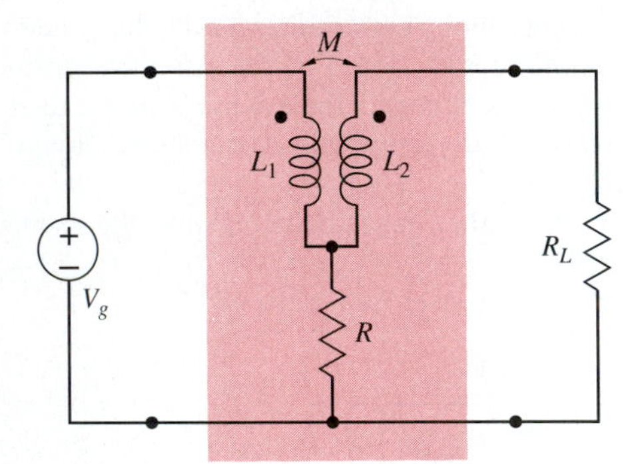

Figure P26

27. Review the discussion of transformer equivalent circuits surrounding Figure 16 in the text. Now use the transformation $V_1 = nV_2'(=nV_1')$ and $I_2' = I_1' = -nI_1$ in the transformer equations in (28a) in the text.

a. From the resulting equations, construct an equivalent model of a transformer and compare it with Figure 16b in the text. Describe the differences to yourself.

b. Draw the equivalent circuit to which your circuit reduces for the special value $n = M/L_2$; note what happens to this circuit when $k = 1$, a perfect transformer.

c. Repeat part (b) for the value $n = L_1/M$.

d. Use these same two values of n in the circuit of Figure 17b in the text and draw the resulting circuits.

e. Compare with the preceding circuits.

28. Each circuit in Figure P28 has been proposed as an equivalent circuit of some transformer. If you agree, find the self and mutual inductances and the coupling coefficients for each case.

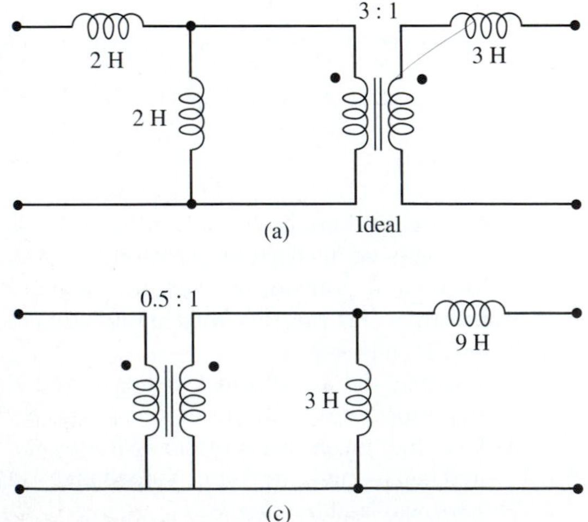

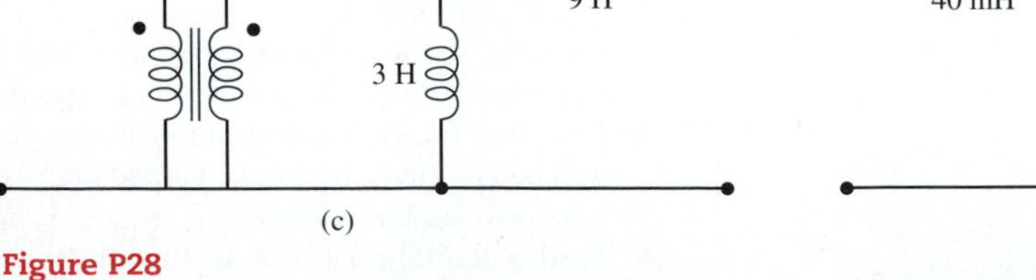

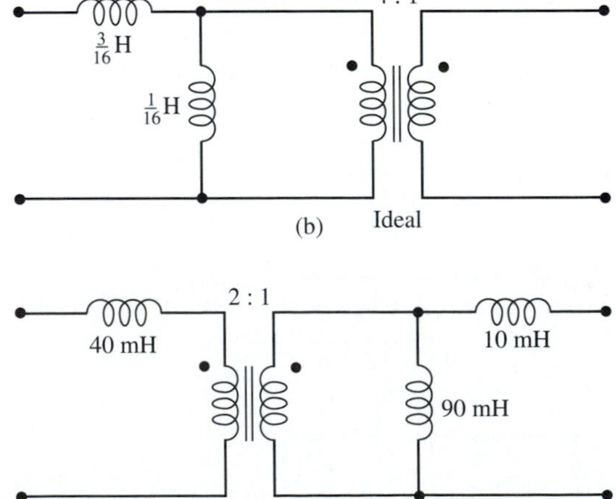

Figure P28

29. A transformer of unknown inductance values is available. The parameters are to be obtained by measurements at the terminals. Assume that the winding resistances and the interwinding capacitances can be neglected. One side of the transformer is excited by a sinusoidal voltage whose frequency is ω rad/s, with the other side open, as shown in Figure P29a. Two measurements—input current and output voltage—are made. Then the other side is excited and the open-circuit voltage on the first side is measured, as in Figure P29b.

a. Determine expressions for L_1, L_2, and M in terms of ω and the measured quantities.

b. Find numerical values for the transformer inductances for $\omega = 400$ rad/s, source voltages $V_{1a} = 120$ V and $V_{2b} = 40$ V, and measured quantities $|I_{1a}| = 1.2$ A, $|V_{2a}| = 48$ V, and $|V_{1b}| = 80$ V.

c. Determine the turns ratio n and the coupling coefficient k.

d. Predict what the measured value of the current in Figure P29b would have been if we had measured it.

e. Find an equivalent circuit of the transformer having only two inductances.

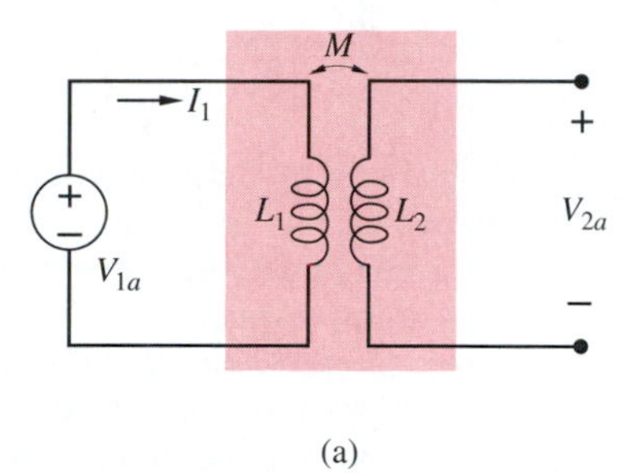

(a)

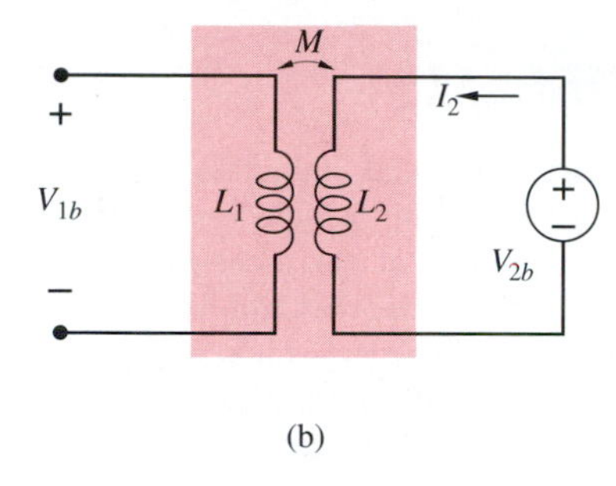

(b)

Figure P29

30. The inductances (self and mutual) of a physical transformer are to be determined by measurements at the terminals. The winding resistances are to be neglected, but the interwinding capacitances are to be accounted for by a capacitor C_x across the primary terminals. Use one or another of the equivalent circuits from Problem 27. For purposes of these measurements, an ammeter, a variable-frequency sinusoidal generator, an AC voltmeter, and two capacitors, C_1 and C_2, are available. Determine L_1, L_2, M, and C_x.

31. The winding resistances of a pair of coupled coils are not negligible and are shown in the circuit of Figure P31.

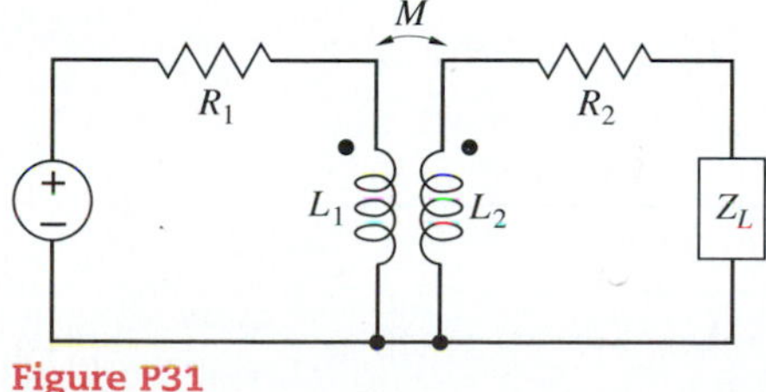

Figure P31

a. Replace the model transformer by an appropriate equivalent circuit.

b. Find the value of the load impedance in terms of frequency and all other parameters in the circuit that will maximize the average power dissipated by the load.

c. For this value of Z_L, find the input impedance presented to the source.

32. Two students had an argument about connecting two identical transformers in series on both their primary and secondary sides, as shown in Figure P32, before connecting them to a load and exciting them with a source. One student (student A), who had worked on the circuit the night before and had gotten as far as solving for the input admittance I_1/V_s, said that it didn't matter how the dots on the transformers were arranged; the results would be the same either with the dots as shown in the diagram or with one of the dots on one of the transformers reversed. The other student (student B) had also worked on the problem the night before and had solved for the voltage gain V_2/V_s. He maintained that the placement of the dots *did* make a difference. It's up to you to settle the argument.

a. After some thought but before any calculations, give a preliminary judgment. (Commit yourself; put it on paper even if nobody else will ever see it.)

b. For both arrangements of the dots, solve for the primary current (proportional to the input admittance) and for the secondary current (proportional to the voltage gain).

c. Solve also for the power delivered to the load and for the power supplied by the source for both sets of dot arrangements.

d. What's the answer to the argument for each of these quantities?

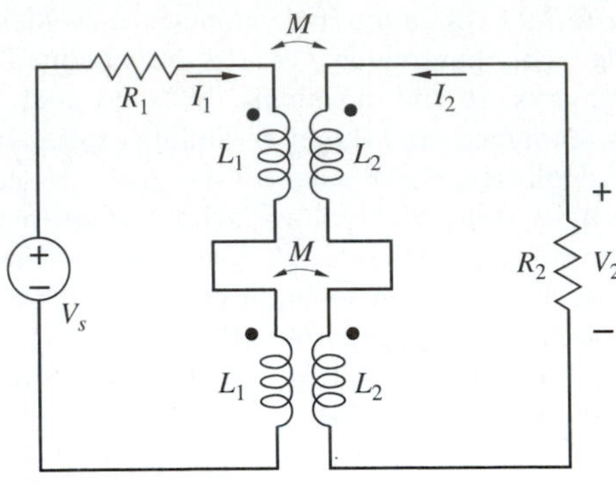

Figure P32

DESIGN PROBLEMS

1. It is desired to measure the self inductances and the mutual inductance of two coils which are wound on the same iron core. Equipment for measuring inductance, essentially an inductance bridge, is available. Design a series of measurements which will give all the desired parameters and the polarities of the coils. Minimize the number of steps in your series of measurements.

2. Figure DP2 represents a detailed model of a transformer. It consists of an ideal transformer with turns ratio $n{:}1$ plus a number of inductances and resistances. The inductance and resistance of the coils are represented by x_1 and x_2, and r_1 and r_2 respectively. X_c and R_c represent the equivalent of the core of the transformer. The purpose of X_c is to represent the current required to magnetize the iron core of the transformer; R_c is to represent the losses in the iron core. The sum of the currents through R_c and X_c is called the *exciting current.* The purpose of this problem is to determine experimentally as many of the coefficients of this equivalent circuit as possible. It is assumed that the turns ratio is 2:1. The following tests are performed.

Open Circuit Test: With the primary (high-voltage) side of the transformer open-circuited, 220 volts are applied to the secondary side. The current at the secondary side is measured to be 1.10A and the average power input to the transformer is measured to be 48.4 W.

Short Circuit Test: With the secondary (low-voltage) side of the transformer short-circuited, 22.8 volts are applied to the primary side. The current at the primary side is measured to be 11.4 A and the average power input to the transformer is measured to be 52.0 W.

Determine as many of the parameters as possible.

Hint: During the short-circuit test, the exciting current can be neglected, because the primary voltage is so much smaller than it is in normal operation.

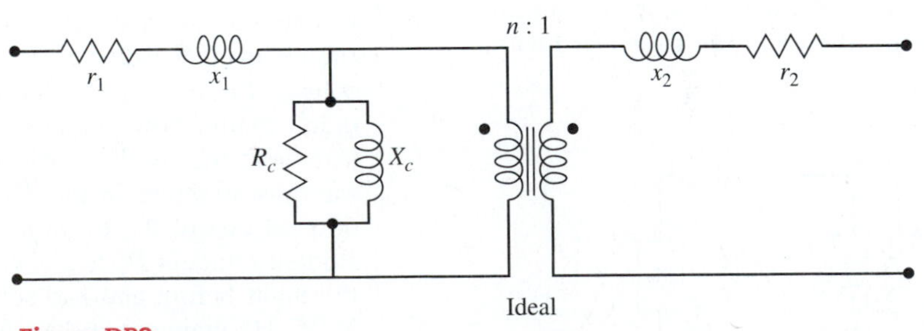

Figure DP2

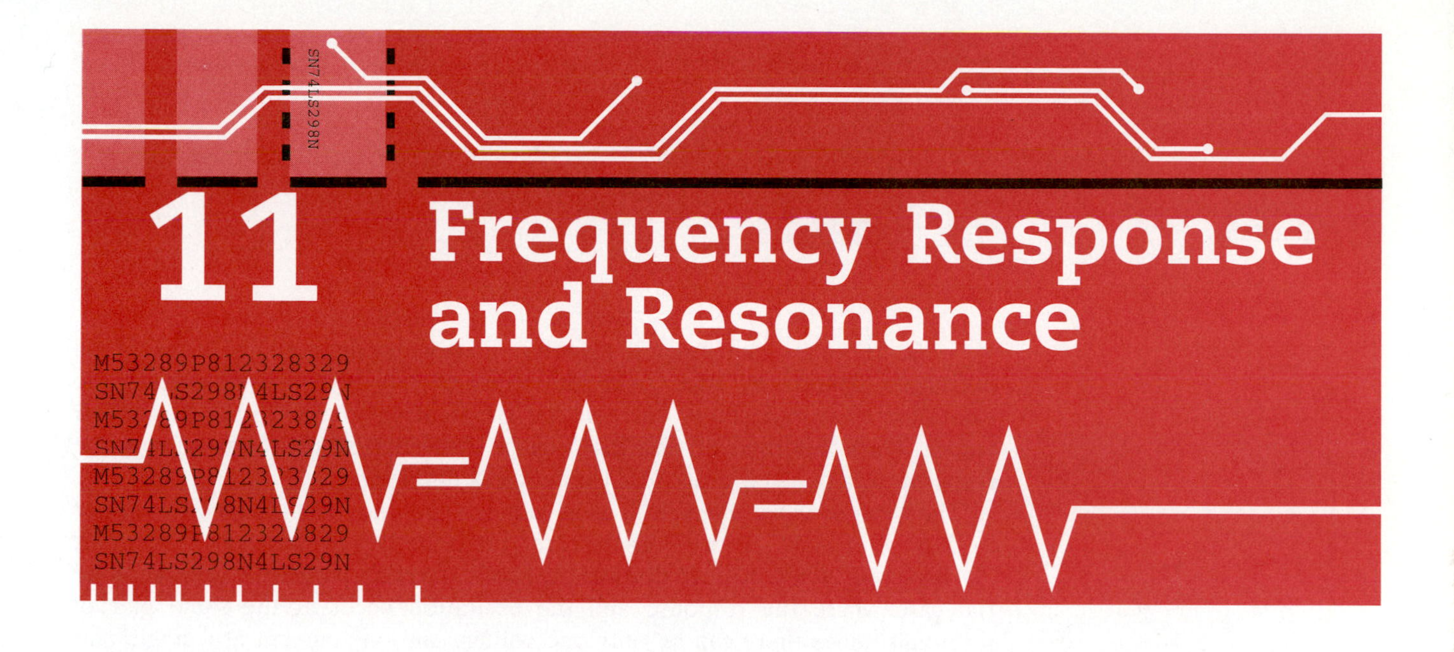

11 Frequency Response and Resonance

In recent chapters we have been dealing with the steady-state response of linear circuits to sinusoidal excitations. All sinusoids were represented by their phasors. For a two-terminal circuit, the response-current phasor is obtained by multiplying the excitation-voltage phasor by the admittance. Or, if the excitation is a current, the response-voltage phasor is obtained by multiplying the excitation-current phasor by the impedance. A similar procedure applies to circuits with more terminals, as we shall more fully develop in this chapter.

The impedance and admittance of a circuit depend on the values of the components in the circuit and on the frequency. In the preceding chapters it was assumed that the circuit components and the frequency were fixed. It is possible, however, for a circuit component or the frequency to vary over some range of values. (A radio is tuned, for example, by varying the value of a component in the radio circuit, usually a capacitor; a voice signal in a telephone conversation is made up of sinusoids from about 100 to 3000 Hz.)

In this chapter we shall study the way in which impedances, admittances, and other network functions of linear networks vary with frequency. Since these functions are complex (as opposed to real, not just complicated), both their magnitudes and their angles will vary with frequency. The concepts introduced in this chapter for simple circuits are applicable to more elaborate networks also.

1 NETWORK FUNCTIONS

If you review Chapter 8, you will find expressions for one voltage or current phasor in terms of another. The simplest case is "Ohm's law for phasors":

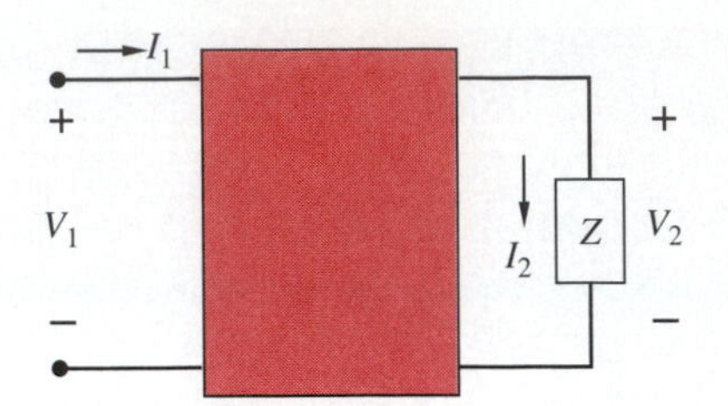

Figure 1
Defining network functions.

$V = ZI$. In each such case, a response phasor is obtained in terms of an excitation phasor. In each case, the ratio of the response phasor to the excitation phasor depends only on the frequency and the circuit parameter values. Once this ratio of response phasor to excitation phasor is known, the response to any given sinusoidal excitation is determined.

The term *network function* is used to mean *the ratio of any response phasor to any excitation phasor.* There are two classes of network functions: *driving-point functions* and *transfer functions.* Driving-point functions refer to the case where the response and the excitation occur at the same pair of terminals. Since there can be only one voltage and one current at a single pair of terminals, the possible driving-point functions are the impedance and the admittance. In Figure 1, for example, if V_1 is the excitation and I_1 is the response, then the appropriate driving-point function is the admittance $Y_1 = I_1/V_1$.

In Figure 1, one branch of the network is explicitly shown. It is assumed that the network includes no independent sources in its interior; the only permissible source, either voltage or current, appears at the left-hand terminals. Four possible transfer functions (ratios of response phasor to excitation phasor) can be defined:

$Z_{21} = V_2/I_1$; transfer impedance
$Y_{21} = I_2/V_1$; transfer admittance
$G_{21} = V_2/V_1$; transfer voltage ratio, or voltage gain
$H_{21} = I_2/I_1$; transfer current ratio, or current gain

Note the order of the subscripts; the first refers to the response, the second to the excitation. The voltage gain and the current gain are dimensionless ratios. The dimensions of Z_{21} and Y_{21} are impedance and admittance, respectively. Note carefully, however, that these two are not each other's reciprocal; indeed, they are defined with different excitations.

Four other such functions can be defined simply by interchanging the locations of the excitations and the responses. The subscripts would then be interchanged.

We have already briefly encountered such functions when studying multiterminal dissipative components in Chapter 4. However, there are two differences between those circuits and the present ones: (1) the circuits in Chapter 4 have no energy-storing devices, so that the variables are not necessarily phasors; and (2) the transfer functions considered there are defined under open-circuit or short-circuit conditions at the ports, whereas here the terminations are more general. The parameter of a voltage-controlled voltage source, for example, is a voltage-gain function.

In this chapter we will be examining driving-point and transfer functions of specific circuits and studying their variation with frequency. The circuits considered serve as examples, but many of them are also important circuits in their own right.

2 FREQUENCY RESPONSE

We will start our study of the frequency response of circuits with the simplest cases and gradually increase circuit complexity.

Single-Component Responses

To start as simply as possible, let's consider the dependence on frequency of the impedances and admittances of the individual components. In the case of resistors and other dissipative components, the impedance is independent of frequency.

For the inductor and capacitor, using the appropriate phasor ratios, we find:

$$Z_L = j\omega L = \omega L e^{j90^\circ} \qquad Y_C = j\omega C = \omega C e^{j90^\circ}$$

$$Y_L = \frac{1}{j\omega L} = \frac{1}{\omega L} e^{-j90^\circ} \qquad Z_C = \frac{1}{j\omega C} = \frac{1}{\omega C} e^{-j90^\circ}$$

$$X_L = \omega L \qquad B_L = -\frac{1}{\omega L} \qquad B_C = \omega C \qquad X_C = -\frac{1}{\omega C}$$

In these cases, the impedance and admittance are purely imaginary. The angle of each is constant at $\pm 90^\circ$: $+90^\circ$ for Z_L and Y_C, and -90° for Y_L and Z_C. The reactances and the susceptances vary with frequency (linearly in one case and hyperbolically in the other), as shown in Figure 2. Identical curves apply for

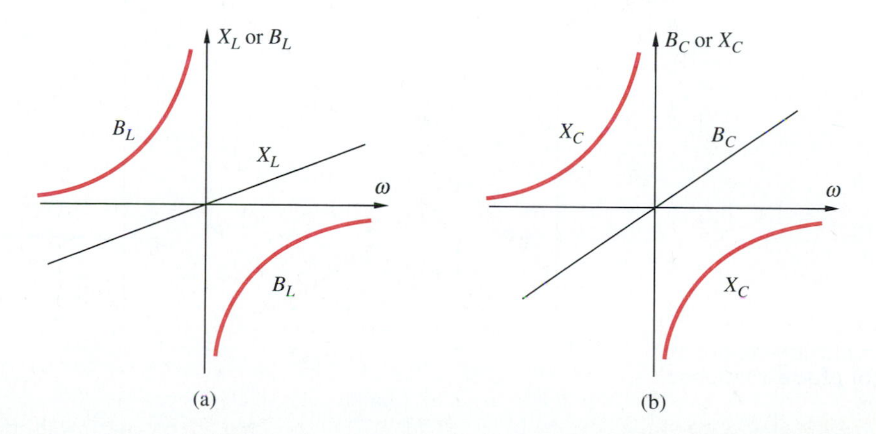

Figure 2 Inductive and capacitive reactance and susceptance variations with frequency.

X_L and B_C; similarly, identical ones apply for B_L and X_C. This is a manifestation of the dual nature of inductance and capacitance.

Frequency Response of Low-Pass RC Circuits

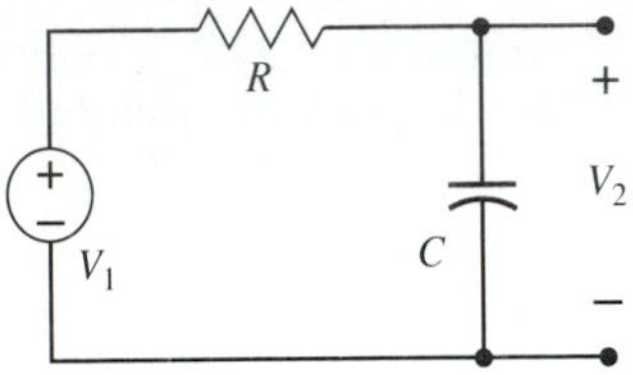

Figure 3 Simple frequency-selective circuit.

A simple, yet significant, circuit is shown in Figure 3. A sinusoidal voltage whose frequency can be varied is applied at the input terminals. The amplitude and phase of this voltage are assumed to remain fixed while only the frequency varies. That is, phasor V_1 is a constant complex number, differing from a DC voltage only in that it is complex rather than real.

We wish to determine the output-voltage phasor V_2. For this purpose, we need the voltage-gain function $G_{21} = V_2/V_1$. Using the voltage-divider relationship, we obtain the voltage ratio as:

$$G_{21} = \frac{V_2}{V_1} = |G_{21}|\, e^{j\phi} = \frac{1/j\omega C}{R + 1/j\omega C} = \frac{1}{RC}\frac{1}{1/RC + j\omega} \tag{1}$$

$$|G_{21}| = \frac{1}{RC}\frac{1}{\sqrt{\omega^2 + (1/RC)^2}}$$
$$\phi = -\tan^{-1}\omega CR \tag{2}$$

The magnitude and angle of G_{21} have been shown explicitly in the two expressions in (2). It is seen that both of these are dependent on the frequency. As the frequency varies, the magnitude and angle both vary.

Sketches of the magnitude of G_{21} and its angle plotted against ω are shown in Figure 4. Either from the figure or from (2), it is seen that the magnitude starts with a value of unity at zero frequency and gradually falls as the frequency increases, approaching 0 as the frequency approaches infinity. For low frequencies, the angle is very close to a linear function of frequency, deviating further and further from a straight line as the frequency increases and reaching $-90°$ as the frequency approaches infinity.

These variations with frequency are also evident from a glance at the circuit. At low frequencies, the magnitude of the capacitor reactance is large. Hence,

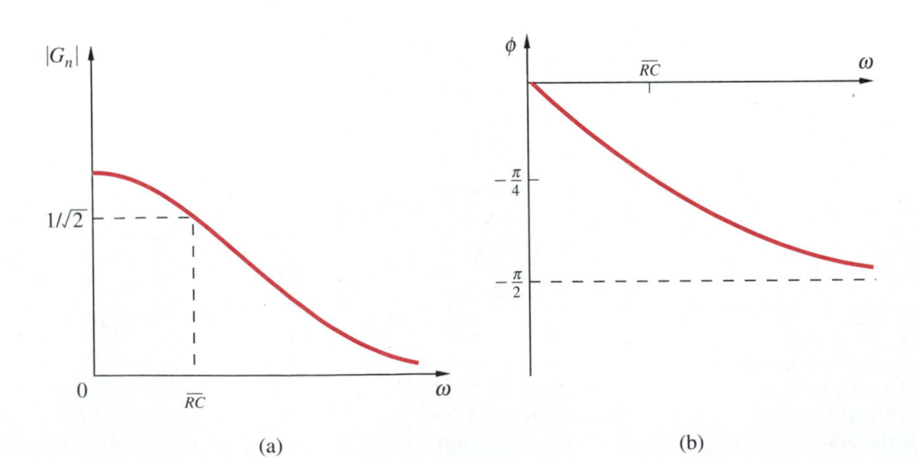

Figure 4 (a) Amplitude-response and (b) phase-response curves.

by voltage division, most of the input voltage appears across the capacitor. As the frequency increases, $1/\omega C$ decreases relative to R, so that more and more of the input voltage appears across the resistor and, thus, less and less of it across the output capacitor.

Plots of the response magnitude and angle (per unit excitation) against ω are called the *frequency-response curves.* Individually, they are called the *magnitude response* and the *phase response,* respectively.

PASSBAND AND BANDWIDTH If we look upon Figure 3 as a signal-transmission circuit, we see that sinusoidal signals having a relatively low frequency are transmitted with not much change in amplitude and with a linear variation in phase. High-frequency sinusoids, on the other hand, have their amplitudes appreciably reduced and their angles reduced by almost 90°. The circuit is, thus, *frequency-selective.* It passes certain sinusoids relatively unchanged and, to a large extent, stops others from passing. Since the signals passed are low-frequency sinusoids, we call this a *low-pass filter circuit.* Sinusoids which are "passed" are said to lie in the *passband,* and those which are "stopped" are said to lie in the *stopband.*

It is clear from Figure 4 that the variation of the magnitude of G_{21} is smooth and continuous; thus there is no particular value of ω to which we can point and say that sinusoids having a lower frequency will be passed while those having a higher frequency will be stopped. Hence, in order to give some quantitative measure of the passband, we arbitrarily pick a value of ω and say that it forms the edge of the passband. It is common practice to choose for this purpose the value of ω at which the square of the magnitude of the response is one-half its maximum value. Thus, we define the *bandwidth* β of the circuit as the *angular frequency interval* over which the square of the magnitude of the response phasor (per unit excitation) is at least one-half its maximum value.

From (2) we note that the maximum value of $|G_{21}|$ is 1 and this occurs at $\omega = 0$. To find the value of ω at which the squared magnitude of G_{21} is $\frac{1}{2}$ (or at which the magnitude itself is $1/\sqrt{2}$), we square (2) and set it equal to $\frac{1}{2}$. The result gives $\omega = 1/RC$. (Confirm this.) Thus, the passband runs from 0 to $\omega = 1/RC$, and the bandwidth is $\beta = 1/RC$ radians per second. Note from (2) that, at the edge of the passband, the angle of the voltage gain is $-45°$. The real and imaginary parts of G_{21} are equal at this frequency.

Two points should be kept in mind. The first is that the definition of bandwidth adopted is not unique. At times, the limits of the passband might be defined differently. Hence, when referring to "the bandwidth," it is necessary to specify exactly what is meant: e.g., the "point seven oh seven bandwidth," since $1/\sqrt{2} = 0.707$.[1] The second point is that we have defined the bandwidth in terms of the angular frequency. Very often, the actual frequency is used instead. This will simply introduce a factor of $1/2\pi$. Thus a bandwidth of 6283 radians per second corresponds to $6283/2\pi = 1000$ Hz.

[1] We will discover later in the chapter that the average power delivered to the circuit at the band edges so defined is one-half the maximum power. So the bandwidth is also called the *half-power* bandwidth. We will use this terminology even though power in resonant circuits will not be discussed until later.

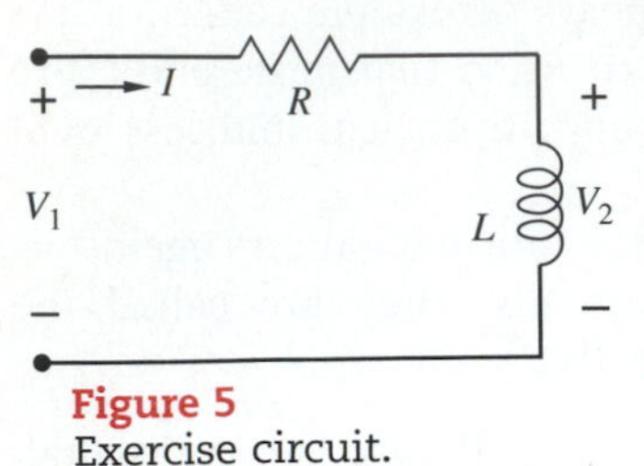

Figure 5
Exercise circuit.

EXERCISE 1

Another simple (first-order) circuit is shown in Figure 5. A sinusoidal source with phasor V_1 is applied, and the desired response phasor is V_2. (a) Find expressions for $|G_{21}|$ and ϕ in terms of the circuit components and ω, where $G_{21} = V_2/V_1 = |G_{21}| e^{j\phi}$. (b) Sketch the corresponding curves against radian frequency. (c) Compare with the response curves in Figure 4, describing the differences. (These might include behavior at low (0), high (∞), and intermediate frequencies, and maxima, minima, and slopes.) □

3 RESONANT CIRCUITS

In your study of physics, you have undoubtedly been exposed to the concept of resonance. An often-used example is a pendulum or a child's swing. If the pendulum is lightly pushed at the correct intervals (with the correct frequency), its motion can be maintained indefinitely. The external force (the excitation), it is said, "resonates" with the natural properties of the pendulum. Such resonating effects exist in electric circuits also.

Just as two properties are needed in a mechanical system for resonance to occur (mass and spring action), so, also, two analogous properties are needed for electrical resonance: inductance and capacitance. But since resistance is inevitably present in a physical inductor, we will explicitly include it. Thus, let's consider the two circuits shown in Figure 6: a series *RLC* circuit and a parallel *RLC* circuit. In the parallel circuit, it is assumed that current is the excitation and voltage the response, the converse being true in the series case. Thus, the appropriate network functions (response phasor over excitation phasor) are the impedance for the parallel circuit and the admittance for the series circuit. In both cases, the excitation is a sinusoid whose amplitude and phase are fixed but whose frequency can be varied.

The two circuits are each other's dual. Hence, the impedance of one can be obtained from the admittance of the other by interchanging L and C and replacing R in the series circuit by G in the parallel circuit. Consequently, we can treat one of the circuits in detail and apply the results to the other one simply by interchanging the appropriate parameters. In what follows, we will treat the series circuit in detail.

The plan is to study the network functions of resonant circuits and the variation of their magnitude and angle in detail as the frequency of a

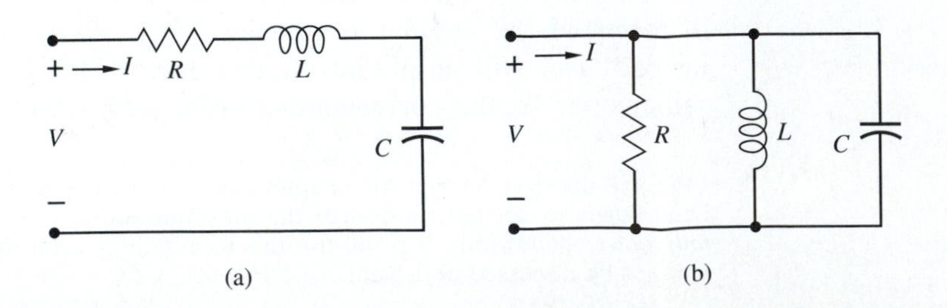

Figure 6
(a) Series and (b) parallel resonant circuits.

sinusoidal source is varied, to determine where the resonating effects are the greatest, to introduce parameters that describe and provide a measure of these effects, and to ascertain how power is distributed among the components.

Resonance in the Series Circuit

The admittance and impedance of the series circuit are:

$$
\begin{aligned}
Y &= \frac{I}{V} = \frac{1}{R + j(\omega L - 1/\omega C)} \\
Z &= \frac{1}{Y} = R + j(\omega L - 1/\omega C)
\end{aligned}
\tag{3}
$$

For our immediate purpose, even though the appropriate network function is the admittance, it is more convenient temporarily to deal with the impedance. As is clear from (3), the real part of the impedance is constant and does not vary with frequency, but the imaginary part does, as sketched in Figure 7.

Shown in this figure are plots of $X_L = \omega L$, $X_C = -1/\omega C$, and their combination. At low frequencies, the capacitive reactance predominates and X is negative. At high frequencies, the inductive reactance predominates and X is positive. Since the two reactances vary in opposite ways with frequency, there will be some frequency at which the two are equal and of opposite sign. At that frequency the two reactances will cancel, making $X = 0$. There, the impedance—and hence also the admittance—is purely real. The voltage and current at the terminals will be in phase. We say that a condition of *resonance* exists, that the circuit is *in resonance*. More precisely, we say:

> A linear circuit is in resonance at a particular frequency, called the *resonant frequency*, if the terminal voltage and current are in phase at that frequency.

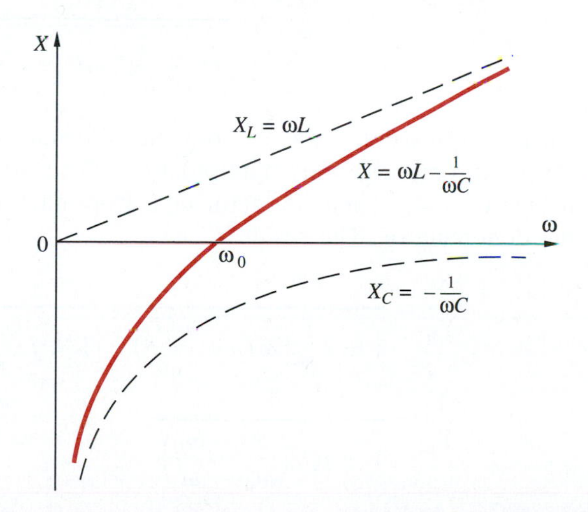

Figure 7
Reactance curve.

EXERCISE 2

(a) It is proposed that an alternative definition be given for the condition of resonance to the effect that it is the condition for which the power factor is unity. Verify that this alternative is equivalent to the original one. (b) It is also proposed that resonance be defined as the condition for which the impedance or admittance is purely real. Verify that this statement, too, is equivalent to the original one. □

Although the resonance concept was introduced via the series RLC circuit, the definition is more inclusive. And although the series circuit is in resonance at just one frequency, other circuits can conceivably be resonant at more than one frequency.

Let the resonant (angular) frequency be labeled ω_0. Then, since the impedance will be real when the voltage and current are in phase, we find from (3) that:

$$\omega_0^2 = \frac{1}{LC} \tag{4}$$

We dealt in detail with the impedance for convenience. Now let's return to the network function itself, the admittance. From (3), the magnitude of the admittance is:

$$|Y| = \frac{1}{\sqrt{R^2 + (\omega L - 1/\omega C)^2}} \tag{5}$$

As the frequency varies, the magnitude of Y will vary and will be a maximum when $\omega L = 1/\omega C$, since the denominator will then have its least value. But this is precisely the frequency at which resonance occurs, and at this frequency, $|Y| = 1/R$. If we let Y_m designate the maximum magnitude of Y, then (3) and (5) will become:

$$\begin{aligned} \frac{Y}{Y_m} &= \frac{1}{1 + j\dfrac{1}{R}\left(\omega L - \dfrac{1}{\omega C}\right)} \\ \frac{|Y|}{Y_m} &= \frac{1}{\sqrt{1 + \dfrac{1}{R^2}\left(\omega L - \dfrac{1}{\omega C}\right)^2}} \end{aligned} \tag{6}$$

These expressions can be put in alternative, more useful, forms by performing some algebraic manipulations. The first of these is to eliminate C using $1/LC = \omega_0^2$. Then we factor $\omega_0 L$ from each term within the parentheses in the denominator. The result is:

$$\begin{aligned} \frac{Y}{Y_m} &= \frac{1}{1 + j\dfrac{1}{R}\left(\omega L - \dfrac{\omega_0^2 L}{\omega}\right)} = \frac{1}{1 + j\dfrac{\omega_0 L}{R}\left(\dfrac{\omega}{\omega_0} - \dfrac{\omega_0}{\omega}\right)} \\ \frac{Y}{Y_m} &= \frac{1}{1 + jQ_0\left(\dfrac{\omega}{\omega_0} - \dfrac{\omega_0}{\omega}\right)}, \qquad Q_0 = \frac{\omega_0 L}{R} \end{aligned} \tag{7}$$

For simplicity, the collection of symbols $\omega_0 L/R$ has been given the name Q_0. This quantity plays an important role in the study of circuits and deserves the further attention we will later give it.

THE QUALITY FACTOR It turns out that what was done for simplicity in (7) has great significance. By using various forms of $\omega_0^2 = 1/LC$, we can write Q_0 in several different forms, as follows:

$$Q_0 = \frac{\omega_0 L}{R} = \frac{1}{\omega_0 CR} = \frac{1}{R}\sqrt{\frac{L}{C}} \tag{8}$$

(Confirm them.) This quantity is called the *quality factor* (or *Q-factor*). Since the frequency in the definition is the resonant frequency, this is the quality factor at resonance. Shortly we shall have much more to say about the general case. For the present, let's think of it as just a shorthand way of expressing the combination of things: $\omega_0 L/R$. Note that, at resonance, Q_0 is the ratio of the reactance of the inductor (and the capacitor, too) to the resistance.[2]

What have all these algebraic manipulations accomplished? In (3) the admittance is written in terms of the circuit elements R, L, and C. But, as the frequency is varied, we are interested in looking at the operation of the circuit as a whole, with emphasis on the behavior at the terminals. For this purpose, the internal circuit parameters are not the most meaningful things to use. Instead of R, L, and C, we have introduced a different set of three parameters in (7) to describe the circuit, namely, Y_m, ω_0, and Q_0. These are all quantities that can be measured at the terminals of the circuit, as we shall discuss shortly.

The magnitude of the relative admittance can now be written in terms of the newly defined parameters as:

$$\frac{|Y|}{Y_m} = \frac{1}{\sqrt{1 + Q_0^2(\omega/\omega_0 - \omega_0/\omega)^2}} \tag{9}$$

Consider plotting this function against frequency. Note first that, since $|Y|$ is divided by its maximum value, the value of the curve at its highest point will always be unity, no matter what the circuit parameters may be. Second, if, instead of ω, we take ω/ω_0 as the variable, then the shape of the curve will be dependent on only one parameter, Q_0. A sketch of (9) is shown in Figure 8a. When Q_0 is low, the curve is relatively broad and has a relatively flat top. But when Q_0 is high, the curve is sharper and narrower.

If the actual magnitude of Y is sketched, instead of the relative value,

[2] It is often the case that, when learners are shown some manipulations of equations that seem to lead to simpler forms, they experience a sense of frustration. "How did you know to make these particular substitutions?" they ask; "I follow what you are doing, but I would never have been able to generate that result if left on my own!" Probably so; but don't for a minute believe that the author or teacher arrived at such simplifications right off the bat. Present ways of developing a line of thought leading to rather simple equations are the result of long study and the cumulative experience of many people over long periods of time. How could you expect to create on first encounter what has taken decades of polishing to accomplish?

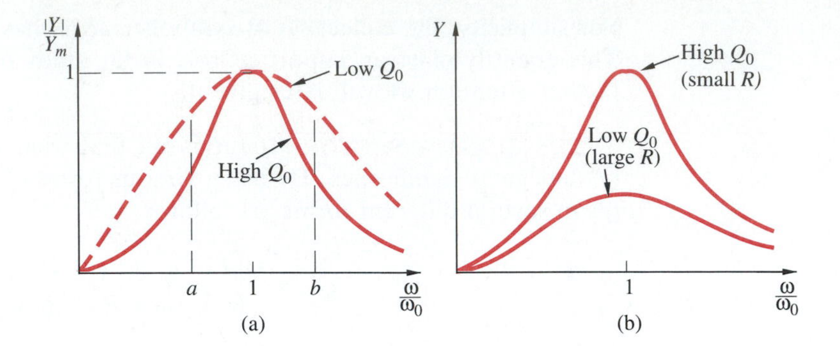

Figure 8
Variation of admittance magnitude with ω: (a) relative; (b) actual.

perhaps a better picture is obtained of the effects of Q_0. This is shown in Figure 8b. For the same inductive reactance, a low Q_0 means a large resistance, and vice versa. In the limiting case of zero resistance, Q_0 will approach infinity and the curve will become very sharp and narrow indeed.

THE UNIVERSAL RESONANCE CURVE The most interesting part of the resonance curve is the immediate vicinity of the resonant frequency. In this vicinity, the curve is controlled by the second term in the denominator of (9). This involves the difference of two terms which are both approximately equal in the vicinity of ω_0. Numerical accuracy is lost when subtracting two quantities which are nearly equal.[3] Hence, near resonance, it would be more useful to use as a variable the *change* in frequency from the resonant value, rather than the frequency itself. Let us, therefore, define the *relative frequency deviation*:

$$\delta = \frac{\omega - \omega_0}{\omega_0} = \frac{\omega}{\omega_0} - 1 \tag{10}$$

This quantity is very close to zero when ω is near the resonant value. When ω is greater than ω_0, δ is positive; when it is less, δ is negative. In terms of the relative frequency deviation, we can now write:

$$\frac{\omega}{\omega_0} - \frac{\omega_0}{\omega} = \delta + 1 - \frac{1}{\delta + 1} = \delta\left(\frac{2 + \delta}{1 + \delta}\right) \tag{11}$$

When ω is near ω_0, δ is small. In this case, the right side is approximately 2δ and the equation can be written:

$$\frac{\omega}{\omega_0} - \frac{\omega_0}{\omega} \approx 2\delta \tag{12}$$

EXERCISE 3

Divide the denominator of the right side of (11) into the numerator (or use the binomial expansion), stopping after the third term. Suppose ω ranges 1%

[3] Thus, suppose 1008 is to be subtracted from 1010 but that only three significant figures are to be retained in the original numbers. The subtraction will give 0. However, if the numbers are expressed as 1000 + 8 and 1000 + 10, emphasis is placed on the last few digits and these are not lost in the subtraction.

above and below ω_0. Compare the values given by the approximation in (12), the three-term expansion you just obtained, and the correct expression in (11), to four decimal places. Percent error?

ANSWER: 2, 1.9901, 1.9901; % error = 0.5. □

Using the preceding approximation, the expressions for the admittance and its magnitude become:

$$\frac{Y}{Y_m} = \frac{1}{1 + j2Q_0\delta}$$
$$\frac{|Y|}{Y_m} = \frac{1}{\sqrt{1 + (2Q_0\delta)^2}} \tag{13}$$

In reviewing these approximate expressions, you might wonder about the conditions under which they are valid. A glance at Figure 8 shows that the variation in the curve for a given relative change of frequency from the resonant value is much greater for the case of high Q_0 than it is for low Q_0. If we let δ vary from zero to some value, say, to the point labeled a or b in the figure, most of the variation in the high-Q_0 curve will be covered, but very little of the variation of the low-Q_0 curve. Saying this in another way, to cover most of the variation of the curve from its maximum value will require a relatively small excursion from the resonant frequency for the case of high Q_0, but it will require a relatively large excursion from the resonant frequency for the case of low Q_0.

Thus, in order to cover the important part of the curve, a larger value of δ will be required for the case of low Q_0 than for high Q_0. Since the error made in the approximation in (12) becomes progressively worse as δ increases, we see that the approximation is best for the case of high Q_0. For this reason, the approximation is called the *high-Q approximation.*

Pursuing the high-Q approximation, let us now plot the relative magnitude of the admittance from (13). If we use $Q_0\delta$ as the abscissa, the resulting curve will be applicable to all high-Q tuned circuits. This curve, called the *universal resonance curve,* is shown in Figure 9. A scale for ω/ω_0 is also shown, but this scale is not linear. It should be repeated that the universal resonance curve applies only to highly resonant circuits.

The Bandwidth

The resonant circuit we have been discussing is another example of a frequency-selective circuit. The magnitude of the current-response phasor (which is proportional to the magnitude of Y) is relatively large over an interval in the vicinity of the resonant frequency. In the present case, both low-frequency and high-frequency sinusoids are "stopped." Frequencies that are "passed" are in the vicinity of ω_0. Such a circuit is called a *bandpass* circuit. Again, a measure of the selectivity is obtained by defining the bandwidth. As before, the bandwidth is defined as the interval of angular frequency over which the magnitude squared of the response phasor exceeds one-half of its maximum value.

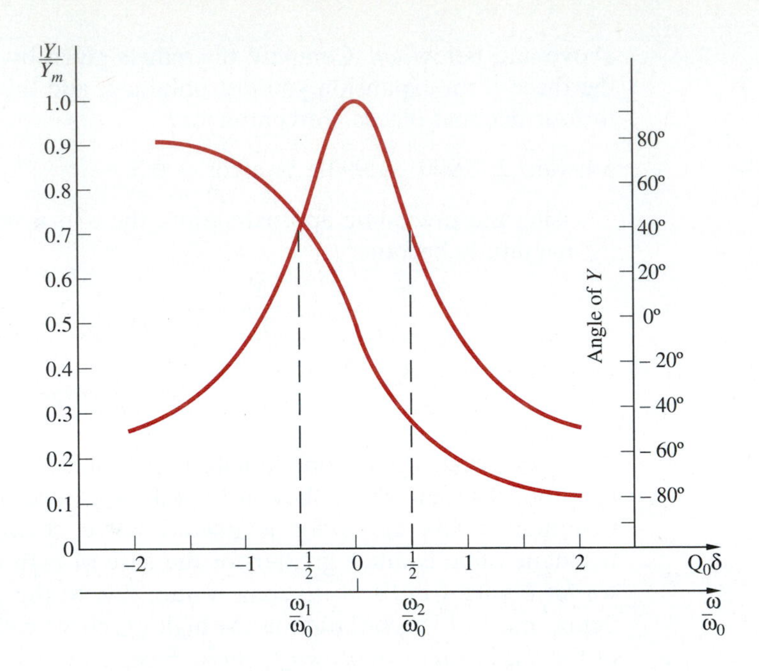

Figure 9
Universal resonance curve.

In order to find the frequency values at the edges of the band, we square $|Y|/Y_m$ and set it equal to $\frac{1}{2}$. For both the exact expression in (9) and the high-Q approximation in (13), this means that the quantity under the radical should equal 2.

Let us first deal with the exact expression in (9). Thus, at the edge of the band:

$$\frac{\omega}{\omega_0} - \frac{\omega_0}{\omega} = \pm\frac{1}{Q_0} \tag{14}$$

This can be rewritten as a quadratic in ω whose roots give two values of frequency, the lower band edge ω_1 and the upper band edge ω_2, as follows:

$$\begin{aligned} \omega_1 &= \omega_0\sqrt{1 + \left(\frac{1}{2Q_0}\right)^2} - \frac{\omega_0}{2Q_0} \\ \omega_2 &= \omega_0\sqrt{1 + \left(\frac{1}{2Q_0}\right)^2} + \frac{\omega_0}{2Q_0} \end{aligned} \tag{15}$$

(Verify these expressions.) The difference of these two frequencies is the bandwidth. Thus:

$$\beta = \omega_2 - \omega_1 = \frac{\omega_0}{Q_0} \tag{16}$$

This is a very useful expression relating the resonant frequency, the bandwidth, and the quality factor. For a given resonant frequency, the higher the Q_0, the smaller the bandwidth.

Now let's form the product of the frequencies at the edges of the passband given in (15). (Do it.) The result is:

$$\omega_1 \omega_2 = \omega_0^2 \tag{17}$$

We find that the resonant frequency is the geometric mean between the two band edges. As a matter of fact, by glancing back at the expression for the magnitude of the admittance in (9), we observe that, if ω is replaced by ω_0^2/ω, the value of Y is not changed. Thus, any two frequencies whose geometric mean equals ω_0 lead to the same value of admittance magnitude. A function possessing this property is said to have *geometric symmetry* around ω_0. It is clear from this discussion that the resonant frequency is *not* in the middle of the passband.

EXERCISE 4

A series resonant circuit has a resonant frequency of $\omega_0 = 10^5$ rad/s, a resistance of 100 Ω, and an inductance of 10 mH. Find the quality factor and the upper and lower band edges. Do the two edges of the band lie at equal distances from the resonant frequency?

ANSWER: $Q_0 = 10$, $\omega_1 = 95.125$ krad/s, $\omega_2 = 105.125$ krad/s. □

BANDWIDTH FOR HIGH-Q APPROXIMATION The preceding development on the subject of bandwidth was carried out using the exact expression for admittance. Let us now repeat this development in terms of the high-Q approximation.

The band edges are found by setting the quantity under the radical in (13) equal to 2. Then:

$$(2Q_0\delta)^2 = 1$$

$$\delta_{1,2} = \pm\frac{1}{2Q_0} \tag{18}$$

In Figure 9, on the $Q_0\delta$ scale, these two points are the ones labeled $+\frac{1}{2}$ and $-\frac{1}{2}$. The corresponding values of ω are found by inserting these values of δ in (10):

$$\omega_2 = \omega_0(1 + \delta_2) = \omega_0 + \frac{\omega_0}{2Q_0}$$

$$\omega_1 = \omega_0(1 + \delta_1) = \omega_0 - \frac{\omega_0}{2Q_0} \tag{19}$$

These are to be compared with (15). The two expressions are identical if $(1/2Q_0)^2$ under the radical can be neglected relative to 1; this is easily the case (within 1% error) for values of Q_0 as low as about 5.

Now let's compute the bandwidth by taking the difference of these two frequencies in the high-Q approximation. The interesting result is the same as (16). That is, the equation relating bandwidth, resonant frequency, and quality factor in (16) holds for the high-Q approximation as well as for the exact relationship. There is now a difference, however. The two edges of the band, ω_1 and ω_2, are centered *arithmetically,* not geometrically, about ω_0, as shown in (19). Thus, the universal resonance curve in Figure 9 has *arithmetic symmetry* about ω_0. If Q_0 is large, the exact curve having geometric symmetry very closely approximates the approximate curve having arithmetic symmetry.

EXERCISE 5

Determine the approximate upper and lower band edges in the resonant circuit given in Exercise 4. Are they equally spaced about the resonant frequency? Find the percent error relative to the resonant frequency in using the approximate band edges compared with the correct band edges. Compare the bandwidths in the two cases.

ANSWER: 105 and 95 krad/s; 0.125% error; same bandwidth. □

Rather than state the bandwidth as so many radians per second, it is more useful to express it as a fraction (or percentage) of the resonant frequency. It is not very informative, for example, to state that the bandwidth of a circuit is 1000 rad/s. This could represent a wide-band circuit if the resonant frequency happened to be 5000 rad/s, or it would be a narrow-band circuit if $\omega_0 = 10^8$ rad/s.

From the basic relationship in (16), the *fractional* (or *relative*) *bandwidth* is:

$$\frac{\beta}{\omega_0} = \frac{1}{Q_0} \tag{20}$$

The relative bandwidth is inversely proportional to the quality factor. The larger Q_0, the narrower the relative bandwidth.

As an afterthought, let's also consider the angle of the admittance, which is $-\tan^{-1} 2Q_0\delta$, shown plotted with the universal resonance curve in Figure 9. This curve is also symmetrical about the point $\delta = 0$. The angle goes from $+90°$ at $\omega = 0$ to $-90°$ at infinite ω. At the lower and upper band edges it is $\pm 45°$, respectively.

The quality factor was defined earlier as a particular relationship among circuit parameter values. From (20), we observe that Q_0 can be measured indirectly in the laboratory by measuring ω_0 and β. The measurements are accomplished by measuring the rms current in response to a sinusoidal voltage excitation of the series resonant circuit as the frequency is varied. The frequency at which the current has its maximum value is ω_0. Then we vary the frequency on either side of the resonant frequency until the current is down to $1/\sqrt{2}$ of its maximum. The difference of these two frequencies is β. Then Q_0 follows from (20).

CAVEAT ABOUT MEASUREMENTS Recall that the description of the circuit in terms of R, L, and C was replaced by a description in terms of Y_m, ω_0, and Q_0. One of the advantages claimed was that these quantities were easily measurable in the laboratory. The preceding paragraph discussed these measurements explicitly, thus confirming the earlier claim.

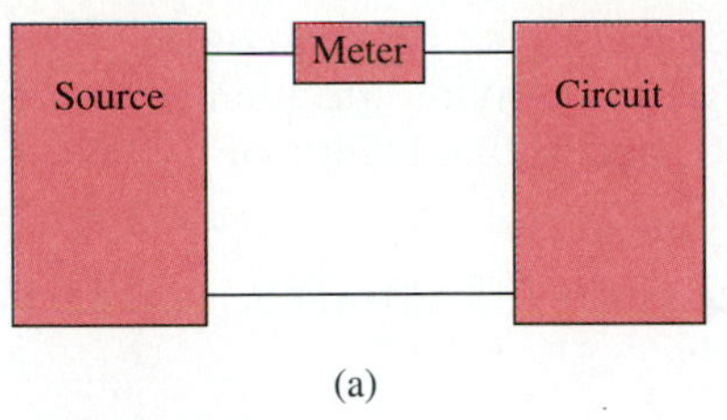

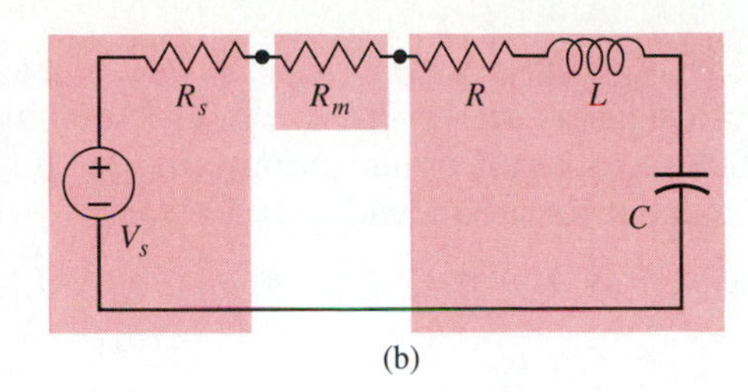

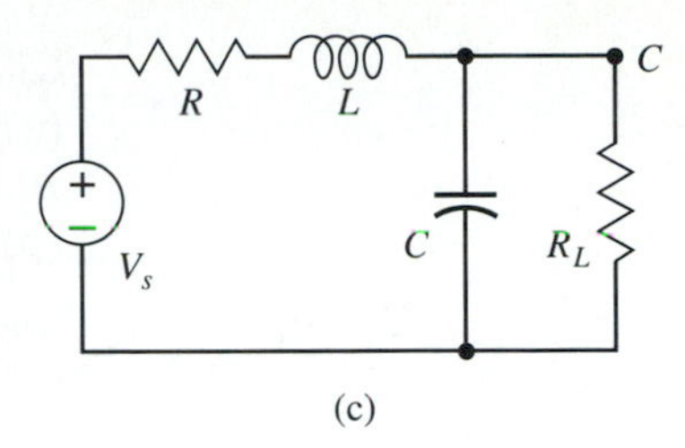

Figure 10 Laboratory measurements can introduce extraneous changes in parameter values.

A word of caution about measurements. The quantities defined in a circuit analysis (such as resonant frequency, quality factor, or maximum admittance magnitude) are always based on a specific circuit model. When measurements are made with real sources and instruments, additional components are introduced that change the parameters we thought were in the model of the circuit.

In the case of the series *RLC* circuit, the experimental arrangement would be as shown in Figure 10a. The simplest model would then take the form of Figure 10b. Both the internal resistance of the source and the resistance of the ammeter "contaminate" the original circuit. In this case the contamination is very simple; it merely adds to the existing circuit resistance. Thus, it will affect Y_m and Q_0 but not the resonant frequency. In other circuits with different experimental arrangements, the theoretical circuit can be modified in still different ways.

EXERCISE 6

The following numerical values are specified for a series resonant circuit: $\omega_0 = 50{,}000$ rad/s, $L = 2$ mH, $R = 10\,\Omega$. (a) Find the values of C, Q_0, β, and Y_m. (b) Find also the upper and lower band edges at which the response is down to $1/\sqrt{2}$ of its maximum value, using both the approximate expressions and the exact expressions. (c) By what percentage do the two values differ?

ANSWER: (a) $C = 0.2\,\mu$F, $Q_0 = 10$; $\beta = 5000$ rad/s; $Y_m = 0.1$ S; (b) $\omega_2 = 52{,}500$ (52,562) rad/s, $\omega_1 = 47{,}500$ (47,562) rad/s; (c) 0.118% (upper), 0.13% (lower). □

EXERCISE 7

The values in the preceding exercise are to be verified through measurements. The experimental setup introduces a 20% increase in the resistance. Determine the new calculated values. By what percentage are the new values different from the original ones?

ANSWER: $Q_0 = 8.33$ (16.7%); $\beta = 6000$ rad/s (20%); $Y_m = 0.0833$ (16.7%); $\omega_2 = 53{,}000$ (0.95%); $\omega_1 = 47{,}000$ (1.05%). □

EXERCISE 8

(a) Design a series *RLC* circuit to have a resonant frequency $\omega_0 =$ 10,000 rad/s, a maximum admittance of 0.1 S, and a bandwidth of 200 rad/s. (b) Repeat for a bandwidth of 500 rad/s.

ANSWER: (a) $R = 10\,\Omega$; $L = 50$ mH; $C = 0.2\,\mu$F. (b) $L = 20$ mH; $C = 0.5\,\mu$F.

□

In the situation illustrated in Figure 10a, a parameter in the resonant circuit became modified, not because we planned it but because it was incidental to some practical application, such as measuring a current and using a nonideal source as excitation. Circuit modifications can occur in other ways also. For example, the capacitor voltage in a tuned circuit may be the input to another circuit which thereby constitutes a load on the tuned circuit. The result might be the appearance of a resistor across the capacitor, as illustrated in Figure 10c. The same thing will occur if we measure the capacitor voltage and take the equivalent resistance of the voltmeter into account. In both these cases, the resistance across the capacitor would be rather large.

Two approaches can be used to deal with this problem. One approach is to deal with the new circuit as it stands, fresh; to find the corresponding impedance, and proceed to analyze it. Since interest is normally in a narrow band of frequencies around ω_0, a second approach would replace the parallel combination of *C* and the new resistor by an approximate series *RC* equivalent (see Chapter 8). This reduces to the circuit in Figure 10b, with changed values of capacitance and resistance. Of course, the validity of this equivalent will impose some limitations on the frequency band, and these must be explored. The details of these approaches will be suggested as problems for you.

Transfer Functions

The admittance is a driving-point function. With the excitation in the series circuit taken to be the voltage, we have so far assumed that the response is the current at the terminals where the voltage is applied. Let us now consider the response to be the voltage across each of the elements. The resistor voltage needs no further comment, since it is proportional to the current.

Note that the ratio of capacitor reactance to the resistance can be rewritten as:

$$\frac{1}{\omega CR} = \frac{1}{\omega_0 CR}\frac{\omega_0}{\omega} = Q_0\frac{\omega_0}{\omega} \tag{21}$$

The ratio of capacitor voltage to the input voltage can be obtained from the voltage-divider relationship as the ratio of the capacitor impedance to the total impedance (with *Z* replaced by $1/Y$), again replacing circuit parameters by their equivalences in terms of ω_0, Q_0, and Y_m:

$$\begin{aligned}\frac{V_C}{V} &= \frac{1}{j\omega C}Y = -j\frac{1}{\omega C}Y = -j\frac{1}{\omega CR}\frac{Y}{Y_m}\\ &= \frac{-jQ_0(\omega_0/\omega)}{1 + jQ_0(\omega/\omega_0 - \omega_0/\omega)}\end{aligned} \tag{22}$$

Confirm this result using equations (7) and (21). The angle of this transfer function *lags* the angle of the admittance by a constant 90°.

Similarly, the transfer voltage ratio of the inductor is again obtained from the voltage-divider relationship; then some algebraic manipulations are performed, to obtain:

$$\frac{V_L}{V} = j\omega LY = j\frac{\omega}{\omega_0^2 C}Y = j\left(\frac{\omega}{\omega_0}\right)^2 \frac{1}{\omega CR}\frac{Y}{Y_m}$$

$$= \frac{jQ_0(\omega/\omega_0)}{1 + jQ_0(\omega/\omega_0 - \omega_0/\omega)} \tag{23}$$

(Don't go on without confirming this.) This time, the angle *leads* that of the admittance by a fixed 90°.

Taking the ratio of these two voltages gives further insight:

$$\frac{V_L}{V_C} = -\left(\frac{\omega}{\omega_0}\right)^2$$

The two voltages differ in phase by 180° at all frequencies. They are equal in magnitude at the resonant frequency, but the inductor voltage has a larger magnitude above resonance and a smaller magnitude below resonance.

Something interesting can be observed in the capacitor voltage and the inductor voltage at resonance. At $\omega = \omega_0$, the transfer functions are:

$$\left.\frac{V_C}{V}\right|_{\omega=\omega_0} = -jQ_0$$

$$\left.\frac{V_L}{V}\right|_{\omega=\omega_0} = jQ_0 \tag{24}$$

In view of the voltage-divider concept, the normal expectation is that the voltage across one of a number of impedances in series is less than the input voltage, since the sum of all the series voltages must add up to the input voltage. However, this notion must be modified when the variables are phasors, which are complex. The sum of two complex numbers can be small even when each one has a large magnitude. An example is $1 + j1000$ and $1 - j1000$, whose sum is only 2. In the series circuit, at the resonant frequency, the magnitudes of the capacitor and inductor voltages are larger than the input voltage magnitude by a factor of Q_0, which can be quite high for high-Q circuits. You can get the shock of your life if you put your hand across a capacitor in a resonant circuit under these circumstances, even for a relatively small input voltage!

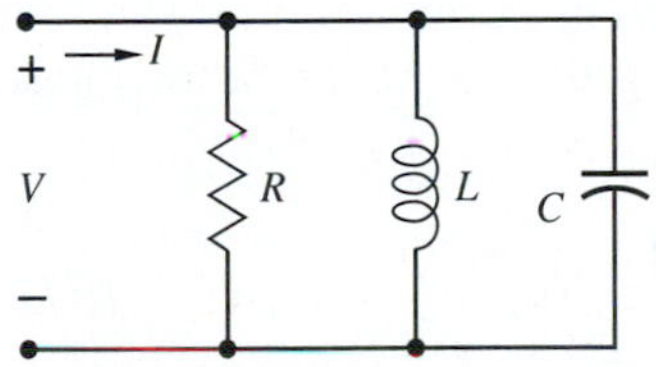

Parallel Resonant Circuits

A parallel *RLC* circuit which is the dual of the series circuit was shown in Figure 6b and is repeated in Figure 11. The excitation here is the current, and the appropriate driving-point function is the impedance. The expression for the impedance will be identical with the expression for the admittance of the series circuit if we interchange *L* and *C* and replace *R* by *G*. Note that, although the

Figure 11
Parallel resonant circuit.

same R, L, and C designations are used, there is no implication that these have the same *values* as in the series circuit.

The entire preceding development of the series resonant circuit will apply to the parallel circuit with an appropriate interchange of voltage and current, admittance and impedance, and so on. The resonant frequency is again defined as the frequency at which the response and excitation are in phase, leading to $\omega_0^2 = 1/LC$, just as before. The response, which is now the voltage, again has a maximum magnitude at this frequency. The bandwidth is defined once again as the frequency interval over which the square of the magnitude is not less than half its maximum value. And so on.

The quality factor—which is now labeled Q_P—is obtained by replacing the element values in the previous definition by their duals: L by C, C by L, and R by G. Thus, instead of $\omega_0 L/R$, we get:

$$Q_P = \frac{\omega_0 C}{G} = \frac{R}{\omega_0 L} = R\sqrt{\frac{C}{L}} \tag{25}$$

By comparing this expression with (8) for Q_0, we find that the quality factor of the parallel circuit in terms of its elements is just the reciprocal of what it is for the series circuit. Note carefully, though, that this does not mean the actual numerical values are reciprocals, since the numerical values of the elements are not necessarily the same in the two circuits.

Letting Z_m be the maximum magnitude of the impedance of the parallel circuit, the expression for the impedance can be written:

$$\frac{Z}{Z_m} = \frac{1}{1 + jQ_P(\omega/\omega_0 - \omega_0/\omega)} \tag{26}$$

(If you aren't convinced by the duality argument, you should derive this result by writing an expression for the impedance in terms of the element values and then use the definitions of ω_0 and Q_P.)

The form of the preceding expression for the impedance is identical with that of the admittance of the series circuit in (7). Hence, the universal resonance curve in Figure 9 will apply to the parallel circuit too, except that the ordinate must now be labeled impedance rather than admittance.

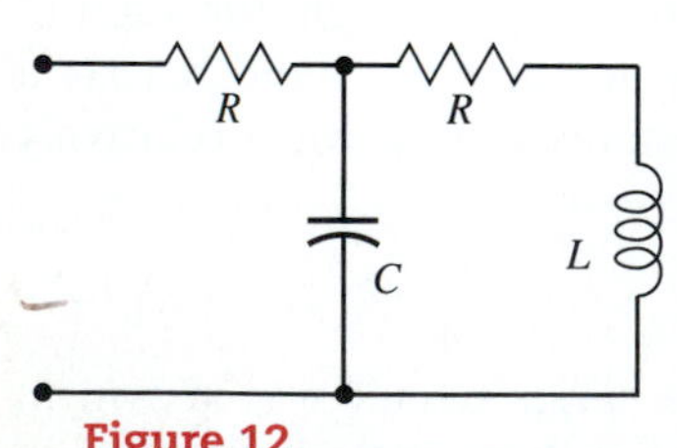

Figure 12
Circuit for finding ω_0.

EXERCISE 9

(a) Find the driving-point impedance in the circuit in Figure 12. (Continued fractions may be the easiest way.) (b) Put the result in the form of two quadratics in ω. (c) Find the resonant frequency in terms of the element values. (d) Show that the terminal voltage and current can never be in phase unless $RC < L/R$.

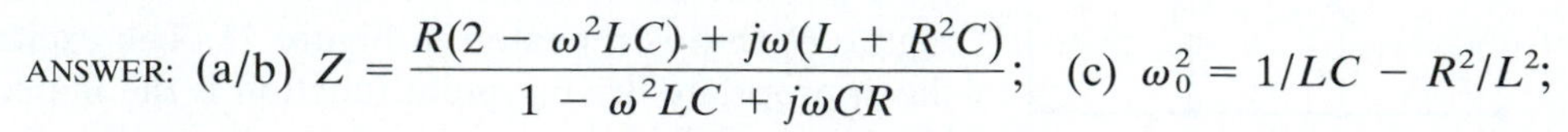

ANSWER: (a/b) $Z = \dfrac{R(2 - \omega^2 LC) + j\omega(L + R^2 C)}{1 - \omega^2 LC + j\omega CR}$; (c) $\omega_0^2 = 1/LC - R^2/L^2$;

(d) $(\omega/\omega_0)^2 > 0$. □

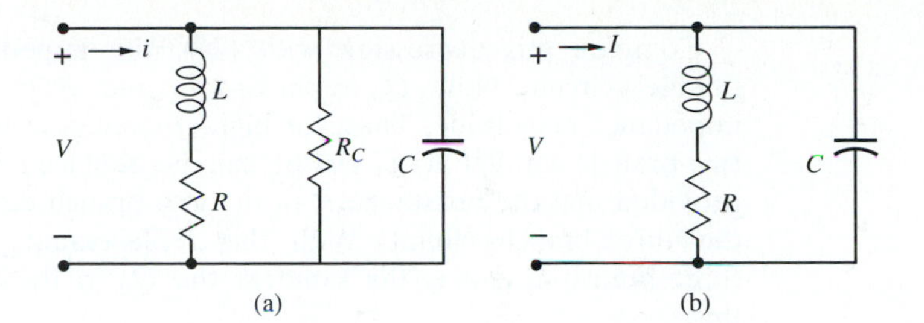

Figure 13
More realistic parallel resonant circuit.

TWO-BRANCH PARALLEL RESONANT CIRCUIT When we considered the natural response of second-order circuits in Chapter 7, the parallel RLC circuit in particular, we commented on the parallel connection of a real coil of wire and a physical capacitor. A model of this arrangement is shown in Figure 13a. The two resistors represent the winding resistance of the coil and the leakage resistance of the capacitor. Ideally, the latter is very large, almost an open circuit; while the winding resistance is very small, almost a short circuit. The effect of neglecting the leakage resistance is much smaller than that of neglecting the winding resistance. Hence, an adequate circuit to represent this physical arrangement is shown in Figure 13b. It might be called a *two-branch parallel resonant circuit.*

The impedance of this circuit is readily found; it can then be manipulated, as follows:

$$\begin{aligned} Z &= \frac{1}{j\omega C + 1/(R + j\omega L)} = \frac{R + j\omega L}{j\omega C(R + j\omega L) + 1} \\ &= \frac{1 + R/j\omega L}{\dfrac{RC}{L} + j\omega C + \dfrac{1}{j\omega L}} = \frac{L}{RC}\frac{1 - j(R/\omega L)}{1 + j\dfrac{1}{R}\left(\omega L - \dfrac{1}{\omega C}\right)} \\ &= \frac{RQ_0^2\left(1 - j\dfrac{\omega_0}{Q_0\omega}\right)}{1 + jQ_0\left(\dfrac{\omega}{\omega_0} - \dfrac{\omega_0}{\omega}\right)} \end{aligned} \tag{27}$$

The right-hand side of the first line is obtained by multiplying both numerator and denominator by $R + j\omega L$. In the second line we first divide numerator and denominator by $j\omega L$ and then factor RC/L from the denominator. The final result is obtained by using the definitions of $\omega_0^2 = 1/LC$ and $Q_0 = \omega_0 L/R$. (Perform each of these steps and confirm the result.)

From this expression we note that the impedance is not purely real when $\omega = \omega_0$, which is the frequency at which the denominator is purely real. At that frequency, the numerator has an imaginary part which equals $-1/Q_0$. For high-Q circuits, this is quite small relative to the real part, so that the impedance is almost real at $\omega = \omega_0$. For high-Q circuits, then, for frequencies in the vicinity of ω_0, the impedance can be written approximately as:

$$Z \approx \frac{RQ_0^2}{1 + jQ_0(\omega/\omega_0 - \omega_0/\omega)} \tag{28}$$

Compare this expression with (26), the impedance of the three-branch parallel circuit. Now Q_0 replaces Q_P and RQ_0^2 is the maximum of the impedance magnitude. Thus, for high-Q circuits in the vicinity of resonance, a two-branch parallel RLC circuit can be replaced by a three-branch circuit, provided that the resistance R in the two-branch circuit is replaced by Q_0^2R in the three-branch circuit. With this replacement, we find that Q_P in the three-branch circuit is the same as the Q_0 in the two-branch circuit. (Show this.)

If the impedance is not real at ω_0, so that the terminal voltage and current are not in phase at that frequency, then what exactly is the resonant frequency in the two-branch circuit? In order to answer that question, we must find the frequency at which the impedance in (27) is real. This will occur when the angle of the numerator equals the angle of the denominator, making the net angle zero. If the angles are equal, then so are their tangents. But the tangents of the angles are the ratios of the imaginary to the real parts in the numerator and denominator. Hence:

For unity power factor:

$$-\frac{\omega_0}{Q_0\omega} = Q_0\left(\frac{\omega}{\omega_0} - \frac{\omega_0}{\omega}\right)$$
$$\omega = \omega_0\sqrt{1 - \frac{1}{Q_0^2}} \tag{29}$$

For high-Q circuits, the frequency for unity power factor is seen to differ very little from ω_0.

How about the frequency at which the magnitude of Z is a maximum? Conceptually, that's easy to find out. We should write an expression for $|Z|$, differentiate it with respect to ω, then set the derivative equal to 0, and solve the resulting equation for ω. Although conceptually simple, the detailed mathematical manipulations are quite tedious. In your spare time you should confirm the following result.

For maximum $|Z|$:

$$\omega = \omega_0\sqrt{\sqrt{1 + \frac{2}{Q_0^2}} - \frac{1}{Q_0^2}} \approx \omega_0\sqrt{1 - \frac{1}{2Q_0^2}} \tag{30}$$

The right side is obtained by expanding the inner radical by the binomial theorem and retaining only the first three terms.

For engineering purposes, the frequency ω_0, the frequency for unity power factor, and the frequency for maximum $|Z|$ can all three be considered practically identical in the case of high-Q circuits. For Q_0 as low as 4, for example, the last two equations give:

For unity power factor: $\omega/\omega_0 = 0.969$
For maximum $|Z|$*:* $\omega/\omega_0 = 0.999$

In the latter case, the error is just 0.1%.

4 NORMALIZATION

If all the numerical computations you would ever carry out in circuit analysis and design were to be done by a digital computer, the size of component and frequency values would be of little concern. However, you will often want to get a feel for the behavior of a circuit, possibly at the initial stages of a design, and will do some hand calculations. On such occasions, it would be very useful if the numerical values involved were convenient, small values rather than many kilohms or megahertz or picofarads.

In the preceding discussion of resonant circuits, we found it useful to refer the admittance magnitude to its maximum value. Thus, the ordinate used in plotting the universal resonance curve in Figure 9 is $|Y|/Y_m$. Similarly, the frequency scale for that curve was taken as ω/ω_0. This process of expressing a quantity relative to some standard value is called *normalizing* or *scaling*.

Normalizing permits dealing with convenient, small numbers in circuit computations. If the frequencies of interest are in the Mrad/s range, define a new frequency variable as the old one over 10^6. Then the new frequency variable will vary near unity rather than near 10^6.

Consider the impedance of a series resonant circuit, say, in (3); now multiply and divide each frequency-dependent term by some standard scaling frequency ω_s:

$$
\begin{aligned}
Z &= R + j\omega L + \frac{1}{j\omega C} \\
&= R + j\left(\frac{\omega}{\omega_s}\right)(\omega_s L) + \frac{1}{j(\omega/\omega_s)(\omega_s C)}
\end{aligned}
\tag{31}
$$

This maneuver of multiplying and dividing by some standard frequency leaves the value of the expression unchanged.

Now suppose both sides of the equation are divided by some constant resistance, say, R_s, which is an impedance-scaling factor. This, too, will leave the expression unchanged:

$$
\begin{aligned}
\frac{Z}{R_s} &= \frac{R}{R_s} + j\left(\frac{\omega}{\omega_s}\right)\left(\frac{\omega_s L}{R_s}\right) + \frac{1}{j(\omega/\omega_s)(\omega_s R_s C)} \\
Z_n &= R_n + j\omega_n L_n + \frac{1}{j\omega_n C_n}
\end{aligned}
\tag{32}
$$

where the subscript n stands for "normalized" and the normalized quantities are:

$$
\begin{aligned}
Z_n &= \frac{Z}{R_s} \qquad L_n = \frac{\omega_s L}{R_s} \\
\omega_n &= \frac{\omega}{\omega_s} \\
R_n &= \frac{R}{R_s} \qquad C_n = \omega_s R_s C
\end{aligned}
\tag{33}
$$

What is the net effect of this normalization? The form of Z_n is the same as that of the original Z. That is, the variation of Z_n with frequency ω_n is the same as the variation of Z with frequency ω. But the scale of values of Z_n is $1/R_s$ times the scale of values of Z. We say that the *impedance level* is scaled, the normalizing (or scaling) factor being R_s. Note that, when we change the impedance level in a circuit by a factor R_s, *all* impedances in the circuit are divided by R_s. For a capacitor, the impedance is inversely proportional to the capacitance. Hence, when we divide the impedance of the capacitor by R_s, this has the effect of *multiplying* the capacitance, as shown in (33). The element values without subscript are the usual values. R_s and ω_s, respectively, are the impedance- and frequency-normalizing factors.

What is the effect of a change in impedance level on a voltage gain or current gain? In the series resonant circuit, for example, consider the ratio of capacitor voltage to source voltage. This equals the ratio of capacitor impedance to the total impedance. Hence, a change of impedance level will have no effect on the voltage gain, since the normalizing constant will cancel from the ratio of impedances. The same is true for the current gain. (Take a specific example, say, the parallel resonant circuit, and confirm this.)

EXAMPLE 1 Let the element values in a series *RLC* circuit be $R = 80\ \Omega$, $L = 0.16$ mH, and $C = 0.025\ \mu$F, leading to a resonant frequency of $1/(LC)^{1/2} = 10^6/2$ rad/s. If we choose an impedance-normalizing factor of $R_s = R = 80$ and a frequency-normalizing factor equal to the resonant frequency $\omega_0 = 10^6/2$, the normalized element values become $R_n = 1$, $L_n = 1$, and $C_n = 1$. Isn't it a lot easier to carry out numerical calculations with such numbers than with the original ones? (Confirm the normalized values.)

It is tempting (and some people succumb to the temptation) to ascribe to these normalized element values the units of ohms, henries, and farads, respectively. That would be incorrect. Looking back at (33) shows that the resistance in ohms has been divided by another resistance, also in ohms; hence, the normalized value is dimensionless. Similarly, the inductive and capacitive reactances in ohms have been divided by a resistance, leaving a dimensionless quantity behind. Indeed, if $R_n = R/R_s$ is to have the dimensions "ohms," then L_n should also have the dimension "ohms," not "henries," since L_n is the normalized value of inductive reactance, normalized by the same scaling factor R_s that R is normalized by.

When realistic parameter values are specified in a given circuit presented for analysis, we *choose* impedance- and frequency-normalizing factors to arrive at normalized parameter values. On the other hand, in the process of synthesis and design, *before* a circuit has been obtained, we might choose a significant frequency (such as a resonant frequency or the edge of a band) as the normalizing factor, so that the normalized value of this frequency becomes 1 (dimensionless). Similarly, a significant impedance (such as a load resistance) is chosen as the normalizing factor, making its normalized value 1. Then, when the circuit has been designed, the resulting parameter values are the normalized ones: R_n, L_n, and C_n. To obtain the proper, realistic component values, we must now *de*normalize. This is the opposite step from normalizing; that is, we invert the relationships in (33). Thus, the denormalized—that is, the actual—values in ohms, henries, and farads are:

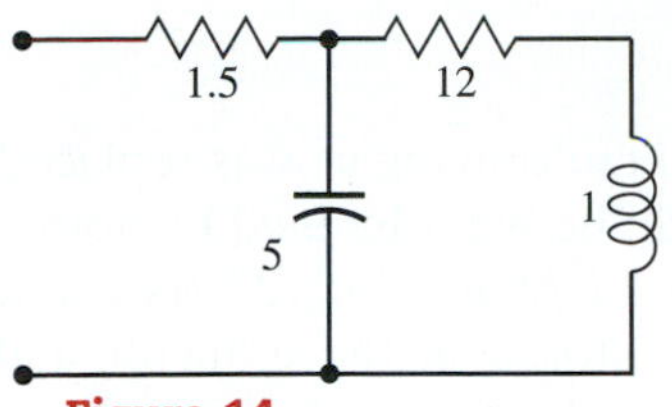

Figure 14
Normalized circuit to be denormalized.

$$R = (R_s)R_n \ \Omega \qquad L = \left(\frac{R_s}{\omega_s}\right)L_n \ \text{H} \qquad C = \left(\frac{1}{\omega_s R_s}\right)C_n \ \text{F} \tag{34}$$

EXERCISE 10

The circuit in Figure 14 has been designed on the basis of impedance- and frequency-normalizing factors of $R_s = 1000\ \Omega$ and $\omega_s = 10^5$ rad/s. Specify the actual component values and give their units.

ANSWER: Left to right: 1.5 kΩ, 0.05 μF, 12 kΩ, 10 mH. □

5 POWER AND ENERGY IN RESONANT CIRCUITS

What we have done in the preceding sections is to introduce the phenomenon of resonance. We used the three-branch series *RLC* circuit as the prototype, but, by duality, the same results apply to the three-branch parallel resonant circuit also. We studied in detail the variation with frequency of the admittance (for the series circuit) and the impedance (for the parallel circuit).

We will now turn to a study of energy storage and dissipation in a resonant circuit. For convenience, we will deal in detail with the series circuit, but, by duality, the same results will apply to the parallel circuit as well. The circuit is repeated in Figure 15. We will initially deal with the variables as functions of time because some of the results apply generally, not only to the sinusoidal steady state.

Power Supplied by Arbitrary Excitation

The immediate objective is to find the power supplied to the circuit by the excitation as a function of time. Contemplate writing KVL around the loop and then eliminating the voltages of the resistor and inductor—but not the capacitor—using the v-i relationships of R and L. The result will be:

$$v = Ri + L\frac{di}{dt} + v_C$$

The power input at any instant of time is the product of v and i; therefore,

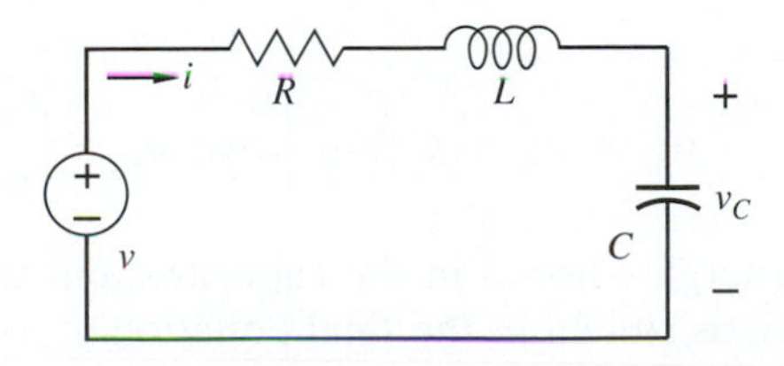

Figure 15
Energy storage and dissipation in a resonant circuit.

multiply both sides of this equation by i, perform some operations, and get:

$$p = vi = Ri^2 + Li\frac{di}{dt} + v_C i = Ri^2 + Li\frac{di}{dt} + Cv_C\frac{dv_C}{dt}$$
$$= Ri^2 + \frac{d}{dt}\left(\frac{1}{2}Li^2 + \frac{1}{2}Cv_C^2\right) \tag{35}$$

In the step before the last one in the first line, the i multiplying v_C was replaced by the i-v relationship of the capacitor. The second line was obtained by noting that $2y\,dy/dx = d(y^2)/dx$. What are the quantities $Li^2/2$ and $Cv_C^2/2$? As you'd expect, they represent the instantaneous energies stored in the inductor and capacitor, respectively. Finally, using w_L and w_C for these energies, we obtain the following:

$$p = vi = Ri^2 + \frac{d}{dt}(w_L + w_C) \tag{36}$$

This equation makes a very important statement, namely, that the power entering the circuit is equal to the power dissipated in the resistor plus the rate of change of the energy stored in the capacitor and the inductor. If we were to integrate this equation over an interval of time, the result would be a statement of the conservation of energy. Although the preceding equation was derived for one particular circuit, it is clearly the statement of a general principle that applies to any linear circuit.

Power and Energy under Sinusoidal Excitation

The preceding section is valid for any excitation waveform. Let us now specify that the excitation is sinusoidal. For convenience, we will choose the capacitor voltage as the reference for phase, so we can write:

$$v_C(t) = \sqrt{2}\,|V_C| \cos \omega t$$
$$i(t) = C\frac{dv_C}{dt} = -\sqrt{2}\omega C\,|V_C| \sin \omega t \tag{37}$$

Using these expressions, the energy stored in the capacitor and that stored in the inductor at any instant of time are as follows:

$$w_C = \tfrac{1}{2}Cv_C^2 = C\,|V_C|^2 \cos^2 \omega t$$
$$w_L = \tfrac{1}{2}Li^2 = L\omega^2C^2\,|V_C|^2 \sin^2 \omega t$$
$$= \left(\frac{\omega}{\omega_0}\right)^2 C\,|V_C|^2 \sin^2 \omega t \tag{38}$$

$$w_C + w_L = C\,|V_C|^2\left[\cos^2 \omega t + \left(\frac{\omega}{\omega_0}\right)^2 \sin^2 \omega t\right] \tag{39}$$

Adding the energies stored in the capacitor and the inductor leads to the total energy stored, as shown in the final equation.

Since $\cos^2 \omega t$ and $\sin^2 \omega t$ have the same maximum value, namely, 1, the maximum value of w_L is $(\omega/\omega_0)^2$ times the maximum value of w_C. Hence, when the frequency is below the resonant frequency, the maximum stored energy in the inductor is less than that in the capacitor, whereas for frequencies higher than the resonant frequency, these maxima are reversed. Exactly at resonance, the energy stored in the inductor has the same maximum value as the energy stored in the capacitor. As a matter of fact, at resonance, the total energy stored is found from the preceding equation to be a constant for all time:

$$(w_C + w_L)\big|_{\omega=\omega_0} = C\,|V_C|^2 = CQ_0^2\,|V|^2 \tag{40}$$

The right-hand side follows from (24).

Since the total stored energy at resonance is a constant but the individual energies stored in L and C vary as the square of a sinusoid, the energy in the circuit surges between the inductor and the capacitor; no energy is supplied by the excitation (except for the energy dissipated in the resistor). This agrees with the fact that the admittance is real at resonance, so that the voltage and current at the terminals are in phase; there is no reactive power at the terminals.

EXERCISE 11

In a series resonant circuit, $R = 100\,\Omega$, $L = 40$ mH, and $C = 0.01\,\mu$F. The source is a 100-V, variable-frequency sinusoidal source. (a) Find the total energy stored in the circuit at the resonant frequency. (b) Find the maximum of the total energy stored at the upper edge of the band, and (c) at the lower edge of the band.

ANSWER: (a) 0.04 J; (b/c) 0.0205 J. □

COMPLEX POWER UNDER SINUSOIDAL CONDITIONS The entire preceding development on power was carried out in the time domain, first for arbitrary time functions and then for sinusoids. Further information can be obtained by repeating the analysis, this time in terms of phasors. The complex power entering the circuit can be found by first writing the loop equation in terms of phasors and then multiplying both sides by the conjugate of the current phasor. Thus:

$$V = RI + j\omega LI - j\frac{1}{\omega C}I$$

$$\begin{aligned} S = VI^* &= R\,|I|^2 + j\left(\omega L\,|I|^2 - \frac{1}{\omega C}|I|^2\right) \\ &= R\,|I|^2 + j\omega(L\,|I|^2 - C\,|V_C|^2) \\ &= P + j2\omega(W_L - W_C) \end{aligned} \tag{41}$$

The third line follows by noting that $|I| = \omega C\,|V_C|$. The last line follows by using equations (13) and (14) from Chapter 9 for the average energy W_L stored in the inductor and W_C in the capacitor. (Confirm each of the steps.)

The last expression tells us that the reactive power is proportional to the difference between the average energy stored in the inductor and the average energy stored in the capacitor. These average stored energies are related by:

$$W_L = \left(\frac{\omega}{\omega_0}\right)^2 W_C \tag{42}$$

This can be verified either by finding the average values from the instantaneous values in (38) or by glancing back at (13) and (14) in Chapter 9 for the average stored energies. Finally, inserting this expression into (41), the complex power becomes:

$$S = P + j2\omega W_C\left[\left(\frac{\omega}{\omega_0}\right)^2 - 1\right] \tag{43}$$

When the frequency is above resonance, the reactive power is positive; below resonance, it is negative. Exactly at resonance, the reactive power vanishes, as we had observed previously.

The expression for the complex power in (41) was derived for the series resonant circuit under study. We will parenthetically note that this is a general relationship, valid for all linear circuits if R is interpreted as the real part of the driving-point impedance, W_L as the average energy stored in all the inductances, and W_C as the average energy stored in all the capacitances.

EXERCISE 12

In a series resonant circuit, the element values are $R = 100\,\Omega$, $L = 40\,\text{mH}$, and $C = 1\,\mu\text{F}$. The steady-state current resulting from a sinusoidal voltage excitation at the resonant frequency is $i(t) = 0.01\sqrt{2}\cos\omega t$ A. (a) Determine the maximum energy stored in the capacitor when the frequency is the resonant frequency. (b) Determine the reactive power supplied to the circuit by the source at the upper and lower band edges. Specify whether the reactive power is inductive or reactive in each case.

ANSWER: (a) $4\,\mu\text{J}$; (b) Upper: $Q = 8.37\,\text{mVAR}$ inductive; lower: $Q = -9.88\,\text{mVAR}$ capacitive. □

AVERAGE POWER DISSIPATED Concentration so far has been on the energy stored in the circuit. Let us now briefly examine the average power dissipated in the resistor. In terms of the phasor current, this is $P = R\,|I|^2$. Using $|I| = \omega C\,|V_C|$ and then $RC = 1/Q_0\omega_0$, the average power becomes:

$$P = R\,|I|^2 = R\omega^2C^2\,|V_C|^2 = \frac{\omega_0}{Q_0}\left(\frac{\omega}{\omega_0}\right)^2 C\,|V_C|^2 \tag{44}$$

Also, note from (38) that $C\,|V_C|^2\,(\omega/\omega_0)^2$ is the maximum value of w_L, the energy stored in the inductor. Using this fact in the preceding equation yields:

$$P = \frac{\omega_0}{Q_0} w_{L\,(\text{max})} \tag{45}$$

Each of the last two equations gives different representations of the average power dissipated in the circuit. But why stop here? Another viewpoint leads to yet another representation. In the expression $P = R\,|I|^2$, let's substitute $I = YV$. Then:

$$P = R\,|I|^2 = R\,|V|^2\,|Y|^2$$
$$P_m = R\,|V|^2\,Y_m^2 \tag{46}$$

The last line follows from the observation that the dissipated power is a maximum (here labeled P_m) at resonance, since the admittance, and so also the current, is a maximum there.

The ratio of the power at any frequency to the power at resonance is then:

$$\frac{P}{P_m} = \frac{|Y|^2}{Y_m^2} \tag{47}$$

This is a very useful expression and gives a new way of specifying the bandwidth. You will recall that the bandwidth was defined as the frequency at which the square of the admittance magnitude is half the square of its maximum value. We now see from this equation that, at the band edges where $|Y| = Y_m/\sqrt{2}$, the average power dissipated in the circuit is half its maximum value, which is the value at resonance. For this reason, the previously defined edges of the band are called the upper and lower *half-power points* and the bandwidth is referred to as the *half-power bandwidth.*

General Definition of Quality Factor

Equation (45) expresses a relationship among Q_0 and some fundamental quantities: the resonant frequency of a circuit, the power dissipated, and the maximum energy stored. Up to this point, remember that Q_0 has simply been a symbol to stand for a combination of other symbols: $Q_0 = \omega_0 L/R$. But (45) permits a more fundamental reinterpretation of this quantity. Solving (45) for Q_0 and using a word description of the result yields:

$$Q_0 = \omega_0 \frac{\text{maximum energy stored in inductor}}{\text{average power dissipated}}$$
$$= 2\pi \frac{\text{maximum energy stored in inductor}}{\text{energy dissipated per cycle}} \tag{48}$$

The last line follows from the first, since ω_0 is 2π divided by the period and the energy dissipated in one period (or one cycle) is the average power dissipated times the period.

Although we have derived these expressions for the series resonant circuit, they are stated in quite general terms and apply to any linear circuit. They give a measure of the quality of a circuit as a storer of energy as compared with its ability to dissipate it. We could have adopted the preceding expression earlier in the chapter as the definition of a quality factor Q at any frequency, of which Q_0 is the value at resonance. We could then have *derived* the expressions for Q_0 with which we started in (8). It's never too late, so let's do it now. Let's make the following fundamental definition of the *quality factor* of a circuit:

$$
\begin{aligned}
Q &= \omega \frac{\text{maximum energy stored}}{\text{average power dissipated}} \\
&= 2\pi \frac{\text{maximum energy stored}}{\text{energy dissipated per cycle}}
\end{aligned}
\tag{49}
$$

From these expressions it is clear that Q is a function of frequency. By substituting expressions for the numerator and denominator valid at $\omega = \omega_0$, we can *derive* $Q_0 = \omega_0 L/R$.[4] Thus, at the resonant frequency, Q takes on the value Q_0.

EXERCISE 13

Use the general definitions of the quality factor in the preceding equations to derive the expression for the Q of a parallel resonant circuit given in (25). □

SUMMARY

- Network functions as ratios of response phasors to excitation phasors; driving-point and transfer functions
- Frequency response: amplitude (magnitude) and phase responses
- Frequency response of low-pass RC circuit
- Half-power or .707 bandwidth
- Series resonant and parallel resonant circuits
- Resonant frequency
- The quality factor Q for series and parallel resonant circuits
- Relationship among bandwidth, Q, and resonant frequency
- Geometric symmetry and arithmetic symmetry
- Universal resonance curve
- Bandwidth for high-Q approximation
- Capacitor voltage in a high-Q resonant circuit
- Resonance in a two-branch parallel resonant circuit
- Interchange of energy between L and C in a resonant circuit
- Power dissipated in terms of the maximum energy stored
- Reactive power in terms of energy stored in L and C
- Magnitude and frequency normalization and denormalization

PROBLEMS

1. A series resonant circuit has the following element values: $R = 5\,\Omega$, $L = 400\,\mu\text{H}$, $C = 100\,\text{pF}$.
   a. Compute the resonant frequency.
   b. Compute the quality factor Q_0.
   c. Compute the magnitude and angle of the impedance of the circuit at a frequency 1% above resonance.
   d. Construct a two-element series circuit equivalent to this circuit at this frequency, including component values.

[4] *Beware*! The symbol Q has been used to stand for quality factor and also for reactive power. That's not all; it has also been used for electric charge. These three quantities don't usually come up in the same discussion, so there should be no reason for excessive confusion on this score.

2. A series RLC circuit with $C = 10\,\text{pF}$ is to be resonant at 2 Mrad/s. Its bandwidth between half-power points is to be 6 krad/s. Find the inductance L and the minimum magnitude of the impedance.

3. Design a series RLC circuit (specify the element values) which is to be resonant at 1 Mrad/s. Its half-power bandwidth is to be 5 krad/s and its impedance at resonance is to be $50\,\Omega$. Would you describe this as a low-Q or high-Q circuit? Specify the exact upper and lower band edges. How are these frequencies related to the resonant frequency? Specify the band edges using the approximate relationship; how are *they* related to ω_0?

4. A series RLC circuit, with $Q_0 = 250$, is resonant at 1.5 Mrad/s.

a. Find the half-power frequencies.

b. Find the frequencies at which the average power supplied to the circuit is 10% of the power supplied at resonance, assuming the input voltage stays constant as the frequency is varied.

5. In a parallel resonant circuit the element values are $C = 0.1\,\mu\text{F}$, $L = 1\,\text{mH}$, and $R = 2\,\text{k}\Omega$.

a. Determine the resonant frequency, the quality factor, and the maximum value of the impedance magnitude.

b. Determine the bandwidth, using approximate methods. How badly do you expect the approximate value of β to differ from the exact value in this circuit?

c. Determine the upper and lower half-power band edges, using approximate methods. How are they related to ω_0? How badly do you expect them to differ from the exact values?

d. Confirm your expectation by finding the exact values of the band edges. How are *these* values related to ω_0?

e. Make a sketch of the impedance magnitude against frequency using the values of ω_0, the band edges, and two other frequencies.

6. The transformer in Figure P6 has a coupling coefficient $k = 0.9995$ (near perfect). The self-inductances are related by $L_1/R_1 = L_2/R_2$.

a. Find the ratio of the two half-power frequencies of the voltage-gain function.

b. If $L_2/R_2 = 1/50\pi$, find the two half-power frequencies.

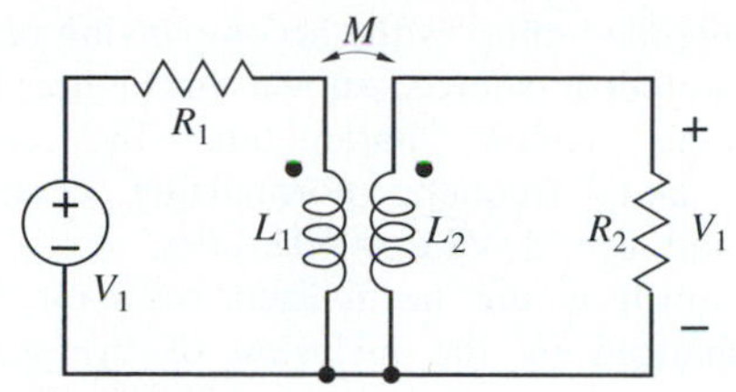

Figure P6

7. A parallel resonant circuit is formed by connecting a real coil of wire with a capacitor (assumed ideal). The simplest model of a coil is just an ideal inductor, but a more realistic model is an inductor in series with a resistor. It is possible to apply the definition of the quality factor in (8) in the text to the coil alone; we say the Q of the coil is $Q = \omega L/R$. Since the coil alone (with a model consisting of L in series with R) has no resonant frequency, the frequency in the definition of the Q-factor of the coil is taken to be any convenient and meaningful frequency.

In a particular case, the Q of a 2-mH coil at 5000 rad/s is found to be 5.

a. Find the capacitance of a capacitor which will resonate with the coil at that frequency when placed in parallel with it, and find the resistance of the coil.

b. Find the bandwidth of the resulting circuit.

8. Going one step further in the model of a coil of wire, the interturn capacitance must be taken into account at high frequencies. A more accurate model of a coil at high frequencies is the one shown in Figure P8, which is identical with the circuit in Figure 13b in the text. In a particular case, it is found that a 2-mH coil self-resonates at a frequency of 10^7 rad/s and has a Q of 100 at that frequency.

a. Find the bandwidth and the approximate band edges.

b. Find the parasitic element values. (They are parasitic because they are unintended effects piggybacking on the host, in this case the inductance.)

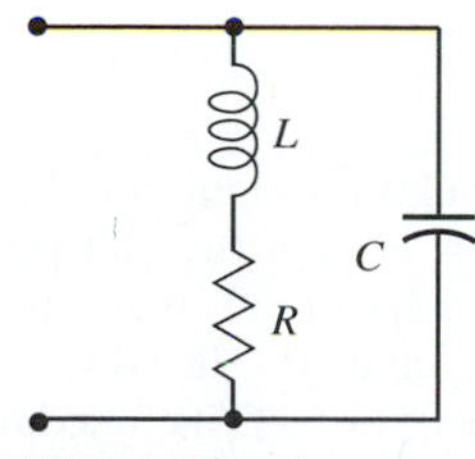

Figure P8

9. Another unintended effect is the accompanying resistance of a nonideal source, shown in Figure P9. Normalize the circuit parameters by using impedance- and frequency-normalizing factors $R_s = 50\ \text{k}\Omega$ and $\omega_s = 1/\sqrt{LC} = 50\ \text{krad/s}$.

a. By how much is the normalized resonant frequency changed by the inclusion of the series resistance?

b. By what percentage does the maximum magnitude of the normalized impedance increase?

c. Determine by how much the normalized half-power bandwidth has changed by the inclusion of the series resistance.

d. How would your answers to these questions change if normalization had not been carried out?

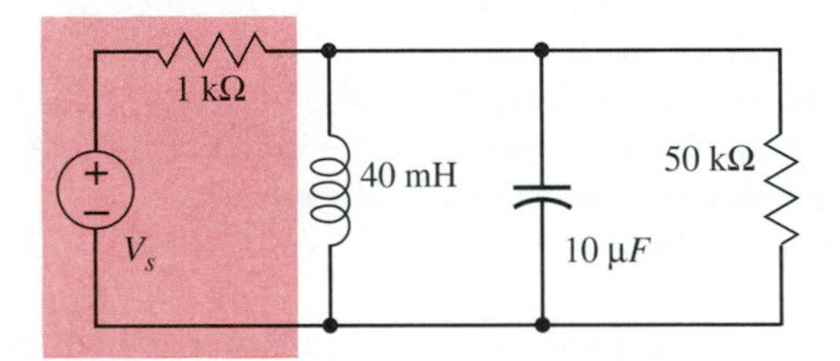

Figure P9

10. At 5000 rad/s, the branch impedances in the circuit of Figure P10 are as follows: $Z_L = j100$, $Z_C = -j100$, and $R = 200$, all in Ω.

a. Find the input admittance at 5000 rad/s.

b. Taking I as the reference for phase, draw a phasor diagram to a convenient scale.

c. Find the resonant frequency and the input admittance at resonance.

d. Find the half-power bandwidth.

e. Make a sketch of the impedance magnitude against frequency.

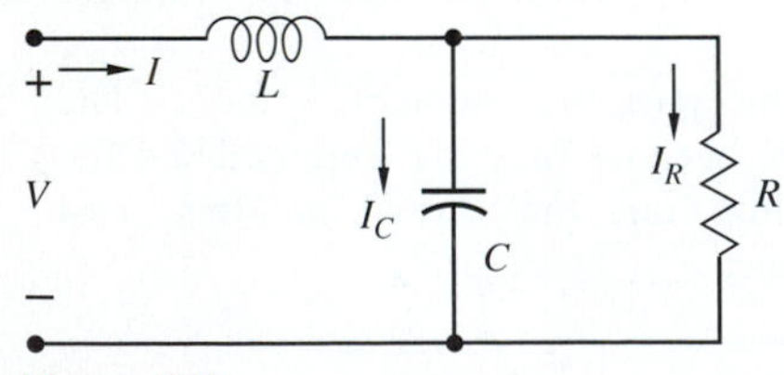

Figure P10

11. In a high-Q two-branch parallel resonant circuit, the resonant angular frequency is ω_0 and the half-power bandwidth is β, both in rad/s. Find the angular frequency, in terms of ω_0 and β, at which the magnitude of the impedance is half its maximum value. Use the high-Q approximation.

12. Using normalizing factors of $\omega_s = 50\ \text{krad/s}$ and $R_s = 10\ \Omega$, the following values are specified for a series resonant circuit: $R_n = 1$, $L_n = 10$, and $\omega_{0n} = 1$.

a. Find the normalized values of C, Q_0, β, and Y_m.

b. Find also the normalized upper and lower band edges at which the response is down to $1/\sqrt{2}$ of its maximum value, using the exact expressions.

c. Repeat part (b), using the approximate expressions. Confirm the way in which these frequencies are related to ω_{0n}.

d. Make sketches of the magnitude and angle of the admittance against frequency.

e. Specify the denormalized values of all quantities, including units.

13. A constant-amplitude sinusoidal voltage $v(t) = 14.14 \cos \omega t$ is used to excite a series RLC circuit whose resonant frequency is $\omega_0 = 1000\ \text{rad/s}$ and whose quality factor is 100. It is found that, at ω_0, $|I| = 2\ \text{A}$.

a. Find the value of the capacitor voltage phasor at the resonant frequency.

b. Find two frequencies at which this phasor magnitude is down to 80% of its value at resonance.

14. The circuit in Figure P14 is used as a frequency-sensing device. The relay closes (turning on the indicator lamp) whenever the magnitude of the current I exceeds $1/\sqrt{5}$ of its value at the resonant frequency. The following information is known: $Q_0 = 100$, $\omega_0 = 1000\ \text{rad/s}$, $|V| = 10\ \text{V}$, and $I = 2\ \text{A}$ at ω_0. Find, within an accuracy of 1 rad/s, the frequency range over which the lamp will be lit.

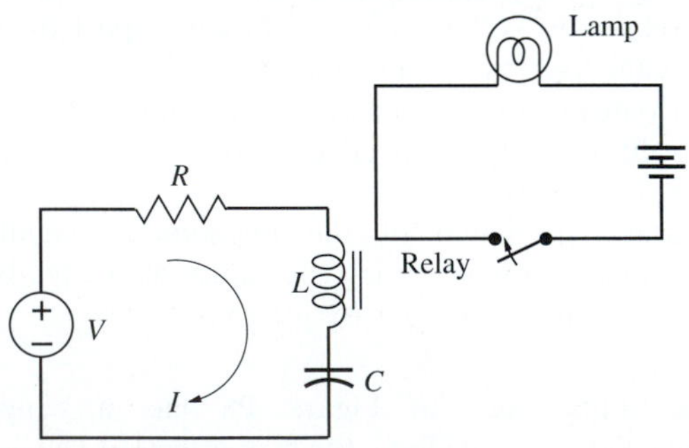

Figure P14

15. In the circuit of Figure P15:

a. Find the resonant frequency.

b. Find the value of the admittance magnitude at that frequency.

c. Find the upper and lower half-power frequencies.
d. Also find the range of frequencies at which the impedance magnitude is no less than 90% of its maximum value.

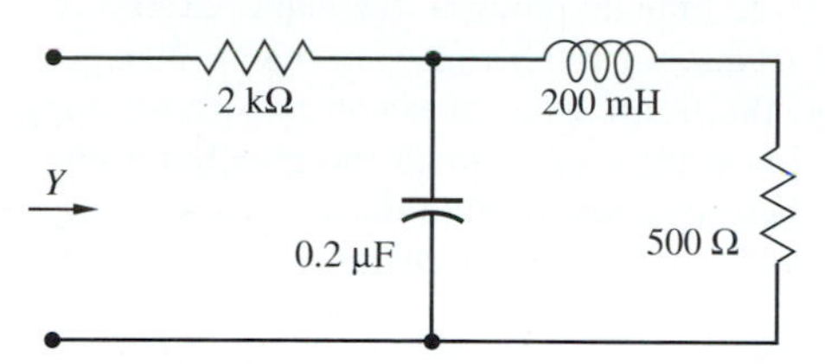

Figure P15

16. In the circuit of Figure P16, the steady-state current resulting from a sinusoidal voltage excitation is $i(t) = 0.01\sqrt{2}\cos\omega t$.
a. Determine the maximum of the energy stored in the capacitor at resonance.
b. Determine the reactive power supplied to the circuit at the upper and lower band edges, and specify if it is inductive or capacitive.

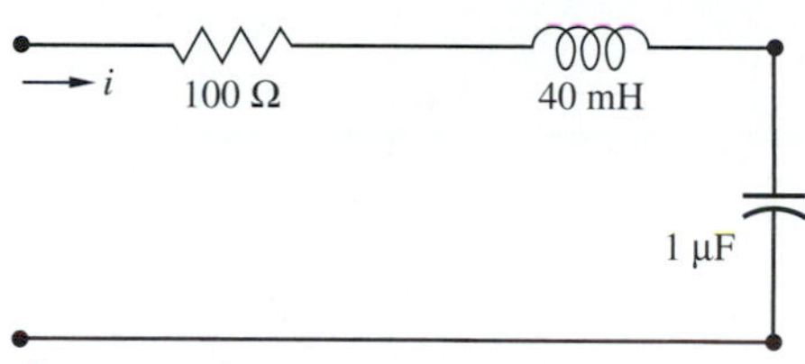

Figure P16

17. A coil of wire whose model is a series L and R has a capacitor $C = 100$ pF connected across its terminals. The circuit is excited by a constant-amplitude, variable-frequency sinusoidal current source. Two measurements of frequency are made in order to determine the parameters of the coil. The frequency is found to be 10 Mrad/s when the voltage has reached its maximum magnitude. When the voltage magnitude is half its maximum, the frequency is found to be 9.9 Mrad/s.
a. Determine the values of L and R.
b. Calculate another frequency at which the magnitude of the voltage is half its maximum.

18. A 1-μF capacitor is connected in series with a coil of wire whose model is a series connection of L and R. The circuit is excited by a variable-frequency, constant-amplitude sinusoidal voltage source. The frequency at which the voltage and the resulting current are in phase is found to be 10 krad/s. The capacitor is then placed in parallel with the coil and the combination is excited by a variable-frequency current source. The frequency at which the voltage and current are now in phase is found to be 9700 rad/s. Determine the unknown parameters.

19. A coil of wire is to be used at such a high frequency that an adequate model to represent it is that shown in Figure P19. To determine the parameters of the model, some measurements similar to the ones in Problems 17 and 18 are proposed, but the highest frequency of the available source is not high enough to reach the peak in the response. The only other equipment available is a few capacitors. Suggest a way to determine the parameter values in the model, and obtain formulas for them in terms of measured frequency, voltage, and current values. Will your scheme always work?

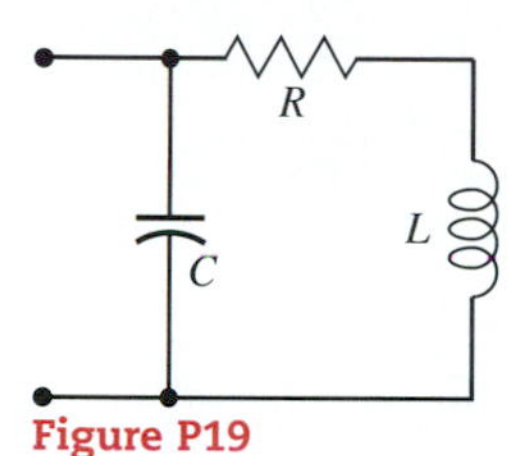

Figure P19

20. A series RLC circuit is excited by a 10-V variable-frequency sinusoidal source. At an angular frequency of 400 rad/s it is found that the magnitudes of the inductor voltage and of the capacitor voltage are both 100 V. At this frequency, the current magnitude is found to be 2 A. Find the element values.

21. The circuit in Figure P21 is excited by a variable-frequency sinusoidal voltage source. The impedance-normalizing factor is taken to be the resistance R, so its normalized value is 1. The current in the resistor is to have a maximum magnitude at a frequency which is taken as the normalizing frequency, so its normalized value is 1. The current in the resistor is to be zero at a normalized frequency of 2.

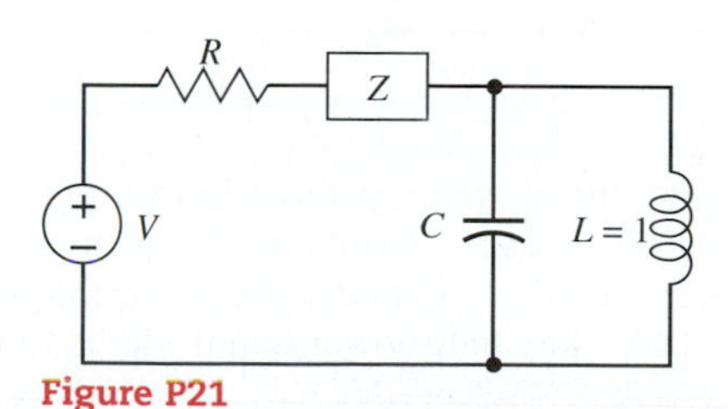

Figure P21

a. Determine the normalized value of C.

b. Specify whether the branch marked Z is a capacitor or an inductor, and determine its normalized value.

22. The circuit in Figure P22 is excited by a variable-frequency sinusoidal voltage source. The current in the resistor is to have its maximum magnitude at a normalized frequency of 3 and a value of zero at a normalized frequency of 2.

a. Determine the normalized value of L.

b. Specify whether the branch marked Z should be a capacitor or an inductor, and determine its normalized value.

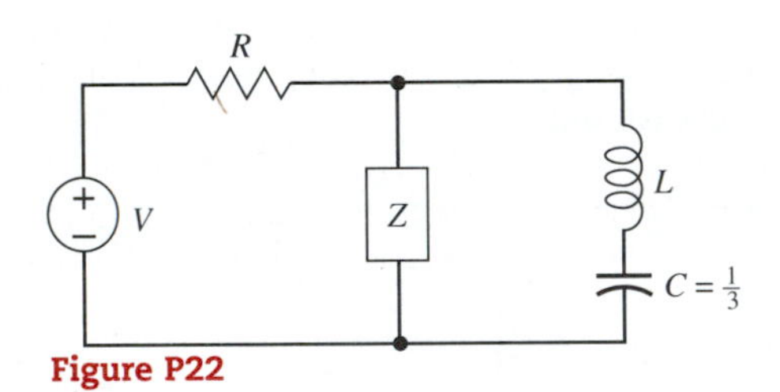

Figure P22

23. The circuit in Figure P23 is excited by a variable-frequency sinusoidal voltage source. The voltage across the capacitor is to have a maximum magnitude at a normalized frequency of 2, and the magnitude of the current supplied by the source is to be a maximum at a normalized frequency of 5.

a. Determine the value of C.

b. The branch marked Z should be a capacitor or an inductor; which is it, and what is its normalized value?

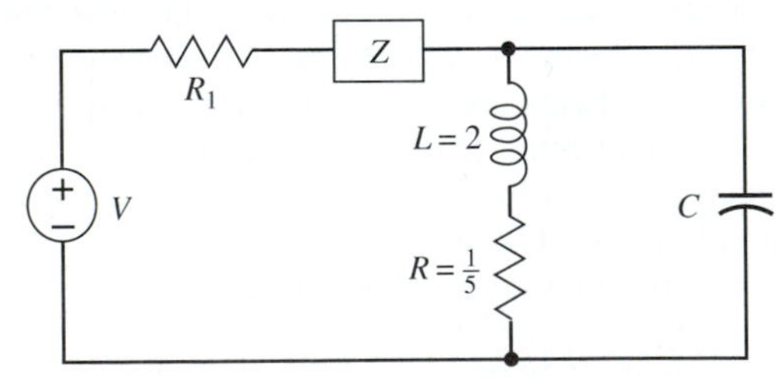

Figure P23

24. A 2-A current source excites a three-branch parallel resonant circuit. The resonant frequency is $\omega_0 = 1$ krad/s. At this frequency, the voltage across the circuit is 10 V. The half-power bandwidth is to be 10 rad/s.

a. Determine the circuit parameters.

b. Determine the band of frequencies over which the voltage exceeds 6 V.

25. In a three-branch parallel resonant circuit the resonant frequency is 5 Mrad/s and $Q_P = 200$.

a. Find the frequencies at which the power supplied to the circuit is one-fourth the maximum power.

b. At these frequencies find the ratio of the capacitor current to the inductor current.

26. The network inside the box in Figure P26 is to be represented by the three-branch parallel circuit shown. C_1 is a known external capacitor. With C_1 disconnected, let the resonant frequency be ω_0 and the half-power bandwidth β. With C_1 connected, the resonant frequency shifts to ω_1. Determine the values of R, L, and C in terms of C_1, ω_0, β, and ω_1.

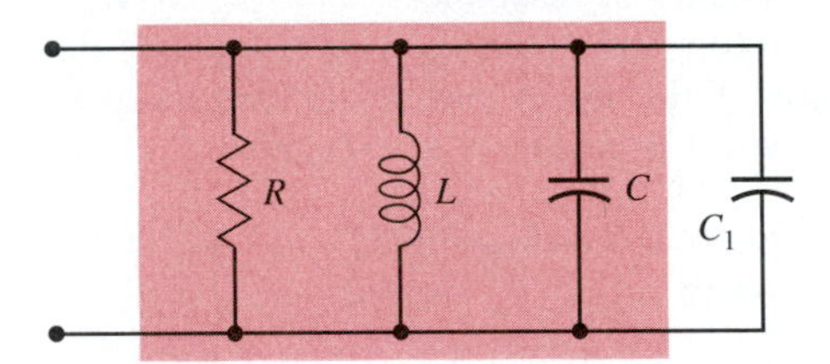

Figure P26

27. A series RLC circuit is excited by a variable-frequency, constant-amplitude voltage source. At a frequency of 20 krad/s it is found that the rms value of the current is 13 A and the power input of 845 W is a maximum. At a frequency of 17.5 krad/s the power input is 125 W. Find the parameter values of the circuit and the source voltage.

28. For each of the circuits in Figure P28 determine an expression for the complex power supplied by the source and verify that it takes the form of (41) in the text.

29. The frequency at which the L and C in Figure P29 resonate is $\omega_0 = 200$ krad/s. Also specified are $R = 400\ \Omega$ and $L = 20$ mH.

a. Assuming a sinusoidal voltage excitation whose rms value is 1 V, find the magnitude of the current at ω_0.

b. By what percent is this value smaller than the maximum magnitude of the current?

c. Find the quality factor and the bandwidth.

d. Find the approximate values of the half-power band edges.

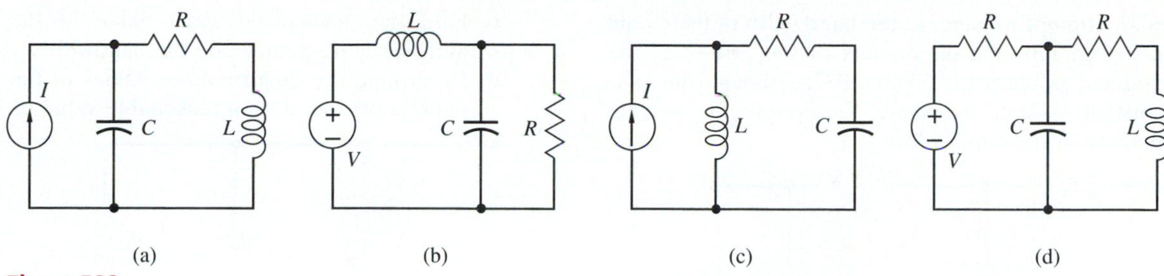

Figure P28

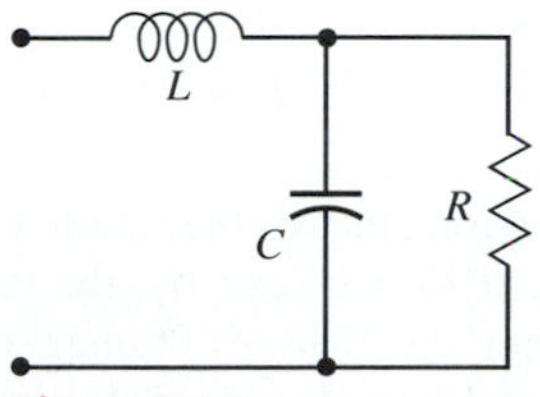

Figure P29

30. In Problem 30 in Chapter 8, you established that the voltage gain in Figure P30a is $V_2/V_1 = -Z_2/Z_1$. (If you didn't do it then, do it now.)

a. Determine the voltage-gain function $G_{21}(j\omega)$ of the circuit in Figure P30b. Note how it is related to the gain of Figure 3 in the text.

b. Sketch the magnitude $|G_{21}(j\omega)|$ and find the value of $RC\omega$ where the magnitude is $1/\sqrt{2}$ of its DC value. Is the bandwidth more or less than the bandwidth of the single RC circuit in Figure 3 in the text?

c. Explore possible changes in the bandwidth by increasing and decreasing the time constant of one of the two stages.

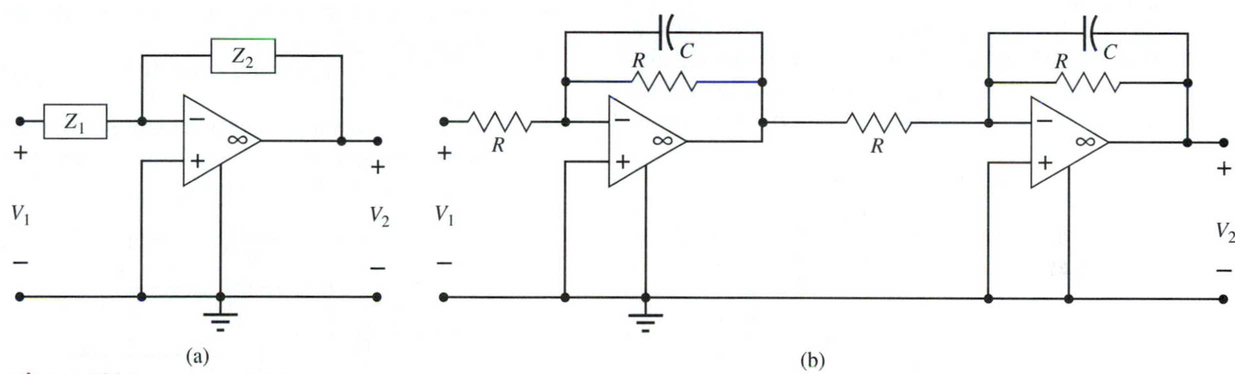

Figure P30

31. To save on the cost of two op amps, someone suggested the circuit of Figure P31 to accomplish the same purpose as Figure P30. Each of the parts within the boxes is referred to as a *stage* of the circuit.

a. Find the voltage gain as a function of frequency. Note the difference between this function, where the presence of one stage interacts with the other, and the one in Problem 30, where the two stages are independent.

b. Sketch the voltage-gain magnitude against frequency. Determine the frequency at which the square of the magnitude is half its maximum value (which occurs at $\omega = 0$, or DC).

c. The time constant of one stage can be varied compared with the time constant of the other stage, possibly to improve the bandwidth. Explore this possibility. Compare with the result of doing this in Problem 30.

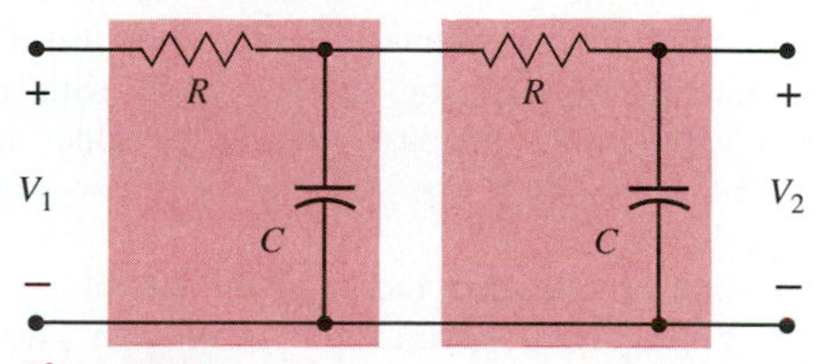

Figure P31

32. In an attempt to increase the bandwidth of the circuit in Figure P31, the parameters of the first stage are modified as shown in Figure P32, although the time constant is kept the same. Determine the percent increase in the bandwidth.

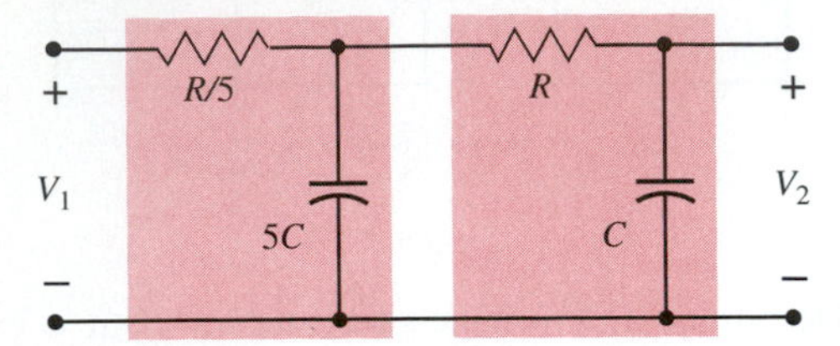

Figure P32

33. A circuit whose voltage gain $G(j\omega) = V_2/V_1$ has the same frequency dependence as the impedance of a parallel resonant circuit (or the admittance of a series resonant circuit) is shown in Figure P33.

a. Determine the voltage-gain function and place it in the form of the admittance in (7) in the text, but with the terms in the denominator expressed in terms of ω and the two time constants a and b.

b. Express ω_0 and Q_0 in terms of the time constants a and b.

c. Specify the range of values of the quality factor in this circuit, and its maximum value, for all possible values of a and b. Would you describe the circuit as high-Q (narrow-band) or low-Q (wide-band)?

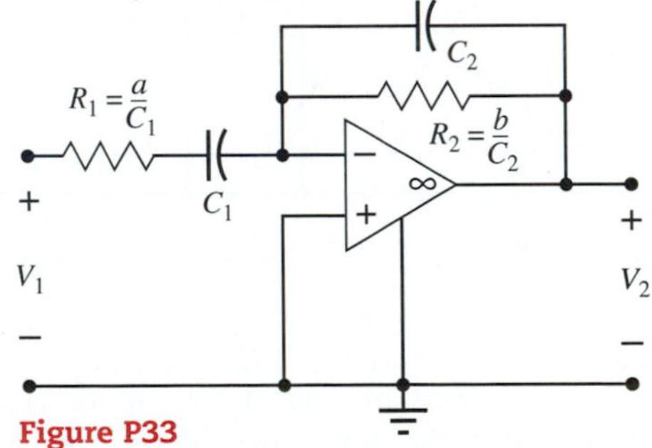

Figure P33

34. The design value of the resonant frequency in the circuit of Figure P34 is specified to be 1.5 Mrad/s. At this frequency, the Q of the coil is 15. Use the design frequency as the normalizing frequency and use R_C as the impedance-normalizing factor. (Deal with normalized values until the last part.) The upper half-power frequency is to be 5% above the resonant frequency.

a. Determine the quality factor of the circuit.

b. Determine the normalized unknown circuit parameters.

c. Find the normalized exact value of the lower half-power frequency and the bandwidth.

d. Determine the denormalized values of the circuit parameters based on a reasonable value of R_C.

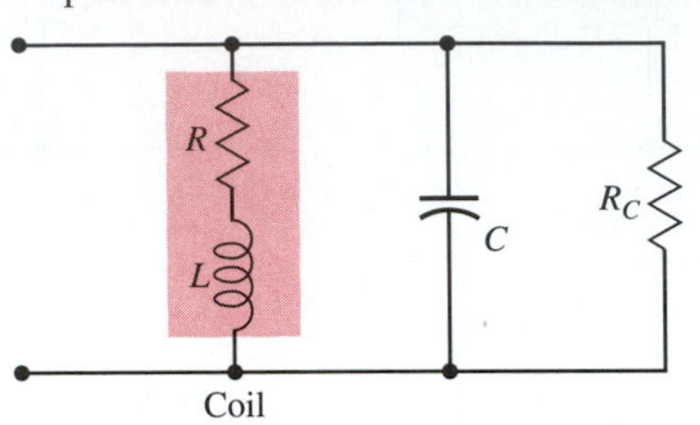

Figure P34

35. Except for a phase inversion, the voltage gain V_2/V_1 in the circuit in Figure P35 has exactly the same dependence on frequency as the admittance of a series RLC circuit or the impedance of a parallel RLC circuit: that is, it is a bandpass circuit. (Neglect this phase inversion in the rest of this problem.)

a. Find an expression for the voltage gain as a function of frequency.

b. Determine the values of R and C which will make the resonant frequency $\omega_0 = 10^5$ and the quality factor $Q_0 = 10$.

c. Find also the maximum magnitude of the voltage gain.

d. Find the maximum energy stored and the average power dissipated at the resonant frequency. From these, confirm the value of Q_0.

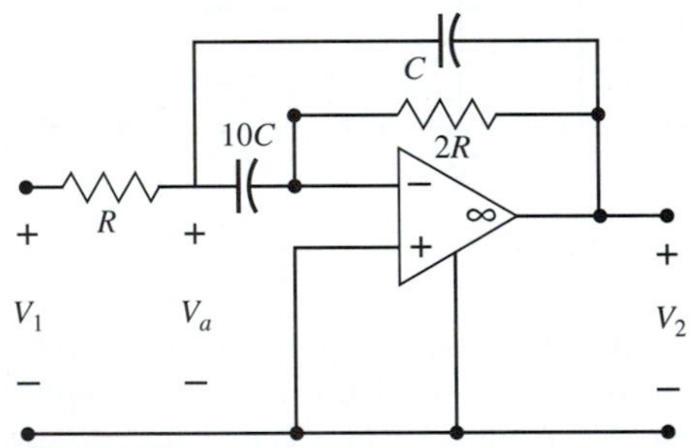

Figure P35

36. At first sight, the circuit in Figure P36 may appear to be complicated. However, recall that an op amp operating linearly acts like an ideal voltage source at its output. Therefore, the various stages operate independently. You are familiar with each of the individual stages.

a. Determine the voltage-gain function V_4/V_1. (Again, in functional dependence on frequency, this is identical with the admittance of a series resonant circuit or the impedance of a parallel resonant circuit, except for a phase inversion.)

b. Determine the resonant frequency, the quality factor, and the maximum voltage gain using the following numerical values: $R = 10$, $R_2 = 100$, $R_1 = 1$, all in kΩ; and $C = 2$ nF.

c. Determine the bandwidth and the frequencies at the band edge from both the exact expression and the approximate expression. Notice the percent error in the latter case and correlate it with the quality factor.

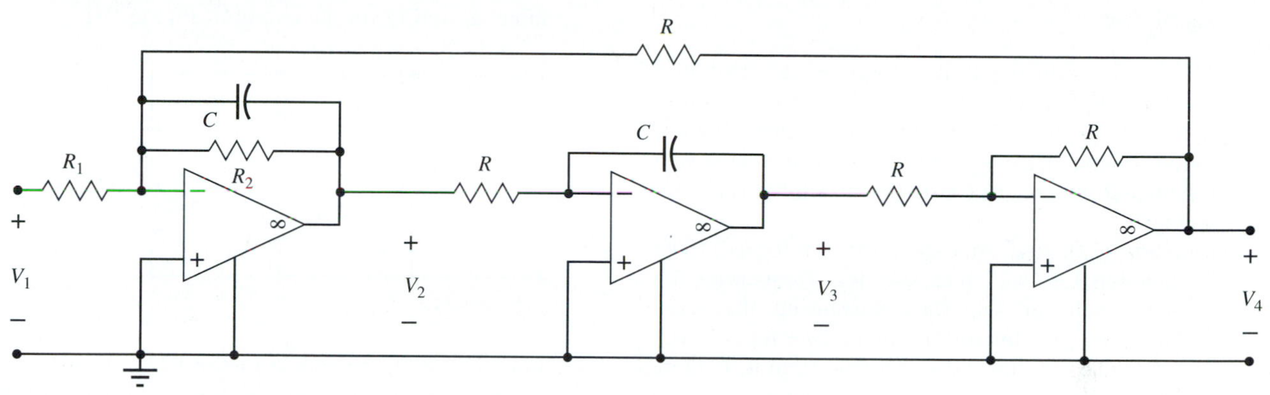

Figure P36

37. A series resonant circuit has a resonant frequency $\omega_0 = 10^4$ rad/s, a quality factor $Q_0 = 6$, and a resistance $R = 50\ \Omega$. In a certain application, a load $R_L = 10\ \text{k}\Omega$ appears across the capacitor, as shown in Figure P37.

a. Determine the values of L and C.

b. Replace the parallel $R_L C$ combination with a series RC circuit equivalent to it at ω_0.

c. Calculate the new resonant frequency and the new quality factor.

d. Calculate the upper and lower band edges in the new circuit and compare with the band edges in the original circuit.

e. Compare the new resonant frequency, the quality factor, and the bandwidth with the corresponding values in the original circuit.

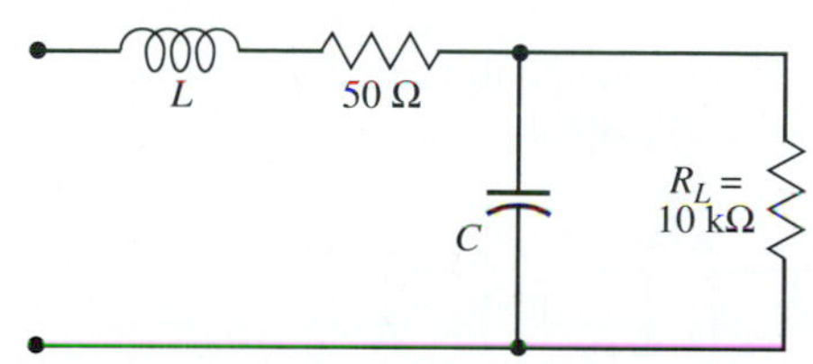

Figure P37

38. A different approach is to be taken in the circuit of Figure P37. The circuit with the load resistance connected is not to be modified.

a. Determine, as a function of frequency, the admittance of the circuit as it stands.

b. From its definition, determine the resonant frequency. (What is the easiest way to determine that the ratio of two complex quantities is real?)

c. Find the half-power band edges and, thus, the bandwidth.

d. Calculate the quality factor. (Does $Q_0 = \omega_0 L/R$ apply?)

e. Compare the resonant frequency, the quality factor, and the bandwidth with the results of Problem 37.

f. Find the maximum energy stored in the circuit and the average power dissipated; from these, confirm the value of the quality factor.

39. a. Find the admittance of the circuit in Figure P39 as a function of frequency.

b. Determine the resonant frequency. At that frequency, determine the magnitude of the admittance.

c. Determine the lower half-power band edge. Why is there no upper one?

d. Find the quality factor at resonance.

e. Determine the maximum energy stored and the average power dissipated at the band edge. How is the value of Q at this frequency related to Q_0?

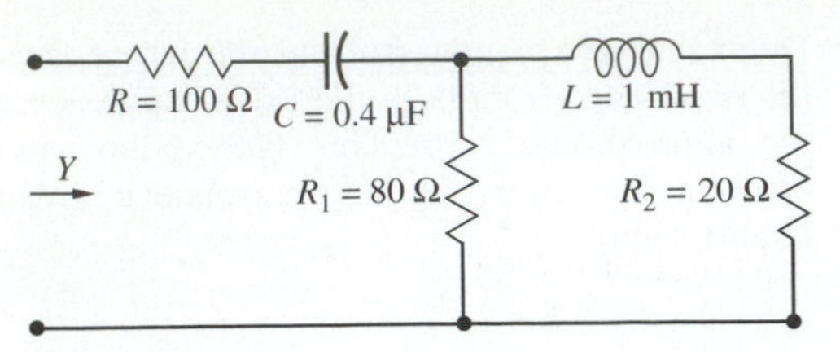

Figure P39

40. The voltage-gain function across the capacitor in a series resonant circuit ($G(j\omega) = V_C/V$) was given in (22) in the text. At the resonant frequency, this was found to have the magnitude Q_0. This is a large gain for high-Q circuits, but is it the maximum capacitor voltage?

a. Using (22), find an expression for $|G(j\omega)|$. The denominator will increase for frequencies on either side of ω_0, thus decreasing the ratio. However, the numerator will *increase* for $\omega > \omega_0$, thus increasing the ratio. The question is, Is there some $\omega > \omega_0$ for which the increase in the numerator is greater than the increase in the denominator? In that case, the gain at ω_0 is not the maximum.

b. If there are such frequencies, find the frequency at which the effect is the greatest; that is, find the frequency for maximum gain magnitude.

c. Find the maximum magnitude of the gain and compare it with Q_0, the magnitude at ω_0.

41. Using an approach like that in Problem 37, determine ω_0 and Q_0 in the circuit of Figure P41.

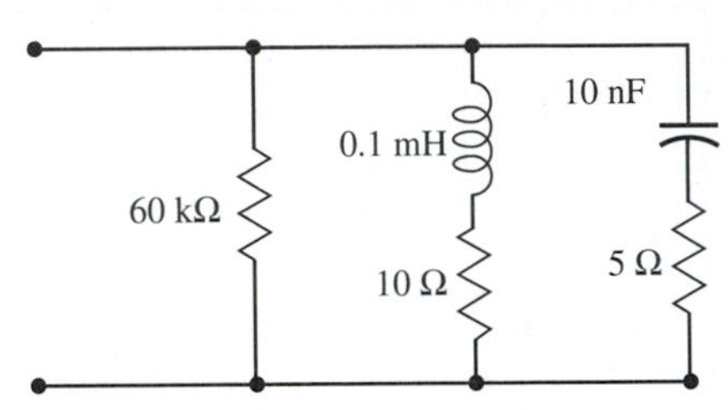

Figure P41

42. For each of the circuits in Figure P42:

a. Write an expression for the admittance of the circuit as a function of frequency.

b. Find the resonant frequency and the half-power frequencies any way you know how.

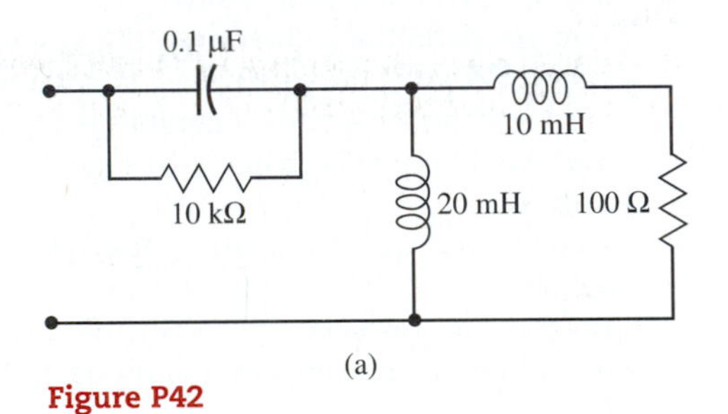

(a)

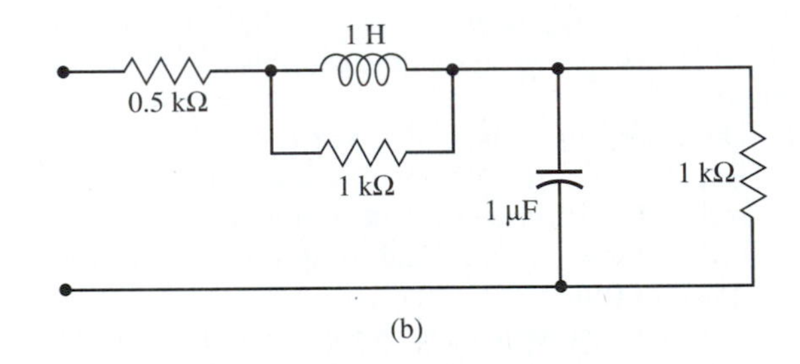

(b)

Figure P42

43. The resonant frequency and the quality factor in a parallel resonant circuit are $\omega_0 = 0.5$ Mrad/s and $Q_P = 10$. The resistance in the circuit is 1 kΩ.

a. Use the values of the resistance and frequency as the normalizing factors R_s and ω_s, and determine L_n and C_n, the normalized values of L and C.

b. Show that the value of Q_P is not dependent on the normalization.

c. Determine the normalized bandwidth and the normalized exact values of the half-power band edges.

d. Determine the magnitude of the impedance at the frequencies 0.95 and 1.05. Are they what you would have expected?

e. Find the average energy stored in L at each of the frequencies in part (d). From this, determine the average energy stored in C; then confirm it by evaluating it directly.

f. Denormalize and specify all values in their proper units.

44. The transformer coupling coefficient in the circuit of Figure P44 is adjustable.

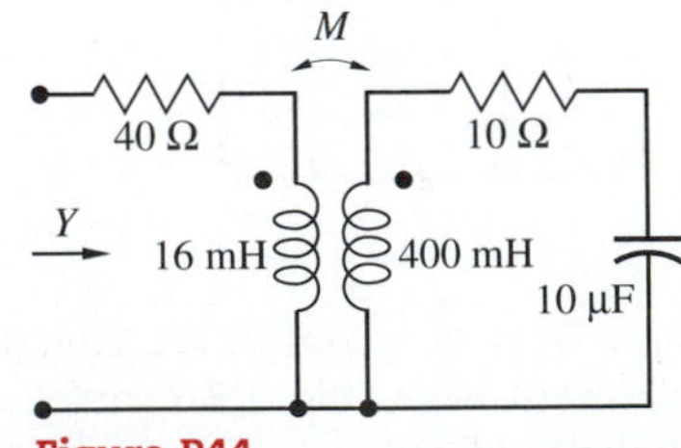

Figure P44

a. Find an expression for the admittance at the terminals of the circuit in terms of k and ω.

b. In terms of the coupling coefficient, find an expression for the frequency at which the admittance is real.

c. What is the range of resonant frequencies if k ranges from 0.5 to 1?

DESIGN PROBLEMS

1. A parallel *RLC* circuit is to be designed as a bandpass filter. A parallel *RLC* circuit is shown in Figure DP1. R_L represents the resistance of the inductor wire.

a. Neglecting the effect of R_L, determine the values of R, L, and C such that the impedance of this circuit is purely resistive at the resonant frequency of $\omega_0 = 10^6$ rad/s and equal to 10 kΩ, and that the magnitude of its impedance at frequencies $\omega = 0.95 \times 10^6$ rad/s and $\omega = 1.05 \times 10^6$ rad/s is equal to 7.07 kΩ.

b. If $R_L = 20\ \Omega$ determine the effect on the resonant frequency and the value of the impedance at this frequency.

c. Reselect the component values to satisfy the original requirements, assuming that $\omega_0 L/R_L = 50$.

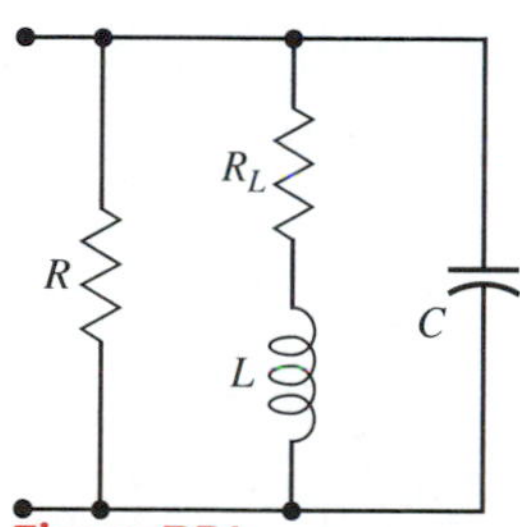

Figure DP1

2. Figure DP2 represents a model of an audio system. The *woofer* is a low-frequency speaker, the *tweeter* is a high-frequency speaker and the *midrange speaker* should respond only to frequencies between the range of frequencies of the other two.

a. Assume that the audio amplifier can be represented as an ideal source, and that the speakers can be represented as purely resistive 8-Ω loads. The design requirements are that the woofer circuit be a low-pass filter with a cut-off frequency of 500 Hz, the tweeter circuit be a high-pass filter with a cut-off frequency of 3500 Hz, and the midrange speaker circuit be a bandpass filter whose 3-db bandwidth extends from 500 to 3500 Hz. Design the system; determine the component values to satisfy the requirements.

b. Determine the effect that an 8-Ω internal resistance of the audio amplifier can have on the frequency response of each of the branches. Use PSpice to determine the ratio of the voltage across each speaker to the voltage of the source.

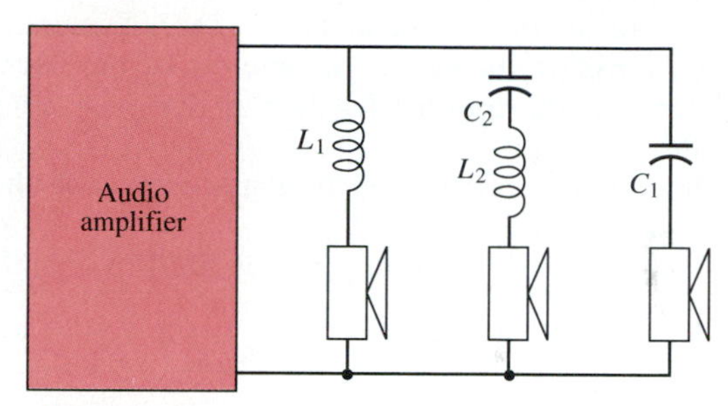

Figure DP2

3. The circuit shown in Figure DP3 is to be used to implement a low-pass filter.

a. Assuming an ideal op amp, determine the transfer function V_0/V_i of the circuit.

b. Determine the value of the circuit elements to implement the following transfer function:

$$\frac{V_0}{V_i} = \frac{1.586 \times 10^8}{-\omega^2 + j2.586\omega + 10^8}$$

c. It is desired to increase the DC gain of this circuit to 2; i.e., for $\omega = 0$, $V_0/V_i = 2$. Suggest additions to this circuit to obtain this desired DC gain, without changing the form of the transfer function.

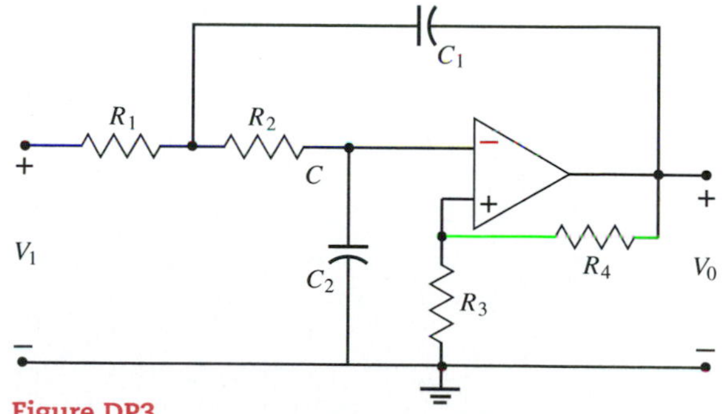

Figure DP3

4. A notch filter is a band elimination filter. Instead of passing a band of frequencies, it eliminates them from the response. The transfer function of a notch filter is given by the following:

$$\frac{V_{\text{out}}}{V_{\text{in}}} = \frac{-\omega^2 + \omega_0^2}{-\omega^2 + j\left(\frac{\omega_0}{Q}\right)\omega + \omega_0^2}$$

where ω_0 is the frequency at which the transfer function is 0. The circuit in Figure DP4 could provide such a transfer function with the proper choices of the parameters.

a. Assuming an ideal op amp, determine the transfer function of the circuit in Figure DP3. Determine relationships among the parameters of the circuit such that the transfer function is that of a notch filter.

b. Determine the values of the parameters such that $\omega_0 = 1000$ and $Q = 3$.

c. Assume that the op amp is not ideal; it has the following parameters:

$$R_i = 500\ \text{k}\Omega, \qquad A = 50{,}000, \quad \text{and} \quad R_0 = 50\ \Omega.$$

Using PSpice, determine the transfer function using the values determined in part (b). Compare this transfer function with the one given in the problem statement.

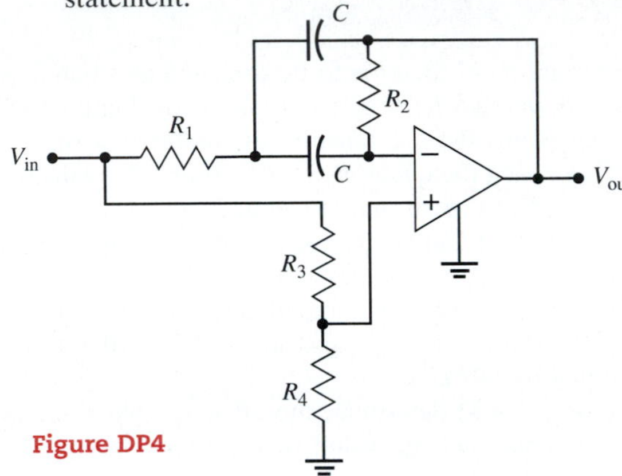

Figure DP4

12 The Complex Frequency and Network Functions

The preceding four chapters have dealt with several different aspects of the steady-state response of linear electrical networks to excitations which are sinusoidal functions of time. In broad outline, the approach for obtaining a solution consists of four parts: (1) First, the problem is formulated in the time domain as a set of differential equations in which the excitation functions are all sinusoids of the same frequency. (2) Then a transformation is carried out so that the voltage and current variables are no longer functions of time but phasors, and the equations are no longer differential but algebraic. (3) The transformed problem is then solved by algebraic methods identical with those used for dissipative networks. (4) Finally, the solution can be converted back into the time domain.

As a matter of fact, the first, second, and last steps are carried out only conceptually (that is, they are really omitted), and the problem is phrased from the beginning in terms of phasors. Basic laws are written in the phasor domain. Portions of networks are represented by impedances or transfer functions and other relationships, such as Thévenin's theorem, applied in the phasor domain.

The fundamental idea on which this approach is based is Euler's theorem, which permits expressing a sinusoid in terms of the exponential $e^{j\omega t}$. In this chapter, we shall exploit this relationship even further by generalizing the exponent.

1 THE COMPLEX FREQUENCY

The exponent in the exponential function $e^{j\omega t}$ is imaginary. Now an imaginary quantity is but a special case of something more general, namely, a *complex* quantity. Consequently, let us introduce a complex quantity s, defined as:

$$s = \sigma + j\omega \tag{1}$$

whose imaginary part is the usual angular frequency ω. We shall call s the *complex-frequency variable,* or simply the *complex frequency.* When the real part of s is 0, then s reverts to just $j\omega$.

The Complex Exponential

What happens if s replaces $j\omega$ in the exponential function $e^{j\omega t}$? By the law of exponents and Euler's theorem, the resulting e^{st} can be written:

$$e^{st} = e^{\sigma t}e^{j\omega t} = e^{\sigma t}(\cos \omega t + j \sin \omega t) \tag{2}$$

A graphical interpretation of this function is readily obtained. We have already interpreted the exponential $e^{j\omega t}$ as a rotating phasor in the complex plane, its projections on the real and imaginary axes generating the cosine and sine functions. Now, however, $e^{j\omega t}$ is multiplied by $e^{\sigma t}$, which plays the role of the magnitude of the phasor. When σ is zero, the phasor magnitude is constant; otherwise, the magnitude will be changing with time.

Figure 1 shows the path traced by the tip of the phasor as it rotates, for both positive and negative values of σ. Both parts of the figure are spirals. When σ is negative, the phasor magnitude is decreasing with time so that the path spirals inward; when σ is positive, the phasor magnitude is increasing with time so that the path spirals outward.

The projections of these spirals onto the real and imaginary axes will be cosine and sine functions multiplied by the real exponential $e^{\sigma t}$. When σ is

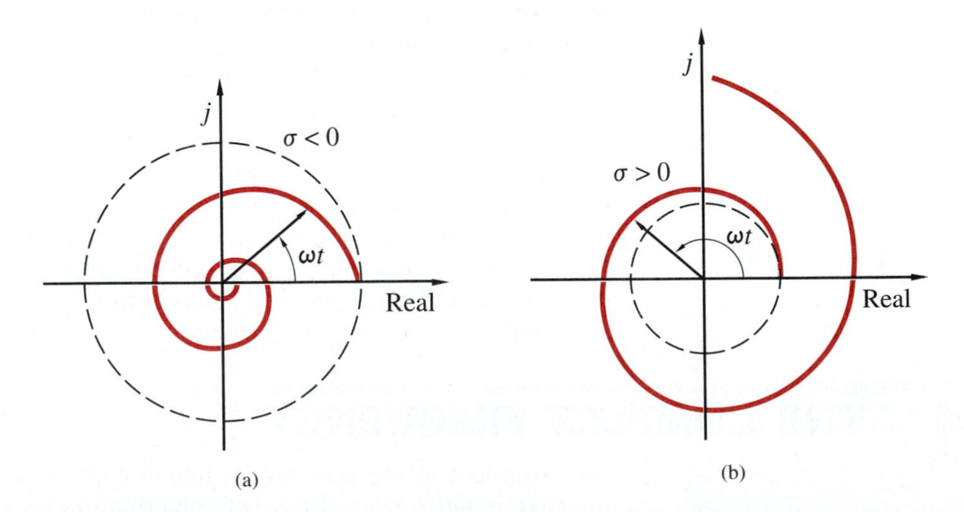

Figure 1.
Spiral paths generated by $e^{\sigma t}e^{j\omega t}$: (a) $\sigma < 0$; (b) $\sigma > 0$.

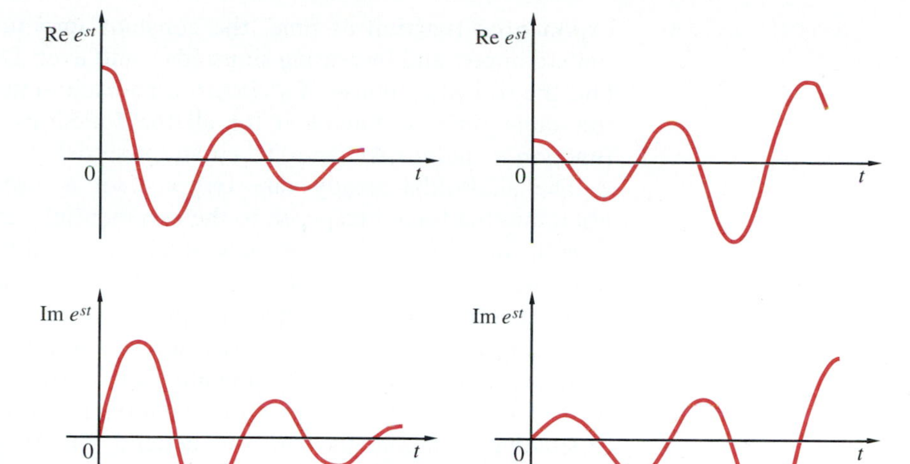

Figure 2
Waveforms of real and imaginary parts of $e^{\sigma t}e^{j\omega t}$: (a) $\sigma < 0$; (b) $\sigma > 0$.

negative, the resulting waveform is a damped sinusoid; when σ is positive, the waveform is an exponentially increasing sinusoid, as in Figure 2.

The same conclusions follow from (2) if the real and imaginary parts are taken; thus:

$$\begin{aligned} \text{Re}\,(e^{st}) &= e^{\sigma t}\cos\omega t \\ \text{Im}\,(e^{st}) &= e^{\sigma t}\sin\omega t \end{aligned} \tag{3}$$

It appears, then, that particular values of s in the exponential function e^{st} lead to sinusoidal waveforms with amplitudes that are constant ($\sigma = 0$), exponentially decreasing ($\sigma < 0$), or exponentially increasing ($\sigma > 0$). But that's not all; when the *imaginary* part of s is zero, so that $s = \sigma$, e^{st} becomes simply a real exponential. The possible waveforms are shown in Figure 3 for positive, negative, and zero values of σ.

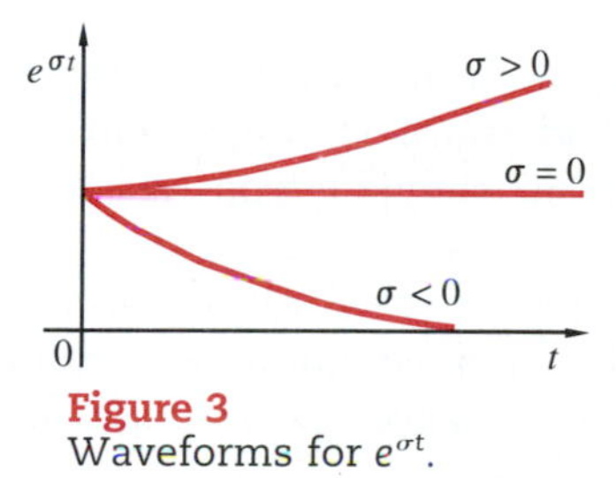

Figure 3
Waveforms for $e^{\sigma t}$.

Now a word about the units of s. The units of ω are expressed as radians per second; but a radian is really dimensionless, since it is defined as the ratio of an arc length of a circle to its radius. Hence, dimensionally, ω is the reciprocal of time. The same should be true of s and of σ. The unit of the exponent of a real exponential has acquired the name *neper,* however, as will be discussed later in the chapter. Thus, if σt has the unit neper, then σ must have the unit *neper per second.* On this basis, the name *neper frequency* has sometimes been given to σ, in contrast to the names *angular frequency* or *radian frequency* for ω.

Phasor Analysis and the Complex Frequency

The concept of the complex exponential introduced in this chapter is a powerful one. The function e^{st} serves to bring under the same roof the real exponential function of time, the constant-amplitude sinusoid, the exponentially damped and increasing sinusoids—and even DC, a steady constant. If we find the forced response of a circuit to an excitation of the form e^{st}, we shall at the same time be finding it for all the functions just mentioned. Thus, the function e^{st} unifies AC and DC circuit analysis!

The sinusoidal steady-state response of a circuit can be found by first obtaining the forced response to the exponential excitation e^{st}, and then setting $\sigma = 0$, or $s = j\omega$. Let us now explore possible changes that we might introduce into the methods of sinusoidal steady-state analysis developed in Chapter 8 by using the complex-frequency variable.

First of all, Kirchhoff's two laws in terms of phasors will remain unchanged. You can demonstrate this by repeating the development in Chapter 8 using e^{st} instead of $e^{j\omega t}$. Next consider the components. Since the v-i relationships of dissipative components do not depend on frequency, they will remain unchanged. The differentiation (and integration) in the v-i relationships of capacitors and inductors will lead to factors of s (or $1/s$) instead of $j\omega$ (or $1/j\omega$). Thus, the impedances of an inductor and a capacitor will be $Z_L = Ls$ and $Z_C = 1/sC$, respectively. All that is necessary is to replace $j\omega$ by s in everything in Chapter 8; the result will be the forced response to the exponential e^{st} as an excitation.

2 NETWORK FUNCTIONS AND THE COMPLEX FREQUENCY

With the concept of complex frequency established, we can use it to determine properties of network functions (driving-point and transfer). Let's start by determining the impedances and admittances of some simple circuits in terms of the complex frequency s.

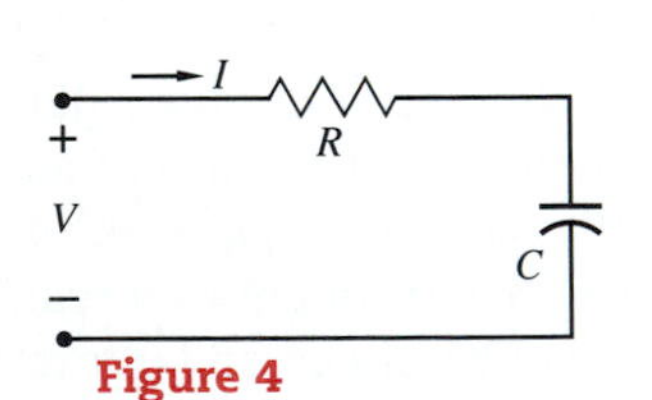

Figure 4
RC series circuit.

A simple series RC circuit excited by a voltage source whose voltage variation is of the form e^{st} is shown in Figure 4. The loop equation in terms of phasors will be $V = RI + I/Cs$. Hence, the impedance (V/I) and the admittance (I/V) are:

$$Z(s) = R + \frac{1}{Cs} = R\frac{s + 1/RC}{s}$$
$$Y(s) = \frac{1}{R}\frac{s}{s + 1/RC} \tag{4}$$

The impedance is in the form of a ratio of two polynomials whose variable is s. By the Fundamental Law of Algebra, nth-degree polynomials have n roots, and they can be written in terms of their factors. In (4) the polynomials are of the first degree and are already in factored form. The root of the polynomial in the denominator of Z is at $s = 0$. Letting $s_1 = -1/RC$ be the root of the numerator, the preceding expressions can be written as follows:

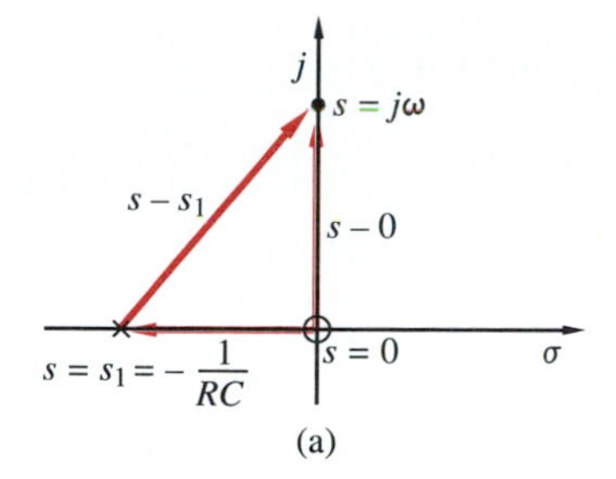

(a)

$$Z(s) = R\frac{s - s_1}{s}$$
$$Y(s) = \frac{1}{R}\frac{s}{s - s_1} \tag{5}$$

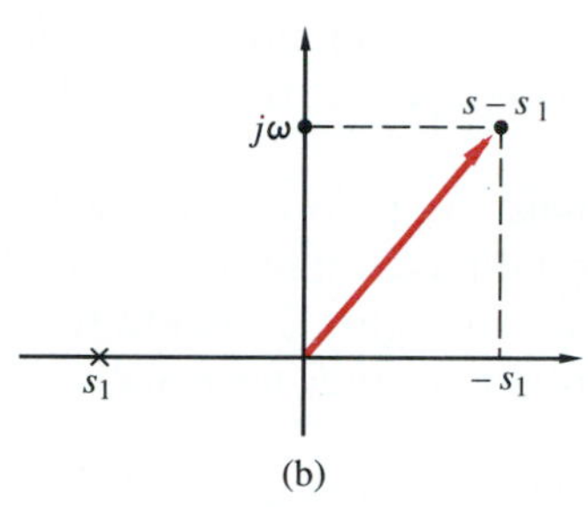

(b)

Figure 5
Representation of the RC impedance in the s-plane.

The quantity s_1 is the particular value of s at which the impedance is zero; it depends only on the parameters of the circuit. Aside from the multiplying constant R, the impedance (or admittance) depends on only two things: the complex-frequency variable, which is contributed by the source, and the roots s_1 and 0, which are contributed by the circuit.

Let's take a slight detour and consider the natural response of this RC circuit, as discussed in Chapter 6. There it was determined that the natural response is characterized by the natural frequency, which, for the series RC circuit, is just $-1/RC$, the same quantity that determines the steady-state response! This is a truly remarkable fact of the utmost significance. We will discuss this at greater length shortly. Now back to the main thread.

It has been established that the impedance of the circuit under consideration is dependent on s, the complex frequency of the source; and on s_1, the complex natural frequency of the circuit. In the present case, these "complex" frequencies are not complex at all. For the source, $s = j\omega$, which is imaginary; for the circuit, s_1 is real and negative. Nevertheless, $j\omega$ and s_1 are simply two particular values of the general complex variable s. This is an extremely productive viewpoint. It permits us to apply to network problems the branch of mathematics known as the theory of functions of a complex variable, with which we can unify a number of aspects of network theory.

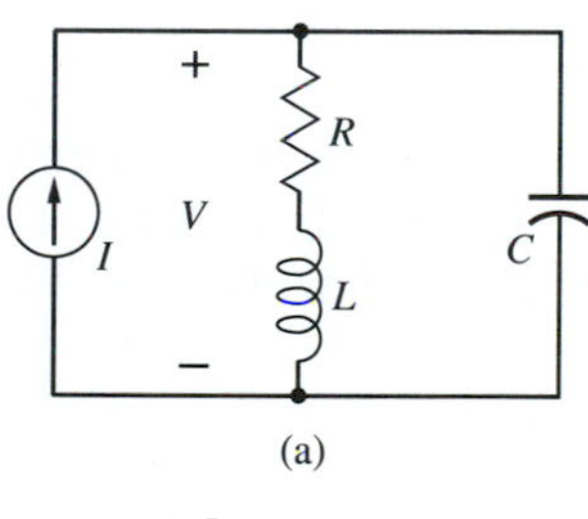

(a)

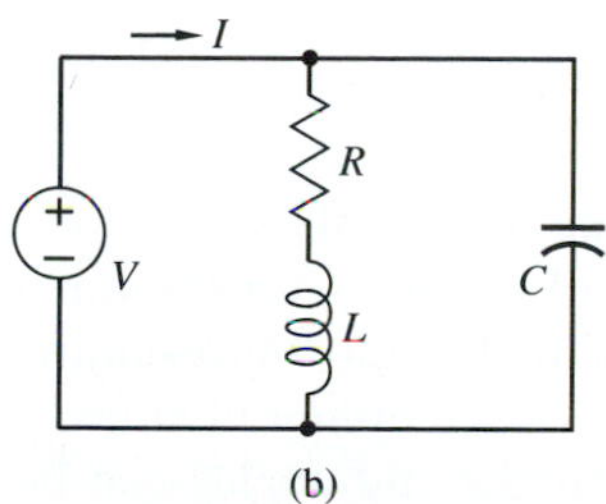

(b)

Figure 6
Two-branch parallel resonant circuit: (a) current-excited and (b) voltage-excited.

Network-Function Representation in the Complex-Frequency Plane

Since s is a complex variable, it can be represented geometrically in the complex plane. The *complex-frequency plane*, or simply the *s-plane*, is shown in Figure 5. The quantities $s = j\omega$ and $s = s_1 = -1/RC$ are represented by directed lines from the origin. The difference $j\omega - s_1$ is found graphically by changing the direction of s_1 and adding. Translating the directed line $j\omega - s_1$ in the first quadrant parallel to itself still represents the same quantity. Thus, the directed line from s_1 to $j\omega$ in the second quadrant still represents $j\omega - s_1$. The factor $j\omega$ (which can be written $j\omega - 0$) in the denominator of Z is also represented by a directed line, from the s-plane origin to a point $j\omega$ on the imaginary axis. Aside from the multiplying factor R, these directed lines completely describe the impedance—and the admittance also.

As another example, consider the two-branch parallel resonant circuit with a current excitation in Figure 6b. The function representing the ratio of response phasor to excitation phasor is the impedance. By combining parallel and series branches, the impedance is:

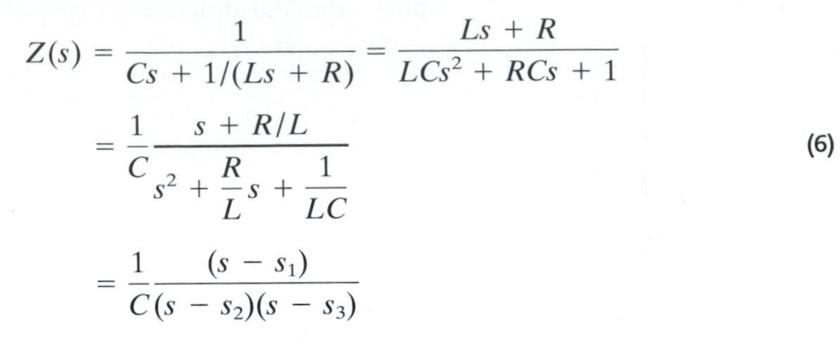

$$Z(s) = \frac{1}{Cs + 1/(Ls + R)} = \frac{Ls + R}{LCs^2 + RCs + 1}$$

$$= \frac{1}{C} \frac{s + R/L}{s^2 + \frac{R}{L}s + \frac{1}{LC}} \tag{6}$$

$$= \frac{1}{C} \frac{(s - s_1)}{(s - s_2)(s - s_3)}$$

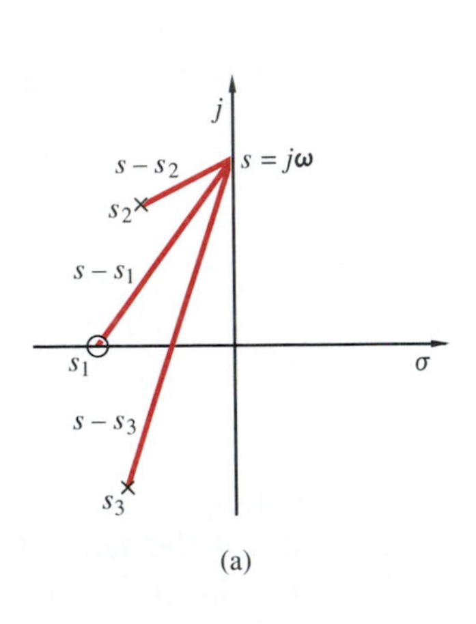

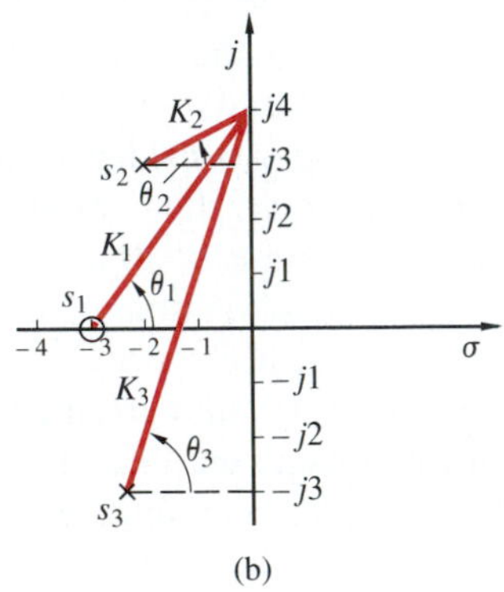

Figure 7 Complex-plane representation of resonant-circuit impedance.

where $s_1 = -R/L$ and s_2 and s_3 are the roots of the denominator quadratic; they are either both real or both complex, one the conjugate of the other. For discussion purposes, assume they are complex.

Again the impedance is a ratio of two polynomials. As before, each of the factors can be represented in the complex plane by a directed line from one of the points s_1, s_2, or s_3 to the point $s = j\omega$, as shown in Figure 7. Aside from the multiplying factor $1/C$ in the preceding equation, the impedance is completely determined by these directed lines.

In a numerical example, it may be required to find the magnitude and angle of Z. This can be done graphically by means of a diagram such as the one in Figure 7. As a first step, let's express each of the factors of (6) in polar form:

$$s - s_1 = K_1 e^{j\theta_1}$$

$$s - s_2 = K_2 e^{j\theta_2} \tag{7}$$

$$s - s_3 = K_3 e^{j\theta_3}$$

The quantities K_1, K_2, and K_3 are the lengths of the directed lines in the diagram, and θ_1, θ_2, and θ_3 are the angles they make with the horizontal, as illustrated in Figure 7b. The magnitude and angle of Z will be obtained by inserting these expressions into (6). Thus:

$$|Z| = \frac{1}{C} \frac{K_1}{K_2 K_3}$$

$$\angle Z = \theta_1 - \theta_2 - \theta_3$$

That is, aside from the factor $1/C$ (which is almost like a normalizing factor), the magnitude of Z is obtained by taking the lengths of the lines representing the denominator factors and dividing into the length of the line representing the numerator factor. Also, the angle of Z is the angle of the numerator factor minus the angles of the denominator factors. These lengths and angles can be measured directly from the diagram.

Each of the preceding examples has dealt with a driving-point function. As a different example, let's consider a transfer function, the one for the circuit shown in Figure 8. (When used in control systems, this is called a *lag network.*) Assuming that the input voltage varies as e^{st}, the desired function is the voltage gain $G_{21} = V_2/V_1$. This is easily obtained from the voltage-divider relationship:

$$G_{21} = \frac{1/C_2 s}{\dfrac{1}{C_2 s} + \dfrac{1}{C_1 s + 1/R}} = \frac{C_1}{C_1 + C_2} \frac{s + 1/RC_1}{s + \dfrac{1}{R(C_1 + C_2)}} = K\frac{s - s_1}{s - s_2} \tag{8}$$

where:

$$K = \frac{C_1}{C_1 + C_2} \qquad s_1 = -\frac{1}{RC_1} \qquad s_2 = -\frac{1}{R(C_1 + C_2)} \tag{9}$$

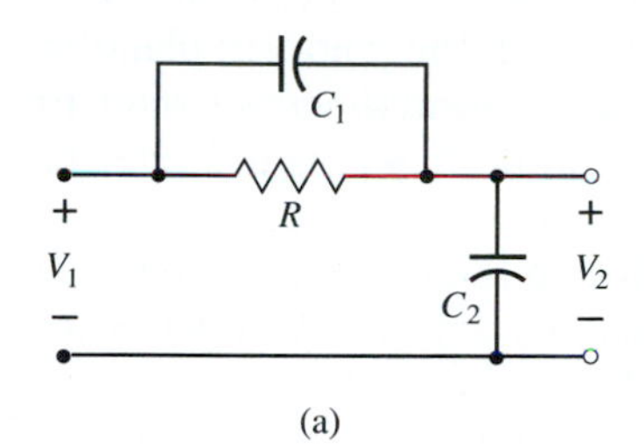

(a)

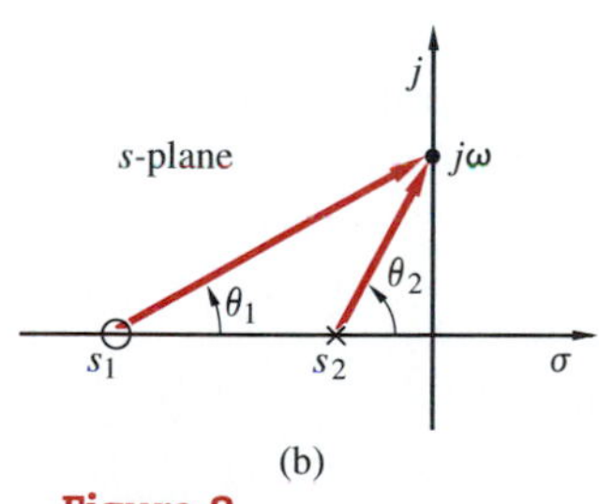

(b)

Figure 8
Lag network and its complex-plane representation.

Again we find that the network function is the ratio of two polynomials. In the present case, both s_1 and s_2 are negative real numbers, s_2 having a smaller magnitude than s_1, as evident from (9). The voltage gain can be determined from the two directed lines from s_1 to $j\omega$ and from s_2 to $j\omega$, as in Figure 8b.

Parenthetically, the origin of the name *lag network* is clear from a consideration of Figure 8. The angle of the voltage-gain function is $\theta_1 - \theta_2$, as you can see from (8). It is clear from the figure that, at any angular frequency ω, the angle θ_1 is smaller than the angle θ_2. Hence, $\theta_1 - \theta_2$ is negative, which means that the angle of the response V_2 is less than that of the excitation V_1, and so the response *lags* the excitation in angle.

POLES AND ZEROS OF NETWORK FUNCTIONS Having examined a number of examples, we are now ready to make some generalizations. Looking back over the examples, we observe that the form of each network function (driving-point or transfer) consists of the ratio of two polynomials in s. Such a function—the ratio of two polynomials—is called a *rational function.* Without examining more complicated circuits in detail, you can easily appreciate that each inductor and capacitor in a circuit will introduce a factor of s into the expression for any network function. This is because the impedance or admittance of these components is proportional to s or $1/s$. The difference between the functions we have explored and the ones corresponding to more complicated networks is that the latter will have higher-order polynomials in their numerator and denominator. In the general case, any network function $H(s)$ of linear, lumped networks will have the following form:

$$H(s) = K\frac{(s - s_{01})(s - s_{02})\cdots(s - s_{0m})}{(s - s_{p1})(s - s_{p2})\cdots(s - s_{pn})} \tag{10}$$

This can be expressed in the form of a theorem:

> Any network function of a linear network is a rational function of the complex-frequency variable s.[1]

When s is purely imaginary ($s = j\omega$), there may be no values of s at which any of the factors in the numerator or denominator of a network function become zero. If s is allowed to take on any complex value whatsoever, however, then both the numerator and denominator will go to zero for some specific values of s. When the denominator goes to zero, the whole function goes to infinity.

Some terminology is in common use to name the values of s for which these things happen. A value of s at which a function vanishes (becomes zero) is called a *zero* of the function. Similarly, a value of s at which the function goes to infinity (the denominator goes to zero) is called a *pole* of the function. Note: Since impedance is the reciprocal of admittance, the poles of the impedance of a network will be zeros of the admittance, and vice versa. In this terminology, it is clear that, aside from a normalizing scale factor such as K in (8), the value of any network function depends only on the locations in the complex plane of its poles and zeros and on the frequency of the source. Sometimes we want to refer collectively to poles and/or zeros. To avoid repetition, we would refer to these as the *critical frequencies.*

When displaying the poles and zeros of a function in the complex plane, it is common practice to use a small circle to identify a zero and a small × to identify a pole. Such a display of the poles and zeros is called a *pole-zero* diagram. (You will notice that we followed this practice in the preceding diagrams.)

Finally, we found a very strong link between the natural frequencies in a circuit and the poles of an appropriate network function. Later in this chapter we will amplify on this.

EXERCISE 1

The transfer impedance of a network, $F(s)$, is given.

$$F(s) = \frac{-2(s-1)}{(s+1)(s+2)}$$

(Assume that the frequency has been normalized.) Draw a complex plane to scale and place the poles and zero at their appropriate positions. (a) Measure the lengths of lines (and the angles they make with the horizontal) from the poles and zeros to the frequency $\omega = 2$ on the imaginary axis. Find the magnitude and the angle of the transfer impedance at this frequency. (b) Confirm your result algebraically by inserting $s = j2$ in $F(s)$. (If you want more practice, repeat the process for other values of ω.) □

[1] This should be taken in context. In this book, *all* the circuits under consideration are lumped and their parameters do not vary with time. Hence, in this context, it isn't necessary to state that, besides being linear, the network must be lumped and time-invariant. However, in a general context, further limitations (lumped and time-invariant) on the validity of the theorem would have to be stated explicitly.

3 POLE-ZERO DIAGRAMS AND RESONANCE

In the preceding section, when discussing graphical methods of determining the steady-state response from pole-zero diagrams, we did not specifically consider that the frequency might be varying; we measured the lengths of the directed lines from the poles and zeros to some point on the $j\omega$-axis. However, it is clear by looking back at Figure 8, say, that the point representing the source frequency can be moved up the axis from zero to infinity. If line lengths and angles are measured at a number of frequencies, then sketches of the magnitude and angle as a function of frequency can be drawn rapidly.

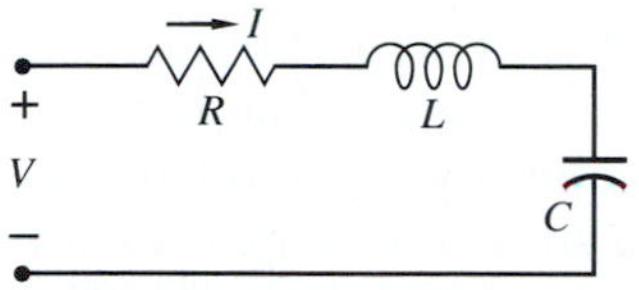

A class of circuits discussed at length in previous chapters is that of resonant circuits, illustrated by the series circuit in Figure 9. In the preceding chapter we plotted the resonance curve against frequency and introduced the concepts of quality factor Q, resonant frequency ω_0, and bandwidth β.

Figure 9
Series resonant circuit.

In Chapter 7 we found the natural frequencies and introduced the concepts of undamped and damped natural frequencies, ω_0 and ω_d, and damping constant α. All these expressions are collected here:

$$
\begin{aligned}
&\omega_0^2 = \frac{1}{LC} \\
&Q_0 = \frac{\omega_0 L}{R} \qquad \alpha = \frac{R}{2L} = \frac{\omega_0}{2Q_0} \\
&\beta = \frac{\omega_0}{Q_0} \qquad \omega_d^2 = \omega_0^2 - \alpha^2
\end{aligned}
\tag{11}
$$

Some observations. First, ω_0 makes its appearance both in the time domain (natural response) and in the frequency domain (resonance). Second, the damping constant (α) in the time domain is just half the bandwidth (β) in the frequency domain.

Let's assume that the excitation is the voltage, so that the appropriate network function representing the response-to-excitation phasor is the admittance. It can be written in several different forms using the preceding parameters as follows:

$$
\begin{aligned}
Y &= \frac{1}{sL + R + \dfrac{1}{sC}} = \frac{1}{L}\frac{s}{s^2 + \dfrac{R}{L}s + \dfrac{1}{LC}} \\
&= \frac{1}{L}\frac{s}{s^2 + 2\alpha s + \omega_0^2} = \frac{1}{L}\frac{s}{s^2 + \dfrac{\omega_0}{Q_0}s + \omega_0^2} \\
&= \frac{1}{L}\frac{s}{s^2 + \beta s + \omega_0^2} = \frac{1}{L}\frac{s}{(s - s_1)(s - s_2)}
\end{aligned}
\tag{12}
$$

The factors in the denominator have been shown explicitly in the last line.[2]

[2] Study these expressions carefully and relate them to each other. This equation will serve as a handy reference for future use.

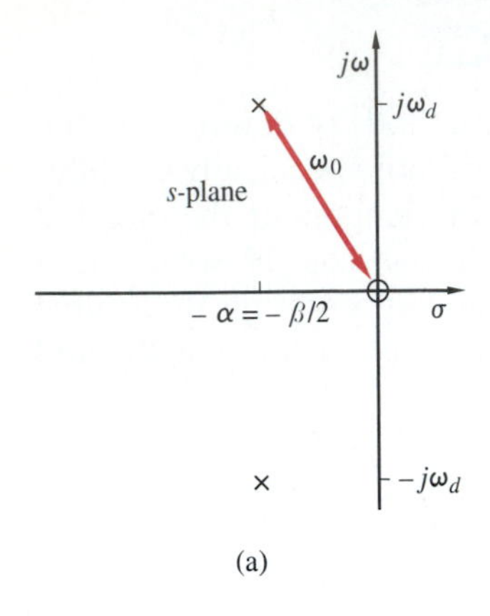

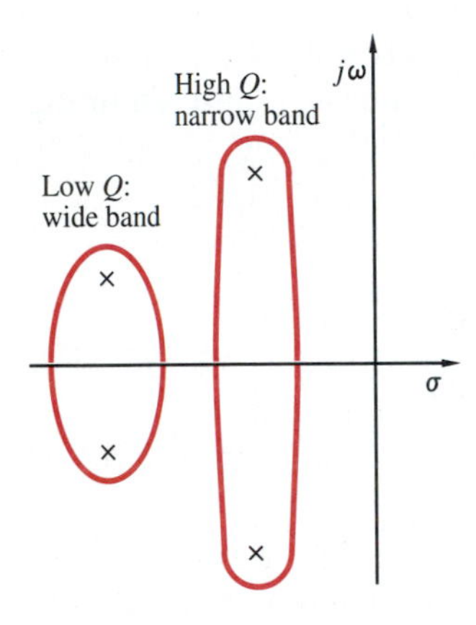

Figure 10
Pole-zero diagram of resonant circuit.

The poles can be expressed in several alternative forms, the most appropriate form depending on whether or not the poles are complex. For the moment, let's assume that the poles are complex. They can be written as:

$$s_{1,2} = -\alpha \pm j\sqrt{\omega_0^2 - \alpha^2} = -\alpha \pm j\omega_d$$
$$s_{1,2} = -\frac{\beta}{2} \pm j\omega_0\sqrt{1 - (1/2Q_0)^2} \tag{13}$$

From the last of these, it is clear that the poles will be complex if $Q_0 > \frac{1}{2}$. The pole-zero diagram is shown in Figure 10.

As noted a few paragraphs back, the magnitude of the real part of the poles (namely, α) is half the half-power bandwidth, and the distance from each pole to the origin (which is the magnitude of the pole) is the resonant frequency ω_0. Remembering that, for a given ω_0, the bandwidth is inversely proportional to Q_0, this means that the closer the poles are to the $j\omega$-axis, the narrower will be the bandwidth and the larger will be the quality factor. Conversely, when the poles are far from the $j\omega$-axis, Q_0 will be small and the bandwidth will be large. These comments are illustrated in Figure 10b.

Graphical Determination of Magnitude and Phase

Let us now estimate from the pole-zero diagram how the magnitude of the admittance will vary as the frequency is increased from zero to some high value. For a source frequency lying somewhere on the imaginary axis, as shown in Figure 11, the normalized magnitude of the admittance will be:

$$|Y| = \frac{|s|}{|s - s_1|\,|s - s_2|} = \frac{K_1}{K_2K_3} \tag{14}$$

where the K's are the lengths of the lines from the poles and the zero to the point on the imaginary axis corresponding to the particular source frequency ω. They can be determined from the diagram as follows:

$$|s| = K_1 = \omega$$
$$|s - s_1| = K_2 = \sqrt{\alpha^2 + (\omega - \omega_d)^2}$$
$$|s - s_2| = K_3 = \sqrt{\alpha^2 + (\omega + \omega_d)^2}$$

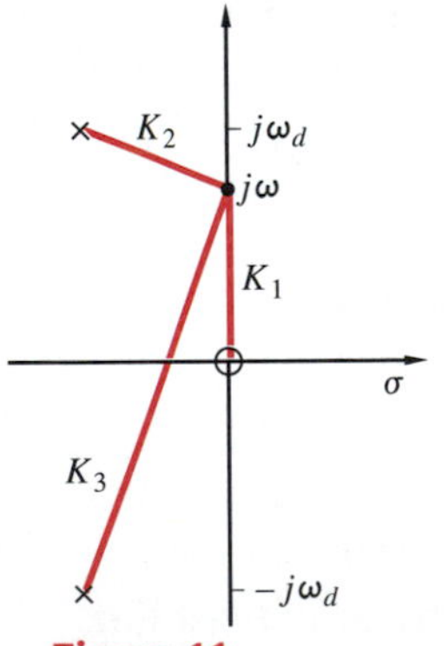

Figure 11
Graphical calculation of resonance curve.

As the frequency increases from zero, K_1 and K_3 increase monotonically. However, K_2 at first decreases until ω equals ω_d, that is, until the varying point takes a position on the imaginary axis just opposite to the pole s_1. From this point on, K_2 begins to increase again. Thus, K_2 has a minimum at $\omega = \omega_d$. Since K_2 is in the denominator of (14), you might think that $|Y|$ must have a maximum at this frequency. But this would not be an accurate conclusion; it isn't necessarily true that a function which depends on more than one factor will reach a maximum when one of the factors ($1/K_2$, in this case) is a maximum.[3]

[3] Suppose that $y = f_1(x)f_2(x)$. Then $y' = f_1f_2' + f_2f_1'$. That is, y' will not generally be zero where f_1' or f_2' is zero.

Nevertheless, if the poles are very close to the imaginary axis (high Q, low α) and we restrict ourselves to the vicinity of ω_d, the lengths of the lines K_1 and K_3 will be varying at approximately the same rate. Hence, the variation of the magnitude of Y will depend almost entirely on the variation of K_2. But, in the case of high Q (low damping), ω_d is approximately the same as ω_0, the frequency at which we know the magnitude of Y is a maximum.

Thus, we see that the visual picture provided by the pole-zero diagram gives a clear interpretation of the phenomenon of resonance and, at least in the high-Q case, it yields quantitative results that are approximately correct to quite a high degree of accuracy. Very often, such an approximate estimate of network behavior is all that is needed.

The lower the Q, the worse will be the approximate result obtained by neglecting the variations of the lines from $s = 0$ and $s = s_2$. When Q_0 becomes less than $\frac{1}{2}$, the poles become real, so the preceding detailed analysis does not apply. The s-plane description in terms of line lengths from poles and zeros, however, is still a valid representation.

Relationship between Steady-State and Natural Responses

The relationship between the steady-state response and the natural response can be demonstrated using the high-Q resonant circuit. With a variable-frequency sinusoidal voltage source, we make measurements of the input current. We determine the frequency at which the current-phasor magnitude reaches a maximum, which we know to be ω_0, and we measure this maximum current. Then we compute $1/\sqrt{2}$ of this maximum as the value at the band edges. Then we vary the frequency of the source and determine the two frequencies ω_1 and ω_2 on either side of ω_0 at which the current magnitude equals this value ($1/\sqrt{2}$ of the maximum). From these frequencies we find the bandwidth $\beta = \omega_2 - \omega_1$ and the damping constant $\alpha = \beta/2$. The damped angular frequency is then found from (13).

EXERCISE 2

A 10-V rms sinusoidal signal source is connected to a resonant circuit. The frequency of the source is varied until a maximum reading (which is 20 mA) is obtained on an rms-reading milliammeter; the frequency at which this happens is 100 kHz. The frequencies at which the current reads 14.14 mA are measured to be 96.15 and 104 kHz, respectively. Calculate the parameters R, L, and C of the circuit.

ANSWER: $R = 500\,\Omega$, $L = 10.14$ mH, $C = 250$ pF. □

Reviewing the discussion of the oscillatory natural response of the series resonant circuit in Chapter 7, you will recall that the *form* of the response, $Ke^{-\alpha t}\cos(\omega_d t + \theta)$, is determined completely by α and ω_d. Thus, except for the constants K and θ (which are the magnitude and angle), the natural response is determined completely from steady-state measurements. We will encounter a universal generalization of this observation in Chapter 14.

4 POLES AND NATURAL FREQUENCIES

When examining each of the circuits discussed in the preceding sections, we found that in each case the admittance function has a pole at the precise point in the s-plane corresponding to the transient (natural) frequency. It is our purpose now to inquire into the relationships among the natural frequencies and the poles and zeros of network functions. We will examine some specific cases in this section and then make some generalizations.

Consider again the two-branch parallel-resonant circuit in Figure 6, once excited by a current source and once by a voltage source, both varying in time as e^{st}. In the first case, the response is the voltage. In this case, the ratio of the response phasor to the excitation phasor is the impedance $Z = V/I_g$:

$$Z = \frac{1}{C}\frac{s + R/L}{s^2 + \frac{R}{L}s + \frac{1}{LC}} \tag{15}$$

The poles of the impedance are the roots of the polynomial in the denominator. (A subscript is used to designate the source both in this case and later for the voltage source so that the variables without subscript can refer to the inductor current and capacitor voltage.)

Next, let's determine the natural frequencies of the same circuit. (Review Chapter 7.) For this purpose, we deal with functions of time and arrive at appropriate differential equations. One equation is obtained by applying KCL at the upper node and eliminating the capacitor current in favor of its voltage by substituting the capacitor i-v relation. A second equation is obtained by applying KVL around the loop formed by R, L, and C and then, by using the v-i relations of R and L, eliminating their branch voltages in favor of the current i. (Carry out these steps and convince yourself that this process uses all the independent equations in the circuit.) The result is:

$$\begin{aligned} C\frac{dv}{dt} + i &= i_g \\ L\frac{di}{dt} + Ri - v &= 0 \end{aligned} \tag{16}$$

Recall that the natural frequencies are found by setting the excitation to zero and assuming that i and v are exponentials $K_i e^{st}$ and $K_v e^{st}$, where s is the unknown natural frequency. When these are substituted into the preceding equations, the result is a pair of homogeneous algebraic equations whose unknowns are K_i and K_v; a nontrivial solution exists only if the determinant of these equations vanishes. Carrying out these steps leads to the following characteristic equation:

$$s^2 + \frac{R}{L}s + \frac{1}{LC} = 0 \tag{17}$$

The roots of this equation are the natural frequencies. (You should go through the details of what was just outlined or carefully review the development in Chapter 7.)

It is immediately evident that this polynomial is identical with the one in the denominator of the impedance in (15). The conclusion is that *the impedance poles are the same as the natural frequencies when the circuit is excited by a current source.*

Let us now store this information, subject to later retrieval, and briefly put aside this aspect of the discussion. Next, let's assume that the same circuit is excited by a voltage source, as in Figure 6b. The current through the capacitor is $C\,dv/dt = C\,dv_g/dt$; with a known source voltage, this current is found by differentiation. The presence of the capacitor will not influence the rest of the circuit, so it might as well be removed. A differential equation for the rest of the circuit is written, then the excitation is set equal to zero, and the current is assumed to have the exponential form Ke^{st}. This process leads to:

$$L\frac{di}{dt} + Ri = v_g$$

$$v_g = 0 \qquad i = Ke^{st}$$

$$K(sL + R)e^{st} = 0$$

$$s + \frac{R}{L} = 0$$

(Again it is for you to carry out the details.) The solitary root of this last equation is the natural frequency under the present conditions. From the equation for the impedance we notice that this root is precisely the zero of the impedance. We conclude that *the zero of the impedance* (*which is the pole of the admittance*) *is the same as the natural frequency when the circuit is excited by a voltage source.*

It is possible to give another interpretation of the two results just obtained. Recall that the natural frequencies of a circuit are obtained by setting the excitation to zero. When the excitation is a current source, setting it to zero means opening the terminals. Under this condition, quite naturally, the natural frequencies are called the *open-circuit natural frequencies.* Recall also that, when the excitation is a current source, it is the impedance which is the ratio of response (voltage) phasor to excitation (current) phasor. Hence, we say that *the poles of the impedance are the open-circuit natural frequencies.*

On the other hand, when the excitation is a voltage source, setting the excitation to zero in order to find the natural frequencies implies short-circuiting the terminals, because that's how a voltage is made to vanish. The natural frequencies under this condition are called the *short-circuit natural frequencies.* Furthermore, when the excitation is a voltage source, it is the admittance which is the ratio of response (current) phasor to excitation (voltage) phasor. Hence, *the poles of the admittance are the short-circuit natural frequencies.*

These results are summarized in the following table:

Natural frequencies	Excitation	Terminal condition
Poles of Z	I-source	Open circuit
Poles of Y	V-source	Short circuit

Although the preceding discussion was based on a single example, the same result applies in general. We will assume the validity of this generalization without further discussion.

EXERCISE 3

The function treated for the lag network in Figure 8 was the voltage-gain function given in (8). Find the short-circuit natural frequencies for this circuit and show that they are the same as the poles of the voltage-gain function in (8). □

Poles of Network Functions for a Given Excitation

Although in the immediately preceding exercise the function treated was the voltage gain, when we considered the open-circuit and short-circuit natural frequencies in the subsection just preceding, the functions treated were the driving-point functions Z and Y. However, the results obtained there apply for any other network function. For the current-source excitation in the resonant circuit of Figure 6b, for example, we might consider also the current ratio I_R/I. It can be obtained from the voltage-divider relationship, as follows:

$$\frac{I_R}{I} = \frac{1/sC}{\dfrac{1}{sC} + R + Ls} = \frac{1/LC}{s^2 + \dfrac{R}{L}s + \dfrac{1}{LC}}$$

We find that the current ratio has the same poles as the impedance in (15). Both of these functions are ratios of a response phasor to an excitation phasor corresponding to open-circuit conditions at the excited terminals.

Indeed, this is a general condition which can be stated as:

In a linear network excited by a sinusoidal source, every network function corresponding to the ratio of a response phasor to the excitation phasor has the same poles.

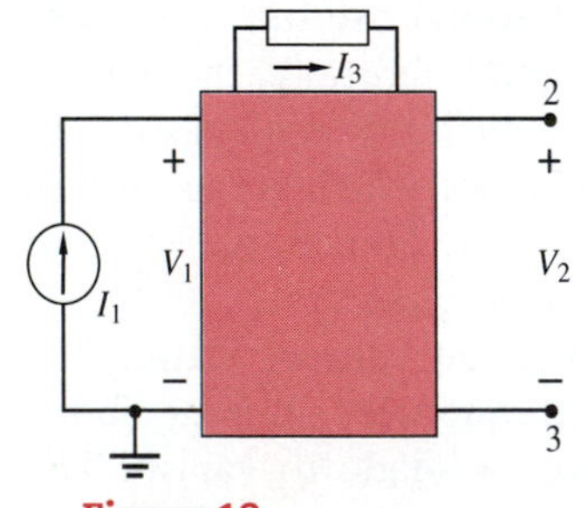

Figure 12
Circuit used to prove that all network functions for a given excitation have the same poles.

A proof of this statement can be outlined as follows. An arbitrary linear network is shown in Figure 12. Let's assume it is excited by a current source I_1 across which the voltage phasor is V_1. Two other response phasors are also shown: a voltage V_2 across nodes 2 and 3 and a current I_3. Assume that a set of node equations has been written in the complex-frequency domain. The solutions of these equations by Cramer's rule are:

$$\frac{V_1}{I_1} = \frac{\Delta_{11}}{\Delta} \qquad \frac{V_2}{I_1} = \frac{\Delta_{21}}{\Delta} \qquad \frac{V_3}{I_1} = \frac{\Delta_{31}}{\Delta} \tag{18}$$

where Δ is the determinant of the node equations and the subscripted Δ's are its cofactors. (See Appendix A.) The zeros of Δ are the poles of each of these functions, and thus, all the functions have the same poles.[4] A similar approach can be taken in finding the current gain I_3/I_1, but we won't pursue the details here.

EXERCISE 4

Assume that the excitation of the network in Figure 12 is a sinusoidal voltage. Using a similar approach, prove that all the functions representing a response phasor to the excitation phasor have the same poles, with the possible caveat discussed in the footnote to the preceding paragraph. □

The Extreme Frequencies: Zero and Infinity

Two specific frequencies are of particular interest. One of these is $s = 0$, corresponding to DC. At this frequency, a network function will have one of three possible values. One possibility is illustrated by the impedance in (6), where $Z \to R$, a constant, as s goes to 0. This is also evident from the circuit in Figure 6; at zero frequency, the capacitor becomes an open circuit and the inductor a short circuit, leaving only the resistor in the circuit.

The other two possibilities are illustrated by the impedance and admittance of Figure 4 given in (5). The admittance has a zero at $s = 0$, while the impedance has a pole there. A pole or a zero at $s = 0$ is detected just as other poles and zeros are: by the presence of a factor s (which can be viewed as $s - 0$).

The other specific frequency is infinity. Again, a function will exhibit three possible behaviors at this frequency. One possibility is illustrated by Z or Y in (5): $Z \to R$ (a constant) as $s \to \infty$. This can also be verified by the circuit in Figure 4, since the capacitor becomes a short circuit at very high frequencies.

A second possibility is illustrated by the impedance in (6). In a rational function, the highest powers of s in the numerator and the denominator dominate at high frequencies. In the case of (6), as $s \to \infty$, $Z \to Ls/LCs^2 = 1/Cs \to 0$. We say that the function has a simple zero at infinity, although this zero is not evident in the form of a factor like one of those in the last line of (6). (*Simple* means "not multiple"; the multiplicity of the zero is the difference of the highest powers in the denominator and the numerator. For example, in the unnumbered equation preceding (19), the zero at infinity is double.)

The third possibility is a rarity. It would occur, for example, in a series RL circuit whose excitation is a current, the network function being the impedance $Z = Ls + R$. As $s \to \infty$, so does Z. We say the impedance has a pole at

[4] Some niceties have to be examined. Is it possible that a zero of Δ will cancel with a zero of one or more of the cofactors but not all of them? In that case, the canceled zero would be a pole of one or more of the responses but not of all of them. The answer is that this is possible. We might say that such canceled factors are *private poles* of those responses from which they do not cancel. This type of discussion will take us too far afield, so we won't pursue it further here. Aside from such possible private poles, the general theorem is true.

infinity. Again, it is the difference in the highest powers of s in the numerator and denominator that determines the behavior at infinity.[5]

EXERCISE 5

Describe the following function completely in terms of its critical frequencies:

$$F(s) = \frac{10s(s + 2)^2}{(s + 3)(s + 4)(s^2 + 2s + 5)}$$

ANSWER: Simple zeros at 0 and ∞; double zero at -2; simple poles at -3, -4, and $-1 \pm j2$. □

Review and Consolidation

Some matters of great significance have been discussed in this chapter. The concept of a complex-frequency variable has revealed an intimate relationship between the natural response of linear networks in the time domain, on the one hand, and the sinusoidal steady-state response, on the other.

Given a linear network, we can obtain a network function in terms of the complex variable s using normal methods of analysis by taking the reactive impedances to be Ls and $1/Cs$, respectively. The poles and zeros of the function are the roots of the denominator and numerator, respectively. What is more, the sinusoidal frequency responses (magnitude and angle) are then obtained by setting $s = j\omega$. A graphical method of making these calculations in the complex plane at different values of ω was described in an earlier section. Another approach will be discussed in the next section.

The poles of the network function in the complex-frequency domain under voltage excitation were found to be the same as the natural frequencies when the terminals at which the voltage excitation appears are short-circuited. Similarly, the poles of the network functions under current excitation are the same as the natural frequencies when the terminals at which the current excitation appears are left open. Thus, to find the natural response, there is no need to write state equations or any other differential equations.

5 LOGARITHMIC MEASURE

Before proceeding, let's review some of the common features of the preceding chapter and this one. Generally speaking, we have been discussing circuits excited by sinusoids of fixed amplitude and phase but variable frequency. In several cases (Figures 4, 7, and 8 in Chapter 11, for example), we plotted

[5] A sinusoidal current source in series with an RL branch is a curiosity. How was it established? Presumably, by switching the source onto the circuit. However, unless this was done at the precise moment that the sinusoidal current has the same value as the initial value of the inductor current, an impulse voltage must occur across the current source.

frequency-response curves—curves of magnitude and angle of network functions (such as voltage gain or admittance) against frequency. In the present chapter (in Figure 11, for example) we described a graphical approach to determining the points on these curves from the pole-zero diagrams of the functions. Glance back at these frequency-response curves; you will notice that the scales on both coordinates are linear.

Linear scales can sometimes be a disadvantage. Suppose the variables cover a large range, say, from 0 to 10^5 or more in frequency and 0 to 1000 in magnitude. A linear scale in such cases can result in the loss of detail. Thus, the interval from 10 to 100 in frequency—which might cover a region of considerable variation in the function—takes up as much of the axis as the interval from 50,000 to 50,090, where the amount of variation of the function might be insignificant. It would be very useful to expand the scale in such a way that the interval from 10 to 100 covered the same length on the axis as the interval from 10,000 to 100,000.

This goal is achieved by the use of a logarithmic scale for either the response variable, the frequency variable, or both. Hence, in this section we shall introduce what amounts to a change of variables from linear to logarithmic ones. The resulting plots will be logarithmic frequency-response curves.

The quantities under consideration are transfer functions in general. In order to convey this generality, let's use the notation $H(j\omega) = |H(j\omega)|\, e^{j\phi(\omega)}$ to represent *any* network function: voltage or current gain, transfer impedance or admittance, even driving-point functions. However, for simplicity, let's refer to it as the gain function. The natural logarithm (logarithm to the base e, designated as ln) of this function is:

$$\begin{aligned} \ln H(j\omega) &= \ln\left(|H(j\omega)|\, e^{j\phi(\omega)}\right) = \ln|H(j\omega)| + \ln e^{j\phi} \\ &= \alpha'(\omega) + j\phi(\omega) \end{aligned} \tag{19}$$

where

$$\alpha'(\omega) = \ln|H(j\omega)| \qquad \text{or} \qquad |H(j\omega)| = e^{\alpha'} \tag{20}$$

and we have used the fact that the log of a product is the sum of the logs of the factors. The quantity α' is the natural logarithm of the gain magnitude. We call it the *logarithmic gain,* to distinguish it from the plain, ordinary gain H; its unit is the *neper.*[6]

The Decibel

For numerical work, the natural logarithm is not as convenient to use as is the common logarithm (to the base 10). In the preceding equation for α', if the common log is used instead of the natural log, the resulting unit will be inconveniently large. Hence, it is common practice to introduce a numerical

[6] This unit is named in honor of the Scottish mathematician John Napier (1550–1617), who invented the natural logarithm and published tables of logarithms which were of such extent and accuracy that no thought was given to recalculation for more than a century. The spelling of the unit for logarithmic gain is a variation on his name.

factor, as in the following definition:

$$\begin{aligned} \alpha(\omega) &= 20 \log |H(j\omega)| \\ |H(j\omega)| &= 10^{\alpha/20} \end{aligned} \tag{21}$$

(To avoid writing the base of logarithms, it is common practice to use "ln" to designate the natural logarithm and "log" for the common logarithm.) Both α and α' are called logarithmic gain, but their units are different. The unit of α is the *decibel* (dB). The relationship between the dB and the neper is found by inserting the second equation in (20) into the first one in (21):

$$\alpha = 20 \log e^{\alpha'} = (20 \log e)\alpha' \approx 8.686\alpha' \tag{22}$$

Thus, the number of decibels is obtained by multiplying the number of nepers by about 8.7.

Note that when mathematical operations involving the entire network function (magnitude and phase) are to be performed, it is the natural logarithm that we use. The common log is used only for numerical convenience when interest is in the magnitude only.

HISTORICAL DIGRESSION It may have occurred to you that if there is a unit called the *deci*bel, there ought to be one called the *bel,* of which the decibel is one-tenth. Historically, this unit arose in the study of sound, where the most meaningful variable is power in a sound wave. Since audio power can range over many powers of 10, it is more convenient to use a logarithmic measure. Thus, the unit adopted is a logarithmic measure of the ratio of two powers.[7]

Thus, if P_1 and P_2 are two values of power, then $\log P_2/P_1$ is a measure of the relative sizes, a logarithmic power ratio with the *bel* as unit.[8] For convenience of size, a unit one-tenth of a bel was chosen and called a decibel. Thus, the original definition was:

$$\alpha = 10 \log \frac{P_2}{P_1} \qquad \text{dB} \tag{23}$$

In the steady state, power is proportional to the square of the magnitude of a voltage or current phasor. Appropriate expressions are:

$$\begin{aligned} P_1 &= |V_1|^2 \operatorname{Re} Y_1 = |I_1|^2 \operatorname{Re} Z_1 \\ P_2 &= |V_2|^2 \operatorname{Re} Y_2 = |I_2|^2 \operatorname{Re} Z_2 \end{aligned} \tag{24}$$

where Z_1 is the impedance (Y_1 the admittance) of the circuit in which power P_1 is dissipated and Z_2 is the impedance (Y_2 the admittance) of the circuit in which P_2 is dissipated. When these equations are substituted into (23) for α,

[7] Similar considerations went into the adoption of the Richter scale in measuring earthquakes. An earthquake measuring 6 on the Richter scale, for example, has an amount of energy 10^6 times that of some standard.

[8] The unit was named after Alexander Graham Bell (1847–1922), an American inventor of Scottish origin whose original profession was teaching the deaf. Bell's initial motivation that led to the invention of the telephone was the desire to create an instrument to help people with impaired hearing.

the result is:

$$\begin{aligned}\alpha &= 10 \log \left|\frac{V_2}{V_1}\right|^2 \frac{\operatorname{Re} Y_2}{\operatorname{Re} Y_1} \\ &= 10 \log \left|\frac{I_2}{I_1}\right|^2 \frac{\operatorname{Re} Z_2}{\operatorname{Re} Z_1} \qquad \text{dB}\end{aligned} \tag{25}$$

In case the two circuits are the same, so that $\operatorname{Re}(Z_2) = \operatorname{Re}(Z_1)$, these expressions reduce to:

$$\alpha = 20 \log \left|\frac{V_2}{V_1}\right| = 20 \log \left|\frac{I_2}{I_1}\right| \tag{26}$$

Only in such a case is the historical definition of the decibel consistent with our definition in (21). However, it is standard practice among electrical engineers to define the dB as in (21), and we shall continue the practice in this book.

Note that, for the simple low-pass circuit of Figure 3 in Chapter 11, for example, the output voltage V_2 is the voltage across a capacitor; hence no average power is dissipated and the historical definition of the decibel in terms of a power ratio would not be applicable. Nevertheless, the electrical engineering definition in terms of a voltage ratio is still valid.

Decades and Octaves

What has been considered so far is a logarithmic measure of the response. Let us now turn to a consideration of a logarithmic measure for the *frequency*. First we normalize the variable by choosing an appropriate reference frequency to which all others are referred. The frequency chosen as reference is arbitrary; call the reference frequency ω_r and assume that all frequencies are divided by it. The logarithmic frequency variable is then defined as:

$$u = \log \frac{\omega}{\omega_r} \qquad \frac{\omega}{\omega_r} = 10^u \tag{27}$$

To get an idea of the unit of u, note that $u = 0$ when $\omega = \omega_r$. To make $u = 1$ requires $\omega = 10\omega_r$. Similarly, when $u = 2$, $\omega = 100\omega_r$. Thus, each unit change in u requires the frequency to change by a factor of 10. For this reason, the unit of u is called a *decade*.

Another unit for the logarithmic frequency in common use is based on the logarithm to the base 2. We write:

$$u' = \log_2 \frac{\omega}{\omega_r} \tag{28}$$

For u' to change from 0 to 1, ω must change from ω_r to $2\omega_r$. Each time the frequency doubles, u' changes by one unit. For this reason, the unit of u' is called the *octave.* Perhaps you are familiar with the octave in connection with music where a musical note is an octave away from another note if it has twice the frequency.

The octave and the decade are related through noting that the log of a quantity to the base 2 equals (the log of the quantity to the base 10) times (the log of 10 to the base 2). Thus:

$$u' = \log_2 \frac{\omega}{\omega_r} = (\log_2 10) \log \frac{\omega}{\omega_r} \approx \frac{10}{3} u \tag{29}$$

That is, the number of octaves in a particular frequency interval is approximately 3.33 times the number of decades.[9]

6 LOGARITHMIC FREQUENCY-RESPONSE PLOTS

Now that logarithmic variables have been defined for both the response and the frequency, let us examine the possible resulting advantages. It is our goal to plot the log magnitude (α) in dB and the angle function ϕ against the suitably normalized logarithmic frequency u. Doing this on ordinary (linear-scale) graph paper is hopeless; it would be very difficult to judge where on the linear scale a particular logarithmic coordinate will fall. (Try it.)

Instead, we use paper whose scales have been laid out logarithmically: "log-log" paper is graph paper both of whose axes have been scaled logarithmically; "log" (or semilog) paper has only one of its axes scaled logarithmically. If the latter paper is used, the frequency is plotted on the logarithmically scaled axis and the gain on a linear scale. Using log-log paper permits plotting both the frequency and the gain on a logarithmic scale.

The idea of plotting α and ϕ against the logarithmic frequency was conceived by Bode; these plots are referred to as *Bode diagrams* or *Bode plots.*[10] Such diagrams have been found to be very useful in the design of feedback amplifiers and filters and in control systems.

[9] In some of your outside reading, you might find the quantity $\log \omega$ being used where $\log \omega/\omega_r$ has been used here. The arguments of logarithms—like the arguments of exponentials and trigonometric functions—*must* be dimensionless. It doesn't make sense to take the logarithm of plain, ordinary frequency. However, a normalized frequency $\omega' = \omega/\omega_r$ might be defined, where ω' is dimensionless so it is OK to take its logarithm. Then, for simplicity, one might go so far as to drop the prime, so that $\log \omega'$ becomes $\log \omega$. If this is done, it must be kept in mind that the ω in this expression is normalized frequency.

[10] Hendryk W. Bode was a twentieth-century mathematician and engineer who spent his professional career at the Bell Telephone Laboratories. (His name has two syllables and rhymes approximately with the bullring cheer "olé!," except that the accent is on the o and not the e.) He made major contributions to the theory of feedback and to the development of feedback amplifiers and filters. He also applied what, in mathematics, are called *Hilbert transforms* to get integral relationships between the real and the imaginary parts of network functions, giving one of these in terms of the other. Electrical engineers refer to these as the *Bode formulas.*

EXAMPLE 1 The low-pass RC circuit from Figure 3 in Chapter 11 is repeated in Figure 13. The voltage-gain function (called H) and its magnitude for $s = j\omega$ are:

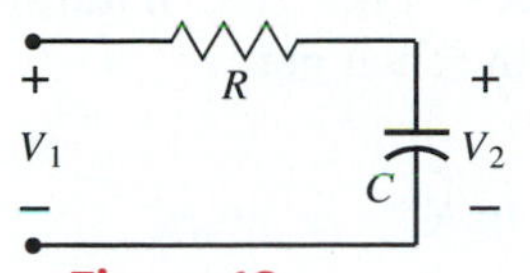

Figure 13
Low-pass RC circuit.

$$H(s) = \frac{V_2}{V_1} = \frac{1/sC}{R + 1/sC} = \frac{1}{RCs + 1} = \frac{1}{s/\omega_r + 1}$$

$$|H(j\omega)| = \frac{1}{\sqrt{(\omega/\omega_r)^2 + 1}} \qquad \omega_r = \frac{1}{RC} \tag{30}$$

The normalizing frequency is chosen as $\omega_r = 1/RC$. A magnitude plot is shown in Figure 14, where, in addition to a scale for $u = RC\omega$ in decades, a scale for ω is also shown. Negative values of u correspond to frequencies below ω_r. As expected, such a plot tends to expand the curve in the vicinity of $\omega = \omega_r$, where the variation of the response is largest. At $u = 0$, $\omega = \omega_r = 1/RC$, the value of $|H(j\omega_r)|$ from (30) is $1/\sqrt{2}$. This frequency was identified in Chapter 11 as the edge of the band. The logarithmic measure at the edge of the band is:

$$\alpha = 20 \log \frac{1}{\sqrt{2}} = -20 \log \sqrt{2} = -10 \log 2 \approx -3 \text{ dB} \tag{31}$$

Thus, here's another way to refer to the bandwidth: the edge of the band is called the *3-dB point* and the bandwidth is the *3-dB bandwidth*.

The variation of the phase with log frequency should also be examined, but this question will be postponed to the next section.

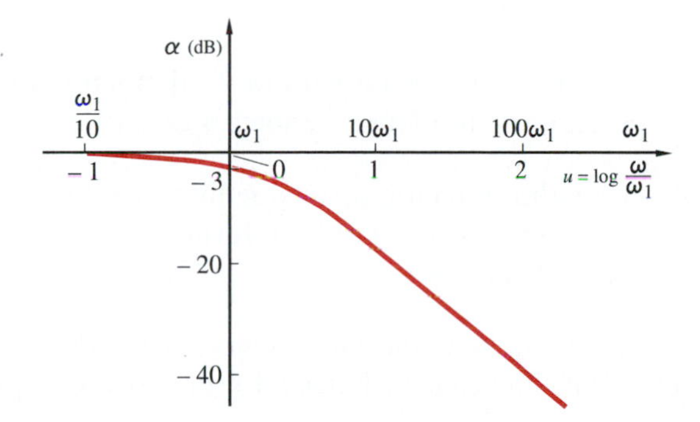

Figure 14
Log-magnitude response curve.

General Case

The example just treated has a single real pole and no finite zeros. The pole is located at $s = -1/RC$; but instead of writing the pole factor as $s + 1/RC$, we normalized the frequency with $\omega_r = 1/RC$ and wrote the pole factor as $RCs + 1$, treating RCs as the normalized frequency.

We plan next to take up the general network function in (10). Let's assume that the frequency has already been normalized and the s in the expression is the normalized frequency. After setting $s = j\omega$, let's take its natural logarithm. Remembering that $\ln(xy/z) = \ln x + \ln y - \ln z$, we get:

$$\ln H(j\omega) = \sum_{k=1}^{m} \ln(j\omega - s_{0k}) - \sum_{k=1}^{n} \ln(j\omega - s_{pk}) + \ln K = \alpha' + j\phi$$

$$\alpha'(\omega) = \sum_{k=1}^{m} \ln|j\omega - s_{0k}| - \sum_{k=1}^{n} \ln|j\omega - s_{pk}| + \ln K \tag{32}$$

$$\phi(\omega) = \sum_{k=1}^{m} \arg(j\omega - s_{0k}) - \sum_{k=1}^{n} \arg(j\omega - s_{pk})$$

where "arg" stands for *argument,* meaning the angle of the function.

The advantage of the change to the logarithm is evident. Multiplication of factors has been replaced by addition, and division has been replaced by subtraction. Furthermore, each zero and pole factor appears separately in a summation. Hence, we need to study the contribution to the frequency response of each separate type of pole and zero factor. To find the variation of the entire function with frequency, we merely add the individual contributions. In the event that the multiplying constant is negative, then an angle π must be added to the phase ϕ.

The distinct types of factors possible in a general network function are as follows:

1. A constant K; assume it is positive.
2. A factor $s + a$ in the numerator or denominator corresponding to a real zero or pole.
3. A factor s in the numerator or denominator corresponding to a zero or pole at $s = 0$. (This is a special case of case 2 with $a = 0$.)
4. A quadratic factor corresponding to a complex conjugate pair of zeros or poles written in standard form: $s^2 + 2\zeta\omega_0 s + \omega_0^2$. (This form will be justified shortly.)

Rather than treat the most general case in (32), let's choose a less general function but one that includes all four possible types of factors listed. Thus:

$$H(s) = K' \frac{s + a}{s(s^2 + 2\zeta\omega_0 s + \omega_0^2)} = \frac{aK'}{\omega_0^2} \frac{s/a + 1}{s\left[\left(\dfrac{s}{\omega_0}\right)^2 + 2\zeta\dfrac{s}{\omega_0} + 1\right]} \tag{33}$$

(Assume that the frequency has been normalized.) On the right-hand side,the constant term of each factor has been factored out, resulting in a new constant multiplier $K = aK'/\omega_0$. Now we construct the logarithmic magnitude (common log) function:

$$\begin{aligned}\alpha &= 20\log|H(j\omega)| = 20\log K + 20\log\left|j\frac{\omega}{a}+1\right| \\ &\quad - 20\log\omega - 20\log\left|1-\left(\frac{\omega}{\omega_0}\right)^2 + j2\zeta\frac{\omega}{\omega_0}\right| \\ &= 20\log K + 20\log\sqrt{\left(\frac{\omega}{a}\right)^2+1} - 20\log\omega \\ &\quad - 20\log\sqrt{\left[1-\left(\frac{\omega}{\omega_0}\right)^2\right]^2 + 4\zeta^2\left(\frac{\omega}{\omega_0}\right)^2} \qquad \text{dB}\end{aligned} \tag{34}$$

This expression puts into evidence the four types of terms that make up the most general gain function. By studying the behavior of these four types, we will have covered every conceivable case.

The constant factor is easy. All that $20\log K$ will do is to shift the logarithmic gain curve vertically by that constant amount.

REAL POLES OR ZEROS Next let's deal with real poles and zeros, starting with a possible pole or zero at $s = 0$. There is a pole at $s = 0$ in (33). The only difference that a zero at $s = 0$ would make is to change the sign of the corresponding term in (34) to +. Remembering that the frequency is normalized, $\log\omega$ is just the logarithmic frequency u. The term of α in (34) corresponding to the pole at $s = 0$ is:

$$\alpha_0 = -20u \qquad \text{dB} \tag{35}$$

That's just the equation of a straight line with a slope of -20 dB per decade passing through the point $u = 0$. This result is shown plotted (in the solid line) in Figure 15. The dashed line with a slope of $+20$ dB per decade shows the corresponding curve if there had been a zero instead of a pole at $s = 0$.

Next let's consider a nonzero real pole or zero; a zero is the case shown in (33). The corresponding term for the logarithmic gain in (34) is:

$$\alpha_r = 20\log\sqrt{\frac{\omega^2}{a^2}+1} = 10\log\left[\left(\frac{\omega}{a}\right)^2+1\right] \tag{36}$$

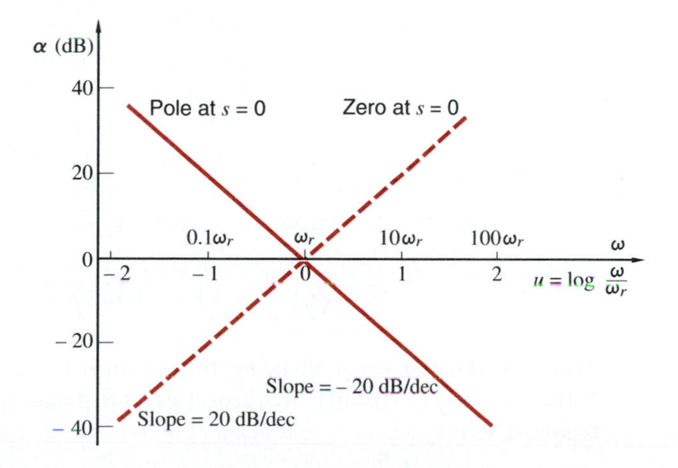

Figure 15
Bode diagrams for a pole or zero of $H(s)$ at $s = 0$.

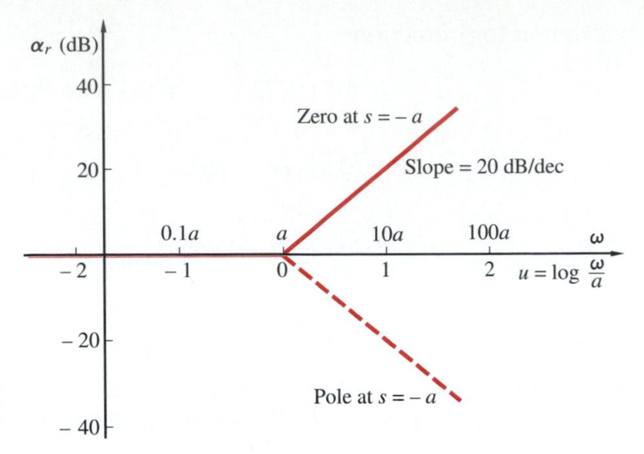

Figure 16
Asymptotic plot of logarithmic gain for real zero (or pole).

The subscript r on α stands for "real."

It would seem that not much is gained if we need to take a succession of values of frequency and then calculate the corresponding value of α from this expression. However, the utility comes from examining the approximation for low and for high frequencies.

At $\omega = 0$, the logarithmic gain in (36) reduces to $20 \log 1 = 0$. At frequencies that are much smaller than a, this function is still very close to 0. For $\omega/a = \frac{1}{10}$, for example, this term's contribution to the logarithmic gain is 0.043. (Confirm this.) On the other hand, for large values of frequency, such that $\omega \gg a$, the dominant term is ω/a, so the function reduces to $20 \log \omega/a = 20\,u$. That is, for high frequencies, α_r becomes a linear function of u with a slope of 20 dB per decade. The gain rises (or falls, if this were a pole factor) at the rate of *20 dB per decade,* or approximately *6 dB per octave.*

Figure 16 shows this asymptotic behavior for high and low frequencies. The low-frequency asymptote is a horizontal line along the u axis, while the high-frequency asymptote is a line with a positive slope of 20 dB per decade passing through the point $u = 0$ (or $\omega = a$). The frequency at the intersection of the two lines is called the *break* frequency. Because of the shape of the curve at this point, it is also called the *corner* frequency. As the frequency increases from low values, or decreases from high values, the curve cannot be distinguished from its asymptotic lines. But the closer the frequency gets to the break point, the farther and farther it deviates from the asymptotes as in Figure 17. According to the asymptotes, the logarithmic gain at the break point is 0. The actual value at this point, where $\omega/a = 1$, is found by inserting this value in the expression for α_r in (36). Thus:

$$\alpha_r\left(\frac{\omega}{a} = 1\right) = 20 \log \sqrt{2} \approx 3 \text{ dB}$$

Thus, the largest error in using the asymptotes instead of the actual values is 3 dB (actually, 3.010 dB, to three decimal places), and it occurs at the break frequency.

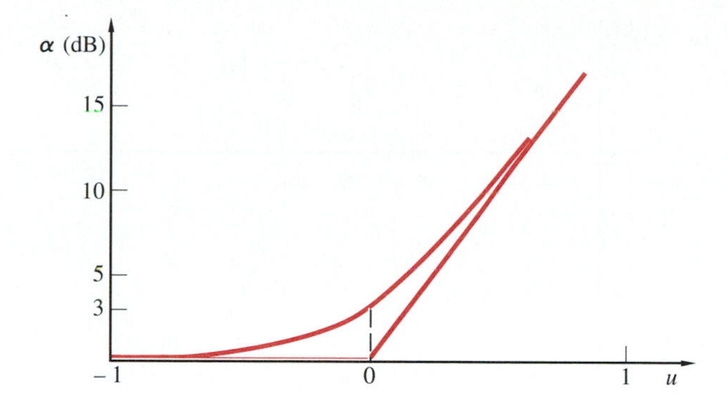

Figure 17
Comparison of actual curve with asymptotes.

COMPLEX POLES OR ZEROS The quadratic factor in (33) is the last one to be considered. At the risk of becoming overwhelmed, you should compare the form of this quadratic with those in (12). The constant term in the quadratic is the same but the coefficient of s is not. By comparison with those in (12), we find that:

$$\zeta = \frac{\beta}{2\omega_0} = \frac{1}{2Q_0} = \frac{\alpha}{\omega_0} \tag{37}$$

Remember that α in this expression is the damping coefficient, the real part of the complex natural frequency—not to be confused with the logarithmic gain. ζ is named the *damping factor*; it is seen to equal the damping coefficient (α) normalized to the undamped frequency (ω_0). The value $\zeta = 1$ corresponds to the critically damped case when the roots of the quadratic are no longer complex, but are real and equal. Thus, the value of the damping factor lies between 0 and 1.

The contribution of a pair of complex poles to the logarithmic gain from (34) is:

$$\alpha_c = -20 \log \sqrt{\left[1 - \left(\frac{\omega}{\omega_0}\right)^2\right]^2 + 4\zeta^2\left(\frac{\omega}{\omega_0}\right)^2} \tag{38}$$

The subscript on α_c stands for "complex." Let's look again at the extreme frequencies. At $\omega/\omega_0 = 0$, the argument of the logarithm reduces to 1, so the logarithmic gain is again 0. As the frequency increases, the gain stays close to zero until ω becomes close to ω_0. When $\omega/\omega_0 = \frac{1}{10}$, for example, even in the worst case of $\zeta = 1$, in this vicinity α_c is only -0.173 dB. As the frequency approaches infinity, the highest power of ω dominates and the gain approaches:

$$\alpha_c = -20 \log \left(\frac{\omega}{\omega_0}\right)^2 = -40u \tag{39}$$

The high-frequency asymptote is again a straight line passing through $u = 0$, but, this time, with a slope of -40 dB per decade. This is true independent of the value of ζ. The break frequency corresponding to $u = 0$ is $\omega = \omega_0$. Although the asymptotes are both independent of the value of ζ, this is not

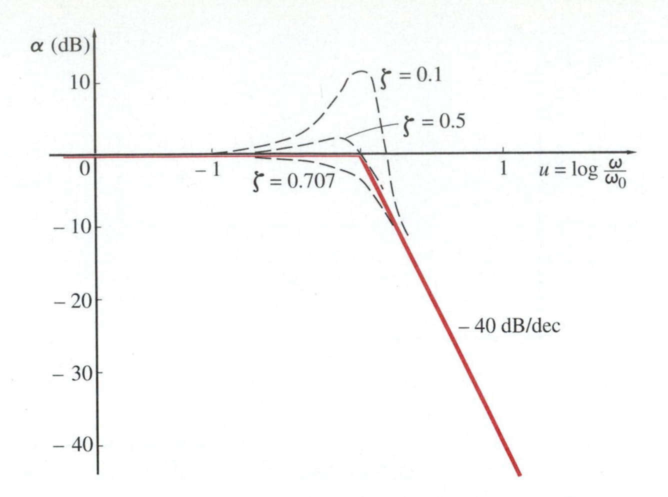

Figure 18
Bode diagram for complex poles.

true of the actual function at intermediate values of frequency. Thus, at the break frequency $\omega/\omega_0 = 1$, the expression for the gain reduces to:

$$\alpha_c = -20 \log 2\zeta \qquad \text{dB} \tag{40}$$

ζ	α_c
1	−6
0.5	0
0.25	6
0.1	14
0.05	20
0	∞

The table gives values of the gain for a range of values of the damping factor from the maximum value on down. The value $\zeta = 0$ corresponds to a pair of poles on the $j\omega$ axis—a pure oscillation with no damping.

The asymptotic plot for complex poles is shown in Figure 18. Also shown are the actual curves for several values of ζ; these differ from the asymptotes over an interval of just a little more than 1 decade around the break frequency. The curve for $\zeta = 1$, the maximum value of the damping factor, corresponds to a double real pole which is the boundary between a complex pair of poles and two real poles.

We see that the curves for some values of ζ demonstrate a peak near the break frequency. By standard methods of calculus (setting the derivative equal to zero) it is possible to find where these peaks occur in frequency and what the values at the peaks are. You will be asked to supply the details in the problems at the end of the chapter. The result of this effort is:

$$\omega_{\max} = \sqrt{1 - 2\zeta^2} \qquad \zeta \leq \frac{1}{\sqrt{2}}$$

$$|F(j\omega)| = 2\zeta\sqrt{1 - \zeta^2}$$

ASYMPTOTIC PHASE Of the four types of factors in the network function in (33), the constant term will not contribute to the phase—unless the constant

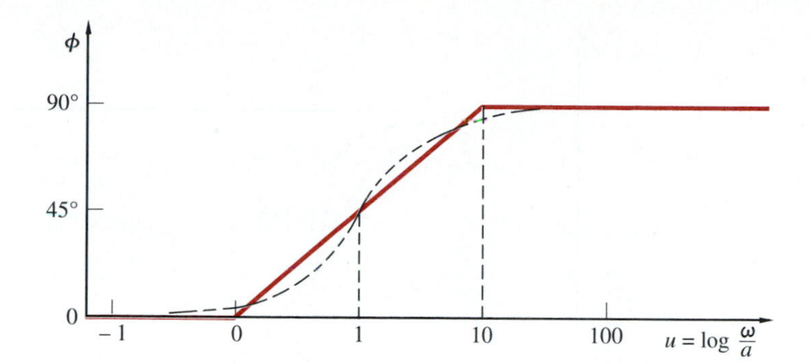

Figure 19
Asymptotic plot of angle contributed by a real zero.

is negative; in that case there will be a contribution of a constant 180°. The factor $s = j\omega$ will contribute a constant angle of $-90°$ if it is a pole and $+90°$ if it is a zero. That leaves a real critical frequency and a complex pair to consider.

The angle contributed by a real zero is:

$$\phi_r = \arg\left(1 + j\frac{\omega}{a}\right) = \tan^{-1}\frac{\omega}{a} = \tan^{-1} 10^u \tag{41}$$

As ω goes from 0 to ∞, ϕ_r goes from 0 to 90°; also, $\phi_r = 45°$ at the break frequency $u = \omega/a = 1$. Most of the angle change takes place over an interval centered at the break frequency $u = 1$.

A plot of (41) is shown in the dashed curve in Figure 19. This can be closely approximated by the solid, piecewise linear plot shown. The low-frequency asymptote is a horizontal line at 0°; The high-frequency asymptote is also a horizontal line but at 90°. In between, from $u = 0$ to $u = 10$, the asymptote is a straight line with a slope of 45° per decade, passing through 45° at $u = 1$ ($\omega = a$). The "corner-frequency" terminology breaks down here, since, for the gain, this frequency corresponds to $u = 1$, whereas for the angle there seem to be two "corners," at $u = \frac{1}{10}$ and 10.

EXERCISE 6

For a real zero, determine the error in the angle at $u = -1$ and $u = 1$ if the asymptotic values (0 and 90°, respectively) are used instead of the actual values.

ANSWER: 5.7°. □

Finally, we examine the angle contributed by a pair of complex poles. Using the quadratic in (33) with $s = j\omega$, the angle is:

$$\phi_c = -\arg\left[1 - \left(\frac{\omega}{\omega_0}\right)^2 + j2\zeta\frac{\omega}{\omega_0}\right] = -\tan^{-1}\frac{2\zeta(\omega/\omega_0)}{1 - (\omega/\omega_0)^2} \tag{42}$$

Not only does this vary with frequency, but it also depends on the parameter ζ. At very low frequencies, the phase is approximately $\tan^{-1}\omega/\omega_0$, which is the same as the variation of the phase for a real pole. At the other extreme, of very high frequencies, the expression reduces to:

$$\phi_c(\omega \gg \omega_0) \rightarrow -\tan^{-1}\frac{-2\zeta}{\omega/\omega_0}$$

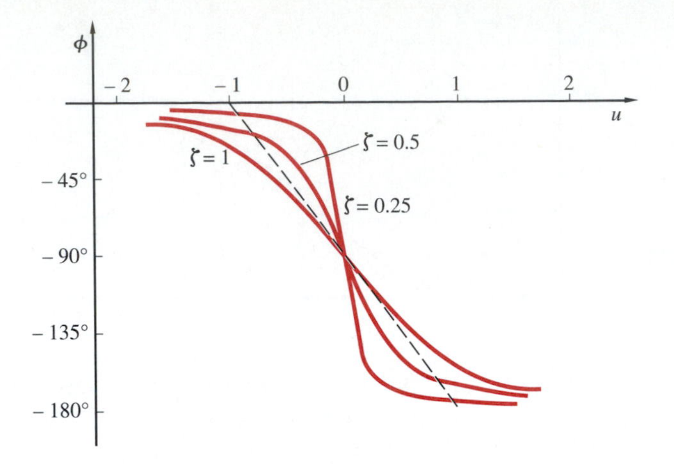

Figure 20
Bode phase diagram for complex poles.

At infinite frequency, this goes to $-180°$, twice as much as for a real pole.

A value of frequency for which the phase is easily found from (42) is the break frequency $10^u = \omega/\omega_0 = 1$. At this frequency, the argument of the arctangent is ∞. Hence $\phi_c = -90°$, halfway between the values at the extreme frequencies, true for all values of ζ.

For a real pole it is not difficult to decide how the low-frequency and high-frequency asymptotes should be connected. The case is not so clear for complex poles. What *is* clear is that the line should pass through $-90°$ at $u = 0$; but what should the slope of the line be? One possibility is to use a single approximating line passing through $-90°$ at $u = 0$ and extending for 1 decade on either side of this corner frequency, as we did for real poles. This gives a very poor approximation to the actual values, especially for low values of ζ. The actual phase curves for several values of ζ are shown in Figure 20, together with the single "approximating" line just described.

If you were asked to approximate the actual curves in Figure 20 by three different lines to give the best approximation near the corner frequency $u = 0$, what lines would you draw? Then what would you do to get each of these lines to meet the low-frequency and high-frequency asymptotes? (Give this some thought before you go on.) I'm sure that, with a little thought, you came up with the answer: (1) find the slope of the actual phase curve evaluated at $u = 1$, with ζ as a parameter; (2) evaluate the slope for the values of ζ in the figure, and maybe a few others between 0 and 1; (3) draw a line with this slope passing through $u = 0$ for each value of ζ; and (4) extend the low-frequency and high-frequency asymptotes to intersect each of the lines.

What is needed is to differentiate with respect to u the expression for ϕ_c in (42) and set $\omega/\omega_0 = 1$. This is suggested for you to carry out at the end of the chapter. The result is:

$$\frac{d\phi}{du} = x\frac{d\phi}{dx} = -2\zeta \frac{x(1 - x^2)}{(1 - x^2)^2 + 4\zeta^2 x^2}\bigg|_{x=1} = -\frac{1}{\zeta}\,\text{rad/dec} \tag{43}$$

where $x = \omega/\omega_0$. Since ζ can range from 1 to 0, this slope can range from -1

to $-\infty$. The intercepts of this line with the low-frequency and high-frequency asymptotes occur at $u = -\zeta\pi/2$ and $+\zeta\pi/2$, respectively.

EXERCISE 7

Using the preceding results, prepare for your own use a set of asymptotic diagrams, one for each of three or four appropriate values of ζ. Label the intercepts of the mid-frequency line with the low-frequency and high-frequency asymptotes.

□

It should be clear to you from the preceding discussion that asymptotic plots of magnitudes give reasonably accurate results, especially for real poles and zeros. For these, relatively small corrections might be needed at (or very near) the break frequencies. The corrections would be greater for complex poles and zeros, but still manageable. On the other hand, for the angle function, the straight-line plot over a fairly wide interval surrounding the break point gives results which are not very satisfactory for the case of complex poles and zeros. In these cases and in that interval, relying on the actual curves for the specific value of damping coefficient involved would be more sensible, unless ζ is very close to 1.

Bode Diagrams

So far, we have investigated the contributions of each type of factor in the numerator and denominator of network functions to the logarithmic gain and the phase. When presented with a network function (or a network from which the network function can be determined) for whch it is desired to draw Bode diagrams, we combine the individual diagrams for each type of factor to obtain the overall diagram. If this plan is to yield a reasonably accurate result, the individual poles and zeros must be sufficiently separated in frequency that the contribution of one pole or zero has already reached its asymptotic value before the next pole or zero comes along.

The function explicitly discussed in the preceding sections is a voltage-gain function, a dimensionless quantity. All procedures would apply equally well to a current-gain function. Furthermore, transfer impedances and transfer admittances—even driving-point functions—could also be included, provided they were normalized to make them dimensionless. The universal resonance curve, for example, can also be plotted logarithmically; the bandwidth would again be called the 3-dB bandwidth.

EXAMPLE 2

The Bode gain and phase diagrams are required for the following transfer function:

$$H(s) = \frac{25 \times 10^9(s^2 + 18{,}000s + 10^8)}{s(s + 1000)(s + 5 \times 10^4)(s + 5 \times 10^5)}$$

SOLUTION The first step in simplifying this expression is to factor out the constant term of each numerator and denominator factor, leaving a 1 for each constant term.

The next step is to normalize the frequency; a convenient value is $\omega_0 = 10^4$. The normalized frequency is then $s' = s/10^4$. The result is:

$$H(s) = \frac{10\left[\left(\frac{s}{10^4}\right)^2 + 1.8 \times \frac{s}{10^4} + 1\right]}{\frac{s}{10^4}\left(\frac{s}{10^3} + 1\right)\left(\frac{s}{5 \times 10^4} + 1\right)\left(\frac{s}{5 \times 10^5} + 1\right)}$$

$$= \frac{10(s'^2 + 1.8s' + 1)}{s'(10s' + 1)(s'/5 + 1)(s'/50 + 1)}$$

The prime on s' is inconvenient and can be confusing. At the risk of even more confusion, let's get rid of the prime, and let s' be renamed s. Thus:

$$H(s) = \frac{10(s^2 + 1.8s + 1)}{s(10s + 1)(s/5 + 1)(s/50 + 1)}$$

Remember that s in this expression is the normalized frequency. (If it seems to you that a "fast one" is being pulled, don't replace the normalized variable s' by s; keep the variable as s' while going through the rest of the problem.)

First we take up the gain diagram. It really isn't necessary to replace s by $j\omega$, take the magnitude, and then take 20 times the logarithm of the result. The only things requiring attention are the following: (1) How does the function behave at the lowest and the highest frequencies? (2) Where are the break frequencies? and (c) Are the breaks up or down? After the frequency-dependent factors are handled, the entire diagram is shifted up by $20 \log 10 = 20$ dB, since 10 is the constant multiplier.

The low frequencies are dominated by the simple pole at $s = 0$. (*Simple* here means first-order, not multiple.) Hence the low-frequency asymptote is a line with a slope of -20 dB/dec. To fix the line totally, we need, in addition to the slope, its intercept on the u axis. But this occurs when $\alpha = 0 = -20 \log \omega$, or $u = 0$. The first break frequency is due to the pole at $10s = -1$ ($\omega = 0.1$, $u = -1$). Hence, there is a down break by another 20 dB/dec at this point, resulting in a line with a slope of -40 dB/dec.

The pair of complex zeros comes next, leading to an upward break of 40 dB/dec at $\omega = 1$, $u = 0$. Thus, the asymptotic curve becomes horizontal. Then come the remaining poles in succession, each causing a break down of 20 dB/dec, the first at $u = \log 5 = 0.7$ and the second at $u = \log 50 = 1.7$. Thus the high-frequency asymptote is falling at -40 dB/dec. Finally, we shift the entire curve up by 20 dB because of the constant. The entire diagram is shown in Figure 21. It is valid because each critical frequency is sufficiently separated from the adjacent ones.

EXERCISE 8

The worst interference of one break on another in the just completed example will occur at the break frequency due to the pole at $s = -5$, caused by its proximity to the pair of zeros. Find the actual gain at $\omega = 5$ and compare with

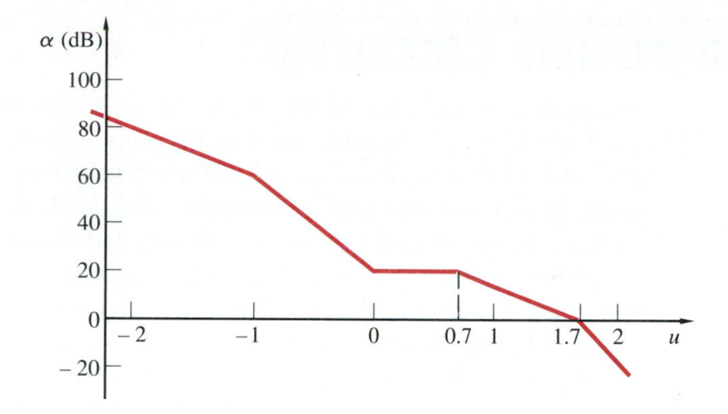

Figure 21
Bode gain diagram for Example 2.

the asymptotic value. How different is it from the expected correction of 3 dB at an isolated break frequency? □

The phase plot in Example 2 is the final task. The pole at the origin contributes a constant angle of $-90°$. Hence, the asymptotic angle curve starts at low frequencies as a horizontal line at $-90°$. Looking at the other extreme, the function approaches $10/s^2$ at very high frequencies. Hence, the high-frequency asymptote is a horizontal line at $-180°$.

In between, we contemplate using the first-order asymptotic lines. The angle would first drop by another 90° because of the first pole, would rise by 180° to 0 because of the pair of complex zeros, and then would drop by 90° because of each of the remaining poles, reaching the predicted $-180°$ at high frequencies.

If we consider the separation between the critical frequencies, however, this approach will be in trouble. In order for the linear rise or fall due to a pole or zero to be valid, there would have to be no interference from the angles of adjacent poles and zeros. This requires a separation of some two decades between adjacent critical frequencies, as we found when examining the angle due to individual poles or zeros. But the actual separations are one decade or less. Hence, the asymptotic approach would not lead to acceptable results, except at the two extreme frequencies. A freehand sketch, guided by the values at the extreme frequencies and at the break points, would lead to no less acceptable results.

EXERCISE 9

Draw a piecewise constant curve for the angle based on the description given in the next-to-last paragraph. The angle function cannot change discontinuously like that. Calculate the angle for some specific frequencies, say, halfway between break frequencies. Then draw a freehand sketch of the angle function, using your best judgment. Finally, calculate the angle for some other specific frequencies and check how close your freehand sketch came to those values. □

7 STAGGER-TUNED CIRCUITS[11]

Physical embodiments of the circuit to be described in this section are widely used in electronic equipment. The purpose in introducing it here, however, is not so much to become acquainted with its properties, as it is to amplify the concept of resonance and to illustrate further the utility of the concept of poles and zeros, the complex-frequency plane, and the frequency response.

The frequency selectivity and filtering properties of the resonant circuits discussed in the preceding and present chapters are useful features. However, for given element values in these circuits, the bandwidth is fixed. In attempting to increase the bandwidth, it may be tempting to interconnect two resonant circuits, say, in parallel. This approach cannot succeed, however, since the natural frequencies of the combined network will be determined not by the parameters of the individual networks separately but by the parameters collectively.

But suppose the two resonant circuits are isolated from each other, say, by amplifiers. We will now follow this line of thought. Figure 22 shows two parallel resonant circuits each of which constitutes the load on an amplifier represented by a controlled current source. The overall voltage gain is determined as follows:

$$V = -gZ_1V_1 \qquad V_2 = -gZ_2V = g^2Z_1Z_2V_1$$
$$G(s) = \frac{V_2}{V_1} = g^2Z_1(s)Z_2(s) \tag{44}$$

$$Z_1(s) = \frac{\dfrac{1}{C_1}s}{s^2 + \dfrac{1}{R_1C_1}s + \dfrac{1}{L_1C_1}} \qquad Z_2(s) = \frac{\dfrac{1}{C_2}s}{s^2 + \dfrac{1}{R_2C_2}s + \dfrac{1}{L_2C_2}} \tag{45}$$

Suppose the two circuits have the same Q and are tuned to the same frequency ω_0. Then, Z_1 and Z_2 will have the same poles and zeros, so the voltage gain function will have double poles and zeros. Hence, the magnitude-

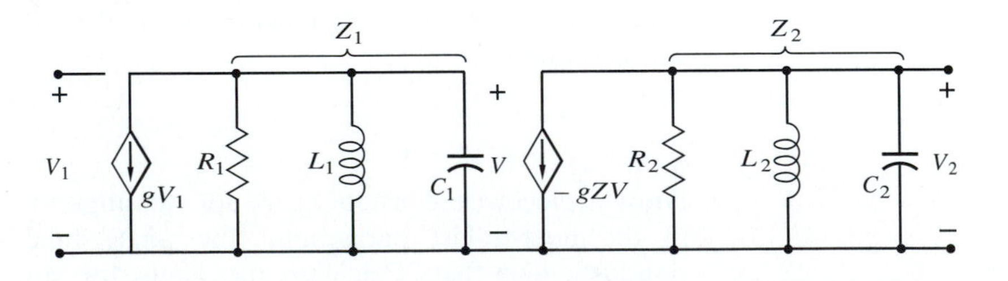

Figure 22
A staggered pair.

[11] This section is intended as a mind stretcher. You won't be handicapped in the material that follows if you omit it, except for two things: (1) you won't be able to carry out some of the design problems suggested at the end of the chapter, and (2) you'll just miss out on a lot of fun!

response curve will resemble that of a simple resonant circuit but will be sharper. (Convince yourself of this by thinking of squaring the universal resonance curve.)

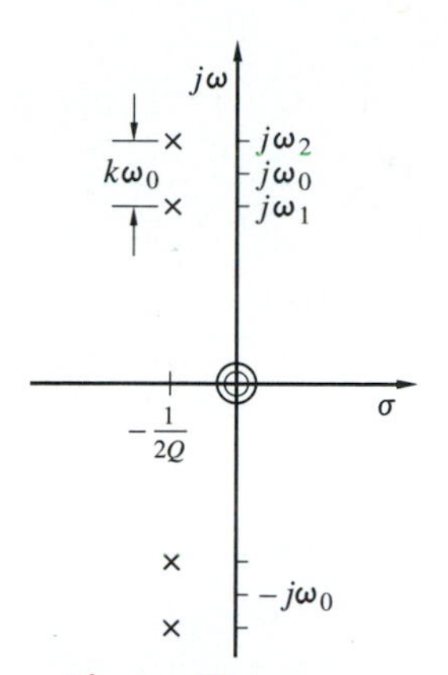

Figure 23
Poles of gain.

Suppose, instead, that the two resonant frequencies are somewhat displaced from each other—we say *staggered*—one above and the other below ω_0: say, $\omega_1 = 1/(L_1C_1)^{1/2} = \omega_0(1 - k/2)$ and $\omega_2 = 1/(L_2C_2)^{1/2} = \omega_0(1 + k/2)$. k is called the *staggering coefficient.* Assuming that the Q's are still the same, the poles of $G(s)$ will be located as shown in Figure 23. In the case of the double pole, the frequency response will have a single peak. As k is increased and the separation between the poles increases far enough, we would expect the frequency-response curve to have two peaks, at points on the $j\omega$ axis opposite each of the poles in the upper half plane. The change from a single peak to a double peak would occur at some intermediate value of k.

Let's confirm the preceding conjectures quantitatively. By using $Q = \omega_0 RC$ and the staggered frequencies just defined, the voltage-gain function in (44) can be written as:

$$G(j\omega) = \frac{g^2R^2}{\left[1 + jQ\left(\dfrac{\omega/\omega_0}{1 - k/2} - \dfrac{1 - k/2}{\omega/\omega_0}\right)\right]\left[1 + jQ\left(\dfrac{\omega/\omega_0}{1 + k/2} - \dfrac{1 + k/2}{\omega/\omega_0}\right)\right]} \tag{46}$$

(Confirm the details.) Let's now define the fractional frequency deviation δ as in (10) in Chapter 11:

$$\delta = \frac{\omega}{\omega_0} - 1 \tag{47}$$

For high-Q circuits and small values of δ (that is, frequencies near ω_0), the following approximations can be made:

$$\begin{aligned} \frac{\omega/\omega_0}{1 - k/2} - \frac{1 - k/2}{\omega/\omega_0} &\approx 2\left(\delta + \frac{k}{2}\right) \\ \frac{\omega/\omega_0}{1 + k/2} - \frac{1 + k/2}{\omega/\omega_0} &\approx 2\left(\delta - \frac{k}{2}\right) \end{aligned} \tag{48}$$

(This can be confirmed by using (47) and then the geometric progression $1/(1 + x) \approx 1 - x$, in which higher-order terms are neglected.) Compare these approximations with the one in (12) in Chapter 11.

With these approximations, the magnitude of $G(j\omega)$ in (46) is:

$$|G(j\omega)| = \frac{g^2R^2}{\sqrt{[1 + (2Q\delta + kQ)^2][1 + (2Q\delta - kQ)^2]}} \tag{49}$$

The locations of the extreme values in this expression can be found by differentiating with respect to δ and setting the derivative equal to zero. You carry out the details and confirm that:

$$\delta[(2Q\delta)^2 - (Q^2k^2 - 1)] = 0 \tag{50}$$

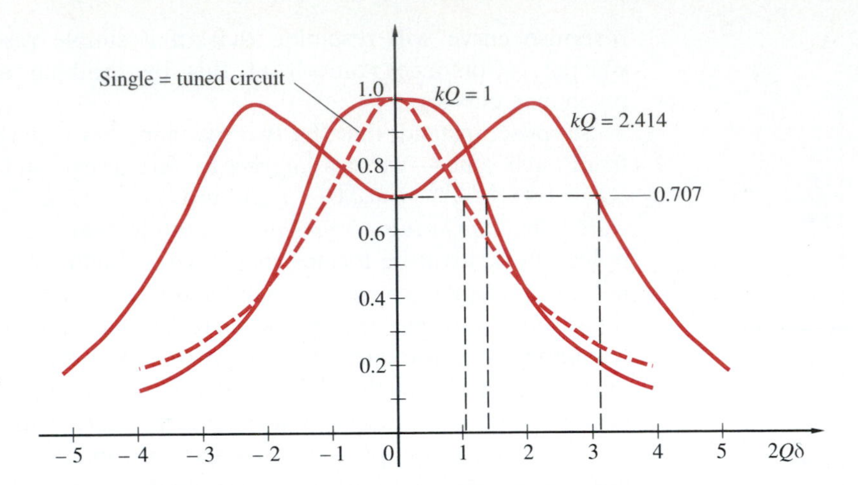

Figure 24
Frequency response of staggered pair.

The derivative always has a zero at $\delta = 0$ ($\omega = \omega_0$), corresponding to either a maximum or a minimum. It has two other zeros at:

$$\delta = \pm\frac{1}{2Q}\sqrt{Q^2k^2 - 1} \tag{51}$$

The condition for these values to be real is $k \geqslant 1/Q$. Thus, $k_c = 1/Q$ is a critical value of staggering. Below this value, the magnitude response has a single peak at $\omega = \omega_0$. Above this critical value, it has a double peak and an intervening valley at $\delta = 0$. At the critical value $k = 1/Q$, the curve is exceptionally flat; it is said to be *maximally flat.* The magnitude response is shown plotted in Figure 24 for two values of the staggering coefficient: the critical value $kQ = 1$ and $kQ = 2.414$.

The heights of the peaks in the magnitude response are computed by inserting the value of δ from (51) into (49). Carry out this process and confirm that:

$$|G(j\omega)|_{\max} = \frac{g^2R^2}{2kQ} \tag{52}$$

Thus, we find that the two peak values are equal and are inversely proportional to the staggering coefficient.

At $\delta = 0$ ($\omega = \omega_0$), the height of the response curve at any value of k is found to be:

$$|G(j\omega)|_{\delta=0} = \frac{g^2R^2}{1 + k^2Q^2} = \frac{2kQ}{1 + k^2Q^2}|G(j\omega)|_{\max} \tag{53}$$

which, at critical staggering, is $g^2R^2/2$. Thus, although the two curves in Figure 24 are drawn with the same peak values, the scale of the ordinate should be divided by the corresponding value of kQ for each curve.

Let us now inquire into the bandwidth of the response. The 3-dB point can be found by setting $|G(j\omega)|$ in (49) equal to $1/\sqrt{2}$ of its maximum value given in (52). Four values of $2Q\delta$ are found at which the response is down 3 dB:

$$\begin{aligned} 2Q\delta &= \pm\sqrt{k^2Q^2 - 1 + 2kQ} \\ 2Q\delta &= \pm\sqrt{k^2Q^2 - 1 - 2kQ} \end{aligned} \tag{54}$$

For critical coupling, the second of these expressions leads to imaginary values. In fact, the quantity under the radial is negative until kQ exceeds 2.414. At this point, the value of $|G|$ at $\delta = 0$ is precisely 3 dB down from the peak value, as seen from (53). For larger values of staggering, the valley in the response curve will dip below the -3-dB level.

The 3-dB points on the outer skirts of the response curve are given in the first expression in (54). Expressing δ in terms of ω, the bandwidth becomes:

$$\begin{aligned} \beta &= \omega_2 - \omega_1 = \omega_0(\delta_2 + 1) - \omega_0(\delta_1 + 1) = \omega_0(\delta_2 - \delta_1) \\ &= \frac{\omega_0}{Q}\sqrt{k^2Q^2 - 1 + 2kQ} \end{aligned} \tag{55}$$

Compare this expression with that of the single-tuned (resonant) circuit, which is $\beta = \omega_0/Q$. The bandwidth is increased by a factor $(k^2Q^2 - 1 + 2kQ)^{1/2}$. At critical coupling, the bandwidth is the same as that of a single-tuned circuit. For $kQ = 2.414 = 1 + \sqrt{2}$, at which the valley in the curve has the same height as the 3-dB band edges, the bandwidth is improved by a factor of 2.7.

This circuit is but one of several different ones which have similar frequency responses. A few of them are described in the problem set and left for you to do.

SUMMARY

- The complex-frequency variable, $s = \sigma + j\omega$
- The complex exponential function in its various forms
- Network analysis in the complex-frequency domain
- Network functions expressed in terms of the complex frequency
- Representation and numerical evaluation of network functions in the complex-frequency plane
- Critical frequencies (poles and zeros) of network functions
- Resonance in terms of critical frequencies of impedance and admittance
- Relationship between steady-state and natural responses
- Relationship between poles and natural frequencies
- Admittance poles: natural frequencies under voltage excitation
- Impedance poles: natural frequencies under current excitation
- Logarithmic measure for network functions: decibels (dB), decades, and octaves
- Frequency-response plots; break, or corner, frequencies
- Asymptotic variation of magnitude and phase functions
- Bode diagrams
- Stagger-tuned circuits; the staggering coefficient
- Maximally flat and oscillatory magnitude responses

PROBLEMS

Some of the circuits to follow were introduced earlier, in Chapter 8 or 11, where questions were posed about the steady state, frequency response, resonance, bandwidth, and related matters. Now we are interested in other matters, such as the locations of the poles and zeros. Check back and review your solutions from Chapter 8 or 11.

1. In the circuit of Figure P1:
   a. Find a second-order differential equation whose variable is i.
   b. Assuming that the source voltage has a variation with time given by Ve^{st}, where $s = \sigma + j\omega$, the forced solution for $i(t)$ will have the form Ie^{st}. Find the "phasor" I of the forced solution of this equation.
   c. From the terminals to which the source is connected, determine the admittance of the circuit in terms of s and compare with the solution for I in part (b).

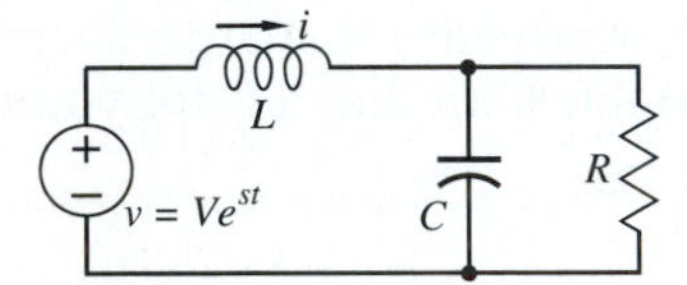

Figure P1

2. A resistor R is to be placed across the terminals of the circuit in Figure P2.
   a. Determine the value of R if the real parts of the impedance poles are to be doubled from their value without R.
   b. Find the percent change in the imaginary part of the impedance poles.
   c. How would you describe, in a general way, the migration of the poles in the complex-frequency plane as a result of R?

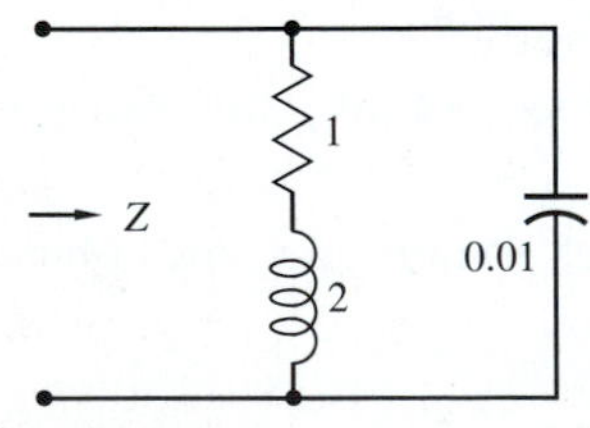

Figure P2

3. In each of the circuits in Figure P3, the given component values are normalized values.
   a. Determine the indicated response-to-excitation functions.
   b. Specify the poles and zeros of these functions.
   c. Explicitly specify the behavior of each function at $s = 0$ and at infinity.
   d. Sketch the poles and zeros in the complex-frequency plane. What are their units?
   e. Sketch the frequency-response curves (magnitude and angle) utilizing the geometry of the pole-zero diagrams, calculating the results for enough frequencies to show the shape of the curve, say, four.

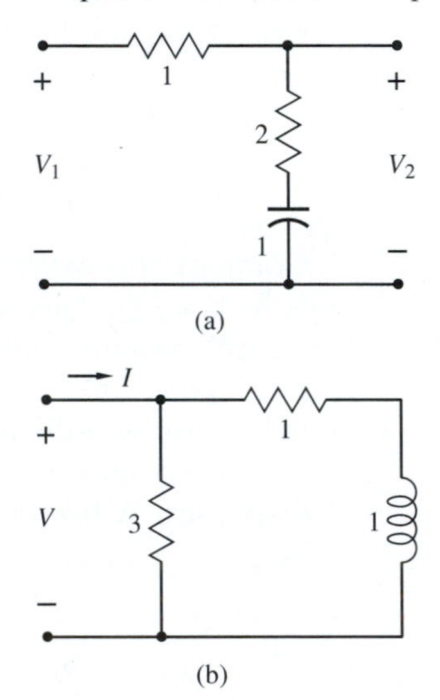

Figure P3

4. The components in Figure P4 are normalized.
   a. Determine expressions for the input impedance V_1/I_1 and the transfer impedance V_2/I_1 and specify their poles and zeros.
   b. Sketch the frequency-response curves from the pole-zero diagram in the complex plane.
   c. Determine the resonant frequency, the quality factor, and the bandwidth.
   d. Let the input current be $i_1(t) = 3\cos 2t$. Determine the steady-state expressions for $v_1(t)$ and $v_2(t)$.

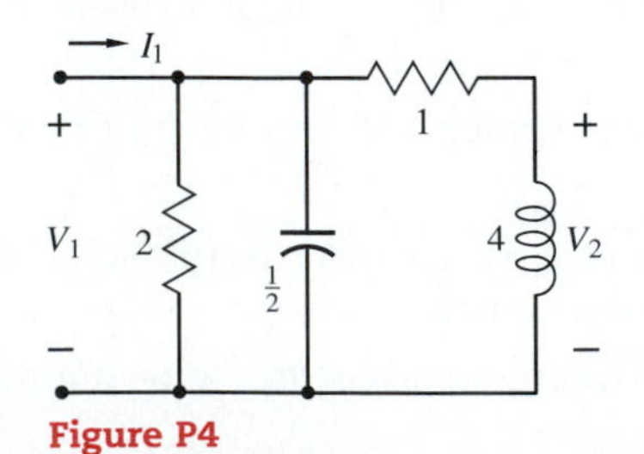

Figure P4

e. Denormalize the parameter values, assuming that the normalizing factors were $R_s = 500\,\Omega$ and $\omega_s = 5000$ rad/s.

5. The components in Figure P5 are normalized.

a. Determine expressions for the input admittance I_1/V_1 and the voltage gain V_2/V_1.

b. Specify their poles and zeros, and sketch them in the complex plane.

c. Determine the resonant frequency, the quality factor, and the bandwidth, and relate them to the locations of the poles.

d. Denormalize the component values, assuming that the normalizing factors were $R_s = 1000\,\Omega$ and $\omega_s = 10{,}000$ rad/s.

e. Write the denormalized values of input-current and output-voltage phasors.

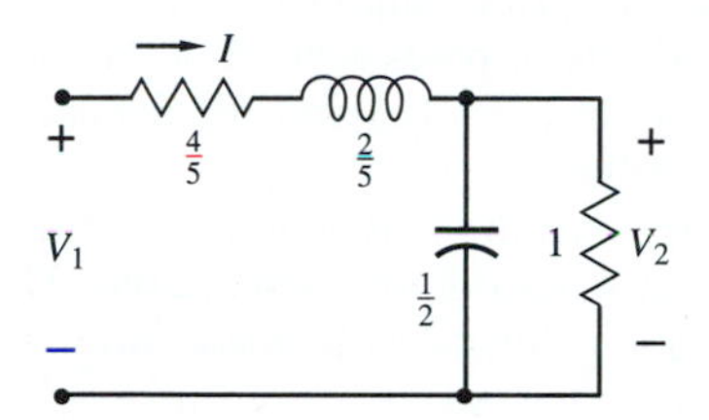

Figure P5

6. The circuit in Figure P6 is the same as that in Figure P5 but with different normalized values.

a. Determine the locations of the poles and zeros of the admittance, and sketch them on the complex-frequency plane.

b. Assuming a voltage excitation v_s, write a pair of differential state equations. From these, determine the natural frequencies and confirm their relationship to the poles of the appropriate function—either impedance or admittance. Assume a current excitation i_s, and again find the resulting natural frequencies; relate them to the poles of the appropriate function.

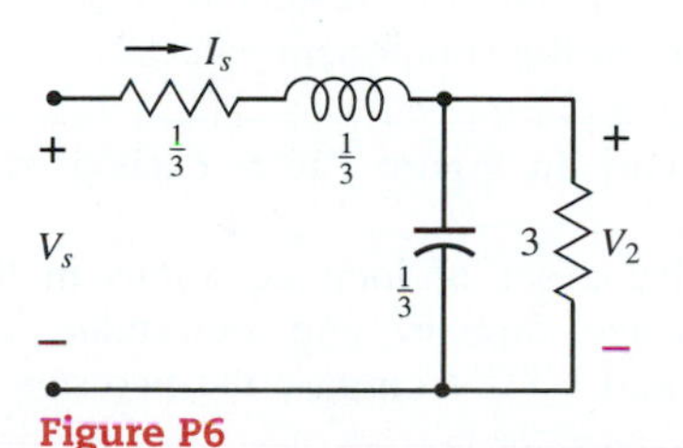

Figure P6

7. a. Determine the poles and zeros of the voltage gain in Figure P7, and locate them in the complex-frequency plane.

b. Sketch the frequency-response curves.

c. Find the steady-state $v_2(t)$ if $v_1(t) = 40 \cos 800t$.

d. If the poles are complex, determine the resonant frequency, the quality factor, and the bandwidth; relate them to the pole locations and the frequency response.

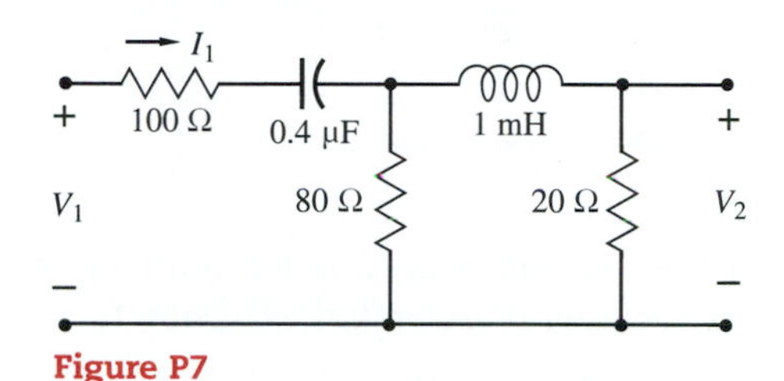

Figure P7

8. Describe how the angle of the voltage-gain function in Figure P3a varies with frequency. Based on this, what name would be appropriate for this circuit?

9. The resistance in a series resonant circuit is $100\,\Omega$. The circuit is to have a half-power bandwidth of 400 rad/s and a quality factor of 10 at resonance.

a. Find the damping constant and the undamped and damped natural frequencies in the natural response.

b. Find the values of L and C.

10. When you are in the middle of working an engineering problem, you sometimes need the dB equivalent of a specific voltage ratio. At such times, you can punch some buttons on a calculator (if one is available) or look it up in a table (if one is available). Although we can't supply you a calculator, this exercise has the objective of making such a table available to you—through your efforts, of course. Construct a table by choosing a number of appropriate numerical values from $x = 0$ to $x = 0.01$ and calculating the corresponding values of $20 \log x$ in dB. Feel free to use a calculator! Keep the table handy for use when you can't find your calculator.

11. The capacitor in Figure P11 represents *parasitic,* or *stray,* capacitance. That is, it is not a component deliberately placed in the circuit but makes its appearance because physical arrangements of physical components inevitably produce capacitive effects. Without C_2, the voltage ratio V_2/V_1 would be independent of frequency.

a. For the circuit in Figure P11, find the poles and zeros of the voltage-gain function.

b. Add an element somewhere in the circuit, and specify its value so as to compensate for the effect of C_2— that is, so that the voltage-gain function is again independent of frequency.

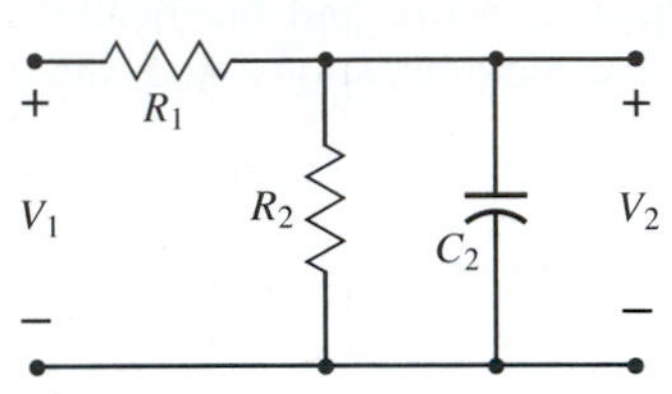

Figure P11

12. **a.** Normalize the following transfer functions and prepare them for drawing Bode diagrams:

1. $F(s) = \dfrac{150(s + 2000)}{s + 16{,}000}$

2. $F(s) = \dfrac{10}{s(s + 1000)}$

3. $F(s) = \dfrac{2s}{s^2 + 20s + 500}$

4. $F(s) = \dfrac{20s^2}{s^2 + 2s + 100}$

5. $F(s) = \dfrac{10(s^2 + 100)}{s^2 + 8s + 10{,}000}$

6. $F(s) = \dfrac{5(s + 80)}{(s + 10)(s + 1000)(s + 7000)}$

b. Draw asymptotic magnitude plots, clearly indicating the break frequencies and the behavior at zero and infinity.

c. Specify the largest errors made at each break frequency compared with the exact values.

d. On the basis of the frequencies for which the gain magnitude is high or low, characterize the functions with such names as low-pass, high-pass, and others.

e. For each such name you attach to a function, study the locations of the poles and zeros and correlate them with the name.

f. Make a stab at drawing asymptotic phase plots.

13. Using what you know about Bode magnitude plots:

a. Find six transfer functions $F(s)$, one corresponding approximately to each of the Bode diagrams shown in Figure P13.

b. Sketch an approximate angle function for each one.

14. For each circuit in Figure P14, determine an expression for the voltage-gain function. Specify the location of the poles and zeros. Notice the distance of the poles from the $j\omega$ axis. Give a descriptive name (for example, low-pass or bandpass) to each circuit.

15. Terminals D and E in Figure P15a are to be joined together (short-circuited) and terminals AB excited by a voltage source V_1e^{st}, as in Figure P15b.

a. Determine the poles and zeros of $Y_1 = I_1/V_1$ and $Y_{21} = I_2/V_1$.

b. How are the two sets of poles related?

c. Now suppose terminals AB are short-circuited and terminals DE are excited by a voltage source V_2e^{st}, as in Figure P15c. Determine the poles and zeros of the functions $Y_2 = I_2/V_2$ and $G_{12} = V_1/V_2$.

d. How are the two sets of poles related in this case?

e. Are your answers to parts (b) and (d) dependent on the element values?

f. The response-to-excitation functions in Figure P15b and c result from different excitations. How are the poles of the functions in these two cases related? Are you surprised?

g. How would your answers be affected if the excitation were a current source, say, in Figure P15b, and the response-to-excitation functions were V_1/I_1 and I_2/I_1?

16. The circuit in Figure P16 is excited by a voltage source V_1e^{st}. Find the poles of the response-to-excitation function V_2/V_1.

a. Do this by writing and solving a set of node equations in the complex-frequency domain. The poles in this case are the short-circuit natural frequencies.

b. Replace the source by a short circuit and write a set of zero-input state equations in the time domain. From these, determine the natural frequencies and compare the results with the poles found in part (a). Do the results of the comparison depend on the component values?

17. Suppose the network in Figure P16 is excited by a current source I_1e^{st}.

a. Write and solve a set of loop equations in the complex-frequency domain and determine the poles of V_1/I_1 and V_2/I_1. Compare the two sets of poles.

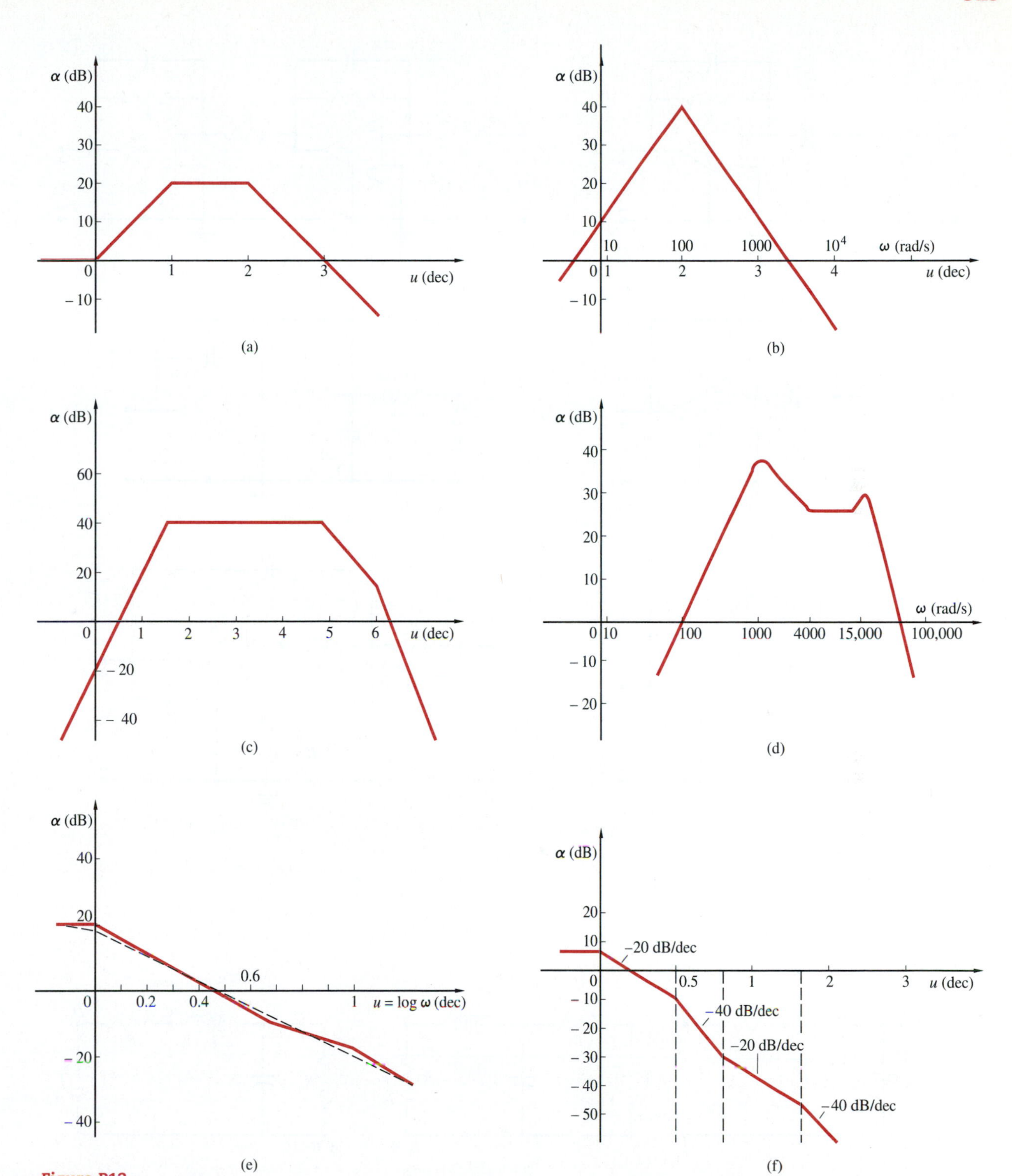
α (dB)
40
30
20
10
0
−10
1
2
3
u (dec)
(a)
α (dB)
40
30
20
10
10
100
1000
10^4
ω (rad/s)
0
1
2
3
4
u (dec)
−10
(b)
α (dB)
60
40
20
0
1
2
3
4
5
6
u (dec)
−20
−40
(c)
α (dB)
40
30
20
10
ω (rad/s)
0
10
100
1000
4000
15,000
100,000
−10
−20
(d)
α (dB)
40
20
0.6
0
0.2
0.4
1
u = log ω (dec)
−20
−40
(e)
α (dB)
20
10
−20 dB/dec
0
0.5
1
2
3
u (dec)
−10
−40 dB/dec
−20
−20 dB/dec
−30
−40
−40 dB/dec
−50
(f)

Figure P13

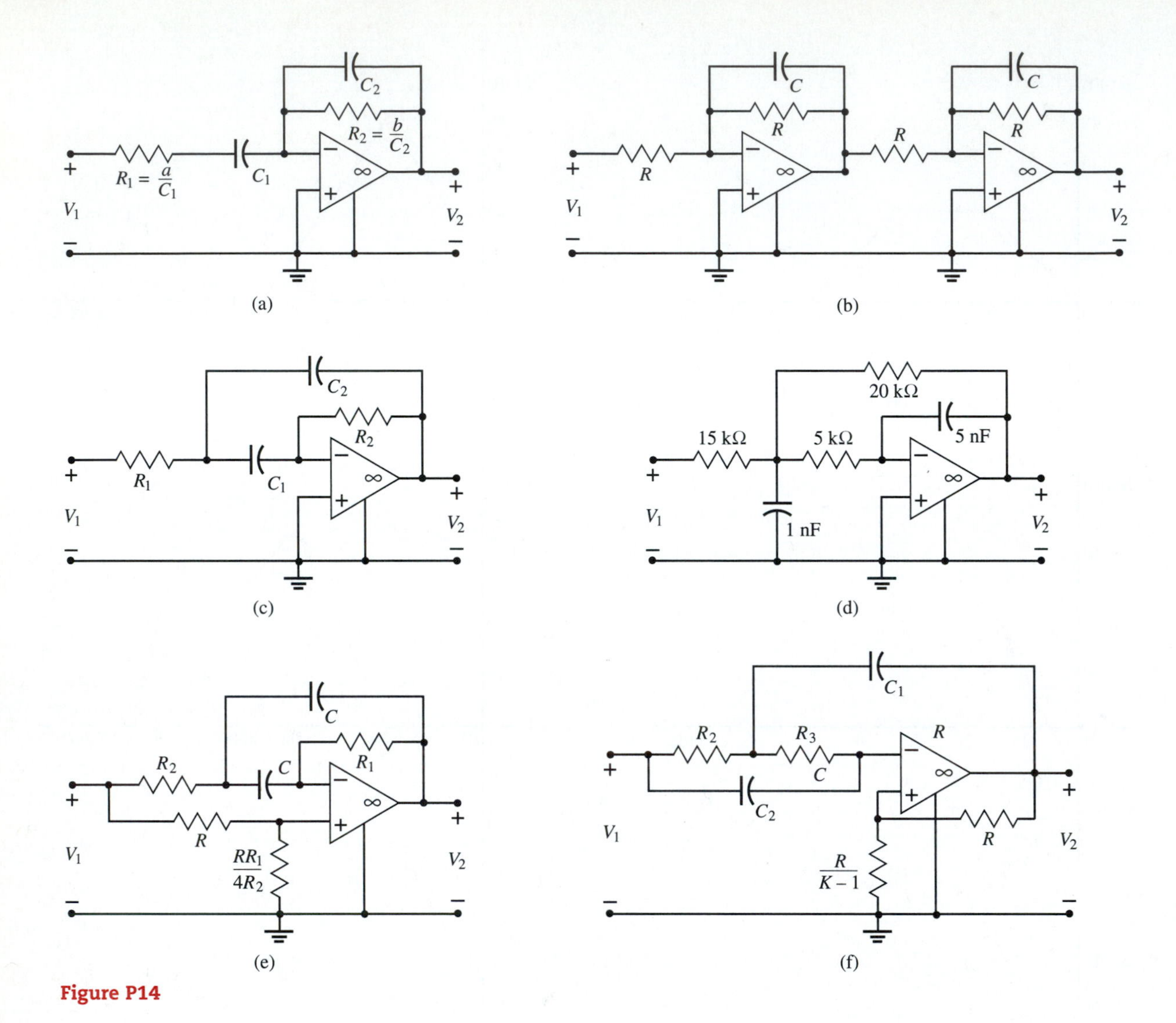

Figure P14

Figure P15

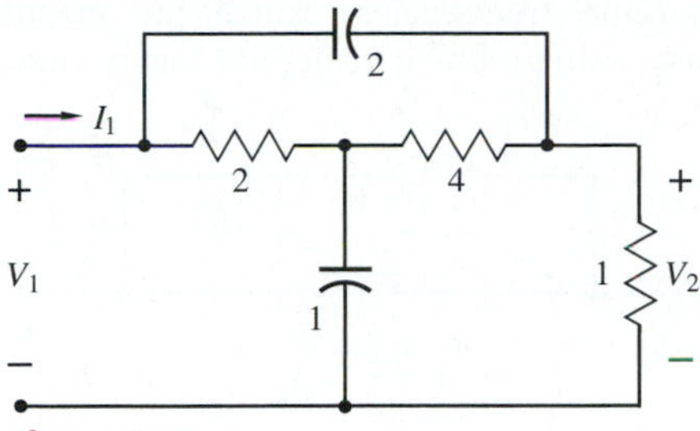

Figure P16

b. Replace the source by an open circuit and write a pair of zero-input state equations in the time domain. Determine the natural frequencies from these equations and compare them with the results of part (a).

18. Repeat Problem 16 for the circuit in Figure P18

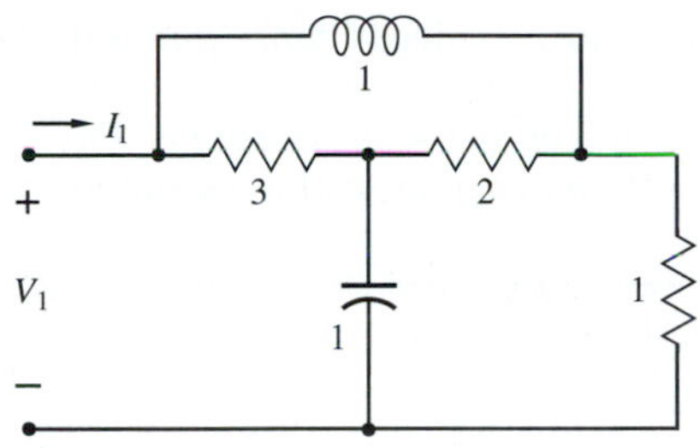

Figure P18

19. Repeat Problem 17 for the circuit in Figure P18.

20. Because of its structure, the network in Figure P20a is called a *lattice*. It is redrawn in Figure P20b as a *bridge*. (By tracing paths through the impedances, confirm that these two forms represent the same network.) The objective is to determine the voltage-gain function in terms of the impedances.
   a. Do this assuming the load impedance is not present. (Perhaps the easiest approach is to use the voltage-divider relationship twice but perhaps you can find an easier one.)
   b. Do it again, this time with the load present; it will take more effort.

21. In Figure P20, let $Z_1 = R_1$, $Z_2 = R_2$, $Z_3 = 1/Cs$, and $Z_4 = Ls$. (Assume these are normalized values.) Find the voltage gain as a function of s. Describe the locations of the poles and zeros and the behavior of the function at infinity.

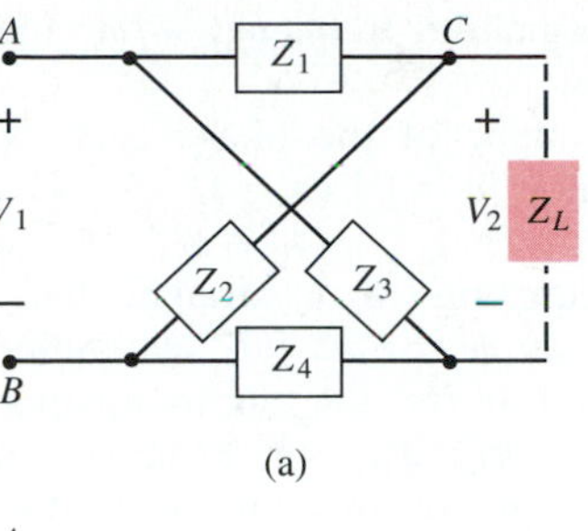

(a)

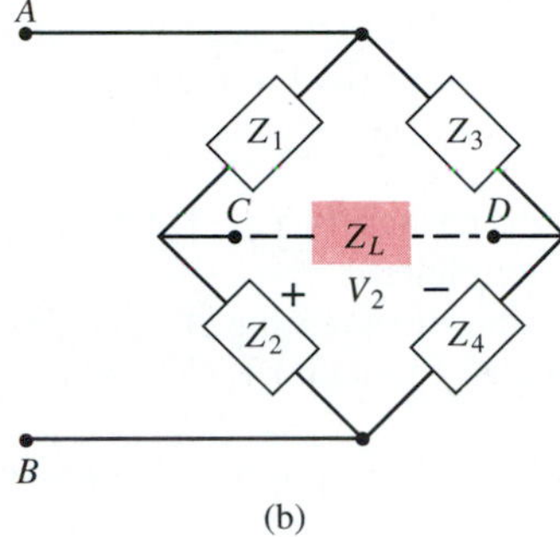

(b)

Figure P20

22. This time, let $Z_1 = 8$, $Z_2 = s$, $Z_3 = 2$, and $Z_4 = 1/s$ (normalized) in Figure P20. Find the voltage gain as a function of s. Sketch the locations of the poles and zeros in the complex plane. One of the zeros is located to the right of the $j\omega$ axis. Suppose you want to sketch graphically the frequency response (magnitude). Describe the difference it makes with this zero on the right rather than in the mirror-image position on the left.

23. Let $F(s) = 1/(s^2 + 2\zeta s + 1)$, where s is a normalized variable. Define $\alpha' = \ln|F(j\omega)|$.
   a. Show that α' has an extreme value at the same frequency where the magnitude of $F(j\omega)$ has an extreme value.
   b. Find the frequency at which α' has a maximum and specify a condition on ζ for a maximum to exist.
   c. Find the maximum value of α' for the permissible values of ζ.
   d. Define α as $20 \log|F(j\omega)|$. Will α have a maximum at the same frequencies as α'? How will its maximum values be related to the maxima of α'?

24. The angle contributed by a pair of complex poles to the total angle of a transfer function is:

$$\phi = -\tan^{-1}\frac{2\zeta x}{1 - x^2}$$

where x is the normalized frequency ω/ω_0 and ζ is the damping factor.

a. Show that the slope of the phase curve at the corner frequency $u = 0$ $(x = 1)$ is $d\phi/du = -\ln 10/\zeta$.

b. Writing the equation of a straight line with negative slope as $\phi = -au + b$, determine the values of a and b for the approximating line passing through $-\pi/2$ at $u = 0$. From this, determine the intercepts of this line with the low-frequency and the high-frequency asymptotes. Specify the separation of the intercepts in number of decades.

25. The circuit inside the dashed box in Figure P25 is a model of a voltage amplifier.

a. Take $1/RC$ as the normalizing frequency and determine the transfer function V_2/V_s as a function of s.

b. Sketch the locations of the poles and zeros in the complex plane; describe also the behavior of the function at infinity.

c. We want a sketch of the magnitude response; will the Bode-diagram approach yield reasonably accurate results?

d. Find the normalized frequency at which the magnitude has a maximum and find this maximum. On the basis of this point and the asymptotic values at the extreme frequencies, sketch the magnitude response, using a linear scale, not logarithmic.

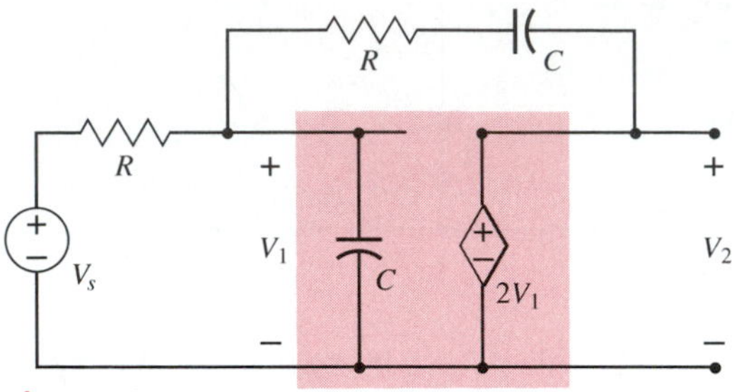

Figure P25

Figure P12-25

26. Two inverting amplifiers are connected as shown in Figure P26.

a. Determine the overall voltage gain as a function of s. The pole of this function contributed by the second stage is to lie at $s = -10{,}000$ units/s. At low frequencies, the logarithmic gain is to equal 20 dB.

b. Determine the unspecified parameter values.

c. Draw the Bode diagram for the magnitude of the gain.

d. At what rate is the logarithmic gain decreasing at high frequencies?

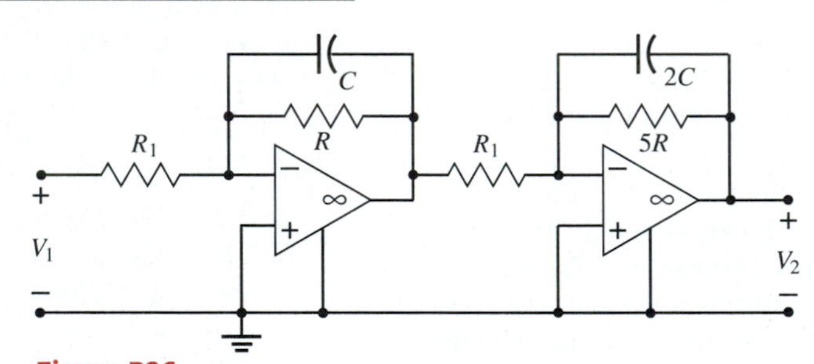

Figure P26

27. Find the poles and zeros of the admittance of the circuit in Figure P27.

a. First assume that the transformer is removed and the components across the secondary side are connected directly to those on the primary side; find the poles and zeros.

b. Then assume the transformer is ideal with a turns ratio approximately determined from the transformer parameters; again find the poles and zeros.

c. Finally, take the transformer to be perfectly coupled, with the primary and secondary inductances shown; again find the poles and zeros.

d. Make a table listing the poles and zeros in these three cases; describe how their locations are shifted or their numbers are increased or reduced by the transformer.

e. Sketch a Bode magnitude diagram and note the separation between break frequencies in each case.

28. The circuit in Figure P28 is called a *phase shifter*. It is to give an output whose phase is controlled by a

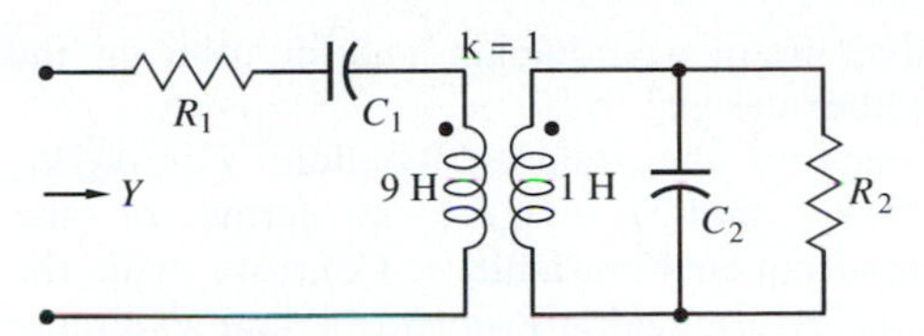

Figure P27

variable resistance R. It is a bridge circuit with a load having such a large impedance that the terminals AB are assumed to be open.

a. Determine the locations of the poles and zeros of the voltage-gain function and locate them on the complex plane.
b. Describe how the gain magnitude varies with frequency. If this were a filter, what would be an appropriate name to describe its passband?
c. Find an expression for the angle of the gain function. Over what range can the phase be varied.
d. Find a value of R for which the phase is $\pi/2$ rad.

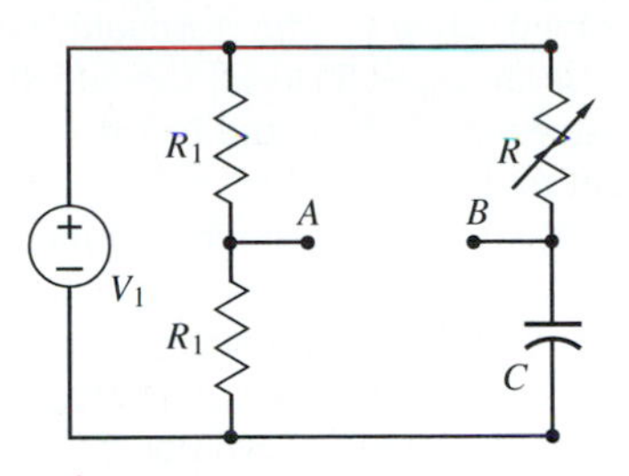

Figure P28

DESIGN PROBLEMS

1. a. Design a two-stage stagger-tuned circuit with a resonant frequency ω_0 krad/s, a quality factor $Q = 15$, and a 3-dB bandwidth that is 8% of ω_0. Assume that the resistors are the same.
   b. Determine the maximum value of the frequency response and the frequency or frequencies at which it occurs.
   c. How many dB below the peak value of the frequency response is its value at ω_0?

2. The purpose of the circuit of Figure DP2 is the same as that of the stagger-tuned circuit in the text. It is to provide a wider-band frequency response than a single-tuned circuit. Two series resonant circuits are formed by the two sides of a transformer. Assume that the resonant frequencies of the self-inductances with the capacitors on their sides are the same, ω_0, as are also the corresponding Q_0's. A parameter that plays the same role in this circuit as the staggering coefficient in the stagger-tuned circuit is the coupling coefficient k.

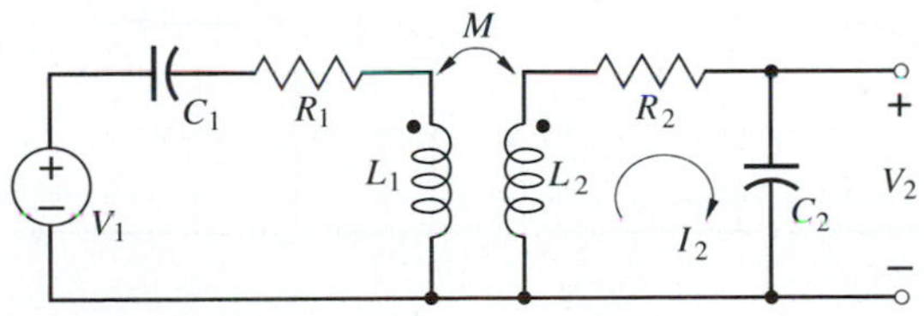

Figure DP2

a. Obtain an expression for the voltage-gain function V_2/V_1 in terms of ω_0, Q_0, and k. (The denominator should be factored into the product of two quadratics.)
b. Find expressions for the poles, simplifying them for large values of Q_0:

$$Q_0 \gg \frac{1}{2\sqrt{1-k}}$$

c. Make a table giving values of Q_0 satisfying this condition for a range of values of k from near 0 to near 1.
d. Sketch the poles and zeros in the complex-frequency plane, showing their real and imaginary parts and showing the trajectory (path) of the poles as the coupling coefficient changes from near 0 to near 1.
e. Express the gain magnitude as a function of frequency, using the approximate expression for the fractional frequency deviation:

$$\delta \approx \frac{1}{2}\left(\frac{\omega}{\omega_0} - \frac{\omega_0}{\omega}\right)$$

f. Carry out an analysis of the magnitude-response curve that parallels the one for the stagger-tuned circuit. In particular: specify a critical coupling coefficient; discuss undercoupling and overcoupling, and how the magnitude peaks change for each. Also draw magnitude-response curves for appropriate coupling coefficients.
g. Determine an expression for the 3-dB bandwidth.
h. Design a double-tuned circuit having capacitor

values of $C_1 = 1\,\mu\text{F}$, $C_2 = 0.4\,\mu\text{F}$, $\omega_0 = 10^4$ rad/s, $Q_0 = 18$, and a bandwidth of 12% of the resonant frequency.

3. This problem is concerned with the design of a three-stage stagger-tuned circuit. The resonant frequency of one stage is at $\omega_0 = 50$ krad/s, while the other two stages are staggered above and below ω_0 by a staggering coefficient k.

a. Show the pole locations in the complex-frequency plane.

b. For high-Q circuits and for frequencies near ω_0, determine the expression for the gain magnitude in terms of the fractional frequency deviation δ.

c. What is the maximum number of peaks and valleys of the magnitude response within the passband?

d. Find the range of staggering coefficients such that the response is not down below 3 dB at the valleys in the response. What is the widest 3-dB bandwidth under this condition, and what is the corresponding value of k?

e. Assume that the quality factors of the impedances are all the same, $Q = 20$, and that the amplifiers have the same g, $g = 1$ S. Determine the circuit parameter values for a value of the coupling coefficient 10% less than the value found in part (d). Find the corresponding 3-dB bandwidth, and draw the frequency-response curve.

4. The circuit in Chapter 11, Figure P36, is repeated in Figure DP4. In Chapter 11 you found its frequency response. We are now interested in its properties in the complex-frequency domain and in utilizing the circuit in filter design.

a. Find each of the transfer functions $F_2 = V_2/V_1$, $F_3 = V_3/V_1$, and $F_4 = V_4/V_1$ in terms of the complex-frequency variable s. Compare with the functions determined in Problem 14, and give these functions appropriate names (low-pass, high-pass, and bandpass).

b. Add the three functions. Describe the general nature of the resulting function. (Do you see why it is called a *biquad*?)

c. Think of a way of obtaining the summation in part (b) by means of a circuit. Would it be any more difficult to obtain a *weighted* sum, that is, a sum in which each function is multiplied by a constant?

d. Your lab has an order to produce a circuit whose normalized transfer function is the following:

$$\frac{V_o}{V_i} = 10\frac{s^2 + 2s + 5}{s^2 + 4s + 13}$$

The size of the order will be 20,000 units if you can design a satisfactory circuit. Using the preceding results, design an appropriate circuit, specifying all normalized component values.

e. The undamped natural frequency is to be 200 krad/s, and the impedance scaling factor is 1 kΩ. Specify the denormalized component values.

f. Without providing detailed calculations, describe how you would design a circuit whose transfer function is the following:

$$\frac{V_o}{V_i} = K\frac{s(s^2 + 16)}{(s^2 + 2s + 2)(s^2 + 6s + 18)}$$

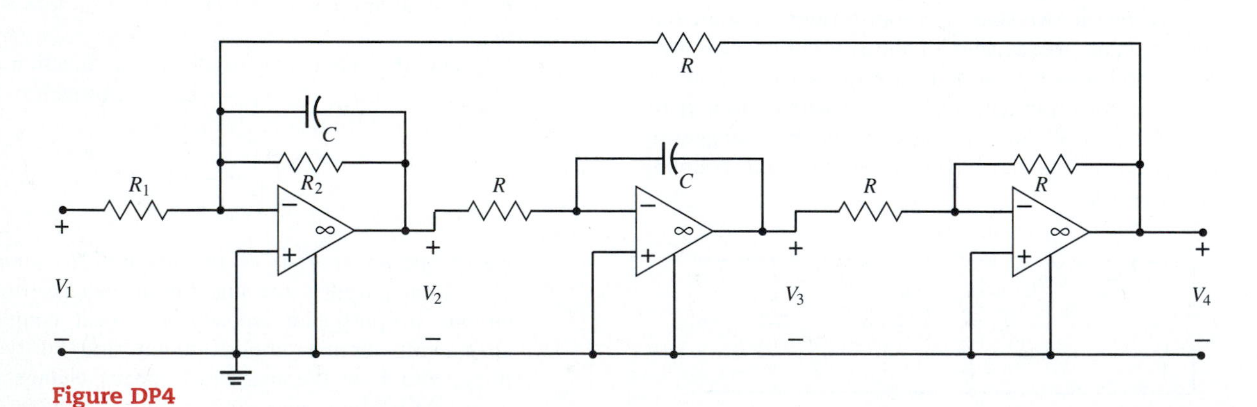

Figure DP4

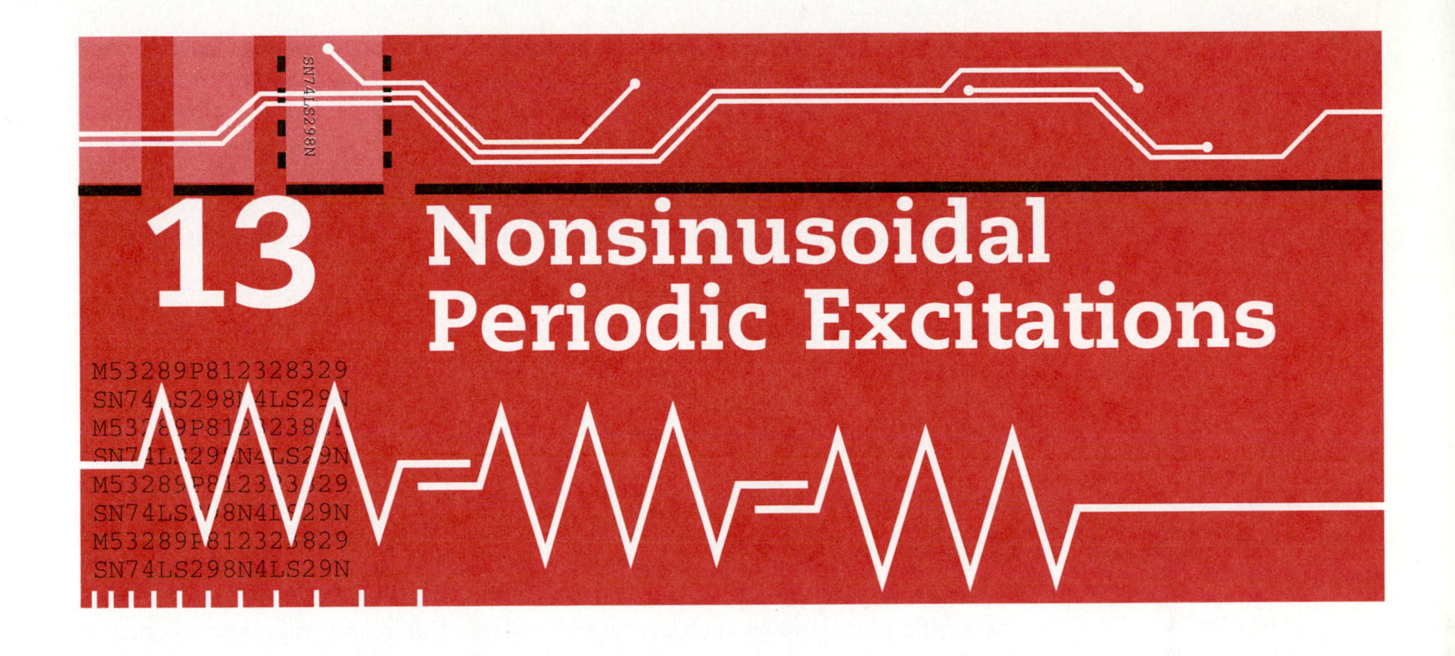

13 Nonsinusoidal Periodic Excitations

You have undoubtedly observed, in your studies so far, that there are two aspects to network theory. One of these, the one to which we have devoted the most attention, has to do with the consequences of interconnecting electrical components to form a topological structure, a network. The second aspect has to do with the signals which are the excitations applied to networks and the resulting responses. The present chapter will be devoted largely to this second aspect. The study of the properties of signals before and after they have been transmitted through a network can be called *signal analysis.*

One of the major goals of linear network analysis is to determine the complete response of an arbitrary (linear) network to an arbitrary excitation. So far we have attacked only parts of this overall problem. For a specific class of networks (dissipative ones) we have solved the problem for arbitrary excitations, but that's because the form of the excitation functions has no influence on the response in such networks; the excitations might as well be simply DC. For another limited class of circuits (first- and second-order), we have indeed found the complete response, but only for a few excitation functions. For arbitrary networks we have solved for just a part of the response—the sinusoidal steady state.

Before we finally tackle the ultimate problem in Chapter 14, we will in this chapter remain in the realm of the steady state, but this time expand the class of functions to include all periodic functions, not just sinusoids. Since the waveforms of periodic functions recur periodically as time goes on, we might intuitively expect that the forced response to such functions will also be periodic. It is appropriate, therefore, to refer to this periodic response as the steady-state response.

One reason for the interest in sinusoids is that the sinusoid itself occurs often in applications. But a much more significant reason is that an arbitrary waveform (subject to certain restrictions to be discussed) can be represented as a combination of sinusoids. The consequences of this fact are of fundamental importance in many branches of science and engineering, including electrical engineering. It has influenced our technical language and mode of thinking to such an extent that, in such areas as communications, we would hardly know how to express ourselves without the concepts it supplies.

In this chapter we will be concerned with exploiting this fact. A certain amount of mathematics is involved; although we will develop some of the mathematical background, you should consult other references for a more complete treatment.

1 FOURIER SERIES

Infinite series of trigonometric functions had been used in the solution of some scientific problems by such mathematicians as Euler and Lagrange as early as 1780. It was the French scientist Jean Baptiste Fourier, however, who recognized certain universal aspects of these series in the representation of periodic functions and in whose honor they are named. (Review the discussion of periodic functions in Chapter 6 before proceeding.)[1]

A periodic function is defined as a function whose form repeats itself in uniform intervals of time $T = 2\pi/\omega_0$, called the *period* of the function. Analytically, this is expressed as:

$$f(t \pm nT) = f(t) \qquad t > 0 \tag{1}$$

Figure 1 shows some examples of periodic functions (see also Chapter 6). Periodic functions have no beginning and no end. Such functions are routine in electrical engineering. The rectifier studied in Chapter 3 produces a periodic function consisting of a half-wave or full-wave rectified sinusoid. Laboratory function generators produce a wide variety of periodic waveforms: square waves, pulse trains, sawtooth waves, and others. A sawtooth wave is also produced by the sweep generator in a TV tube. (Such practical waveforms obviously *do* have a beginning and an end. However, for practical purposes they behave like periodic functions.)

Cosine-and-Sine Form of Series

Fourier showed that a periodic function $f(t)$, subject to certain restrictions, which we shall shortly describe, can be represented by means of an infinite

[1] Jean Baptiste Joseph Fourier (1768–1830) was a mathematical physicist whose main contribution came in the area of heat propagation. His famous book on the subject was published in 1822, fifteen years after he started work in the area. Like Oliver Heaviside a century later, he ignored mathematical rigor and treated cavalierly such niceties as convergence, limits, and continuity.

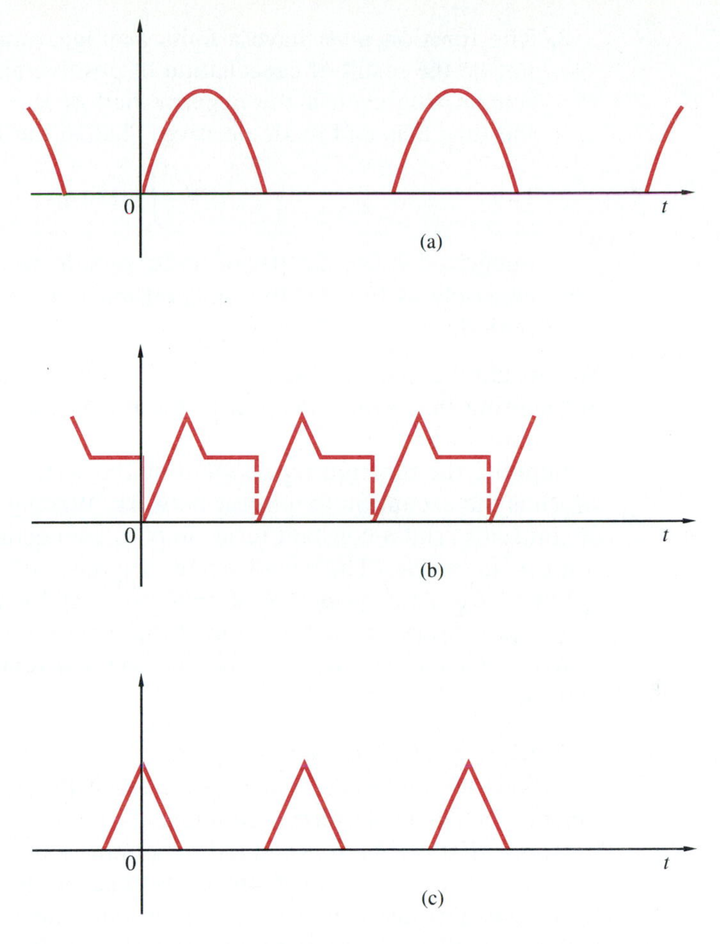

Figure 1 Some periodic functions.

series of trigonometric terms, now called a *Fourier series,* as follows:

$$\begin{aligned} f(t) &= a_0 + a_1 \cos \omega_0 t + a_2 \cos 2\omega_0 t + \cdots + a_n \cos n\omega_0 t + \cdots \\ &\quad + b_1 \sin \omega_0 t + b_2 \sin 2\omega_0 t + \cdots + b_n \sin n\omega_0 t + \cdots \\ &= a_0 + \sum_{n=1}^{\infty} (a_n \cos n\omega_0 t + b_n \sin n\omega_0 t) \end{aligned} \tag{2}$$

The quantity $\omega_0 = 2\pi/T$, the angular frequency of the periodic function, is called the *fundamental* frequency. Each trigonometric function in the series has a frequency which is an integral multiple of the fundamental. These frequencies are called the *harmonics.*

The conditions which a periodic function must satisfy if it is to be represented by a Fourier series, called the *Dirichlet conditions,* are the following:

1. The function may have discontinuities, but there must be only a finite number of them in one period.
2. The function may be oscillatory, but there must be a finite number of maxima and minima in one period.

3. The function must have a finite average value; furthermore, this should not be the result of cancellation of positive and negative areas—it should remain true even if the negative half of the function is inverted around the time axis and made positive. That is, the integral

$$\int |f(t)|\, dt$$

integrated over one period must remain finite. This condition is stated concisely as follows: the function must be *absolutely integrable* over one period.

We would be hard-pressed to find any function normally encountered in engineering that *would not* satisfy these conditions. (Try hard and see if you can create one.)

Suppose the function represented by the series comes from a voltage source which is the excitation to a linear network. Writing the source voltage as a sum of sinusoids (and a constant term) implies connecting a set of these component sources in series. The steady-state response of the network can then be obtained from the principle of superposition by treating each source alone. (The same discussion would apply if the function were the current of a current source.) That's one of the reasons the existence of Fourier series is so appealing.

Finite Sums of Sinusoids

Before we undertake the major task of determining for a given periodic function what its harmonic components might be, let's first consider a few functions that can be represented by a finite sum of sinusoids.

Figure 2 shows a plot of sin ωt, its cube, its fourth power, and the sum of these two. The last (Figure 2d) is a periodic function which, although smooth, looks more like a triangular wave train than a sinusoid. But where's the sum of sinusoids, you might ask? To answer that, we have to do analytically what was done graphically in the figure—that is, analytically add the two powers of sin ωt. For convenience, let's temporarily set $\omega t = x$ and then use appropriate trigonometric identities to reduce the powers of sin x. One identity that can be applied immediately is $2 \sin^2 x = 1 - \cos 2x$. The result becomes:

$$\begin{aligned}
\sin^4 x &= (\sin^2 x)^2 = \left(\frac{1 - \cos 2x}{2}\right)^2 \\
&= \tfrac{3}{8} - \tfrac{1}{2}\cos 2x + \tfrac{1}{8}\cos 4x \\
\sin^3 x &= \sin x \sin^2 x = \sin x\left(\frac{1 - \cos 2x}{2}\right) \\
&= \tfrac{3}{4}\sin x - \tfrac{1}{4}\sin 3x \\
\sin^3 x + \sin^4 x &= \tfrac{3}{8} + \tfrac{3}{4}\sin x \\
&\quad - \tfrac{1}{2}\cos 2x - \tfrac{1}{4}\sin 3x + \tfrac{1}{8}\cos 4x
\end{aligned}$$

Not every step in arriving at the final result is shown. In the expansion of the fourth power, after using the aforementioned identity once, we use it again. In

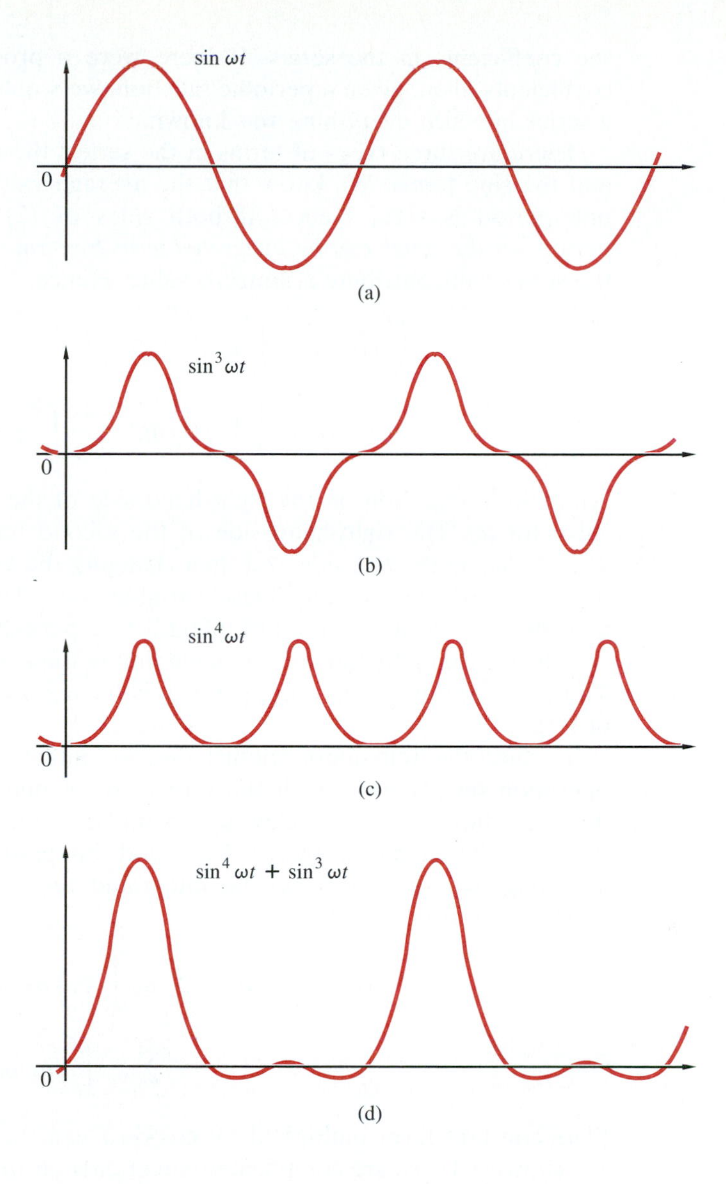

Figure 2
Plots of $f(t) = \sin \omega t$, f^3, f^4, and $f^3 + f^4$.

the expansion of the cubic term, the identities relating to the sine of the sum and difference of two angles are used, in addition to the previous identity. The result is a sum of sines and cosines up to the fourth harmonic. (You should supply the missing steps.)

What is seen in this example is that a finite sum of sines and cosines, each with specific coefficients, is equivalent to a periodic function which does not resemble a sinusoid. Now back to the main thread.

Determination of the Coefficients

For a given periodic function, the fundamental frequency ω_0 will be known. Looking over (2), you will note that the only unknown things remaining are the coefficients in the series. If there were a procedure for calculating the coefficients, then, given a periodic function, we would be able to represent it by a series in which everything was known.

There are three types of terms in the series: the constant, the cosine terms, and the sine terms. We know that the average value of a sine or cosine over one period is zero. Hence, if both sides of (2) are integrated over one period—*if the series can be integrated term by term*—only the constant term in the series will contribute a nonzero value. Hence:

$$\int_0^T f(t)\,dt = \int_0^T a_0\,dt + 0$$

$$a_0 = \frac{1}{T}\int_0^T f(t)\,dt = \frac{1}{2\pi}\int_0^{2\pi} f(t)\,d(\omega_0 t) \tag{3}$$

After the integration on the right-hand side of the first line is performed, we solve for a_0. The right-hand side of the second line is obtained by inserting $T = 2\pi/\omega_0$ in the left side and then changing the variable of integration from the time variable t to the phase variable $\omega_0 t$. (The limits are also changed accordingly.)[2] Now the result of integrating a periodic function over one period and then dividing by the period yields simply the average value of the function. Thus, the constant term a_0 in the series is the average value, or *DC component,* of $f(t)$.

To find the remaining coefficients, we might think of performing some operation on (2) so that all the terms except one will vanish, just as in the determination of a_0. To follow up on this thought, suppose we multiply both sides of the equation by $\cos k\omega_0 t$ and integrate over one period, again assuming that the series *can* be integrated term by term; k is any positive integer. The result will be:

$$\int_0^T f(t)\cos k\omega_0 t\,dt = \sum_{n=1}^{\infty} a_n \int_0^T \cos n\omega_0 t \cos k\omega_0 t\,dt + \sum_{n=1}^{\infty} b_n \int_0^T \sin n\omega_0 t \cos k\omega_0 t\,dt \tag{4}$$

(The constant term multiplied by $\cos k\omega_0 t$ vanishes when integrated, so it is not shown.) These are complicated integrands on the right. However, each one reduces to the sum of two terms by means of the following identities:

$$\begin{aligned} 2\cos n\omega_0 t \cos k\omega_0 t &= \cos(n+k)\omega_0 t + \cos(n-k)\omega_0 t \\ 2\sin n\omega_0 t \cos k\omega_0 t &= \sin(n+k)\omega_0 t + \sin(n-k)\omega_0 t \end{aligned} \tag{5}$$

[2] If you consult other references on Fourier series, you will find a number of different systems of notation. In mathematics texts the variable x is usually used for $\omega_0 t$. The mathematics of Fourier series is probably best developed in terms of such a dimensionless variable. Our purpose, however, is not to give the neatest possible mathematical development but, concurrently, to use variables suggestive of engineering interpretations. To be mathematically neat, for example, we would have had to express $f(t)$ on the right-hand side in (3) as a function of $\omega_0 t$, not of t, but then the name of the function would have had to be changed from f to something else.

When n is any integer different from k, each term on the right in the two identities in (5) is a sinusoid. Now recall that the integral of a sinusoid over one period is zero. Hence, the integrals in (4) will both vanish. However, when $k = n$, something significant happens in (5): the second term on the right becomes $\sin 0 = 0$ in the second line and $\cos 0 = 1$ in the first line. Nothing different from before happens when the second line is integrated, but now, in the first line, the integral in (4) over one period yields $T/2 = \pi/\omega_0$. (Write out the details.) Since the coefficient of this term is a_n, the final result becomes:

$$a_n = \frac{2}{T}\int_0^T f(t)\cos n\omega_0 t\, dt = \frac{1}{\pi}\int_0^{2\pi} f(t)\cos n\omega_0 t\, d(\omega_0 t) \tag{6}$$

This is a formula for calculating the coefficients of all the cosine terms.

In a completely similar manner, the b_n coefficients can be found by first multiplying both sides of (2) by $\sin k\omega_0 t$ and then integrating term by term, assuming it is possible to do so. The integrals will again all vanish except when $n = k$.[3] The result will give an expression for b_n.

COEFFICIENT FORMULAS For easy reference, the formulas for all the coefficients will be collected here:

$$a_0 = \frac{1}{T}\int_0^T f(t)\, dt = \frac{1}{2\pi}\int_0^{2\pi} f(t)\, d(\omega_0 t) \tag{7a}$$

$$a_n = \frac{2}{T}\int_0^T f(t)\cos n\omega_0 t\, dt = \frac{1}{\pi}\int_0^{2\pi} f(t)\cos n\omega_0 t\, d(\omega_0 t) \tag{7b}$$

$$b_n = \frac{2}{T}\int_0^T f(t)\sin n\omega_0 t\, dt = \frac{1}{\pi}\int_0^{2\pi} f(t)\sin n\omega_0 t\, d(\omega_0 t) \tag{7c}$$

(If you haven't done so already, you are urged to carry out each step and confirm the results.)

EXAMPLE 1 The waveform shown in Figure 3 is the periodic voltage of a signal generator. Find the Fourier coefficients and write out the terms of the series through the sixth harmonic.

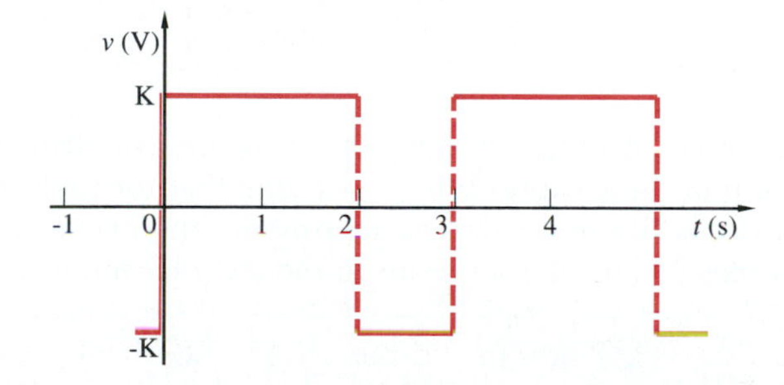

Figure 3
Rectangular waveform of Example 1.

[3] This property of sinusoidal functions—namely, the vanishing of the integral over one period of the product of two sines or cosines with unequal harmonic frequencies—is called the *orthogonality* property. The terminology comes by analogy with two vectors that are at right angles, or orthogonal, so that their product (dot product) vanishes.

SOLUTION In this example, the period is $T = 3$ s. An integration is not needed to find the average value of the function. The area under the positive part of the curve ($2K$) is twice that of the negative part; hence, the net area is half that of the positive part (K). Thus the DC component of the wave is $K/3$. If you want some practice, confirm this answer by using the formula for the constant term.

The convenient interval of integration to use in this example is the period from 0 to 3 s. In this range, the voltage function is a simple function (just K) from 0 to 2 and another simple function ($-K$) from 2 to 3. Thus, a_n is calculated as follows:

$$a_n = \frac{2}{T}\int_0^T v(t)\cos n\omega_0 t\, dt$$

$$= \frac{2}{3}\int_0^2 K\cos n\omega_0 t\, dt + \frac{2}{3}\int_2^3 -K\cos n\omega_0 t\, dt$$

$$= \frac{K}{n\pi}\left(\sin\frac{2\pi n}{3}t\bigg|_0^2 - \sin\frac{2\pi n}{3}t\bigg|_2^3\right)$$

$$a_n = \frac{2K}{n\pi}\sin\frac{4\pi n}{3}$$

Similarly, b_n is calculated as follows:

$$b_n = \frac{2}{T}\int_0^T v(t)\sin n\omega_0 t\, dt$$

$$= \frac{2}{3}\int_0^2 K\sin n\omega_0 t\, dt + \frac{2}{3}\int_2^3 -K\sin n\omega_0 t\, dt$$

$$= -\frac{K}{n\pi}\cos\frac{2n\pi}{3}t\bigg|_0^2 + \frac{K}{n\pi}\cos\frac{2n\pi}{3}t\bigg|_2^3$$

$$b_n = \frac{2K}{n\pi}\left(1 - \cos\frac{4n\pi}{3}\right)$$

Numerical values of the coefficients are established by inserting integer values of n in these expressions. It is found that for both the cosine and the sine terms the coefficients of the third harmonic and every multiple of the third harmonic vanish. Thus, the series up to the sixth harmonic can be written as follows:

$$v(t) = \frac{K}{3} + \frac{K\sqrt{3}}{\pi}\left(-\cos\frac{2\pi}{3}t + \frac{1}{2}\cos\frac{4\pi}{3}t - \frac{1}{4}\cos\frac{8\pi}{3}t + \frac{1}{5}\cos\frac{10\pi}{3}t + \cdots\right)$$

$$+ \frac{3K}{\pi}\left(\sin\frac{2\pi}{3}t + \frac{1}{2}\sin\frac{4\pi}{3}t + \frac{1}{4}\sin\frac{8\pi}{3}t + \frac{1}{5}\sin\frac{10\pi}{3}t + \cdots\right)$$

One other observation should be made before leaving this example: the coefficients (the harmonic amplitudes) are inversely proportional to the order of the harmonics.

An important conclusion should be observed from the coefficient formulas in (7). It might be thought that the formula for a_0 can be obtained from that for a_n simply by setting $n = 0$. It is clear from (7), however, that this is incorrect, since, for $n = 0$, the expression for a_n reduces to twice the expression for a_0. Thus, the DC component must always be calculated separately.

EXERCISE 1

The rectangular wave in Figure 3 is modified as follows. The positive part of the wave now extends from 0 to 1 and the negative part from 1 to 3 s, and this pattern is repeated. In addition, the positive amplitude becomes $3K/2$ and the negative amplitude, $-K/2$. (a) Find the coefficients of the Fourier series. (b) Write out the terms of the series through the third harmonic.

ANSWER: $a_0 = \dfrac{K}{2}; \qquad a_n = \dfrac{2K}{n\pi}\sin\dfrac{2n\pi}{3}; \qquad b_n = \dfrac{2K}{n\pi}\left(1 - \cos\dfrac{2n\pi}{3}\right).$ □

Fourier Series Theorem

A few points still remain to be clarified in the preceding development. First, note that all the integrations cover one period, ranging from 0 to T (or 0 to 2π). It is not necessary that this particular range be covered, so long as one period is covered. That is, the same result will be obtained if the range of integration is from t_1 to $T + t_1$, where t_1 is an arbitrary value of time chosen in each case to make life easier. We have chosen $t_1 = 0$ so far.

Next, the procedure for calculating the *Fourier coefficients* (the a's and b's) does not constitute a proof that the series exists and converges to the specified periodic function $f(t)$. The proof requires a fairly extensive development which, needless to say, we shall not indulge in here. Assuming that the series representation of $f(t)$ exists, however, then (7) will give the correct values of the coefficients.

Finally, to arrive at the expressions for the coefficients, we assumed that the series could be integrated term by term. It is a mathematical theorem that term-by-term integration of a series representing a function is possible only if the series converges *uniformly* in the integration interval. (See any calculus book.) As part of proving the Fourier theorem, it has to be proved that the series does indeed converge uniformly—except at discontinuous points. At a point of discontinuity, it turns out that the Fourier series converges to the mean of the two values of the function on either side of the discontinuity. (If a function jumps from 2 to 8, for example, its Fourier series converges to 5 at that point.)

For reference purposes, the preceding discussion is stated as the following *Fourier series theorem*:

> If a periodic function with period T satisfies the Dirichlet conditions, then it can be represented by a trigonometric series as given in (2) with the coefficients specified by (7). The series converges to the function (and converges uniformly) at all continuous points; at a point of discontinuity t_0, the series converges to $[f(t_0-) + f(t_0+)]/2$.

Shifting the Time Origin

Since a periodic function goes through its periodic variations for all time, the specific point which is chosen as the origin of time should not be of great significance. Let's now examine what the effect will be on the Fourier coefficients if the origin of time is shifted by an amount, say, t_1. That is, given a function with a particular time origin, a Fourier series has been obtained. Following this, the origin is moved to a point whose time value was t_1 with the old origin. What happens to the Fourier series?

A specific example, a triangular wave, is shown in Figure 4a. Its Fourier series will not have a constant term, since its average value is zero. Someone came up with the following coefficient values, with the time origin selected as

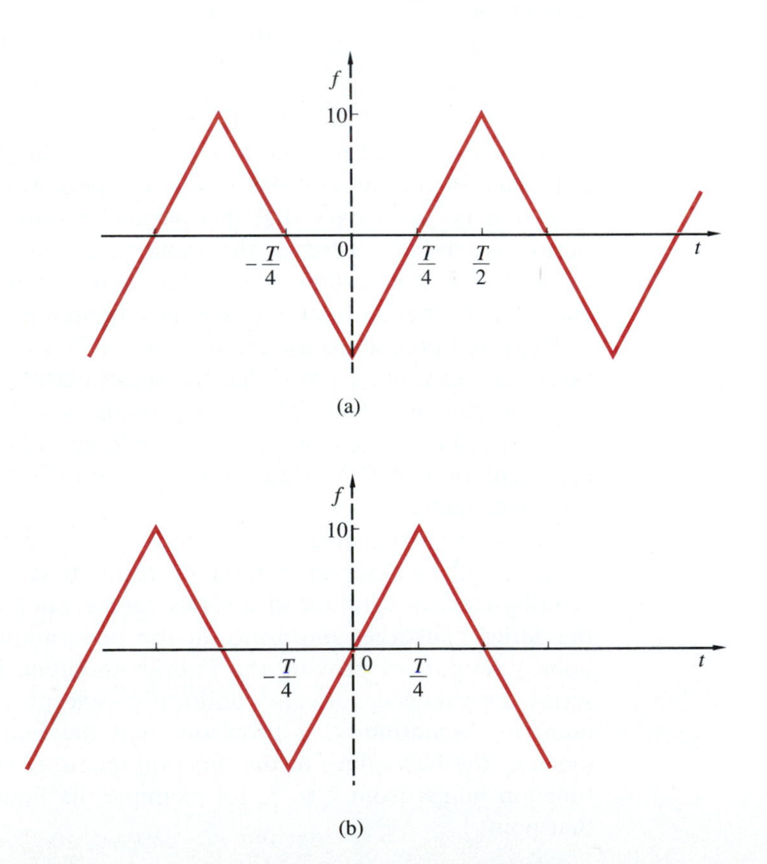

Figure 4
Triangular wave.

shown:

$$a_n = \frac{20}{\pi^2 n^2}(\cos n\pi - 1)$$

$$= \begin{cases} -\dfrac{40K}{\pi^2 n^2} & (n \text{ odd}) \\ 0 & (n \text{ even}) \end{cases}$$

$$b_n = 0$$

$$f(t) = -\frac{40}{\pi^2}\left(\cos \omega_0 t + \frac{1}{9}\cos 3\omega_0 t + \frac{1}{25}\cos 5\omega_0 t + \cdots\right)$$

(Go through the calculation and confirm, please.) The first few harmonics in the series are also shown. Now suppose the time origin is shifted by a quarter period to the right, as indicated in Figure 4b. For any time t relative to the old origin, the new time will be $t - T/4$. The new series is obtained by replacing t by this value in the old series, resulting in:

$$f(t) = -\frac{40}{\pi^2}\left[\cos \omega_0\left(t - \frac{T}{4}\right) + \frac{1}{9}\cos 3\omega_0\left(t - \frac{T}{4}\right) + \frac{1}{25}\cos 5\omega_0\left(t - \frac{T}{4}\right) + \cdots\right]$$

$$= -\frac{40}{\pi^2}\left[\cos\left(\omega_0 t - \frac{\pi}{2}\right) + \frac{1}{9}\cos\left(3\omega_0 t - \frac{3\pi}{2}\right) + \frac{1}{25}\cos\left(5\omega_0 t - \frac{5\pi}{2}\right) + \cdots\right]$$

$$= -\frac{40}{\pi^2}\left(\sin \omega_0 t - \frac{1}{9}\sin 3\omega_0 t + \frac{1}{25}\sin 5\omega_0 t - \cdots\right)$$

We confirm that the magnitudes of the coefficients have not been affected by the time shifting. We find that the angle of each harmonic has been increased by $\pi/2$ times the order of the harmonic. (Take note that $\pi/2$ corresponds to a quarter period in terms of the phase variable.) Using the trigonometric identity relating to the cosine of the sum of the two angles leads to the final form. The result of shifting the time origin by a quarter period is quite interesting; what started as a series of cosines has been converted to a series of sines. This phenomenon will be further discussed in the next section.

As you go along in this chapter, looking over examples and working out exercises, you should make note of the features you observe in the Fourier series of various types of functions. For example, in the earlier example of a rectangular wave, we called attention to the fact that the coefficients were inversely proportional to the order n of the harmonic. For the triangular wave just discussed, this isn't true. In this case, the nonzero coefficients are inversely proportional to the *square* of the harmonic order. That means that the size of the harmonic amplitudes will decrease faster as more and more harmonics are added. This is a significant matter. Suppose we take, say, 5% as a cutoff; that is, if a harmonic contributes less than 5% of the fundamental to the overall series value, we will neglect it in comparison with the others. How many harmonics would have to be kept for the rectangular wave, and how many for the triangular wave?

2 SIMPLIFICATIONS RESULTING FROM SYMMETRIES[4]

In determining the Fourier series of the rectangular wave in Figure 3, it was found that some of the harmonics are missing (every third one). This must have something to do with the shape of the function. Similarly, in the triangular wave in the immediately preceding section (Figure 4), there were no sine terms with one choice of origin and no cosine terms with another choice of origin. Perhaps by studying how the shape of a waveform or the choice of its time origin influences the harmonics in its Fourier series (perhaps making some of the harmonics vanish, or perhaps just the cosine terms or just the sine terms), we can avoid the effort of calculating the coefficients only to find that some of them are zero.

Even Symmetry

This line of thought leads us to examine the cosine and sine functions for some of their properties. Let's first concentrate on the cosine. Besides its being periodic, we know that $\cos(-t) = \cos t$. This property illustrates a general class of functions:

Even function: $$f(-t) = f(t) \tag{8}$$

Thus, along the time axis, an even function is the mirror image of itself reflected about the origin. Whatever its value at a given distance from the time origin in one direction, its value at the same distance in the other direction is the same.

If two even functions are added together, their sum will be even. If an even function is added to another function which is not even, their sum cannot be even. (Demonstrate this by replacing t by $-t$ in the sum of the two functions and then using the property of evenness.)

Given an even periodic function, is it possible for its Fourier series to contain sine terms? The answer is, clearly, no; because the sine function is not even, the series would not satisfy (8). Hence, given an even periodic function,

[4] This section is devoted to the exploration of simplifications in the formulas of the Fourier coefficients for functions possessing certain symmetry properties. The amount of computational labor involved in evaluating the coefficients for such functions is reduced. However, there is not much of fundamental significance here, aside from the identification of functions having the properties in question. If you have the desire for a lot of practice on the coefficient formulas, here's the place to get it. Otherwise, skim it to get a feel and go on to the next section.

half the work of obtaining a Fourier series is already done, just by recognizing its even symmetry. However, as noted many times, the choice of the origin of time is arbitrary; we choose it for convenience. It might be that a certain function is not an even function for some particular choice of the time origin but can become one if the time origin is shifted.

An example was the triangular wave described in the preceding section in Figure 4. With the origin chosen as in Figure 4a, the function is even; and indeed, we found the series representing the function with that origin to contain only cosine terms.

EXERCISE 2

A square wave is shown in Figure 5 without an origin of time. Select a time origin to make the resulting function even, and determine the Fourier coefficients. Choose the one-period integration interval in such a way that the evaluation of the integrals is simplified. Note how the harmonic amplitudes vary with n.

ANSWER: Choose the origin in the center of one of the "bumps," say, the positive one. Choose a symmetrical integration interval, from $-T/2$ to $+T/2$. $a_{2k-1} = 10/(2k - 1)\pi$, for $k = 1$ to ∞. $b_n = 0$. □

In evaluating the coefficients in this exercise, you may have noticed that the integration over the negative half period gives the same value as the integration over the positive half period. This is a general result, as we shall now demonstrate.

For an even function $f(t) = f(-t)$, the Fourier a_n coefficients are found by integrating from $-T/2$ to $T/2$. Consider writing the integral as the sum of two integrals, one from $-T/2$ to 0 and the other from 0 to $T/2$. By carrying out a sequence of steps, we can show that the first of these is equal to the second:

$$\int_{-T/2}^{0} f(t) \cos n\omega_0 t \, dt = -\int_{0}^{-T/2} f(t) \cos n\omega_0 t \, dt$$

$$= \int_{0}^{-T/2} f(-t) \cos n\omega_0(-t) \, d(-t) = \int_{0}^{T/2} f(t) \cos n\omega_0 t \, dt$$

In the second step, the order of integration is reversed, resulting in a reversal of sign. In the next step, this negative sign is associated with the t in dt; also, the sign of t in $f(t)$ and in the cosine has been reversed—permitted because these are even functions. Finally, the variable $-t$ in the preceding step is changed to t, requiring a change in the upper limit to $+T/2$. The end result is

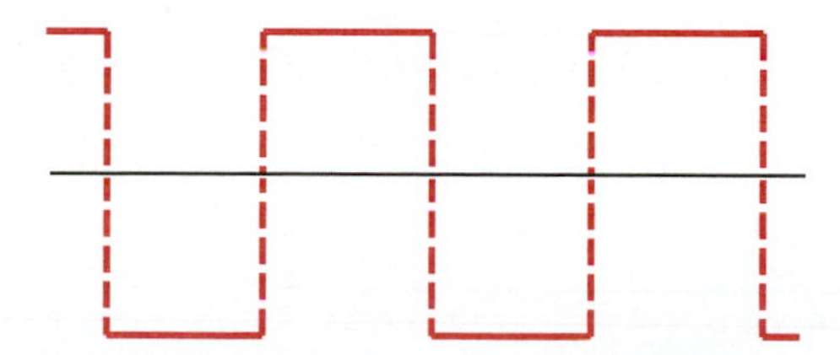

Figure 5
Square wave.

that the integral over the negative half period is identical with the one covering the positive half period. Thus, to find the a_n coefficients for an even function, it is necessary to integrate only over the positive half period and to multiply by 2.

For periodic functions with even symmetry

$$a_n = \frac{4}{T}\int_0^{T/2} f(t)\cos n\omega_0 t\,dt \tag{9}$$

Odd Symmetry

A second type of symmetry is exemplified by a sine function, which has the property that $\sin(-x) = -\sin x$. Again, this is an example of a general class of functions having the following property:

Odd function: $$f(-t) = -f(t) \tag{10}$$

Thinking in terms of a graph of the function, this time the reflection is not only about the vertical axis but about the horizontal axis also. That is, the value of an odd function at some time is the negative of its value at the same value of negative time.

Let's explore the implications of the preceding definition of an odd function for $t = 0$. By definition, $f(-0) = -f(0)$. But 0 and -0 represent the same time, so this result requires the value of the function at $t = 0$ to equal its own negative. For a continuous function, the only way that is possible is for $f(0) = 0$. Thus, the graph of a *continuous* odd function must pass through $t = 0$. (Convince yourself by drawing some examples that the same is true for functions discontinuous at $t = 0$.)

EXERCISE 3

For the square wave in Figure 5, choose the origin of time in order to make the function an odd function. Complete the discussion of the preceding paragraph for odd functions that have a discontinuity at the origin. □

EXERCISE 4

The graph sketched in Figure 6 represents half of a periodic function. Sketch the second half if the function is to be (a) even and (b) odd. □

Given an odd function, it is not possible for the Fourier series representing this function to contain any cosine terms. Since the location of the time origin is intimately related to whether or not the function is odd, it is sometimes possible to make a given function odd by appropriate choice of the origin. In

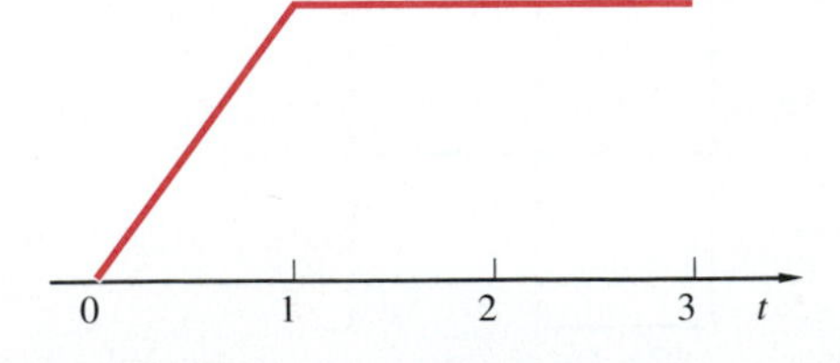

Figure 6
Half a periodic function; to be completed for even or odd result.

that case, there is no need to evaluate the a_n in the series, because we know they must be zero. Can the series representing an odd function have a DC component? The answer is again no, because the sign of the constant term would not change if t were replaced by $-t$.

EXERCISE 5

Carry out a development similar to the one carried out for an even function to show that the b_n coefficients in the Fourier series representation of an odd function can be determined by integrating over only the positive half period and multiplying by 2. You should obtain the following result:

For functions with odd symmetry:

$$a_n = 0 \qquad b_n = \frac{4}{T}\int_0^{T/2} f(t)\sin n\omega_0 t\,dt \tag{11}$$

□

EXERCISE 6

(a) Determine the coefficients of a series of sines representing the square wave in Figure 5 with a proper choice of time origin. (b) Confirm your result, using the cosine series found in Exercise 2 with an appropriate shift of the time origin. □

Half-Wave Symmetry

In the preceding sections we explored Fourier series representing functions possessing certain symmetries. We found that cosines are absent from series representing functions with odd symmetry and sines are absent from series representing functions with even symmetry. But in some examples, we also found that no even harmonics were present in the series. Is there some particular property that leads to this result?

The series for the triangular wave in Figure 4 had this property. With the origin shown in Figure 4, this function has even symmetry. It is redrawn in Figure 7 in order to display another feature that is observable. The two points marked A and B are half a period apart, and this fact does not depend on the location of the time origin. We notice that the value of the function at t_B is the negative of its value at t_A, a half period away. Let's generalize this property and give it a name. A periodic function is said to have *half-wave symmetry* if

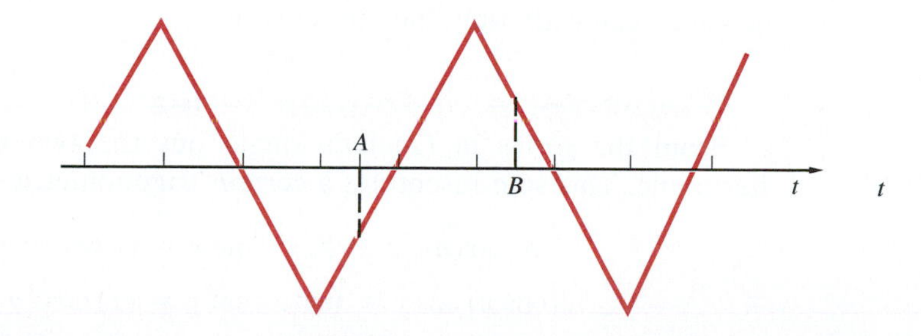

Figure 7 Function with half-wave symmetry.

it satisfies the following condition:

$$f\left(t + \frac{T}{2}\right) = -f(t) \tag{12}$$

Let's now use this property to find a simplification of the coefficient formulas. The first step is to use the period centered at $t = 0$, from $-T/2$ to $T/2$, as the integration interval. Then, following a procedure similar to the one we used in the case of functions with even symmetry:

1. We find that the integral over one period is the sum of two integrals each over half a period, one from $-T/2$ to 0 and the other from 0 to $T/2$.
2. We change variables, but this time replacing t by $t + T/2$, and we change the limits of integration accordingly: when $t = -T/2$, $t + T/2 = 0$; and when $t = 0$, $t + T/2 = T/2$.
3. Finally, we use the half-wave symmetry property in (12).

Carry out the details of the outlined steps to confirm the following results:

For periodic functions with half-wave symmetry:

$$\begin{aligned} a_{2k-1} &= \frac{4}{T}\int_0^{T/2} f(t)\cos(2k-1)\omega_0 t\,dt \\ b_{2k-1} &= \frac{4}{T}\int_0^{T/2} f(t)\sin(2k-1)\omega_0 t\,dt \end{aligned} \tag{13}$$

Writing the index n as $2k - 1$, with k ranging over all positive integers, automatically leads to odd harmonics only.

3 ALTERNATIVE FORMS OF THE FOURIER SERIES

The form of the Fourier series considered so far contains both cosine and sine terms generally. For each harmonic, these terms can be combined using certain trigonometric identities, and the series can be rewritten in terms of either cosines alone, with initial angle, or sines alone, with initial angle. We shall consider explicitly only the cosine case.

Cosine Form of Fourier Series

From the series in (2) let's single out the two terms involving the nth harmonic. This sum resembles a certain trigonometric identity as follows:

$$a_n \cos n\omega_0 t + b_n \sin n\omega_0 t = c_n \cos(n\omega_0 t - \phi_n)$$

$$(\cos x)\cos y + (\sin x)\sin y = (1)\cos(y - x)$$

In order to compare these two expressions, the coefficient on the right side of the top line should be the same as the corresponding coefficient of the second line. Thus, the top equation should be divided by c_n. From the right-hand sides, we identify y with $n\omega_0 t$ and x with ϕ_n. By comparing the coefficients of the two cosines and the two sines (after dividing the top equation by c_n), we find $a_n/c_n = \cos\phi_n$ and $b_n/c_n = \sin\phi_n$. Squaring and adding these expressions, and dividing the second by the first, result in:

$$c_n = \sqrt{a_n^2 + b_n^2} \qquad \phi_n = \tan^{-1}\frac{b_n}{a_n} \tag{14}$$

Using these results, the original series in (2) can be rewritten as follows:

$$f(t) = c_0 + \sum_{n=1}^{\infty} c_n \cos(n\omega_0 t - \phi_n) \tag{15}$$

In this form, it is clearer that the function $f(t)$ is made up of a DC component and an AC component which consists of a fundamental and its harmonics. The quantities c_n and $-\phi_n$ are the *harmonic amplitudes* and *phase angles,* respectively. Note, however, that c_n is the peak value of the harmonic, not its rms value. For uniformity of notation, the DC value has been renamed $c_0\,(= a_0)$.

Given a periodic function, in order to obtain a Fourier series representation in the form of (15), the coefficients are first calculated from (7). From these, the harmonic amplitudes and phase angles are calculated using (14). (How about the DC component?)

EXERCISE 7

a. Express in the form of (15), from DC through the fifth harmonic, the Fourier series representing the rectangular wave of Figure 3 obtained in Example 1.

b. Repeat for the modified wave whose series you obtained in Exercise 1.

ANSWER:

a. $$v(t) = \frac{K}{3} + \frac{2\sqrt{3}K}{\pi}\left[\cos\left(\frac{2\pi}{3}t + 120°\right) + \frac{1}{2}\cos\left(\frac{4\pi}{3}t - 60°\right) + \frac{1}{4}\cos\left(\frac{8\pi}{3}t + 120°\right) + \frac{1}{5}\cos\left(\frac{10\pi}{3}t - 60°\right) + \cdots\right]$$

b. $$v(t) = \frac{K}{2} + \frac{2\sqrt{3}K}{\pi}\left[\cos\left(\frac{2\pi}{3}t - 60°\right) + \frac{1}{2}\cos\left(\frac{4\pi}{3}t + 120°\right) + \frac{1}{4}\cos\left(\frac{8\pi}{3}t - 60°\right) + \frac{1}{5}\cos\left(\frac{10\pi}{3}t + 120°\right) + \cdots\right]$$ □

You may be wondering about a seeming inconsistency in the case of functions possessing odd symmetry. Their series are supposed to contain sines only, not cosines, yet the series described in this section have exclusively cosine terms. This is not really an inconsistency, since $\sin x = \cos(x - \pi/2)$; hence, sines can be expressed in terms of cosines at nonzero phase angles.

Indeed, that same trigonometric identity can be used to convert the cosine terms in the series being discussed in this section to sine terms; only the phase angles will be different, by 90°.

EXERCISE 8

Write a series similar to (15) but with sine functions instead of cosines, using d_n for the harmonic amplitudes and ψ_n for the phases. Express the amplitudes and phases in terms of c_n and ϕ_n in (14).

ANSWER: $$f(t) = d_0 + \sum_{n=1}^{\infty} d_n \sin(n\omega_0 t + \psi_n)$$

$$d_0 = c_0, \qquad d_n = c_n, \qquad \psi_n = -\phi_n + \pi/2$$ □

Exponential Form of Fourier Series

Since, using Euler's theorem, sinusoids can be expressed in terms of exponentials, it should be possible to place in exponential form the Fourier series described in the preceding section. Thus, for the nth harmonic, we get:

$$\begin{aligned} c_n \cos(n\omega_0 t - \phi_n) &= \frac{c_n}{2}(e^{j(n\omega_0 t - \phi_n)} + e^{-j(n\omega_0 t - \phi_n)}) \\ &= F_n e^{jn\omega_0 t} + F_n^* e^{-jn\omega_0 t} \end{aligned} \tag{16}$$

A new coefficient, F_n, has been introduced.[5] The new coefficient is related to the previously defined ones by:

$$\begin{aligned} F_n &= \frac{c_n}{2} e^{-j\phi_n} = \frac{1}{2}\sqrt{a_n^2 + b_n^2}(\cos\phi_n - j\sin\phi_n) \\ &= \tfrac{1}{2}(a_n - jb_n) \end{aligned} \tag{17}$$

[5] When we first introduced the Fourier series, we first used lowercase a_n and b_n for the coefficients. Then came lowercase c_n. Now suddenly, we jumped over d and e and came up with capital F_n. Why? One possible explanation is that d_n is being reserved for the series involving sines with initial angle. But the real reason is the notational symmetry that results when, for a function of time $f(t)$, we use the capital form of the same letter for the Fourier coefficient of the exponential form. Then, if the signal is a voltage $v(t)$, the Fourier coefficient will be V_n; if the signal is $i(t)$, the coefficient will be I_n; and so on. Furthermore, as we shall see later in this chapter and in the next, the Fourier series can be generalized in such a way that the notational symmetry becomes even more useful, so bear with it.

The last line comes from the discussion preceding (14). Now, putting (16) for the nth harmonic into the series in (15), a new form is obtained:

$$f(t) = F_0 + \sum_{n=1}^{\infty} F_n e^{jn\omega_0 t} + \sum_{n=1}^{\infty} F_n^* e^{-jn\omega_0 t}$$

$$= \sum_{n=-\infty}^{\infty} F_n e^{jn\omega_0 t} \quad (18)$$

A number of steps were taken to arrive at the second line. First, the constant term c_0 was relabeled F_0. Why? You guessed it: for convenience, for uniformity of notation. Then, note in the second summation of the first line that n appears in three places: in the exponent, in the coefficient F_n^*, and as the summation index. Let's replace n by $-n$ in the second summation. The exponent now becomes positive, the same as the exponent in the first summation; the coefficient becomes F_{-n}^*, and the summation extends over negative values of n. But if n is replaced by $-n$ in the expression for F_n in (17), this becomes:

$$F_{-n} = \tfrac{1}{2}(a_{-n} - jb_{-n}) = \tfrac{1}{2}(a_n + jb_n) = F_n^* \quad (19)$$

The intermediate results $a_{-n} = a_n$ and $b_{-n} = -b_n$ follow from the coefficient formulas in (7). The end result is that the two summations running over positive values of n can be replaced by a single summation running over both positive and negative values, as shown in the second line in (18).

The expression in (18) is called (can you guess?) the *exponential form* of the Fourier series. To find the F_n coefficients, we can find the a_n and b_n coefficients from (7) and insert them into (17). After some manipulation, the result becomes:

$$F_n = \frac{1}{T}\int_0^T f(t)e^{-jn\omega_0 t}\,dt = \frac{1}{2\pi}\int_0^{2\pi} f(t)e^{-jn\omega_0 t}\,d(\omega_0 t) \quad (20)$$

To confirm this expression, multiply both sides of the exponential form of the series by $e^{-jk\omega_0 t}$ and integrate over one period.

The coefficients of the exponential form of the series are complex; they contain information that includes both the harmonic amplitude and the phase. Note carefully that the magnitude of F_n is just one-half of the nth harmonic amplitude (or $1/\sqrt{2}$ of the rms value). Except for this scale factor, we can look upon F_n as the phasor representing the nth harmonic.

In the preceding section, attention was devoted to functions possessing certain kinds of symmetry. Since the exponential is neither even nor odd, we cannot use simplified formulas for finding the coefficients of the exponential series for even or odd functions. However, half-wave symmetry is not origin-dependent. Hence, for functions with half-wave symmetry, it should be valid to integrate over half the period and to double the coefficient in the exponential form as well. Thus:

For functions with half-wave symmetry:

$$F_n = \frac{2}{T}\int_0^{T/2} f(t)e^{-jn\omega_0 t}\,dt \qquad n \text{ odd} \quad (21)$$

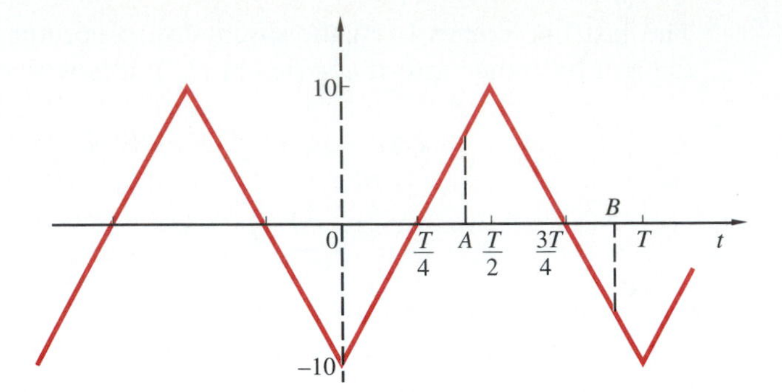

Figure 8
Triangular wave displaying half-wave symmetry.

EXAMPLE 2 Calculate the F_n coefficients of the exponential form of the series representing the triangular wave shown in Figure 8. This function has half-wave symmetry. The origin is chosen to simplify the integrand, as shown in Figure 8. Integrating from 0 to $T/2$ results in a single expression representing the function over the entire half period, namely, $f(t) = 40t/T - 10$. Hence, the coefficients will be:

$$F_n = \frac{2}{T}\int_0^{T/2}\left(\frac{40}{T}t - 10\right)e^{-jn\omega_0 t}\,dt$$

$$= \frac{2}{T}\left\{\frac{40}{T}\left[\frac{-e^{-jn\omega_0 t}(1 + jn\omega_0 t)}{-4n^2\pi^2/T^2}\right] + \frac{10e^{-jn\omega_0 t}}{j2n\pi/T}\right\}\Bigg|_0^{T/2}$$

$$= -\frac{40}{n^2\pi^2} \qquad n \text{ odd}$$

Notice that, for the exponential form, a single coefficient formula applies for all values of n, including $n = 0$. This is a distinct difference from the trigonometric form of the series, where a_0 cannot be obtained from the formula for a_n. However, even in the case of the exponential form, where the same formula applies for all the coefficients, it is not possible to set $n = 0$ in an expression for F_n obtained by evaluating the formula. Thus, in the preceding example, setting $n = 0$ in the result obtained for F_n gives infinity. Clearly, that is not the DC component.

4 THE FREQUENCY SPECTRUM

When discussing the properties of a time function, we are accustomed to emphasizing certain aspects of its waveform, such as its amplitude and its duration. These aspects describe the signal in the *time domain.* We have now found that a periodic signal can be described another way, as a summation of sinusoids with *harmonically* related frequencies, each an integral multiple of a lowest frequency, the *fundamental.* This is a frequency-domain description.

In other words, a signal is completely described if a statement, either analytical or graphical, is made of its dependence on time. We now find that it is also completely described in terms of the harmonic amplitudes and phases of the sinusoids (plus DC) that make up the signal. All information about the signal conveyed by specifying $f(t)$ is also conveyed by specifying the Fourier coefficients F_n. You will note from (20) that F_n is dependent on ω_0 and n. Just as a graphical portrayal of the signal can be obtained by plotting $f(t)$ against t, so also a graphical portrayal can be obtained by plotting the magnitude and angle of F_n (the harmonic amplitudes and angles) against $n\omega_0$. Since n is an integer, however, the abscissa does not take on all continuous values, but only discrete values. This is a description in the *frequency domain.*

A Pulse Train

The preceding ideas will be illustrated by means of examples, one of which is the pulse train shown in Figure 9. The pulse amplitude is V, its width is w, and the repetition rate is $1/T$ (radian frequency $\omega_0 = 2\pi/T$). To make the function even, the origin of time is chosen at the center of a pulse. Analytically, the function can be expressed as:

$$v(t) = \begin{cases} V & -\dfrac{w}{2} < t < \dfrac{w}{2} \\ 0 & \dfrac{w}{2} < t < T - \dfrac{w}{2} \end{cases}$$

$$v(t + T) = v(t)$$

First, let's calculate the coefficients of the exponential form, using (20) and choosing a period with symmetrical limits around the origin. Since the function is 0 over part of this range, the only contribution to the integral will come from the interval $-w/2$ to $+w/2$. Hence the coefficients will be:

$$F_n = \frac{1}{T}\int_{-w/2}^{w/2} Ve^{-jn\omega_0 t}\,dt = \frac{2V}{n\omega_0 T}\sin\frac{n\omega_0 w}{2}$$

$$= V\frac{w}{T}\frac{1}{n\omega_0 w/2}\sin\frac{n\omega_0 w}{2} \tag{22}$$

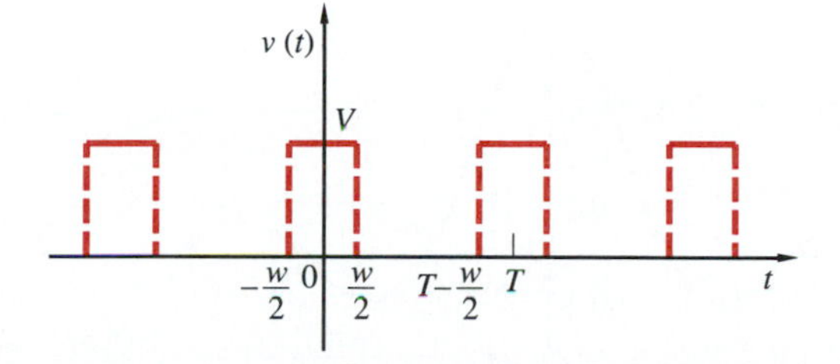

Figure 9
Pulse train.

The final form is obtained by multiplying both numerator and denominator by $w/2$. Note in the example that F_n is real and that $F_{-n} = F_n$ (for the same value of n, of course).

With this F_n, the series representing the pulse train becomes:

$$v(t) = \frac{2V}{\omega_0 T}\left[\frac{\omega_0 w}{2} + \left(\sin\frac{\omega_0 w}{2}\right)e^{j\omega_0 t} + \frac{1}{2}\left(\sin \omega_0 w\right)e^{j2\omega_0 t} + \cdots \right.$$

$$\left. + \left(\sin\frac{\omega_0 w}{2}\right)e^{-j\omega_0 t} + \frac{1}{2}\left(\sin \omega_0 w\right)e^{-j2\omega_0 t} + \cdots\right]$$

$$v(t) = \frac{w}{T}V\left(1 + 2\sum_{n=1}^{\infty}\frac{2}{n\omega_0 w}\sin\frac{n\omega_0 w}{2}\cos n\omega_0 t\right) \tag{23}$$

To arrive at the final form, the negative and positive exponentials for each n have been combined, leading to the trigonometric form that displays the individual harmonics. (You should confirm these results; in particular, find the constant term by finding the average value geometrically.)

In this example, we find that the F_n coefficients are purely real. This means that the harmonic phases should all be 0 or 180°, a fact borne out by the series in (23). Let's now examine the expression for F_n in (22); it can be rewritten in the following form:

$$F_n = \frac{wV}{T}\frac{\sin(n\omega_0 w/2)}{n\omega_0 w/2} \tag{24}$$

If we think of $n\omega_0 w/2$ as a continuous variable x, then the right-hand side (apart from a multiplying constant) has the form of $(\sin x)/x$. This function arises very often in signal analysis in different contexts and is called the *sampling function*; thus:

$$\text{Sa}(x) \equiv \frac{\sin x}{x} \tag{25}$$

The sampling function is plotted in Figure 10. (Show by l'Hôpital's rule that the function approaches 1 when x goes to 0.)

We will now use the preceding ideas to arrive at a graphical portrayal of the harmonic amplitudes and phases. Recall that c_n is the amplitude of the nth harmonic and that the magnitude of F_n is just half of c_n. There is a choice, then,

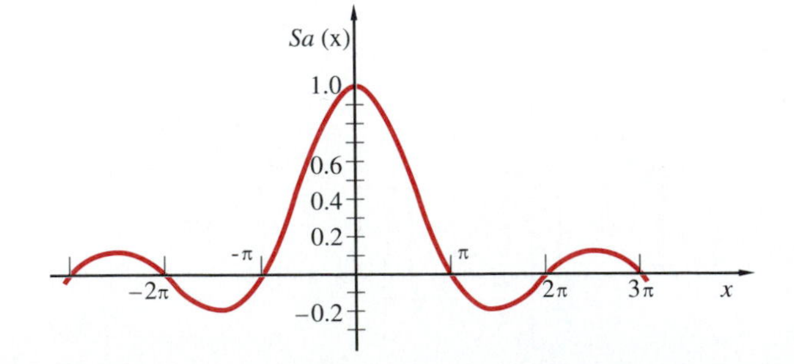

Figure 10 The sampling function $(\sin x)/x$.

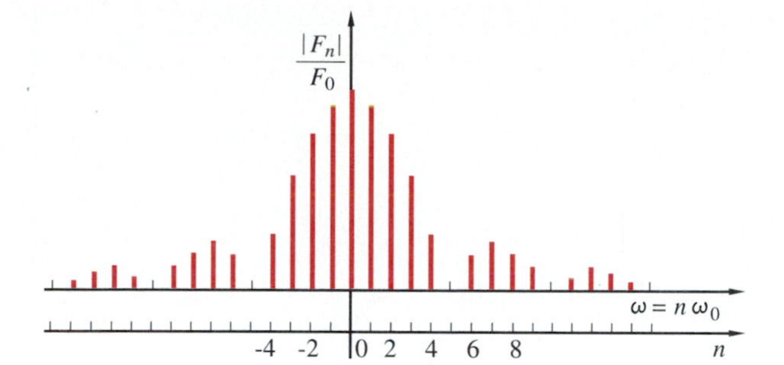

Figure 11
Amplitude spectrum of pulse train.

of portraying c_n or $|F_n|$. For future notational convenience, let's deal with the magnitude of F_n while speaking of it as the harmonic amplitude; we'll be off by a factor of $\frac{1}{2}$ for each coefficient, but the *relative* sizes of the harmonic amplitudes will be maintained.

In order to display a graph, it will be necessary to give a numerical value to $\omega_0 w$ or, alternatively, to the ratio of the pulse width w to the period T. This quantity, w/T, is called the *duty cycle,* a measure of the fraction of a period covered by the pulse. Small duty cycle means narrow pulse. Figure 11 shows a plot of $|F_n|$ for the duty cycle $w/T = \frac{1}{5}$. The graph is normalized by plotting $|F_n|/F_0$ instead of F_n. In the present case, $F_0 = wV/T = V/5$. The abscissa can be taken as either the angular frequency, which takes on the discrete values ω_0, $2\omega_0$, etc., or as the index n, which takes on integer values. Note that the amplitude of every multiple of the fifth harmonic is zero.

The harmonic phase can be portrayed in the same way. In the present example, the phase is 0 whenever F_n is positive and π radians whenever F_n is negative. That is, $\phi_n = 0$ for n between 0 and 5, between 10 and 15, between 20 and 25, etc.; and it is π for n between 5 and 10, between 15 and 20, and so on. (Sketch it for yourself.)

Collectively, these plots of $|F_n|$ and ϕ_n as functions of the discrete frequency are called *frequency spectra*; individually, they are called the *amplitude spectrum* and the *phase spectrum,* respectively. Because the plots consist of vertical lines at discrete frequencies, they are also called *line spectra* or *discrete spectra.*

Rectified Sinusoid

In an earlier section we took note of the fact that the magnitudes of the Fourier coefficients representing a function decreased with increasing n. (If this did not happen, indeed, how could the series converge?) We will now give further consideration to this question with the help of another waveform. Figure 12a shows a full-wave rectified sinusoid whose period is half the period of the original sine function of which this is the rectified version. Analytically, the function can be expressed as:

$$v(t) = V_m \sin\frac{\pi t}{T} \qquad 0 \leq t \leq T$$

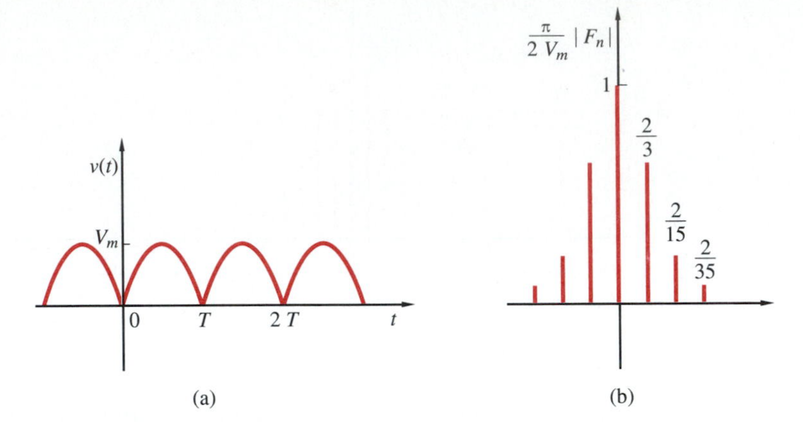

Figure 12
Full-wave rectified sinusoid and its amplitude spectrum.

Since this is an even function, there will be no sine terms in the series, so let's determine the trigonometric form of the series. The coefficients will be:

$$a_0 = \frac{1}{T}\int_0^T V_m \sin\frac{\pi}{T}t\,dt = \frac{2V_m}{\pi}$$

$$a_n = \frac{2}{T}\int_0^T V_m \sin\frac{\pi}{T}t \cos\frac{2\pi n}{T}t\,dt$$

$$= \frac{V_m}{T}\int_0^T \left[\sin\frac{(2n+1)\pi}{T}t - \sin\frac{(2n-1)\pi}{T}t\right]dt$$

$$= \frac{-4V_m}{\pi(4n^2-1)}$$

which, with $\omega_0 = 2\pi/T$, leads to:

$$v(t) = \frac{2V_m}{\pi}\left(1 - \frac{2}{3}\cos\omega_0 t - \frac{2}{15}\cos 2\omega_0 t - \frac{2}{35}\cos 3\omega_0 t - \cdots\right)$$

The amplitude spectrum is shown in Figure 12b. Notice that, in comparison with the pulse train, the harmonic amplitudes of the rectified sinusoid decrease more rapidly as the order of the harmonic increases. This is an illustration of a general property of Fourier coefficients which we shall now briefly describe.

Variation of Coefficients with Harmonic Order

The pulse train (and the rectangular wave shown in Figure 3) have discontinuities, and we found that their harmonic amplitudes decrease as $1/n$. But for the rectified sinusoid (and the triangular wave shown in Figure 4), we found that the harmonic amplitudes decrease approximately as $1/n^2$. These latter waveforms have no discontinuities, but if we differentiate the functions, we will find that their first derivatives have discontinuities.

The "smoothness" of a function can be described as the number of times it can be differentiated without encountering a discontinuity. For example, a sinusoid can be differentiated an infinite number of times; it *is* smoooooth. It can be represented by a Fourier series of only one term: itself. On the other hand, a function with a discontinuity is not smooth at all; it would require a large number of terms in a Fourier series to approximate it. It is a property of the Fourier coefficients that the smoother the function represented by the series, the faster the coefficients decrease with increasing n. We have seen this to be the case for discontinuous functions and continuous functions with discontinuous first derivatives. We could test the proposition with a continuous function with continuous first derivative but discontinuous second derivative.

EXERCISE 9

The periodic function in Figure 13 is made up of arcs of parabolas, as shown. Hence both the first and second derivatives will be continuous. Extrapolating from the preceding discussion, you might surmise that the Fourier coefficients will decrease as $1/n^3$. Determine the Fourier coefficients and find out if you are right. Use any and all symmetries to simplify the calculation.

ANSWER: $a_0 = 0 = b_n$; $a_n = -128/n^3\pi^3$ $(n = 1, 5, 9 \ldots)$, $a_n = 128/n^3\pi^3$ $(n = 3, 7, 11 \ldots)$. □

This discussion leads to another interesting result also. If a function of time is differentiated and the exponential form of its Fourier series is differentiated term by term, each exponential is multiplied by $jn\omega_0$. That is, each harmonic phase is increased by $\pi/2$, and $|F_n|$ is multiplied by $n\omega_0$. If $f(t)$, like the pulse, has discontinuities, its derivative does not exist at those points. In that case, F_n decreases as $1/n$. When this is multiplied by n, the coefficients no longer vanish as n goes to infinity, so the Fourier series of the derivative will not converge. However, if $f(t)$ is continuous and differentiable, then the Fourier coefficients will decrease at least as $1/n^2$. Thus, multiplying these F_n's by n still leaves a converging Fourier series. The upshot is that the Fourier series of the

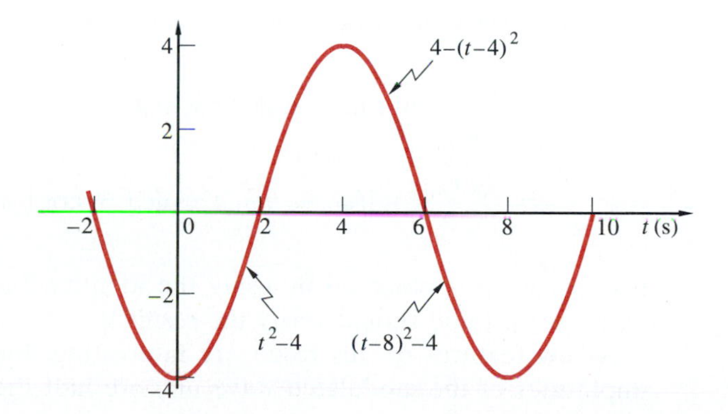

Figure 13
A "smoother" periodic function.

derivative of a differentiable function can be obtained from the Fourier series of the function itself by the differentiating term by term

EXERCISE 10

Sketch the graph of a full-wave rectified sinusoid and, on a set of axes just under this graph, sketch its derivative. Differentiate term by term the Fourier series of the rectified sinusoid. Finally, determine the Fourier series of the differentiated function from the formula for the coefficients and confirm that the two sets of Fourier coefficients are the same. □

Whenever you are presented with a periodic function—say, as the excitation of a circuit—and asked about its Fourier series, before doing anything else, describe its features to yourself. Does it lie mainly above the horizontal axis or below it? That is, what is its average value? Can the average value be found geometrically? Does the function exhibit any kind of symmetry to simplify calculations? How "smooth" does it appear to be? That is, how fast will its Fourier coefficients decrease with n? This familiarity with the function may not help in evaluating integrals, but it will surely help to reduce the effort, to detect possible errors, and to give you a feel for what you are doing.

Spectrum of an Amplitude-Modulated Wave: Example

An amplitude-modulated wave is a sinusoid whose amplitude is a function of time. It can be expressed analytically as follows:

$$v(t) = v_m(t) \cos \omega_0 t \tag{26}$$

The cosine is the *carrier* (its frequency is the *carrier frequency*) and $v_m(t)$ is the *modulation.* Suppose the modulating voltage is a periodic function with fundamental frequency ω_1. Then it can be represented by a Fourier series, in, say, the following form:

$$v_m(t) = \sum_{n=0}^{\infty} c_n \cos (n\omega_1 t + \phi_n) \tag{27}$$

Assuming that the series can be multiplied term by term by the carrier, the modulated wave $v(t)$ can be written as follows:

$$\begin{aligned} v(t) &= \sum_{n=0}^{\infty} c_n \cos (n\omega_1 t + \phi_n) \cos \omega_0 t \\ &= \sum_{n=0}^{\infty} \frac{c_n}{2} \{\cos [(\omega_0 + n\omega_1)t + \phi_n] + \cos [(\omega_0 - n\omega_1)t - \phi_n]\} \end{aligned} \tag{28}$$

The last form is obtained by using the identity $2 \cos x \cos y = \cos (x + y) + \cos (x - y)$. (You should verify the result.)

A few features of this result are interesting. First, note that the harmonic amplitudes of the modulated wave $v(t)$ are half the corresponding amplitudes

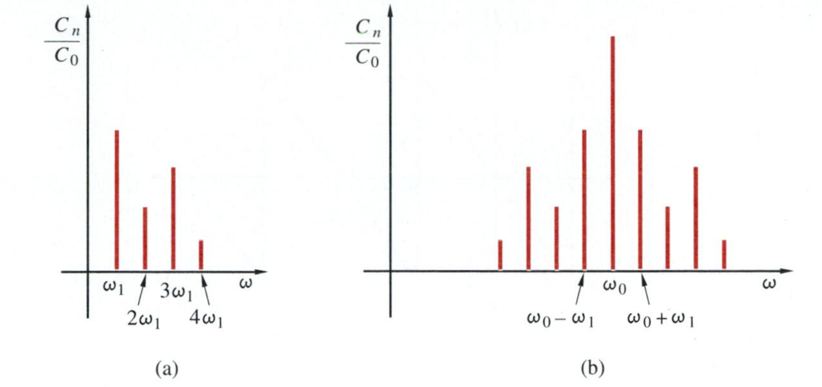

Figure 14 Spectrum of amplitude-modulated wave: (a) modulating spectrum; (b) modulated spectrum.

of the modulating wave $v_m(t)$. More important, the frequency spectrum of the modulated wave is a replica of the modulating spectrum, but moved up the frequency axis so that the spectrum is centered at the carrier frequency ω_0. A possible amplitude spectrum of $v_m(t)$ and the resulting spectrum of $v(t)$ are shown in Figure 14.

Notice that what has been plotted is the *normalized* harmonic amplitudes and not the normalized values of $|F_n|$. Does this make any difference? The only difference is that F_n takes on values for negative n, as well as positive n, but c_n does not. The harmonics of the modulating wave are clustered on either side of the carrier frequency. These two frequency intervals are called the *sidebands*—the *upper sideband* and the *lower sideband.*

Band-Limited Functions

When writing expressions for the Fourier series representing an arbitrary periodic function, we have assumed that the index n ranges up to infinity—that is, that the series has an infinite number of terms. While an infinite number of frequency components will indeed be necessary to represent an arbitrary periodic function, this will not be true of *every* periodic function. For example, the function:

$$f(t) = \cos^2 \omega_0 t = \tfrac{1}{2} + \tfrac{1}{2}\cos 2\omega_0 t$$

is a periodic function which has only a DC component and a second harmonic. (Sketch it.) In Section 1, the functions $\sin^3 x$ and $\sin^4 x$ were discussed; they, also, are periodic functions with a small number of harmonics. In fact, any function represented by the sum of a finite number of terms from a Fourier series will be a periodic function, so this periodic function will have a finite number of harmonics.

A function whose frequency components don't extend to infinity is called a *band-limited* function. In engineering, we often approximate a given periodic function by the first few terms of its Fourier series. The resulting approximating function is band-limited.

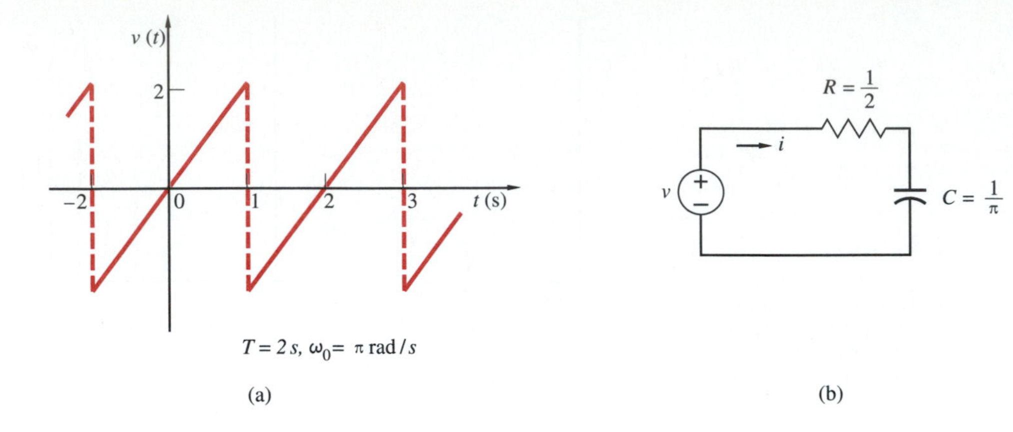

Figure 15 Sawtooth sweep voltage in an RC circuit.

5 STEADY-STATE RESPONSE TO PERIODIC WAVEFORMS

So far in this chapter our concern has been with the analysis of periodic signals; networks have not made an appearance. We are now ready to discuss the steady-state response of networks when the excitations are periodic functions. The stage will be set by considering some examples, from which the generalization will follow.

Sawtooth Sweep Voltage Applied to an RC Circuit

The sawtooth voltage shown in Figure 15 is applied to the *RC* circuit. Our goal is to find the steady-state current.[6] The approach is to determine the Fourier series representing the sawtooth excitation, find the steady-state response to each term in the series, and then add these partial responses.

The Fourier series of the sawtooth is found to be:

$$a_n = 0 \qquad b_n = (-1)^n \frac{4}{n\pi}$$

$$v(t) = \frac{4}{\pi}\left(-\sin \pi t + \frac{1}{2}\sin 2\pi t - \frac{1}{3}\sin 3\pi t + \cdots\right) \tag{29}$$

(Carry out the details of determining the coefficients.) The series expresses the source voltage as a sum of voltages, each of which can be taken as the voltage of a source that is connected in series with all the others. Since the circuit is linear, the response to the sum of these excitations is the sum of the responses to the individual excitations.

Each of the component excitations is a sinusoid. The steady-state response to each of them can be determined using phasors and network functions. The

[6] You will recall that a similar problem was tackled in Chapter 6 using a different approach. There we used the natural response of a first-order circuit together with the periodicity of the excitation to arrive at the steady-state response. That approach would be difficult to generalize to higher-order circuits.

only constraint is that each sinusoid has a different frequency which must be taken into account when evaluating the network function, in this case an impedance. The only other caution is that the Fourier coefficients are harmonic amplitudes, not rms values.

Remembering the relationships among the coefficients in the different forms of a Fourier series, as in (17), and combining this with the preceding equation, we get:

$$V_n = -j\frac{b_n}{2} = \frac{b_n}{2}e^{-j90^\circ} = (-1)^n\frac{2}{n\pi}e^{-j90^\circ} \tag{30}$$

Note the use of the specific notation V_n rather than the general F_n.

Now let's find the steady-state response to each sinusoid. The current phasor at each harmonic frequency will be the voltage phasor at that frequency multiplied by the admittance of the circuit evaluated at that frequency. The admittance of the *RC* circuit is:

$$Y_n(jn\omega_0) = \frac{1}{R - j\dfrac{1}{n\omega_0 C}} = \frac{1}{\dfrac{1}{2} - j\dfrac{1}{n}}$$

Notice that as n increases, the admittance approaches a constant value of 2. Numerical values for each harmonic are obtained by inserting the appropriate value of n. Thus:

Fundamental: $$Y(j\omega_0) = \frac{1}{0.5 - j1.0} = 0.895e^{j63.4^\circ}$$

2d harmonic: $$Y(j2\omega_0) = \frac{1}{0.5 - j0.5} = 1.414e^{j45^\circ}$$

3d harmonic: $$Y(j3\omega_0) = \frac{1}{0.5 - j0.333} = 1.66e^{j33.6^\circ}$$

4th harmonic: $$Y(j4\omega_0) = \frac{1}{0.5 - j0.25} = 1.79e^{j26.5^\circ}$$

$$\vdots$$

The I_n coefficients of the current series, proportional to the current phasors, are obtained by multiplying each voltage V_n coefficient from (30) by the admittance for the corresponding harmonic from the preceding values. The result becomes:

Fundamental: $$\frac{2}{\pi}e^{-j90^\circ}(0.895e^{j63.4^\circ}) = \frac{2}{\pi}(0.895e^{-j26.6^\circ})$$

2d harmonic: $$\frac{1}{\pi}e^{-j90^\circ}(1.414e^{j45^\circ}) = \frac{2}{\pi}(0.707e^{-j45^\circ})$$

3d harmonic: $$\frac{2}{3\pi}e^{-j90^\circ}(1.66e^{j33.6^\circ}) = \frac{2}{\pi}(0.553e^{-j56.4^\circ})$$

4th harmonic: $$\frac{1}{2\pi}e^{-j90^\circ}(1.79e^{j26.5^\circ}) = \frac{2}{\pi}(0.448e^{-j63.5^\circ})$$

$$\vdots$$

Finally, remembering that the harmonic amplitudes c_n are twice the magnitudes of the corresponding F_n, the trigonometric form of the series can now be written:

$$\frac{\pi}{4} i(t) = 0.895 \cos(\pi t - 26.6°) + 0.707 \cos(2\pi t - 45°)$$
$$+ 0.553 \cos(3\pi t - 56.4°) + 0.448 \cos(4\pi t - 63.5°) + \cdots$$

To find the steady-state current exactly requires calculating all harmonic components. An acceptable approximation, however, is obtained by retaining a relatively few terms. The number of terms necessary in this example will be relatively large because this series does not converge rapidly, the fourth-harmonic amplitude still being half the amplitude of the fundamental.

Rectifier Filter Circuit

Let us now consider an example resulting in a more rapidly converging output series. Figure 16 shows a full-wave rectified voltage applied to a load R through an intermediate LC combination which acts as a filter. The box on the left in Figure 16a—which is the output of a rectifier—can be viewed as a voltage source. The objective is to determine the steady-state output voltage.

The Fourier series representing the input waveform was previously determined in connection with Figure 12. The appropriate network function is the transfer voltage ratio. From the voltage-divider relationship, this is easily found to be:

$$G_{21}(jn\omega_0) = \frac{V_2}{V_1} = \frac{\dfrac{R}{jn\omega_0 CR + 1}}{\dfrac{R}{jn\omega_0 CR + 1} + jn\omega_0 L}$$
$$= \frac{1}{1 - n^2\omega_0^2 LC + jn\omega_0 L/R}$$

If we multiply the F_n coefficients of the input-voltage series by the value of G_{21} for corresponding values of n, the result will be the coefficients of the exponential form of the series for the output voltage. We assume that the frequency of the sinusoid that was rectified is the power-line frequency ($f = 60$ Hz, $\omega_p = 377$ rad/s), so ω_0 will be twice the power-line angular frequency, or $\omega_0 = 2\omega_p = 754$ rad/s, as evident from the waveform in Figure

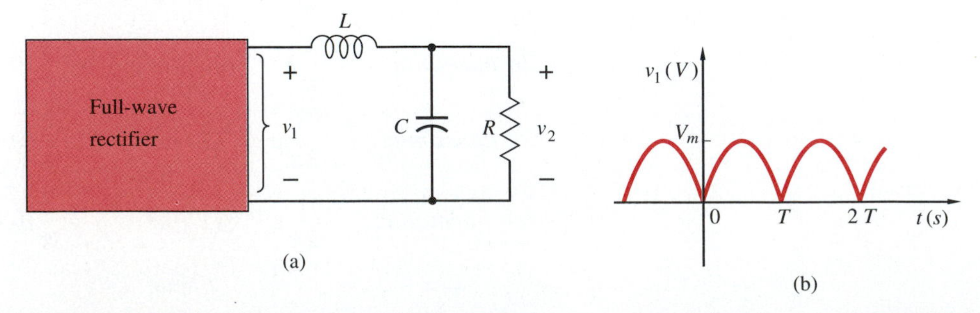

Figure 16
Rectifier-filter circuit.

16, where T is half the period of the unrectified sinusoid. Finally, remembering that the magnitude of F_n is half the harmonic amplitude, the trigonometric series for the output voltage becomes:

$$v_2(t) = 2\frac{V_m}{\pi}[1 + 0.0553 \cos(2\omega_p t + 38.8°)$$
$$+ 0.0029 \cos(4\omega_p t + 19.1°) + \cdots$$

Only two harmonics of the power frequency ω_p have been shown, because the coefficients decrease so rapidly with n that the next nonzero harmonic (the sixth) will contribute a negligible amount to the total. (Confirm that the coefficient of the next nonzero cosine term will be 0.00054.)[7] Not only do the coefficients of the rectified sinusoid decrease approximately as $1/n^2$, but in the present case the magnitude of the network function also decreases rapidly with n, unlike the admittance in the previous example. The amplitude of the fourth harmonic of the power frequency is only about 0.25% of the DC component. Because the entire AC component is so small, it constitutes only a *ripple* on the DC component. Do you see why, then, this circuit is called a *filter*? The major AC component, the second harmonic, has an amplitude which is only 5.5% of the DC component, in contrast with the rectified sinusoid, where this proportion is about 67%.

General Procedure

Having examined some examples, we can now describe the general procedure for finding the steady-state response of a linear network to a periodic excitation. First we find the Fourier spectrum of the input waveform, calculating as many terms as are needed for an acceptable approximation. The most appropriate coefficients to calculate are probably the F_n coefficients of the exponential form, because these include both the harmonic amplitude information and the phase.

What is an "acceptable" approximation will depend on the circumstances. In power-frequency filters, for example, it is common practice to retain only the second harmonic and to describe the ripple as the amplitude of this harmonic relative to the DC component as a percentage. In other applications, where the details of the waveform have more significance, more terms will be needed in order for the approximation to be acceptable.

Next, we determine the network function appropriate for the problem, whether driving-point or transfer. The magnitude and angle of this function are evaluated at each of the harmonic frequencies retained in the series approximating the input function. Finally, the product of each F_n coefficient of

[7] There is a possibility of confusion in referring to the harmonics of the rectified sinusoid. Normally, harmonics refer to multiples of the frequency of the periodic waveform in question—in this case, the full-wave rectified sinusoid. However, it would be useful in the case of a power supply (part of which is the rectifier) to speak of harmonics of the *power* frequency. Thus, the fundamental of the full-wave rectified wave is the second harmonic of the power frequency; the next term in the preceding expression is the fourth harmonic of the power frequency, and so on. We will use this terminology here, even at the expense of possible confusion.

the input with the complex value of the network function at the corresponding harmonic frequency gives the corresponding coefficient in the Fourier series of the steady-state output response.

The effect of the network is simply to modify the amplitudes and phases of the harmonics already present in the input. No frequency components exist in the output waveform which were not present in the input. This is a direct result of the linearity of the network. By contrast, in a nonlinear network in general, the output waveform will contain frequency components which not only are absent in the input waveform, but are not even harmonics of its frequency.

A note of pessimism should be injected here. Although the process just described is straightforward and the numerical calculations are not too tedious, yet the waveform of the output is not generally clear from its Fourier series. If the actual form of the output wave is important, then a point-by-point plot of the output will be required. In some cases, however, the exact form of the output wave is not important, as in the rectifier-filter problem. In that case, it is required only to know the approximate amplitude of the ripple.

A FREQUENCY MULTIPLIER The preceding comments accentuate and amplify some of the remarks made in the introduction of this chapter. There it was stated that the *conceptual framework* of Fourier, or frequency, analysis may be as important as a means of thought as the quantitative results—at least at the introductory level.

We will further illustrate this idea by considering the network shown in Figure 17a. A sinusoidal source v_1 with a frequency ω_0 is available, but a

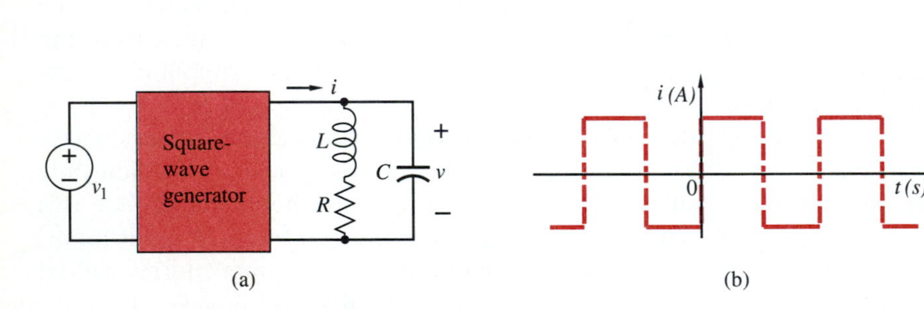

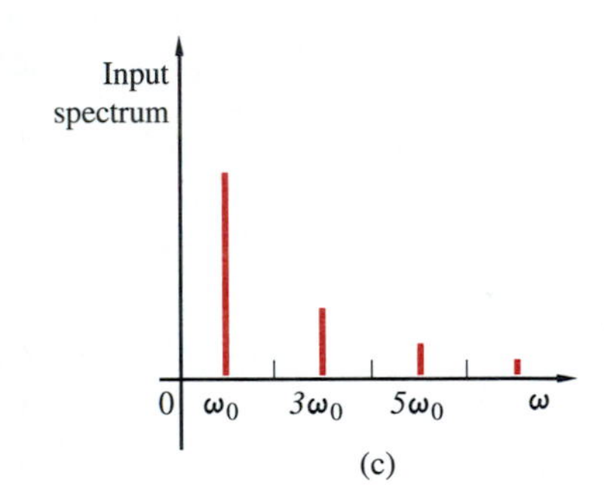

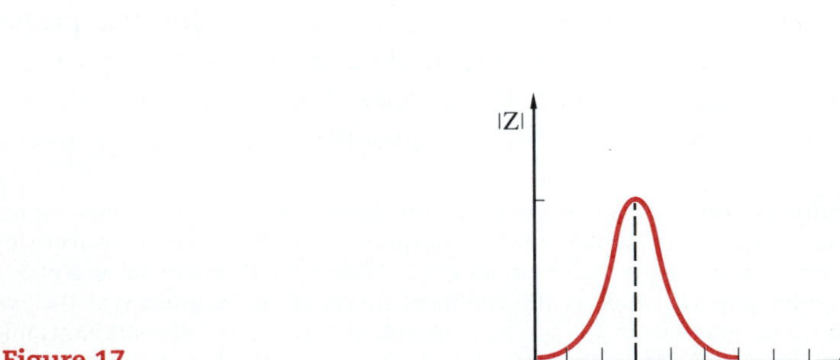

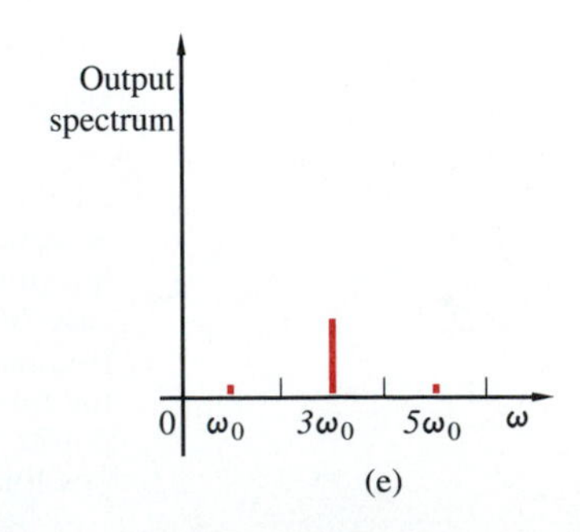

Figure 17
A frequency multiplier and its spectrum.

sinusoid of 3 times the frequency is desired. With v_1 as a source, it is possible to generate a square wave by means of the (obviously nonlinear) circuit labeled "square-wave generator" in the figure.

For our present purposes, let's assume that the the current i supplied by the box is a square wave, as shown in Figure 17b, with an amplitude spectrum as shown in Figure 17c. The current is supplied to a two-branch parallel resonant circuit which is assumed to have a high Q and to be tuned to the third-harmonic frequency $3\omega_0$. The resonance curve (a plot of the impedance magnitude of the resonant circuit) is shown in Figure 17d. The amplitude spectrum of the output is simply the product of the input spectrum and the resonance curve.

Because the response of the tuned circuit is low at the fundamental frequency, this frequency component will have a low amplitude in the output, and the higher the Q, the lower it will be. The same will be true of the fifth- and higher-harmonic amplitudes. The third-harmonic amplitude in the output will be relatively high, since the peak of the amplitude response of the tuned circuit occurs at that frequency. The resulting output spectrum is shown in the last part of the figure. Because of the nonzero amplitudes of the fundamental and fifth harmonics, the output is not a pure sinusoid. Nevertheless, these *distortion components* are small, and the output is approximately a sinusoid. The overall network—for obvious reasons—is a *frequency multiplier.* In this case, it is a tripler, although, by proper tuning, other multiples can be obtained.

Qualitative Aspects of Fourier Analysis

The major point to be made here is that, although quantitative results are indeed involved in the preceding example, this aspect is less significant than the conceptual, qualitative analysis in terms of frequency spectra. This same approach, a consequence of Fourier analysis, can be applied in many specific engineering situations.

Very often, it may be desired to obtain approximate relationships concerning the response which can be expected from a network. In such a case, the frequency response of the network can be idealized so as to emphasize the first-order effects and to neglect less significant results. To illustrate this point, note that the essential features of the resonance curve in Figure 17d are its relatively large values near $\omega = 3\omega_0$ and the relatively low values for all frequencies outside a range centered at $3\omega_0$. These features can be represented approximately by the idealized bandpass amplitude response shown in Figure 18. If the square wave in Figure 17a had been applied to a network having this idealized response (there is no such real network), then the output would have been a pure sinusoid, without any distortion components. Thus, a first-order approximation to the correct result would be obtained from an idealized frequency response.

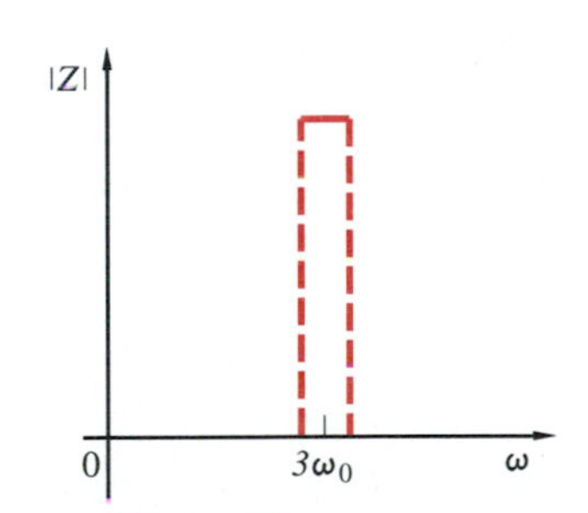

Figure 18
Idealized bandpass response.

Another consideration similar to the preceding one may be answered by means of Fourier concepts. In a signal-transmission network, the fidelity with which the signal is transmitted is an important property. The only modifications of the signal which can be tolerated are a possible reduction in its amplitude (which can always be restored by an amplifier) and a possible delay in time. That is, if the input signal is $f(t)$, then the output signal is $Kf(t - t_d)$, where K is a constant and t_d is a time delay; any other modification would distort the signal.

The effect on the Fourier coefficients of a shift in the time origin was discussed in Section 1. From that discussion, recall that the function $Kf(t - t_d)$ will have Fourier (F_n) coefficients that are $Ke^{-jn\omega_0 t_d}$ times the coefficients of the function $f(t)$. This means that the transfer function of the network should be $Ke^{-jn\omega_0 t_d}$. That is, the amplitude response should be constant for all frequencies and the phase response should be proportional to frequency. A network having these properties (*flat amplitude, linear phase*) is called a *distortionless* transmission network. It is not possible for a real network to have such a transfer function; we can only approximate it.

6 NATURE OF APPROXIMATION BY A FINITE FOURIER SERIES

In the preceding section we examined approximations of input and output waveforms by finite numbers of terms in a Fourier series. There we considered the *degree* of approximation—what percentage the nth harmonic constitutes of the first harmonic or the DC component. We did not examine the *nature* of the approximation. This is the question that we shall now pursue.

One period of a sweep voltage is shown by the dashed lines in Figure 19. Let S_k designate a finite sum of the terms in a Fourier series up to (and including) the kth harmonic. Also shown in the diagram in a solid curve is S_3, where:

$$S_3 = \frac{4}{\pi}\left(\sin \omega_0 t - \frac{1}{2}\sin 2\omega_0 t + \frac{1}{3}\sin 3\omega_0 t\right)$$

The error at any time can be defined as the difference between the given function and the finite sum at that instant. It is clear that the error is not the same at each instant; indeed, it does not always have the same sign, being sometimes positive and sometimes negative.

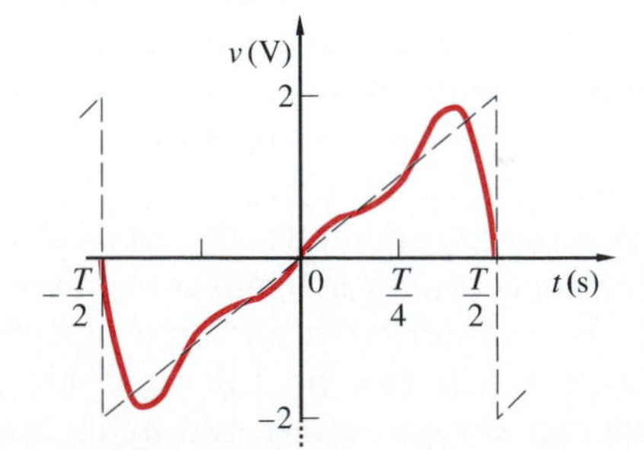

Figure 19
Finite-sum approximation of a sweep signal.

Generally, several different varieties of approximation (and resulting error) can be distinguished. One is the *equal-ripple* approximation. Here, the approximating function oscillates about the desired function in such a manner that the maximum deviations are all equal. (In that case, it turns out, these error peaks are minimized.) Another type of approximation results in zero error in the center of the approximating time interval, with increasing error toward the ends of the interval, a so-called *maximally flat* approximation.[8]

Least-Mean-Square Approximation

In both specific cases in the preceding paragraph, the error varies with time over the interval in question. What we will consider here is another type of error, which yields a single numerical value. First the error function is squared (guaranteeing a positive value); then its average value is taken over the interval. This is the *mean-square* error (mean = average). The square root of the result can also be taken, resulting in the *root-mean-square* (or rms) error, although this step does not give anything fundamentally different.

Suppose $S_k(t)$ is a finite trigonometric series whose coefficients are to be determined so that this finite sum will be a *least*-mean-square approximation to a function $f(t)$. It turns out that if the coefficients in the finite sum are chosen in accordance with the Fourier formulas, then the approximation will, indeed, yield the minimum mean-square error. That is, for a fixed number of harmonics, no other choice of coefficients in a trigonometric series will give an approximation to $f(t)$ having a smaller mean-square error than the Fourier coefficients will.

Furthermore, the Fourier coefficients have another property. Suppose a finite Fourier series S_k has been determined for a particular function but now we want to increase the number of harmonics in the approximation to a finite sum S_{k+m}. Again the harmonic coefficients of the new sum are to be determined so that the mean-square error is minimized. It turns out that the mean-square (or the rms) error will again be minimized if the coefficients are chosen in accordance with the Fourier coefficients. The first k harmonic coefficients don't have to be recalculated; only the new ones have to be calculated, and the Fourier formulas are the ones that will lead to the least rms error.

The preceding discussion is not a proof that a finite sum with coefficients determined by the Fourier formulas approximates a given function in a least-mean-square sense. We shall now undertake this proof.[9] For convenience, we will use the phase variable $x = \omega_0 t$ and the exponential form of the Fourier series.

[8] The stagger-tuned circuit discussed in Chapter 11 exhibits each of these properties for different degrees of staggering. Thus, the maximally flat approximation (to a rectangular magnitude function) results from critical staggering. For larger values of k, the approximation is equal-ripple—the twin peaks have the same amplitude.

[9] Although there isn't anything here beyond your abilities, you may skip this proof if you wish. If you do, though, you will be missing a neat little process that can be a joy to experience.

Let $S_k(x)$ be a finite sum of a Fourier series representing a function $f(x)$. The exponential form can be written as:

$$S_k(x) = \sum_{n=-k}^{k} F_n e^{jnx} \tag{31}$$

The error is $e(x) = f(x) - S_k(x)$.[10] The square of the error is now easily written as:

$$\begin{aligned} e^2(x) &= [f(x) - S_k(x)]^2 = \left[f(x) - \sum_{n=-k}^{k} F_n e^{jnx}\right]^2 \\ &= f^2(x) - 2\sum_{n=-k}^{k} F_n f(x) e^{knx} + \left(\sum_{n=-k}^{k} F_n e^{jnx}\right)^2 \end{aligned} \tag{32}$$

This is the expression whose average we are to find. There are three terms in the second line that we will examine individually. If we call F_{rms} the rms value of $f(x)$, taking the average value of the first of the three terms gives F_{rms}^2.

In the last term, we will be squaring a sum of exponentials. There will be two classes of terms in the product:

$$(F_n e^{jnx})(F_{-n} e^{-jnx}) = F_n F_{-n} \tag{33a}$$

$$(F_n e^{jnx})(F_m e^{jmx}) = F_n F_m e^{j(n+m)x} \tag{33b}$$

In the second of these, n and m take on any integer values between $-k$ and $+k$, except for $m = -n$. But the average value of an exponential is zero over a range of 2π. Hence, only the terms in (33a) will contribute to the average value.

Taking the mean value of $e^2(x)$ in (32) and using the observations just completed, the mean-square error, designated $\langle e^2 \rangle$, becomes:

$$\langle e^2 \rangle = F_{\text{rms}}^2 + \sum_{n=-k}^{k} F_n F_{-n} - 2\sum_{n=-k}^{k} F_n \frac{1}{2\pi}\int_0^{2\pi} f(x) e^{jnx}\, dx \tag{34}$$

This is the quantity to be minimized by appropriate selection of the coefficients, which, here, play the role of variables. For a function of several variables to have a minimum, the partial derivatives of the function with respect to each of the variables must be zero. Treating $n = 0$ separately, this

[10] Again there is a notational problem, as we encountered with the symbol Q which stands for several different quantities. The symbol e, for error, has the same difficulty since e is also the base of natural logarithms and is used for excitation. Stay alert to this potential source of confusion.

condition results in:

$$\frac{\partial\langle e^2\rangle}{\partial F_0} = 2F_0 - \frac{1}{\pi}\int_0^{2\pi} f(x)\,dx = 0$$
$$\frac{\partial\langle e^2\rangle}{\partial F_n} = F_{-n} - \frac{1}{\pi}\int_0^{2\pi} f(x)e^{jnx}\,dx = 0 \tag{35}$$

Finally, solving for the "variables"—the coefficients—yields:

$$F_0 = \frac{1}{2\pi}\int_0^{2\pi} f(x)\,dx$$
$$F_n = \frac{1}{\pi}\int_0^{2\pi} f(x)e^{-jnx}\,dx \tag{36}$$

(The last one was obtained by replacing $-n$ by n on both sides.) *These are exactly the formulas for the Fourier coefficients.* Since the second derivative of the mean-square error is positive (show this), the extreme value just found is, indeed, a minimum.

Here is what we have established: for a fixed number of harmonics, the mean-square error between the function and its approximation is minimized if the coefficients are chosen in accordance with the Fourier formulas.

Discontinuities and the Gibbs Phenomenon

Whenever a periodic function has discontinuities, its Fourier series representation displays a phenomenon that was first investigated by the American mathematician Willard Gibbs. This section will provide a brief discussion of this phenomenon.

Look back at the sweep function and its partial-sum approximation in Figure 19. The value of the function at the discontinuity is indeterminate, but its values just before and just after the discontinuity are -2 and $+2$, respectively. From the diagram, it is evident that the *approximation* has a value of 0 at the discontinuity of the function; this is the average of the values on either side of the discontinuity. Indeed, we previously remarked that, at a discontinuity of a function, a Fourier series representing the function converges to the mean of the values on either side of the discontinuity. Here, we observe an example of a general phenomenon, that even a partial Fourier sum takes on this same value.

Furthermore, we find in Figure 19 that the approximation oscillates around the actual function, with the greatest deviation from the function occurring just before and just after the discontinuity. This deviation is called the *overshoot.* If the number of harmonics in the partial sum is increased, the number of oscillations will increase, but the maximum overshoot still occurs on either side of the discontinuity. It turns out that, even in the limit as the number of harmonics in the partial sum approaches infinity, the overshoot remains. It will simply be a vertical line in the limit and is found to amount to about 9% of the discontinuity above it and below it. In honor of Willard Gibbs, this behavior of

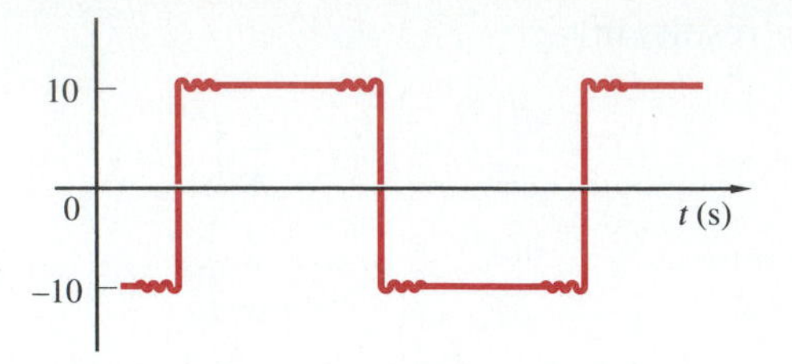

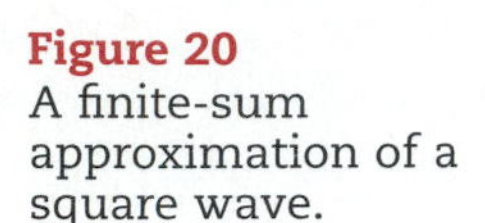
Figure 20
A finite-sum approximation of a square wave.

the overshoot is called the *Gibbs phenomenon.* Obviously, we haven't proved the Gibbs phenomenon here.

The diagram in Figure 20 illustrates a finite Fourier series approximation to a square wave. Assuming a large number of harmonics, what value (accurate to two decimal places) will the finite sum take on at the peak occurring just after a discontinuity?

7 POWER AND ENERGY

One of the problems of importance in network theory, no matter what the waveform of the excitation, is the determination of the energy dissipated in a network and the energy stored in it. Specifically, the power delivered to a resistor and the energy stored in an inductor and a capacitor at each instant are:

$$\begin{aligned} p(t) &= Ri^2(t) = Gv^2(t) \\ w_L(t) &= \tfrac{1}{2}Li^2(t) \\ w_C(t) &= \tfrac{1}{2}Cv^2(t) \end{aligned} \tag{37}$$

Each case involves the square of a voltage or current.

Of even greater interest than the instantaneous variation of these quantities is their average value over a period of the function. For *sinusoidal* v and i, these average values are:

$$\begin{aligned} P &= V_{\text{rms}}I_{\text{rms}} \cos\theta \\ W_L &= \tfrac{1}{2}LI_{\text{rms}}^2 \\ W_C &= \tfrac{1}{2}CV_{\text{rms}}^2 \end{aligned} \tag{38}$$

(The subscript rms is used for emphasis.)

Average Power into a Network in Terms of Harmonics

Now suppose the excitation applied to a network and the resulting steady-state response are periodic, though not sinusoidal. Let both the voltage and current be expanded in a Fourier series:

$$\begin{aligned} v(t) &= V_0 + V_{m1}\cos(\omega_0 t + \alpha_1) + V_{m2}\cos(2\omega_0 t + \alpha_2) + \cdots \\ i(t) &= I_0 + I_{m1}\cos(\omega_0 t + \beta_1) + I_{m2}\cos(2\omega_0 t + \beta_2) + \cdots \end{aligned} \tag{39}$$

In the notation used here, V_0 and I_0 represent the DC components; V_{mk} and I_{mk} represent the amplitudes of the kth harmonics.

The instantaneous power is the product of v and i, assuming standard references. If the two series in the preceding expressions are multiplied term by term (this is permissible for all functions that can be represented by Fourier series), the result will be a rather extensive expression. After multiplication, we plan to find the average power. For this purpose, four types of terms can be distinguished:

1. The product of the two DC components
2. The product of a DC component and a harmonic
3. Products involving the same harmonic frequency
4. Products involving different harmonic frequencies

Finding the average involves integrating over a period and dividing by the period.

The first term is easy; the average value of a constant equals that constant. The terms of type 2 are also easy; the average value of a sinusoid is zero. The cross-product terms of the fourth type also contribute nothing to the average, because of the orthogonality property of sinusoids discussed in Section 1. That leaves the terms involving like harmonics. If we let θ_k be the phase angle between the kth-harmonic voltage and current, the contribution of the kth harmonic to the average becomes:

$$\begin{aligned} \frac{1}{T}\int_0^T v_k i_k\,dt &= \frac{1}{T}\int_0^T V_{mk}I_{mk}\cos[(k\omega_0 t + \beta_k) + \theta_k]\cos(k\omega_0 t + \beta_k)\,dt \\ &= \frac{V_{mk}I_{mk}}{2}\cos\theta_k \end{aligned} \tag{40}$$

The last step follows from the trigonometric identity $2\cos x \cos y = \cos(x + y) + \cos(x - y)$. In the last term here, the difference of the arguments is just θ_k. The cosine of the sum of the two arguments is $\cos(2k\omega_0 t + \beta_k)$, which contributes zero to the average value. (Carry out the details and confirm the result.) Taking all of this into account, the average power becomes:

$$\begin{aligned} P &= \frac{1}{T}\int_0^T vi\,dt = P_0 + \sum_{k=1}^{\infty} P_k \\ P_k &= \tfrac{1}{2}V_{mk}I_{mk}\cos\theta_k \end{aligned} \tag{41}$$

P_k is the average power of the kth harmonic.

EXERCISE 11

The preceding result was derived using the trigonometric form of the Fourier series in (39). Repeat the process using the exponential form of the two series and confirm the result. □

The preceding equation is a very interesting result. It states that, when the excitation is periodic, the average power delivered to a network is the sum of the average powers of the individual harmonics, including the DC component. This result is very significant; it is a special case of a more general result to be discussed in a later section, called *Parseval's theorem.*

Thus the average power delivered to a network by a periodic excitation can be computed in either of two ways. If the waveforms of the voltage and current are known functions of time, then a time integration can be performed. On the other hand, if the frequency spectra of the voltage and current are known, then the average power can be computed as a summation of the average powers of the frequency components. These two procedures are completely equivalent, since there is a one-to-one correspondence between a waveform and its frequency spectra.

Nevertheless, determination of the average power from the frequency spectra permits the acquisition of additional information which is not evident in the time domain.

Suppose, for example, that the network to which the power is delivered is frequency-selective, like the tuned circuit in the frequency-multiplier example of Figure 17. That is, certain frequencies are "passed" while others are "stopped." It is of importance to know what fraction of the average power delivered is contained in the frequency components that fall within the passband of the network. For this purpose, the summation in the first line of (41) is carried out only over the appropriate frequency components.

The development leading to the average power delivered to a network can be repeated for the average energy stored in a capacitor or an inductor, starting with (37) and the Fourier series in (39). All the steps carried out before are now repeated, with some small differences. Since only the voltage or the current is involved, for example, instead of $V_{mk}I_{mk}$ we will have V_{mk}^2 or I_{mk}^2. Furthermore, $\theta_k = 0$ (and $\cos\theta_k = 1$), since the phase difference between a current and itself, or a voltage and itself, is zero. Thus, the counterparts of (41) become:

$$\begin{aligned} W_L &= \sum_{k=0}^{\infty} W_{Lk} \qquad W_{Lk} = \tfrac{1}{2}LI_{\text{rms}\,k}^2 \\ W_C &= \sum_{k=0}^{\infty} W_{Ck} \qquad W_{Ck} = \tfrac{1}{2}CV_{\text{rms}\,k}^2 \end{aligned} \tag{42}$$

RMS Value of a Periodic Function

Another particularly useful relationship can be derived from the expression in (41) by considering the network to be simply a resistor. Then $v = Ri$, $V_{mk} = RI_{mk}$, and $\theta_k = 0$. In this case, (41) becomes:

$$P_R = R\left(I_0^2 + \tfrac{1}{2}\sum_{k=1}^{\infty} I_{mk}^2\right) = R\left(I_0^2 + \sum_{k=0}^{\infty} I_{\text{rms}\,k}^2\right) \tag{43}$$

Now recall the definition of the rms value of a periodic function in (3) in Chapter 6. We see that the quantity within parentheses in the preceding equation is the square of the rms value of the current. Hence:

$$I_{\text{rms}} = \sqrt{I_0^2 + \sum_{k=1}^{\infty} I_{\text{rms}\,k}^2} \tag{44}$$

A similar result applies for the voltage or any other function, of course. What this tells us is that there are two ways to calculate the rms value of a periodic function. One way is to use the basic definition in the time domain. Another is to obtain the amplitude spectrum from its Fourier series and use (44), the number of terms to be used depending on the desired accuracy.

EXERCISE 12

The rms value of the sawtooth voltage waveform one cycle of which was shown in Figure 4 in Chapter 6 was found there to be 5.7735. **a.** Determine the number of terms needed to find this rms value by the expression in (44) (with a suitable change in notation) if only a 1% error is acceptable. **b.** Repeat for 5% permissible error. **c.** Find the error if only the DC term is used.

ANSWER: $v(t) = 5 - \dfrac{10}{\pi}\sum \dfrac{1}{n}\sin n\pi t$

a. DC and eight harmonic terms.
b. Only DC and fundamental yield an error of 5.03%.
c. 13.4%. □

Combining the last equation with the immediately preceding two for the average power in a resistor and the average energy stored in a capacitor and an inductor, we can write the latter in terms of rms values, as follows:

$$P_R = RI_{\text{rms}}^2$$
$$W_C = \tfrac{1}{2}CV_{\text{rms}}^2 \qquad W_L = \tfrac{1}{2}LI_{\text{rms}}^2 \tag{45}$$

An interesting conclusion follows from these expressions. Referring back to Chapter 9, we observe that, for the average power dissipated in R and the average energy stored in L and C, the same expressions, in terms of rms values, apply both for sinusoids and for arbitrary periodic functions.

8 FOURIER INTEGRAL AND FOURIER TRANSFORM

So far in this chapter we have been concerned with periodic functions. Such functions, we have found, can be represented by an infinite series of harmonic components. In this section we will briefly introduce a generalization of the Fourier series which applies to transient (aperiodic) waveforms.[11] We will not present an exhaustive treatment—just enough to whet your appetite.

[11] It applies even to periodic waveforms, after appropriate modification.

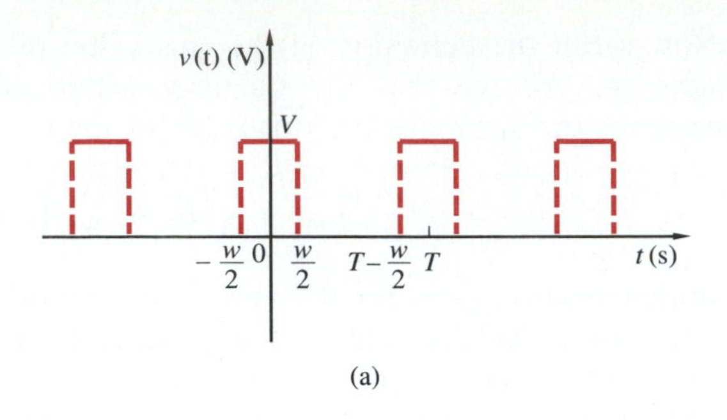

(a)

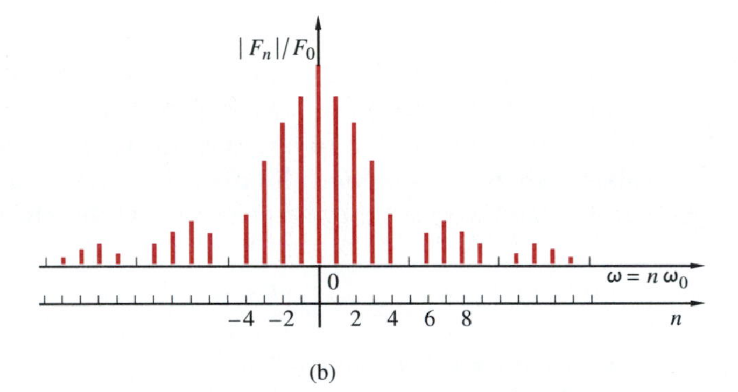

(b)

Figure 21 A pulse train and its amplitude spectrum.

A Single Pulse

The subject will be introduced with the help of the pulse train treated earlier and redrawn in Figure 21a. The Fourier coefficients and the series were given in (22) and (23) and are rewritten here:

$$F_n = V\frac{w}{T}\frac{1}{n\omega_0 w/2}\sin\frac{n\omega_0 w}{2} \tag{46a}$$

$$\begin{aligned} v(t) &= V\frac{w}{T}\sum_{n=-\infty}^{\infty}\left(\frac{2}{n\omega_0 w}\sin\frac{n\omega_0 w}{2}\right)e^{jn\omega_0 t} \\ &= V\frac{w}{T}\left[1 + 2\sum_{n=1}^{\infty}\left(\frac{2}{n\omega_0 w}\sin\frac{n\omega_0 w}{2}\right)\cos n\omega_0 t\right] \end{aligned} \tag{46b}$$

The normalized amplitude spectrum for a duty cycle of $w/T = \frac{1}{5}$ is also redrawn, in Figure 21b.

Now let's inquire what would happen to the frequency spectrum if the duty cycle (the value of w/T) were reduced by increasing the period. For example, suppose that the period is increased by a factor of 4 over what it was, thereby making $w/T = \frac{1}{20}$, instead of $\frac{1}{5}$. The argument of $\sin n\omega_0 w/2$ becomes $n\pi/20$; this will equal π when $n = 20$. In this case, it will be the amplitude of the twentieth, instead of the fifth, harmonic that becomes zero. But since the fundamental frequency is now 4 times less than before, the actual frequency

$\omega = n\omega_0$ at which the spectral amplitude is zero remains the same. This means that, keeping the scale in Figure 21b the same, there will be 4 times as many lines in the spectrum, and no other effect. (Keep in mind that the discrete spectrum in Figure 21b is normalized. While increasing the period reduces the size of each line in the spectrum, the *relative* harmonic amplitudes will not change, since the plot is normalized.)

You can intuitively appreciate that, if the period is further increased and made to approach infinity, the lines in the spectrum will become denser and denser until, in the limit, no lines can be distinguished, only a continuous spectrum. The waveform itself in this case is no longer a pulse train but a single pulse—since an infinite time must pass before the next pulse in the train comes along. In other words, it is really an aperiodic function.

It is possible to verify this intuitive argument rigorously. The concept of representing a periodic function by a superposition of frequency components can be generalized and extended to aperiodic waveforms. Instead of a series of discrete components there will be an integral of continuous frequency components. The frequency spectrum will now be a *continuous,* as opposed to a discrete, spectrum. The discrete frequencies $n\omega_0$ merge into the continuous frequency ω.

A General Aperiodic Function

By intuitive reasoning for a particular function (a pulse), the preceding section demonstrated that the discrete spectrum for a pulse train becomes a continuous spectrum for a single pulse. We shall now consider this process in the general case.

To begin, let's see if any patterns can be observed by scrutinizing the exponential form of the Fourier series and the formula for the corresponding coefficients when the latter is multiplied by the period T:

$$\begin{aligned} TF_n &= \int_{-T/2}^{T/2} f(t)e^{-jn\omega_0 t}\,dt \\ f(t) &= \sum_{n=-\infty}^{\infty} F_n e^{jn\omega_0 t} \end{aligned} \tag{47}$$

As in the pulse train, the plan is to let the period go to infinity, so the limits on the integral in the expression for the coefficients become $-\infty$ and ∞. The harmonic frequencies will no longer be discrete values but continuous; that is, $n\omega_0 \to \omega$. Since $T = 2\pi/\omega_0$, some manipulation permits writing the Fourier coefficients as:

$$F_n = \frac{1}{T}(TF_n) = \frac{\omega_0}{2\pi}(TF_n) = \frac{\Delta\omega}{2\pi}(TF_n) \tag{48}$$

Note that ω_0 has two meanings. It is the fundamental frequency, but it is also the *frequency interval* between one harmonic and the next; that is, it is the *change* in frequency, $\Delta\omega$, in going from one harmonic to the next one. This meaning has been used in the right-hand side of (48).

Next we substitute the expression for F_n in (48) into the Fourier series in (47), resulting in:

$$f(t) = \frac{1}{2\pi} \sum_{n=-\infty}^{\infty} (TF_n) e^{jn\omega_0 t} \Delta\omega \tag{49}$$

As T goes to infinity, the interval between harmonics ($\Delta\omega$) goes to zero. The summation in the series then becomes, by definition, an integral. Finally, we need a notational change. T refers to the period of a periodic function, and F_n refers to the discrete Fourier coefficients. We need a different symbol for TF_n to reflect the change from a series of discrete components to a continuous integral. We will use $F(j\omega)$ for this purpose.

The Fourier Transform Pair

The final step is to let T go to infinity; then the discrete harmonics $n\omega_0$ become ω, and TF_n becomes $F(j\omega)$. Using these in the summation in (49) and the integral in (47) results in:

$$F(j\omega) = \int_{-\infty}^{\infty} f(t) e^{-j\omega t}\, dt \tag{50a}$$

$$f(t) = \frac{1}{2\pi} \int_{-\infty}^{\infty} F(j\omega) e^{j\omega t}\, d\omega \tag{50b}$$

This is a remarkable pair of formulas with a remarkable degree of symmetry. Both integrands contain an exponential—one positive, the other negative—and this is the only substantial difference between the two expressions.[12] They are extremely important in engineering. Look them over carefully; study them; compare them; describe their details to yourself out loud; become thoroughly familiar with them.

Given a function of time $f(t)$, it is *transformed* by the *Fourier integral* in (50a) to a function of frequency $F(j\omega)$, called the *Fourier transform* of $f(t)$, which is generally complex. Conversely, given a function of frequency $F(j\omega)$, it can be *inverse-transformed* by the process of (50b) into a function of time $f(t)$ which is, thus, known as the *inverse Fourier transform* of $F(j\omega)$. There is a unique, one-to-one correspondence between a function $f(t)$ and its Fourier transform, $F(j\omega)$. That is, if $F(j\omega)$ is found as the Fourier transform of a function $f(t)$, then taking the inverse transform of $F(j\omega)$ will lead to no other function of t but the original $f(t)$.

The symbol $\mathscr{F}$ (which is a script eff) is used to designate the Fourier transform. The inverse transform is designated $\mathscr{F}^{-1}$. This unique relationship between a function $f(t)$ and its Fourier transform is indicated by:

$$\mathscr{F}^{-1}[F(j\omega)] = f(t) \qquad \leftrightarrow \qquad F(j\omega) \equiv \mathscr{F}[f(t)] \tag{51}$$

This particular $f(t)$ and $F(j\omega)$ constitute a *transform pair.*

[12] The 2π factor in one of the equations can easily be redistributed between the two. Thus, if we define a new function $g(t) = \sqrt{2\pi}\, f(t)$ and then replace $f(t)$ by its equivalent in terms of $g(t)$, there will be a $1/\sqrt{2\pi}$ multiplier in front of each integral. But this is a mathematical artificiality that has no physical significance.

The notation for a Fourier transform pair is convenient for another reason. The Fourier transform of a current $i(t)$, for example, can be written as $I(j\omega)$; that of a voltage $v(t)$ as $V(j\omega)$.

The mathematical operation of integration has the following property. If the sum of two functions is integrated, the result is the same as integrating the individual functions and then adding the integrals. Thus integration is a linear operation. It follows that the Fourier transform is linear; the transform of a sum of functions can be obtained by finding the transforms individually and adding.

EXAMPLE 3 Find the Fourier transform of the exponential voltage $v(t) = e^{-at}u(t)$ for $t \geqslant 0$, where a is real and positive. Direct evaluation of the Fourier integral is what it takes. Thus:

$$\begin{aligned} V(j\omega) &= \int_{-\infty}^{\infty} e^{-at}u(t)e^{-j\omega t}\,dt = \int_{0}^{\infty} e^{-(a+j\omega)t}\,dt \\ &= \left.\frac{e^{-(a+j\omega)t}}{-(a+j\omega)}\right|_{0}^{\infty} = \frac{1}{a+j\omega} \end{aligned} \tag{52}$$

Hence, in the preceding notation:

$$\mathcal{F}[e^{-at}u(t)] = \frac{1}{a+j\omega} \tag{53}$$

Relevance of the Fourier Transform

Let's pause for a brief review. We started the chapter by determining that a periodic function can be expanded in a trigonometric series whose frequencies are integral multiples of a fundamental. If this periodic function is the input to a linear network, the *steady-state* response of the network can be found harmonic by harmonic; we multiply the appropriate network function, evaluated at each harmonic frequency, by the Fourier coefficient of that harmonic. The result is the harmonic coefficient of the output.

Later we extended these notions to aperiodic functions of time. Now the Fourier series is replaced by a Fourier transform. By analogy, we would multiply the Fourier transform of a given input function by the appropriate network function, which is a function of frequency. The result is the transform of the output function. The output function itself is found by taking the inverse transform. It sounds simple enough.

To carry out this plan with some degree of competence, however, requires some familiarity with the properties of Fourier transforms and some expertise in evaluating the transform of a given excitation function. While this sounds straightforward, we will not pursue this plan to its conclusion. The reason is that another transform will be introduced in the next chapter which has some advantages over the Fourier transform for this purpose. Nevertheless, we will proceed sufficiently along the path to gain some familiarity with Fourier transforms and their properties.

Certain Properties of Fourier Transforms

Being complex, the Fourier transform can be written in terms of its magnitude and angle and its real and imaginary parts as:

$$F(j\omega) = |F(j\omega)|\, e^{j\phi(\omega)} = A(\omega) + jB(\omega) \tag{54}$$

Both the magnitude and the angle are real functions of the continuous frequency variable. They are the amplitude (or magnitude) spectrum and the phase spectrum, respectively. The real and imaginary parts are each real functions of ω (no j in these functions); hence their functional dependence is shown to be on ω, not $j\omega$.

EXERCISE 13

By applying Euler's theorem in (50a), the expression for the Fourier integral, find integral expressions for the real and imaginary parts of a Fourier transform. Make a note whether they are even or odd functions.

ANSWER: $A(\omega)$ even, $B(\omega)$ odd. □

Using these results and replacing ω by $-\omega$ in (54) leads to:

$$\begin{aligned} F(-j\omega) &= F^*(j\omega) \\ F(j\omega)F(-j\omega) &= |F(j\omega)|^2 \end{aligned} \tag{55}$$

That is, the conjugate of a Fourier transform function is obtained by replacing ω by $-\omega$.

EXERCISE 14

Expressing the magnitude and angle functions in terms of the real and imaginary parts, and using the preceding results, show that the magnitude and angle are even and odd functions, respectively. □

EXERCISE 15

Let $f(t)$ be a single pulse of height K and width w. Find its Fourier transform by evaluating the Fourier integral in (50). Compare with the value of TF_n, given in (46), for the pulse train in the limit as T approaches infinity. Sketch the amplitude and phase spectra. Confirm that they satisfy the properties just discussed. □

EXERCISE 16

Confirm the even and odd properties of the Fourier transform described in this section, using the transform of the exponential in (53). □

Parseval's Theorem

In an earlier section we found the power dissipated in a network in terms of the harmonic amplitudes (or rms values) of the input voltage and current, when these were periodic. We shall now undertake a similar endeavor for aperiodic waveforms.

Here's how the problem will be framed. The energy dissipated in a resistor R carrying a current $i(t)$ is the integral over all time of $Ri^2(t)$ (or $Gv^2(t)$). The R (or G) will be eliminated if we normalize and take the energy dissipated per unit resistance (or per unit conductance). We can use $f(t)$ to stand for either $i(t)$ or $v(t)$. Hence, the *normalized* energy W will be:

$$W = \int_{-\infty}^{\infty} f^2(t)\,dt \tag{56}$$

It is our objective now to find an equivalent expression involving the frequency domain, just as we did in (40) and (41) where we inserted the Fourier series of the voltage and current. This time, we'll insert the equivalent of the series, namely, the inverse transform. We will not replace both of the f factors in f^2, however; only one of them will be replaced by the inverse transform of $F(j\omega)$ as given in (50b). The result will be:

$$W = \int_{-\infty}^{\infty} \left[\frac{1}{2\pi}\int_{-\infty}^{\infty} F(j\omega)e^{j\omega t}\,d\omega\right] f(t)\,dt \tag{57}$$

There are two integrals in this expression; the inside one, to be performed first, is with respect to ω, and the second one is with respect to t. Let's interchange these two operations and note that, in the integral with respect to ω, $f(t)$ can be considered a constant and moved outside the integral sign. Similarly, in the integral with respect to t, $F(j\omega)$ can be taken as a constant and moved outside the integral sign. The results of these steps are:

$$\begin{aligned} W &= \frac{1}{2\pi}\int_{-\infty}^{\infty} F(j\omega)\left[\int_{-\infty}^{\infty} f(t)e^{j\omega t}\,dt\right] d\omega \\ &= \frac{1}{2\pi}\int_{-\infty}^{\infty} F(j\omega)F(-j\omega)\,d\omega = \frac{1}{2\pi}\int_{-\infty}^{\infty} |F(j\omega)|^2\,d\omega \end{aligned} \tag{58}$$

The quantity in brackets in the first line would be the Fourier integral of $f(t)$ if ω were replaced by $-\omega$. Hence, it is $F(-j\omega)$, as shown in the first term of the second line.

Finally, combining the last expression with (56), we get:

$$\int_{-\infty}^{\infty} f^2(t)\,dt = \frac{1}{2\pi}\int_{-\infty}^{\infty} |F(j\omega)|^2\,d\omega \tag{59}$$

This is *Parseval's theorem*. It states that not only can the normalized energy associated with a signal $f(t)$ be determined by an integration of the square of $f(t)$ in the time domain, but it can also be determined by an integration of the magnitude squared of its Fourier transform in the frequency domain (divided by 2π).

EXAMPLE 4 As an example that does not illustrate the power of this theorem but provides some acquaintance with the mechanics, let us find the normalized energy contained in the exponential whose Fourier transform was obtained in (53). For simplicity, let us assume $a = 1$. Using the $F(j\omega)$ from (53), we find:

$$\frac{2}{2\pi}\int_0^\infty \frac{1}{(1+\omega^2)^2}\,d\omega = \frac{1}{\pi}\frac{1}{2}\left(\frac{\omega}{\omega^2+1} + \tan^{-1}\omega\right)\Bigg|_0^\infty = \frac{1}{2}$$

(Supply the missing steps and confirm the result.) Note that the magnitude squared is an even function of frequency; hence, we can integrate over half the range and multiply by 2.

EXERCISE 17

Confirm Parseval's theorem for this example by integrating in the time domain. □

Energy Content of Signal

Let's now examine the frequency-domain integration in Parseval's theorem in (59), concentrating on the dimensions. The entire expression represents energy per unit resistance or conductance, if $f(t)$ represents current or voltage, respectively; it is called the 1-Ω energy by some, though still treated as energy, in joules. We say it is the *energy content* of the signal. Since $|F|^2\,d\omega$ is energy, $|F(j\omega)|^2$ is energy per unit angular frequency, commonly called the *energy density*. (By the same reasoning, $f^2(t)$ is also an energy density but with respect to time.)

If it were only a matter of identifying dimensions, the idea of energy density would not be very significant. The significance arises from the following considerations. If the entire energy content lies in the entire frequency range from $-\infty$ to ∞, then a fraction of the energy content contained in a certain band of frequencies is obtained by integrating over that particular band.

To illustrate, a network is excited as illustrated in Figure 22. Let the excitation signal be $v_1(t) = 10e^{-t/1000}$ and the network be the low-pass RC circuit shown in Figure 22b. Then, the Fourier transform of the input function

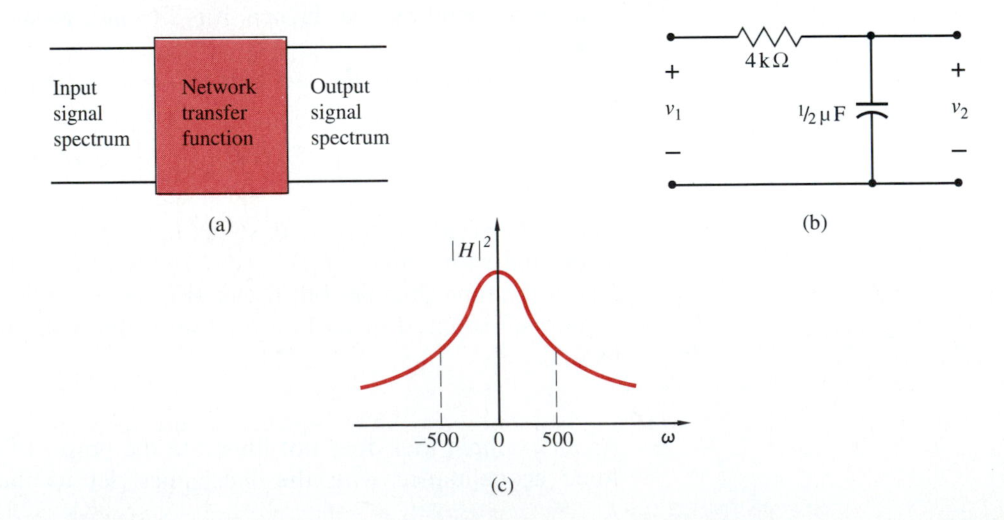

Figure 22
Energy-density calculation for a low-pass filter.

$v_1(t)$ and the network function $H(j\omega)$ are:

$$V_1(j\omega) = \mathscr{F}(10e^{-t/1000}) = \frac{10}{10^{-3} + j\omega}$$

$$H(j\omega) = \frac{1}{1 + j\omega RC} = \frac{1}{1 + j0.002\omega}$$

The product of these two functions is $V_2(j\omega)$, the Fourier transform of the output voltage.

Whereas the input is known as a function of time, the output function is not, although its Fourier transform is known. Even if the output signal as a function of time is not known, its energy content can be found over any frequency band. Because of the even nature of the magnitude-squared function, we integrate over the positive frequency band only and multiply by 2. The relevant results are as follows:

$$V_2(j\omega) = \frac{10^4}{(1 + j10^{-3}\omega)(1 + j2 \cdot 10^{-3}\omega)}$$

$$|V_2(j\omega)|^2 = \frac{10^8}{(1 + 10^{-6}\omega^2)(1 + 4 \cdot 10^{-6}\omega^2)} \tag{60}$$

$$2\int_{\omega_1}^{\omega_2} |V_2(j\omega)|^2\, d\omega = \int_{\omega_1}^{\omega_2} \frac{2 \times 10^8\, d\omega}{(1 + 10^{-6}\omega^2)(1 + 4 \cdot 10^{-6}\omega^2)}$$

EXERCISE 18

The half-power frequency of the preceding filter is $\omega = 500$ rad/s. (Confirm this.) The frequency band of interest is shown in Figure 22c. Integrate and determine the fraction of the energy content within the filter passband.

ANSWER: 70.5%. □

Fourier Transform of Functions with No Fourier Integral[13]

In order for the Fourier integral of a function to exist, the function must satisfy the Dirichlet conditions, given for periodic functions in Section 1. However, those conditions now apply for all time, from $-\infty$ to $+\infty$. In particular, $f(t)$ must be absolutely integrable over this infinite interval. This condition is somewhat more stringent for aperiodic functions than it is for periodic functions.

If a function is to be absolutely integrable (have a finite area under the curve), it cannot approach anything but zero at infinity. Hence, a *necessary condition* for the existence of its Fourier integral is that $f(t)$ approach 0 as t approaches infinity. This condition is not sufficient, however, since some functions which satisfy it do not have a Fourier integral. That is, it is possible

[13] This section can be skipped without incurring any liabilities in the rest of the book. It is included only to expand understanding; once found, no use will be made of the transforms studied here. You might want to read it out of intellectual curiosity and compare it with the material in the study of Laplace transforms in the next chapter.

for a curve to go to zero at infinity in such a way that the area under the curve becomes infinite. Such a function has no Fourier integral.

As an example, consider the step function with discontinuity at t_0: $f(t) = Ku(t - t_0)$. The area under the entire step function is infinite, which means that the function is not absolutely integrable. (Sketch the function and verify.) Thus, *the step function does not have a Fourier integral.* (Attempt to evaluate the Fourier integral for the step function to confirm this result.) Similarly, the absolute value of a sinusoid is a rectified sinusoid. As time goes to infinity, the area under this curve grows without limit. Hence, this function, too, has no Fourier integral.

We see that some simple functions which are very important in engineering—such as a constant (DC), cosine and sine functions (AC), a unit step, and an exponential with an imaginary exponent—are not absolutely integrable, and so they have no Fourier integral. This is distressing; what can we do?

What we will do is not take no for an answer! We *want* all functions of interest in engineering to have a Fourier *transform*; if they don't have a Fourier *integral,* we'll redefine the transform in these cases so that they will at least have a transform. In these cases, the Fourier transform is not the same as the Fourier integral.[14] With that in mind let's look at each of these significant functions in turn.

FOURIER TRANSFORM OF THE IMPULSE First let's consider the impulse, first introduced in Chapter 6. (Review that material if you feel the need.) Recall the mutual relationship between the step function and the impulse:

$$\int_{-\infty}^{t} \delta(t - t_1)\, dt = u(t - t_1) \qquad t \geqslant t_1 \tag{61}$$

Another significant property of the impulse is the following:

$$\int_{-\infty}^{t} f(t)\delta(t - t_1)\, dt = f(t_1)u(t - t_1) \qquad t > t_1 \tag{62}$$

This is easily proved by noting that the impulse (and, therefore, the entire integrand) is 0 everywhere except at $t = t_1$. The only value of f which is not multiplied by 0 is $f(t_1)$, which is just a constant. This constant can be taken outside the integral; hence, what's left is just the integral in (61). The result follows. It's almost as if the impulse scans, or samples, the function and picks out only this particular value. That's why (62) is called the *sampling* property of the impulse, and sometimes also the *sifting* property. This property will be very useful in the evaluation of the Fourier transform of the impulse.

Let's now evaluate the Fourier integral of a unit impulse:

$$\mathscr{F}[\delta(t - t_1)] = \int_{-\infty}^{\infty} e^{-j\omega t}\delta(t - t_1)\, dt = e^{-j\omega t_1}u(t - t_1) \tag{63}$$

[14] That's the difference between engineering and mathematics (or science, in general). We want to make something "work," whatever it takes.

(You may worry that the upper limit of the integral in the sampling property is not the same as the one in the Fourier integral. But note that the only requirement on the upper limit t in (62) is that it exceed t_1; surely ∞ satisfies that condition.) Although you may be surprised that something that goes to infinity at some time has a Fourier integral and, hence, a Fourier transform, the impulse *is* absolutely integrable, so there should be no surprise. Specifically, the Fourier transform of an impulse occurring at a specific time t_1 is an exponential function of $j\omega$: $e^{-j\omega t_1}$ starting at time t_1.

Something interesting results if we look for the energy content of the impulse. The magnitude squared of an exponential with an imaginary exponent is 1. This is to be integrated over an infinite range of frequencies. Hence, *the energy content of an impulse is infinite.*

Another way of saying this is that the Fourier transform of the impulse is not absolutely integrable. Hence, we should not expect that the inverse Fourier *integral* exists—and we just found that it doesn't. Nevertheless, since the Fourier *transform* of the impulse occurring at t_1 is $e^{-j\omega t_1}$, the inverse Fourier *transform* of $e^{-j\omega t_1}$ is $\delta(t - t_1)$; these two functions constitute a transform pair:

$$\delta(t - t_1) \leftrightarrow e^{-j\omega t_1} \tag{64}$$

FOURIER TRANSFORM OF AN IMAGINARY EXPONENTIAL The objective in this section is to find the Fourier transform of an exponential function of time $e^{j\omega_1 t}$. Since the absolute value of this function is 1 over all time, it is not absolutely integrable, so its Fourier *integral* does not exist. We will proceed to define the Fourier *transform* by interchanging the roles of time and frequency in the preceding transform pair.

Look back at the defining integrals in (50). As noted back then, these are highly symmetric integrals—aside from a trivial factor of 2π in one of them and an interchange of sign in the exponentials. If those two points are appropriately handled, time and frequency can be interchanged in the two integrals and the direct transform becomes the inverse, and vice versa. That is, if the inverse transform in (50) is to become the direct transform, both sides must be multiplied by 2π. Thus, $f(t)$ is replaced by $2\pi f(\omega)$, the signs in the exponents are interchanged, and then $F(j\omega)$ is replaced by $F(jt)$. Performing these steps on the transform pair in (64) leads to:

$$2\pi\delta(\omega - \omega_1) \leftrightarrow e^{j\omega_1 t}$$

or

$$e^{j\omega_1 t} \leftrightarrow 2\pi\delta(\omega - \omega_1) \tag{65}$$

The first line is written in accordance with the way it was constructed, but the second line is the normal way of showing the transform pair, with the function of time first.

The impulse in frequency has the same meaning as the impulse in time. It can be depicted as a vertical arrow located at ω_1 along the ω axis, with its strength (in this case 2π) written alongside. (Sketch it.)

FOURIER TRANSFORM OF CONSTANT (DC) AND COSINE AND SINE (AC) FUNCTIONS From the transform of the imaginary exponential it is a simple matter to find the Fourier transform of a constant. We let ω_1 go to zero in

(65). The exponential reduces to 1 and its Fourier transform remains an impulse but located at the frequency origin. If the constant is K instead of 1, the Fourier transform pair becomes:

$$K \leftrightarrow 2\pi K\delta(\omega) \tag{66}$$

We see that the magnitude spectrum of a DC signal K is an impulse with a strength of $2\pi K$ located at the frequency origin. The phase spectrum is zero.

The transforms of the cosine and sine functions can be obtained from (65) using Euler's theorem. Thus:

$$\begin{aligned}\mathscr{F}(\cos \omega_1 t) &= \mathscr{F}[\tfrac{1}{2}(e^{j\omega_1 t} + e^{-j\omega_1 t})] \\ &= \pi\delta(\omega - \omega_1) + \pi\delta(\omega + \omega_1]\end{aligned} \tag{67}$$

$$\begin{aligned}\mathscr{F}(\sin \omega_1 t) &= \mathscr{F}\left[\frac{1}{j2}(e^{j\omega_1 t} - e^{-j\omega_1 t})\right] \\ &= -j[\pi\delta(\omega - \omega_1) - \pi\delta(\omega + \omega_1)]\end{aligned} \tag{68}$$

Several interesting bits of information can be obtained from these two equations. First, the magnitude spectra of both the sine and the cosine functions are the same. They consist of two impulses of strength π at $\pm\omega_1$ on the frequency axis, corresponding to the frequency of the sinusoids. Second, the phase spectra are different; the cosine has a phase spectrum of 0° and the sine a phase spectrum of −90°.

CONSOLIDATION

This chapter started by concerning itself with the steady-state response of linear circuits to periodic excitations. The first step was to convert the periodic functions to a series of sinusoids with harmonic frequencies. Then, since the steady-state response of a circuit to a sinusoid is known, the response to the periodic function is obtained as a linear combination of the harmonic responses.

But much more than this was made possible. A whole new way of looking at signals was opened up. Not only can they be described by their variation with time, but also by their frequency spectra. The magnitude and phase spectra are discrete for periodic functions, but they become continuous when the Fourier concepts are extended to aperiodic functions. In this case, the sums of discrete harmonics become integrals of continuous frequency components. Parseval's theorem introduces the concept of the frequency content of signals over various frequency intervals. We can now determine how much of the energy in a signal is contained within the passband of a circuit.

It would be logical to use the Fourier transform extension of the Fourier series to aperiodic functions in order to find the response of circuits to such functions. However, we didn't pursue that line, promising to introduce a different transform in the next chapter which can also perform this function and has certain advantages in doing so.

SUMMARY

- The Dirichlet conditions for representation by a Fourier series
- Cosine and sine form of Fourier series
- The fundamental frequency and its harmonics

- ormulas for the Fourier coefficients
- Symmetries: even, odd, and half-wave
- Cosine-with-angle form of Fourier series
- Exponential form of Fourier series and coefficient formulas
- Frequency spectra: magnitude and phase spectra
- The sampling function
- Variation of Fourier coefficients with harmonic order
- Band-limited functions
- Steady-state response of a linear circuit to a periodic function
- Fourier series as the least-mean-square approximation
- The Gibbs phenomenon
- rms value of a periodic function in terms of harmonic rms values
- The Fourier integral: generalization of Fourier series to aperiodic functions
- The Fourier transform concept
- Properties of magnitude and angle, and real and imaginary parts of a Fourier transform
- Energy density and Parseval's theorem
- Energy content of a signal over a band of frequencies
- Fourier transform of functions with no Fourier integral: exponential, step, cosine, sine

PROBLEMS

Some of the problems in this chapter will involve some of the following waveforms. Reference will be made to them by number when they are used. Each might be a voltage or a current; which one it is will be specified in each problem. Draw and label each of these waveforms for use in the problems.

1. Square wave: $T = 20\pi$ ms, peak-to-peak amplitude = 40 SI units, average value = 0
2. Sawtooth: period = 20 ms, height = 10 SI units, lowest point rests on the horizontal axis
3. Symmetrical triangular wave: period = 100 μs, average value = 0, peak-to-peak amplitude = 20 SI units
4. Waveform 1 with every other cycle missing
5. Waveform 2 with every other cycle missing
6. Waveform 3 with every other cycle missing
7. Waveform 3 with every other cycle inverted
8. Waveform 1 with two consecutive cycles missing for each cycle present
9. Waveform 1 with three consecutive cycles missing for each cycle present, and with a constant 20 units added to the value of the function
10. Waveform 3 with the lower half cycles missing
11. Waveform 1 with three half cycles missing after an upper half cycle
12. Waveform 2 shifted down 5 units, with three half cycles missing after an upper half cycle
13. Waveform 3 with three half cycles missing after an upper half cycle
14. Sinusoid: $T = 2\pi$ ms, peak-to-peak amplitude = 10 SI units, every other cycle missing

Save the results of any work you perform with these waveforms for use in later problems.

1. For the functions in Figure P1, specify all the types of symmetry each can exhibit with proper choice of the origin of time and/or the location of the time axis. (Specify these choices.)

2. The function in Figure P2 forms half of one period of a periodic function.
   a. Complete the sketch for functions having even symmetry.
   b. Complete the sketch for functions having odd symmetry.
   c. For each of these functions, find the Fourier coefficients and write out the series up to the third nonzero harmonic. Note the terms that are absent in each case.

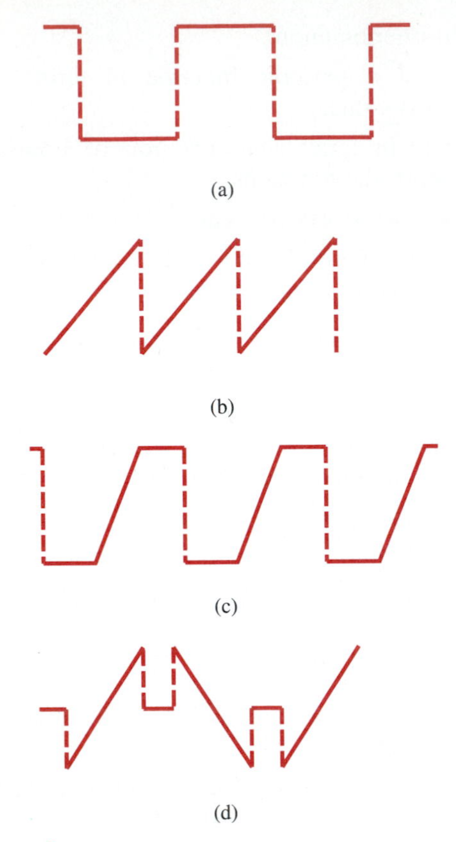

Figure P1

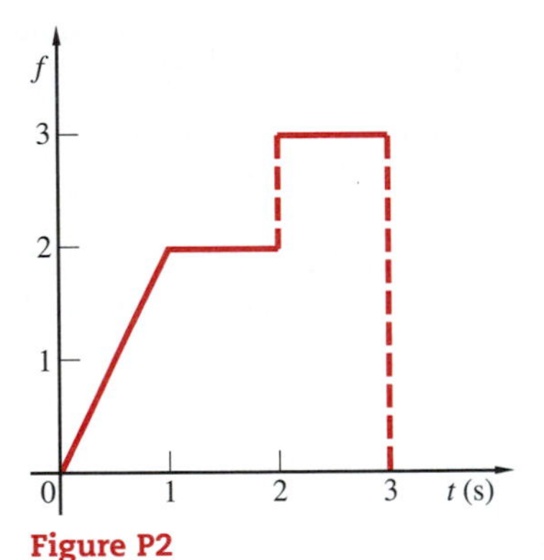

Figure P2

3. Suppose the graph in Figure P2 constitutes one quarter of one period of a periodic function.

a. Complete the sketch for functions having even and half-wave symmetry.

b. Repeat for functions having odd and half-wave symmetry.

c. For each function, find the Fourier coefficients and write out the series up to the third nonzero harmonic. Note the terms that are absent in each case.

4. **a.** Find the rms value of each of the periodic functions generated in Problem 2.

b. Repeat for each of the functions in Problem 3.

5. Find the coefficients in the exponential form of the Fourier series of a half-wave rectified sinusoid whose period is 1 ms. (Select the origin of time to reduce your effort.) Write out the terms of the series for n from -3 to $+3$.

6. Assume that waveform 10 is a voltage.

a. Draw the waveform.

b. Determine the coefficients in the trigonometric form of its Fourier series. (Do whatever is needed to simplify the calculation.)

c. Write out the terms of the series up to $n = 3$.

d. Determine the rms value of the function.

7. Assume that waveform 11 is a current. Repeat Problem 6.

8. The periodic function in Figure P8 represents a current.

a. Find the coefficients in the trigonometric form of its Fourier series and write out the first few terms in the series.

b. Move the time origin 2 s (one-fourth of a period) to the right and substitute the new time variable in the series obtained in part (a) to obtain a new series.

c. With that new time origin, determine the coefficients of the new Fourier series, using the coefficient formulas; compare with the result in part (b).

d. Determine the rms value. Does it depend on the choice of origin?

9. The coefficients F_n of the exponential form of a certain Fourier series are found to have the following property: $F_{-n} = F_n$. Giving a persuasive argument, specify the terms that will be absent in the trigonometric form of the series.

10. In a particular case, the coefficients in the exponential form of a Fourier series are found to have the following property: $F_{-n} = -F_n$. Derive expressions

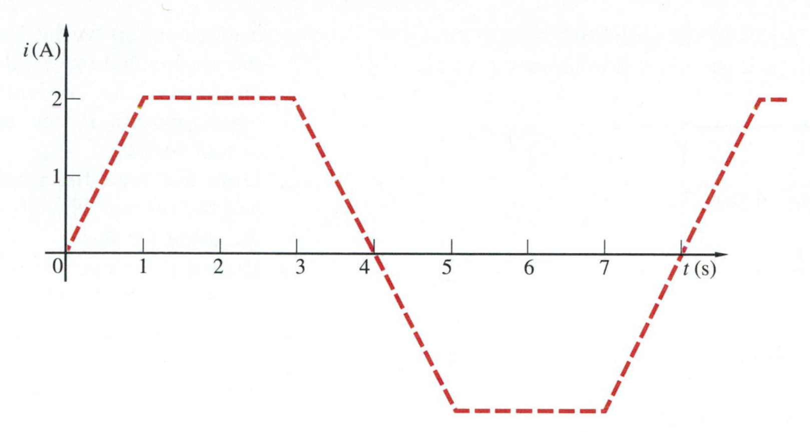

Figure P8

for the coefficients of the trigonometric form, a_n and b_n, in terms of F_n.

11. The waveform in Figure P11 is a sinusoid whose peaks have been truncated. (Would a linear circuit be able to do that?)

a. Find the coefficients of the trigonometric form of its Fourier series.

b. Using the results of the preceding two problems, determine the coefficients of the exponential form.

c. Find the rms value of the wave on the basis of its definition.

d. Find the approximate rms value from the first few Fourier coefficients.

e. How many terms in the series will you need to include so that the approximate rms value is at least 96% of its correct value?

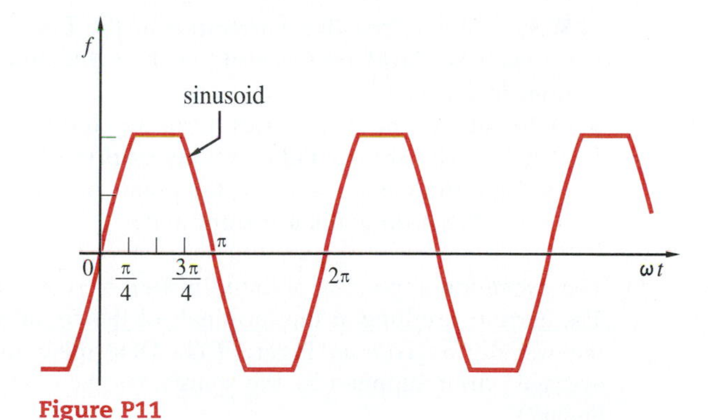

Figure P11

12. a. Draw and label waveform 7 and specify the symmetries that it has, if any.

b. Determine the coefficients in its trigonometric Fourier series.

c. Specify the harmonic amplitudes and angles up to the third nonzero harmonic.

d. Write out the exponential form of the series from $n = -3$ to $+3$.

13. Repeat Problem 12 for waveform 4.

14. Waveform 5 is the excitation voltage v_1 in Figure P14.

a. Draw and label the waveform.

b. Find the first few terms in the Fourier series of the output voltage. Comment on how fast the coefficients in the output series decrease with increasing n.

c. Repeat all the preceding if the excitation is waveform 7.

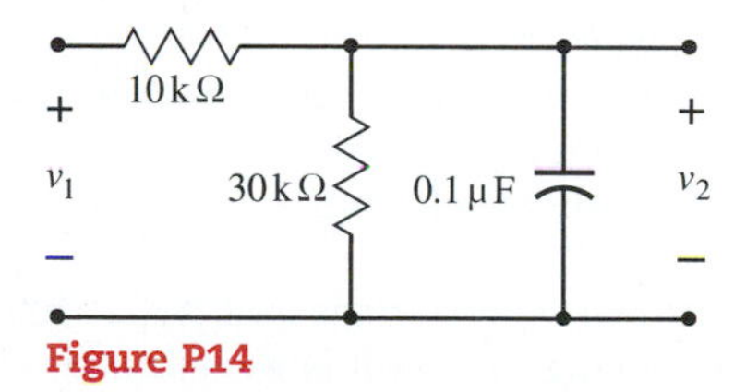

Figure P14

15. The excitation in Figure P15 is waveform 6.

a. Draw and label the waveform.

b. Find the general expression for the Fourier coefficients.

c. Find the numerical values of the first few terms in the Fourier series of the output voltage. Note how fast the coefficients go down with n.

d. Convert the series to the cosine-with-angle form.
e. Repeat all steps if the excitation is waveform 12.

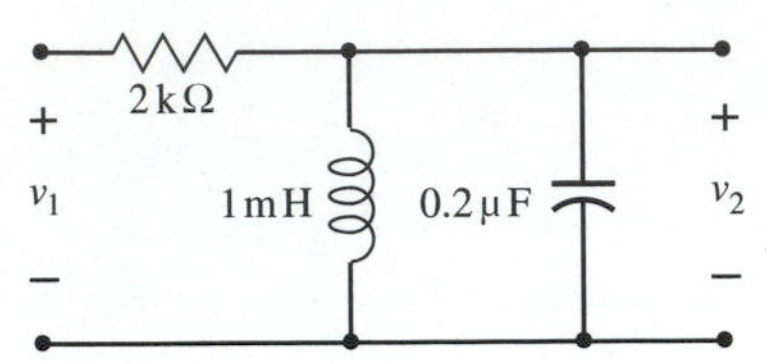

Figure P15

16. The voltage exciting the circuit of Figure P16 is waveform 8.
a. Draw and label the waveform.
b. Find the general expression for the Fourier coefficients of the input.
c. Find the numerical values of the first few terms in the Fourier series of output voltage v_2. How fast are the coefficients dropping with n? Repeat if the output voltage is v_3.
d. Repeat all steps if the input is waveform 10.

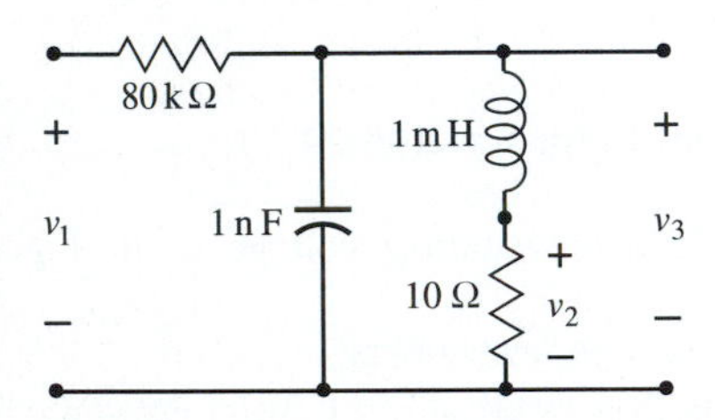

Figure P16

17. a. Using the coefficients of the exponential form of its Fourier series, find an approximate rms value of waveform 6.
b. Confirm your result by finding the rms value from the waveform itself.
c. How many terms of the series are required if the error is to be no more than 3%.
d. Repeat all steps for waveform 2.

18. a. Find the average power delivered to a 100-Ω resistor if the voltage across it is waveform 5. Use enough terms so that the first neglected harmonic would contribute less than 5% to the result.
b. Find the power directly from the waveform and compare.
c. Repeat all steps for waveform 1.

19. The input voltage in Figure P19 is waveform 10.
a. Using the exponential form of the Fourier coefficients up to the fourth harmonic, determine the approximate rms value of the output.
b. Determine the percentage improvement in the approximation if one more term is used in the output series.
c. Draw and label the amplitude spectra of the input and the output. (Specify the form of the series you are using for this.)
d. Repeat if the excitation is waveform 13.

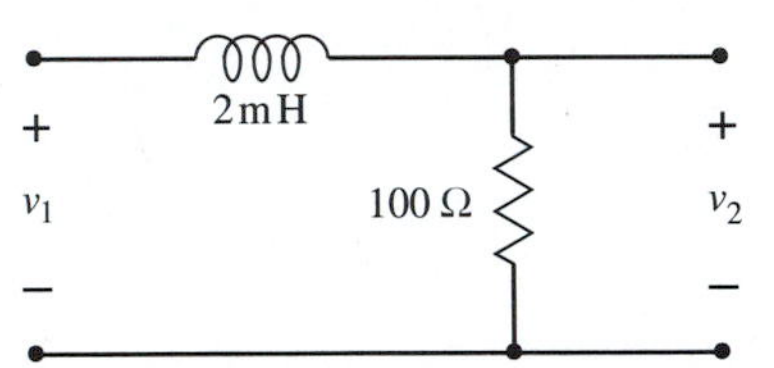

Figure P19

20. The excitation v_1 of the circuit in Figure P20 is waveform 3.
a. Using the exponential form of the Fourier series of the excitation, find the average power dissipated in the resistor, within 2%. (How is this related to the rms value of the output?)
b. Find the energy content in the output signal i within the 3-dB passband of the circuit.
c. Repeat all steps if the excitation is waveform 9.

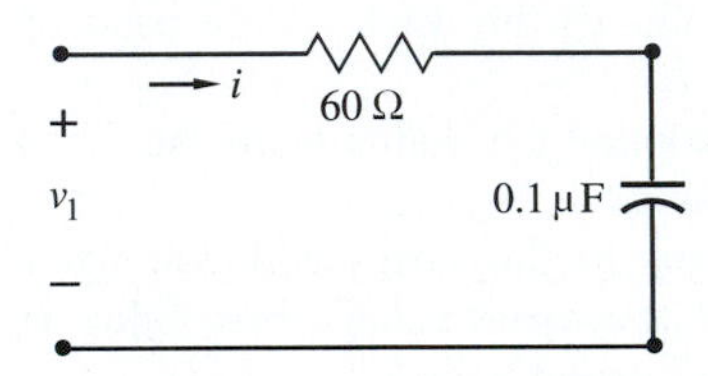

Figure P20

21. a. Write out the first five harmonics in the Fourier series (any form) representing each of the functions in Figure P21.
b. Differentiate the first series term by term and compare with the second one. Any surprises?
c. Evaluate the second series at the points of discontinuity. What do you learn from that?

22. The excitation applied to a circuit is $v(t) = 3\cos^3 \omega t$. The current resulting at the terminals of the circuit is the waveform given in Figure P11. Determine the average power supplied by the source. (Is the circuit linear?)

23. A periodic voltage waveform applied to a circuit consisting of a 100-Ω resistor in series with a 10-mH

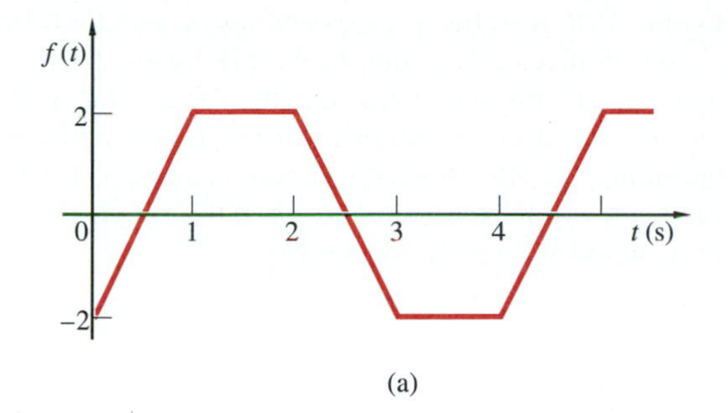

(a)

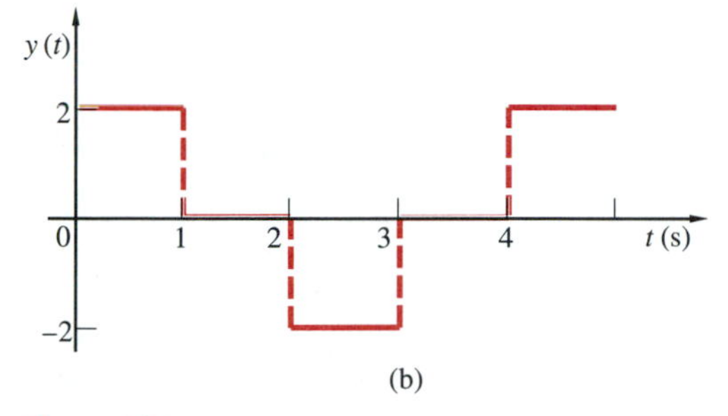

(b)

Figure P21

inductor has a period $T = 100\ \mu$s. The only nonzero coefficients in its exponential Fourier series are $F_0 = 8$, $F_{1,-1} = 5 \pm j6$, $F_{2,-2} = 3 \pm j1$, $F_{3,-3} = 1$, $F_{4,-4} = \pm j0.3$. Determine the average power dissipated in the circuit.

24. From the Fourier series of the listed waveforms you have worked out in the preceding problems construct a table whose columns list:

a. The waveform number
b. Its rms value (to within 5%)
c. The rms value of the fundamental
d. The ratio of the last column to the preceding one; that is, how big a part of the effective value of the waveform is made up by the fundamental

Are there any patterns?

25. The square wave in the frequency-multiplier circuit of Figure 17 in the text has a period of $\pi/10^5$ s.

a. Select the time origin to make the function odd, and determine the values of the first three nonzero terms of its Fourier series, confirming that they are odd harmonics.
b. A 50-mH coil is available, with a Q of 100 at the frequency of the function. Connect a capacitor C in series with the coil so that the resonant circuit is tuned to the third harmonic of the square wave.
c. Determine the magnitude of the first- and fifth-harmonic distortion components in the output as a percentage of the third-harmonic magnitude. (These fractions could be called the *rejection ratios.*)
d. Suppose the capacitor is adjusted to tune the circuit to a frequency slightly higher than the third harmonic in order to make both distortion components in the output have the same magnitude. This will mean that the third harmonic does not fall at the maximum of the resonance curve, so that its magnitude also will be reduced. Again determine the ratio of the first and fifth harmonic in the output to the desired third harmonic. Is there an improvement?

26. The square wave in the frequency-multiplier circuit in Figure 17 in the text is to be a voltage. For its Fourier series, use the coefficients determined in Problem 25. The resonant circuit is to be replaced by the stagger-tuned circuit discussed in Chapter 12, Section 7.

a. Determine the rejection ratios of the distortion components for critical staggering, assuming ω_0 is the frequency of the third harmonic of the wave.
b. See if you can improve this by increasing the staggering coefficient.
c. Design the circuit.

27. The steady-state response of a first-order RC circuit to a rectangular-wave excitation was discussed in Chapter 6, Section 3, using a time-domain approach. (Review that discussion.) There we used the total time response appropriate to each interval in the wave and matched the solutions at the boundaries of the intervals to determine the unspecified constants.

A different approach is taken in this chapter. The circuit in Chapter 6, Figure 20, is repeated in Figure P27a. The rectangular wave is the same as the one in Chapter 6, Figure 21, except that the average value of that wave (12 V) is subtracted, leaving the average value of 0 here. The solution given in Chapter 6 applies except that 12 V should be subtracted in each interval. Use RC as a frequency-normalizing standard. (That is, the normalized value of RC is 1.)

a. Based on the results in Chapter 6, sketch the output-voltage wave on the same axes as the source voltage.
b. Find the Fourier coefficients (exponential form) of the rectangular input-voltage wave in Figure P27b.
c. Find the coefficients of the output voltage up to the fifth harmonic.

d. From the series, make a point-by-point sketch of the output voltage over one period and place it on the same axes as the previous sketch in part (a).

Comment on your findings about the relative ease of the two approaches. Which approach would increase more in difficulty for higher-order circuits?

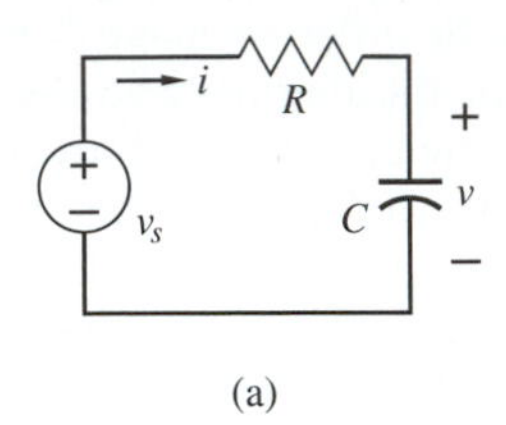

(a)

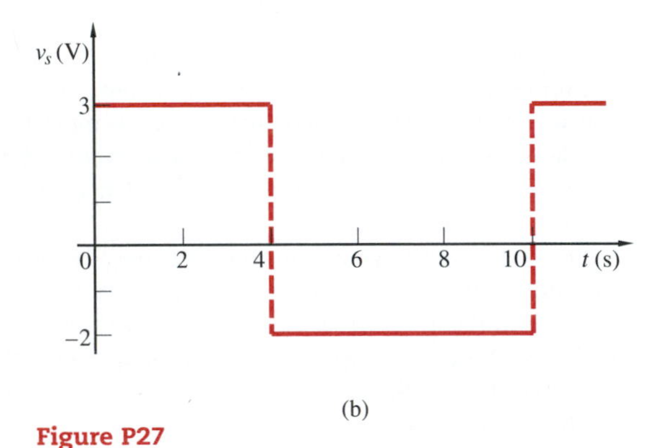

(b)

Figure P27

28. In an effort to reduce the second- and higher-harmonic amplitudes even further, the circuit in Figure P28 has been proposed as a rectifier-filter circuit, in place of the inverted ell in Figure 16 in the text. The added inductance has the same value as the old one. Find the amplitudes of the fourth and sixth harmonics (of the power frequency) relative to that of the second harmonic. Is the filtering action improved and the ripple reduced?

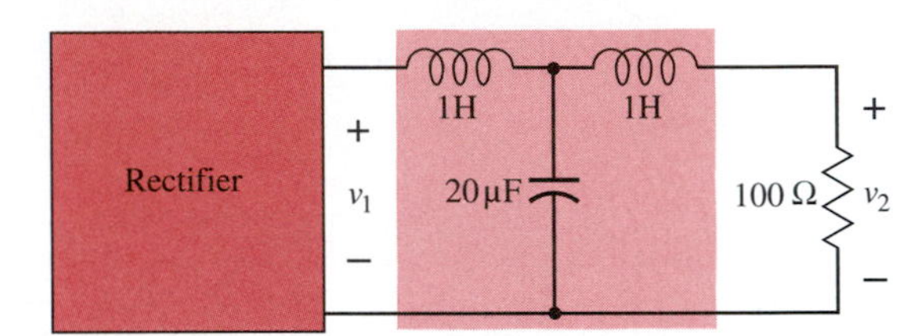

Figure P28

29. Instead of the tee filter in Figure P28, someone suggested the pi filter shown in Figure P29. The current excitation is the full-wave rectified sine function given in Figure 12 in the text; its Fourier series was given there also.

a. Determine the ratios of the (power-frequency) second-harmonic and fourth-harmonic amplitudes to the DC component in the filter output. Is there an improvement over the inverted-ell filter in the text?

b. Repeat the calculation if the input is a half-wave rectified sine at the power frequency.

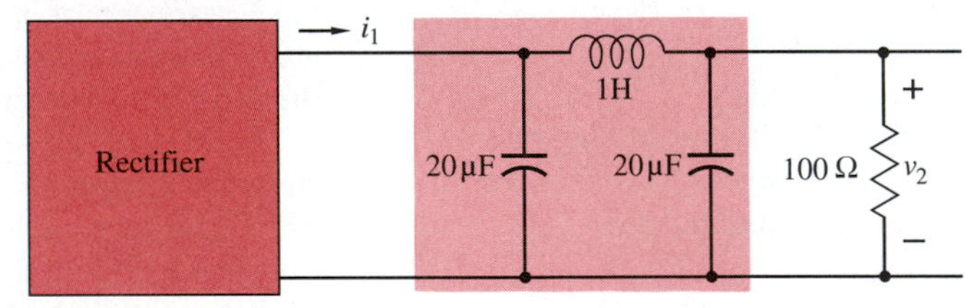

Figure P29

30. The parameter values given in the bridge circuit in Figure P30 are normalized. The excitation is the truncated sinusoid shown. The variable RL branch is adjusted so that the bridge is balanced at the fundamental frequency ω_0, which is also normalized. Find the amplitude of the largest harmonic in the output voltage $v_2(t)$.

31. Waveform 3 is a voltage excitation applied to the circuit in Figure P16. Change the value of C so that the resonant frequency of the circuit equals the third harmonic of the excitation.

a. Find the harmonic of the output voltage $v_3(t)$ that has the largest amplitude.

b. Compare the energy content of this harmonic with the total energy content of the wave. What is your conclusion about letting this harmonic represent the entire wave?

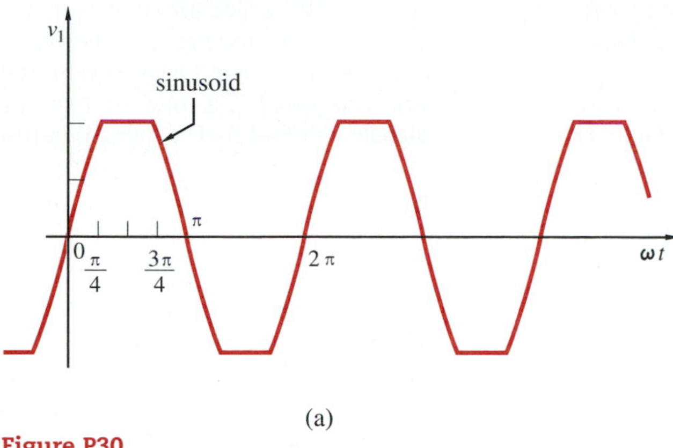

(a)

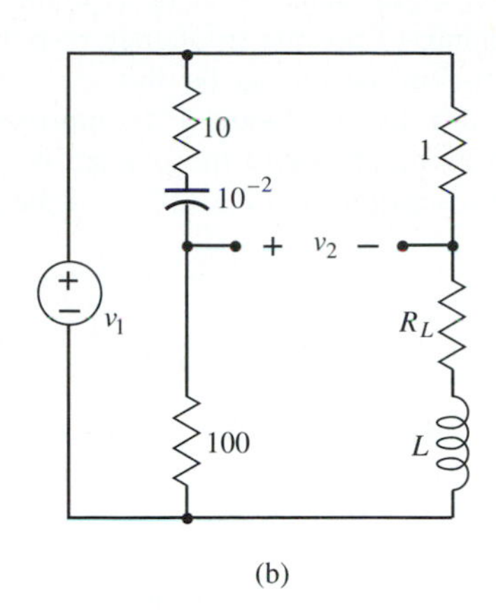

(b)

Figure P30

32. A train of positive pulses has a pulse repetition rate of 500 pulses per second and a pulse width of 0.1 μs.

a. Find the duty cycle of the pulse train.

b. Find the lowest frequency for which the spectral amplitude is zero.

c. How many lines are there in the amplitude spectrum from $\omega = 0$ to the frequency in part (b)?

d. Find the maximum amplitude of a frequency component that is at a higher frequency than the one in part (b).

e. Find the energy content in the frequency components lying in the band of frequencies extending up to the frequency in part (b).

33. The parameter values in Figure P33 are normalized values. The input voltage v_1 is waveform 3.

a. Find the coefficients in the exponential form of the Fourier series in general form.

b. Determine the first few terms of the output series.

c. The next order of business is to look for the steady-state response as a function of time from the complete solution of the appropriate differential equation. Write the differential equation.

d. Find the complete solution.

e. Describe in detail how you would determine any unknown quantities.

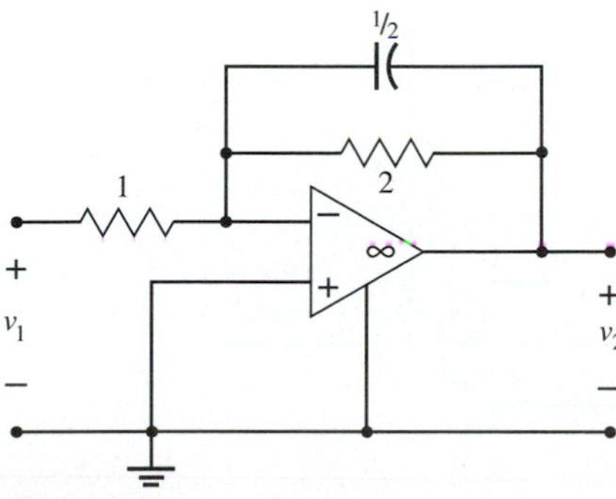

Figure P33

34. A 3-bit digital signal can be represented by a periodic waveform consisting, within each period, of a series of up to three rectangular pulses located at regular intervals. The number 5 (digital 101), for example, is represented by a pulse in the first and third positions and an absent pulse (no pulse) in the second position. After an interval where the function is zero, the pattern repeats itself.

Provision is to be made for digital signals up to 8 bits. If each pulse width is 1 μs and each space between two pulses is also 1 μs, an interval of 15 μs is needed to accommodate 8 bits. To indicate that the pattern has ended, an additional "dead" interval of 4 μs is included. Thus, for an 8-bit signal, each 19-μs cycle starts with a 4-μs interval of no pulses followed by 15 μs which can include up to eight 1-μs pulses separated by seven 1-μs spaces.

A communications channel for transmitting digital signals of up to 8 bits is to be designed. As a first step (the only one to be carried out here), the bandwidth required for transmitting such signals is to be determined. Let's deal here with 3-bit digital signals; each cycle, then, starts with 14 μs of zero signal followed by a 5-μs interval that can have up to three pulses. Assume a digital signal 111 (decimal 7).

a. Determine the bandwidth required to transmit this signal if at least 90% of the energy content of the input signal is to be contained in the output signal. (Take the frequency response of the transmission system to be flat over the entire bandwidth; this is obviously an approximation.)

b. How much would the needed bandwidth increase if we required at least 95% of the energy content?

c. The signal 000 requires no bandwidth to transmit. You might conjecture that signals between 000 and 111 will require an intermediate width of band between 0 and what is required for 111. But are you sure? Try one or two additional 3-bit signals and find their needed bandwidth.

DESIGN PROBLEMS

1. The input to the vertical amplifier of an oscilloscope can be modeled as shown in Figure DP1. The op amp in the figure is assumed ideal, $R = 1\,\text{M}\Omega \pm 2\%$ and $C = 20\,\text{pF} \pm 2\%$. The probe is used to attenuate the input signals by a desired factor.
   a. Determine the transfer function of the attenuator/scope system V_2/V_1 as a function of frequency.
   b. Determine the values of R_p and C_p such that signals are attenuated by a factor of 10, i.e., $V_2/V_1 = 1/10$, over a frequency range from DC (0 Hz) to 5 MHz.
   c. Determine the acceptable tolerances on R_p and C_p such that the difference in the attenuation of signals over the frequency range of interest is less than 1%.

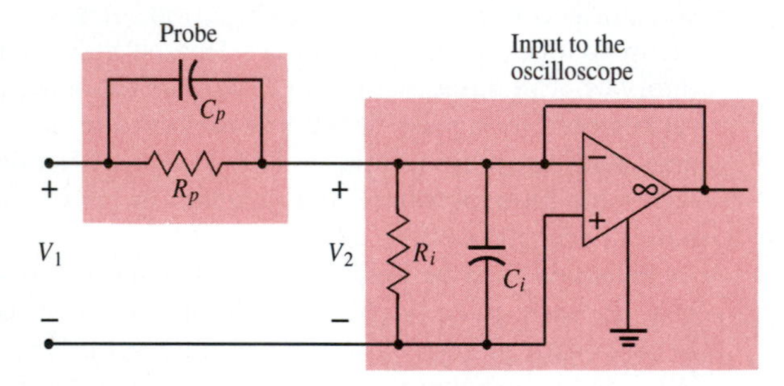

Figure DP1

2. a. Using the definition of the Fourier transform, determine the Fourier transform of a signal $g(t) = f(t - \tau)$ given that the Fourier transform of $f(t)$ is $F(\omega)$. The signal $g(t)$ is a delayed version of $f(t)$.
   b. A problem arising in high-freqency communication systems is the presence of echo signals in the received signal. This problem can be characterized as follows. If a signal $f(t)$ is sent from the transmitter, this signal, plus delayed and attenuated copies of it, can be received at the receiver. In this problem it is assumed that the received signal is of the form $g(t) = f(t) + \alpha f(t - \tau)$. If the Fourier transform of $f(t)$ is $F(\omega)$ determine the Fourier transform $G(\omega)$ of $g(t)$.
   c. The diagram in Figure DP2 represents a *transversal* filter. It is a schematic diagram of signal flow with "boxes" representing operations performed on the signals. The boxes marked delay cause the signal passing through them to be delayed by a time T without any other change. The circles with letters a_i cause an incoming signal to be multiplied by a coefficient a_i. As you might have guessed, the circle with a summation symbol yields an output consisting of the sum of the signals entering it. Find the transfer function of this filter. Assume the input is some signal $f(t)$ with Fourier transform $F(\omega)$, and,

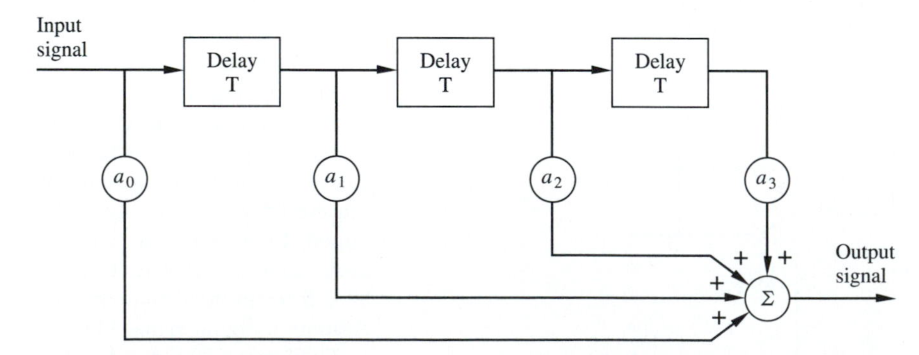

Figure DP2
A four-tap transversal filter.

using the result of part (a) of this problem and the linearity of the Fourier transform, determine the Fourier transform of the output signal. The ratio of the two gives the transfer function.

d. Assuming that α in the signal of part (b) is very small compared to 1, approximate the Fourier transform $G(\omega)$. Design a transversal filter to recover the original signal $f(t)$ from the received signal $g(t)$. Specify the number of taps and the constants of the filter.

14 Laplace Transforms and the Complete Response to Arbitrary Waveforms

This chapter has been a long time arriving. It has been a major objective in this book to determine the complete response of an arbitrary linear network to excitations of arbitrary waveform. Until now we have achieved only parts of this objective. Either we have limited the type and number of components present in a circuit, or we have limited the type of excitation function, or both.

The most general networks *were* treated—but only for periodic excitations in the steady state, especially sinusoids. The relative simplicity of the phasor method for finding the sinusoidal steady-state solution is a direct consequence of the fact that differential equations are transformed into algebraic equations of transformed variables, even though the variables and the coefficients in the resulting equations are complex. If you think about this a little, you might begin to wonder whether a similar result cannot, perhaps, be accomplished by some sort of transformation, even when the excitation is an arbitrary one and when we are interested in the complex response, not just in the steady state. A brief introduction to this possibility was treated in connection with Fourier transforms in Chapter 13. Pursuing this thought and exploiting its consequences are the objectives of this chapter.

Historically, it was the English engineer Oliver Heaviside who first used transform methods (he called them *operational methods*) in the solution of network problems.[1] The methods that he introduced were a product of his

[1] Oliver Heaviside (1850–1925) lacked a formal university education; he was self-taught. He lived in poverty most of his life; he couldn't even pay the annual dues of the Institution of Electrical Engineers, so he was dropped from membership until 1908, when he was made an honorary member. He made contributions also to telegraphy and radio-wave propagation. One of the ionized layers in the ionosphere is named after him: the *Heaviside layer.*

keen insight and inspiration. Since he was mainly interested in the solution of practical problems, he was indifferent to rigorous mathematical justification of his techniques. This led to strong antagonisms between Heaviside and the mathematicians of his day, and to bitter criticisms of his heuristic methods. Since then, his methods have been improved and others have supplied mathematical justification.

Oliver Heaviside, like Michael Faraday before him, was one of those rare individuals who seem to possess an intuitive insight into the workings of nature. With bold leaps of the imagination, they introduce new concepts and new ideas which seem strange and unorthodox to their contemporaries, and so not easily acceptable. When these ideas are refined, however, and made familiar, often by lesser individuals, they become the warp and woof of our knowledge.

1 THE LAPLACE TRANSFORM

It will be very helpful to recall here the basic features of the steady-state response to a sinusoidal excitation. The first step is to replace the sinusoids by their phasors. We think of this step as a transformation of variables from one domain to another—from the time domain to the phasor domain. The differential equations resulting from the application of the fundamental laws are transformed into algebraic equations with the phasors as variables, since the operation of differentiation becomes replaced by multiplication by $j\omega$ (generalized to s). These equations are then solved for the phasors. Finally, the phasors can be replaced by the sinusoidal functions of time which they represent. The last step can be thought of as the inverse of the first step, the *inverse transformation.*

Parenthetically, you yourself have, in the past, used a mathematical transformation from one domain into another where the resulting mathematical operations are easier to carry out: namely, the logarithm function. When two numbers are to be multiplied together, we turn to the logarithm function. The logarithm of the product of the two numbers is the sum of the logarithms of the individual numbers. Hence, after we take the logarithms of the two numbers and we add them, we then take the antilogarithm of the result to arrive at the final product. The multiplication operation has been transformed into the simpler operation of addition.

An Integral Transform in General

It would be very useful to have a similar sequence of steps that permits the calculation of a network response when the excitation is any function of time, not just sinusoidal; and when the desired response is not just the steady state but the complete response—so that the initial conditions must be introduced somehow.

In Chapter 13, the concept of a frequency spectrum was discussed. There we represented a periodic function by an infinite series of sinusoids. For a periodic function $f(t)$:

$$F_n = \frac{1}{T}\int_0^T f(t)e^{-jn\omega_0 t}\,dt \tag{1a}$$

$$f(t) = \sum_{n=-\infty}^{\infty} F_n e^{jn\omega_0 t} \tag{1b}$$

In these expressions F_n is the coefficient of the exponential form of the Fourier series representing $f(t)$.

This was then generalized to aperiodic functions, again extending over all time, just like periodic functions. The summation of discrete frequency components in (1b) became an integral of continuous frequency components. We obtained the Fourier integral and its inverse as:

$$F(j\omega) = \int_{-\infty}^{\infty} f(t)e^{-j\omega t}\,dt \tag{2a}$$

$$f(t) = \frac{1}{2\pi}\int_{-\infty}^{\infty} F(j\omega)e^{j\omega t}\,d\omega \tag{2b}$$

The equations in this pair have a mutual relationship. One of them converts a function of time $f(t)$ to a quantity $F(j\omega)$ which is a function of ω. The other reconverts $F(j\omega)$ back to a function of time. We can think of them as a pair of *transforms,* the first one being a *direct* transform and the second one an *inverse* transform.

Neglecting the details, let's examine the basic features of the expression in (1a). First of all, the right-hand side contains an integral; although the limits of integration here are specific ones, there are basically two limits: a lower one and an upper one, say, a and b. Next, the integrand is the product of two functions, one of which is a function of time $f(t)$; although the second function is a specific one here, its basic feature is that it is a function of both t and another variable n (or $\omega = n\omega_0$). In general form, the relationship can, therefore, be expressed as follows:

$$F(x) = \int_a^b f(t)K(x,t)\,dt \tag{3}$$

Such expressions arise quite often in science and engineering. In general terms, such an expression is an *integral transform.* The function $K(x, t)$ is called the *kernel* of the transform. After integration with respect to t and the substitution of the limits, the resulting function $F(x)$ is no longer a function of t; $F(x)$ is called the *transform* of $f(t)$. Many mathematical properties of integral transforms can be studied independently of the specific form of the kernel and independently of the specific physical interpretation. Obviously we will not pursue such studies here. Using a Fourier interpetation, though, $F(x)$ is a kind of "spectrum" function of $f(t)$.

Development of the Laplace Transform

Now let's consider the possibility of using an integral transform to solve the problem we're undertaking. If $f(t)$ is the variation with time of a network variable, using the integral transform in (3) certainly eliminates the time t but introduces another variable x. We don't know yet what kind of equations the differential equations in t are converted into under such a transformation. It is conceivable that the answer may depend on the kernel and on the limits of integration, so let's decide what these might be.

The integration in (1a) is over one period of a periodic function. The period of an aperiodic function, however, is infinite, so it would seem reasonable to choose the upper limit as ∞. As for the lower limit, the origin of time is arbitrary. Usually, the time after which a network response is to be found, such as an instant of switching, is chosen as $t = 0$, and so it would be reasonable to choose the lower limit as $t = 0$. Finally, as for the kernel, it has already been observed that, for aperiodic functions, the Fourier spectrum loses its discreteness. Thus, the nth-harmonic frequency $n\omega_0$ of the Fourier series becomes just any frequency ω. It might be reasonable, therefore, to take the kernel as $e^{-j\omega t}$. This is the kernel used in the Fourier integral in (2a).

This choice, however, is unsatisfactory, because, with that kernel, the integral (Fourier) transforms of a number of functions that are important in engineering (such as DC and AC) do not exist, as noted in the preceding chapter. Hence we ought to look for a different kernel.

DEFINITION OF THE LAPLACE TRANSFORM Suppose that, instead of $j\omega$ in the exponent, we choose the complex-frequency variable $s = \sigma + j\omega$ so that the real part in the exponent can serve as a convergence factor to remove the difficulty brought about by the upper limit. Hence, the desired integral transform becomes:

$$F(s) = \int_0^\infty f(t)e^{-st}\,dt \tag{4}$$

For any given function $f(t)$, the real part σ of the complex variable s must be large enough for the integral to converge. This means that σ can be different for different $f(t)$ functions.

This relationship defines an integral transform that is a function of the complex variable s called the *Laplace transform* of $f(t)$. It is convenient to use a symbolic notation to identify this transform. We write:

$$F(s) = \mathscr{L}[f(t)] = \int_0^\infty f(t)e^{-st}\,dt \tag{5}$$

The symbol for this transform is a script ell. We read the equation as "$F(s)$ is the Laplace transform of $f(t)$."[2]

[2] This is named for French mathematician Pierre Simon Laplace (1749–1827), who ranks next to Newton in importance of scientific contributions, primarily in astronomy, celestial mechanics, and mathematics. By the age of 18 Laplace had mastered applied mathematics. He remained active in research throughout his life and published prodigiously. But he also published popular works (without mathematics) for the general public. The last volume of his five-volume work on celestial mechanics was published when he was 76.

THE INVERSION INTEGRAL The counterpart of the inverse Fourier transform in (2b) is obtained by replacing $j\omega$ in that expression by s and noting that σ is a constant in $s = \sigma + j\omega$; the only restriction on σ is that it be large enough for convergence. Hence, $ds = j\,d\omega$. With those changes in (2b), the result is:

$$f(t) = \mathcal{L}^{-1}[F(s)] = \frac{1}{2\pi j}\int_{\sigma-j\infty}^{\sigma+j\infty} F(s)e^{st}\,ds \tag{6}$$

This is called the *inverse Laplace transform*; the integral on the right is the *inversion integral.*

Note that, even though the direct Laplace transform is "one-sided"—it extends in only one direction in time—nevertheless, its "spectrum" function $F(s)$ is two-sided, extending over both positive and negative frequencies.[3] Whereas carrying out the Laplace integration in the time domain is relatively straightforward, evaluating the inverse transform in the complex domain requires knowledge of mathematics beyond what is assumed in this book. We will discover an alternative, easier way of finding the inverse transform of functions relevant to linear lumped circuits with constant parameters.

A function of time and its Laplace transform constitute a transform pair. Knowing either of them uniquely fixes the other through the Laplace integral or its converse, the inversion integral. We indicate this mutual relationship symbolically by a double-sided arrow:

$$f(t) \leftrightarrow F(s)$$

This notation states that the Laplace transform of a function $f(t)$ is the function $F(s)$ whose inverse transform, in turn, is the same function $f(t)$. This expression stands for the combination of (5) and (6).

Review and Preview

We have now established the following. Given a function of time $f(t)$, it can be transformed into a function of the complex variable s. We still need to determine the transforms of such functions of time as are likely to be excitations applied to circuits, such as pulses, steps, sinusoids, ramps, exponentials, and combinations of such functions.

Our general plan in the solution of circuits of arbitrary complexity, whether initially relaxed or not, when excited by arbitrary functions of time is the following:

1. Determine the Laplace transforms of certain common functions of time. Place these in a table of transform pairs for future reference. We assume

[3] It is possible to define a *two-sided* Laplace transform also. However, since we are interested in functions of time which start after the occurrence of some event—such as the activation of a switch—which is taken as the origin of time, the one-sided transform is more meaningful for us.

that an arbitrary function can be expressed as a combination of such common functions.

2. Study the properties of Laplace transforms, both for the insights such study may provide and for help in carrying out the previous step.

3. Carry out circuit analysis in the complex s domain to arrive at the Laplace transform of any desired circuit voltage or current.

4. Determine the actual currents and voltages from their transforms using the table of transform pairs.

This appears to be an excellent plan—with one possible drawback. While some of you might want to get to the circuit solutions in steps 3 and 4 quickly, we have to spend some considerable effort in the preliminaries in steps 1 and 2 before we have the needed competence to tackle 3 and 4. Grit your teeth and bear it—if you must. Better yet, dig in and enjoy it—for it can be fun.

2 EVALUATING LAPLACE TRANSFORMS

In accordance with the plan just described, before studying the use of Laplace transforms in the solution of network problems, we need to determine the transforms of some simple functions of time which might be the excitations applied to networks. It would be useful also to examine some of the properties of transforms.

When seeking to evaluate Laplace transforms, a question arises how to interpret the point $t = 0$ in the lower limit of the Laplace integral: $t = 0+$, or $t = 0-$? Clearly, it will make no difference if the integrand is continuous. No problem results even if there is a step function at the origin. Integration is a smoothing operation; the integral of a discontinuous function is continuous. Hence, in the case of a step function also, both possible lower limits will yield the same value.

Note that:

$$\int_{0-}^{\infty} f(t)e^{-st}\,dt = \int_{0-}^{0+} f(t)e^{-st}\,dt + \int_{0+}^{\infty} f(t)e^{-st}\,dt$$

The only difference between the two choices of lower limit is the integral from $0-$ to $0+$. This integral vanishes if $f(t)$ is an ordinary point function, with or without discontinuities. The only nonzero contribution to this integral will come from a possible impulse (or an even weirder "function") in $f(t)$. Thus, if $f(t)$ includes an impulse $K\delta(t)$, then, from the sampling property of an impulse, its contribution to the integral will be $Ku(t)$. It seems that if we want to handle the possibility of an impulse at $t = 0$, we should choose $t = 0-$ as the lower limit. However, a method for handling such an impulse that permits the opposite choice will be discussed in the next section.

Laplace Transforms of Some Specific Functions

We now embark on a project to determine and tabulate the Laplace transforms of a set of basic functions that are likely to be excitations of linear networks. This is a straightforward task of applying the basic definition of the Laplace transform in (5) to each such function. Besides yielding the required transforms, this project will provide some practice in evaluating the Laplace integral.

THE IMPULSE The first "function" to be considered is a *displaced* unit impulse $\delta(t - t_1)$, translated t_1 seconds on the positive t axis:

$$\mathscr{L}[\delta(t - t_1)] = \int_0^\infty \delta(t - t_1)e^{-st}\,dt = e^{-st_1}u(t - t_1) \tag{7}$$

The final result follows from the sampling property of the impulse. Thus, the Laplace transform of an impulse displaced by t_1 is an exponential function of complex frequency s, starting at $t = t_1$.

Suppose $t_1 \to 0$ in the preceding result. The impulse is now located at the origin, and the exponential reduces to 1 for $t > 0$. Hence:

$$\mathscr{L}[\delta(t)] = u(t) \tag{8}$$

We see that the presence of an impulse at the origin can be handled by taking it to be a displaced impulse and then reducing the displacement to zero. We will henceforth assume that by $t = 0$ we mean $0+$.

THE EXPONENTIAL FUNCTION Take as the next example the exponential function of time e^{xt}, where x may be real (positive or negative) or complex. The Laplace transform is found by inserting this function as $f(t)$ in (5):

$$\mathscr{L}(e^{xt}) = \int_0^\infty e^{xt}e^{-st}\,dt = \left.\frac{e^{-(s-x)t}}{s - x}\right|_0^\infty \tag{9}$$

If the upper limit is to lead to a finite result, then σ (the real part of s) should be greater than x (or its real part, if x is complex). Hence:

$$\mathscr{L}(e^{xt}) = \frac{1}{s - x} \qquad \text{Re}\,(s) \geqslant \text{Re}(x) \tag{10}$$

Even an *increasing* exponential has a Laplace transform; it is necessary only to choose a large enough convergence factor σ.

The exponential is a transcendental function, whereas its Laplace transform, $1/(s - x)$, is an algebraic function, a member of the class of rational functions.

Thus, the Laplace transform has converted a function high in the hierarchy of functions to one which is lower down. As we go along, you should observe how this is a general property of the Laplace transform.

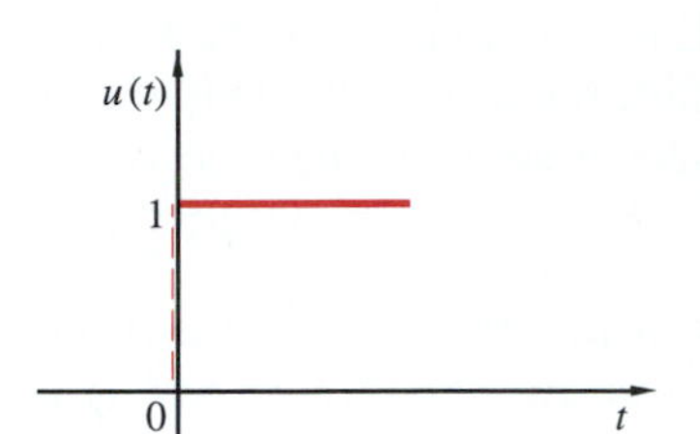

(a)

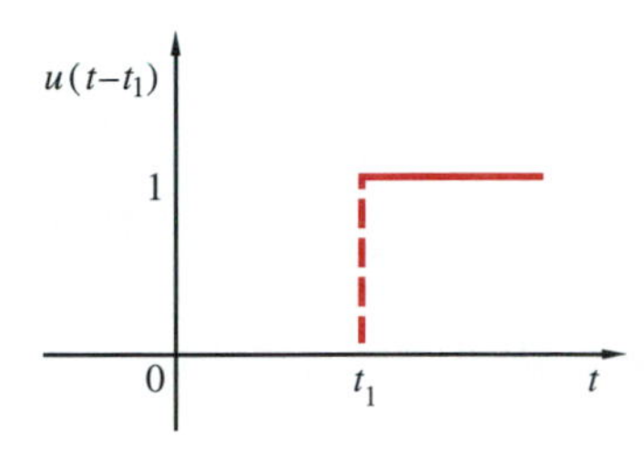

(b)

Figure 1
Unit step (a) at $t = 0$ and (b) shifted to t_1.

THE STEP FUNCTION The Laplace transform of several functions related to the exponential can be obtained directly from the Laplace transform of the exponential without the need to evaluate the corresponding Laplace integrals.

Take the unit step in Figure 1a. It can be regarded as an exponential function e^{xt}, defined for positive values of t, in which x goes to zero. Symbolically:

$$\mathscr{L}[u(t)] = \mathscr{L}\left[\lim_{x\to 0} e^{xt}\right] \tag{11}$$

Suppose we let x go to zero in the Laplace transform of the exponential function in (10). The result will be:

$$\lim_{x\to 0} \mathscr{L}(e^{xt}) = \lim_{x\to 0} \frac{1}{s - x} = \frac{1}{s} \tag{12}$$

There is a problem here. The question is whether finding the Laplace transform of the exponential and *then* taking the limit, as in this expression, leads to the same result as *first* taking the limit of e^{xt} as t goes to zero and *then* taking the Laplace transform, as in the previous equation. A similar question arose in Chapter 13 when we wanted to know whether a Fourier series could be integrated term by term—that is, whether we had to add the terms of the series *first* and *then* integrate or could integrate each term first and then add the terms. The answer is again the same; it involves the uniform convergence of the Laplace integral in (5). That is, it is a theorem of mathematics that the two operations can be interchanged if the integral is *uniformly* convergent. In the present case, this condition is known to be satisfied whenever the integral converges at all, namely, for Re $(s) \geqslant$ Re (x). Hence, the two operations can be interchanged. The conclusion is:

$$\mathscr{L}[u(t)] = \frac{1}{s} \tag{13}$$

You should confirm this result by direct evaluation of the Laplace integral of the unit step.[4]

The "step" in the step function in Figure 1a occurs at $t = 0$. The Laplace transform of a displaced step $u(t - t_1)$, can be found easily by integration of the Laplace integral. Since the range from 0 to t_1 contributes nothing:

$$\mathscr{L}[u(t - t_1)] = \int_{t_1}^{\infty} (1)e^{-st}\, dt = -\frac{1}{s}e^{-st}\Big|_{t_1}^{\infty} = \frac{1}{s}e^{-t_1 s} \tag{14}$$

[4] The interchange of two such mathematical operations is not always justified, and we will have occasion to discuss a case later where it isn't. That's why it's necessary to devote as much attention to this subject as we did.

Again, in order for the integral to converge, the real part of s must be sufficiently large; in this case, $\sigma > 0$ is enough.

We find that the transform of the unit step shifted t_1 units of time to the right is the same as that of the undisplaced step multiplied by $e^{-t_1 s}$. This result is of considerable significance and will be generalized soon, to include other functions displaced along the time axis.

THE SINE AND COSINE FUNCTIONS Let us next find the transforms of the trigonometric functions $\sin \omega t$ and $\cos \omega t$, where t is the variable and ω is a constant. Each is written as a linear combination of two exponentials, as follows:

$$\sin \omega t = \frac{e^{j\omega t} - e^{-j\omega t}}{2j} \tag{15a}$$

$$\cos \omega t = \frac{e^{j\omega t} + e^{-j\omega t}}{2} \tag{15b}$$

For the sine function, the Laplace transform can be written:

$$\begin{aligned} \mathscr{L}(\sin \omega t) &= \mathscr{L}\left(\frac{e^{j\omega t} - e^{-j\omega t}}{2j}\right) = \mathscr{L}\left(\frac{e^{j\omega t}}{2j}\right) - \mathscr{L}\left(\frac{e^{-j\omega t}}{2j}\right) \\ &= \frac{1}{2j}\left(\frac{1}{s - j\omega} - \frac{1}{s + j\omega}\right) = \frac{\omega}{s^2 + \omega^2} \end{aligned} \tag{16}$$

The Laplace transform of the difference of two functions was written as the difference of the transforms. This is OK since integration (taking the Laplace transform) is commutative with subtraction (or addition). In the second line, we used (9) for the transform of an exponential, first with $j\omega$, then $-j\omega$, replacing x.

The Laplace transform of the cosine function can be found in the same way. The details will be left for you to carry out. The result will be:

$$\mathscr{L}(\cos \omega t) = \frac{s}{s^2 + \omega^2} \tag{17}$$

Note that, for the transforms of both the sine and the cosine functions, the "x" in the exponent is the purely imaginary $\pm j\omega$. Hence, the convergence condition $\operatorname{Re} s > \operatorname{Re} x$ requires only that $\operatorname{Re} s > 0$.

EXERCISE 1

Since the hyperbolic sine and cosine functions can also be expressed in terms of exponentials, use the Laplace transform of exponentials to find those of the hyperbolic sine and cosine. Confirm the results by evaluating the Laplace integral.

ANSWER: See Table 1. □

Table 1
Laplace transform pairs

$f(t)$	$F(s) = \mathscr{L}[f(t)]$	$f(t)$	$F(s) = \mathscr{L}[f(t)]$
$\delta(t)$	1	$t^n e^{-at}$	$\dfrac{n!}{(s+a)^{n+1}}$
$u(t)$	$\dfrac{1}{s}$	$\sin \omega t$	$\dfrac{\omega}{(s^2+\omega^2)}$
t	$\dfrac{1}{s^2}$	$\cos \omega t$	$\dfrac{s}{(s^2+\omega^2)}$
t^n	$\dfrac{n!}{s^{n+1}}$	$e^{-at}\sin \omega t$	$\dfrac{\omega}{[(s+a)^2+\omega^2]}$
e^{-at}	$\dfrac{1}{(s+a)}$	$e^{-at}\cos \omega t$	$\dfrac{(s+a)}{[(s+a)^2+\omega^2]}$

DAMPED SINUSOIDS Damped sinusoids constitute a class of functions that are very significant in engineering. In Chapter 7 we found them to be part of the natural response whenever the natural frequencies are complex. These are the functions:

$$e^{-at}\cos \omega_1 t \qquad \text{and} \qquad e^{-at}\sin \omega_1 t$$

(We have placed a subscript on ω to emphasize that this is some particular constant frequency.) To find the Laplace transform of such functions, we note that the sine and cosine can both be written in terms of exponentials, as in (15). The real and imaginary exponentials can then be combined.

The Laplace transform of the damped cosine function is:

$$\begin{aligned}
\mathscr{L}(e^{-at}\cos \omega_1 t) &= \mathscr{L}\left(\frac{e^{-(a-j\omega_1)t} + e^{-(a+j\omega_1)t}}{2}\right) \\
&= \frac{1}{2}\left(\frac{1}{s+a-j\omega_1} + \frac{1}{s+a+j\omega_1}\right) \\
&= \frac{s+a}{(s+a)^2+\omega_1^2} = \frac{s+a}{s^2+2as+a^2+\omega_1^2}
\end{aligned} \tag{18}$$

This is a rational function with two complex poles at $s = -a \pm j\omega_1$. The convergence condition on s is more relaxed than it is for the cosine alone, namely, $\operatorname{Re} s > -a$.

EXERCISE 2

The Laplace transform of the damped sine function can be obtained in the same way. Carry out the details to show that:

$$\mathscr{L}(e^{-at}\sin \omega_1 t) = \frac{\omega_1}{(s+a)^2+\omega_1^2} = \frac{\omega_1}{s^2+2as+a^2+\omega_1^2} \tag{19}$$

□

POWERS OF t Another function introduced in Chapter 6 was the ramp function $f(t) = tu(t)$. The Laplace transform of this function can be found from the Laplace integral using integration by parts. Thus:

$$\mathscr{L}(t) = \int_0^\infty te^{-st}\,dt = -\frac{t}{s}e^{-st}\Big|_0^\infty + \frac{1}{s}\int_0^\infty e^{-st}\,dt \tag{20}$$

(Verify this.) At the lower limit, the first term on the right disappears. At the upper limit, one factor (t) of the product te^{-st} becomes infinite; the other becomes zero, assuming that the real part of s is positive. Application of l'Hôpital's rule, however, shows that the limit approached by xe^{-x} is zero as x approaches infinity as long as its real part is positive. Thus the first term on the right in the preceding equation disappears for appropriate values of s. The integration in the second term on the right is routine; the final result for the Laplace transform of t is:

$$\mathscr{L}(t) = \frac{1}{s^2} \tag{21}$$

The Laplace transform of higher powers of t can also be found by integrating by parts. Carry out the details for $f(t) = t^2$. For t^n the result is:

$$\mathscr{L}(t^n) = \frac{n!}{s^{n+1}} \tag{22}$$

In a later section we will take a different approach to arrive at these same results.

Transform-Pair Tables

The functions and their Laplace transforms that were just determined are collected in Table 1. The first column gives a function of time, the second column its Laplace transform. Notice that all but one of the transforms determined so far are rational functions of the variable s, that is, ratios of two polynomials.

Does the Laplace transform exist for all possible functions $f(t)$? A function may approach a constant as time increases (like a step function) or oscillate (like the sine and cosine), or even increase with time (like an exponential with a positive real exponent), and still have a Laplace transform. We have seen from examples that the value of s for which the Laplace integral will converge can vary for different functions. The exponential e^{-st} acts as a convergence factor. But there may be functions that increase faster with time than such an exponential decrease, so that e^{-st} becomes ineffective as a convergence factor. Such functions are few and far between.

One such function which is not Laplace-transformable is an exponential with a power of t greater than 1 in the exponent. Such functions, however, though mathematically interesting, are of little engineering interest. The linear models we have been using are valid for "small" signals. If network variables become too large, then the linear model will no longer be valid. Furthermore, very large voltages and currents will cause components to be destroyed, and so

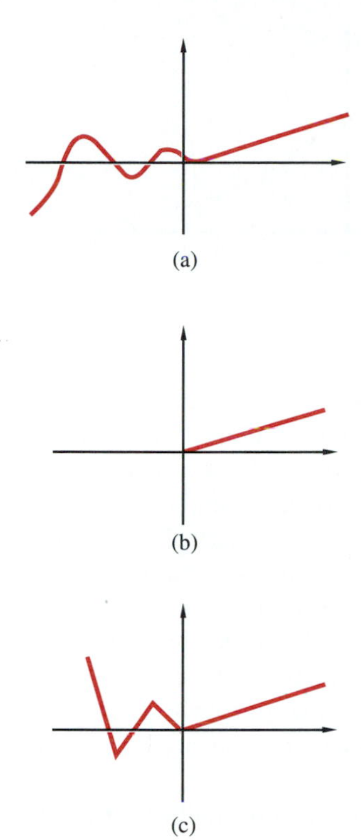

Figure 2
Three time functions with the same Laplace transform.

we have no engineering interest in functions which are not Laplace-transformable.

In more advanced studies in Laplace transform theory, it is shown that a function is transformable if it is (1) *piecewise continuous* and (2) of *exponential order.* A piecewise continuous function is continuous except possibly for a finite number of discontinuities. Being of exponential order means that the function should rise no faster for large t than an exponential (with a linear exponent). The Laplace integral will converge for any such function.

Behavior at the Origin of Time

Since the range of integration of the Laplace integral starts at $t = 0$, the Laplace transform is quite indifferent to the behavior of a function of time for negative t. Thus, all the functions in Figure 2 have the same Laplace transform, even though their behavior is radically different for negative t.

Since negative values of t are without interest, we shall henceforth assume that all functions we deal with are zero for negative t. To indicate this assumption analytically, we could multiply a function $f(t)$ by a unit step $u(t)$. Thus, $f(t)u(t)$ is simply $f(t)$ for positive values of t, since $u(t)$ is just 1 there; and it is 0 for negative values of t, since $u(t)$ is 0 there.

This approach causes any function which is not zero at the origin to have a discontinuity there. Thus, the function $\cos \omega t$ has the value 1 at $t = 0$ whether the origin is approached from negative values or positive values. However, $u(t) \cos \omega t$ is identically zero for negative t and jumps to the value 1 at $t = 0$. As noted earlier, the choice of $0-$ or $0+$ for the point $t = 0$ will make no difference for a function with a discontinuity at $t = 0$. We have chosen the simpler notation (0) to stand for $(0+)$, as previously noted.

3 SIGNIFICANT GENERAL PROPERTIES

It is possible to continue evaluating the Laplace integral for more and more functions of time to build up a table of transform pairs. However, we shall now turn to an examination of a number of important general properties of Laplace transforms. These can provide insight into the relationships between operations performed on a function in the time domain and the corresponding operation in the complex s domain. As a collateral benefit, there is the possibility that by applying general properties to certain specific cases, additional transform pairs can be generated.

Real Shifting Theorem

One of the functions whose Laplace transform was determined in the preceding section was the displaced unit step. Indeed, this shifting of other functions (besides the step) along the time axis is a common occurrence. It would be useful, therefore, if a simple relationship could be found between the transform of a time function and that of the same function displaced by a certain value of time along the time axis. This is now the subject of discussion.

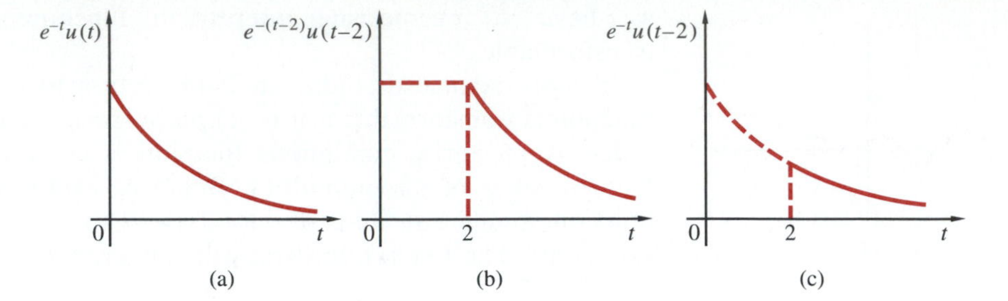

Figure 3
Shifted function of time:
(a) $f(t)u(t)$,
(b) $f(t - t_1)u(t - t_1)$,
(c) $f(t)u(t - t_1)$.

An arbitrary function of time which is identically zero for negative time by virtue of the multiplier $u(t)$ is exemplified by the specific function in Figure 3a. In Figure 3b the same function is shown shifted t_1 units of time to the right. Equivalently, we can think of the origin as being moved t_1 units to the left. This amounts to a change of variable from t to $t - t_1$. For example, the value of the function at $t = 0$ before is the same as the value at $t = t_1$ now.

Observe that this is quite different from the function $f(t)u(t - t_1)$, shown in Figure 3c. The function in this figure is *not* shifted; it's the same $f(t)$ except that the multiplier $u(t - t_1)$ causes the product to vanish for all values of t up to $t = t_1$. The function $f(t)$ remains fixed, but we slide the unit step to the right, thereby annihilating everything in its path up to the occurrence of the step.

Let's now determine the Laplace transform of the shifted function in Figure 3b by evaluating the Laplace integral:

$$\mathscr{L}[f(t - t_1)u(t - t_1)] = \int_0^\infty f(t - t_1)u(t - t_1)e^{-st}\,dt$$

$$= \int_{t_1}^\infty f(t - t_1)e^{-st}\,dt$$

Since the integrand is zero up to $t = t_1$, the lower limit can be moved up to t_1, as in the last step. Now make a change of variable $t - t_1 = x$; then $t = x + t_1$ and $dt = dx$, so that when $t = t_1$, $x = 0$. Hence, $e^{-st} = e^{-(x+t_1)} = e^{-st_1}e^{-sx}$. Using these, the preceding expression becomes:

$$\mathscr{L}[f(t - t_1)u(t - t_1)] = e^{-st_1}\int_0^\infty f(x)e^{-sx}\,dx = e^{-st_1}F(s) \quad (23)$$

Note that the integral in the middle expression is the Laplace integral with x as a dummy variable.

The result expressed by this equation is called the *time-shift theorem* or the *real shifting theorem* (*real* in contrast with *complex,* which will be discussed later). The theorem states that the effect of shifting a function t_1 units on the time axis is to multiply its Laplace transform by e^{-st_1}. It has many useful applications. You should note that the transform of the shifted unit step in (14) is a special case of the general case treated here.

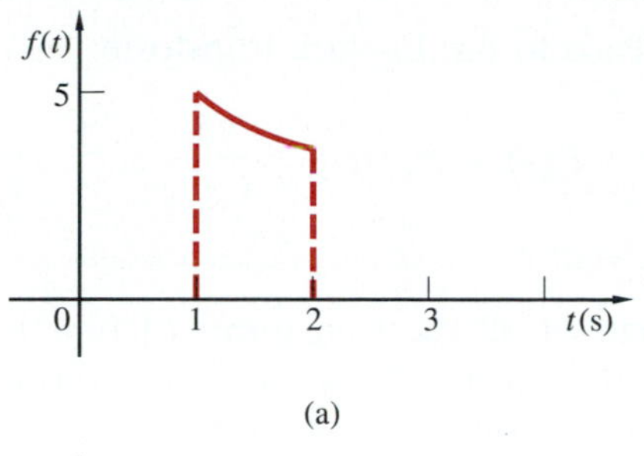

(a)

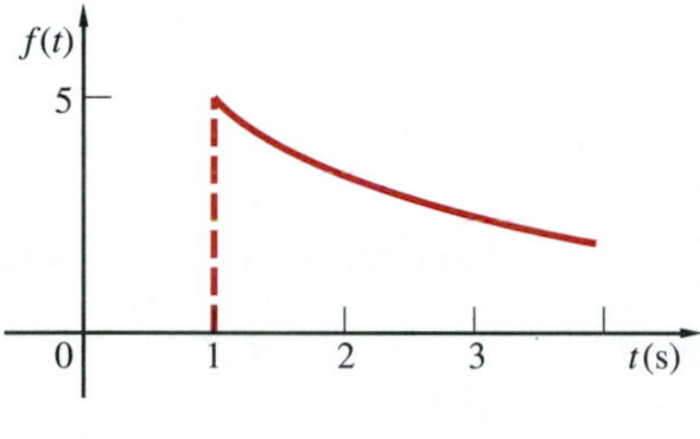

(b)

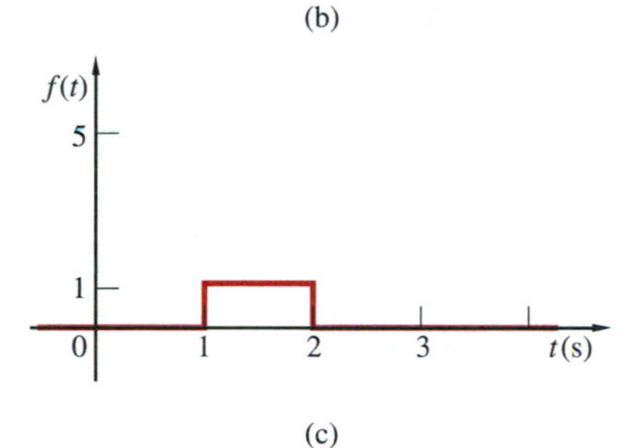

(c)

Figure 4
A shaped pulse (a) formed by the product of a shifted function (b) and a gate function (c).

EXAMPLE 1 Let's use the shifting theorem to determine the Laplace transform of the pulse with an exponential top shown in Figure 4a. The pulse starts at $t = 1$ s and ends at $t = 2$ s. This function can be written as the product of the shifted exponential function $5e^{-(t-1)}$ in Figure 4b and the rectangular pulse in Figure 4c:

$$f(t) = 5e^{-(t-1)}[u(t-1) - u(t-2)]$$

The rectangular pulse can be imagined as a "gate" or a "window" that blocks any part of the function except the part within the gate. It might thus be called a *gate function,* which, in this case, would be labeled $g_{1,2}$ and, more generally, g_{t_1,t_2}.

The shifting theorem can be applied to the first term in the preceding equation, but the second term is not in the right form. This can be corrected, however, by adding and subtracting 1 in the exponent. Thus:

$$f(t) = 5e^{-(t-1)}u(t-1) - 5e^{-1}e^{-(t-2)}u(t-2)$$

Since the transform of e^{-t} is $1/(s+1)$, application of the shifting theorem to this function leads to the Laplace transform:

$$F(s) = \mathscr{L}[f(t)] = \frac{5}{s+1}e^{-s} - \frac{5}{s+1}e^{-(2s+1)}$$

Observe that, of all the transforms of functions treated so far, only those resulting from the shifting theorem are not rational functions.

It would be useful to have a table—just like the table of transform pairs—that collects the general properties of the Laplace transform in one place for easy reference. Table 2 is such a table. Each of the properties, or operations, will be derived and discussed in the sections that follow.

Table 2
Transform properties

Operation	Time function	Transform
Linearity	$K_1 f_1(t) + K_2 f_2(t)$	$K_1 F_1(s) + K_2 F_2(s)$
Time derivative	$\frac{df}{dt}$	$sF(s) - f(0)$
	$\frac{d^2f}{dt^2}$	$s^2F(s) - sf(0) - f'(0)$
Time integral	$\int_0^t f(x)\,dx$	$\frac{1}{s}F(s)$
Time shift	$f(t-a)u(t-a)$	$F(s)e^{-as}$ $\quad a \geqslant 0$
Frequency shift	$f(t)e^{-at}$	$F(s+a)$
Frequency derivative	$-tf(t)$	$\frac{dF(s)}{ds}$
Scaling	$f\left(\frac{t}{a}\right)$	$aF(as)$
Initial value	$f(0)$	$\lim_{s\to\infty} sF(s)$
Final value	$\lim_{t\to\infty} f(t)$	$\lim_{s\to 0} sF(s)$ Poles of $sF(s)$ left half plane only
Periodic function	$f(t) = f(t-nT)$	$\frac{F_a(s)}{1-e^{-Ts}}$ where $F_a(s) = \mathscr{L}[f_a(t)]$ and $f_a(t) = f(t)[u(t) - u(t-T)]$

Scaling (Normalization)

It is sometimes convenient to change the time scale. That is, if the processes of interest are taking place in microsecond or nanosecond intervals, we might want to slow down the time uniformly so that the processes take place in seconds or even minutes. This would be especially useful if laboratory confirmation is needed or an analog computer is to be used, where component response times are more sluggish.

Suppose time is normalized by a factor a; that is, divide t by a. We would like to know what happens to the Laplace transform. To find out, we evaluate the Laplace integral of $f(t/a)$:

$$\mathcal{L}\left[f\left(\frac{t}{a}\right)\right] = \int_0^\infty f\left(\frac{t}{a}\right)e^{-st}\,dt$$

$$= a\int_0^\infty f\left(\frac{t}{a}\right)e^{-ast/a}\,d\left(\frac{t}{a}\right) = aF(as) \tag{24}$$

To obtain the second line, we multiply and divide by a in the exponent and under the integral sign. The resulting combination $a\,dt/a$ can be written as $a\,d(t/a)$. If you think of t/a as a new time variable t', then the integral looks just like the Laplace integral, with t' instead of t, *except* that s is replaced by as. Hence, the final result follows.

The factor a on the right can be put on either side of the equation, so the preceding result can be written in two forms:

$$\mathcal{L}\left[f\left(\frac{t}{a}\right)\right] = aF(as)$$

$$\frac{1}{a}\mathcal{L}\left[f\left(\frac{t}{a}\right)\right] = F(as) \tag{25}$$

It seems that, if the time scale is changed by dividing t by a, the Laplace transform of the function with the new time scale is obtained from the old Laplace transform by making the inverse change in the scale of s and then multiplying by a. Thus, if a is greater than 1, the time scale is expanded. A particular point on the t scale represents a smaller value on the t' scale. That means that more t' is needed to reach the same value of time as t. The scale is compressed if $a < 1$. If the time scale is compressed, the s scale is expanded, and vice versa. Not only that, but the multiplier of the tranform function is correspondingly changed.

EXAMPLE 2 The time constant in an RC circuit is 55 μs. The natural response will be $Ke^{-t/0.000055}$. Take $a = 0.000055$; then $t' = t/0.000055$. The Laplace transform of $Ke^{-t'}$ is just $K/(s + 1)$. Hence, using the preceding scale-change result, we find that:

$$\mathcal{L}(Ke^{-t/0.000055}) = (0.000055)\frac{K}{(0.000055)s + 1} = \frac{K}{s + 1/0.000055}$$

This example is so simple that no benefit of using the theorem is evident. Obviously, the benefit would be more evident with more complicated examples.

Indeed, the theorem may be of greater benefit when going from the s to the time domain. Although we referred to the preceding process as a change in time scale, the frequency scale is also changed, but in the opposite direction.

As mentioned earlier, the overall objective is to arrive at the Laplace transform of a circuit response, by methods being developed here. From this we must obtain the time response itself. The function of s may be of some numerical complexity. If we could carry out a normalization in frequency (that is, in s), the complexity of this function might be reduced. (When starting from the s domain like this, the second form in (25) would be the more appropriate to use.) The inverse transform obtained from this simplified function will be in normalized time. To arrive at $f(t)$ from $(1/a)f(t/a)$, we multiply both t and the function by a.

Laplace Transforms of Derivatives and Integrals

The equations resulting from an application of the fundamental circuit laws are generally differential equations. It is possible also that they are integrodifferential equations, depending on the order in which Kirchhoff's laws and the component relationships are applied (see Chapter 7). If the Laplace transform is to be used to solve circuit equations, we will need to know the transforms of the derivative and of the integral of a function.

TRANSFORM OF A DERIVATIVE Let's first consider the derivative of a function of time $f(t)$. (It might be a voltage or a current.) Its derivative is $f'(t) = df/dt$. Assume that $f(t)$ and its derivative are both Laplace-transformable functions. The Laplace transform of the derivative, by definition, is:

$$\mathscr{L}\left(\frac{df}{dt}\right) = \int_0^\infty \frac{df}{dt} e^{-st}\,dt \tag{26}$$

To evaluate the integral, let's integrate by parts. Define:

$$\begin{aligned} u &= e^{-st} & dv &= \frac{df}{dt}dt \\ du &= -se^{-st}\,dt & v &= f(t) \end{aligned} \tag{27}$$

Then insert in the formula:

$$\int_0^\infty u\,dv = uv\big|_0^\infty - \int_0^\infty v\,du \tag{28}$$

The result will be:

$$\mathscr{L}\left(\frac{df}{dt}\right) = f(t)e^{-st}\big|_0^\infty + s\int_0^\infty f(t)e^{-st}\,dt \tag{29}$$

At the lower limit, the first term reduces to $-f(0)$. At the upper limit, there must be some value of s which will cause the first term to vanish as $t \to \infty$, since $f(t)$ is assumed to be transformable. Hence, the first term vanishes at the upper limit. The integral in the last term is, by definition, $F(s)$. The final result is:

$$\mathcal{L}\left(\frac{df}{dt}\right) = \int_0^\infty \frac{df}{dt} e^{-st}\, dt = sF(s) - f(0) \tag{30}$$

(Remember that $f(0)$ is the value of $f(t)$ just after $t = 0$.)

This is the desired expression. Aside from the constant $f(0)$, which is the initial value of the function $f(t)$, differentiation in time has been replaced by multiplication by s in the s domain.

The Laplace transform of higher-order derivatives can be found by repeated application of this formula. For the second derivative, for example:

$$\begin{aligned}\mathcal{L}[f''(t)] &= s\mathcal{L}[f'(t)] - f'(0) = s[sF(s) - f(0)] - f'(0) \\ &= s^2F(s) - sf(0) - f'(0)\end{aligned} \tag{31}$$

where $f'(0)$ is the value of the derivative at $t = 0$.

EXERCISE 3

Find an expression for the Laplace transform of the third derivative of a function $f(t)$. □

EXERCISE 4

From the transform of cos ωt, confirm the expression for the Laplace transform of sin ωt, using the fact that a sine function is proportional to the derivative of a cosine function. □

EXERCISE 5

Confirm the Laplace transform of $f(t) = t^2$ knowing the transform of t^3. □

TRANSFORM OF AN INTEGRAL Now let's turn to the definite integral of a function $f(t)$:

$$g(t) = \int_0^\infty f(x)\, dx \tag{32}$$

If g is the integral of f, then f must be the derivative of g. Thus:

$$f(t) = \frac{dg}{dt} \tag{33}$$

But now the previous result about the transform of a derivative can be used. Assuming that $g(t)$ and its derivative (that is to say, $f(t)$ and its integral) are transformable, we can use (30) to get:

$$\mathcal{L}[f(t)] = \mathcal{L}\left(\frac{dg}{dt}\right) = sG(s) - g(0) \tag{34}$$

where $G(s) = \mathcal{L}[g(t)]$ is the transform of the definite integral we are seeking. To find $g(0)$, put $t = 0$ for the upper limit in the definite integral in (32); this results in $g(0) = 0$ and leads to:

$$\mathscr{L}\left[\int_0^\infty f(x)\,dx\right] = \frac{1}{s}F(s) \tag{35}$$

This is again a gratifyingly simple result. Integration in the time domain has been converted into division by s in the s domain. Just as is true for derivatives, the result can be extended to higher-order integrals.

EXERCISE 6
Find the transform of the second-order integral of $f(t)$. □

EXERCISE 7
Confirm the transform of cos ωt from that of sin ωt. □

EXERCISE 8
Confirm $\mathscr{L}(t^n)$ in (22) knowing that $\mathscr{L}(t^{n-1}) = (n-1)!/s^n$. □

One of our ultimate objectives is the solving of differential (or integro-differential) equations using Laplace transforms. The theorems discussed in this section, relating the transform of the derivative and integral of a function to the transform of the function itself, can be directly applied toward this goal, as we shall more fully explore in a later section. But we have also observed here that these theorems can be used to find the Laplace transforms of some functions if those of their derivatives or integrals are known. In other words, to find a Laplace transform pair, it isn't always necessary to evaluate the Laplace integral; it may be possible to find the transform by using general properties of Laplace transforms or theorems related to them. That's one of the reasons (besides intellectual curiosity) why we will be taking up other such general properties or theorems.

Initial-Value and Final-Value Theorems

The Laplace and inversion integrals provide an association between a given function in the time domain and some particular function in the complex s domain. To get the function of s, we must evaluate an integral involving the time function over all positive values of t. Hence, we cannot, in general, pick out some point in time and say that this is related to some particular point or region in the complex s plane. Although such a point-to-point correspondence between the two domains does not exist *in general,* nevertheless, there are two *specific* points in time that have some kind of correspondence with two specific points in the complex plane.

INITIAL-VALUE THEOREM Glance back at (30), which gives the Laplace transform of the derivative of a function. Since the derivative is assumed to be transformable, there will be some positive value of the real part of s for which $e^{-st}\,df/dt$ will approach zero as t goes to infinity. For any larger positive value of Re (s), the same result will be true a fortiori; in particular, for $s \to \infty$. Let's take the limit of both sides of (30) as $s \to \infty$:

$$\lim_{s\to\infty} \int_0^\infty \frac{df}{dt} e^{-st}\, dt = \lim_{s\to\infty} [sF(s) - f(0)] \tag{36}$$

On the left side, the explicit limit on s and the limit on t implied by the integral can be interchanged, because the integral is uniformly convergent. When this is done, the limit of the integrand will vanish as $s \to \infty$, as just noted. Hence, the left side approaches 0. The final result will be

$$\lim_{s\to\infty} sF(s) = f(0) \tag{37}$$

This result is called the *initial-value theorem.* It expresses the initial value of a function of time in terms of the limit approached by the Laplace transform of the function (multiplied by s), as s approaches infinity. Hence, if the transform of a function is known, the initial-value theorem permits finding the initial value of the function without actually finding the inverse transform. Furthermore, the theorem can be applied repeatedly, assuming that higher derivatives are Laplace-transformable, to find the initial slope of a function and initial values of higher derivatives.

EXAMPLE 3 To illustrate, let's find both the initial value and the initial slope of a function whose transform is:

$$F(s) = \frac{(s+2)(s+4)}{(s+1)(s+3)(s+5)}$$

From the initial-value theorem, we get:

$$f(0) = \lim_{s\to\infty} sF(s) = \lim_{s\to\infty} \frac{s(s+2)(s+4)}{(s+1)(s+3)(s+5)} = 1$$

When going to the limit, the highest powers of s in the numerator and denominator dominate; hence, the other powers can be neglected.

Knowing the initial value of $f(t)$, we can find the Laplace transform of the derivative:

$$\mathcal{L}\left(\frac{df}{dt}\right) = sF(s) - f(0) = \frac{s(s+2)(s+4)}{(s+1)(s+3)(s+5)} - 1$$

$$= \frac{-(3s^2 + 15s + 15)}{(s+1)(s+3)(s+5)}$$

Again applying the initial-value theorem, this time to the preceding function:

$$\frac{df}{dt}(0) = \lim_{s\to\infty} s\mathcal{L}\left(\frac{df}{dt}\right) = \lim_{s\to\infty} \frac{-s(3s^2 + 15s + 15)}{(s+1)(s+3)(s+5)} = -3$$

Thus, we find that the function $f(t)$ starts out at a positive value of 1 but with a negative slope of -3.

EXERCISE 9

Find the initial value and the initial slope of a function $f(t)$ whose Laplace transform is:

$$F(s) = \frac{2(s+1)}{s^2 + 2s + 2}$$

ANSWER: $f(0) = 2$; $f'(0) = -2$. □

As noted in Example 3, the highest powers of s in the numerator and denominator of $F(s)$ predominate when s goes to infinity. Hence, the initial value $f(0)$ will be zero whenever the highest power of s in the denominator is 2 or more higher than that in the numerator. Furthermore, if the two highest powers are the same, then $sF(s) \to \infty$ when s goes to infinity. (The implications of this will be explored subsequently.) Thus, a nonzero, noninfinite value of $sF(s)$ in the limit as s goes to infinity will occur only when the highest power of s in the denominator is exactly 1 more than the highest power in the numerator—that is, if $F(s)$ has a simple zero at infinity.

FINAL-VALUE THEOREM The initial-value theorem relates zero time and infinite s. Curiosity would naturally suggest that we seek the converse relationship, if one exists, relating infinite time and zero s. Look again at the expression for the Laplace transform of the derivative of a function given in (30). This time we seek the limit as s approaches 0. Thus:

$$\lim_{s\to 0} \int_0^\infty \frac{df}{dt} e^{-st}\, dt = \lim_{s\to 0} [sF(s) - f(0)] \tag{38}$$

Again we contemplate interchanging the integration and the limit as s goes to zero. This interchange will be permitted only if the integral converges for the value of s under consideration, namely, $s = 0$. If $df/dt = e^{5t}$, for example, then s must have a real part greater than 5 if the integral is to converge. In this case, it is clearly not possible to let s go to 0 before integrating. The conclusion is that, in order to interchange the two operations, the smallest value of s for which the Laplace integral of df/dt converges must be less than zero.

With that restriction, we interchange the two operations and allow s to approach zero under the integral. Then:

$$\lim_{s\to 0} \int_0^\infty \frac{df}{dt} e^{-st}\, dt = \int_0^\infty \frac{df}{dt}\, dt = \lim_{t\to\infty} f(t) - f(0) \tag{39}$$

Using this result in (38), the final result becomes:

$$\lim_{t\to\infty} f(t) = \lim_{s\to 0} sF(s) \tag{40}$$

This is the *final-value theorem*. It specifies the limit approached by a function of time $f(t)$ as t goes to infinity in terms of the limit approached by $sF(s)$ for zero s. The conditions of applicability of this theorem are more restrictive than those of the initial-value theorem. It is actually easier to state the condition in terms of $F(s)$ than in terms of $f(t)$. This condition is:

> The final-value theorem applies to a function $f(t)$ only if its Laplace transform $F(s)$ has no poles in the right half plane or on the $j\omega$ axis, except for a possible simple pole at $s = 0$.

(That last exception comes about because a possible factor s in the denominator corresponding to such a pole is canceled when $F(s)$ is multiplied by s.) Confirm for yourself that the s-domain condition on the poles of $F(s)$ is equivalent to the convergence of the Laplace integral in (38).

The final-value theorem cannot be applied to such simple functions as sinusoids or to increasing exponentials:

$$\mathcal{L}(\sin \omega t) = \frac{\omega}{s^2 + \omega^2}$$

$$\mathcal{L}(e^{at}) = \frac{1}{s - a} \qquad a > 0$$

The first has poles on the imaginary axis, the second in the right half plane.

EXERCISE 10

A function of time and its Laplace transform are given as follows:

$$f(t) = (1 - 5e^{-2t} + 2e^{-4t})u(t)$$

$$F(s) = \mathcal{L}f(t) = \frac{-2s^2 + 6s + 8}{s(s + 2)(s + 4)}$$

Confirm that the validity condition of the final-value theorem is satisfied; then verify the theorem.

ANSWER: $f(t)|_{t\to\infty} = 1 = sF(s)|_{s\to 0}$. □

Complex-Domain Theorems

In the preceding section, a number of general properties or theorems were introduced. They were based on operations performed in the time domain, such as the real shifting theorem and derivatives and integrals in the time domain. What happens if some of these operations are performed in the complex s domain instead? We will now explore such matters.

COMPLEX SHIFTING THEOREM Look back at the real shifting theorem in (23). There we found that when a function of time is shifted by a time t_1, its transform is multiplied by an exponential. We might take a wild stab and conjecture by analogy that if a transform function $F(s)$ is shifted by some value of s, then the time function whose transform it is will be multiplied by an exponential involving this value of s. Let's see.

Suppose $\mathscr{L}[f(t)] = F(s)$. What will happen if we shift s by $-a$, where a is a real constant—that is, if we consider $F(s + a)$? Let's see if we can get some information from the inversion integral in (6) if the function to be inverted is $F(s + a)$. There are two other places where s appears in the integral. One is in the exponent; by adding and subtracting a, the exponential can be written $e^{-at+(s+a)t}$. By the law of exponents, this can then be written as the product of two exponentials. The other place that s appears is in the differential ds; but since a is constant, $ds = d(s + a)$. Finally, the limits of integration would have to be changed appropriately if the variable of integration is to be $s + a$. With these changes, the inversion integral becomes:

$$\mathscr{L}^{-1}[F(s + a)] = \frac{1}{2\pi j}\int_{\sigma - j\omega}^{\sigma + j\omega} F(s + a)e^{st}\, ds$$

$$= e^{-at}\left[\frac{1}{2\pi j}\int_{\sigma + a - j\omega}^{\sigma + a + j\omega} F(s + a)e^{(s+a)t}\, d(s + a)\right]$$

Note that the exponential e^{-at} does not depend on s, so it is moved outside the integral. The quantity inside the brackets in the last line is just the inversion integral. (This might be clearer if you make the change of variable $z = s + a$.) The final result is:

$$\mathscr{L}^{-1}[F(s + a)] = e^{-at}\mathscr{L}^{-1}[F(s)] = e^{-at}f(t)$$
$$\mathscr{L}[e^{-at}f(t)] = F(s + a) \tag{41}$$

(Go through the details to confirm this.)

The wild guess turned out to be correct! If a function of time is multiplied by an exponential e^{-at}, then s in its transform is shifted by a. In contrast with the real shifting theorem, this might be called the *complex shifting theorem*. It might also be called the *complex frequency-shift* (or s-shift) theorem, in contrast with the time-shift theorem.

EXAMPLE 4 We earlier found the Laplace transform of $e^{-at}\cos\omega t$ by direct evaluation of the Laplace integral. Instead, we could have used the s-shift theorem. Thus:

$$\mathscr{L}(e^{-at}\cos\omega t) = \mathscr{L}(\cos\omega t)\big|_{s+a} = \left.\frac{s}{s^2 + \omega^2}\right|_{s+a} = \frac{s + a}{(s + a)^2 + \omega^2}$$

It's certainly gratifying that this gives the same result as previously found in (18).

EXERCISE 11

Use the s-shift theorem to find the Laplace transform of $e^{-at}\sin\omega t$ from the Laplace transform of $\sin\omega t$. □

Perhaps a better appreciation of the frequency shift can be obtained by considering the movement of poles and zeros in the complex s-plane resulting from an s-shift. A transform function and the function resulting from a 5-unit frequency shift are given:

Original Function	**Shifted Function**
$F(s) = \dfrac{s}{(s + 20)(s^2 + 100)}$	$F(s + 5) = \dfrac{s + 5}{(s + 25)[(s + 5)^2 + 100]}$
$\mathscr{L}^{-1}[F(s)] = f(t)$	$\mathscr{L}^{-1}[F(s + 5)] = e^{-5t}f(t)$

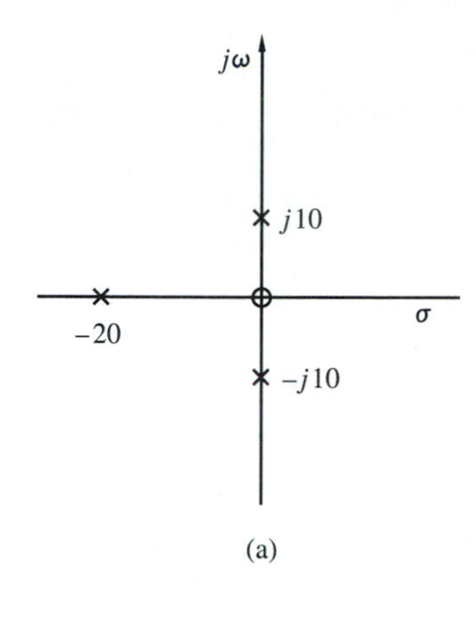

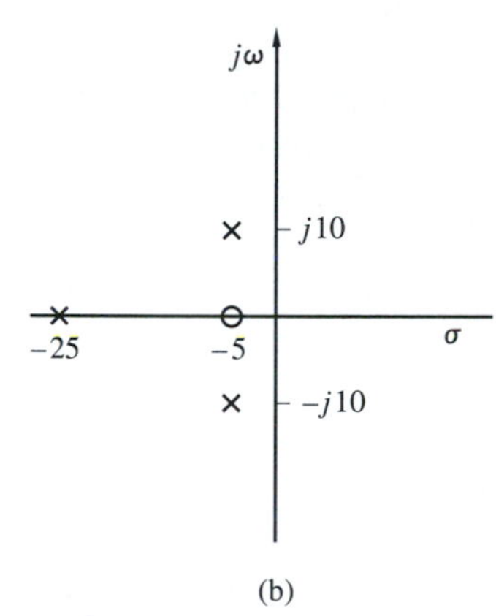

Figure 5
Shift of zeros and poles of the transform in the complex plane when $f(t)$ is multiplied by e^{-5t}.

The poles and zeros of both the original function and the frequency-shifted function are shown in Figure 5. We see that the s-shifting moves all the poles and zeros 5 units horizontally to the left in the complex plane. Study these diagrams carefully.

Every example examined so far involves shifting along the *real* axis in the complex plane. Let's now consider a different variety. Specifically, consider the following amplitude-modulated wave:

$$f_1(t) = f(t)\cos\omega_0 t = \tfrac{1}{2}f(t)e^{j\omega_0 t} + \tfrac{1}{2}f(t)e^{-j\omega_0 t}$$

where $f(t)$ is the modulating function (whose transform is $F(s)$) and ω_0 is the carrier frequency. On the right side, the cosine has been expressed in terms of exponentials. Now take the Laplace transform of $f_1(t)$ and apply the frequency-shift theorem:

$$F_1(s) = \mathscr{L}[f_1(t)] = \tfrac{1}{2}F(s - j\omega_0) + \tfrac{1}{2}F(s + j\omega_0)$$

Remember that $F(s)$ is the spectrum function of the modulating function $f(t)$. Hence, by $F(s - j\omega_0)$, we understand the spectrum of the modulating function shifted vertically upward a distance ω_0 in the complex plane. Similarly, $F(s + j\omega_0)$ is the spectrum of the modulating wave shifted vertically downward a distance ω_0.

EXERCISE 12

Suppose the frequency components of the modulating signal $f(t)$ lies in the audio band up to 10 kHz. The carrier has a frequency of 1 MHz. In the complex plane draw three sets of axes, showing the carrier poles on one, the modulating frequency poles on the other, and the poles of the modulated carrier on the third. Pay particular attention to the scale. □

DERIVATIVE IN THE FREQUENCY DOMAIN In the time domain, a very significant theorem is that $\mathscr{L}(df/dt) = sF(s) - f(0)$ if $\mathscr{L}[f(t)] = F(s)$. Aside from the initial-value term, differentiation in the time domain is transformed into multiplication by s in the frequency domain. On this basis, we might conjecture that differentiation in the s domain will be replaced by multiplication by t in the time domain.

Before we attempt to confirm this conjecture, let's try an example. Let $f(t)$ be an impulse $\delta(t)$ whose transform we know to be $F(s) = 1$. Then the derivative of $F(s)$ is zero. But the inverse transform of 0 is nothing but 0, *not* $t\delta(t)$, and so the conjecture is *not true* for *this* "function" of time.

To determine the reality, let's use the inversion integral on the derivative

$dF(s)/ds$ and then integrate by parts:

$$\mathscr{L}^{-1}\left[\frac{d}{ds}F(s)\right] = \frac{1}{2\pi j}\int_{\sigma-j\infty}^{\sigma+j\infty}\frac{dF(s)}{ds}e^{st}\,ds$$

$$= \frac{1}{2\pi j}F(s)e^{st}\Big|_{\sigma-j\infty}^{\sigma+j\infty} - tf(t) \tag{42}$$

(Letting $u = e^{st}$ and $dv = (dF/ds)\,ds$, supply the missing steps.) Sure enough, there is a term proportional to $tf(t)$. If only the rest of the right-hand side would go to zero, we would have the answer. Note that the integration is carried out with respect to s, not t. When the upper and lower limits are substituted for s, this term (not including the coefficient) becomes:

$$F(\sigma + j\infty)e^{(\sigma+j\infty)t} - F(\sigma - j\infty)e^{(\sigma-j\infty)t} = e^{\sigma t}[F(\sigma + j\infty)e^{j\infty} - F(\sigma - j\infty)e^{-j\infty}]$$

(You should interpret $e^{j\infty}$ as the limit of $e^{j\omega t}$ as $\omega \to \infty$.)

In this expression, σ is just a constant which might be 0 or not. In any case, $e^{\sigma t}$ is not zero. Furthermore, $e^{\pm j\infty}$ is indeterminate. We know at least that the magnitude of an exponential with an imaginary exponent is 1 for finite values of the exponent. Hence, the only way the whole expression can go to zero is for $F(s) = F(\sigma + j\omega) \to 0$ in the limit as $\omega \to \infty$. *That's it*; that's the condition that must be satisfied by $F(s)$ if the inverse transform of its derivative is to equal $-tf(t)$. The theorem we have arrived at is:

$$\mathscr{L}^{-1}\left[\frac{dF(s)}{ds}\right] = -tf(t) \qquad F(s)\big|_{s\to\infty} \to 0 \tag{43}$$

Does that condition on $F(s)$ check out for $F(s) = 1$ (that is, for $f(t) = \delta(t)$)? It certainly doesn't; $F(s)$ does not go to zero as $s \to \infty$, and so the theorem is not valid for this case, as already noted.

Unhappy with dealing with the inverse transform, and seeking an approach that might work better, a student suggested trying the following: write $F(s)$ as the direct Laplace integral of $f(t)$, and then take the derivative with respect to s. Let's follow that idea:

$$\frac{d}{ds}F(s) = \frac{d}{ds}\int_0^\infty f(t)e^{-st}\,dt = \int_0^\infty f(t)\frac{d}{ds}e^{-st}\,dt$$

$$= \int_0^\infty [-tf(t)]e^{-st}\,dt = \mathscr{L}[-tf(t)]$$

That seems to give the answer without any conditions. The crucial step on the right side of the top line, however, is the interchange of the operations of (1) differentiation with respect to s and (2) integration with respect to t. But such an interchange is permitted only if the integral is uniformly convergent. This condition is not satisfied when $f(t)$ is an impulse, or an even weirder "function" such as the derivative of an impulse. You should never interchange such operations without confirming its permissibility. The conclusion is that the preceding equation is not valid generally; it is valid only for the uniform-convergence condition.

Consider the following sequence of $F(s)$ functions:

$$\cdots s^2 \quad s \quad 1 \quad \frac{1}{s} \quad \frac{1}{s^2} \quad \frac{1}{s^3} \cdots$$

Any one is proportional to the derivative of the one immediately to its left. Take $F(s) = s$, for example; in this case, $dF(s)/ds = 1$, which is on the right of s. We would like to apply the theorem in (43) to these functions. Each function in the sequence from $1/s$ to the right satisfies the condition of the theorem. Thus, applying the theorem to $F(s) = 1/s$ (that is, $f(t)$ a step function):

$$\mathscr{L}^{-1}[F(s)] = \mathscr{L}^{-1}\left(\frac{1}{s}\right) = u(t)$$

$$\mathscr{L}^{-1}\left[\frac{dF(s)}{ds}\right] = \mathscr{L}^{-1}\left(-\frac{1}{s^2}\right) = -tu(t)$$

The last step follows from (43). This confirms the transform of the ramp given in (21). The theorem in (43) does not apply to the functions to the left of $1/s$.

EXERCISE 13

Use the result just obtained and the complex shifting theorem to determine the function of time whose Laplace transform is $1/(s + 3)^2$. □

EXERCISE 14

Use the theorem in (43) to establish the result (22). □

Laplace Transform of Periodic Functions

The overall approach taken in this chapter is the following. When a circuit is excited by a function of time, we transform the problem to the complex-frequency domain. This requires, among other things, finding the Laplace transform of the excitation. Until now, we have not explicitly considered periodic functions, so let's correct that deficiency.

A function that is periodic but multiplied by a unit step so that it doesn't start till $t = 0$ is illustrated in Figure 6. By definition:

$$f(t) = f(t - nT) \tag{44}$$

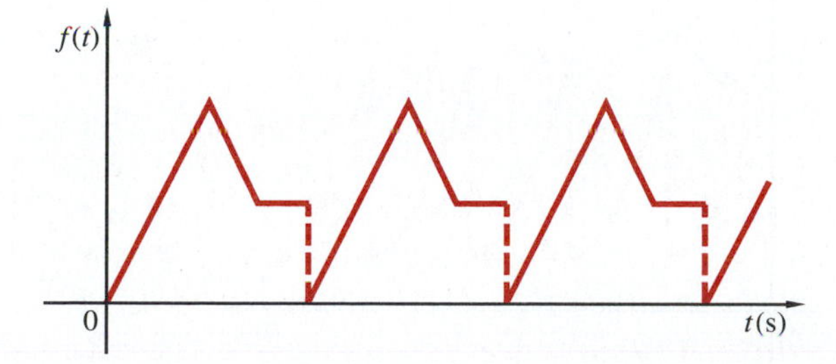

Figure 6 Periodic function from t = 0 on.

where T is the period. Let $f_a(t)$ be the same as $f(t)$ over the first period but 0 everywhere else. Then the kth cycle of the function is the same as the first one but displaced k periods to the right. Thus, each cycle can be written as the first cycle displaced by an appropriate number of periods. The entire function can then be expressed as an infinite sum of such displaced cycles. Thus:

$$\begin{aligned} f(t) &= f_a(t) + f_a(t - T) + \cdots + f_a(t - nT) + \cdots \\ &= \sum_{n=0}^{\infty} f_a(t - nT) \end{aligned} \tag{45}$$

Now we take the $\mathscr{L}$ transform. You can see what's coming: we plan to integrate the infinite sum and then interchange the summation with the integral. But does the integral of the sum equal the sum of the integrals? The answer is yes. Note from the real shifting theorem that:

$$\mathscr{L}[f_a(t - nT)] = F_a(s)e^{-nTs} \tag{46}$$

Hence:

$$\begin{aligned} F(s) &= \mathscr{L}[f(t)] = \sum_{n=0}^{\infty} \mathscr{L}[f_a(t - nT)] = \sum_{n=0}^{\infty} F_a(s)e^{-nTs} \\ &= F_a(s) \sum_{n=0}^{\infty} e^{-nTs} = F_a(s)(1 + e^{-Ts} + e^{-2Ts} + \cdots) \end{aligned} \tag{47}$$

You might recognize the series in the last line as a geometrical progression: $1 + x + x^2 + \cdots$; its sum is $1/(1 + x)$. Using this, the Laplace transform of a periodic function $f(t)$ with period T and first cycle $f_a(t)$ is:

$$F(s) = \frac{F_a(s)}{1 - e^{-Ts}} \tag{48}$$

EXAMPLE 5 An example of a periodic function is shown in Figure 7. The first cycle can be expressed as:

$$f_a(t) = 10e^{-2t}[u(t) - u(t - 1)]$$

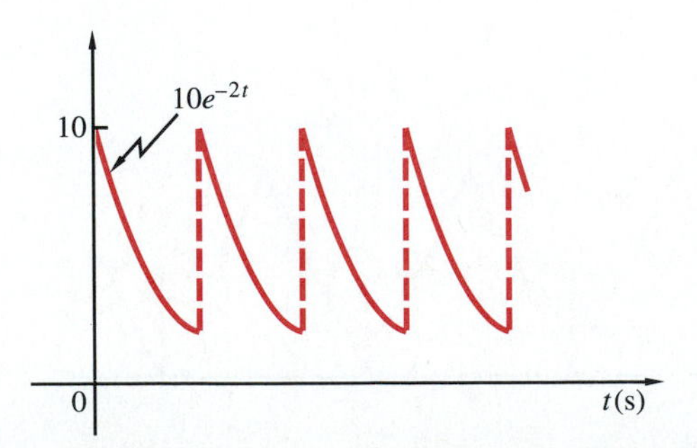

Figure 7
Example of a periodic function.

Since the function extends only up to $t = 1$ s, the Laplace transform is obtained by integrating only from 0 to 1. Thus:

$$F_a(s) = \int_0^1 10e^{-2t}e^{-st}\,dt = \frac{10}{-(s+2)}e^{-(s+2)t}\Big|_0^1$$

$$= \frac{10}{s+2}(1 - e^{-(s+2)})$$

This is the transform of the first period. To find the Laplace transform of the entire periodic function, we use (48) to get:

$$F(s) = \frac{10}{s+2}\frac{1 - e^{-(s+2)}}{1 - e^{-s}}$$

Here we experience something new. Whereas most Laplace transforms we have met were rational functions—possibly with exponentials in s, resulting from time shifting, in some numerators—here we have some rather complex function of s in the denominator.

If the preceding function is a circuit excitation, we would find the output transform (by methods to be discussed shortly) and then we would take the inverse transform. The output transform will have the form of (48) except that $F_a(s)$ will be the transform of the first period of the *output*, not of the input. We would find the inverse transform of the part of the output transform corresponding to $F_a(s)$ in (48); this would be the first period of the output. The rest of the output would simply be copies of this first period shifted by multiples of the period.

EXERCISE 15

A sawtooth waveform starting at $t = 0$ is shown in Figure 8. Find its Laplace transform.

ANSWER:

$$F(s) = \frac{10}{s^2}\frac{1 - e^{-2s}(2s + 1)}{1 - e^{-2s}}$$

□

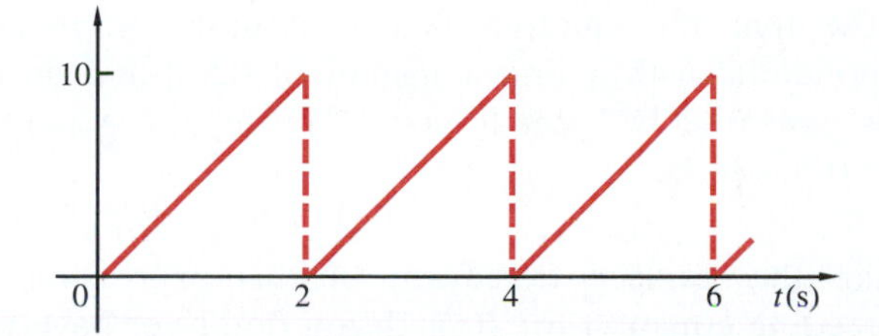

Figure 8
Waveform for Exercise 15.

4 FINDING INVERSE TRANSFORMS

The preliminaries are over. We are ready for the final two steps in the general plan for finding the complete response of a circuit to arbitrary excitations: to find the Laplace transform of any response voltage or current, and then to determine the corresponding functions of time. We will start with a simple example.

Simple Prototype Circuit

The topic will be introduced with the differential equation of a simple circuit. Suppose a voltage step is applied to a series RL circuit in which there is an initial current of 1 A. (Draw the circuit for yourself.) Assume the parameter values are such that the appropriate differential equation is:

$$\frac{di}{dt} + 2i = 5u(t) \qquad i(0) = 1 \text{ A}$$

It is our plan to Laplace-transform the entire equation. We are immediately confronted with a big question. When introducing the Laplace transform of a derivative (and an integral), we *assumed* that the transform of the function and its derivative (and its integral) exist. But at the start of a problem, we don't know the response function. How can we be sure that it and its derivative (and its integral) have transforms? That's easy: we really don't have to know; we just *assume they exist.* After we obtain what we think is a solution, we can confirm it by inserting it into the original equation and verifying that the equation is satisfied.

Back to the example. We transform the preceding equation and solve for the transform of the unknown current:

$$\mathscr{L}\left(\frac{di}{dt}\right) + 2\mathscr{L}(i) = \mathscr{L}[5u(t)]$$

$$sI(s) - 1 + 2I(s) = \frac{5}{s}$$

$$I(s) = \frac{s + 5}{s(s + 2)}$$

(Confirm this result by supplying the details.) The general plan is to use the table of transform pairs to find the inverse transform. If we had a more extensive table, it might contain the inverse transform of this particular function. Failing that, let's rely on ingenuity. (That's *always* a good plan.) We know that the solution has a natural component, in this case, a single exponential with a time constant of 0.5 s; and an equilibrium component, in this case, the DC steady state. Hence, the general form of the solution for $t > 0$ has to be:

$$i(t) = K_1 + K_2e^{-2t}$$

Take the Laplace transform of this expression and compare it with the preceding equation for $I(s)$. If you don't yet have the transform of a step and an exponential memorized, look them up in the table:

$$I(s) = \mathscr{L}[i(t)] = \frac{K_1}{s} + \frac{K_2}{s+2}$$

To find the coefficients K_1 and K_2, we equate this expression for $I(s)$ with the preceding one. Then we clear the fractions and equate the coefficients of like powers of s. The result is:

$$I(s) = \frac{K_1}{s} + \frac{K_2}{s+2} = \frac{s+5}{s(s+2)}$$

$$K_1(s+2) + K_2 s = s + 5$$

$$K_1 + K_2 = 1 \qquad 2K_1 = 5$$

$$K_1 = 2.5 \qquad K_2 = -1.5$$

$$I(s) = \frac{2.5}{s} - \frac{1.5}{s+2}$$

$$i(t) = (2.5 - 1.5e^{-2t})u(t)$$

At $t = 0$ this expression reduces to $i(0) = 1$, which is the correct initial value of the current.

What was just done yields a solution of the problem well enough, but it isn't totally satisfactory, because we started by *assuming* the form of the solution. Before proceeding to other circuits, let's pause briefly to deal with just part of the problem. Without writing circuit equations to arrive at the transforms of circuit variables, let's assume that this step has been carried out and we are *given* a function of s which is the Laplace transform of a voltage or current. How do we find the inverse transform? That's what we'll take up now.

Partial-Fraction Expansion

When the inversion integral was first introduced in (6), it was noted that the direct evaluation of this integral requires mathematics that is not assumed in this book. When presented with a transform function $F(s)$, it is our plan to use the transform-pair tables to find the corresponding function of time. However, it is unlikely that all given transform functions are simple enough to appear in the table. What can we do?

There is a neat way to deal with the problem just posed. The idea is to break up the given function into a sum of simpler ones, as done in the preceding example, such that the inverse transform of each term is recognizable or appears in the table. Except for transforms of periodic functions, all the transforms we have seen so far have been rational functions, possibly with exponentials in s resulting from time shifting.

Rational functions are characterized by their poles and zeros. The poles may be real or complex. (Complex poles in a real circuit excited by real functions of time always come in complex conjugate pairs.) They may be distinct (simple) or multiple; "simple" means that no two poles are the same. The details of the procedures to be developed are slightly different for each of these categories of poles, so we'll deal with them one at a time, starting with the simplest case.

SIMPLE REAL POLES Without considering how it was obtained, suppose the following function is the Laplace transform of a circuit output:

$$F(s) = \frac{s + 2}{(s + 1)(s + 3)}$$

This is a rational function having two simple real poles and one (finite) zero. A function having this form does not appear in the transform-pair table, so we plan to break it up as a sum of one-pole functions. Guided by the previous example, let's try:

$$F(s) = \frac{s + 2}{(s + 1)(s + 3)} = \frac{K_1}{s + 1} + \frac{K_2}{s + 3}$$

K_1 and K_2 are constants whose values we don't know and so must determine. Such an expression in which a rational function is expanded as a sum of terms, each one of which involves just one of the poles of the function, is referred to as a *partial-fraction expansion.*

Either from the expression for the Laplace transform of an exponential in (10) or from the table of transform pairs, we see that the inverse transform of each term is an exponential. Hence, the inverse transform of the original expression can be obtained as a sum of exponentials if the coefficients in the partial-fraction expansion can be found. A little thought will show that the approach used in the previous example is inadequate in general. It required solving two simultaneous equations for two constants. More unknowns would mean solving more equations simultaneously. Something better is needed.

To this end, suppose each side of the preceding expression is multiplied by one of the two pole factors in turn. The resulting expressions will be:

$$(s + 1)\left[\frac{s + 2}{(s + 1)(s + 3)}\right] = K_1 + \frac{K_2(s + 1)}{s + 3}$$

$$(s + 3)\left[\frac{s + 2}{(s + 1)(s + 3)}\right] = \frac{K_1(s + 3)}{s + 1} + K_2$$

In each of these expressions, one of the two constants appears alone on the right. On the left side, the multiplying factor is the same as one of the factors in the denominator, so it cancels in each case. Now evaluate each expression on the left at the corresponding value of s; $s = -1$ in the first expression and $s = -3$ in the second:

$$K_1 = \tfrac{1}{2} = K_2$$

You can confirm these values by inserting them in the partial-fraction expansion and recombining the terms to verify that the original rational function is obtained.[5]

[5] Another check, used by some, might be to evaluate both the original function and the partial-fraction expansion at some convenient value of s, such as $s = 0$ or $s = -2$, at which the function is 0 in this case. (But can you be sure that the two quantities are the same in general if they happen to be equal at one particular frequency?)

Summarizing the steps in these calculations, a simple general expression can be written, giving a formula for calculating the values of the constants in a partial-fraction expansion:

$$K_i = (s - s_i)F(s)\big|_{s=s_i} \tag{49}$$

Each K_i value is called the *residue* of the function at the corresponding pole. To find the residue of $F(s)$ at pole s_i, we multiply $F(s)$ by $s - s_i$, cancel this with the corresponding factor in the denominator, and then evaluate the result at s_i. Although established only for an example, this result holds generally for any rational function whose poles are *simple*—that is, for factors $(s - s_i)$ in the denominator that appear to the first power only.

EXERCISE 16

Find the inverse Laplace transform of the function:

$$F(s) = \frac{s + 3}{(s + 1)(s + 2)}$$

ANSWER: $(2e^{-t} - e^{-2t})u(t)$. □

SIMPLE COMPLEX POLES In the preceding example and in Exercise 16, the poles were all real (and negative). The same approach is valid if the poles are complex, except that the numerical work would involve complex numbers. After such a partial-fraction expansion for a pair of complex poles, the inverse transform would be a pair of exponentials with complex exponents. By Euler's theorem, this pair of exponentials can then be combined and rewritten as damped sinusoids. This case requires no further elaboration.

For complex poles in some cases, however, there is another approach. Instead of separate partial fractions for each pole of a complex pair, followed by a combination of the inverse transforms by Euler's theorem, we can think of first combining the two partial fractions in the transform and then taking an inverse transform of the combination. An example will illustrate. The following transform function with three poles is given. The residue at the real pole $s = -1$ is found to equal 1. The partial fraction corresponding to this pole is subtracted from the original function to leave a remainder, as shown:

$$F(s) = \frac{2(s + 2)^2}{(s + 1)(s^2 + 4s + 5)}$$

$$\text{Res}\big|_{s=-1} = \frac{2(s + 2)^2}{s^2 + 4s + 5}\bigg|_{s=-1} = 1$$

$$G(s) = F(s) - \frac{1}{s + 1} = \frac{s + 3}{s^2 + 4s + 5}$$

$G(s)$, the function remaining after the partial fraction at the pole $s = -1$ is subtracted, has a pair of complex poles at $s = -2 \pm j1$. However, rather than concentrate on the individual poles, let's take the entire remainder function as a unit and compare it with (18) or (19) (see also the transform-pair table). It doesn't match either of them exactly. However, simply subtracting and adding a constant in the numerator leads to the following:

$$G(s) = \frac{s+3}{(s+2)^2+1} = \frac{s+2}{(s+2)^2+1} + \frac{1}{(s+2)^2+1}$$

Every time s appears on the right-hand side of $G(s)$, it appears as $s + 2$. This should ring a bell and remind you of the complex frequency-shifting theorem: $\mathscr{L}[e^{-at}f(t)] = F(s+a)$. That is, (1) replace $s + 2$ by s, (2) find the inverse transform of the result, and then (3) multiply by e^{-2t}. You can also find these transform pairs in the table. Thus, the inverse transform of the entire function is:

$$f(t) = \mathscr{L}^{-1}[F(s)] = (e^{-t} + e^{-2t}\cos t + e^{-2t}\sin t)u(t)$$

EXERCISE 17

The following transform function is given:

$$F(s) = \frac{s^3 + 4s^2 - 2s - 3}{(s+2)(s+3)(s^2+2s+3)}$$

(a) Find the residues of $F(s)$ at $s = -2$ and $s = -3$.

ANSWER: 3, −2.

(b) Determine the partial fraction corresponding to the complex poles.

ANSWER: Numerator = −3.

(c) Find the inverse transform of part (b).

ANSWER: $2.12e^{-t}\sin\sqrt{2}\,tu(t)$. □

MULTIPLE REAL POLES Some complications arise when multiple poles are present. The problem will be illustrated by the following transform function, the form of whose partial-fraction expansion is given on the right:

$$F(s) = \frac{18(s+5)}{s(s+2)(s+3)^2} = \frac{K_0}{s} + \frac{K_1}{s+2} + \frac{K_2}{s+3} + \frac{K_3}{(s+3)^2}$$

In addition to simple poles, this function has a double pole at $s = -3$, so the partial-fraction expansion *must* contain a term accounting for this double pole; it *might* also contain a term with $s + 3$ to the first power in the denominator. (Compare with a second-order polynomial in x which *must* have an x^2 term, but possibly also x^1 and x^0 terms.) A higher-order pole would lead to more terms in the partial-fraction expansion, with higher powers of $s + 3$ in the denominators.

The residues at the simple poles (0 and −2) can be found in the same way as before. (Do it, and get $K_0 = 5$ and $K_1 = -27$.) If we try finding K_2 in the same way (multiplying both sides by $s + 3$ and then letting $s = -3$), nothing but trouble will result, because a factor $s + 3$ will still remain in the denominator. You have probably already concluded that, to avoid this difficulty, we ought to multiply both sides by $(s + 3)^2$, instead of $s + 3$, and then set $s = -3$. If that is done, the result will be:

$$\begin{aligned}(s + 3)^2 F(s)\big|_{s=-3} &= 18\frac{(s + 5)}{s(s + 2)}\bigg|_{s=-3} \\ &= \left[(s + 3)^2\left(\frac{K_0}{s} + \frac{K_1}{s + 2}\right) + (s + 3)K_2 + K_3\right]\bigg|_{s=-3} \\ &= 12 \\ K_3 &= 12\end{aligned}$$

We have successfully found all the constants except K_2; now what do we do?

Look over the second line in the preceding expression, *before* $s = -3$ has been substituted. We see a factor $s + 3$ multiplying K_2. One way to get rid of that factor is to differentiate both sides of the equation. Then, when we put $s = -3$, everything on the right will disappear except K_2. Thus:

$$K_2 = \frac{d}{ds}[(s + 3)^2 F(s)]\bigg|_{s=-3} = 22$$

With all the constants determined, collect all the terms in the partial-fraction expansion and confirm the original function.

The next step in the process would be to find the inverse transform. All the terms are familiar, from the previous examples if not before, except for $K_3/(s + 3)^2$. This is in the general form $F(s + 3)$. That should remind you of a shifting theorem; use it to find the inverse transform.

For higher-order poles, we would have to differentiate more often. This is a straightforward procedure, but writing the expression in general terms for a pole of arbitrary order makes it look formidable, and so we won't do it.

EXERCISE 18

Determine the partial-fraction expansion and then the inverse Laplace transform of the following transform function. Confirm that your expansion is correct by (1) recombining the terms to arrive at the original expression and (2) evaluating both the original expression and your expansion for some particular value of s; they ought to be the same. (Choose a value of s that minimizes numerical calculations.)

$$F(s) = \frac{5(s + 8)}{(s + 4)^2(s^2 + 2s + 2)}$$

ANSWER: $f(t) = [e^{-4t}(1.7 - 2t) + e^{-t}(1.7\cos t - 3.1\sin t)]u(t)$. □

DOUBLE COMPLEX POLES[6] Complex poles come in conjugate pairs; hence, the denominator of a transform function with a double complex pair of poles

[6] This subsection deals with a problem not often encountered. The topic is not a prerequisite for the further development. Some of you may wish to skip it.

(and possibly other poles as well) will be at least of the fourth degree. You are unlikely to run into functions of greater complexity than this, so we'll stick to just a double pair of poles. Actually, the function might contain simple (real or complex) poles as well. The partial fractions corresponding to these poles can be calculated and then subtracted from the given function to leave the function being treated here.

One approach is to follow the procedure just outlined for multiple real poles. The only complication will be that, because the poles are complex, the resulting exponentials in the response will be complex. This is no big problem, since Euler's theorem can be invoked to convert the complex exponential functions of time to damped sinusoids. Just as for simple complex poles, another approach keeps the partial fractions corresponding to a pair of complex poles together.

To avoid getting bogged down in a mass of algebra, we'll use a mixture of numerical and literal values, with numerical values whenever the notation will be simplified without destroying generality. The most general case we will treat will be:

$$F_0(s) = \frac{N_0(s)}{(s^2 + 4s + 13)^2} = \frac{N_0(s)}{[(s + 2)^2 + 9]^2} \tag{50}$$

The denominator can always be written in the form on the right, no matter what the coefficients may be. The numerator is a polynomial of order at most 3. In all cases, its powers of s can always be written as $(s + 2)^k$ by adding and subtracting appropriate terms. Thus, if $N_0(s) = s^2 + s + 8$, we note that $(s + 2)^2 = s^2 + 4s + 4$. The coefficient of s in the original numerator is only 1, so we must add and subtract $3s$. Pursuing this process to the end leads to:

$$\begin{aligned} N_0(s) &= s^2 + s + 8 = (s + 2)^2 - 3s + 4 \\ &= (s + 2)^2 - 3(s + 2) + 10 \end{aligned}$$

In this manner, all powers of s in the original function can be written as powers of $s + 2$ with changed coefficients. (Note that, although the numerical value 2 is used here, this can be any number; similarly, 9 in the denominator can be any number.)

What we have done should remind you of the complex shift theorem. If we replace $s + 2$ in the given function by s and find the inverse transform of the new function—say, $f(t)$—the inverse transform of the original function will be $e^{-2t}f(t)$. Let's assume we have replaced $s + 2$ by s and we now have the function:

$$F_1(s) = \frac{N_1(s)}{(s^2 + 9)^2} \tag{51}$$

Now make one more observation. The squared denominator like the one here can come about by differentiating a rational function having a denominator $s^2 + 9$ to the first power. Thus, suppose:

$$F(s) = \frac{As + B}{s^2 + 9} \tag{52}$$

Then,

$$F_2(s) = \frac{dF(s)}{ds} = \frac{A(s^2 + 9) - (As + B)2s}{(s^2 + 9)^2}$$

$$= \frac{-As^2 - 2Bs + 9A}{(s^2 + 9)^2} \tag{53}$$

Suppose the numerator of $F_1(s)$ in (51)—the original numerator after $s + 2$ has been replaced by s—has exactly the form of this numerator. Then, by the theorem about the derivative with respect to s:

$$\mathscr{L}^{-1}[F_2(s)] = -t\mathscr{L}^{-1}[F(s)] \tag{54}$$

The function whose inverse transform is needed is the one in (52). Depending on the numerator, this will be a sine, a cosine, or their combination. Then the inverse transform of the original function is this sine-cosine combination multiplied by $-t$ and then by e^{-2t}.

The numerator in (51), however, will not generally have the exact form of the numerator of (53). When it does not, $F_1(s)$ will not be the derivative of a function like the one in (52). Now what? Maybe there's a clue in the case of a double *real* pole. In that case, remember, there was a term $1/(s + a)$ in addition to $1/(s + a)^2$. Therefore, to $F_2(s)$ let's add a function $F_3(s)$ having the same complex pair of poles but to the first power, as follows:

$$F_2(s) + F_3(s) = \frac{-As^2 - 2Bs + 9A}{(s^2 + 9)^2} + \frac{Cs + D}{s^2 + 9}\left(= \frac{(Cs + D)(s^2 + 9)}{(s^2 + 9)^2}\right)$$

$$= \frac{Cs^3 + (D - A)s^2 + (9C - 2B)s + 9(A + D)}{(s^2 + 9)^2} \tag{55}$$

Both terms in the first line have been put over the same denominator.

We're almost home. We now compare this with $F_1(s)$ in (51), the given function with $s + 2$ replaced by s. Comparing coefficients of like powers of s, the four constants $(ABCD)$ are now identified and the inverse transform is easily found.

To conclude, let's choose some specific numerical values. Let the function with a double pair of complex poles be:

$$F_0(s) = \frac{4s + 2}{[(s^2 + 2)^2 + 9]^2} = \frac{4(s + 2) - 6}{[(s + 2)^2 + 9]^2} \quad \rightarrow \quad F_1(s) = \frac{4s - 6}{(s^2 + 9)^2}$$

Comparing this with the general function in (55), we find:

$$\begin{aligned} C &= 0 & A &= -\tfrac{1}{3} \\ D - A &= 0 & B &= -2 \\ 9C - 2B &= 4 & C &= 0 \\ 9(A + D) &= -6 & D &= -\tfrac{1}{3} \end{aligned}$$

With these numerical values, the function $F(s)$ is:

$$F(s) = \frac{As + B}{s^2 + 9} = -\frac{\frac{1}{3}s + 2}{s^2 + 9}$$

$$f(t) = \mathcal{L}^{-1}[F(s)] = -(\tfrac{1}{3}\cos 3t + 2\sin 3t)u(t)$$

By the complex derivative theorem, this is to be multiplied by $-t$.

Now turn to the function with the simple pole:

$$F_3(s) = \frac{Cs + D}{s^2 + 9} = \frac{-\frac{1}{3}}{s^2 + 9} \qquad \leftrightarrow \qquad -\tfrac{1}{9}\sin 3t\, u(t)$$

The two time functions are now added; then, by the complex shifting theorem, they are multiplied by e^{-2t}. The final result is:

$$f_1(t) = e^{-2t}[(\tfrac{2}{3}t - \tfrac{1}{9})\sin 3t + \tfrac{1}{3}t\cos 3t]u(t)$$

That *was* a lot of work. Was it worth it? Although you are unlikely to meet a double complex pole often, this development illustrates how you might tackle an analytical task.

REVIEW Let us now review the process of finding a partial-fraction expansion and then an inverse transform of a rational function with a double pair of complex conjugate poles.

1. Let $F_0(s)$ be the given function:

$$F_0(s) = \left.\frac{N_0(s)}{[(s + a)^2 + b^2]^2}\right|_{s+a\to s} \qquad \to \qquad F_1(s) = \frac{N_1(s)}{(s^2 + b^2)^2}$$

2. Define two functions $F(s)$ and $F_3(s)$ according to (52) and (55), having the same poles as F_1 but with numerator coefficients (A, B, C, D) that are unknown; then define $F_2(s)$ as the derivative of $F(s)$:

$$F(s) = \frac{As + B}{s^2 + b^2} \qquad F_3(s) = \frac{Cs + D}{s^2 + b^2}$$

$$F_2(s) = \frac{d}{ds}F(s) = \frac{-As^2 - 2Bs + b^2A}{(s^2 + b^2)^2}$$

3. Form $F_2(s) + F_3(s)$ as the desired partial-fraction expansion. Place the entire expression over the same denominator.

4. Equate the resulting numerator with $N_1(s)$:

$$Cs^3 + (D - A)s^2 + (b^2C - 2B)s + b^2(A + D) = N_1(s)$$

5. Determine the coefficients A, B, C, D by equating coefficients of like powers of s on both sides of this equation.

6. Finally, the desired inverse transform is:

$$\mathcal{L}^{-1}[F_0(s)] = e^{-at}\left[-t\mathcal{L}^{-1}\left(\frac{As + B}{s^2 + b^2}\right) + \mathcal{L}^{-1}\left(\frac{Cs + D}{s^2 + b^2}\right)\right]u(t)$$

This has been a rather lengthy development for this one particular case. The result alone may not be worth the effort. However, participating in the thought processes carried out in the analysis may make up for it.

5 TRANSFORM NETWORKS

Now that the groundwork has been laid, we are ready to deal with the problem of finding the complete response of a network when it is subjected to any (Laplace-transformable) excitations. The steps in the process can be summarized as follows:

1. Apply the fundamental laws (Kirchhoff's and the component relationships) to the network, obtaining a set of equations which generally contain derivatives (and possibly integrals) of the variables.

2. Transform these equations, converting them to algebraic equations in which the variables are Laplace transforms. Initial conditions are introduced automatically.

3. Solve the algebraic equations for the transform variables.

4. Reconvert to functions of time via the inverse transform. This is the *complete* solution, and the initial conditions are already incorporated.

You are by now well versed in each of these steps. The preceding section dealt at length with step 4. As for solving the algebraic equations in step 3, this was treated in Chapter 5 for real coefficients and in Chapter 8 for complex coefficients in the sinusoidal steady state.

We have met a variation of steps 1 and 2 before. In the study of the sinusoidal steady state in Chapter 8, we introduced another kind of transformed variable, namely, a phasor, to represent sinusoids. There, too, we contemplated applying fundamental laws to a given network to obtain the relevant equations, and *then* replacing the variables by their phasors; *but we rejected the idea.* Instead, we noted that Kirchhoff's laws are satisfied by the phasors themselves and that the component equations can be written directly in terms of phasors, which are the transform variables in the sinusoidal steady state. It should be clear that the same approach can be taken here.

First, recall that Kirchhoff's laws are simply linear algebraic combinations of voltages and currents. It is certainly possible to take the Laplace transform of such linear combinations, the result being the same linear combinations, but now with Laplace-transformed variables.

Dissipative components (resistors and multiterminal components) have v-i relationships that are algebraic. Taking the Laplace transform of such a relationship leads to an identical relationship but with variables that are Laplace transforms. Thus, $v(t) = Ri(t)$, when transformed, becomes $V(s) = RI(s)$.

The inductor and the capacitor have v-i relationships that include a derivative (or an integral). The transforms are:

$$
\begin{aligned}
i(t) = C\frac{dv(t)}{dt} \quad &\rightarrow \quad
\begin{cases} I(s) = CsV(s) - Cv(0) \\ V(s) = \dfrac{1}{Cs}I(s) + \dfrac{1}{s}v(0) \end{cases} \\
v(t) = L\frac{di(t)}{dt} \quad &\rightarrow \quad
\begin{cases} V(s) = LsI(s) - Li(0) \\ I(s) = \dfrac{1}{Ls}V(s) + \dfrac{1}{s}i(0) \end{cases}
\end{aligned}
\tag{56}
$$

Note that the initial values automatically show up in the transformed relationships.

Initially Relaxed Networks

In the initially relaxed case, the transformed V-I relationships for the capacitor and inductor are identical with what in Chapter 8 was called Ohm's law for phasors:

$$V(s) = Z(s)I(s) \qquad I(s) = Y(s)V(s)$$

where $Z(s) = Ls$ or $1/Cs$, except that the variables are now Laplace transforms instead of phasors.

There's a glimmer here of something very significant. All the concepts and procedures developed for the sinusoidal steady state were based on these two results: that Kirchhoff's laws apply to the transformed variables (phasors, in that case) and that the component V-I relationships take the form of Ohm's law. Later, in Chapter 12, we generalized the concept of frequency by introducing a complex-frequency variable s (what foresight!) and we interpreted the procedures of phasor analysis to consist of finding the forced response to an excitation of the form e^{st}.

Now we find that, for an initially relaxed network, the same V-I relationships apply—except that the variables are Laplace transforms instead of phasors. Instead of defining network functions (such as impedance, admittance, and voltage gain) as ratios of phasors, we define them as ratios of Laplace transforms. When written in terms of s, both definitions lead to identical functions.

As a consequence of all this, it is now clear that all the formal analytic procedures carried out in terms of phasors—such as series and parallel combinations, solutions of loop or node equations, and Thévenin's theorem—are directly applicable to the Laplace transform solution *whenever the network is initially relaxed.*

EXAMPLE 6 The switch in the circuit of Figure 9 has been open for a long time when it is closed at $t = 0$. The component values are normalized. The objective is to find the output voltage $v(t)$ and the input current $i(t)$. Prior to switching, there was obviously no current in the inductor, and whatever charge there may have been on the capacitor in the past had been discharged by the time of the switching. Hence, the circuit is initially relaxed.

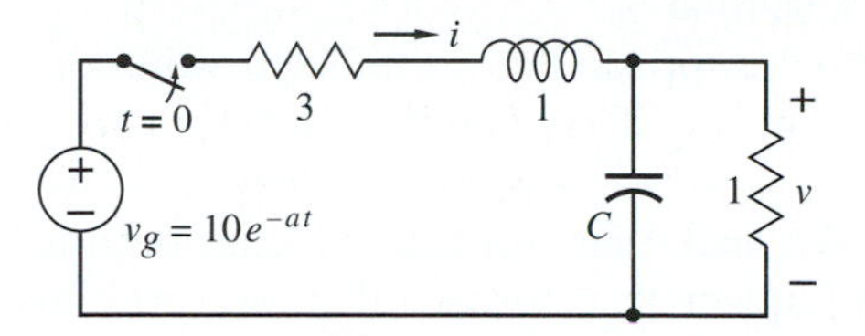

Figure 9 Circuit with double natural frequencies for C = 1.

SOLUTION The previously utilized approach was to derive appropriate differential equations and to solve them, subject to the initial conditions. This approach is still valid; and Laplace transforms can then be used in the solution of the differential equations. But there is now a simpler approach. There is no need to write differential equations and *then* to Laplace transform them. The circuit variables themselves are taken as transformed variables so that the relevant equations are already algebraic. Thus, the output-voltage transform in Figure 9 can be obtained by the voltage-divider relationship as follows:

$$V(s) = \frac{1/(Cs+1)}{1/(Cs+1)+s+3}V_g(s) = \frac{1/C}{s^2+(3+1/C)s+4/C}V_g(s)$$

$$= \frac{10/C}{[s^2+(3+1/C)s+4/C](s+a)}$$

Similarly, the input-current transform is obtained by multiplying the excitation transform by the admittance seen from the source terminals—or dividing by the impedance:

$$I(s) = \frac{Vg(s)}{Z(s)} = \frac{V_g(s)}{s+3+\dfrac{1}{Cs+1}} = \frac{s+1/C}{s^2+(3+1/C)s+4/C}V_g(s)$$

$$= \frac{10(s+1/C)}{[s^2+(3+1/C)s+4/C](s+a)}$$

The quadratic in the denominator of each of the response functions $V(s)$ and $I(s)$ is the characteristic polynomial; its roots are the poles of the voltage gain and also of the input admittance. The other pole, at $s = -a$, is contributed by the excitation.

From here on the solution procedure consists of expanding each of the transform functions into partial fractions. The inverse transforms (obtained using the transform table if necessary) then give the output functions of time. We will pursue this matter in the following section.

THE LONG-DEFERRED CASE OF CRITICAL DAMPING Long, long ago, in our first study of second-order circuits in Chapter 7, there was one particular case that we deferred. We noted then that, based on the parameter values, the two natural frequencies of second-order circuits could take on two real or complex values. At the boundary between these two cases, for one particular combination of the parameter values, the two real natural frequencies are equal—the case of critical damping. We will now take up this case using the preceding Example 6.

In Figure 9, suppose the normalized value of C is 1. Then the characteristic values (the poles of $V(s)$ and $I(s)$ in the expressions in Example 6) are found to be equal ($s = -2$)—what in Chapter 7 was called the case of critical damping. We had then postponed consideration of this case. Now we find that—with Laplace transforms—this case presents no great difficulty. Assuming the pole contributed by the excitation ($s = -a$) is not the same as -2, the partial-fraction expansion of $V(s)$ and its inverse transform are easily found to be:

$$V(s) = \frac{10}{(s+2)^2(s+a)}$$

$$= 10\left(\frac{1/(2-a)^2}{s+a} + \frac{1/(a-2)}{(s+2)^2} - \frac{1/(a-2)^2}{s+2}\right)$$

$$v(t) = \mathscr{L}^{-1}[V(s)] = \frac{10}{(a-2)^2}[e^{-at} + ((a-2)t - 1)e^{-2t}]$$

(You find the partial fraction expansion of $I(s)$.) The first term is the equilibrium solution resulting from the excitation. The remainder is the natural response, the transient.

Not only does the double pole cause no difficulty, there is no difficulty even if the excitation pole is the same as the natural frequency: $a = -2$. In this case, $V(s)$ will have a triple pole at $s = -2$ but its partial-fraction expansion can be easily determined. The solution for $v(t)$ will have the form: $v(t) = (At^2 + Bt + C)e^{-2t}$. Carry out the details and determine the A, B, and C coefficients.

EXERCISE 19

Following the same procedure, carry out the solution for the input current $i(t)$ for the same capacitance value and for both $a \neq -2$ and $a = -2$. □

EXERCISE 20

Write the state equations in the time domain for the circuit in Figure 9. Confirm that Laplace transforming these equations and solving for $V(s)$ leads to the same expression obtained in Example 6. □

Initially Active Networks

In Example 6 the circuit was assumed to be initially relaxed. How can we handle nonzero initial conditions? Glance back at Chapter 2 where the capacitor and inductor were first introduced. There it was noted that an initially charged capacitor is equivalent to a relaxed capacitor in series with a battery. Similarly, an inductor with an initial current is equivalent to a relaxed inductor in parallel with a constant-current source. This is evident in the Laplace transform domain from (56). Consequently, initial capacitor voltages and inductor currents can be treated as sources, everything else in the analysis remaining unaltered.

Recall that the definition of network functions (driving-point and transfer) specifically excludes independent sources within the network. Hence, when a network is not initially relaxed, such network functions cannot be defined. That is to say, for example, that if a voltage source is applied to an initially *active* network—not relaxed—and the current in the source is to be found, this cannot be done by multiplying the voltage transform by the "admittance," because no admittance exists for this network. Instead, other methods (node equations or Thévenin's theorem, for example) must be used.

EXAMPLE 7 The switch in Figure 10a has been open a long time when it is closed at $t = 0$. The current in the inductor is to be found following the closing of the switch.

Since the circuit has been active for a long time with the switch open, it can be assumed that no changes in voltage or current are taking place at the time of closing the switch at $t = 0$. Specifically, the current in the inductor is not changing so there is no voltage across it; the battery voltage appears across the two resistors in the loop with the inductor. Hence, $i(0) = 24/12 = 2$ A. We intend to deal with Laplace transforms after the switch is closed. The initially active inductor can be replaced by a relaxed inductor in parallel with a 2-A DC current source. In the transform domain, this combination can be replaced by its Thévenin equivalent, shown in Figure 10b. The voltage source has a transform which is constant; so the source itself is an impulse whose strength is $4\ \text{V}\cdot\text{s}$ (volt-seconds).

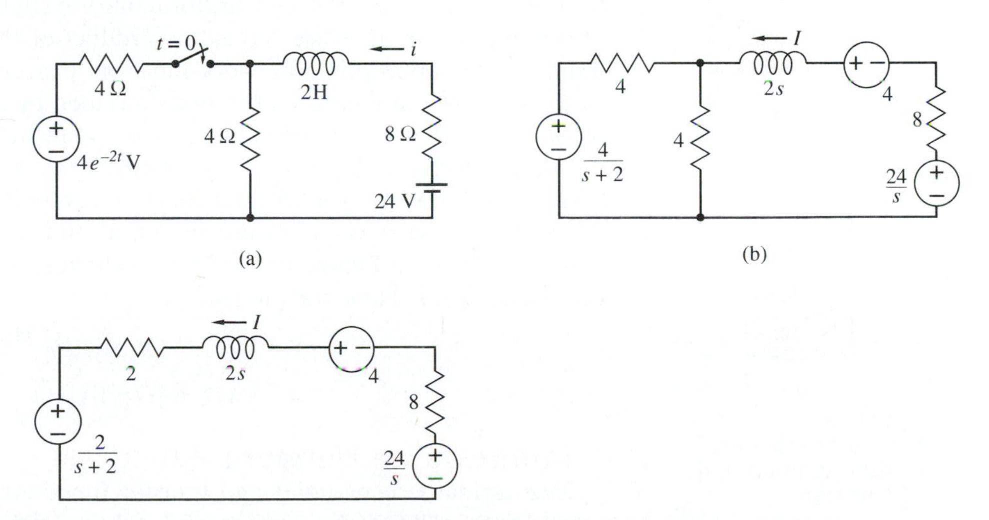

Figure 10
Circuit of Example 7.

After the switch is closed, the transform circuit whose variables are Laplace transforms has the form of Figure 10b; the battery and the initial-condition source are in series. The two resistors and the source on the left can be replaced by their Thévenin equivalent, leading to the single-loop circuit in Figure 10c. The transform current in this loop is obtained by dividing the equivalent transform source voltage by the total impedance. Then follows a partial-fraction expansion and its inverse transform:

$$I(s) = \frac{24/s + 4 - 2/(s+2)}{2s + 10} = \frac{2s^2 + 15s + 24}{s(s+2)(s+5)}$$

$$= \frac{12/5}{s} - \frac{1/3}{s+2} - \frac{1/15}{s+5}$$

$$i(t) = [\tfrac{12}{5} - \tfrac{1}{3}e^{-2t} - \tfrac{1}{15}e^{-5t}]u(t)$$

(Confirm the results.) This includes a constant and two exponentials. However, the latter are different in kind. The one with time constant 1/2 is part of the equilibrium response resulting from the exponential source. The other, with time constant $L/R = 1/5$, is the natural response. Even though $e^{-2t}/3$ is part of the equilibrium response, we can't say it is part of the steady state like the constant term, since there is nothing steady about it.

6 TRANSFER FUNCTIONS AND THE IMPULSE RESPONSE

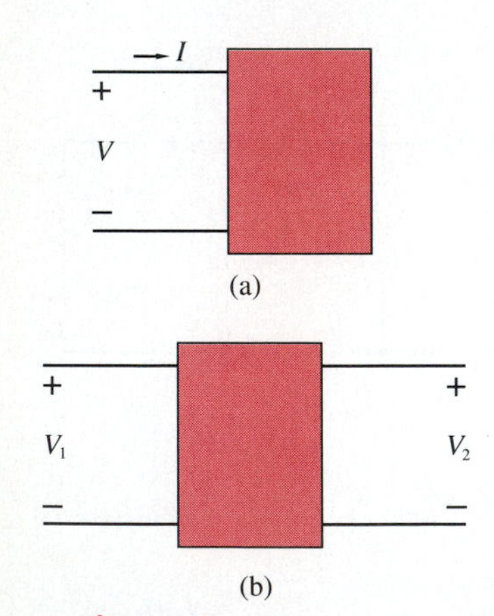

Figure 11
Network functions: (a) driving-point and (b) transfer.

We have established that, when an initially-relaxed linear network is subjected to an arbitrary (but Laplace-transformable) excitation, the Laplace transform of the output can be expressed as the product of the Laplace transform of the input and the appropriate network function, expressed as a function of s. Two cases are shown in Figure 11, the first described by a driving-point function, the second by a transfer function. The driving-point function can be an impedance or an admittance, depending on whether the excitation is the voltage or current, respectively. The transfer function can be any one of four possibilities, depending on what variables are the input and output. The output and input variables shown in Figure 10b are both voltages, so the appropriate function is the voltage gain. Thus, for the two cases:

$$V(s) = [Z(s)]I(s)$$

$$V_2(s) = [G(s)]V_1(s)$$

Determining Network Functions

The various driving-point and transfer functions can be determined by any appropriate method: by writing and solving simultaneous equations in the frequency domain or by such techniques as series and parallel combinations, voltage dividers, or Thévenin's theorem. We will illustrate with an example.

EXAMPLE 8 The goal in the initially relaxed circuit in Figure 12 is to find $v_2(t)$ in response to an excitation $v_g(t)$. (Before you continue, think up and describe to yourself a number of different ways of attacking this problem.) One approach would be to derive the state equations, carry out their Laplace transform, and then solve for the current $I(s)$. The desired output transform is then $R_2I(s)$; its inverse transform is the desired $v_2(t)$. Carry out this process to compare with what follows.

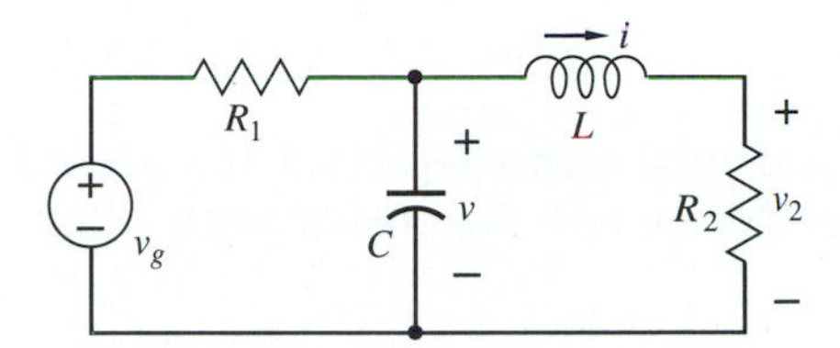

Figure 12
Circuit of Example 8.

A simpler approach is to carry out standard circuit analysis procedures to determine $V_2(s)$. One possibility is to convert the combination (C, R_1, V_g) to a Thévenin equivalent; then $V_2(s)$ is obtained from the voltage-divider relationship. Another possibility is to write a pair of mesh equations, then solve for $I(s)$, and then get $V_2(s) = R_2I(s)$. These, too, are left for you to do.

Instead, let's write a single node equation, choosing V as the node voltage, with L and R_2 taken as a single branch. After solving for $V(s)$, the output voltage follows from the voltage divider relationship:

$$CsV + \frac{V - V_g}{R_1} + \frac{V}{Ls + R_2} = 0$$

$$V(s) = \frac{Ls + R_2}{(Ls + R_2)(R_1Cs + 1) + R_1} V_g(s)$$

$$V_2(s) = \frac{R_2}{Ls + R_2} V(s)$$

$$= \frac{R_2}{R_1LCs^2 + (R_1R_2C + L)s + R_1 + R_2} V_g(s)$$

Suppose the normalized component values in the preceding circuit are $R_1 = C = 1$, $L = 1/3$, $R_2 = 5/3$. Then, $V_2(s)$ is:

$$V_2(s) = \frac{5}{(s + 2)(s + 4)} V_g(s)$$

This is in the form "output transform equals transfer function × input transform." Thus, the poles of the output are those of the transfer function *and* those of the input transform. Suppose the input voltage is $v_g(t) = 10\cos 2t$. Its Laplace transform can be obtained from the table, if you don't remember it. Then a partial-fraction expansion can be carried out, with the following result:

$$V_2(s) = \frac{50s}{(s+2)(s+4)(s^2+4)}$$
$$= \frac{-6.25}{s+2} + \frac{5}{s+4} + \frac{1.25s + 7.5}{s^2+4}$$

(Confirm each line.) Finally, the inverse transform is:

$$v_2(t) = -6.25e^{-2t} + 5e^{-4t} + 1.25(\cos 2t + 3\sin 2t)$$

The two exponential terms represent the natural response; the sinusoids, the steady state. (Compare this result with what you found from the state equations.)

EXERCISE 21

Find the phasor representing the sinusoidal source in just completed Example 8 and evaluate the transfer function at the frequency of the source. Determine the phasor output voltage and confirm the steady-state component of the preceding solution. □

EXAMPLE 9 This example deals again with the circuit in Figure 12 but with somewhat different parameter values. For the numerical values used in Example 8, the natural frequencies had two distinct real values: −2 and −4. Suppose the normalized component values are changed to $R_1 = C = L = 1$ and $R_2 = 3$. Then

$$V_2(s) = \frac{3}{(s+2)^2} V_g(s)$$

This is the critically damped case; the transfer function has a double real pole (coincident natural frequencies). As seen in Example 6, with our present knowledge there would be no difficulty in dealing with this case. To determine the inverse transform of the output voltage, the input voltage must also be specified. We will consider several possibilities.

Let's initially assume that the input voltage is a unit impulse; its Laplace transform is 1. The result for the output-voltage transform is $V_2(s) = 3/(s+2)^2$; it is already in the form of a partial-fraction expansion. Hence

$$v_2(t) = \mathscr{L}^{-1}\left(\frac{3}{(s+2)^2}\right) = 3te^{-2t}$$

This is the impulse response of the network.

Next let the input be $v_g(t) = 2e^{-t}$; then $V_g(s) = 2/(s + 1)$. Hence

$$V_2(s) = \frac{6}{(s + 1)(s + 2)^2} = \frac{6}{s + 1} - 6\left(\frac{1}{s + 2} + \frac{1}{(s + 2)^2}\right)$$

$$v_2(t) = 6e^{-t} - 6(1 + t)e^{-2t}$$

There are no surprises here. The term in the form $(A + Bt)e^{-at}$ is what we would expect from a double pole. Sometimes, as in the case of the impulse excitation just treated, the coefficient A is zero.

Finally, let's return to the original component values in Example 8 which led to the output transform $V_2(s) = 5V_g(s)/(s + 2)(s + 4)$. Now suppose that the excitation is $v_g(t) = 4e^{-2t}$. Then

$$V_2(s) = \frac{20}{(s + 2)^2(s + 4)} = \frac{10}{(s + 2)^2} - \frac{5}{s + 2} + \frac{5}{s + 4}$$

$$v_2(t) = 5(2t - 1)e^{-2t} + 5e^{-4t}$$

The form of this function of time is exactly the same as that of the preceding one, even though the natural frequencies in this case are distinct. It is clear that if the output transform has a double pole, it doesn't matter whether this results from the transfer function having a double pole, or the input transform having a pole at the same location as a pole of the transfer function.

A MORE EXTENSIVE NETWORK The circuit shown in Figure 13, being of third order, is somewhat more complex than the ones so far considered. The circuit is assumed to be initially relaxed. The output voltage is to be determined for any (Laplace transformable) input voltage. Once the transfer function is found, the desired result will follow. Again a number of approaches can be taken. One possibility is to observe that (1) $V_2(s) = R_2I_2(s)$; (2) $I_2(s)$ can be obtained from I_1 by the current-divider expression; and (3) $I_1(s)$ is obtained as the input-voltage transform divided by the input impedance (or times the input admittance). Only steps (2) and (3) require some effort.

Perhaps the simplest way to find the input impedance is by continued fractions starting from the input terminals. (Refresh your memory from Chapter 3.) Then we carry out the steps just listed, but not necessarily in that order:

$$Z_1 = R_1 + \frac{1}{C_1s} + \frac{1}{C_2s + 1/(Ls + R_2)}$$

$$= \frac{(R_1C_1s + 1)(LC_2s^2 + C_2R_2s + 1) + C_1s(Ls + R_2)}{C_1s(LC_2s^2 + C_2R_2s + 1)}$$

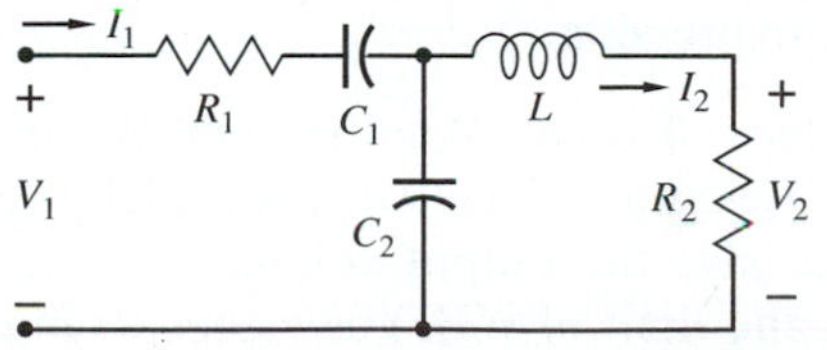

Figure 13
Needed: voltage gain in initially relaxed circuit.

(Don't forget: confirm each step.) Then the two response transforms become:

$$I_1(s) = Y_1(s)V_1(s)$$

$$= \frac{C_1 s(LC_2 s^2 + C_2 R_2 s + 1)V_1(s)}{(R_1 C_1 s + 1)(LC_2 s^2 + C_2 R_2 + 1) + C_1 s(Ls + R_2)}$$

$$V_2(s) = R_2 I_2(s) = \frac{R_2(1/C_2 s)I_1}{Ls + R_2 + 1/C_2 s}$$

$$= \frac{R_2}{LC_2 s^2 + C_2 R_2 s + 1} Y_1(s)V_1(s)$$

Finally, we form V_2/V_1, multiply out all parentheses, collect terms, and divide both numerator and denominator by the coefficient of the highest power of s in the denominator:

$$\frac{V_2(s)}{V_1(s)} = \frac{R_2 C_1 s}{(R_1 C_1 s + 1)(LC_2 s^2 + C_2 R_2 s + 1) + C_1 s(Ls + R_2)}$$

$$= \frac{R_2 C_1 s}{R_1 C_1 C_2 L s^3 + (LC_1 + LC_2 + R_1 R_2 C_1 C_2)s^2 + (R_1 C_1 + R_2 C_2 + R_2 C_1)s + 1}$$

$$= \frac{(R_2/R_1 LC_2)s}{s^3 + \left[\frac{R_2}{L} + \frac{1}{R_1}\left(\frac{1}{C_1} + \frac{1}{C_2}\right)\right]s^2 + \left[\frac{1}{LC_2} + \frac{R_2}{LR_1}\left(\frac{1}{C_1} + \frac{1}{C_2}\right)\right]s + \frac{1}{LC_1 C_2 R_1}}$$

Note that this is a rational function with three poles and a single (finite) zero. Also note that the admittance, the reciprocal of the impedance in the previous equation, has exactly the same poles but three finite zeros. We will expand on this thought shortly.

CHECKING THE WORK In the process of carrying out the algebraic details of establishing the Laplace transform responses in a given circuit problem, how do we know that no errors are made? Retracing each step is one way to confirm algebraic work. However, other checks are possible:

1. *Dimensional check.* One of these is a dimensional check. A voltage gain is dimensionless; at each step in the process, including especially the final form, we must check each term to verify the proper dimensions. Thus, s has the dimensions of reciprocal time. Hence, a coefficient multiplying s to the first power must have the dimension of time; the coefficient of higher powers of s must have the dimension of higher powers of time. Thus, in $R_2 C_1 s$ in the numerator of the second line in the preceding equation, $R_2 C_1$ must have the dimension of time and, sure enough, it does. Make a dimensional check on all the terms in the denominator of the same expression.

2. *Initial-value theorem.* Another check is to apply the initial-value theorem. Because the network is initially relaxed, the inductor current—and hence also the output voltage—is initially zero. By the initial value theorem, the limit of $sG(s)$ as s goes to infinity must, therefore, be zero,

where $G(s) = V_2(s)/V_1(s)$. The highest power in the numerator of $sG(s)$ is 2, and in the denominator, 3. Thus $sG(s)$ does indeed go to zero as s goes to infinity.

3. *Final-value theorem.* This theorem constitutes a third check. However, first we must ensure that the conditions for the validity of the theorem are satisfied, namely, that there are no poles in the right half plane or on the j-axis. Right-half-plane poles, of course, mean increasing exponential response functions. Even if this could occur, in the real world it would mean the circuit would soon be destroyed, since components will break down when voltages are too high. For this example, let's defer consideration of the final-value theorem until after numerical values have been introduced.

4. *Network check at extreme frequencies.* Another check results from an examination of the network at the extreme frequencies of 0 and infinity. At 0 frequency (DC), inductors behave like short circuits (zero impedance) and capacitors like open circuits (infinite impedance)—and vice versa at infinite frequency. The inductors and capacitors in the circuit can be replaced by these open and short circuits, leaving only a dissipative circuit. Thus, a transfer function reduces to a constant: zero or nonzero. However, in addition, a driving-point function might go to infinity (caused by a capacitor across a voltage-source input, for example).

Although these four ways of checking for errors have been introduced for an example, they are obviously general and can be used in all cases.

ILLUSTRATION CONTINUED The voltage gain in the preceding illustration was in literal terms. In order to pursue the illustration (Figure 13) any further, we will resort to numerical values. Assume the following: $R_1 = 1.5\,\text{k}\Omega$, $R_2 = 20\,\Omega$, $L = 0.2\,\text{H}$, $C_1 = 1\,\mu\text{F}$, and $C_2 = 0.5\,\mu\text{F}$. When these values are inserted (using the appropriate SI units), the coefficients will be encumbered by high powers of 10. To avoid this, we can normalize the frequency. To see what a convenient normalizing value might be, the value of the constant term in the denominator is found to have a factor 10^9. Since the highest power of s in the denominator is 3, a normalizing value of 10^3 should do the trick. That is, let the old s be divided by 1000 to get a new, normalized s. Of course, in order not to change the value of the function, each term divided by 10^3 must also be multiplied by 10^3.

Inserting the numerical values and normalizing as described leads to the following:

$$G(s) = \frac{V_2(s)}{V_1(s)} = \frac{0.385s}{(s + 0.1)(s^2 + 2s + 10)}$$

$$= \frac{0.385s}{(s + 0.1)[(s + 1)^2 + 9]}$$

Notice how the normalization has resulted in convenient numerical values. If you were skeptical earlier when time constants had such numerical values as 1 or 2 and frequencies had such nice, round values as 3 or 4, you can now relax. Those were not 1 or 2 *seconds,* nor were they 3 or 4 *radians per second*; they were dimensionless, normalized values, used just for convenience.

Now that the transfer function (voltage gain, in this case) has been obtained, we can find the transform of the output voltage:

$$V_2(s) = G(s)V_1(s)$$

The poles of $G(s)$ constitute the natural frequencies (short-circuit) of the network. Consider a partial-fraction expansion of $V_2(s)$. There will be two categories of poles: those belonging to $G(s)$—the natural frequencies—and those belonging to V_1. When the inverse transform is taken, the latter will constitute the forced response.

Something very significant happens if $V_1(s) = 1$. What is the "function" of time whose Laplace transform is 1? It is none other than a unit impulse. Under these circumstances, there is no forced response; the entire response is the natural response. All the impulse input does is to "stimulate" the network, to give it . . . an impulse.

This example has demonstrated that the transfer function of a network is simply the Laplace transform of its *impulse response.* In the present case, the "transfer function" is the voltage gain. In other cases it can be any other network function, including a driving-point function, such as an admittance. In that case, the "output" will be the input current in response to the input voltage.

Let's pursue the numerical example further. With the input voltage an impulse, the output transform is the previously found transfer function. A partial-fraction expansion can be obtained by finding the residue at the real pole and subtracting the corresponding partial fraction from the original function to obtain the remainder, pertinent to the complex poles. You supply the details:

$$\begin{aligned} V_2(s) = G(s) &= \frac{-0.00392}{s + 0.1} + \frac{0.00392(s + 100)}{(s + 1)^2 + 3^2} \\ &= \frac{-0.00392}{s + 0.1} + 0.00392\frac{(s + 1)}{(s + 1)^2 + 3^2} \\ &\quad + 0.129\frac{3}{(s + 1)^2 + 3^2} \end{aligned}$$

If you don't recognize the inverse transforms, look them up in the table. The impulse response is:

$$v_2(t) = (-0.00392e^{-0.1t} + 0.00392e^{-2t}\cos 3t + 0.129e^{-2t}\sin 3t)u(t)$$

Improper Rational Functions

All of the rational functions explicitly treated so far have been proper. That is to say, the degree of the denominator has exceeded that of the numerator. This is perhaps better described in terms of the behavior of the function at infinity. As s goes to infinity, the dominant terms in the numerator and denominator of a rational function are the highest powers of s; all other terms become negligible compared with these. In the preceding example (the circuit of Figure 13), the highest powers of s in the numerator and denominator of the voltage-gain function are 1 and 3, respectively. Hence, $G(s)|_{s\to\infty} \to 1/s^2 \to 0$. (That also means that the inverse transform of such a function has a 0 initial value, as well as a 0 final value if it has no poles in the right half plane.)

However, look back at the input admittance in that example. It has the same poles as the voltage gain. This is not surprising, since these poles represent the natural frequencies; for a given input, we would expect the same natural frequencies for any output response: the output voltage, input current, or any other response. The admittance is a cubic polynomial over a cubic. As s goes to infinity, the admittance (and so the initial value of its inverse transform) approaches a nonzero constant. With the given numerical values, the admittance and its partial-fraction expansion become:

$$Y_1(s) = 0.1\frac{s(s^2 + 0.1s + 10)}{(s + 0.1)(s^2 + 2s + 10)}$$

$$= 0.1 + \frac{0.01}{s + 0.1} - \frac{s - 0.1}{(s + 1)^2 + 3^2}$$

For improper rational functions, the partial-fraction expansion is obtained by dividing the denominator into the numerator until such time as, in the remainder, the degree of the denominator exceeds that of the numerator. The remainder is then expanded as before. Carry out this process to arrive at the preceding expansion.

If we again assume a unit-impulse input voltage, the current at the input terminals (the impulse response there) will be the inverse transform of the admittance. The constant term in the partial-fraction expansion (corresponding to the value of the admittance at infinite frequency) will result in an impulse in the impulse response. Is this reasonable?

It can be confirmed from the circuit. As previously discussed, at infinite frequency inductors behave like open circuits (infinite impedance) and capacitors like short circuits (zero impedance). Thus, at infinite frequency, the circuit in Figure 13 reduces to a resistor R_1 across the input terminals. Therefore, an impulse voltage input will indeed result in an impulse input current.

7 THE CONVOLUTION CONCEPT

What has been achieved up to this point is the following. Given a linear network and an arbitrary (but transformable) excitation, the response can be determined by the methods described. What we would like to do now is to determine the response of a network to a given excitation, not when the network is given, but when the response to some *standard* excitation is given.

We know, for instance, that the steady-state response to a periodic excitation can be obtained as a Fourier series from the steady-state response to a sinusoid. The standard function in this case is a sinusoid. The standard functions we will consider now are an impulse and a step.

To carry out this plan, some mathematical manipulations are needed. The final result requires a certain mathematical step with which you are probably unfamiliar. It is straightforward, however, and we will carry it out here.

The Convolution Theorem[7]

Suppose a voltage or current excitation function $e(t)$, whose transform is $E(s)$, is applied to an initially relaxed network; it is desired to find a specific response $r(t)$ with transform $R(s)$. The response transform is the transfer function times the excitation transform; therefore, letting the relevant transfer function be $H(s)$:

$$R(s) = H(s)E(s) \tag{57}$$

In particular, if the excitation is an impulse, $e(t) = \delta(t)$, its transform is $E(s) = 1$. The response and its transform in this case can be labeled $r_\delta(t)$ and $R_\delta(s)$. Then:

$$\begin{aligned} R_\delta(s) &= H(s) \\ r_\delta(t) &= \mathscr{L}^{-1}[H(s)] \end{aligned} \tag{58}$$

After that slight digression to the impulse response, return to the general case in (57). In the preceding parts of this chapter, the response $r(t)$ is obtained by expanding the right side in partial fractions and then taking the inverse transform. The goal now is to express both $H(s)$ and $E(s)$ in terms of the time functions of which they are the transforms using the Laplace transform definition. This assumes that $H(s)$ is an ordinary transform function. Suppose the result can be manipulated into the form:

$$R(s) = \int_0^\infty (\cdot)e^{-st}\,dt \tag{59}$$

Then, from the definition of the Laplace transform, the desired time response $r(t)$ is whatever lies between the parentheses.

Although the preceding discussion was phrased in terms of an excitation and its response, the mathematics does not depend on these interpretations. Hence, in developing the result, a more general notation will be used; we will return to the physically more meaningful notation later. Let $F_1(s)$ and $F_2(s)$ be two transform functions whose inverse transforms are $f_1(t)$ and $f_2(t)$. That is:

$$F_1(s) = \int_0^\infty f_1(u)e^{-su}\,du \tag{60a}$$

$$F_2(s) = \int_0^\infty f_2(y)e^{-sy}\,dy \tag{60b}$$

[7] You can omit this subsection and go on to the next one, if you dont mind accepting as true, without corroboration, what you are told is true.

(Dummy variables other than t are used to avoid later confusion.) Now consider the product of these two functions, motivated by the product on the right side of (57). The product of integrals on the right can be converted to a double integral:

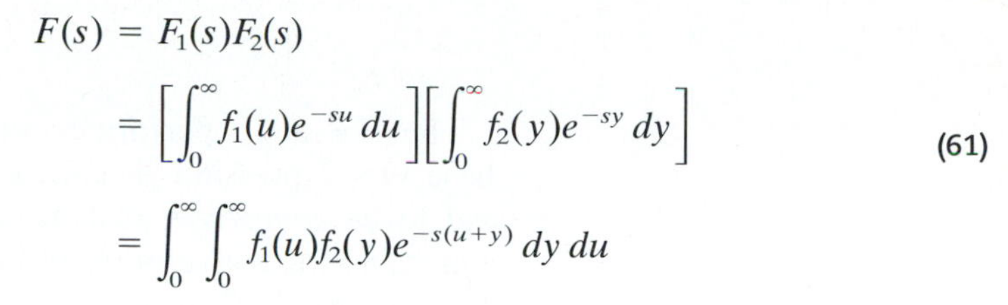

$$F(s) = F_1(s)F_2(s) = \left[\int_0^\infty f_1(u)e^{-su}\,du\right]\left[\int_0^\infty f_2(y)e^{-sy}\,dy\right] = \int_0^\infty \int_0^\infty f_1(u)f_2(y)e^{-s(u+y)}\,dy\,du \tag{61}$$

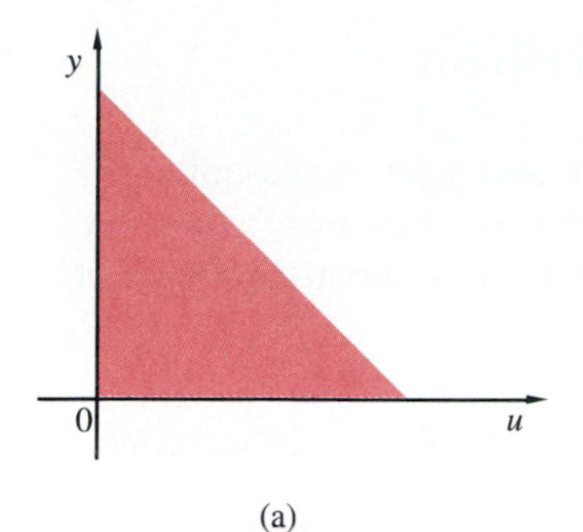

(a)

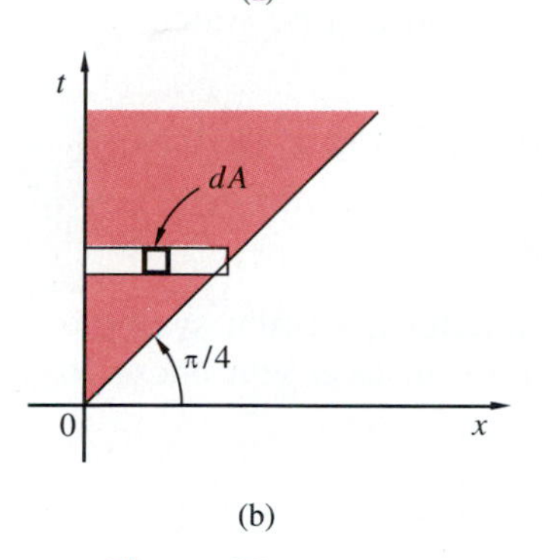

(b)

Figure 14
Regions of integration.

Converting to a double integral is justified, since each integral in the product is a constant with respect to the other variable of integration. The double integral is carried out over an area whose coordinate axes are u and y. The integration is to be performed over the entire first quadrant, as indicated in Figure 14a.

To facilitate the interpretation of this integral, let's make a transformation to a new set of variables, as follows:

$$\begin{aligned} t &= u + y \\ x &= u \end{aligned} \tag{62}$$

The second of these is actually an identity transformation and is not needed; it is included only for clarity. The double integral must now be expressed in terms of the new variables. It turns out that the element of area $du\,dy$ in the old variables is the same as the element $dx\,dt$ in the new variables.[8]

To complete the transformation of variables, the new limits of integration must be determined. Since $t = u + y = x + y$, and since y takes on only positive values, t can be no less than x. The line $t = x$ in the x-t plane bisects the first quadrant, as shown in Figure 14b; thus the desired area of integration is the area between this line and the t axis, as shown. In order to cover this area, we first integrate with respect to x from $x = 0$ to $x = t$; then we integrate with respect to t from zero to infinity.

With the change of variables in (62) and with the corresponding limits, as just discussed, (61) now becomes:

$$F(s) = \int_0^\infty \left[\int_0^t f_1(x)f_2(t-x)\,dx\right]e^{-st}\,dt \tag{63}$$

This is exactly in the form of (59); hence, we can identify the quantity in the parentheses of (59) as $f(t) = \mathscr{L}^{-1}[F_1(s)F_2(s)]$. It should be clear that if $F_1(s)$ in (60) is written in terms of the dummy variable y and $F_2(s)$ is written in terms of u, then in the preceding result the arguments of f_1 and f_2 will be interchanged.

[8] If you look up a calculus book, you will find that the two elements of area are related through the Jacobian, which, in this case, happens to equal 1.

$$du\,dy = \left|\frac{\partial u}{\partial t}\frac{\partial y}{\partial x} - \frac{\partial u}{\partial x}\frac{\partial y}{\partial t}\right| dx\,dt = dx\,dt$$

Hence, the final result can be written in the following two alternative forms:

$$
\begin{aligned}
f(t) = \mathcal{L}^{-1}[F_1(s)F_2(s)] &= \int_0^t f_1(x)f_2(t-x)\,dx \\
&= \int_0^t f_1(t-x)f_2(x)\,dx
\end{aligned}
\tag{64}
$$

The operation performed on two functions $f_1(t)$ and $f_2(t)$ represented by these two expressions is called the *convolution* of the two functions; they are said to be *convolved.* Instead of the expressions in (64), a shorthand way of expressing the convolution of two functions is $f_1 * f_2$.

Response to Arbitrary Excitation in Terms of Impulse Response

The result derived in the immediately preceding section can be stated as a theorem:

The convolution theorem: Given two transformable functions $f_1(t)$ and $f_2(t)$ with transforms $F_1(s)$ and $F_2(s)$, the inverse transform of the product $F_1(s)F_2(s)$ is the convolution $f_1(t) * f_2(t)$ of the functions.

The mathematical form of the convolution of two functions, written symbolically as $f_1 * f_2$, was given in (64), and is repeated here, in case you chose to omit that section.

$$
\begin{aligned}
f(t) = \mathcal{L}^{-1}[F_1(s)F_2(s)] &= \int_0^t f_1(x)f_2(t-x)\,dx \\
&= \int_0^t f_1(t-x)f_2(x)\,dx
\end{aligned}
\tag{64}
$$

To accentuate the mathematical generality, the preceding development was carried out in a general notation. Our primary interest, however, is to determine the response $r(t)$ of an initially relaxed network, with a transfer function $H(s)$, to an excitation $e(t)$. The convolution theorem gives us the answer.

We have already noted that the transfer function $H(s)$ is the Laplace transform $R_\delta(s)$ of the impulse response $r_\delta(t)$. Let us initially assume that $H(s)$ has a zero at infinity. (The opposite case will be treated later.) In the convolution theorem let f_1 be the impulse response r_δ and let f_2 be the excitation e. Then, the response $r(t) = \mathcal{L}^{-1}[H(s)E(s)]$ is the convolution $r_\delta(t) * e(t)$:

$$
r(t) = \int_0^t r_\delta(x)e(t-x)\,dx = \int_0^t r_\delta(t-x)e(x)\,dx \tag{65}
$$

This is a very valuable result which permits us to detemine the response of a network to an arbitrary (transformable) excitation $e(t)$ knowing only the impulse response of the network. The latter is the inverse transform of the relevant transfer function.

Although the convolution theorem applies to any transformable excitation function, the excitations of greatest utility with this approach are "pulselike" functions. These can take many different forms: rectangular, triangular (any shape), half a sinusoidal cycle, or any composite shape. These are the types of excitation functions we will concentrate on in what follows.

One way of looking at the result in (65) is to say that the output is obtained from the input, but that the values of the input at each point are "weighted" by the impulse response. For this reason the impulse response is sometimes called the *weighting function.*

Consider the first form of the integral. The process to be carried out is a little tricky to interpret. Keep referring to (65). The response at time t is the area from $x = 0$ to t under some curve. This curve is obtained by multiplying two curves together. One of these two is the impulse response $r_\delta(x)$. The second one, $e(t - x)$, needs interpreting. The excitation is $e(x)$. What is $e(-x)$? It is obtained by *folding* $e(x)$ about the vertical axis so that it runs to the left. Then $e(t - x)$ is obtained from $e(-x)$ by adding t to the argument. This amounts to sliding or shifting $e(x)$ to the right by an amount t. These two curves, $r_\delta(x)$ and $e(t - x)$, are then multiplied point by point and the area under the result from 0 to t is found. All of this work gives the value of the response at only one point t. The process must be repeated for each value of t. An example is badly needed to clarify this matter.

EXAMPLE 10

The simple high-pass circuit in Figure 15a is excited by the voltage pulse in Figure 15b. All values, including time, are normalized. The objective is to find the output voltage by means of convolution.

SOLUTION The transfer function in this case is:

$$\frac{V_2(s)}{V_1(s)} = \frac{s}{s+1} = 1 - \frac{1}{s+1}$$

The transfer function is nonzero (it has the value 1) at infinity. Hence, if an impulse is applied, the response will also contain an impulse. This part of the transfer function can be treated separately, dealing with the remaining transfer function $(-1/(s + 1))$ after the nonzero value at infinity (namely 1) has been removed. Carry out this approach in your spare time. Here, we'll deal with the entire transfer function together.

$$H(s) = \frac{s}{s+1} = 1 - \frac{1}{s+1}$$

$$r_\delta(t) = \mathscr{L}^{-1}\left(\frac{s}{s+1}\right) = \delta(t) - e^{-t}$$

To apply the preceding interpretation of the convolution integral to this example, we will use the first form of the integral in (65):

$$r(t) = \int_0^t r_\delta(x)e(t - x)\,dx$$

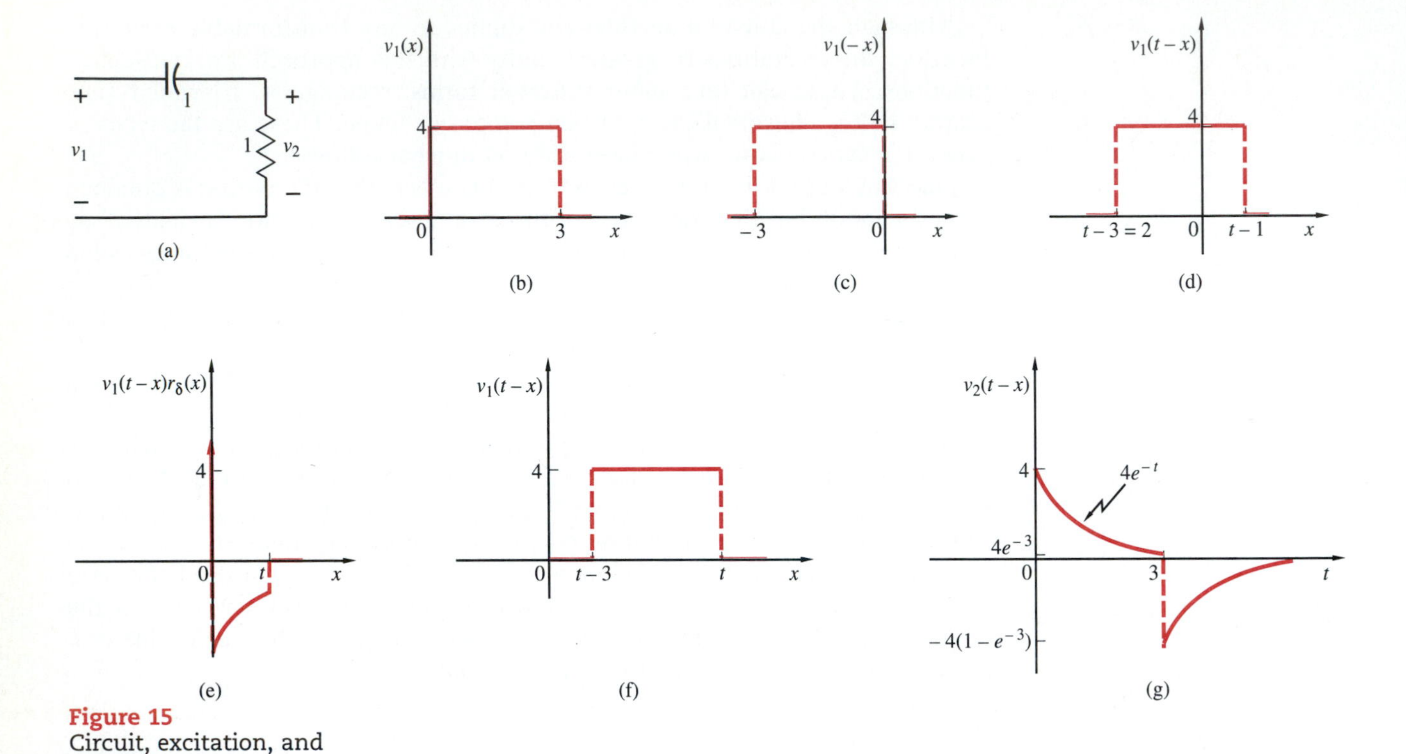

Figure 15
Circuit, excitation, and impulse response in Example 10.

The excitation, $e(x) = 4$ from $x = 0$ to 3 and zero everywhere else, couldn't be much simpler. Folding the function to $e(-x)$ amounts to plotting the function backward from $x = 0$, as shown in Figure 15c. Sliding the function to the right by an amount t is illustrated in Figure 15d, with $t = 1$. The next step is to multiply the folded and shifted function in Figure 15d by the impulse response point by point. The result is shown in Figure 15e; it has two parts, consisting of the impulse and the exponential which is truncated by the pulse. Note that the product is zero for $x < 0$. The final step is to find the area under the curve, including the contribution from the impulse. The result from all this is the value of the response at just one point $t = 1$. The process is repeated for each value of t.

Note that this process can be carried out numerically, even if no formula is known for the excitation or for the impulse response. They might be known only graphically or numerically, for example. But when analytical expressions are known for the excitation and the impulse response, as in this case, the convolution integral can be carried out analytically.

In this example, there are two distinct intervals in $t > 0$ to be considered, one for t between 0 and 3, the other from 3 on. Even though the excitation has ceased in the latter case, the pulse serves to charge the capacitor, which discharges after the pulse has ended. From the convolution integral in the first range:

$$v_2(t) = \int_0^t 4[\delta(x) - e^{-x}]\,dx = 4u(t) + 4e^{-x}\Big|_0^t$$
$$= 4e^{-t} \qquad 0 \leqslant t \leqslant 3$$

The impulse response doesn't start until $t = 0$, even though the folded and shifted excitation starts before $t = 0$, so the response does not start until $t = 0$.

For $t > 3$, the integrands are the same. But now the folded and shifted excitation in Figure 15f does not start until $t - 3$. Hence, the convolution integral has $t - 3$ as its lower limit:

$$v_2(t) = \int_{t-3}^t 4[\delta(x) - e^{-x}]\,dx = 4e^{-x}\Big|_{t-3}^t$$
$$= 4[e^{-t} - e^{-(t-3)}] = 4(1 - e^3)e^{-t} \qquad t \geqslant 3$$

Note that the impulse occurs outside the range of integration and so it contributes nothing to the result in this range. The complete response is shown in Figure 15g.

Rather than simply accepting the final result, we ought to check it for reasonableness. When the pulse is first applied to the circuit, the capacitor voltage cannot change instantly. So, the entire pulse first appears across the output resistor. As the capacitor charges exponentially, the output voltage must drop exponentially since the sum of the two voltages must add up to the constant pulse voltage. When the excitation turns off, the capacitor voltage again cannot change suddenly, so the entire 4-V drop must appear across the output. The capacitor now discharges; the output voltage (now equal to the capacitor voltage) must follow it exponentially to a completely discharged condition. The output curve is, thus, eminently reasonable. We would have had to look for an error if it had not been exactly as obtained.

The excitation in the problem just completed was too simple to cover the range of complications that might be expected in the convolution integral. A slightly more complicated example will now be considered.

EXAMPLE 11 The low-pass *RC* circuit shown in Figure 16a is excited by the voltage pulse shown in Figure 16b. Parameter values and the time are normalized. The goal here is to find $v_2(t)$ using the convolution idea. The impulse response shown in Figure 16c, is a simple exponential:

$$r_\delta(x) = \tfrac{1}{4}e^{-x/4}$$

SOLUTION Let's apply the preceding description of the convolution integral to this example. The steps are shown in Figure 17 for the specific value $t = 0.25$. First (Figure 17a) there is a plot of the impulse response $r_\delta(x)$, followed by the excitation $e(x)$ in Figure 17b. Note that the vertical scale for r_δ is expanded for clarity. The folded version $e(-x)$ comes next, in Figure 17c. Then we slide this

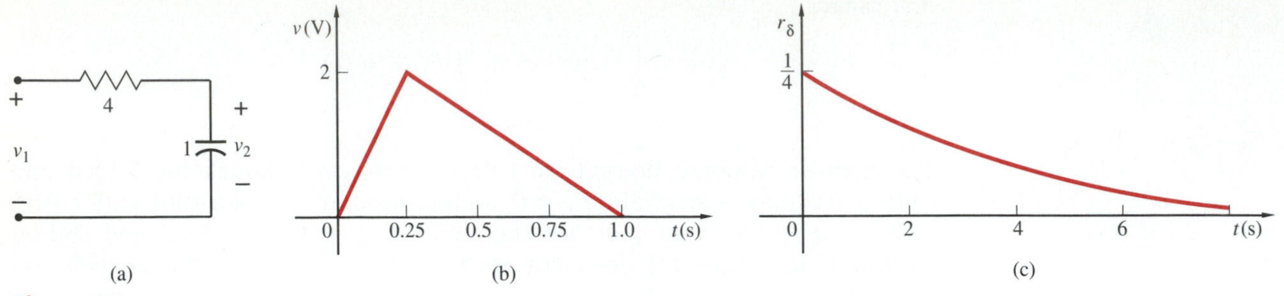

Figure 16
Finding the response by convolution.

curve to the right an amount $t = 0.25$ (Figure 17d). Although this takes a lot of describing, very little in the way of effort has been required so far. Now the work begins: the folded curve shifted to the right by $t = 0.25$ has to be multiplied point by point by the first curve (the impulse response) to yield the last curve, which is the integrand of the convolution integral (Figure 17e). The response at $t = 0.25$ is the area under this curve.

Note that the part of the folded and shifted excitation to the left of the vertical axis contributes nothing to the response, since the impulse response is 0 there. For other values of t, the first three curves in this set are the same; the fourth curve is shifted to the right by a different value of t, so that the final curve, and hence the area under it, will be different.

In the present example, as well as the previous one, an analytical

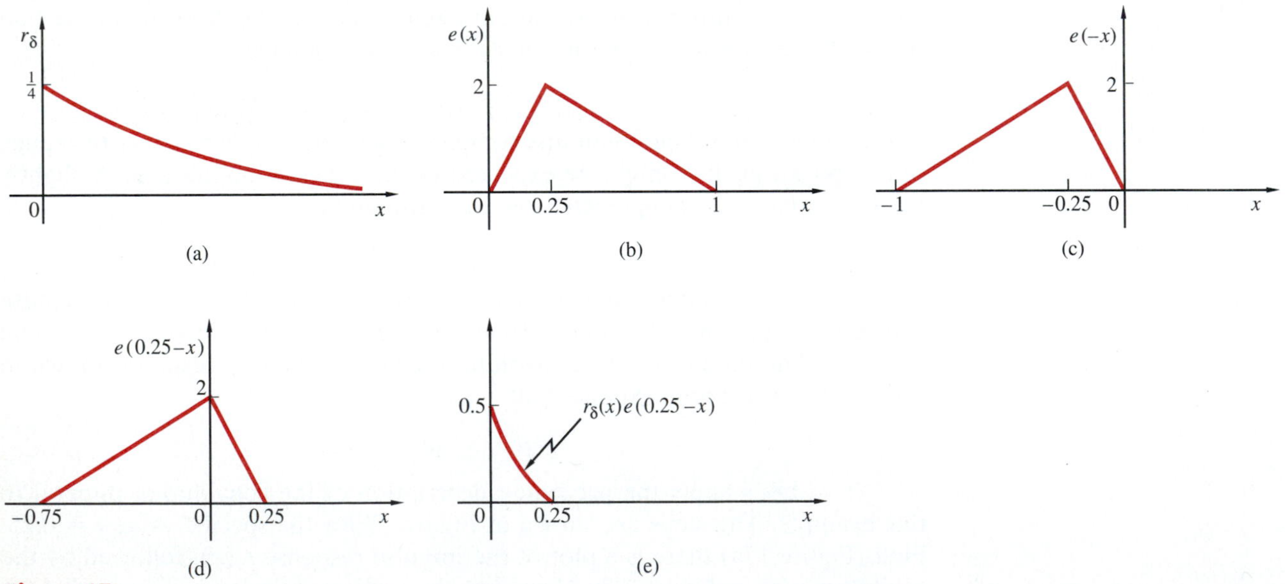

Figure 17
Interpretation of convolution.

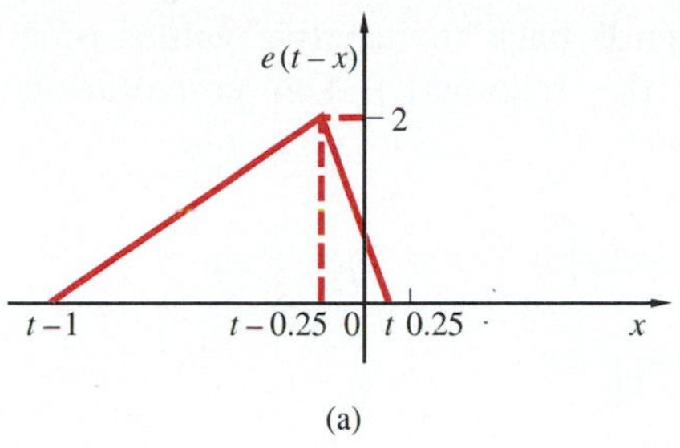

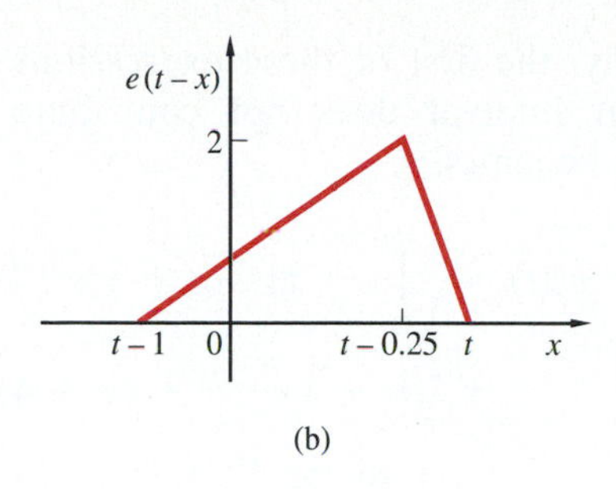

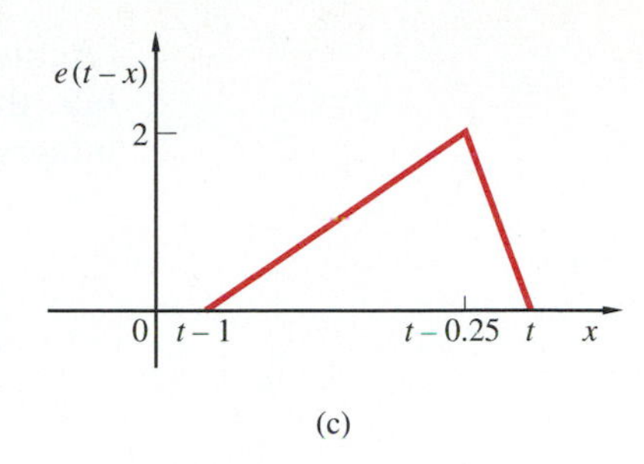

Figure 18
Folding and shifting.

expression can be written for the excitation, and so it is possible to carry out the convolution integral analytically. Not counting $t < 0$, there are three distinct intervals of time over which different analytical expressions apply: from 0 to 0.25, from 0.25 to 1, and from 1 on. In the latter interval, although the excitation is 0, there will still be a response, the natural response, as a consequence of the preceding pulse.

We will now evaluate the convolution integral in these three ranges. Figure 18 shows the folded excitation shifted to the right by the three different values of t lying in the three ranges. The presence of two variables might be a source of confusion, so let's clear it up. Both t and x represent time, but x is a dummy variable of integration. After we integrate, x vanishes. On the other hand, during the integration t is fixed. After the integration, however, t takes on all values of time in the specified ranges.

The excitation and its folded and shifted version in the first range are:

$$v_1(x) = 8x \qquad 0 \leqslant x \leqslant 0.25$$
$$v_1(t - x) = 8(t - x) \qquad t - 0.25 \leqslant x \leqslant t$$

Note that, although the folded and shifted excitation extends forward from $x = t - 0.25$, the impulse response does not start until $t = 0$. So the response in the first range is:

$$v_2(t) = 2\int_0^t (t - x)e^{-x/4}\,dx = -8te^{-x/4}\Big|_0^t + 8(x + 4)e^{-x/4}\Big|_0^t$$
$$= 8(t - 4 + 4e^{-t/4}) \qquad 0 \leqslant t \leqslant 0.25$$

Carry out the details of the integration to confirm the result.

In the second range, $0.25 \leqslant t \leqslant 1$, the integration is still to be carried out from 0 to t. Now, however, the integrand is expressed by different functions over different parts of this range. Study Figure 18b; in the two ranges, the folded and shifted excitation is:

$$v_1(t - x) = \tfrac{8}{3}[1 - (t - x)] \qquad 0 \leqslant x \leqslant t - 0.25$$
$$v_1(t - x) = 8(t - x) \qquad t - 0.25 \leqslant x \leqslant t$$

(Actually, the first of these expressions extends back to negative values of x but that interval does not contribute to the response.) The convolution integral becomes:

$$\begin{aligned} v_2(t) &= \frac{2}{3}\int_0^{t-0.25}(1-t+x)e^{-x/4}\,dx + 2\int_{t-0.25}^{t}(t-x)e^{-x/4}\,dx \\ &= \tfrac{8}{3}[(t-1)e^{-x/4} - (x+4)e^{-x/4}]\big|_0^{t-0.25} \\ &\quad + 8[-te^{-x/4} + (x+4)e^{-x/4}]\big|_{t-0.25}^{t} \\ &= \tfrac{8}{3}[5 - t - 16e^{-(t-0.25)/4} + 12e^{-t/4}] \qquad 0.25 \leq t \leq 1 \end{aligned}$$

(Confirm the details. Should $v_2(0.25)$, obtained from the lower limit in the second interval, be the same as $v_2(0.25)$ obtained from the upper limit in the first interval? Is it?)

In the third range, when $t > 1$, the integrands are the same as they were in the preceding equation. Now, however, the folded and shifted excitation is 0 from $x = 0$ to $t - 1$. Hence, the lower limit of integration for this range is changed from 0 to $t - 1$; that is the only thing different between this interval and the preceding one. In the third range, then, the convolution integral becomes:

$$\begin{aligned} v_2(t) &= \frac{2}{3}\int_{t-1}^{t-0.25}(1-t+x)e^{-x/4}\,dx + 2\int_{t-0.25}^{t}(t-x)e^{-x/4}\,dx \\ &= \tfrac{8}{3}[(t-1)e^{-x/4} - (x+4)e^{-x/4}]\big|_{t-1}^{t-0.25} \\ &\quad + 8[-te^{-x/4} + (x+4)e^{-x/4}\big|_{t-0.25}^{t} \\ &= \tfrac{8}{3}[4e^{-(t-1)/4} - 16e^{-(t-0.25)/4} + 12e^{-t/4}] \qquad t \geq 1 \end{aligned}$$

(Confirm this result. Should the value of $v_2(1)$ obtained from the lower limit in the third interval be the same as the value obtained from the upper limit in the second interval? Is it?)

From the last line it is clear that the response in this range is simply a decaying exponential. (Is that reasonable?) The entire response over all ranges is sketched in Figure 19. The excitation function is sketched on the same axes.

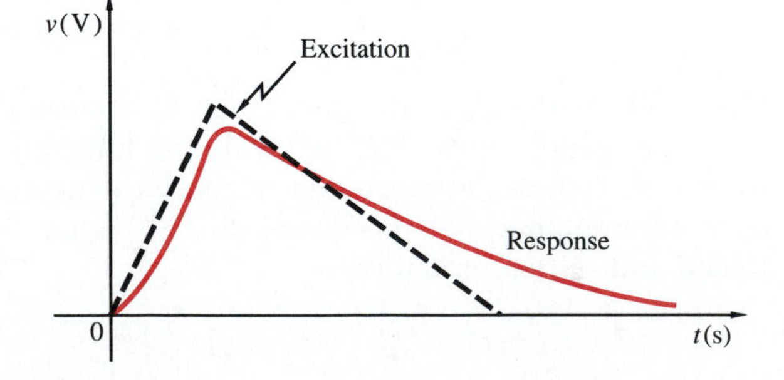

Figure 19 Response obtained by convolution.

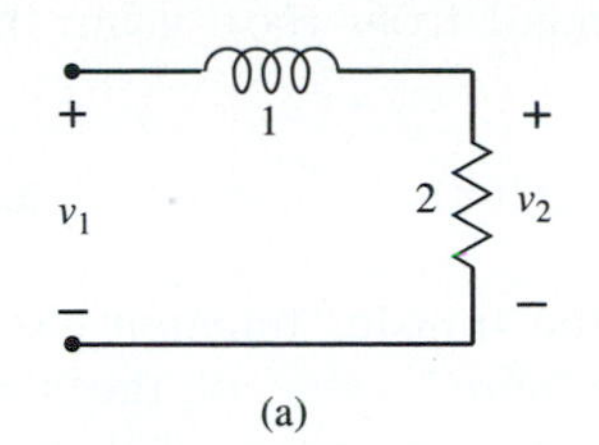

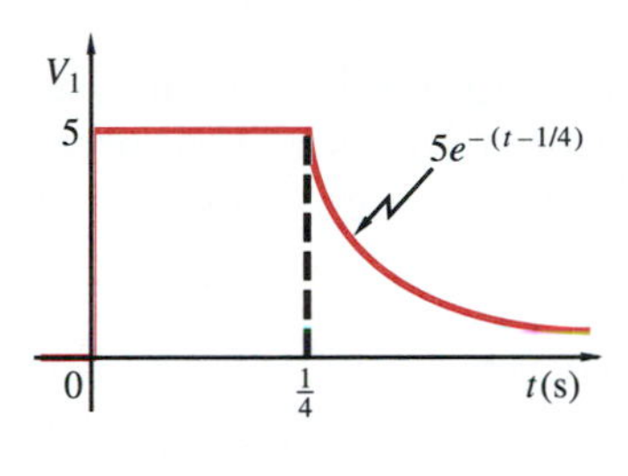

Figure 20
Circuit and excitation for Exercise 22.

EXERCISE 22

The circuit shown in Figure 20 is excited by the input shown. Find the response $v_2(t)$ using convolution, doing the folding and sliding on the impulse response.

ANSWER: $v_2(t) = 5(1 - e^{-2t})$ for $0 \leq t \leq \frac{1}{4}s$. $v_2(t) = 10e^{-(t-1/4)}(1 - e^{-(t-1/4)}) + 5(e^{1/2} - 1)e^{-2t}$ for $t \geq \frac{1}{4}s$. □

One final comment. The preceding development explicitly assumed that $H(s) = 0$ at infinity. What happens if it isn't? Then $H(s)$ will include a constant term $H(\infty)$ in its partial-fraction expansion. The inverse transform of $H(s)$, which is the impulse response, will include an impulse with weight $H(\infty)$. Thus, no insurmountable problem occurs when $H(s)$ is not zero at infinity.

Look at the convolution integral in (65). There it was assumed that $H(\infty) = 0$; hence, for a rational $H(s)$, $r_\delta(x)$ consists of exponentials with real or complex exponents. Now if $H(\infty)$ is a nonzero constant, the impulse response will include a term $H(\infty)e(t)$. (Confirm that.) It is a simple matter to give an interpretation for such a result; it's as if there were a straight-through resistive connection from the input to the output at infinite frequency. Glance back at Figure 15a. At infinite frequency, the capacitor acts as a short circuit; there is a direct connection from input to output and $H(\infty) = 1$.

The conclusion from all this is the following. When the transfer function is 0 at infinity, the convolution integral in (65) gives the response to an excitation $e(t)$. If $H(\infty)$ is not 0, then a term $H(\infty)e(t)$ must be added to this response.

Convolution and the Step Response

It has been established by the previous development that, using the operation called convolution, the response of an initially relaxed network to an arbitrary excitation is determined from a knowledge of the response of the network to a standard function. The standard function considered so far is the impulse. A similar conclusion is true if the standard function is a step function. That's the subject of this section.

Let $r_u(t)$, with Laplace transform $R_u(s)$, denote the response to a unit step of a network whose transfer function is $H(s)$. Naturally, r_u is called the *step response*. Since the transform of a unit step is $1/s$, (57) leads to:

$$R_u(s) = \frac{1}{s}H(s) = \frac{1}{s}R_\delta(s) \tag{66}$$

This gives a relationship between the *transforms* of the impulse and step responses. To get the relationship between the *time* responses, we take the inverse transform, either as this expression stands or after multiplying both sides by s. The results are:

$$r_u(t) = \int_0^t r_\delta(x)\,dx \tag{67a}$$

$$r_\delta(t) = \frac{d}{dt}r_u(t) + r_u(0)\delta(t) \tag{67b}$$

The initial value of the step response is readily found from (66), using the initial-value theorem:

$$r_u(0) = \lim_{s\to\infty}[sR_u(s)] = \lim_{s\to\infty} H(s) = H(\infty) \tag{68}$$

The step response will have a zero initial value if the transfer function has a zero at infinity. If $H(\infty)$ is nonzero, however, the initial value of the step response will be nonzero and, from (67b), the impulse response will itself contain an impulse.

That was by way of introduction. Glance again at (57), which relates the transforms of a network response to that of an arbitrary excitation. Multiplying and dividing on the right by s, we can put this equation into a number of different forms:

$$R(s) = s\left[\frac{H(s)}{s}E(s)\right] = s[R_u(s)E(s)] \tag{69a}$$

$$R(s) = \left[s\frac{H(s)}{s}\right]E(s) = [sR_u(s)]E(s) \tag{69b}$$

$$R(s) = [sE(s)]\left[\frac{H(s)}{s}\right] = [sE(s)]R_u(s) \tag{69c}$$

In each case, (66) was used to obtain the right-hand side.

Only the first of these will be considered here. The response $r(t)$ is obtained by taking the inverse transform of this expression. The right side is the product of two transforms multiplied by s. But the product of two transforms is the transform of the convolution of the corresponding time functions. If you also recall how the transform of the derivative of a function is related to the transform of the function, you will anticipate where we are heading. Let's do it gradually. Let the function whose transform is $R_u(s)E(s)$ be $f(t)$. The convolution theorem tells us:

$$\begin{aligned} f(t) = \mathscr{L}^{-1}[R_u(s)E(s)] &= \int_0^t r_u(x)e(t-x)\,dx \\ &= \int_0^t r_u(t-x)e(x)\,dx \end{aligned} \tag{70}$$

The value of the integral at $t = 0$ will be zero unless the step response contains an impulse at $t = 0$.[9] Is that possible? From (68) we see that the step response is finite at $t = 0$ if $H(\infty)$ is finite. That is, the transfer function of the network should not have a pole at infinity if the convolution is to be zero at $t = 0$. We assume this to be the case.[10]

The result of all this is that, when the last equation is combined with (69a),

[9] Or unless the excitation contains an impulse. But the response to an impulse in the excitation can be determined without going through all this; it is the impulse response.
[10] For $H(s)$ to have a pole at infinity, the highest power of s in its numerator should be greater than the highest power in the denominator. You would be hard put to find a significant network for which that is true.

we have:

$$r(t) = \frac{d}{dt} f(t) = \frac{d}{dt} \int_0^t r_u(x) e(t - x)\, dx$$
$$= \frac{d}{dt} \int_0^t r_u(t - x) e(x)\, dx \tag{71}$$

This result ranks in importance with the convolution theorem. It provides an alternative way for obtaining the response of an initially relaxed network to an excitation, this time in terms of the step response.

EXAMPLE 12 The circuit and excitation in this example are the same as the ones in Example 10. The objective this time is to find the response $v_2(t)$ using the step response in (71). As your part of the effort draw any diagrams needed, showing the necessary folding and shifting for the appropriate intervals.

SOLUTION The first step is to find the step response. Its transform and the resulting step response are:

$$R_u(s) = \frac{H(s)}{s} = \frac{1}{s + 1}$$
$$r_u(t) = e^{-t}$$

Using (70), the function whose derivative is the step response is given by the following convolution integral in the two ranges:

$$f(t) = \int_0^t 4e^{-x}\, dx = 4(1 - e^{-t}) \qquad 0 \leqslant t \leqslant 3$$
$$f(t) = \int_{t-3}^t 4e^{-x}\, dx = 2(e^{-(t-3)} - e^{-t})$$
$$= 4(1 - e^{-3})e^{-(t-3)} \qquad t \geqslant 3$$

(The value of $f(3)$ ought to be the same whether obtained from the upper limit of the first expression or the lower limit of the second one. Is it? Such a confirmation provides a check on our numerical work.) Finally, the response to the pulse is the derivative of $f(t)$:

$$r(t) = \frac{d}{dt} f(t) = 4e^{-t} \qquad 0 \leqslant t \leqslant 3$$
$$= -4(1 - e^{-3})e^{-(t-3)} \qquad t \geqslant 3$$

A comparison with the solution obtained in Example 10 is gratifying. We must have done *something* right.

One further thought before we quit. Glance back at (69). We used the first equation of that group to find the response to an arbitrary function in terms of the step response. The other two equations lead to alternative expressions for finding a response in terms of the step response. We will not pursue the matter any further here, but you are free to do so in your spare time.

SUMMARY

- Integral transforms
- The Laplace transform and the inversion integral
- Evaluating Laplace transforms: impulse, step, exponential, sinusoids, damped sinusoids, powers of t
- Real shifting theorem
- Scaling and time normalization
- Laplace transform of a derivative
- Laplace transform of an integral
- Initial-value theorem
- Final-value theorem and its conditions of validity
- Complex shifting theorem
- Inverse transform of frequency-domain derivatives
- Laplace transform of a periodic function
- Partial-fraction expansion of a rational function
- Expansion with simple real and complex poles
- Expansion with multiple real poles
- Expansion with double complex poles
- Solving circuit problems with the Laplace transform
- Solving intially relaxed circuit problems
- Solving initially active circuit problems
- The impulse response and its relationship to a transfer function
- Checks for detecting errors: dimensional, initial-value theorem, final-value theorem, circuit behavior at extreme frequencies
- Partial-fraction expansion of improper rational functions
- The convolution theorem
- Interpretation of the convolution integral
- Evaluation of the convolution integral in circuit problems
- The step response and convolution

PROBLEMS

1. Find the Laplace transforms of each of the following functions, either by direct integration or by other means:

a. $f(t) = te^{-2t}$
b. $f(t) = t \sin 10t$
c. $f(t) = 10 \cos (\omega t - \pi/3)$
d. $f(t) = at^3 + bt^2 + ct + d$
e. $f(t) = te^{-2t}u(t - 2)$
f. $f(t) = (t - 2)e^{-2(t-2)}u(t)$
g. $f(t) = \sin 10t\, u(t - 1)$
h. $f(t) = \sin (10t - 10)\, u(t - 1)$

2. Sketch the following functions and then find their Laplace transforms:

a. One cycle of $\sin \omega t$
b. A curve that rises linearly from 0 at $t = 0$ to 5 at $t = 1$ s; stays constant at 5 from $t = 1$ to 3 s; and then drops to 0 thereafter
c. An isosceles triangle rising from 0 to 2 at 1 s, then dropping to 0 at 2 s and remaining 0

3. The function of time sketched in Figure P3 can be formed by a linear combination of displaced functions.

a. Derive such a linear combination.
b. Find its Laplace transform.

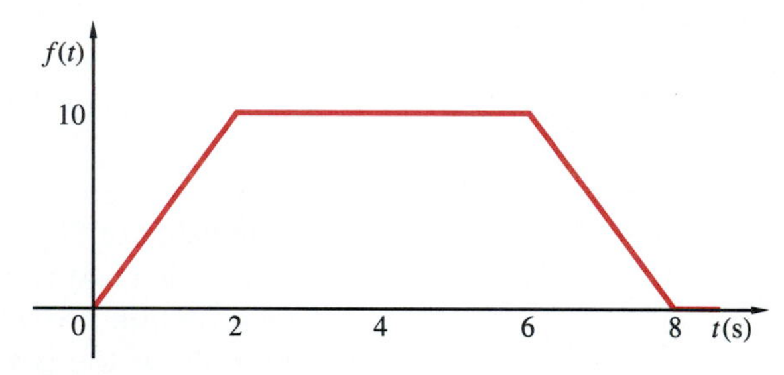

Figure P3

4. Using only the time-differentiation property (repeatedly, if necessary), find the Laplace transforms of the following functions:

$$e^{at}, \quad \sin \omega t, \quad \cos \omega t, \quad \sinh bt, \quad \cosh bt$$

5. Using the time-differentiation property on the function $f(t) = e^{-10t} \sin 10t$, find the Laplace transform of $e^{-10t} \sin (10t - 45°)$.

6. The Laplace transform of a function can be determined directly or found in a table of transform pairs. Now suppose the time scale is expanded so that each old unit of time is expanded to 4 new units. Use the scaling property to determine the Laplace transform of the new function.

7. The Laplace transform of a certain function $f(t)$ is $F(s) = 5(s + 2)/(s + 1)(s + 3)$. Now a sketch of $f(t)$ is moved 2 s to the right along the time axis. Determine the Laplace transform of the new function.

8. The Laplace transform of a function is $F(s) = s/(s^2 + 2s + 2)$. The poles and zero are now moved horizontally to the left by one unit. Find the inverse transform of the new function of s.

9. Assuming $F(s)$ is the Laplace transform of $f(t)$:
a. Use the complex differentiation property to determine the inverse transform of $d^2F(s)/ds^2$.
b. Apply this result to find the inverse transforms of the following functions of s:

1. $F(s) = \dfrac{5}{(s + 2)^3}$

2. $F(s) = \dfrac{2(s + 4)}{(s^2 + 1)^4}$

10. Use the complex differentiation property of Laplace transforms to find the Laplace transforms of the following functions:
a. $f(t) = 2t^3$
b. $f(t) = t \sin \omega t$

11. a. Use the time-differentiation property and a knowledge of the transforms of $\sin \omega t$ and $\cos \omega t$ to show that:

1. $\mathscr{L}(t \cos \omega t) = \dfrac{(s^2 - \omega^2)}{(s^2 + \omega^2)^2}$

2. $\mathscr{L}(t \sin \omega t) = \dfrac{2\omega s}{(s^2 + \omega^2)^2}$

b. Verify these results by direct evaluation.
c. Also verify them by expressing the sinusoids in terms of exponentials and using the complex shifting theorem.

12. a. Find the Laplace transforms of the functions that follow in the left column, using the table of transforms (Table 1) and also using appropriate frequency-domain properties
b. With the left-column transforms given, find the Laplace transforms of the functions in the right column, using the real differentiation property.

$f_1(t) = t \cos 2t \qquad g_1(t) = 5 \cos 2t - 10t \sin 2t$

$f_2(t) = te^{-t/10} \qquad g_2(t) = e^{-t/10}(10 - t)$

$f_3(t) = e^{-3t/4} \sin 2t \qquad g_3(t) = 5e^{-3t/4} \cos (t + 20.6°)$

13. Find the Laplace transforms of the following periodic functions:
a. A half-wave rectified sine function
b. A full-wave rectified sine function
c. A sawtooth function
d. A symmetrical square wave
e. An isosceles triangular wave whose average value is half the peak-to-peak value

14. Find the initial values and initial slopes of the inverse transforms of the following functions. When possible, find the final values also.

a. $F(s) = \dfrac{3s + 2}{s^2 + 3s + 6}$

b. $F(s) = \dfrac{2s + 5}{(s + 1)(s^2 + 4)}$

c. $F(s) = \dfrac{2s^2 + 10}{s^3 + 2s^2 + 5s}$

d. $F(s) = \dfrac{(2s + 7)(1 - e^{-s})}{s^2 + 2s + 3}$

15. Find the partial-fraction expansions of the following functions in s:

a. $F(s) = \dfrac{2s^2 + 6s + 10}{s^2 + 4s + 3}$

b. $F(s) = \dfrac{s^3 + 4s^2 + 3s + 8}{(s + 2)(s + 4)}$

16. Find the inverse transforms of the following functions of s. Brute force will always work, but, wherever possible, use simplifications and the transform properties discussed in the text. Confirm the initial-value theorem and the final-value theorem whenever applicable.

a. $F(s) = \dfrac{3s(s + 1)}{(s + 2)(s + 5)}$

b. $F(s) = \dfrac{6(s^2 - s + 2)}{s(s + 1)(s + 3)}$

c. $F(s) = \dfrac{s + 20}{s(s^2 + 2s + 10)}$

d. $F(s) = \dfrac{3s + 1}{(s + 2)(s^2 + 4s + 8)}$

e. $F(s) = \dfrac{s^2 + 2s + 2}{s(s + 2)^2}$

f. $F(s) = \dfrac{24s^2 + 48s + 48}{(s + 1)(s^2 + 4s + 29)}$

g. $F(s) = \dfrac{2se^{-2s}}{(s + 1)(s + 3)}$

h. $F(s) = \dfrac{12s^2 + 33s + 20}{(s + 2)(s + 4)(s^2 + 2s + 2)}$

i. $F(s) = \dfrac{s(1 - e^{-2s})}{s^2 + 2s + 2}$

j. $F(s) = \dfrac{2s^2 + s + 3}{s^2(s^2 + 4)^2}$

k. $F(s) = \dfrac{(s + 3)e^{-s}}{(s^2 + 2s + 10)^2}$

l. $F(s) = \dfrac{s(s + 2)e^{-3s}}{(s + 1)^2(s + 4)}$

17. Use Laplace transform methods to find the solutions of the following problems from Chapter 7:

10 12 13 16 17 19 20 22
23 24 25 28 29 33 35

18. The circuit in Figure P18 is in equilibrium when the switch is closed at $t = 0$.

a. Find the capacitor voltage just before the switch is closed.

b. Find $v(t)$ for $t \geqslant 0$.

c. Find $v_C(t)$ for $t \geqslant 0$.

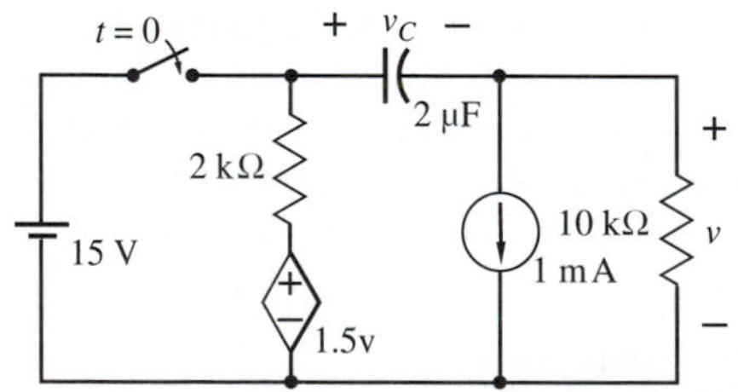

Figure P18

19. The circuit in Figure P19 is in equilibrium when the switch is opened at $t = 0$.

a. Draw the transform-domain circuit valid for $t \geqslant 0$.

b. Find the Laplace transform of v and then its inverse transform.

c. Find the Laplace transform of i and then its inverse transform.

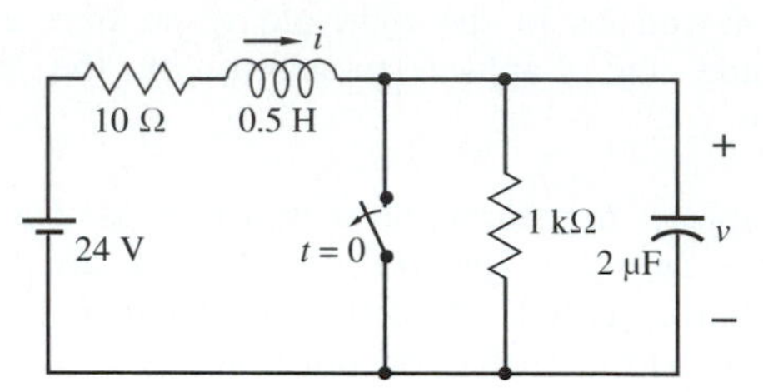

Figure P19

20. The circuit in Figure P20 is in equilbrium when the switch is closed at $t = 0$.

a. Find the Laplace transform functions $V_1(s)$ and $V_2(s)$. (Confirm your results by using more than one approach: node equations, mesh equations, Thévenin's theorem, for example).

b. Find the inverse transforms $v_1(t)$ and $v_2(t)$ for $t \geqslant 0$.

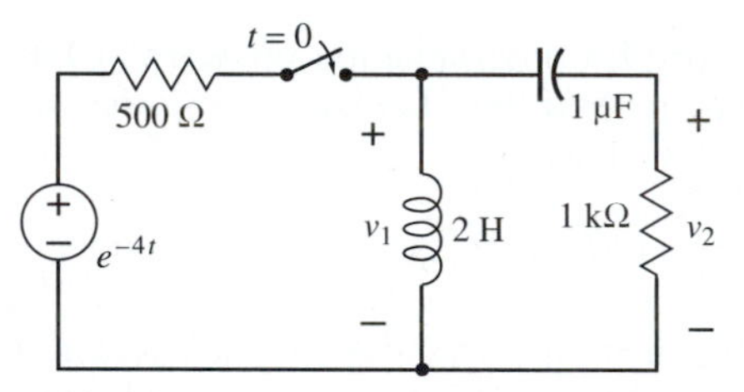

Figure P20

21. The circuit in Figure P21 has been in equilibrium when the make-before-break switch is moved from position A to B at $t = 0$. The parameter values are $L_1 = 500$ mH, $L_2 = 125$ mH, $k = 0.8$, $R_1 = 4\,\Omega$, $R_2 = 2\,\Omega$, and $v_s = 10\cos 400t$. Find $v_2(t)$ for $t \geqslant 0$ by first finding its Laplace transform.

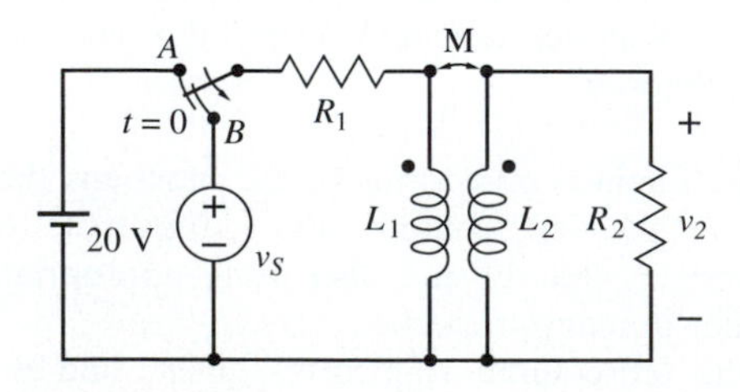

Figure P21

22. The excitation and the resulting response of an initially relaxed circuit are the following:

$$e(t) = \sin 2t$$
$$r(t) = -\tfrac{1}{2}e^{-2t} + \tfrac{1}{8}\cos 2t + \tfrac{1}{8}\sin 2t$$

a. Determine the network function $H(s)$.
b. From this, find the time response when the excitation is the sawtooth wave in Figure P22.

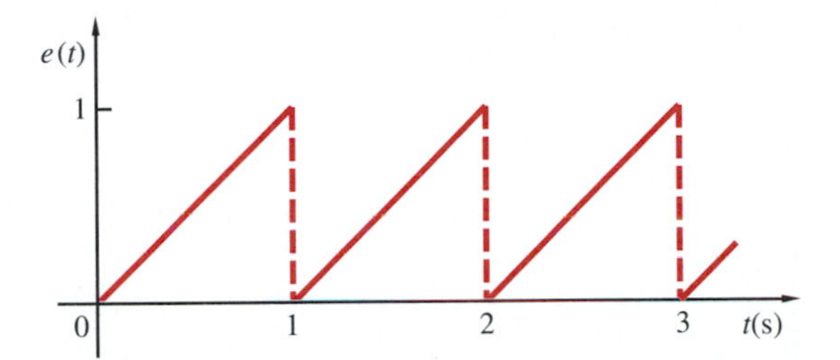

Figure P22

23. An initially relaxed circuit is shown in Figure P23. The location of the excitation and the desired response are indicated.
a. Find the impulse response.
b. Find the step response.
c. Confirm that (67) in the text relating the impulse and step responses is satisfied.

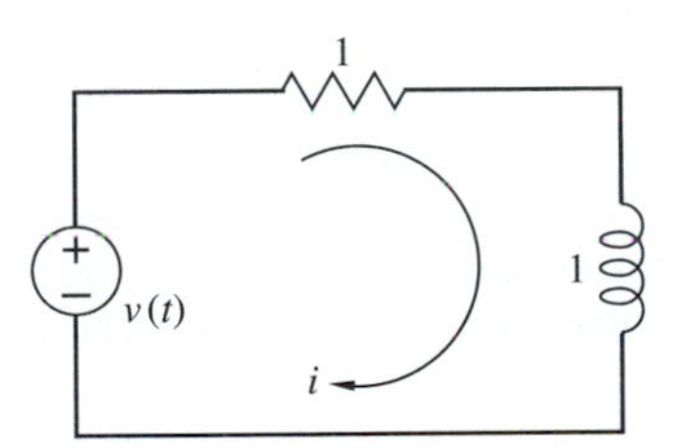

Figure P23

24. Repeat Problem 23 for the circuit in Figure P24.

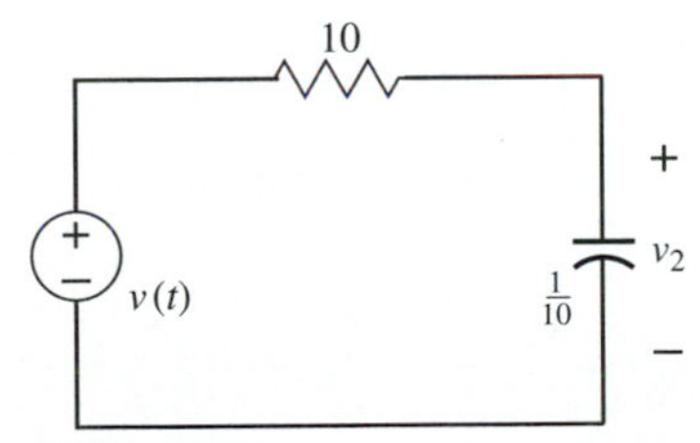

Figure P24

25. Repeat Problem 23 for the circuit in Figure P25.

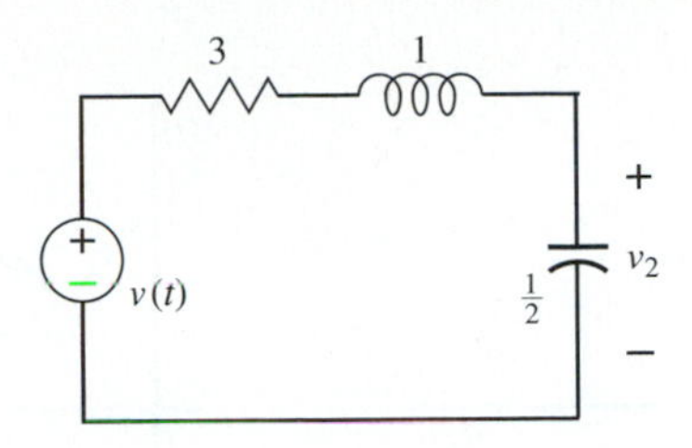

Figure P25

26. Repeat Problem 23 for the circuit in Figure P26.

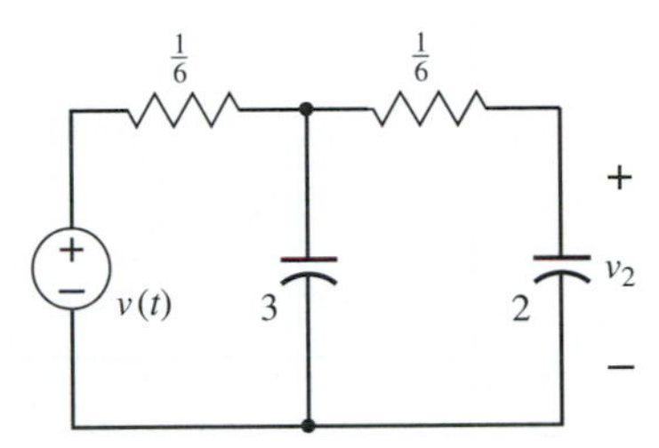

Figure P26

27. Repeat Problem 23 for the circuit in Figure P27.

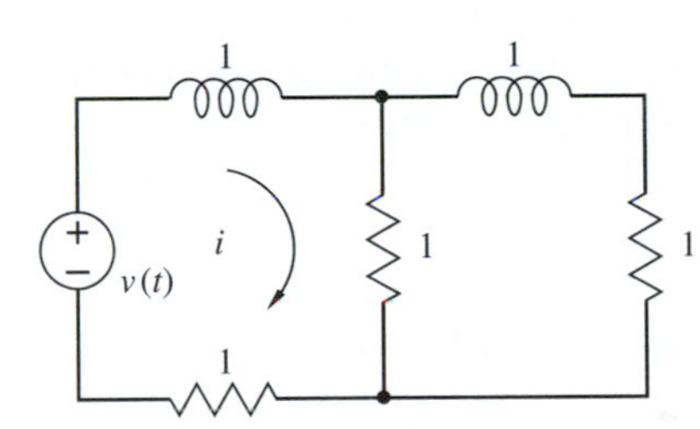

Figure P27

28. A number of excitation functions are shown in Figure P28a through e.
a. For any one of these functions, find the indicated response in the circuit shown in Figure P23, using the impulse response and convolution.
b. Repeat, using the step response and (71) in the text.

29. Repeat Problem 28 for the circuit in Figure P24.

30. Repeat Problem 28 for the circuit in Figure P25.

31. Repeat Problem 28 for the circuit in Figure P26.

32. Repeat Problem 28 for the circuit in Figure P27.

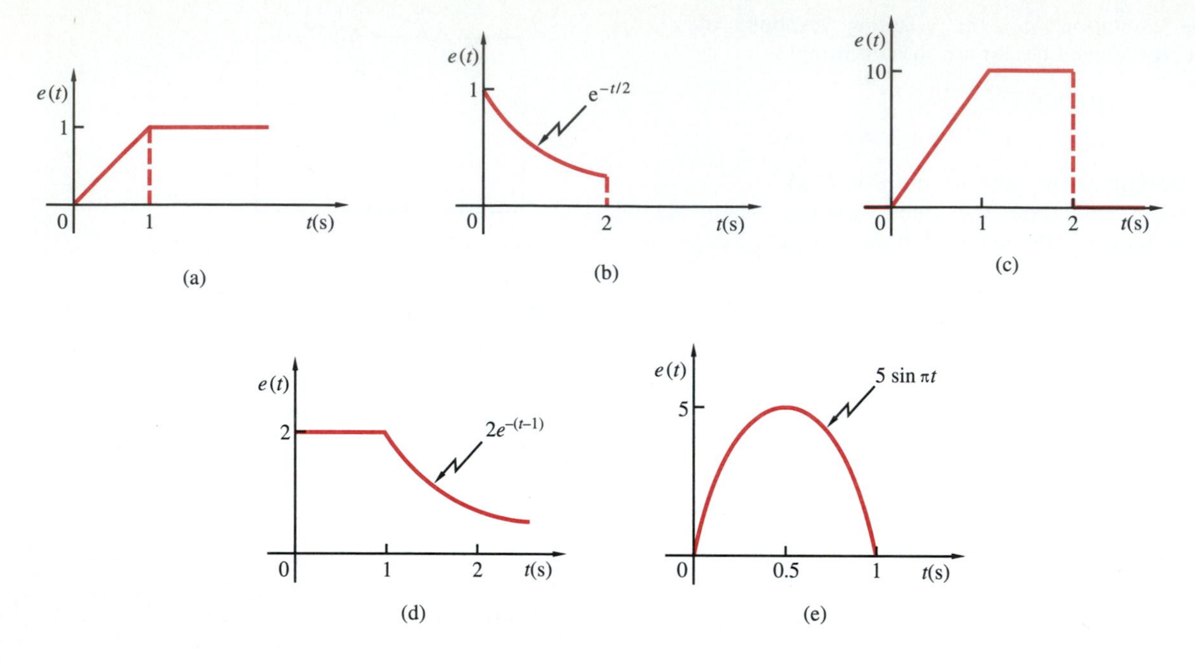

Figure P28

33. The circuit in Figure P33 is initially relaxed. Assuming an impulse voltage excitation, find the current response at the input terminals.

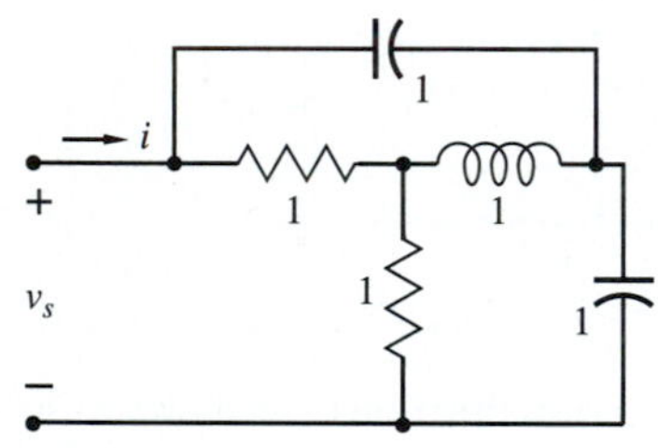

Figure P33

34. The excitation to an initially relaxed network is $e(t) = \cos 2t$. The Laplace transform of the response is found to be:

$$R(s) = \frac{s^2 + 2s + 5}{(s^2 + 2s + 4)(s^2 + 4)}$$

a. Determine the steady-state response.
b. Determine the natural response.
c. The cosine input is replaced by a step. Determine the step response.

35 An *integral equation* is like a differential equation except that it contains integrals of the unknown function instead of (or in addition to) its derivatives. Use Laplace transforms and convolution to work out a procedure and solve the following integral equations for $f(t)$:

a.
$$f(t) + \int_0^t f(x)e^{-2(t-x)}\,dx = 2t$$

b.
$$f(t) + \int_0^t f(x)e^{-(t-x)}\sin(t-x)\,dx = 5$$

c.
$$\frac{df}{dt} + 2f(t) + 9\int_0^t f(x)(t-x)\,dx = 1 - e^{-2t}$$
$$f(0) = 0$$

DESIGN PROBLEMS

1. The rise time is defined as the time required for a signal to rise from 10% to 90% of its steady-state value in response to a step excitation. In Figure DP1a the rise time is $t_2 - t_1$. The overshoot is the peak excursion above the steady-state value.

Select values of R and L in the circuit shown in

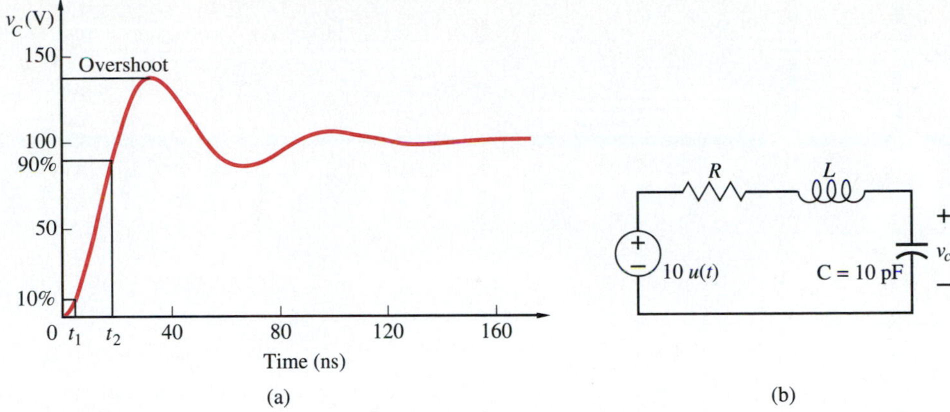

Figure DP1

Figure DP1b, such that the rise time of the output voltage is less than or equal to 12.5 ns and the overshoot is less than 50%.

2. Design a circuit, as in Figure DP2, consisting of only passive elements, such that $v_0 = 1 - 1.01 \times e^{-6\times10^4 t} \cos(3.105 \times 10^5 t + 0.128)$ when the input is a unit step. (The angle is in radians.)

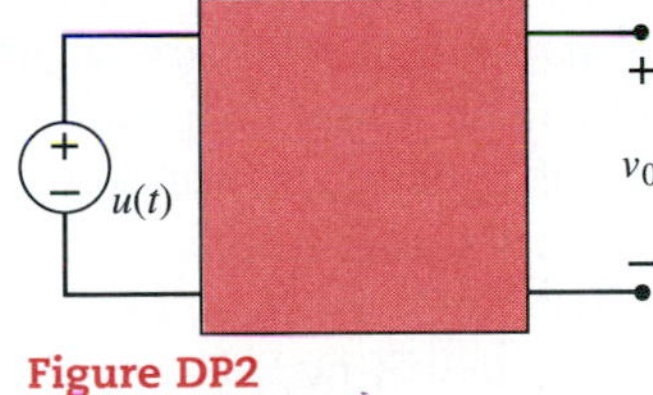

Figure DP2

3. It is common to draw block diagrams representing the flow of signals in many systems. The block diagram in Figure DP3 is an example. The signals at different points in the system are indicated. Let $R(s)$, $E(s)$, $F(s)$, and $G(s)$ represent the Laplace transforms of the signals $r(t)$, $e(t)$, $f(t)$, and $g(t)$.
   a. From the block diagram $e(t) = r(t) - g(t)$, determine a relationship between $E(s)$, $R(s)$, and $G(s)$.
   b. The transfer function of the amplifier $\frac{F(s)}{E(s)}$ is a frequency independent constant K, and the transfer function of the servomotor is given by $\frac{G(s)}{F(s)} = \frac{0.8}{s(0.15s + 1)}$. Determine the overall transfer function $\frac{G(s)}{R(s)}$.
   c. Determine the value of K, for which $g(t)$ will be critically damped when the input $r(t)$ is a unit step. Determine $g(t)$ for this value of K.

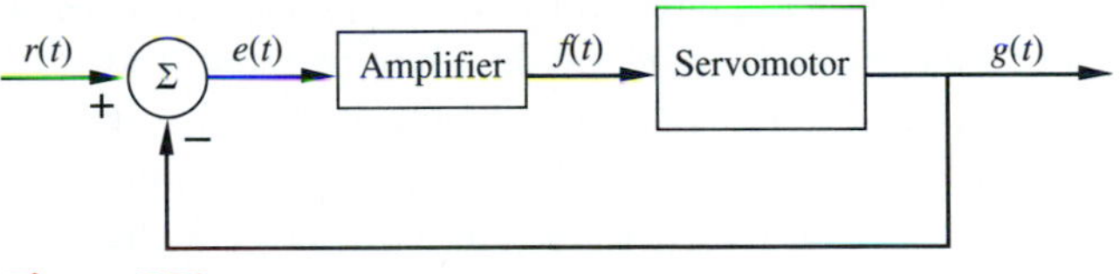

Figure DP3
Block diagram of a control system.

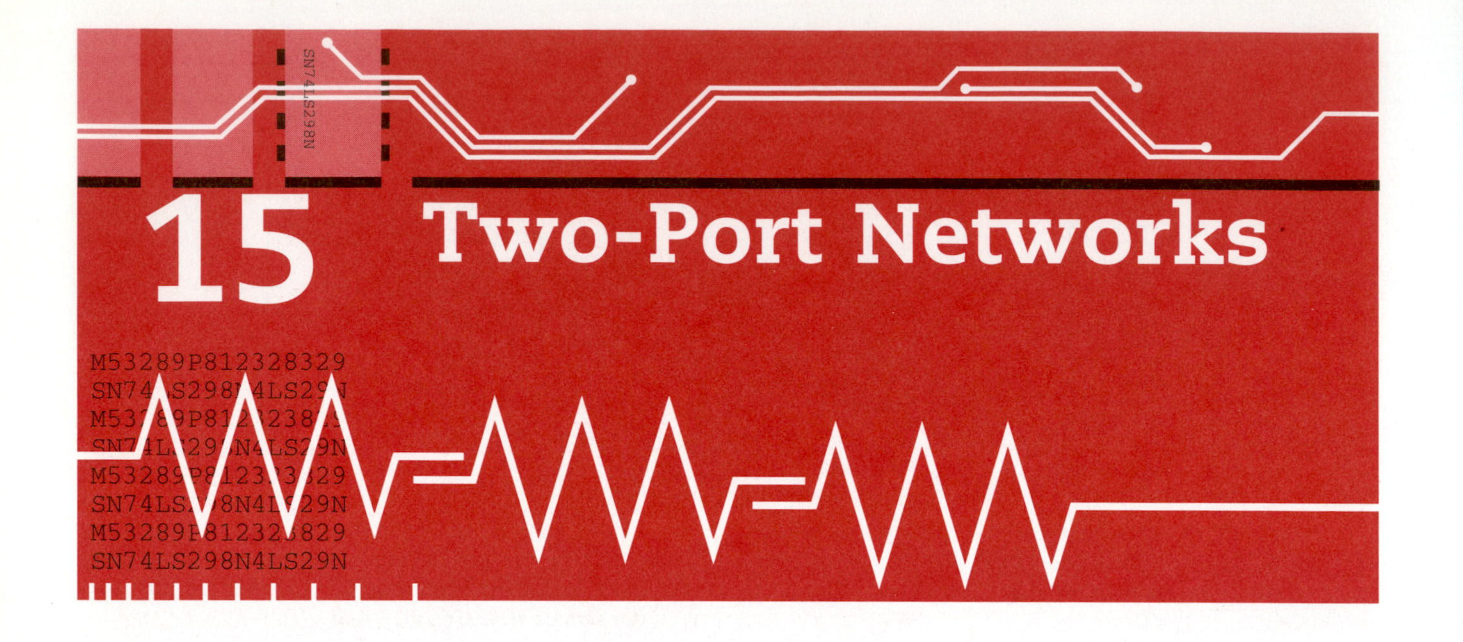

15 Two-Port Networks

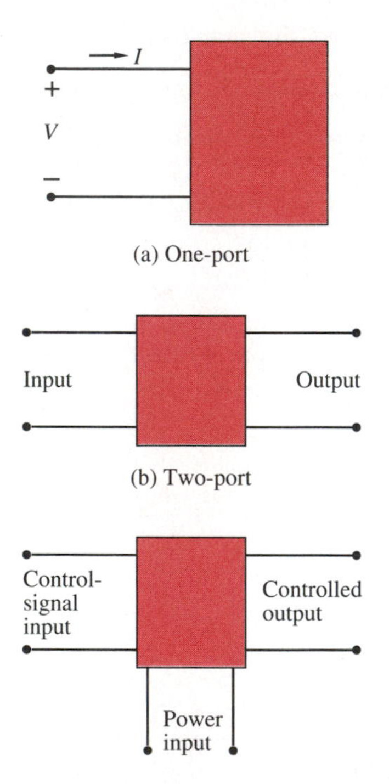

Figure 1
Various multiports.

So far in the discussion of networks, we have been concerned with determining currents or voltages—and sometimes power dissipated and energy stored—after being given the sources and the details of the network structure. We have not been restricted as to the permissible locations of the sources or the permissible places where an output could be taken. In many applications, however, we have no interest in the internal parts of a network; we want to know only the relationships among voltages and currents at the locations in a network where external connections can be made. It is only at those locations where an exciting source will be connected or an output will be taken. Figure 1a shows the simplest network from this point of view; it has just one pair of terminals, a single port. Either the voltage or the current is the excitation, the other being the response. In the complex-frequency domain, the network function is a driving-point function, either impedance or admittance.

Other networks may have more pairs of terminals to which external connections can be made; they are multiports. Figure 1b shows a two-port. Signals are fed into one of the ports (the *input* port) and extracted from the other (the *output* port). (Such a description requires that there be no sources internal to the ports.) An example might be an amplifier, a filter, or a transmission line (power or telephone).

In other systems, different varieties of signals may enter the system at different ports. A three-port is shown in Figure 1c. The power is supplied at one of the ports while a signal which is to control the power enters a second port. The controlled output signal then appears at a third port. An op-amp circuit fits this description exactly. A modulator can also be described as a three-port. The so-called carrier enters at one port, the modulating signal at another; the modulated carrier is extracted from a third port. The modulating process is not a linear one, however, so that this particular three-port must contain nonlinear components.

Although multiport concepts are not all restricted to *linear* electrical networks—as illustrated in the preceding paragraph—we are limited in this book to linear networks only. The concepts to be developed can easily be extended to other types of systems—mechanical, electromechanical, thermal, chemical—which can all be described in terms of the concepts to be discussed in this chapter. Of course, the variables used to describe such systems will not be voltages and currrents.

In this chapter we will limit the discussion to two-ports, although the ideas can easily be generalized to higher-order multiports. We will study (1) the different ways in which two-ports can be described in the complex-frequency domain (which includes the sinusoidal steady state) and (2) the properties of the network functions that characterize a two-port. An introduction to the subject was briefly discussed for dissipative circuits in Chapter 4. Review that material before you go on.

1 ADMITTANCE AND IMPEDANCE REPRESENTATIONS

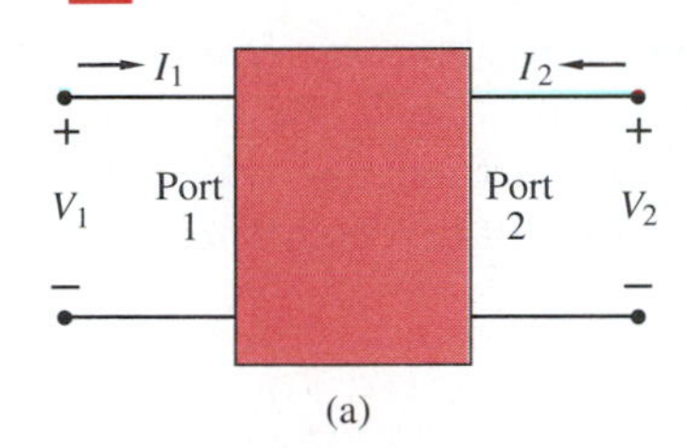

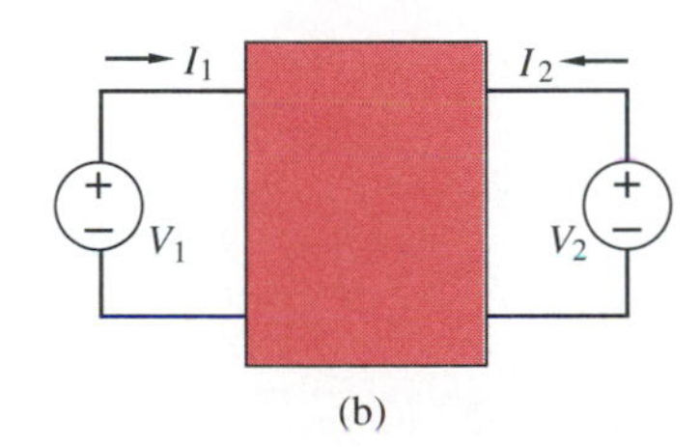

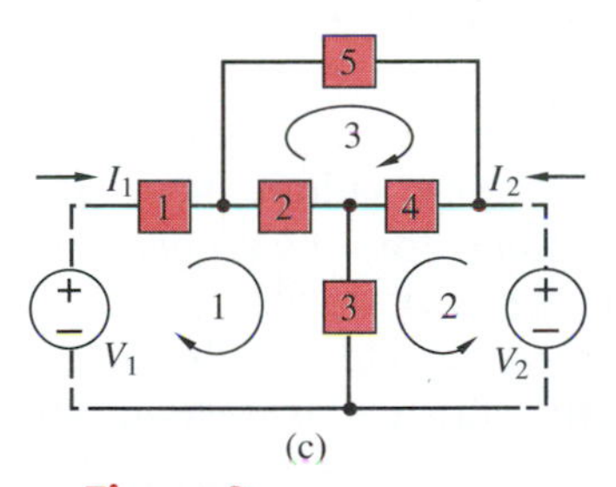

Figure 2
Conventional two-port references and example circuit.

A two-port structure is shown in Figure 2a. Standard references are universally used for the port variables, as shown, even though a source might be connected at one port and a load at the other. (Remember, there are no independent sources inside.) In the case of a one-port, there is a single voltage and a single current, the ratio of whose Laplace transforms (or phasors) is the impedance (or the admittance); and this quantity completely characterizes the network. But in the case of a two-port, there are two voltages and two currents; how should we describe the behavior of the network at its terminals? There are different possibilities depending on which specific variables are considered to be excitations and which ones responses.

Short-Circuit Admittance Parameters

In Figure 2b the excitations are the voltages, so the responses are the currents. Without knowing anything about the network lying between the two ports (other than that it is linear and contains no sources), we know that a set of loop equations can be written and solved for all the loop currents. In this process, we make sure to choose the loops so that each of the sources appears in only one loop. We number the loops so that loops 1 and 2 are the ones containing the corresponding sources, and we choose those loop currents so that they coincide with the port currents. (All of this can always be done.)

A nontrivial example is shown in Figure 2c. The loops are chosen and numbered as described. You can confirm that the loop equations for the loops shown are:

$$\begin{bmatrix} Z_1 + Z_2 + Z_3 & Z_3 & -Z_2 \\ Z_3 & Z_3 + Z_4 & Z_4 \\ -Z_2 & Z_4 & Z_2 + Z_4 + Z_5 \end{bmatrix} \begin{bmatrix} I_1 \\ I_2 \\ I_3 \end{bmatrix} = \begin{bmatrix} V_1 \\ V_2 \\ 0 \end{bmatrix}$$

Study this equation. Note that all elements of the coefficient matrix are impedances. Exactly two of the entries in the voltage vector, corresponding to the two-port excitations, are nonzero.

In the most general linear network, with the same approach in choosing and numbering the loops, a matrix equation of the same form will be obtained. For variety, let's write out the equations in scalar form; thus:

$$\begin{aligned} a_{11}I_1 + a_{12}I_2 + a_{13}I_3 + \cdots + a_{1n}I_n &= V_1 \\ a_{21}I_1 + a_{22}I_2 + a_{23}I_3 + \cdots + a_{2n}I_n &= V_2 \\ a_{31}I_1 + a_{32}I_2 + a_{33}I_3 + \cdots + a_{3n}I_n &= 0 \\ \cdot \quad \cdot \quad \cdot \quad \cdot \quad \cdot \quad & \cdot \\ a_{n1}I_1 + a_{n2}I_2 + a_{n3}I_3 + \cdots + a_{nn}I_n &= 0 \end{aligned} \tag{1}$$

Dimensionally, the coefficients of the loop equations are impedances, although impedance symbols were not used to represent them, for reasons that will become clear as we proceed. It is not our goal to find *all* the currents in the network, only those at the ports, chosen to be the same as loop currents 1 and 2.

The solution for these two currents will have contributions from each source, and since the network is linear, each current will be a linear combination of the two excitation voltages. Thus the port-current transforms (or phasors) can be expressed in terms of the port-voltage transforms (or phasors) as follows:

$$\begin{aligned} I_1 &= y_{11}V_1 + y_{12}V_2 \\ I_2 &= y_{21}V_1 + y_{22}V_2 \end{aligned} \tag{2a}$$

or, in matrix form:

$$\mathbf{I} = \mathbf{YV} \tag{2b}$$

where

$$\mathbf{I} = \begin{bmatrix} I_1 \\ I_2 \end{bmatrix} \qquad \mathbf{V} = \begin{bmatrix} V_1 \\ V_2 \end{bmatrix} \qquad \mathbf{Y} = \begin{bmatrix} y_{11} & y_{12} \\ y_{21} & y_{22} \end{bmatrix} \tag{3}$$

Dimensionally, the coefficients multiplying the voltages are admittances. Even though capitals are normally used for (one-port) admittances and impedances, it is standard to use lowercase letters for these two-port parameters. A boldface capital letter (**Y**) is used, however, to designate the matrix of the coefficients. The matrix form, indeed, would be the same, independent of the number of rows and columns. Thus, the matrix form in (2b) would remain valid for a multiport of any order.

For a one-port, only one quantity (admittance or impedance) is needed to completely describe the network at its terminals. For a general linear two-port, we find that four quantities are needed. Once these four quantities are known, the description of the *I-V* relationships at the terminals is complete. The four *admittance parameters,* or *y-parameters,* as they are referred to, can be determined, at least conceptually, by solving the equations in (1), say, by Cramer's rule. Thus:

$$I_1 = \frac{\Delta_{11}}{\Delta} V_1 + \frac{\Delta_{21}}{\Delta} V_2$$
$$I_2 = \frac{\Delta_{12}}{\Delta} V_1 + \frac{\Delta_{22}}{\Delta} V_2 \tag{4}$$

where Δ is the determinant of the equations in (1) and the subscripted Δ's are its cofactors. (Review Appendix A, if you need to.)

The y-parameters in (2) can now be immediately recognized by comparing the preceding expression with (2a). They will be:

$$y_{11} = \frac{\Delta_{11}}{\Delta} \qquad y_{12} = \frac{\Delta_{21}}{\Delta}$$
$$y_{21} = \frac{\Delta_{12}}{\Delta} \qquad y_{22} = \frac{\Delta_{22}}{\Delta} \tag{5}$$

While setting up the determinant and its cofactors in an extensive network may involve a nontrivial amount of work, the process is straghtforward.

EXERCISE 1

The loop equations for the circuit in Figure 2c were provided just prior to the general loop equations in (1). Assume that $Z_1 = R_1$, $Z_2 = sL_2$, $Z_3 = R_3$, $Z_4 = R_4$, and $Z_5 = 1/5Cs$. Using (5), find the y-parameters of this two-port. Use dimensional checks along the way to avoid errors. Specify the highest power of s in numerator and denominator of each y-parameter if written as a rational function. □

y-PARAMETERS AS SHORT-CIRCUIT ADMITTANCES While the immediately preceding discussion gives formulas for calculating the y-parameters, it does not yield a simple interpretation for them. Such an interpretation can be easily obtained.

Suppose that only port 1 is excited and port 2 is short-circuited, as shown in Figure 3a, so that V_2 is zero. Setting $V_2 = 0$ in (2a) reduces these equations to $I_1 = y_{11}V_1$ and $I_2 = y_{21}V_1$. From these, we find:

$$y_{11} = \left.\frac{I_1}{V_1}\right|_{V_2=0} \tag{6a}$$

$$y_{21} = \left.\frac{I_2}{V_1}\right|_{V_2=0} \tag{6b}$$

From the manner in which these expressions were obtained, it is natural that they would be called the *short-circuit admittances.*

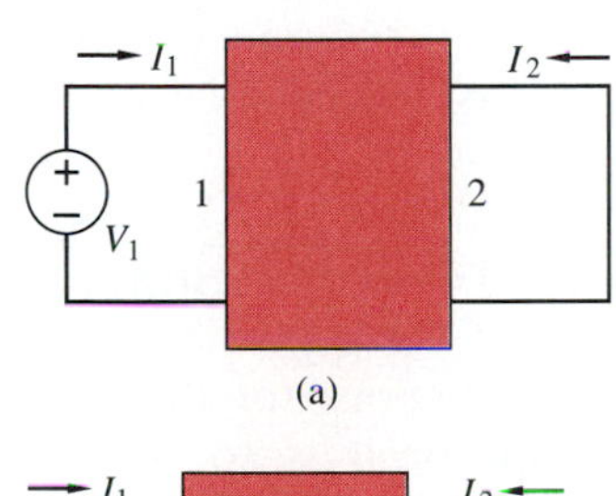

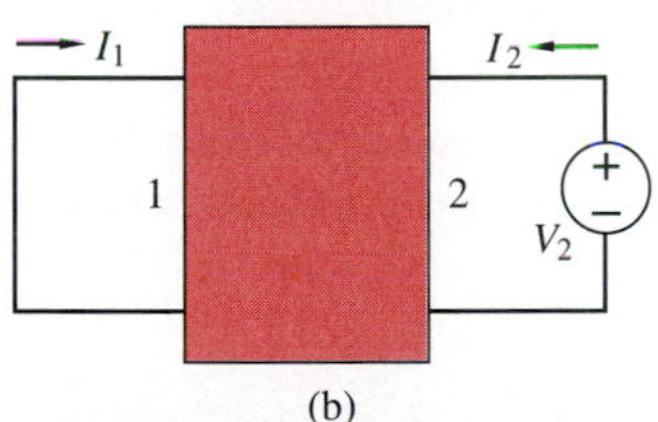

Figure 3
Interpretation of the short-circuit parameters.

The quantity y_{11} is the input admittance at port 1 with port 2 short-circuited. On the other hand, y_{21} is a transfer admittance; it is the *short-circuit transfer admittance* from port 1 to port 2. The order of the subscripts is consistent with the order previously defined for transfer functions. (Because the transmission of signals is normally taken to be from left to right in Figure 3a, y_{21} is also called the *forward* short-circuit admittance parameter.)

Now suppose that the source and short circuit are interchanged, as in Figure 3b. Setting $v_1 = 0$ in (2a) leads to:

$$y_{22} = \left.\frac{I_2}{V_2}\right|_{V_1=0} \tag{7a}$$

$$y_{12} = \left.\frac{I_1}{V_2}\right|_{V_1=0} \tag{7b}$$

In this case V_2 is the only source in the network. By definition, then, y_{22} is the input admittance at port 2 when port 1 is short-circuited. Since port 2 is normally the output port, this admittance is also called the *short-circuit output admittance.* Also, y_{12} is the *transfer admittance* from port 2 to port 1 when port 1 is short-circuited; that is, it is the *reverse short-circuit transfer admittance.*

Collectively, these y parameters are called the *short-circuit admittance* parameters. Two of them (y_{11} and y_{22}) are driving-point admittances, and two (y_{21} and y_{12}) are transfer admittances. The expressions in (6) and (7) can be looked upon as the definitions of the short-circuit admittance parameters.

In the preceding development, since one port is short-circuited and the other excited by a voltage source, it might appear that the condition at one port is different in nature from the condition at the other port. This is not the case, however. A voltage source whose voltage is zero is indistinguishable from a short circuit. Therefore, in the defining relationships in (6) and (7), both ports carry the same kind of excitation (a voltage source), except that one of the sources has the value zero.

DETERMINING y-PARAMETERS FROM THEIR DEFINITION Given a two-port, how do we go about determining the y-parameters? One possibility has already been discussed: writing and solving the loop equations for the given two-port, assuming it is voltage-excited. Another approach is to use the definitions of the parameters just discussed. That is, excite the ports with voltage sources at both ports, one of which has the value zero; two of the parameters follow. Then interchange the nonzero and the zero excitations, and determine the other two parameters.

EXAMPLE 1 Find the y parameters of the pi circuit in Figure 4a.

SOLUTION We assume that the branches of the circuit represent self-contained components, without coupling from one branch to another. For example, there

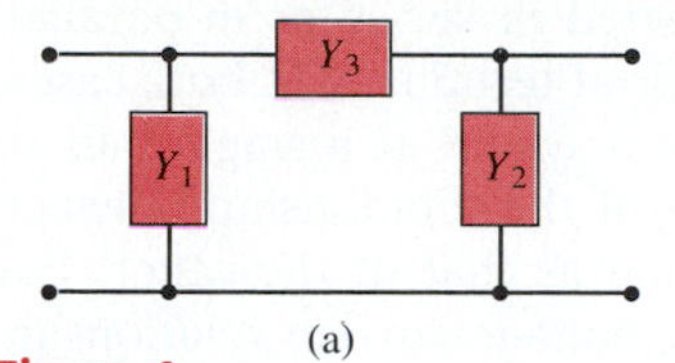

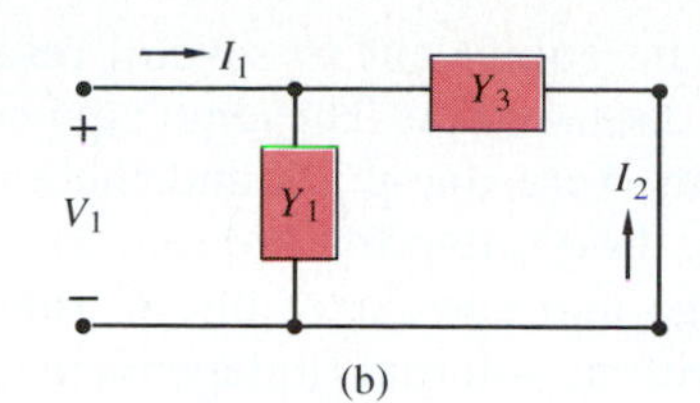

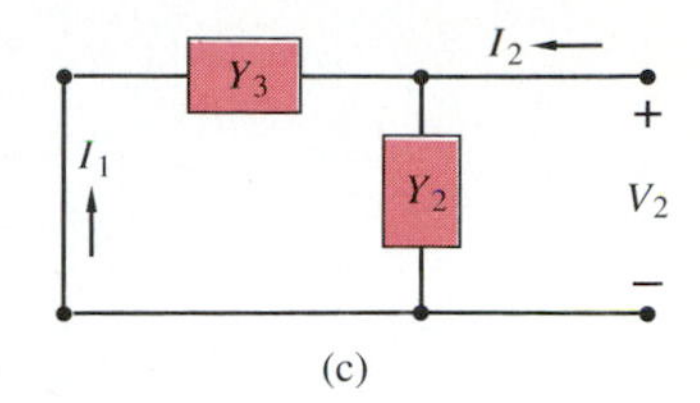

Figure 4
Determining the short-circuit parameters of the pi.

cannot be a controlled source in the circuit whose controlling branch and controlled branch are in different arms of the pi. Can there be controlled sources at all, even if both controlling and controlled branches are in the same arm of the pi? The answer is yes; the only requirement is that each branch in the pi be a one-port.

In the circuit of Figure 4a, the results can be obtained simply by inspection. With port 2 short-circuited, as in Figure 4b, Y_1 and Y_3 are in parallel. Hence, y_{11} is just their sum. Note that I_2 is the current in a branch whose voltage is V_1; taking the references into account, $I_2 = -Y_3V_1$. Hence:

$$\begin{aligned} y_{11} &= Y_1 + Y_3 \\ y_{21} &= -Y_3 \end{aligned} \tag{8}$$

Next, with port 1 short-circuited, as in Figure 4c, Y_2 and Y_3 are in parallel, so y_{22} is just their sum. Now I_1 is the current in a branch whose voltage is V_2, so taking the reference into account, $I_1 = -Y_3V_2$. Hence:

$$\begin{aligned} y_{22} &= Y_2 + Y_3 \\ y_{12} &= -Y_3 \end{aligned} \tag{9}$$

The procedure of short-circuiting a port in the definitions of the y-parameters makes clear why we had to stipulate that there could be no coupling between branches. If there is a coupling between branches, then short- or open-circuiting a port can influence the component values in other branches. Hence, results obtained under these circumstances would not be applicable to the original two-port. Note, however, that these considerations are not general; they apply only in those cases where short- or open-circuiting a port will cause a branch in the circuit to become short- or open-circuited.

THE PI NETWORK AS A TWO-PORT EQUIVALENT The concept of the equivalence of two circuits at a pair of terminals was introduced early in this book. The first appearance of the idea was in Chapter 3, where a single resistor was found to be equivalent to several resistors connected in series or in parallel. Later we discussed the Thévenin equivalent at a pair of terminals. In both cases, the circuits were one-ports and the equivalence was found at a single pair of terminals. Two one-ports are said to be equivalent if the relationship between the voltage and current of one of them is the same as that of the other. For circuits with no internal (independent) sources, the voltage-current relationship in terms of Laplace transforms (or phasors) can be expressed as either an impedance or an admittance.

We will now discuss the equivalence of two-ports in general. The vehicle used to discuss equivalence will be the y-parameters. We make the following definition:

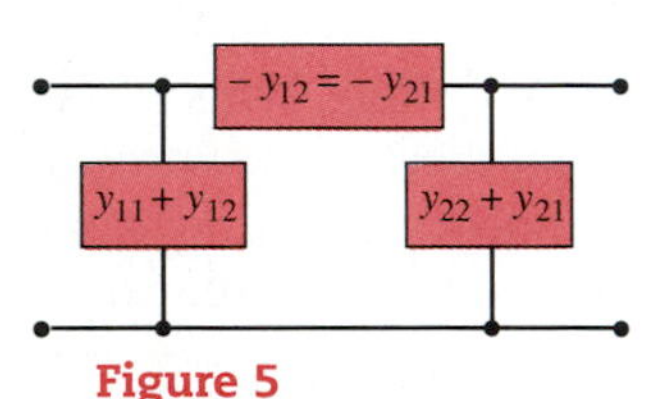

Figure 5
Pi equivalent of a set of y-parameters with $y_{12} = y_{21}$.

> Two two-ports are said to be equivalent if the equations relating the port voltages and currents for one two-port are the same as those of the other two-port.

The equivalence of two-ports says nothing about their internal construction; it refers only to the ports.[1]

Let's return to the pi network in Figure 4a, whose short-circuit parameters were given in (8) and (9) in terms of the three branch admittances. In particular, we observe in this case that $y_{12} = y_{21}$. It is possible to invert these equations and solve for the three branch admittances in terms of the y-parameters. Thus:

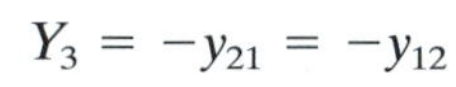

$$Y_3 = -y_{21} = -y_{12}$$

$$Y_1 = y_{11} + y_{21}$$

$$Y_2 = y_{22} + y_{21}$$

(Confirm these.) Hence, given any set of y-parameters for which $y_{12} = y_{21}$, a pi network can be found which has the same y-parameters, as shown in Figure 5. Hence, the pi network is equivalent to the network represented by those y-parameters. We shall have more to say about this subject later.

The equivalence just discussed is possible only because of the special condition $y_{12} = y_{21}$. This is an example of a general property which we shall now discuss before continuing.

THE RECIPROCITY CONDITION IN TWO-PORTS As noted, the condition $y_{12} = y_{21}$ in the preceding example is an instance of a property possessed by a class of two-port networks. An arbitrary linear two-port is shown in Figure 6a. The equations relating the port voltage and current phasors are those given in (2). In Figure 6b, the exciting voltage source at the input port is set equal to

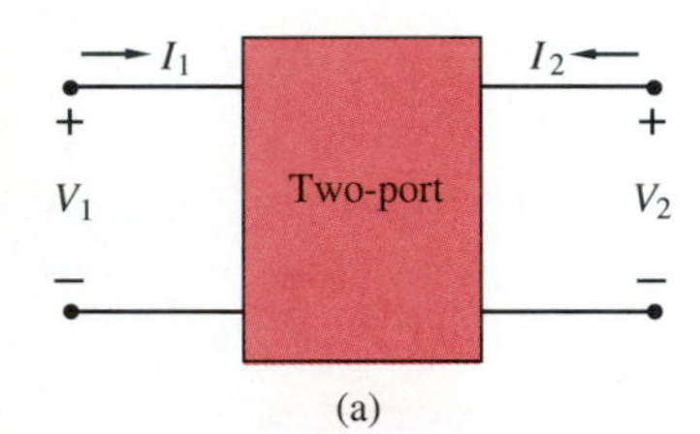

(a)

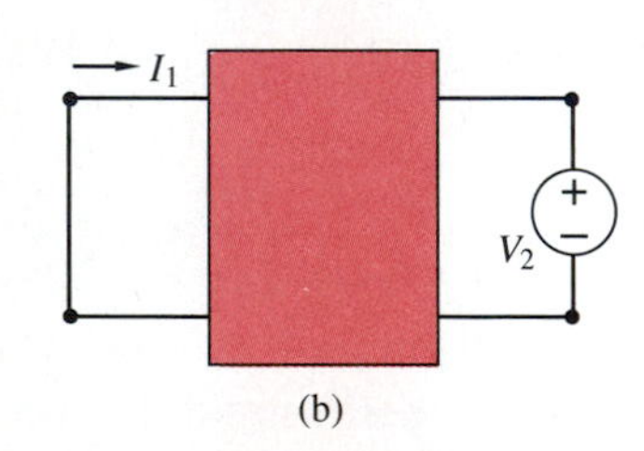

(b)

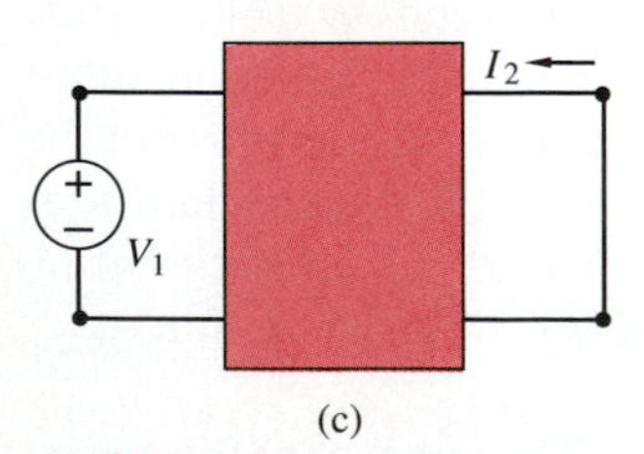

(c)

Figure 6
Development of the concept of reciprocity.

[1] Twice before, we danced around the idea of two-port equivalent circuits without acknowledging the generality of what was being done. One of the occasions was when we dealt with the equivalence of the tee and pi structures in Chapter 8. The other was when we discussed the tee equivalent of a transformer in Chapter 10. Strictly speaking, we should have deferred those discussions until now.

zero, so $V_1 = 0$. The same two-port appears in Figure 6c, but now the converse is true: the source voltage at the output port is short-circuited, making $V_2 = 0$.

Now let's consider the equations in (2a) under these conditions. The first equation under the first condition in Figure 6 and the second equation under the second condition in Figure 6 become:

$$I_1 = y_{12}V_2$$
$$I_2 = y_{21}V_1 \tag{10}$$

Now let's assume that the voltage phasors in the two parts of the experiment in Figure 6 are identical. Then, either the two current phasors are equal or they are not. This is the basis for identifying two classes of two-ports.

Let the voltages in Figure 6 be considered *inputs* (independent variables), and let the currents be considered the resulting *outputs* (dependent variables).

With the voltage in port 1 set equal to zero, let I_1 be the phasor current in port 1 in response to a voltage in port 2; then, with the voltage in port 2 set equal to zero, let I_2 be the phasor current in port 2 in response to a voltage in port 1. We say that the two-port is *reciprocal* if I_1 is identical with I_2 whenever the two voltage phasors are equal; otherwise, the two-port is *nonreciprocal.*

On the basis of the preceding definition, we can now describe the pi network in Figure 4a as a *reciprocal two-port.* In terms of the y parameters, the reciprocity condition is:

$$y_{12} = y_{21} \tag{11}$$

Open-Circuit Impedance Parameters

Upon initiating the development of this chapter, we assumed in Figure 2 that the excitations at the ports were voltage sources and the responses were currents. The excitations and responses can equally well be reversed. Then we write a set of node equations, selecting the nodes in such a way that node 1 is one of the nodes (the upper one) at port 1 and node 2 is the upper node at port 2.[2]

Then we solve for the node voltages. For linear networks, each node-voltage phasor is a linear combination of source-current phasors. We want only node

[2] With one possible difference in an arbitrary network, the general form of the node equations will be the same as that of the loop equations given in (1), except that the voltage and current variables will be interchanged. For the node equations, the dimensions of the a_{ij} coefficients will be admittance instead of impedance. The only difference in form will be that up to two additional equations, besides the first two, can have nonzero currents on the right. This comes about because the second terminal of each source in the network may not be the datum node. In such a case, the node equations at these nodes will carry a nonzero current-source term. However, although there can be current-source entries in up to two more equations, these entries will be only $-I_1$ or $-I_2$. When the node equations are solved, only two terms will result, as in (12).

voltages 1 and 2, so:

$$V_1 = z_{11}I_1 + z_{12}I_2$$
$$V_2 = z_{21}I_1 + z_{22}I_2 \qquad (12)$$
$$\mathbf{V} = \mathbf{Z}\mathbf{I}$$

$$\mathbf{V} = \begin{bmatrix} V_1 \\ V_2 \end{bmatrix} \qquad \mathbf{I} = \begin{bmatrix} I_1 \\ I_2 \end{bmatrix} \qquad \mathbf{Z} = \begin{bmatrix} z_{11} & z_{12} \\ z_{21} & z_{22} \end{bmatrix} \qquad (13)$$

Dimensionally, each of the the coefficients in these equations is an impedance. Collectively, then, they are called the *z-parameters.*

Given a two-port network, one method of determining the z-parameters is to write a set of node equations, assuming the two-port is excited by current sources at the ports. Solving these equations for V_1 and V_2 yields the z-parameters. Examples will be suggested for you to carry out in the problem set.

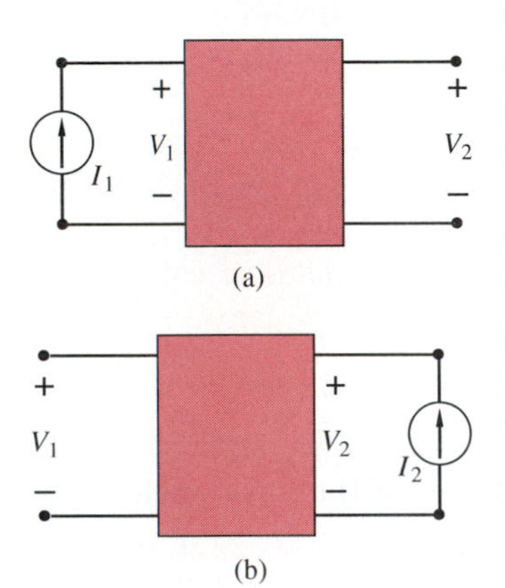

Figure 7 Open-circuit parameters.

Just as we did for the y-parameters, it is possible to give an interpretation for the z-parameters by letting first one, then the other, of the two source currents go to zero, as in Figure 7, effectively open-circuiting one of the two ports. Actually, a current source with zero current is indistinguishable from an open circuit. Therefore, the conditions at the ports are identical; both ports are excited by current sources, one of whose current happens to be zero.

By noting what each of the equations in (12) becomes under each of the two port conditions, we arrive at the following:

$$z_{11} = \left.\frac{V_1}{I_1}\right|_{I_2=0} \qquad z_{12} = \left.\frac{V_1}{I_2}\right|_{I_1=0}$$
$$z_{21} = \left.\frac{V_2}{I_1}\right|_{I_2=0} \qquad z_{22} = \left.\frac{V_2}{I_2}\right|_{I_1=0} \qquad (14)$$

(You should write the equations resulting from each of the circuits in Figure 7 and confirm the preceding results.) Since these definitions apply under open-circuit conditions at the ports, the parameters are called the *open-circuit impedance parameters.*

Two of the parameters are again driving-point functions and two are transfer functions. Thus, z_{11} is the *open-circuit input impedance* (the impedance at port 1 when port 2 is open) and z_{22} is the *open-circuit output impedance* (the impedance at port 2 when port 1 is open).[3] Also, z_{21} is the *open-circuit forward transfer impedance* from port 1 to port 2, and z_{12} is the *open-circuit reverse transfer impedance* from port 2 to port 1.

[3] There is some ambiguity in this terminology. Port 1 is normally taken as the input port and port 2 the output port. Yet in Figure 7b there is only one excitation, or "input" quantity, and that is at the "output" port. The context usually makes clear whether the terms *input* and *output* refer to the location of a port, or to a particular voltage or current. In the case under discussion, z_{22} is the driving-point impedance at what is considered the output port. It is the impedance at the output; hence, the *output impedance.*

CALCULATION OF z-PARAMETERS OF A NETWORK In determining the impedance parameters of a given network, a number of different approaches are possible. One approach is to use the defining relationships in (14), especially in simple cases.

EXAMPLE 2 Given the tee network in Figure 8, the object is to find the open-circuit impedances.

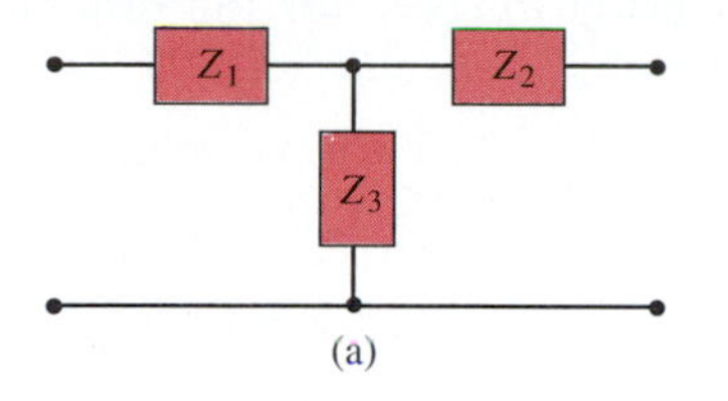

(a)

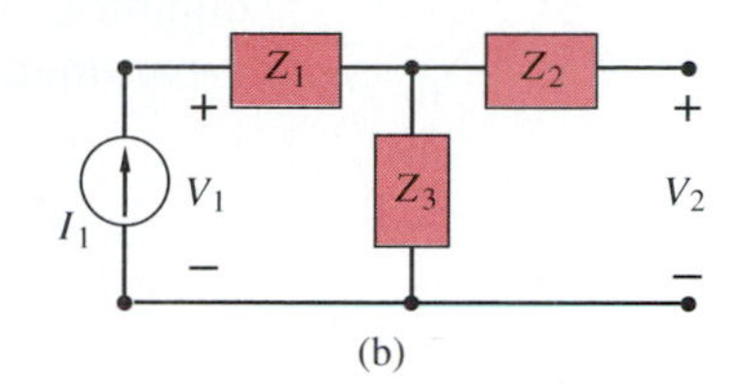

(b)

(c)

(d)

Figure 8 Open-circuit parameters of a tee.

SOLUTION The simplest approach in this case is to use the relationships in (14). Two of the parameters are determined by applying a current source at one port and leaving the other open, as in Figure 8b. The other two are obtained by interchanging the source and the open circuit, as in Figure 8c. The results will be:

With $I_2 = 0$	**With $I_1 = 0$**
$z_{11} = Z_1 + Z_3$	$z_{22} = Z_2 + Z_3$
$z_{21} = Z_3$	$z_{12} = Z_3$

We find that the impedance parameters of a tee are related in a rather simple way to the branch impedances. We find also that z_{12} and z_{21} are both equal to the same impedance, so they are equal to each other.

It is tempting to conclude that any two-port for which $z_{12} = z_{21}$ must be reciprocal. However, reciprocity was defined for voltage-source excitations, whereas here we have current sources. The following question needs an answer: If the reciprocity condition is satisfied under the conditions of voltage-source excitations in Figure 6, will it also be satisfied under the current-source excitation conditions in Figure 7? To answer the question, we need to examine the relationships among the z- and the y-parameters. However, we will take up one other matter before doing that.

THE TEE NETWORK AS A TWO-PORT EQUIVALENT In an earlier section, we defined the equivalence of two two-ports based on the port equations being the same. On the basis of equal short-circuit admittance equations, we had established that a pi network is equivalent to any two-port for which the forward and reverse transfer admittances are equal.

It should be possible to establish equivalence on the basis of the impedance equations also. The preceding equations specified the open-circuit impedance parameters in terms of the branch impedances of the tee in Figure 8a. These equations can be inverted to solve for the branch impedances in terms of the z-parameters:

$$Z_3 = z_{21} = z_{12}$$

$$Z_1 = z_{11} + z_{21}$$

$$Z_2 = z_{22} + z_{21}$$

A tee with these branch impedances is shown in Figure 8d. It shows that a tee network is equivalent to any two-port for which $z_{21} = z_{12}$. In the next section, we'll see that this condition implies reciprocity.

Relationships Between Impedance and Admittance Parameters

The admittance equations of a linear two-port were given in (2) and (3); the matrix form is repeated here:

$$\mathbf{I} = \mathbf{YV} \qquad \mathbf{Y} = \begin{bmatrix} y_{11} & y_{12} \\ y_{21} & y_{22} \end{bmatrix} \tag{15}$$

Assuming that the admittance matrix $\mathbf{Y}$ is nonsingular (meaning that it has an inverse), the preceding equation in matrix form can be solved by multiplying both sides by the inverse $\mathbf{Y}^{-1}$, leading to:

$$\mathbf{V} = \mathbf{Y}^{-1}\mathbf{I}$$

$$\begin{bmatrix} V_1 \\ V_2 \end{bmatrix} = \begin{bmatrix} \dfrac{y_{22}}{\Delta_y} & \dfrac{-y_{21}}{\Delta_y} \\ \dfrac{-y_{12}}{\Delta_y} & \dfrac{y_{11}}{\Delta_y} \end{bmatrix} \begin{bmatrix} I_1 \\ I_2 \end{bmatrix} \tag{16}$$

$$\Delta_y = \det \mathbf{Y} = y_{11}y_{22} - y_{12}y_{21}$$

Δ_y is the determinant of the matrix of y-parameters.

By comparing this equation with (12), a number of interesting conclusions are obtained. First, from a comparison of coefficients, the impedance parameters can be written in terms of the admittance parameters as follows:

$$z_{11} = \frac{y_{22}}{\Delta_y} \qquad z_{12} = \frac{-y_{12}}{\Delta_y}$$

$$z_{21} = \frac{-y_{21}}{\Delta_y} \qquad z_{22} = \frac{y_{11}}{\Delta_y} \tag{17}$$

It is clear from these expressions that, once the admittance parameters of a network have been determined, the impedance parameters can be obtained from them. This constitutes another method for determining the z-parameters of a given two-port.

EXAMPLE 3 Use (17) to determine the impedance parameters of the pi network in Figure 4a.

SOLUTION It's simply a matter of substituting the values of the y-parameters of the pi from (8) and (9) into (17); the results are:

$$z_{11} = \frac{Y_2 + Y_3}{\Delta_y} \qquad z_{12} = z_{21} = \frac{Y_3}{\Delta_y}$$
$$z_{22} = \frac{Y_1 + Y_3}{\Delta_y} \qquad \Delta_y = Y_1Y_2 + Y_1Y_3 + Y_2Y_3 \tag{18}$$

A second significant conclusion results from taking the ratio of z_{21} to z_{12} in (17):

$$z_{21} = \frac{y_{21}}{y_{12}} z_{12} \tag{19}$$

For a reciprocal network, we know that $y_{12} = y_{21}$. Using this in the preceding equation leads to the conclusion that, for a reciprocal network, $z_{12} = z_{21}$ also.

A third conclusion from a comparison of (12) and (16) involves the matrix forms. Since $\mathbf{V} = \mathbf{ZI}$ from (12) and $\mathbf{V} = \mathbf{Y}^{-1}\mathbf{I}$ from (16), the open-circuit impedance matrix and the short-circuit admittance matrix are each other's inverses: $\mathbf{Z} = \mathbf{Y}^{-1}$ and $\mathbf{Y} = \mathbf{Z}^{-1}$.

EXERCISE 2

Use the last paragraph to determine the y-parameters in terms of the z-parameters. (The results should be duals of (17).) □

Since the short-circuit admittance matrix is the inverse of the open-circuit impedance matrix, given the latter ($\mathbf{Z}$), we can determine the former ($\mathbf{Y}$). This constitutes a third approach for the determination of the y-parameters.

Parallel and Series Connections of Two-Ports

Complex networks and transmission systems can often be made up by interconnecting simpler two-ports. Conversely, when confronted with the need to analyze a large network, it may be possible to identify subnetworks as two-ports whose interconnections make up the larger network. It may be easier to analyze the smaller subnetworks. Two specific forms of interconnection, the series and parallel, will be explored in this section.

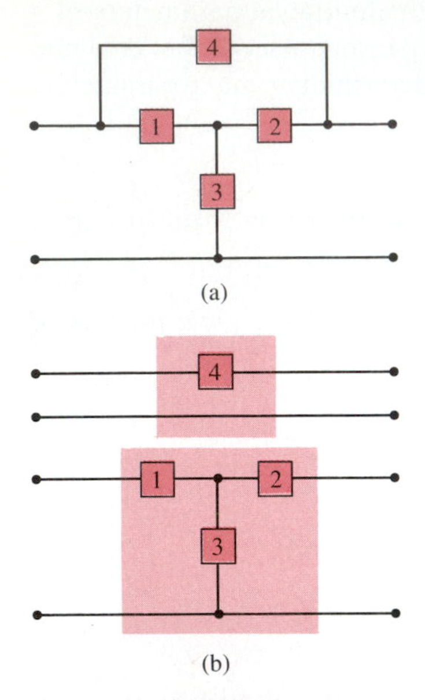

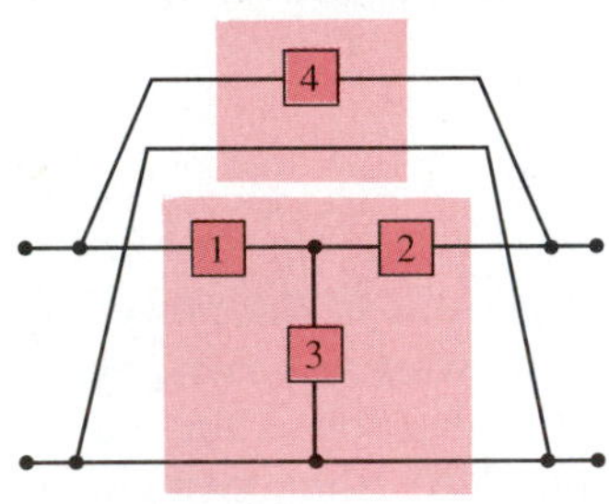

Figure 9
Bridged tee as parallel-connected two-ports.

TWO-PORTS CONNECTED IN PARALLEL To introduce the subject, let's consider the bridged-tee network shown in Figure 9a. It's easy to see that this two-port can be viewed as an interconnection of the two simpler two-ports shown in Figure 9b. The manner of interconnecting these simpler two-ports to give the original one is shown in Figure 9c. The two input ports are connected in parallel, as are the two output ports.

The idea introduced in the preceding paragraph is generalized in Figure 10, which shows two general two-ports whose input ports and output ports are both connected in parallel. The input voltages of both two-ports are the same, and so are their output voltages. As for the currents, the total input current is made up of the sum of the two individual input currents; and the overall output current is made up of the sum of the two individual output currents. If each two-port is described by its admittance representation, these facts lead to the following sequence of equations in matrix form:

$$\mathbf{I}_a = \begin{bmatrix} I_{1a} \\ I_{2a} \end{bmatrix} = \begin{bmatrix} y_{11a} & y_{12a} \\ y_{21a} & y_{22a} \end{bmatrix} \begin{bmatrix} V_{1a} \\ V_{2a} \end{bmatrix} = \mathbf{Y}_a \mathbf{V} \qquad \mathbf{I} = \mathbf{I}_a + \mathbf{I}_b = \mathbf{YV}$$

$$\mathbf{I}_b = \begin{bmatrix} I_{1b} \\ I_{2b} \end{bmatrix} = \begin{bmatrix} y_{11b} & y_{12b} \\ y_{21b} & y_{22b} \end{bmatrix} \begin{bmatrix} V_{1b} \\ V_{2b} \end{bmatrix} = \mathbf{Y}_b \mathbf{V} \tag{20}$$

$$\mathbf{Y} = \mathbf{Y}_a + \mathbf{Y}_b = \begin{bmatrix} y_{11a} + y_{11b} & y_{12a} + y_{12b} \\ y_{21a} + y_{21b} & y_{22a} + y_{22b} \end{bmatrix}$$

The conclusion is:

> The short-circuit admittance matrix of two two-ports connected in parallel is the sum of the individual short-circuit admittance matrices.

EXERCISE 3

Find the y-parameters of the bridged-tee in Figure 9, first by finding those of the tee and the bridging two-port, and then by using the preceding equations.

□

A word of caution here. Two two-ports are to be connected in parallel, as in Figure 10. Before they are so connected, they are individually described by

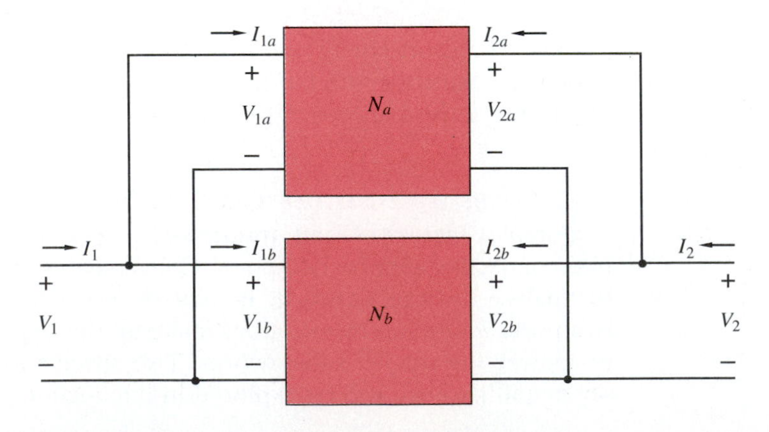

Figure 10
Parallel connection of two two-ports.

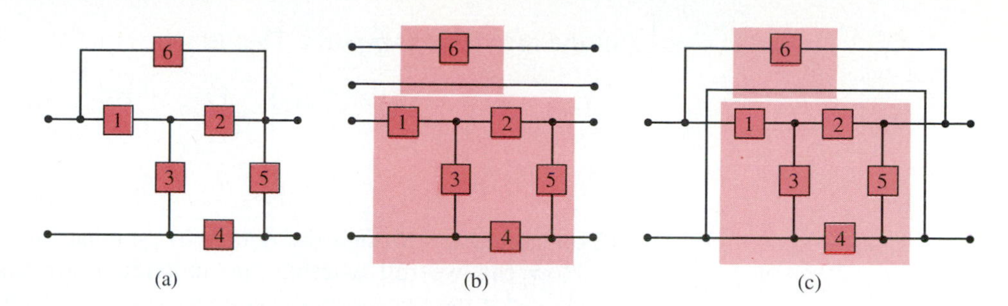

Figure 11
Improper parallel connection of two-ports.

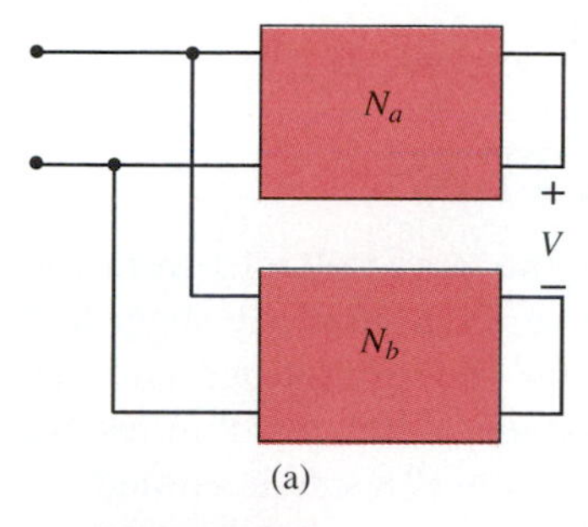

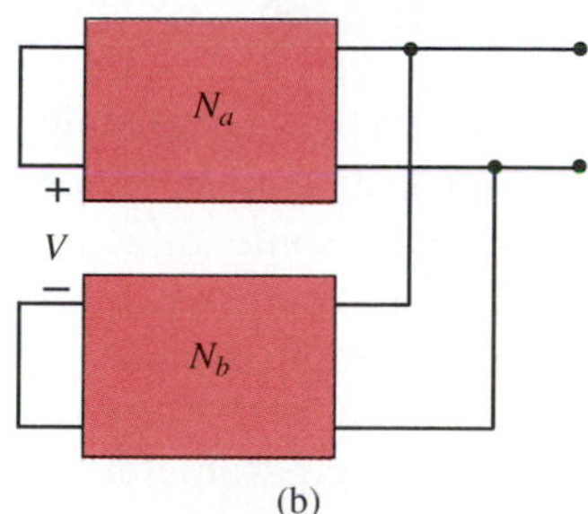

Figure 12
Test for parallel connection of two-ports.

their y-parameters, which are determined by the specific constituents of the two-ports. Do you think that the result in (20) will remain valid if the act of connecting the two-ports in parallel causes one or more of their components to become open-circuited or short-circuited?

It shouldn't be expected that the overall y-parameters can be obtained by adding the individual ones—unless the individual ones are not modified by the act of making the interconnection. For example, consider the two-port in Figure 11a. To simplify, suppose we were contemplating decomposing it into the two two-ports shown in Figure 11b and wanted to find the overall y-parameters by adding the individual ones. Our hopes would be dashed. The problem is that the straight-through connection in the two-port containing branch 6 shorts out branch 4 in the other two-port when the parallel connection is made.

A test can be devised to determine if the parallel interconnection of two two-ports can legitimately be made. The two-ports are connected in parallel at only one of the ports, either the input or the output port, and excited there by a voltage source, the other ports being short-circuited, as in Figure 12. In order for the I-V relationships of the individual two-ports to remain unaltered when connected in parallel, the voltage marked V in Figure 12 must be zero in both cases. It is not difficult to prove this result, so we'll leave it up to you. (Hint: Determine the voltage across the two bottom terminals in each part of the figure.)

Clearly, a class of two-ports for which a parallel connection is always possible is the class of common-terminal two-ports, where there is a direct connection from one of the terminals of one port to one of the terminals of the other port. This class includes the tee, the pi, the bridged tee, and ladder networks in general.

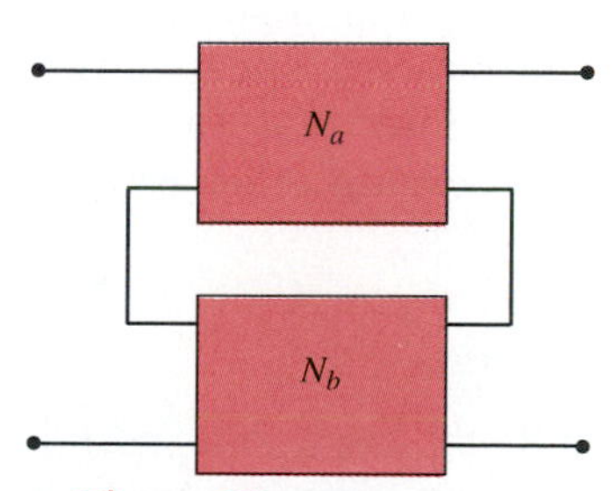

Figure 13
Series connection of two-ports.

TWO-PORTS CONNECTED IN SERIES Without considering an example, simply by analogy with the parallel connection of two-ports, we can conceive of two two-ports connected in series at each of the pairs of ports, as shown in Figure 13. In the present case, the current vectors of the two networks are the same, whereas the overall voltage vector is the sum of the individual voltage vectors. Assume, as before, that the V-I relationships of the individual two-ports are not altered when the two-ports are interconnected. Then, by a process which is the dual of the equations in (20), we find that the z-parameters of the composite two-port consist of the sum of the z-parameters of the individual two-ports. That is:

$$\begin{bmatrix} z_{11} & z_{12} \\ z_{21} & z_{22} \end{bmatrix} = \begin{bmatrix} z_{11a} + z_{11b} & z_{12a} + z_{12b} \\ z_{21a} + z_{21b} & z_{22a} + z_{22b} \end{bmatrix} \tag{21}$$

$$\mathbf{Z} = \mathbf{Z}_a + \mathbf{Z}_b$$

(You should go through the details to establish this result.)

How can we tell whether the individual two-port parameters will remain unaltered if the two-ports are connected in series? An answer will be suggested for you to work out in the problem set.

2 HYBRID AND TRANSMISSION PARAMETERS

In our study of two-ports so far, we assumed that the port voltages were the excitations and the currents the responses, or vice versa. Somebody suggested that those aren't the only possibilities for selecting the excitations and the responses out of the four voltage and current variables; other possibilities are illustrated in Figure 14. But, you might ask, why bother? Isn't it enough to have two methods? Won't it be redundant to waste time on one more approach to doing the same thing?

You might have a point. On the other hand, having different ways of doing things is useful because it's possible that a certain approach cannot handle some particular problem at all, or can handle it only with some difficulty. Another approach might do the job and do it better. If only the open-circuit impedance representation of two-ports were available, for example, how would we handle the parallel connection of two-ports? It might be that there are some two-ports which can absolutely not be described by y-parameters or z-parameters; how would we deal with such two-ports? Consider, for example, the very simple two-port shown in Figure 15. The only equations that can be written for this two-port are:

$$V_1 = 0 \qquad I_2 = 5I_1$$

This is neither an open-circuit impedance representation nor a short-circuit admittance representation. (The two-port is a current-controlled current source; its two-port representation was given in Chapter 4.) Therefore, although it may appear to be tedious, we'll go on to develop other ways of representing two-ports.

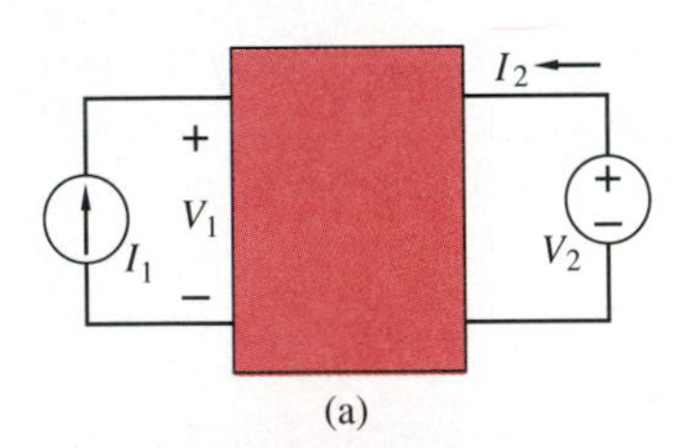

(a)

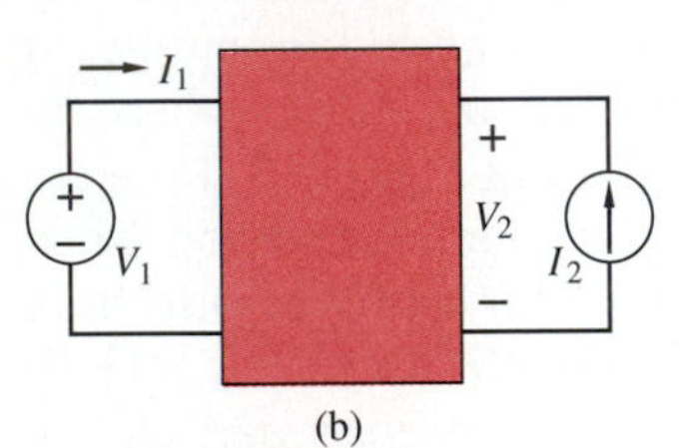

(b)

Figure 14
Excitations for hybrid-parameter descriptions of a two-port.

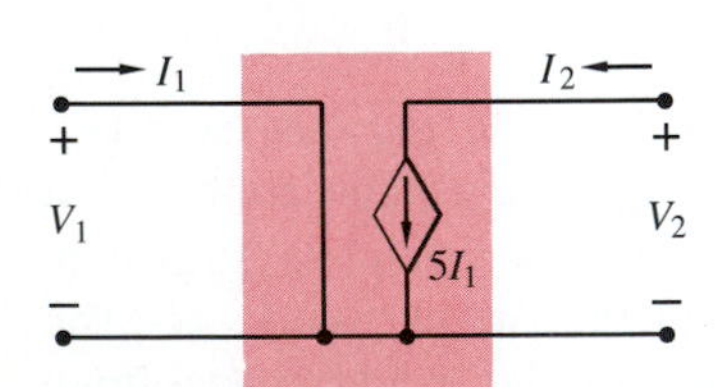

Figure 15
Example for hybrid-parameter description of a two-port.

The Hybrid Parameters

Refer again to the two cases in Figure 14. In both cases, one port is current-excited and the other port is voltage-excited. Again, since the network is linear, any response phasor will be a linear combination of the excitation phasors.

HYBRID *h*-PARAMETERS Let's start with the case in Figure 14a; in this case, the responses are V_1 and I_2, so the response-to-excitation equations become:

$$\begin{aligned} V_1 &= h_{11}I_1 + h_{12}V_2 \\ I_2 &= h_{21}I_1 + h_{22}V_2 \end{aligned} \tag{22}$$

The coefficients in these equations, called the *h*-parameters, are not all dimensionally the same. For this reason they are called *hybrid* parameters. As before, interpretations of the hybrid parameters can be obtained by setting $I_1 = 0$ (open-circuit input) or $V_2 = 0$ (short-circuit output):

$$\begin{aligned} h_{11} &= \left.\frac{V_1}{I_1}\right|_{V_2=0} = \frac{1}{y_{11}} && \text{short-circuit input impedance} \\ h_{12} &= \left.\frac{V_1}{V_2}\right|_{I_1=0} && \text{open-circuit reverse-voltage gain} \\ h_{21} &= \left.\frac{I_2}{I_1}\right|_{V_2=0} && \text{short-circuit forward-current gain} \\ h_{22} &= \left.\frac{I_2}{V_2}\right|_{I_1=0} = \frac{1}{z_{11}} && \text{open-circuit output admittance} \end{aligned} \tag{23}$$

Given a two-port, the hybrid *h*-parameters can be obtained from the preceding expressions, which can be taken as their definitions.[4]

EXAMPLE 4 The objective in this example is to find the hybrid *h*-parameters of the two-port in Figure 16. The parameter values are normalized.

SOLUTION Taking the ground node as datum, KCL equations can be written at the other two nodes in the Laplace transform domain or the phasor domain. The *I*-*V* relations of the components are then used appropriately to arrive at the following two equations:

$$\begin{aligned} I_1 &= 2V_1 + 2s(V_1 - V_2) \\ I_2 &= \frac{V_2}{5} + 3I_1 + 2s(V_2 - V_1) \end{aligned}$$

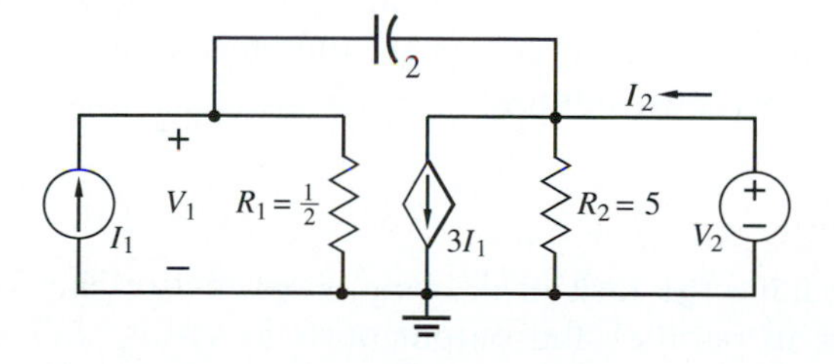

Figure 16
Circuit of Example 4.

[4] The hybrid parameters for multiterminal components were introduced briefly in Chapter 4.

It is now simply a matter of algebra to rearrange these equations to express V_1 and I_2 in terms of I_1 and V_2. The only thing needed in the first equation is to solve for V_1; this will give one of the two-port equations. When this is substituted into the second of the two preceding equations, the result—after rearrangement—is the second two-port equation. You supply the details; the result will be:

$$V_1 = \frac{1}{2(s+1)} I_1 + \frac{s}{s+1} V_2$$

$$I_2 = \frac{2s+3}{s+1} I_1 + \frac{11s+1}{5(s+1)} V_2$$

HYBRID g-PARAMETERS The inverse of the preceding hybrid h representation relates to Figure 14b. The responses here are I_1 and V_2. The corresponding equations are written:

$$\begin{aligned} I_1 &= g_{11} V_1 + g_{12} I_2 \\ V_2 &= g_{21} V_1 + g_{22} I_2 \end{aligned} \tag{24}$$

These are also hybrid parameters—*the hybrid g-parameters.*

EXERCISE 4

By successively setting each excitation to zero in Figure 14b, formulate interpretations for each of the hybrid g-parameters similar to those in (23) for the h-parameters. □

The hybrid parameters find their greatest utility in two-ports containing controlled sources (transistor models). Often, the hybrid parameters are the most convenient to use for such circuits; sometimes they are the only possible ones.

We found earlier that, for two two-ports connected in parallel at both ports, the y-parameters come in handy; and for two two-ports connected in series at both ports, the z-parameters are very useful. How about the case where two ports, one from each of two two-ports, are connected in parallel and the other ports of the same two-ports are connected in series? A little thought will show that the hybrid parameters are the ones to use in these cases, one type for each of the two possibilities.

EXERCISE 5

You have the task of drawing diagrams for the following two cases: (1) input ports in parallel and output ports in series, and (2) input ports in series and output ports in parallel. Show the relationship of the overall voltage and current vectors to the individual current and voltage vectors. From these,

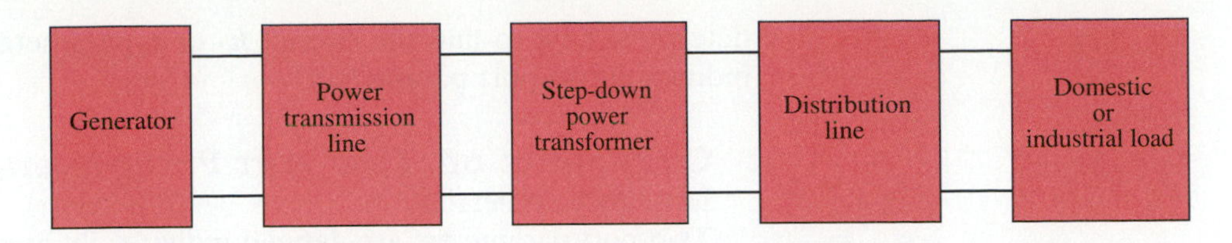

Figure 17
Cascaded two-ports.

the chapter. A thorough appreciation of this discussion will require you to solve several of these problems.

The Transmission Parameters

Historically, one of the earliest electrical systems was a power system. The structure of such a system resembles the diagram in Figure 17. The power is transmitted from the generator (after being stepped up) over a transmission line at a high voltage, through a transformer that reduces the voltage, through a shorter distribution line, to the ultimate user. The two-ports in such a connection are said to be in *cascade.* They are also said to be connected in a *chain.* At each step, both the voltage and the current on one side of a two-port are to be determined in terms of those on the other side, which are presumed known. This situation is different from all four of the preceding interconnections (parallel, series, series-parallel, and vice versa). Thus, we should not be surprised if there were no simple relationship for the parameters of the overall interconnection in terms of the component two-port parameters using any of the previously defined sets of parameters.

Figure 18
Two-port for defining transmission parameters.

What we need is a set of equations that expresses the voltage and current at one port in terms of the voltage and current at the other port. Using the standard notation in Figure 18, there are obviously two possibilities, as follows:

$$\begin{bmatrix} V_1 \\ I_1 \end{bmatrix} = \begin{bmatrix} a_{11} & a_{12} \\ a_{21} & a_{22} \end{bmatrix} \begin{bmatrix} V_2 \\ -I_2 \end{bmatrix}$$
$$\begin{bmatrix} V_2 \\ I_2 \end{bmatrix} = \begin{bmatrix} b_{11} & b_{12} \\ b_{21} & b_{22} \end{bmatrix} \begin{bmatrix} V_1 \\ -I_1 \end{bmatrix} \quad (25)$$

(Actually, the second of these relationships is redundant; we can always renumber the ports if transmission is to occur from right to left.)

On the basis of the *function* they perform, the parameters in these equations are called the *transmission* parameters, since they involve transmission from one port to the other. But the structure of a cascade connection of two-ports resembles a chain, each of the component two-ports being a link in the chain. Because of the *structure,* then, they are also called *chain* parameters. In the early days, the a_{ij} parameters were labeled $ABCD$, so the set was called the $ABCD$ parameters.[6]

[6] As noted, these parameters were first introduced in the study of power transmission lines in the late nineteenth century and were called *general circuit constants* at that time—"constants" because the frequency was fixed. Later, since behavior with variable frequency became important, some people called them general circuit *functions.* There is, however, nothing more "general" about them than the other sets of parameters under consideration in this chapter. They take their place as one of several ways of describing a two-port.

determine how to find the overall h- or g-parameters from the correspon individual two-port parameters.[5]

Comments on Two-Port Parameters from Measurements

Two-port parameters are defined individually by setting one of the four p variables equal to zero (by short- or open-circuiting a port), applying excitation at the other port, and measuring the response at either port. T case of the y-parameters is given in (6) and (7), the z-parameters in (14), an the h-parameters in (23). (Drawing a two-port diagram with the port variable labeled will be helpful in pursuing this development; then, as the discussio proceeds, you can open- and short-circuit the ports, and pursue th consequences.)

With each excitation, two responses can be measured. Hence, two parameters are determined by measuring an excitation and its two responses. Determining four two-port parameters thus requires six measured quantities. By properly selecting the excitations and the corresponding responses, a complete set of two-port parameters can be obtained in this way. For example, if one port at a time is short-circuited, the other port is voltage-excited, and both currents are measured each time, the result is the set of y-parameters, as in (6) and (7).

It is also possible, however, that a pair of parameters from one set and a pair from a different set are determined in this way. Then some further manipulations will be needed to find either full set of parameters. These manipulations are straightforward.

There is yet another possibility. For each port condition (open- or short-circuited), it is possible to apply one excitation and measure only one response. Since there are a total of four such port conditions, this process will require making four excitation and four response measurements.

Finally, in case it is known that the two-port in question is reciprocal, then one fewer response needs to be measured. Thus, a set of three parameters is obtained by two excitation and three response measurements. In the case of the approach taken in the preceding paragraph, only three excitations and three responses need be measured.

A number of problems based on this discussion are included at the end of

[5] By now you might be complaining that too much of this effort is left for you. You might question why you aren't just given a table showing all the possible two-port representations, the relationships among all the types of parameters, and conditions satisfied by the appropriate set of parameters for each possible type of interconnection of two two-ports. The reason is that there would be little educational value to this. You would be much better off if you spent some time *discovering* these relationships. The point is that the results themselves are not as important as the process you go through in order to discover them. A few years from now you may not remember the specific results, but if you have learned how to pursue an investigation, you will be a far better engineer. There are books available called handbooks in which one can look up all kinds of known information. A handbook is handy when you need specific pieces of information or numerical values in order to pursue a project. That information is useful, but it cannot substitute for the knowledge of *how* you should pursue your project.

One distinction from the previous sets of two-port equations is the minus sign appearing before the "output" current. (If transmission is from right to left, the "output" is I_1.) That's because the idea of transmission in one direction implies that both current references ought to be in the same sense with respect to the flow, rather than in the same sense with respect to the ports. Anyway, that's how the references were chosen in the early days. At the present time, there are two choices: either we redefine the transmission parameters on the basis of the usual choice of port reference for the "output" current, causing a change in sign of the parameters having subscripts 12 and 22; or we effectively change the reference for the "output" current by placing a minus sign in front of it in the equations. That's what we now do, rather than cause a change in previously calculated parameters for often-used transmission networks.

As we did for the other sets of parameters, interpretations for the transmission parameters can be obtained by open-circuiting and short-circuiting each of the ports in turn. The results are:

$$\frac{1}{a_{11}} = \left.\frac{V_2}{V_1}\right|_{I_2=0} \qquad \text{open-circuit voltage gain}$$

$$\frac{1}{a_{12}} = -\left.\frac{I_2}{V_1}\right|_{V_2=0} = -y_{21} \qquad -(\text{short-circuit transfer admittance})$$

$$\frac{1}{a_{21}} = \left.\frac{V_2}{I_1}\right|_{I_2=0} = z_{21} \qquad \text{open-circuit transfer impedance} \tag{26}$$

$$\frac{1}{a_{22}} = -\left.\frac{I_2}{I_1}\right|_{V_2=0} \qquad -(\text{short-circuit current gain})$$

It appears that the reciprocals of the transmission parameters are all transfer functions of some kind, with transmission in the forward direction, from port 1 to port 2. The chain parameters are not all the same dimensionally; a_{12} is an impedance, a_{21} is an admittance, and a_{11} and a_{22} are dimensionless.

EXAMPLE 5 Find the chain parameters of the tee in Figure 19a.

SOLUTION First we open-circuit the output port (set $I_2 = 0$), as in Figure 19b, and we voltage-excite the input. From the voltage gain and the input admittance of the resulting circuit, we find:

$$V_2 = \frac{Z_3}{Z_1 + Z_3} V_1 \qquad I_1 = \frac{V_1}{Z_1 + Z_3}$$

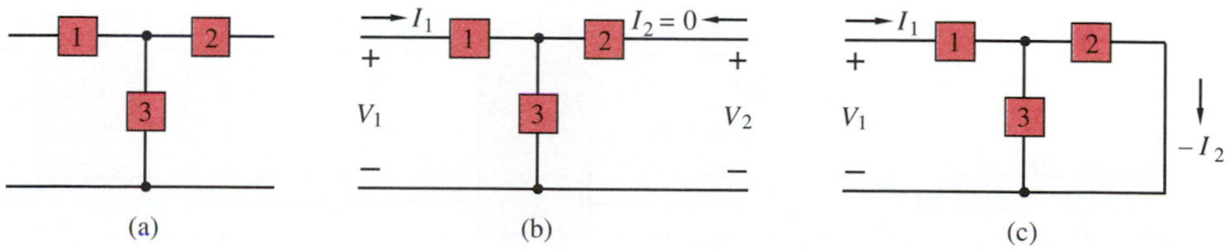

Figure 19 Finding the chain parameters of the tee.

With the output open, the first set of equations in (25) reduce to $V_1 = a_{11}V_2$ and $I_1 = a_{21}V_2$. Combine with the preceding equations and solve:

$$a_{11} = \frac{V_1}{V_2} = \frac{Z_1 + Z_3}{Z_3} = 1 + \frac{Z_1}{Z_3}$$

$$a_{21} = \frac{I_1}{V_2} = \frac{I_1}{V_1}\frac{V_1}{V_2} = \frac{1}{Z_3}$$

Next we short-circuit the output port (set $V_2 = 0$), as in Figure 19c. From the input impedance and the current-divider relationship of the resulting circuit, we find:

$$V_1 = \left(Z_1 + \frac{Z_2Z_3}{Z_2 + Z_3}\right)I_1$$

$$-I_2 = \frac{Z_3}{Z_2 + Z_3}I_1$$

The first set of transmission equations in (25) under this port condition reduce to $V_1 = -a_{12}I_2$ and $I_1 = -a_{22}I_2$. Equating these with the preceding equations and solving for a_{12} and a_{22} results in:

$$a_{22} = -\frac{I_1}{I_2} = \frac{Z_2 + Z_3}{Z_3}$$

$$a_{12} = -\frac{V_1}{I_2} = \frac{V_1}{I_1}\left(-\frac{I_1}{I_2}\right) = \left(Z_1 + \frac{Z_2Z_3}{Z_2 + Z_3}\right)\left(\frac{Z_2 + Z_3}{Z_3}\right)$$

Cascade Connection of Two-Ports

As previously noted, the greatest utility of the transmission parameters lies in describing the port relationships of cascaded two-ports. Two such two-ports are shown in Figure 20. We use the first set of equations in (25) to express the relationship among the port variables for each of the individual two-ports, as follows:

$$\begin{bmatrix} V_1 \\ I_1 \end{bmatrix} = \begin{bmatrix} V_{1A} \\ I_{1A} \end{bmatrix} = \begin{bmatrix} a_{11A} & a_{12A} \\ a_{21A} & a_{22A} \end{bmatrix}\begin{bmatrix} V_{2A} \\ -I_{2A} \end{bmatrix}$$
$$\begin{bmatrix} V_{2A} \\ -I_{2A} \end{bmatrix} = \begin{bmatrix} V_{1B} \\ I_{1B} \end{bmatrix} = \begin{bmatrix} a_{11B} & a_{12B} \\ a_{21B} & a_{22B} \end{bmatrix}\begin{bmatrix} V_{2B} \\ -I_{2B} \end{bmatrix} \tag{27}$$

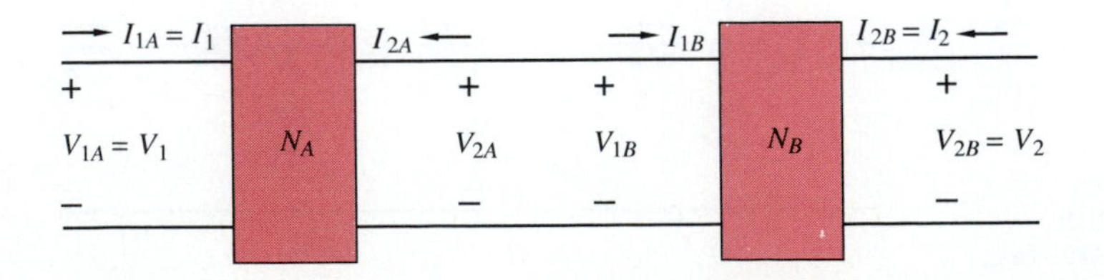

Figure 20
Two two-ports in cascade.

We observe that $V_{1B} = V_{2A}$ and $I_{1B} = -I_{2A}$; that is, the vector of input variables for network B equals the vector of output variables for network A, as shown in the last equation. Thus, starting from the transmission equations of network A and successively inserting the parts of (27) yields:

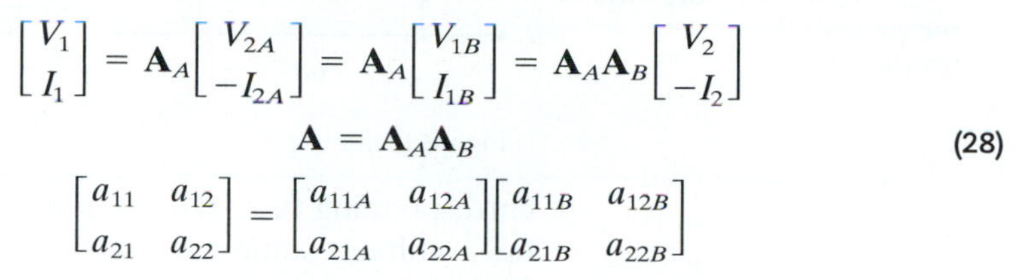

$$\begin{bmatrix} V_1 \\ I_1 \end{bmatrix} = \mathbf{A}_A \begin{bmatrix} V_{2A} \\ -I_{2A} \end{bmatrix} = \mathbf{A}_A \begin{bmatrix} V_{1B} \\ I_{1B} \end{bmatrix} = \mathbf{A}_A \mathbf{A}_B \begin{bmatrix} V_2 \\ -I_2 \end{bmatrix}$$

$$\mathbf{A} = \mathbf{A}_A \mathbf{A}_B \tag{28}$$

$$\begin{bmatrix} a_{11} & a_{12} \\ a_{21} & a_{22} \end{bmatrix} = \begin{bmatrix} a_{11A} & a_{12A} \\ a_{21A} & a_{22A} \end{bmatrix} \begin{bmatrix} a_{11B} & a_{12B} \\ a_{21B} & a_{22B} \end{bmatrix}$$

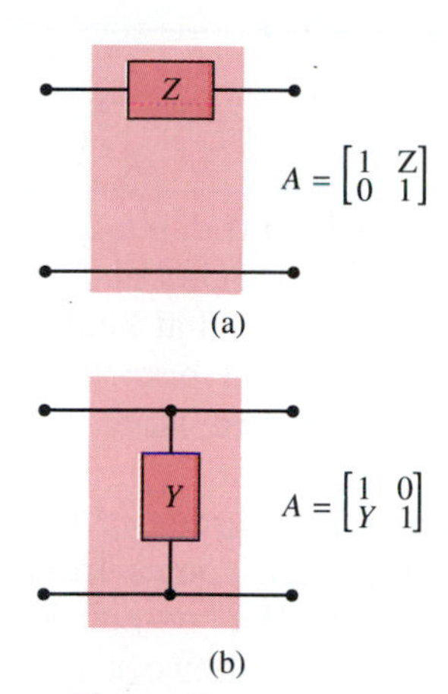

Figure 21 Elementary two-ports and their chain parameters.

This result demonstrates that the transmission matrix of two two-ports connected in cascade is the product of the transmission matrices of the individual two-ports.

EXERCISE 6

The tee network in Figure 19 can be considered the cascade connection of the elementary two-ports in Figure 21: one of the type in Figure 21a followed by one of the type in Figure 21b, followed by another of the sort in part (a). It is claimed that the chain parameters of these elementary two-ports are the ones given in the figure. (a) Verify these and (b) use them to confirm the chain parameters previously found for the tee of Figure 19. □

3 EQUIVALENT TWO-PORTS

In the preceding sections you became acquainted with six different sets of equations, all of which express relationships among the port variables. This is unlike the case of one-ports, where there are only two relations and one of them is the inverse of the other. Which set of equations should be used in the determination of equivalence? From the definition of equivalence, any one of them can be used. In previous sections we used the y-parameters to find a pi equivalent and we used the z-parameters to find a tee equivalent. However, these were for the special case of reciprocal two-ports.

Only one parameter, the impedance (or admittance), is needed to completely describe a one-port. For the general two-port, four parameters are needed—only three for the special case of a reciprocal two-port. Therefore, it might be conjectured that, in general, four components would be needed in the equivalent circuit of a two-port. Let's see.

Let's initially look at the y-parameters. The admittance equations are:

$$\begin{aligned} I_1 &= y_{11} V_1 + y_{12} V_2 \\ I_2 &= y_{21} V_1 + y_{22} V_2 \end{aligned} \tag{29}$$

The terms involving the driving-point admittances are easy to represent. Each would be represented by an admittance across the corresponding port. But, then, what do we do with the transfer admittances? Each of the terms

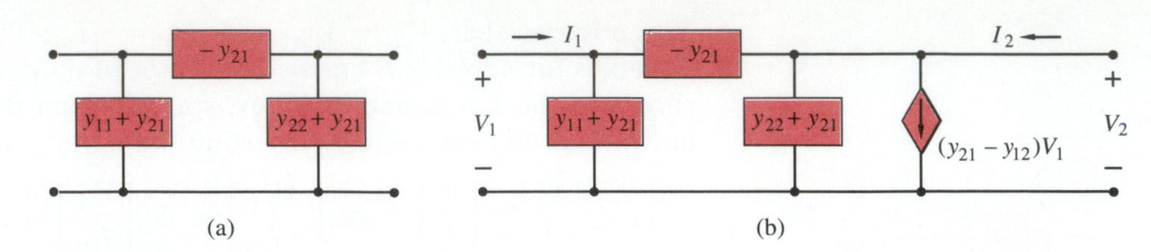

Figure 22 y-parameter equivalent two-port: (a) reciprocal; (b) nonreciprocal.

Figure 15-22

corresponding to a transfer admittance represents a current which is dependent on a voltage somewhere else. Of course, that way of stating it immediately brings to mind something we've seen before: a controlled source. (What kind?) Hence, a possible equivalent two-port consists of a two-terminal branch in parallel with a VCCS across each port. (Draw the diagram and label the components and the variables.)

The preceding is not incorrect, but it places too much of the burden of the equivalence on the controlled sources. An alternative approach comes from noting that if $y_{21} = y_{12}$, then the two-port is reciprocal. We noted, in connection with the discussion surrounding Figure 5, that a pi network can represent the y-parameters of a reciprocal two-port. This representation is shown in Figure 22a. When the two-port is nonreciprocal, however, y_{21} and y_{12} are not equal. In that case, by adding and subtracting $y_{12}V_1$ in the second y-parameter equation, we can write:

$$\begin{aligned} I_1 &= y_{11}V_1 + y_{12}V_2 \\ I_2 &= y_{12}V_1 + y_{22}V_2 + (y_{21} - y_{12})V_1 \end{aligned} \tag{30}$$

The last term on the right in the second equation can be represented by a voltage-controlled current source across the output port, since this term contributes a current to the output which is proportional to the voltage somewhere else. All the other terms in the two equations are identical to ones for a reciprocal two-port which resulted in the pi network in Figure 22a. Hence, an equivalent circuit representing the equations in (30) will be that pi augmented by a VCCS across the output port, as shown in Figure 22b.

EXERCISE 7

A variation on the equivalent circuit just obtained can be found by adding and subtracting a term $y_{21}V_2$ in the first equation in (29). (a) Determine this equivalent two-port. (b) For the following admittance matrix, specify the admittance values in both the circuit you found and the one in Figure 22b:

$$Y = \begin{bmatrix} s + 0.4 & 0.05 \\ 0.1 & 0.2 \end{bmatrix}$$ □

Other equivalent circuits can be worked out starting with other parameters. For the impedance parameters, an approach similar to the one just carried out for the admittance parameters will lead to two other equivalent two-ports. These will be suggested as problems whose details will be left for you to work out.

4 TERMINATED TWO-PORTS

Although the admittance, impedance, and hybrid parameters are *defined* with certain ports short-circuited or open-circuited, the corresponding two-port equations represent the V-I relationships of the two-port no matter what is connected at the ports. A typical situation is illustrated in Figure 23a, where a source network terminates one of the ports and a load the other. In Figure 23b, the source network is replaced by its Norton equivalent, but it could equally well be the Thévenin equivalent, described in terms of $V_s = Y_s I_s$ and $Z_s = 1/Y_s$.

Given a circuit in the form of Figure 23b (or a set of two-port parameters and the source and load), we might be called upon to find, as problems in network analysis, any of the following functions: transfer impedance, transfer admittance, voltage gain, current gain, and power gain. For each of these functions, there are two possibilities: one specifies the ratio of two-port output to *two-port* input variable; the other specifies the ratio of two-port output to *excitation* input. Both of these are of interest in different situations. We might also be called on to find (1) the input impedance at the input port with the load in place and (2) the Thévenin voltage and impedance at the output port with the source in place.

If this were a handbook, we would give you worked-out expressions for all of those functions in terms of the two-port parameters of the network and the source and load impedances. Since it is not a handbook, we'll work out some of them to give you the flavor of how to proceed; then you can work out some of the others as problems.

Refer to Figure 23b. In addition to the two-port equations, the port variables will be constrained by the source and load terminations. The relevant equations are:

$$\begin{aligned} V_2 &= -Z_L I_2 \qquad & V_1 &= V_s - Z_s I_1 \\ I_2 &= -Y_L V_2 \qquad & I_1 &= I_s - Y_s V_1 \end{aligned} \tag{31}$$

where $V_s = Z_s I_s$. The equations have been written twice, once with the voltages in terms of the currents and once with the currents in terms of the voltages.

The preceding equations, together with the two-port equations, constitute a set of four equations in four unknowns, the only known quantity being the source voltage or current. These four equations can be solved for any one of

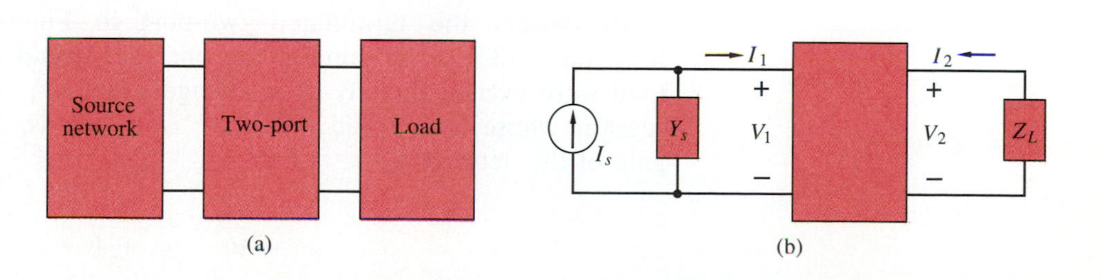

Figure 23 Two-port terminated in a source network at one port and a load at the other.

the port variables. Any transfer function from the source to an output can be obtained from these solutions. Any transfer function from the two-port input to the output can be obtained as a ratio of two of these solutions. The only remaining question is, Which two-port parameters should we use? Convenience and relevance, is the answer. If the open-circuit impedance parameters have already been determined for the two-port, for example, then use them, by all means.

As an illustration, let's deal with the y-parameters. The relevant two-port equations are repeated here:

$$\begin{aligned} I_1 &= y_{11}V_1 + y_{12}V_2 \\ I_2 &= y_{21}V_1 + y_{22}V_2 \end{aligned} \tag{32}$$

When the current equations in (31) are substituted into these, the resulting equations can be solved for V_1 and V_2 and their ratio taken. The results are as follows:

$$\begin{gathered} V_1 = \frac{(y_{22} + Y_L)}{\Delta} I_s \qquad V_2 = \frac{-y_{21}}{\Delta} I_s \\ \frac{V_2}{V_1} = \frac{-y_{21}}{y_{22} + Y_L} \\ \Delta = y_{11}y_{22} - y_{12}y_{21} + y_{11}Y_L + y_{22}Y_s + Y_L Y_s \end{gathered} \tag{33}$$

The output-to-input voltage gain has been explicitly shown. In addition, these equations give V_1/I_s, the impedance faced by the current source; and V_2/I_s, the output-to-excitation transfer impedance.

5 LAST WORD: SYNTHESIS

This chapter has presented a number of different ways of representing a two-port. The real benefit of all this algebra might not be clearly evident to you when the arena is limited to the analysis of networks. The benefits become much more evident when the domain of activity is expanded to include the synthesis of networks as well. That is an extensive subject beyond the scope of this book. Nevertheless, an example will be considered which, if nothing else, might justify to you the mass of detail in this chapter.

Reconsider the terminated two-port in Figure 23b. The voltage-gain function in (33) is the same as the one that would result if the two-port with load were excited directly by a voltage source V_1. Assume that the load is a resistor whose normalized resistance is 1; that is, $Z_L = 1$. Then the voltage gain of the network is:

$$\frac{V_2}{V_1} = \frac{-y_{21}}{y_{22} + 1} \tag{34}$$

Now suppose the following synthesis problem is posed. A normalized function of s is specified as the voltage gain of an unknown two-port terminated in a unit (normalized) resistance:

$$\frac{V_2}{V_1} = K\frac{s}{s^2 + 4s + 3} \tag{35}$$

The objective is to create a two-port network which has this specified function as its voltage gain. Any value of the multiplying constant is acceptable, but the larger the better because then less amplification might be needed in the system of which this network is a part to provide an adequate level of gain.

First we notice that the poles are real and negative. We also know from previous chapters that the natural frequencies of *RC* networks are real and negative. Hence, it might be reasonable to look for an *RC* two-port to fill the bill. By direct comparison of the last two equations, you might be tempted to identify $-y_{21} = s$ and $y_{22} = s^2 + 4s + 2$; but that won't work. Such a function has a double pole at infinity and no negative real poles; it can't be the admittance of an *RC* two-port. We need another approach.

Suppose we try dividing both numerator and denominator of the given V_2/V_1 function in (35) by $s + a$, where a is still to be determined. Then, by comparing the resulting function with (34), we identify:

$$-y_{21} = K\frac{s}{s + a}$$

$$y_{22} = \frac{s^2 + 4s + 3}{s + a} - 1$$

$$= \frac{Ks}{s + a} + \frac{s^2 + (3 - K)s + (3 - a)}{s + a}$$

$$y_{22} + y_{21} = \frac{s^2 + (3 - K)s + (3 - a)}{s + a}$$

The motivation for the last step and the maneuver on the right-hand side of the previous step is the pi structure of Figure 5, where one of the branches is $y_{22} + y_{21}$. The desired two-port will indeed have this structure if each of the functions $-y_{21}$ and $y_{22} + y_{21}$ can be implemented as an *RC* admittance.

For reasons beyond your present knowledge, the value of $-a$ (the pole of these functions) must lie between the zeros of the last function for an *RC* network. Doing some exploratory calculations involving K and a, let's choose $K = \frac{1}{2}$ and $a = 1.6$. Then:

$$-y_{21} = \frac{1}{2}\frac{s}{s + 1.6} \qquad y_{22} + y_{21} = s + \frac{7}{8} + \frac{0.025s}{s + 1.6}$$

On the right-hand side a modified partial-fraction expansion has been carried out. The reciprocal of $-y_{21}$ is $2 + 3.2/s$, which is easily recognized as the impedance of a 2-unit resistance in series with a capacitance $1/3.2$.

Now turn to $y_{22} + y_{21}$. This admittance is seen to consist of the parallel connection of a unit capacitance, an $\frac{8}{7}$-unit resistance, and another branch

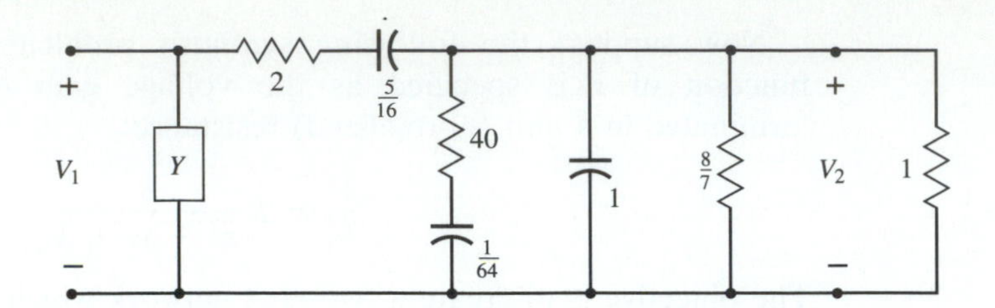

Figure 24 Network realization of a given voltage-gain function.

which has the same structure as $-y_{21}$. The entire resulting network is shown in Figure 24.

The admittance Y across the input (the input branch of the pi) has no influence on the gain function V_2/V_1. Omitting it would lead to the simplest two-port.

This illustration has given just a small taste of network synthesis. The overall problem is that of creating a network to realize a specified transfer function with specified load and source impedances. Being well versed in the many ways of representing two-ports goes a long way toward your eventual ability to carry out this process.

Final Last Word

You may have noticed that, at various points in the preceding development, we made choices among alternatives. One choice had to do with the value of a; another, with the value of K. Such choices generally influence the numerical values of the parameters but have no other significant effect. A major choice was in the decision to look for both an *RC* circuit and one that has a pi structure. That was just one option; there are others. One other option will be suggested for you to do as a problem.

Unlike analysis, synthesis and design always involve options and choices. There are many criteria for making a choice. Some of them, as in the preceding example, are technical (for example, network structure, or component types). Such choices are the easiest to make. Other criteria—aesthetic, economic, ethical, environmental—are much more difficult to apply. As you proceed in your studies and in your profession, you will have occasion to design some things. Your standing will depend on how careful you are in applying such other design criteria.

Good Luck!

SUMMARY

- Representation of two-port networks
- Short-circuit admittance representation
- Definition of short-circuit admittance parameters
- Open-circuit impedance representation
- Definition of open-circuit impedance parameters
- Reciprocity property
- Pi and tee equivalents of reciprocal two-ports
- Equivalent circuits of nonreciprocal two-ports

- Relationships between y-parameters and z-parameters
- Parallel connections of two-ports
- Series connections of two-ports
- Hybrid h- and hybrid g-parameters
- Transmission (chain) parameters
- Cascade connections of two-ports
- Transfer functions of terminated two-ports
- Synthesis example

PROBLEMS

Numerical values of all parameters in this set of problems are normalized.

1. A number of two-port circuits are shown in Figure P1. From their basic definitions, find the short-circuit admittance parameters for each circuit.

2. Find a relationship between the determinant Δ_z of the set of z-parameter equations and the determinant Δ_y of the y-parameter equations of a two-port.

3. By determining z_{12} and z_{21} in the two-port of Figure P3, confirm that it is electrically symmetrical although not structurally symmetrical. Structural symmetry is sufficient but not necessary for electrical symmetry.

4. **a.** Find y_{11} as a function of s in the two-port in Figure P4.
   b. Locate the poles and zeros on the complex-frequency plane.
   c. Repeat parts (a) and (b) for z_{11}.

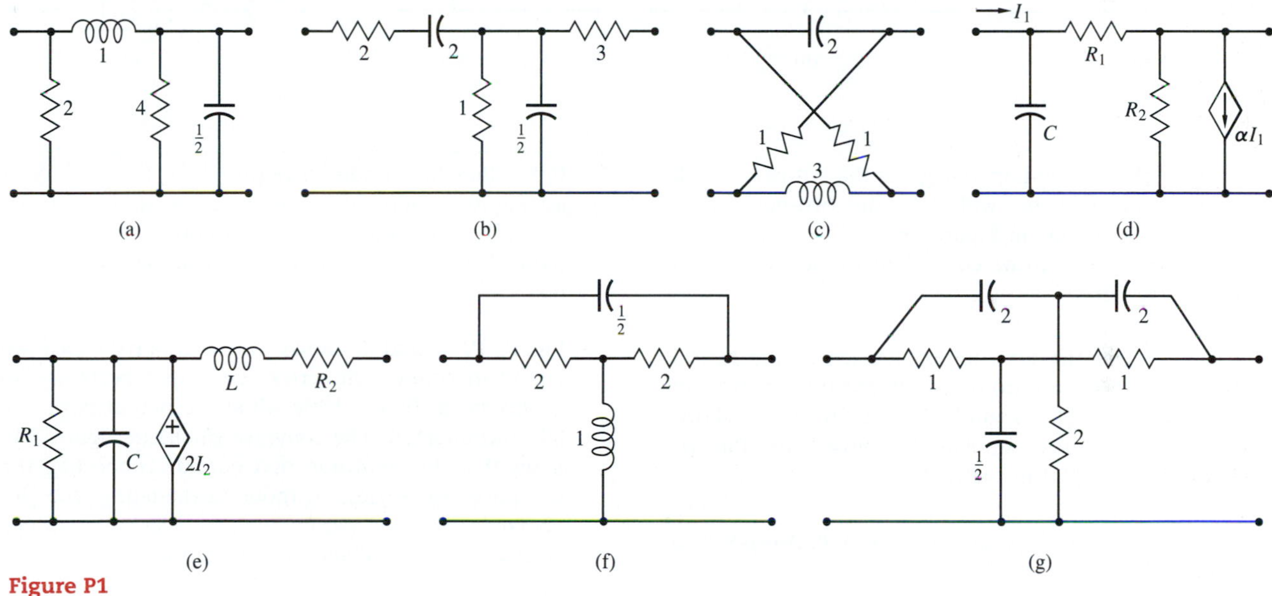

Figure P1

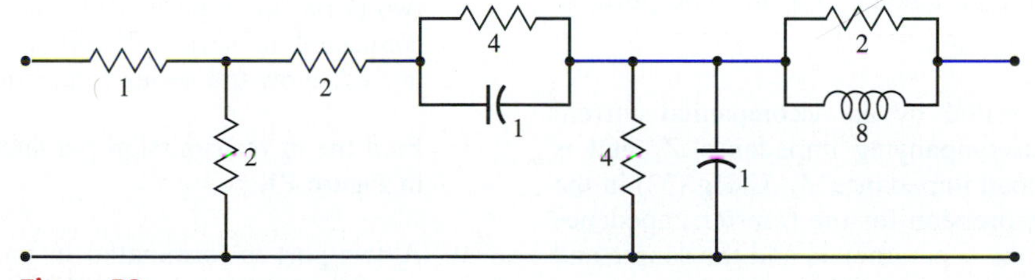

Figure P3

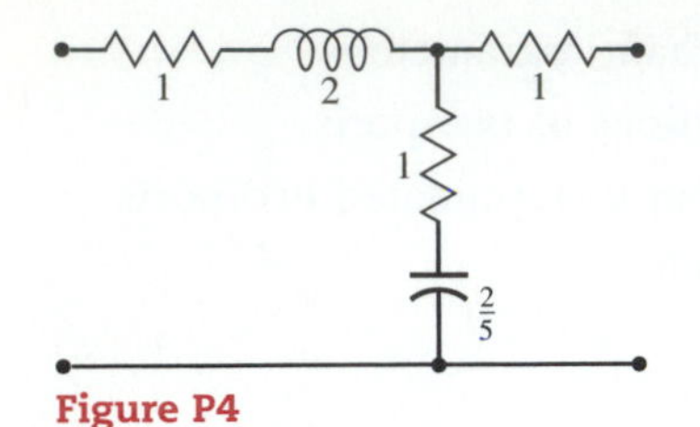

Figure P4

5. Find the open-circuit impedance parameters of the two-ports in Figure P1 from their basic definitions.

6. Two two-ports are equivalent if they have the same two-port parameters (any set).
   a. Find the branch admittances of the bridged tee in Figure P6c that will make it equivalent to the two-port in Figure P6a.
   b. Repeat if the bridged tee is to be equivalent to the two-port in Figure P6b.

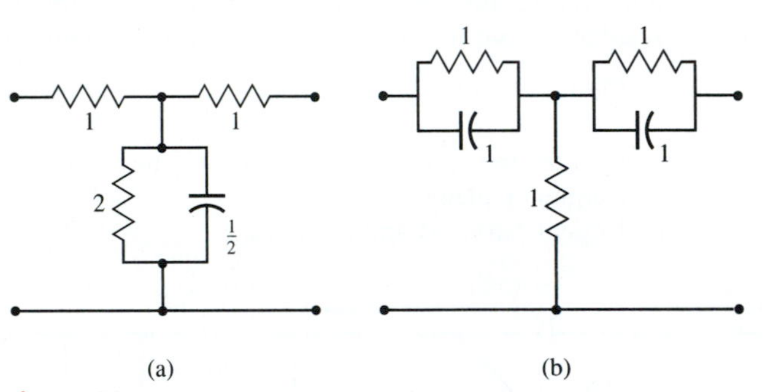

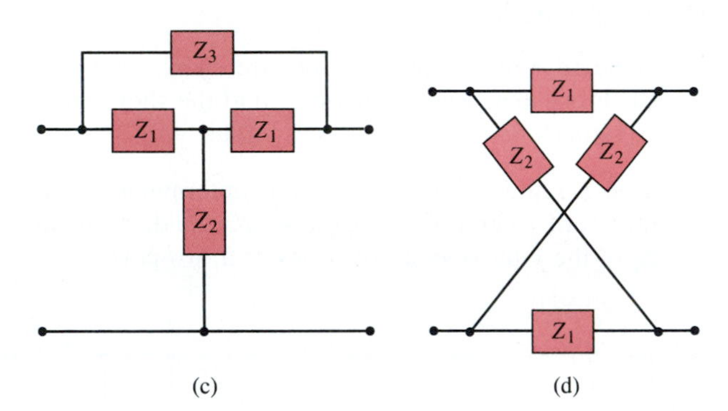

Figure P6

7. a. Find the branch impedances of the lattice network in Figure P6d that will make this two-port equivalent to the one in Figure P6a.
   b. Repeat if it is to be equivalent to the two-port in Figure P6b.

8. From (24) in the text, interpret each of the hybrid g-parameters as a ratio of a response transform to an excitation transform under open-circuit or short-circuit conditions, similar to those for the h-parameters in (23) in the text.

9. Find the hybrid h-parameters for the two-ports in Figure P1.

10. Find the hybrid g-parameters for the two-ports in Figure P1.

11. A two-port is excited by an accompanied current source I_s with accompanying impedance Z_1 and is terminated in a load impedance Z_2. Using (32) in the text, derive an expression for the transfer impedance V_2/I_s in terms of the y-parameters and the source and load impedances.

12. Prove that the condition required for the validity of adding the individual y-parameters to obtain the overall y-parameters of two two-ports connected in parallel is $V = 0$ in both networks in Figure 12 in the text.

13. Figure P13a and b shows two two-ports which we intend to connect in series. The input ports are so connected in Figure P13a while the output ports are left unconnected. The converse is true in Figure 13b. Show that the condition that permits connecting the two-ports in series, without invalidating the individual two-port parameters, is that the voltage labeled V in each diagram equal zero.

14. On the basis of Problem 13, decide if each pair of two-ports in Figure P14a and b can be validly connected in series. If either pair can be so connected, show the series connection.

15. Find the a_{ij} transmission parameters of the two-ports in Figure P1.

16. A two-port is terminated in an impedance Z_2. The two-port itself is described by one of the following

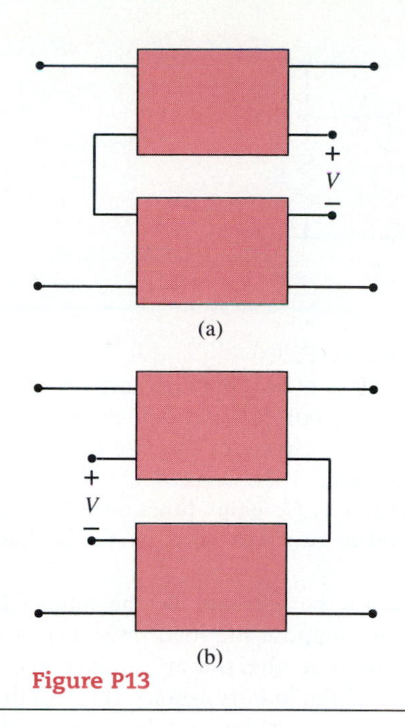

Figure P13

transmission equations:

$$\begin{bmatrix} V_1 \\ I_1 \end{bmatrix} = \begin{bmatrix} -n & 0 \\ 0 & \dfrac{-1}{n} \end{bmatrix} \begin{bmatrix} V_2 \\ -I_2 \end{bmatrix} \quad \begin{bmatrix} V_1 \\ I_1 \end{bmatrix} = \begin{bmatrix} n & 0 \\ 0 & \dfrac{1}{n} \end{bmatrix} \begin{bmatrix} V_2 \\ -I_2 \end{bmatrix}$$

a. Determine the input impedance $Z_1 = V_1/I_1$ in each case in terms of n and Z_2. What would you call such a two-port?

b. Repeat part (a) if the transmission matrix is:

$$\mathbf{A} = \begin{bmatrix} 0 & n \\ \dfrac{1}{n} & 0 \end{bmatrix}$$

17. Assume that the g-parameter equations are given.

a. By solving them for V_1 and I_2 in terms of I_1 and V_2, find the h-parameters in terms of the g-parameters.

b. Find a relationship between the h-parameter matrix $\mathbf{H}$ and the g-parameter matrix $\mathbf{G}$.

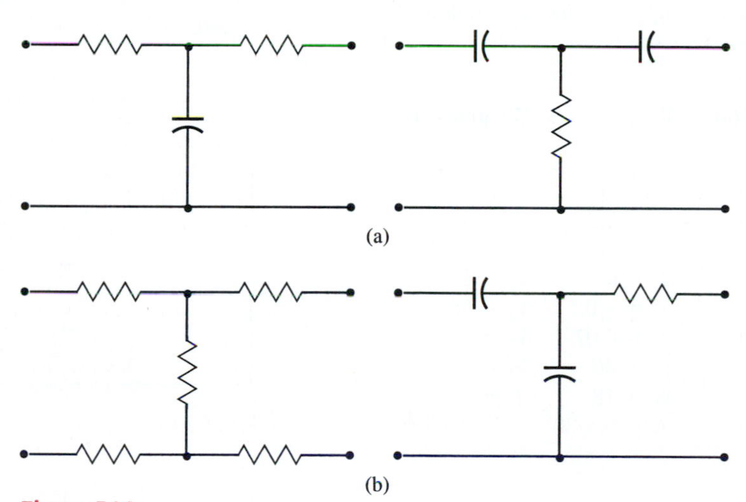

Figure P14

c. How are the determinants, Δ_h and Δ_g, of these matrices related?

18. The equations defining the transmission parameters are:

$$V_1 = a_{11}V_2 + a_{12}(-I_2)$$
$$I_1 = a_{21}V_2 + a_{22}(-I_2)$$

These equations are to be rearranged to express each voltage in terms of the two currents.

a. Carry out that rearrangement and identify the z-parameters in terms of the transmission parameters.

b. Repeat part (a) but this time express each current in terms of both voltages to arrive at the y-parameters in terms of the transmission parameters.

19. Given one set of two-port parameters, the objective of this problem is to find the parameters of any other set in terms of them. The z-parameters, for example, were given in terms of the y-parameters in (17) in the text. Also, in the preceding two problems you worked out relationships between two other sets. Continue with this approach.

Construct a table showing the relationship of each two-port parameter in one set to the parameter in each of the four other sets. The table will contain 20 rows (five sets of four rows each) and 5 columns. The columns corresponding to a given set of parameters will have the same entries as the rows corresponding to that set.

20. This problem requires the determination of a set of two-port parameters from a set of measurements at the ports. Several cases are given in the following table. One of the ports is open- or short-circuited and an excitation applied at the other port. Two responses are then measured. Excitations and responses are phasors; when the two-port is only dissipative, the phasors all have the same angle, taken as 0. The units of all voltages are volts and the units of all currents are amperes. Find a complete set of two-port parameters from each set of measurements.

Port Condition	Excitation	Response A	Response B
a. $V_2 = 0$	$I_1 = 0.1$	$V_1 = 10$	$I_2 = 4$
$I_1 = 0$	$V_2 = 20$	$V_1 = 0.2$	$I_2 = 0.01$
b. $V_2 = 0$	$V_1 = 10$	$I_1 = 1$	$I_2 = -0.5$
$V_1 = 0$	$I_2 = 0.5$	$I_1 = 0.5$	$V_2 = 25$
c. $V_1 = 0$	$I_2 = 1$	$I_1 = -0.1$	$V_2 = 50$
$I_2 = 0$	$V_1 = 20$	$I_1 = 0.02$	$V_2 = 8$
d. $I_2 = 0$	$I_1 = 2$	$V_1 = 40$	$V_2 = 40$
$V_2 = 0$	$I_1 = 1$	$V_1 = 18$	$I_2 = 5$
e. $V_2 = 0$	$V_1 = 10$	$I_1 = 0.5\angle 30°$	$I_2 = 0.5\angle 135°$
$I_1 = 0$	$V_2 = 100\angle 60°$	$I_2 = 1\angle 45°$	$V_1 = 200\angle 30°$

21. Using as models (30) in the text and Exercise 7, which follows it, derive two different equivalent circuits of a nonreciprocal two-port based on the open-circuit impedance equations.

22. An amplifier circuit is shown in Figure P22.

a. Find one of the sets of two-port parameters of the circuit—you choose. (If you want more practice, find other sets of parameters.)

b. Draw an equivalent circuit representing these parameters. (It might be useful to write the relevant two-port equations.)

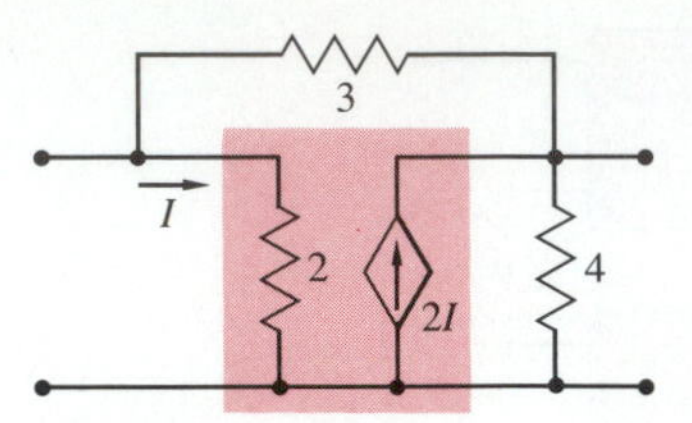

Figure P22

Figure 15 P-22

23. In the circuit of Figure P23:

a. Find one of the sets of two-port parameters of the two-port within the box (any set you choose).

b. Draw an equivalent circuit of the two-port based on these parameters.

c. Determine the voltage-gain function V_2/V_1 in terms of the set of two-port parameters you have chosen.

d. Let the variables be phasors in the sinusoidal steady state, and suppose the load resistance is to be varied. Determine the power supplied by V_s when the value of the load resistance is set so that the maximum power is being delivered to the load.

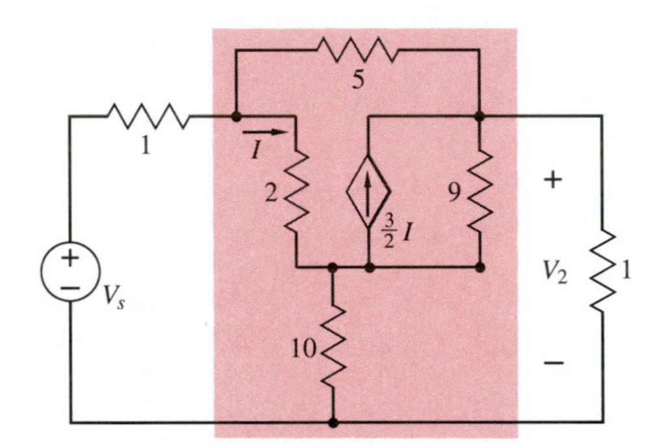

Figure P23

24. The diagram in Figure P24 represents a linear two-port—either reciprocal or nonreciprocal—with a load Z_L (or Y_L) and a source circuit represented here by its Thévenin equivalent. Either from the two-port equations or from an appropriate equivalent circuit:

a. Derive an expression for the transfer impedance V_2/I_1 in terms of the z-parameters and the terminations.

b. Derive a similar expression for V_2/V_s.

25. Repeat Problem 24, this time in terms of the y-parameters.

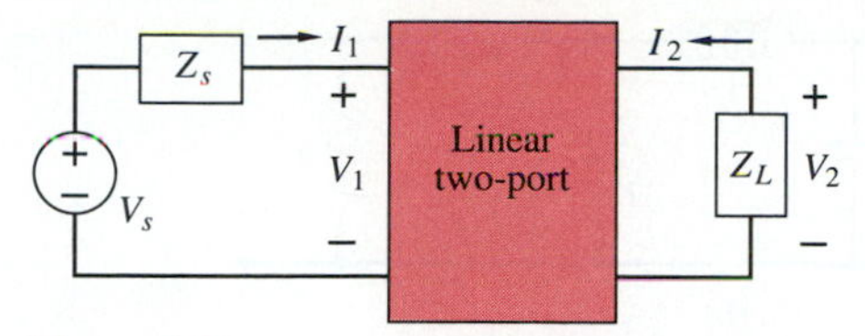

Figure P24

26. Repeat Problem 24, this time in terms of the h-parameters.

27. Repeat Problem 24, this time in terms of the g-parameters.

28. In the circuit in Figure P28:
   a. Find the set of hybrid g-parameters or h-parameters of the network within the box.
   b. From the results of Problem 26 or Problem 27, determine V_2/V_s as a function of s.

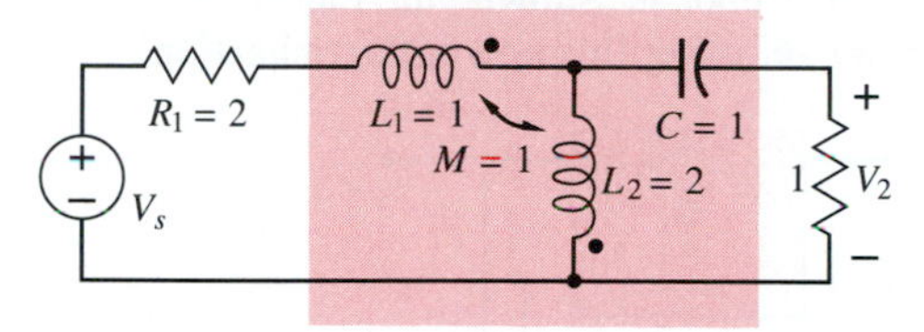

Figure P28

Figure P 15-28

29. Repeat Problem 28 for the circuit in Figure P29.

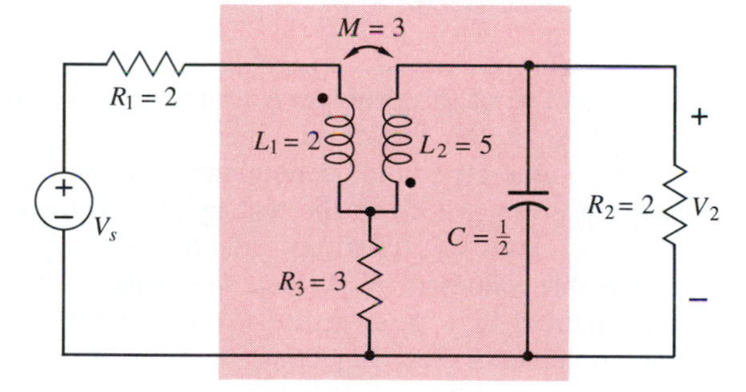

Figure P29

30. A certain transformer is characterized by the primary inductance L_1, the turns ratio n, and the coupling coefficient k. Find its transmission parameters in terms of these quantities.

31. Two identical transformers are connected as shown in Figure P31. There is no coupling from one transformer to the other.
   a. Find the open-circuit impedance parameters of the entire two-port.
   b. Draw an equivalent circuit based on these parameters.
   c. Repeat parts (a) and (b) for the short-circuit admittance parameters.

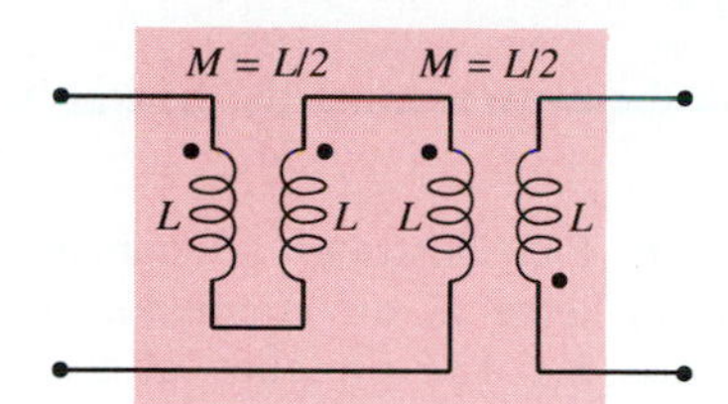

Figure P31

32. The nonreciprocal two-port in Figure P32 is represented by its y-parameters.
   a. Using the equivalent circuit you derived in Exercise 7 in the text, determine an expression for the voltage gain V_2/V_1 in the terms of the load R and the y-parameters. (The excitation is assumed to be V_1.)
   b. Using this equivalent circuit once more, replace the network to the left of R by a Thévenin equivalent and again determine V_2/V_1. (It had better be the same!)
   c. Confirm this result from the original equations, using the I-V relationship of the load.

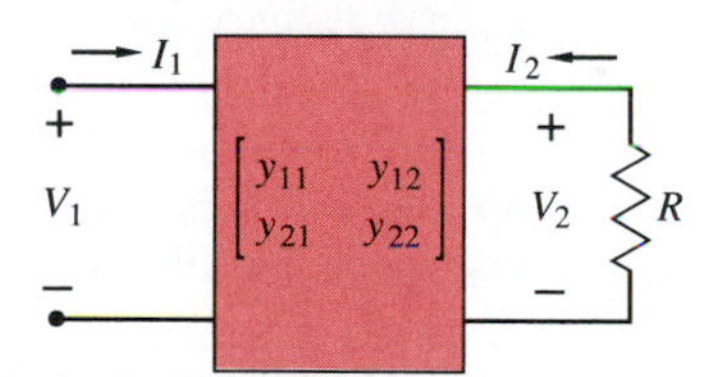

Figure P32

33. a. Starting from the z-parameter equations of a nonreciprocal two-port, add and subtract an appropriate term in the first equation and find an equivalent circuit described by the resulting equations.
   b. Repeat part (a) but this time add and subtract an appropriate term in the second equation.

34. a. Use the concept explored in Exercise 6 in the text to find the transmission parameters of each of the low-pass filter two-ports in Figure P34.
   b. From these results, determine the y-parameters of the two-ports.

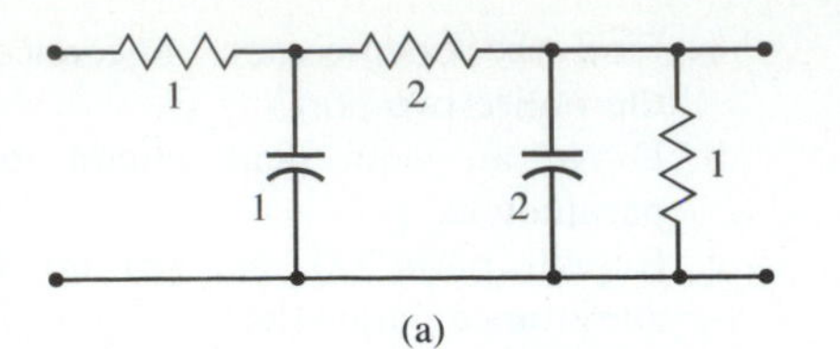

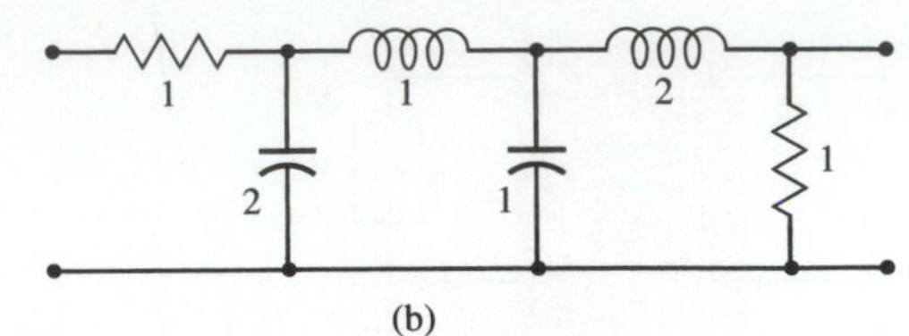

Figure P34

c. Determine also the z-parameters the easiest way you know.

35. The circuit in Figure P35 illustrates a symmetrical two-port which is to serve as an attenuator. It is terminated in a load R_0. The presence of the attenuator is to have no influence on the load presented to the source. That is, the impedance presented to the source should be R_0 with or without the attenuator. The voltage "gain" function is $G_{21} = V_2/V_1$.
 a. Derive design equations giving the resistances R_1 and R_2 of the attenuator in terms of R_0 and G_{21}.
 b. Design an attenuator for a voltage reduction of 10 to 1 and a load resistance $R_0 = 770\ \Omega$.
 c. Repeat part (b) for a 40-dB attenuation working into a 500-Ω load.

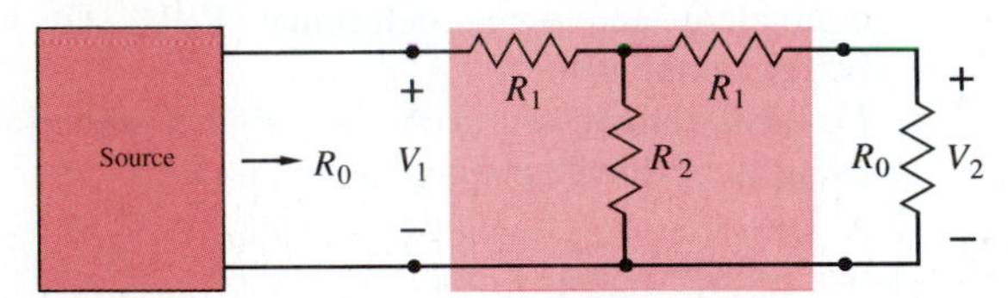

Figure P35

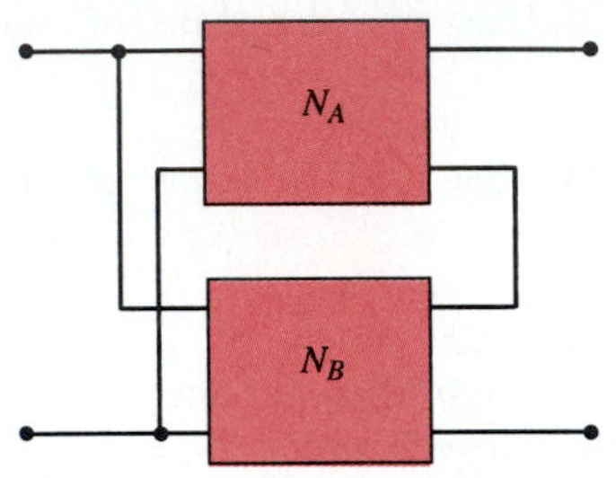

Figure P36

36. An interconnection of two two-ports is shown in Figure P36. The input ports are in parallel and the output ports in series.
 a. Find the relationship of the overall port variables to the port variables of the individual two-ports.
 b. Find the overall set of hybrid parameters (you determine which set) in terms of the hybrid parameters of the individual two-ports.

37. The converse of the interconnection of two two-ports in Figure P36 is shown in Figure P37. Repeat Problem 36, using this new interconnection.

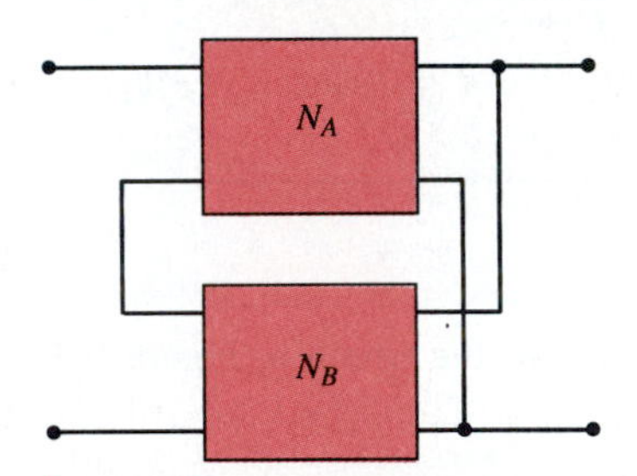

Figure P37

38. The objective of this problem is to find an alternative realization of the circuit in Figure 24 in the text. Divide numerator and denominator of the gain function given in (35) in the text by $s^2 + 3$. Identify the $-y_{21}$ and y_{22} functions. identify a range of permissible values of K if these functions are to be implemented in a pi structure. Construct such a pi that implements these functions.

DESIGN PROBLEMS

1. Design an L-type two-port network as shown in Figure DP1. This network is to match a 550-Ω resistive load to a source with an internal impedance of 50 Ω. The frequency of the source is 500 kHz. The configuration of the two-port should help eliminate a 1-kHz interfering signal. Assume that Z_a and Z_b consist of only pure reactive elements.

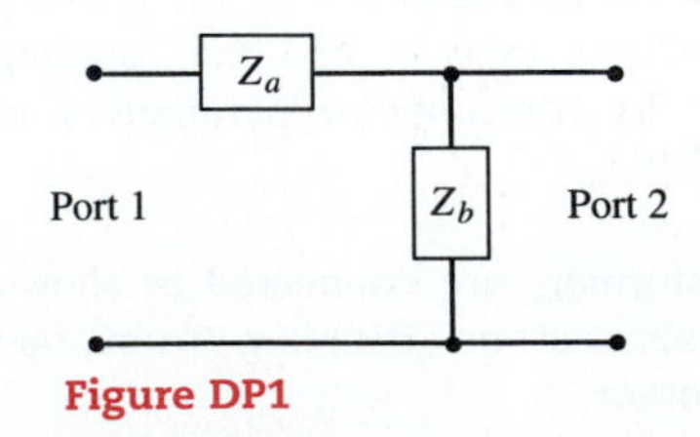

Figure DP1

2. Design a two-port with the following hybrid parameters:

$$h_{11} = \frac{0.5s^3 + s^2 + 2s + 2}{s(0.5s^2 + 2)} \qquad h_{21} = \frac{1}{0.5s^2 + 2}$$

$$h_{12} = \frac{1}{0.5s^2 + 1} \qquad h_{22} = \frac{0.5s(0.5s^2 + 2)}{s + 1}$$

You can design this two-port with two capacitors, an inductor and a resistor.

3. The part of the circuit of Figure DP3 shown as a colored box represents the high frequency equivalent of a transistor. The value of the transistor parameters are as follows: $r_{bb'} = 100\ \Omega$, $r_{b'e} = 1\ \text{k}\Omega$, $C_e = 100\ \text{pF}$, $r_{b'c} = 4\ \text{M}\Omega$, $r_{ce} = 80\ \text{k}\Omega$, and $g_m = 50\ \text{mA/V}$. Determine values of R_L and R_E such that the current gain $|i_L|/|i_{in}| > 47$ and the input impedance $|Z_{in}| > 20\ \text{k}\Omega$ for the range of frequencies between 100 Hz and 100 kHz.

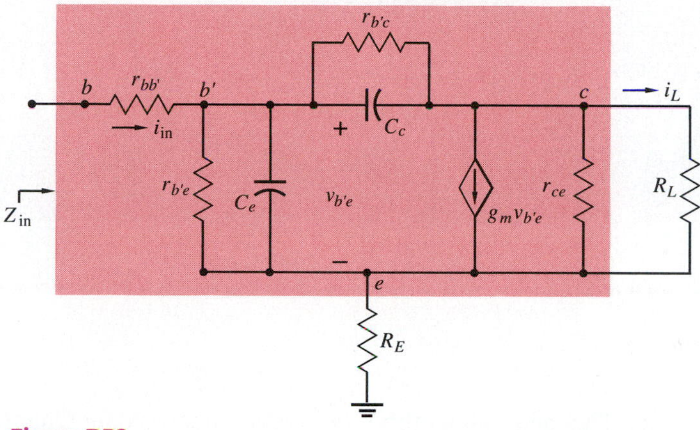

Figure DP3
A transistor circuit.

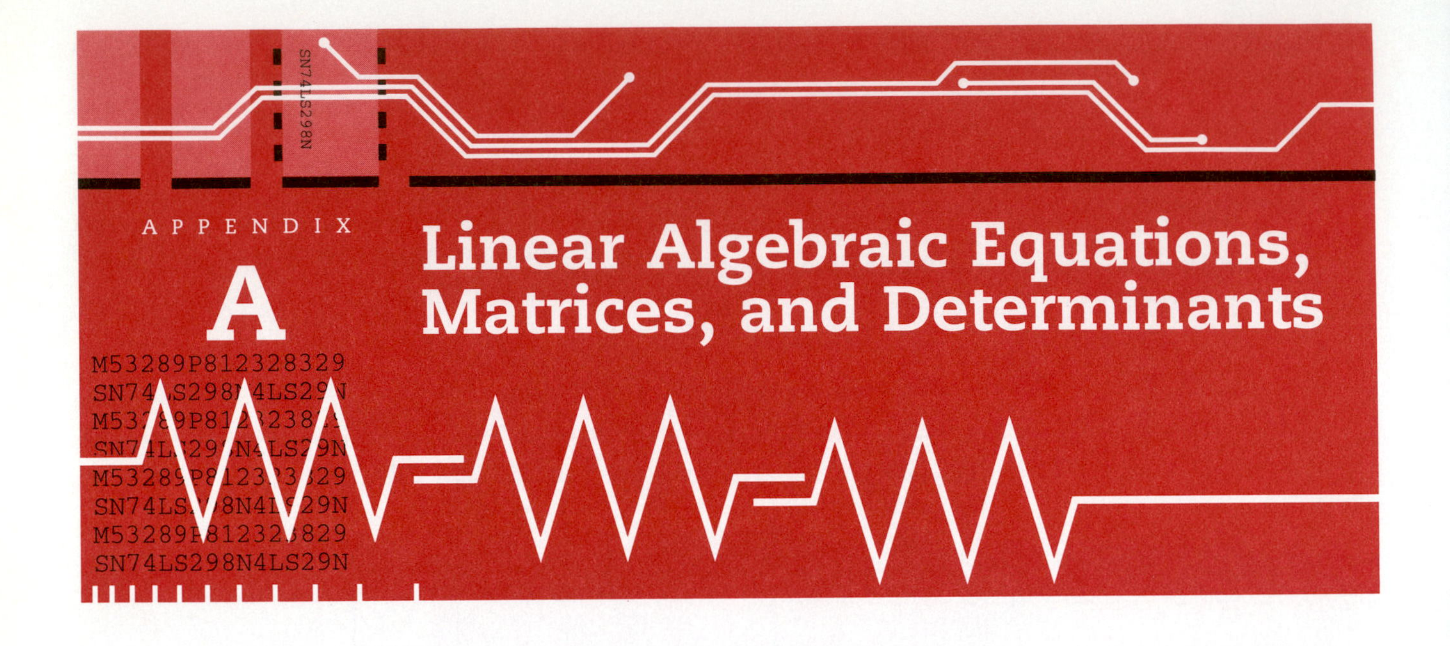

APPENDIX A Linear Algebraic Equations, Matrices, and Determinants

The general methods of circuit analysis in Chapter 5 require constructing, manipulating, and solving linear algebraic equations. In this appendix, we will briefly describe some mathematical concepts and procedures that facilitate this process. The discussion is not exhaustive. You will need a better understanding of these concepts than presented here, but this presentation can serve (1) as a review and (2) as a handy reference when you want to refresh your memory on a particular point.

The origins of the relevant equations are the application of the basic circuit laws (KCL, KVL, and the *v-i* relationships of components) and the combinations of these by node analysis, loop analysis, or other methods. The resulting equations for some example might be as follows:

$$\begin{aligned} 3x_1 - 2x_2 \qquad\quad &= y_1 \\ -2x_1 + 5x_2 + 4x_3 &= y_2 \\ 3x_1 \qquad\quad + 8x_3 &= y_3 \end{aligned} \tag{1}$$

where the y's are the *independent* variables and the x's are the *dependent* variables. In addition to the variables, the equations include *coefficients* which multiply the independent variables.

It would be extremely useful if such a set of equations (no matter how many equations there were) could be written compactly so as to display the three entities (two sets of variables and the coefficients) separately. This is what matrices permit us to do.

1 MATRICES

A *matrix* is defined as a rectangular array of *elements* arranged in *rows* and *columns* within brackets. It is identified by a **boldface** name. It can have any number of rows and any number of columns:

$$\mathbf{A} = \begin{bmatrix} a_{11} & a_{12} & \cdots & a_{1n} \\ a_{21} & a_{22} & \cdots & a_{2n} \\ \cdot & \cdot & \cdots & \cdot \\ a_{m1} & a_{m2} & \cdots & a_{mn} \end{bmatrix} = [a_{ij}]_{mn} \tag{2}$$

The *order* of a matrix is a pair of numbers designating the number of its rows (first) and of its columns (second). The order of $\mathbf{A}$ in (2) is mn (or m by n), meaning it has m rows and n columns. When the number of rows is 1, the matrix is called a *row* matrix. For example, $\mathbf{A}_j$ in (3), which follows, is the jth row of $\mathbf{A}$ in (2). When the number of columns is 1, as in $\mathbf{X}$ in (3), the matrix is called a *column* matrix; a column matrix is also called a *vector.*

$$\mathbf{A}_j = [A_{j1} \quad A_{j2} \quad \cdots \quad A_{jn}] = [A_{jk}]_{1n} \qquad \mathbf{X} = \begin{bmatrix} x_1 \\ x_2 \\ \vdots \\ x_k \end{bmatrix} = [x_i]_{k1} \tag{3}$$

Square, Symmetrical, and Unit Matrices

When the number of rows and columns is the same ($m = n$), the matrix is *square.* The *main diagonal* of a square matrix consists of all the elements whose two subscripts are the same: $a_{11}, a_{22}, \ldots, a_{nn}$. An important class of square matrices is the *symmetrical* matrix, in which $a_{ij} = a_{ji}$, meaning that an element on one side of the main diagonal is equal to its similarly positioned counterpart on the other side. (A more mathematically satisfying definition will be given subsequently.)

An important subclass of symmetrical matrices is the *diagonal* matrix, in which the only nonzero elements lie on the main diagonal, all other (off-diagonal) elements being zero. Furthermore, if all the diagonal elements are 1, the matrix is a *unit* (or *identity*) matrix. Thus:

$$\mathbf{D} = \begin{bmatrix} d_{11} & 0 & 0 & \cdots & 0 \\ 0 & d_{22} & 0 & \cdots & 0 \\ 0 & 0 & d_{33} & \cdots & 0 \\ \cdot & \cdot & \cdot & \cdots & \cdot \\ 0 & 0 & 0 & \cdots & d_{nn} \end{bmatrix} \qquad \mathbf{U} = \begin{bmatrix} 1 & & & & \\ & 1 & & \mathbf{0} & \\ & & 1 & & \\ & \mathbf{0} & & \ddots & \\ & & & & 1 \end{bmatrix} \tag{4}$$

(Note how it can be indicated that the off-diagonal elements are all zero in the second of these, instead of showing each element as zero.) One other special matrix is the *triangular* matrix, in which all elements on one side of the main diagonal are zero:

$$\mathbf{T}_u = \begin{bmatrix} a_{11} & a_{12} & a_{13} & \cdots & a_{1n} \\ 0 & a_{22} & a_{23} & \cdots & a_{2n} \\ 0 & 0 & a_{33} & \cdots & a_{3n} \\ \cdot & \cdot & \cdot & \cdots & \cdot \\ 0 & 0 & 0 & \cdots & a_{nn} \end{bmatrix} \qquad \mathbf{T}_l = \begin{bmatrix} a_{11} & & & & \\ a_{21} & a_{22} & & & \mathbf{0} \\ a_{31} & a_{32} & a_{33} & & \\ \cdot & \cdot & \cdot & \ddots & \\ a_{n1} & a_{n2} & a_{n3} & \cdots & a_{nn} \end{bmatrix} \tag{5}$$

The first is the *upper triangular* and the second the *lower triangular* matrix.

Matrix Algebra

This section will describe the rules of matrix algebra, pointing out along the way how they differ from the rules of ordinary (scalar) algebra.

EQUALITY Two matrices $\mathbf{A} = [a_{ij}]$ and $\mathbf{B} = [b_{ij}]$ of the same order are defined to be equal if all corresponding elements of the two matrices are equal: $a_{ij} = b_{ij}$ for all rows i and all columns j.

ADDITION AND SUBTRACTION The sum and difference of two matrices $\mathbf{A} = [a_{ij}]$ and $\mathbf{B} = [b_{ij}]$ of the *same order* are defined as follows:

$$\mathbf{C} = [c_{ij}] = \mathbf{A} \pm \mathbf{B} \quad \Rightarrow \quad c_{ij} = a_{ij} \pm b_{ij} \qquad \text{all } i \text{ and } j$$

Matrix addition is *associative*: $\mathbf{A} + (\mathbf{B} + \mathbf{C}) = (\mathbf{A} + \mathbf{B}) + \mathbf{C}$.

MULTIPLICATION The product $\mathbf{C} = [c_{ij}]$ of two matrices $\mathbf{A} = [a_{ij}]$ and $\mathbf{B} = [b_{ij}]$ is defined as follows:

$$\mathbf{C} = \mathbf{AB} \quad \Rightarrow \quad c_{ij} = \sum_{k=1}^{r} a_{ik} b_{kj} \tag{6}$$

That is, each element of the product matrix $\mathbf{C}$ is obtained by multiplying together, one-by-one, the elements of the appropriate row i of the first matrix $\mathbf{A}$ by the corresponding elements of the appropriate column j of the second matrix $\mathbf{B}$ and then adding the products. Thus, to obtain c_{23}, one multiplies each element of the second row of $\mathbf{A}$ with the corresponding element of the third column of $\mathbf{B}$. An example might be as follows:

$$\mathbf{A}_{\text{row 2}} = [3 \quad -1 \quad 5 \quad 2] \qquad \mathbf{B}_{\text{col 3}} = \begin{bmatrix} 0 \\ 1 \\ 4 \\ -3 \end{bmatrix}$$

Then: $c_{23} = 3(0) + (-1)(1) + 5(4) + 2(-3) = 13$.

It was fortunate in this example that we didn't run out of elements in the second row of $\mathbf{A}$ before we reached the last element in the third column in $\mathbf{B}$. That good fortune is not an accident. The definition of multiplication *requires* the number of *columns* in the first matrix in a product (that is, the number of elements in each row) to equal the number of *rows* of the second matrix (the number of elements in each column). Such matrices are said to be *conformable* for multiplication.

Matrix multiplication has some properties which are the same as the corresponding ones for scalar multiplication. Thus matrix multiplication is *associative*: $\mathbf{A}(\mathbf{BC}) = (\mathbf{AB})\mathbf{C}$; and it is *distributive* over addition: $\mathbf{A}(\mathbf{B} + \mathbf{C}) = \mathbf{AB} + \mathbf{AC}$. But matrix multiplication has a number of properties which are not the same and even seem to be counterintuitive:

1. Matrix multiplication is *not commutative.* That is, it is possible that $\mathbf{AB} \neq \mathbf{BA}$. This possibility can occur for two reasons. In the first place, the two matrices may be nonconformable for one of the two possible products. If the orders of the two matrices are $\mathbf{A}_{mn}$ and $\mathbf{B}_{rp}$, then conformability requires $n = r$ for product $\mathbf{AB}$ and $p = m$ for product $\mathbf{BA}$. It is possible for one of these requirements to be satisfied but not the other. And second, even if the matrices are conformable for both orders of multiplication, it is possible for the two product matrices to be unequal.

2. Division or cancellation does not exist in matrix algebra. In scalar algebra, if $ky = kx$, the k can be canceled, yielding $y = x$. A similar result does not hold true for matrices. Thus, if $\mathbf{AB} = \mathbf{AC}$, we can't just cancel the $\mathbf{A}$ and get $\mathbf{B} = \mathbf{C}$. This is readily apparent without considering a specific example by observing the possible orders of the three matrices. Let $\mathbf{A}$ be a matrix having n columns. Then, for conformability, $\mathbf{B}$ and $\mathbf{C}$ must have n rows. But that places no limitation on their number of columns—they can have different numbers of columns. In that case, $\mathbf{B}$ and $\mathbf{C}$ cannot be equal.

3. In scalar algebra, if the product $xy = 0$, then either $x = 0$ or $y = 0$. This property is not true for matrices. That is, if $\mathbf{AB} = \mathbf{0}$ it does not follow that either $\mathbf{A} = \mathbf{0}$ (all elements of $\mathbf{A} = 0$) or $\mathbf{B} = \mathbf{0}$.

From the rules for multiplication, the following important property of the unit matrix can be verified. Given a matrix $\mathbf{M}$:

$$\mathbf{UM} = \mathbf{M} \qquad \text{and} \qquad \mathbf{MU} = \mathbf{M} \tag{7}$$

where the order of $\mathbf{U}$ is chosen so that $\mathbf{M}$ and $\mathbf{U}$ are conformable.

PARTITIONING Sometimes it is convenient to rewrite a given matrix so that parts of it, called *submatrices,* are themselves treated as matrices. As an example, the following 3×5 matrix can be partitioned into four submatrices, as shown by the dashed lines:

$$\mathbf{A} = \left[\begin{array}{cc:ccc} a_{11} & a_{12} & a_{13} & a_{14} & a_{15} \\ \hdashline a_{21} & a_{22} & a_{23} & a_{24} & a_{25} \\ a_{31} & a_{32} & a_{33} & a_{34} & a_{35} \end{array}\right] = \begin{bmatrix} \mathbf{A}_{11} & \mathbf{A}_{12} \\ \mathbf{A}_{21} & \mathbf{A}_{22} \end{bmatrix} \tag{8}$$

where

$$\mathbf{A}_{11} = [a_{11} \quad a_{12}] \qquad \mathbf{A}_{12} = [a_{13} \quad a_{14} \quad a_{15}]$$

$$\mathbf{A}_{21} = \begin{bmatrix} a_{21} & a_{22} \\ a_{31} & a_{32} \end{bmatrix} \qquad \mathbf{A}_{22} = \begin{bmatrix} a_{23} & a_{24} & a_{25} \\ a_{33} & a_{34} & a_{35} \end{bmatrix}$$

The submatrices can have any number of rows and columns except that each submatrix in a row must have the same number of rows of the original matrix and each submatrix in a column must have the same number of columns of the original matrix. In particular, a single element of the given matrix can form one of the submatrices in the partition.

TRANSPOSE Given a matrix $\mathbf{A}$, the *transpose* of $\mathbf{A}$, labeled $\mathbf{A}'$, is obtained from $\mathbf{A}$ by interchanging its rows and columns. Thus, if $\mathbf{A} = [a_{ij}]$, then $\mathbf{B} = \mathbf{A}' = [b_{ij} = a_{ji}]$. Unless the matrix is square, the transpose of $\mathbf{A}$ will have a different order from $\mathbf{A}$.

What would it mean for the transpose of a square matrix $\mathbf{A}$ to equal $\mathbf{A}$? That is, suppose:

$$\mathbf{A} = [a_{ij}] = \mathbf{A}' = [a_{ji}]$$

Then, it follows that $a_{ij} = a_{ji}$ for all rows i and columns j. This is the condition for $\mathbf{A}$ to be a *symmetrical* matrix, so a symmetrical matrix is characterized by $\mathbf{A}' = \mathbf{A}$.

2 DETERMINANTS

The *determinant* of a square matrix $\mathbf{A}$ is a *scalar* indicated by the following notation:

$$\Delta = \det \mathbf{A} = \begin{vmatrix} a_{11} & a_{12} & \cdots & a_{1n} \\ a_{21} & a_{22} & \cdots & a_{2n} \\ \cdot & \cdot & \cdots & \cdot \\ a_{n1} & a_{n2} & \cdots & a_{nn} \end{vmatrix} \tag{9}$$

(Note that, for a determinant, the array of elements is between vertical bars, not brackets.)

The *cofactor* of element a_{ij} of $\Delta = \det \mathbf{A}$, designated Δ_{ij}, is the determinant of an array obtained from $\mathbf{A}$ by removing row i and column j from it and then multiplying the result by $(-1)^{i+j}$. Thus, for a square matrix of order 3, Δ_{13} is obtained by removing the first row and third column, giving:

$$\Delta_{13} = (-1)^{1+3} \begin{vmatrix} a_{21} & a_{22} \\ a_{31} & a_{32} \end{vmatrix} = \begin{vmatrix} a_{21} & a_{22} \\ a_{31} & a_{32} \end{vmatrix}$$

Properties of Determinants

Some of the properties of determinants useful in evaluating them are as follows.

1. A square matrix and its transpose have the same determinant: $\det \mathbf{A} = \det \mathbf{A}'$.
2. If all elements of any row of a determinant are multiplied by k, then the determinant is multiplied by k. Likewise for columns.

3. If any two rows of a determinant are interchanged, the sign of the determinant is changed. Likewise for columns.

4. A determinant is zero if any two rows are identical. Likewise for columns.

5. If all elements of a row are zero, then the determinant is zero. Likewise for columns.

6. Adding to each element of a row the same multiple of the corresponding element of any other row leaves the determinant unchanged. Likewise for columns.

7. The determinant of the product of two matrices equals the product of the two determinants.

8. The determinant of a unit matrix is 1.

9. The determinant of both a diagonal matrix and a triangular matrix equals the product of the diagonal elements.

Cofactor Expansions

The scalar value of a determinant can be evaluated as follows:

$$\begin{aligned} \Delta = \det \mathbf{A} &= \sum_{i=1}^{n} a_{ij}\Delta_{ij} \qquad \text{for any column } j \\ &= \sum_{j=1}^{n} a_{ij}\Delta_{ij} \qquad \text{for any row } i \end{aligned} \tag{10}$$

These are referred to as *cofactor expansions* along a row and along a column, respectively.

Each cofactor, in turn, can be expanded and the process continued until the order is reduced to 2. For a matrix of order 2, each cofactor is $\pm$ a single element, so the evaluation of a second-order determinant can be easily remembered. It is:

$$\det \mathbf{A} = \begin{vmatrix} a_{11} & a_{12} \\ a_{21} & a_{22} \end{vmatrix} = a_{11}a_{22} - a_{12}a_{21} \tag{11}$$

Thus, in evaluating a high-order determinant, the expansion process is used until second-order determinants are reached. Then, the preceding expression for a second-order determinant is used to complete the evaluation. Using such expansions, it is easy to show for a diagonal matrix $\mathbf{D}$ and a unit matrix $\mathbf{U}$ that:

$$\det \mathbf{D} = d_{11}d_{22}\cdots d_{nn} \qquad \text{and} \qquad \det \mathbf{U} = 1$$

Thus, the determinant of a diagonal matrix is the product of its diagonal elements.

Pivotal Condensation

While the cofactor expansion of determinants in (10) is very useful symbolically, it has a disadvantage numerically. For computational purposes, reducing the number of arithmetic operations (especially multiplications) is important. (Division is multiplication by the reciprocal.) In the numerical evaluation of determinants by cofactor expansion, the number of arithmetic operations increases excessively with increasing order.

Another method for the numerical evaluation of determinants that requires significantly fewer arithmetic operations is called *pivotal condensation.* Given a square matrix $\mathbf{A}$ of order n, the first step is to partition it, taking a_{11} as a submatrix:

$$\mathbf{A} = \begin{bmatrix} a_{11} & \mathbf{A}_{12} \\ \mathbf{A}_{21} & \mathbf{A}_{22} \end{bmatrix} \tag{12}$$

Note that, whereas $\mathbf{A}_{22}$ is of order $n - 1$, $\mathbf{A}_{12}$ is a row matrix of order 1 by $(n - 1)$ and $\mathbf{A}_{21}$ is a column matrix of order $(n - 1)$ by 1. The determinant of $\mathbf{A}$ is found as follows:

$$\det \mathbf{A} = \frac{1}{a_{11}^{n-2}} \det (a_{11}\mathbf{A}_{22} - \mathbf{A}_{21}\mathbf{A}_{12}) \tag{13}$$

Obviously, a_{11} cannot equal zero. If it is zero in the given matrix, the first column can be interchanged with any other column whose first element is not zero. According to property 3 of determinants, this would simply introduce a minus sign. A *condensation* has taken place around the pivot a_{11}, since the order of the determinant to be evaluated has been reduced by 1. The process can be repeated until the order has been reduced to 2.

3 LINEAR ALGEBRAIC EQUATIONS

Matrices provide a neat, compact way of expressing sets of linear algebraic equations and of solving them. Let's use the set of equations in (1) at the beginning of this appendix. In matrix form, this set can be written as follows:

$$\begin{bmatrix} 3 & -2 & 0 \\ -2 & 5 & 4 \\ 3 & 0 & 8 \end{bmatrix} \begin{bmatrix} x_1 \\ x_2 \\ x_3 \end{bmatrix} = \begin{bmatrix} y_1 \\ y_2 \\ y_3 \end{bmatrix} \tag{14}$$

$$\mathbf{AX} = \mathbf{Y}$$

where $\mathbf{A}$ is the square coefficient matrix and $\mathbf{X}$ and $\mathbf{Y}$ are the vectors of dependent and independent variables, respectively. (Applying the rules of matrix multiplication and equality, you should verify that this expression expands into the original equations.)

Solution

A solution for vector $\mathbf{X}$ would result if both sides of (14) are divided by matrix $\mathbf{A}$, but division of matrices is not permissible. Suppose, instead, that both sides of $\mathbf{AX} = \mathbf{Y}$ are multiplied by a square matrix (temporarily called $\mathbf{I}$) of the same order as $\mathbf{A}$, yielding $\mathbf{IAX} = \mathbf{IY}$. That's certainly a permissible operation, but what does it gain? It would gain a great deal if **IAX** were to equal **X**; the result would then be the solution for **X**. It happens that **IAX** = **X** if **IA** = **U**, a unit matrix. Thus, solving the set of equations amounts to finding such a matrix **I**. So much for the preliminaries.

Inverting a Matrix

Given a square matrix **A**, we define another matrix, designated $\mathbf{A}^{-1}$ and called the *inverse* of **A**, by the following:

$$\mathbf{A}^{-1}\mathbf{A} = \mathbf{A}\mathbf{A}^{-1} = \mathbf{U} \tag{15}$$

(The temporary name **I** used in the preceding paragraph stood for *inverse.*) Thus, given a matrix equation **AX** = **Y**, we multiply both sides by the inverse of **A** and get a solution for **X** as $\mathbf{A}^{-1}\mathbf{Y}$. The only remaining problem is a means for finding the inverse. For this purpose, we need one other definition.

The Adjoint of a Matrix

Given a square matrix **A**, we define the *adjoint* of **A**:

$$\text{adj}\,\mathbf{A} = \begin{bmatrix} \Delta_{11} & \Delta_{21} & \cdots & \Delta_{n1} \\ \Delta_{12} & \Delta_{22} & \cdots & \Delta_{n2} \\ \cdot & \cdot & \cdots & \cdot \\ \Delta_{1n} & \Delta_{2n} & \cdots & \Delta_{nn} \end{bmatrix} \tag{16}$$

in which the element in position ij is the cofactor of a_{ji}. That is, the elements in a *row* of adj **A** are the cofactors of the elements in the corresponding *column* of **A**.

Solving a Set of Linear Equations

A number of different methods can be used to solve a set of linear algebraic equations. One method is based on cofactor expansions. Using the adjoint of a matrix defined in (16), the solution for the set of equations **AX** = **Y** is given by:

$$\mathbf{X} = \mathbf{A}^{-1}\mathbf{Y} = \frac{1}{\det \mathbf{A}}(\text{adj}\,\mathbf{A})\mathbf{Y} \tag{17}$$

In scalar form, using (16), this can be written as follows:

$$x_j = \frac{\Delta_{1j}}{\Delta}y_1 + \frac{\Delta_{2j}}{\Delta}y_2 + \cdots + \frac{\Delta_{nj}}{\Delta}y_n \tag{18}$$

where $\Delta = \det \mathbf{A}$. The last equation is referred to as *Cramer's rule.*

As an illustration, the preceding relationships will be used to find the solution of the equations in (1). First, the cofactors are calculated; then, using one of the possible expansions (say, along the first row), the determinant is found using (10):

$$\Delta_{11} = \begin{vmatrix} 5 & 4 \\ 0 & 8 \end{vmatrix} = 40 \qquad \Delta_{12} = -\begin{vmatrix} -2 & 4 \\ 3 & 8 \end{vmatrix} = 28 \qquad \Delta_{13} = \begin{vmatrix} -2 & 5 \\ 3 & 0 \end{vmatrix} = -15$$

$$\Delta_{21} = -\begin{vmatrix} -2 & 0 \\ 0 & 8 \end{vmatrix} = 16 \qquad \Delta_{22} = \begin{vmatrix} 3 & 0 \\ 3 & 8 \end{vmatrix} = 24 \qquad \Delta_{23} = -\begin{vmatrix} 3 & -2 \\ 3 & 0 \end{vmatrix} = -6$$

$$\Delta_{31} = \begin{vmatrix} -2 & 0 \\ 5 & 4 \end{vmatrix} = -8 \qquad \Delta_{32} = -\begin{vmatrix} 3 & 0 \\ -2 & 4 \end{vmatrix} = -12 \qquad \Delta_{33} = \begin{vmatrix} 3 & -2 \\ -2 & 5 \end{vmatrix} = 11$$

$$\Delta = 3\Delta_{11} + (-2)\Delta_{12} = 3(40) - 2(28) = 64$$

Finally, using the expression for the solution $\mathbf{X} = \mathbf{A}^{-1}\mathbf{Y}$, we get:

$$\begin{bmatrix} x_1 \\ x_2 \\ x_3 \end{bmatrix} = \frac{1}{64}\begin{bmatrix} 40 & 16 & -8 \\ 28 & 24 & -12 \\ -15 & -6 & 11 \end{bmatrix}\begin{bmatrix} y_1 \\ y_2 \\ y_3 \end{bmatrix}$$

The number of arithmetic operations in the evaluation of the determinant and its cofactors can be reduced by the use of pivotal condensation. In the preceding example the cofactors are trivial. As for the determinant, the required partitions of $\mathbf{A}$ for pivotal condensation are:

$$a_{11} = 3 \qquad \mathbf{A}_{12} = [-2 \quad 0]$$

$$\mathbf{A}_{21} = \begin{bmatrix} -2 \\ 3 \end{bmatrix} \qquad \mathbf{A}_{22} = \begin{bmatrix} 5 & 4 \\ 0 & 8 \end{bmatrix}$$

These submatrices are now inserted into (13):

$$\begin{aligned} \det \mathbf{A} &= \frac{1}{3^{3-2}} \det\left(3\begin{bmatrix} 5 & 4 \\ 0 & 8 \end{bmatrix} - \begin{bmatrix} -2 \\ 3 \end{bmatrix}[-2 \quad 0]\right) \\ &= \frac{1}{3}\det\left(\begin{bmatrix} 15 & 12 \\ 0 & 24 \end{bmatrix} - \begin{bmatrix} 4 & 0 \\ -6 & 0 \end{bmatrix}\right) = \frac{1}{3}\det\begin{bmatrix} 11 & 12 \\ 6 & 24 \end{bmatrix} \\ &= \tfrac{1}{3}(264 - 72) = \tfrac{1}{3}(192) = 64 \end{aligned}$$

The value of the determinant is the same. Count the number of arithmetic operations by both methods and compare.

Gaussian Elimination

A numerically much more efficient method for solving a set of linear equations is the method of *Gaussian elimination.* It is based on number 6 of the properties of determinants previously listed. Starting with n equations in n unknowns, the process eliminates one of the variables from $n - 1$ of the equations (all but the first one). Holding this one equation aside, it next eliminates another of the variables from all but one of the remaining $n - 1$ equations. It continues in this fashion, with one fewer equation each time, eliminating one variable after another until, after the $(n - 1)$th step, only one equation is left in one unknown. This sequence is the *forward elimination* step in the process.

The last equation is solved for the one remaining variable. This is then substituted into all other equations. The next-to-last equation then has just a single unknown for which it can be solved, and the process continued. The complete solution is obtained by proceeding in a backward sequence in this fashion, one equation at a time, in which a single unknown remains, successively substituting the solution for this one variable into all remaining equations. This is the *back substitution* step.

The Gaussian elimination procedure will be illustrated on the system of linear equations in (1). To eliminate x_1 from the second equation, the first equation is multiplied by $\frac{2}{3}$ and is added to the second. Similarly, to eliminate x_1 from the third equation, the first equation is multiplied by -1 and added to the third. The result of this step is:

$$\begin{aligned} 3x_1 - 2x_2 \qquad\quad &= y_1 \\ \tfrac{11}{3}x_2 + 4x_3 &= \tfrac{2}{3}y_1 + y_2 \\ 2x_2 + 8x_3 &= -y_1 + y_3 \end{aligned}$$

The forward elimination process continues by eliminating x_2 from the last equation. This is done by multiplying the second equation by $-\frac{6}{11}$ and adding it to the last. The result is:

$$x_3 = \tfrac{1}{64}(-15y_1 - 6y_2 + 11y_3)$$

This completes the forward substitution step. Now x_3 is substituted into the equations remaining from the forward substitution step (the first two of the preceding set of three equations). The only unknown remaining in the last equation of that pair will be x_2, so the equation is solved for x_2:

$$x_2 = \tfrac{1}{16}(7y_1 + 6y_2 - 3y_3)$$

Finally, x_2 is inserted into the first equation (x_3 is missing in that equation), resulting in the solution for x_1:

$$x_1 = \tfrac{1}{8}(5y_1 + 2y_2 - y_3)$$

Comparing these answers with those resulting from Cramer's rule, we see that they are the same.

These low-order equations do not illustrate well the computational efficiency of the Gaussian elimination method. To solve a set of n equations by Cramer's rule requires approximately $2(n + 1)!$ multiplications.[1] For $n = 3$, this number is 48. But for $n = 5$, it is 1440, and for $n = 10$, it is almost 80 million! In Gaussian elimination, the number of multiplications is approximately $n^3/3$. For $n = 10$, this is only 333 compared with almost 8 million!

[1] Most of the time spent by a computer to perform arithmetic operations is spent on multiplication. (Division is counted as a multiplication by the reciprocal.) Hence, a measure of the computational efficiency is the number of multiplications.

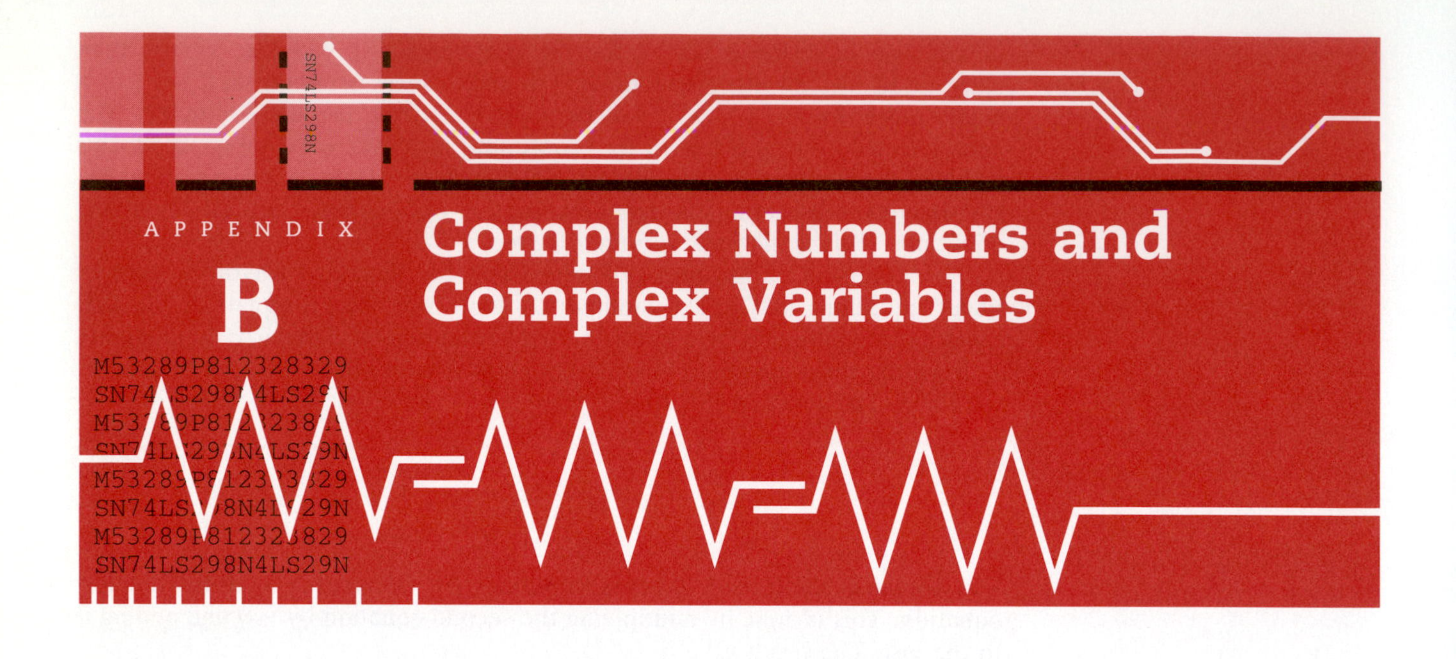

APPENDIX B Complex Numbers and Complex Variables

This appendix will provide an introduction to a subject which is of major importance in circuit analysis—not to mention other areas, such as control systems and communications theory. The arithmetic of complex numbers facilitates the sinusoidal steady-state solution of electric networks. The algebra and calculus of complex variables permit insights into the behavior of networks which would be unavailable otherwise.

1 THE ARITHMETIC OF COMPLEX NUMBERS

Definition: A complex number is a particular combination of two real numbers: an *ordered pair,* written (a, b). When we say that the pair of real numbers is *ordered,* we mean that the sequence (or order) of writing the two numbers is important. By definition, then, a complex number has two parts, each of which is a real number. The first member of the pair is called the *real part* of the complex number, the second member being the *imaginary part.* This terminology is unfortunate, since there is nothing more imaginary about the "imaginary" part of a complex number than there is about the real part.

A real number is a special case of a complex number, one whose imaginary part is zero. That is, complex numbers of the form $(a, 0)$ are real. Another special case is the class of complex numbers whose real part is zero. Such numbers are called *imaginary* numbers; they have the form: $(0, b)$.

Arithmetic Operations

We now turn to a definition of the elementary arithmetic operations for complex numbers. Since real numbers are special cases of complex numbers, any operation defined for complex numbers must reduce to the corresponding operation for real numbers, whenever the complex numbers involved are real.

EQUALITY Two complex numbers $A = (a_1, a_2)$ and $B = (b_1, b_2)$ are equal if, and only if, $a_1 = b_1$ and $a_2 = b_2$; that is, the two real parts must be equal and so must the two imaginary parts.

ADDITION The sum of two complex numbers $A = (a_1, a_2)$ and $B = (b_1, b_2)$ is defined as a complex number $C = (c_1, c_2)$ such that:

$$C = A + B = (a_1 + b_1, a_2 + b_2)$$

That is, the sum has a real part which is the sum of the real parts of the two numbers and it has an imaginary part which is the sum of the imaginary parts.

SUBTRACTION The difference of two complex numbers A and B is defined as $A - B = A + (-B)$; that is, the difference is obtained by changing the sign of the subtrahend and adding.

MULTIPLICATION The product of two complex numbers $A = (a_1, a_2)$ and $B = (b_1, b_2)$ is defined as:

$$C = AB = ([a_1b_1 - a_2b_2], [a_2b_1 + a_1b_2])$$

You might wonder how in the world somebody thought up such a definition, and not some other combination of the two ordered pairs! One criterion has already been mentioned: the result must be compatible with multiplication of real numbers when the imaginary parts a_2 and b_2 are zero. (Verify that it is.) Another criterion is utility, but you are not yet in a position to judge whether the definition is useful (it is).

RECTANGULAR FORM From the definition of a complex number as an ordered pair, the only way to distinguish the second member of the pair is from its position, its order. It would be useful if a "tag" could be attached to the imaginary part in order to distinguish it no matter where it appeared. This can be done by observing that, using the definitions of addition and multiplication, any complex number can be written in the following forms:

$$\begin{aligned}(a_1, a_2) &= (a_1, 0) + (0, a_2)\\ &= (a_1, 0) + (0, 1)(a_2, 0)\end{aligned}$$

In the first step, the complex number has been written as the sum of two complex numbers, the first having a zero imaginary part and the second a zero real part. In the next step, this second complex number has been written as the product of two other complex numbers, the second of which is simply a real number a_2. The first number in this product is a special imaginary number. It arises so often that it is identified with a special symbol:

$$(0, 1) \equiv j$$

With this notation, the preceding complex number can be written:

$$(a_1, a_2) = a_1 + ja_2$$

The form of the complex number on the right-hand side is referred to as the *rectangular form.*[1]

THE UNIT IMAGINARY NUMBER Something significant about the unit imaginary number j can be discovered by multiplying j by itself:

$$j^2 = (0, 1)(0, 1) = (-1, 0) = -1$$
$$j = \sqrt{-1}$$

That is, j is the square root of -1.[2]

Using the symbol j, it is possible to perform multiplication in a more convenient way than indicated in the original definition. Thus, using the ordinary rules of algebra, we find:

$$\begin{array}{r} a_1 + ja_2 \\ \times \quad b_1 + jb_2 \\ \hline a_1b_1 + ja_2b_1 + ja_1b_2 + j^2a_2b_2 \\ = (a_1b_1 - a_2b_2) + j(a_2b_1 + a_1b_2) \end{array}$$

This agrees with the original definition, of course.

CONJUGATE The conjugate of a complex number $A = a_1 + ja_2$ is the complex number $A^* = a_1 - ja_2$. The conjugate of A, designated by an asterisk, has the same real part as A but an imaginary part which is of opposite sign. Something interesting occurs if a complex number is multiplied by its conjugate:

$$AA^* = (a_1 + ja_2)(a_1 - ja_2) = a_1^2 + a_2^2$$

Thus, the product of A and its conjugate is a real number consisting of the sum of the squares of the real and imaginary parts of A.

[1] Mathematicians use the symbol i instead of j for the unit imaginary number. However, since i stands for current in electrical engineering, j is the symbol universally used by electrical engineers.

[2] Let's wax philosophical. Undoubtedly, the first contact of humans with numbers came from counting—their ears, toes, children, goats. Thus, the first set of numbers people became acquainted with were the positive integers. Fractions followed soon after, since objects had to be divided among two or more people on occasion. (When Solomon suggested dividing a child evenly between two women, each of whom claimed the child as her own, the real mother objected and Solomon's wisdom was proved.) The introducton of negative numbers was a milestone. In some respects, negative numbers are counterintuitive; who ever heard of a negative toe? Thus, -1, the unit of negative numbers, is a major conceptual advance. Similarly, the unit imaginary number j represents another major conceptual leap. Finding that one of these is the square root of the other is a mathematical delight.

DIVISION The quotient of two complex numbers $A = a_1 + ja_2$ and $B = b_1 + jb_2$ is defined as the complex number $C = c_1 + jc_2$ which, when multiplied by B, gives A. As it should be, this definition is compatible with the definition of division of real numbers. Thus, $C = A/B$ means $BC = A$, or:

$$(b_1 + jb_2)(c_1 + jc_2) = a_1 + ja_2$$

From the definitions of multiplication and equality, we find:

$$b_1c_1 - b_2c_2 = a_1$$

$$b_2c_1 + b_1c_2 = a_2$$

This pair of equations can now be solved for the unknowns c_1 and c_2 (by means of determinants, for example). You should make no attempt to try to remember the resulting expressions. Instead, a different approach will be described.

RATIONALIZING A clue to a possible approach is obtained by recalling that the product of a complex number and its conjugate is a real number. Hence, given a quotient of two complex numbers, A/B, if both numerator and denominator are multiplied by the conjugate of B, the value of the quotient will be unchanged but the resulting denominator will be a real number. Thus:

$$\frac{A}{B} = \frac{a_1 + ja_2}{b_1 + jb_2} = \frac{AB^*}{BB^*} = \frac{(a_1 + ja_2)(b_1 - jb_2)}{b_1^2 + b_2^2}$$

When the indicated multiplication in the numerator is carried out, the real and imaginary parts of the complex number will be placed in evidence. This process is called *rationalizing* the fraction.

Graphical Representation

The set of all real numbers, we know, can be put in a one-to-one correspondence with points on a straight line. Any real number can be represented by the distance of a point on the line from another point called the *origin.* Now a complex number consists of a *pair* of real numbers, so one line alone will be insufficient to represent it. Instead of a line, we will need a *plane* to represent a complex number.

Figure 1 shows a pair of axes at right angles to each other. Each part (real and imaginary) of a complex number can be put in a one-to-one correspondence with points on one of these lines. The horizontal axis is called the *real axis,* while the vertical one is called the *imaginary,* or *j,* axis. Any point in the plane can be described in terms of its horizontal and vertical (real and imaginary) coordinates. Hence, any point in the plane can represent a complex number. This description of a complex number is said to be in *rectangular coordinates.*

As an example, consider the complex number $A = 3 + j4$. This is represented in the figure by the point marked A having a horizontal (real) coordinate 3 and a vertical (imaginary) coordinate 4.

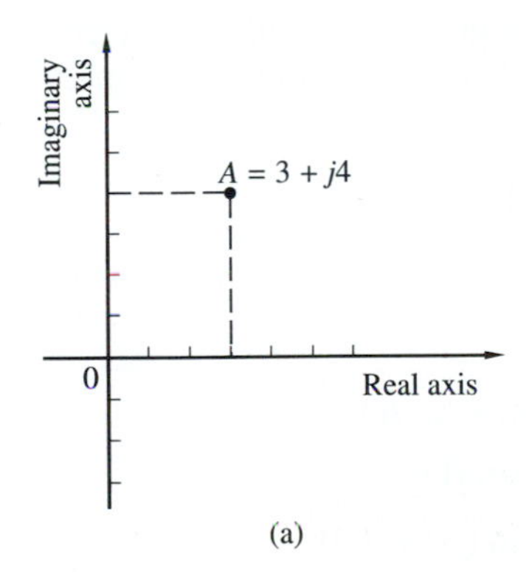

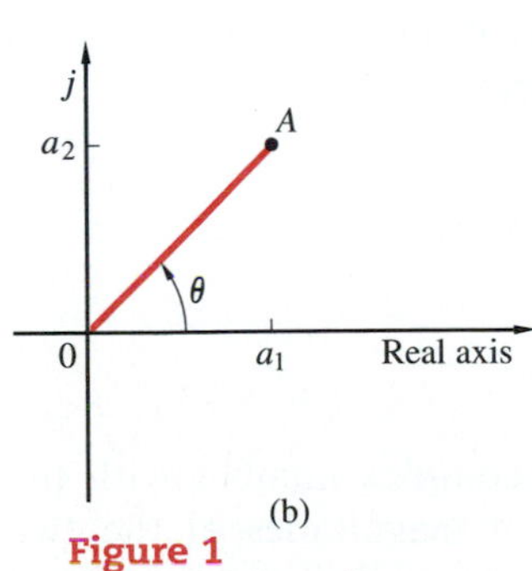

Figure 1
The complex plane.

Polar Coordinates

Besides rectangular coordinates, a point in a plane can be described in terms of *polar* coordinates also. The polar coordinates of a point are (1) the distance of the point from the origin and (2) the angle which the line from the origin to the point makes with the real axis. This is illustrated in Figure 1b.

Thus a complex number A can be described in terms not only of its real and imaginary parts, but of its *magnitude* $|A|$ and its *angle* θ. From the geometry of the figure, it is clear that:

$$a_1 = |A| \cos \theta$$
$$a_2 = |A| \sin \theta$$

Hence, a complex number can also be expressed in what is called the *trigonometric* form:

$$A = |A| \cos \theta + j\,|A| \sin \theta$$

The essential information in this expression is the magnitude $|A|$ and the angle θ. Thus, instead of the preceding expression, the following symbolic notation is used to express a complex number:

$$A = |A| \underline{/\theta}$$

(We read this as "magnitude of A at an angle θ.") This notation is simply symbolic; mathematical operations are not to be performed on it.

MULTIPLICATION AND DIVISION IN POLAR COORDINATES Multiplication and division of complex numbers are greatly simplified in terms of the magnitude and angle. Let us write two complex numbers in trigonometric form:

$$A = |A| (\cos \alpha + j \sin \alpha)$$
$$B = |B| (\cos \beta + j \sin \beta)$$

Their product will become:

$$\begin{aligned} C = AB &= |A|\,|B|\,[(\cos \alpha \cos \beta - \sin \alpha \sin \beta) \\ &\quad + j(\sin \alpha \cos \beta + \cos \alpha \sin \beta)] \\ &= |A|\,|B|\,[\cos (\alpha + \beta) + j \sin (\alpha + \beta)] \end{aligned}$$

The last step follows from trigonometric identities involving the sine and cosine of the sum of two angles. The final result is in the form:

$$C = |C| (\cos \theta + j \sin \theta)$$

where

$$|C| = |A|\,|B|$$
$$\theta = \alpha + \beta$$

That is, the product of two complex numbers is a complex number with the following attributes: its magnitude is the product of magnitudes of the two numbers and its angle is the sum of the two angles. For computational purposes, this is the form in which multiplication is most easily carried out. Symbolically:

$$A = |A|\underline{/\alpha} \qquad B = |B|\underline{/\beta}$$
$$C = |A|\,|B|\underline{/\alpha + \beta}$$

The same approach can be used to find the quotient of two complex numbers. If $C = A/B$, then $A = BC$, and so, using the same notation for magnitudes and angles as before:

$$|A|\underline{/\alpha} = |B|\underline{/\beta}\,|C|\underline{/\theta} = |B|\,|C|\underline{/\beta + \theta}$$
$$C = \frac{A}{B} = \frac{|A|}{|B|}\underline{/\alpha - \beta}$$

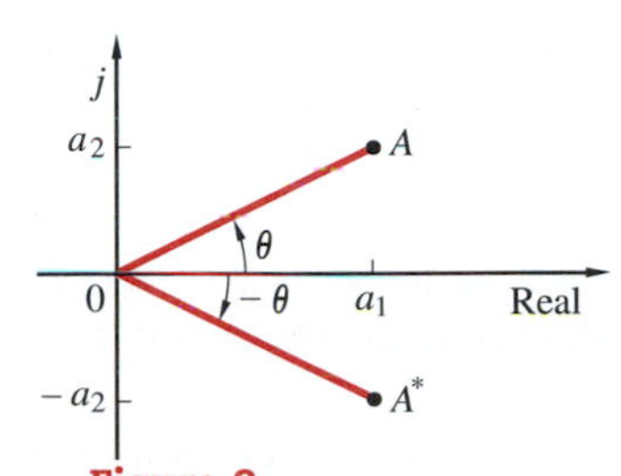

Figure 2
A number and its conjugate in the complex plane.

That is, the quotient of two complex numbers is a complex number with the following attributes: its magnitude is the quotient of the magnitudes of the two numbers and its angle is the difference of the angles. In numerical computations, this form for division is easy to carry out.

THE CONJUGATE IN POLAR COORDINATES As a final task, let's examine the conjugate of a complex number in polar coordinates. Figure 2 illustrates a complex number $A = a_1 + ja_2 = |A|\underline{/\alpha}$ located in the complex plane. The conjugate of A has the same real part but a negative imaginary part. Hence, its position in the plane is the image of A about the real axis, as shown in the figure.

It is clear that the conjugate of A has the same magnitude as A but an angle which is the negative of the angle of A. In symbolic notation, A^* can be written:

$$A^* = |A|\underline{/-\alpha}$$

Now take the product of A by its conjugate. Symbolically:

$$AA^* = |A|\underline{/\alpha}\,|A|\underline{/-\alpha} = |A|^2$$

That is, a complex number multiplied by its conjugate is a real number equal to the square of the magnitude of the complex number.

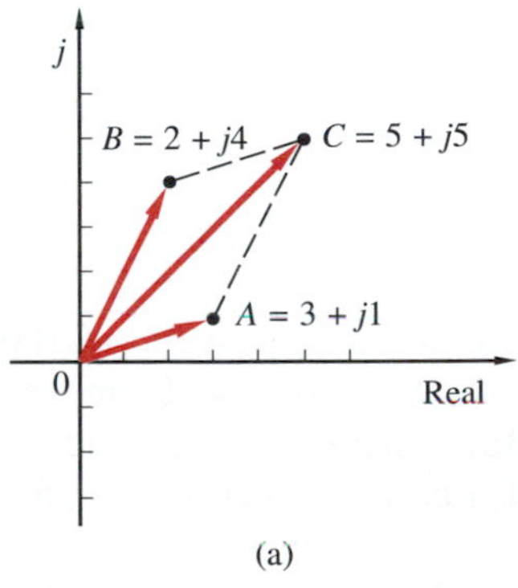

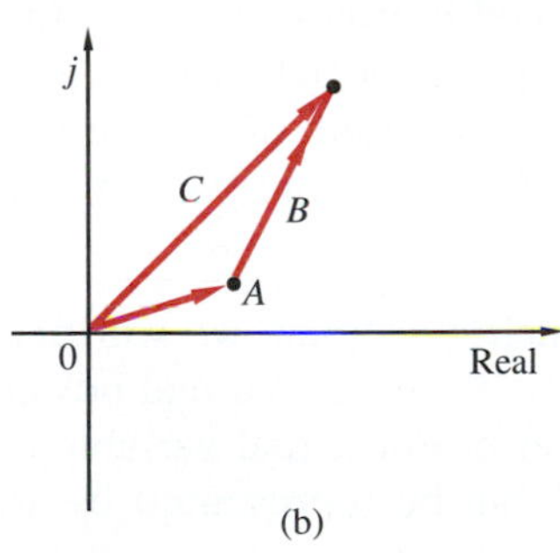

Figure 3
Graphical addition of complex numbers.

Similarity with Two-Dimensional Vectors

The preceding discussion should have brought to your consciousness that complex numbers have many properties in common with two-dimensional vectors. Thus, when represented graphically, a complex number can be regarded as a directed line from the origin to a point in a plane. The magnitude of the complex number is the same as the length of the vector, and the angle specifies the direction of the vector. Furthermore, if the vector from the origin is translated parallel to itself anywhere in the plane, it still represents the complex number, since its length and the angle it makes with the horizontal axis do not change in the translation.

The operations of addition and subtraction can be performed in a graphical manner very easily in the complex plane. As an example, consider Figure 3a. The parallelogram law for adding vectors is valid, as the figure illustrates.

Studying Figure 3b shows that the polygon rule for the addition of vectors is also valid for complex numbers. That is, the tail of one vector in a sum is joined to the head of the preceding one. The sum is then the vector joining the tail of the first vector to the head of the last one. Although the example shows only two vectors, this result holds for any number of vectors in a sum.

2 COMPLEX VARIABLES AND FUNCTIONS

So far, we have been discussing the arithmetic of complex numbers. We have been considering complex numbers—A, B, and so on—that are constant. If a complex quantity takes on a sequence of values, it is considered a *variable*. If the values that one complex variable takes on are dependent on the values of a second complex variable, the first complex variable is said to be a *function* of the second. Thus, just as there are functions of real variables, there are also functions of complex variables.

The simplest functions to think of are powers. Thus, let us designate a complex variable as $z = x + jy$. Then, by z^2 we mean the variable z multiplied once by itself. Higher powers of z have the obvious meaning that z is multiplied by itself more than once.

A linear combination of various powers of a variable is a function of the form:

$$f(z) = z^n + a_1 z^{n-1} + \cdots + a_{n-1} z + a_n$$

Polynomial is the obvious term for this function. Again obviously, a *rational function* is a ratio of two polynomials of a complex variable.

Transcendental Functions

For a complex variable, there is no difficulty in defining powers of the variable, polynomials, or rational functions, because they follow from the meanings of multiplication, addition, and division. But difficulty is encountered in the attempt to define transcendental functions of a complex variable—such as trigonometric functions and exponentials.

The tangent of a real number, for example, is defined in trigonometry as the ratio of two sides of a right triangle, one of the angles of which is the given number. Now what could possibly be meant by the tangent of a complex number? There is no such thing as a complex angle of a triangle in the physical world.

THE EXPONENTIAL FUNCTION To deal with such questions, we will first consider the exponential. The meaning of an exponential raised to a real power is clear; for example, e^2 means multiply e by itself once. For a real variable x, e^x has a similar meaning. But we also know that e^x can be represented by an infinite series, as follows:

$$e^x = 1 + x + \frac{x^2}{2!} + \frac{x^3}{3!} + \cdots$$

In the case of a complex variable $z = x + jy$, the original meaning of e^z as multiplying e by itself a complex number of times is meaningless. However, we *define* the exponential e^z as the same as the preceding series expansion, but with x replaced by z. Thus, *by definition*:

$$e^z = 1 + z + \frac{z^2}{2!} + \frac{z^3}{3!} + \cdots$$

This is meaningful since the only operations involved on the right are multiplication and addition.

In the special, but very important, case in which the real part of z is zero, so that $z = jy$ is purely imaginary, the preceding infinite series reduces to:

$$\begin{aligned} e^{jy} &= 1 + jy + \frac{(jy)^2}{2!} + \frac{(jy)^3}{3!} + \frac{(jy)^4}{4!} + \cdots \\ &= \left(1 - \frac{y^2}{2!} + \frac{y^4}{4!} - \cdots\right) \\ &\quad + j\left(y - \frac{y^3}{3!} + \frac{y^5}{5!} - \cdots\right) \end{aligned}$$

Using the fact that $j^2 = -1$, all terms involving even powers in the first line will be real and all terms involving odd powers will be imaginary. In the second line, all the real terms have been collected together and all the imaginary terms likewise.

EULER'S THEOREM Just as the exponential function of a real variable can be represented by an infinite series, so also can trigonometric functions of real variables. The series for $\sin y$ and $\cos y$ are:

$$\cos y = 1 - \frac{y^2}{2!} + \frac{y^4}{4!} - \cdots$$

$$\sin y = y - \frac{y^3}{3!} + \frac{y^5}{5!} - \cdots$$

Comparing these expressions with the preceding one, the important conclusion is reached that:

$$e^{jy} = \cos y + j \sin y$$

This expression is known as *Euler's theorem*; it is of major importance in network theory. This theorem expresses an exponential in terms of trigonometric functions.

It would be extremely useful to invert this relationship and express trigonometric functions in terms of exponentials. This goal is facilitated by replacing y by $-y$ in Euler's theorem; thus:

$$e^{-jy} = \cos(-y) + j\sin(-y) = \cos y - j \sin y$$

The right side follows from the fact that the cosine is an even function $[f(-y) = f(y)]$ and the sine is an odd function $[f(-y) = -f(y)]$. Now, by adding and subtracting the last two equations, the following result is obtained:

$$\cos y = \tfrac{1}{2}(e^{jy} + e^{-jy})$$

$$\sin y = \frac{1}{2j}(e^{jy} - e^{-jy})$$

These expressions, inversions of Euler's theorem, also arise very often in the study of electric networks.

One addititonal form for expressing the sine and cosine in terms of an exponential can be found. This follows by noting that the exponential in Euler's theorem is a complex quantity whose real part is a cosine and whose imaginary part is a sine. Hence:

$$\cos y = \text{Re}\,(e^{jy})$$

$$\sin y = \text{Im}\,(e^{jy})$$

The notation $\text{Re}\,(A)$ is read "real part of A" and $\text{Im}\,(A)$ is read "imaginary part of A." These results will be found to be very useful in the determination of the sinusoidal steady-state solution of circuit problems.

THE LAW OF EXPONENTS The "law of exponents" is valid for complex exponentials, just as it is for real exponentials. This can be demonstrated by means of the infinite-series definition, but we shall not take the trouble to do so here. Thus, if z and w are two complex variables, then:

$$e^{z+w} = e^z e^w$$

In particular, if $z = x$ and $w = jy$:

$$e^{x+jy} = e^x e^{jy} = e^x(\cos y + j \sin y)$$

The last step follows from Euler's theorem. This result is very important in the study of the natural response of second-order (and higher) circuits.

As a final point, let's compare Euler's theorem with the trigonometric form for expressing complex numbers discussed in Section 1. From the comparison, it is clear that a complex number can be expressed in still one other form; namely:

$$A = |A|\, e^{j\alpha}$$

This form also serves to put into evidence the magnitude and angle of a complex number, just as does the symbolic form $|A|\underline{/\alpha}$ for writing a complex number. It has the added benefit that mathematical operations can be performed on it.

INDEX